AF327412

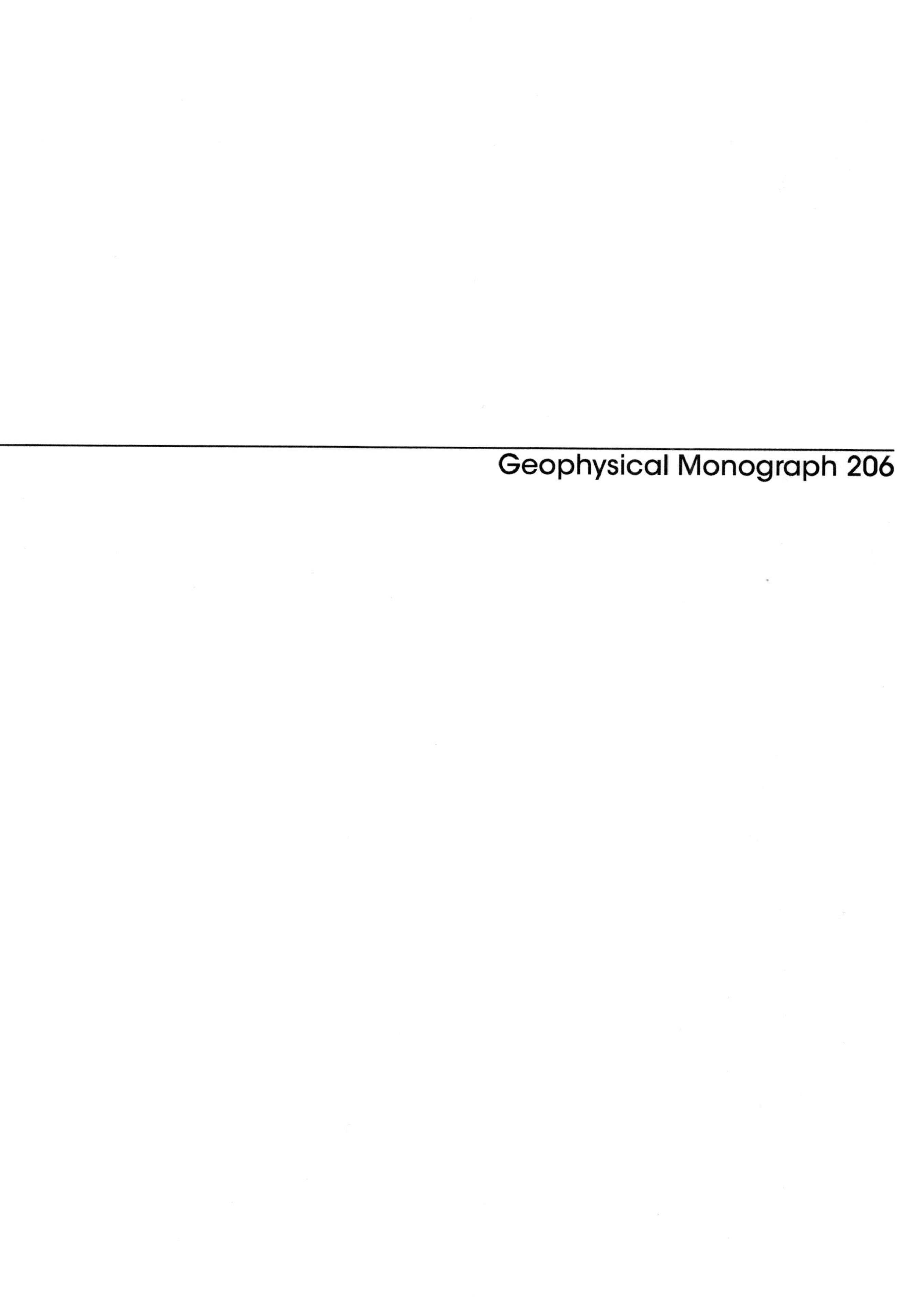

Geophysical Monograph 206

Geophysical Monograph Series

Geophysical Monograph 206

Remote Sensing of the Terrestrial Water Cycle

Venkat Lakshmi
Douglas Alsdorf
Martha Anderson
Sylvain Biancamaria
Michael Cosh
Jared Entin
George Huffman
William Kustas
Peter van Oevelen
Thomas Painter
Juraj Parajka
Matthew Rodell
Christoph Rüdiger
Editors

This work is a co-publication between the American Geophysical Union & John Wiley & Sons, Inc.

WILEY

CONTENTS

CONTRIBUTORS

Karim C. Abbaspour
Eawag, Swiss Federal Institute of Aquatic Science
and Technology
Dübendorf, Switzerland

Mohamed Ahmed
Department of Geosciences
Western Michigan University
Kalamazoo, Michigan, USA
and
Department of Geology
Faculty of Science
Suez Canal University
Ismailia, Egypt

George H. Allen
Department of Geological Sciences
University of North Carolina
Chapel Hill, North Carolina, USA

C. A. Anderson
School of Sustainable Engineering and
Built Environment
Arizona State University
Tempe, Arizona, USA

Jeremy D. Apgar
New York–New Jersey Trail Conference
Mahwah, New Jersey, USA

Igor Appel
IMSG NOAA Center for Satellite Applications
and Research
Washington, DC, USA

J. L. Awange
Western Australian Centre for Geodesy and
The Institute for Geoscience Research
Curtin University
Perth, Australia

Gregory S. Babonis
Department of Geology
State University of New York at Buffalo
Buffalo, New York, USA

Roger G. Barry
National Snow and Ice Data Center (NSIDC)
University of Colorado at Boulder
Boulder, Colorado, USA

Matthew W. Becker
Department of Geology
California State University at Long Beach
Long Beach, California, USA

R. Edward Beighley
Department of Civil and Environmental Engineering
Northeastern University
Boston, Massachusetts, USA

Katrina E. Bennett
International Arctic Research Center and
Water and Environmental Research Center
University of Alaska
Fairbanks, Alaska, USA

Gopal Bhatt
Department of Civil & Environmental Engineering
The Pennsylvania State University
University Park, Pennsylvania, USA

G. Robert Brakenridge
CSDMS, INSTAAR,
University of Colorado
Boulder, Colorado, USA

Qing Cao
Advanced Radar Research Center
National Weather Center
Norman, Oklahoma, USA

Lexi P. Coons
USDA Forest Service
Helena National Forest
Helena, Montana, USA

Michael Cosh
USDA-ARS Hydrology and Remote Sensing Laboratory
Beltsville, Maryland, USA

Wade Crow
USDA-ARS Hydrology and
Remote Sensing Laboratory
Beltsville, Maryland, USA

Robert E. Davis
U.S. Army Cold Regions Research and Engineering
Laboratory
Hanover, New Hampshire, USA

Elias Deeb
U.S. Army Cold Regions Research and
Engineering Laboratory
New Hanover, New Hampshire, USA

Tom De Groeve
Joint Research Centre of the European Commission
Ispra, Italy

Richard de Jeu
Vrije Universiteit Amsterdam
The Netherlands

Jianbin Duan
Division of Geodetic Science
School of Earth Sciences
Ohio State University
Columbus, Ohio, USA

Christopher Duffy
Department of Civil &
Environmental Engineering
The Pennsylvania State University
University Park, Pennsylvania, USA

Michael T. Durand
School of Earth Sciences and
Byrd Polar Research Center
Ohio State University
Columbus, Ohio, USA

Ammar Rafiei Emam
Department of Cartography
GIS and Remote Sensing
University of Göttingen Germany
and
Department of Desert Management
Yazd University
Yazd, Iran

Mustafa Kemal Emil
Department of Geosciences
Western Michigan University
Kalamazoo, Michigan, USA

Daniel Esteban-Fernandez
Jet Propulsion Laboratory
California Institute of Technology
Pasadena, California, USA

Bin Fang
Department of Earth and Ocean Sciences
University of South Carolina
Columbia, South Carolina, USA

Fan Fang
NASA Goddard Space Flight Center
Goddard Earth Sciences Data and
Information Services Center (GES DISC)
Greenbelt, Maryland, USA
and
ADNET Systems, Inc.
Rockville, Maryland, USA

Tom G. Farr
Jet Propulsion Laboratory
California Institute of Technology
Pasadena, California, USA

K. Fleming
Physics of Earthquakes and Volcanoes
Centre Potsdam GFZ German Research Centre
for Geosciences
Potsdam, Germany

E. Forootan
Institute of Geodesy and Geoinformation
Bonn University, Bonn, Germany

Thian Yew Gan
National Snow and Ice Data Center (NSIDC)
University of Colorado at Boulder
Boulder, Colorado, USA
and
Department of Civil and
Environmental Engineering
University of Alberta
Edmonton, Canada

Mekonnen Gebremichael
Department of Civil & Environmental Engineering
University of Connecticut
Storrs, Connecticut, USA

Kelly E. Gleason
College of Earth, Ocean and Atmospheric Sciences
Oregon State University
Corvallis, Oregon, USA

Adam K. Gobena
Department of Civil & Environmental Engineering
University of Alberta
Canada

Jonathan J. Gourley
NOAA/National Severe Storms Laboratory
Norman, Oklahoma, USA

Bin Guan
Joint Institute for Regional Earth System
Science & Engineering
University of California
Los Angeles, California, USA

Feyera A. Hirpa
Department of Civil & Environmental Engineering
University of Connecticut
Storrs, Connecticut, USA

Yang Hong
School of Civil Engineering and
Environmental Sciences
Water Technology for Emerging Region (WaTER) Center
University of Oklahoma
Oklahoma, USA
and
HyDROS Lab and Advanced Radar Research Center
National Weather Center
Norman, Oklahoma, USA

Thomas M. Hopson
Research Applications Laboratory
National Center for Atmospheric Research
Boulder, Colorado, USA

J. Indu
Department of Civil Engineering
Indian Institute of Science
Bangalore, India

Thomas J. Jackson
USDA-ARS Hydrology and
Remote Sensing Laboratory
Beltsville, Maryland, USA

Nasreen Jahan
Department of Civil and Environmental
Engineering
University of Alberta
Edmonton, Canada

Robert J. Joyce
NOAA Climate Prediction Center
College Park, Maryland, USA
and
Innovim, LLC
Greenbelt, Maryland, USA

Hahn Chul Jung
Hydrological Sciences Laboratory
NASA Goddard Space Flight Center
Greenbelt, Maryland, USA

Martin Kappas
Department of Cartography
GIS and Remote Sensing
University of Göttingen
Göttingen, Germany

Jason Kaye
Department of Crop & Soil Sciences
The Pennsylvania State University
University Park, Pennsylvania, USA

Pierre-Emmanuel Kirstetter
School of Civil Engineering and Environmental Sciences
University of Oklahoma
NOAA/National Severe Storms Laboratory
Norman, Oklahoma, USA
and
Advanced Radar Research Center
National Weather Center
Norman, Oklahoma, USA

D. Nagesh Kumar
Department of Civil Engineering
Indian Institute of Science
Bangalore, India
and
Centre for Earth Sciences
Indian Institute of Science
Bangalore, India

Venkat Lakshmi
Department of Earth and Ocean Sciences
University of South Carolina
Columbia, South Carolina, USA

A. S. Laliberte
Earthmetrics
Brownsville, Oregon, USA

Haksu Lee
NOAA National Weather Service
Office of Climate, Weather, and Water Services
Silver Spring, Maryland, USA
and
Len Technologies
Oak Hill, Virginia, USA

Hyongki Lee
Department of Civil and Environmental Engineering
University of Houston
Houston, Texas, USA
and
National Center for Airborne Laser Mapping
University of Houston
Houston, Texas, USA

Guang-Dih Lei
NASA Goddard Space Flight Center
Goddard Earth Sciences Data and
Information Services Center (GES DISC)
Greenbelt, Maryland, USA
and
ADNET Systems, Inc.
Rockville, Maryland, USA

Glen E. Liston
Cooperative Institute for Research in the Atmosphere
Colorado State University
Fort Collins, Colorado, USA

Zhen Liu
Jet Propulsion Laboratory
California Institute of Technology
Pasadena, California, USA

Eugene J. Mar
College of Earth, Ocean, and Atmospheric Sciences
Oregon State University
Corvallis, Oregon, USA

Steven A. Margulis
Department of Civil and Environmental Engineering
University of California
Los Angeles, California, USA

Zachary F. Miller
Department of Geological Sciences
University of North Carolina
Chapel Hill, North Carolina, USA

David Mocko
NASA Goddard Space Flight Center
Goddard Earth Sciences Data and
Information Services Center (GES DISC)
Greenbelt, Maryland, USA
and
Science Applications International Corp.
Greenbelt, Maryland, USA

Delwyn Moller
Jet Propulsion Laboratory
California Institute of Technology
Pasadena, California, USA
and
Remote Sensing Solutions Inc.
Barnstable, Massachusetts, USA

Noah P. Molotch
Department of Geography
Institute of Arctic and Alpine Research
University of Colorado at Boulder
Boulder, Colorado, USA

and
Jet Propulsion Laboratory
California Institute of Technology
Pasadena, California, USA

Tadanobu Nakayama
Center for Global Environmental Research
National Institute for Environmental Studies (NIES)
Ibaraki, Japan

Christopher M. U. Neale
Water for Food Institute
University of Nebraska–Lincoln
Nebraska, USA

Anne W. Nolin
College of Earth, Ocean,
and Atmospheric Sciences
Oregon State University
Corvallis, Oregon, USA

G. Odhiambo
Department of Geography and Urban Planning
UAE University
Al Ain, United Arab Emirates

P. Omondi
IGAD Climate Prediction and Applications
Centre (ICPAC)
Nairobi, Kenya

John Osterberg
U.S. Bureau of Reclamation
Denver, Colorado, USA

Thomas H. Painter
Jet Propulsion Laboratory
California Institute of Technology
Pasadena, California, USA

Robert Parinussa
Vrije Universiteit
Amsterdam, The Netherlands

Tamlin M. Pavelsky
Department of Geological Sciences
University of North Carolina
Chapel Hill, North Carolina, USA

W. Petersen
NASA Wallops Flight Facility
Wallops Island, Virginia, USA

N. A. Pierini
School of Sustainable Engineering and
Built Environment
Arizona State University
Tempe, Arizona, USA

R. C. Pipunic
Department of Infrastructure Engineering
University of Melbourne
Parkville, Victoria, Australia

Jessica R. Price
Department of Earth and Ocean Sciences
University of South Carolina
Columbia, South Carolina, USA

Zhangcai Qin
Department of Earth, Atmospheric and
Planetary Sciences
Purdue University
West Lafayette, Indiana, USA

Guo Yu Qiu
Key Laboratory for Urban Habitat Environmental
Science and Technology
School of Environment and Energy
Peking University
Shenzhen, Guangdong, China

Joan Ramage
Earth and Environmental Sciences Department
Lehigh University
Bethlehem, Pennsylvania, USA

A. Rango
USDA-ARS Jornada Experimental Range
Las Cruces, New Mexico, USA

Karl Rittger
Jet Propulsion Laboratory
California Institute of Technology
Pasadena, California, USA

Travis R. Roth
College of Earth, Ocean, and Atmospheric Sciences
Oregon State University
Corvallis, Oregon, USA

Hualan Rui
NASA Goddard Space Flight Center
Goddard Earth Sciences Data and
Information Services Center (GES DISC)
Greenbelt, Maryland, USA
and
ADNET Systems, Inc.
Rockville, Maryland, USA

D. Ryu
Department of Infrastructure Engineering
University of Melbourne
Parkville, Victoria, Australia

S. Saripalli
School of Earth and Space Exploration
Arizona State University
Tempe, Arizona, USA

M. Schwaller
NASA Goddard Space Flight Center
Greenbelt, Maryland, USA

A. Schreiner-McGraw
School of Earth and Space Exploration
Arizona State University
Tempe, Arizona, USA

Kathryn Alese Semmens
USDA-Agricultural Research Service
Hydrology and Remote Sensing Lab
Beltsville, Maryland, USA
and
Earth and Environmental Sciences Department
Lehigh University
Bethlehem, Pennsylvania, USA

Yuning Shi
Earth and Environmental Systems Institute
The Pennsylvania State University
University Park, Pennsylvania, USA

Shoichi Shige
Graduate School of Science
Kyoto University
Kyoto, Japan

A. Slaughter
USDA-ARS Jornada Experimental Range
Las Cruces, New Mexico, USA

Subramania I. Sritharan
International Center for Water Resources Management
Central State University
Wilberforce, Ohio, USA

Mohamed Sultan
Department of Geosciences
Western Michigan University
Kalamazoo, Michigan, USA

Saleh Taghvaeian
Department of Biosystems and Agricultural
Engineering
Oklahoma State University
Stillwater, Oklahoma, USA

Aina Taniguchi
Graduate School of Science
Kyoto University
Kyoto, Japan

Bill Teng
NASA Goddard Space Flight Center
Goddard Earth Sciences Data and
Information Services Center (GES DISC)
Greenbelt, Maryland, USA
and
ADNET Systems, Inc.
Rockville, Maryland, USA

William Teng
NASA Goddard Space Flight Center
Goddard Earth Sciences Data and
Information Services Center (GES DISC)
Greenbelt, Maryland, USA
and
ADNET Systems, Inc.
Rockville, Maryland, USA

Fei Tian
State Key Laboratory of Urban and Regional Ecology
Research Center for Eco-Environmental Sciences
Chinese Academy of Sciences
Beijing, China

Kevin E. Trenberth
National Center for Atmospheric Research (NCAR)
Boulder, Colorado, USA

Humberto Vergara
School of Civil Engineering and
Environmental Science
University of Oklahoma
Norman, Oklahoma, USA
and
Advanced Radar Research Center
University of Oklahoma
Norman, Oklahoma, USA

E. R. Vivoni
School of Earth and Space Exploration
Arizona State University
Tempe, Arizona, USA
and
School of Sustainable Engineering and
Built Environment
Arizona State University
Tempe, Arizona, USA

Bruce Vollmer
NASA Goddard Space Flight Center
Goddard Earth Sciences Data and
Information Services (GES DISC)
Greenbelt, Maryland, USA

John Wahr
Department of Physics
University of Colorado at Boulder
Boulder, Colorado, USA

J. P. Walker
Department of Civil Engineering
Monash University
Clayton, Victoria, Australia

Xuguang Wang
Center for Analysis and Prediction of Storms
University of Oklahoma
Norman, Oklahoma, USA
and
School of Meteorology
University of Oklahoma
Norman, Oklahoma, USA

Doyle R. Watts
Department of Earth and Environmental Sciences
Wright State University
Dayton, Ohio, USA

Rafal Wojick
Department of Civil & Environmental Engineering
Massachusetts Institute of Technology
Cambridge, Massachusetts, USA

Pingping Xie
NOAA Climate Prediction Center
College Park, Maryland, USA

Yu Jiu Xiong
Department of Water Resources and
Environments
School of Geography and Planning
Sun Yat-Sen University
Guangzhou, Guangdong, China
and
Key Laboratory of Water Cycle and
Water Security in Southern China of
Guangdong High Education Institute
Sun Yat-Sen University
Guangzhou, Guangdong, China

Munehisa K. Yamamoto
Graduate School of Science
Kyoto University
Kyoto, Japan

Eugene Yan
Environmental Science Davison
Argonne National Laboratory
Argonne, Illinois, USA

Xuan Yu
Department of Civil & Environmental
Engineering
The Pennsylvania State University
University Park, Pennsylvania, USA

Ting Yuan
Department of Civil and Environmental Engineering
University of Houston
Houston, Texas, USA
and
National Center for Airborne Laser Mapping
University of Houston
Houston, Texas, USA

Yu Zhang
School of Civil Engineering and Environmental Sciences
University of Oklahoma
Norman, Oklahoma, USA
and
Advanced Radar Research Center
University of Oklahoma
Norman, Oklahoma, USA
and
Center for Analysis and Prediction of Storms
University of Oklahoma
Norman, Oklahoma, USA

Shao Hua Zhao
Environmental Satellite Center
Ministry of Environmental Protection
Beijing, China

Qianlai Zhuang
Department of Earth, Atmospheric and
Planetary Sciences
Purdue University
West Lafayette, Indiana
and
Department of Agronomy
Purdue University
West Lafayette, Indiana, USA

PREFACE

The terrestrial hydrologic cycle involves both land surface and atmospheric transport of water in all phases (solid/liquid/water vapor) via liquid and frozen precipitation, infiltration and recharge, surface runoff and snowmelt, stream/river flow, and evapotranspiration. These different transport mechanisms are interconnected and strongly affected by the land-atmosphere dynamics and surface heterogeneity in soil type, topography, and vegetation. Similarly, meteorological forcing of precipitation and radiation are also variable in space and time. In situ systems cannot capture entirely the variability at the surface, as these are often point measurements. In addition, many of these long-term in situ measurements are being discontinued due to budgetary and programmatic constraints. Large-scale field campaigns to measure multiple geophysical variables and properties over different landscapes have been critical to scientific progress, but are limited in space and time. Many of these experiments have provided valuable data sets for the atmospheric, hydrologic, and remote sensing communities. These data sets have been used to construct algorithms that scale between point measurements on the ground and measurements from aircraft and satellites, and providing superior spatial coverage over single-point measurements. Numerous international programs, such as Prediction of Ungauged Basins (PUB) and Hydrology for the Environment, Life and Policy (HELP), which focus on the lack of rain gauge data within single watersheds, have been instrumental in bringing worldwide attention to the lack of ground observations. With the advent of the Earth Observing System (EOS) era in the 1990s, NASA, ESA, and JAXA have launched numerous spaceborne sensors to study the various components of the terrestrial water cycle. These include sensor/missions, to estimate soil moisture—Advanced Microwave Scanning Radiometer (AMSR), Soil Moisture and Ocean Salinity (SMOS); precipitation—Tropical Rainfall Measuring Mission (TRMM); vegetation—Moderate Resolution Imaging Spectroradiometer (MODIS); surface water level—JASON-1 and JASON-2 and TOPEX-POSEIDON; and groundwater—Gravity Recovery and Climate Experiment (GRACE). Currently as a result of the National Research Council Decadal Survey in 2007 a whole new suite of sensors have been recommended to study the water cycle and are scheduled to be launched within the next 10 years.

Therefore, this is a very opportune time for the geosciences community in general and the hydrological community in specific to focus on the use of satellite data for estimating and monitoring the components of the hydrological cycle. *This monograph resulting from the AGU Chapman Conference on Remote Sensing of the Terrestrial Water Cycle held in February 2012 examines the use of available satellite data and data from future missions that can be used to expand our knowledge in quantifying the spatial and temporal variations in the terrestrial water cycle.*

This monograph has seven distinct sections, viz. remote sensing of precipitation, evapotranspiration, surface water, snow, soil moisture, groundwater, and modeling and data assimilation.

PRECIPITATION

Satellite estimates of precipitation are key to understanding the global water cycle with high temporal resolution. These chapters address particular challenges in analyzing precipitation over land, including understanding rain/no-rain classification, making retrievals over complex terrain, integration of model information (which is important in areas where satellites have trouble making retrievals), and intercalibration of all precipitation-relevant satellites.

EVAPOTRANSPIRATION

Evapotranspiration (ET) is required at many different spatial and temporal scales in climate, weather, [...] hydrology, and agricultural research and applications. Satellite remote sensing is viewed as one of the only technologies that can be used with land surface models to derive ET from field to global scales, particularly in regions with little or no ground resources available. The two chapters in this monograph utilize satellite data to estimate spatially distributed ET without the need for detailed ground data, comparing output to more traditional methods including water balance and micrometeorological methods.

SURFACE WATER

Water flux throughout the world's rivers, lakes, and wetlands is one of the longest monitored components in terrestrial water balance science and more generally in the global water cycle. Knowing the surface water flux places a constraint on the other components in the water balance. Yet, numerous basins still remain poorly observed with conventional in situ measurements. The vast diversity of satellite sensors now available provide invaluable complementary observations, such as water storage change estimates [...] at the basin [...] scale or spatially distributed geomorphological parameters. Expected future satellite missions will increase our understanding of surface water fluxes.

SNOW

Water stored in snow cover is one of the main components of water balance in many parts of the world. The large spatial variability of snow characteristics, particularly in mountains, makes remote sensing a very important alternative to ground snow observations. The snow cover chapters in this book present examples of integrating remote sensing snow observations into hydrologic modeling, improvements in satellite snow cover retrieval and snowmelt and melt-refreeze estimation, advances in snow estimation under vegetation cover, and assessing impacts of climatic factors on snow pack changes in the past.

SOIL MOISTURE

Soil moisture near the surface is a relatively new field of study for remote sensing. This surface state influences both energy and water fluxes at the land surface-atmosphere interface, and is the key state in the partitioning between water infiltration and runoff. Recent advances in satellite remote sensing have pushed the envelope in the science of soil moisture estimation, improving estimates for modeling and applications. Spanning from SMMR, SSM/I, and AMSR-E heritage data to the upcoming SMAP mission, and with the advent of very high resolution products from ESA's Sentinel-1, the dramatic increase in data availability is allowing new and innovative explorations of soil moisture [...] utilization, as presented in the chapters of this section.

GROUNDWATER

Groundwater comprises more than half of the world's drinking water and a large portion of the water used to irrigate crops. However, groundwater data are scarce in most of the world, making remotely sensed groundwater information that much more valuable. The chapters on groundwater present examples of remote sensing techniques that have the potential to transform our ability to monitor groundwater variability globally.

MODELING

Global observations and modeling of the terrestrial water cycle have advanced considerably over recent years. Many processes related to changes due to human and natural causes can be better understood and even predicted. Yet, with the uncertainty related to climate change and the highly variable nature of many of the terrestrial processes, there is still much research needed to address the challenges related to freshwater availability. Continuation of existing Earth observational capabilities is as paramount as the development of new capabilities related to monitoring, modeling, and prediction of the terrestrial water cycle.

Venkat Lakshmi
University of South Carolina

Douglas Alsdorf
Ohio State University

Martha Anderson
USDA-Agricultural Research Service

Sylvain Biancamaria
Centre National de la Recherche Scientifique
Laboratoire d'Etudes en Geophysique et
Oceanographie spatiales

Michael H. Cosh
USDA-Agricultural Research Service

Jared Entin
NASA Headquarters

George J. Huffman
NASA Goddard Space Flight Center

William Kustas
USDA-Agricultural Research Service

Peter van Oevelen
International GEWEX Project Office

Thomas H. Painter
Jet Propulsion Laboratory

Juraj Parajka
Vienna University of Technology

Matthew Rodell
NASA Goddard Space Flight Center

Christoph Rüdiger
Monash University

Section I: Precipitation

1

Rain/No-Rain Classification Using Passive Microwave Radiometers

J. Indu[1] and D. Nagesh Kumar[1,2]

1.1. INTRODUCTION

Precipitation is a critical variable driving the atmosphere's general circulation through latent heat release. As such, accurate quantification of the spatiotemporal variability of precipitation is essential for applications involving environmental, atmospheric, water resource, and related science and engineering disciplines. The increased availability of data products from microwave (passive and active) remote sensing has contributed toward our understanding of the spatiotemporal distribution of precipitation by providing near-real-time spatially continuous precipitation estimates at smaller temporal sampling intervals [*Petty*, 1994; *Ferraro*, 1997; *Bauer*, 2001; *Grecu and Anagnostou*, 2001; *Kummerow et al.*, 2001; *Turk et al.*, 2002; *McCollum and Ferraro*, 2003; *Wilheit et al.*, 2003; *Ferraro et al.*, 2005; *Levizzani and Gruber*, 2007]. These include data products from the Special Sensor Microwave Imager (SSM/I) on Defense Meteorological Satellite Program satellites [*Ferraro*, 1997], Advanced Microwave Sounding Unit (AMSU) on National Oceanic and Atmospheric Agency (NOAA) Polar Orbiting environmental satellites [*Ferraro et al.*, 2005], Tropical Rainfall Measuring Mission (TRMM) microwave imager (TMI) and precipitation radar (PR) [*Kummerow et al.*, 2001; *Wang et al.*, 2009], Advanced Microwave Scanning Radiometer-Earth Observing System (AMSR-E) [*Wilheit et al.*, 2003] on National Aeronautics and Space Administration (NASA) and Japan Aerospace and Exploration Agency (JAXA) joint satellites, etc. Along with the widespread acceptance of microwave-based precipitation products, it has also been recognized that these products contain large uncertainties [*Petty*, 1994; *Smith et al.*, 1998; *Kummerow et al.*, 1998, 2005; *Coppens et al.*, 2000]. Studies quantifying global uncertainty offered by microwave rainfall algorithms show climatologically distinct space/time domains that contribute approximately 25% uncertainty to rainfall product that goes undetected by a microwave radiometer [*Kummerow et al.*, 2005]. Of these, nearly 20% is attributed to changes in cloud morphology and microphysics and 5% to changes in the rain/no-rain thresholds. The purpose of this chapter is to describe the foundations of rain/no-rain classification (RNC) based on passive microwave brightness temperatures, outstanding issues, areas of future research, and a comprehensive review of the existing RNC algorithms, based on the works by *Grody* [1991], *Adler et al.* [1993], *Ferraro et al.* [1998], *Seto et al.* [2005, 2009], *Kida et al.* [2009], and *Kubota et al.* [2007].

The physically based overland rainfall retrieval algorithms incorporate rainfall screening as an integral part, without which the succeeding overland rain retrieval technique gets corrupted easily. From the work by *Grody* [1991], *"the physics of rain detection and screening are every bit as important as those of conversion."* Studies by rainfall intercomparison projects including algorithm intercomparison projects sponsored by the Global Precipitation Climatology Project and NASA WetNet Precipitation Intercomparison Projects conclude that inadequate screening of nonraining pixels complicates the simplest to the most complex of retrieval algorithms.

To date, various approaches exist to detect raining areas within a radiometer footprint. While some of these techniques are easy to implement, some others involve sophisticated programming logic for correct implementation.

[1] *Department of Civil Engineering, Indian Institute of Science, Bangalore, India*

[2] *Centre for Earth Sciences, Indian Institute of Science, Bangalore, India*

Remote Sensing of the Terrestrial Water Cycle, Geophysical Monograph 206. First Edition. Edited by Venkat Lakshmi.

Currently, there exist two schools of thoughts for describing rainfall screening methodologies. One approach addresses screening as part of the rainfall retrieval problem. The other approach considers RNC as an essential preprocessing step for proper identification of potential rain measurements before the actual retrieval process. Regardless of which philosophy is followed, typical RNC classification algorithms should accurately identify rainfall signatures over surfaces covered with snow/ice that offer difficulty in uniquely separating rainfall signature from the surface conditions. This implies that an algorithm should either "dynamically" determine nonraining pixels or it should depend on suitable surface masks based upon climatology (e.g., for snow and ice) or geography (e.g., for deserts) [*Ferraro et al.*, 1996]. The organization of this chapter is as follows: Section 1.2. presents a discussion on the fundamental principle of passive microwave data and radiative transfer model. Atmospheric attenuation (i.e., reduction of a signal due to atmospheric gases, hydrometeors) is a critical factor affecting radiometer brightness temperature. Hence, Section 1.3. discusses the complex interactions of atmospheric hydrometeors (like water vapor, ice, precipitation) with different microwave frequencies. Section 1.4. describes the fundamentals of the RNC classification technique and highlights prominent RNC algorithms that are embedded in the Goddard profiling (GPROF), the global satellite mapping (GSMaP) of precipitation, and the Goddard scattering (GSCAT) algorithms. Section 1.5. summarizes various indices used for performance evaluation of a typical RNC classification. Compared to RNC classification over oceans, overland classification offers a myriad of complications as land presents itself as a radiometrically warm background with highly varying surface emissivities [*Spencer et al.*, 1989; *Grody*, 1991; *Adler et al.*, 1994; *Ferraro*, 1997]. The fairly complex atmospheric attenuation in the scattering regime complicates rainfall delineation even further. For these reasons, overland RNC warrants separate attention. There are several open questions that need to be addressed. These have been discussed in Section 1.6. followed by the conclusions in Section 1.7.

1.2. PRINCIPLES OF PASSIVE MICROWAVE SATELLITE MEASUREMENTS

Radiometry is the field of science related to measurement of incoherent electromagnetic radiation. According to thermodynamic principles, all materials (gases, liquids, solids) both emit and absorb incoherent electromagnetic energy. The magnitude of thermal emission I can be expressed as a product of emissivity (ε) and the Planck (blackbody) function $B(T)$ as

$$I_\lambda = \varepsilon_\lambda B_\lambda(T) = [\varepsilon_\lambda (2hc^2\lambda^{-5})]/(e^{hc/\lambda kT} - 1) \quad (1.1)$$

where h is Planck's constant, k is Boltzmann's constant, c is the speed of light, and T is thermal temperature [*Elachi*, 1987]. By approximating the thermal emission from the Planck function using Rayleigh-Jeans formula, the microwave brightness temperature can be conveniently expressed as a linear function of physical temperature and emissivity (ε) as

$$Tb = \varepsilon T_{\text{PhysicalTemperature}}, \quad (1.2)$$

where ε is a complex function of the dielectric constant whose values are quite well known for gases and calm water but not so well understood for the complicated case of rough water and land surfaces [*Elachi*, 1987].

A downward-viewing spaceborne radiometer is built to sense the upwelling electromagnetic energy emanating from the surface, which reaches the top of the atmosphere after attenuation. The brightness temperatures registered by this radiometer depends on absorption and scattering properties of atmosphere and background emissivity, which vary with frequency and polarization. The intensity of brightness temperature (T_b) incident on a spaceborne microwave radiometer (directed toward Earth), indicates radiation received by the spaceborne antenna from regions of space, which are defined by the antenna pattern. "*The antenna pattern is usually strongly peaked along its beam axis. And when pointing towards the ground, its spatial resolution or footprint size is defined by the angular region over which the antenna power pattern is less than 3 dB down from its value at beam center*" [*Njoku*, 1982]. The total noise power resulting from the thermal radiation incident on the antenna, also known as "antenna temperature" is expressed as a function of the antenna gain pattern [$G(\theta, \phi)$] and the brightness temperature distribution incident [TB(θ, ϕ)] as

$$T_a = \frac{1}{4\Pi} \iint_{4\Pi} T_b(\theta, \phi) G(\theta, \phi) \, d\Omega. \quad (1.3)$$

As shown in Figure 1.1, the distribution of T_b is composed of self-emitted radiation from land/sea, upward emission from the atmosphere, and downward atmospheric emission that is rescattered by the surface toward the antenna coupled with atmospheric attenuation. Therefore, an interpretation of T_b will essentially reveal the physical properties of the media that produce them. Knowing the atmosphere, surface environmental parameters, and radiometer characteristics, radiative transfer models (RTMs) can be used to normalize the measured T_b to a common reference for comparison. This implies that RTMs can interpret T_b from radiometers with different characteristics having different viewing geometries (incidence angles) and operating at different frequencies [*Chandrasekhar*,

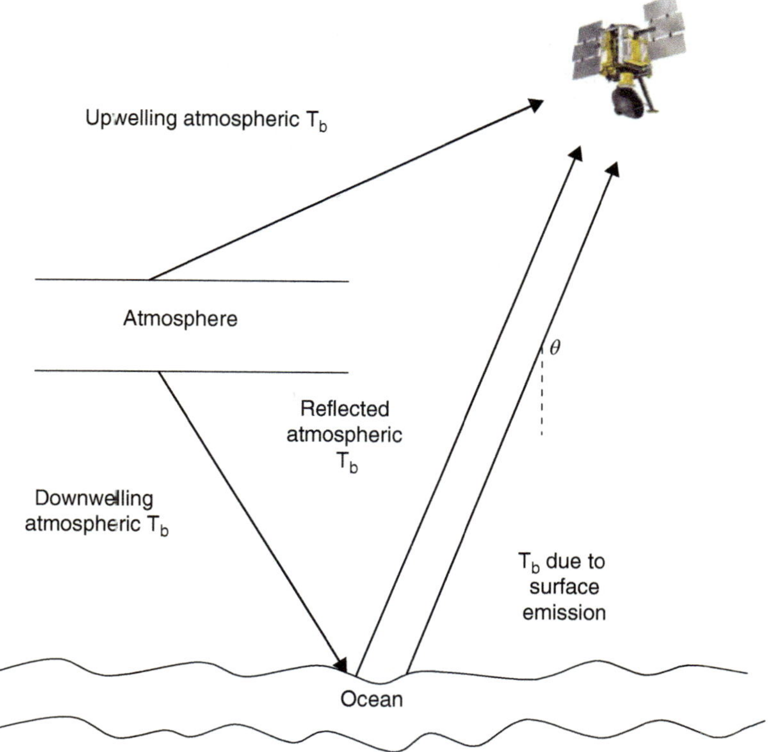

Figure 1.1 Spaceborne radiometer observing the ocean at a nadir angle θ.

1960; *Volchok and Chernyak*, 1968; *Paris*, 1971; *Snider and Westwater*, 1972; *Fraser*, 1975; *Savage*, 1976; *Wilheit et al.*, 1977]. Several factors contribute to the quantitative accuracy of RTMs such as the realism in specifying hydrometeor shape, size, and phases encountered in real rain clouds, knowledge of vertical hydrometeor profiles, and proper generation of local radiative interaction properties (like single scatter albedo, extinction coefficient, etc.).

The general problem of radiative transfer was solved by *Chandrasekhar* [1960] and further extended for microwaves in a cloudy atmosphere by *Volchok and Chernyak* [1968], *Paris* [1971], *Snider and Westwater* [1972], *Fraser* [1975], *Savage* [1976], and *Wilheit et al.* [1977]. For a medium in thermodynamic equilibrium, the change in microwave intensity I_λ over a distance ds in the direction (θ, ϕ) is given by the expression:

$$\frac{dI_\lambda}{ds} = -\left(k_{ab} + k_{sc}\right) I_\lambda + k_{ab} B_\lambda + \frac{k_{sc}}{4\Pi}$$

$$\int_0^{2\Pi} \int_0^{\Pi} P_\lambda^{'}\left(\theta_s, \phi_s, \theta, \phi\right) I_\lambda \left(\theta_s, \phi_s\right) \sin\theta_s \, d\theta_s \, d\phi_s, \qquad (1.4)$$

where, $P_\lambda^{'}\left(\theta_s, \phi_s, \theta, \phi\right)$ is an inverse scattering phase function that describes the relative contribution of each polar angle θ_s and each azimuthal angle ϕ_s to the energy scattered in the direction (θ, ϕ); and k_{ab}, k_{sc} denote the absorption and scattering coefficients, respectively. In other words, the gradient of I_λ along s is determined by the balance of energy lost by absorption and by scattering out of the direction of I and energy gained by thermal emission and by scattering into the direction of I. The literature presents several solutions for RTM based on the assumptions of Marshall Palmer drop size distribution, horizontally homogeneous atmosphere in thermal equilibrium [*Savage et al.*, 1995; *Lovejoy and Austin*, 1980]. Generally, forward radiative transfer equations found in the literature are either scalar or vector models. While the scalar model solves radiative transfer equations with a single Stokes' parameter and considers only the polarization effects caused by the surface, vector models consider polarization effects caused by both surface and hydrometeors. Vector models consider four Stokes' parameters, which make the calculations complex and difficult to implement. Hence, vector models require comparatively larger amount of computational time when compared with scalar models. These will not be discussed in this chapter. Comprehensive details regarding microwave radiative transfer are given in *Liou* [1980] and *Janssen* [1993].

Uncertainty in effectively interpreting microwave T_b stems from several factors, like antenna pattern, deviation of spacecraft attitude parameters (roll, pitch, yaw), alteration of land surface emissivity during rainfall, nonhomogeneity of land surfaces that results in high and variable surface emissivity, atmospheric attenuation, etc. Attenuation results from the complex interaction of electromagnetic waves with ice, water vapor, oxygen, and other precipitation sized hydrometeors aloft, which can be liquid and/or solid and which may precipitate to surface as rainfall or snowfall depending on the temperature in the subcloud layer. As atmospheric attenuation complicates detection of rainfall signature within T_b, knowledge regarding the sources and sinks of microwave radiation within the atmosphere is crucial to fully understand these uncertainties.

1.3. ATMOSPHERIC ATTENUATION OF MICROWAVES

With the advent of microwave radiometers on board satellites like Defence Meteorological Satellite Program (DMSP), TRMM, Global Precipitation Mission (GPM), Megha Tropiques (MT), etc., microwave rainfall products have become an indispensable source for precipitation information and for real-time applications in flood forecasting. A choice of frequency channels on board these satellites are made based on the geophysical parameter to be studied and its sensitivity to major atmospheric constituents. Several studies have examined the response of microwave frequency channels due to precipitation-sized particles in the atmosphere [*Weinman and Guetter*, 1977; *Spencer*, 1986; *Wu and Weinman*, 1984]. Works have also been conducted to estimate sensitivity of T_b to variations in atmospheric and precipitation parameters using cloud radiative models such as those by *Weinman and Guetter* [1977], *Wilheit et al.* [1982], *Wu and Weinman* [1984], *Szejwach et al.* [1986], *Olson* [1987], and *Kummerow and Weinman* [1988]. This section describes the field of spectroscopy, an age-old science explained by quantum mechanics during the first half of the twentieth century involving the study of absorption and emission by gases. The five possible ways in which radiation interacts with atmospheric gases are ionization-dissociation interaction, electronic transition, vibrational transition, rotational transition, and forbidden transition [*Kidder and Vonder Haar*, 1995]. Among these, vibrational and rotational transitions are important for satellite meteorology as they occur mostly in the infrared and microwave portion of the electromagnetic spectrum. Some of the prominent sources causing atmospheric attenuation of microwaves are discussed below.

1.3.1. Absorption by Gaseous Atmosphere

An extensive study of microwave absorption of atmospheric gases (both theoretically and experimentally) shows that, emission/absorption in gaseous atmosphere is dominated by the presence of water vapor and oxygen [*Waters*, 1976; *Ulaby and Stiles*, 1981]. Absorption characteristics of these gases are summarized by *Staelin* [1969], *Paris* [1971], *Derr* [1972], *Waters* [1976], and *Fraser* [1975]. Microwaves undergo resonant absorption and emission at certain frequencies due to the quantum energy states of the water vapor/oxygen molecules. Within microwave spectrum, these molecules are subjected to rotational transition wherein a molecule changes rotational energy states. This causes a peak in T_b measured by a radiometer. The magnitude of increase in T_b depends on the total number of water vapor/oxygen molecules along the propagation path through the atmosphere. At higher altitudes there is a decrease in the number of water vapor/oxygen molecules per unit volume. This in turn reduces the bandwidth of water vapor/oxygen emission (absorption) leading to an increase in absorption at the peak of resonance. The rotational lines of water and oxygen are "pressure broadened" in the atmosphere owing to the presence of other gases; there is also a slight dependence on temperature [*Kidder and Vonder Haar*, 1995]. Water vapor has a weak absorption line at 22.235 GHz and a strong line at 183 GHz. All sensors currently used for precipitation make a measurement near 22.235 GHz like TMI at 21.3 GHz and AMSR at 23.8 GHz. Oxygen has two major peaks, one near 60 GHz and another at 118.75 GHz. More details regarding the absorption characteristics of atmospheric water vapor and oxygen can be obtained from *Paris* [1971] and *Fraser* [1975].

1.3.2. Cloud Liquid Water

In an atmosphere with cloud particles, the prominent sources and sinks of microwave energy are local emission and absorption. In the scattering regime, cloud droplets interact weakly with microwave radiation. As the cloud liquid water particles are usually less than 100 μm in diameter, much smaller than the wavelength of radiation, for this Rayleigh region, the scattering effect is negligible. Generally, for the Rayleigh regime, the effect of cloud particles on microwave radiation depends on liquid water content, cloud temperature, and wavelength of radiation. When microwave radiation interacts with rain clouds, the phenomenon is similar to an ensemble of drops with no coherence from drop to drop in the phase of scattered light. Within a rain volume, the usual practice is to assume the raindrops to be randomly distributed. Once we calculate the scattering and absorption for a single drop of spherical dielectric, it is possible by integration to determine corresponding coefficients for a rain cloud that is an ensemble of drops. The particle sizes of raindrops within a rain-bearing cloud are usually described by a continuous function known as drop size distribution (DSD). This function is responsible for defining the concentration of rain particles per unit volume per unit increment of the drop radius.

Studies by *Fraser* [1975] estimated the single-scatter albedo as a function of wavelength for DSD representing very thin fair weather cumulus and very dense cumulonimbus clouds. Generally, for cumulus clouds, scattering remains negligible at all wavelengths. For cumulonimbus clouds raining at 150 mm/h, scattering will be small only for all values of wavelength >~3 cm. Studies by *Lovejoy and Austin* [1980] concluded that scattering can be neglected for all clouds if wavelength is >0.5 cm, and for rain rate <10 mm/h if wavelength is >1 cm [*Stepanenko*, 1968; *Wilheit et al.*, 1977]. A complete and published summary of extinction, scattering, and absorption coefficients and scattering phase functions is available in *Savage* [1978]. Savage approximated the scattering phase functions by Legendre polynomials, and expressed the Legendre coefficients (for the phase function) and extinction, scattering, and absorption coefficients as power law relations in liquid water content [*Barrett and Martin*, 1981]. *Lovejoy and Austin* [1980] assessed the relative contribution of cloud droplets and raindrops to total cloud layer absorption and came out with the conclusions that at rain rates = 10 mm/h, cloud absorption was 30–40% as large as rain absorption. This conclusion was consistent with the observations by *Gorelik et al.* [1971]. In 1976, Savage stated that addition of a non-scattering cloud layer above the rain layer counteracted scattering and resulted in T_b increase by an amount proportional to the cloud layer thickness. It must be noted that as cloud droplets, water vapor, and oxygen all absorb (but do not scatter) microwave radiation, they have the potential to confuse precipitation estimates based on absorption [*Barrett and Martin*, 1981]. Authoritative treatment of this subject may be found in *Gunn and East* [1954], *Shifrin and Chernyak* [1968], *Paris* [1971], *Schwiesow* [1972], *Hansen and Travis* [1974], *Savage* [1976], and *Fraser* [1975].

1.3.3. Surface Emission

In the microwave spectrum, an emitting surface must be considered as a gray body so that its emissivity value stays lower than unity. For homogeneous land surfaces, the variability in microwave radiances depends on surface skin temperature and surface emissivity, while the variability for open water bodies is attributed to the atmospheric constituents such as columnar water vapor, temperature profiles, and presence of cloud liquid water. Unlike the oceans, it is very difficult to model land surface properties in the microwave spectrum due to the spatiotemporal variations of soil features like roughness, vegetation cover, and moisture content. The response of different surface types on the temperature and humidity retrievals has been quantified by *English* [1999]; in these studies microwave emission errors for different continental surfaces were evaluated by using a mathematical technique to potentially extend the low-altitude sounding information over solid surfaces. Microwave land surface emissivity for various surface conditions on a global scale was attempted by *Prigent et al.* [1998], *Weng et al.* [2001], and *Pellerin et al.* [2003].

Different surfaces contribute varying amounts of emission to a microwave radiometer footprint. Figure 1.2 shows the T_b variations for different microwave frequency channels over land and ocean surfaces [*Ferraro et al.* 1998].

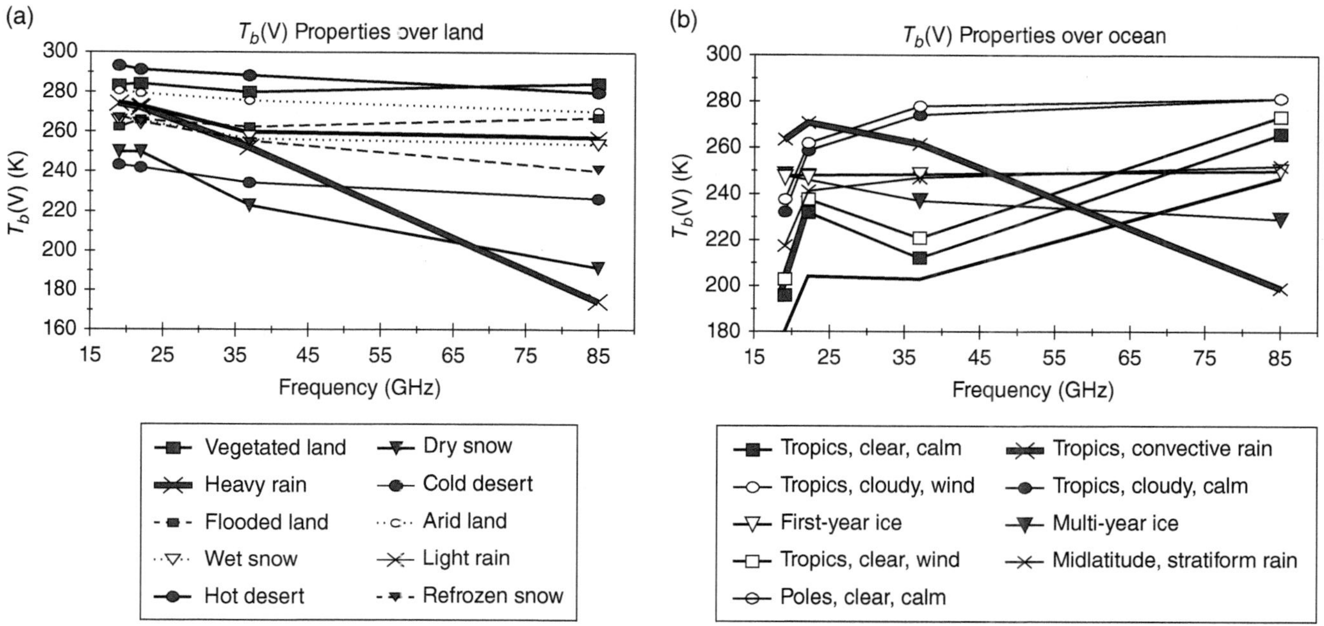

Figure 1.2 Emissivity characteristics of various surface types represented by T_b as a function of frequency (a) over land and (b) over ocean [*Ferraro et al.*, 1998].

Oceans provide a stable and uniformly "cold" background for a radiometer, emphasizing more the extinction of upwelling radiation by atmospheric constituents. Emissivity of sea surface is dependent on the dielectric properties of seawater through the Fresnel equation. Studies were conducted to predict the dielectric constant of seawater with an aim to improve the retrieval of atmospheric parameters [*Klein and Swift*, 1977]. Over snow-covered soil, the emissivity values depend on the dielectric constant of frozen soil (~3), thickness, water equivalent, and liquid water distribution. If snow is dry, T_b decreases with an increase in snow water equivalent. In the case of wet snow, even a small increase in the amount of liquid water causes T_b to rise due to volume scattering. The dependence of snow on T_b is more prominent at microwave frequencies >8–12 GHz. In the presence of vegetation, microwave radiation gets emitted, absorbed, and scattered with the radiative properties mostly controlled by the vegetation density, dielectric properties, and relative size of vegetation components with respect to wavelength. Increasing vegetation density increases the emissivity in horizontal polarization and reduces the emissivity polarization difference [*Prigent et al.*, 1997]. In effect, the presence of vegetation reduces the radiometric sensitivity to soil moisture. Studies have developed computational schemes to improve the mathematical description of surface emissivity for several land types: bare soil, vegetation canopy, and snow-covered terrain [*Shi et al.*, 2002; *Ferrazoli et al.*, 2000; *Fung*, 1994].

Theoretical models for microwave emission from soils have been presented by many studies [*Njoku and Kong*, 1977; *Wilheit*, 1978; *Burke et al.*, 1979; *England*, 1976] by considering emission from soil for a range of moisture and temperature profiles. At low microwave frequencies, T_b is strongly affected by soil moisture content. This strong dependence is owing to the comparatively high dielectric constant of water (~80) compared to that of dry soil. The dielectric constant of wet soil can reach 20 or more, resulting in an emissivity change at 1.4 GHz from about 0.95 for dry soil to 0.6 for wet soils. Despite a strong sensitivity of the emissivity on soil moisture, microwave remote sensing of soil moisture from space is complicated due to highly varying surface roughness and vegetation cover, which is aggravated by the presence of mixed surfaces within satellite field of view (FOV). To summarize, varying emissivity values from spatiotemporal variations of land surface types can deeply affect radiometer observations, often leading to rainfall retrieval errors.

a large variety of shapes and modes depending on atmospheric temperature and humidity conditions. If we simplify the domain of shapes that the ice nucleation process can create, they can be considered as columns and plates. The usual practice is to adopt a Marshall-Palmer size distribution for ice particle sizes in all theoretical treatments.

Sensitivity of T_b values to the integrated mass of ice/rain depends on frequency, until the optical depth reaches the saturation level [*Evans et al.*, 1995]. *Fulton and Heymsfield* [1991] studied the response of T_b (18, 37, 92, 183 GHz) to hydrometeors due to intense convection and suggested that even the lowest microwave frequency channel (18 GHz) is significantly obscured by deep convective ice mass. Generally, with an increase in microwave frequency (>60 GHz), scattering signatures become more pronounced and a dramatic increase is observed in volume scattering (k_s), absorption coefficients (k_a), and single-scatter albedo. This is because the aging of ice results in internal voids that tend to scatter microwave radiation. And, an increase in microwave frequencies is accompanied by an increase in the scattering cross section of inhomogeneities, thereby causing a decrease in radiation emanating from them. Thus, at higher frequencies, scattering dominates with microwave radiation acting relatively transparently to the rain below freezing level. In satellite meteorology, the high single-scatter albedo produced by ice is crucial, as it heavily depresses high-frequency microwave channels. Ice has much smaller absorption coefficients than water that result in high albedos at all SSM/I frequencies. A single scatter albedo approaching unity indicates that any thermal radiation upwelling from below an ice layer that is attenuated by the ice will be scattered out of the radiometer's field of view, with very little ice-emitted radiation to replace it [*Spencer et al.*, 1989]. When the scattering coefficient is large, and since there is very little (2.7 K) downwelling radiation from cosmic background, very low values of T_b will be recorded at 85.5 GHz frequency. This extremely low T_b observed is usually attributed to convective rainfall. Due to the high sensitivity of 85.5 GHz frequency to frozen ice, RNC algorithms for land regions are essentially based on ice scattering at this frequency. Studies have also been conducted by *Anagnostou and Kummerow* [1997] that suggest that as T_b at 85.5 GHz (V) frequency is more variable in raining than in nonraining area, studies of rainfall screening can utilize even the standard deviation of T_b at 85.5 GHz (V) frequency in a 5 × 5 pixel window [*Biscaro and Morales*, 2007].

1.3.4. Ice

Ice particle shapes are crucial in scattering regimes, as they significantly affect the emerging radiance field [*Mugnai and Wiscombe*, 1986; *Bohren*, 1986]. Crystals of ice exhibit

1.3.5. Precipitation

At microwave wavelengths, precipitation-sized drops interact strongly with microwave radiation [*Kidder and Vonder Haar*, 1995]. Interaction of electromagnetic (EM)

waves with a spherical dielectric causes scattering (redirecting) or absorption (conversion to mechanical energy) of radiation depending on the size of precipitation particles [*Barrett and Martin*, 1981]. One of the earlier studies by *Mie* [1908] introduced the general mathematical solution for scattering and absorption of EM waves by a dielectric sphere of arbitrary radius. Later on, this was applied to the context of rain by *Gunn and East* [1954]. The expressions for Mie efficiency factors are given by

$$Q_{ex}\left(n_c,\lambda\right) = \frac{\sigma_{ex}}{\Pi r^2}, \qquad (1.5)$$

$$Q_{sc}\left(n_c,\lambda\right) = \frac{\sigma_{sc}}{\Pi r^2}. \qquad (1.6)$$

In equations (1.5) and (1.6), r denotes the radius of the rain drop, λ stands for the wavelength, and n_c represents the complex index of refraction; Q_{ex} and Q_{sc} refer to the Mie efficiency factors of extinction coefficient and scattering coefficient for a single drop. The symbols σ_{ex} and σ_{sc} represent the effective cross sections for extinction and scattering.

Fraser [1975] has calculated the Mie efficiency factors for extinction and scattering and the Rayleigh extinction coefficient for a range of drop sizes [*Barrett and Martin*, 1981]. If we consider a single raindrop particle whose size is much smaller than the wavelength of EM waves, Rayleigh approximation to the exact Mie expression applies. The absorption cross section will then be proportional to the cube of particle diameter and hence proportional to the volume and mass of the raindrop while scattering cross section will be negligible. When cloud drops coalesce into raindrops with dimensions comparable to microwave wavelengths, absorption per unit mass increases and scattering can no longer be ignored. Based on the theoretical calculations by *Savage* [1976], rain rates of even a few millimeters per hour cause depression (below 260 K) in T_b for microwave frequencies close to 100 GHz. Studies by *Kidder and Vonder Haar* [1977] used T_b threshold values to discriminate raining from nonraining pixels. Although attempts to use measurements at 37 GHz for land regions [*Weinman and Guetter*, 1977; *Spencer et al.*, 1983; *Spencer*, 1986] met with partial success, mainly in cases of heavier convective rainfall, more reliable microwave rainfall monitoring was made possible only after the launch of SSM/I (1987) [*Barrett et al.*, 1988; *Spencer et al.*, 1989]. *Spencer et al.* [1989] calculated the scattering and absorption properties of rain for the three main wavelengths (19.35, 37, 85.5 GHz) that have been used to measure precipitation (Figure 1.3). Their study came out with the conclusions that liquid drops both absorb and scatter microwaves of which absorption dominates [*Kidder and Vonder Haar*, 1995], especially in the frequency range below 22 GHz. This implies that, in this frequency range,

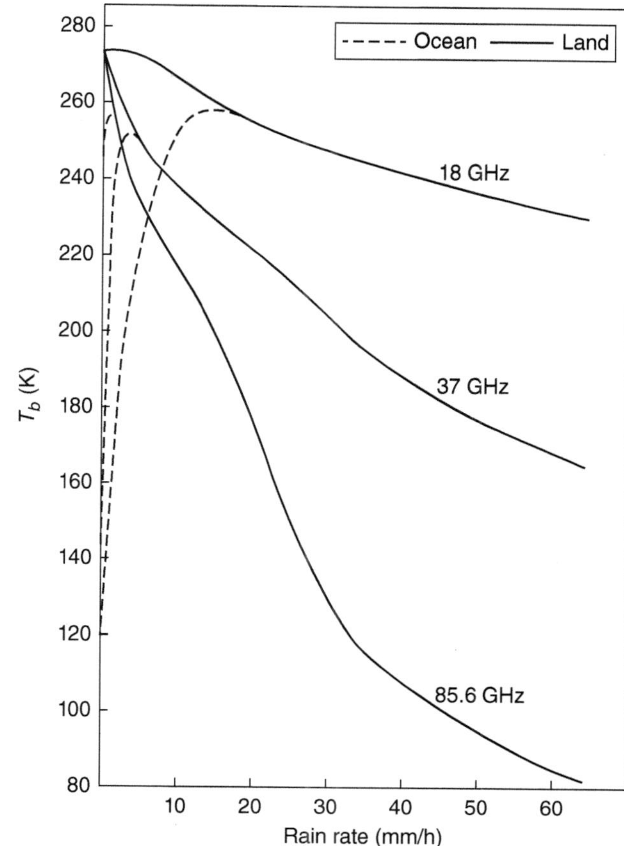

Figure 1.3 Relationship of T_b rain rate at 18, 37, and 85.6 GHz. [Reprinted from radiative transfer modeling of *Wu and Weinman*, 1984.]

scattering does not occur and ice particles above rain are nearly transparent. Another prominent result of their study is that with an increasing rain rate, scattering and absorption both increase with microwave frequencies.

1.4. RAIN/NO-RAIN CLASSIFICATION METHODS

An ideal approach toward understanding the fundamentals of RNC classification is based on the emissivity characteristics of background surface. RNC algorithms follow different principles when the underlying surface is land or ocean. Ocean surfaces, which appear "cold" to a radiometer operating in the microwave region, offer good contrast for the detection of rain drops, which appear radiometrically warm. As this phenomenon utilizes the strong physical relationship between low-frequency (6–37 GHz) T_b and liquid rainfall, overocean techniques are essentially emission based. Land, however, offers a radiometrically warm background, which tends to hide emission from raindrops. Overland RNC techniques solely rely on ice scattering at a high-frequency (85 GHz) microwave channel and are ambiguous in nature [*Wilheit*,

1986; *Spencer et al.,* 1989; *Grody,* 1991; *Adler et al.,* 1993; *Ferraro and Marks,* 1995; *Lin and Hou,* 2008; *Wang et al.,* 2009; *Gopalan et al.,* 2010]. This is mainly due to the highly varying emissivity from the land surface background that clutters rainfall signature. The complicated nature of high-frequency microwave scattering with ice crystals adds to the uncertainty, thereby rendering the use of radiative transfer models extremely difficult. It should be noted that the microwave frequencies are utilized to retrieve several geophysical parameters over ocean; for example, 6.8 and 37 GHz are used for wind speed retrieval [*Wentz,* 1983], 7 GHz is used for retrieving sea surface temperature [*Chelton and Freilich,* 2005], and so forth.

Some of the first rainfall screening studies were by *Ferraro* et al. [1986] and *Wentz* and Cavalieri [1995]. They proposed using multiple channels to identify a rainfall signature from radiometer-received T_b. Their study, involved an intensive analysis of Scanning Multichannel Microwave Radiometer (SMMR) passes over central North America on 20 January, 1979. The study region was chosen to represent a wide variety of surface and atmospheric features ranging from harsh winter conditions and deep snow cover in the northern latitudes to heavy convective rains.

Based on the criteria of spatial resolution and sensitivity to surface parameters, differences between 18 and 37 GHz channels were selected as the optimum channel combination for rain/no-rain discrimination. Their study observed that the presence of clouds and precipitation led to an increase in T_b. *Ferraro et al.* [1994] expanded on these ideas and developed a set of geographical screens for land, ocean, semiarid land, coastlines, and sea ice. Using SSM/I data over ocean surfaces, *Wentz and Cavalieri* [1995] proposed the no-rain algorithm based on the fundamental principles of radiative transfer and explicitly showed the physical relationships between the input (T_b) and the output (wind speed, columnar water vapor, columnar cloud liquid water, rain rate, and effective radiating temperature for upwelling radiation). Later on, this algorithm was extended to include the effects of rainfall. Comprehensive details can be found in *Wentz* [1997]. The screening methodology that has evolved continuously throughout the years is the Grody-Ferraro screening methodology (discussed in Section 1.4.1). This is currently built in to the GPROF algorithm [*Kummerow et al.,* 2001]. Various versions of GPROF have been applied in SSM/I, TMI, and AMSR-E missions [*McCollum and Ferraro,* 2003; *Wang and Wolff,* 2010]. The purpose of this section is to describe the major indices used for demarcating the rainfall signature within a microwave FOV, through which we discuss some of the prominent RNC algorithms adopted for land, ocean, and coastlines, using data from satellites like SSM/I, TRMM, etc.

1.4.1. Scattering Index

The technique of using a scattering index (SI) to delineate raining pixels originated from the studies by *Grody* [1991]. Initially, the idea was proposed to create geographical masks for eliminating T_b cluttering due to desert sand and ice-capped land surfaces. This was essential, as uncertainty in detecting scattering caused by desert/ice-capped surfaces led to false estimates of rainfall over these regions. RNC algorithms based on SI by *Grody* [1991] largely relied on regression relationships of microwave low-frequency channels, especially the 19 and 22 GHz. Vertical polarization measurements were preferred, as they resulted in smaller aliasing effects in the presence of mixed boundaries (e.g., coastlines). *Adler et al.* [1993] devised a global empirical relation for SSM/I to calculate the estimated value of T_b at 85 GHz (V) under nonrainy conditions ($T_{b,Estimated}$), using a fixed quantity of 243 K. Later on *Ferraro et al.* [1994] and *Ferraro and Marks* [1995] introduced the concept of using low-frequency channel combinations (10–37 GHz) to represent $T_{b,Estimated}$. Since the introduction of Grody-Ferraro screening methodology [*Ferraro et al.,* 1986, 1998; *Grody,* 1991], it has been the most applied technique for use in microwave land precipitation algorithms. The key idea in this technique is that radiation emitted from land surfaces is affected by ice particles and raindrops at high frequency 85 GHz T_b. Calculation of $T_{b,Estimated}$ involves simulation of 85 GHz T_b values for clear sky conditions (i.e., nonscattering condition). The difference between $T_{b,Estimated}$ and the observed 85 GHz T_b ($T_{b,Observed}$) gave a measure quantifying the degree of scattering by ice particles and raindrops, wherein the rain rate is proportional to the amount of scattering. As rainfall SI models offered an indirect and nonunique relation that varied from region to region, empirical relationships were largely employed between precipitation and SI to map rainfall over land surfaces [*Spencer et al.,* 1989; *Kidd and Barrett,* 1990; *Conner and Petty,* 1998; *Adler et al.,* 1994; *Dinku and Anagnostou* 2005].

As experience with SI-based studies grew, it became increasingly clear that a new suite of algorithms was necessary that efficiently modeled the value of $T_{b,Estimated}$ to suit the highly varying emissivity from the background land surface. Results of the ensuing development for overland regions are summarized in Table 1.1. Approaches involved using 85 GHz (H) channel instead of 85 GHz (V) channel to depict $T_{b,Observed}$ [owing to the failure of the first of SSM/I's 85.5 GHz (V) channel T_b] [*Adler et al.,* 1994; *Kummerow and Giglio* 1994], using channels of 19 GHz (V) and 22 GHz (V) to represent $TB_{Estimated}$ owing to their increased sensitivity to land surface emissivity. SI-based RNC classification techniques are currently being used for overland RNC classification embedded in

Table 1.1 Prominent scattering index based RNC methods

SI No:	Algorithm Proposed by	Observed T_b (K)	Estimated T_b (K)	SI Threshold (K)
1.	*Grody* [1991]	85(V)	$450.2 - 0.506 \times T_{b,19V} - 1.874 \times T_{b,22V} + 0.006 \times T_{b,22V}^2$	SI > 10
2.	*Adler et al.* [1994] (GSCAT)	85(H)	251	SI > 4
3.	*Kummerow and Giglio* [1994]	85(H)	$\text{Min}[T_{b,37H}, 265]$	SI > 0
4.	*Ferraro* [1997]	85(V)	$451.9 - 0.44 \times T_{b,19V} - 1.775 \times T_{b,22V} + 0.005 \times T_{b,22V}^2$	SI > 10
5.	*Ferraro* [1997]	37(V)	$62.18 + 0.773 \times T_{b,19V}$	SI > 5
6.	*Kummerow et al.* [2001] (GPROF)	85(V)	$T_{b,22V}$	SI > 8
7.	M1 [*Seto et al.,* 2005]	85(V)	μ	SI > $k_0\sigma$
8.	M2 [*Seto et al.,* 2005]	85(V)	$a + b \times T_{b,22V}$	SI > $k_0\sigma_e$

Source: Modified and adapted from *Seto et al.* [2005].

prominent algorithms such as GSCAT and GPROF algorithms, which are discussed in Sections 1.4.3 and 1.4.4.

1.4.2. Polarization-Corrected Temperature

Atmospheric hydrometeors have a depolarizing effect on microwave radiation that is emitted and reflected from a highly polarized surface [*Wu and Weinman*, 1984; *Huang and Liou*, 1983]. Therefore, polarization offers a great deal of information for separating the highly polarized radiances of the ocean from the essentially unpolarized radiances due to precipitation volume scattering [*Weinman and Guetter*, 1977]. *Spencer et al.* [1989] proposed an index comprised of linear combinations of vertical and horizontal polarizations to eliminate contrast between land and water/wet surfaces to yield a precipitation signal whose interpretation does not vary much depending on the background surface. The conceptual diagram of this index, known as polarization-corrected temperature (PCT) is shown in Figure 1.4.

The PCT relates the vertically and horizontally polarized T_b, and *Spencer et al.* [1989] described it as a measure of the distance from the no-scattering line. Earlier studies by *Spencer* [1986] noted that upon addition of nonscattering materials to the atmosphere above a nonraining, oceanic scan spot, the observed T_b will tend to move along the no-scattering line. When scattering materials (e.g., precipitation) gets introduced into the atmosphere, the point moves off the no-scattering line. As scattering lowers the T_b values, observations in which precipitation occurs will essentially fall between the no-scattering line and the no-polarization line. If $T_{b,\text{HCLF}}$ and $T_{b,\text{VCLF}}$ refer to the horizontally and vertically polarized cloud free ocean T_b respectively, $T_{b,\text{H}}$ and $T_{b,\text{V}}$ are the horizontally and vertically polarized T_b that are at least partially affected by any combination of clouds and precipitation, $T_{b,\text{VOLA}}$ and $T_{b,\text{HOLA}}$ are the vertically

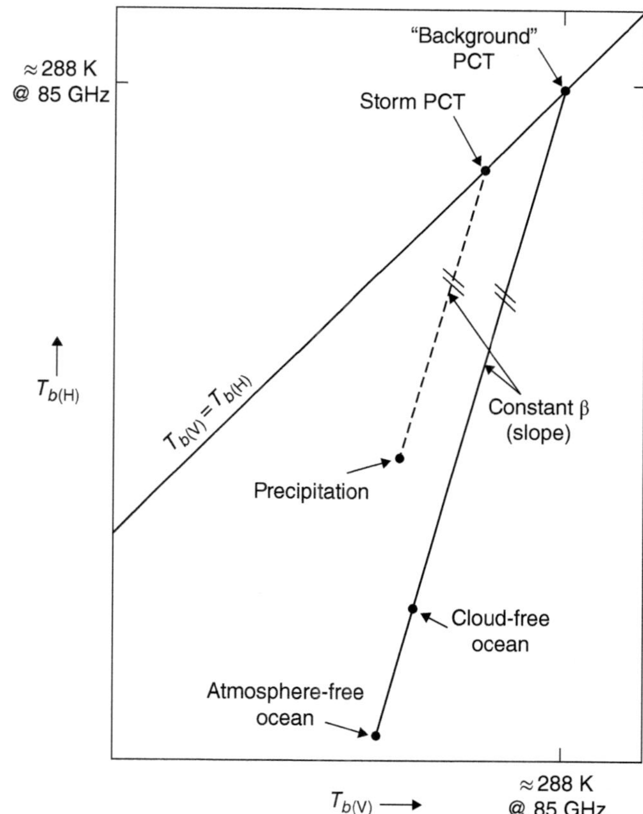

Figure 1.4 Schematic diagram of vertically and horizontally polarized T_b of ocean with and without an overlying atmosphere [*Spencer et al.*, 1989].

and horizontally polarized T_b of the ocean with no overlying atmosphere, then the expression for PCT is given by

$$\text{PCT} = \frac{\beta T_{b,\text{H}} - T_{b,\text{V}}}{\beta - 1}, \qquad (1.7)$$

where

$$\beta = \frac{T_{b,\text{VCLF}} - T_{b,\text{VOLA}}}{T_{b,\text{HCLF}} - T_{b,\text{HOLA}}}. \tag{1.8}$$

Calculation of PCT at any frequency requires a nearly constant value of β. Experiments using SSM/I observations of global cloud-free oceanic areas show that to obtain a physically meaningful value of PCT (between 275 and 290 K) the value of β should be 0.45. Absolute accuracy for β is not as important as keeping it constant in all subsequent calculations. Equation 1.7 for PCT can be rewritten as

$$PCT = 1.818T_{b,v} - 0.818T_{b,h} \tag{1.9}$$

PCT values using 85.5 GHz frequency channels will generally be lower than the background PCT, although its T_b depression will be much less than that due to the strong volume scattering effects of precipitation. Hence, this property of 85.5 GHz PCT is employed to detect cloud liquid water. Studies by *Mugnai et al.* [1993] using numerical model simulations demonstrated that 85 GHz signals represent emissions from upper-level liquid and ice scattering in the upper reaches of tall precipitation clouds. Therefore, PCT using 85.5 GHz channel is sensitive to precipitation top height for tall convection and surface rainfall for moderate convection. A modified relation for PCT was proposed by *Kidd and Barret* [1990] using SSM/I's 85.5 GHz channels as the basis for estimation of precipitation over both land and water. PCT is currently being used for RNC classification and the succeeding rainfall retrieval oceans and coastlines in the algorithm of the GSMaP, which is discussed in Section 1.4.4

1.4.3. Goddard Scattering Algorithm

The Goddard scattering algorithm (GSCAT) was first proposed by *Adler et al.* [1993]. Their study detected the existence of rainfall signature using an empirical logic tree applied to multiple channels. The algorithm relied on frequencies of 86 and 37 GHz (both in the horizontal polarization) to eliminate nonraining areas. This technique worked well over ocean and land areas but suffered from the inability to detect rain from clouds below the freezing level. Efforts were undertaken to modify the GSCAT RNC algorithm by including geographical screens (for deserts and snow-covered surfaces) similar to the work by *Grody* [1991]. *Adler et al.* [1994] modified this algorithm and included better quality control and use of lower frequency channels to differentiate cold surface and desert from precipitation. This differentiation was

essential when rainfall retrieval is to be made globally. *Adler et al.* [1994] and *Kummerow and Giglio* [1994] created GSCAT2, which used 85 GHz (H) instead of 85 GHz (V), to represent the value of $T_{b,\text{Observed}}$ and a constant value (251 K) for $T_{b,\text{Estimated}}$ without using any regression equations such as *Grody* [1991]. It was developed using channel information from SSM/I sensors. The methodology employed several checks, including the existence of cold ocean, coastline, desert, ice-covered regions, and ambiguous cold surface possible precipitation checks. These checks prevented surface effects that might lead to false identification of rain regions. GSCAT-2 proved to perform successfully in the SSM/I era to demarcate rain/no-rain regions, but not without some false rain identifications. The overall procedure for identifying raining pixels was not all that dissimilar from the scattering index by *Grody* [1991]. In an intercomparison study involving seven microwave techniques over Japan, *Lee et al.* [1991] showed that the GSCAT had the highest correlation with the Grody scheme during the convective regime in July–August 1989. The scattering signatures in GSCAT were used to retrieve rain intensity in proportion to the amount of scattering by ice and graupel aloft based on radiative transfer calculations applied to numerical cloud model results.

1.4.4. Goddard Profiling Algorithm

The GPROF algorithm is considered the established algorithm framework for microwave rainfall products from TRMM (launched in November 1997), Aqua (satellite of AMSR-E launched in May 2002), and included in the initial plans for the proposed Global Precipitation Measurement (GPM) mission (to be launched in 2014). GPROF follows separate sets of algorithms for RNC classification over land, ocean, and coastlines. Various versions of GPROF screening methodology have been implemented in SSM/I, TMI, and AMSR-E missions with an improved version to be applied in GPM mission [*McCollum and Ferraro*, 2003; *Wang et al.*, 2009; *Gopalan et al.*, 2010].

1.4.4.1. RNC Over Land

The RNC classification algorithm of GPROF for land regions [*Kummerow et al.*, 2001] assumes that T_b at 21.3 GHz (V) represents the nonscattering portion of T_b from 85 GHz (V). A scattering index threshold of 8 K is fixed to judge rainfall signature from a pixel/footprint. All pixels that exceed this threshold were identified as "possible rain" and were then processed using the full Bayesian algorithm to quantify the rain rate, which could be zero or nonzero. GPROF version 4 used the screening methodology of GSCAT 2 [*Adler et al.*, 1994]. Version 5 of GPROF [*Petty* 1994] employed polarization-based emission and scattering indices that could isolate signal coming from rain clouds with the background variability.

1.4.4.2. RNC Over Oceans

Over oceans, the predictable ocean surface emissivity offers contrast to the signals emanating from liquid hydrometeors over the range of microwave frequencies. Yet, the RNC detection technique of TRMM TMI usually fails over oceans to detect shallow rain observed by PR owing to the small scale of shallow rain when compared with the resolution of channels used in the emission-based algorithm. As clouds are optically thick at 85 GHz, it becomes very difficult to use the emission-based algorithm to detect shallow rains. Owing to the contrast between atmospheric liquid and low emissivity ocean surface, screening rainfall pixels over oceans relies on estimation of the liquid water path (LWP). The screening of GPROF over the ocean consists of two processes: checking the LWP and screening out clear ocean pixels and ice surface pixels. The flowchart for GPROF method over the ocean is shown in Figure 1.5.

In the first process, based on the study of *Karstens et al.* [1994], the LWP is checked using TMI low-resolution channels of 22 GHz (V) and 37 GHz (V) using the relation

$$LWP = 0.39 \log\left(285 - T_{b,22V}\right) - 1.40 \log\left(285 - T_{b,37V}\right) + 4.29. \tag{1.9}$$

All the footprints having LWP values less than the maximum LWP was classified under "no rain," where the value of maximum LWP (kg /m²) is based on the following relation:

$$LWP\_max = 0.25 * \left(\frac{FLH}{4000}\right). \tag{1.10}$$

Here the value of freezing level height (FLH) is derived from the work by *Wilheit et al.* [1991] and the values of 0.25 and 4000 represent the liquid water content and a typical FLH [*Wilheit et al.*, 1991].

The second process was based on the GSCAT algorithm [*Adler et al.*, 1994], which employed three checks. The first check employed threshold values for T_b from 22 GHz (V) and 85 GHz (H) channels to identify the target pixel as "possibly rain." This check was originally used to screen ice surfaces and "possible rain," but in the GPROF version 6 algorithm, this was utilized to detect rainfall signature. The second check aimed to distinguish ice surfaces from "rain" using T_b at 22 GHz (V). The third check was to identify clear ocean using T_b at 37 GHz (H) and 85 GHz (H). If the target pixel was not identified as ice surface/clear ocean by these checks, it was flagged as "possible rain." After the screening process, the footprints identified as "possible rain" are processed using the Bayesian algorithm to quantify rain rate.

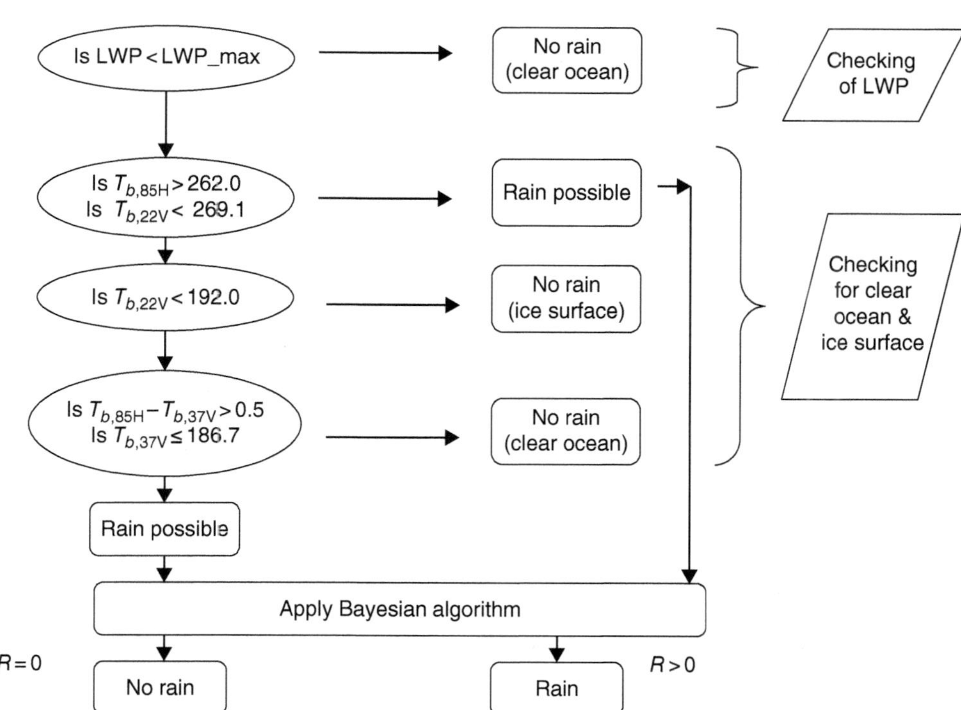

Figure 1.5 Flowchart for RNC method of GPROF over ocean [*Kida et al.*, 2009].

1.4.4.3. RNC Over Coasts

Rain identification over land or water involves checks for snow/desert surfaces, which tend to depress the high-frequency microwave channels. Coastlines are much more difficult as, for either water or land surfaces, adding the opposite surface into the footprint will have the same effect as rainfall. Over land, adding surface water to the footprint will reduce the T_b's, as does scattering caused by rain, and adding land to a water footprint will increase T_b's, similarly to rain over water, resulting in emission [*McCollum and Ferraro*, 2005]. Over coasts, microwave footprint is a combination of radiometrically warm land surface and cold ocean surface. One of the very first RNC algorithms developed for coasts was using SSM/I channels by *Adler et al.* [1994]. They proposed a complex decision tree method as shown in Figure 1.6, to isolate rainfall signature without using the SI method by *Grody* [1991]. More details about this method are available in *Huffman and Adler* [1993], which will hereinafter be referred to as HA93. The HA93 algorithm was implemented in GPROF and has remained in use in successive GPROF versions [*Wilheit et al.*, 2003].

Bennartz [1999] provided a technique to account for the coastline complexity, which involved use of effective antenna pattern function and scan geometry of the microwave instrument and high-resolution land-water mask to analyze the fraction of land versus water within each radiometer footprint. The study represented the atmospheric contribution from rain assuming a constant land T_b, while the land versus water fraction was used to weight the relative contribution of land-ocean surfaces to background signal. This technique was implemented to retrieve column water vapor for noncloud conditions using SSM/I data for the Baltic Sea region of western Europe. The study concluded that satellite navigation uncertainty created a dominant source of error. The assumption of a constant value to represent T_b from highly varying land surfaces fails in raining situations, for which cases the method of *Bennartz* [1999] cannot be implemented. The land surface emissivities have not yet been incorporated in the land component of GPROF rainfall algorithms. And, use of a straight cutoff for T_b values also cannot be implemented over coasts as water within the footprint reduces the high-frequency

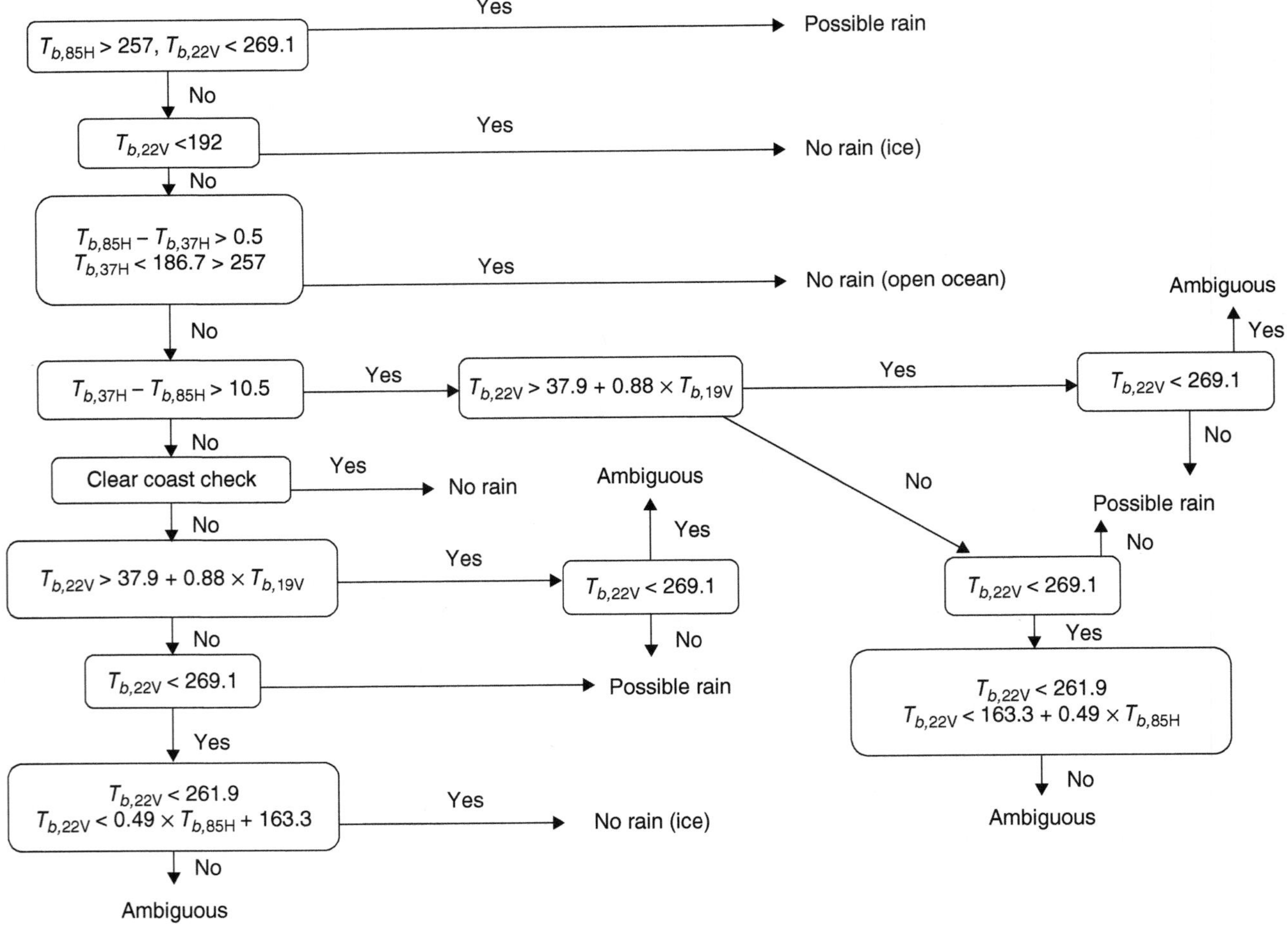

Figure 1.6 Decision tree for demarcating rainfall over coasts (HA93) [*McCollum and Ferraro*, 2005].

T_b's. As a result, combination of several criteria were examined to classify a footprint as having "no rain," "possibly rain," or "ambiguous." The criteria were mostly determined from the studies of *Grody* [1991] and *Adler et al.* [1993].

The HA93 decision tree method for RNC classification of coastline is summarized in figure 1.6. In this figure, the "clear coast check" by *Adler et al.* [1993] involves the following checks for T_b from 85 GHz (H) and 37 GHz (H):

$$\sigma\left(T_{b,85\mathrm{H}}\right) > 10K, \tag{1.11}$$

$$\rho\left(T_{b,37\mathrm{H}}, T_{b,85\mathrm{H}}\right) > 0.5, \tag{1.12}$$

$$\text{Slope} < 1.2, \tag{1.13}$$

where

$$\text{Slope} = \rho\left(T_{b,37\mathrm{H}}, T_{b,85\mathrm{H}}\right)\left[\frac{\sigma\left(T_{b,85\mathrm{H}}\right)}{\sigma\left(T_{b,37\mathrm{H}}\right)}\right],$$

and σ (standard deviation) and ρ (cross correlation) are computed on a 5×5 footprint array centered on the footprint of interest. This test identifies cases in which low humidity allows the (similar) surface emission signals from $T_{b,37\mathrm{H}}$ and $T_{b,85\mathrm{H}} / T_{b,89\mathrm{H}}$ to dominate the microwave signal [*McCollum and Ferraro*, 2005].

The HA93 algorithm added ambiguous classes for TB combinations for which rainfall rate was retrieved based on the requirement that another scheme be applied to estimate whether the retrieval was useful or an artifact. In case this could not be done, no estimate was made for such footprints in the "ambiguous" class, leaving "holes" in the resulting rainfall map. Footprints that were classified under "possible rain" were required to satisfy a cutoff threshold using higher frequency channels to be flagged as "rain" over land regions. The HA93 algorithm used $T_{b,85\mathrm{H}} < 257\,K$ criterion to assign a positive rain classification. The study conducted by *McCollum and Ferraro* [2005] using AMSR-E data suggested that a threshold using $T_{b,85\mathrm{H}}$ fails to provide a clear cutoff. The availability of TMI T_b's collocated with TRMM PR rainfall rates enabled choosing cutoff criteria that could efficiently separate raining from nonraining footprints. To summarize, the study by *McCollum and Ferraro* [2005] provided two major improvements to the existing RNC classification algorithm for coasts. The first step was to estimate conditions where positive rain rates should be estimated rather than leaving the areas without estimates as in the previous algorithm. Owing to the high correlation among the various TMI microwave channels, principal component analysis often provides a useful technique to separate signals of geophysical variables by the creation of mutually orthogonal statistically uncorrelated eigenvectors [*Conner and Petty*, 1998]. Therefore, the second step modified the cut-off threshold for rain/no-rain classification by using a PCT criterion instead of a straight T_b cut-off. These modifications were implemented in 2004 for the version 6 TMI product and third release of AMSR-E products with a slight difference for each product. The significant changes implemented for the latest version (version 7) of the GPROF TMI ocean algorithm involves addition of the probability of precipitation parameter, wherein pixels are not screened before Bayesian scheme. The algorithm developers recommend using 50% probability of rainfall threshold within the FOV when comparing with instantaneous PR and TMI rain rates. For the TMI coastal algorithm, a change in the land/ocean classification has been implemented [*Zagrodnik and Jiang*, 2013].

1.4.5. Global Satellite Mapping of Precipitation (GSMaP) Algorithm

The GSMap algorithm was developed by the Earth Observation Research Center, Japan Aerospace Exploration Agency (JAXA/EORC), and has been further improved with the use of PR measurements. Comparative studies by *Kummerow et al.* [2001] evaluated the performance of TRMM monthly rainfall estimates from both its sensors TMI and PR, which revealed a bias between both the rainfall products of nearly 30% over ocean and 26% over land (using version 5 of data products). The TRMM version 6 algorithms display improvements within level 2 surface rain retrieval algorithms based on physical principles. Results of intercomparison studies between version 5 and version 6 algorithms are presented in *Chiu et al.* [2006]. GSMaP is drawn up to the highest levels of precision and resolution with temporal resolution of 1 h and spatial resolution of 0.1°. The RNC algorithms implemented in GSMaP for over land, over ocean, and coastal regions are discussed below.

1.4.5.1. RNC Over Land

Seto et al. [2005] developed the RNC classification algorithm (version 4.5) for GSMaP that was employed in TRMM. Their study involved statistically summarizing all the TMI T_b values under no-rainfall conditions of PR 2A25, for the land regions into a database that represented both the spatial and temporal variations of T_b. This "land surface brightness temperature database" contained the spatiotemporal variations of T_b including the effects of sand and fallen snow [*Seto et al.*, 2005]. Due to the varied spatial resolutions of TMI channels among themselves as well as with PR, footprint size was defined by means of effective field of view (EFOV). PR footprints, the center of which lie within a

TMI footprint, were chosen as reference. In their study, all the PR observations within a TMI footprint that had a "no-rain" or "rain-possible" flag were adjudged to be in no-rain conditions. Their study summarized TMI observations under no-rain conditions in a database with resolution of 1 month and 1° latitude × 1° longitude. The distribution of T_b values under no-rain conditions was represented using a Gaussian distribution. The mean (μ) and standard deviation (σ) of 85 GHz (V) T_b were calculated to represent the distribution and stored in the database.

Seto et al. [2005] proposed two RNC methods (named as M1 and M2) for real-time use. The first method (M1) used the parameters estimated from the database of TMI T_b under no-rain conditions. The value for $T_{b,\text{Estimated}}$ was fixed as equivalent to μ and the threshold of scattering was judged at $k_0\sigma$ where k_0 was a constant in space and time. The thresholds for M1 differed with month and grid. This was an improvement over the threshold of *Adler et al.* [1994], which remained fixed at a constant value of 251 K. M2 considered a linear regression fit using least mean square error, between T_b (21.3 V) and T_b (85.5 V), both under no-rain conditions, using the database.

$$\left(T_{b,85.5V}\right)_{\text{NoRain}} \sim a + b\left(T_{b,21.3V}\right)_{\text{NoRain}} \qquad (1.14)$$

The subscript denotes observations conducted under no-rain / clear sky conditions. If σ_e is the standard deviation of residuals of equation and k_0 is a constant in both space and time, the pixel fulfilling the criterion of equation (1.11) is adjudged as containing a rainfall signature:

$$\left(T_{b,85.5V}\right)_{\text{Estimated}} - \left(T_{b,85.5V}\right)_{\text{Observed}} > k_0\sigma_e \qquad (1.15)$$

The number of rain pixels increased or decreased depending on the value of k_0, which varied with regions and seasons. The usual practice was to affix a constant value for k_0 for simplicity reasons. For GSMaP, the value of k_0 adopted was 3.5 and no desert/snow masks were employed as in *Grody* [1991]. The proposed RNC for version 4.7 was the same as that of version 4.5 with the only difference being in the retrieval part. These RNC methods are also known as PR-dependent methods as they cannot be applied to other microwave radiometers not accompanied by spaceborne precipitation radar. The methods (M1 and M2) were modified with an aim to make the RNC methods independent of PR so that these could be applied to data from other microwave radiometers as well. Comprehensive details regarding these can be found in *Seto et al.* [2009].

1.4.5.2. RNC Over Oceans

Over oceans, GSMaP adopted the method of *Kida et al.* [2009]. Their study employed two stages for detection of rain and no-rain footprints, as shown in Figure 1.7. In the first stage, deep rain pixels were

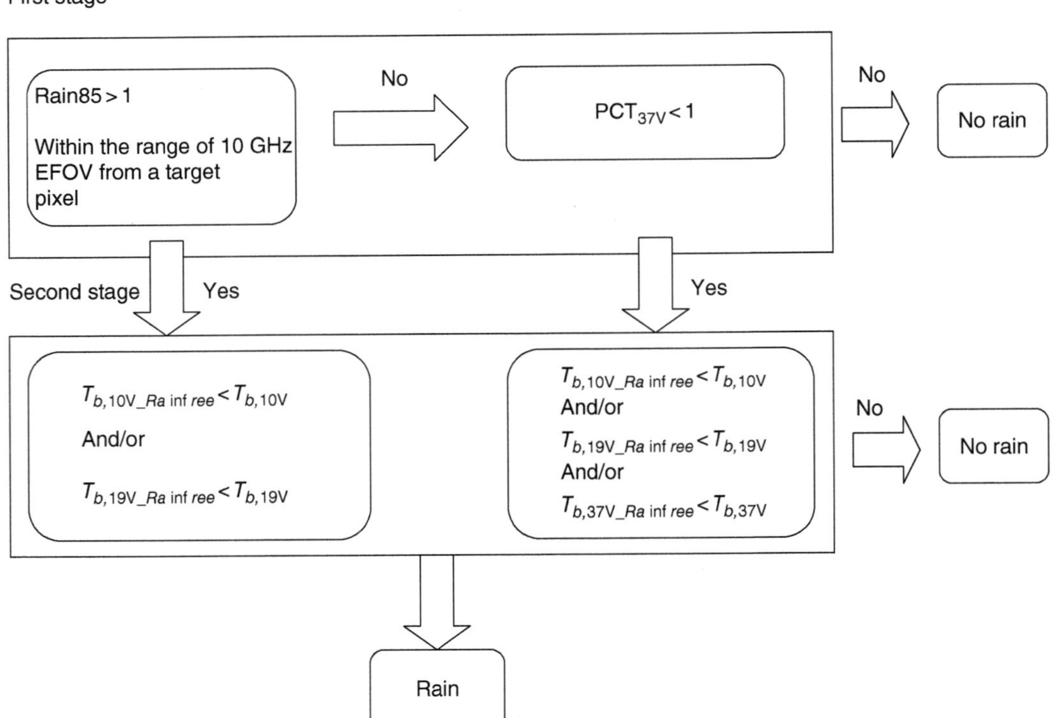

Figure 1.7 Flowchart for RNC classification used in GSMaP [*Kida et al.*, 2009].

determined by T_b from 85 GHz (V) scattering signature, and shallow rain pixels were determined with normalized polarization difference at 37 GHz (V). The study considered all the pixels of 85 GHz (V) lying within the EFOV of 10 GHz (V) pixel. The condition for detection of deep rain pixel was then fixed as the existence of one or more pixels of 85 GHz (V) within the EFOV of 10 GHz (V), having a rain rate > 1mm/h (rain 85 pixels). The study classified the target pixel (central pixel) as deep rain pixel upon fulfilling the above condition. If not, the normalized PCT [*Petty*, 1994] as given by equation 1.12 was used to check the existence of shallow rain:

$$ \mathrm{PCT}_{37V} = \frac{T_{b,37V} - T_{b,37H}}{T_{b,37V\_Ra\inf ree} - T_{b,37H\_Ra\inf ree}}. \qquad (1.16) $$

Based on the results of the first stage, their study checked emission signatures from raindrops in the second stage. The three checks used in the second stage are shown in Figure 1.7 from which it can be seen that T_b values of channels 19 GHz (V), 10 GHz (V), and 37 GHz (V) were used.

Kida et al. [2009] proposed modifications for the RNC method of GSMaP in order to use T_b from 37 GHz (V) more efficiently. Their study was essentially a PR-dependent method wherein the level 2 standard product 2A25 [*Iguchi*, 2007] was used as the validation product. Their study modified two conditions used in the first stage of GSMaP. In the original GSMaP algorithm,

for the first stage, in the presence of pixels whose rain 85 > 1 mm/h within the EFOV of 10 GHz (V), the central target pixel was identified as a deep rain pixel, even if it may actually be a shallow rain pixel. This led to misclassification of most of the shallow rain pixels as no-rain pixels (Figure 1.8). *Kida et al.* [2009] modified the first-stage algorithm by checking the rain rate of 85 GHz (V) T_b pixel just for the target pixel, to avoid misclassification of shallow rain pixels as deep rain pixels. The second modification was use of 37 GHz (V) T_b instead of normalized PCT to detect shallow rain. This was because T_b at 37 GHz (H) was known to be more sensitive to wind speed than T_b at 37 GHz (V). With increase in wind speed, T_b at 37 GHz (H) increases more than T_b at 37 GHz (V). And in the case of an extraordinary event such as a typhoon, characterized by strong wind speed, PCT (using 37 GHz) will be less than 1, leading to misclassification of shallow rain in windy regions. Hence, their study preferred Tb from 37 GHz (V) channel, which was less sensitive to wind speed variations.

1.4.5.3. RNC Over Coasts

Over coastal areas, GSMaP used the RNC algorithm proposed by *Kubota et al.* [2007]. Their study was an improvement over the RNC detection method of *McCollum and Ferraro* [2005], which detected precipitating areas using PCT index at 85 GHz (V) and a decision tree of several empirical conditions for TMI T_b's. *Kubota et al.* [2007] used the condition of surface temperature < 273.2 K for flagging no-rain pixels over coastal areas.

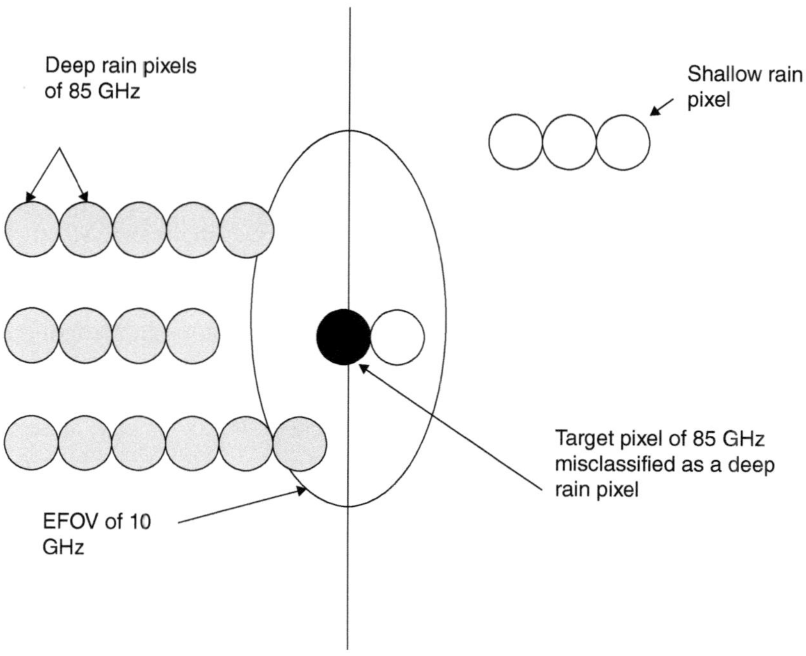

Figure 1.8 Example of a shallow rain pixel being misclassified as a deep rain pixel [*Kida et al.*, 2009].

The previously employed condition of $T_{b,85H} > 257$, $T_{b,22V} < 269.1$ leads to false rainfalls during the winter in midlatitude coastal areas and hence is avoided. *Kubota et al.* [2007] classified ambiguous class into "possible rain" and "no-rain" classes. Instead of using a PCT cutoff threshold, their study applied a scattering index threshold of RainPCT85 > 1, to possible rain cases. Here, rainPCT85 stands for rain rate from PCT index calculated using 85 GHz (V) channel frequency. Selection of a suitable threshold for RNC classification is as important as deriving the algorithm itself. For the GMSaP algorithm, *Kida et al.* [2008] proposed a parameterization of rain/no-rain threshold value of cloud liquid water as a function of storm height based on CloudSat precipitation product and the cloud liquid water derived from Aqua/AMSRE.

1.5. RNC PERFORMANCE ANALYSIS

RNC classification is a typical example for dichotomous classification having just two probabilities of either zero (denoting "no") or unity (denoting "yes") whose result can be expressed in 2×2 contingency matrix as shown in Table 1.2. In Table 1.2, the element a denotes the number of correct (or yes) forecasts of an observed event, c refers to the number of events that occurred but were not forecast, b is the number of forecasts of events that did not occur, and d is the number of correct forecasts of events that did not occur. It should be noted that Table 1.2 depicts dichotomous classification wherein collocated TMI and PR observations are analyzed assuming PR observations to be "true" or near perfect. Some studies increase the resolution of low-frequency channels by linear interpolation to match the resolution of 85 GHz (V) channels. Data collocation can then be performed using the geolocation information from TRMM PR and TMI data set to assign a TMI pixel at the 85 GHz (V) resolution as the nearest neighbor for every PR pixel in an orbit [*Gopalan et al.*, 2010; *Indu and Kumar*, 2013]. Data collocation has also been carried out by aggregating all of the PR observations within a TMI footprint before skill score calculations [*Seto et al.*, 2005]. This section discusses some of the statistical descriptors used to analyze performance of this binary classification. The commonly used indices for RNC accuracy assessment are summarized in Table 1.3.

The most commonly used indices are the probability of detection (POD) or prefigurance [*Panofsky et al.*, 1965] and the false alarm ratio (FAR). POD is the likelihood that an event would be estimated, given that it occurred, whereas FAR is an element of the conditional distribution of events given the estimate. Due to the negative orientation of FAR, smaller values indicate better estimates [*Wilks*, 1995]. Another measure used to compare the average estimate with the average observations is the frequency bias (B). B signifies the ratio of number of yes estimates to the number of yes observations. In this measure less than one indicates underestimation and greater than one indicates overestimation. The threat score (TS) or critical success index (CSI) indicates the number of correct yes estimates divided by the total number of occasions on which the event was estimated or observed [*Wilks*, 1995]. CSI has been widely used as a performance measure for rare events (as rainfall extremes) as it does not use the content of null events, as in POD and FAR [*Montero-Martinez et al.*, 2012]. These indices help to answer questions regarding meteorological aspects such as: (1) How reliable is a product in detecting precipitation? (obtained using POD); (2) How to quantify overall bias of satellite estimates using ground truth? (obtained using B); (3) How often does a product indicate precipitation during nonprecipitating scenarios? (obtained using FAR).

Performance analysis of binary classification can also be characterized using relative accuracy measures or skill scores. Skill scores quantify the agreement between forecast and observations [*Tartaglione*, 2010]. A skill score is the ratio of differences [*Stanski et al.*, 1989; *Wilks*, 1995] of scalar representations of the classification performance. An estimating scheme cannot be useful if it yields skill scores that can be obtained by less sophisticated estimating procedures [*Storch and Zwiers*, 1999]. It is to be noted that no single skill score can be used to indicate forecast skill. Hence, various skill scores or relative accuracy measures are derived from the contingency table. Different skill scores perform differently. Some of the skill scores used are Heidke skill score (HSS), Kuiper skill score (KSS), Gilbert skill score (GSS), and odd's ratio skill score (ORSS). The reference accuracy measure in the HSS [*Heidke*, 1926] is the hit rate of random estimates, subject to the constraint that the marginal distributions of estimates and observations characterizing the contingency table for the random estimates are the same as the marginal distributions in the actual verification data set [*Wilks*, 1995]. HSS is a generalized skill score that tends to eliminate classifications occurring purely due to chance. Thus, perfect classification yields HSS value of 1, which implies

Table 1.2 Layout of contingency matrix[a]

	Event Observed	
	Rain Judged by PR (Yes)	No Rain Judged by PR (No)
Event Forecast		
Rain judged by TMI (Yes)	a	b
No rain judged by TMI (No)	c	d

[a]Assuming that the rainflag by PR is near perfect and can be used as validation data.

Table 1.3 List of performance statistics used in the study

Serial No.	Performance Measure	Formula	Range
1.	Probability of detection (POD)	$\dfrac{a}{a+c}$	[0, 1]
2.	Threat score (TS) of Critical success index (CSI)	$\dfrac{a}{a+b+c}$	[0, 1]
3.	Miss rate (M)	$1-H$	[0, 1]
4.	False alarm ratio (FAR)	$\dfrac{b}{a+b}$	[0, 1]
5.	Heidke skill score (HSS)	$\dfrac{2(ad-bc)}{(a+c)(c+d)+(a+b)(b+d)}$	$[-\infty, 1]$
6.	Kuiper skill score (KSS)	$\dfrac{ad-bc}{(a+c)(b+d)}$	[−1, 1]
7.	Percent correct (PC) or hit rate (H)	$\dfrac{a+d}{n}$	[0, 1]
8.	Bias (B)	$\dfrac{a+b}{a+c}$	$[0, \infty]$
9.	Gilbert skill score (GSS)	$\dfrac{a-a_r}{a+b+c-a_r} \qquad a_r=\dfrac{(a+b)(a+c)}{n}$	[0, 1]
10.	Odds ratio skill score (ORSS)	$\dfrac{ad-bc}{ad+bc}$	[−1, 1]
11.	Log odds ratio	$\ln(a)+\ln(d)-\ln(b)-\ln(c)$	$[-\infty, \infty]$

that the performance of the classification is 100%. Another popular skill score, which was proposed by Pierce [1884] is the Kuiper skill score, known by many names such as true skill statistic [*Flueck, 1987*], Pierce skill score, Kuiper's performance index [*Murphy and Daan, 1985*], and so forth. KSS does not depend on climatological event frequency. Perfect classification results in a KSS value of 1, whereas inferior classification can even take negative values. The highlight of KSS is that both random and constant estimates receive a zero score. It also tends to zero if the event under consideration is rare in nature. Both HSS and KSS can be extended to multicategory estimate as well. Gilbert skill score, known as equitable threat score (ETS), was proposed by *Gilbert* [1884]. He proposed that, when the expected number of hits obtained by a random estimating system with the same estimate rate of occurrence as the actual estimating system is subtracted from both the numerator and denominator, it results in a measure known popularly as ETS or Gilbert skill score (GSS) [*Hogan et al.*, 2010]. GSS is widely used to operationally assess the performance in forecasting events over a range of thresholds [*Tartaglione*, 2010]. A perfect estimate gives GSS a value of 1, whereas a worst estimate gives GSS a value of 0. *Doswell et al.* [1990] has proved that ETS is monotonically but nonlinearly related to the

truly equitable Heidke skill score with the relation ETS = HSS/(2-HSS). However, GSS can be a useful index only if supplemented with additional information such as the frequency of occurrence of the event [*Mason, 1989*].

The odds or risk of an event happening, denoted by ORSS, is the performance index that can be subjected to statistical significance testing. ORSS is the ratio of probability that the event will happen to the probability of it not happening. This measure depends solely on the conditional joint probabilities and not on the marginal probabilities, and is therefore independent of any bias between observations and estimates. When the hit and false alarm rates are identical, odds ratio becomes unity. ORSS varies between +1 and −1, where a score of 1 indicates perfect skill and a score of 0 represents no skill. Negative values imply that the estimate was opposite to what was originally observed. *Stephenson* [2000] stated that associated variables that give odds ratio larger than unity can be tested for significance by considering the natural logarithm of the odds ratio referred also as "log odds," which is asymptotically Gaussian distributed. Agresti [1996] stated that when any one of the cell counts is zero, the asymptotic standard error in log odds becomes infinite and the odds ratio can no longer be meaningfully tested for significance. More information regarding validation of satellite-based precipitation estimation can be obtained from the research by the International Precipitation Working Group (IPWG) (*http://www.isac.cnr.it/~ipwg/IPWG.html*).

It should be noted that the performance of an RNC algorithm is strongly affected by the spatial and temporal scales for rainfall detection and the rainfall detection threshold. Identification of an optimum threshold for RNC classification depends on the missed rain amount, false rain amount, and most importantly on the user requirements. If the end product is to generate rain maps for flood warning, a lower threshold should be fixed for safety reasons. A larger threshold value is preferred if the resulting rain map is to be used for drought warning [*Seto et al.*, 2005]. Difference in spatial resolutions of multifrequency satellite microwave observations is a common problem in RNC classification and subsequent rainfall retrieval process. Usage of channels with larger FOV will incorporate more noise from outside the test area, thereby resulting in decreased accuracy. When the radiometer FOV is partially filled with cloud or rain, it results in beam filling effects that have been well documented in many studies [*Melitta and Katsaros*, 1995; *Greenwald et al.*, 1997; *Bremen et al.*, 2002; *Chiu et al.*, 1990; *Short and North*, 1990; *Kummerow et al.*, 1998]. Some studies have used techniques to deconvolve the TMI T_b's at 10 GHz and convolve at 21–85 GHz to a common FOV corresponding to 19 GHz (H) FOV [*Backus and Gilbert*, 1970]. A resampling technique is applied to the TMI data to perform this operation using a linear combination of nearby observations using knowledge regarding TMI antenna patterns and scan geometry. The crucial points to be considered are the parallax effects between the instruments (e.g., TMI and PR) and their antenna pattern, the inclusion of signals from outside the EFOV if footprint is defined using EFOV, beam filling effects, and the like [*Bauer et al.*, 2002]. Studies showing temporal variations of FAR and POD for various RNC algorithms were presented in *Seto et al.* [2005]. Their study revealed that diurnal variations in the physical temperature of land surface cause T_b values of no-rain pixels to be lower at night. This directly implies that at night, a constant rainfall detection threshold for T_b will tend to produce a higher FAR.

1.6. OPEN QUESTIONS

The ultimate goal of RNC classification is to produce the most accurate detection of rainfall signature from microwave footprints over any background surface (land, ocean, coast). Since the accuracy of RNC algorithms is tied to the characteristic properties of microwave data in multiple frequency channels, the need to understand and improve quality of information retrieved from various channels is obvious. There remain several open questions that need to be addressed such as uncertainty due to RNC, inability of present algorithms to detect light rain, and so forth. Here, we outline some important questions and possible future research directions.

1.6.1. Resampling/Matchup Errors

The distribution of microphysical parameters, precipitation cells, cloud cover, and associated volume absorption and scattering coefficients are subjected to large variability within a microwave radiometer footprint [*Smith and Kidder*, 1978; *Austin and Geotis*, 1978; *Mugnai and Smith*, 1988]. For quantitative assessment of rain/no-rain classification, by far the biggest issue is the nonuniform spatial resolution wherein diffraction limits the attainable resolution. Microwave footprints are characterized with large horizontal resolution (FOV). And when information from multiple frequencies having differing resolutions need to be combined to retrieve geophysical parameters, which are not necessarily homogeneous over the footprint area, the problem of resolution mismatch becomes critical. The literature presents several approaches to solve this issue in the form of simple averaging of high-resolution measurements to match the low-resolution data, matching the resolution of all the frequencies to a common 25 km spatial resolution. *Backus and Gilbert* [1970] proposed a technique for inverting seismic data in order to retrieve Earth density profiles. This technique reconstructed measurements at different resolutions from those originally sampled using a combination of oversampled measurements. Theory by *Backus and Gilbert* [1970] is being largely employed in atmospheric remote sensing applications to convolve microwave measurements to a common resolution [*Hollinger et al.*, 1987]. Another common approach being followed is averaging of inhomogeneous retrieved parameters from data of differing resolutions. However, the accuracy of retrieval is strongly influenced by input data resolution, which if not done carefully tends to incorporate errors. Further, due to the varying FOV sizes and spectral bands used by different microwave radiometers, some of the thresholds used for RNC screening are unreliable, which tends to get aggravated at the interfaces between surface types (i.e., coastline) and results in "ambiguous" classification in the resulting rainfall product. Before undertaking experiments using the plethora of available microwave products, it would be worthwhile to understand these effects.

1.6.2. Drawback of Modeling RNC Using Database

Many authors utilize a statistical database of T_b under no-rain conditions to model the nonscattering portion of 85 GHz (V) instead of using a set of sensitive channel combinations. The creation of such a consistent database requires TRMM-like satellites without which the databases cannot be produced globally. The issue is that a database built on TRMM observations cannot be applied to areas beyond TRMM coverage area. Moreover, satellites such as GPM and MT have slightly different channel frequencies than TRMM TMI (85.5 GHz for TMI, 89.0 GHz for AMSR-E) and therefore a previously created database would need to be adjusted to suit each sensor. To overcome this problem, radiative transfer modeling of land surfaces is essential.

1.6.3. Assumption of Marshall Palmer Size Distribution for Ice

Atmospheric hydrometeors are generally nonspherical [*Pruppacher and Klett*, 1978]. The alteration of T_b measurements due to hydrometeor size, phase, layer depth, and other factors is quantified by *Mie* [1908] theory. Ice particles are complicated with unknown particle sizes, densities, and shapes, all affecting T_b. The assumption of Marshall-Palmer size distribution for ice is a result of convenience, due to the unavailability of direct measurements of ice particle sizes within convective storms. Even if the measurements existed, modeling using radiative transfer becomes highly difficult. This is why we rely on empirical relationships. For overland regions, scattering-based modeling is not very well understood. Ice particles actually found within a rainstorm have highly varying shapes and sizes that depend on atmospheric temperature and humidity. Ice particle shapes offer a crucial parameter for single scattering calculations that affects emerging radiance field [*Mugnai and Wiscombe*, 1986; *Bohren*, 1986]. Yet they remain modeled as an ensemble of spheres, despite the evidence that oriented ice particles can cause polarization difference of up to 10 K in stratiform regions associated with strong convection. When the particles causing scattering are small in size compared to the wavelength of observation, scattering is more or less assumed to be well approximated by that of a sphere of equal mass, but the actual electromagnetic solutions are completely intractable for all but a very few shapes. This area still needs to be addressed.

1.6.4. Additional Factors

Microwave sensors are relatively insensitive to low-altitude liquid-water-bearing clouds. There are no specific screens developed to reliably separate nonraining and raining clouds. Moreover, defining a suitable threshold to detect raining clouds so that it suits every climatic regime is nearly impossible due to the lack of corresponding validation data. The usual procedure for selecting a threshold is to either detect even the lightest rain at the cost of misclassifying nonraining clouds or to detect rain only above a certain threshold at the expense of eliminating light-rain pixels. The choice of either a liberal or a conservative threshold affects the accuracy of RNC and this area needs to be researched further.

Also, most of the overland precipitation retrieval algorithms rely on a fixed land-sea-coast database that might not exactly match up with what the instrument (i.e., radiometer) is viewing as it enters the decision tree classification scheme. Ambiguity is further offered to the retrieval process due to the existence of inland water bodies [*Sudradjat et al.*, 2010]. Over oceans, RNC classification is affected by strong winds that tend to impact the emitting temperature and polarization signature especially at low microwave frequencies. This in turn causes misclassification of the area as rain where no rain is present. When ocean surface is clear and calm, the emissivity will be at its lowest. In the presence of a dry atmosphere, very low T_b values can result that may again cause erroneous screening of raining area. These pose serious problems for 85 GHz scattering based techniques using single polarization temperature thresholds to delineate rain areas.

The scattering properties of each surface type are different. A database consisting of scattering properties of different land surface covers can be generated to create different SI-based RNC algorithms sensitive for each surface type. Change in land use land cover (such as deforestation) that affect scattering signatures of land covers need to be accounted. Usage of ancillary information for land surface types might improve the rain detection over problematic surfaces such as desert, semiarid regions, coastal, and inland regions. In order to develop more physically based approaches for overland precipitation retrieval, all these problematic issues need to be addressed.

1.7. CONCLUSIONS

Rainfall screening methodologies have progressed with the evolution of new microwave sensors. Over the years, RNC studies have evolved from simple relations formulated for computational expedience to more elaborate computer-intensive schemes that effectively discriminate rainfall occurrence. This chapter discussed the principles of rain/no-rain (RNC) algorithms that have been developed through the years processing data from satellite-borne passive microwave radiometers. Any screening algorithm would depend on the characteristic interaction of atmospheric constituents with the microwave frequency energy. Hence, we summarized the interaction of microwaves with major sources and sinks of microwave frequency energy such as water vapor, cloud water, cloud ice, precipitation, and the like. Two different approaches to RNC classification were presented: one based on scattering signatures of atmospheric constituents at high microwave frequencies and the other based on emission-based techniques for over ocean rainfall delineation. We focused on the major rainfall screening techniques used for the different backgrounds of land, ocean, and coastlines. Quantifying the uncertainty caused in overland rainfall retrieval due to improper screening of raining pixels is not a trivial task owing to cluttering of microwave brightness temperature by varying overland emissivity, beam filling errors, resolution matchup errors, and the like. Often the rainfall products suffer from underestimation of warm rainfall due to low-level liquid-water-bearing clouds that go undetected by a microwave radiometer. Over the years, overland RNC algorithms have been continuously evolving to bridge the gap in bias caused between the passive and active microwave rainfall products. Such studies have made significant contributions to improve the existing RNC techniques used in prominent algorithms such as the global satellite mapping (GSMaP) and Goddard profiling algorithm (GPROF). Many recent studies have propagated the idea of RNC classification to land surfaces of small areal extent. Challenges for RNC classification include detecting shallow rainfall, improving accuracy through derivation of global threshold values, addressing the nonhomogeneity of geophysical parameters within the large field of view of microwave footprints, and using RNC screening techniques independent of PR that can be applied to satellites outside the field of view of TRMM. Future work in these directions will aim to create global RNC classification algorithms using scattering properties of different land cover types, which can be globally applied for sensors of slightly differing fields of view.

ACKNOWLEDGMENTS

The authors thank Dr. G. Huffman from NASA Goddard Space Flight Center and Dr. V. Lakshmi from University of South Carolina for invitation of this study. The authors wish to thank the two anonymous reviewers of this manuscript for their insightful suggestions, which significantly improved the text.

REFERENCES

Adler, R. F., H.-Y. M. Yeh., N. Prasad, W.-K. Tao, and J. Simpson (1991), Microwave simulations of a tropical rainfall system with a three dimensional cloud model, *J. Appl. Meteor.*, *30*, 924–953.

Adler, R., A. Negri, P. Keehn, and I. Hakkarinen (1993), Estimation of monthly rainfall over Japan and surrounding waters from a combination of low-orbit microwave and geosynchronous IR data, *J. Appl. Meteorol.*, *32*, 335–356, doi:10.1175/1520-0450(1993)032<0335:EOMROJ>2.0.CO;2.

Adler, R., G. J. Huffman, and R. R Keehn (1994), Global tropical rain estimates from microwave-adjusted geosynchronous IR data, *Remote Sens. Rev.*, *11*, 125–152.

Agresti, A. (1996), An Introduction to Categorical Data Analysis, New York, USA, Wiley Interscience Publication.

Anagnostou, E. N., and C. Kummerow (1997), Stratiform and convective classification of rainfall using SSM/I 85-GHz brightness temperature observations., *J. Atmos. Oceanic Technol.*, *14*, 570–575.

Austin, P. M., and S. G. Geotis (1978), Evaluation of the quality of precipitation data from a satellite borne radiometer. Final Report, NASA Contract NSG-5024., MIT, Cambridge, MA.

Backus, G., and F. Gilbert (1970), Uniqueness in the inversion of inaccurate gross earth data, *Philos. Trans. Roy. Soc. London*, *A266*, 123–192.

Barrett, E. C., and D. W. Martin, (1981), *The Use of Satellite Data in Rainfall Monitoring*, Academic, San Diego.

Barrett, E. C., C. Kidd, and Bailey J. O., (1988), The special sensor microwave imager: a new intstrument with rainfall monitoring potential, *Int. J. Remote Sensing*, *9*, 1943–1950.

Bauer, P (2001), Including a melting layer in microwave radiative transfer simulation for clouds., *Atmos. Res.*, *57*, 9–30.

Bauer, P., J. F. Mahfouf., W. S. Olson., F. S. Marzano, F. S., S. Di Michele., A. Tassa., and A. Mugnai (2002), Error analysis of TMI rainfall estimates over ocean for variational data assimilation, *Q. J. R. Meteorol. Soc.*, *120*, 1367–1388.

Bauer, P., and P. Schluessel (1993), Rainfall, total water, ice water and water vapor over the sea from polarized microwave simulations and SSM/I data., *J. Geophys. Res.*, *98*(D11), 737–759.

Bennartz, R. (1999), On the use of SSM/I measurements in coastal regions, *J. Atmos. Oceanic Technol.*, *18*, 1838–1855.

Biscaro, T. S., and C. A. Morales (2007), Continental passive microwave based rainfall estimation algorithm: Application to the Amazon basin., *J. Appl. Meteor. Climatol.*, *47*, 1962–1981.

Bohren, C. F (1986), Absorption and scattering of light by non-spherical particles. Extended abstracts, Sixth Conf. on Atmos. Rad., Williamsburg, VA, Amer. Meteor. Soc., pp. 1–7.

Bremen, L. V., E. Ruprecht, and A. Macke (2002), Errors in liquid water path retrieval arising from cloud inhomogeneities: The beam-filling effect. *Meteor. Z.*, *11*, 13–19.

Burke, W. J., T. Schmugge, and J. F. Paris (1979), Comparison of 2.8 and 28 cm microwave radiometer observations over soils with emission model calculation., *J. Geophys. Res.*, doi:10.10 29/0JGREA0000840000C1000287000001.issn:0148-0227.

Chandrasekhar, S. (1960), *Radiative Transfer*, Oxford University Press, Oxford.

Chelton, D. B., and M. H. Freilich (2005), Scatterometer-based assessment of 10m wind analyses from the ECMWF and NCEP numerical weather prediction models, *Mon. Wea. Rev.*, *133*, 409–429.

Chiu, L. S., G. R. North, D. A. Short, and A. McConnell (1990), Rain estimation from satellites: Effect of finite field of view, *J. Geophys. Res.*, *95*, 2177–2185.

Chiu, L. S., D.-B. Shin, and J. Kwaitkowski (2006), *Surface rain rates from Tropical Rainfall Measuring Mission satellite algorithms, Earth Science Satellite Remote Sensing*, Vol. I, edited by J. Qu, *et al.*, Springer-Tsinghua University Press, Berlin, pp. 317–336.

Conner, M. D., and G. W. Petty (1998), Validation and intercomparison of SSM/I rain-rate retrieval methods over the continental United States, *J. Appl. Meteor.*, *37*, 679–700.

Coppens, D., Z. S. Haddad, and E. Im (2000), Estimating the uncertainty in passive microwave rain retrieval. *J. Atm. Oceanic. Technol.*, *17*, 1618–1629.

Derr, V. E. (1972), *Remote Sensing of the Troposphere*, U.S. Govt. Printing Office, Washington, D.C.

Dinku, T., and E. N. Anagnostou (2005), Regional differences in overland rainfall estimation from PR-calibrated TMI algorithm, *J. Appl. Meteor.*, *44*, 189–205.

Doswell, C. A. III., R. Davies-Jones., and D. L. Keller (1990), On summary measures of skill in rare event forecasting based on contingency tables, *Weath. Forecasting.*, *5*, 576–586.

Elachi. C. (1987), *Introduction to the Physics and Techniques of Remote Sensing*, pp. 32–37, Wiley, New York.

England, A. W. (1976), Relative influence upon microwave emissivity of fine-scale stratigraphy, internal scattering and dielectric properties, *Pure Appl. Geophys.*, *114*, 287–299.

English, S. J. (1999), Estimation of temperature and humidity profile information from microwave radiances over different surface types, *J. Appl. Meteor.*, *38*(10), 1526–1541.

Evans. K. F., J. Turk, T. Wong, and G. L. Stephens (1995), A Bayesian approach to microwave precipitation profile retrieval, *J. Appl. Meteor.*, *34*, 260–279.

Ferraro, R. R. (1997), Special sensor microwave imager derived global rainfall estimates for climatological applications, *J. Geophys. Res.*, *102*, 16715–16735.

Ferraro, R. R., and G. F. Marks (1995), The development of SSM/I rain rate retrieval algorithms using ground based radar measurements, *J. Atmos. Oceanic Technol.*, *12*, 755–770, doi:10.1175/1520-0426(1995)012<0755:TDOSRR>2.0.CO;2.

Ferraro, R. R., N. C. Grody, and J. A. Kogut (1986), Classification of geophysical parameters using passive microwave satellite measurements, *IEEE Trans. Geosci. Remote Sens.*, *24*, 1008–1013.

Ferraro, R. R., N. C. Grody, and G. G. Marks (1994), Effects of surface conditions on rain identification using the DMSP-SSM/I, *Remote Sens. Rev.*, *11*, 195–209.

Ferraro, R. R., F. Weng, N. C. Grody, and A. Basist (1996), An eight-year (1987–1994) time series of rainfall, clouds, water vapor, snow cover, and sea-ice derived from SSM/I measurements, *BAMS*, *77*, 891–905.

Ferraro, R. R., E. A. Smith, W. Berg, and G. J. Huffman (1998), A screening methodology for passive microwave precipitation retrieval algorithms, *J. Atmos. Sci.*, *55*, 1583–1600. doi:10.1175/1520-

Ferraro, R. R., E. A. Smith, W. Berg, and G. Huffman (2005), NOAA operational hydrological products derived from the AMSU, *IEEE Trans. Geosci. Remote Sens.*, *43*, 1036–1049.

Ferrazoli, P., J. P. Wigneron, L. Guerriero, and A. Chanzy (2000), Multifrequency emission of wheat modeling and application, *IEEE Trans. Geosci. Remote Sens.*, *38*, 2598–2607.

Fraser, R. S. (1975), Interaction mechanisms—within the atmosphere, in *Manual of Remote Sensing*, Vol. 1, pp. 181–233, edited by F. J. Janza, *Theory, Instruments and Techniques*, American Society of Photogrammetry, Falls Church, Va.

Fulton, R., and G. M. Heymsfield (1991), Microphysical and radiative characteristics of convective clouds during COHMEX, *J. Appl. Meteor.*, *30*, 98–116.

Fung, A. K. (1994), *Microwave Scattering and Emission Models and Their Applications*, Artech House, pp. 277–303, Boston.

Gilbert, G. F. (1884). Finley's tornado predictions. *Amer. Meteor. J.*, *1*, 166–172.

Gopalan, K., N.-Y. Wang, R. Ferraro, and C. Liu (2010), Status of the TRMM 2A12 land precipitation algorithm, *J. Atmos. Oceanic Technol.*, *27*, 1343–1354.

Gorelik, A. G., V. V. Kalashnikov, B. G. Kutuza, and V. I. Semiletov (1971), Measurements of space distribution of brightness temperatures of clouds and rain at 0.8 and 1.35 centimeter wavelengths, *Advances in Satellite Meteorology*, Wiley, New York.

Grecu, M., and E. N. Anagnostou (2001), Overland precipitation estimation from the TRMM passive microwave observations, *J. Appl. Meteor.*, *40*, 1367–1380.

Greenwald, T. J., S. A. Christopher, and J. Chou, (1997), Cloud liquid water path comparisons from passive microwave and solar reflectance satellite measurements: Assessment of sub-field-of-view cloud effects in microwave retrievals, *J. Geophys. Res.*, *102*, 19,585–19,596.

Grody, N. C. (1991), Classification of snow cover and precipitation using the Special Sensor Microwave Imager, *J. Geophys. Res.*, *96*, 7423–7435.

Guillou, C., S. J. English, C. Prigent, and D. C. Jones (1996), Passive microwave airborne measurements of the sea surface response at 89 and 157 GHz, *J. Geophys. Res.*, *101*(C5), 3775–3788.

Gunn, K. L. S., and T. W. R. East (1954), The microwave properties of precipitation particles, *Q. J. Roy. Meteorol. Soc.*, *80*, 522–545.

Hansen, J. E., and L. D. Travis (1974), Light scattering in planetary atmospheres, *Spa. Sci. Rev.*, *16*, 527–610.

Heidke, P. (1926), Berechnung des Erfolges und der Gute der Windstarkevorhersagen in Sturnwarnungsdienst, *Geografiska Annaler*, *8*, 301–349.

Hogan, R. J., C. A. T. Ferro, I. T. Jolliffe, and D. B. Stephenson (2010), Equitability revisited: Why the "equitable threat score" is not equitable, *Weather Estimating*, *25*(2), 710–726, doi:10.1175/2009WAF2222350.1.

Hollinger, J., R. Lo., G. Peo., R. Savage., and J. Pierce (1987), *The Special Sensor Microwave/Imager's User's Guide, Naval Research Laboratory Technical report*, pp. 1–120, Nav. Res. Lab., Washington, D. C.

Huang, R., and K. N. Liou (1983), Polarized microwave radiation transfer in precipitating cloudy atmospheres: Applications to window frequencies, *J. Geophys. Res.*, *88*, 3885–3893.

Huffman, G. J., and R. F. Adler (1993), Precipitation estimation from SSM/I data with the Goddard Scattering Algorithm. Proc. Shared Proceeding Network SSM/I Algorithm Symp., Monterey, CA, Fleet Numerical Oceanography Center [available from Dudley Knox Library, Naval Postgraduate School, 411 Dyer Rd., Monterey, CA 93943].

Iguchi, T. (2007), Space-borne radar algorithms, in *Measuring Precipitation from Space-EURAINSAT and the Future*, edited by V. Levizzani, P. Bauer, and F. J. Turk, pp. 199–212, Springer, New York.

Indu, J. and D. N. Kumar (2013), Copula based modeling of TRMM TMI brightness temperature with rainfall type, *IEEE Trans. Geosci. Remote Sens*, doi:10.1109/TGRS.2013.2285225.

Janssen, M. A. (1993), *Atmospheric Remote Sensing by Microwave Radiometry*, Wiley, New York.

Karstens, U., C. Simmer, and E. Ruprecht (1994), Remote sensing of cloud liquid water, *Meteor. Atmos. Phys.*, *54*, 157–171.

Kida, S., S. Shige, T. Manabe, T. S. L'Ecuyer, and G, Liu (2008), Validation of rain/no-rain threshold value of cloud liquid water for microwave precipitation retrieval algorithm using CloudSat precipitation product, *Proc. SPIE7152, Remote Sensing of Atmosphere and Clouds II*, 715209, doi:10.1117/12.804924.

Kida, S., S. Shige, T. Kubota, A. Aonashi, and K. Okamoto (2009), Improvements on rain/no-rain classification methods for microwave radiometer observations over the ocean using a 37 GHz emission signature, *J. Meteor. Soc. Jpn.*, *87A*, 165–181.

Kidd, C. and E. C. Barrett (1990), The use of passive microwave imagery in rainfall monitoring, *Remote Sens. Rev.*, *4*, 415–450.

Kidder, S. Q., and T. H. Vonder Haar (1995), *Satellite Meteorology-An Introduction, Academic*, San Diego.

Kidder, S. Q., and T. H. Vonder Haar (1977), Seasonal oceanic precipitation frequencies from Nimbus 5 microwave data, *J. Geophys. Res.*, *82*, 2083–3086.

Klein, L. A., and C. T. Swift (1977), An improved model for the dielectric constant of sea water at microwave frequencies, *IEEE J. Oceanic Eng.*, *2*, 104–111.

Kubota, T., et al. (2007), Global precipitation map using satellite-borne microwave radiometers by the GSMaP project: production and validation, *IEEE Trans. Geosci. Remote Sens.*, *45*, 2259–2275.

Kummerow, C. (2001), The evolution of the Goddard Profiling Algorithm (GPROF) for rainfall estimation from passive microwave sensors, *J. Appl. Meteorol.*, *40*, 1801–1820, doi:10.1175/1520-0450(2001)040<1801:TEOTGP>2.0.CO;2.

Kummerow, C., and J. A. Weinman (1988), Determining microwave brightness temperatures from precipitating horizontally finite and vertically structured clouds, *J. Geophys. Res.*, *93*, doi: 10.1029/88JD01623.

Kummerow, C., and L. Giglio (1994), A passive microwave technique for estimating rainfall and vertical structure information from space. Part I: Algorithm description, *J. Appl. Meteor.*, *33*, 3–18.

Kummerow, C., W. S. Olson., and L. Giglio (1996), A simplified scheme for obtaining precipitation and vertical hydrometeor profiles from passive microwave sensors, *IEEE Trans. Geosci. Remote Sens.*, *34*, 1213–1232, doi:10.1109/36.536538.

Kummerow C., W. Barnes T. Kozu J. Shiue, and J. Simpson (1998), The Tropical Rainfall Measuring Mission (TRMM) sensor package, *J. Atmos. Oceanic Technol.*, *15*, 809–817, doi:10.1175/1520-0426(1998)015<0809:TTRMMT>2.0.CO;2.

Kummerow, C., W. S. Olson., and L. Giglio (2001), The evolution of the Goddard Profiling Algorithm (GPROF) for rainfall estimation from passive microwave sensors, *J. Appl. Meteor.*, *40*, 1801–1820.

Kummerow, C., W. Berg, J. T. Stahle, and H. Masunaga (2006), Quantifying global uncertainties in a simple microwave rainfall algorithm, *J. Atmos. Oceanic. Technol.*, *23*, 23–37.

Lee, T. H., J. E. Janowiak., and P. A. Arkin (1991), *Atlas of Products from the Algorithm Intercomparison Project 1: Japan and Surrounding Oceanic Regions*, June - August 1989, University Corporation for Atmospheric Research, Washington, DC.

Levizzani, V., and A. Gruber (2007), The International Precipitation Working Group: A bridge towards operational applications, in *Measuring Precipitation from Space: EURAINSAT and the Future*, edited by Levizzani, Turk and Bauer, pp. 705–712, Springer, New York.

Lin, X., and A. Y. Hou (2008), Evaluation of coincident passive microwave rainfall estimates using TRMM PR and ground measurements as references, *J. Appl. Meteor. Climatol.*, *47*, 3170–3187.

Liou, K.-N (1980), *An Introduction to Atmospheric Radiation*, Academic, New York.

Lovejoy, S., and G. L. Austin (1980), The estimation of rain from satellite-borne microwave radiometers, *Quart. J. Roy. Meteor. Soc.*, *106*, 255–276.

Marzano F. S., A. Mugnai, E. A. Smith, X. Xiang, J. Turk, and J. Vivekanandanan (1994), Active and passive remote sensing of precipitating storms during CaPE. Pan II: Intercomparison of precipitation retrievals from AMPR radiometer and CP-2 radar, *Meteorol. Atmos. Phys.*, *10*, 29–54.

Mason I. B (1989), Dependence of the critical success index on sample climate and threshold probability, *Aust. Meteor. Mag.*, *37*, 75–81.

McCollum, J., and R. R. Ferraro (2003), The next generation of NOAA/NESDIS SSM/I, TMI and AMSR-E microwave land rainfall algorithms, *J. Geophys. Res.*, *108*, 8382–8404.

McCollum, J., and R. R. Ferraro (2005), Microwave rainfall estimation over coasts, *J. Atmos. Ocean. Technol*, *22*, 497–512.

Melitta, J., and K. B. Katsaros (1995), Using coincident multi-spectral satellite data to assess the accuracy of special sensor microwave imager liquid water path estimates, *J. Geophys. Res.*, *100*, 16,333–16,339.

Mie, G. (1908), Beitrage zur optic truber median, speziell kolloidaler metallosunger (Contribution on the optics of scattering media, special colloidal metal solutions), *Ann. Phys.*, *25*, 377–445.

Montero-Martinez, G., V. Zarraluqui-Such, and F. Garcia-Garcia (2012), Evaluation of 2B31 TRMM product rain estimates for single precipitation events over a region with complex topographic features, *J. Geophys. Res.*, *117*, D02101, doi:10.1029/2011JD16280.

Mugnai A., and E. A. Smith (1988), Radiative transfer to space through a precipitation cloud at multiple microwave frequencies. Part I: Model description, *Am. Meteor. Soc.*, 1055–1073.

Mugnai A., and W. J. Wiscombe (1986), Scattering from non-spherical Chebyshev particles. Part I: Cross sections, single-scattering albedo, asymmetry factor, and backscattered fraction, *Appl. Opt.*, *25*, 1235–1244.

Mugnai A., E. A. Smith, and G. J. Tripoli (1993), Foundations for statistical-physical precipitation retrieval from passive microwave satellite measurements. Part II: Emission source and generalized weighing function properties of a time-dependent cloud-radiation model, *J. Appl. Meteor.*, *32*, 17–39.

Murphy, A. H., and H. Daan (1985), Forecast evaluation. *Probability Statistics, and Decision Making in the Atmospheric Sciences*, edited by A. H. Murphy and R. W. Katz, Boulder, CO, pp. 379–437, Westview Press.

Njoku, E. G. (1982), Passive microwave remote sensing of the earth from space—A review, *Proc. IEEE, 70*, 728–750.

Njoku, E.G., and J. A. Kong (1977), Theory for passive microwave remote sensing of near-surface soil moisture, *J. Geophys. Res., 82*, 3108–118.

Olson, W. S. (1987), Estimation of rainfall rates in tropical cyclones by passive microwave radiometry, Ph.D. thesis, Univ. of Wisc., Madison.

Panofsky, H. A., and G. W. Brier (1965), Some applications of statistics to meteorology, The Pennsylvania State University, University Park, PA, 224.

Paris, J. F. (1971), Transfer of thermal microwaves in the atmosphere, Dept. of Meteorology, Texas A and M Univ., College Station, Tex.

Panofsky, H. A., and G. W. Brier (1965), Some applications of statistics to meteorology, The Pennsylvania State University, University Park, PA, 224.

Pierce C. S. (1884), The numerical measure of the success of predictions, *Science, 4*, 453–454.

Pellerin, T., J.-P. Wigneror, J.-C. Calvet, and P. Waldteufel (2003), Global soil moisture retrieval from a synthetic L-band brightness temperature data set, *J. Geophys. Res., 108*, 4364, doi:10.1029/2002JD003086.

Petty, G. (1994), Physical retrievals of over-ocean rain rate from multichannel imagery. Part I: Theoretical characteristics of normalized polarization and scattering indices, *Meteor. Atmos. Phys., 54*, 79–99.

Prigent, C., W. B. Rossow, and E. Matthews (1998), Global maps of microwave land surface emissivities: Potential for land surface characterization, *Radio Sci., 33*, 745–751.

Pruppacher, H. R., and J. D. Klet (1978), *Microphysics of Clouds and Precipitation, Reidel*, Dordrecht.

Savage, R. C. (1976), The transfer of thermal microwaves through hydrometeors, Ph.D. thesis, Dept. of Meteorology, University of Wisc., Madison.

Savage, R. C. (1978), The radiative properties of hydrometeors at microwave frequencies, *J. Appl. Meteorol, 17*, 904–911.

Savage, R. C., E. A. Smith, and A. Mugnai (1995), Concepts for a geostationary imaging sounder (GeoMIS), in *Preprints of the International Geoscience and Remote Sensing Symposium* (10–14 July, Firenze, Italy), Vol. *1*, pp. 652–654.

Schwiesow, R. L (1972), Atomic, molecular, particulate, and collective generalized scattering, in *Remote Sensing of the Troposphere. Environmental Research Laboratories*, NOAA, Boulder, Col., Chapter 10.

Seto, S., N. Takahashi, and T. Iguchi (2005), Rain/no-rain classification methods for microwave radiometer observations overland using statistical information for brightness temperatures under no-rain conditions, *J. Appl. Meteor., 44*, 1243–1259.

Seto, S., T. Kubota, T. Iguchi, N. Takahashi, and T. Oki (2009), An evaluation of over-land rain rate estimates by the GSMaP and GPROF algorithms: The role of lower-frequency channels, *J. Meteorol. Soc. Jpn., 87A*, 183–202, doi:10.2151/jmsj.87A.183.

Shi, J., K. S. Chen, Q. Lin, T. J. Jackson, P. E. O'Neill, and L. Tsang (2002), A parameterized surface reflectivity model and estimation of bare-surface soil moisture with L-band radiometers, *IEEE Trans. Geosci. Remote Sens., 40*, 2674–2686.

Shifrin, K. S., and M. M. Chernyak (1968), Microwave absorption and scattering by precipitation, in *Transfer of Microwave Radiation in the Atmosphere*, pp. 69–78. Jerusalem, Israel Program for Scientific Translations.

Short, D. A., and G. R. North (1990), The beam filling error in Nimbus-5 ESMR observations of GATE rainfall, *J. Geophys. Res., 95*, 2187–2193.

Smith, E. A., and S. Q. Kidder (1978), A multispectral satellite approach to rainfall estimates. Presented at the 18th AMS Conf. on Radar Meteorology (28–31 March, Atlanta, GA) *The Use of Satellite Data in Rainfall Monitoring*, edited by E.C. Barrett and D. W. Martin, pp. 160–163, Acadamic, New York.

Smith, E. A., et al. (1998), Results of WetNet PIP-2 projects, *J. Atmos. Sci., 55*, 1483–1536.

Snider, J. B., and E. R. Westwater (1972), *Radiometry, in Remote Sensing of the Atmosphere, Environmental Research Laboratories*, NOAA, Boulder, Col., Chapter 15.

Spencer, R. W. (1986), A satellite passive 37 GHz scattering based method for measuring oceanic rain rates, *J. Climate Appl. Meteor., 25*, 754–766.

Spencer, R. W., W. S. Olson., Wu Rongzhang., D. W. Martin., J. A. Weinman., and D. A. Santek, (1983), Heavy thunderstorms observed over land by the Nimbus-7 scanning multichannel microwave radiometer, *J. Climat. Appl. Meteorol., 22*, 1041–1046, 1983.

Spencer, R. W., H. M. Goodman, and R. E. Hood (1989), Precipitation retrieval over land and ocean with the SSM/I, Part I: Identification and characteristics of the scattering signal, *J. Atmos. Oceanic Technol., 6*, 254–273.

Staelin, D. H. (1969), Passive remote sensing at microwave wavelengths, *Proc. IEEE, 57*, 427–439.

Stanski, H. R., L. J. Wilson, and W. R. Burrows (1989), *Survey of Common Verification Methods in Meteorology*, WMO/TD-No.359, World Meteorological Organization, Geneva, Switzerland.

Stepanenko, V. D. (1968), Contrasts of radio brightness temperatures in clouds and precipitation, *Transfer of Microwave Radiation in the Atmosphere*, US Dept. of Commerce, Springfield, Va.

Stephenson, D. B (2000), Use of the "Ødds ratio" for diagnosing estimate skill, *Wea. Estimating, 15*, 221–232.

Storch, H. V., and F. W. Zwiers, (1999), *Statistical Analysis in Climate Research*, Cambridge University Press, Cambridge.

Sudradjat, A., N. A. Wang, K. Gopalan, and R. R. Ferraro (2010), Prototyping a generic, unified land surface classification and screening methodology for GPM-Era microwave land precipitation retrieval algorithms, *J. Appl. Meteor. Climatol., 50*, 1200–1211.

Szejwach, G., R. F. Adler, I. Jobard, and R. Mack (1986), A cloud model radiative model combination for determining microwave Tb-Rain rate relations, presented at the Second Conference on Satellite Meteorology/Remote Sensing and Applications, *Am. Meteorol. Soc.*, Williamsburg, Va., May 13–16.

Tartaglione, N. (2010), Relationship between precipitation estimate errors and skill scores of dichotomous estimates, *Weather Estimate., 25*, 355–365.

Turk, J., E. Ebert, H.-J. Oh, B.-J. Sohn, B. Levizzani, E. Smith, and R. Ferraro (2002), Verification of an operational global precipitation analysis at short time scales, 1st Intl. Precipitation Working Group (IPWG) Workshop, Madrid, Spain, 23–27 Spet.

Ulaby, F. T., and W. H. Stiles (1981), Microwave response of snow, *Adv. Space Res.*, *1*, 131–149.

Volchok, B. A., and M. M. Chernyak (1968), Transfer of microwave radiation in clouds and precipitation, Transfer of Microwave Radiation in the Atmosphere, NASA TT F-590, pp. 90–97.

Wang, J., and D. B. Wolff (2010), Evaluation of TRMM ground-validation radar-rain errors using rain gauge measurements, *J. Appl. Meteor. Climatol.*, *49*, 310–324.

Wang, N. Y., C. Liu., R. Ferraro, E. Zipser, and C. Kummerow (2009), TRMM 2A12 land precipitation product status and future plans, *J. Meteor. Soc. Jpn*, *87A*, 237–253.

Waters, J. W. (1976), Absorption and emission by atmospheric gases, in *Methods of Experimental Physics*, Vol. 12: *Astrophysics, Part B*, edited by M. L. Meeks, Academic, New York.

Weinman, J. A., and P. J. Guetter (1977), Determination of rainfall distribution from microwave radiation measured by the Nimbus-6 ESMR., *J. Appl. Meteor.*, *16*, 437–442.

Weng, F., B. Yan, and N. C. Grody (2001), A microwave land emissivity model, *J. Geophy. Res.*, *106*, 20, 115–20 123.

Wentz, F. J (1983), A model function for ocean microwave brightness temperatures, *J. Geophys. Res.*, *88*(C3), 1892–1908.

Wentz, F. J. (1997) A well calibrated ocean algorithm for special sensor microwave/Imager, *J. Geophys. Res.*, *102*(C4), 8703–8718.

Wentz, F. J., and D. J. Cavalieri (1995), *A 20-year Geophysical Data Set from Window-Frequency Microwave Radiometers*, Remote Sensing Systems, Santa Rosa, calif.

Wilheit, T. T (1978), A review of applications of microwave radiometry to oceanography, *Boundary Layer Meteorol.*, *13*, 277–293.

Wilheit T.T (1986), Some comments on passive microwave measurement of rain, *BAMS*, *67*, 1226–1232.

Wilheit, T. T., A. T. C., Chang, M. S. V., Rao, E. B. Rodgers, and J. S. Theon, (1977), A satellite technique for quantitatively mapping rainfall rates over the oceans, *J. Appl. Meteorol.*, *16*, 551–560.

Wilheit, T. T., A. T. C. Chang, M. S. V. Rao, E. B. Rodgers, and J. S. Theon (1982), Microwave radiometric observations near 19.35, 92 and 183 GHz of precipitation in Tropical Storm cora., *J. Appl. Meteor.*, *21*, 1137–1145.

Wilheit, T. T., A. T. C. Chang, and L. S. Chiu (1991), Retrieval of monthly rainfall indices from microwave radiometric measurements using probability distribution functions, *J. Atmos. Oceanic Technol.*, *8*, 118–136.

Wilheit, T. T., C. D. Kummerow, and R. Ferraro (2003), Rainfall algorithms for AMSR-E, *IEEE Trans. Geosci. Remote Sens.*, *41*, 204–214.

Wilks, D. S. (1995), *Statistical Methods in the Atmospheric Sciences*, Academic, San Diego.

Wu, R., and J. A. Weinman (1984), Microwave radiances from precipitating clouds containing aspherical ice, combined phase, and liquid hydrometeors, *J. Geophys. Res.*, *89*, 7170–7178.

Zagrodnik, J., and H. Jiang (2013), Investigation of PR and TMI Version 6 and Version 7 rainfall algorithms in landfalling tropical cyclones relative to the NEXRAD stage-IV multi-sensor precipitation estimate dataset, *J. Appl. Meteor. Climatol*, *52*, 2809–2827, doi:http://dx.doi.org/10.1175/JAMC-D-12-0274.1.

2

Improvement of TMI Rain Retrieval Over the Indian Subcontinent

Shoichi Shige, Munehisa K. Yamamoto, and Aina Taniguchi

2.1. INTRODUCTION

It is difficult to measure precipitation in mountainous regions. Observation networks of rain gauges are too sparse to resolve important structures in the precipitation field. Ground-based radar is blocked by topography, preventing measurement of near-surface precipitation rates. The launch of the Tropical Rainfall Measuring Mission (TRMM) satellite [*Kummerow et al.*, 1998] brought a new age of active remote sensing of precipitation from space, with the first spaceborne precipitation radar [PR; *Kozu et al.*, 2001; *Okamoto*, 2003]. The PR makes accurate observations of precipitation with high spatial resolution over both ocean and land via the well-developed 2A25 retrieval algorithm [*Iguchi et al.*, 2000, 2009]. Using PR data, *Xie et al.* [2006] found orographic rainbands over narrow mountain ranges of Asia, which are not local phenomena but have far-reaching effects on the continental-scale monsoon. *Nesbitt and Anders* [2009] produced high-resolution precipitation 10-year climatologies from PR data, revealing strong coupling between precipitation and topography.

Following the great success of the TRMM, the development of high-resolution satellite rainfall products (0.1°–0.25° latitude/longitude and 0.5–3 hourly) has accelerated, by combining data from microwave radiometers (MWRs) in low Earth orbit and infrared radiometers (IRs) [see review in *Gebremichael and Hossain*, 2010]. However, poorly performing high-resolution satellite rainfall products over mountainous areas were identified by *Kubota et al.* [2009] for Japan, and by *Dinku et al.* [2010] for Africa and South America. *Kubota et al.* [2009] suggested that a major reason for such errors is that MWR algorithms might underestimate heavy rainfall associated with shallow orographic rainfall systems.

The dominant view that heavy rainfall results from deep clouds was formed based on observational studies in the United States, such as the pioneering work of *Byers and Braham* [1949]. Under this paradigm, some studies use "deep convection" as synonymous with "heavy precipitation" [e.g., *Gray and Jacobson*, 1977]. MWR algorithms, which also assume that heavy rainfall results from deep clouds, have been developed under this paradigm [e.g., *Spencer*, 1984; *McCollum and Ferraro*, 2003; *Weng et al.*, 2003; *Ferraro et al.*, 2005]. However, heavy rainfall can be caused by shallow orographic convection, especially in moist Asian monsoon regions [e.g., *Takeda et al.*, 1976; *Takeda and Takase*, 1980; *Sakakibara*, 1981]. Warm-rain processes are enhanced by low-level orographic lifting of moist air, so that rainfall develops in the lower portions of growing convective elements and precipitates while the clouds are still growing.

There have been few studies on improvements to MWR rain retrieval for mountainous regions. Following the study of *Vicente et al.* [2002] on improvements to IR-based rain retrieval, *Kwon et al.* [2008] developed topographic correction factors as functions of terrain slope, low-level wind, and moisture parameters for the terrain of the Korean Peninsula, within the Goddard profiling (GPROF) algorithm [*Kummerow et al.*, 2001; *McCollum and Ferraro*, 2003; *Olson et al.*, 2006; *Wang et al.*, 2009]. GPROF is the TRMM microwave imager (TMI) facility algorithm.

Graduate School of Science, Kyoto University, Kyoto, Japan

Remote Sensing of the Terrestrial Water Cycle, Geophysical Monograph 206. First Edition. Edited by Venkat Lakshmi.
© 2015 American Geophysical Union. Published 2015 by John Wiley & Sons, Inc.

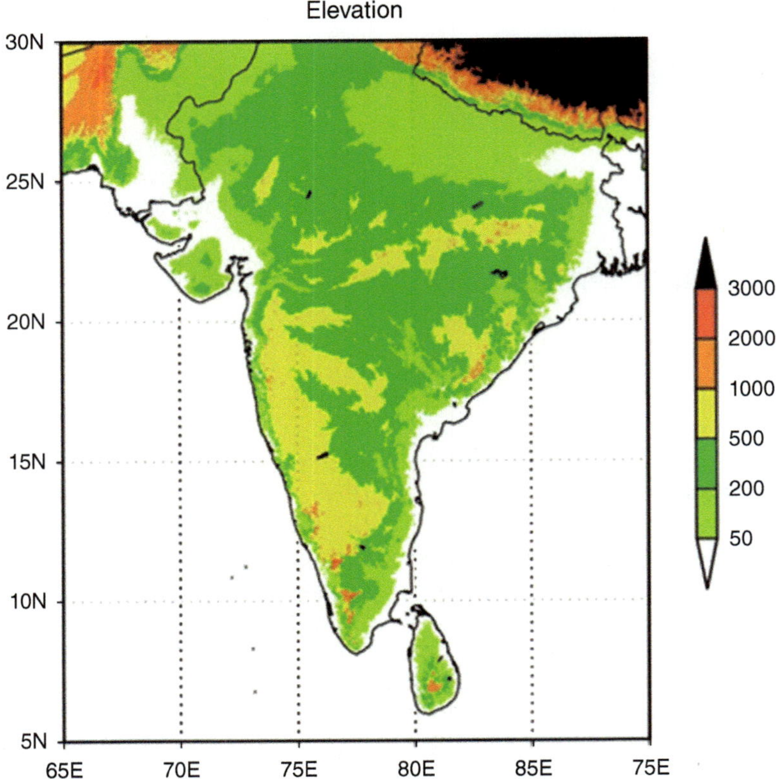

Figure 2.1 SRTM30 elevation (m).

Shige et al. [2013] (hereafter S13) recently improved the performance of rainfall estimates from the global satellite mapping of precipitation (GSMaP) MWR algorithm [*Aonashi et al.*, 2009; *Shige et al.*, 2009] from TMI data of the Kii Peninsula. This peninsula is a region of Japan with heavy precipitation, where satellite methods of estimating maximum rainfall amounts have been shown to be poor [*Negri and Adler*, 1993; *Kubota et al.*, 2009]. S13 did not introduce orographic correction factors in the GSMaP MWR algorithm but rather incorporated a dynamic selection of lookup tables (LUTs) appropriate for heavy orographic rainfall, based on an orographic/nonorographic rainfall classification scheme. LUTs derived from orographic precipitation profiles are used to estimate rainfall for an orographic rainfall pixel determined by conditions of orographically forced vertical motion and convergence of surface moisture flux. LUTs derived from original precipitation profiles are used to estimate rainfall for a nonorographic rainfall pixel. Rainfall estimates using the revised GSMaP MWR algorithm from TMI data agree more with the TRMM PR and gauge-calibrated ground radar data than estimates from the original algorithm. A follow-up study by *Taniguchi et al.* [2013] (hereafter T13) applied the GSMaP MWR algorithm with the orographic/nonorographic rainfall classification scheme to Typhoon Morakot (2009), which produced a huge amount of rainfall and had a catastrophic impact on Taiwan [*Wu and Yang*, 2011; *Wu*, 2013]. That classification scheme is extended to the GSMaP MWR algorithm for all passive microwave radiometers in orbit, including both microwave imagers and microwave sounders.[1] Retrieved rainfall rates, together with IR images, are used by the GSMaP Kalman filter algorithm [*Ushio et al.*, 2009], and the results agree more with rain gauge observations.

As reviewed by *Lin* [2007] and *Houze* [2012], topography affects precipitating clouds through extremely complex mechanisms, which are strongly dependent on numerous factors. Appropriate thresholds of orographic/nonorographic rainfall classification schemes for the occurrence of orographic rainfall may vary with region due to regional variability in precipitation processes. In this study, the GSMaP MWR algorithm with the orographic/nonorographic rainfall classification scheme is applied to rain retrieval over the western coast of India (Figure 2.1), which is one of the heaviest rainfall areas of the southwest monsoon [*Grossman and Durran*, 1984; *Ogura and Yoshizaki*, 1988].

[1] The orographic/nonorographic rainfall classification scheme has been used by the version of GSMaP products introduced on 2 September 2014, which are available in near real time (about 4 h after observation) via the Internet [*Kachi et al.*, 2011].

We have made two major improvements to the GSMaP MWR algorithm. The first is to use the orographic/nonorographic rainfall classification scheme with appropriate thresholds for orographic rainfall. The second is to distinguish precipitation-sized ice particle density in the radiative transfer model calculations between orographic rainfall and nonorographic rainfall types. Estimates using the revised algorithm agree more with PR estimates, not only over the western coast but alsoover nearly the entire Indian subcontinent.

2.2. GSMaP MWR ALGORITHM

We focused on GSMaP MWR (hereafter GSMaP) from TMI data because estimates from this algorithm can be compared with PR observations. The GSMaP algorithm consists of a forward-calculation section that calculates LUTs, showing the relationship between rainfall rate and brightness temperature, T_b, via a radiative transfer model (RTM), and a retrieval section that estimates the precipitation rate from the observed T_b using the LUTs. Here, the GSMaP overland algorithm is described. For further details, refer to *Aonashi et al.* [2009].

LUTs showing the relationship between rainfall rate and T_b were computed daily for $5.0° \times 5.0°$ latitude-longitude boxes by incorporating atmospheric, surface, and precipitation-related variables into the four-stream RTM developed by *Liu* [1998]. Atmospheric temperature, freezing-level height (FLH), and surface temperature were adapted from the Japan Meteorological Agency (JMA) global analysis (GANAL). Surface emissivity is set at 0.9 for all frequencies.

We constructed convective and stratiform precipitation models for precipitation-related variables (such as hydrometeor profiles) for six land precipitation types (severe thunderstorm, afternoon shower, shallow rain, extratropical frontal systems, organized rain, or high land rainfall). These types were determined from the stratiform pixel ratio, the stratiform rain ratio, the precipitation area, the precipitation top height, the rain intensity, the diurnal cycle, and the rain yield perflash (RPF) [*Williams et al.*, 1992; *Takayabu*, 2006] obtained from PR and TRMM lightning imaging sensor (LIS) in ~100 km mesoscale boxes by *Takayabu* [2008]. High land rainfall corresponds to rainfall as observed over the Tibetan Plateau, but we have not included the orographic types considered in S13 or T13. We averaged the convective and stratiform precipitation profiles of PR data over prescribed surface precipitation ranges[2] for each precipitation type. In this

averaging, profiles relative to the FLH were used to exclude the effect of atmospheric temperature variations [*Kubota et al.*, 2007]. For the RTM calculations, we used the precipitation profiles associated with the dominant precipitation types in $2.5° \times 2.5°$ latitude-longitude, which are statistically classified trimonthly.

For the raindrop-size distribution, we used statistical models for each precipitation type that were constructed from PR data [*Kozu et al.*, 2009]. They influence the over-ocean algorithm through the emission signals but not the overland algorithm [*Aonashi et al.*, 2009]. We used conventional models for frozen and mixed-phase particle distributions that could not be estimated from PR data. For the frozen-particle size distribution, we applied the Marshall-Palmer distribution [*Marshall and Palmer*, 1948] to both convective and stratiform precipitation. We calculated the refractivity of convective and stratiform frozen particles, assuming it to be a mixture of ice and air with an empirically prescribed constant density $(0.2 \, g/cm^3)$. We parameterized the size distribution and refractivity of mixed-phase particles for stratiform precipitation in terms of atmospheric temperature [*Nishituji et al.*, 1983] and neglected mixed-phase particles for convective precipitation. In the RTM calculations, we assumed all precipitation particles were spherical and computed the absorption and scattering coefficients and the phase functions using the Mie theory [*Mie*, 1908].

For the overland algorithm, we assumed that precipitation is lognormally distributed in the horizontal and the standard deviation of the natural logarithm of the precipitation is 1. Using this assumption, we derived the convective and stratiform LUTs with a horizontal in homogeneity from those calculated for horizontally homogenous precipitation. The LUTs used in the retrieval section were weighted averages of the above convective and stratiform LUTs. We determined the weights using the statistical frequency distribution of the PR convective and stratiform precipitation for each precipitation type and surface precipitation rate.

We used the rain/no-rain classification method of *Seto et al.* [2005] over land and that of *Kubota et al.* [2007] (an improvement of the *McCollum and Ferraro* [2005] method) over coasts. We obtained surface rainfall estimates by using LUTs to combine estimates from polarization-corrected temperatures (PCT) [*Spencer et al.*, 1989] at 85 GHz (PCT85) and 37 GHz (PCT37).

2.3. IMPROVEMENT OF OROGRAPHIC/ NONOROGRAPHIC CLASSIFICATION SCHEME

We acquired rain gauge observation network data from the Asian Precipitation—Highly Resolved Observational Data Integration Towards Evaluation of Water Resources project [APHRODITE; *Yatagai et al.*, 2012] (Figure 2.2a).

---

[2]The prescribed surface precipitation ranges are 0.475–0.525, 0.95–1.05, 1.9–2.1, 2.85–3.15, 3.8–4.2, 5.7–6.3, 7.6–8.4, 9.5–10.5, 14.25–15.75, 19.0–21.0, 28.5–31.5, 38–42, 57–63, 76–84, 114–126, 152–168, and 190–210 mm/h.

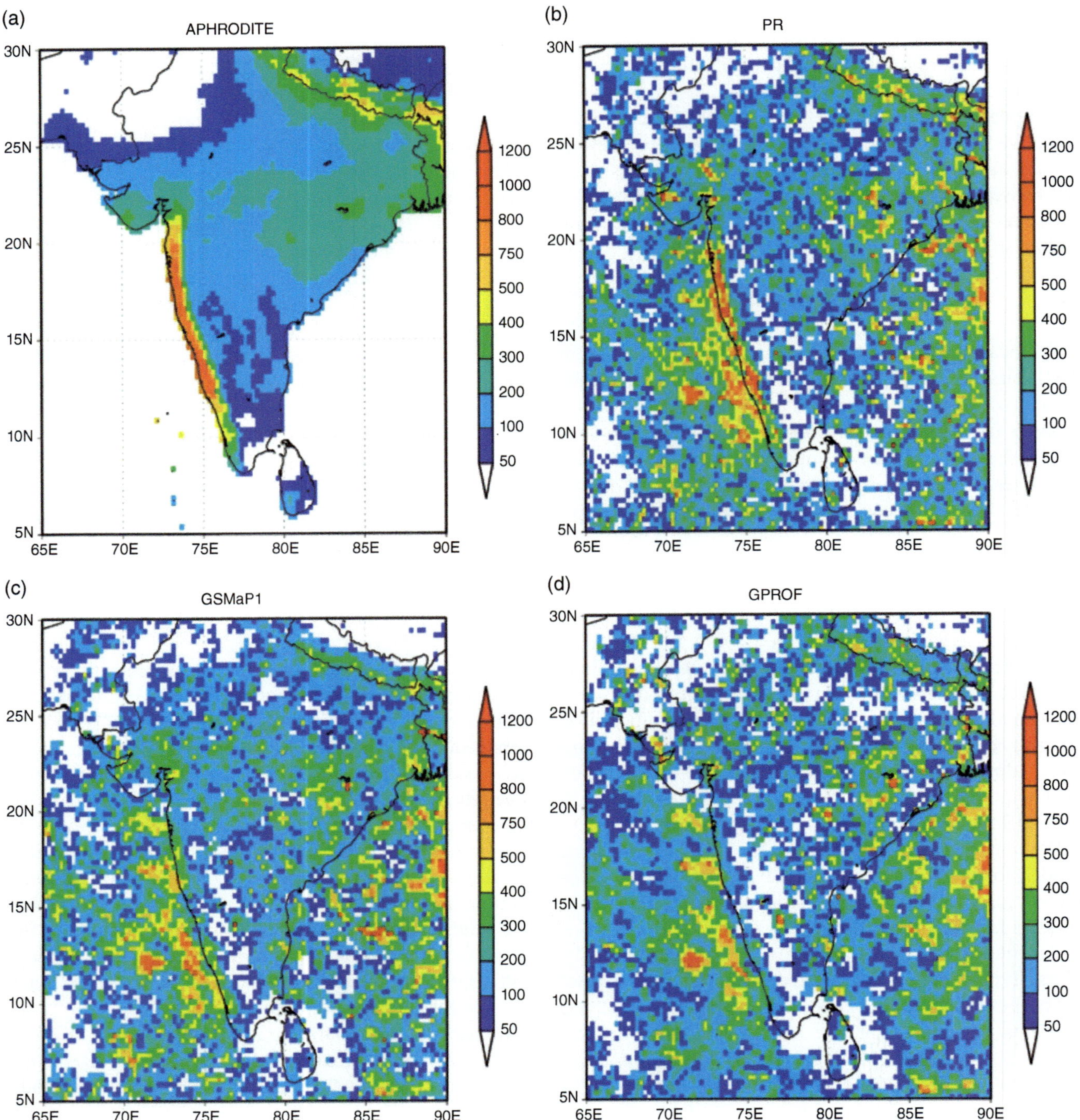

Figure 2.2 Average rain rates (mm/month) for summer (June–August 2007) based on (a) APHRODITE, (b) PR, (c) GSMaP1 from TMI data, and (d) GPROF from TMI data.

From the daily 0.25° × 0.25° grid point precipitation data set, there were striking spatial gradients of summer average rain rates over the Indian coast, west of the Western Ghats. This feature was well captured by the PR rain rate with heavy rainfall offshore (Figure 2.2b), as previously noted by *Xie et al.* [2006]. The original GSMaP (hereafter GSMaP1) rain rates agreed well with the PR on broad scales (Figure 2.2c). However, there were underestimations exceeding 600 mm/month for the Indian coast, west of the Western Ghats (Figure 2.2e), similar to results for the Kii Peninsula of Japan (S13, Figure 9a). GPROF obtained similar results (Figures 2.2d, and 2.2f).

Here, the GSMaP algorithm with orographic/nonorographic rainfall classification scheme (hereafter

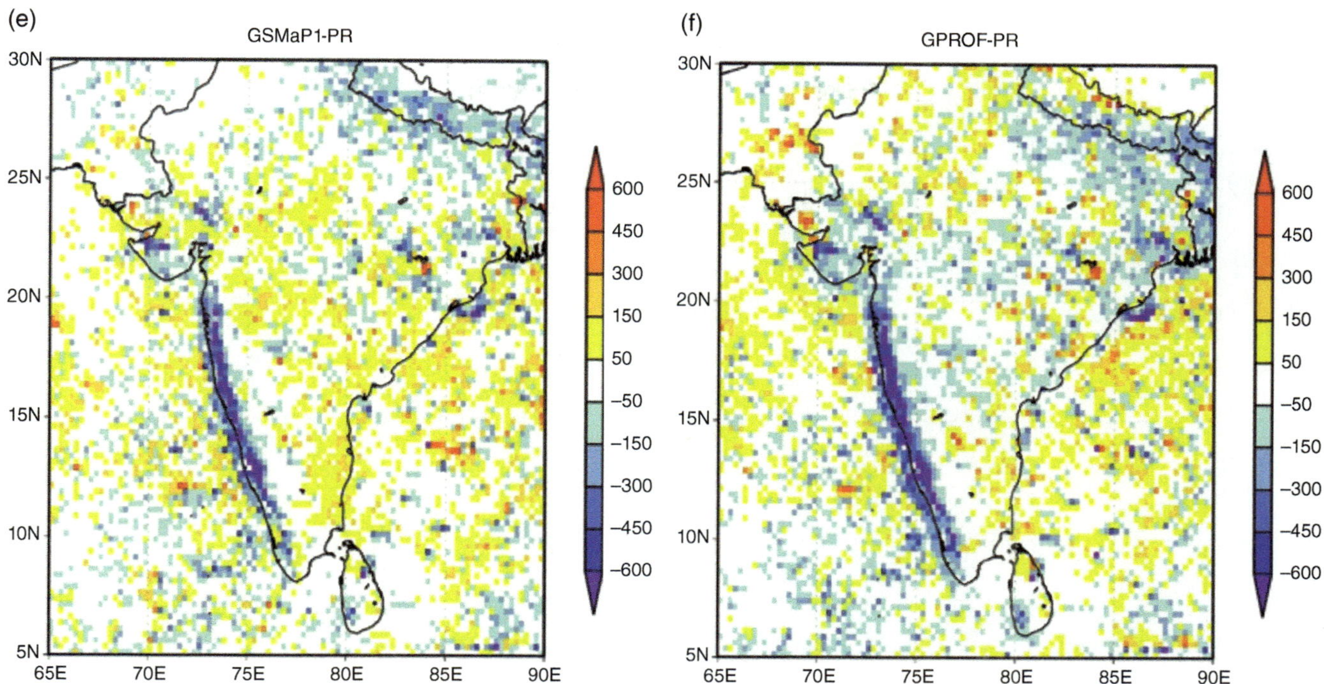

Figure 2.2 (Continued) Rain rate differences between (e) GSMaP1 from TMI data and PR (i.e., GSMaP1 – PR) and (f) GPROF from TMI data and PR (i.e., GPROF – PR) for summer (June–August 2007).

GSMaP2), developed by S13 and T13, was applied to retrieval over the Indian subcontinent. That classification scheme is based on orographically forced upward motion w (m/s) and moisture flux convergence Q (s), as follows:

$$w = \frac{Dh}{Dt} = \mathbf{V}_H \cdot \nabla h, \qquad (2.1)$$

$$Q = -\nabla \cdot \mathbf{V}_H q, \qquad (2.2)$$

where $\mathbf{V}_H$ is the horizontal surface wind (m/s) from GANAL, and q is the water vapor mixing ratio (kg/kg) calculated from the GANAL surface-specific humidity; and $h\,(x, y)$ is the elevation (m) derived from the Shuttle Radar Topography Mission [SRTM30; *Werner*, 2001; *Farr et al.*, 2007], with horizontal grid spacing approximately 1 km (Figure 1) and averaged over a 50 km horizontal length scale in the calculation of w, based on T13.

An orographic rainfall pixel is determined using the same condition as in S13 and T13:

$$w > 0.1\,(\mathrm{m/s}) \quad \text{and} \quad Q > 0.5 \times 10^{-6}\,(\mathrm{s}). \qquad (2.3)$$

For an orographic rainfall pixel determined by equation (2.3), rainfall rates were estimated using LUTs calculated from orographic precipitation profiles. To preserve the simplicity, we used the same orographic precipitation profiles observed by the PR over the Kii Peninsula as S13 and T13. Conversely, for a nonorographic rainfall pixel, rainfall rates were estimated using LUTs calculated from the

original precipitation profiles. In the GSMaP2 algorithm, LUTs calculated from the appropriate precipitation profiles are dynamically selected according to atmospheric conditions derived from 6 hourly GANAL data. The original algorithm (GSMaP1) only deals with trimonthly variations of typical hydrometeor profiles. However, GSMaP2 rain rates (Figure 2.3a) and differences between GSMaP2 and PR (Figure 2.3b) are nearly the same as those for GSMaP1 (Figures 2.2c and 2.2e). The results suggest that orographic LUTs are not used in GSMaP2.

Figure 2.4 shows observations derived from a TRMM satellite overpass of a mesoscale convective system (MCS) over the Indian coast west of the Western Ghats, on 30 June 2007. Surface rain rates > 30 mm/h were detected in the PR data (Figure 2.4a) but not in the GSMaP2 data (Figure 2.4b), which are computed mainly according to TMI PCT85 (Figure 2.4c). GSMaP2 rain rates are the same as those of GSMaP1 (not shown). Figure 2.4d shows orographically forced vertical motion estimated from equation (2.1). Westerly wind impinges on the mountains, and there is upward motion stronger than 0.01 m/s along the coast where strong rain rates are estimated by the PR; but this motion is weaker than the threshold in equation (2.3) (i.e., 0.1 m/s). Therefore, orographic LUTs are not used in GSMaP2.

As discussed in S13, thresholds in equation (2.3) can be selected liberally (i.e., detection of areas with weak upward motion, at the expense of misclassifying nonorographic rain pixels) or conservatively (i.e., detection of

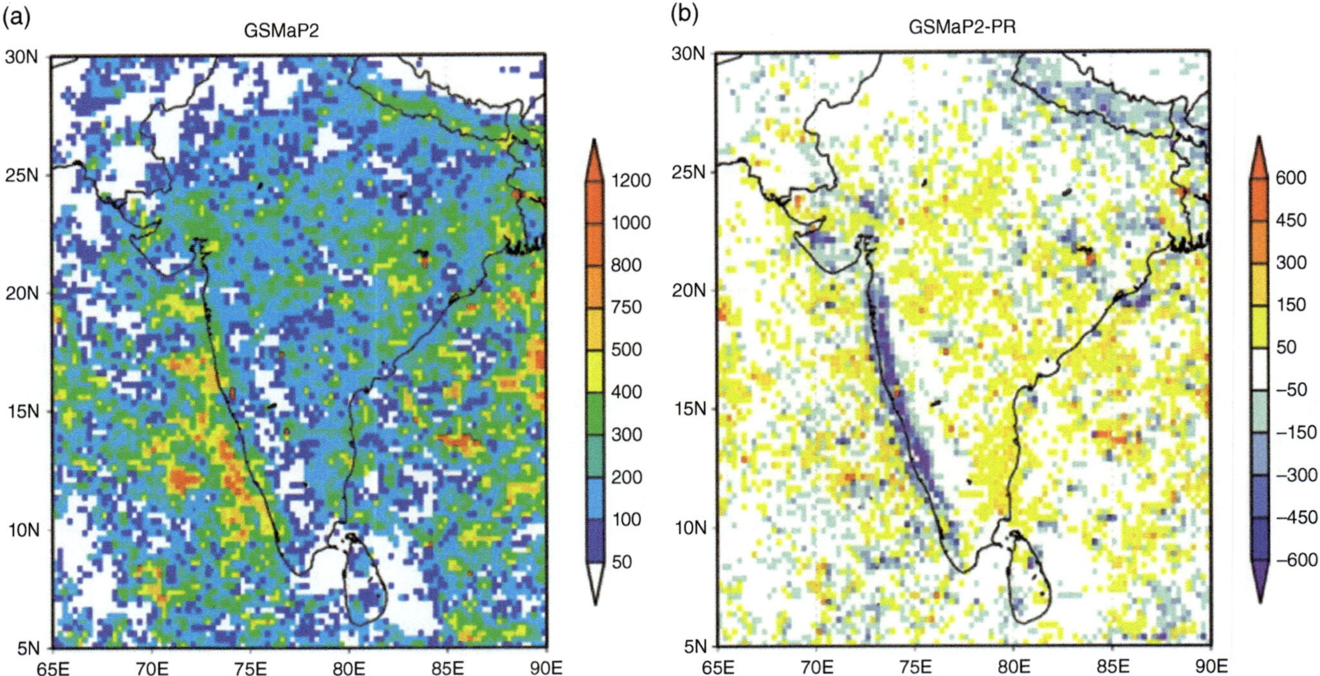

Figure 2.3 As in Figures 2.2c and 2.2e but for (a) average rain rates (mm/month^1) estimated by GSMaP2 algorithm from TMI data, and (b) rain rate differences between GSMaP2 and PR (GSMaP2 − PR).

only upward motion above a certain threshold, at the expense of eliminating some orographic rain pixels). Even for the Kii Peninsula case in S13, orographic/nonorographic rainfall classification involved incorrectly eliminating many orographic rainfall pixels. Upward motions in our case (Figure 2.4d) are weaker than those of the S13 case (see Figure 2.7a of S13), resulting in elimination of all orographic rainfall pixels.

Here, the thresholds in equation (2.3) were selected liberally as

$$w > 0.01\,(\text{m/s}) \quad \text{and} \quad Q > 0.3 \times 10^{-6}\,(\text{s}). \qquad (2.4)$$

For the case shown in Figure 2.4, we do not need to change the Q threshold in equation (2.3); however, for many other cases over the western coast of India, the Q threshold in this equation eliminates many orographic rainfall pixels, as later shown. Figure 2.5 shows rain rates estimated by GSMaP with the thresholds in equation (2.4) (hereafter GSMaP3). The thresholds produce orographic rain pixels, but the GSMaP3 estimates for orographic rain pixels are uniformly greater than 30 mm/h.

Figure 2.6a depicts GSMaP3 rain rates averaged over summer (June–August 2007). The algorithm captured the rain maximum over the west coast of India. However, there are large positive values above 600 mm/month over that coast (Figure 2.6b) that indicate GSMaP3 overestimation there, as expected from the result of the case shown in Figure 2.5. LUTs derived from convective and

stratiform precipitation profiles for a case of orographic rainfall over the Kii Peninsula resulted in overestimation by GSMaP3.

2.4. IMPROVEMENT OF PRECIPITATION-RELATED VARIABLE MODELS IN RTM CALCULATIONS

As discussed in S13 and T13, the PCT85 in LUTs strongly depends on the precipitation profile input to the RTM calculations. The orographic precipitation profiles for the Kii Peninsula case (Figures 2.7c and 2.7d) had much lower precipitation top heights than those of the original profiles for organized rain type (Figures 2.7a and 2.7b) used by GSMaP2 (GSMaP3) to produce LUT for nonorographic rain retrieval in the case shown by Figure 2.4b (Figure 2.5). PCT85 in the LUT calculated from the orographic precipitation profiles for the Kii Peninsula case (blue line in Figure 2.8) decreased more slowly with rainfall rate than that from the original precipitation profiles (thin black line in Figure 2.8). This is because the ice layer thickness in the orographic precipitation profiles was smaller than in the original profiles. Therefore, for a given PCT85, the LUT obtained from the orographic precipitation profiles gives a higher rainfall rate than that obtained from the original.

Figures 2.7e and 2.7f show average precipitation profiles obtained from PR data, within the box over the west coast of India (15°−20°N × 70°−75°E) during summer (June August) 2003–2011. We only use PR data matching

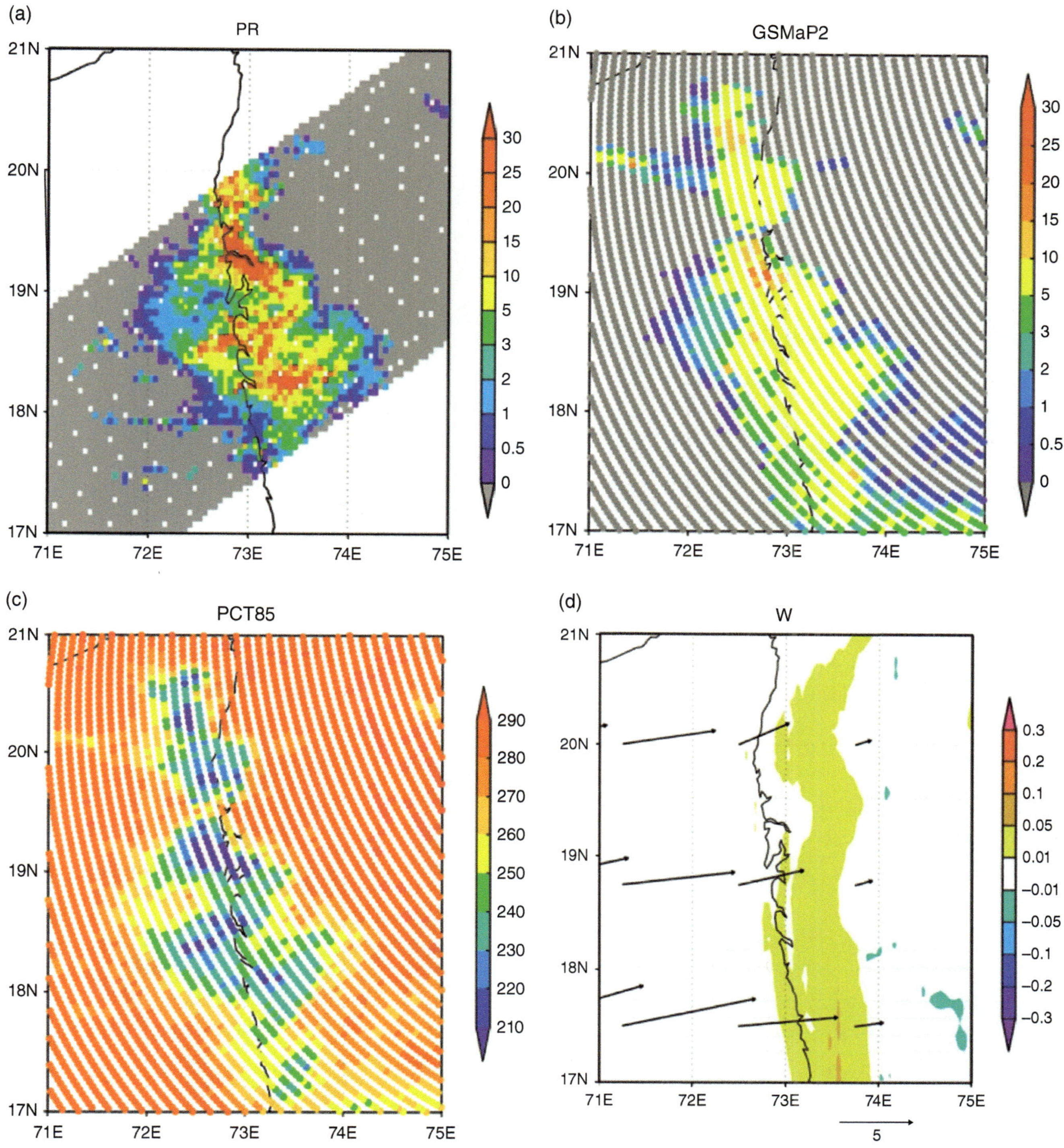

Figure 2.4 Case study of a mesoscale convective system (MCS) over Indian coast, 30 June 2007 (TRMM orbit No. 54821): (a) PR 2A25 near-surface rain rate (mm/h¹), (b) GSMaP2 surface rain (mm/h¹), (c) TMI PCT85 (K), and (d) orographically forced vertical motion (m/s¹), estimated by equation 2.1 with surface horizontal winds from GANAL data at 0000 UTC (Coordinated Universal Time) 30 June 2007.

of the conditions in equation (2.4) for this averaging. While precipitation top heights of stratiform precipitation profiles over the Indian coast (Figure 2.7f) are almost the same as those of original stratiform precipitation profiles (Figure 2.7b), precipitation top heights of convective precipitation profiles over India (Figure 2.7e) are lower than those of the original convective precipitation profiles (Figure 2.7a). This demonstrates weaker cold-rain

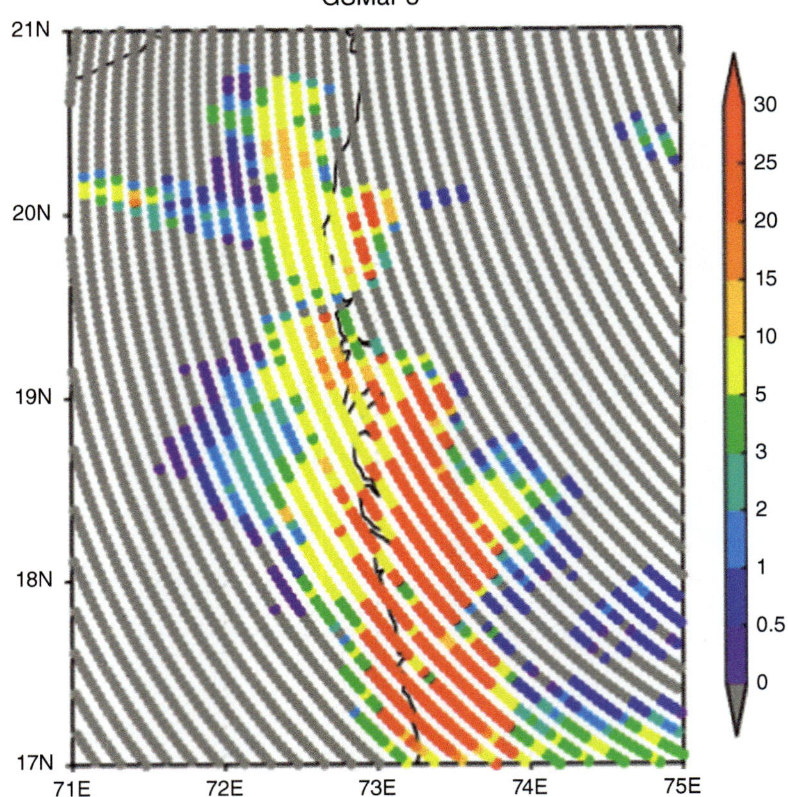

Figure 2.5 As in Figure 2.4b but for surface rain rates estimated by GSMaP3 algorithm from TMI data.

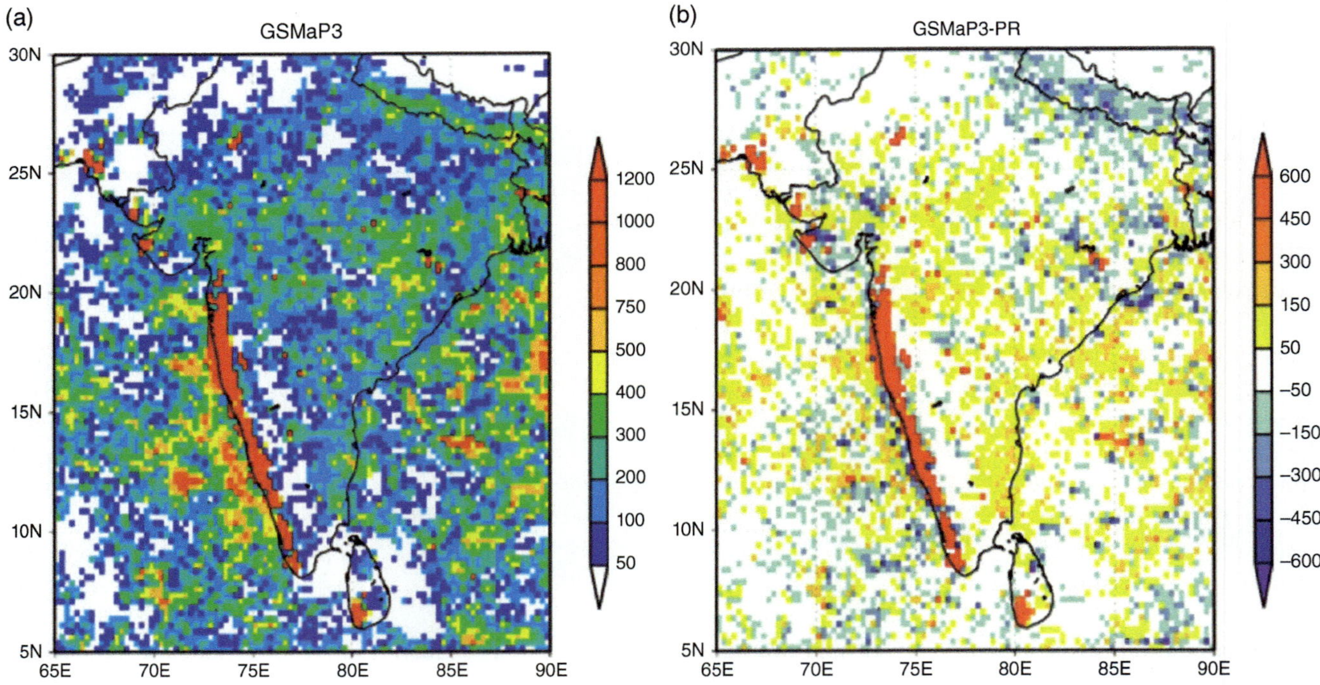

Figure 2.6 As in Figures 2.2c and 2.2e but for (a) average rain rates (mm/month) estimated by GSMaP3 algorithm from TMI data and (b) rain rate differences between GSMaP3 and PR (GSMaP3 – PR).

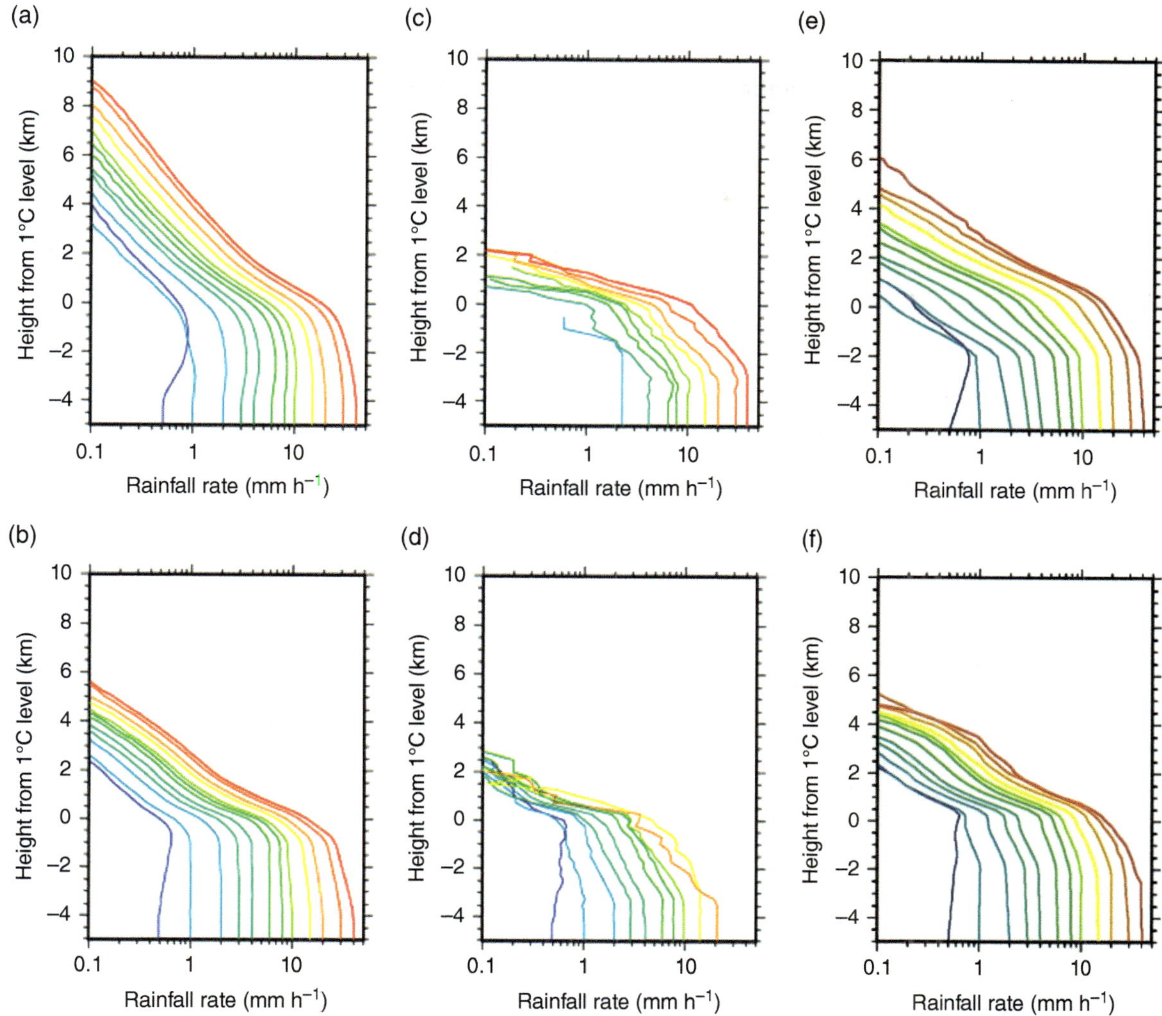

Figure 2.7 Precipitation profile models for (a) convective and (b) stratiform rainfall, used by GSMaP2 (GSMaP3) to produce LUT for nonorographic rain retrieval in the case shown by Figure 2.4b (Figure 2.5); (c) convective and (d) stratiform rainfall used by GSMaP2 (GSMaP3) to produce LUT for orographic rain retrieval in the case shown by Figure 2.4b (Figure 2.5); (e) convective and (f) stratiform rainfall obtained by averaging convective and stratiform precipitation profiles observed by PR within the box over the west coast of India ($15°$–$20°$N × $70°$–$75°$E) for summer (June–August) 2003–2011. Ordinate is height difference from FLH.

processes, consistent with earlier studies using data from the LIS and PR [*Petersen and Rutledge*, 2001; *Takayabu*, 2006, 2008]. *Hirose and Nakamura* [2002] also showed that the precipitation profiles over the Western Ghats have a downward-increasing precipitation rate feature, which is the manifestation of maritime profiles with weaker cold-rain processes [*Zipser and Lutz* 1994; *Shige et al.* 2008]. The southwest monsoon wind impinges on the coastal mountains of Western Ghats [*Xie et al.*, 2006], and low-level orographic lifting of maritime air enhances warm-rain processes.

Despite the differences of convective precipitation profiles, PCT85 in the LUT calculated from the orographic precipitation profiles over the western coast of India (thin red line in Figure 2.8) is nearly the same as that in the LUT calculated from original (thin black line in Figure 2.8). Therefore, the LUTs calculated from the orographic precipitation profiles over the western coast of India provide almost the same estimates as those calculated from original profiles (not shown). The LUTs in the GSMaP algorithm are weighted averages of the convective and stratiform LUTs. We determined the weights from the statistical frequency distributions of the PR convective and stratiform rain for each precipitation type and surface rainfall rate. Stratiform rain is much more common than convective rain, and therefore the weighted LUTs are biased toward stratiform. *Houze* [2012] demonstrated the increased frequency of occurrence of broad stratiform radar echo toward the mountain range of Western Ghats. Thus, the LUT calculated from the orographic

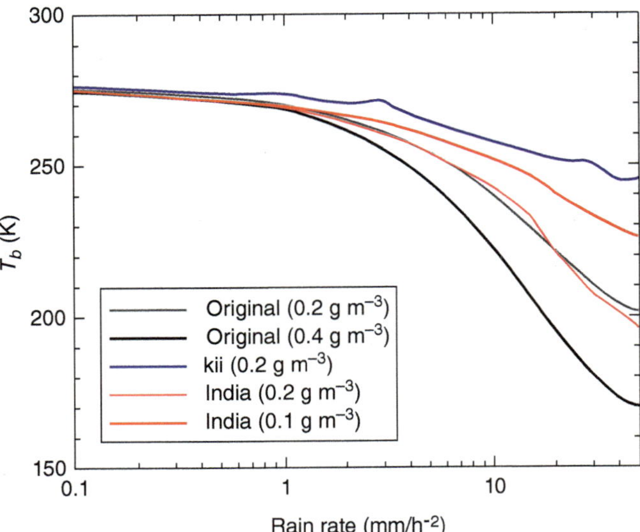

Figure 2.8 LUTs for PCT85 calculated from precipitation profiles shown in Figures 2.7a and 2.7b assuming precipitation-sized ice particle densities of 0.2 g/cm³ (thin black line) and 0.4 g/cm³ (thick black line), those in Figures 2.7c and 2.7d assuming density of 0.2 g/cm³ (thick blue line), and those in Figures 2.7e and 2.7f assuming densities of 0.2 g/cm³ (thin red line) and of 0.1 g/cm³ (thick red line).

precipitation profiles over the west coast of India is nearly the same as that in the LUT calculated from original precipitation profiles because precipitation top heights of stratiform precipitation profiles over the Indian coast (Figure 2.7f) are almost the same as those of original stratiform precipitation profiles (Figure 2.7b).

Highly uncertain parameters of precipitation-related variable models in the RTM calculations are precipitation size ice particle properties, such as size distribution and density. These affect the microwave scattering signature [*Weng and Grody*, 2000; *Bennartz and Petty*, 2001; *Skofronkick-Jackson et al.*, 2002]. Extensive research on this topic is beyond the scope of this chapter, so we focus the effect of precipitation size ice particle density on the LUTs and thus rain estimates.

Precipitation size ice particle density was empirically assumed to be 0.2 g/cm³ for all precipitation types in the GSMaP algorithm [*Aonashi et al.*, 2009]. It was empirically assumed to be 0.3 g/cm³ in the *Aonashi and Liu* [2000] algorithm, which was the originator of GSMaP. To simulate a realistic large horizontal extent of stratiform cloud and the contribution of stratiform rain in the tropics, graupel and snow are widely used as precipitation particles in cloud-resolving models [e.g., *Tao and Simpson*, 1989, 1993; *McCumber et al.*, 1991; *Tao et al.*, 1991]. The density of graupel is 0.4 g/cm³, while that of snow is 0.1 g/cm³ [*Rutledge and Hobbs*, 1984]. The density depends on precipitation regime; high-density precipitation size ice particles (graupel) are dominant in stronger

cold-rain processes, while low-density precipitation size ice particles (snow) are dominant in weaker cold-rain processes. Therefore, we assume that precipitation size ice particle density for the original profiles is 0.4 g/cm³, and that for orographic precipitation profiles over the western Indian coast is 0.1 g/cm³. PCT85 in the LUT calculated from the orographic precipitation profiles over the west coast of India, with precipitation-size ice particle density of 0.1 g/cm³ (thick red line in Figure 2.8), decreased more rapidly than that in the LUT calculated from the orographic precipitation profiles for the Kii Peninsula case (blue line in Figure 2.8). However, it decreased more slowly with rainfall rate than that in the LUT calculated from the orographic precipitation profiles over the west coast of India using precipitation-size ice particle density of 0.2 g/cm³ (thin red line in Figure 2.8). Conversely, PCT85 in the LUT calculated from the original precipitation profiles with precipitation size ice particle density of 0.4 g/cm³ (thick black line in Figure 2.8) decreased more rapidly with rainfall rate than that in the LUT calculated from the original precipitation profiles with precipitation size ice particle density of 0.2 g/cm³ (thin black line in Figure 2.8).

We applied an experimental algorithm (hereafter GSMaP4) to the TMI data. In this algorithm, the thresholds in equation (2.4) are used for orographic/nonorographic classification, as in the GSMaP3 algorithm. However, we estimated rainfall rates using LUTs calculated from the orographic precipitation profiles over the west coast of India, using a precipitation size ice particle density of 0.1 g/m³ for an orographic rainfall pixel. We estimated rainfall rates using LUTs calculated from the original precipitation profiles, using a precipitation size ice particle density of 0.4 g/m³ for a nonorographic rainfall pixel.

Figure 2.9 shows rain rates estimated by GSMaP4 for the MCS case over the coast west of the Western Ghats on 30 June 2007. The GSMaP4 algorithm mitigates the overestimation of GSMaP3 for this case. The GSMaP4 algorithm also differs from GSMaP3 in that for an orographic rainfall pixel, only PCT85 scattering signals were used for rain retrieval. We did not use the adjustments of rain estimates from those signals using PCT37 scattering signals, which were introduced by *Aonashi et al.* [2009] for rain retrieval of tall precipitation. This is because application of the scattering correction of tall precipitation is not appropriate for shallow orographic rain and results in underestimation, similar to the GSMaP2 estimates (Figure 2.4b).

Figure 2.10a depicts GSMaP4 rain rates averaged over summer (June–August 2007). GSMaP4 captured the rain maximum over the west coast of India. Although there remained large values (Figure 2.10b), overestimation by GSMaP3 was mitigated by GSMaP4. Small overestimations by GSMaP1 (Figure 2.1d), GSMaP2 (Figure 2.3b), and GSMaP3 (Figure 2.6b) over the Indian subcontinent

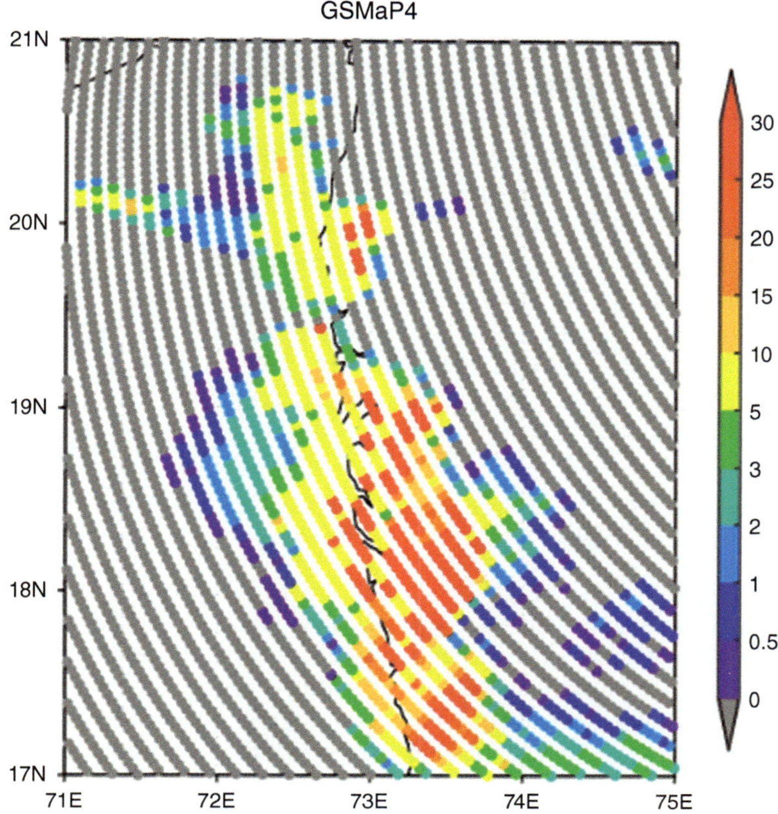

Figure 2.9 As in Figure 2.4b but for surface rain rates estimated by GSMaP4 algorithm from TMI data.

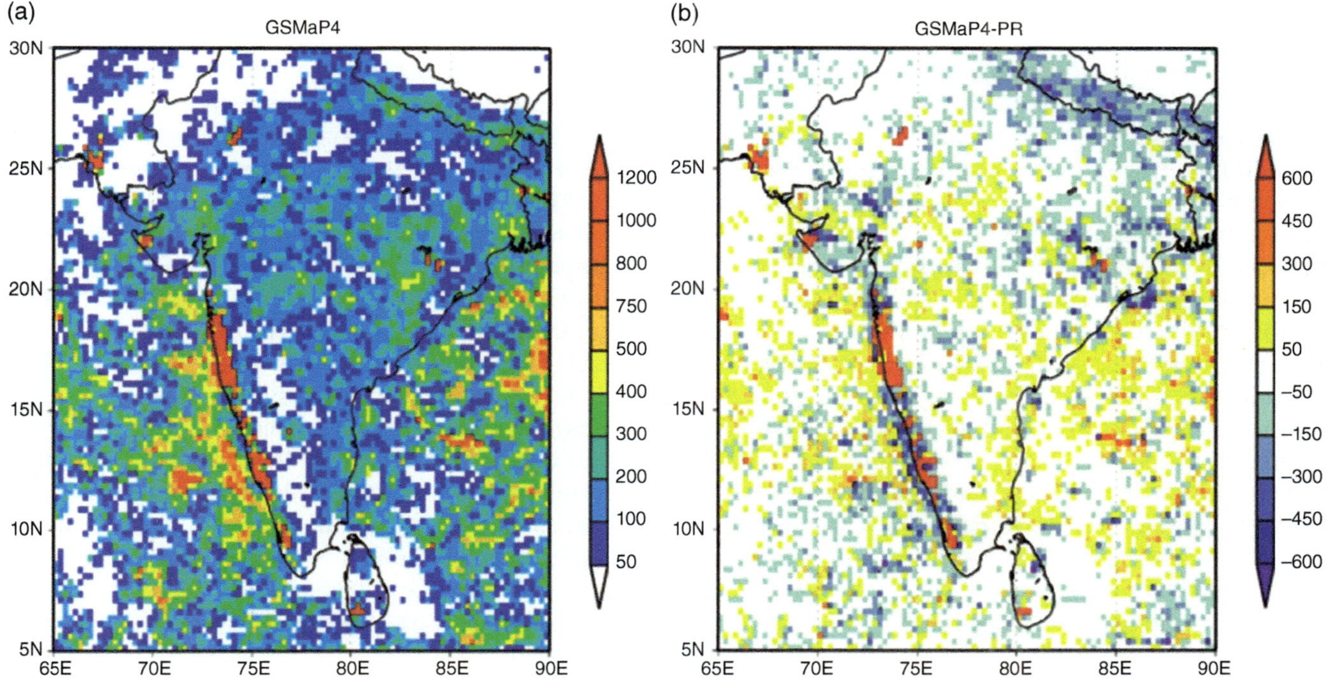

Figure 2.10 As in Figures 2.2c and 2.2e but for (a) average rain rates (mm/month) estimated by GSMaP4 algorithm from TMI data and (b) rain rate differences between GSMaP4 and PR (GSMaP4 – PR).

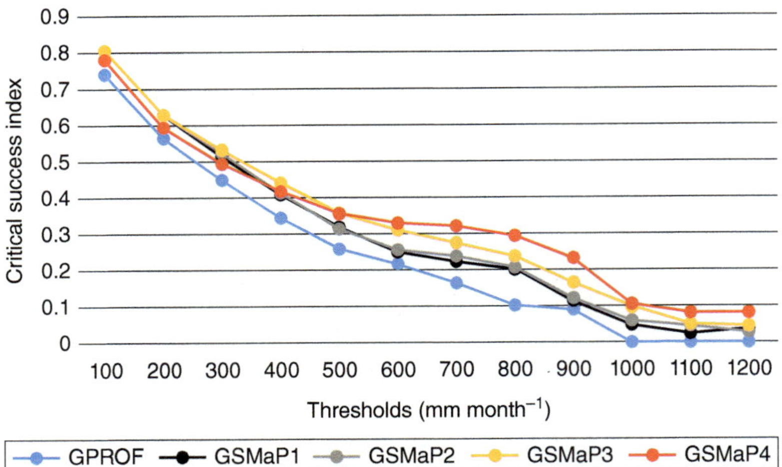

Figure 2.11 Comparisons of CSI for summer (June–August 2007) based on different rainfall thresholds for GPROF, GSMaP1, GSMaP2, GSMaP3, and GSMaP4.

(except for its western coast) are slightly mitigated by GSMaP4. Previous studies using data from the LIS and PR [*Takayabu*, 2006, 2008] suggested a continental precipitation regime (i.e., stronger cold-rain processes) over the Indian subcontinent, except for its western coast. This result suggests that the precipitation size ice particle density of $0.4\,kg/m^3$ is more appropriate than $0.2\,kg/m^3$. Conversely, underestimation by GSMaP over the Tibetan Plateau was not mitigated. Retrieval of the GSMaP algorithm consists of a rain/no-rain classification and estimation of rain rates over a delineated rain area. *Yamamoto et al.* [2011] showed that this classification failed to detect areas of rain on the plateau. Dynamic selection of LUTs based on orographic/nonorographic rainfall classification helps improve estimations of rain rates over the delineated rain area, and therefore underestimation by the GSMaP algorithm was not mitigated.

Figure 2.11 shows the comparison of the critical success index (CSI) for different rainfall thresholds. GPROF has lower CSI values than all versions of GSMaP for all thresholds and has zero values of CSI for rainfall thresholds greater than 1000 mm/month because it does not have rainfall estimates that exceed that threshold. However, GSMaP2 has almost the same values of CSI as GSMaP1, but GSMaP3 has higher values of CSI than GSMaP1 and GSMaP2. This supports our previous statement that orographic LUTs were not used in GSMaP2 because of the threshold in equation (2.3). GSMaP4 has the highest CSI values for thresholds greater than 600 mm/month.

Finally, we revisit our discussion on the threshold for Q. Figure 2.12 shows the comparisons of CSI from GSMaP4 for Q thresholds of 0.1×10^{-6}, 0.3×10^{-6}, and 0.5×10^{-6}. All three thresholds have similar CSI values for both tails of rainfall thresholds. For rainfall thresholds of 600–900, the CSI is slightly higher when $Q = 0.3 \times 10^{-6}$, which supports the selection of Q in equation (2.4). The GSMaP4 results, however, are not very sensitive to the threshold for Q. To preserve the simplicity and transparency, we have used the fixed thresholds for w and Q in equation (2.4) globally.

2.5. SUMMARY AND FUTURE WORK

In this chapter, we improved the performance of rainfall estimates by the GSMaP algorithm [*Aonashi et al.*, 2009] from TMI data over the Indian subcontinent. This area has heavy rainfall over its western coast and is one of the heaviest rainfall areas of the southwest monsoon. We have made improvements to the orographic/nonorographic rainfall classification scheme developed by S13 and T13 and to precipitation-related variable models in the RTM calculations.

The new rainfall classification eliminated nearly all orographic rainfall pixels over the west Indian coast. We lowered thresholds used in the classification so that orographic rainfall pixels were detected. However, we used LUTs calculated from the orographic precipitation profiles for the Kii Peninsula case of orographic rainfall for orographic rainfall pixels, which overestimated the rainfall when compared with PR estimates.

Precipitation top heights of convective precipitation profiles over the Indian coast were higher than those of convective precipitation profiles for the Kii Peninsula case of orographic rainfall used in GSMaP3, demonstrating stronger cold-rain processes. Conversely, precipitation top heights of convective precipitation profiles over India (Figure 2.7e) were lower than those of original convective precipitation profiles (Figure 2.7a), indicating weaker cold-rain processes, consistent with studies using

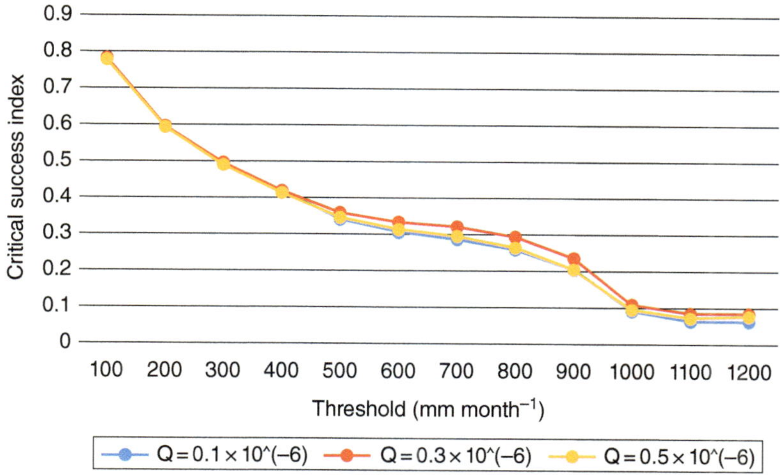

Figure 2.12 As in Figure 2.11 but for GSMaP4 using the thresholds for Q of 0.1×10^{-6}, 0.3×10^{-6} and 0.5×10^{-6}.

data from the LIS and PR [*Petersen and Rutledge*, 2001; *Takayabu*, 2006, 2008]. Despite differences of convective precipitation profiles, PCT85 in the LUT calculated from the orographic precipitation profiles over the west coast of India is nearly the same as that in the LUT calculated from original precipitation profiles. This is because the LUTs in the GSMaP algorithm are weighted averages of convective and stratiform LUTs, calculated according to the statistical frequency distribution of the PR convective and stratiform rain for each precipitation type and surface rainfall rate; therefore, the weighted LUTs are biased toward the stratiform LUTs.

A precipitation size ice particle density was empirically determined for the GSMaP algorithm. Differences in convective precipitation profiles reflect varying precipitation processes, resulting in precipitation size ice particle density variations. We applied an experimental algorithm to the TMI data. In this algorithm, LUTs are calculated from the orographic precipitation profiles over the west coast of India, using precipitation size ice particle density of $0.1 \, \text{g/m}^3$ for an orographic rainfall pixel. LUTs are calculated from the original precipitation profiles, using precipitation size ice particle density of $0.4 \, \text{g/m}^3$ for a nonorographic rainfall pixel. The former corresponds to the density of snow and the latter to the density of graupel, which are widely used in cloud-resolving models. The experimental algorithm estimates are in better agreement with PR estimates, not only over India's west coast but also over nearly the entire Indian subcontinent.

Based on these results, it appears necessary to distinguish precipitation size ice particle properties, such as particle density and size distribution, between different precipitation types (beyond precipitation profiles). To accomplish this, combined radar and radiometer analyses of precipitation profiles [*Viltard et al.*, 2000; *Bennartz and Petty*, 2001; *Masunaga and Kummerow*, 2005; *Shige*

et al., 2006, 2008; *Munchak and Kummerow*, 2011] are very important for improvements of not only rainfall retrieval algorithms but also cloud-resolving models, in which incomplete and uncertain parameterizations of ice microphysical processes persist [e.g., *Eito and Aonashi*, 2009].

The results obtained in this study are only for a short period. A longer term examination of biases during varying monsoon conditions (dry/wet) may reveal different results owing to associated varying precipitation structures.

2.6. ACKNOWLEDGMENTS

This study was supported by a Grant-in-Aid for Scientific Research (KAKENHI). The authors thank G. Huffman from NASA Goddard Space Flight Center for his invitation of this chapter. We also thank S. Nesbitt from the University of Illinois at Urbana-Champaign and an-anonymous reviewer for their constructive comments that helped improve the clarity of the chapter. The authors also thank K. Aonashi from the Meteorological Research Institute, Japan Meteorological Agency, and T. Kubota and S. Kida from the Earth Observation Research Center, Japan Aerospace Exploration Agency, for their discussions.

REFERENCES

Aonashi, K., and G. Liu (2000), Passive microwave precipitation retrievals using TMI during the Baiu period of 1999. Part I: Algorithm description and validation, *J. Appl. Meteorol.*, *39*, 2024–2037.

Aonashi, K., et al. (2009), GSMaP passive microwave precipitation retrieval algorithm: Algorithm description and validation, *J. Meteor. Soc. Japan*, *87A*, 119–136.

Bennartz, R., and G. W. Petty (2001), The sensitivity of microwave remote sensing observations of precipitation to ice particle size distributions, *J. Appl. Meteorol.*, *40*, 345–364.

Byers, H. R., and R. R. Braham, Jr. (1949), *The Thunderstorm*, U.S. Government Printing Office, Washington, D.C.

Dinku, T., S. J. Connor, and P. Ceccato (2010), Comparison of CMORPH and TRMM-3B42 over mountainous regions of Africa and South America, in *Satellite Rainfall Applications for Surface Hydrology*, edited by M. Gebremichael and F. Hossain, pp. 193–204, Springer, New York.

Eito, H., and K. Aonashi (2009), Verification of hydrometeor properties simulated by a cloud-resolving model using a passive microwave satellite and ground-based radar observations for a rainfall system associated with the Baiu front, *J. Meteor. Soc. Japan*, *87A*, 425–446.

Farr, T. G., et al. (2007), The Shuttle Radar Topography Mission, *Rev. Geophys.*, *45*, RG2004, doi:10.1029/2005RG000183.

Ferraro, R., F. Weng, N. Grody, L. Zhao, H. Meng, C. Kongoli, P. Pellegrino, S. Qiu, and C. Dean (2005), NOAA operational hydrological products derived from the Advanced Microwave Sounding Unit, *IEEE Trans. Geosci. Remote Sens.*, *43*(5), 1036–1049.

Gebremichael, M., and F. Hossain (Eds) (2010), *Satellite Rainfall Applications for Surface Hydrology*, Springer, New York.

Gray, W. M., and R. W. Jacobson, Jr. (1977), Diurnal variation of deep cumulus convection, *Monthly Weather Rev.*, *105*, 1171–1188.

Grossman, R. L., and D. R. Durran (1984), Interaction of low-level flow with the Western Ghat Mountains and offshore convection in the summer monsoon, *Monthly Weather Rev.*, *112*, 652–672.

Hirose, M., and K. Nakamura (2002), Spatial and seasonal variation of rain profiles over Asia observed by spaceborne precipitation radar, *J. Climate*, *15*, 3443–3458.

Houze, R. A., Jr. (2012), Orographic effects on precipitating clouds, *Rev. Geophys.*, *50*, RG1001, doi:10.1029/2011RG000365.

Iguchi, T., T. Kozu, R. Meneghini, J. Awaka, and K. Okamoto (2000), Rain-profiling algorithm for the TRMM precipitation radar, *J. Appl. Meteorol.*, *39*, 2038–2052.

Iguchi, T., T. Kozu, J. Kwiatkowski, R. Meneghini, J. Awaka, and K. Okamoto (2009), Uncertainties in the rain profiling algorithm for the TRMM precipitation radar, *J. Meteor. Soc. Japan*, *87A*, 1–30.

Kachi, M., T. Kubota, T. Ushio, S. Shige, S. Kida, K. Aonashi, K. Okamoto, and R. Oki, (2011), Development and utilization of "JAXA global rainfall watch" system based on combined microwave and infrared radiometers aboard satellites (in Japanese). *IEEJ Trans. Fund. Mater.*, *131*, 729–737, doi:10.1541/ieejfms.131.729.

Kozu, T., et al. (2001), Development of precipitation radar onboard the Tropical Rainfall Measuring Mission (TRMM) satellite, *IEEE Trans. Geosci. Remote Sens.*, *39*, 102–116.

Kozu, T., T. Iguchi, T. Kubota, N. Yoshida, S. Seto, J. Kwiatkowski, and Y. N. Takayabu (2009), Feasibility of raindrop size distribution parameter estimation with TRMM precipitation radar, *J. Meteorol. Soc. Japan*, *87A*, 53–66.

Kubota, T., et al. (2007), Global precipitation map using satelliteborne microwave radiometers by the GSMaP project: Production and validation, *IEEE Trans. Geosci. Remote Sens.*, *45*, 2259–2275.

Kubota, T., T. Ushio, S. Shige, S. Kida, M. Kachi, and K. Okamoto (2009), Verification of high-resolution satellite-based rainfall estimates around Japan using a gauge-calibrated ground-radar dataset, *J. Meteor. Soc. Japan*, *87*, 203–222.

Kummerow, C., W. Barnes, T. Kozu, J. Shiue, and J. Simpson (1998), The Tropical Rainfall Measuring Mission (TRMM) sensor package, *J. Atmos. Oceanic Technol.*, *15*, 808–816.

Kummerow, C., et al. (2001), The evolution of the Goddard profiling algorithm (GPROF) for rainfall estimation from passive microwave sensors, *J. Appl. Meteorol.*, *40*, 1801–1820.

Kwon, E.-H., B.-J. Sohn, D.-E. Chang, M.-H. Ahn, and S. Yang (2008), Use of numerical forecasts for improving TMI rain retrievals over the mountainous area in Korea, *J. Appl. Meteor. Climatol.*, *47*, 1995–2007.

Lin, Y.-L. (2007), *Mesoscale Dynamics*, Cambridge.

Liu, G. (1998), A fast and accurate model for microwave radiance calculations, *J. Meteor. Soc. Japan*, *76*, 335–343.

Marshall, J. S., and W. M. Palmer (1948), The distribution of raindrops with size, *J. Meteor.*, *5*, 165–166.

Masunaga, H., and C. D. Kummerow (2005), Combined radar and radiometer analysis of precipitation profiles for a parametric retrieval algorithm, *J. Atmos. Oceanic Technol.*, *22*, 909–929.

McCollum, J. R., and R. R. Ferraro (2003), Next generation of NOAA/NESDIS TMI, SSM/I, and AMSR-E microwave land rainfall algorithms, *J. Geophys. Res.*, *108*, 8382, doi:10.1029/2001JD001512.

McCollum, J. R., and R. R. Ferraro (2005), Microwave rainfall estimation over coasts, *J. Atmos. Oceanic Technol.*, *22*, 497–512.

McCumber, M., W.-K. Tao, J. Simpson, R. Penc, and S.-T. Soong (1991), Comparison of ice-phase microphysical parameterization schemes using numerical simulations of convection, *J. Appl. Meteor.*, *30*, 985–1004.

Mie, G. (1908), Beiträge zur Optik trüber Medien, speziell kolloidaler Metallösungen [Contributions to the optics of turbid media, particularly of colloidal metal solutions], *Ann. Phys.*, *25*, 377–445.

Munchak, S. J., and C. Kummerow (2011), A modular optimal estimation method for combined radar-radiometer precipitation profiling, *J. Appl. Meteor. Climatol.*, *50*, 433–448, doi:10.1175/2010JAMC2535.1.

Negri, A. J., and R. F. Adler (1993), An intercomparison of three satellite infrared rainfall techniques over Japan and surrounding waters, *J. Appl. Meteor.*, *32*, 357–373.

Nesbitt, S. W., and A. M. Anders (2009), Very high resolution precipitation climatologies from the Tropical Rainfall Measuring Mission precipitation radar, *Geophys. Res. Lett.*, *36*, L15815, doi:10.1029/2009GL038026.

Nishitsuji, A., M. Hoshiyama, J. Awaka, and Y. Furuhama (1983), An analysis of propagative character at 34.5 GHz and 11.5 GHz between ETS-II satellite and Kasima station—On the precipitation model from stratus, *IEICE Trans. Commun.*, *J66-B*, 1163–1170 (in Japanese).

Ogura, Y., and M. Yoshizaki (1988), Numerical study of orographic-convective precipitation over the eastern Arabian Sea and the Ghat Mountains during the summer monsoon, *J. Atmos. Sci.*, *45*, 2097–2122, doi:10.1175/1520-0469 (1988)0 45<2097:NSOOCP>2.0.CO;2.

Okamoto, K. (2003), A short history of the TRMM precipitation radar, in Cloud Systems, Hurricanes and the Tropical Rainfall Measurement Mission (TRMM): A Tribute to Dr. Joanne Simpson, *Meteorol. Monogr.*, *51*, edited by W.-K. Tao and R. Adler, 187–195.

Olson, W. S., et al. (2006), Precipitation and latent heating distributions from satellite passive microwave radiometry. Part I: Method and uncertainty estimates, *J. Appl. Meteorol. Climatol.*, *40*, 702–720.

Petersen, W. A., and S. A. Rutledge (2001), Regional variability in tropical convection: Observations from TRMM, *J. Clim.*, *14*, 3566–3586.

Rutledge, S. A., and P. V. Hobbs (1984),The mesoscale and microscale structure and organization of clouds and precipitation in midlatitude cyclones. XII: A diagnostic modeling study of precipitation development in narrow cold-frontal rainbands, *J. Atmos. Sci.*, *41*, 2949–2972.

Sakakibara, H. (1981), Heavy rainfall from very shallow convective clouds, *J. Meteor. Soc. Japan*, *59*, 387–394.

Seto, S., N. Takahashi, and T. Iguchi (2005), Rain/no-rain classification methods for microwave radiometer observations over land using statistical information for brightness temperatures under no-rain conditions, *J. Appl. Meteor.*, *44*, 1243–1259.

Shige, S., H. Sasaki, K. Okamoto, and T. Iguchi (2006), Validation of rainfall estimates from the TRMM precipitation radar and microwave imager using a radiative transfer model: 1. Comparison of the version-5 and -6 products, *Geophys. Res. Lett.*, *33*, L13803, doi:10.1029/2006GL026350.

Shige, S., T. Watanabe, H. Sasaki, T. Kubota, S. Kida, and K. Okamoto (2008), Validation of western and eastern Pacific rainfall estimates from the TRMM PR using a radiative transfer model, *J. Geophys. Res.*, *113*, D15116, doi:10.1029/2007JD009002.

Shige, S., T. Yamamoto, T. Tsukiyama, S. Kida, H. Ashiwake, T. Kubota, S. Seto, K. Aonashi, and K. Okamoto (2009), The GSMaP precipitation retrieval algorithm for microwave sounders. Part I: Over-ocean algorithm. *IEEE Trans. Geosci. Remote Sens.*, *47*, 3084–3097.

Shige, S., S. Kida, H. Ashiwake, T. Kubota, and K. Aonashi (2013), Improvement of TMI rain retrievals in mountainous areas, *J. Appl. Meteor. Climatol.*, *52*, 242–254.

Skofronick-Jackson, G. M., A. J. Gasiewski, and J. R. Wang (2002), Influence of microphysical cloud parameterizations on microwave brightness temperatures, *IEEE Trans. Geosci. Remote Sens.*,, *40*, 187–196.

Spencer, R. W. (1984), Satellite passive microwave rain rate measurement over croplands during spring, summer and fall, *J. Climate Appl. Meteor.*, *23*, 1553–1562.

Spencer, R. W., H. M. Goodman, and R. E. Hood (1989), Precipitation retrieval over land and ocean with SSM/I: Identification and characteristics of the scattering signal, *J. Atmos. Oceanic Technol.*, *6*, 254–273.

Takayabu, Y. N. (2006), Rain-yield per flash calculated from TRMM PR and LIS data and its relationship to the contribution of tall convective rain, *Geophys. Res. Lett.*, *33*, L18705, doi:10.1029/2006GL027531.

Takayabu, Y. N. (2008), Observing rainfall regimes using TRMM PR and LIS data, *GEWEX News*, *18*, 9–10.

Takeda, T., and K. Takase (1980), Radar observation of rainfall system modified by orographic effects, *J. Meteor. Soc. Japan*, *58*, 500–516.

Takeda, T., N. Moriyama, and Y. Iwasaka (1976), A case study of heavy rain in Owase area, *J. Meteor. Soc. Japan*, *54*, 32–41.

Taniguchi, A. S. Shige, M. K. Yamamoto, T. Mega, S. Kida, T. Kubota, M. Kachi, T. Ushio, and K. Aonashi (2013), Improvement of high-resolution satellite rainfall product for Typhoon Morakot (2009) over Taiwan, *J. Hydrometeor, 14*, 1859–1871.

Tao, W.-K., and J. Simpson (1989), Modeling study of a tropical squall-type convective line, *J. Atmos. Sci.*, *46*, 177–202.

Tao, W.-K., and J. Simpson (1993), Goddard cumulus ensemble model. Part I: Model description, *Terr. Atmos. Oceanic Sci.*, *4*, 35–72.

Tao, W.-K., J. Simpson, and S.-T. Soong (1991), Numerical simulation of a sub-tropical squall line over Taiwan Strait, *Monthly Weather Rev.*, *119*, 2699–2723.

Ushio, T., et al. (2009), A Kalman filter approach to the Global Satellite Mapping of Precipitation (GSMaP) from combined passive microwave and infrared radiometric data. *J. Meteor. Soc. Japan*, *87A*, 137–151.

Vicente, G. A., J. C. Davenport, and R. A. Scofield (2002), The role of orographic and parallax corrections on real-time high-resolution satellite rainfall rate distribution, *Int. J. Remote Sens.*, *23*, 221–230.

Viltard, N., C. Kummerow, W. S. Olson, and Y. Hong (2000), Combined use of the radar and radiometer of TRMM to estimate the influence of drop size distribution on rain retrievals, *J. Appl. Meteorol.*, *39*, 2103–2114.

Wang, N. Y., C. Liu, R. Ferraro, D. Wolff, E. Zipser, and C. Kummerow (2009), TRMM 2A12 land precipitation product-status and future plans, *J. Meteor. Soc. Japan*, *87A*, 237–253.

Weng, F., and N. C. Grody (2000), Retrieval of ice cloud parameters using a microwave imaging radiometer, *J. Atmos. Sci.*, *57*, 1069–1081.

Weng, F., L. Zhao, R. R. Ferraro, G. Poe, X. Li, and N. C. Grody (2003), Advanced microwave sounding unit cloud and precipitation algorithms, *Radio Sci.*, *38*(4), 8068, doi:10.1029/2002RS002679.

Werner, M. (2001), Shuttle Radar Topography Mission (SRTM), Mission overview, *J. Telecom. (Frequenz)*, *55*, 75–79.

Williams, E. R., S. A. Rutledge, S. G. Geotis, N. Renno, E. Rasmussen, and T. Rickenbach (1992), A radar and electrical study of tropical "hot towers", *J. Atmos. Sci.*, *49*, 1386–1395.

Wu, C.-C. (2013), Typhoon Morakot (2009): Key findings from the journal TAO for improving prediction of extreme rains at landfall., *Bull. Amer. Meteor. Soc.*, *94*, 155–160.

Wu, C.-C., and M. J. Yang (2011), Preface to the special issue on "Typhoon Morakot (2009): Observation, modeling, and forecasting", *Terr. Atmos. Ocean. Sci.*, *22*, I, doi:10.3319/TAO.2011.10.01.01(TM).

Xie, S. P., H. M. Xu, N. H. Saji, Y. Q. Wang, and W. T. Liu (2006), Role of narrow mountains in large-scale organization of Asian monsoon convection, *J. Clim.*, *19*, 3420–3429, doi:10.1175/JCLI3777.1.

Yamamoto, M. K., K. Ueno, and K. Nakamura (2011), Comparison of satellite precipitation products with rain gauge data for the Khumb Region, Nepal Himalayas, *J. Meteor. Soc. Japan*, *89*, 597–610.

Yatagai, A., K. Kamiguchi, O. Arakawa, A. Hamada, N. Yasutomi, and A. Kitoh (2012), APHRODITE: Constructing a long-term daily gridded precipitation dataset for Asia based on a dense network of rain gauges, *Bull. Amec. Meteorol. Soc.*, *93*, 1401–1415, doi:10.1175/BAMS-D-11-00122.1.

Zipser, E. J., and K. R. Lutz (1994), The vertical profile of radar reflectivity of convective cells: A strong indicator of storm intensity and lightning probability? *Monthly Weather Rev.*, *122*, 1751–1759.

3

Integrating Information from Satellite Observations and Numerical Models for Improved Global Precipitation Analyses: Exploring for an Optimal Strategy

Pingping Xie[1] and Robert J. Joyce[1,2]

3.1. INTRODUCTION

Two categories of techniques have been developed in the last 15 years or so to generate precipitation estimates at high spatial and temporal resolution by fusing infrared (IR) and passive microwave (PMW) observations from multiple satellite platforms. In the first category, estimates of instantaneous precipitation rates are derived from IR brightness temperature observed by geostationary (GEO) satellites using an empirical relationship established through comparison against retrievals from concurrent PMW measurements by low-earth-orbit (LEO) platforms [*Hsu et al.*, 1997; *Kuligowski*, 2002; *Turk et al.*, 2004; *Huffman et al.*, 2007]. In some techniques, the IR-based estimates are further combined with PMW-based precipitation retrievals to improve quantitative accuracy [*Turk et al.*, 2004; *Huffman et al.*, 2007]. Common to these so-called Euler approach techniques is the use of the IR- and PMW-based estimates only at the target grid box to define the final fused precipitation analysis.

The other category of techniques, called the Lagrangian approach, produces precipitation analysis through the propagation of PMW-derived instantaneous precipitation rates along cloud motion vectors from their observation times to the target analysis time (*Joyce et al.*, 2004; *Ushio et al.*, 2009; *Huffman et al.*, 2011]. The motion vectors are defined through comparison of the location and patterns of cloud systems depicted on two consecutive GEO IR images that are often 30 min apart. Recent versions of this technique further blend these propagated

PMW retrievals with IR-based estimates to ensure stable performance over periods of sparse PMW sampling [*Ushio et al.*, 2009; *Joyce and Xie*, 2011]. The Climate Prediction Center (CPC) Morphing technique (CMORPH) of *Joyce et al.* [2004] and *Joyce and Xie* [2011] and the global satellite mapping of precipitation (GsMAP) of *Ushio et al.* [2009] are two Lagrangian-based algorithms that produce high-resolution global precipitation estimates on an operational basis. Meanwhile, a processing system, called Integrated Multi-satellite Retrievals for GPM (IMERG; *Huffman et al.*, 2011], is being developed for the production of Global Precipitation Mission [GPM; *Hou et al.*, 2014] Level 3 blended satellite precipitation estimates. Development of the IMERG takes advantage of the propagation and morphing techniques of *Joyce et al.* [2004] and *Joyce and Xie* [2011] enhanced by the PMW intercalibration components of *Huffman et al.* [2007] and the IR-based technique of *Hsu et al.* [1997]. Evaluation studies exhibit superior performance of the Lagrangian-based techniques in the production of precipitation analyses, especially when PMW retrievals of frequent sampling are available [*Xie et al.*, 2007; *Ebert et al.*, 2007].

None of the integration techniques described above produce precipitation estimates that cover global regions of high latitudes beyond the 60°N/S parallels. Even within this latitude band, no accurate precipitation estimation can be made over snow-covered land areas or over oceanic sea ice. In general, performance of satellite precipitation estimates is compromised over the mid- and high latitudes and during cold seasons [*Ebert et al.*, 2007]. Limitations and shortcomings of the current generation integrated satellite precipitation estimates are largely attributable to the two types of inputs to the systems:

[1]*NOAA Climate Prediction Center, College Park, Maryland, USA*

[2]*Innovim, LLC, Greenbelt, Maryland, USA*

Remote Sensing of the Terrestrial Water Cycle, Geophysical Monograph 206. First Edition. Edited by Venkat Lakshmi.
© 2015 American Geophysical Union. Published 2015 by John Wiley & Sons, Inc.

GEO IR data and LEO PMW precipitation retrievals. The GEO IR data are used to derive precipitation estimation at a high space and time resolution in virtually all of the techniques. In addition, in CMORPH and other Lagrangian approach-based techniques, the GEO IR data are also utilized to compute cloud motion vectors to propagate instantaneous PMW precipitation fields. The GEO IR data, however, is available only over a latitude band from ~60°S to 60°N. Current generation techniques to derive precipitation estimates from satellite PMW and IR observations exhibit less than desirable skills over mid- and high latitudes, especially during cold seasons, and over snow and sea ice cover.

The objective of this work is to examine potential information sources for mid- and high latitude precipitation and to explore optimal strategies to expand integrated satellite precipitation estimates to cover the entire globe from pole to pole. While solutions may differ based on the various technology frameworks used in integration, the CMORPH technique developed at NOAA Climate Prediction Center (CPC) is utilized as the testbed for this investigation. The results achieved in this work, however, are expected to provide useful insights for the developments of other techniques to produce pole-to-pole coverage of global precipitation estimates. As we will discuss in this chapter, we explore the possibility of creating two versions of the integrated precipitation analyses: one using all available information from both satellite observations and numerical model forecasts/simulations to achieve best possible coverage and accuracy and the other using satellite observations only for applications in model validations.

This chapter is organized into seven sections. Section 3.2 describes the current generation CMORPH technique and its limitations; Section 3.3 presents the Kalman filter (KF)-based CMORPH developed by *Joyce and Xie* [2011] to be used here as the integration framework for the pole-to-pole CMORPH; Sections 3.4 and 3.5 examine, respectively, potential information sources for the definition of precipitation and cloud motion vectors over the globe covering the entire latitude band from the equator to the poles; Section 3.6 explores strategies and demonstrates technical feasibility to integrate information of different sources for the optimal definition of global precipitation analyses. Finally, conclusions will be given in Section 3.7.

3.2. CURRENT GENERATION CMORPH AND ITS LIMITATIONS

CMORPH [*Joyce et al.*, 2004] derives high-resolution (8 km × 8 km) half-hourly precipitation analyses over a quasi global domain (60°S–60°N) through the integration of PMW estimates from all available LEO satellites in a Lagrangian framework (Figure 3.1, left). The CMORPH processing system is composed of three parts: (a) defining cloud motion vectors, (b) preprocessing of instantaneous PMW-based precipitation retrievals from individual satellites, and (c) constructing global precipitation estimates through propagating and morphing the individual PMW retrievals.

First, motion vectors of precipitating cloud systems are defined on a 2.5° lat/lon grid over the globe using a cross-correlation technique applied to the full-resolution IR data observed from GEO satellites. Spatially lagged pattern correlation between two consecutive GEO IR images is computed for each grid point using data over a 5° lat/lon domain centering at the target grid point. Proxy precipitation motion vectors are then derived from the cloud motion vectors using a model derived from comparisons between cloud motion vectors and radar precipitation motion [*Joyce et al.*, 2004]. The proxy precipitation motion vectors defined at 2.5° lat/lon grid are then interpolated onto a global grid of 8 km × 8 km and used to propagate the PMW-based instantaneous precipitation retrievals as described below.

The CMORPH preprocessing subsystem generates gridded maps of combined PMW precipitation rates, called MWCOMB, from raw level 2 PMW precipitation retrieval packages received from respective sources. Level 2 precipitation rate retrievals from individual platform/sensors are decoded and quality controlled to eliminate suspicious data. These quality-controlled level 2 retrievals of instantaneous precipitation rate are then mapped from their respective observation times and field of view (FOV) onto a global grid field of 0.0727° lat/lon (8 km × 8 km, at the equator) in a temporal interval of 30 min. The gridded level 2 precipitation retrievals from different sensors are then calibrated against a common reference through matching a probability density function (PDF) of the precipitation retrievals from the target sensor against that for a reference precipitation field. Over the tropics and subtropics, level 2 precipitation retrievals from the TRMM Microwave Imager (TMI) are used as the reference, while over the extra-tropical areas beyond the 40°N/S parallels, those from the Advanced Microwave Scanner Radiometer (AMSR) aboard NASA's Earth Observing System (EOS) *Aqua* satellite are utilized as the PDF calibration standard.

The gridded maps of the combined PMW retrievals (MWCOMB) are then propagated from their respective observations to the target analysis time along the radar-adjusted cloud motion vectors. The propagation is separately performed from both forward and backward directions to take full advantage of the PMW observations available around the target scan analysis time. The final CMORPH precipitation estimates are achieved through "morphing" the forward and backward-propagated PMW

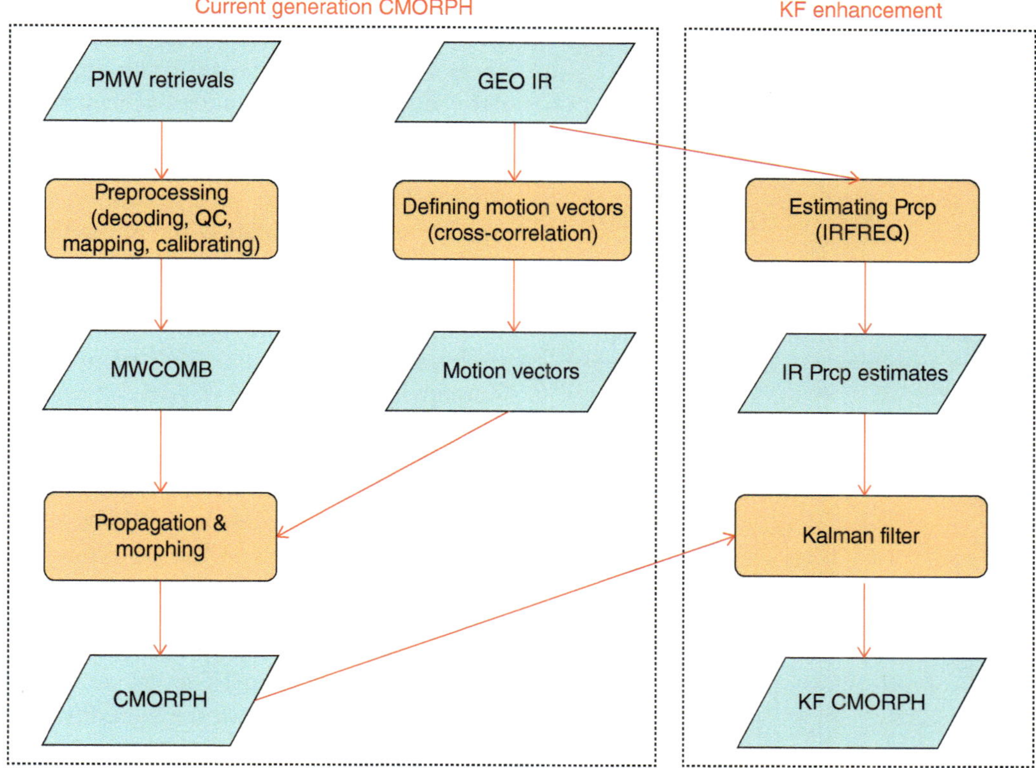

Figure 3.1 Flowchart of the current generation CMORPH (dashed framework on the left) and the Kalman filter (KF) enhancement (dashed framework on the right). Blue parallelograms indicate input/output data while brown rectangles are analysis procedures. The two parts together comprise the Kalman-filter-based CMORPH [*Joyce and Xie*, 2011].

retrievals, in which a linear combination is taken with the weighting coefficients inversely proportional to the temporal length of propagation. Through this Lagrangian approach, the CMORPH technique defines high-resolution precipitation estimates by interpolating PMW retrievals in a combined time-space domain with consideration of the cloud system movements. Precipitation intensity of the cloud systems, however, is assumed to be unchanged throughout the propagation period.

The operational version CMORPH-integrated satellite precipitation estimates started real-time production from December 2002. A retrospective CMORPH reprocessing has recently been completed that generates precipitation estimates using a fixed integration algorithm and PMW level 2 precipitation retrievals of the most recent identical version from all available sensors. As a result, a consistent data set of integrated satellite analysis of 30-min precipitation has been constructed on an 8 km × 8 km grid over the globe from 60°S to 60°N covering a 15-year period from 1998 to the present [*Xie et al.*, 2013].

While CMORPH consistently presents superior performance in the estimation of the spatial distribution and temporal variations of precipitation over most of the global regions and for all seasons compared to other similar techniques, shortcomings exists in the current operational version CMORPH precipitation products. One such limitation is the incomplete global coverage of the CMORPH precipitation estimates. The current CMORPH, and all other high-resolution satellite precipitation analyses, only cover from latitudes 60°S to 60°N, largely because of the geographical restriction of the GEO IR images used to derive the propagation vectors.

Even within the latitude coverage of 60°S–60°N, the quality of CMORPH precipitation is compromised over extra-tropical areas and during cold seasons. As shown in Figure 3.2, performance of the current version CMORPH is relatively stable over regions with surface air temperature above the freezing point. However, the performance degrades quickly as the temperature decreases from 0 °C to −15 °C when CMORPH fails to capture any precipitation with correlation and relative bias close to 0 and −100%, respectively. The compromised performance of CMORPH is attributable largely to the fact that the current generation PMW precipitation retrievals used as

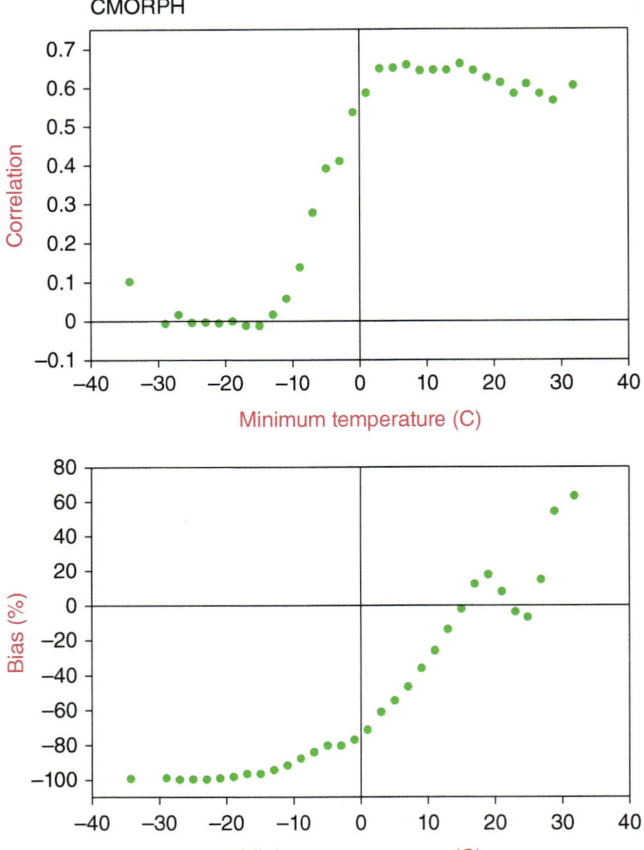

Figure 3.2 (Top) Correlation and relative bias (% bottom) of the CMORPH daily precipitation estimates against the CPC gauge-based analysis, calculated as a function of minimum surface air temperature. Calculations are based on daily data on a 0.25° lat/lon grid over the global land for 2009. Only data over a 0.25° lat/lon grid with one or more reporting gauge are included in the comparison.

because of the assumptions that PMW retrievals propagated along cloud motion vectors exhibit superior skills in representing precipitation variations than estimates derived from IR observations at their target time. Error of the propagated PMW precipitation retrievals, however, increases with temporal distance from original scan time. GEO IR-based precipitation estimates therefore provide useful information for filling the temporal gaps between LEO PMW observations.

A modified CMORPH integration algorithm is developed to combine the precipitation information from both the LEO PMW and GEO IR observations under a Kalman filter framework [Figure 3.1; *Joyce and Xie*, 2011]. To this end, a technique, called IR frequency (IRFREQ), is developed to estimate precipitation from the GEO IR data at an 8 km × 8 km spatial resolution through a PDF matching against colocated PMW retrievals. PMW retrievals derived from LEO platforms are first propagated from their respective observation times to the target analysis time in both the forward and backward directions. A weighted mean of the forward, and backward-propagated PMW, and IR-based precipitation estimates (IRFREQ) are computed with the PMW-based weights defined as a function of sensor type, length of propagation time, location and season, and used as the first guess of the integrated precipitation analysis. The IR-based precipitation estimates (IRFREQ) are used when/where PMW observations are not available within a time period of 90 min as the IR-based weights become relevant. This represents an improvement upon the morphing process in the current operational version CMORPH where PMW weights are computed only as a linear function of propagation time, and IR information was not used at all for precipitation derivation or no contribution to the final estimate. Comprehensive examinations demonstrated improved quality of the KF-based CMORPH-integrated precipitation estimates in capturing precipitation and its variations over the globe from 60°S to 60°N [*Joyce and Xie*, 2011].

Different from the operational CMORPH [*Joyce et al.*, 2004], the KF-CMORPH [*Joyce and Xie*, 2011] is capable of taking in precipitation fields from multiple sources to enhance the PMW retrievals, enabling the construction of integrated precipitation analysis over regions and for seasons where/when quality PMW retrievals are not available.

inputs to the CMORPH present an error in detecting and quantifying precipitation (both rainfall and snowfall) over land areas covered by snow or ice over the ground [*Wang et al.*, 2009; *Gopalan et al.*, 2010). Encouraging progress is being made by the GPM level 2 PMW precipitation retrieval product developers to improve retrievals of rainfall in association with warm clouds and orographic forcing and snowfall during cold seasons.

3.3. KALMAN-FILTER-BASED CMORPH INTEGRATION ALGORITHM

The current operational version CMORPH defines precipitation analysis by integrating PMW precipitation retrievals from LEO platforms. Precipitation information from IR-based precipitation estimates and other sources is not utilized as inputs to its processing system, largely

3.4. POTENTIAL INFORMATION SOURCES FOR GLOBAL PRECIPITATION DEFINITION

One primary cause of the limited spatial coverage and degraded performance of CMORPH and all other current generation high-resolution integrated precipitation products is the insufficient use of available information. As described in the previous sections, two categories of information are needed for the production of high-quality, high-resolution global precipitation under a Lagrangian framework, that is, information on the instantaneous precipitation distribution at the moment of instrument observation and accurate motion vectors of the precipitating cloud systems. Development of the KF-CMORPH makes it possible to integrate precipitation information from multiple sources. In this section, we will examine the potential sources for the definition of precipitation distribution over the entire globe from pole to pole.

Retrievals of instantaneous precipitation rates derived from PMW measurements aboard LEO satellites have demonstrated good results, especially over tropics and during warm seasons [*Ebert et al.*, 2007; *Xie et al.*, 2007], and used as the primary source for the precipitation distribution by CMORPH [*Joyce et al.*, 2004; *Joyce and Xie*, 2011] and other published high-resolution integrated satellite precipitation estimates [*Hsu et al.*, 1997; *Turk et al.*, 2004; *Joyce et al.*, 2004; *Huffman et al.*, 2007; *Ushio et al.*, 2009). The current generation PMW-based precipitation retrieval techniques, however, present compromised skills in estimating precipitation over extra-tropical areas and for cold seasons and are unable to accurately detect precipitation (rainfall or snowfall) over surfaces covered by ice or snow. As shown in Figure 3.3 (*top*), huge gaps exist in valid (i.e., not flagged) PMW precipitation retrievals even during summer months over high-latitude areas (e.g., the Arctic Ocean and Greenland), suggesting a need to include precipitation information from other sources in constructing an integrated precipitation analysis with complete spatial coverage from pole to pole.

Another source of precipitation information is estimation derived from satellite-observed IR data. While earlier IR-based efforts aim to derive time/space-averaged precipitation through calibration against ground truth [e.g., *Arkin*, 1979; *Negri et al.*, 1984], IR precipitation estimation techniques developed in the last 15 years or so assign precipitation rates at the raw GEO IR pixel resolution (~4 km) through empirical cloud-precipitation relationship established from comparison with concurrent PMW retrievals [*Hsu et al.*, 1997; *Kuligowski*, 2002; *Turk et al.*, 2004; *Huffman et al.*, 2007; *Joyce and Xie*, 2011]. One potential source for mid- and high-latitude precipitation is the estimates derived from IR window channel observations made by the Advanced Very High Resolution Radiometers (AVHRR) aboard NOAA polar-orbiting satellites. As shown in Figure 3.2 (bottom), AVHRR observations from operational LEO satellites combined provide frequent coverage over the middle and high latitudes, filling the gaps of valid PMW retrievals from LEO satellites. More than 40% of the 30 min temporal intervals are covered by one or more scans from the AVHRR over high latitudes poleward of 60°N, with the maximum reporting rate of 50% or higher observed valid retrievals over the polar cap due to the

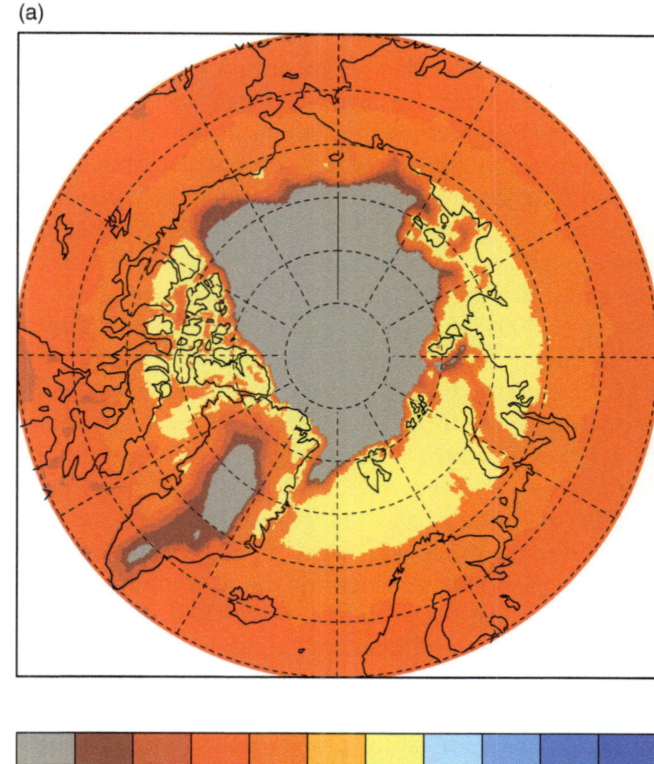

(a)

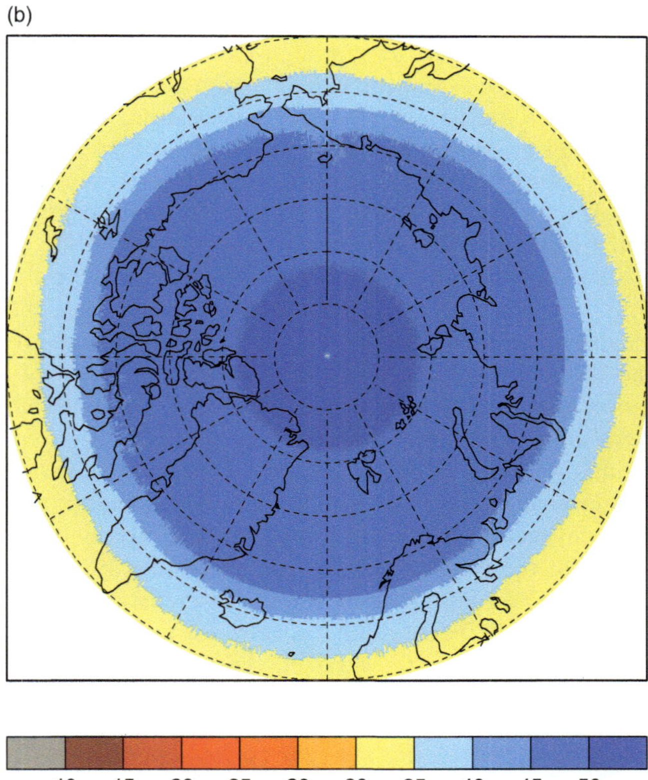

(b)

Figure 3.3 Percentage of 30 min time slots covered (a) by PMW retrievals from all nine LEO satellites and (b) by AVHRR IR observations from five LEO satellites over a one-month period from 1 July to 31 July 2009.

orbit overlap sampling pattern of the polar-orbiting satellites.

A key question here is whether or how well an AVHRR IR-based technique can estimate precipitation over the middle and high latitudes. Although IR-based precipitation estimates in general presents degraded skill in deriving precipitation from nonconvective clouds, a carefully calibrated technique may be able to extract information for extra-tropical precipitation with reasonable accuracy [*Xie and Arkin*, 1998; *Hsu et al.*, 1997; *Kuligowski*, 2002; *Turk et al.*, 2004; *Huffman et al.*, 2007]. To this end, a simple technique is developed, following the practice for GEO IR-based IRFREQ, to derive precipitation estimates through matching the PDF of the AVHRR window channel brightness temperature (T_b) against colocated PMW precipitation retrievals. The PDF matching is performed for different latitude band and for land and ocean, respectively. Over the polar caps where no PMW retrievals are available, PDF tables created for the next lower latitude band are used in the precipitation estimation, assuming that the cloud-precipitation relationship is similar over the polar caps and their neighboring regions.

AVHRR IR-based precipitation estimates are then produced on a 0.05° lat/lon grid over the entire globe for a 2 month period from 1 July to 31 August 2009. Despite the limited tuning for the PDF matching technique, the AVHRR IR-based estimates exhibit reasonable spatial distribution patterns for the high-latitude precipitation (Figure 3.4). To examine the quantitative accuracy of the AVHRR IR-based and the PMW-retrieved precipitation over high latitudes, the two types of satellite precipitation estimates are compared to radar-derived precipitation over Finland and its adjacent oceans for July and August of 2009. Both the PMW- and the AVHRR IR-based precipitation estimates to be compared here are from the same polar orbiting satellite, the NOAA 18, yielding statistics reflecting purely the performance of the two techniques in estimating instantaneous precipitation rates. Not a surprise at all, the PMW precipitation retrievals exhibit better skills than the AVHRR IR-based estimates (Table 3.1). Correlation of satellite-derived 30 min precipitation over a 0.25° lat/lon grid against the Finnish radar observations is 0.349/0.317 over land/ocean for the PMW retrievals, compared to 0.262/0.260 for the IR-based estimates. The performance statistics for the AVHRR IR-based estimates, however, are reasonable for precipitation of the targeted time/space scale, suggesting the feasibility to fill in temporal gaps of PMW satellite-based precipitation observations with the estimates derived from the AVHRR IR data aboard polar-orbiting satellites.

Another potential source of global precipitation is the forecasts and simulations made by state-of-the-art

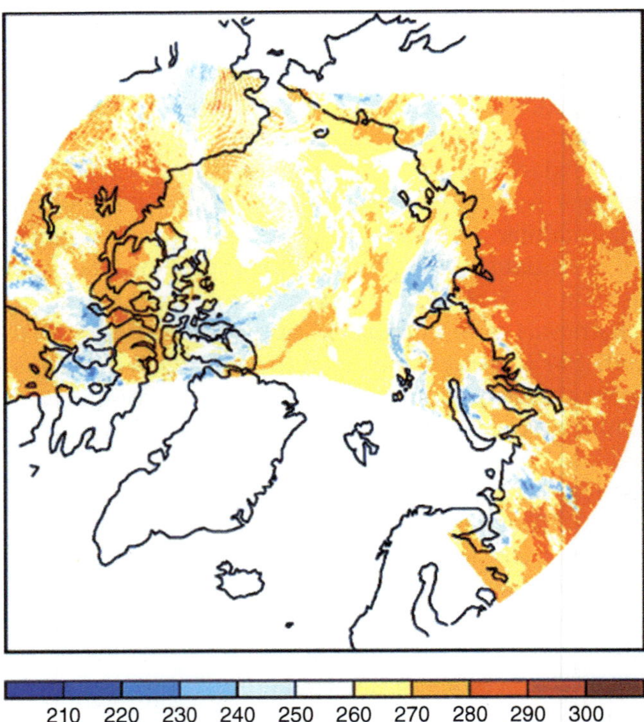

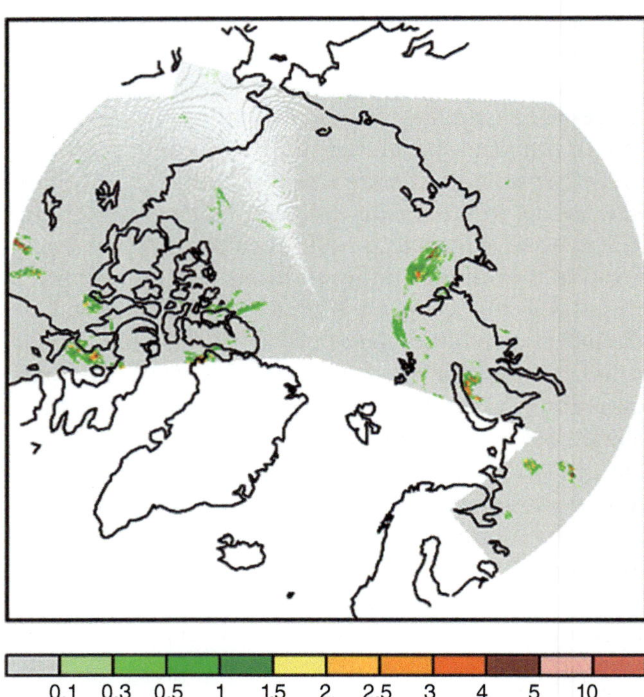

Figure 3.4 Sample AVHRR infrared (IR) window channel brightness temperature (*top*, °K) and precipitation estimates (mm/h) derived from the AVHRR IR data (*bottom*) for 00:00 UTC, 1 August 2009. The AVHRR IR data are from NOAA 18 polar orbiting satellite.

Table 3.1 Correlation between the PMW retrieved and the AVHRR IR-based precipitation with radar observations over Finland and adjacent oceans[a]

Satellite Estimates	Land	Ocean
PMW retrievals from NOAA 18	0.349	0.317
IR-based estimates from NOAA18	0.262	0.260

[a]The comparisons are for 30 min precipitation averaged over a grid box of 0.25° lat/lon for July and August 2009.

numerical models. These numerical model forecast and analysis systems assimilate information from *in situ* measurements, satellite observations, and many other sources and define the intensity and distribution of precipitation as a result of the simulated dynamic, thermodynamic, and physical processes. While these numerical models may eventually generate precipitation fields with superior performance than those achieved through a pure statistical analysis technique such as CMORPH, recent examinations have shown that model precipitation fields present good skills over extra tropics and for cold seasons while their performance is less than desirable for meso-scale cloud systems especially over the tropics [*Ebert et al.*, 2007].

Although inclusion of model-based information into an integrated precipitation analysis makes it inappropriate as an "observation" to evaluate other models, optimal blending of model precipitation with observations extends the spatial coverage and improves the quantitative accuracy of precipitation analyses for many other applications such as weather/climate monitoring and water cycle research. *Xie and Arkin* [1996, 1997] included precipitation fields generated by the National Centers for Environmental Prediction (NCEP) / National Center for Atmospheric Research (NCAR) reanalysis [*Kalnay et al.*, 1996] as an optional input in constructing their CPC merged analysis of precipitation (CMAP). The model produced precipitation fields examined and used in these studies, however, are available for a relatively coarse spatial and temporal resolution. For example, the NCEP/NCAR reanalysis is implemented on a spatial grid of ~220 km and the precipitation fields are generated for output every 6 h.

A new generation of reanalyses have been constructed by three numerical modeling centers, including the Climate Forecast System reanalysis [CFSR, *Saha et al.*, 2010] by the NOAA National Centers for Environmental Prediction Center (NCEP), the Modern Era Retrospective Analysis for Research and Applications [MERRA; *Bosilovich*, 2008] by the NASA Goddard Space Flight Center (GSFC) Global Modeling and Assimilation Office (GMAO), and the European Center for Medium-Range Weather Forecasts (ECMWF) interim reanalysis [*Dee et al.*, 2011]. Generated at a refined time/space resolution, precipitation fields from these reanalyses may be utilized as a supplementary source for the global information to be used as inputs to our integrated global precipitation analysis.

To examine the feasibility and optimal strategy to utilize the reanalysis precipitation in our integrated global precipitation analysis, daily precipitation fields from the above mentioned three reanalyses and those from the operational version CMORPH satellite estimates are compared against the daily gauge analysis of *Xie et al.* (2010) for a 12 year period from 1999 to 2010. To this end, precipitation fields for the three sets of reanalyses and the CMORPH are integrated from their respective native time/space resolution to 0.5° lat/lon and daily resolution and compared against the daily gauge analysis over the global land. Comparisons are performed for four latitude bands (60°N–90°N, 40°N–60°N, 20°N–40°N, and 20°S–20°N) to examine the latitude dependence of the performance.

The operational version CMORPH-integrated satellite estimates present superior performance in capturing precipitation and their variations over tropics (20°S–20°N) throughout the annual cycle and over subtropics (20°N–40°N) during warm seasons (Figure 3.5). The daily precipitation fields from the three sets of reanalyses, meanwhile, exhibit correlation of similar magnitude with that for the CMORPH over middle latitudes (40°N–60°N) during warm seasons and outperform the satellite estimates over middle latitudes (40°N–60°N) during cold seasons. Over high latitudes (60°N–90°N), where no data are available from CMORPH or any other integrated satellite precipitation estimates, the three sets of reanalyses show consistently high performance in estimating daily precipitation, with a pattern correlation of 0.6 or higher observed throughout the annual cycle. The complementarity of the performance for the satellite estimated and reanalysis precipitation shown in Figure 3.5 strongly suggests the necessity of including precipitation information from the state-of-the-art reanalyses to enhance our CMORPH-integrated global precipitation analysis.

3.5. POTENTIAL INFORMATION SOURCES FOR THE DEFINITION OF CLOUD MOTION VECTORS

Precipitation fields of extensive spatial coverage and successive time steps are needed to define the cloud motion vectors for use in a Lagrangian approach integration algorithm such as CMORPH. Over the middle and high latitudes where GEO IR data are not available, at least three categories of data might be utilized to provide information on precipitation and its movements, that is, PMW retrievals, AVHRR IR-based estimates, and the precipitation fields produced by the state-of-the-art reanalyses. PMW retrievals and AVHRR IR precipitation

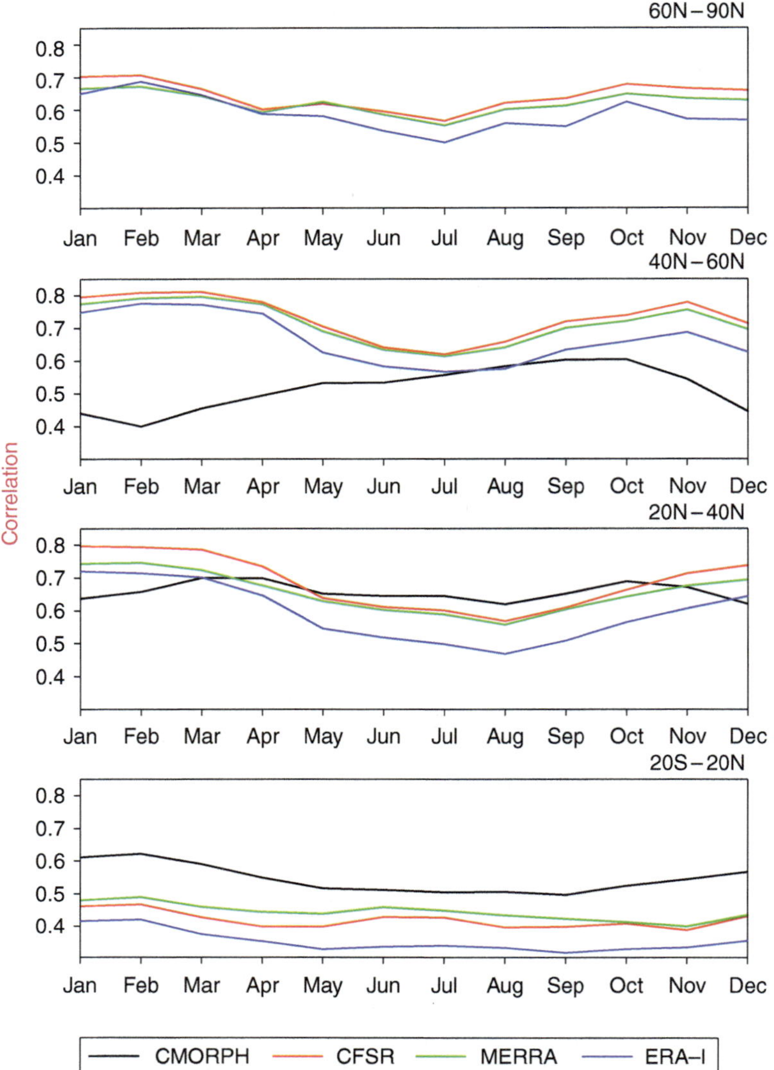

Figure 3.5 Seasonal variations of the pattern correlation between the daily precipitation from CMORPH satellite estimates (black), CFS reanalysis (CFSR, red), Modern Era Reanalysis for Research and Applications (MERRA, green), and the ECMWF Interim reanalysis (ERA-I, blue) against daily gauge analysis. Results for four different latitude bands (60°N–90°N, 40°N–60°N, 20°N–40°N, and 20°S–20°N) are calculated separately to examine the latitude dependence of the performance. Statistics are calculated for daily precipitation averaged over a 0.5° lat/lon grid. Only data over a 0.5° lat/lon grid with one or more reporting gauges are included in the calculations.

estimates aboard LEO satellites provide an observation-based source of information. Temporal/spatial coverage of the existing LEO satellite observations, however, need to be examined for their applications in deriving the motion vectors. *Key et al.* [2003] demonstrated that cloud drift and water vapor winds can be derived in the polar regions through the use of Moderate Resolution Imaging Spectroradiometer (MODIS) observations when successive overpasses are available within 100 min from the *Terra* and *Aqua* satellites. In our practice, consecutive fields of a shorter time interval (at least less than 60 min) are desirable to capture the evolution of precipitating cloud systems [*Joyce and Xie*, 2011]. As shown in

Figure 3.6, even for July 2009 when PMW and AVHRR observations are available from nine and six polar LEO satellites, respectively, only about 20–30% of the grid boxes over high latitudes are covered by PMW or AVHRR observations from two consecutive observations of 30 min intervals. This means that precipitation fields from LEO satellites, especially when the PMW retrievals and the AVHRR IR-based estimates are combined, can be utilized to define precipitating cloud motion vectors only over part of the region and for a fraction of the time over the high latitudes.

By definition, precipitation fields produced by operational numerical models and modern reanalysis systems

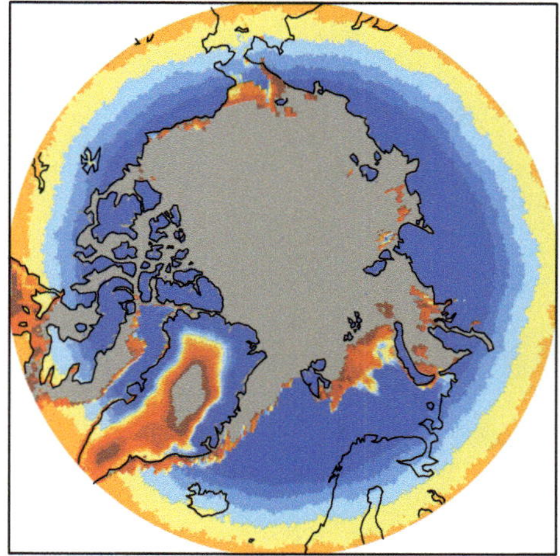

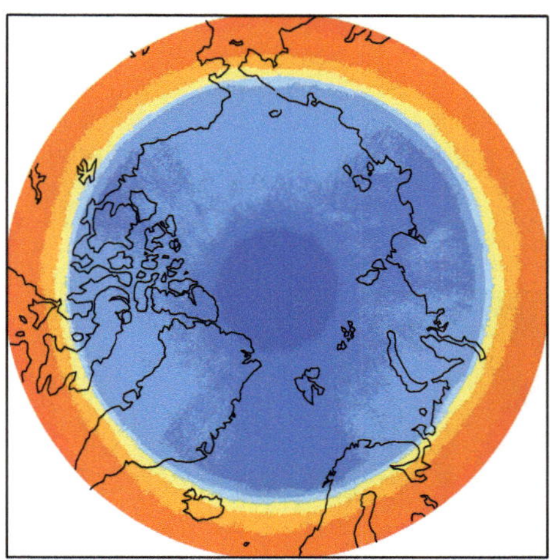

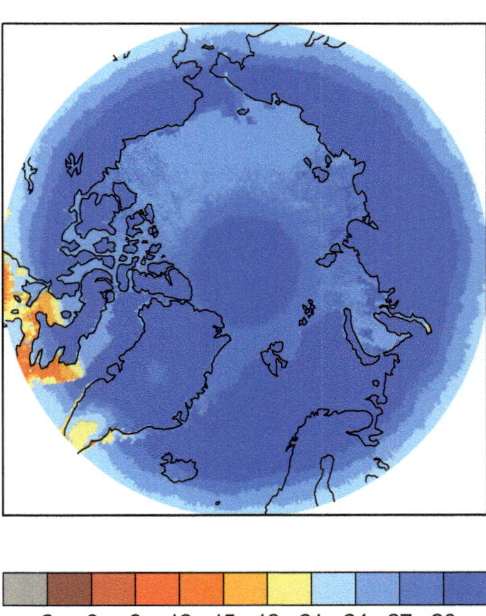

Figure 3.6 Percentage of successive 30 min intervals during July 2009 covered with precipitation data from (*top*) PMW retrievals alone, (*middle*) AVHRR IR precipitation estimates, and (*bottom*) combined PMW and AVHRR precipitation estimates.

present complete spatial coverage over the entire globe. Precipitation fields generated by an operational weather forecast model such as the NOAA/NCEP Global Forecast System (GFS) often exhibit refined accuracy and spatial resolution compared to those produced by a reanalysis system that is based on an older version of a numerical model and a fixed set of inputs. Frequent updates of the operational forecast model (new observations, physics package, grid system, etc.) yield discontinuities of significant magnitude in the resulting precipitation time series. Among the three state-of-the-art reanalyses, the CFSR is implemented on a T382 (~35 km) grid for hourly outputs. The precipitation fields from MERRA and European Center for Medium-range Weather Forecasting (ECMWF) Reanalysis - Interim (ERA-I), meanwhile, are available at a coarser space/time resolution of 0.5° lat/lon hourly and 0.5° lat/lon 3 hourly, respectively. In this section, we examine the possibility to derive precipitating motion vectors from the CFSR hourly precipitation fields on its native T382 grid.

Following the practice for the definition of cloud motion vectors using the GEO IR data, a cross-correlation technique is applied to the CFSR hourly precipitation fields at its native T382 grid for July 2009. Figure 3.7 illustrates an example of the resulting cloud motion vectors for 00 UTC, 1 July 2009, superimposed on the precipitation patterns over a part of the Southern Ocean covered by a cyclone at the time. The precipitation motion vectors derived from the hourly CFSR precipitation fields look reasonable, compared to the 700 hPa wind fields that are often used as a proxy for synoptic system movements (not shown) and to expected cloud system movements associated with a cyclone.

CFSR-based precipitation motion are constructed for all hours of July 2009. To examine the performance of deriving precipitation motion vectors, CFSR global precipitation field for each hour is propagated for an hour along the vectors and compared against the next hour CFSR precipitation. Figure 3.8 shows a comparison of the global hourly precipitation for 01 UTC, July 1, 2009, as defined in the CFSR simulation (*top*) as well as fields derived through propagation of the previous hour precipitation using reanalysis-derived motion vectors (bottom). The propagated hourly precipitation presents excellent agreement in both overall pattern and position of the precipitation associated with major weather systems, relative to hourly fields simulated by the CFSR. To quantitatively assess the accuracy of the CFSR-based cloud motion vectors, correlation between the propagated and simulated hourly fields are computed for July 2009. As shown in Figure 3.9 propagated CFSR hourly precipitation presents good agreement with the simulated fields. Correlation is relatively low (~0.7) over tropics where precipitation is dominated by rapidly evolving mesoscale cloud systems whose movements are difficult to capture through the application of a cross-correlation technique using data

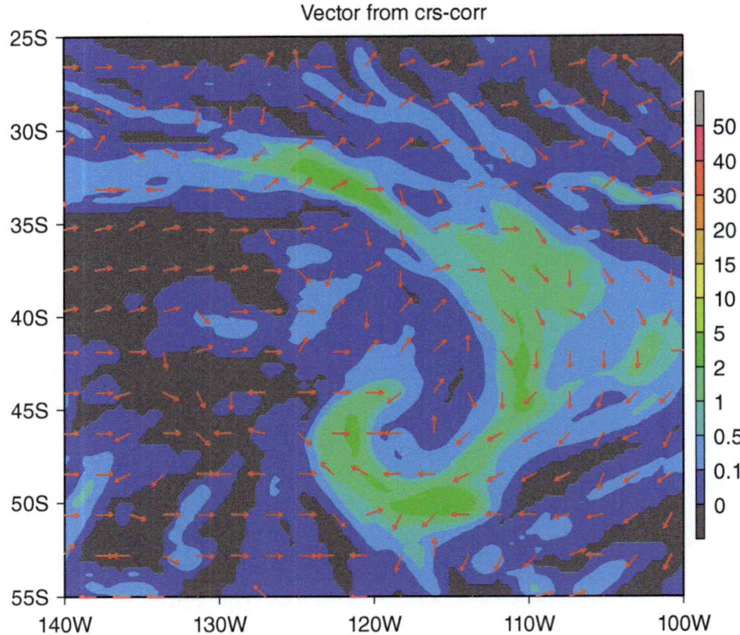

Figure 3.7 Direction of the precipitating cloud motion vectors over a section of Southern Ocean for 00:00 UTC, 1 July 2009, derived through applying the cross-correlation technique with the CFSR hourly precipitation fields, superimposed on the CFSR hourly precipitation. Precipitation rates are mm/h.

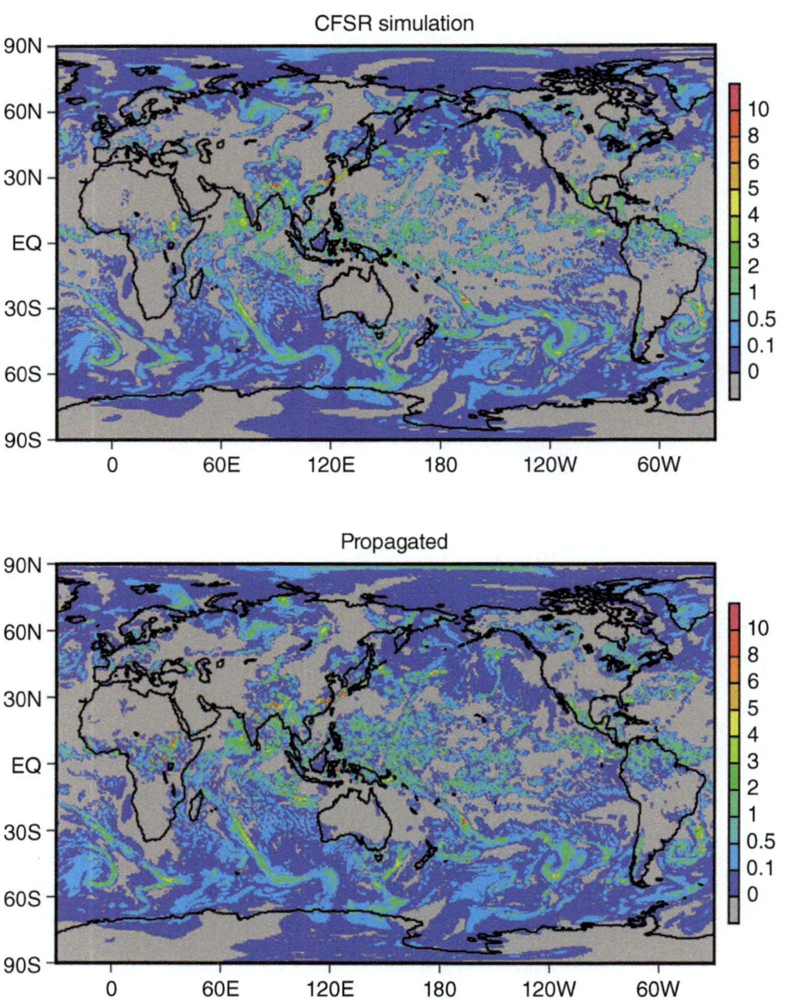

Figure 3.8 Global precipitation distribution (mm/h) for 01 UTC, 1 July 2009, as defined in the CFSR (*top*) and derived by propagating the precipitation field for the previous time step to the target hour (*bottom*).

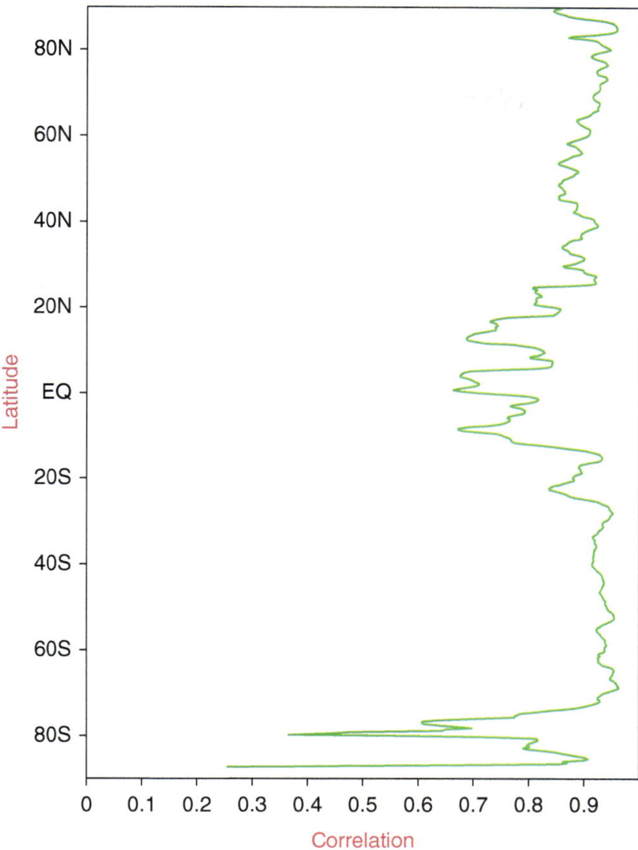

Figure 3.9 Latitudinal profile of the correlation between the hourly precipitation in the CFSR and that derived by propagating the precipitation field for the previous hour along the precipitation motion vectors. Statistics is calculated using hourly data for 1 July 2009.

over a wide domain. The correlation improves over middle and high latitudes (~0.9), a reflection of the relatively large spatial scale of the precipitating cloud systems detected by the cross-correlation technique. Low and unstable correlation over the Antarctic is caused by the lack of extensive and long-lasting precipitation systems over the region during the season.

3.6. STRATEGY FOR CONSTRUCTING POLE-TO-POLE CMORPH

The results reported in Sections 3.4 and 3.5 illustrate clearly the importance of combined information use from LEO PMW, GEO/LEO IR, and numerical model forecasts/reanalyses in defining instantaneous precipitation fields as well as cloud motion vectors. An objective analysis system for the second-generation CMORPH is designed to construct precipitation analysis on a high spatial and temporal (0.05° lat/lon; 30 min) resolution over the entire globe from pole to pole through the integration of precipitation information derived from geostationary

and LEO satellites and from state-of-the-art numerical models (reanalyses). The second-generation CMORPH, as illustrated in Figure 3.10, takes in level 2 data sets of LEO PMW retrievals, gridded IR data from GEO and LEO platforms, and precipitation fields from numerical models (reanalyses) as inputs to the system to derive both the precipitation patterns and their movements.

Level 2 PMW precipitation retrievals from individual LEO satellites are first mapped on to a 0.05° lat/lon grid field of 30 min interval. These gridded precipitation estimates are then intercalibrated against a common reference and combined in time and space to form a combined field of PMW retrievals of instantaneous precipitation rate (MWCOMB). Precipitation estimates are derived from the GEO/LEO IR temperature data through PDF matching against the MWCOMB.

To define the motion vectors for the precipitating clouds, precipitation fields derived from the numerical models and from the satellite observations (including combined LEO PMW retrievals MWCOMB and the GEO/LEO IR precipitation estimates) are subjected to cross correlation separately to get the reanalysis-based and satellite observation-based motion vectors, respectively. These vectors are then combined through an objective analysis technique such as the optimal interpolation (OI) method of *Gandin* [1965] in which the vector analysis for the previous time step is used as the first guess, while the satellite and model (reanalysis) derived vectors are utilized as the observations to update the first guess.

Information on precipitation and precipitating cloud motion vectors derived from the above described procedures are then integrated into a high-quality, high-resolution pole-to-pole global precipitation analysis through the KF framework developed by *Joyce and Xie* [2011]. In this procedure, error for the reanalysis precipitation fields is defined as a function of latitude, season, underlying surface type (land/ocean), while error for the estimates derived from the satellite PMW and IR-derived precipitation changes with season, latitude, underlying surface type, sensor type, and temporal length of propagation.

Two versions of the integrated precipitation analyses will be generated by the second-generation CMORPH. First, global precipitation analysis will be generated using data from all available sources from both the satellites and the model (reanalysis) to ensure the best possible quality. A variation of the analysis will be then produced using only the satellite observation data following the procedures listed inside the left frame of Figure 3.10. Although the resulting observation-only version analysis will not be up to the same quality level as the full version analysis over middle and high latitudes, especially over cold season, it provides an independent source of information to verify precipitation fields generated by the reanalyses and other numerical models.

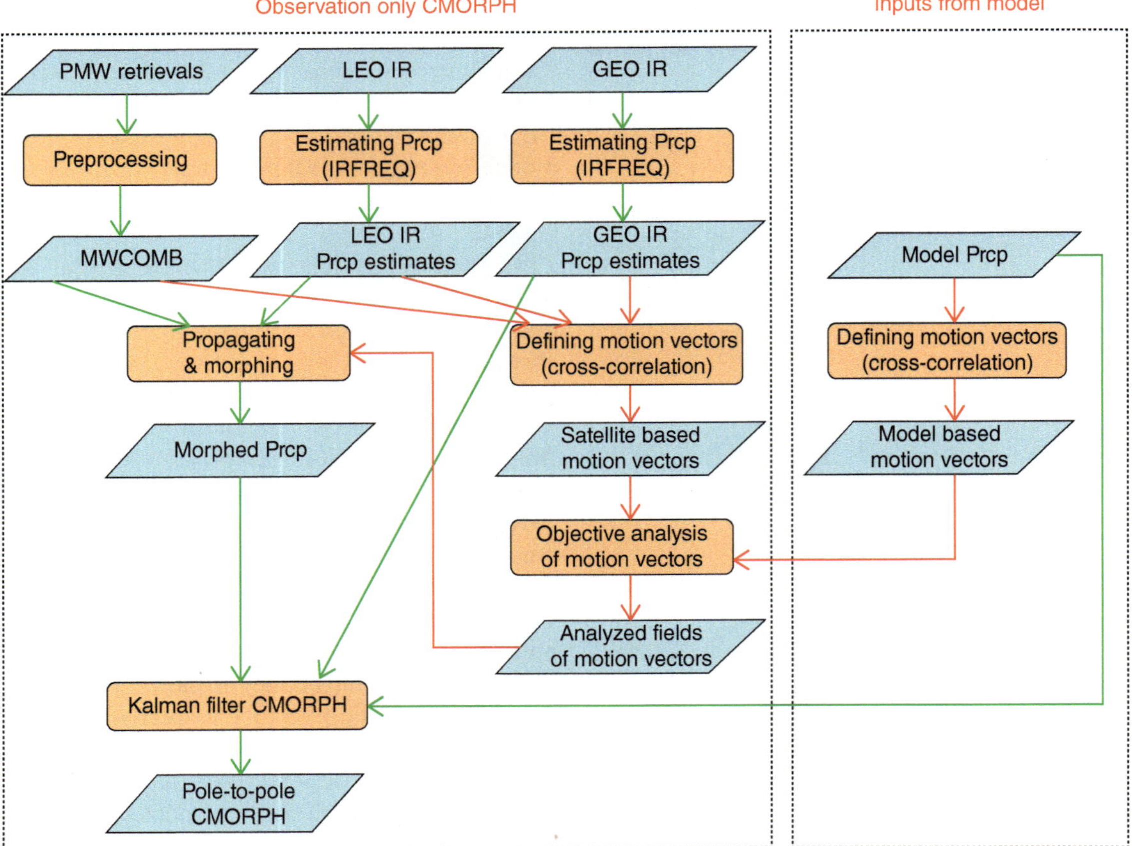

Figure 3.10 Flowchart of the second-generation CMORPH-integrated high-resolution global precipitation analysis system. Block diagrams enclosed in the dashed line frame on the left side are processing for the observation-only CMORPH, while those in the dashed frame on the right are for the incorporation of model-based information. Blue parallelograms indicate input/output data while brown rectangles are analysis procedures. Green and red lines indicate data flow directions for the precipitation and cloud motion vectors, respectively.

To demonstrate the technical feasibility of the conceptual model for the second-generation CMORPH, a simplified version of the above-described analysis system is developed consisting of only primary components shown in Figure 3.10. The simplified test system takes input precipitation information from level 2 PMW retrievals and the precipitation estimates derived from LEO IR data over regions over the entire globe. Since the LEO platforms carrying AVHRR instruments also have PMW sensors aboard, AVHRR-based precipitation estimates are utilized to fill in the gaps where/when no valid PMW retrievals are available, often over regions poleward of 50°S and 70°N. As shown in Figure 3.11, even with PMW retrievals available from nine LEO satellites and IR-based estimates from six polar orbiting platforms, a large part of the globe is still left uncovered with any valid satellite observations in a 30 min sampling period (top panels). The satellite coverage is reduced with the increasing period of sampling and reaches to more than 90% over a 3 h period over most part of the high latitudes.

In this simplified test version of the second-generation global CMORPH, the precipitation motion vectors are defined by combining those derived from the CFSR hourly precipitation over the entire globe and those from the GEO IR-based precipitation estimates over the geostationary coverage from 60°S to 60°N. Instead of a full-version OI-based integration that presents improved quantitative accuracy but requires a set of comprehensive procedures for error modeling and parameter tuning, a weighted mean of the CFSR and GEO IR-based motion vectors are taken to define the combined motion vectors over regions of geostationary coverage. The weight is defined as a function of season and latitude so that over tropics and subtropics the blended motion vectors are dominated by those derived from GEO IR precipitation fields, while the CFSR-based motion vectors receive increasing weights toward the high latitude and over the cold seasons to reflect the examination results shown in Section 3.4 and to ensure smooth transition to regions of higher latitudes where only CFSR-based vectors are used.

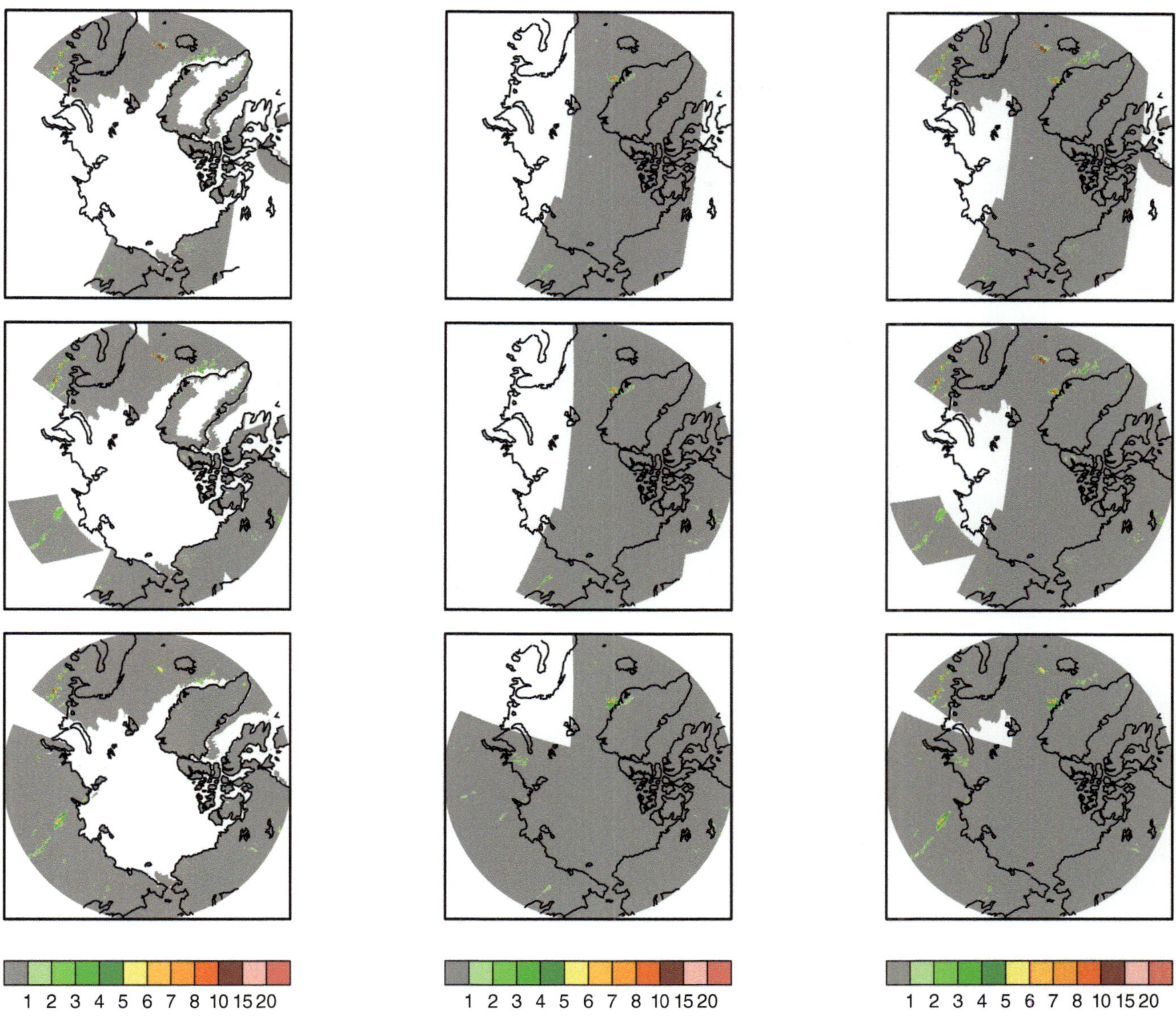

Figure 3.11 Sample satellite precipitation data coverage (mm/h) over; (*top*) a 30 min period from 18:00 to 18:30 UTC, 1 July 2009; (*middle*) a 60 min period from 18:00 to 19:00 UTC, 1 July 2009; and (*bottom*) a 180 min period from 18:00 to 21:00 UTC, 1 July 2009. Results for using PMW retrievals only, LEO IR only, and PMW and LEO IR combined are plotted on the left, middle, and right columns, respectively.

Precipitation information from individual sources is integrated in this test version through the current operational version of CMORPH [*Joyce et al.*, 2004]. The KF-based CMORPH framework of *Joyce and Xie* [2011] is not used here to avoid potentially complicated technical procedures especially in its applications to the analysis of high-latitude regions. This practice enables us to focus our efforts on the examination of technical feasibility to propagate the observed precipitation fields through the CFSR-satellite blended motion vectors.

The preliminary test version second-generation CMORPH is produced on a 0.05° lat/lon grid over the entire globe from pole to pole in 30 min intervals for a 24 h period from 00:00 to 23:30 UTC, 1 July 2009. As illustrated in Figure 3.12, the test version CMORPH presents complete global coverage of precipitation with reasonable distribution patterns associated with major synoptic systems. No discontinuities are visible around the 60° parallels across which the motion vectors are defined using data from different sources, or at 50°S and 70°N parallels, from which toward the poles the input precipitation transitions from LEO PMW based to LEO AVHRR based in this northern hemisphere summertime case. This is reconfirmed from the latitudinal profiles of zonal mean precipitation computed for the 48 half-hourly fields of integrated precipitation fields for the test period (Figure 3.13). The

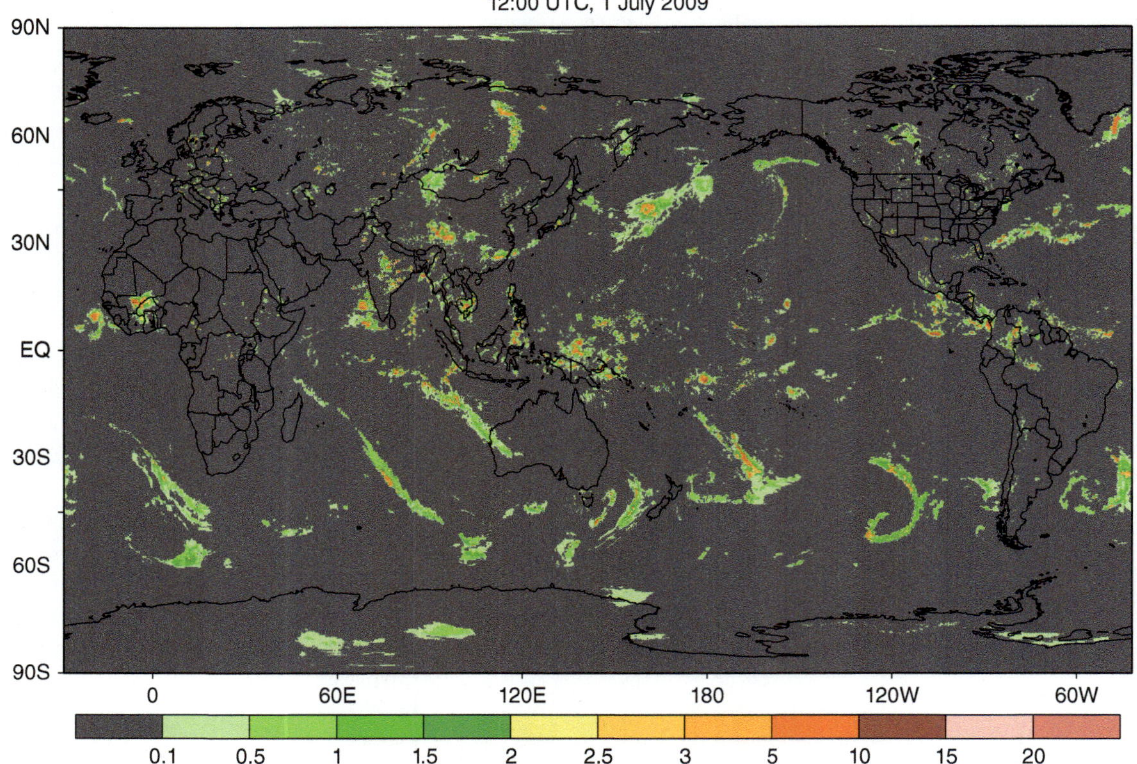

Figure 3.12 Sample precipitation analysis (mm/h) for 12:00 UTC, 1 July 2009, generated by the test version pole-to-pole global CMORPH.

latitudinal profile of zonal mean precipitation derived from the test version pole-to-pole CMORPH (Figure 3.13, black line) exhibits no discontinuities around the 60°N/S where inputs used to define the analysis differ greatly. In addition, it agrees very well with those based on the operational version CMORPH (red line) and the TMPA V7 (green line) over latitude bands where the two other products are available. Furthermore, animations of the resulting half-hourly global fields (not shown) reveal smooth evolution of the global precipitation in association with major synoptic weather systems over various parts of the world.

3.7. CONCLUSIONS AND DISCUSSIONS

The primary goal of this work is to explore an optimal strategy to expand the data domain of a high-quality, high-resolution precipitation analysis such as CMORPH to cover the entire globe from pole to pole. To this end, potential information sources for the definition of precipitation and the motion vectors of precipitating cloud systems are investigated and examined. In addition to retrievals derived from PMW measurements aboard LEO satellites and estimates based on IR observations from GEO platforms that are utilized as inputs to the current generation of high-resolution satellite precipitation analyses covering a latitude band up to 60°S/N, estimates

derived from IR data from polar LEOs and the precipitation analyses produced by the state-of-the-art numerical models (reanalyses) are needed to fill in PMW retrieval coverage gaps and to ensure the quantitative accuracy of the resulting precipitation analysis over middle and high latitudes.

Motion vectors are another category of information needed to integrate the precipitation fields from individual sources into a global analysis of complete spatial coverage and refined quality through a Lagrangian approach. To ensure complete spatial coverage and stable quality of the motion vectors for the precipitating cloud systems, vectors need to be computed first from the movements of precipitating cloud systems captured by individual sources including the GEO IR-based precipitation, PMW retrievals, and IR-based estimates from polar LEOs, as well as the precipitation fields generated by numerical models (reanalyses). Vectors derived from individual sources are then combined through an objective analysis technique to form a global analysis of the cloud motion vectors.

In this work a simplified version of the processing system is explored to demonstrate the technical feasibility to expand the integrated precipitation analysis to cover the entire globe from pole to pole for a summer month, July 2009. In this exercise, PMW retrievals from nine LEO satellites and precipitation estimates derived from AVHRR

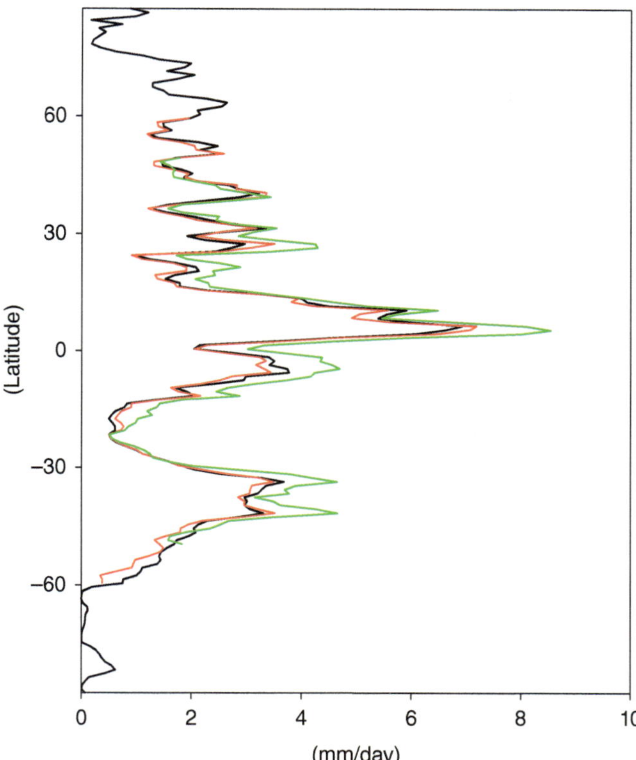

Figure 3.13 Latitudinal profile of zonal mean precipitation (mm/h) for 1 July 2009, derived from the test version pole-to-pole global CMORPH (black), current operational version CMORPH (red) and the Tropical Rainfall Measurement Mission (TRMM) Multi-satellite Precipitation Analysis (TMPA) Version 7 data set (green).

IR window channel brightness temperatures from six LEO satellites are utilized as input precipitation information. Motion vectors are defined by combining those computed, respectively, from the precipitating cloud system movements captured by the GEO IR-based precipitation estimates and by the precipitation fields produced by the CFS reanalysis. Information on the precipitation systems and their movements are integrated through the current operational version of the CMORPH [*Joyce et al.*, 2004], producing a high-resolution precipitation analysis at a 0.05° lat/lon grid over the entire globe from pole to pole.

Tests described in this chapter demonstrate the technical feasibility of constructing a high-quality, high-resolution precipitation analysis over the entire globe from pole to pole through the integration of information from multiple sources, including PMW retrievals from LEO platforms, IR-based estimates from both the GEO and LEO satellites, and precipitation fields generated by the state-of-the-art numerical models (reanalyses). Continued work is needed to complete the development of the objective analyses techniques, to establish an operational processing system to produce the pole-to-pole global integrated precipitation analysis, and to examine the performance of

the pole-to-pole CMORPH in representing precipitation over different regions and for various seasons. Key techniques to be developed and refined as components to the second-generation CMORPH include the following:

• Estimating precipitation from LEO AVHRR IR observations over various parts of the globe and for all seasons;

• Deriving motion vectors from individual sources, especially those from the combined LEO PMW retrievals and AVHRR IR precipitation estimates over the middle and high latitudes; and

• Blending motion vectors from individual sources into a combined field through an objective analysis technique such as OI.

One logistical and computational issue encountered in this preliminary test is the definition of the grid system covering the high-latitude regions. In the current exercise, the pole-to-pole CMORPH was executed on a global equal angle grid of 0.05° lat/lon. This practice requires the processing of precipitation fields and cloud motion vectors on a very large number of grid boxes of very small nominal size over the polar caps, consuming huge computer time and memory space. Experiments are needed to explore for a more spatially representative and computationally efficient grid system. Potential candidates include a Gaussian grid used widely in the numerical global modeling [e.g., *Saha et al.*, 2010] and polar stereographic projection grid [e.g., *Key et al.*, 2003]. Of particular interest is the two-step approach developed successfully by *Turk et al.* [2010] in constructing their global satellite cloud product (GSCP). Gridded fields of satellite IR-based cloudiness over the polar caps are first registered on two separate grid systems of polar stereo-projection and then converted onto an equal angle grid of 0.04° lat/lon. Potential impacts of such methods, especially the alias in the PDF of the precipitation intensity caused by the grid system conversion, need to be examined in the designing of the grid system for our applications in precipitation analysis.

Two remaining issues left untouched in this study include the definition of orographic precipitation and the incorporation of snowfall estimates as part of the precipitation estimates. Both issues are very important to improving the quantitative accuracy of the resulting precipitation analysis, especially over the mountainous regions and for cold seasons. Recent developments indicate that orographic rainfall not detected by PMW retrievals may be estimated through combining information from the satellite observations and numerical model simulations. Algorithms to derive snowfall estimates from PMW observations have been developed and put into operations recently at NOAA/NESDIS [*Wang et al.*, 2011; *Meng et al.*, 2011]. New techniques need to be developed to incorporate satellite estimated snowfall with information from other sources (e.g., reanalyses) to provide an estimation with improved quality.

While the current chapter describes the exploration for an optimal strategy to expand the CMORPH to cover the entire globe, comprehensive examinations of the integrated satellite precipitation analysis need to be conducted in lieu of its potential applications in weather, climate, hydrometeorology, and other fields. In particular, inhomogeneities in the resulting precipitation time series and the alias in the PDF structure of the precipitation intensity caused by the different input satellite data used, and the propagation/morphing processing need to be examined and quantified in future work.

ACKNOWLEDGMENTS

The work reported in this chapter is carried out at the NOAA Climate Prediction Center (CPC) under joint supports from NOAA Climate Program Office (CPO) Climate Monitoring and Observations Program, NOAA United States Weather Research Program (USWRP) Hydrometeorology Testbed (HMT), NASA's Precipitation Measurement Missions (PMM) Global Precipitation Measurement (GPM) international satellite mission program, and the NOAA National Environmental Satellite Data and Information Service (NESDIS) Joint Polar Satellite System (JPSS) Risk Reduction Program (listed in order of receiving supports). The authors would like to express their thanks to G. J. Huffman, R. Ferraro, R. Cifelli, Song Yang, Joe Turk, Craig Long, and Wei Shi for their encouragements and valuable discussions throughout this work.

REFERENCES

Arkin, P. A. (1979), The relationship between fractional coverage of high cloud and rainfall accumulations during GATE over the B-Scale Array, *Mon. Wea. Rev., 107*, 1382–1387.

Bosilovich, M. (2008), NASA's modern-era retrospective analysis for research and applications: Integrating Earth observations, *Earthzine*, available online at http://www.earthzine.org/2008/09/26/nasas-modern-era-retrospective-analysis.

Dee, D. P., et al., (2011), The ERA-Interim reanalysis: Configuration and performance of the data assimilation system, *Q. J. R. Met. Soc., 137*, 553–597.

Ebert, E. E., J. E. Janowiak, and C. Kidd (2007), Comparison of near real time precipitation estimates from satellite observations and numerical models, *Bull. Am. Meteor. Soc., 88*, 47–64, doi:10.1175/BAMS-88-1-47.

Gandin L. S. (1965), Objective Analysis of Meteorological Fields, Israel Program for Sceintific Translations, 242pp.

Gopalan, K., N.-Y. Wang, R. Ferraro, and C. Liu (2010), Status of the TRMM 2A12 Land Precipitation Algorithm, *J. Atmos. Oceanic Technol., 27*, 1343–1354, doi:10.1175/2010JTE CHA1454.1.

Hou, A. Y., et al. (2014), The Global Precipitation Measurement (GPM) Mission, *BAMS*, doi:10.1175/BAMS-D-13-00164.1.

Hsu, K-L, X. Gao, S. Sorooshian, and V. Gupta (1997), Precipitation estimation from remotely sensed information using artificial neural networks, *J. Appl. Meteor., 36*, 1176–1190.

Huffman, G. J., R. F. Adler, D. T. Bolvin, G. Gu, E. J. Nelkin, K. P. Bowman, E. F. Stocker, and D. B. Wolff (2007), The TRMM multi-satellite precipitation analysis: Quasi-global, multi-year, combined-sensor precipitation estimates at fine scale, *J. Hydrometeor., 8*, 38–55.

Huffman, G. J., et al. (2011), GPM day-1 multi-satellite algorithms (iMERG), 2011 PMM Science Team Meeting, Nov. 7–11, 2011, Denver, CO.

Joyce, R. J., and P. Xie (2011), Kalman filter based CMORPH, *J. Hydrometeor, 12*, 1547–1563.

Joyce, R. J., J. E. Janowiak, P. A. Arkin, and P. Xie (2004), CMORPH: A method that produces global precipitation estimates from passive microwave and infrared data at high spatial and temporal resolution, *J. Hydrometeor., 3*, 487–503.

Kalnay, E., et al. (1996), The NCEP/NCAR 40-year reanalysis project, *BAMS, 77*, 437–471.

Key, J. R., D. S Santek, C. S. Velden, N. Bormann, J.-N. Thépaut, L. P. Riishojgaard, Y. Zhu, and W. P. Menzel (2003), Cloud-drift and water vapor winds in the polar regions from MODIS, *IEEE Trans. Geosci. Remote Sensing, 41*, 482–492.

Kuligowski, R. J. (2002), A self-calibrating real-time GOES rainfall algorithm for short-term rainfall estimates, *J. Hydrometeor., 3*, 112–130.

Meng, H., R. F. Ferraro, and B. Yan (2011), Satellite remote sensing of snowfall rate, 2011 PMM Science Team Meeting, Nov. 7–11, Denver, CO.

Negri, A. J., R. F. Adler, and P. J. Wetzel (1984), Rainfall estimation from satellites: An examination of the Griffith–Woodley technique, *J. Appl. Meteor., 23*, 102–106.

Saha, A., et al., (2010), The NCEP climate forecast system reanalysis, *BAMS, 91*, 1015–1057.

Turk, F. J., E. E. Ebert, B.-J. Sohn, H.-J. Oh, V. Levizzani, E. A. Smith, and R. Ferraro (2004), Validation of an operational global precipitation analysis at short time scales, 12thAMS Conf. on Satellite Meteorology and Oceanography, January 11–15, Seattle, WA.

Turk, F. J., S. Miller, and C. Castello (2010), A dynamic global cloud layer for virtual globes, *Int. J. Remote Sensing, 31*, 1897–1914.

Ushio, T., et al. (2009), A Kalman filter approach to the global satellite mapping of precipitation (GSMaP) from combined passive microwave and infrared radiometric data, *J. Meteor. Soc. Jpn. 87A*, 137–151.

Wang, N-Y., C. Liu, R. Ferraro, D. Wolff, E. Zipser, and C. Kummerow (2009), TRMM 2A12 land precipitation product—Status and future plans, *J. Meteor. Soc. Jpn., 87A*, 237–253.

Wang, N.-Y., K. Gepalan, and R. F. Ferraro (2011), Developing winter precipitation algorithm over land from satellite observations and C3VP field campaign, 2011 PMM Science Team Meeting, Nov. 7–11, Denver, CO.

Xie, P., and P. A. Arkin (1996), Analyses of global monthly precipitation using gauge observations, satellite estimates, and numerical model predictions, *J. Climate, 9*, 840–858.

Xie, P., and P. A. Arkin (1997), Global precipitation: A 17-year monthly analyses based on gauge observations, satellite estimates, and numerical model outputs, *BAMS*, *78*, 2539–2558.

Xie, P., and P. A. Arkin (1998), Global monthly precipitation estimates from satellite-observed outgoing longwave radiation, *J. Climate, 11*, 137–164.

Xie, P., M. Chen, A. Yatagai, T. Hayasaka, Y. Fukushima, and S. Yang (2007), A gauge-based analysis of daily precipitation over East Asia, *J. Hydrometeor.*, *8*, 607–626.

Xie, P., M. Chen, and W. Shi (2010), CPC unified gauge-based analysis of global daily precipitation, 24th AMS Conf. on Hydrology, 17–22 January, Atlanta, GA.

Xie, P., R. Joyce, and S. Wu (2013), A 15-year high-resolution gauge-satellite merged analysis of precipitation, 27th AMS Conf. on Hydrology, 7–11 January, Austin, Tx.

4

Research Framework to Bridge from the Global Precipitation Measurement Mission Core Satellite to the Constellation Sensors Using Ground-Radar-Based National Mosaic QPE

Pierre-Emmanuel Kirstetter,[1,2,3] Yang Hong,[1,3] Jonathan J. Gourley,[2] Qing Cao,[3] M. Schwaller,[4] and W. Petersen[5]

4.1. INTRODUCTION

Rainfall is central to the hydrologic cycle, and its high variability in space and time makes its measurement a challenge for hydrologic, meteorological, and climatic applications. Given their quasi global coverage, satellite-based quantitative precipitation estimates (QPEs) are becoming widely used for such purposes, and a major interest arises for their detailed error characterization and quantification in data assimilation and climate analysis [*Stephens and Kummerow*, 2007] and more specifically over land in hydrological modeling of natural hazards and budgeting water resources [*Grimes and Diop*, 2003; *Lebel et al.*, 2009] as highlighted by the International Precipitation Working Group (IPWG; see http://www.isac.cnr.it/~ipwg/) [*Turk et al.*, 2008; *Yang et al.*, 2006; *Zeweldi and Gebremichael*, 2009; *Sapiano and Arkin*, 2009; *Wolff and Fisher*, 2009]. Comprehensive sensor calibration and ground validation research needs to be conducted to ensure proper accuracy and precision of the spaceborne QPE missions [*Petersen and Schwaller*, 2008] like the current Tropical Rainfall Measurement Mission (TRMM; http://trmm.gsfc.nasa.gov) and the Global Precipitation Measurement Mission (GPM; http://gpm.gsfc.nasa.gov).

Current high-resolution, satellite-based (QPEs) are based on low-Earth-orbiting (LEO) passive/active microwave (PMW) measurements, used in combination with geostationary data to provide a frequent (1–3 h) refresh of gridded precipitation accumulations [*Joyce et al.*, 2004; *Huffman et al.*, 2007; *Ushio et al.*, 2006; *Ebert*, 2007]. How uncertainties from LEO platforms propagate through such combined products is not totally understood, and there is a need for comparison of platforms (Figure 4.1). Issues have been identified for both active and passive satellite-based precipitation retrieval algorithms and motivate ongoing and future research: variability of the precipitation inside the sampling volumes and physical properties of hydrometeors like their phase state. A quantitative and detailed characterization of these uncertainties is therefore needed at the original QPE resolution. This task is often impaired by the difficulty of obtaining a reference rainfall product commensurate with the retrieval scale of the instantaneous satellite products. We propose an original framework to tackle these issues. One possible approach is to examine all sources of errors separately and evaluate their cumulative effects [*L'Ecuyer and Stephens*, 2002; *Kummerow et al.*, 2006]. However, the underconstrained nature of remote sensing of precipitation impairs the error separation methods, partly because the models used to perform the retrievals are very sensitive to unobserved atmospheric parameters [*Yang et al.*, 2006; *Stephens and Kummerow*, 2007], particularly over land. The problem is addressed here by comparing the satellite QPE overall accuracy with respect to an external, independent reference precipitation product.

[1]*School of Civil Engineering and Environmental Sciences, Water Technology for Emerging Region (WaTER) Center, University of Oklahoma, Norman, Oklahoma, USA*

[2]*NOAA/National Severe Storms Laboratory, Norman, Oklahoma, USA*

[3]*HyDROS Lab and Advanced Radar Research Center, National Weather Center, Norman, Oklahoma, USA*

[4]*NASA Goddard Space Flight Center, Greenbelt, Maryland, USA*

[5]*NASA Wallops Flight Facility, Wallops Island, Virgina, USA*

Remote Sensing of the Terrestrial Water Cycle, Geophysical Monograph 206. First Edition. Edited by Venkat Lakshmi.
© 2015 American Geophysical Union. Published 2015 by John Wiley & Sons, Inc.

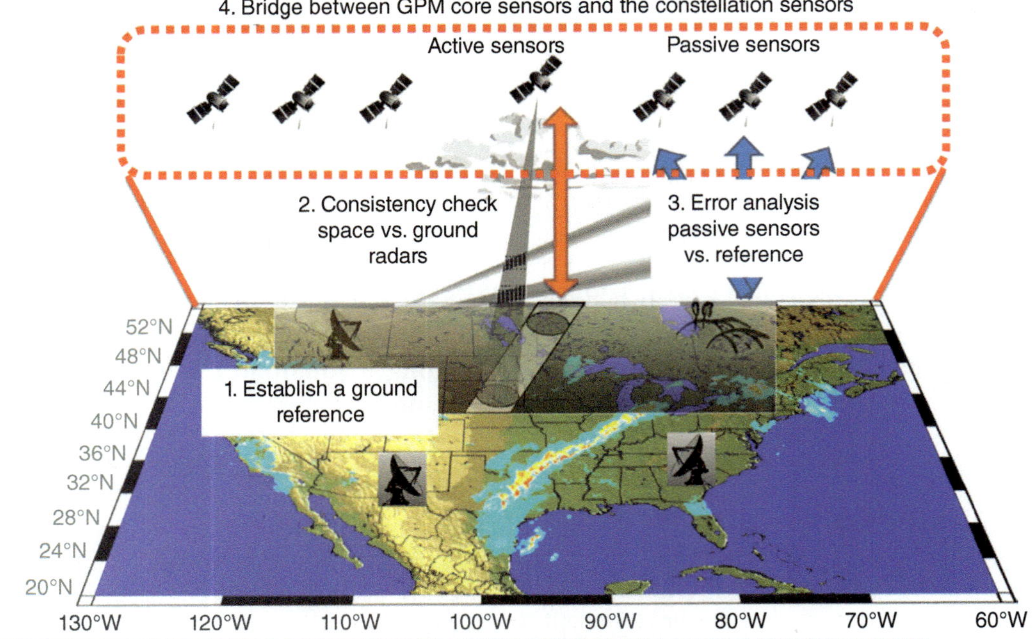

Figure 4.1 Research framework and overview flowchart to bridge from the future Global Precipitation Measurement mission core satellite to the constellation sensors using ground-radar-based National Mosaic QPE. An example of NMQ/Q2 instantaneous precipitation rates at 0725 UTC on 11 April 2011 is shown.

Estimating surface rain rates from satellite measurements poses challenges because of the indirect nature of the observations, the calibration and operating protocol of the instrument itself, the rainfall variability, and the retrieval algorithm used. Joint precipitation observations with the NEXRAD ground-based radar network and from space sensors provide unique opportunities for comparison of QPE as estimated from various sensors. Although measurements from NEXRAD and TRMM/GPM do not match in terms of frequency, polarization state, beam geometry, and incidence angle, they are physically consistent through the identification of hydrometeors and estimation of particle size distribution (PSD). Given the significant research already performed on the ground-based National Mosaic and QPE system (NMQ) data at unprecedented high resolution and accuracy [*Zhang et al.*, 2011a] and the high-resolution spaceborne precipitation measuring techniques, an opportunity exists to compare NASA's space-based precipitation products such as current TRMM precipitation radar (PR), TRMM microwave imager (TMI), the dual-frequency phased array radar (DPR) and other passive sensors of GPM (e.g., GPM microwave imager, GMI) for robust comparison to ground radar products and evaluation over the CONUS (Figure 4.1). We focus here on the TRMM-PR (2A25 products) and TRMM-TMI (2A12 products) version 7.

Specific aspects need to be accounted for when performing such comparisons. Issues have been identified for both active and passive satellite-based precipitation retrieval algorithms and motivate ongoing and future research: variability of the precipitation inside the sampling volumes [non uniform beam filling (NUBF) effects] and physical properties of hydrometeors like their phase state. The comparison procedure needs to be adapted according to the type of space sensors. While some consistency between space-based radars and ground-based radars regarding characterization and quantification of precipitation is expected, overland passive sensor retrievals are specifically sensitive to the vertical water distributions (especially the scattering properties of the ice-phase precipitation) and the surface conditions. The reference has to perform better than the space sensors regarding precipitation detection to ensure proper evaluation, and a correct reference precipitation classification (rain type) is required to target some of the physical factors contributing to erroneous satellite precipitation retrievals.

NMQ is an independent reference for space-based precipitation products regarding resolution, accuracy, and sample size. It provides a consistent database in time and space up to 55° in latitude including various geographical (plains, mountains) and meteorological (subtropical to midlatitudes) conditions for robust comparison (Figure 4.1). Precipitation error characterization is conditioned on microphysics (precipitation rates, types, phases) and, by taking full advantage of the NMQ insights into precipitation, across scales (from subsatellite pixel to mesoscale). This work is anticipated to aid in the development of GPM core and constellations' retrieval algorithms for both active and passive sensors and provide cross-platform, consistent error characterizations for use in the combined products (Figure 4.1).

The PR, TMI data, and steps required to refine the Q2 ground-based rainfall to arrive at the reference rainfall used for comparisons are presented in Section 4.2. Section 4.3 assesses the ability of PR rain retrievals to detect, classify, and quantify rainfall. Section 4.4 details the abilities of TMI with focus on specific factors like soil moisture and compares its capabilities to the PR. The chapter closes with concluding remarks in Section 4.5.

4.2. REFERENCE FOR EVALUATION OF LEVEL 2 SATELLITE-BASED PRECIPITATION RETRIEVALS

The satellite-ground precipitation comparison focuses on three complementary aspects: precipitation detection, characterization (e.g., convective, stratiform), and quantification. The satellite rainfall estimate $R(A)$ is compared with a reference rainfall $R_{ref}(A)$ over a spatial domain A (a satellite pixel) at the snapshot timescale. The reference rainfall $R_{ref}(A)$ is a proxy of the true (and unknown) area-averaged rainfall rate over the same area A. While we do not know the truth at ground, we need to correctly assess the reference's uncertainties for a reliable quantitative comparison of precipitation products as in *Chen et al.* [in press].

4.2.1. Original Ground-Based Products and Preprocessing

Although the quantitative interpretation of the weather radar signal in terms of rainfall may be complex, radars enable a reliable evaluation of area-averaged rainfall estimates. The NOAA/NSSL and University of Oklahoma (OU) NMQ system (http://nmq.ou.edu) incorporates data from all WSR-88D radars and automated rain gauge networks in the conterminous United States (CONUS) [*Zhang et al.*, 2005, 2011a; *Vasiloff et al.*, 2007; *Kitzmiller et al.*, 2011]. The system generates high-resolution national three-dimensional (3D) reflectivity mosaic grids and a suite of severe weather and QPE products at a 1 km horizontal resolution and 5 min update cycle: rain types, precipitation phase, rain rate, freezing level height, etc. The geographical extension and time period covered by the NMQ compiles a broad sample of precipitation situations (stratiform, convective, orographically enhanced precipitation, etc.) over different surfaces (plains vs. mountain) from a stable and uniform observation system (see Figure 4.2). At hourly time step, Q2 adjusts radar estimates with automated rain gauge networks using a spatially variable bias multiplicative factor. A radar quality index (RQI) is produced to represent the radar QPE uncertainty associated with reflectivity changes with

height and near the melting layer [*Zhang et al.*, 2011b; Figure 4.2a).

One should note that it is not possible to "validate" the PR estimates in a strict sense because independent rainfall estimates with no uncertainty do not exist. Yet trustworthy values of the Q2 rainfall estimates within the satellite pixel are needed to evaluate the satellite estimates. To refine the reference data set as much as possible, several quality control steps are applied in order to use ground radar data only where confidence in skill is very high and artifacts (beam blockage due to mountain, range effects, etc.) are essentially minimized and deemed negligible. Blending techniques build on the locally reliable rain gauge measurement and the space-time resolution of the radar [*Amitai et al.*, 2009; *Kirstetter et al.*, 2012a, 2013]. A conservative approach is followed by (i) filtering out instances when the radar and gauge have significant quantitative disagreement [i.e., radar-rain gauge ratios outside of the range (0.1–10)] and (ii) by retaining only the best measurement conditions (i.e., no beam blockage and radar beam below the melting layer) using the RQI product as described in *Kirstetter et al.* [2012a, 2013]. One must keep in mind these improvements may not screen out all possible errors in ground-based radar estimates. The NEXRAD radars' sensitivity (about −5 dBZ at a distance of 50 km) allows generally high performance for rainfall detection compared to satellite sensors. A simplified rain-type classification is elaborated from the vertically integrated liquid (VIL) content derived from the NMQ 3D mosaics at the original resolution (1 km, 5 min). The ice water content VIL_{ice} is computed from the 3D reflectivity mosaics between the freezing level height and the echo top, and a two-step approach similar to *Steiner et al.* [1995] is applied to identify convective areas [*Kirstetter et al.*, 2013]. *Kirstetter et al.* [2012a] showed the increased consistency between TRMM-PR and the Q2-based reference following sequential NMQ data quality control steps including bias correction using rain gauges and filtering using the RQI product. This finding highlights the importance of matching the scales and refining the accuracy of the reference data set as much as possible before reaching meaningful conclusions about the satellite sensor's accuracy.

4.2.2. Building the Reference Rainfall

Because of the highly variable precipitation processes, the resolution of the ground reference should match the satellite products resolution to reduce noise when comparing satellite retrievals to R_{ref}. The spatial resolution of NMQ (1 km) is finer than any satellite sensor, allowing the resolution of the reference to be specifically adapted to each sensor using spatial sampling techniques [*Kirstetter et al.*, 2012a]. Note that the time difference

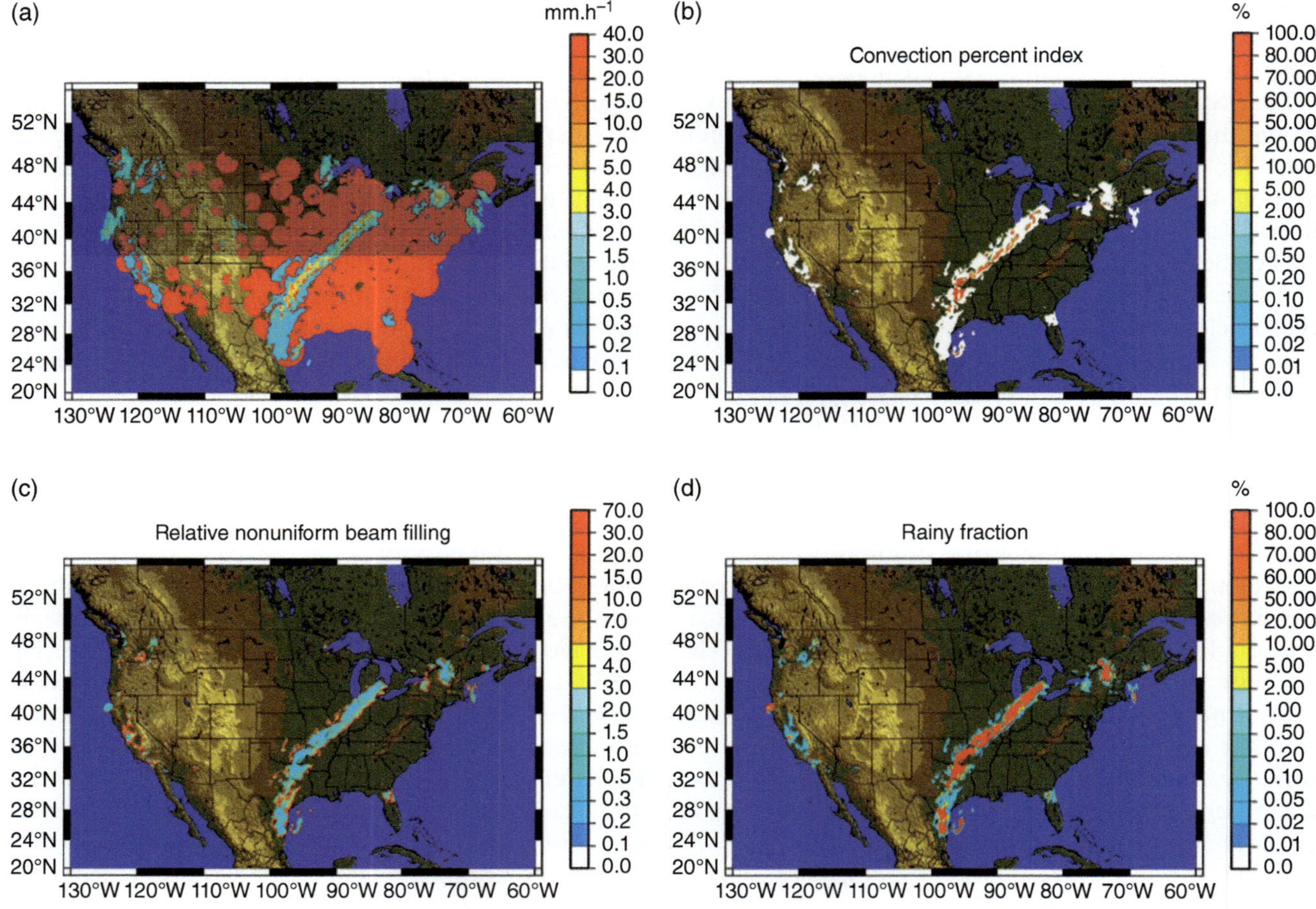

Figure 4.2 (a) Map of CONUS area with NMQ/Q2 instantaneous rain rates at 0725 UTC on 11 April 2011. The red area shows the good quality radar coverage corresponding to radar quality index equal to 1. The shaded area is not sampled by the TRMM-PR; (b) the convective percent index (CPI) computed at the PR footprint resolution; (c) the relative nonuniform beam filling (RNUBF); and (d) the rainy fraction (RF).

between the Q2 data and satellite data (max 2.5 min) could add random noise (but not bias) in the comparison, although other factors like the difference in resolution may have more significant impacts. All significant rain fields observed coincidentally by TRMM overpasses and the NEXRAD network from March to October 2011 are collected. Given that the maximum latitude coverage by TRMM is 38°, the data set is dominated by deep convection common to the southern portion of the United States during the warm season. However, the CONUS-wide coverage includes a diversity of weather systems in the database such as tropical storms, orographic rainfall in mountainous regions, and stratiform precipitation. The Q2 products closest in time to the TRMM satellite local overpass schedule time are used. To determine the reference rainfall R_{ref} (A) over the sensor pixel A, a block-Q2 rainfall pixel matches each sensor pixel. All of the Q2 pixels (rainy and nonrainy) found within a footprint sensor are located to compute unconditional mean rain rates

for the Q2 at the sensor pixel scale. When more than 25% of Q2 pixels have missing values, the data pair is discarded from the comparison. Matched PR/TMI and R_{ref} (A) estimates only exist at locations where both the PR/TMI and ground radars have taken actual observations. The advantages of the current technique over gridded approaches are that the satellite products remain untouched, hence preserving their characteristics: the total rainfall amount, the total rainy area, the convective/stratiform contribution, and the probability density function (PDF) shapes. All of these properties may therefore be compared to the reference at once.

An extended description of the reference rainfall regarding occurrence, variability, and types of rainfall within the satellite sensor's field of view (FOV) is assessed through several products derived at the footprint resolution:

1. The rainfall type through a convective percent index (CPI, Figure 4.2b) quantifying the volume contribution of

convective rainfall to R_{ref} as follows. The CPI is expressed in percent between 0% (purely stratiform rainfall within the PR FOV) to 100% (purely convective rainfall).

2. The quantitative variability of the Q2 values $\sigma_{footprint}$ [*Kirstetter et al.*, 2012a]. Because $\sigma_{footprint}$ presents some correlation with the reference rainfall rate, we use the relative nonuniform beam filling quantity as $RNUBF = \sigma_{footprint}/R_{ref}$ (unitless, Figure 4.2c) to assess the impact of the variability of the Q2 rain rate relative to R_{ref}; typical values range from 0.1 (homogenous Q2 values within the FOV) to 50 (highly variable Q2 values).

3. The rainy fraction (RF; Figure 4.2d) characterizes the rainfall occurrence within the PR pixel. It represents the FOV filling with the proportion of positive Q2 values inside the PR pixel and is expressed in percent between 0% (no rainfall within the FOV) and 100% (FOV completely filled).

Given its spatial extension and native resolution, the NMQ products provide large samples of matched ground-satellite comparison pairs. This allows judicious selections of data pairs to monitor the comparison quality.

4.2.3. TRMM Active and Passive Sensors

TRMM is a joint mission between NASA and the Japan Aerospace Exploration Agency (JAXA) designed to monitor and study tropical precipitation since its launch in 1997. It comprises the first space-based PR measuring reflectivity profiles at Ku and and the TRMM TMI measuring brightness temperatures [*Kummerow et al.*, 1998].

With the PR, surface rain rates are estimated over the southern United States up to a latitude of 37°N (see Figure 4.2a). The scan geometry and sampling rate of the PR lead to footprints spaced approximately 5.1 km in the horizontal and along track, over a 215 km wide swath centered within the 760 km wide TMI swath. The TRMM product used in this work is the PR 2A25 product (version 7) described in *Iguchi et al.* [2000, 2009], which provides 3D reflectivity and 2D rain rate fields at ground. The 2A23 product classifies rain into stratiform and convective [*Awaka et al.*, 2007]. The 2A25 algorithm relies on a hybrid attenuation correction method that combines the surface reference technique and Hitschfeld-Bordan method [*Iguchi et al.*, 2000; *Meneghini et al.*, 2000, 2004; *Takahashi et al.*, 2006]. It uses models to describe the hydrometeor drop size distributions (DSDs) according to the 2A23 products, which are adjusted to match the observed path integrated attenuation (PIA). The PR is a well-calibrated and very stable radar. The PR detection of rainfall depends primarily on the instrument sensitivity around 17 dBZ, or 0.5 mm/h. The various conditions of filling of the FOV by rainfall may impact the detection capabilities of the PR. Primary errors in rainfall retrievals

have mainly been attributed to attenuation correction of the radar signal (involving the incorrect physical assumptions related to convective versus stratiform rainfall classification and assumed drop size distribution), NUBF, and conversion from reflectivity-to-rainfall intensity [*Iguchi et al.*, 2000, 2009; *Yang et al.*, 2006; *Wolff and Fisher*, 2008]. PR observations provide a more direct measurement of the rain rate than TMI, and PR may be regarded as a "calibrator" for the passive microwave precipitation estimates. Our data set covers 7 months (March–October 2011) of satellite overpasses over the lower CONUS (up to a 37°N in latitude). The variable eSurfRain (estimated surface rain) was extracted from the 2A25 files as the PR surface QPE. Despite the seemingly short period for evaluation, the use of gridded Q2 data for reference provided a large sample size totaling 1,625,942 nonzero PR-reference pairs including 468,921 pixels classified as stratiform by the PR and 185,860 convective pixels.

The TMI measures brightness temperatures (T_b) at five microwave frequencies, which are used by the Goddard profiling algorithm (GPROF). This algorithm uses a Bayesian formulation to estimate surface rain rates and vertical hydrometeor profiles [*Kummerow et al.*, 2011]. These physical retrievals rely upon a large set of simulated brightness temperatures at each microwave channel, using clouds from numerical models or actual satellite observations to build the database. An up-front screen flags rain before attempting to quantify the precipitation rate. It is important to note some specific issues for the precipitation retrieval over land. The main information used for the retrieval comes from the scattering channels because of the radiometrically warm land surface and variable surface emissivity. As the brightness temperature at these frequencies is mostly sensitive to the scattering processes in the higher regions of the cloud [*Wilheit et al.*, 2003], the available information for the retrieval is not directly related to surface precipitation. The overland algorithms are generally empirical (the correlation between ice and rainfall is variable). Hence, there are particular needs for a detailed assessment of GPROF performances, especially regarding the rain/no-rain mask and the quantitative retrieval [*Gopalan et al.*, 2010]. We consider two data sets for TMI. One covers the same period as for PR (March–October 2011) of satellite overpasses over the lower CONUS to focus on the QPE at ground. We refer to this data set as the "CONUS data set" hereafter.

The other data set is used to focus on the influence of integrated information of the vertical cloud structure and surface conditions on the microwave-based precipitation overland retrievals. It includes: (i) surface soil moisture simulated using the NASA catchment land surface model forced with stage IV WSR-88D radar-rainfall data and

(ii) the normalized difference vegetation index (NDVI) derived from the Moderate resolution Imaging Spectroradiometer (MODIS) aboard NASA's *Terra* satellite. This data set covers two years (2009–2010) of satellite overpasses over the Oklahoma region, which is well instrumented. The climate in this region varies from west to east with the western areas being drier and as a result much less vegetated. Most of the precipitation falls as rain during the spring months (i.e., April–June) predominantly from supercell thunderstorms and mesoscale convective systems. Tropical systems are rare in this region, but storms with active warm-rain processes occur later in the warm season. Snow and ice storms occur occasionally during winter and are much more prevalent in the northwest part of the domain. We refer to this data set as the "Oklahoma data set" hereafter. In this data set, particular situations like frozen surfaces (the freezing level height is below the ground level) or warm rain (positive reference rainfall with VIL_{ice} equal to zero) are specifically addressed. The TMI resolution is considered to be 15 km [*Lin and Hou*, 2012].

The PR-calibrated TMI is, in turn, used as the calibrator for other passive microwave sensors, which collectively enable the creation of quasi-global-scale combined precipitation products [*Huffman et al.*, 2007]. Thus, PR/TMI have fundamental impacts on satellite-based precipitation estimates from other LEO passive microwave measurements and a number of satellite-based, high-resolution precipitation products [*Ebert*, 2007; *Hong et al.*, 2007].

4.3. COMPARISON BETWEEN GROUND-BASED AND SPACE-BASED RADARS

Ongoing and future research for active satellite-based precipitation retrieval algorithms focus on issues such as horizontal variability of the precipitation inside the sampling volumes (NUBF), sampling geometric effects, attenuation of the radar signal, and physical properties of hydrometeors like their phase state. The comparison with independent references can further guide the development and improvement of space radar algorithms, which includes attenuation correction, melting layer detection, hydrometeor classification, and precipitation estimation.

4.3.1. Precipitation Detection

Satellite sensor detection capabilities impact the understanding of the rain rate spectra as seen from space and on the climatic variability of light rain [e.g., *Lin and Hou*, 2012]. NMQ provides generally better detection performance than PR [*Kirstetter et al.*, 2012a, Fig. 6] except in specific situations like at great distance from the ground

radar (>150 km) and in mountainous areas when the PR benefits from sampling of precipitation closer to the surface [*Wen et al.*, 2013]. We consider cases where the reference rainfall is positive and evaluate PR's rainfall detection capabilities and the percentage of rain occurrence and volume likely missed for various precipitation conditions.

It is commonly noted that PR misses weaker echoes probably due to its sensitivity [*Schumacher and Houze*, 2000]. Satellite rainfall retrievals are often characterized by a rainfall rate threshold, under which the detection capabilities degrade [*Petty*, 1995, 1997; *Lin and Hou*, 2008, 2012; *Berg et al.*, 2010]. In such cases, the probability of detection (POD) as a function of rainfall rate would be a step function at the threshold. Figure 4.3a shows the cumulative distributions of rain occurrence and volume as functions of the rain intensity. Light precipitation (rain rates < 1 mm/h) dominates at 78% the rain intensity spectra in terms of the fraction of rain occurrence, while the fraction of intermediate rain intensity (1 mm/h < R_{ref} < 10 mm/h) is 19%, and the fraction of the heavy rain intensities (R_{ref} > 10 mm/h) is only 3%. If the PR detection threshold is 0.5 mm/h (as assumed in *Lin and Hou* [2012]), then it misses 68% of the rain occurrence. However, intermediate and heavy rain intensities dominate the total rainfall volume: heavy rainfall contributes as much as 58%, while 35% for intermediate rain intensities, and only 7% for light rain intensities. With the same detection threshold, the PR would miss 8% of the rainfall volume. Figure 4.3a depicts the behavior of the POD of PR as a function of the reference rain rate. The POD consistently improves with rain rate with a sharp increase between 0.1 and 1.5 mm/h. Yet the POD is neither null for low rain rates nor equal to unity for high rain rates, and the transition between the two sills is not a step function, which suggests that the rain intensity is not the only factor driving the detection capabilities of PR. In finding a reference rainfall threshold defining the detection capabilities of the PR, a trade-off between increasing the percentage of correct detections and minimizing misses is found by using the Heidke skill score [HSS; *Wolff and Fisher*, 2009; *Wilks*, 2011]. Maximizing the HSS enables us to identify the threshold and to evaluate this model: 1 indicates a perfect delineation between PR detection above some threshold and missed precipitation below it; 0 indicates no skill (the detected occurrence of precipitation is the same as the PR without correlation); negative values indicate a model no better than random guess. Figure 4.3b shows the HSS computed for various rain rate thresholds over the entire data set. The maximum (HSS = 0.65) occurs at 0.53 mm/h, a value very close to the 0.5 mm/h one assumed by *Lin and Hou* [2012]. The threshold model involving a minimum detectable rain rate from PR is an approximation, and the

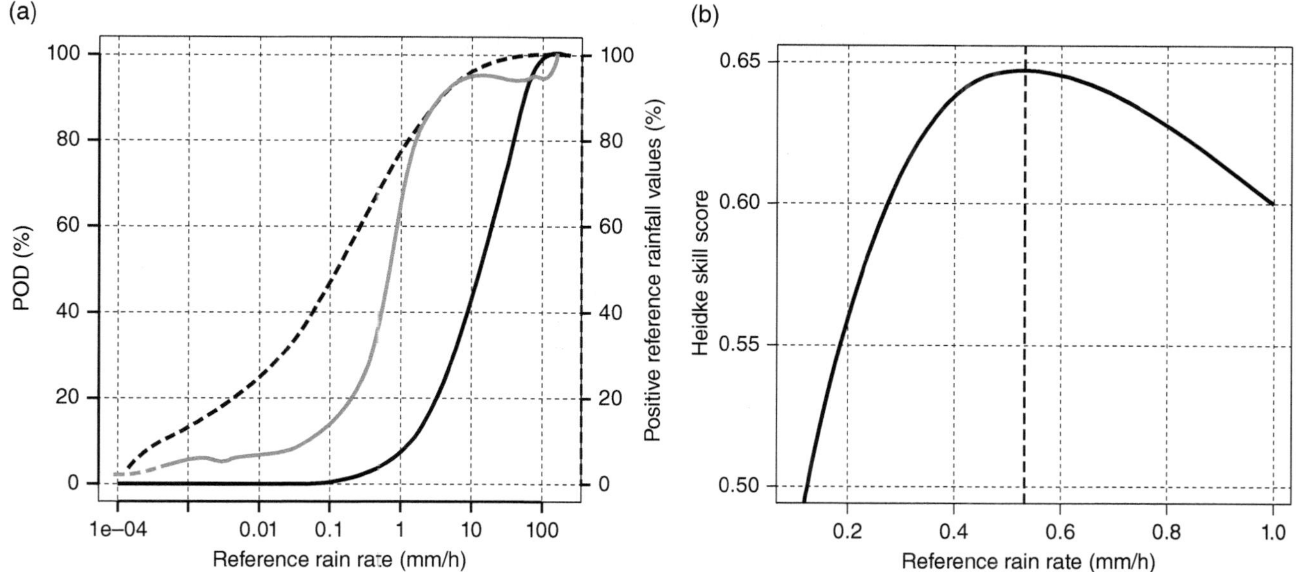

Figure 4.3 (a) Probability of detection of the reference rainfall by the PR (gray) as a function of the reference rain intensity. The accumulated fractions of rain occurrence (dotted line) and volume (plain line) are shown in black and plotted against the secondary ordinate. (b) Heidke skill score for the threshold model of detection for PR (Section 4.3.1) as a function of the minimum detected reference rain intensity. The dotted line denotes the threshold (0.53 mm/h) identified by the maximum of the HSS [from *Kirstetter et al.*, 2013].

capabilities vary with the horizontal variability within the field of view.

In investigating factors driving the detection capabilities of PR, Figures 4.4a and 4.4b, respectively, show how the reference rainfall R_{ref} and reference rainfall type (CPI) evolve as functions of the rain fraction (RF) and the relative NUBF (RNUBF). The convolution of the subpixel rainfall rate (type) distributions in the FOV by the PR antenna pattern causes the same reference rain rate (CPI) to be characterized by various conditions of rain occurrence and quantitative variability. As an example, moderate and homogeneous precipitation filling the FOV (e.g., RF = 95% and RNUBF = 0.2) results in the same footprint-averaged reference rainfall rate as a scene with inhomogeneous but heavy precipitation (e.g., RF = 55% and RNUBF = 2). Lighter rain rates are found for lower beam filling conditions (RF < 60%) with various quantitative variability, whereas the stratiform type rainfall is associated with lower RNUBF values. The highest mean rain rates are found for filled FOV and RNUBF within the interval (0.5–2). The highest convective contributions are consistently found for higher RNUBF values (the convective rainfall is associated with higher variability). The impact of rainfall variability within the PR field of view is assessed with the rain fraction (occurrence variability of rainfall) and the RNUBF (quantitative variability of rainfall). The reference data are separated into various subsamples according to these two variables. All coincident and collocated satellite values are considered

and sorted according to the reference samples. PR sampled robust rainfall only 33% of the time (80% of total volume of rainfall). The PR FOV is at least 80% filled only 26% of the time (84% of total volume of rainfall). Situations where the FOV filling is low and/or the variability of the rainfall is high occur frequently and require a detailed assessment. Figure 4.4c shows the behavior of the POD by the PR relative to the reference as a function of the RF and the RNUBF. Various conditions of variability lead to equivalent detection performances by the PR. The POD degrades (i) with lower RNUBF values and (ii) with low rainfall filling of the PR pixel. More specifically, the detection capabilities degrade significantly (i.e., POD < 50%) for RF < 40%. The highest POD rates (>90%) are found for the conditions RNUBF >1, RF > 90%, and 90% of rainfall is properly classified for RF > 95%. Misses are associated with high intermittency and/or the "rain/no-rain" limits of rain fields. A significant filling of the FOV (70%) is required to detect rainfall and a high quantitative variability (RNUBF >1) has a positive impact from the PR perspective, probably because it enables the raining signal to exceed the 17 dBZ threshold, which limits PR detection. The rainfall variability at fine spatial scale within the PR FOV impacts the PR detection performance. Segregating rain from no-rain boundaries is also a driving contributor to the PR rain rate errors, probably linked to the lack of sensitivity in the most inhomogeneous and light parts of the edges of rainy regions [*Schumacher and Houze*, 2000].

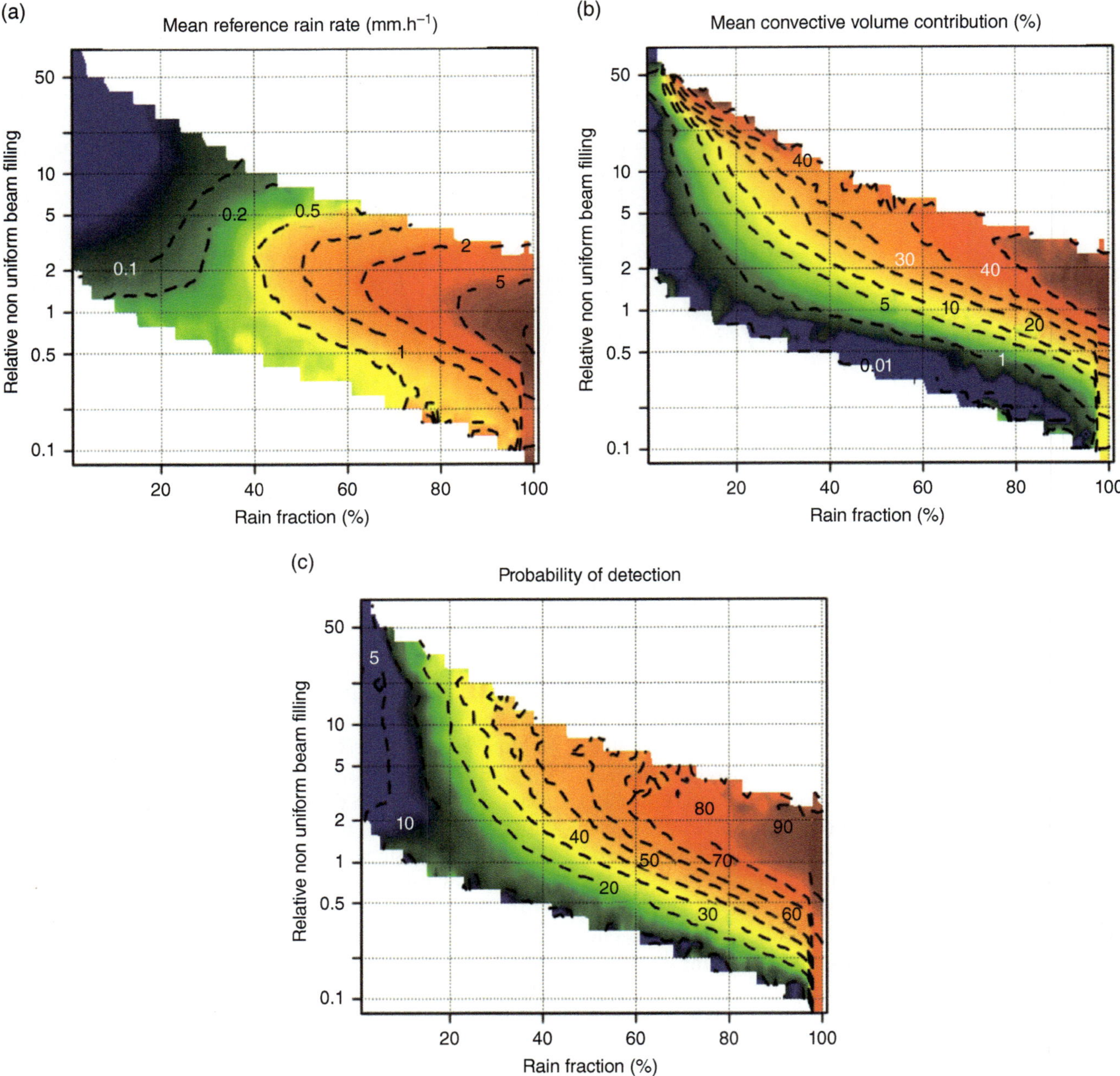

Figure 4.4 (a) Mean reference rainfall R_{ref}; (b) mean reference rainfall convective volume contribution (CPI); (c) probability of detection by the PR as functions of the rain fraction (RF) and the relative nonuniform beam filling (RNUBF) [from *Kirstetter et al., 2014*].

4.3.2. Precipitation Characterization

Precipitation classifications from satellite algorithms have profound impacts on the accuracy of the quantitative retrievals. As an example, the classification drives the vertical model of rainfall used to correct for the attenuation of the PR signal, to estimate the vertical profile of reflectivity, and the rainfall rate at ground [*Iguchi et al.,* 2009]. Satellite rainfall classification capabilities,

the conditions driving potential misclassification, and the proportion of rain likely to be misclassified have been largely unknown. This is partly due to the fact that classification of precipitation type from nonpolarimetric active remote sensing relies partly on subjective analysis based on interpretation of precipitation spatial variability. We investigate relevant factors (occurrence, type, and rate of rain within the PR FOV) driving the convergence or divergence of the PR classification relative to the reference. We consider cases where both the PR and the reference rainfall are positive, so that precipitation detectability is not a factor. The PR classification is a binary decision (i.e., either convective or stratiform), while the ground reference provides a volume contribution of convective rainfall varying continuously from 0% (purely stratiform reference rainfall) to 100% (purely convective reference rainfall). The reference rainfall is separated into purely stratiform (59% of the data sample) and convective when the convective volume contribution is positive. Again, all coincident and collocated PR values are considered and sorted according to the reference sample.

Gray lines in Figure 4.5a show the cumulative distributions of CPI occurrence and convective volume contribution as functions of the CPI. The CPI is quite evenly distributed with light convective contribution (CPI < 25%) representing 30% of the population, the fraction of intermediate convective contribution (25% < CPI < 75%) being ~30%, and the fraction of the significantly convective reference rainfall (CPI > 75%) is 40%. The last class dominates the total convective rainfall volume with a contribution of more than 90%. Figure 4.5a also depicts the POD of the PR as a function of the convective volume contribution into the reference rainfall. The POD improves with CPI with a sharp increase from 13% to nearly 40% at low CPI values, then increases more slowly yet regularly with the convective contribution to reach ~80% for a purely convective reference rainfall. Therefore the PR shows consistent classification performances relative to the reference. Yet the POD is neither null for low CPI nor equal to unity for high CPI values, and the transition between the two sills is not a step function as would be the case for a perfect classification of rainfall relative to the reference. This result suggests that the CPI is not the only factor driving the classification capabilities of PR. In evaluating the convergence between the PR and reference classifications, we identify a reference rainfall threshold defining the classification capabilities of the PR. Similarly with the detection aspect, we evaluate the convective contribution threshold describing the stratiform/convective separation by the PR with a trade-off between increasing the percentage of convective and stratiform classification. We use the HSS to quantify the accuracy of precipitation classification (stratiform vs. convective) relative to random chance at a given reference convective contribution. By computing the HSS for various CPI thresholds over the entire data set, a maximum value (HSS = 0.56) is found for CPI = 6%; this indicates the 2A25 product presents overall agreement with the independent classification provided by the reference.

Figures 4.5b and 4.5c provide an in-depth view of the 2A25 classification relative to the reference by showing the proportion of convective classification as a function of the variability within the PR field of view (rain fraction

and RNUBF) for both sensors. The reference classification is regularly distributed and depends primarily on the RNUBF with values ranging from ~3% to more than 95%. This is expected since convective rainfall involves more quantitative variability than with stratiform rainfall. The 2A25 classification shows different patterns than the reference (Figure 4.5c). There is less dependence on RNUBF, and the proportion of convective detection does not exceed 85%. While the classification dependence on the RNUBF shows the same trend as the reference for RF > 50%, the 2A25 classification shows two detection maxima for extreme RNUBF values in the domain RF < 50%. More investigation is necessary to identify the reason why the PR is prone to incorrectly detect convective activity under these last conditions.

4.3.3. Precipitation Quantification

The uncertainties associated with satellite estimates of rainfall include systematic errors as well as random effects from several sources [*Yang et al.*, 2006; *Kirstetter et al.*, 2012a, 2012b, 2013]. There is a fundamental issue in segregating the proportion of the scatter due to purely random error and the proportion due to conditional biases of the satellite estimates that may be either positive or negative, producing additional scatter. *Kirstetter et al.* [2012a, 2012b] showed that PR presents a tendency to overestimate light rain rates and underestimate higher rain rates. The most significant errors are most likely due to a combination of inaccurate Z-R relationship, nonuniform beam filling, and/or attenuation of the PR radar signal. It is difficult to distinguish between these different influences by comparing solely rain rates at ground.

The 2A25 algorithm uses different Z-R relationships in the convective/stratiform profiling components. The rain type impacts the bias of the PR relative to the reference. In order to provide some insight into the influence of rain typing on the PR error, the departures of PR estimates from the Q2 reference values are analyzed as functions of the convective contribution on a point-to-point basis. The residuals are defined as the difference between the reference rainfall (R_{ref}) and the satellite estimates (R): $\varepsilon = (R - R_{ref})$. Only pairs for which R_{ref} and R are both nonzero are considered in the calculations, so as to remove any discrepancies related to detectability. Figures 4.6a and 4.6b show the residuals as a function of the convective contribution, CPI. All coincident and collocated PR values are considered and sorted according to the reference sample. The data set is also segregated according to the PR-based rainfall-type classification. The conditional PDFs of residuals ε present a high conditional shift from the 0 line and a large conditional spread. Figures 4.6a and 4.6b show a shift toward lower rainfall rates as CPI increases. These features are more pronounced with PR-indicated convective echoes

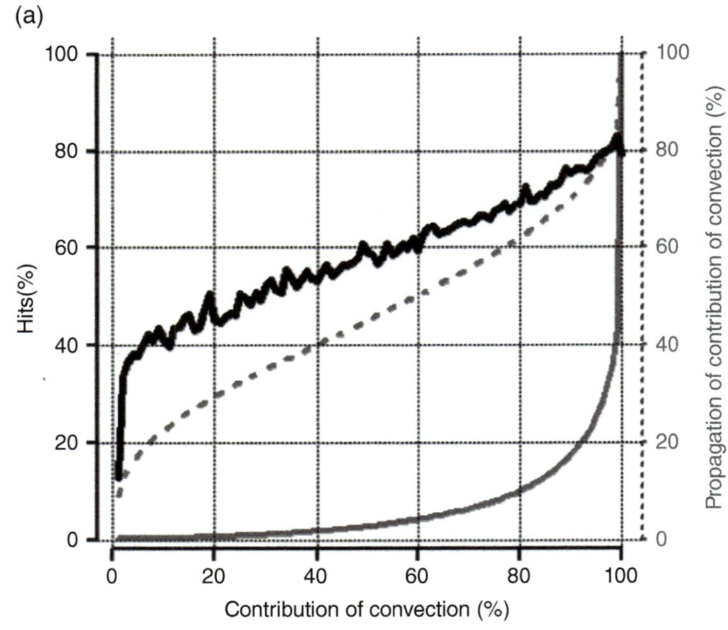

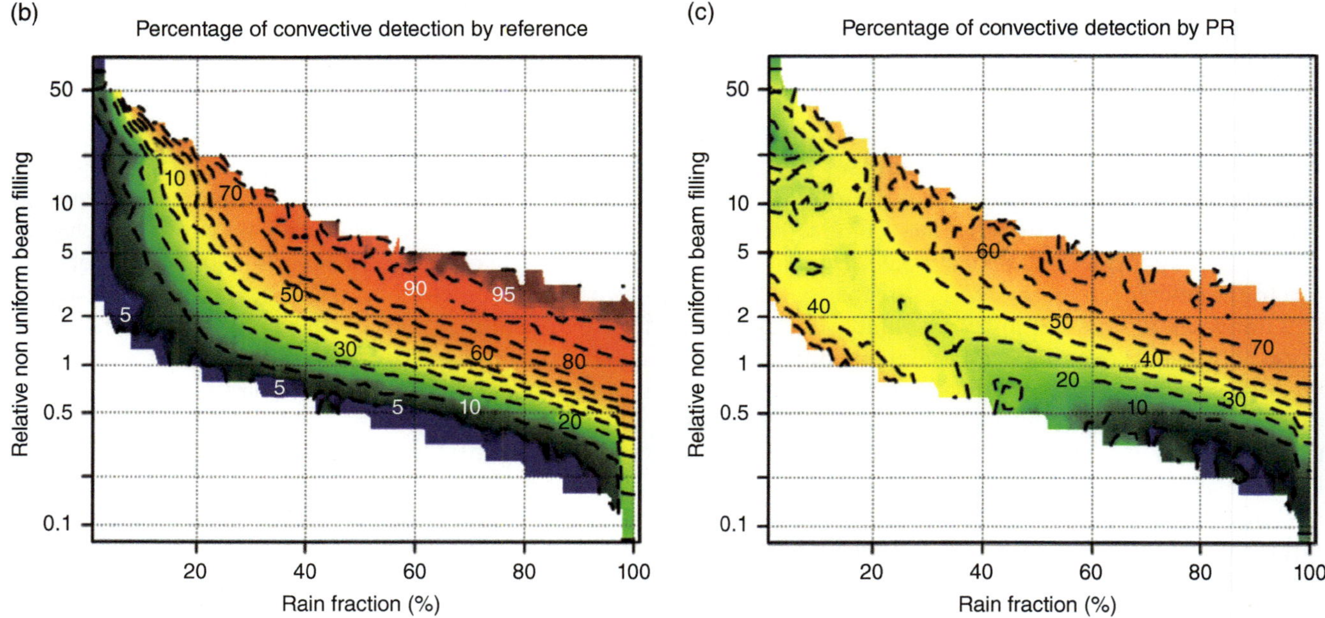

Figure 4.5 (a) Proportion of convective detection by the PR as a function of the reference rainfall convective volume contribution. The proportion of convective contribution by occurrence (dotted line) and volume (plain line) are figured in gray. Proportion of convective detection by (b) the reference and (c) the PR [from *Kirstetter et al.*, 2014].

(Figure 4.6a). Both 2A25 classifications present a tendency to overestimate rain rates for low CPI values (CPI < 20%, the conditional median of residuals are positive) and underestimate rain rates for CPI values higher than 90% (negative median of residuals). As an example, the convective PR model overestimates at CPI = 20% (mainly stratiform) rain rates with an occurrence of 80% and with a representative bias of +2 mm/h and underestimates at CPI = 100% (convective) rain rates with an occurrence of 70% and with a representative bias of −10 mm/h. The stratiform PR model overestimates at CPI = 0% (stratiform) rain rates with an occurrence of 55% and with a representative bias of +0.5 mm/h and underestimates at CPI = 100% (convective) rain rates with an occurrence of 75% and with a representative bias of −3 mm/h.

Because of the asymmetric density of residuals and for a better representativity, we consider the conditional median of the residuals to compare the systematic error component

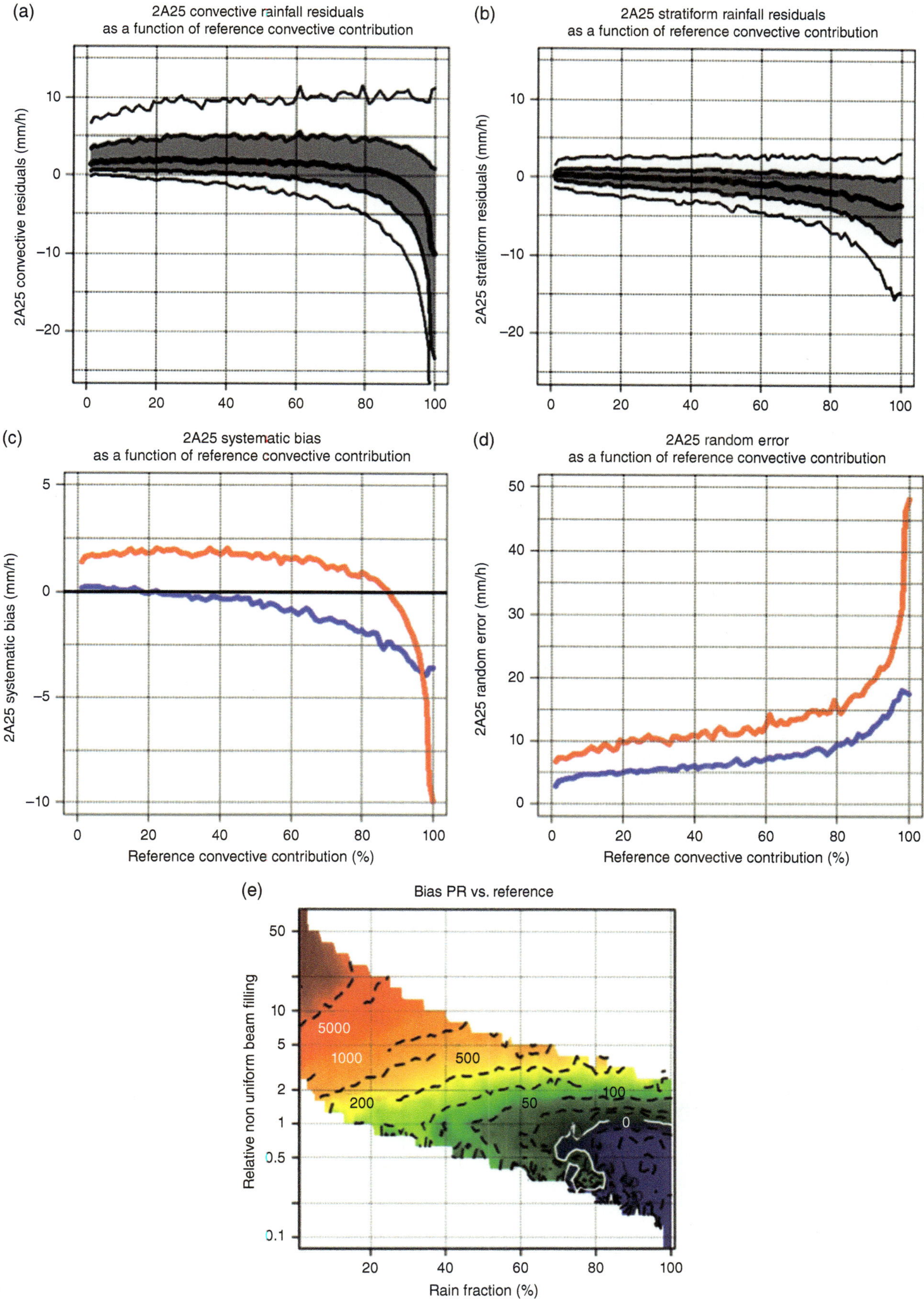

Figure 4.6 Density plots of residuals (PR—reference rainfall) versus CPI for (a) convective and (b) stratiform rainfall. The thick black line represents the median (50% quantile), the dark gray-shaded region represents the area between the 25% and 75% quantiles, and the light gray-shaded region represents the area between the 10% and 90% quantiles. The (c) systematic part and (d) random part of error are represented for stratiform (blues lines) and convective (red lines) types; and (e) bias (%) of the PR relative to the reference as a function of the rain fraction (RF) and the relative nonuniform beam filling (RNUBF).

for both 2A25 classifications as well as the interquantile (90%–10%) value to assess the random part of the error. Figures 4.6c and 4.6d show the conditional biases and random errors of both classifications of 2A25 relative to the reference. The conditional biases are very distinct according to the PR rainfall types. The PR convective systematic biases present a shift toward higher values (+2 mm/h) compared to stratiform biases for CPI < 80%. Biases cover a much broader range for the PR convective type (from −10 mm/h to 2 mm/h) than for the stratiform type (from −3 to 0.5 mm/h). They are both decreasing functions of the convective contribution CPI, with overestimation at values < 20% and underestimation for convective contribution > 90%. Apparently, both stratiform and convective profiling algorithms in 2A25 lack sufficient dynamics to deal with extreme rainfall amounts. In addition to the Z-R relationships, the consistent underestimation at high rainfall rates could also be related to the attenuation correction algorithm, or even total loss of the radar signal. The random discrepancies increase consistently with CPI for both types, and it is worth noting that the spread of the PR convective rainfall spectra in Figure 4.6d is greater (~ +5 mm/h) than for the stratiform PR rainfall whatever the convective contribution. Such features could be related to the 2A25 Z-R relationships and potentially indicate greater uncertainties in quantifying the convective rainfall.

Insight into the impact of the fine-scale variability of rainfall on the bias of the PR relative to the reference is provided in Figure 4.6e. The bias of PR relative to the reference is computed as a function of the variability within the PR field of view (rain fraction and RNUBF). The bias is regularly distributed along the RNUBF and RF axes with highest underestimation (−30%) around RF ~ 100% and RNUBF ~ 0.5. More generally, the PR underestimates relative to the reference for RF > 80% and RNUBF < 1. For lower FOV filling conditions and higher quantitative variability, 2A25 overestimates relative to the reference with bias values exceeding 1000%. In these conditions the reference rain rates are low (Figure 4.4a) and the NUBF correction is almost nonexistent in the 2A25 algorithm.

4.4. COMPARISON BETWEEN GROUND-BASED RADARS AND SPACE-BASED PASSIVE SENSORS

The GPM constellation will include a number of satellites and passive sensors with various effective resolutions for rainfall retrievals. Issues have been identified for passive satellite-based precipitation retrieval algorithms and motivate ongoing research: variability of the precipitation inside the sampling volumes (i.e., NUBF), geometric effects (parallax; the surface location of the TRMM-TMI footprint may be displaced from the footprint at cloud top), physical properties of hydrometeors, and surface conditions. Over land, the parallax effects cause the effective position of the TMI footprint to depend on the vertical extent and altitude of frozen hydrometeors in storms, which varies with the 0°C height, the top of the storm, etc. The errors in the effective footprint position in the TMI data are not accounted for, which could add random noise in the comparison between TMI and the reference rainfall. The detection issue is particularly significant with passive sensors when warm clouds are considered. Warm clouds are described as those that develop droplets and yield rainfall at temperatures greater than 0°C and/or have not developed the ice phase. In essence, they produce rainfall via warm rain processes. They can be associated to orographic clouds, early stages of cumulonimbus, stratus, stratocumulus, or even cumulus congestus that have their entire life cycle within the warm phase. These clouds do not lead to a significant brightness temperature depression and are often classified as "no rain" by the screening procedure. Overland surface microwave emissivity affects the extraction of key information about precipitating water from brightness temperatures. The interaction between microwave radiation and the land surface is complex, being dependent on a large number of highly variable surface characteristics, such as soil humidity and roughness, vegetation properties, or snow cover [*Aires et al.*, 2011; *Ferraro et al.*, 2013]. In this section we will compare the performances of TMI with the PR using the 2011 data sample (Section 4.2.3.). For this purpose all products (reference, PR, and TMI) are resampled to the TMI footprint resolution in the CONUS data set.

4.4.1. Precipitation Detection

Because variable land surface conditions (soil moisture, vegetation) yield detectable signals in the microwave spectrum, radiometers face issues with light precipitation detection [*Gopalan et al.*, 2010]. Using the Oklahoma data set, surface impacts on precipitation detection using satellite-based passive sensors are investigated from soil moisture simulations using the NASA catchment land surface model [*Koster et al.*, 2000; *Ducharne et al.*, 2000; *Maggioni et al.*, 2011] forced with radar-rainfall fields and from the MODIS normalized difference vegetation index (Table 4.1).

Table 4.1 Sample definition according to different classes of soil wetness and vegetation index

	Dry Soil Wetness Value <0.178	Medium	Wet Soil Wetness Value >0.309
Low vegetation NDVI < 0.411	2183	3357	2215
Higher vegetation NDVI > 0.411	5575	12158	5543

Table 4.2 Contingency table for satellite precipitation detection relative to reference[a]

			Reference		
			> 0.	= 0.	Σ Estimates
Satellite estimates	> 0.	Frozen	2.50	76.6	2797
		Warm rain	0.60	—	25
		General	46.4	—	14413
	=0.	Frozen	20.9	—	738
		Warm rain	99.4	—	4429
		General	53.6	—	16618
	Σ reference	Frozen	826	2709	3535
		Warm rain	4454	0	4454
		General	31031	0	31031

[a] The results are provided for frozen precipitation, warm rain and general (other) cases.

Table 4.2 shows the contingency tables for satellite rain/no-rain occurrence relative to the reference with percentile of hits (H; both reference and satellite detect rain), misses (M; satellite does not detect rain while reference does), false alarms (F; satellite detects rain while reference does not), and correct rejections (C; both reference and satellite do not detect rain). The reference data are separated into three subsamples: frozen surface, warm rain, and the other (general) cases. All coincident and colocated satellite values are considered and sorted according to the reference samples.

The detection behavior is very distinct for the three subsamples, with high proportion (at least 75%) of false detection (M+F) in case of warm rain and frozen surface while limited to 54% in the general case. In case of frozen surface, the radiometrically cold emissivity impacts the T_b so that the 2A12 is triggered to false alarms (77%) and still misses a significant part (21%) of the frozen precipitation. Warm cloud precipitation does not produce significant ice content and would likely escape detection as the 2A12 algorithm depends on the ice-scattering signal to detect rain. This finding is noted from previous 2A12 studies [*Liu and Zipser*, 2009; *Wang et al.*, 2009] and is consistent with a study over Africa with a similar algorithm [*Kirstetter et al.*, 2012b]. The misses (M) are generally the main contributors (at least 53%) to the false detection (M+F) population except for the frozen surface. The 2A12 algorithm retrieves rain only when a clear scattering signature is detected in the T_b's, and this effect is expected to be particularly sensitive in the no-rain to light-rain transition regions and also for all the events that do not develop ice phase (like in the case of warm rain). Table 4.3 shows the discarded rain volumes in question. The misses of 2A12 represent more than 90% of the reference rainfall volume in the frozen and warm rain cases while around 12% of the reference rainfall volume in the general cases.

Figure 4.7a brings additional insight into the detection performances of the 2A12 algorithm for the general cases.

Table 4.3 Discarded rain volumes due to misses from 2A12 relative to reference[a]

Frozen	90.6%
Warm rain	99.6%
General	12.1%

[a] The results are provided for frozen precipitation, warm rain, and general (other) cases.

The probability of detection of the algorithm consistently increases with the ice water content from less than 10% for $VIL_{ice} < 0.1 \, g/m^2$ to more than 70% for $VIL_{ice} > 5 \, g/m$. This confirms that 2A12 tends to miss the lightest intensities and the rain/no rain-transition areas. Detectability in 2A12 is dictated by signals from the 21V and 85V GHz channels, both of which respond differently to surface conditions and storm microphysical properties. Thus, it is difficult to disentangle and quantify the respective impacts from the surface and atmospheric factors. Performances vary according to the surface conditions, that is, better over vegetated areas (POD rate 15–20% higher than over less vegetated areas), and POD systematically improves over less vegetated areas for all VIL_{ice} values from wet surfaces to dry surfaces. Note that because of the correlation between the surface properties and the climatological rainfall features in Oklahoma (not shown here), it is difficult to uniquely distinguish the impact of the surface conditions on the 2A12 retrievals from the impact of the vertical structure of the precipitating clouds. More research is needed to investigate those aspects.

Figure 4.7b shows the cumulative distributions of rain occurrence and volume as functions of the rain intensity at the TMI resolution for the CONUS data set. It depicts the behavior of the POD of the PR (resampled at the TMI resolution) and TMI as a function of the reference rain rate. The POD consistently improves for both sensors with rain rate with a sharp increase between 0.1 and 1.5 mm/h. Similar to the PR, the TMI

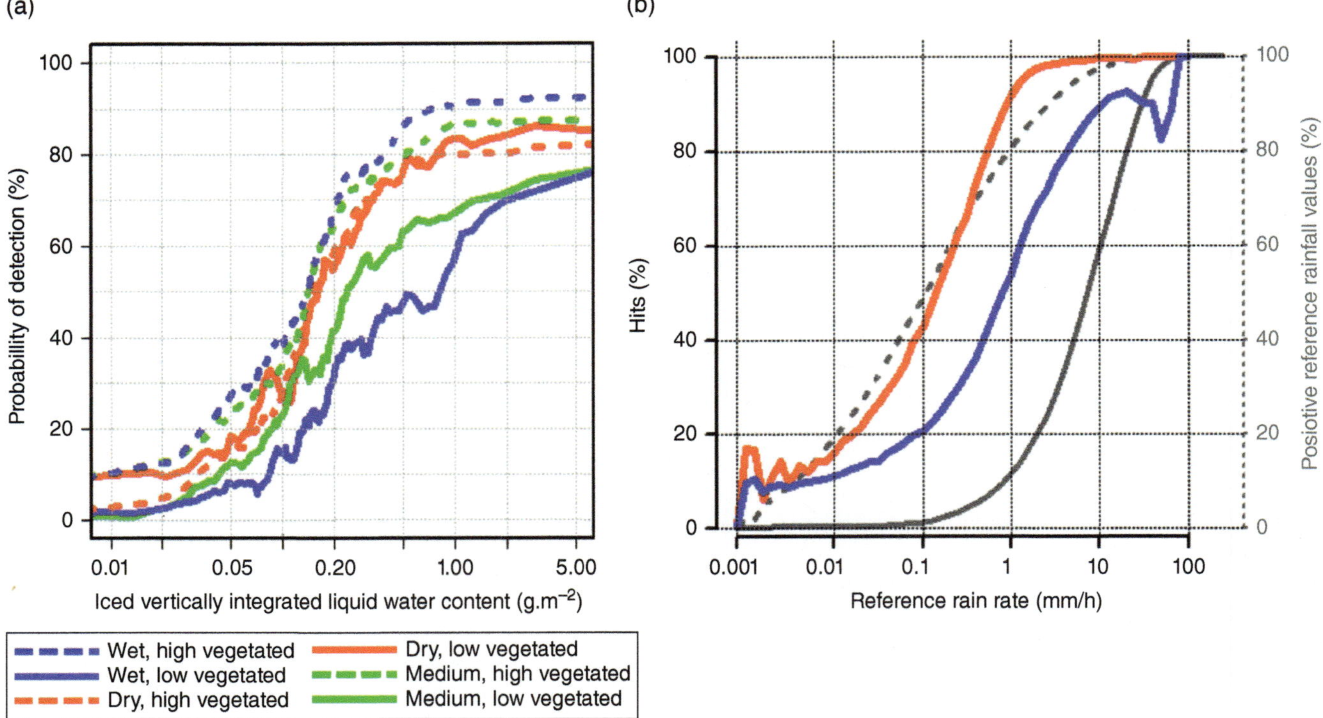

Figure 4.7 (a) Probability of detection of 2A12 as function of ice water content for various soil moisture and vegetation conditions; (b) probability of detection of the reference rainfall by the TMI (blue) and the PR (at the TMI footprint resolution, red) as a function of the reference rain intensity. The accumulated fractions of rain occurrence (dotted line) and volume (plain line) are plotted against the secondary ordinate (in gray).

POD is neither null for low rain rates nor equal to unity for high rain rates, and the transition between the two sills is not a step function, which suggests that the rain intensity is not the only factor driving the detection capabilities of TMI. The POD of TMI is below the PR one at all rain intensities, which is consistent with the fact that the 2A12 algorithm retrieves rain only when a clear scattering signature is detected in the brightness temperatures. The PR at this resolution misses rain rate occurrences at a rate of 48%, which represents 4% in volume, while TMI misses 68% occurrences of rainfall for a volume of 20%.

4.4.2. Precipitation Characterization

The radiometer algorithms quantify the convective contribution in the rain rate retrievals. We address this aspect focusing on cases when both 2A12 and the reference rainfall are greater than zero over the CONUS data set so that precipitation detection is not a factor. The PR rainfall is resampled to the TMI footprint resolution to compare the classification performances of TMI with the PR.

Figure 4.8 compares the conditional PDFs of the PR and TMI convective contribution (CPI) relative to the reference. All panels in Figure 4.8 show a shift toward higher CPI values as the reference CPI increases, as we would expect. As an example, at the reference CPI = 20% the interquantile 25%–50% of the PR CPI distribution is between 0% and 72% and between 65% and 97% at the reference CPI = 80%. At the reference CPI = 20%, the interquantile 25%–50% of the TMI CPI distribution is between 0% and 72% and between 38% and 85% at the reference CPI = 80%. Both space sensors show consistent classification performances relative to the reference.

At lower reference CPI values (<10%) the PR and TMI show different behavior, with the PR conditional median becoming positive for reference CPI > 9%, while the TMI spectra show a sharp increase up to 7% resulting in an overestimation of the convective contribution. Over the same reference CPI domain, the 75% quantile of the PR CPI distribution shows a sharp increase from 0% to 60%, while the TMI equivalent is higher between 52% and 67%. As an active sensor, the PR better separates convective from stratiform echoes because on average closer to the reference (Section 4.3.2.) and presenting less uncertainties than TMI.

For CPI > 10%, the dynamic ranges of convective contribution spectra are greater for the PR than for TMI. The conditional median of the PDF for the PR is closer

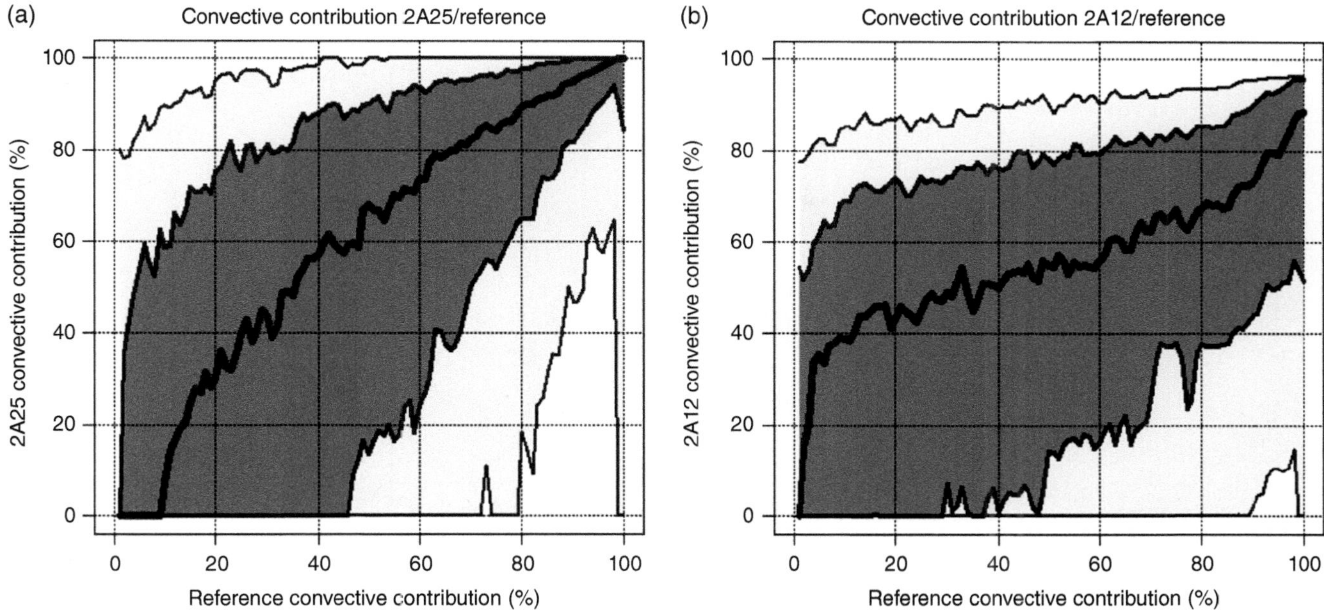

Figure 4.8 Density plots of CPI for (a) PR (at the TMI footprint resolution) and (b) TMI versus reference CPI. The thick black line represents the median (50% quantile), the dark gray-shaded region represents the area between the 25% and 75% quantiles, and the light gray-shaded region represents the area between the 10% and 90% quantiles.

to the 1 : 1 line than TMI, which shows general overestimation for lower reference CPI values (between 10% and 50%) and underestimation of higher CPI values. The 2A12 algorithm's skill in detecting and quantifying convection depends on the ice content, ice density, or spatial homogeneity of the ice field that impacts T_b's at 85 GHz. The uncertainties in convective contribution remain significant at all levels. It is worth noting that the spread of the PR convective contribution spectra decreases for higher reference CPI values. The uncertainties of the PR classification decrease faster than TMI with higher reference CPI.

4.4.3. Precipitation Quantification

To investigate the rainfall quantification from 2A12, Figure 4.9a shows the behavior of the mean relative error (MRE) (expressed in percentage and defined as $MRE = 100\left(\overline{R_{ref}} - \overline{R}\right)/\overline{R_{ref}}$) as a function of VIL_{ice} for various conditions of soil moisture and vegetation with the Oklahoma data set. A rainy pixel is included in the statistics if both satellite and the reference are nonzero to emphasize the satellite sensor's ability to quantify precipitation when it is raining (the case of satellite having zero rainfall when it is raining has been addressed in Section 4.4.1.). This is relevant given the significant misses of the satellite. The 2A12 algorithm significantly overestimates the reference mean values for dry and medium soil moisture conditions over the range of ice water content values. For wet conditions,

2A12 overestimates the reference values for $VIL_{ice} < 0.3\,g/m^2$ and underestimates the reference values for $VIL_{ice} > 1\,g/m^2$. The degree of overestimation is significantly related to the soil moisture with greater MRE values (up to 250%) over dry surface conditions and lower MRE values (up to 150%) over wet surface conditions. The overestimation decreases with increasing VIL_{ice} with the strongest decrease being at VIL_{ice} below $1\,g/m^2$.

Some insight into the influence of rain typing on the PR and TMI errors is provided by Figures 4.9b and 4.9c computed with the 2011 data set. The residuals for PR and TMI are shown as a function of the reference convective contribution. The conditional PDFs of residuals ε present a conditional shift from the 0 line and a conditional spread. Figures 4.9b and 4.9c show a shift toward lower rainfall rates as CPI increases. These features are more pronounced with TMI (Figure 4.9c). Both 2A25 and 2A12 products present a tendency to overestimate rain rates for light and moderate CPI values (CPI < 60%) and underestimate rain rates for CPI values higher than 80%. As an example, the 2A25 product overestimates at CPI = 20% (mainly stratiform) rain rates with an occurrence of 55% and with a representative bias of +0.2 mm/h and underestimates at CPI = 100% (convective) rain rates with an occurrence of 75% and with a representative bias of −7 mm/h. The stratiform PR model overestimates at CPI = 20% rain rates with an occurrence of 70% and with a representative bias of +1.5 mm/h and underestimates at CPI = 100% (convective) rain rates with an occurrence of 65% and with a representative bias of −6 mm/h.

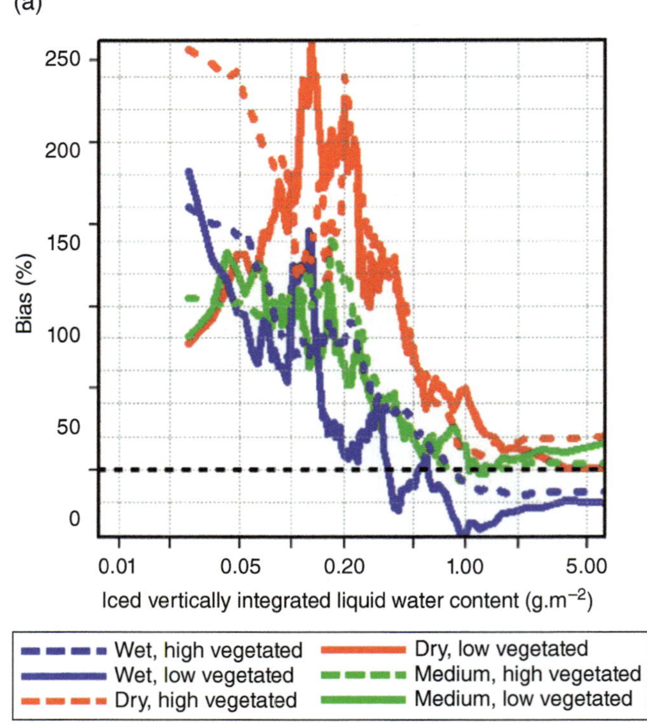

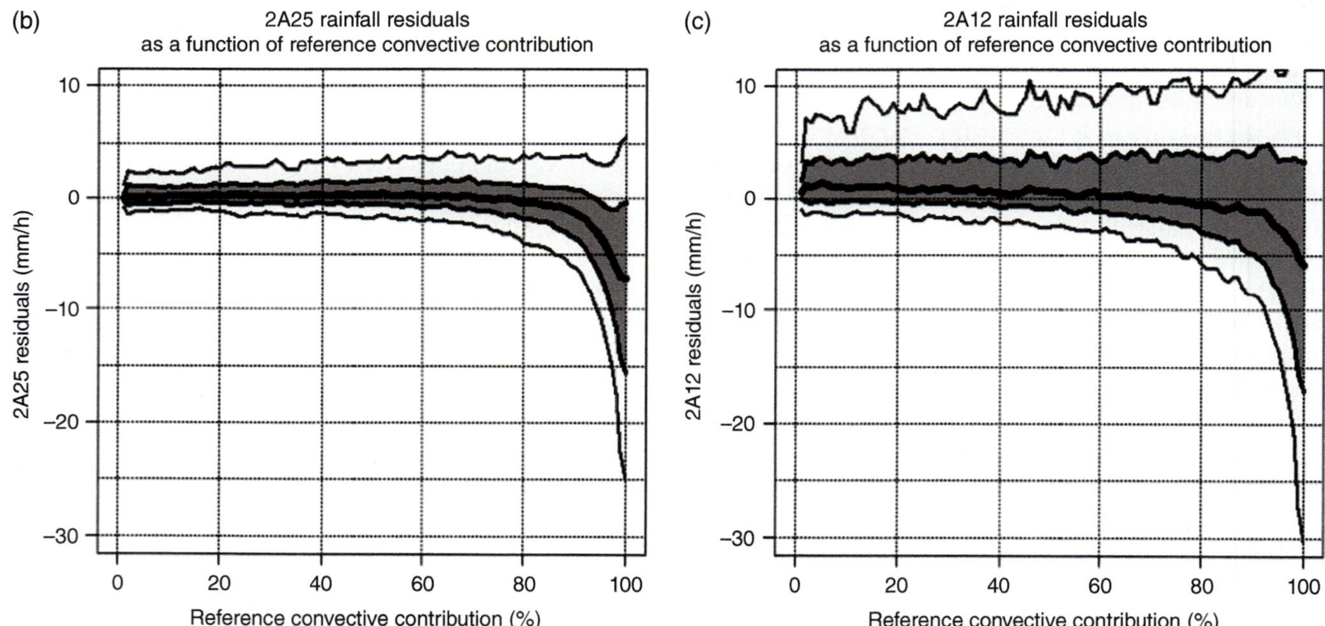

Figure 4.9 (a) Mean relative error of 2A12 versus reference as functions of ice water content for various soil moisture and vegetation conditions; density plots of residuals (satellite—reference rainfall) versus reference CPI for (a) PR and (b) TMI . The thick black line represents the median (50% quantile), the dark gray-shaded region represents the area between the 25% and 75% quantiles, and the light gray-shaded region represents the area between the 10% and 90% quantiles.

The conditional biases are both decreasing functions of the convective contribution CPI and present more dynamics for the 2A12 product from +2mm/h for CPI < 20% to −6mm/h for CPI = 100%. This is consistent with the 2A12 overestimation noted on Figure 4.9a. The 2A12 algorithm apparently lacks sufficient dynamics to deal with extreme rainfall amounts compared to the 2A25 algorithm. The random discrepancies increase consistently with CPI for both products, and it is worth noting that the spread of the TMI convective rainfall spectra (uncertainties) is greater than for the PR rainfall whatever the convective contribution. While the Bayesian methodology in the 2A12 product yields estimates that are well constrained as designed from the a priori database, this approach apparently does not accommodate extreme rainfall amounts.

4.5. CONCLUSIONS AND PERSPECTIVES

Rainfall variability affectes satellite-based rainfall retrieval algorithms in terms of detection capabilities, characterization of rainfall types, and quantification of rainfall rates. A 7 month data sample of TRMM PR and TRMM TMI-based rainfall products was analyzed using gauge-adjusted and quality-filtered surface rainfall estimates derived from NMQ/Q2. Several high-resolution Q2 products were used to characterize the reference rainfall in terms of occurrence, types, and rate at both sensors' pixel resolution to evaluate the PR and TMI detection, classification, and quantification performances.

For TRMM PR, primary errors due to incorrect physical assumptions related to convective versus stratiform rainfall classification, NUBF, and conversion from reflectivity-to-rainfall intensity were investigated. Rainfall detection and classification capabilities vary with the horizontal variability within the instrument's field of view (nonuniform beam filling). The PR detects rainfall when the FOV is significantly filled with rainfall (at least 70%). While segregating rain from no-rain boundaries is a driving contributor to the PR rain rate errors, probably linked to the lack of sensitivity in the most inhomogeneous and light parts of the edges of rainy regions, simple rain rate threshold-based detection models are not accurate in case of high rainfall variability. We thus recommend caution when using them for evaluating the PR detection performances. The PR classification generally agrees with the reference, however, misclassification is shown to occur with variability of the rainfall within the FOV not resolved by the 2A23 algorithm, with false convective detection associated with low filling of the FOV and low RNUBF values. Regarding quantification, significant error is most likely due to a combination of inaccurate Z-R relationship, nonuniform beam filling, and/or attenuation (even total loss) of the PR radar signal. Both stratiform and convective profiling algorithms in 2A25 seem to lack sufficient dynamics to deal with extreme rainfall amounts. For lower FOV filling conditions and higher quantitative variability, 2A25 overestimates relative to the reference and underestimates otherwise. Results from the error model presented herein provide insights into the most significant characteristics of PR rainfall retrieval errors that need to be taken into account when such data are used in applications.

As a passive sensor, 2A12 rainfall estimates over land rely on the scattering signature of precipitation in ice phase aloft and are also sensitive to soil moisture, vegetation, and vertical cloud structure. Significant ice water content (>0.2 g/m^2) is required for the 2A12 algorithm to detect rainfall and its detection capabilities are lower than PR for all rainfall intensities. As an active sensor, PR better separates convective from stratiform echoes and presents better dynamics in quantifying the convective contribution to the rainfall. The 2A12 products overestimate the reference mean values over the range of ice water content values. Once again, the 2A12 algorithm is inadequate to deal with extreme rainfall amounts compared to the 2A25 algorithm.

In terms of perspectives, the same framework and reference rainfall data sets can be readily applied to rainfall and even snowfall retrievals from other sensors onboard LEO satellites (i.e., TMI, AMSR-E, SSMI, MADRAS). This framework will also be applied to GPM rainfall estimates following its launch in 2014. The NEXRAD network has been upgraded with dual-polarization technology. Compared to single-polarized radars, the polarization technology will improve the identification of hydrometeor phase (liquid, melting, or frozen) and types (e.g., hail, snow) and changes in particle size distribution within individual storm systems and between meteorological regimes. The advent of coincidental upgrades to the NEXRAD network with dual polarization and the launch of GPM with a dual-frequency radar provides unique opportunities for comparison of cloud microphysical properties as seen by various sensors. Another important issue to study is how the various error sources in PR and TMI, which is often used as a calibrator, propagate when merging with geostationary infrared data for a number of satellite-based, high-resolution precipitation products. Current needs exist for error information for the development of level 3 integrated satellite precipitation products, in particular on the detection aspect (likelihood of rain) and uncertainty on the quantification of precipitation rates. The GPM core satellite sensors will be used as a reference to intercalibrate the constellation radiometers to provide self-consistent radiometric observations across the constellation. However, because satellite coincidences within a few minutes are rare, a CONUS-wide consistent reference from NMQ will be the ideal tool to compare and intercalibrate the constellation satellites to the GPM core sensors.

REFERENCES

Aires, F., C. Prigent, F. Bernardo, C. Jiménez, R. Saunders, and P. Brunel (2011), A tool to estimate land-surface emissivities at microwave frequencies (TELSEM) for use in numerical weather prediction. *Q. J. R. Meteorol. Soc.*, *137*, 690–699.

Amitai, E., X. Llort, and D. Sempere-Torres (2009), Comparison of TRMM radar rainfall estimates with NOAA next-generation QPE, *J. Meteorol. Soc. Japan*, *87A*, 109–118.

Awaka, J., T. Iguchi, and K. Okamoto (2007), Rain type classification algorithm, *Adv. Global Change Res.*, *28*, 213–224.

Berg, W., T. L'Ecuyer, and J. M. Haynes (2010), The distribution of rainfall over oceans from spaceborne radars, *J. Appl. Meteor. Climatol.*, *49*, 535–543.

Chen, S., J. J. Gourley, Y. Hong, P. E. Kirstetter, J. Zhang, K. Howard, Z. Flamig, J. Hu, and Y. Qi (2013), Evaluation and uncertainty estimation of NOAA/NSSL Next Generation National Mosaic QPE (Q2) over the Continental United States, *Journal of Hydrometeorology,14*, 1308–1322. doi:10.1175/JHM-D-12-0150.1

Ducharne, A., R. D. Koster, M. J. Suarez, M. Stieglitz, and P. Kumar (2000), A catchment-based approach to modeling land surface processes in a general circulation model: 2. Parameter estimation and model demonstration, *J. Geophys. Res.*, *105*(20), 24,809–24,822.

Ebert, E. (2007), Methods for verifying satellite precipitation estimates, in *Measuring Precipitation from Space: EURAINSAT and The future*, edited by V. Levizzani, P. Bauer, and F. J. Turk, pp. 345–356, Springer.

Ferraro, R., et al. (2013), An evaluation of microwave land surface emissivities over the continental UnitedStates to benefit GPM-era precipitation algorithms, *IEEE Trans. Geosci. Remote Sens.*, *51*(1), 378–398.

Gopalan, K., N. Y. Wang, R. Ferraro, and C. Liu (2010), Status of the TRMM 2A12 Land Precipitation Algorithm, *J. Atmos. Oceanic Technol.*, *27*, 1343–1354.

Grimes, D. I. F., and M. Diop (2003), Satellite-based rainfall estimation for river flow forecasting in Africa. I: Rainfall estimates and hydrological forecasts. *Hydrol. Sci. J.*, *48*, 567–584.

Hong, Y., R. F. Adler, F. Hossain, S. Curtis, and G. J. Huffman (2007), A first approach to global runoff simulation using satellite rainfall estimation, *Water Resour. Res* , *43*(8).

Huffman, G. J., R. F. Adler, D. T. Bolvin, G. Gu, E. J. Nelkin, K. P. Bowman, Y. Hong, E. F. Stocker, and D. B. Wolff (2007), The TRMM multi-satellite precipitation analysis: Quasi-global, multi-year, combined-sensor precipitation estimates at fine scale, *J. Hydrometeorol.*, *8*(1), 38–55.

Iguchi, T., T. Kozu, R. Meneghini, J. Awaka, and K. Okamoto (2000), Rain-profiling algorithm for the TRMM precipitation radar, *J. Appl. Meteor.*, *39*, 2038–2052.

Iguchi, T., T. Kozu, J. Kwiatkowski, R. Meneghini, J. Awaka, and K. Okamoto (2009), Uncertainties in the rain profiling algorithm for the TRMM precipitation radar, *J. Meteor. Soc. Japan*, *87A*, 1–30.

Joyce, R. J., J. E. Janowiak, P. A. Arkin, and P. Xie (2004), CMORPH: A method that produces global precipitation estimates from passive microwave and infrared data at high spatial and temporal resolution, *J. Hydrometeorol.*, *5*, 487–503.

Kirstetter, P. E., Y. Hong, J. J. Gourley, S. Chen, Z. L. Flamig, J. Zhang, M. Schwaller, W. Petersen, and E. Amitai (2012a), Toward a framework for systematic error modeling of NASA spaceborne radar with NOAA/NSSL ground radar-based National Mosaic QPE, *J. Hydrometeorol.*, doi:10.1175/JHM-D-11-0139.1.

Kirstetter, P. E., N. Viltard, and M. Gosset (2012b), An error model for instantaneous satellite rainfall estimates: Evaluation of BRAIN-TMI over West Africa, *Q. J. R. Meteorol. Soc.*, doi:10.1002/qj.1964.

Kirstetter, P. E., Y. Hong, J. J. Gourley, M. Schwaller, W. Petersen, and J. Zhang (2013), Comparison of TRMM 2A25 products Version 6 and Version 7 with NOAA/NSSL ground radar-based national mosaic QPE, *J. Hydrometeorol.*, *14*(2), 661–669, doi:10.1175/JHM-D-12-030.1.

Kirstetter, P. E., Y. Hong, J. J. Gourley, M. Schwaller, W. Petersen, and Q. Cao 2014: Impact of sub-pixel rainfall variability on spaceborne precipitation estimation: evaluating the TRMM 2A25 product, *Quarterly Journal of the Royal Meteorological Society, in press.* doi:10.1002/qj.2416

Kitzmiller, K. et al. (2011), Evolving multisensor precipitation estimation methods: Their impacts on flow prediction using a distributed hydrologic model, *J. Hydrometeorol.*, *12*, 1414–1431.

Koster, R. D., J. Suarez, A. Ducharne, M. Stieglitz, and P. Kumar (2000), A catchment-based approach to modeling land surface processes in a general circulation model: 1. Model structure, *J. Geophys. Res.*, *105*(20), 24,809–24,822.

Kummerow, C., W. Barnes, T. Kozu, J. Shiue, and J. Simpson (1998), The Tropical Rainfall Measuring Mission (TRMM) sensor package, *J. Atmos. Oceanic Technol.*, *15*, 809–817.

Kummerow, C., W. Berg, J. Thomas-Stahle, and H. Masunaga (2006), Quantifying global uncertainties in a simple microwave rainfall algorithm, *J. Atmos. Oceanic Technol.*, *23*, 23–37.

Kummerow, C. D., S. Ringerud, J. Crook, D. Randel, and W. Berg (2011), An observationally generated a priori database for microwave rainfall retrievals, *J. Atmos. Oceanic Technol.*, *28*(2), 113–130.

Lebel, T., C. Cappelaere, S. Galle, N. Hanan, L. Kergoat, S. Levis, B. Vieux, L. Descroix, M. Gosset, and E. Mougin (2009), AMMA-CATCH studies in the Sahelian region of West-Africa: An overview, *J. Hydrol.*, *375*(1–2), 3–13.

L'Ecuyer, T. S., and G. L. Stephens (2002), An uncertainty model for Bayesian Monte Carlo retrieval algorithms: Application to the TRMM observing system, *Q. J. Meteorol. Soc.*, *128*, 1713–1737.

Lin, X., and A. Y. Hou (2008), Evaluation of coincident passive microwave rainfall estimates using TRMM PR and ground measurements as references, *J. Appl. Meteor. Climatol.*, *47*, 3170–3187.

Lin, X., A. Y. Hou (2012), Estimation of rain intensity spectra over the Continental United States using ground radar–gauge measurements, *J. Climate*, *25*, 1901–1915.

Liu, C., and E. J. Zipser (2009), "Warm rain" in the tropics: Seasonal and regional distribution based on 9 years of TRMM data, *J. Climate*, *22*, 767–779.

Maggioni, V., R. H. Reichle, and E. N. Anagnostou (2011), The effect of satellite-rainfall error modeling on soil moisture prediction uncertainty, *J. Hydrometeorol.*, *12*, 413–428.

Menegheni, R., T. Iguchi, T. Kozu, L. Liao, K. Okamoto, J. A. Jones, and J. R. Kwiatkowski (2000), Use of the surface reference technique for path attenuation estimates from the TRMM precipitation radar, *J. Atmos. Ocean. Technol.*, *40*, 2053–2070.

Menegheni, R., J. A. Jones, T. Iguchi, K. Okamoto, and J. R. Kwiatkowski (2004), A hybrid surface reference technique and its application to the TRMM precipitation radar, *J. Atmos. Ocean. Technol.*, *21*, 1645–1658.

Petersen, W. A., and M. R. Schwaller (2008), NASA GPM Ground Validation Science Implementation Plan, NASA Rep., available at http://pmm.nasa.gov/sites/default/files/document_files/GPM_GVS_imp_plan_Jul08.pdf.

Petty, G. W. (1995), Frequencies and characteristics of global oceanic precipitation from shipboard present-weather reports, *Bull. Amer. Meteorol. Soc.*, *76*, 1593–1616.

Petty, G. W. (1997), An intercomparison of oceanic precipitation frequencies from 10 special sensor microwave/imager rain rate algorithms and shipboard present weather reports, *J. Geophys. Res.*, *102*, 1757–1777.

Sapiano, M. R. P., and P. A. Arkin (2009), An intercomparison and validation of high-resolution satellite precipitation estimates with 3-hourly gauge data, *J. Hydrometeorol.*, *10*, 149–166.

Schumacher, C., and R. A. Houze, Jr. (2000), Comparison of radar data from the TRMM satellite and Kwajalein oceanic validation site, *J. Appl. Meteorol.*, *39*, 2151–2164.

Steiner, M., R. A. Houze, and S. E. Yuter (1995), Climatological characterization of three-dimensional storm structure from operational radar and rain gauge data, *J. Appl. Meteorol.*, *34*(9), 1978–2007.

Stephens, G. L., and C. D. Kummerow (2007), The remote sensing of clouds and precipitation from space: A review, *J. Atmos. Sci.* 64(11), 3742–3765.

Takahashi, N., H. Hanado, and T. Iguchi (2006), Estimation of path-integrated attenuation and its nonuniformity from TRMM/PR range profile data, *IEEE Trans. Geosci. Remote Sens.*, *44*(11), 3276–3283.

Turk, F. J., P. Arkin, E. E. Ebert, and M. R. P. Sapiano (2008), Evaluating high resolution precipitation products. *Bull. Amer. Meteorol. Soc.*, *89*, 1911–1916.

Ushio, T., et al. (2006), A combined microwave and infrared radiometer approach for a high resolution global precipitation mapping in the GSMAP project Japan, paper presented at the 3rd Int. Precipitation Working Group Workshop on Precipitation Measurements, Melbourne, Australia.

Vasiloff, S., et al. (2007), Improving QPE and very short term QPF: An initiative for a community-wide integrated approach, *Bull. Amer. Meteorol. Soc.*, *88*, 1899–1911.

Wang, N. Y., C. Liu, R. Ferraro, D. Wolff, E. Zipser, and C. Kummerow (2009), The TRMM 2A12 land precipitation product: Status and future plans, *J. Meteorol. Soc. Japan*, *87A*, TRMM special issue, 237–253.

Wen, Y., Q. Cao, P.E. Kirstetter, Y. Hong, J. J. Gourley, J. Zhang, G. Zhang, and B. Yong (2013), Incorporating NASA space-borne radar data into NOAA National Mosaic QPE system for improved precipitation measurement: a physically based VPR identification and enhancement method. *J. Hydrometeorol.*, *14*, 1293–1307. doi:10.1175/JHM-D-12-0106.1.

Wilheit, T., C. D. Kummerow, and R. Ferraro (2003), Rainfall algorithms for AMSR-E, *IEEE Trans. Geosci. Remote Sens.*, *41*, 204–214.

Wilks, D. S. (2011), *Statistical Methods in the Atmospheric Sciences*, Elsevier.

Wolff, D. B., and B. L. Fisher (2008), Comparisons of instantaneous TRMM ground validation and satellite rain-rate estimates at different spatial scales, *J. Appl. Meteorol. Climatol.*, *47*, 2215–2237.

Wolff, D. B., and B. L. Fisher (2009), Assessing the relative performance of microwave-based satellite rain-rate retrievals using TRMM ground validation data, *J. Appl. Meteorol. Climatol.*, *48*(6), 1069–1099.

Yang, S., W. S. Olson, J. J. Wang, T. L. Bell, E. A. Smith, and C. D. Kummerow (2006), Precipitation and latent heating distributions from satellite passive microwave radiometry. Part II: Evaluation of estimates using independent data, *J. Appl. Meteorol. Climatol.*, *45*(5), 721–739.

Zeweldi, D. A., and M. Gebremichael (2009), Sub-daily scale validation of satellite-based high-resolution rainfall products, *Atmos. Res.*, *92*(4), 427–433.

Zhang, J., K. Howard, and J. J. Gourley (2005), Constructing three-dimensional multiple radar reflectivity mosaics: Examples of convective storms and stratiform rain echoes, *J. Atmos. Ocean. Technol.*, *22*, 30–42.

Zhang, J., et al. (2011a), National mosaic and multi-sensor QPE (NMQ) system: Description, results and future plans, *Bull. Amer. Metereal. Soc.*, *92*, 1321–1338.

Zhang, J., Y. QI, K. Howard, C. Langston, and B. Kaney (2011b), Radar Quality Index (RQI): A combined measure of beam blockage and VPR effects in a national network, in *Proceedings, International Symposium on Weather Radar and Hydrology*, IAHS Publ. 351.

Section II: Evapotranspiration

5

Estimating Regional Evapotranspiration Using a Three-Temperature Model and MODIS Products

Yu Jiu Xiong,[1,2] Guo Yu Qiu,[3] Shao Hua Zhao,[4] and Fei Tian[5]

5.1. INTRODUCTION

Evapotranspiration (ET) is a critical component of both water and energy balance and is also of crucial importance in the disciplines of hydrology [*Trenberth et al.*, 2006], meteorology [*Teuling et al.*, 2009], agriculture [*Oki and Kanae*, 2006], and in water-related areas when considering water shortages and global climate change. Large-scale ET measurements with spatially continuous features are required in water and energy-related applications. However, it is difficult to accurately quantify ET at the regional scale owing to the heterogeneity of the land surface and the large number of controlling factors involved, e.g., plant biophysics, soil properties, relative humidity, and topography [*Bastiaanssen et al.*, 1998; *Glenn et al.*, 2007; *Monteith*, 1965; *Shuttleworth and Wallace*, 1985; *Su*, 2002].

Spatially distributed remote sensing data can be used as proxies for important controlling variables, and a large body of research has indicated that satellite remote sensing is the most feasible and economical method for mapping the spatial distribution of ET at scales ranging from the regional [e.g., *Allen et al.*, 2007; *Bastiaanssen et al.*, 1998; *Idso et al.*, 1975; *Jackson et al.*, 1983; *Jiang and Islam*, 2003; *Kustas and Daughtry*, 1990; *Moran et al.*, 1989; *Norman et al.*, 1995] to the global [e.g., *Jung et al.*, 2010; *Mu et al.*, 2007; *Vinukollu et al.*, 2011; *Zhang et al.*, 2010]. Remote-sensing-based methods fall into two categories: (1) empirical/statistical methods that estimate ET using an empirical equation deduced from remotely sensed variables, e.g., land surface temperature [*Carlson et al.*, 1995; *Jackson et al.*, 1977]; and (2) analytical methods, which estimate ET on the basis of the Priestley-Taylor approach [*Priestley and Taylor*, 1972], the Penman-Monteith equation [*Cleugh et al.*, 2007; *Mu et al.*, 2007, 2011], or the residual method of the energy balance equation [*Anderson et al.*, 1997; *Bastiaanssen et al.*, 1998; *Kustas and Anderson*, 2009; *Norman et al.*, 1995, 2003]. The three-temperature (3 T) model is a dual-source model that addresses the energy (e.g., latent heat) exchanges between the soil and vegetation and the overlying atmosphere and is deduced from the energy balance equation, but sensible heat is eliminated by introducing reference sites (dry soil without evaporation and dry canopy without transpiration) [*Qiu et al.*, 1996, 1998]. This constitutes the biggest difference with other ET estimation methods such as SEBAL (surface energy balance algorithm for land) [*Bastiaanssen et al.*, 1998], SEBS (surface energy balance system) [*Su*, 2002; *Jia et al.*, 2003], the Penman-Monteith equation [*Cleugh et al.*, 2007], and N95 [*Norman et al.*, 1995] because surface resistance is necessary for estimating the sensible heat flux, but its reproducibility has at times

[1] *Department of Water Resource and Environments, School of Geography and Planning, Sun Yat-Sen University, Guangzhou, Guangdong, China*

[2] *Key Laboratory of Water Cycle and Water Security in Southern China of Guangdong High Education Institute, Sun Yat-Sen University, Guangzhou, Guangdong, China*

[3] *Key Laboratory for Urban Habitat Environmental Science and Technology, School of Environment and Energy, Peking University, Shenzhen, Guangdong, China*

[4] *Environmental Satellite Center, Ministry of Environmental Protection, Beijing, China*

[5] *State Key Laboratory of Urban and Regional Ecology, Research Center for Eco-Environmental Sciences, Chinese Academy of Sciences, Beijing, China*

Remote Sensing of the Terrestrial Water Cycle, Geophysical Monograph 206. First Edition. Edited by Venkat Lakshmi.
© 2015 American Geophysical Union. Published 2015 by John Wiley & Sons, Inc.

limited the remote sensing applications [*Li et al.*, 2009; *Matsushita and Fukushima*, 2009]. Field tests show that the 3 T model is accurate with a mean absolute error (MAE) of less than 0.20 mm/d [*Qiu et al.*, 1998, 2006]; in addition, a previous application of the 3 T model based on the Enhanced Thematic Mapper (ETM+) imagery in a semiarid grassland in northern China showed that MAE was only 0.23 mm/d between the modeled ET and the observational data [*Xiong and Qiu*, 2011]. However, how well the 3 T model performs for remote sensing applications at a regional scale is still unclear.

The Moderate Resolution Imaging Spectroradiometer (MODIS), onboard NASA's *Terra* and *Aqua* satellites, has a high temporal resolution and covers the Earth multiple times per day, depending on the spectral characteristics of interest. MODIS data have spatial resolutions of 250, 500, and 1000 m. Furthermore, the MODIS science team has developed a series of products for global research, which are designed to represent certain parameters of the land surface, such as the land surface temperature (LST). Because some of the products contain fundamental variables in commonly adopted ET retrieving algorithms and are freely available, MODIS products have been used to estimate ET [*Cleugh et al.*, 2007; *Mu et al.*, 2007, 2011]. The MODIS-based estimations, while unable to discriminate the influence of land surface heterogeneity at the field scale, effectively reproduce watershed average responses, thereby illustrating the utility of this sensor for estimating regional-scale ET [*Mu et al.*, 2011; *Velpuri et al.*, 2013].

The objectives of this study are (1) to apply the 3 T model to estimate regional ET based on MODIS products, (2) to test the performance of the method at the basin scale, and (3) to analyze spatiotemporal trends in ET over the study basin.

5.2. MATERIALS AND METHODS

5.2.1. Description of the Study Area

The Jinghe River Basin, covering an area of approximately 45,421 km², is located in the southern part of the Loess Plateau, China, at approximately 34°46′–37°19′ N and 106°14′–108°42′E (Figure 5.1). Topographically high in the northern part and low in the southern part, this area is characterized by hills and syncline valleys, with Liupan Mountain on the west side and Ziwu Mountain on the east side. Elevation ranges from 360 to 2900 m above sea level, with a mean value of approximately 1390 m. The continental climate varies from subhumid to semiarid, with an annual average temperature, precipitation, and pan evaporation of 8–13 °C, 390–560 mm, and 1000–1300 mm, respectively. Precipitation decreases from the southeast to the northwest and mainly occurs from July to September, accounting for 50–70% of the total annual rainfall. Vegetation cover is relatively low, and the main vegetation types are prairie and desert grass, with some forest and shrubs distributed in the Liupan and Ziwu mountains.

5.2.2. Data Preparation

Data sets used for this study included hydrological data, meteorological data, remote sensing data, and geospatial data. The hydrological data represented the annual runoff at the Zhangjiashan gauge station, the outlet of the Jinghe River Basin. The meteorological data consisted of daily observational values of air temperature, precipitation, and cloud cover, obtained from 16 national meteorological stations (triangles in Figure 5.1).

Four types of remote sensing data were collected from five MODIS products with vertical and horizontal tiles of 26 and 5, respectively. These included 8 day MODIS land surface temperature/emissivity data (MOD11A2), leaf area index (LAI) and fraction of photosynthetically active radiation (FPAR) data (MOD15A2), 16 day global MODIS vegetation indices (MOD13A2), and albedo (MCD43B3). All the MODIS products were recorded between 2004 and 2006. MODIS land use and land cover (LULC) products (MCD12Q1, Figure 5.1) recorded in the same periods were also used. All the data used were provided by NASA at http://e4ftl01. cr.usgs.gov/.

The geospatial data consisted of vector data for the administrative boundaries and digital elevation model (DEM), i.e., the Shuttle Radar Topography Mission (SRTM) 90 m DEM [*Jarvis et al.*, 2008] provided by NASA at http://srtm.csi.cgiar.org/.

The data processing steps were as follows: (1) the MODIS products were projected to a spatial resolution of 1 km using the universal transverse mercator (UTM) projection (WGS84, UTM zone 48 N) with the MODIS conversion toolkit; (2) two consecutive 8 day MODIS products in each year were averaged to retrieve the value over 16 days, e.g., the average of day 1 and day 9 and the average of day 353 and day 361; (3) all meteorological data were averaged for every 16 day period according to the composition dates of the MODIS product in each year (except for precipitation, which was totaled on a daily basis), and then were interpolated to 1 km resolution for the study area by applying the inverse distance weight (IDW) interpolation method; and (4) the DEM was preprocessed and projected onto the coordinate system with the same spatial resolution as that of the processed MODIS product.

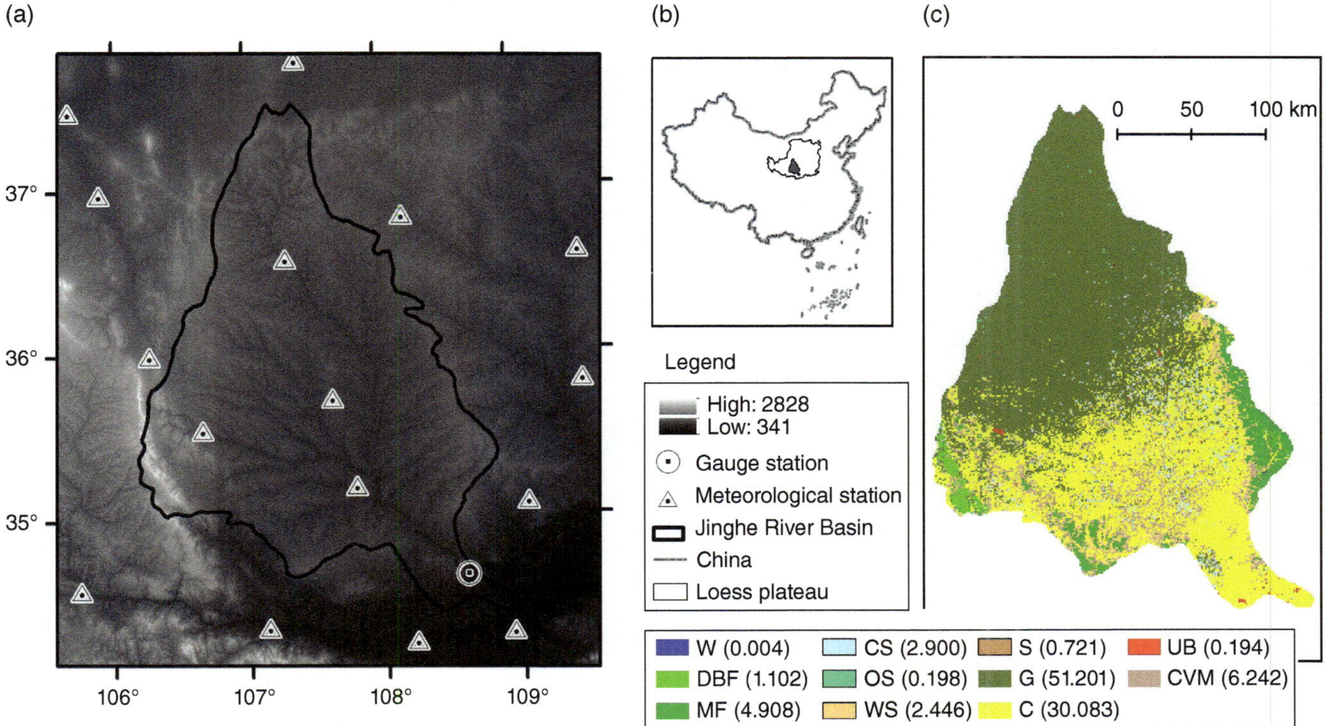

Figure 5.1 Geographic information of the study area: (a) location and SRTM DEM of the Jinghe River Basin; (b) schematic of the selected study area in China; and (c) land use and land cover (LULC) map of the study area based on the MODIS land use product MCD12Q1: (a) and (c) share the same scale bar; in LULC legends, W, DBF, MF, CS, OS, WS, S, G, C, UB, and CVM represent water, deciduous broadleaf forest, mixed forest, closed shrubland, open shrublands, woody savannas, savannas, grasslands, croplands, urban and builtup, and cropland/natural vegetation mosaic, respectively; the figure inside of the parentheses is percentage of the total land cover of each area; the triangles in (a) are meteorological stations, and the dot is the Zhangjiashan gauge station.

5.2.3. Theory of the 3T Model and ET Estimation

The 3T model, deduced from the energy balance equation, is a dual-source model with the advantage that aerodynamic resistance is eliminated by introducing reference sites (dry soil without evaporation and dry canopy without transpiration) [*Qiu et al.*, 1996, 1998]. Later, the 3T model was revised for remote sensing applications and includes equations (5.1)–(5.3) [*Xiong and Qiu*, 2011, 2014]. The 3T model can be applied to soil, vegetation, and a mixture of both, but it may not be suitable for estimating ET over other land surfaces, such as water or snow-covered areas.

$$LE_s = R_{n,s} - G_s - (R_{n,sd} - G_{sd})\frac{T_s - T_a}{T_{sd} - T_a} \quad \text{soil} \quad (5.1)$$

$$LE_c = R_{n,c} - R_{n,cp}\frac{T_c - T_a}{T_{cp} - T_a} \quad \text{vegetation} \quad (5.2)$$

$$L(\text{ET}) = LE_s' + LE_c' \quad \text{mixture of soil and vegetation} \quad (5.3)$$

where E_s, E_c, and ET in millimeters are soil evaporation, vegetation transpiration, and evapotranspiration, respectively; L is the latent heat of vaporization, approximately $2.49 \times 10^6\,\text{W/m}^2/\text{mm}$. The subscripts s, c, a, sd, and cp represent soil, vegetation, atmosphere, reference soil, and reference canopy, respectively; R_n is net radiation and G is soil heat flux in W/m²; T is temperature in Kelvins; and E_s' and E_c' are the ET values for the soil and vegetation components within a given pixel, respectively.

In addition to air temperature, which is usually obtained from meteorological stations, other input variables in the 3T model can be retrieved from remote sensing data. Figure 5.2 shows the flowchart of the ET estimations based on the 3T model and MODIS products. First, the status of a pixel is distinguished by the normalized difference vegetation index (NDVI). When a pixel's NDVI is less than the NDVI$_{\text{min}}$, it is assumed to be a soil pixel. Pixels with an NDVI larger than the NDVI$_{\text{max}}$ are assumed to be vegetated pixels. When the NDVI is between NDVI$_{\text{min}}$ and NDVI$_{\text{max}}$, it is considered to be a mixture of soil and vegetation. The NDVI was retrieved from the 8 day MOD13A2 product. NDVI thresholds (NDVI$_{\text{min}}$

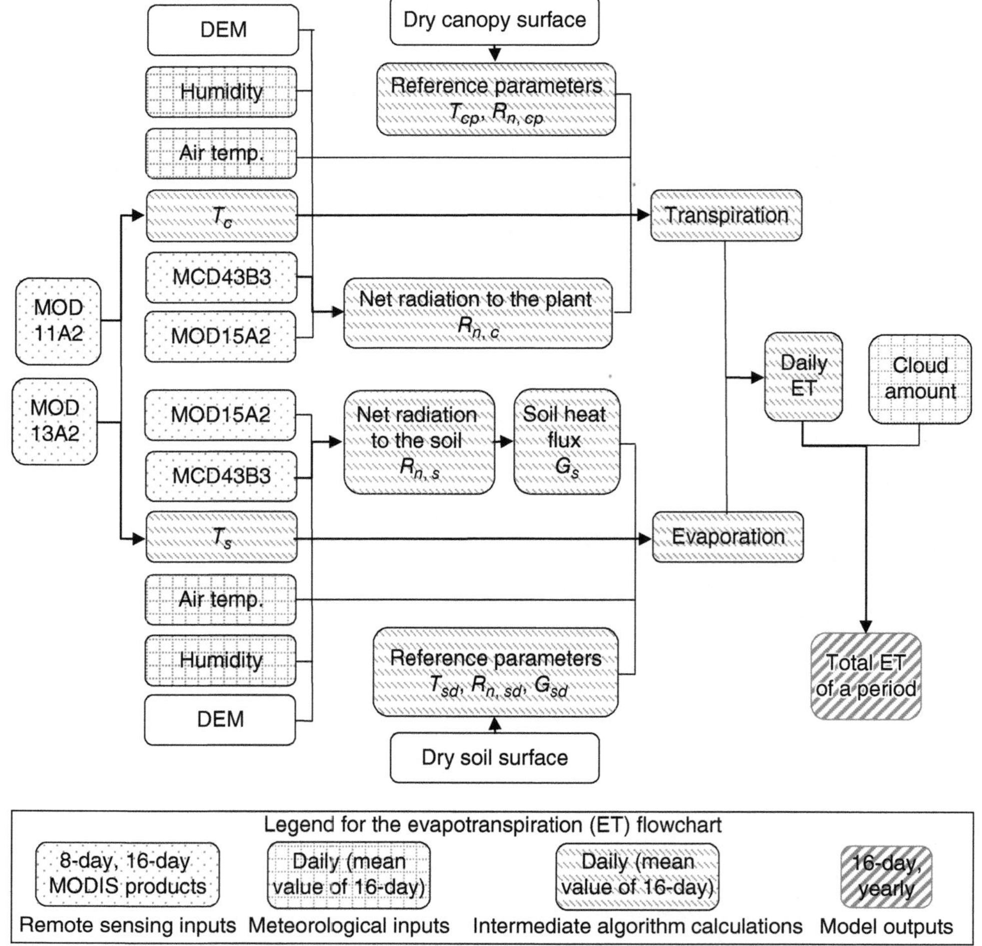

Figure 5.2 Flowchart of ET estimation based on the 3T model and MODIS products.

and NDVI$_{max}$) can be determined from the lower and upper 3% tails of the NDVI distribution. In this study, NDVI$_{min}$ = 0.2 and NDVI$_{max}$ = 0.5 were used, according to *Sobrino et al.* [2001].

T_s and T_c are the temperatures for the soil and vegetation components within a given pixel, respectively, which were estimated from the 8 day land surface temperature data ($T_{MOD11A2}$) using equation (5.4) [*Lhomme et al.*, 1994]. R_n was estimated using equation (5.5) and G was estimated using equation (5.6). The reference temperatures (T_{sd} and T_{cp}) were assumed to be a regional maximum value for the soil surface and vegetation temperatures [*Zhao et al.*, 2010; *Xiong and Qiu*, 2014]. Reference net radiation ($R_{n,sd}$ or $R_{n,cp}$) and reference soil heat flux (G_{sd}) were estimated using equations (5.7) and (5.8), respectively [*Xiong and Qiu*, 2011].

$$fT_c + (1-f)T_s = T_{MOD11A2}$$

$$T_s - T_c = a(T_{MOD11A2} - T_a)^m \quad (5.4)$$

where f is the fractional vegetation cover and is related to NDVI [*Carlson et al.*, 1995]. a and m are empirical coefficients: $a = 0.1$ and $m = 2$ [*Lhomme et al.*, 1994]. $T_{MOD11A2}$ is the land surface temperature data provided by MOD11A2.

$$R_n = (1-\alpha)R_{swd} + \varepsilon_0\varepsilon_a\sigma T_a^4 - \varepsilon_0\sigma T_{MOD11A2}^4$$
$$= (1-\alpha)\tau SE_0\cos\theta + \varepsilon_0\varepsilon_a\sigma T_a^4 - \varepsilon_0\sigma T_{MOD11A2}^4 \quad (5.5)$$

where R_{swd} represents the incoming short wave radiation, which can be calculated by using the atmospheric transmittance coefficient (τ), the solar constant (S), the eccentricity correction factor (E_0), and the solar zenith angle (θ) [*Xiong and Qiu*, 2011]; α is the surface albedo estimated from the 8 day MCD43B3 product; σ is the Stefan-Boltzmann constant (5.67×10^{-8} W/m²/K⁴); and ε_0 and ε_a are the land surface and atmospheric emissivity, respectively. ε_0 was calculated by considering NDVI in three different cases with respect to the maximum and minimum

NDVI values [*Sobrino et al.*, 2001]. ε_a was estimated from air temperature according to *Swinbank* [1963]:

$$G = R_n [\Gamma_c + (1-f)(\Gamma_s - \Gamma_c)] \tag{5.6}$$

where Γ_c and Γ_s are empirical coefficients: $\Gamma_s = 0.315$ is valid for full vegetation cover [*Kustas and Daughtry*, 1990] and $\Gamma_c = 0.05$ is valid for bare soil [*Monteith*, 1973]:

$$R_{n,r} = f(R_{swd}, \alpha_r, \varepsilon_{0r}, T_r, T_a) \tag{5.7}$$

$$G_{sd} = f(R_{n,sd}) = 0.315 R_{n,sd} \tag{5.8}$$

where the subscript r refers to the reference site, i.e., the dry soil or dry canopy. For the dry soil, $\alpha_r = 0.275$ [*Qiu et al.*, 1998], $\varepsilon_r = 0.970$ [*Sobrino et al.*, 2004], and $T_r = T_{sd}$, while $\alpha_r = 0.225$ [*Qiu et al.*, 1996], $\varepsilon_r = 0.990$ [*Sobrino et al.*, 2004], and $T_r = T_{cp}$ are used for the dry canopy. These four constants (0.275, 0.970, 0.225, and 0.990) used for the dry condition may not be suitable for all places; however, our previous results showed that they had little impact on the modeled ET results (see *Xiong and Qiu* [2011] for details).

According to the model input, an ET value estimated using the 3 T model was instantaneous (a mean value for a 16 day period in this study). This remotely sensed instantaneous ET can be scaled to a daily value using equation (5.9) [*Jackson et al.*, 1983] and is similar to the sinusoidal variation of solar radiation in the daytime on cloudless days. Equation (5.10) was proposed to obtain the total ET value for n days by summing the mean daily ET and considering cloud effects for each day because the MODIS products represent a mean for mostly cloudless conditions during that time interval:

$$ET_{MOD} = ET_i \frac{2N}{\pi \sin(\pi \cdot t / N)} \tag{5.9}$$

$$ET_{ndays} = a_C (n \cdot ET_{MOD})$$

$$a_C = \frac{D_{clear} + (n - D_{clear})(1-C)}{n} \tag{5.10}$$

where ET_i, ET_{MOD}, and ET_{ndays} in millimeters are the instantaneous, daily, and n days' total ET rate, respectively. N is the time duration of ET in the daytime, which was assumed to be the daily sunshine hours minus 2 h, and t is the time interval between sunrise and the data collection time point of the passing satellite sensor; n is the temporal granularity of the MODIS product and is here equal to 16; a_C is the cloud coefficient, D_{clear} is the number of sunny days in n days, and C is the mean total cloud cover for n days. These are routine observational items by meteorological stations, and we adopted their observations.

5.2.4. Model Evaluation

The modeled ET was evaluated in two ways. First, the modeled yearly ET was validated with a value estimated using the water balance method [equation 5.11]:

$$ET = P - R \tag{5.11}$$

where P in millimeters is the annual precipitation interpolated from the 16 national meteorological stations data, and R in millimeters is the runoff measured at the outlet gauge station of the Jinghe River Basin.

Because validation of the annual ET using the water balance method does not capture seasonal and spatial variations efficiently, the modeled ET was also compared with an ET product (MOD16A2), which was estimated on the basis of MODIS products and the Penman-Monteith equation [*Mu et al.*, 2007, 2011]. This method was adopted for validation because both modeled ET and MOD16A2 used MODIS products as inputs, have similar temporal resolutions, and share the same spatial resolution.

The MAE [equation 5.12], widely used in other studies [e.g., *Anderson et al.*, 2004; *Timmermans et al.*, 2007], was used to assess the model performance:

$$MAE = \frac{1}{n} \sum_{i=1}^{n} |E_i - M_i| \tag{5.12}$$

where E_i and M_i are paired model estimates/predictions and measured/observed variables, respectively; n is the number of data points.

5.3. RESULTS AND DISCUSSIONS

5.3.1. Comparison of Observed and Estimated ET

Following the procedures listed in Section 5.2.3., mean ET for each 16 day period and the yearly ET were calculated, as shown in Figures 5.3 and 5.4, respectively. Statistical results show that (1) the modeled mean daily ET of each 16-day period varied from 1.43 to 4.26 mm over the 3 year period, with mean values of 2.99, 2.94, and 2.89 mm/d for 2004, 2005, and 2006, respectively; and (2) the mean annual ET of the catchment was 487, 418, and 403 mm for 2004, 2005, and 2006, respectively.

The actual ET values of the study area from 2004 to 2006, measured using the water balance equation, were 400, 427, and 448 mm, respectively, as shown in Table 5.1. A comparison of the modeled ET and the water balance ET shows that the difference between them was small, with absolute errors of 87, 9, and 45 mm/yr for 2004, 2005, and 2006, respectively. These results indicate that the modeled ET was close to the water balance ET on a yearly scale.

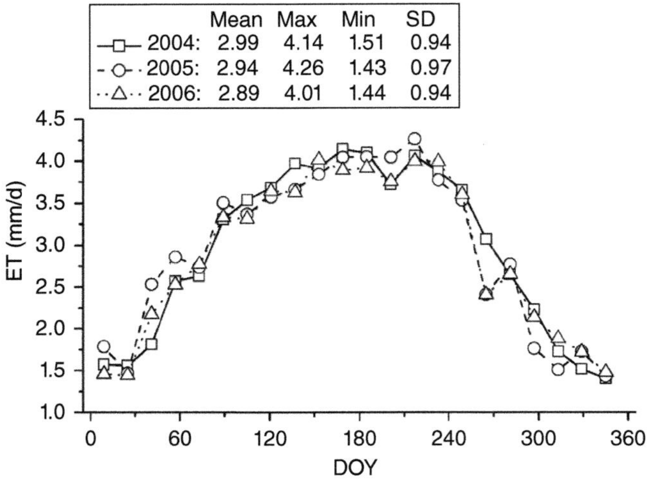

Figure 5.3 Mean daily ET of the Jinghe River Basin for each 16-day period estimated by the 3T model and equation (5.9).

A pixel-by-pixel comparison between the modeled daily ET and MOD16A2 maps for the same period in 2004 shows that although some of the randomly paired ET values are distributed closely along the 1:1 line, the majority of the modeled ET values are higher than those for MOD16A2, especially for MOD16A2 values lower than 4 mm/d (Figure 5.5). The MAE between the modeled ET and MOD16A2 was 1.18 mm/d, with minimum and maximum absolute errors of 0.00 and 9.24 mm/d, respectively. The differences can also be seen in Figure 5.4. The annual ET distribution of the two estimations shows a similar trend, i.e., high ET values are often present in areas with abundant vegetation coverage and low ET values occur in sparsely vegetated areas; however, the ET values estimated by the 3T model are generally higher than the MOD16A2 values (Figure 5.4). We chose data in 2004 because validation results on a yearly scale showed that the maximum MAE occurred in 2004, and we wanted to show the daily comparison for the worst case. At the same time, we realized that MOD16A2 itself carries uncertainties of up to 20% on a basin scale [*Mu et al.*, 2011], and it underestimated ET compared to that estimated by the water balance equation (Table 5.1), especially in 2005 and 2006, with absolute errors of 68 and 98 mm/yr, respectively. The ET estimated by the 3 T model was much closer to the water balance ET (Table 5.1). In addition, we found that upstream of the Jinghe River Basin (including Liupan Mountain), the highest ET occurred in forest land during 1997–2003, with a mean value of 484 mm/yr, followed by cropland and grassland, which varied from 399 to 450 mm/yr and 348–408 mm/yr, respectively [*Zhang et al.*, 2011]. Figure 5.6a indicates that our modeled ET for different LULCs was close to the results of *Zhang*

et al. [2011], whereas the MOD16A2 ET value was much lower, and the highest ET occurred in cropland (Figure 5.6b). *Feng et al.* [2012] also reported that MOD16A2 can be over-underestimated ET in China's Loess Plateau region. These results suggest that the ET estimated by the 3 T model shows a similar spatial distribution as the MOD16A2, but at basin scale, the 3 T model ET is much closer to the water balance ET than is the MOD16A2 value.

5.3.2. Spatiotemporal Characteristics of ET in the Jinghe River Basin

Using the 3 T model, we retrieved the ET in each 16 day period from 2004 to 2006 at the pixel scale in the Jinghe River Basin. Figure 5.3 shows the mean daily ET for each 16 day period from 2004 to 2006 at the basin scale, obtained by averaging the value of all pixels. The mean ET value showed little difference over the 3 years, with a maximum margin of 0.10 mm/d. The change trend in each year was also the same, i.e., the intraannual ET increased from January to July and then decreased throughout the remaining months. Maximum ET was observed during the summer. The ET seasonal changes indicate that the modeled results are highly correlated with the study area climate. In winter, the temperature is very low within the basin, and vegetation is dormant; thus, ET is at the annual minimum value. In spring, the temperature rises gradually, vegetation begins to grow, and ET is thus higher than that in winter. In the summer rainy season, with higher temperatures and better light, heat, and water conditions, vegetation grows rapidly and significantly, yielding the year's best growth period. Thus, ET reaches a peak value in summer, which is significantly higher than the ET values in other seasons. During autumn, as temperatures drop, there is less precipitation, vegetation grows slowly, and finally stops growing, resulting in decreased ET.

Spatially, ET exhibits a gradual decreasing trend from south to north, as shown in Figure 5.4, and this change trend accords well with the regional climate and vegetation distributions. The southern part of the study area has subhumid climate, while the northern part is semiarid. Therefore, precipitation in the north is significantly lower than in the south, and the south has higher ET values. However, some low values were observed around the outlet of the basin in the south. This was caused by null value pixels (e.g., no cloud-free data) in the MODIS products in these areas, especially in MCD43B3, leading to underestimation when summing periodical ET values on a yearly scale. The highest ET value (>600 mm/yr) occurred in the mountains where there is abundant vegetation coverage. In contrast, the

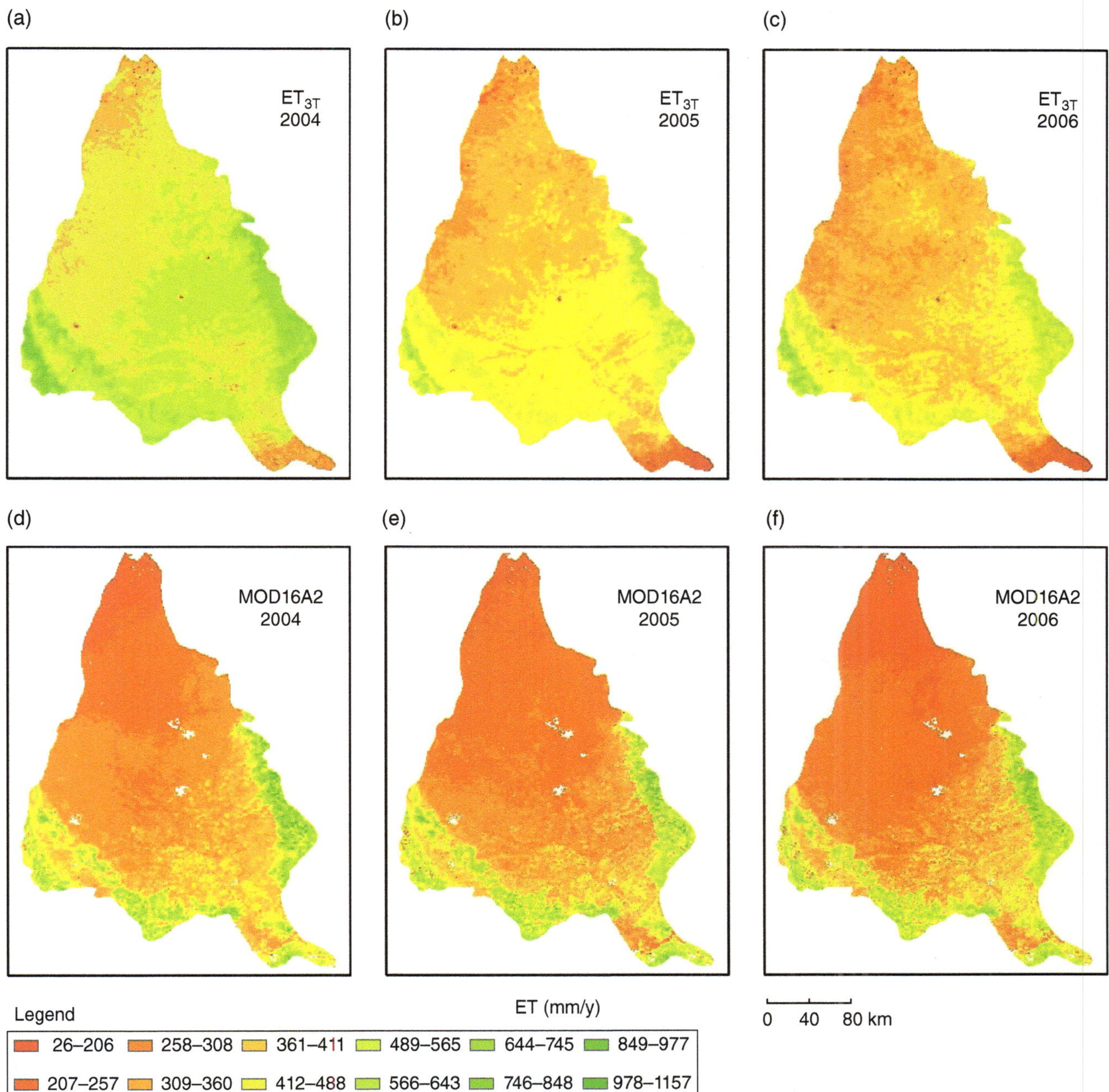

Figure 5.4 Annual ET map of the study area: (a)–(f) share the same scale bar and legend; (a), (b), and (c) are the annual ET (ET_{3T}) estimated by the 3T model and equation 5.10, and (d), (e), and (f) are from the MODIS ET product (MOD16A2). The white areas in MOD16A2 indicate that there is no ET estimation.

Table 5.1 Comparison of ET estimation between the 3T model and other methods at yearly scale[a]

	Observation					Absolute Error	
Year	P	R	ET_0	ET_{3T}	MOD16A2	$|ET_0 - ET_{3T}|$	$|ET_0 - MOD16A2|$
2004	422	22	400	487	372	87	28
2005	449	22	427	418	359	9	68
2006	467	19	448	403	350	45	98
Mean						47	65

[a] P, R, ET, and absolute error are in mm/yr; ET_0 means ET calculated from water balance method; ET_{3T} means ET modeled by the 3T model; MOD16A2 is MODIS ET product.

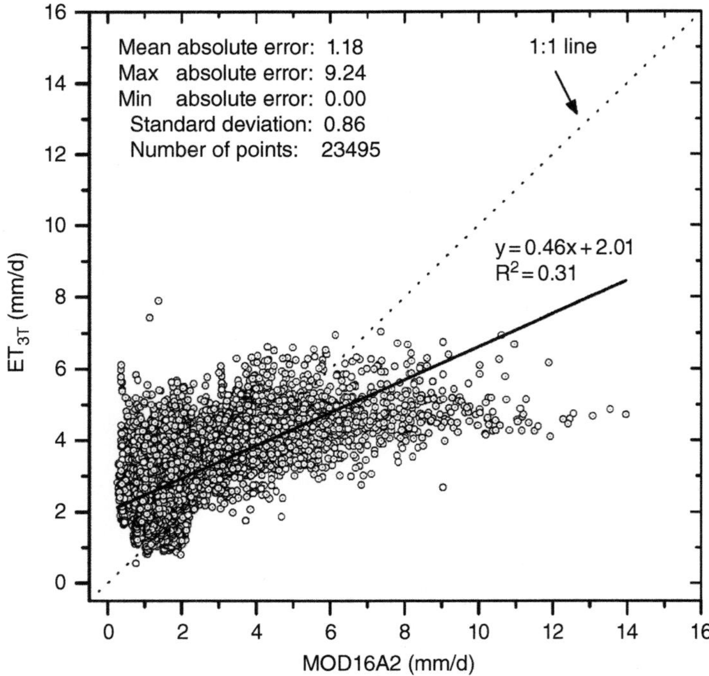

Figure 5.5 Pixel-by-pixel comparison between the modeled daily ET and the MOD16A2 products in 2004.

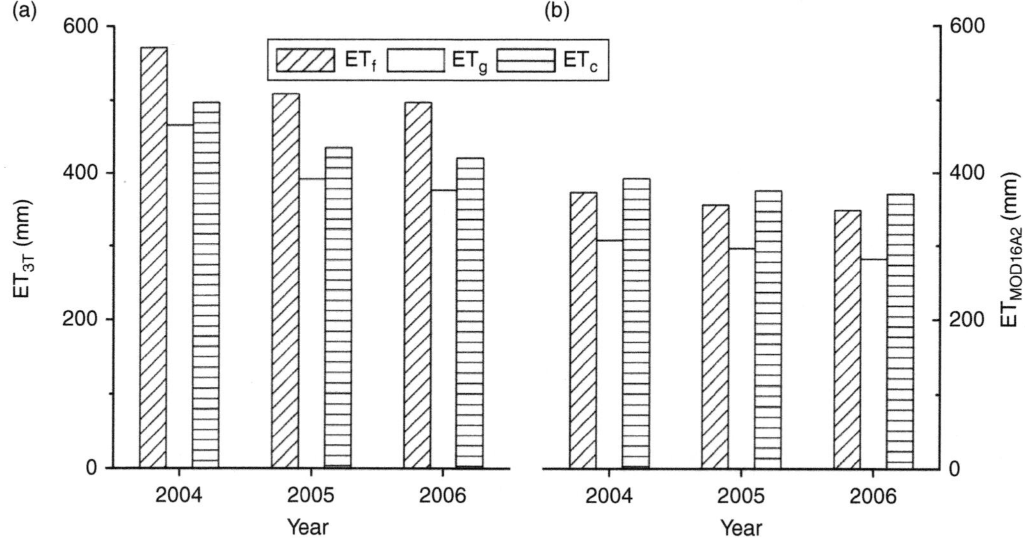

Figure 5.6 Mean annual ET from different land cover and land use types: (a) ET modeled by the 3T model (ET_{3T}); and (b) the MODIS ET product (MOD16A2): the subscripts f, g, and c represent forest, grassland, and cropland, respectively.

lowest ET value (<300 mm/yr) occurred in the sparsely vegetated areas of the north.

The ET values for areas with different land coverage were also investigated on a yearly scale. On the basis of the MODIS land use product (MCD12Q1, Figure 5.1), the LULC map was merged into three categories: forest (including deciduous broadleaf forest, mixed forest, and open and closed shrublands), grassland (including grassland, savannas, and woody savannas), and cropland. Figure 5.6a shows that the forest had the highest ET, with a mean value of 526 mm over the 3 years, followed by cropland and grassland. The ET for cropland, with a

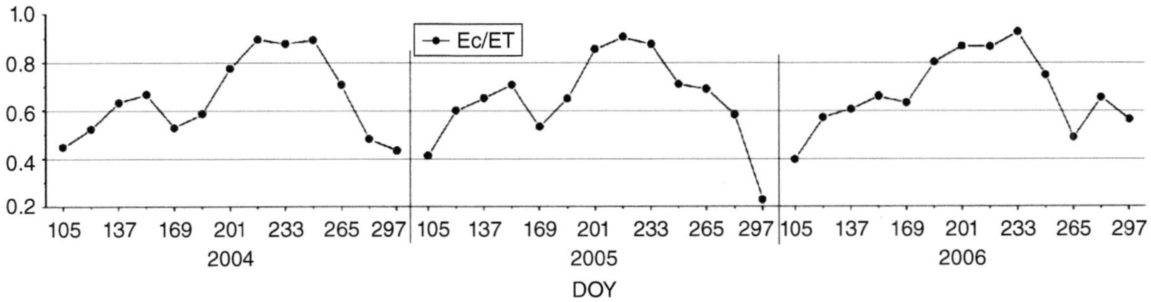

Figure 5.7 Water use efficiency of cropland from April to October (from DOY 105 to 297) in the Jinghe River Basin: E_c is the mean daily crop transpiration and ET is the mean daily evapotranspiration of the same cropland, both estimated by the 3T model; the cropland is retrieved from MCD12Q1, as shown in Figure 5.1.

mean value of 450 mm, was higher than that for grassland with a mean value of 411 mm. These results were close to the findings of *Zhang et al.* [2011] (see Section 5.3.1. for details). The ET for different LULC types also indicates that the modeled ET corresponded to the study area's climate, especially in its spatial distribution of water resources and solar energy.

5.3.3. Water Use Efficiency of Cropland in the Jinghe River Basin

In arid and semiarid regions where water resources are scarce, it is vitally important to improve water use efficiency (WUE) by reducing soil evaporation (E_s) and increasing crop transpiration (E_c) [*Howell*, 2001; *Kang et al.*, 2002; *Zhang et al.*, 2012]. In the Jinghe River Basin, water resources come mainly from surface water supplemented by precipitation, and irrigation consumes most of the water resources; *Li* [2008] reported that irrigation consumed 0.55 billion m^3 of water on average during 1956–2004, accounting for 31.37% of the mean annual runoff of the basin in that same period. WUE was very low not only in the study area but also in the Loess Plateau [*Chen et al.*, 2003]. It is essential that irrigation water be saved and WUE be improved in the croplands in the study area. Using the 3T model, we evaluated the E_s and E_c of the cropland in the growing season (from April to October). Statistical results show that the ratio of E_c to ET (E_c/ET) ranged from 22.8% to 92.9% over the 3 year period, with a mean value of 65.8% (Figure 5.7). In other words, approximately 34.2% of the water was returned to the atmosphere in the form of soil evaporation, and the WUE was low. Although the E_c/ET value showed some differences from year to year, the change trend in each year was nearly the same, i.e., E_c/ET increased gradually before DOY 233 and thereafter decreased gradually. This indicates that WUE might be improved at the beginning and the end of the growing season by reducing soil evaporation.

5.3.4. Uncertainties in the ET Estimation Based on the 3T Model and MODIS Data

On account of this study's lack of flux tower observational data, the modeled ET could only be compared to the water balance ET on a yearly scale, and although the difference between them was small (with a mean absolute error of 47 mm/yr; see Section 5.3.1. for details), the validation does not capture the seasonal variation of the modeled ET. The 3T model should be tested thoroughly by using ground observational data in future studies.

In addition, the ET process is affected by many factors [*Bastiaanssen et al.*, 1998; *Glenn et al.*, 2007; *Monteith*, 1965; *Shuttleworth and Wallace*, 1985; *Su*, 2002], so some uncertainty characterizes estimated ET values [e.g., *Cleugh et al.*, 2007; *Mu et al.*, 2011; *Senay et al.*, 2013; *Velpuri et al.*, 2013]. As shown by *Mu et al.* [2011] and *Velpuri et al.* [2013], factors causing uncertainty in ET estimation are mainly model/algorithm limitations, model parameterization, errors in ground-based observation (e.g., flux tower measurements), spatial heterogeneity within the MODIS pixel, and mismatch between pixel to flux tower footprint. We will address the algorithm limitation and model parameterization issues here. First, only cloud cover and the number of sunny days were considered in the algorithm proposed to scale a daily ET value to the total value for a giver period (here 16 days), and the lack of other important parameters, such as wind speed, may contribute to the uncertainties in the modeled ET value. Second, the interpolation of meteorological parameters (i.e., air temperature and cloud cover) using IDW may lead to error and uncertainty, because unlike the topography of this study area, IDW is known to work well over relatively flat, homogeneous terrain [*Dodson and Marks*, 1997]. Third, uncertainties occurred in the model parameterization. In addition to the uncertainties from the MODIS products (e.g., LST, LAI and FPAR) as summarized in *Mu et al.* [2011], the algorithm adopted to parameterize T_s and T_c—the soil and canopy component temperatures, respectively—may cause error because there were two empirical coefficients in equation 5.4, which may not be suitable for all land surfaces. Although the uncertainty caused by the above factors was likely to have been counteracted by the difference ratio form in equation 5.1 or 5.2 [e.g., $(T_s - T_a)/(T_{sd} - T_a)$], as shown in our previous study [*Qiu et al.*, 1998] and the study of *Sun et al.* [2009] using a revised 3T model (Sim-ReSET model), the methods adopted to parameterize the model input should be carefully chosen depending on the study requirements and data availability.

5.4. CONCLUSION

A method was proposed for estimating periodical (16 day or yearly) ET by using the 3T model and MODIS products, i.e., MODIS land surface temperature/emissivity data (MOD11A2), LAI and FPAR data (MOD15A2), global MODIS vegetation indices (MOD13A2), and albedo (MCD43B3). Except for air temperature, these MODIS products could provide all the model input required for the 3T model either directly or indirectly. Combined with the advantages of the 3T model, i.e., not needing to calculate resistance and less dependence on meteorological observations, the method proposed in this study makes the remote estimation of ET more viable. A case study performed in the Jinghe River Basin between 2004 and 2006 showed the following results: (1) when comparing the modeled ET with values obtained from the water balance equation on a yearly scale, the estimated annual ET was close to the water balance ET, with a mean absolute error of 47 mm/yr for the 3 year period; (2) the intraannual ET increased from January to July and then decreased throughout the remaining months, with a mean value of 2.94 mm/d for the 3 year period; (3) ET declined along a gradient starting in the southern mountain areas (>600 mm/yr) and moving down into the sparsely vegetated regions in the north (<300 mm/yr); and (4) the ratio of crop transpiration to ET ranged from 22.8% to 92.9% over the 3 year period, with a mean value of 65.8%, indicating that 34.2% of the water returned to the atmosphere in the form of soil evaporation, and that water use efficiency could be improved by reducing soil evaporation. All in all, the results of this study indicated that the proposed method and the MODIS products are useful for practical estimation of regional ET. In combination with more ground observational data, we plan to thoroughly test the 3T model in future studies.

ACKNOWLEDGMENTS

This study was supported by the National Natural Science Foundation of China (No. 41201433); the Specialized Research Fund for the Doctoral Program of Higher Education of China (No. 20110171120001), the National Natural Science Foundation of China (No. 91025008 and 41101313); and the National Basic Research Program (973 Program) (No. 2009CB421303). Thanks to the China Meteorological Administration for providing meteorological data, to NASA for providing MODIS products, to Priscilla Lynne Young for improving the quality of the English, and to anonymous reviewers for their valuable comments.

REFERENCES

Allen, R. G., M. Tasumi, and R. Trezza (2007), Satellite-based energy balance for mapping evapotranspiration with internalized calibration (METRIC)-Model, *J. Irrigation Drainage Eng. 133*(4), 395–406.

Anderson, M. C., J. M. Norman, G. R. Diak, W. P. Kustas, and J. R. Mecikalski (1997), A two-source time-integrated model for estimating surface fluxes using thermal infrared remote sensing, *Remote Sens. Environ., 60*, 195–216.

Anderson, M. C., J. M. Norman, J. R. Mechikalski, R. D. Torn, W. P. Kustas, and J. B. Basara (2004), A multiscale remote sensing model for disaggregating regional fluxes to micrometeorological scales, *J. Hydrometeorol., 5*, 343–363.

Bastiaanssen, W. G. M., M. Menenti, R. A. Feddes, and A. A. M. Holtslag (1998), A remote sensing surface energy balance algorithm for land (SEBAL) 1. Formulation, *J. Hydrol., 212–213*, 198–212.

Carlson, T. N., W. J. Capehart, and R. R. Gillies (1995), A new look at the simplified method for remote sensing of daily evapotranspiration, *Remote Sens. Environ., 54*, 161–167.

Chen, J., D. He, and S. Cui (2003), The response of river water quality and quantity to the development of irrigated agriculture in the last 4 decades in the Yellow River Basin, China, *Water Resour. Res., 39*(3), 1047, doi:10.1029/2001WR001234.

Cleugh, H. A., R. Leuning, Q. Mu, and S. W. Running (2007), Regional evaporation estimates from flux tower and MODIS satellite data, *Remote Sens. Environ., 106*, 285–304.

Dodson, R., and D. Marks (1997), Daily air temperature interpolated at high spatial resolution over a large mountainous region, *Climate Res., 8*, 1–20.

Feng, X. M., G. Sun, B. J. Fu, C. H. Su, Y. Liu, and H. Lamparski (2012), Regional effects of vegetation restoration on water yield across the Loess Plateau, China, *Hydrol. Earth Syst. Sci., 16*, 2617–2628.

Glenn, E. P., A. R. Huete, P. L. Nagler, K. K. Hirschboeck, and P. Brown (2007), Integrating remote sensing and ground methods to estimate evapotranspiration, *Crit. Rev. Plant Sci., 26*(3), 139–168.

Howell, T. A. (2001), Enhancing water use efficiency in irrigated agriculture, *Agron. J., 93*, 182–289.

Idso, S., T. Schmugge, R. Jackson, and R. Reginato (1975), The utility of surface temperature measurements for the remote sensing of surface soil water status, *J. Geophys. Res., 80*(21), 3044–3049.

Jackson, R. D., R. J. Reginato, S. B. Idso, and P. J. Pinter (1977), Wheat canopy temperature: A practical tool for evaluating water requirements, *Water Resourc. Res.*, *13*, 651–656.

Jackson, R. D., J. L. Hatfield, R. J. Reginato, S. B. Idso, and P. J. Pinter (1983), Estimation of daily evapotranspiration from one time-of-day measurements, *Agric. Water Manag.*, *7*, 351–362.

Jarvis, A., H. I. Reuter, A. Nelson, and E. Guevara (2008), Hole-filled SRTM for the globe Version 4, available from the CGIAR-CSI SRTM 90m database at http://srtm.csi.cgiar.org.

Jia, L., Z. Su, B. van den Hurk, M. Menenti, A. Moene, H. A. R de Bruin, J. J. B. Yrisarry, M. Ibanez and A. Cuesta (2003), Estimation of sensible heat flux using the Surface Energy Balance System (SEBS) and ATSR measurements, *Physics and Chemistry of the Earth, Parts A/B/C*, *28*(1), 75–88.

Jiang, L., and S. Islam (2003), An intercomparison of regional latent heat flux estimation using remote sensing data, *Int. J. Remote Sens.*, *24*, 2221–2236.

Jung, M., (2010), A recent slowdown in global land evapotranspiration due to limited moisture supply, *Nature*, *467*, 951.

Kang, S. Z., L. Zhang, Y. L. Liang, X. T. Hu, H. J. Cai, and B. J. Gu (2002), Effects of limited irrigation on yield and water use efficiency of winter wheat in the Loess Plateau of China, *Agric. Water Manag.*, *55*, 203–216.

Kustas, W. P., and M. Anderson (2009), Advances in thermal infrared remote sensing for land surface modeling, *Agric. Forest Meteorol.*, *149*, 2071–2081.

Kustas, W. P., and C. S. T. Daughtry (1990), Estimation of the soil heat flux/net radiation ratio from spectral data, *Agric. Forest Meteorol.*, *49*, 205–223.

Lhomme, J. -P., B. Monteny, and M. Amadou (1994), Estimating sensible heat flux from radiometric temperature over sparse millet, *Agric. Forest Meteorol.*, *68*, 77–91.

Li, X. G. (2008), Theory of water resources system coupling and its application to study of water resources and hydrology in Jinghe River, Ph.D. thesis, Chang'an University, Xi'an, China.

Li, Z. L., R. Tang, Z. Wan, Y. Bi, C. Zhou, B. Tang, G. Yan, and X. Zhang (2009), A review of current methodologies for regional evapotranspiration estimation from remotely sensed data, *Sensors*, *9*(5), 3801–3853.

Matsushita, B., and T. Fukushima (2009), Methods for retrieving hydrologically significant surface parameters from remote sensing: A review for applications to East Asia region, *Hydrol. Process.*, *23*(4), 524–533.

Monteith, J. L. (1965), Evaporation and environment, *Symposia Soc. Exper. Biol.*, *19*, 205–234.

Monteith, J. L. (1973), *Principles of Environmental Physics*, 2nd ed., Edward Arnold Press, London.

Moran, M. S., R. D. Jackson, L. H. Raymond, L. W. Gay, and P. N. Slater (1989), Mapping surface energy balance components by combining Landsat thematic mapper and ground-based meteorological data, *Remote Sens. Environ.*, *30*, 77–87.

Mu, Q., F. A. Heinsch, M. Zhao, and S. W. Running (2007), Development of a global evapotranspiration algorithm based on MODIS and global meteorology data, *Remote Sens. Environ.*, *111*, 519–536.

Mu, Q., M. Zhao, and S. W. Running (2011), Improvements to a MODIS global terrestrial evapotranspiration algorithm, *Remote Sens. Environ.*, *115*, 1781–1800.

Norman, J. M., W. P. Kustas, and K. S. Humes (1995), Source approach for estimating soil and vegetation energy fluxes in observations of directional radiometric surface temperature, *Agric. Forest Meteorol.*, *77*, 263–293.

Norman, J. M., M. C. Anderson, W. P. Kustas, A. N. French, J. Mecikalski, R. Torn, G. R. Diak, T. J. Schmugge, and B. C. W. Tanner (2003), Remote sensing of surface energy fluxes at 10^{1}-m pixel resolutions, *Water Resourc. Res.*, *39*(8), 1221.

Oki, T., and S. Kanae (2006), Global hydrological cycles and world water resources, *Science*, *313*(5790), 1068–1072.

Priestley, C. H. B., and R. J. Taylor (1972), On the assessment of surface heat flux and evaporation using large-scale parameters, *Monthly Weather Rev.*, *100*, 81–92.

Qiu, G. Y., T. Yano, and K. Momii (1996), Estimation of plant transpiration by imitation leaf temperature. I. Theoretical consideration and field verification, *Trans. Japanese Soci. irrigation Drainage Reclam. Engi.*, *64*, 401–410.

Qiu, G. Y., T. Yano, and K. Momii (1998), An improved methodology to measure evaporation from bare soil based on comparison of surface temperature with a dry soil, *J. Hydrol.*, *210*, 93–105.

Qiu, G. Y., P. J. Shi, and L. M. Wang (2006), Theoretical analysis of a soil evaporation transfer coefficient, *Remote Sens. Environ.*, *101*, 390–398.

Senay, G. B., S. Bohms, R. K Singh, P. H. Gowda, N. M. Velpuri, H. Alemu, and J. P. Verdin (2003), Operational evapotranspiration mapping using remote sensing and weather datasets: A new parameterization for the SSEB approach, *Journal of the American Water Resources Association.*, *49*(3), 577–591.

Shuttleworth, W. J., and J. S. Wallace (1985), Evaporation from sparse crops—An energy combination theory, *Q. J. R. Meteorol. Soc.*, *111*, 839–855.

Sobrino, J. A., N. Raissouni, and Z. L. Li (2001), A comparative study of land surface emissivity retrieval from NOAA data, *Remote Sens. Environ.*, *75*, 256–266.

Sobrino, J. A., J. C. Jimenez-Munoz, and L. Paolini (2004), Land surface temperature retrieval from LANDSAT TM 5, *Remote Sens. Environ.*, *90*, 434–440.

Su, Z. (2002), The surface energy balance system (SEBS) for estimation of the turbulent heat fluxes, *Hydrol. Earth Sci.*, *6*(1), 85–99.

Sun, Z., Q. Wang, B. Matsushita, T. Fukushima, Z. Ouyang, and M. Watanabe (2009), Development of a Simple Remote Sensing EvapoTranspiration model (Sim-ReSET): Algorithm and model test, *J. Hydrol.*, *376*(3–4), 476–485.

Swinbank, W. C. (1963), Long-wave radiation from clear skies, *Q. J. Royal Meteorol. Soc.*, *89*, 339–348.

Teuling, A. J., et al. (2009), A regional perspective on trends in continental evaporation, *Geophys. Res. Lett.*, *36*, L02404.

Timmermans, W. J., W. P. Kustas, M. C. Anderson, and A. N. French (2007), An intercomparison of the Surface Energy Balance Algorithm for Land SEBAL and the Two-Source Energy Balance TSEB modeling schemes, *Remote Sens. Environ.*, *108*(4), 369–384.

Trenberth, K. E., L. Smith, T. Qian, A. Dai, and J. Fasullo (2006), Estimates of the global water budget and its annual cycle using observational and model data, *J. Hydrometeorol.*, *8*, 758–769.

Velpuri, N. M., G. B. Senay, R. K. Singh, S. Bohms, and J. Verdin (2013), A comprehensive evaluation of two MODIS evapotranspiration products over the conterminous United States: Using point and gridded FLUXNET and water balance ET, *Remote Sens. Environ.*, *139*, 35–49.

Vinukollu, R. K., E. F. Wood, C. R. Ferguson, and J. B. Fisher (2011), Global estimates of evapotranspiration for climate studies using multi-sensor remote sensing data: Evaluation of three process-based approaches, *Remote Sens. Environ.*, *115*(3), 801–823.

Xiong, Y. J., and G. Y. Qiu (2011), Estimation of evapotranspiration using remotely sensed land surface temperature and the revised three-temperature model, *Int. J. Remote Sens.*, *32*(20), 5853–5874.

Xiong, Y. J., and G. Y. Qiu (2014), Simplifying the revised three-temperature model for remotely estimating regional evapotranspiration and its application to a semi-arid steppe, *Int. J. Remote Sens.*, *35*(6), 2003–2027.

Zhang, K., J. S. Kimball, R. R. Nemani, and S. W. Running (2010), A continuous satellite-derived global record of land surface evapotranspiration from 1983–2006, *Water Resourc. Res.*, *46*, W09522.

Zhang, S. L., P. T. Yu, Y. H. Wang, H. J. Zhang, K. Valentina, S. C. Huang, W. Xiong, and L. H. Xu (2011), Estimation of actual evapotranspiration and its component in the upstream of Jinghe Basin, *Acta Geograph. Sinica*, *66*(3), 385–395 (in Chinese with English abstract).

Zhang, Y., J. Ma, X. Chang, J. van Wonderen, L. Yan, and J. Han (2012), Water resources assessment in the Minqin Basin: An arid inland river basin under intensive irrigation in Northwest China, *Environ. Earth Sci.*, *65*(6), 1831–1839.

Zhao, S. H., Y. H. Yang, G. Y. Qiu, Q. M. Qin, Y. J. Yao, Y. J. Xiong, and C. Q. Li (2010), Remote detection of bare soil moisture using a surface-temperature-based soil evaporation transfer coefficient, *Int. J. Appl. Earth Observ. Geoinform.*, *12*, 351–358.

6

Water Use and Stream-Aquifer-Phreatophyte Interaction Along a Tamarisk-Dominated Segment of the Lower Colorado River

Saleh Taghvaeian,[1] Christopher M. U. Neale,[2] John Osterberg,[3]
Subramania I. Sritharan,[4] and Doyle R. Watts[5]

6.1. INTRODUCTION

Invasive vegetation species such as saltcedar or tamarisk (*Tamarix* spp.) and Russian olive (*Eleagnus angustifolia*) have spread throughout riparian ecosystems of the western United States, out-competing and replacing native species such as cottonwoods (*Populus* spp.), willows (*Salix* spp.), and mesquite (*Prosopis* spp.). Tamarisk, in particular, is one of the most dominant invasive species in the Lower Colorado River Basin that has a high tolerance to drought [*Cleverly et al.*, 1997] and salinity [*Glenn et al.*, 1998, *Vandersande et al.*, 2001] and grows in medium to dense stands, covering large areas of the generally wider floodplains. For decision makers and water managers in the semiarid western United States with scarce water resources, it is of crucial importance to accurately estimate tamarisk evapotranspiration (ET) in order to better estimate releases from reservoirs along the Colorado River system as well as the potential amount of water that can be salvaged by its removal. However, tamarisk ET rates reported in the literature are inconsistent, ranging from very low to unrealistically high values [*Owens and Moore*, 2007]. For example, the U.S. Bureau of Reclamation (USBR) estimates riparian water consumption along the Lower Colorado River by assuming that annual tamarisk ET is about 86% of the reference ET, while *Murray et al.* [2009] estimated a ratio of only 42% over the same area.

Such differences have resulted in contrasting opinions on effectiveness of Tamarisk control efforts for water conservation purposes. Fostering an aggressive eradication program, *Zavaleta* [2000] reported that the negative effects of tamarisk water consumption on agricultural and municipal water supplies, hydropower generation, and flood control reach an annual value as high as US$285 million. On the other hand, *Vandersande et al.* [2001] found that water use of tamarisk is similar to other native species, and *Murray et al.* [2009] concluded that water salvage from tamarisk removal in the Lower Colorado River would be negligible. Most of the methods that have been developed for quantifying tamarisk water use are based on point measurements, representing the very local condition of the site where measurements take place. Given the high level of heterogeneity in the hydroclimatological conditions of riparian communities, extrapolating the results of point measurements to catchment and basin scales may fail to provide a comprehensive picture of actual riparian water consumption. Air and spaceborne remotely sensed imagery provide spatially distributed information that can significantly assist decision makers and water managers. Existing remote sensing methods for mapping riparian ET fall into two main categories: empirical approaches based on vegetation indices (VI) and physically based models for quantifying surface energy balance components.

[1] *Department of Biosystems and Agricultural Engineering, Oklahoma State University, Stillwater, Oklahoma, USA*

[2] *Water for Food Institute, University of Nebraska–Lincoln, Nebraska, USA*

[3] *U.S. Bureau of Reclamation, Denver, Colorado, USA*

[4] *International Center for Water Resources Management, Central State University, Wilberforce, Ohio, USA*

[5] *Department of Earth and Environmental Sciences, Wright State University, Dayton, Ohio, USA*

Remote Sensing of the Terrestrial Water Cycle, Geophysical Monograph 206. First Edition. Edited by Venkat Lakshmi.
© 2015 American Geophysical Union. Published 2015 by John Wiley & Sons, Inc.

Several VI-based methods have been previously proposed. *Nagler et al.* [2005] developed a bi-parameter regression equation that related ET measurements of energy-flux towers (Bowen ratio and eddy covariance) to the point measurement of air temperature and spatially distributed enhanced vegetation index (EVI; *Huete et al.*, 2002], obtained from the Moderate Resolution Imaging Spectroradiometer (MODIS) instrument. Applying this method over a tamarisk-dominated corridor in the Upper Colorado River Basin resulted in annual ET of about 700 mm [*Dennison et al.*, 2009]; while *Hultine et al.* [2010] measured only 260–270 mm over the same area, using sap-flux sensors that were specifically calibrated for tamarisk studies. This significant overestimation error was attributed to the fact that in model development, ET measurements from energy-flux towers were plotted against EVI of a single MODIS pixel containing that tower. However, tower upwind footprints are highly variable in size and direction and sometimes fall over surfaces other than the narrow riparian corridors [*Hultine et al.*, 2010]. *Nagler et al.* [2009a] modified the EVI method by making two adjustments. The first adjustment was the replacement of air temperature with grass-based reference ET (ET_0), estimated at a standard weather station. The second adjustment was the use of sap flux technique rather than energy balance towers in estimating actual ET rates. Both EVI approaches (original and modified) were applied over the Cibola National Wildlife Refuge (the same site in this study), where the original equation produced 20% larger estimates than the modified one. This difference was attributed to evaporation, since sap-flux sensors measure only transpiration (compared to flux towers that measure evapotranspiration). However, such a contribution from soil evaporation seems to be too high for a semiarid area with annual precipitation of less than 100 mm and average depth to groundwater of more than 2 m. In addition, most rainfall events in this area happen during winter, while the data were collected over summer months (June–September).

A major drawback of the VI approach is that vegetation indices are not efficient in capturing stress development, unless stress factors prolong enough to cause detectable changes in plants vegetative conditions (*Nagler et al.*, 2005, 2009a). In addition, the inherent empiricism in VI approach limits its application to sites with hydroclimatological conditions similar to the one where they were developed [*Scott et al.*, 2008]. In the case of MODIS-EVI, another limitation arises from the fact that high temporal resolution of MODIS imagery comes at the cost of a spatial resolution that is rather coarse (250 m for visible bands) for mapping ET of heterogeneous riparian communities. This could be problematic especially in differentiating between water consumption of different species in mixed stands, as well as in estimating ET along the edges of riparian corridors, where MODIS pixels may partially cover water bodies or bare soils. Pixel contamination could have significant effects, given that many riparian corridors along western rivers are only a few hundred meters wide. *Scott et al.* [2008] improved the EVI method by incorporating MODIS-derived nighttime land surface temperature (LST) maps. Although this new model was successfully validated over the same area it was developed, it should be noted that the pixel size of MODIS thermal band is even larger than its visible bands (1 km). *Groeneveld et al.* [2007] developed a linear regression equation that approximated the ratio of actual to reference ET based on scaled normalized difference vegetation index (NDVI) estimates derived from Landsat imagery. The high spatial resolution of Landsat visible bands was a great advantage in capturing riparian heterogeneity (64 Landsat pixels can easily fit in one MODIS pixel), but a source of error was the mismatch between the highly variable footprint of energy flux towers (used in model parameterization) and fixed pixels used in NDVI extraction.

Unlike the VI approach, remotely sensed energy balance (RSEB) models take advantage of the ability of air- and spaceborne imagery to quantify net radiation and sensible and soil heat fluxes. Latent heat flux is then estimated as the residual of the energy balance equation. Due to recent improvements in estimating sensible heat flux [*Norman et al.*, 1995; *Bastiaanssen et al.*, 1998a], the RSEB models perform satisfactorily, with accuracies that range from 67% to 97%, and above 94% for daily and seasonal temporal scales, respectively [*Gowda et al.*, 2008]. However, it should be noted that these models are originally developed to be applied over agricultural areas. Therefore, modifications may be required before they can be applied over riparian ecosystems, where biophysical characteristics of surface vegetation are significantly different than agricultural areas.

To the best of our knowledge, only a few riparian applications of RSEB models have been reported. One example is the research carried out by *Bawazir et al.* [2009] in the Middle Rio Grande Basin in New Mexico. The implemented model was very similar to the surface energy balance algorithm for land (SEBAL), developed by *Bastiaanssen et al.* (1998a). A major modification was in the selection of the wet pixel, which is used in interpolating sensible heat flux between two known extremes. In newer versions of SEBAL, a well-irrigated pixel at full-cover and low temperature is selected as the wet pixel, and it is assumed that temperature gradient and consequently sensible heat flux over this pixel is negligible. In the study by *Bawazir et al.* [2009], however, the wet pixel was selected from the footprint of an eddy-covariance tower over a dense tamarisk canopy. Instead of assuming a negligible value, associated temperature gradients were obtained from the measurements of the same tower. Although this modified RSEB model was successful in accurately estimating ET over tamarisk and cottonwood thickets of the Middle Rio Grande Basin, its application is limited to areas where energy flux towers are available. Another RSEB model known as METRIC has been also applied to estimate riparian ET at the Middle Rio Grande Basin [*Allen et al.*, 2007b] and at the Nebraska Panhandle [*Kamble et al.*, 2013].

This research experiment was conducted at a tamarisk-dominated riparian forest along the Lower Colorado River in southern California to achieve the following objectives:

• Study the poorly understood connection between surface water flows and subsurface fluxes into the floodplain groundwater system, resulting from riparian water uptake.

• Estimate riparian ET using a RSEB model and another method based on high-frequency, point measurements of groundwater fluctuations.

• Investigate possible sources of error in applying RSEB models over riparian communities and suggesting appropriate modifications.

• Evaluate the performance of each method through a comparison with the measurements of a Bowen ratio energy flux tower.

The need to conduct this study stemmed from the fact that the lack of a thorough knowledge on the complex stream-aquifer-phreatophyte interaction poses operational challenges for the management of the Lower Colorado River. In addition, applying and modifying RSEB models is important for studying the effect of the saltcedar leaf beetle (*Diorhabda carinulata*) defoliation on tamarisk water consumption, an effect that is largely unknown [*Hultine et al.*, 2010]. The leaf beetles have been released in a few locations along the Upper Colorado River, and it is likely that they would travel southward to the Lower Colorado River, since recent studies have shown that their spread rate is rather fast [*Hultine et al.*, 2009]. Hence, it is critical to accurately identify current ET rates, so the water salvage from future beetle defoliation can be estimated by comparison with existing estimates.

6.2. METHODS AND MATERIALS

6.2.1. Study Area

The study area was located in the lower portion (978 ha) of the Cibola National Wildlife Refuge (CNWR) in southern California. Established in 1964, CNWR occupies about 70 km² of floodplains on the west bank of the Lower Colorado River. Figure 6.1 demonstrates the entire CNWR and the new engineered river channel on the east side. To the west is the old Colorado River channel. It currently carries small regulated flows to support

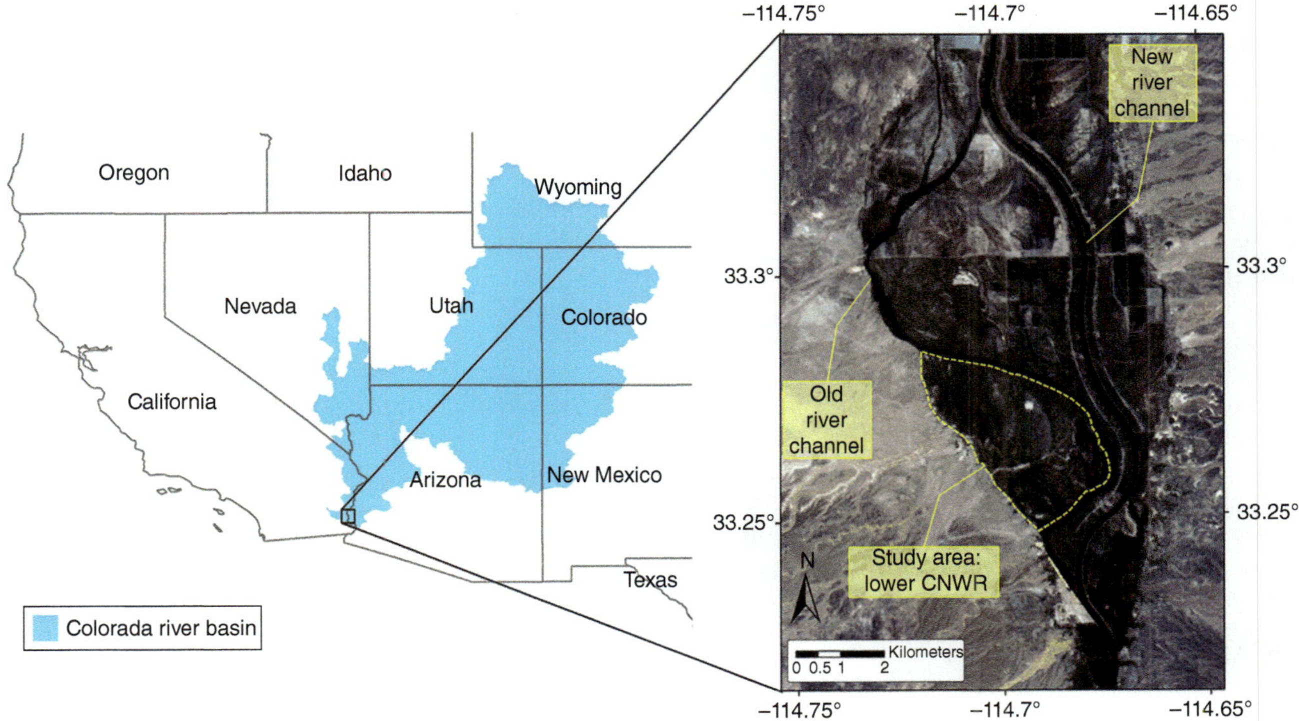

Figure 6.1 Location of CNWR in the Lower Colorado River Basin.

the wildlife, as well as the agricultural return flows from the Palo Verde Irrigation District (PVID) upstream. Ground elevation ranges from 66 m at the eastern boundary (the river) to about 70 m at the western edge (desert hills). The CNWR is home to more than 280 bird species and several phreatophytes. Over 90% of the area is covered by tamarisk [*Nagler et al.*, 2009b] with an average age of about 20 years [*Godaire and Klinger*, 2007]. Mesquite (*Prosopis velutina*) and arrowweed (*Pluchea sericea*) are the next most prevalent species.

Three measuring stations, named Slitherin, Diablo, and Swamp, were installed within the lower CNWR at sites that were variable in tamarisk density, groundwater availability, and groundwater quality. Five observation wells were installed at each of these stations to an approximate depth of 7.5 m. Table 6.1 summarizes different characteristics of

these sites, with geohydrological parameters being averaged for the study period: January 2008 to January 2009. The distances were measured within the ArcGIS software, using georeferenced stations and river layers.

Figure 6.2 illustrates two high-resolution airborne images of the lower portion of CNWR, where the present study was conducted. The plot on the left is a false-color image acquired with the USU (Utah State University) airborne multispectral system [*Cai and Neale*, 1999; *Neale and Crowther*, 1994], while the plot on the right is a map of canopy height, developed based on LiDAR (Light Detection and Ranging) data from the USU airborne LASSI (Lidar-assisted stereo imager) system [*Geli et al.*, 2012]. The average tamarisk height was 2.5, 4.0, and 6.0 m around Swamp, Diablo, and Slitherin stations, respectively.

6.2.2. Groundwater Characteristics

As mentioned before, five observation wells were installed at each measuring station (total of 15 wells) to monitor groundwater dynamics. One well was in a central position and the rest of the wells were located at about 100 m from the central well in NW, NE, SE, and SW directions [*Zhu et al.*, 2011]. Groundwater electrical conductivity (EC) was measured at each well on a monthly basis. In addition, automatic pressure transducers (model HOBO U20, Onset Computer Corporation, Bourne, MA, USA) were installed below groundwater level in each well to monitor water head and temperature at high frequency. A HOBO barometric pressure transducer was also installed above groundwater level at one of the Slitherin wells to monitor changes in

Table 6.1 Characteristics of measuring stations

Characteristics	Swamp	Slitherin	Diablo
Tamarisk density	Medium-high	High	Medium-low
Distance from old river channel (m)	100	650	1500
Distance from new river channel (m)	850	2900	2400
Groundwater temperature (°C)	22.2	21.7	22.3
Depth to groundwater (m)	2.9	3.5	3.4
Groundwater electrical conductivity (dS/m)	4.1	8.0	17.0

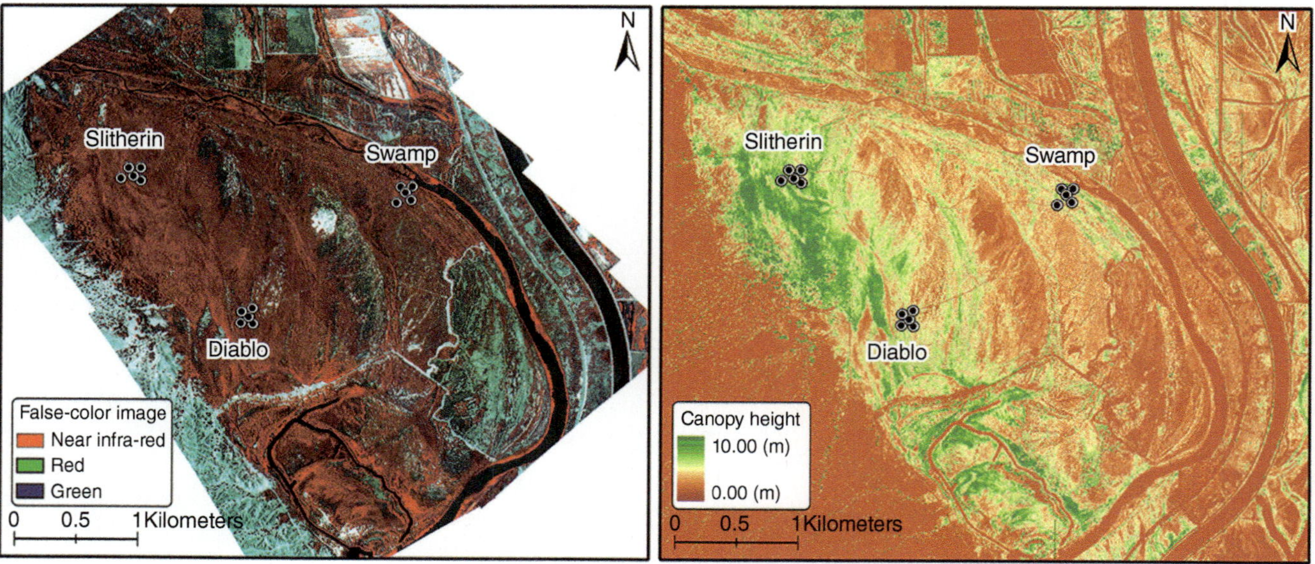

Figure 6.2 False-color multispectral airborne image (*left*) and a LiDAR-derived canopy height (*right*) of the lower CNWR and the location of measuring stations.

atmospheric pressure. Recorded atmospheric pressure was subtracted from all water pressure measurements of the submerged sensors to obtain the pressure that was exerted only by the water column above the sensors. Water head data were then converted to groundwater depth, using the measured distance between sensors and ground surface at each observation well. Daily and seasonal patterns in groundwater depth were analyzed to study the effect of water availability on tamarisk ET.

Groundwater depth needed to be converted to groundwater elevation before any analysis of subsurface flow and stream-aquifer interactions could be performed. In order to do so, soil surface elevation at each observation well was extracted from the high-resolution digital elevation models (DEM) derived from the LiDAR data. Then, groundwater depth at each well was subtracted from the corresponding soil surface elevation to obtain groundwater elevation.

An important aspect of a stream-aquifer interaction study is the identification of subsurface flow direction to determine if the stream is acting as a source or a sink. In this study, groundwater elevation data were analyzed in conjunction with river stage data. The USBR owns and operates stage gauges at two sites within a few kilometers of CNWR, where river stage is measured on an hourly basis. These sites are Taylor Ferry and Cibola, located upstream and downstream of CNWR, respectively. However, we needed to select only one point on the river to be able to compare river stage data with groundwater elevation. Therefore, a line passing through the Swamp site and perpendicular to the river was drawn in ArcGIS, and the intersection of this line and the river was selected as the river stage reference point. This point was about 24 river kilometers downstream of Taylor Ferry and 7 km upstream of Cibola sites. It was assumed that the river stage varies between these two sites in a linear fashion, so the stage value at the reference point was interpolated proportional to its distances from the two sites. Assuming a linear change in the river stage may introduce some error into analysis. However, this error does not seem to be significant since the new river channel has an engineered path and slope. In addition, no flow-regulating structures were located between the two monitoring sites. The drop in river stage from Taylor Ferry to Cibola had small variations during the study period (2008), with an average and standard deviation of 6.91 and 0.23 m, respectively.

6.2.3. Tamarisk Water Use

Tamarisk water use was estimated using three independent methods. The first was the White method, which is based on high-frequency (measured every 5 s and averaged every 15 min) point measurements of groundwater

fluctuations. The second method was SEBAL, which is a RSEB model, and the third method was based on the measurements of a Bowen ratio energy flux tower.

6.2.3.1. White Method

After a comprehensive study of groundwater dynamics in the Escalante Valley in southeastern Utah, *White* [1932] proposed a method for estimating riparian ET from water table fluctuations:

$$\mathrm{ET} = S_y(24r \pm s), \tag{6.1}$$

where ET is daily evapotranspiration (mm), S_y is the specific yield of the aquifer (dimensionless), r is the average rate of groundwater recharge between midnight and 4:00 A.M. (mm/h), and s is the net change in water table over a 24 h period (mm). The White method is based on several key assumptions. The first assumption is that r represents daily average rate of groundwater influx. Another important assumption is that diurnal decline and the following nocturnal incline in groundwater level is a result of the presence and absence of water extraction by taproots of phreatophytes, respectively. However, other factors such as barometric pressure changes, freeze-thaw processes, significant rainfall and infiltration, and anthropogenic factors may induce rapid groundwater fluctuation and disturb the natural discharge-recharge process in the study area [*Gribovszki et al.*, 2010]. In this study the first factor was accounted for by measuring atmospheric pressure on the same time scale and subtracting it from groundwater pressure measurements to obtain a pressure that is exerted only by the head of water above sensors. The next two factors were ruled out since CNWR is located in an arid environment with annual precipitation of less than 100 mm and minimum air temperatures that rarely fall below zero. In other words, there was no significant diffuse discharge during tamarisk growing season from precipitation events or from thawing of frozen soil moisture. The Diablo and Slitherin sites were also far enough from any anthropogenic activity and from the Colorado River; thus, the fourth factor was not an issue either. However, the Swamp site was close to the old Colorado River channel that currently carries agricultural drainage water from the upstream PVID and some regulated flow rates for supporting the riparian ecosystem. It is also directly influenced by the fluctuating river stage of the Colorado River. Because of this, this site was excluded from ET estimation by the White method.

The White method is particularly sensitive to the value of specific yield. Previous studies have shown that except for clean sand, laboratory-derived values of this parameter result in a significant overestimation error [*Gribovszki et al.*, 2010]. This is due to the fact that in nature, the rise and fall of water tables do not happen instantaneously. To avoid this error, *Meyboom* [1996] suggested a 50% reduction in S_y values and called it "readily available specific yield." *Loheide et al.* [2005] defined guidelines for estimating the readily available specific yield based on sediment texture. In applying the White method to groundwater fluctuation data collected at Diablo, Slitherin, and Swamp stations, *Zhu et al.* [2011] obtained a constant S_y value from *Loheide et al.* [2005] and used it for all observation wells at these three stations. The same reference [*Loheide et al.*, 2005] was used in this study. Unlike *Zhu et al.* [2011], however, a unique S_y value was assigned to each observation well, using the actual subsurface soil properties from borehole logging information collected during drilling the wells.

Since the White method is solely based on diurnal fluctuations in groundwater, any water extraction by phreatophytes from parts of the vadose zone that are not hydraulically connected to the aquifer is neglected. Therefore, the results are usually considered to only represent that part of the total daily ET that is provided by the aquifer. Over the CNWR, however, annual precipitation rarely exceeds 100 mm, with usually less than a quarter of this amount falling during the growing season of tamarisk. Due to such a low precipitation and the aridity of this region, the contribution of vadose zone water content to ET is negligible, and most of riparian water consumption is provided by the aquifer. In addition, a previous study on isotope signatures of water samples obtained from tamarisk stem, soil vadose zone, and aquifer revealed that vadose zone moisture had a negligible contribution to tamarisk water uptake [*Nagler et al.*, 2008].

6.2.3.2. SEBAL Model

In this study, the Idaho implementation of SEBAL model was used for mapping spatially distributed ET over the study area. SEBAL has been successfully applied over agricultural ecosystems in more than 30 countries, producing accurate estimates of crop ET [*Bastiaanssen et al.*, 1998b, 2005; *Ramos et al.*, 2009], but its applications to riparian ecosystems has remained very limited. In this model, net radiation (R_n) is estimated through quantifying all of the incoming and outgoing components of short- and long-wave radiation. Soil heat flux (G) is then modeled as a ratio of net radiation. Finally, sensible heat flux (H) is mapped using an innovative approach that interpolates H between two extreme conditions, representing minimum and maximum sensible heat fluxes. Knowing R_n, G, and H, the latent heat flux (LE) can be calculated as the residual of the energy balance equation, assuming that the energy consumed in photosynthesis and the canopy storage of energy are both insignificant:

$$LE = R_n - G - H. \qquad (6.2)$$

In the Idaho implementation of SEBAL model, the minimum H flux is calculated as the remainder of the energy balance equation [equation 6.2], where R_n and G are obtained from a cold and well-irrigated agricultural field at full cover. The LE flux is assumed to be 5% larger that the reference alfalfa-based ET, measured at a standard weather station at the hour of satellite overpass. The maximum condition occurs over a hot agricultural bare soil, where there is little or no water to evaporate and the available energy is translated to sensible heat flux. After identifying these two extreme limits, H is interpolated for all other pixels, using the Monin-Obukhov similarity theory. In this study, hot and cold pixels were selected over the upstream PVID, which has numerous fallow fields (for hot pixel), as well as well-irrigated alfalfa and cotton fields (for cold pixel) during the long growing season of this region.

The LE flux estimated from equation 6.2 is only an instantaneous estimate at the time of data acquisition. Thus, it has limited applications for practical purposes such as managing water resources. Several methods have been proposed for extrapolating instantaneous to daily values [*Colaizzi et al.*, 2006; *Chávez et al.*, 2008]. The original extrapolation method in SEBAL is based on the evaporative fraction (EF), which is the ratio of instantaneous LE to instantaneous available energy ($R_n - G$). Assuming that instantaneous EF is equal to daily EF [*Brutsaert and Sugita*, 1992; *Zhang and Lemeur*, 1995; *Crago*, 1996], daily LE is calculated by multiplying instantaneous EF and daily available energy. Daily available energy is estimated as a function of surface albedo, 24 h solar radiation, and atmospheric transmittance, as explained in *Bastiaanssen et al.* [2005]. Although this technique has provided reliable results in many studies [*Gowda et al.*, 2008; *Bastiaanssen et al.*, 2005; *Ramos et al.*, 2009], its accuracy is hampered over agricultural areas in arid/semiarid regions, where afternoon advection can substantially enhance the ET of well-irrigated crops.

Trezza [2002] proposed to replace EF with the ratio of instantaneous ET to instantaneous alfalfa-based reference ET (ET_r), estimated at a standard weather station. This new method (ET_rF) has shown to yield improved estimates of daily ET [*Allen et al.*, 2007a, 2007b]. Similarly, grass-based reference ET (ETo) could be used in upscaling instantaneous ET values (EToF method). An important question in the application of RSEB models in riparian areas is which one of existing upscaling methods performs better for extrapolating instantaneous to daily ET estimates. For agricultural crops with ample access to water, ET_rF and ET_0F methods that are based on water use of nonstressed reference crops may be more appropriate. The ET process over some riparian ecosystems, however, is more water limited than energy limited. In addition, phreatophytes adapted to arid/semiarid climates have different mechanisms for controlling their stomatal conductance and transpiration rate. For tamarisk at CNWR, *Glenn et al.* [2013] found that both parameters decreased in response to increasing groundwater depth and salinity. *Devitt et al.* [1997] also observed that when water table was about 3.0 m from soil surface, tamarisk individuals were not able to meet increased atmospheric demand under advective conditions. Hence, EF may be a more appropriate upscaling technique for mapping ET over riparian ecosystems such as the CNWR as it is less sensitive to advection. In the present study, the EF and ET_0F methods were implemented and the performance of each method in extrapolating instantaneous tamarisk ET was evaluated by using expert knowledge and by comparing the results with White approach approximations and the estimates from Bowen ratio towers.

Another major concern in extrapolating instantaneous ET is the validity of a key assumption that is made in both techniques. According to this assumption, the instantaneous $ET_rF/ET_0F/EF$ at the time of satellite overpass is equal to the daily value. This may not be the case if phreatophytes experience afternoon reduction in transpiration. Using sap flux sensors, *Nagler et al.* [2009a] measured the hourly transpiration rate at several sites in the CNWR and found that tamarisk at Slitherin had a rather constant EF throughout the day. However, measurements near the Diablo tower showed signs of a midday (early afternoon) reduction due to stomata closure, but it was compensated by nocturnal transpiration, resulting in the validity of assuming instantaneous EF is similar to daily EF.

6.2.3.3. *Bowen Ratio*

Three Bowen ratio energy flux towers (Radiation and Energy Balance Inc., Seattle, WA) were installed at each station to obtain an independent estimate of energy balance components. In order to estimate soil heat flux, soil moisture sensors, soil heat flux plates, and soil temperature probes were installed at three locations (sunny, partial shade, and shaded), representing the footprint of the net radiometer and the upwind ecosystem. At each tower, temperature and vapor pressure gradients were measured at two different heights, namely 2.0 and 3.0 m above the tamarisk canopy. The location of the instrumentation at these two heights was switched every 15 min, using an automatic exchange mechanism, to minimize biases caused from nonsimilar instrumental errors. Measured variables were averaged over every two consecutive 15 min periods. Hence, energy balance components were estimated on a 30 min basis. Collected data were transmitted via cell phone connection to a center at the Wright State University, where they were downloaded and examined for quality control. A detailed description of the installed Bowen ratio systems and

data processing can be found in *Chatterjee* [2010]. Several filters were applied to raw data before estimating the latent heat flux. For example, data were excluded from further analysis when G was larger than 500 W/m², which is about the average value of R_n measured at study area. In addition, data were rejected when the Bowen ratio was between -0.5 and -1.5 to avoid obtaining unreasonable results [*Payero et al.*, 2003].

In this study, only the estimates based on the measurements of the Slitherin Bowen ratio tower were used for comparison against the estimates of other methods. As explained before, the Swamp station was excluded from analysis due to its proximity to the old river channels, which made it impossible to apply the White method. The Bowen ratio measurements at the Diablo tower were also excluded from further analysis since about half of the data were missing due to power failure at different times during the study period. The available data showed that, in general, the ET was lower than the measured values at the Slitherin tower. Detailed explanation on Bowen ratio instrumentation, data quality control, and data processing are provided in *Chatterjee* [2010].

6.2.4. Other Required Data

Daily ET_0 estimates and other required weather parameters were obtained from a nearby weather station located in Blythe, California, and operated by The California Irrigation Management Information System (CIMIS). In addition, 21 cloud-free satellite (Landsat TM5) images acquired between January 2008 and January 2009, were downloaded from the website of the U.S. Geological Survey (USGS) Global Visualization Viewer, (http://glovis.usgs.gov/). This ample cloud-free imagery was available because the study area is located in the overlap zone of two Landsat paths (paths 38 and 39, row 37).

6.3. RESULTS AND DISCUSSION

6.3.1. Borehole Logging

The borehole logging provided basic data on soil physical characteristics to a depth of 6.0 m for each observation well. The most dominant soil textures at Diablo were loamy-sand and sand, occupying from 74% to 95% of the total studied depth among the five observation wells. The soil layers were mainly structureless, rarely having massive or blocky structures with loose to very friable consistency. Soil textures were similar for Slitherin wells, with 75%–91% of the studied layers being classified as loamy-sand, sandy-loam, and sand. Soil structure was slightly improved at this station, with weak blocky and platy structures being occasionally observed. Soil consistency ranged from loose to friable, fairly similar to the condition at Diablo. Based on these data, an S_y value was selected from *Loheide et al.* [2005] for each observation well. Selected values ranged from 0.085 to 0.100 for the Diablo wells and from 0.115 to 0.130 for the Slitherin wells. These values were all smaller than the constant value of 0.140 used by *Zhu et al.* [2011] for the same wells. It should be noted that overestimation in S_y value directly translates into overestimation of ET when applying the White method. Borehole logging results for Swamp wells are not presented here as this station was excluded from White method analysis due to its proximity to the outflow drain.

6.3.2. Groundwater Characteristics

Groundwater electrical conductivity (EC) was measured at every observation well and then averaged for all five wells of each measuring station. The EC did not have significant temporal variations, but it increased with distance from the surface water (old and new river channels). The annual average EC was 17.0, 8.1, and 4.1 dS/m for Diablo, Slitherin, and Swamp, respectively (Figure 6.3). The high levels of groundwater EC at Diablo promote the dominance of tamarisk, as previous studies have shown that native species have a maximum salt tolerance that is significantly smaller than that of tamarisk [*Glenn et al.*, 1998; *Vandersande et al.*, 2001]. *Glenn et al.* [2012] conducted a comprehensive study on salt concentration in CNWR soil and aquifer and concluded that the Swamp site is the only station that may support the growth of native species. Although tamarisk can survive high groundwater salinity levels at Slitherin and Diablo sites, *Glenn et al.* [2013] argued that the rate of tamarisk water use is negatively affected by aquifer salinity. Another study conducted along the Gila River in Arizona reported a 77 mm decrease in tamarisk annual water use per each unit (dS/m) of increase in EC of soil saturation extract [*Van Hylckama*, 1970].

The groundwater depth was determined for each measuring stations by averaging the measurements of the five observation wells in that station. Slitherin and Swamp had the greatest and smallest groundwater depths, with annual averages of 3.5 and 2.9 m from soil surface, respectively. The water table at Diablo was slightly closer to the surface than at Slitherin, with annual average of 3.4 m. This means that overall tamarisk individuals at Swamp had a better access to groundwater than the other two stations, which is due to the closer distance between this station and the river. The day-to-day groundwater fluctuation was also investigated during the study period. Groundwater level had a distinct seasonal pattern under Slitherin and Diablo, with its deepest level occurring in mid to late summer. This is when atmospheric demand and riparian water consumption are substantially high (Figure 6.4). As the air

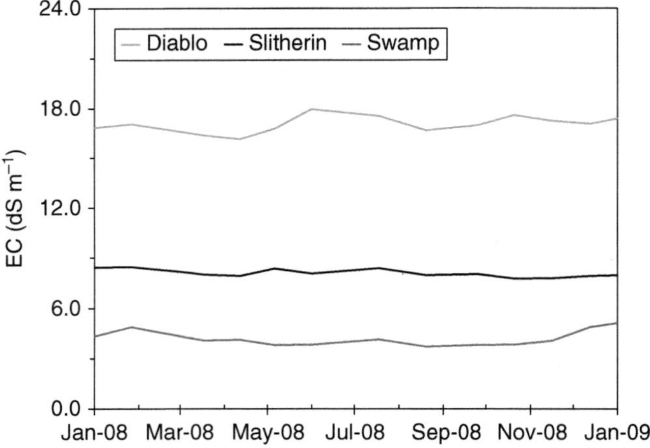

Figure 6.3 Average groundwater EC during the study period at all measuring stations.

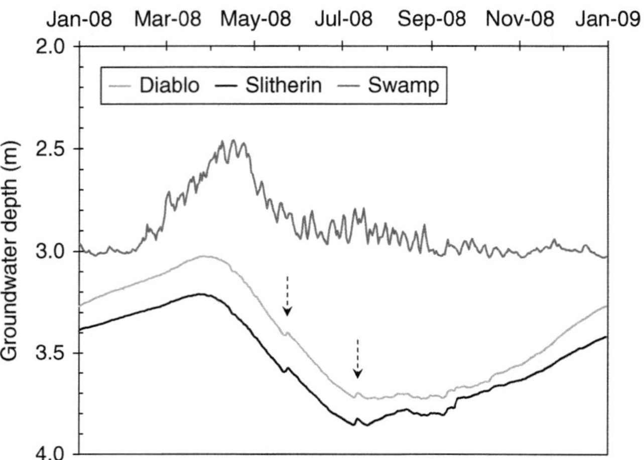

Figure 6.4 Average daily groundwater depths during the study period at all measuring stations.

temperature started to decrease in October and tamarisk began to senesce, the aquifer recharge rate became greater than water extraction by tamarisk. Consequently, groundwater levels rose until they reached their shallowest level in April. The water table at Swamp had different behavior, with a peak in spring, and an approximately constant level of about 3.0 m from the surface during the rest of the year. Higher frequency fluctuations observed at this station were influenced by stage variations in both old and new Colorado River channels. The annual magnitude of water level variation was 0.70, 0.64, and 0.56 m under Diablo, Slitherin, and Swamp, respectively. For all three sites, groundwater returned to about the same level over a period of one year.

The two indents in Slitherin and Diablo groundwater depth curves (marked by dashed arrows in Figure 6.4)

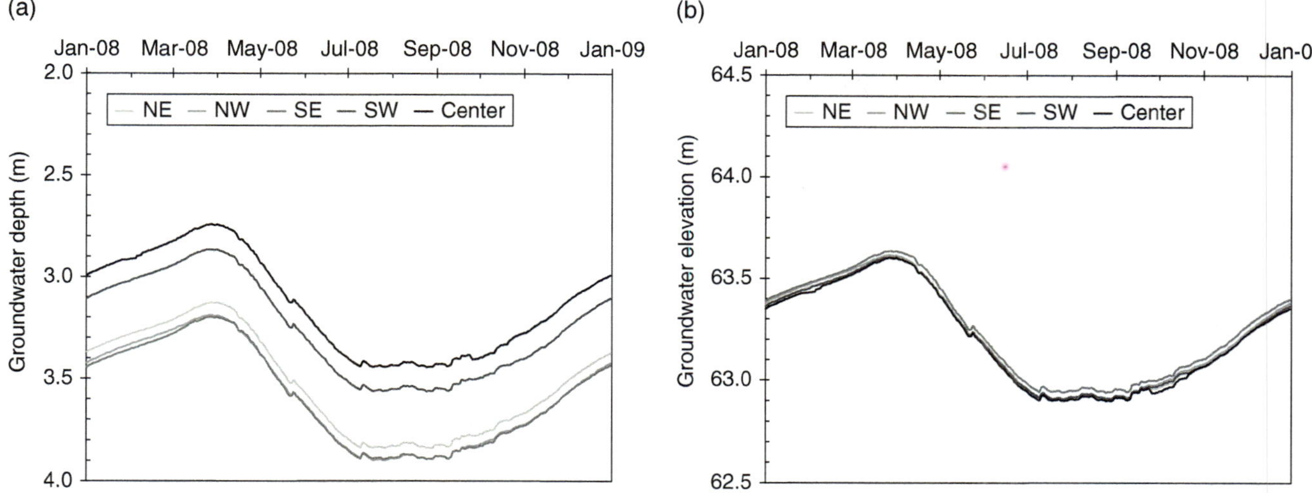

Figure 6.5 Daily average: (a) groundwater depths and (b) groundwater elevation, at five observation wells of the Diablo station.

coincided with two monsoon rain events in late May and mid-July with cumulative depth of 12.2 and 8.6 mm, respectively. Since the magnitude of these events was fairly small, the observed rise in groundwater level is most probably due to the loading effect of precipitated water on the aquifer. In other words, the CNWR aquifer was acting as a geological weighing lysimeter, meaning that it did not have hydraulic connection with precipitated water and the reason behind the observed rise in water table was the pressure exerted on aquifer formation by the weight of water [*Rasmussen and Mote*, 2007]. These two events together consisted 29% of the total rainfall during the study period (71 mm). The rest of precipitations happened during the period when groundwater was gradually rising, therefore the effect on water table was not detectable.

Since the ground elevation varied from one observation well to another, the direction of subsurface flow could not be determined just by evaluating groundwater depth. Hence, a LiDAR-derived, high-resolution DEM ground elevation map was used in the ArcGIS environment to extract the ground elevation at each observation well. The groundwater depth was then subtracted from ground elevation to obtain groundwater elevation. This conversion removed almost all of the variability among the five observation wells of each station. As an example, Figure 6.5 shows daily groundwater depth and elevation for the five wells at Diablo. The range of the ordinate is the same (2.0 m) for both plots.

Daily groundwater elevation data were then averaged over all five wells at each station and compared with the river stage to study the stream-aquifer interaction (all elevation data were based on the same datum). This comparison confirmed that groundwater fluctuations at

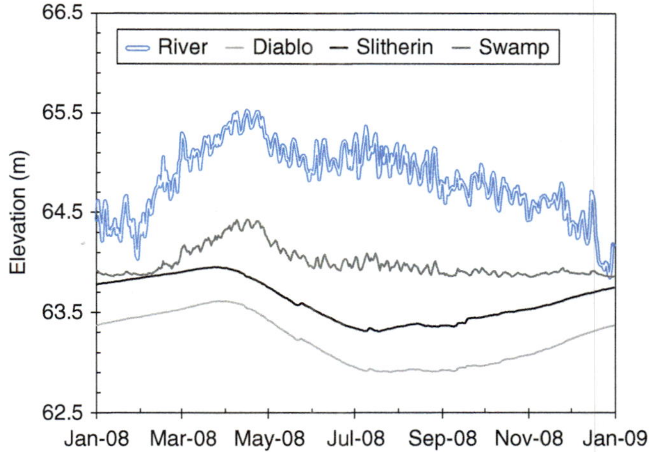

Figure 6.6 Daily average groundwater elevation at measuring stations, along with the Colorado River stage.

Swamp were strongly influenced by the variations in the Colorado River stage (Figure 6.6). Groundwater elevation at Diablo was always lower than the two other stations and the river stage, ranging from 62.9 m in late summer to 63.6 m in April. The fact that water table elevation at Diablo was the lowest reveals that the direction of subsurface flow was from surface water bodies toward the heart of the CNWR. On average, the water table at Slitherin was 0.4 m higher than Diablo, with a minimum level of 63.3 m and a maximum of 64.0 m, occurring at about the same times as Diablo. The Colorado River stage reached a maximum level of 65.5 m in mid-April and a minimum level of 63.9 m in late December. During the entire study period, river stage was always at a higher level than the CNWR aquifer, except for only one date (28 December 2008), when it was equal to the aquifer head

(a)

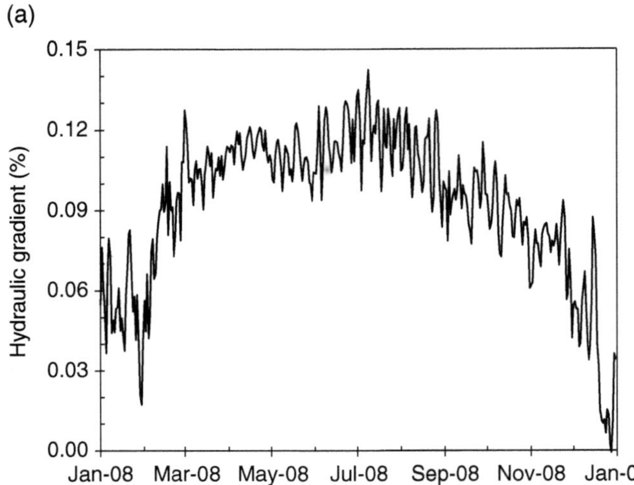

(b)

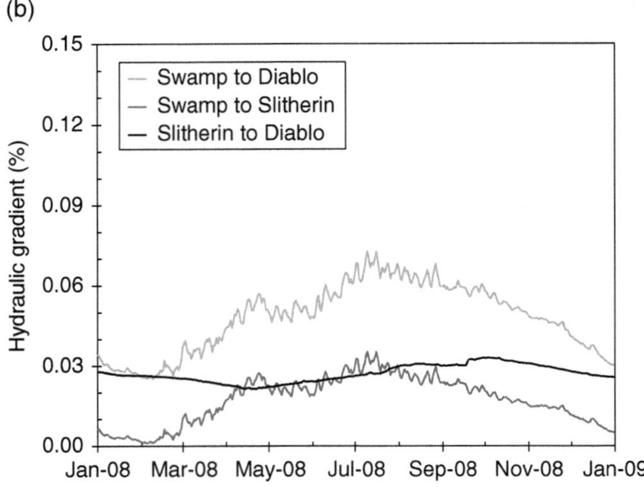

Figure 6.7 Daily average hydraulic gradients (a) from the river to Swamp and (b) between the stations.

level at Swamp. This finding suggests that the hydrologic interaction between the Colorado River and the CNWR aquifer is a source-sink interaction. Water flows from the river toward the riparian forest.

The hydraulic gradients were also estimated based on measured elevation of water table and river stage. The hydraulic gradient between the river and Swamp varied from zero to a maximum of 0.14% (1.4 mm/m) in July, reaffirming the presence of a westward subsurface flow (Figure 6.7a). In comparison with river to Swamp gradient, interstation gradients were smaller. Swamp-Diablo and Swamp-Slitherin gradients had similar patterns (Figure 6.7b) following the greening and senescence of tamarisk, which occur in March and November at the CNWR [*Nagler et al.*, 2009b]. However, the magnitudes were different. The Swamp-Diablo hydraulic gradient reached a maximum of 0.07%, which was half of the river Swamp gradient and two times larger than the Swamp-Slitherin gradient. Interestingly, the Slitherin-Diablo gradient did not pose a significant seasonal pattern, mainly due to the similarity in (i) seasonal pattern of tamarisk water use and (ii) the rate of aquifer recharge by the Colorado River. This suggests that a rather constant southward flow existed within the lower CNWR, which was most probably fed by the old river channel on the north side of the Slitherin site. As mentioned before, this channel carries a significant amount of irrigation return flow from the upstream PVID, as well as a fraction of the river flow.

6.3.3. Tamarisk Water Use

6.3.3.1. White Method

Figure 6.8 demonstrates daily water use estimates using the White method, overlaid on corresponding reference ET (ET_0) and groundwater depth for Slitherin and Diablo.

For the sake of a clear representation, a 5 day moving average was applied to daily water use estimates (both tamarisk water use and ET_0) to smooth some of the day-to-day variations. The results revealed that tamarisk growing season is about 240 days long in CNWR, with emerging new leaves occurring in mid-March and full senescence occurring by mid-November. Some negative estimates were obtained beyond this period, due to the absence of tamarisk-induced diurnal fluctuations. The negative values can be dealt with in two ways: (i) by limiting the analysis period to tamarisk growing season and (ii) by averaging them over longer periods (weekly and monthly), when the errors cancel each other out and yield a near-zero estimate.

According to Figure 6.8, groundwater depth and tamarisk ET were strongly coupled. The increase in tamarisk water use coincided with the decline in aquifer level and *vice versa*. At Slitherin, water use reached a maximum of 8.0 mm/d in early July, when water table fell to its deepest level of 3.85 m from the soil surface. It seems that the rapid reduction in water use that happened soon after was a result of this deep water level. This reduced rate of water uptake was probably smaller than the rate of aquifer recharge, as water level started to rise until it reached 3.78 m from the surface in mid-August. At this point, the tamarisk seems to have regained better access to water, as water use increased and caused a second drop in water table. The ET_0 graph confirms that variations in tamarisk water use were influenced by variations in water availability (groundwater depth) and not changes in atmospheric demand. While small-scale variations in ET_0 and tamarisk water use were similar, ET_0 values were not unusually low or out of their seasonal trend during the period with a significant reduction in water uptake by tamarisk. After reaching a depth of 3.81 m, aquifer level remained

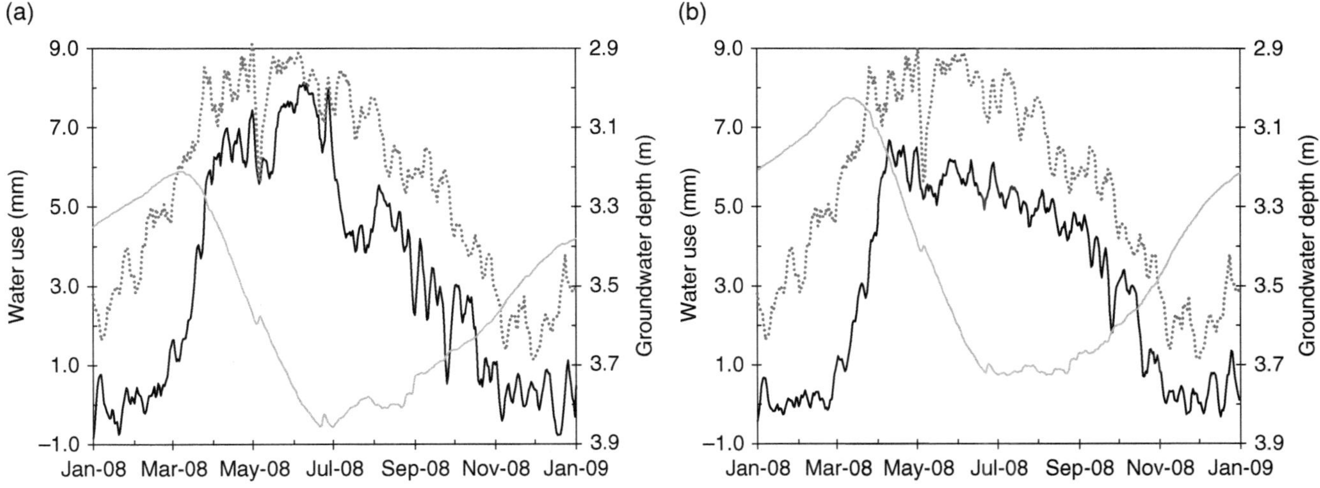

Figure 6.8 Daily 2008 tamarisk water use estimated by the White method (solid black), overlaid on groundwater depth (solid gray) and ET_0 (dotted gray) at (a) Slitherin and (b) Diablo.

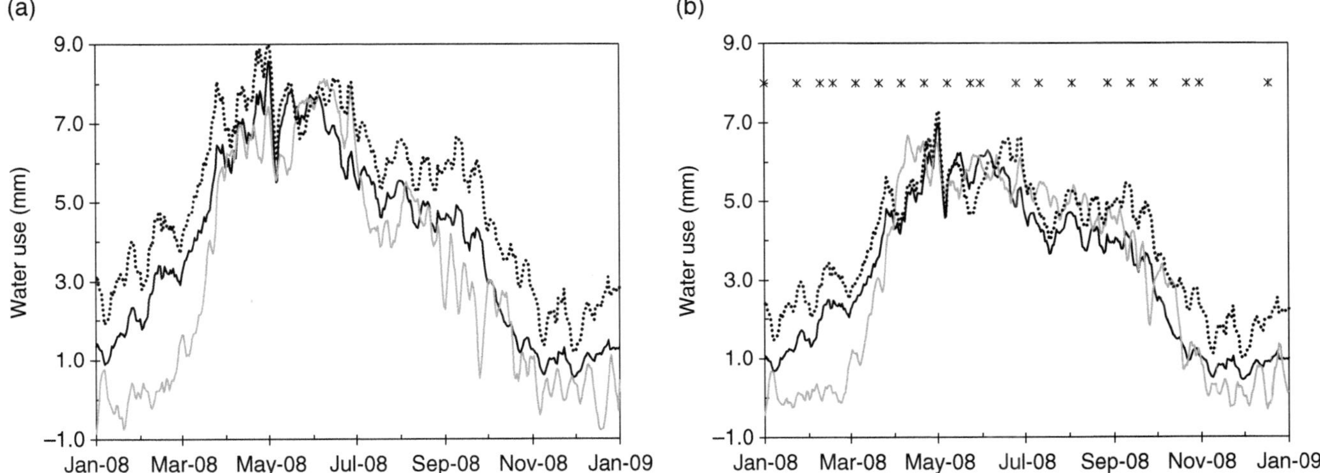

Figure 6.9 Tamarisk water use estimates of SEBAL-EF (solid black), SEBAL-ET_0F (dashed black), and White (solid gray) methods for (a) Slitherin and (b) Diablo sites. The asterisks in (b) represent the dates of satellite overpass.

approximately constant for 2 weeks and then started to increase as tamarisk water uptake decreased due to lower plant and atmospheric demands. Such a distinct aquifer-phreatophyte interaction was not observed at Diablo, probably because groundwater at this station was always closer to the surface compared to Slitherin. The interaction between groundwater depth and groundwater uptake by tamarisk suggests that dropping the water table may be regarded as a method of controlling tamarisk water use. Previous studies have also reported that tamarisk transpiration has an inverse relationship with water table depth [*Van Hylckama*, 1970; *Devitt et al.*, 1997; *Neale et al.*, 2011; *Glenn et al.*, 2013]. In the Hassayampa River floodplain in Arizona, for example, *Horton et al.* [2001] observed tamarisk dieback when groundwater depth fell below 2–3 m from the surface.

6.3.3.2. SEBAL Model

Among the ET_0F and EF methods of upscaling ET, the former resulted in higher water use estimates. At Slitherin, the difference between the results of these methods reached about 2 mm/d (Figure 6.9). Water use estimates were generally smaller at Diablo, but they showed the same behavior, with ET_0F results being larger than those based on the EF method. Our field observations were more consistent with the lower estimates from the EF method. In addition, EF-based results were closer to the White method estimates. However, despite their smaller values in comparison to the ET_0F results, the EF estimates were still significantly larger than the White method results during the first 3 months of 2008. For example, while the White method estimated no significant water use in early March, SEBAL-EF predicted 2.0–3.0 mm of ET per day (Figure 6.9).

Part of this difference can be attributed to the effect of evaporation from soil and vegetation surfaces after precipitation events that occurred in January 2008 (4.0 mm on 7 January and 14 mm during the last week of January). While precipitation events influence SEBAL results, they are not accounted for in the White method as it is solely based on water uptake from aquifer. However, the difference between SEBAL-EF and White estimates was also significant in February and March, a period when no measurable precipitation occurred. It appears that the White estimate of near-zero water use was more reasonable during this period since tamarisk was still dormant and precipitation was negligible.

One hypothesis for explaining the over estimation of SEBAL is that the empirical equations for estimating surface roughness length in this model are developed for agricultural crops, which have a relatively short and homogeneous height compared to tamarisk trees that were several meters tall over Slitherin. To investigate this hypothesis, a high-resolution, LiDAR-derived map of tamarisk canopy height was used to estimate actual roughness length, based on *Prueger and Kustas* [2005]. A comparison of results using the actual roughness length layer with the estimates using empirical equations showed no significant difference (results not presented here). This was not surprising as any effect from an underestimated canopy height would have been projected over the entire study period, not only the first few months of the year. In addition, *Wang et al.* [2009] evaluated the sensitivity of SEBAL estimates over pecan orchards in New Mexico and found that changing the value of roughness length from 0 to 2.5 (representing canopy heights from 0 to 20.8 m) did not have any significant effect on the results, especially when canopy cover was larger than 50%. Similar results were reported by *Tasumi* [2005] and *Long et al.* [2011].

Further investigation of all the steps in running the SEBAL model revealed that the probable reason behind the overestimation of ET during the first 3 months of the year was the dominant presence of shadows in the tamarisk forest that artificially lowered the canopy temperatures detected by satellite sensors. The overpass time for all Landsat scenes used in this study was within 7 min of 10:00 A.M. Pacific standard time. This fixed overpass time resulted in a sun elevation angle that varied between 30.8° on 19 January and 66.2° on 11 June 2008. The values of sun elevation and azimuth angles reported in the header file of the Landsat images and the LiDAR-derived map of top-of-canopy elevation were used as input data to the "hill-shade" function in ArcGIS. Analyzing the generated maps showed that the percentage of the shaded pixels (pixels with shaded relief value of zero) ranged from 33.4% on 19 January to 2.5% on 26 May 2008. This extensive presence of shadows lowered the apparent canopy temperature in the 120-m by 120-m pixels of the Landsat

thermal band. Contaminated pixels tend to shift more toward the selected cold extreme in the image, resulting in a lower sensible heat flux estimate from the model and consequently a higher latent heat flux.

6.3.3.3. Modified SEBAL

In order to modify the SEBAL results to better represent the water use of tamarisk during its dormant period, the soil adjusted vegetation index [SAVI: *Huete*, 1988] was used. Since this VI is sensitive to green vegetation biomass, its evolution closely follows the growing season of tamarisk. The remotely sensed SAVI would have a minimum value during tamarisk dormancy, but it would increase as new leaves emerge until it reaches a maximum when the canopy is fully developed. The value of SAVI would then remain fairly constant until the onset of senescence, when it starts to decrease. Hence, a normalized SAVI could serve as an adjusting coefficient to reduce SEBAL estimates when tamarisk transpiration is reduced and/or stopped. The normalized SAVI was estimated on a distributed basis by dividing the SAVI of each pixel in every satellite image by the maximum SAVI of the same pixel among all 21 images. Multiplying the normalized SAVI by the results of SEBAL-EF model was effective in reducing the remotely sensed tamarisk water use during dormancy to levels similar to that estimated by the White method (Figure 6.10).

Using the normalized SAVI to correct for the effect of shaded pixels has two main advantages. First, its estimation does not require additional data, as it can be mapped based on the same remotely sensed data used in running the SEBAL model. Second, the spatially distributed nature of normalized SAVI makes it appropriate in dealing with a high level of heterogeneity that is usually observed in riparian communities. For example, the effect of different greening times would be accounted for if thickets of different species are present in a riparian forest. The main caveat of this method is the underestimation of ET after precipitation events that occur during vegetation dormancy, as the evaporation would be erroneously reduced by a low value of normalized SAVI. This limitation, however, would not impose significant challenges for applying this modification to riparian ecosystems of arid/semiarid regions.

6.3.3.4. Bowen Ratio

The ET measurements of the Slitherin Bowen ratio tower had a pattern that was similar to that of the White and modified SEBAL-EF methods (Figure 6.11). The values were close to zero until mid-March, when they started a rapid increase and reached about 7.0 mm/d in mid-June. A strong coupling between tamarisk water use and groundwater depth was observed in Bowen ratio data as well, especially during the period when groundwater

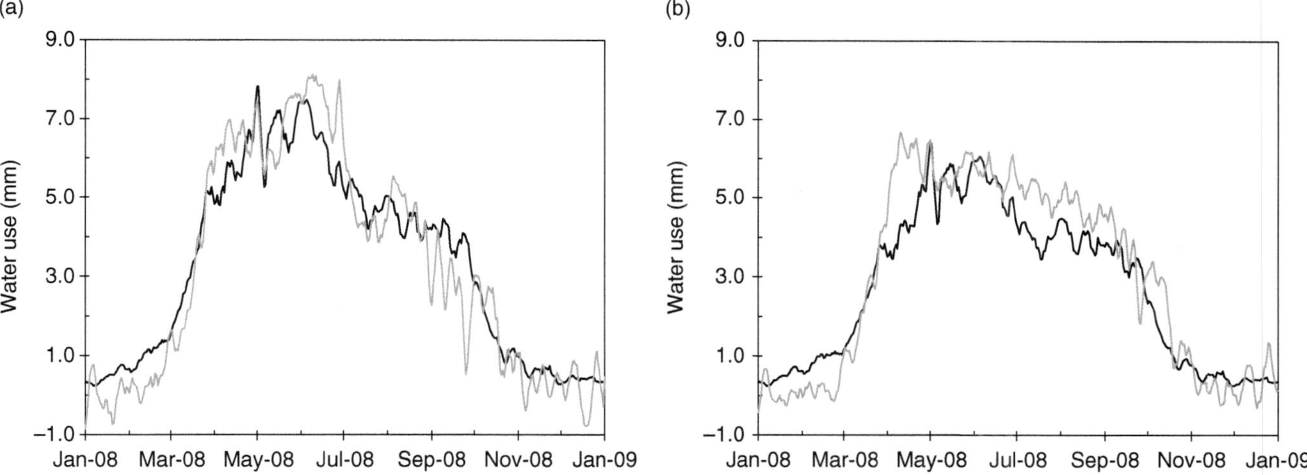

Figure 6.10 Tamarisk water use estimates of SAVI-adjusted SEBAL-EF (black) and White (gray) methods over (a) Slitherin and (b) Diablo.

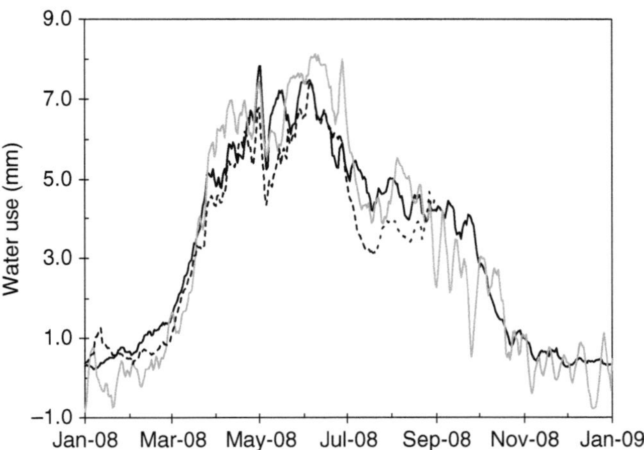

Figure 6.11 Tamarisk water use estimates of SEBAL-EF (solid black), Bowen ratio (dashed black), and White (solid gray) methods at Slitherin.

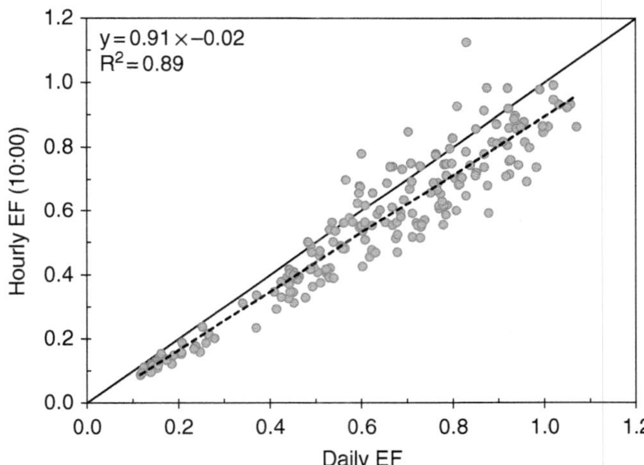

Figure 6.12 Hourly EF values at the time of Landsat overpass (10:00 A.M.) plotted against daily EF values, both estimated based on Bowen ratio measurements. The solid black line is the one-to-one line and the dashed black line is the regressed line.

reached its lowest levels from the soil surface in early July. Over a 6 month period (mid-March to mid-September) during the tamarisk growth season, the White and modified SEBAL-EF results had a mean bias error (MBE) of 0.86 and 0.62 mm/d, respectively, when compared to Bowen ratio results. The values of root mean square error (RMSE) were 1.22 and 1.01 mm/d, respectively, for the same two methods. The small errors suggest that the implemented modifications to the White and SEBAL methods were satisfactory. These modifications included the selection of S_y parameters based on borehole logging data and *Loheide et al.* [2005] guidelines for the White method and the use of SAVI-adjusted EF technique for the SEBAL model. Without SAVI adjustment, the SEBAL-EF results had MBE and RMSE of 1.33 and 1.61 mm/d, respectively. The SEBAL-ET$_0$F results had

MBE and RMSE of 2.20 and 2.44 mm/d, respectively, indicating that the EF method performed better than EToF in daily ET extrapolation over riparian canopies.

Several previous studies have proved the validity of the assumption that the EF at the hour of satellite overpass is similar to the daily (24 h) EF value [*Brutsaert and Sugita*, 1992; *Zhang and Lemeur*, 1995; *Crago*, 1996]. To further examine the adequacy of the EF for extrapolating instantaneous values of ET, Bowen ratio data collected at Slitherin were analyzed to examine the EF conservation assumption under the specific hydroclimatological conditions of CNWR. Figure 6.12 presents a scatterplot of EF values for the hour of Landsat overpass (10:00 A.M.) against daily EF values for 200 clear-sky days during the study period. The regression line fitted to the data had a slope of 0.91, a negligible intercept, and a coefficient of determination of 0.89. In other words, hourly EF was 9% smaller than daily EF on average. This amount of difference is smaller than the expected error of daily SEBAL estimates (±15%) reported by *Bastiaanssen et al.* [2005].

6.3.3.5. Monthly and Annual Tamarisk Water Use

Monthly and annual tamarisk water use rates were estimated based on the White method and adjusted SEBAL-EF model. These estimates are reported in Table 6.2, along with the corresponding estimates of grass-based reference ET. In general, both methods provided fairly similar monthly estimates, with an average absolute difference of 12.3 and 16.7 mm/month for Slitherin and Diablo, respectively. Besides the uncertainties associated with each method, the possible mismatch between the footprints of SEBAL and White methods may have contributed to the observed differences in monthly and annual estimates. To extract SEBAL results for each station, a circular footprint encompassing all five observation wells at each station was used. The center of this circle was located on the central well and its radius was 230 m (enclosing an approximate area of 0.17 km²). However, groundwater fluctuations at each observation well may be influenced by phreatophytes at farther distances. A possible mismatch of footprints would be less important at Slitherin, where the canopy is more homogeneous and at full cover, while at Diablo, tamarisk thickets are interspaced with bare soil. For Slitherin, the tamarisk water use estimates based on Bowen ratio measurements are also presented in Table 6.2 for the 9 months with available data. The estimates were closer to SEBAL results, with average absolute difference of 11.9 mm/month.

Table 6.2 Monthly reference ET and tamarisk water use over slitherin and diablo (all in mm)

| Period | ET$_0$ | Slitherin | | | Diablo | |
		White	SEBAL	Bowen ratio	White	SEBAL
Jan	78	0	11	14	0	11
Feb	96	0	19	17	1	18
Mar	163	30	51	35	20	39
Apr	224	152	139	126	129	102
May	241	198	195	177	181	156
Jun	257	221	208	194	173	168
Jul	247	196	176	167	167	146
Aug	219	143	142	110	156	123
Sep	182	110	129	129	131	116
Oct	152	77	96	NA	99	82
Nov	92	24	29	NA	27	24
Dec	60	11	15	NA	6	13
Total	2,011	1,162	1,210	NA	1,090	998

The monthly water use estimates of this study were significantly lower than those reported by *Zhu et al.* [2011]. This is mainly due to the fact that they used a constant readily available specific yield value from *Loheide et al.* [2005], assuming that the aquifer is composed of sandy material at all observation wells. The same reference [*Loheide et al.*, 2005] was used in this study, but a unique value was assigned to each well, using the soil texture data identified in the borehole logs. The annual water use at Slitherin was close to, but lower than, the estimates of *Nagler et al.* [2008], who applied the MODIS-derived EVI approach and obtained a value of 1300 mm, averaged over a 6 year period (2000–2006). However, their estimate for Diablo was significantly larger at 1430 mm. This was 43% and 31% greater than SEBAL and White results in this study, respectively. Our results are also similar to the annual tamarisk ET of 1076 mm estimated by Bowen ratio towers at the Havasu National Wildlife Refuge, which is about 250 river kilometers upstream of CNWR [*Westenburg et al.* 2006].

6.3.3.6. Tamarisk Vegetation Coefficients

In managing water deliveries on the Lower Colorado River, USBR utilizes an approach that is known as the Lower Colorado River Accounting System (LCRAS). The LCRAS is based on the crop coefficient (K_c) concept, which is widely used in water management of irrigated agriculture. The K_c represents the ratio of water used by agricultural crops in different stages of growth to that of a reference surface (grass or alfalfa). Once these ratios are developed for any specific type of crop and climatic conditions, they can be multiplied by the reference ET, estimated at standard weather stations, to provide an approximation of water consumption by that crop. Similar to agricultural crops, crop coefficients (or better called vegetation coefficients) are developed and applied for riparian species in LCRAS. Evaluating the accuracy of tamarisk K_c values used in LCRAS based on the results of this study would be useful to river managers.

In order to be consistent with the traditional four-stage K_c curves of LCRAS, piecewise linear relationships were fitted to the daily K_c values based on both the adjusted SEBAL-EF and White methods (Figure 6.13). The maximum midseason K_c in this study was 0.76, which was estimated at Slitherin station based on the White method. The midseason K_c value assigned to tamarisk in LCRAS is significantly larger at 1.10 (Marvin Jensen, personal communication). On an annual basis, SEBAL and White estimates of tamarisk K_c ranged from 0.50 to 0.60 for the two stations, while LCRAS assumes a value of 0.86. The K_c values obtained in this study were consistent with findings of *Nagler et al.* [2009b], who reported that annual water use at CNWR was about half of the reference ET. Over the Mojave River floodplain in California, *Neale*

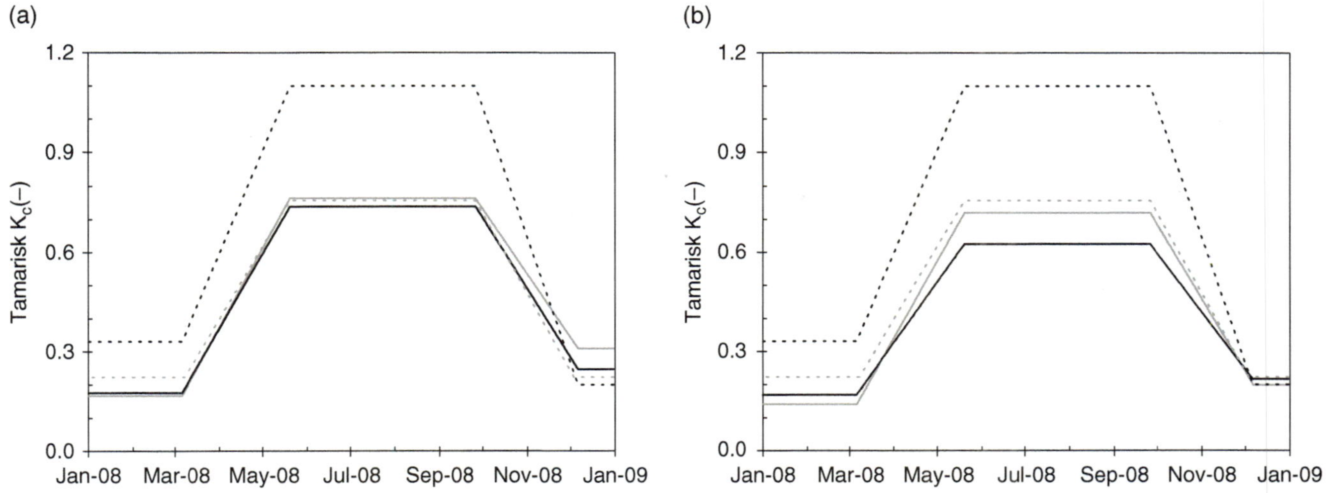

Figure 6.13 Piecewise K_c curves for (a) Slitherin and (b) Diablo based on adjusted SEBAL-EF (solid black), White (solid gray), LCRAS (dashed black), and *Westenburg et al.* [2006] (dashed gray).

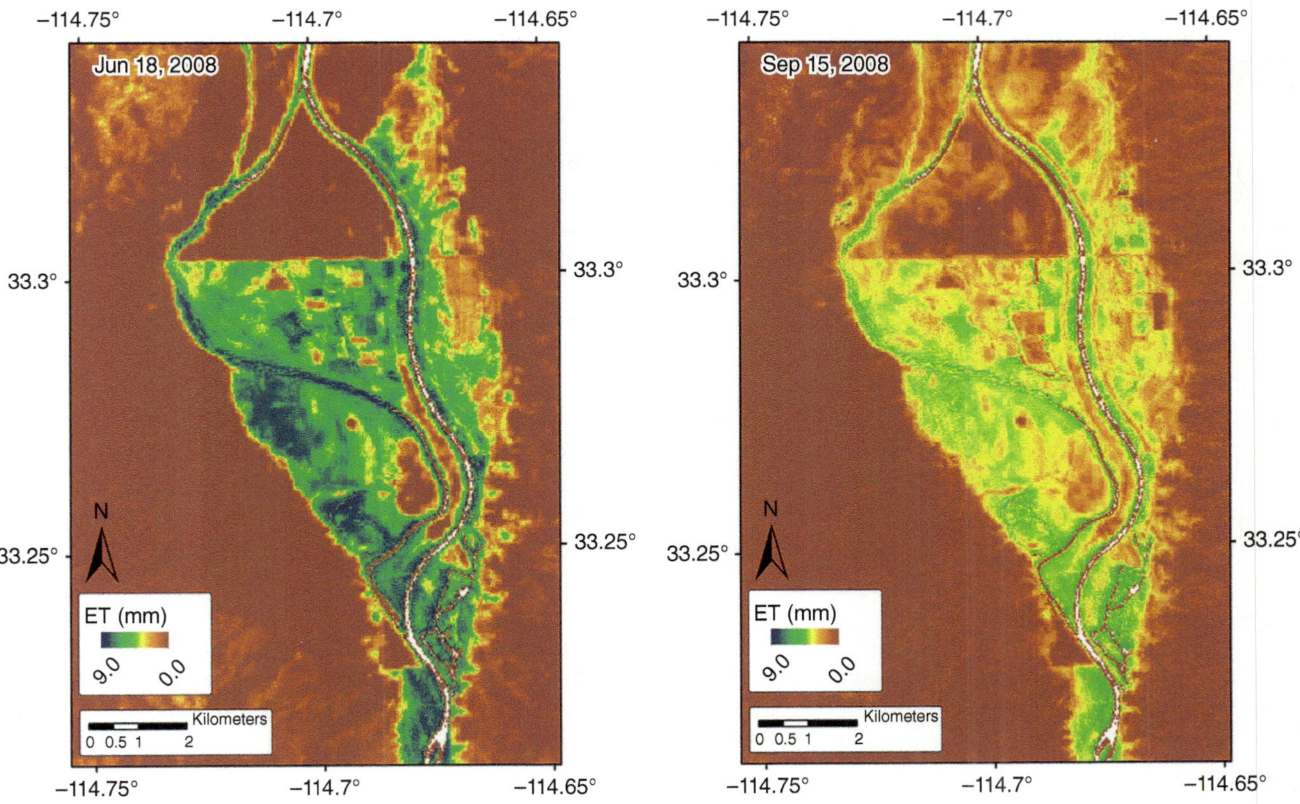

Figure 6.14 Spatially distributed ET rates modeled by the adjusted SEBAL-EF model.

et al. [2011] also estimated a total tamarisk ET that was 48% of the annual ET_0 ($K_c = 0.48$). Using Bowen ratio towers, *Westenburg et al.* [2006] found that tamarisk ET was 60% of the annual ET_0 at Havasu National Wildlife Refuge. This suggests that LCRAS K_c will most probably result in an overestimation of tamarisk water use.

6.3.3.7. Wide-Area Water Use

The potential of RSEB models is in their ability to provide spatially distributed estimates of riparian water use. Figure 6.14 illustrates ET maps generated by the adjusted SEBAL-EF model for two dates in 2008: 18 June, which is during the peak water use period, and

Table 6.3 Monthly reference ET and tamarisk water use over the entire study area (all in mm)

Period	ET_0	SEBAL
Jan	78	11
Feb	96	17
Mar	163	38
Apr	224	90
May	241	139
Jun	257	155
Jul	247	142
Aug	219	116
Sep	182	111
Oct	152	78
Nov	92	22
Dec	60	13
Total	2,011	932

15 September, when ET rates are reduced according to both SEBAL-EF and White methods. Table 6.3 reports the monthly and annual water use estimates of the SEBAL-EF model, averaged over the entire study area (lower CNWR).

The spatially averaged daily ET over CNWR ranged from 0.2 mm in January to 6.8 mm in June. This peak daily ET rate was 15% lower than the rates reported by *Nagler et al.* [2009b] over the same area. Based on their EVI approach, average daily ET rates had reached a maximum of about 8.0 mm in every year during the 2000–2007 period. The annual water use averaged over the entire lower CNWR (978 ha) was 932 mm. This was 46% of ET_0 and 72% of the average ET over the PVID (1286 mm), estimated using the SEBAL model during the same period [*Taghvaeian and Neale*, 2011]. The most dominant crop at PVID is alfalfa, occupying over 60% of the total area under field crops.

The total area of tamarisk monocultures along the Lower Colorado River (below Davis Dam to the U.S.-Mexico border) is about 18,200 ha [*Nagler et al.*, 2008]. Assuming that these regions are similar to the CNWR in terms of water consumption, total volume of monthly water loss by tamarisk for the entire Lower Colorado Basin varies from 2.0 Mm³ in January to 28.2 Mm³ in June. These estimates are 0.3% and 2.3% of the corresponding river flow below the Davis Dam. For a 6 month period (mid-March to mid-September) during peak tamarisk water use, the total water consumption adds up to 130.4 Mm³, which is 1.8% of river flow below the Davis dam during the same period. On an annual basis, tamarisk water use was estimated as 169.6 Mm³, which is 5% smaller than the 178.4 Mm³ estimated by *Nagler et al.* [2008] and 1.4% of annual river flow below the Davis Dam. The total area of riparian ecosystems in the Lower Colorado River Basin is 34,000 ha. If this entire area becomes occupied by tamarisk stands similar to those at CNWR, the total annual riparian water use would be about 316.9 Mm³, which is again 5% smaller than the 333.2 Mm³ estimated by *Nagler et al.* [2008] and only 43% of the 748.1 Mm³ reported in the 2008 LCRAS report [*U.S. Bureau of Reclamation*, 2009b]. This estimate is 2.6% of the total river flow below the Davis Dam in 2008 [*U.S. Bureau of Reclamation*, 2009a] and increases to 3.3% if the analysis is performed over the 6 month period (mid-March to mid-September) during peak tamarisk water use.

6.4. CONCLUSIONS

A strong aquifer-phreatophyte interaction was observed at further distances from the river, as the increase and decrease in tamarisk water use coincided with the falling and rising of the water table, respectively. At closer distances to the river, groundwater fluctuations were influenced by variations in river stage. Converting groundwater depth to groundwater elevation using the high-resolution, LiDAR-derived ground elevation map was effective in removing the variability among the five observation wells at each of the three measuring stations. Analyzing the elevation data in conjunction with river stage information revealed that the Colorado River acted as a source to the riparian aquifer during the study period. The hydraulic gradients were smallest during tamarisk dormancy and largest during the peak water use period.

In applying the SEBAL model over CNWR, it was found that the EF method resulted in a lower error than the ET_0F method in extrapolating instantaneous ET to daily estimates, mainly due to the fact that riparian transpiration is more water limited than energy limited. During the tamarisk growing season, the results of SEBAL-EF approach were very similar to the estimates using the White method, which is solely based on diurnal groundwater fluctuations. During tamarisk dormancy, however, SEBAL-EF results were significantly larger. This appeared to be caused by the presence of shaded soil and vegetation surfaces when the sun elevation angle was low, resulting in the contamination of detected surface temperature. Applying a distributed, normalized vegetation index was found to be effective in adjusting for this error. In comparison with the estimates of a Bowen ratio energy flux tower at one of the measuring stations, the White and modified SEBAL models had satisfactory performances, as evidenced in their small errors. The results of the modified SEBAL model were used to provide water use estimates over the study area and the entire tamarisk monocultures of the Lower Colorado River basin. The results were compared with grass-based reference ET, average ET of an upstream irrigation district, and the river flows below the Davis Dam.

ACKNOWLEDGMENTS

This research was partially funded by the U.S. Bureau of Reclamation under a contract with the Alliance Universities with subcontract to Utah State University. Additional funding was provided by Utah Agricultural Experiment Station projects UTA00793 and UTA01140 and the Remote Sensing Services Laboratory, Department of Civil and Environmental Engineering, Utah State University. The authors would like to express their appreciation to Mr. John Weiss (Blythe Hydrographic Office, River Operations Group, USBR) for kindly providing us with river flow and stage databases and to Dr. Marvin Jensen for his valuable comments on the results.

REFERENCES

R. G., Allen, M. Tasumi, and Trezza R. (2007a), Satellite-based energy balance for mapping evapotranspiration with internalized calibration (METRIC)-model, *ASCE J. Irr. Drain. Eng*, *133*(4), 380–394.

Allen, R. G., M. Tasumi, A. Morse, A. Trezza, J. L. Wright, W. Bastiaanssen, W. Kramber, I. Lorite-Torres, and C. W. Robison, (2007b), Satellite-based energy balance for mapping evapotranspiration with internalized calibration (METRIC)– applications, *ASCE J. Irr. Drain Eng.*, *133*(4), 395–406.

Bastiaanssen, W. G. M., M. Menenti, R. A. Feddes, and A. A. Holtslang (1998a), A remote sensing surface energy balance algorithm for land (SEBAL): 1. Formulation, *J. Hydrol.*, 212–213, 198–212.

Bastiaanssen, W. G. M., H. Pelgrum, J. Wang, Y. Ma, J. F. Moreno, G. J. Roerink, and T. van der Wal, (1998b), A remote sensing surface energy balance algorithm for land (SEBAL): 2. Validation, *J. Hydrol.*, 212–213: 213–229.

Bastiaanssen, W. G. M., E. J. M. Noordman, H. Pelgrum, G. Davids, B. P. Thoreson, and R. G. Allen, (2005), SEBAL model with remotely sensed data to improve water-resources management under actual field conditions, *J. Irr. Drain Eng.*, *131*(1), 85–93.

Bawazir, A. S., Z. Samani, M. Bleiweiss, R. Skaggs, and T. Schmugge (2009), Using ASTER satellite data to calculate riparian evapotranspiration in the Middle Rio Grande, New Mexico, *Int. J. Remote Sens.*, *30*(21), 5593–5603.

Brutsaert, W., and M. Sugita, (1992), Application of self-preservation in the diurnal evolution of the surface energy budget to determine daily evaporation, *J. Geophys. Res.*, *97*, 18,377–18,382.

Cai, B., and C. M. U. Neale (1999), A method for constructing three dimensional models from airborne imagery, paper presented at the 17th Biennial Workshop on Color Photography and Videography in Resource Assessment, Reno, NV, 5–7 May 1999.

Chatterjee, S. (2010), Estimating evapotranspiration using remote sensing: A hybrid approach between MODIS derived enhanced vegetation index, Bowen ratio system, and ground based micro-meteorological data, Ph.D. dissertation, Wright State University, Dayton, Ohio.

Chávez, J. L., C. M. U. Neale, J. H. Prueger, and W. P. Kustas, (2008), Daily evapotranspiration estimates from extrapolating instantaneous airborne remote sensing ET values, *Irrig. Sci. 27*, 67–81.

Cleverly, J. R., S. D., Smith, A., Sala, and D. A. Devitt (1997), Invasive capacity of *Tamarix ramosissima* in a Mojave Desert floodplain: The role of drought, *Oecologia*, *111*, 12–18.

Colaizzi, P. D., S. R. Evett, T. A. Howell, and J. A. Tolk, (2006), Comparison of five models to scale daily evapotranspiration from one-time-of day measurements, *Trans. ASABE*, *49*(5), 1409–1417.

Crago, R. D. (1996), Conservation and variability of the evaporative fraction during the daytime, *J. Hydrol.*, *180*(1–4), 173–194.

Dennison, P. E., P. L., Nagler, K. R. Hultine, E. P. Glenn, and J. R. Ehleringer (2009), Remote monitoring of tamarisk defoliation and evapotranspiration following saltcedar leaf beetle attack, *Remote Sens. Environ.*, *113*, 1462–1472.

Devitt, D. A., A. Sala, K. A. Mace, and S. D. Smith (1997), The effect of applied water on the water use of saltcedar in a desert riparian environment, *J. Hydrol.*, *192*, 233–246.

Geli, H. M. E., C. M. U. Neale, D. Watts, J. Osterberg, H. A. R. De Bruin, W. Kohsiek, R. T. Pack, and L. E. Hipps (2012), Scintillometer-based estimates of sensible heat flux using lidar-derived surface roughness, *J. Hydrometeor.*, *13*, 1317–1331.

Glenn, E. P., R. Tanner, S. Mendez, T. Kehret, D. Moore, J. Garcia, and C. Valdes (1998), Growth rates, salt tolerance and water use characteristics of native and invasive riparian plants from the delta of the Colorado River, Mexico, *J. Arid Environ.*, *40*, 281–294.

Glenn, E. P., K. Morino, P. L. Nagler, R. S. Murray, S. Pearlstein, and K. R. Hultine (2012), Roles of saltcedar (*Tamarix* spp.) and capillary rise in salinizing a non-flooding terrace on a flow-regulated desert river, *J. Arid Environ.*, *79*, 56–65.

Glenn, E. P., P. L. Nagler, K. Morino, and K. R. Hultine (2013), Phreatophytes under stress: Transpiration and stomatal conductance of saltcedar (*Tamarix* spp.) in a high-salinity environment, *Plant Soil*, doi:10.1007/s11104-013-1803-0.

Godaire, J. E., and R. E. Klinger, (2007), Tree-ring dating of Tamarisk (*Tamarix ramosissima*), Cibola Nat. Wildlife Refuge, Calif., U.S. Bur. of Reclamation, Denver, CO, USA.

Gowda, P. H., J. L. Chávez, P. D. Colaizzi, S. R. Evett, T. A. Howell, and J. A. Tolk (2008), ET mapping for agric water manage: Present status and challenges, *Irrig. Sci.*, *26*, 223–237.

Gribovszki, Z., J. Szilágyi, and P. Kalicz (2010), Diurnal fluctuations in shallow groundwater levels and streamflow rates and their interpretation — A review, *J Hydrol.*, *385*, 371–383.

Groeneveld, D. P., W. M. Baugh, J. S. Sanderson, and D. J. Cooper (2007), Annual groundwater evapotranspiration mapped from single satellite scenes, *J. Hydrol.*, *344*, 146–156.

Horton, J. L., T. E. Kolb, and S. C. Hart (2001), Responses of riparian trees to interannual variation in ground water depth in a semi-arid river basin, *Plant Cell Environ.*, *24*, 293–304.

Huete, A. R. (1988), A soil-adjusted vegetation index (SAVI), *Remote Sens. Environ.*, *25*, 295–309.

Huete, A. R., K. Didan, T. Miura, E. Rodriquez, X. Gao, and L. Ferreira (2002), Overview of the radiometric and biophysical performance of the MODIS vegetation indices, *Remote Sens. Environ.*, *83*, 195–213.

Hultine, K. R., J. Belnap, C. van Riper III , J. R. Ehleringer, P. E. Dennison, M. E. Lee, P. L. Nagler, K. A. Snyder, S. E. Uselman, and J. B. West (2009), Tamarisk biocontrol in the western United States: Ecological and societal implications, *Front. Ecol. Environ.*, *8*(9), 467–474.

Hultine, K. R., P. L. Nagler, K. Morino, S. E. Bush, K. G. Burtch, P. E. Dennison, E. P. Glenn, and J. R. Ehleringer (2010), Sap flux-scaled transpiration by tamarisk (*Tamarix* spp.) before, during and after episodic defoliation by the saltcedar leaf beetle (*Diorhabda carinulata*), *Agric. Forest Meteorol. 150*, 1467–1475.

Kamble, B., A. Irmak, D. L. Martin, K. G. Hubbard, I. Ratcliffe, G. Hergert, S. Narumalani, and R. J. Oglesby (2013), Satellite-based energy balance approach to assess riparian water use, in *Evapotranspiration — An Overview*, edited by S. G. Alexandris, InTech, doi:10.5772/52929.

Loheide II, S. P., J. J. Butler, Jr., and S. M. Gorelick (2005), Use of diurnal water table fluctuations to estimate groundwater consumption by phreatophytes: A saturated-unsaturated flow assessment, *Water Resourc. Res.*, *41*, W07030.

Long, D., V. P. Singh, and Z.-L. Li (2011), How sensitive is SEBAL to changes in input variables, domain size and satellite sensor? *J. Geophys. Res.*, *116*(D21), D21107.

Meyboom, P. (1966), Groundwater studies in the Assiniboine River drainage basin—Part I: Evaluation of a flow system in south-central Saskatchewan, *Bull. Geol. Surv. Can.*, *139*, 65p.

Murray, R. S., P. L. Nagler, K. Morino, and E. P. Glenn (2009), An empirical algorithm for estimating agricultural and riparian evapotranspiration using MODIS enhanced vegetation index and ground measurements of ET. II. Application to the Lower Colorado River, *U.S. Remote Sens.*, *1*, 1125–1138.

Nagler, P. L., R. L. Scott, C. Westenberg, J. R. Cleverly, E. P. Glenn, and A. R. Huete (2005), Evapotranspiration on western U.S. rivers estimated using enhanced vegetation index from MODIS and data from eddy covariance and Bowen ratio flux towers, *Remote Sens. Environ.*, *97*, 337–351.

Nagler, P. L., E. P. Glenn, K. Didan, J. Osterberg, F. Jordan, and J. Cunningham (2008), Wide-area estimates of stand structure and water use of *Tamarix* spp. on the Lower Colorado River: Implications for restoration and water management projects, *Restor. Ecol.*, *16*(1), 136–145.

Nagler, P. L., K. Morino, R. S. Murray, J. Osterberg, and E. P. Glenn (2009a), An empirical algorithm for estimating agricultural and riparian evapotranspiration using MODIS enhanced vegetation index and ground measurements of ET. I. Description of method, *Remote Sens.*, *1*, 1273–1297.

Nagler, P. L., K. Morino, K. Didan, J. Erker, J. Osterberg, K. R. Hultine, and E. P. Glenn (2009b), Wide-area estimates of saltcedar (*Tamarix* spp.) evapotranspiration on the lower Colorado River measured by heat balance and remote sensing methods, *Ecohydrology*, *2*, 18–33.

Neale, C. M., and B. G. Crowther (1994), An Airborne Multi-Spectral Video/Radiometer Remote Sensing System: Development and Calibration, *Remote Sens. Environ.*, *49*, 187–94.

Neale, C. M. U., et al. (2011), Evapotranspiration water use analysis of saltcedar and other vegetation in the Mojave River Floodplain, 2007 and 2010, Mojave Water Agency Water Supply Management Study, Phase 1 Rep., U. S. Bur. of Reclamation, August 2011, available at http://www.usbr.gov/lc/socal/reports/MWAStudy/Phase1.pdf.

Norman, J. M., W. P. Kustas, and K. S. Humes (1995), A two-source approach for estimating soil and vegetation energy fluxes from observations of directional radiometric surface temperature, *Agric. Meteorol.*, *77*, 263–293.

Owens, M. K., and G. W. Moore (2007), Saltcedar water use: Realistic and unrealistic expectations, *Rangeland Ecol. Manag.*, *60*(5), 553–557.

Payero, J. O., C. M. U. Neale, J. L. Wright, and R. G. Allen (2003), Guidelines for validating Bowen ratio data, Trans. *ASAE*, *46*(4), 1051–1060.

Prueger, J. H., and W. P. Kustas (2005), Aerodynamic methods for estimating turbulent fluxes. in *Micrometeorology in Agricultural Systems*, Agronomy Monograph No. 47, edited by J. L. Hatfield, and J. M. Baker, American Society of Agronomy, Crop Science Society of America, Soil Science Society of America, Madison, Wisconsin.

Ramos, J. G., C. R. Cratchley, J. A. Kay, M. A. Casterad, A. Martinez-Cob, and R. Dominguez (2009), Evaluation of satellite evapotranspiration estimates using ground-meteorological data available for the Flumen District into the Ebro Valley of N.E. Spain, *Agric. Water Manag.*, *96*, 638–652.

Rasmussen, T. C., and T. L. Mote (2007), Monitoring surface and subsurface water storage using confined aquifer water levels at the Savannah River Site, *USA, Vadose Zone J.*, *6*, 327–335.

Scott, R. L., W. L. Cable, T. E. Huxman, P. L. Nagler, M. Hernandez, and D. C. Goodrich (2008), Multiyear riparian evapotranspiration and groundwater use for a semiarid watershed, *J. Arid Environ.*, *72*, 1232–1246.

Taghvaeian, S., and C. M. U. Neale (2011), Water balance of irrigated areas: A remote sensing approach, *Hydrol. Process.*, *25*, 4132–4141, doi:10.1002/hyp.8371.

Tasumi, M., R. Trezza, R. G. Allen, and J. L. Wright (2005), Operational aspects of satellite-based energy balance models for irrigated crops in the semi-arid U.S., *Irrig. Drain. Syst.*, *19*, 355–376.

Trezza, R. (2002), Evapotranspiration using a satellite-based surface energy balance with standardized ground control, Ph.D. dissertation, Utah State University. Logan, Utah.

U.S. Bureau of Reclamation (2009a), Colorado River accounting and water use report; Arizona, California, and Nevada; Calendar year 2008, U.S. Dep. of the Interior, Bur. of Reclamation, Lower Colorado Region, Boulder Canyon Operations Office, Water Conservation & Accounting Group, available at http://www.usbr.gov/lc/region/g4000/4200Rpts/DecreeRpt/2008/2008.pdf.

U.S. Bureau of Reclamation (2009b), Lower Colorado River accounting system, evapotranspiration and evaporation calculations; Calendar year 2008, U.S. Dep. of the Interior, Bur. of Reclamation, Lower Colorado Regional Office, Boulder City, Nev., available at http://www.usbr.gov/lc/region/g4000/4200Rpts/LCRASRpt/2008/2008LCRAS.pdf.

Vandersande, M. W., E. P. Glenn, and J. L. Walworth (2001), Tolerance of five riparian plants from the lower Colorado

River to salinity drought and inundation, *J. Arid Environ.*, *49*, 147–159.

Van Hylckama, T. E. A. (1970), Water use by salt cedar, *Water Resourc. Res. 6*(3), 728–735.

Wang, J., T. W. Sammis, V. P. Gutschick, M. Gebremichael, and D. R. Miller (2009), Sensitivity analysis of the surface energy balance algorithm for land (SEBAL), *Trans. ASABE, 52*(3), 801–811.

Westenburg, C. L., D. P. Harper, and G. A. DeMeo (2006), Evapotranspiration by phreatophytes along the lower Colorado River at Havasu National Wildlife Refuge, Arizona: U.S. Geol. Surv. Sci. Investig. Rep. 2006–5043, available at http://pubs.water.usgs.gov/sir20065043.

White, W. N. (1932), A method of estimating ground-water supplies based on discharge by plants and evaporation from soil: Results of investigations in Escalante Valley, Utah, U.S. Geol. Surv. Water Supply Pap. 659-A.

Zavaleta, E. (2000), The economic value of controlling an invasive shrub, *AMBIO, 29*(8), 462–467.

Zhang, L., and R. Lemeur (1995), Evaluation of daily ET estimates from instantaneous measurements, *Agric. Meteorol., 74*, 139–154.

Zhu, J., M. Young, J. Healey, R. Jasoni, and J. Osterberg (2011), Interference of river level changes on riparian zone evapotranspiration estimates from diurnal groundwater level fluctuations, *J. Hydrol., 403*(3–4), 381–389.

Section III: Surface Water

7

Controls of Terrestrial Water Storage Changes Over the Central Congo Basin Determined by Integrating PALSAR ScanSAR, Envisat Altimetry, and GRACE Data

Hyongki Lee,[1,2] Hahn Chul Jung,[3] Ting Yuan,[1,2] R. Edward Beighley,[4] and Jianbin Duan[5]

7.1. INTRODUCTION

Interseasonal and interannual variations in surface water storage volume, a key term in terrestrial water balance at any scale, are poorly known. Therefore, their impact on precipitation, evaporation, infiltration, and runoff cannot be well known [*Alsdorf et al.*, 2007]. Fundamentally, the land surface water budget represented by hydrologic models is in great error due to the lack of consistent hydrological observations with uniform spatial density globally. Consequently, the amount of water stored on wetlands and floodplains and subsequently exchanged with their main channels is largely unknown as well. This opens a science question related to the dynamics of the global water cycle: "What is the spatial and temporal variability in the world's terrestrial surface water storage?" [*Durand et al.*, 2010].

Various satellite geodetic instruments have been used with their own advantages and disadvantages to quantify terrestrial water dynamics. For example, the satellite radar altimeter, which was originally developed for ocean circulation studies, has also been successfully used to

---

[1]*Department of Civil and Environmental Engineering, University of Houston, Houston, Texas, USA*

[2]*National Center for Airborne Laser Mapping, University of Houston, Houston, Texas, USA*

[3]*Hydrological Sciences Laboratory, NASA Goddard Space Flight Center, Greenbelt, Maryland, USA*

[4]*Department of Civil and Environmental Engineering, Northeastern University, Boston, Massachusetts, USA*

[5]*Divion of Geodetic Science, School of Earth Sciences, Ohio State University, Columbus, Ohio, USA*

observe water height changes over rivers, lakes, and wetlands [e.g., *Birkett et al.*, 2002; *Crétaux and Birkett*, 2006; *Calmant et al.*, 2008; *Lee et al.*, 2009, 2011]. The radar altimeter is currently the only instrument that can provide regularly repeated observation of water heights from space. However, it is a nadir-looking one-dimensional profiling instrument, and thus the satellite ground track must intersect with the water bodies. Interferometric synthetic aperture radar (InSAR) has been a unique technique to map two-dimensional water height changes beneath flooded forests between SAR acquisition dates with high spatial resolution (~40 m) [e.g., *Alsdorf et al.*, 2000; *Wdowinski et al.*, 2008; *Kim et al.*, 2009; *Jung et al.*, 2010]. InSAR can provide only spatially relative water height changes, and thus it requires a vertical reference to retrieve absolute water height changes [*Kim et al.*, 2009]. The method's apparent limitation would be that it requires a tree trunk that can play a role as a corner reflector, enabling us to achieve the so-called double-bounce backscattering. The Gravity Recovery And Climate Experiment (GRACE) mission, launched in 2002 with objectives including mapping of the global gravity field and its temporal variation, has also been extensively used in many areas of hydrologic studies [e.g., *Swenson and Wahr*, 2006; *Rodell et al.*, 2007; *Syed et al.*, 2009; *Chen et al.*, 2009; *Alsdorf et al.*, 2010; *Lee et al.*, 2011]. GRACE has been successfully used to reveal terrestrial water storage (TWS) changes, in terms of total changes in water stored in surface, soil, and groundwater reservoirs, over large river basins. However, GRACE also has its own limitations, including its coarse spatial resolution (> ~300 km half-wavelength) and its inability to separate a vertical structure of a TWS change into individual components. In addition, the growing availability of

---

Remote Sensing of the Terrestrial Water Cycle, Geophysical Monograph 206. First Edition. Edited by Venkat Lakshmi.
© 2015 American Geophysical Union. Published 2015 by John Wiley & Sons, Inc.

satellite data has also increased the opportunities to study wetland flooded area classification. The Congo and Amazon wetland mappings were carried out during both high and low water seasons using mosaics of SAR images acquired by the Japanese Earth Resource Satellite-1 (JERS-1) as part of the Global Rain Forest Mapping (GRFM) project [*De Grandi et al.*, 2000; *Hess et al.*, 2003]. Wetland probability mapping in the Congo Basin was also developed using optical and radar remotely sensed data [*Bwangoy et al.*, 2010]. In addition, *Betbeder et al.* [2013] generated detailed spatial distribution of forested wetland types in the Cuvette Centrale of the Congo Basin using Moderate Resolution Imaging Spectroradiometer (MODIS) and advanced land observing satellite (ALOS) phased array type L-band synthetic aperture radar (PALSAR) imagery. But these maps are limited to little temporal resolution as to provide seasonal variations in inundation extent. Although global satellite-derived inundation maps were generated on a monthly basis for 1993–2000 [*Prigent et al.*, 2007], the spatial resolution of 773 km^2 (i.e., equal area grid of 0.25° × 0.25° at the equator) is not fine enough to study the regional scale of wetland hydrology and hydrodynamics.

In this study, we intend to integrate different types of satellite measurements over the central Congo Basin to determine the controls of its TWS changes, which cannot be revealed if applied independently. The Congo Basin is the world's third largest in size (~3.7 million km^2), second only to the Amazon River in discharge (~40,200 m^3/s annual average). The Congo River is the only major river that crosses the equator twice and receives year-round rainfall from the migration of the Inter Tropical Convergence Zone (ITCZ), which leads to a bimodal discharge pattern [*Campbell*, 2005]. Despite its enormous size, the Congo Basin has not experienced the same degree of new research compared to the Amazon because the lack of in situ data has limited our hydrologic understanding of the basin. A few studies have only recently investigated the Congo's terrestrial water dynamics using remote sensing measurements and modeling [*Jung et al.*, 2010; *Beighley et al.*, 2011; *Lee et al.*, 2011; *O'Loughlin et al.*, 2013]. In this study, we investigate the controls of TWS changes by tracking

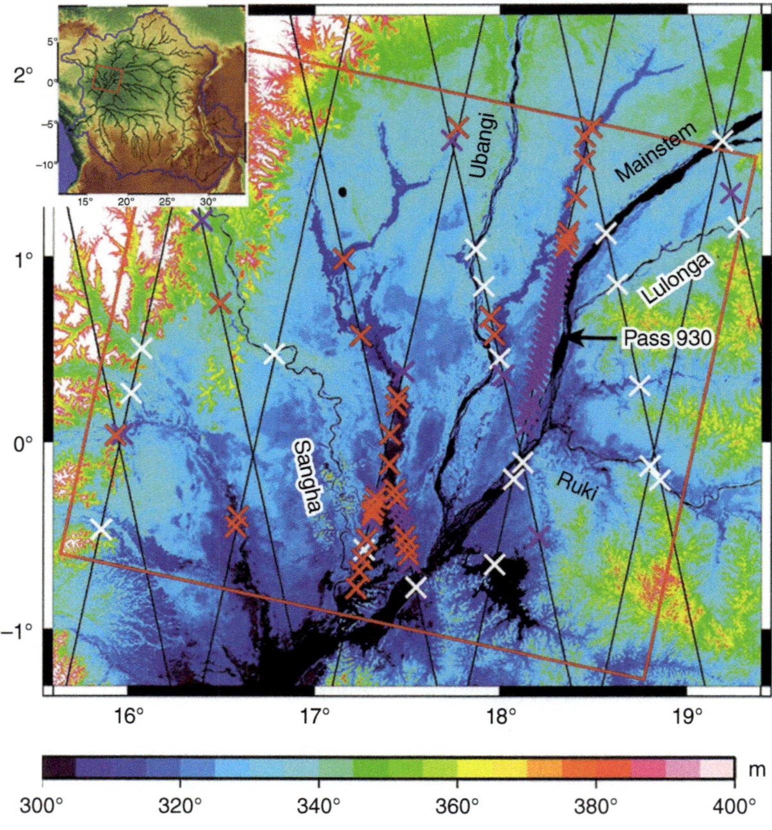

Figure 7.1 Location of the central Congo Basin with the ScanSAR coverage (red box). Background shows topography from the Shuttle Radar Topography Mission (SRTM) C-band Digital Elevation Model (DEM). Intersections between Envisat altimeter and the nonvegetated, herbaceous, and forest surfaces are indicated with white, red, and purple 'x's, respectively.

surface (channels and wetlands) and subsurface (soil moisture and groundwater) components of the water balance over the central Congo where most of the low-lying interfluvial wetlands are located around the confluence of the Congo mainstem and its major tributaries including the Ubangi (~3800 m^3/s annual discharge [*Laraque et al.*, 2001], which is the largest right-bank tributary of the Congo mainstem), the Sangha (~1600 m^3/s annual discharge [*Laraque et al.*, 2001]), and the Lulonga and Ruki rivers (Figure 7.1). We use total TWS changes from GRACE, nonvegetated/herbaceous/forest surface water height changes from Envisat radar altimetry, and inundation extents from ALOS PALSAR ScanSAR data with higher spatial and temporal resolutions in order to quantify the following: (1) surface (nonvegetated, herbaceous, and forest) storage change from the combination of water height changes and inundated areas and (2) subsurface storage change as the difference between GRACE TWS changes and surface storage changes, over our study region (red box in Figure 7.1).

7.2. DATA SETS

7.2.1. Envisat Radar Altimetry

In this study, we use Envisat Geophysical Data Record (GDR), which contains 18 Hz (~350 m along-track sampling) retracked measurements from OCEAN, ICE-1, ICE-2, and SEA ICE retrackers from the periods of October 2002 to September 2010 with a 35 day repeat period. We use ICE-1 [*Wingham et al.*, 1986] retracked measurements, which have been demonstrated to be a suitable retracker for inland water bodies [*Frappart et al.*, 2006a; *Lee et al.*, 2010]. The instrument corrections, media corrections (dry and wet troposphere corrections from the European Centre for Medium-Range Weather Forecasts model, and the ionosphere correction based on global ionosphere maps), and geophysical corrections (solid Earth and pole tides) are applied. In addition, the 5.6 m level ultra stable oscillator anomalies for cycles 44–85 are corrected using the European Space Agency's correction table.

7.2.2. PALSAR ScanSAR

PALSAR (L-band), an enhanced version of JERS-1 SAR (L-band, HH polarization), was jointly developed by Japan Aerospace Exploration Agency (JAXA) and the Japan Resources Observation Systems Organization (JAROS) [*Rosenqvist et al.*, 2004]. PALSAR is carried by ALOS with a 46 day repeat cycle. PALSAR is a fully polarimetric instrument, operating in fine-beam mode with

Table 7.1 List of ALOS PALSAR ScanSAR scenes used in this study

Date	Path	Polarization
5 December 2006	274	HH
23 July 2007	274	HH
7 September 2007	274	HH
8 December 2007	274	HH
23 January 2008	274	HH
9 March 2008	274	HH
25 October 2008	274	HH
10 December 2008	274	HH
25 January 2009	274	HH
12 March 2009	274	HH
27 April 2009	274	HH
13 December 2009	274	HH
15 March 2010	274	HH
15 June 2010	274	HH
16 December 2010	274	HH

single polarization (HH or VV), dual polarization (HH + HV or VV + VH), or full polarimetry (HH + HV + VH + VV). It also features wide-swath ScanSAR mode, with single polarization (HH or VV). The ScanSAR mode data is usually comprised of 5 subswaths. The ScanSAR beam mode has 350 km swath width and 100 m ground resolution [*Shimada*, 2008]. ScanSAR has been used for intensive monitoring of seasonal inundation dynamics in major wetlands and river basins [*Rosenqvist*, 2008]. In this study, we have collected 15 scenes of PALSAR ScanSAR mode data set (L1.5) to map the spatial extent and temporal dynamics of flooding in the central Congo (see Table 7.1). The products in level L1.5 provided by the Alaska SAR Facility (ASF) are processed by the JAXA PALSAR processor in ground range geometry and radiometric calibration. Speckle noise was removed by a 3 × 3 median filter.

7.2.3. TWS Anomalies from GRACE

This study uses Release 5 (RL05) GRACE data processed by Center for Space Research (CSR), Jet Propulsion Laboratory (JPL), and GeoForschungs Zentrum (GFZ) for the period of November 2006–June 2010 (44 months). To reduce the GRACE south–north striping [*Swenson & Wahr*, 2006], we use a decorrelation method by *Duan et al.* [2009], and apply a 3° Gaussian filter [*Guo et al.*, 2010]. From each of the monthly gravity fields, the surface mass changes in equivalent water height (EWH) changes are computed as in *Wahr et al.* [1998] at 1° × 1° grid spacings and spatially averaged over the study region. Finally, total TWS changes are obtained by multiplying the EWH changes by the study area (350 km × 350 km).

7.3. RESULTS

7.3.1. Water Height Anomalies from Envisat Altimetry

To identify "water-covered" surface, backscattering coefficients (σ_0) from radar altimetry (ALT σ_0 hereinafter to distinguish it from SAR σ_0 in Section 7.3.2.) have been used [*Birkett*, 1998; *Lee et al.*, 2009]. Typically, a simple threshold (typically ~20 dB) has been identified to determine whether the radar pulse returns from the water surface, based on the fact that stronger return is obtained from the water-covered surface. However, we find that the water surface beneath the flooded forest may exhibit lower ALT σ_0 (<20 dB; e.g., red oval in Figure 7.2). It is also noted that mean and standard deviation of the TOPEX Ku-band σ_0 are 12.28 and 0.74 dB, respectively [*Papa et al.*, 2003], although these numbers represent basin-scale values. In this study, we examined individual height profiles along Envisat passes and established a total of 91 so-called virtual stations over nonvegetated (18), herbaceous (35), and forest (38) (Figure 7.1), following the land cover map in *Hansen et al.* [2008]. We generated time series of surface elevation changes and ALT σ_0 changes (from ICE-1 retracker) over each virtual station

and computed correlation coefficients (CCs) between them. Most of the forest virtual stations are obtained along Envisat Pass 930, which flies along the forest located between the mainstem and the Ubangi River (Figure 7.1), by spatially averaging ~15 high-rate (18 Hz) measurements (~5 km interval) along the tracks. This interval enables us to generate smooth enough time series, and at the same time to obtain enough number of forest virtual stations. As can be seen from examples of time series in Figure 7.3, nonvegetated (i.e., open water) and herbaceous surfaces exhibit larger water height variation and stronger ALT σ_0 (<20 dB) than forest surface. More interestingly, we see that the time series of elevation and ALT σ_0 changes over herbaceous and forest surfaces reveal stronger correlations between them. This is summarized in Figure 7.4 showing that most of the virtual stations in herbaceous and forest surfaces have CCs higher than 0.6, while the virtual stations in nonvegetated surface have CCs lower than 0.6. This indicates that ALT σ_0 is more sensitive to the water height changes over vegetated wetland that is seasonally flooded and becomes less sensitive to the water height changes over the open river channel, which is permanently flooded. Finally, we discarded 1 virtual station and 2 virtual stations in herbaceous and

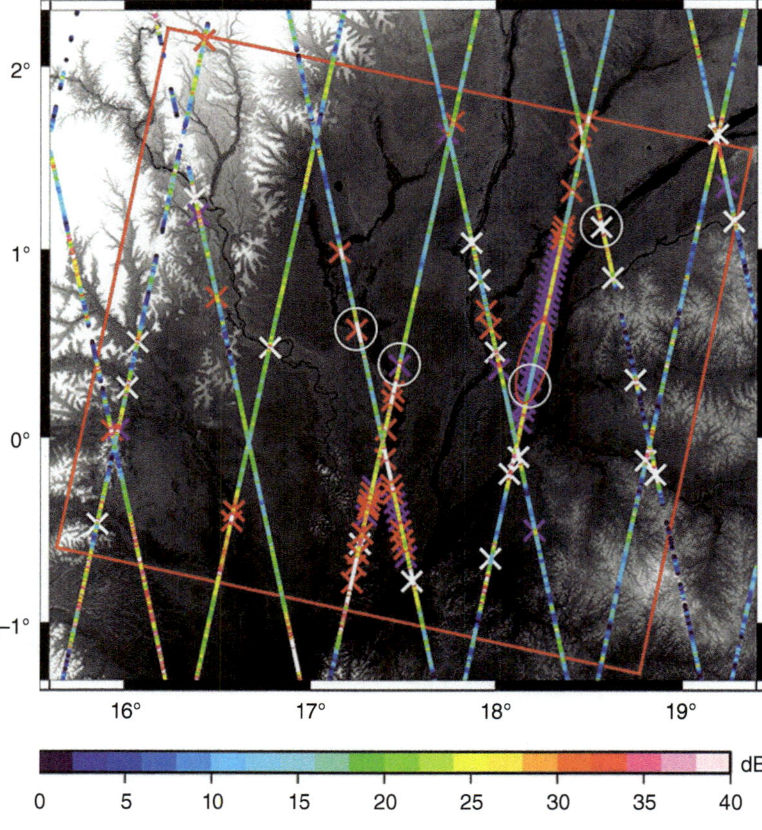

Figure 7.2 Envisat 18-Hz σ_0 (dB) from the ICE-1 retracker over the study area from cycle 12 and 43 obtained in December (high-water season). Inundated forest areas with relatively lower σ_0 are shown in a red oval. The white circles represent the locations of the time series in Figure 7.3.

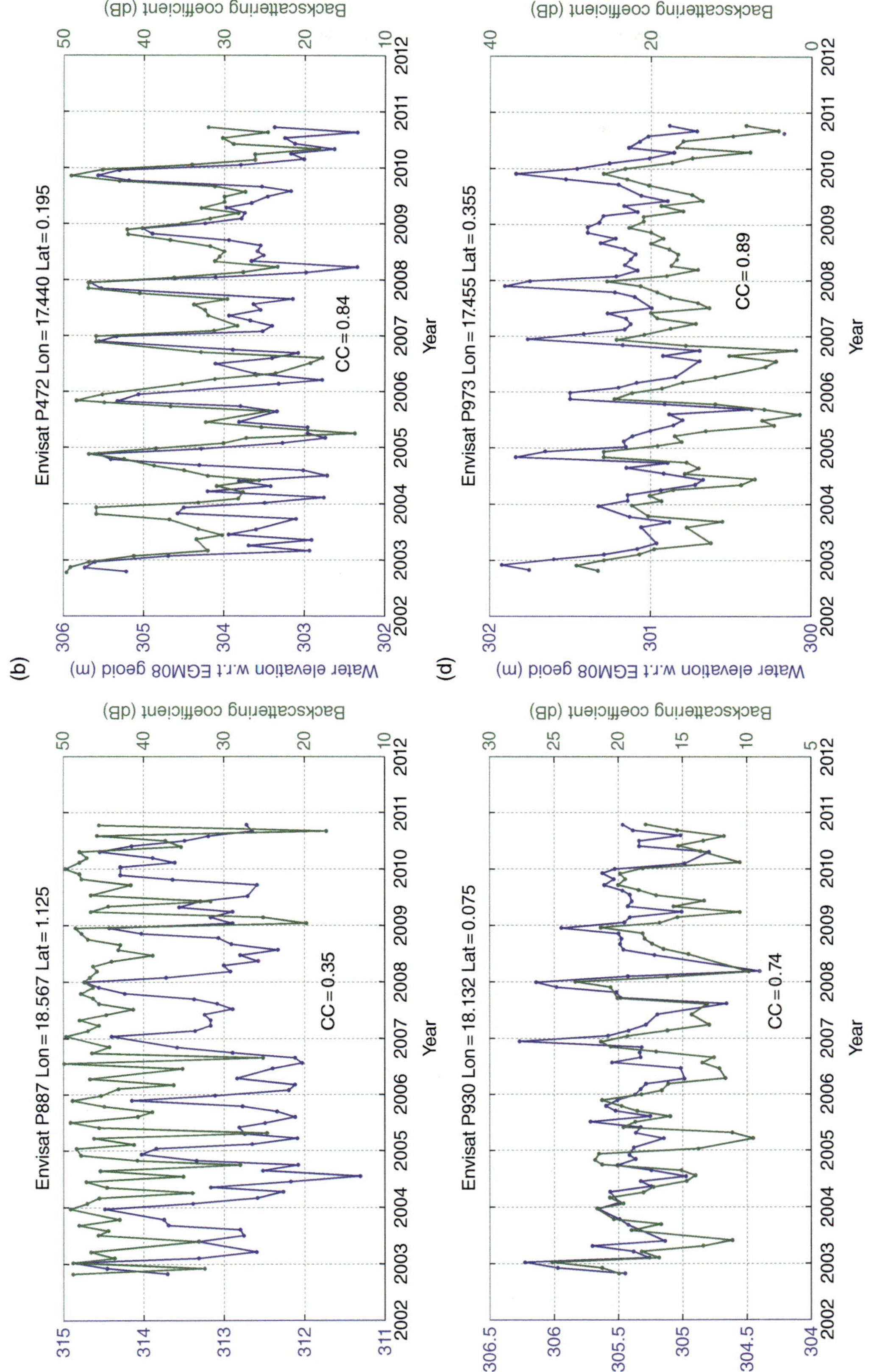

Figure 7.3 Examples of time series of water level and σ_0 changes from Envisat altimeter over (a) nonvegetated, (b) herbaceous, and (c)–(d) forest surfaces. We observe high correlation between the water level and σ_0 changes over the herbaceous and forest surfaces.

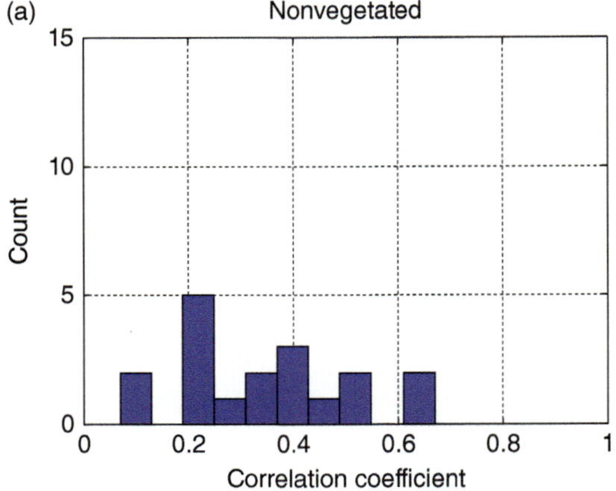

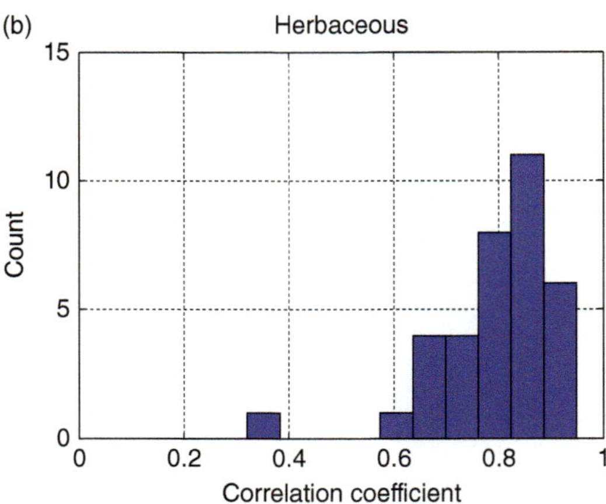

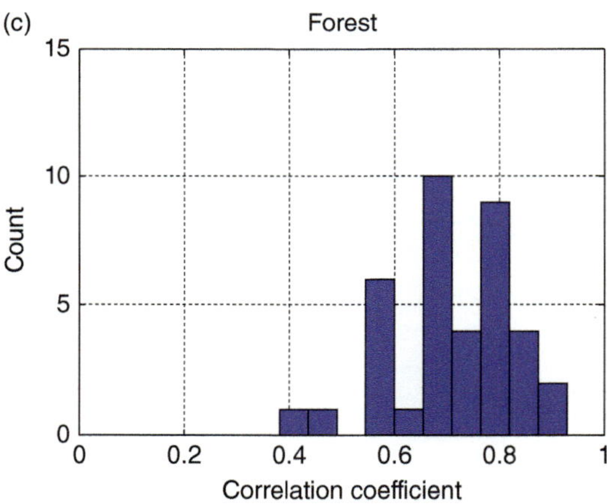

Figure 7.4 Histograms of correlation coefficients between the time series of water level and σ_0 changes over the virtual stations in (a) nonvegetated, (b) herbaceous, and (c) forest surfaces.

forest areas, respectively, which have CCs lower than 0.6, assuming that these virtual stations may not be located over water-covered surfaces, and accepted the others as virtual stations in seasonally or permanently flooded wetlands. We use all of the virtual stations over the nonvegetated surface regardless of their CCs. Those 36 virtual stations in forest will also be used as an indicator to classify flooded forest areas from the ScanSAR images in Section 7.3.2.

7.3.2. Inundated Areas from PALSAR ScanSAR

Because radar pulse interactions with flooded forest typically follow a double-bounce travel path that returns energy to the antenna, flooded forest in SAR imagery shows greater amplitude (i.e., higher SAR σ_0) relative to nonflooded forest [*Jung & Alsdorf*, 2010]. This indicates that SAR σ_0 can be used to distinguish between flooded and nonflooded forests. Figure 7.5 illustrates this spatiotemporal variation of SAR σ_0 as an example. While we observe highest SAR σ_0 in December, especially over the interfluvial wetlands around the Congo mainstem, lowest SAR σ_0 is observed in March among these. In addition, flooded nonvegetated (i.e., open water) and flooded herbaceous (e.g., savannah) areas show very low amplitude due to specular backscattering. First of all, we mask out flooded nonvegetated surface using MODIS-derived land cover maps [*Hansen et al.*, 2008]. Next, we classify areas with less than $-14\,$dB SAR σ_0 as flooded herbaceous as verified in the Amazon [*Hess et al.*, 2003]. To classify flooded and nonflooded forests with lacking ground truth data, a simple threshold method using SAR σ_0 is implemented in this study. In central Amazon, a threshold of $-6.5\,$dB was used to separate flooded from nonflooded forests in the GRFM project mosaics generated from L-band JERS-1 SAR, which is a predecessor of ALOS PALSAR [*Hess et al.*, 2003]. In a different study by *Rosenqvist* [2008], $-4.6\,$dB was used as a threshold to classify flooded forest in central Congo using a series of PALSAR ScanSAR images (Frame 271). It turns out that these two different thresholds lead to significantly different estimates of flooded forest areas as can be seen from the top and bottom panels of Figure 7.6. For example, using the ScanSAR image obtained on 8 December 2007 (high-water season), threshold of $-4.6\,$dB leads us to obtain flooded forest mostly between the mainstem and the Lulonga River. However, if we use $-6.5\,$dB, most of the interfluvial wetlands in the study area become flooded. Moreover, we also observe notable differences in the flooded area using the ScanSAR image obtained on 9 March 2008 (low-water season).

In this study, we attempt to find a site-specific threshold by utilizing the Envisat virtual stations set up in the forest. We first extract SAR σ_0 from five ScanSAR images obtained in December (5 December 2006, 8 December 2007, 10 December 2008, 9 December 2009, 16 December 2010) at the virtual station locations based on the fact that all of the virtual stations become flooded during high-water season. We also extract SAR σ_0 from the five ScanSAR images over nonflooded (*terra firma*) forest along the Envisat ground tracks. From the normalized histogram showing their distributions in Figure 7.7, we clearly observe a distinction between flooded (mean: $-4.8\,$dB) and

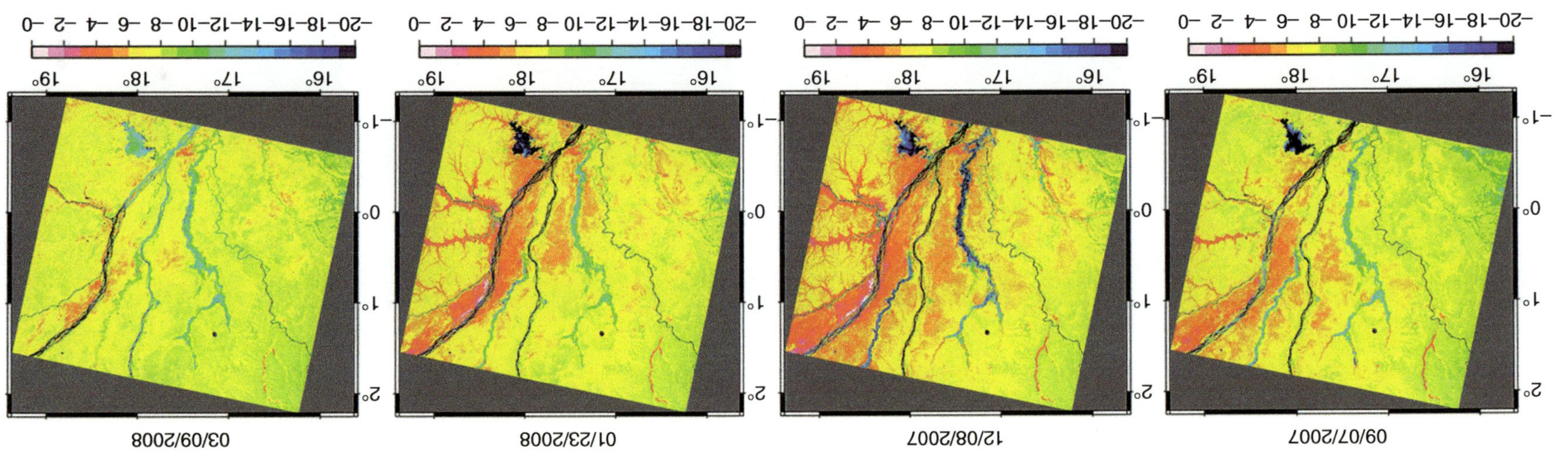

Figure 7.5 Examples of maps of PALSAR ScanSAR σ_0 showing its seasonal variation, especially over the interfluvial wetlands.

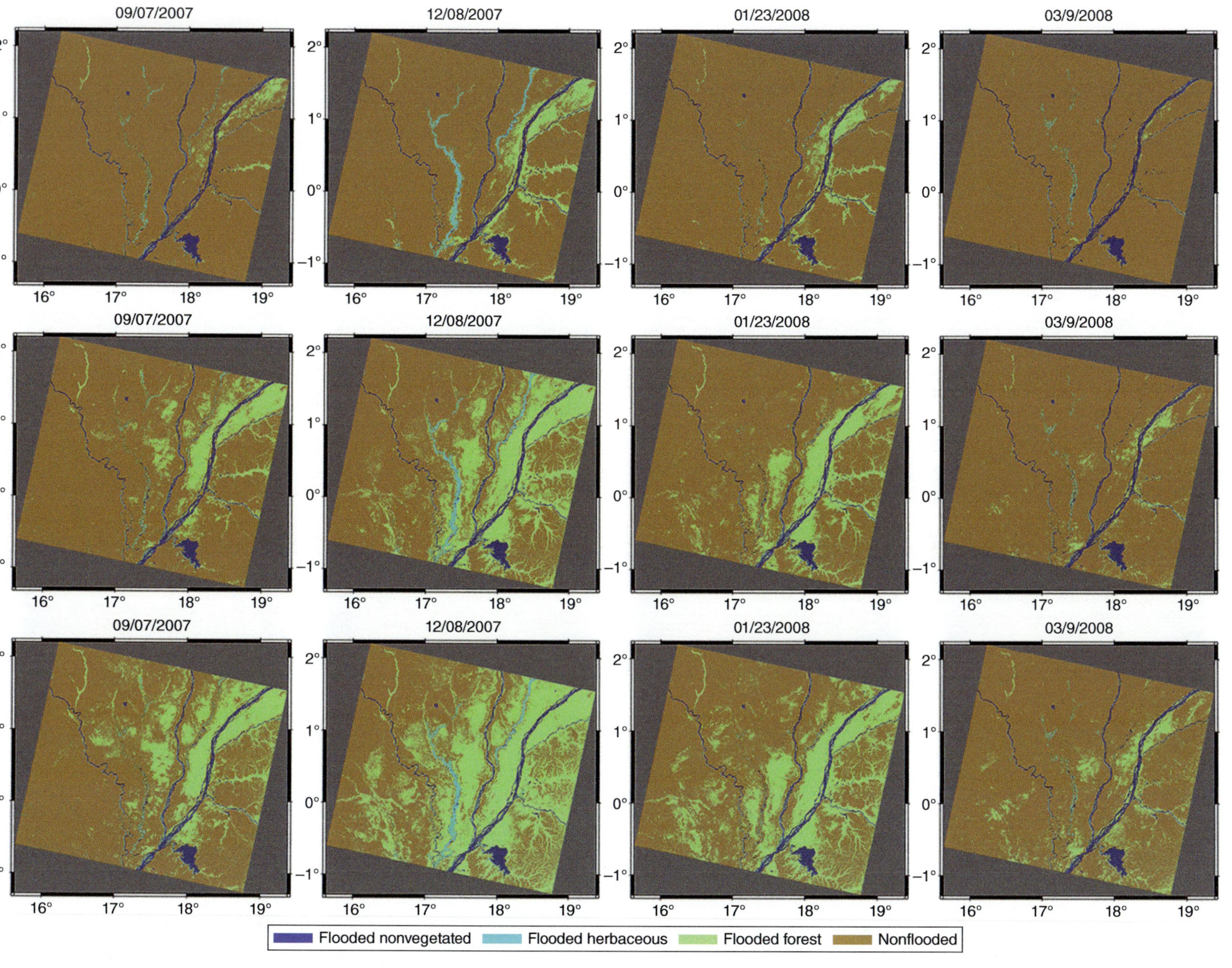

Figure 7.6 Examples of classification maps of inundated extent over nonvegetated, herbaceous, and forest surfaces using −4.6 dB (*top*), −6.0 dB (*middle*), and −6.5 dB (*bottom*) as a threshold for the flooded forest classification.

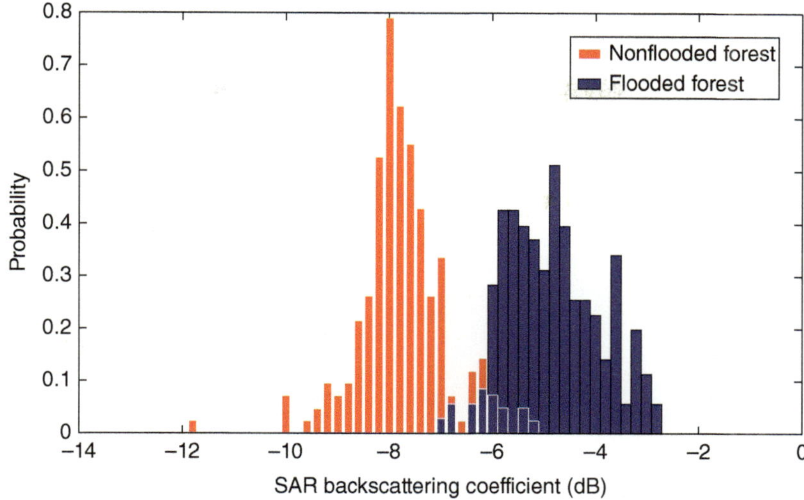

Figure 7.7 Normalized histogram of SAR σ_0 over flooded (blue) and nonflooded (red) forests. The distribution of SAR σ_0 for nonflooded forest higher than $-6.0\,$dB is shown with bars with white boundaries.

nonflooded (mean: $-7.6\,$dB) forests. We also observe an overlapped distribution range between $-7.0\,$dB (minimum σ_0 of flooded) and $-5.1\,$dB (maximum σ_0 of nonflooded). However, it demonstrates that the flooded forest shows higher probability than the nonflooded forest above $-6.0\,$dB, which is a threshold implemented in this study. The flood classification maps generated using $-6.0\,$dB threshold are shown in the middle panel of Figure 7.6. It is noted that we have no available contemporaneous in situ data to validate it, but we attempt to use this statistically reasonable estimate based on another independent radar instrument to estimate the flooded forest areas.

7.3.3. TWS Anomalies over the Central Congo

There have been several studies to estimate surface water storage changes by combining water level time series from radar altimetry and in situ gauges and inundation extent maps generated from radar or optical imaging sensors over the Rio Negro, Mekong, and Lower Ob' river basins [*Frappart et al.*, 2005, 2006b, 2008, 2010, 2011]. Maps of spatially interpolated surface water levels are constructed using averages of the water levels for each altimetry or in situ stations [*Frappart et al.*, 2008]. The inundated area maps determine the temporal and spatial resolution of the water level maps.

Considering the limited number of virtual stations in cross-track direction of radar altimetry and number of ScanSAR images used for inundated extent maps, we temporally interpolate Envisat water levels at ScanSAR dates and simply average them over nonvegetated, herbaceous, and forest surfaces to generate their water level change time series over the central Congo. As can be seen from Figure 7.8a, among the three land cover types, the

nonvegetated areas have the largest water level changes of about 2 m, and the forest areas have the smallest changes of about 0.5 m. The error bars are computed as 1-σ standard errors (standard deviations of the averaged water level anomalies divided by square roots of number of virtual stations for each of nonvegetated, herbaceous, and forest areas). Figure 7.8b shows the seasonal variation of inundated areas in forest and herbaceous surfaces. The maximum is observed in December, whereas the minimum is observed in March–April. The maximum flooded forest area ranges from 29,400 km^2 (December 2007) to 17,712 km^2 (December 2010), and the minimum flooded forest area varies from 7600 km^2 (March 2009) to 3400 km^2 (March 2010).

The surface water storage changes are then computed by multiplying the inundated areas with the mean water level changes over nonvegetated, herbaceous, and forest areas (Figure 7.8c). Timings of minima and maxima agree well with each other. The annual storage changes over nonvegetated and herbaceous areas are about 3.4 and 6.9 km^3, respectively. The annual storage changes over forest areas reveal larger interannual variation, decreasing from 14 km^3 in 2007 to 8.4 km^3 in 2009. Finally, total surface storage changes are calculated by summing the contribution of single compartments and compared with TWS changes from GRACE in Figure 7.8d. The error bars of the total surface storage changes represent the variances among the nonvegetated, herbaceous, and forest storage changes. It is noted that total surface storage changes agree reasonably well with TWS changes in both timings and amplitudes. This indicates that the TWS signal is governed by surface water, and thus the subsurface storage changes, including changes in root-zone soil moisture and groundwater, may be insignificant over the central Congo. Although we have no in situ measurements

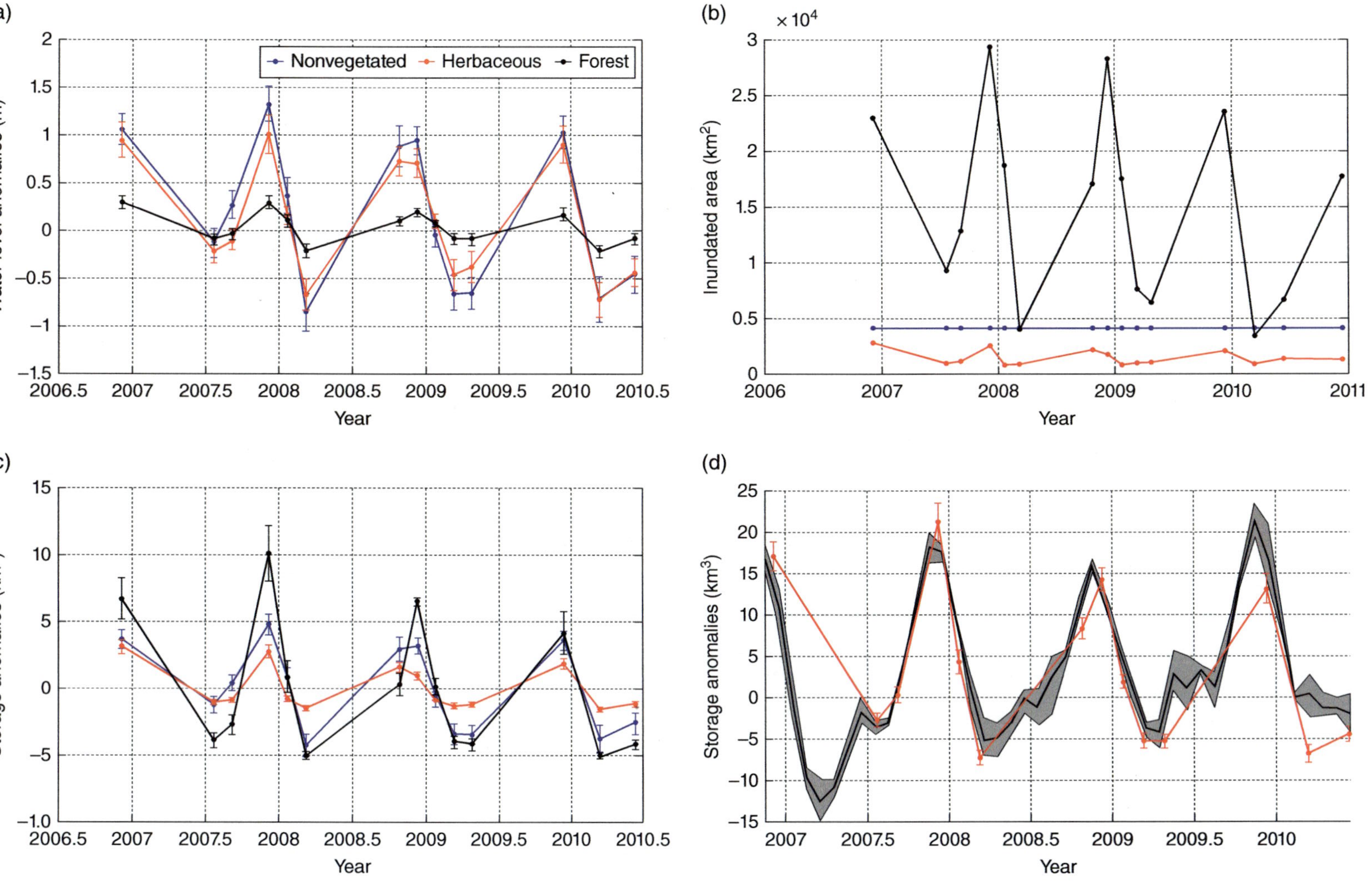

Figure 7.8 (a) Mean water level changes from Envisat altimeter averaged over nonvegetated (blue), herbaceous (red), and forest (green), interpolated at ScanSAR dates; (b) variations of the inundated areas over nonvegetated, herbaceous, and forest surfaces; (c) water storage changes over nonvegetated, herbaceous, and forest surfaces by multiplying (a) and (b); (d) the shading illustrates the range of TWS changes estimated using CSR, JPL, GFZ GRACE solutions. The red dots indicate the total surface storage changes computed by the summation of water storage changes over nonvegetated, herbaceous, and forest surfaces.

to validate it, this is in agreement with *Laraque et al.* [2001], one of the few Congo studies that states that the central Congo has no important aquifer. Accordingly, the authors showed that the hydrologic regime in the central Congo is similar to the regional rainfall rhythm revealed by comparing historic in situ discharge and rainfall data.

7.4. CONCLUSION AND DISCUSSIONS

In this study, we integrate multiple satellite measurements, water level changes from radar altimeter, inundated extent from SAR imagery, and TWS changes from GRACE to characterize and quantify TWS change and its surface and subsurface components over the central Congo Basin. The annual variations of the TWS changes during the period of 2007–2010 range between 21 and 31 km^3 and are mostly controlled by surface storage changes. In contrast to the central Congo, which has insignificant subsurface storage changes, the amplitude of the subsurface storage changes represents more than a third of the amplitude of TWS changes over the Negro River Basin, the second largest tributary to the Amazon River [*Frappart et al.*, 2011]. Our findings will contribute to provide a basis for determining and predicting the impacts of climate change and deforestations on the distribution of terrestrial water stores and fluxes in the Congo Basin.

Error sources on the water storage calculation include classification error of flooded extent, error in mean water level changes obtained from limited number of virtual stations, and altimeter measurement error itself. In addition, insufficient number of ScanSAR images and the satellite's long (44 day) repeat period also prevent us from obtaining the flood extent map with more frequent sampling.

The Surface Water and Ocean Topography (SWOT) mission, scheduled to be launched in 2019, is a Ka-band swath mapping radar interferometer that will provide simultaneous measurements of water elevation and inundated area for inland water bodies. SWOT will directly measure fluvial inundated area as well as water elevation, with spatial sampling in the order of tens of meters [*Biancamaria et al.*, 2010; *Lee et al.*, 2010]. For the purpose of monitoring surface water storage variations in wetlands, lakes, or reservoirs, the key measurements retrieved by SWOT will be repeat observations of surface water elevation and inundated area from which storage change can be readily calculated. SWOT measurements will thus enable the first global characterization of freshwater storage changes, providing the means to address the open science questions mentioned above.

ACKNOWLEDGMENTS

This research was supported by grants from NASA's GRACE Program (NNX12AJ95G), New Investigator Program (NNX14AI01G) and Terrestrial Hydrology Program (NNX12AQ36G). The second author was supported by NASA's Postdoctoral Program (NPP) at the Goddard Space Flight Center (GSFC). We thank two anonymous reviewers for providing constructive comments. GRACE Release-05 Level-2 data were provided by Center for Space Research (CSR), GeoForschungsZentrum (GFZ), and Jet Propulsion Laboratory (JPL). ALOS PALSAR data were provided by Alaska Satellite Facility (ASF), and Envisat altimetry data were provided by European Space Agency (ESA). Some of the figures were prepared using the Generic Mapping Tool (GMT) graphics package.

REFERENCES

Alsdorf, D. E., J. M. Melack, T. Dunne, L. A. K. Mertes, L. L. Hess, and L. C. Smith (2000), Interferometric radar measurements of water level changes on the Amazon floodplain, *Nature*, *404*, 174–177.

Alsdorf, D. E., E. Rodriguez, and D. P. Lettenmaier (2007), Measuring surface water from space, *Rev. Geophys.*, *45*, doi:10.1029/2006RG000197.

Alsdorf, D., S.-C. Han, P. Bates, and J. Melack (2010), Seasonal water storage on the Amazon floodplain measured from satellites, *Remote Sensing Environ.*, *114*, 2448–2456.

Beighley, R.E., et al. (2011), Comparing satellite derived precipitation datasets using the Hillslope River Routing (HRR) model in the Cong. River Basin, *Hydrol. Proc.*, doi:10.1002/hyp.8045.

Betbeder, J., V. Gond, F. Frappart, N. Baghdadi, G. Briant, and E. Bartholome (2014), Mapping of central Africa forested wetlands using remote sensing, *IEEE J. Select. Topics Appl. Earth Obser. Remote Sensing*, *7*, 531–542.

Biancamaria, S., et al. (2010), Preliminary characterization of SWOT hydrology error budget and global capabilities, *IEEE J. Selec. Topics Appl. Earth Obser. Remote Sensing*, *3*, 6–19.

Birkett, C. (1998), Contribution of the TOPEX NASA radar altimeter to the global monitoring of large rivers and wetlands, *Water Resources Res.*, *34*, 1223–1239.

Birkett, C., L. A. K. Mertes, T. Dunne, M. H. Costa, and M. J. Jasinski (2002), Surface water dynamics in the Amazon Basin: Application of satellite radar altimetry, *J. Geophys. Res.*, *107*, D20, doi:10.1029/2001JD000609.

Bwangoy, J.-R. B., M. C. Hansen, D. P. Roy, G. D. Grandi, and C. O. Justice (2010), Wetland mapping in the Congo Basin using optical and radar remotely sensed data and derived topographical indices, *Remote Sensing Environ.*, *114*, 73–86.

Calmant, S., F. Seyler, and J.-F. Cretaux (2008), Monitoring continental surface waters by satellite altimetry, *Surv. Geophys.*, *29*, 247–269.

Campbell, D. (2005), The Congo River basin, The World's Largest Wetlands: Ecology and Conservation, pp. 149–165. Cambridge University Press, New york.

Chen, J. L., C. R. Wilson, B. D. Tapley, Z. L. Yang, and G. Y. Niu (2009), 2005 Drought event in the Amazon River basin as measured by GRACE and estimated by climate models, *J. Geophys. Res.*, *114*, B05404, doi:10.1029/2008JB006056.

Crétaux, J.-F., and B. Charon (2006), Lake studies from satellite radar altimetry, *Comp. Rendus Geosci.*, *338*, 1098–1112.

De Grandi, G., P. Mayaux, Y. Rauste, A. Rosenqvist, M. Simard, and S. S. Saatchi (2000), The Global Rain Forest Mapping Project JERS-1 Radar Mosaic of Tropical Africa: Development and product characterization aspects, *IEEE Trans. Geosci Remote Sensing*, *38*, 2218–2233.

Duan, X. J., J. Y. Guo, C. Shum, and W. van der Wal (2009), On the postprocessing removal of correlated errors in GRACE temporal gravity field solutions, *J.Geodesy*, *83*, 1095–1106.

Durand, M., L.-L. Fu, D. P. Lettenmaier, D. E. Alsdorf, E. Rodriguez, and D. Esteban-Fernandez (2010), The Surface Water and Ocean Topography Mission: Observing terrestrial surface water and oceanic submesoscale eddies, *Proc.IEEE*, *98*, 766–779.

Frappart, F., F. Seyler, J.-M. Martinez, J. G. León, and A. Cazenave (2005), Floodplain water storage in the Negro River basin estimated from microwave remote sensing of inundation area and water levels, *Remote Sensing Environ.*, *99*, 387–399.

Frappart, F., S. Calmant, M. Cauhopé, F. Seyler, and A. Cazenave (2006a), Preliminary results of ENVISAT RA-2-derived water levels validation over the Amazon basin, *Remote Sensing Environ.*, *100*, 252–264.

Frappart, F., et al. (2006b), Water volume change in the lower Mekong from satellite altimetry and imagery data, *Geophys. J. Int.*, *167*, 570–584.

Frappart, F., et al. (2008), Interannual variations of river water storage from a multiple satellite approach: A case study for the Rio Negro River basin, *J. Geophys. Res.*, *113*, D21104, doi:10.1029/2007JD009438.

Frappart, F., et al. (2010), Interannual variations of the terrestrial water storage in the Lower Ob' Basin from a multisatellite approach, *Hydrol. Earth Syst. Sci.*, *14*, 2443–2453.

Frappart, F., et al. (2011), Satellite-based estimates of groundwater storage variations in large drainage basins with extensive floodplains, *Remote Sensing Environ.*, *115*, 1588–1594.

Guo, J. Y., X. J. Duan, and C. Shum (2010), Non-isotropic filtering and leakage reduction for determining mass changes over land and ocean using GRACE data, *Geophys. J. Int.*, *181*, 290–302.

Hansen, M. C., D. P. Roy, E. Lindquist, B. Adusei, C. O. Justice, and A. Altstatt (2008), A method for integrating MODIS and Landsat data for systematic monitoring of forest cover and change in the Congo Basin, *Remote Sensing Environ.*, *112*, 2945–2513.

Hess, L. L, J. M. Melack, E. M. L. M. Novo, C. C. F. Barbosa, and M. Gastil (2003), Dual-season mapping of wetland inundation and vegetation for the central Amazon basin, *Remote Sensing Environ.*, *87*, 404–428.

Jung, H. C., and D. Alsdorf (2010), Repeat-pass multi-temporal interferometric SAR coherence variations with Amazon floodplain and lake habitats, *Int. J Remote Sensing*, *31*, 881–901.

Jung, H., et al. (2010), Characterization of complex fluvial systems via remote sensing of spatial and temporal water level variations, *Earth Surface Processes and Landforms*, *35*, 294–304.

Kim, J. W., Z. Lu, H. Lee, C. Shum, C. M. Swarzenski, T. W. Doyle, and S. H. Baek (2009), Integrated analysis of PALSAR/Radarsat-1 InSAR and ENVISAT altimeter data for mapping absolute water level changes in Louisiana wetlands, *Remote Sensing Environ.*, *113*, 2356–2365.

Laraque, A., G. Mahé, D. Orange, and B. Marieu (2001), Spatiotemporal variations in hydrological regimes within Central Africa during the XXth century, *J. Hydrol.*, *245*, 104–117.

Lee, H., et al. (2009), Lousiana wetland water level monitoring using retracked TOPEX/POSEIDON altimetry, *Marine Geodesy*, *32*, 284–302.

Lee, H., M. Durand, H. Jung, D. Alsdorf, C. Shum, and Y. Sheng (2010), Characterization of surface water storage change in Arctic lakes using simulated SWOT measurements, *Int. J. Remote Sensing*, *31*, 3931–3953.

Lee, H., et al. (2011), Characterization of terrestrial water dynamics in the Congo Basin using GRACE and satellite radar altimetry, *Remote Sensing Environ.*, *115*, 3530–3538.

O'Loughlin, F., M. A. Trigg, G. J.-P. Schumann, and P. D. Bates (2013), Hydraulic characterization of the middle reach of the Congo River, *Water Resour. Res.*, *49*, 5059–5070.

Papa, F., B. Legresy, and R. Remy (2003), Use of the Topex-Poseidon dual-frequency radar altimeter over land surfaces, *Remote Sensing Environ.*, *87*, 136–147.

Prigent, C., F. Papa, F. Aires, W. B. Rossow, and E. Matthews (2007), Global inundation dynamics inferred from multiple satellite observations, 1993–2000, *J. Geophys. Res.*, *112*, D12107, doi:10.1029/2006JD007847.

Rodell, M., J. Chen, H. Kato, J. Famiglietti, J. Nigro, and C. Wilson (2007), Estimating ground water storage changes in the Mississippi River basin (USA) using GRACE, *Hydrogeol. J.*, *15*, 159–166.

Rosenqvist, A. (2008), Mapping of seasonal inundation in the Congo river basin using PALSAR ScanSAR, ALOS PI Symposium, Rhodes, Greece, Nov. 3–7.

Rosenqvist, A., M. Shimada, and M. Watanabe (2004), ALOS PALSAR: Technical outline and mission concepts, 4th International Symposium on Retrieval of Bio- and Geophysical Parameters from SAR Data for Land Applications, Innsbruck, Austria, November 16–19.

Shimada, M. (2008), PALSAR ScanSAR ScanSAR Interferometry, IGARSS 2008, Boston, July 6–11.

Swenson, S., and J. Wahr (2006), Post-processing removal of correlated errors in GRACE data, *Geophys. Res. Lett.*, *33*, L08402, doi:10.1029/2005GL025285.

Syed, T. H, J. S. Famiglietti, and D. P. Chambers (2009), GRACE-based estimates of terrestrial freshwater discharge from basin to continental scales, *J. Hydrometeor.*, *524*, 22–40.

Wahr, J., M. Molenaar, and F. Bryan (1998), Time variablility of the Earth's gravity field: Hydrological and oceanic effects and their possible detection using GRACE, *J. Geophys. Res.*, *103*, 30,205–30,229.

Wingham, D., C. G. Rapley, and H. Griffiths (1986), New techniques in satellite altimeter tracking systems, Proceedings of IGARSS' 86 Symposium, Zürich, 8–11 Sept.

Wdowinski, S., S. W. Kim, F. Amelung, T. H. Dixon, F. Miralles-Wilhelm, and R. Sonenshein (2008), Space-based detection of wetlands' surface water level changes from L-band SAR interferometry, *Remote Sensing Environ.*, *112*, 681–696.

8

Spatial Patterns of River Width in the Yukon River Basin

Tamlin M. Pavelsky, George H. Allen, and Zachary F. Miller

8.1. INTRODUCTION

River width is among the most fundamental parameters governing river hydrology, geomorphology, biogeochemistry, and ecology. The width of a river reflects the amount of water flowing through it, the sediment load it is carrying, the composition of its bed and banks, and the influence of local topography and tectonics [*Leopold and Wolman*, 1957; *Finnegan et al.*, 2005]. As one of the three primary dimensions of stream flow, width is integral to the calculation of river discharge both directly and through slope area scaling methods such as the Manning and Chézy formulas [*Chanson*, 2004]. Accurate hydraulic and hydrologic modeling of rivers consequently requires the input of spatially distributed widths as a primary parameter [*Yamazaki et al.*, 2011; *Neal et al.*, 2012]. Fundamentally, width and its ratio with depth are among the most basic parameters that encapsulate variations in river morphology [*Nicholas*, 2013]. In bedrock channels, river width (along with slope) acts as a primary indicator of the incision potential and sediment transport capacity of a river [*Allen et al.*, 2013]. Recognition that river width reflects difficult-to-measure landscape characteristics (e.g., tectonic deformation) has generated considerable study of river width in the field of theoretical geomorphology [*Kirby and Whipple*, 2012; *Yanites et al.*, 2010].

The flux of nutrients such as carbon and nitrogen between subsurface reservoirs, rivers, and the surrounding atmosphere is mediated in part by river width [*Butman and Raymond*, 2011; *Donner et al.*, 2002; *Tank et al.*, 2008]. Ecologically, variations in width affect the ways in which rivers are barriers to dispersion of terrestrial [*Yeager*, 1991] and avian [*Hayes and Sewlal*, 2004] species, while also acting as habitat for many aquatic species [*Jenkins et al.*, 2004; *Kemp et al.*, 1999]. River width holds significance for humans by influencing the feasibility and cost of building bridges [*Troitsky*, 1994], transporting goods [*McCartney*, 1986], and developing and managing fisheries [*Prévost et al.*, 2003]. As such, knowledge of spatial variations in river width is of primary importance for scientists and engineers in a broad range of fields.

Despite this fact, to date there are no uniform, high-resolution observational data sets of spatial variations in river width available for any major river basin, globally. Instead, outside of localized study areas widths are generally estimated using long-recognized relationships with discharge and, by proxy, upstream drainage area. These relationships were summarized more than half a century ago in the pioneering work of *Leopold and Maddock* [1953] on downstream hydraulic geometry (DHG). They recognized that river width, depth, and velocity exhibit consistent power-law relationships with discharge of the following form:

$$w = aQ^b, \qquad (8.1a)$$

$$d = cQ^f, \qquad (8.1b)$$

$$v = kQ^m, \qquad (8.1c)$$

where w, d, and v are width, depth, and velocity, respectively, Q is discharge, and a, c, k, b, f, and m are empirically derived coefficients and exponents that vary from

Department of Geological Sciences, University of North Carolina, Chapel Hill, North Carolina, USA

Remote Sensing of the Terrestrial Water Cycle, Geophysical Monograph 206. First Edition. Edited by Venkat Lakshmi.
© 2015 American Geophysical Union. Published 2015 by John Wiley & Sons, Inc.

basin to basin. By definition, the product of the coefficients and the sum of the exponents must be equal to one.

Since the original development of DHG theory there has been substantial work devoted to exploring variations in the coefficients and exponents and understanding the strengths and limitations of the overall approach. While DHG relationships are generally empirically derived, there is some evidence that they can also be explained theoretically based on conservation of mass and momentum, channel resistance, sediment transport laws, and continuity of flow [*Smith*, 1974; *Yang et al.*, 1981; *Molnar and Ramirez*, 2002; *Griffiths*, 2003; *Singh et al.*, 2003; *Eaton and Church*, 2007]. Considerable progress has been made in understanding variability in b, f, and m, which are most commonly between ~0.4–0.5 (b), ~0.3–0.4 (f), and ~0.1–0.2 (m), respectively [*Leopold and Maddock*, 1953; *Park*, 1977]. In other words, as discharge decreases downstream, rivers generally adjust most by widening, then by deepening, while changes in velocity are relatively small. In contrast, DHG coefficients are much less well understood and are generally assumed to reflect the idiosyncrasies of individual river systems [*Park*, 1977].

While many rivers appear to exhibit DHG-type relationships, there are also cases in which channel geometry variations cannot be explained by discharge alone. For example, variations in lithology, sediment load, channel slope, and differential tectonic uplift may all perturb river form in such a way that width, depth, and velocity do not increase predictably with added discharge [*Wohl*, 2004; *Allen et al.*, 2013]. In addition, human impacts on flow such as water withdrawals and diversions can perturb natural DHG relationships [*Eaton and Church*, 2007]. Finally, all studies of DHG to date have relied on either high-density measurements in a particular study site or more widely dispersed measurements, such as from stream gauges, over a larger region. The most extensive studies of generalized DHG outside of a specific location have used no more than ~1500 measurement locations [*Park*, 1977; *Moody and Troutman*, 2002; *Lee and Julien*, 2006]. DHG relationships developed at small scales are often applied to much larger river basins, an approach that often produces significant errors [*Klein*, 1981]. Further, measurements used to construct DHG relationships are often relatively sparse and usually not randomly distributed in space, making it is difficult to fully assess the true sources of downstream variability in width, depth, and velocity.

Here, we use software designed to measure river widths from satellite imagery [*Pavelsky and Smith*, 2008] to measure the widths of all rivers in the Yukon River Basin wider than ~100 m (and many narrower rivers) from water masks derived from Landsat imagery. This effort complements other recent work focused on measuring river width from satellite imagery [*O'Loughlin et al.*, 2013; *Allen et al.*, 2013; *Pavelsky and Smith*, 2008; *Yamazaki et al.*, in review; *Miller et al.*, in review]. The Yukon Basin was selected because it is among the largest basins globally without significant human alteration of the river network. We link these widths to drainage areas derived from a digital elevation model (DEM) and establish a drainage area-discharge relationship within the Yukon Basin to estimate mean summertime discharge at all ~7.5×10^5 measurement locations. We then use this data set to (a) understand the distribution of widths in the basin and (b) obtain the first uniform, high-resolution DHG estimates for the entire basin and, separately, its major tributaries.

8.2. STUDY AREA

The Yukon River Basin encompasses ~854,700 km² in Alaska and Northwestern Canada, making it the fourth largest drainage basin in North America (Figure 8.1). It is largely bounded on the south by major mountain ranges, including the Alaska Range, the Wrangell–St. Elias mountains, and smaller ranges in British Columbia. The northern and western boundaries are defined by lower elevation mountain ranges, including the Brooks Range in northern Alaska. The characteristics of the landscape vary widely throughout the basin. Many southern tributaries, including the White and Tanana rivers, are fed by glacial melt and exhibit characteristic patterns of glacial streams including high sediment load and extensive braiding [*Brabets et al.*, 2000]. In contrast, tributaries draining the northern and eastern portions of the basin, including the Pelly, Stewart, Porcupine, and Koyukuk, are largely free of glaciers and exhibit more meandering planforms. The entire basin is within the permafrost zone, ranging from sporadic (10%–30%) permafrost in the southwest to continuous permafrost in the north (Figure 8.1) [*Brown et al.*, 1997].

Hydrologically, the Yukon Basin is dominated by snowmelt-driven (nival) and glacial flow regimes. As a result, the mainstem and most tributaries exhibit wintertime low flows and springtime and summertime maxima associated with snowmelt and glacial ablation. The mean annual discharge of the Yukon at Pilot Station (the downstream-most gauge) is ~6430 m³/s. There is substantial evidence that the hydrology of the Yukon River Basin is changing in response to anthropogenic climate change [*Hinzman et al.*, 2005], including patterns of increased permafrost thaw [*Osterkamp*, 2005; *Walvoord and Striegl*, 2007; *Walvoord et al.*, 2012], earlier river ice breakup [*Sagarin and Micheli*, 2001], and accelerating glacial recession [*Kaser et al.*, 2010]. Many areas exhibit increasing wintertime stream flow associated with larger groundwater contributions, but mean annual stream flow within the basin has not changed significantly over the

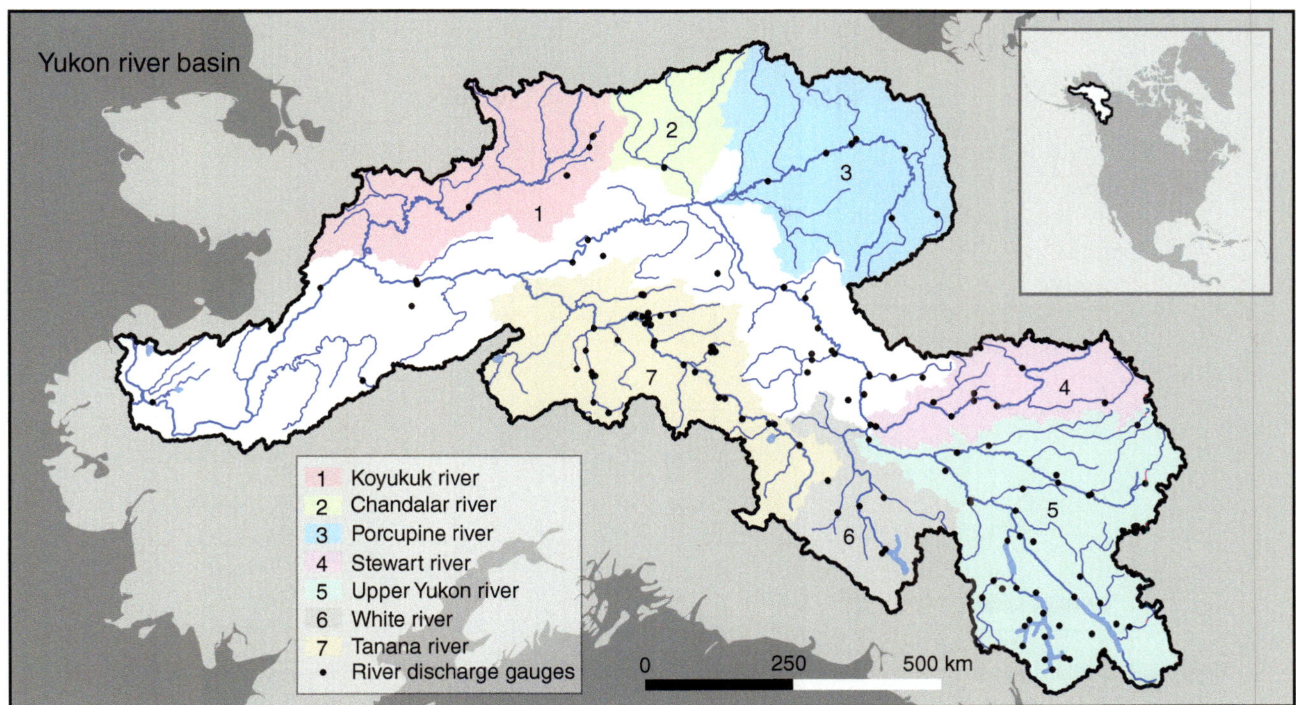

Figure 8.1 Location map showing the Yukon Basin, its principal river channels, extent of major subbasins, and locations of river discharge gauges used in this study.

period of record [*Walvoord and Striegl*, 2007]. There are just 101 current or historical gauging stations located within the basin, and major tributaries including the Porcupine and the Koyukuk are no longer gauged. This gauge density is far lower than in most other North American basins of comparable size. There are also few long-term, reliable measurements of precipitation within the basin [*Brabets et al.*, 2000]. As a result, hydrology is now poorly constrained except for measurements of aggregated flow near the mouth of the basin and in a few large subbasins.

8.3. DATA AND METHODS

8.3.1. Calculating River Widths

The first step in calculating river widths for the Yukon Basin is to develop a mask differentiating water from all other land cover types. For this purpose, we use a combination of Landsat-derived products at 30 m spatial resolution. Landsat was selected because of its extensive data availability, suitable spectral bands, and high spatial resolution. For the portion of the Yukon Basin within the United States, we extracted the water class from the National Land Cover Database (NLCD) 2001 Alaska land cover classification [*Selkowitz and Stehman*, 2011]. This product, derived from Landsat Thematic Mapper (TM) and Enhanced Thematic Mapper + (ETM+) imagery

acquired largely between 1999 and 2002, was developed by the U.S. Geological Survey (USGS) to provide a consistent land cover classification scheme for Alaska, matching the product already available for the conterminous United States [*Homer et al.*, 2004]. All imagery used for the classification was acquired during the summer, between June and early September [*Selkowtiz and Stehman*, 2011]. No attempt was made to specifically calibrate the classification to a common river discharge frequency [*D. Selkowitz*, personal communication], but this would have been very difficult in any case due to lack of discharge data for many major tributaries and limited cloud-free Landsat imagery.

No land cover classification comparable to the NLCD is available for the Canadian portion of the Yukon Basin. To develop a water mask, cloud-free Landsat TM and ETM + images acquired between June and early September were downloaded from the Global Land Cover Facility (GLCF) at the University of Maryland (www.landcover.org) and the USGS (*glovis.usgs.gov*). Often only a few suitable images were available for each Landsat tile, so it was not possible to calibrate image selection to match a consistent discharge frequency. While it would have been ideal to apply the NLCD classification methods to match the Alaska product, these methods are largely focused on differentiating vegetation types and are far more complex than required for the relatively simple task of differentiating water from land. As such, we instead use simple thresholds in band 5

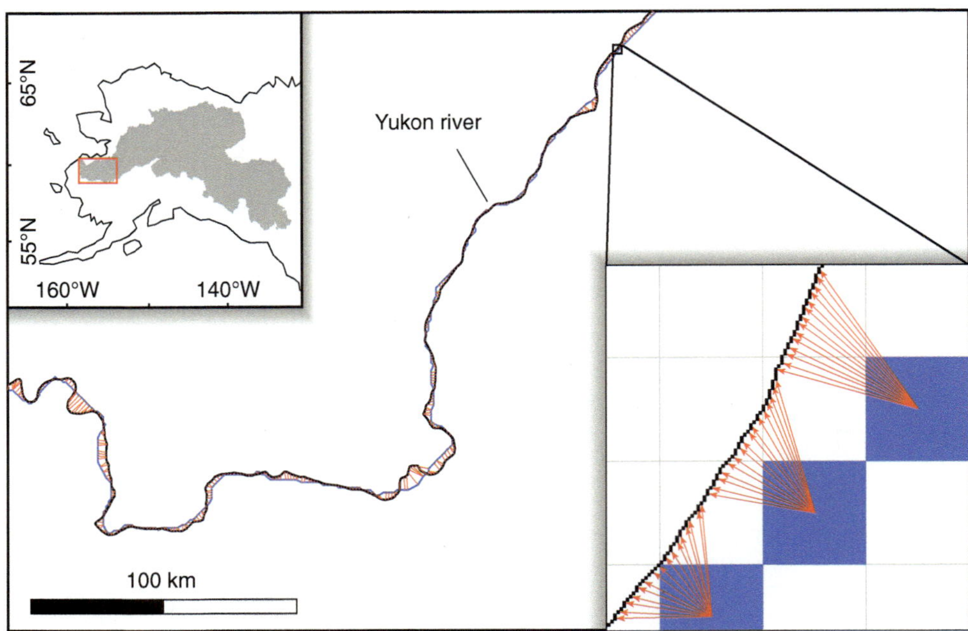

Figure 8.2 Method for combining river width and upstream drainage area data sets. Red arrows, enlarged in the inset, show data transfer path of drainage area from the DEM flow line (blue) to the nearest pixel in the RivWidth centerline (black).

(near-infrared, 1.55–1.75 μm), which previous studies suggest can be used to accurately differentiate water from land [*Sheng et al.*, 2001; *Overton*, 2005]. The resulting binary masks were then mosaicked with the NLCD-derived mask for Alaska to produce a basin-wide map of inundation extent. This map was manually corrected to remove bridges and other hydrologic discontinuities from rivers.

The resulting map of inundation extent was used as input to RivWidth, software developed by *Pavelsky and Smith* [2008] to automate the extraction of river flow width. RivWidth takes in a binary water mask, removes water bodies other than the river network, derives centerlines for all rivers, determines the direction orthogonal to the river centerline at each centerline pixel, and calculates the total flow width along each orthogonal. More detailed discussion of RivWidth is provided by *Pavelsky and Smith* [2008], *Smith and Pavelsky* [2008], and at http://www.unc.edu/~pavelsky/. RivWidth has been substantially improved since its original 2008 release, and using the latest version (v0.4) on a fast workstation it is possible to calculate widths for the entire Yukon Basin at 30 m spatial resolution in a few hours. The final output of RivWidth is an image showing the width at each centerline pixel and a text file containing the image coordinates, Universal Transverse Mercator (UTM) coordinates, river width, and number of channels for each centerline pixel. There are several large lakes connected to the river network in the upstream portions of the Yukon Basin, and widths of these lakes have been removed in this analysis.

8.3.2. Estimating Continuous River Discharge

In order to develop DHG relationships between width and discharge, it is necessary to match each width measurement with an estimate of river discharge. As discharge has been measured at relatively few locations in the Yukon Basin, we exploit the well-known relationship between discharge and drainage area to estimate discharge continuously [*Mosley and McKerchar*, 1993]. To determine basin area for each centerline pixel, we use a flow accumulation map and flow lines extracted from the Hydro1k database of digital topography [*Verdin and Greenlee*, 1998]. While Hydro 1k is available only at the relatively coarse resolution of 1 km, it is the only hydrologically corrected digital elevation model (DEM) available for the Yukon Basin. Other DEMs are either unavailable at high latitudes (Hydrosheds, Shuttle Radar Topography Mission) or are not hydrologically conditioned (ASTER, National Elevation Dataset, Canadian DEMs). We combined the RivWidth and DEM-based data sets by assigning drainage area to each RivWidth centerline pixel from the nearest DEM flow line pixel (Figure 8.2). To produce an estimate of discharge from a drainage area, we use a basin-wide linear fit between Q and A using data from 101 stream gauges (Figure 8.3). We constrain the best-fit regression to pass through the origin using a regression through the origin method [*Eisenhauer*, 2003], although the slope of the regression line is similar, using an ordinary least-squares method. Mean annual discharge values were obtained for all gauges with at least 2 years of continuous

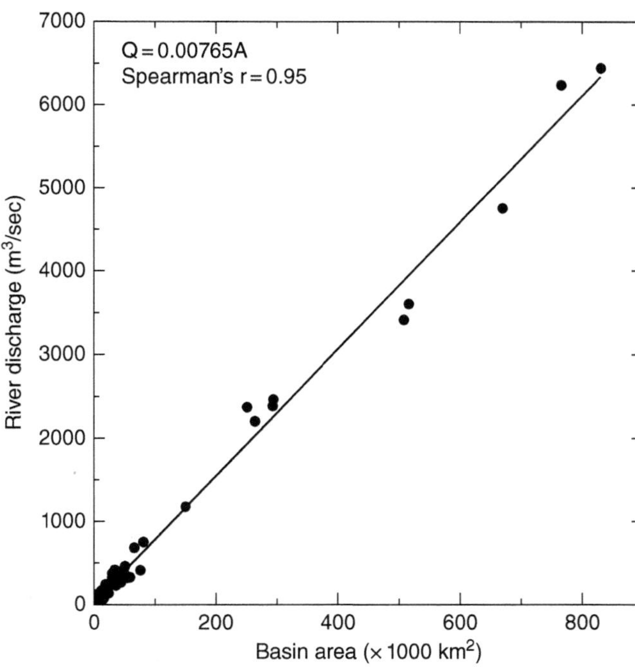

Figure 8.3 Relationship between drainage area and mean annual river discharge for the Yukon River Basin, derived for gauge locations operated by the USGS and Water Survey of Canada.

discharge measurements from the USGS and Water Survey of Canada websites. Recent work by *Brabets and Walvoord* [2009] suggests that annual discharge in the Yukon Basin has remained relatively constant over the past ~70 years. In principle, hydraulic geometry relationships should be constructed using the channel-shaping flow, usually bankfull discharge. As measurements are rarely available in bankfull conditions, however, several studies [*Leopold and Maddock*, 1953; *Leopold et al.*, 1964; *Griffiths*, 1980; *Ibbitt et al.*, 1999] have compared the results of using mean annual discharge instead and have found that DHG exponents are similar over a range of discharge frequencies. We use Spearman's ρ [*Spearman*, 1904], a nonparametric goodness-of-fit test, rather than a parametric correlation metric because the data are not normally distributed. The strong relationship between discharge and drainage area (Spearman's $\rho = 0.95$) suggests that basin-wide discharge estimates are likely to be comparatively accurate. However, the relatively small number of gauges in some tributary basins, especially those draining glaciers, suggests that we may not fully capture the discharge-drainage area relationship.

8.3.3. Assessing Width-Discharge Relationships

To determine the strength and variability of the relationship between width and discharge in the Yukon Basin, we construct DHG relationships between width

and discharge for the entire basin and for major subbasins, including the Koyukuk, Tanana, Chandalar, Porcupine, White, Stewart, and Upper Yukon (Figure 8.1). For the entire basin and for each subbasin, we regress log-transformed width against log-transformed discharge. The resulting linear regression equation can be transformed to yield the coefficient (a) and the exponent (b) in equation (1). For each relationship, we also assess goodness of fit using a simple coefficient of determination (r^2) between the log-transformed width and discharge.

RivWidth widths are validated against a subset of 19 USGS gauges for which at least 10 years of continuous discharge measurements are available and which have channel geometry data from manual discharge surveys available on the USGS website. We compare the mean of the five geographically closest RivWidth-derived widths against the width value from the manual discharge measurement that most closely matches the mean discharge for the June–September season, the same period of Landsat imagery used in creation of the water mask. As with the discharge-drainage area relationship, we assess goodness of fit using Spearman's ρ.

8.4. RESULTS

8.4.1. Measurement of River Widths

We measured total flow width of rivers at 566,829 cross sections in the Yukon Basin, representing ~19,700 km of channel (Figure 8.4). To our knowledge, this is the first major river basin in which width is measured continuously for all principal rivers. The largest measured width in the basin, 7251 m, is located in the Yukon Flats in central Alaska, a tectonically controlled basin in which the Yukon River is highly braided. The narrowest width measured is 30 m, which is dictated by the limits of the Landsat spatial resolution. There are a number of error sources that impact width measurements. Mixed pixels along river banks and islands produce uncertainty of one pixel resolution (or 30 m) for each channel in the cross section [*Pavelsky and Smith*, 2008]. In addition, the water classification used here represents a snapshot in time and likely does not perfectly represent the river extent. The fact that our classification is derived from the NLCD for Alaska and raw Landsat imagery for Canada introduces additional uncertainty. To test the magnitude of this uncertainty, we calculated widths from both data sets for ~70 km of the Porcupine River downstream of the international border. We found a mean bias of 2.9 m and a mean absolute error of 29 m. These values suggest that there is little difference between the two classifications that cannot be explained by uncertainty in classification of water-land boundary pixels. Because RivWidth calculates width along a cross section orthogonal to the centerline of the

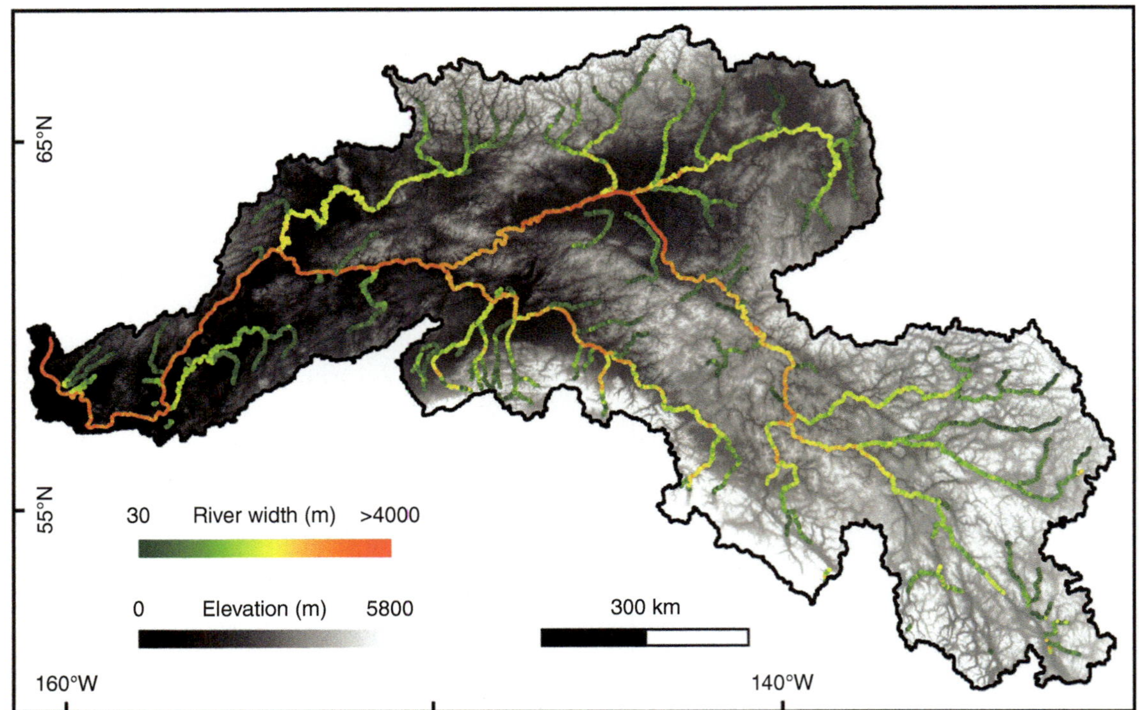

Figure 8.4 Map of river widthrs for the Yukon Basin superimposed on the Hydro1k DEM. Widths are scaled geometrically instead of arithmetically to better convey width variability.

river as a whole, in braided reaches individual river channels are not always perpendicular to this overall orthogonal and width is sometimes overestimated. For example, the widest width measurement in our data set occurs at a location on the Yukon where several channels diverge in different directions, and the measured width represents a substantial overestimate.

Despite these potential sources of error, validation of the widths against a limited set of in situ width measurements at mean summertime discharge from USGS measured cross-sections shows a strong correspondence ($\rho = 0.92$, $p < 0.0001$) between ground-based and remotely sensed widths (Figure 8.5). Remotely sensed widths exhibit a mean bias of 30 m, suggesting that they are, on average, slightly higher than the width at mean discharge for the June–September period. The mean absolute error is 40 m (25%), which compares favorably with the minimum possible uncertainty associated with remotely sensed width measurements from Landsat of 30 m [*Pavelsky and Smith*, 2008].

8.4.2. Distribution of River Widths

Using the width data set developed here, it is possible for the first time to calculate the frequency distribution of observed widths for a large river basin. The width distribution in the Yukon Basin closely follows a power law distribution, with an exponent of -1.72 and $r^2 = 0.98$

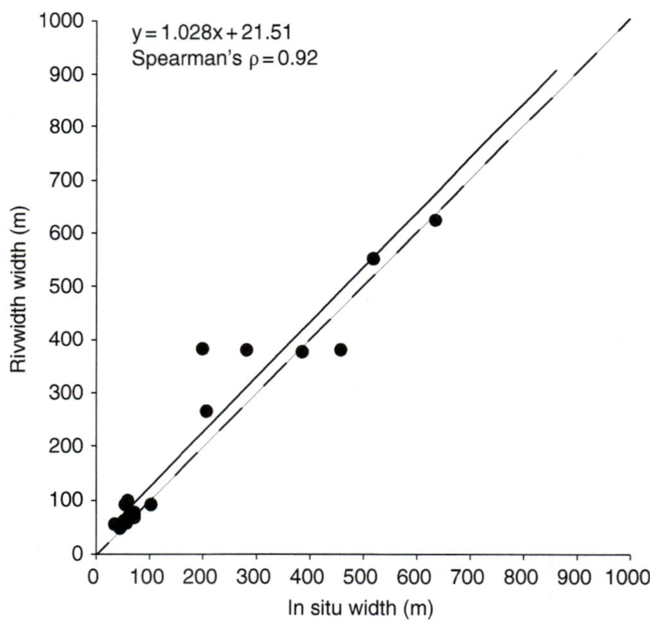

Figure 8.5 Comparison of widths measured at USGS gauging stations close to mean summertime discharge and widths measured using RivWidth. The dashed line is a 1-to-1 line.

(Figure 8.6a). We do not include width values below 100 m in this portion of the analysis because we measure only a portion of rivers with widths below this threshold. Many previous studies that required measurements of

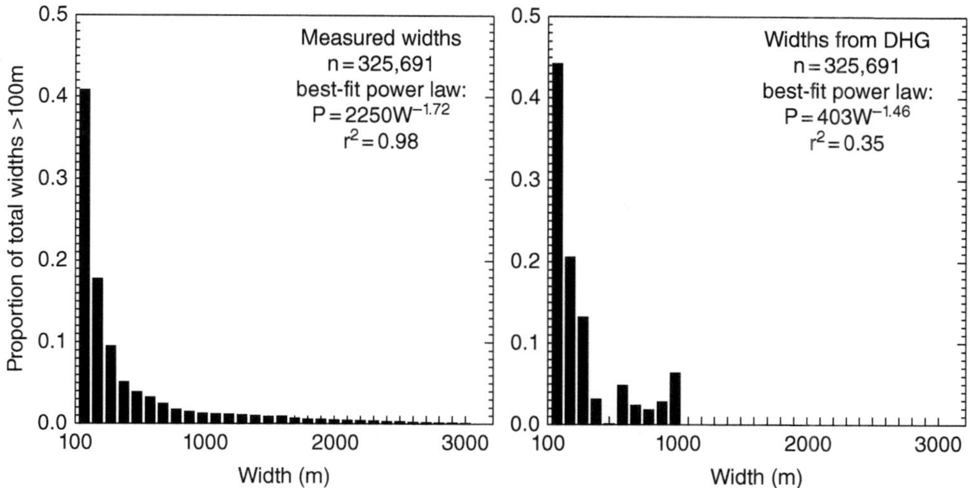

Figure 8.6 Histograms of width observation frequency for the Yukon Basin derived from (a) RivWidth observations and (b) a downstream hydraulic geometry relationship.

river width have relied on estimates based on DHG relationships [e.g., *Butman and Raymond*, 2011; *Paiva et al.*, 2013]. The distribution of Yukon Basin widths calculated using the width-discharge relationship discussed in Section 8.4.3. are shown in Figure 8.6b. Although the same general pattern of width is evident in the observed and DHG-based data sets, there are several key differences. The DHG-based data set, with a maximum width of 1082 m, does not include any of the widest widths measured in the observed data set. In addition, observed widths match a power law relationship much less closely with a somewhat different exponent (−1.46) and a much weaker overall fit ($r^2 = 0.35$). The apparent clustering of predicted widths is due to the fact that some ranges of drainage area values do not occur in the Yukon Basin, and a one-to-one match between width and drainage area is assumed. Finally, there are considerably more widths in the 100–200 m range in the DHG-based data set compared to observations. All of these differences suggest that DHG-based relationships are unlikely to produce fully realistic distributions of river widths for large basins. Instead, it may be possible to model such distributions using simple power law relationships like the one shown in Figure 8.6a.

8.4.3. Width-Discharge Relationships

Using basin areas derived from Hydro1k and discharge measurements from USGS and Water Survey of Canada gauges, we develop a width-discharge relationship for the entire Yukon Basin (Figure 8.7). Because Hydro1k poorly captures stream networks in some lowland environments, it was impossible to accurately link 21,715 (3.8%) of the width observations to accurate estimates of basin area. The resulting width-discharge relationship calculated from the remaining 545,114 data points can be expressed via the power law equation:

$$W = 20.6Q^{0.46} \qquad (8.2)$$

However, some of the discharge values apparent in Figure 8.7 are implausibly low for rivers that are >30 m wide, and these may be skewing the best-fit relationship. Nearly all of the discharge estimates of less than 10 m³/s are found in the Tanana, White, and Upper Yukon subbasins immediately downstream of glaciers, where basin area is quite small, but summertime discharge from the glaciers can be quite large. Because we estimate discharge based on drainage area, we substantially underestimate the summertime discharge for these rivers. If we remove from consideration all observations with $Q \le 10$ m³/s (~8% of all observations), the resulting power law relationship is

$$W = 14.2Q^{0.52} \qquad (8.3)$$

We believe that this relationship better reflects the actual width-discharge relationship than does equation (8.2), which almost certainly contains erroneous discharge estimates downstream from glaciers. Variability in discharge accounts for more than two thirds of width variability ($r^2 = 0.70$), implying that it is the dominant control on river width as suggested by *Leopold and Maddock* [1953], *Park* [1977], and many others based on more limited data sets. The *b* exponent of 0.52 is quite similar to the global width-discharge exponent of 0.50 identified by *Moody and Troutman* [2002]. It is also identical to the exponent from the width-discharge relationship developed using in situ width measurements from 21 USGS discharge gauges. That equation is

$$W = 6.2Q^{0.52} \qquad (8.4)$$

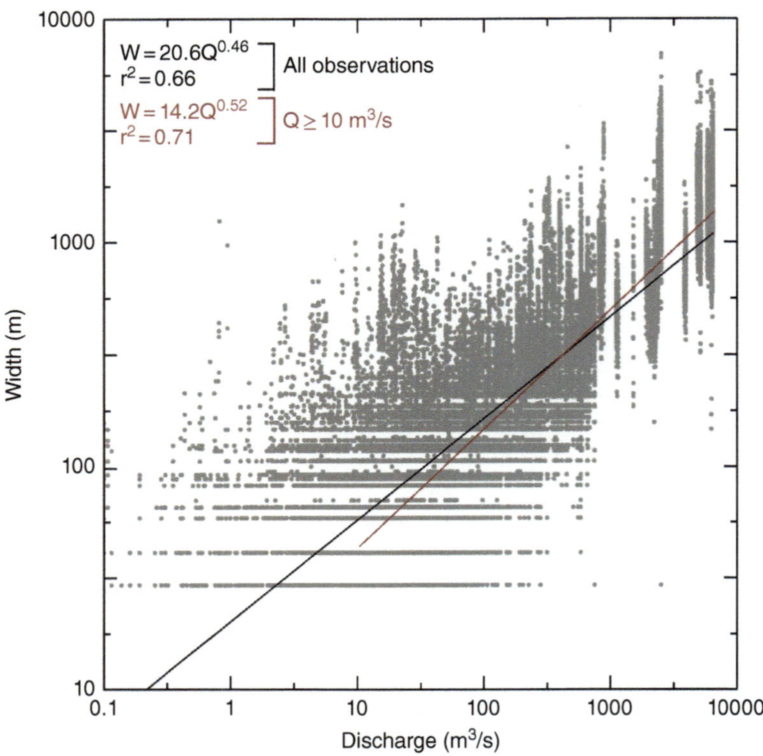

Figure 8.7 Scatterplot of log-transformed widths against log-transformed discharge for the Yukon Basin. Because of the very large number of widths measured, a random subsample of 50,000 points is shown here.

It is unclear why the coefficient is so much smaller in equation (8.4) than equation (8.3), but it may be related to the small number of available in situ width measurements.

Width-discharge relationships for subbasins of the Yukon vary somewhat from the equation for the basin as a whole (Table 8.1). We include values both with and without $Q \leq 10\,m^3/s$, although the only subbasins for which the resulting relationships are significantly different are those influenced by substantial glaciation: the Tanana, White, and, to a lesser extent, the Upper Yukon. For the remainder of this analysis, we refer only to the relationships with $Q \geq 10\,m^3/s$. Values of b range from 0.43 for the Upper Yukon and Porcupine to 0.65 for the White. Goodness of fit ranges from relatively low ($r^2 = 0.45$) for the Porcupine to comparatively high ($r^2 = 0.74$) for the Koyukuk.

8.5. DISCUSSION AND CONCLUSIONS

There are two principal conclusions to be drawn from this analysis. The first conclusion is that the width-discharge relationship for the Yukon Basin developed here with >500,000 data points is consistent with previous studies of downstream hydraulic geometry using many fewer observations [*Leopold and Maddock*, 1953; *Knighton*, 1974; *Park*, 1977; *Rhoads*, 1991; *Wohl*, 2004].

The hydraulic geometry coefficient a has been studied relatively little and the factors controlling its variation are not well understood. Values of a in the Yukon Basin and its subbasins are consistent with the range of previous values for other watersheds [*Rhoads*, 1991; *Moody and Troutman*, 2002]. Values of the hydraulic geometry coefficient b are related primarily to bank strength, which is primarily determined by sediment and vegetation characteristics [*Eaton and Church*, 2007]. Unfortunately, we are not aware of continuous measurements of these characteristics in the Yukon Basin to compare against DHG exponents from the various subbasins. However, we do have access to gauge data near the mouth of each major tributary that can be used to calculate variations in mean annual runoff among the subbasins (Table 8.1). There is a statistically significant positive relationship ($p < 0.001$) between runoff and the b exponent (Figure 8.8). This relationship is unlikely to arise solely from errors in discharge estimation from drainage area, since variations in the discharge-drainage area relationship primarily affect the a coefficient instead. Instead, we speculate that subbasins of the Yukon with high runoff are likely to also have relatively high sediment flux and more braided planforms because they are more likely to contain glaciers and rivers with more erosive potential. Past studies suggest that the b exponent varies inversely with bank strength,

Table 8.1 Downstream Hydraulic Geometry Characteristics of Major Subbasins of Yukon Basin[a]

River	Basin Area (km²)	Mean Q (m³/s)	Runoff (cm/yr)	n	a (≥10 m³/s)	b (≥10 m³/s)	r² (≥10 m³/s)
Porcupine	117,000	623	17	72,734	19.9(18.6)	0.42(0.43)	0.49(0.45)
Koyukuk	83,000	770	29	54,693	14.2(14.1)	0.49(0.49)	0.74(0.74)
Upper Yukon	150,400	1178	25	81,523	20.7(15.6)	0.39(0.45)	0.58(0.58)
Stewart	52,000	464	28	28,897	11.3(11.2)	0.53(0.53)	0.66(0.66)
White	46,900	566	38	14,908	16.9(7.9)	0.50(0.65)	0.36(0.49)
Tanana	114,000	1246	34	100,882	42.1(20.9)	0.38(0.53)	0.44(0.46)
Chandalar	35,500	210	19	23,258	19.0(18.6)	0.47(0.48)	0.42(0.46)
Entire Yukon	*854,700*	*6430*	*24*	*545,114*	*20.6(14.2)*	*0.46(0.52)*	*0.66(0.71)*

[a]Q is the mean annual discharge, runoff is calculated by dividing the mean discharge by the basin area, and n represents the total number of river width measurements made in that subbasin. a and b refer to the coefficient and exponent in equation (1) and are shown both for the full range of data and for only observations with estimated discharge ≥10 m³/s. The r² value represents the goodness of fit of these power law equations to the width and discharge data.

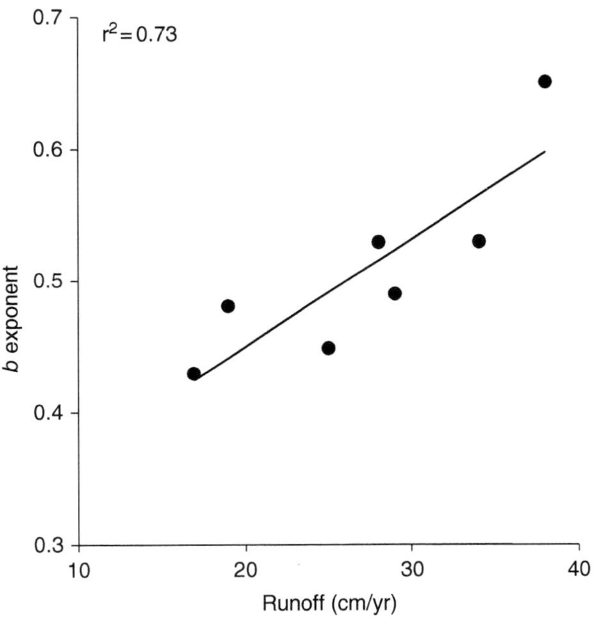

Figure 8.8 Comparison of b exponents for estimated discharge ≥10 m³/s and runoff for seven subbasins of the Yukon River summarized in Table 8.1.

with discharge variations as a secondary factor [*Eaton and Church*, 2007]. Rivers with a high sediment load to transport capacity ratio tend to have faster rates of lateral migration [*Finnegan et al.*, 2007], plausibly decreasing the ratio of bank strength to lateral incision potential of a river. In addition, aggradational rivers (such as portions of the Tanana [*Yarie et al.*, 1998]) are likely to have higher b exponents than will primarily incising rivers. However, this preliminary finding merits additional future testing and may not be extensible to other basins.

The second primary conclusion of this study is that it is now practical to continuously map river widths across large river basins like the Yukon. Many previous studies have used downstream hydraulic geometry to estimate variations in river width for large rivers [e.g., *Yamazaki et al.*, 2011; *Butman and Raymond*, 2011; *Paiva et al.*, 2013]. However, widths produced from these relationships are often unreliable, both because frequency distributions are not fully captured (e.g., Figure 8.6) and because, absent local width and discharge data, studies are often constrained to using global- or continental-scale relationships over smaller areas. Already, hydrologic and hydrodynamic modeling studies using continuous widths measured from satellite imagery using RivWidth are more accurately capturing variability in flood wave dynamics and inundation extent [e.g., *Neal et al.*, 2012], and multitemporal observations of river width change are being used to track temporal variations in discharge [*Smith and Pavelsky*, 2008; *Brackenridge et al.*, 2005]. In the future, studies of carbon evasion and nutrient fluxes influenced by inundation extent are also likely to benefit from improved knowledge of river widths. Given recent advances in the availability of high-resolution satellite imagery and computing, it will soon be fully feasible to observe variations in river form on global scales rather than relying on statistical relationships extrapolated from sparse measurements.

ACKNOWLEDGMENTS

This study was funded by NASA New Investigator Program Grant # NNX12AQ77G, managed by Ming-Ying Wei.

REFERENCES

Allen, G. H., J. B. Barnes, T. M. Pavelsky, and E. Kirby (2013), Lithologic and tectonic controls on bedrock channel form at the northwest Himalayan front, *J. Geophys. Res, 118*, 1–20.

Brabets, T. P., and M. A. Walvoord (2009), Trends in stream-flow in the Yukon River Basin from 1944 to 2005 and the influence of the Pacific Decadal Oscillation, *J. Hydrol.*, *371*, 108–119.

Brabets, T. P., B. Wang, and R. H. Meade (2000), Environmental and hydrologic overview of the Yukon River Basin, Alaska, U.S. Geological Survey Water Resources Investigations, Rep. 99-4204, U.S. Geol. Surv., Anchorage, AK.

Brakenridge, G. R., S. V. Nghiem, E. Anderson, and S. Chien (2005), Space-based measurement of river runoff, *Eos Trans. AGU*, *86*(19), 185–188.

Brown, J., O. J. Ferrians, Jr., J. A. Heginbottom, and E. S. Melnikov (1997), Circum-Arctic map of permafrost and ground-ice conditions, Circum-Pacific Map Series, U.S. Geol. Surv, Reston, VA.

Butman, D., and P. A. Raymond (2011), Significant efflux of carbon dioxide from streams and rivers in the United States, *Nature Geosci.*, *4*, 839–942.

Chanson, H. (2004), *Hydraulics of Open Channel Flow*, Butterworth-Heinemann, Burlington, MA.

Paiva, R. C. D., D. C. Buarque, W. Collichson, M-P. Bonnet, F. Frappart, S. Calmant, and C. A. B. Mendes (2013), Large-scale hydrologic and hydrodynamic modeling of the Amazon River basin, *Water Resourc. Res.*, *49*, 1226–1243.

Donner, S. D., M. T. Coe, J. D. Lenters, T. E. Twine, and J. A. Foley (2002), Modeling the impact of hydrological changes on nitrate transport in the Mississippi River Basin from 1955 to 1994, *Global Biogeochem. Cycles*, *16*(3), 1043.

Eaton, B. C., and M. Church (2007), Predicting downstream hydraulic geometry: A test of rational regime theory, *J. Geophys. Res.*, *112*, F03025.

Eisenhauer, J. G. (2003), Regression through the Origin, *Teaching Statistics*, *25*(3), 76–80.

Finnegan, N. J., G. Roe, D. R. Montgomery, and B. Hallet (2005), Controls on the channel width of rivers: Implications for modeling fluvial incision of bedrock, *Geology*, *33*(3), 229–232.

Finnegan, N. J., L. S. Sklar, and T. K. Fuller (2007), Interplay of sediment supply, river incision, and channel morphology revealed by the transient evolution of an experimental bedrock channel, *J. Geophys. Res.*, *112*(F3), F03S11, doi:10.1029/2006jf000569.

Griffiths, G. A. (1980), Hydraulic geometry relationships of some New Zealand gravel bed rivers. *J. Hydrol. (NZ)*, *19*, 106–118.

Griffiths, G. A. (2003), Downstream hydraulic geometry and hydraulic similitude, *Water Resourc. Res.*, *39*(4), 1094.

Hayes, F. E., and J-A. N. Sewlal (2004), The Amazon River as a dispersal barrier to passerine birds: Effects of river width, habitat and taxonomy, *J. Biogeogr.*, *31*, 1809–1818.

Hinzman, L. D., et al. (2005), Evidence and implications of recent climate change in Northern Alaska and other Arctic regions, *Climatic Change*, *72*, 251–298.

Homer, C., C. Huang, L. Yang, B. Wylie, and M. Coan (2004), Development of a 2011 national landcover database for the United States, *Photogram. Eng. Remote Sens.*, *70*, 829–840.

Ibbitt, R. P., G. R. Willgoose, and M. J. Duncan (1999), Channel network simulation models compared with data from the Ashley River, New Zealand, *Water Resourc. Res.*, *35*(12), 3875–3890.

Jenkins, R. A., K. R. Wade, and E. Pugh (2004), Macroinvertebrate-habitat relationships in the River Teifi catchment and the significance to conservation *Freshwater Biology*, *14*, 23–42.

Kaser, G., M. Groshauser, and B. Marzeion (2010), Contribution potential of glaciers to water availability in different climate regimes, *Proc. Natl. Acad. Sci.*, *107*(47), 20,223–20,227.

Kemp, J. L., D. M. Harper, and G. A. Crosa (1999), Use of "functional habitats" to link ecology with morphology and hydrology in river rehabilitation, *Aquatic Conserv. Marine Freshwater Ecosyst.*, *9*, 159–178.

Kirby, E., and K. X. Whipple (2012), Expression of active tectonics in erosional landscapes, *J. Struct. Geol.*, doi:10.1016/j.jsg.2012.07.009.

Klein, M. (1981), Drainage area and the variation of channel geometry downstream, *Earth Surf. Process. Landforms*, *6*, 589–593.

Knighton, A. D. (1974), Variation in width-discharge relation and some implications for hydraulic geometry, *Geol. Soc. Am. Bull.*, *85*, 1069–1076.

Lee, J.-S., and P. Y Julien (2006), Downstream hydraulic geometry of alluvial channels, *J. Hydraul. Eng.*, *132*(12), 1347–1352.

Leopold, L. B., and T. Maddock (1953), The hydraulic geometry of stream channels and physiographic implications, U.S. Geological Survey Professional Paper.

Leopold, L. B., and M. G. Wolman (1957), River channel patterns: braided, meandering and straight, Physiographic and Hydraulic Studies of Rivers Geological Survey Professional Paper 282-B.

Leopold, L. B., M. G. Wolman, and J. P. Miller (1964), *Fluvial Processes in Geomorphology*, W. H. Freeman, San Francisco.

McCartney, B. (1986), Inland waterway navigation project design, *J. Waterway Port Coastal Ocean Eng.*, *112*(6), 645–657.

Miller, Z. F., T. M. Pavelsky, and G. H. Allen (in review), Quantifying river form variations in the Mississippi Basin using remotely sensed imagery, *Hydrology and Earth System Science*.

Molnar, P., and J. A. Ramirez (2002), On downstream hydraulic geometry and optimal energy expenditure: Case study of the Ashley and Taieri Rivers, *J. Hydrol.*, *259*, 105–115.

Moody, J. A., and B. M. Troutman (2002), Characterization of the spatial variability of channel morphology, *Earth Surf. Proc. Landforms*, *21*, 1251–1266.

Mosley, M. P., and A. I. McKerchar (1993), Streamflow, in *Handbook of Hydrology*, edited by D. R. Maidment, pp. 8.1–8.39, McGraw-Hill, New York.

Neal, J., G. Schumann, and P. Bates (2012), A subgrid channel model for simulating river hydraulics and floodplain inundation over large and data sparse areas, *Water Resourc. Res.*, *48*, W11506.

Nicholas, A. (2013), Morphodynamic diversity of the world's largest rivers, *Geology*, *47*(4), 475–478.

O'Loughlin, F., M. A. Trigg, G. J.-P. Schumann, and P. D. Bates (2013), Hydraulic characterization of the middle reach of the Congo River, *Water Resourc. Res.*, *49*(8), 5059–5070, doi:10.1002/wrcr.20398.

Osterkamp, T. E. (2005), The recent warming of permafrost in Alaska, *Global Planet. Change, 49*, 187–202.

Overton, I. C. (2005), Modeling floodplain inundation on a regulated river: Integrating GIS, remote sensing, and hydrological models, *River Res. Applic., 21*, 991–1001.

Park, C. C. (1977), World-wide variations in hydraulic geometry exponents of stream channels: An analysis and some observations, *J. Hydrol., 33*, 133–146.

Pavelsky, T. M., and L. C. Smith (2008), RivWidth: A software tool for the calculation of river widths from remotely sensed imagery, *IEEE Geoscience and Remote Sens. Lett., 5*(1), 70–73.

Prévost, E. E. Parent, W. Crozier, I. Davidson, J. Dumas, G. Gudbergsson, K. Hindar, P. McGinnity, J. MacLean, and L. M. Sættem (2003), Setting biological reference points for Atlantic salmon stocks: Transfer of information from data-rich to sparse-data situations by Bayesian hierarchical modelling, *ICES J. Marine Sci. Journal du Conseil, 60*(6), 1177–1193.

Rhoads, B.L. (1991), A continuously varying parameter model of downstream hydraulic geometry, *Water Resources Research, 27*(8), 1865–1872.

Sagarin, R., and F. Micheli (2001), Climate change in nontraditional data sets, *Science, 294*, 811.

Selkowitz, D. J., and S. V. Stehman (2011), Thematic accuracy of the National Land Cover Database (NLCD) 2001 land cover for Alaska, *Remote Sens. Environ., 115*, 1401–1407.

Sheng, Y., P. Gong, and Q. Ziao (2001), Quantitative dynamic flood monitoring with NOAA AVHRR, *Int. J. Remote Sens, 22*(9), 1709–1724.

Singh, V. P., C. T. Yang, and Z.Q. Deng (2003), Downstream hydraulic geometry relations: 1. Theoretical development, *Water Resourc. Res., 39*(12), 1337.

Smith, T. R. (1974), A derivation of the hydraulic geometry of steady-state channels from conservation principles and sediment transport laws, *J. Geol., 82*, 98–104.

Smith, L. C., and T. M. Pavelsky (2008), Estimation of river discharge, propagation speed and hydraulic geometry from space: Lena River, Siberia, *Water Resourc. Res., 44*, W03427.

Spearman, C. (1904), The proof and measurement of association between two things, *Am. J. Psych., 15*(1), 72–101.

Tank, J. L., E. J. Rosi-Marshall, M. A. Baker, and R. O. Hall, Jr. (2008), Are rivers just big streams? A pulse method to quantify nitrogen demand in a large river, *Ecology, 89*(10), 2935–2945.

Troitsky, M. S. (1994), *Planning and Design of Bridges*, New York, Wiley.

Verdin, K. L., and S. K. Greenlee (1998), HYDRO1k documentation, U.S. Geol. Surv., available at https://lta.cr.usgs.gov/sites/default/files/GTOPO30_README.doc.

Walvoord, M. A. and R. G. Striegl (2007), Increased groundwater to stream discharge from permafrost thawing in the Yukon River basin: Potential impacts on lateral export of carbon and nitrogen, *Geophys. Res. Lett., 34*, L12402.

Walvoord, M. A., C. I. Voss, and T. P. Wellman (2012), Influence of permafrost distribution on groundwater flow in the context of climate-driven permafrost thaw: Example from Yukon Flats Basin, Alaska, United States, *Water Resourc. Res., 48*, W07524.

Wohl, E. (2004), Limits of downstream hydraulic geometry, *Geology, 32*(10), 897–900.

Yamazaki, D., S. Kanae, H. Kim, and T. Oki (2011), A physically based description of floodplain inundation dynamics in a global river routing model, *Water Resourc. Res., 47*, W04501.

Yamazaki, D., F. O'Loughlin, M. A. Trigg, Z. F. Miller, T. M. Pavelsky, and P. D. Bates (2014), Development of the Global River Width Database, *Water Resourc. Res, 50*, 3467–3480.

Yang, C. T., C. C. Song, and M. J. Woldenberg (1981), Hydraulic geometry and minimum rate of energy dissipation, *Water Reourc. Res., 17*(4), 1014–1018.

Yanites, B. J., G. E. Tucker, K. J. Mueller, Y. G. Chen, T. Wilcox, S.-Y. Huang, and K.-W. Shi (2010), Incision and channel morphology across active structures along the Peikang River, central Taiwan: Implications for the importance of channel width, *Geol. Soc. Am. Bull., 22*(7/8), 1192–1208.

Yarie, J., L. Viereck, K. Van Cleve, and P. Adams (1998), Flooding and ecosystem dynamics along the Tanana River, *BioScience, 48*(9), 690–695.

Yeager, C. P. (1991), Possible antipredator behavior associated with river corssings by proboscis monkeys (*Nasalis iarvatus*), *Am. J. Primatol., 24*, 61–66.

9

Near-Nadir Ka-band Field Observations of Freshwater Bodies

Delwyn Moller[1,2] and Daniel Esteban-Fernandez[1]

9.1. INTRODUCTION

The ability to observe and monitor the volume of water stored and flowing in rivers, lakes, and wetlands globally is of paramount importance. Although in situ measurements can be very accurate in quantifying river discharge and stage, they are globally sparse (both temporally and spatially), with the exception of the United States and other industrialized regions. On the other hand, satellite measurements could allow basin-wide measurements of discharge and water storage globally.

The Surface Water Ocean Topography (SWOT) satellite mission concept has been recommended by the National Academy of Sciences Decadal Review of Earth Science and Applications from Space (ESAS) [*National Research Council*, 2007]. SWOT is projected to launch in 2020 and would be a mission that partners National Aeronautics and Space Administration NASA, Centre national d'études spatiales CNES and the Canadian Space Agency CSA. The SWOT mission would provide measurements of water surface elevation in both the oceanic and terrestrial, with its key instrument payload being a Ka-band radar inteferometer (KaRIN) capable of making high-resolution wide-swath height measurements [*Fu et al.*, 2009]. In particular, by looking at a range of near-nadir incidence angles (from $0.5° < \theta_i < 4.5°$ on both sides of the spacecraft) the coverage of KaRIN would be substantially greater than that of conventional altimetry, which only provides a nadir track measurement. A Ka-band center frequency, in combination

with a relatively large separation between interferometric antennas (hereafter termed baseline) of 10 m, provides a high level of accuracy, at a technologically feasible baseline, while still penetrating clouds and light rain.

Fundamental to the measurement concept are the assumptions that (1) there would be sufficient backscatter from the water at the off-nadir incidence angles, and (2) that the scene would remain sufficiently correlated over the inland water bodies during the aperture synthesis time to acheive the required resolution (the ocean resolution requirement is relatively coarse and so would not be a driving issue). However there currently exists little supporting data, especially for inland water bodies. Over the past 20 years, there is a wealth of Ku-band radar data over the ocean (e.g., Tropical Rainfall Measuring Mission TRMM, Topex, Jason series), which has allowed the determination of the near-nadir behavior of the cross section over the ocean. In the case of Ka-band, near-nadir ocean scattering observations have been published by *Walsh et al.* [1998], *Vandemark et al.* [2004], and *Tanelli et al.* [2006]. From measurements obtained with the NASA scanning radar altimeter, *Walsh et al.* [1998] characterize the dependence of the ocean surface $\sigma_0 (\theta_i)$ for $0 < \theta_i < 25°$ for low to gale force ocean surface wind speed conditions. *Chapman et al.* [1994] estimated a decorrelation time of just 3 ms for the ocean using their Ka-band scatterometer.

Radar observations of freshwater bodies are even more limited both at Ka-band frequencies and for near-nadir incidence angles. In particular, *Vandemark et al.* [2004] presented limited nadir σ_0-derived surface slope observations of a single inland water body as a point of comparison for their more substantial ocean observations. There are indications from the study of Ku-band nadir

[1] *Jet Propulsion Laboratory, California Institute of Technology, Pasadena, California, USA*

[2] *Remote Sensing Solutions, Inc., Barnstable, Massachusetts, USA*

Remote Sensing of the Terrestrial Water Cycle, Geophysical Monograph 206. First Edition. Edited by Venkat Lakshmi.
© 2015 American Geophysical Union. Published 2015 by John Wiley & Sons, Inc.

altimeter waveforms that the nadir cross section exhibits a peak that is much brighter than the near-nadir cross section [*Berry et al.*, 2004]. However the functional form of $\sigma_0(\theta_i)$ is unknown due to the limited dynamic range of altimeters. Recently, *Fjortoft et al.* [2013] showed near-field, near-nadir σ_0 measurements as a function of wind speed for a focused wave-tank study, as well as interferometric airborne Ka-band synthetic aperture radar SAR-derived σ_0 for several water bodies, the ocean, and a variety of land cover types. There are no currently available data sets that measure near-nadir τ_c at Ka-band frequencies. Fortunately, the launch of Indian Space Research Organization ISRO-CNES SARAL/AltiKa promises to contribute to the state of the knowledge in the near future [*Vincent et al.*, 2006].

The research in this chapter provides some initial observations and validating data in support of Ka-band near-nadir altimetry over freshwater bodies. In particular, we present field results for measurements of the two key parameters of interest: (1) the angular decay of the water scattering cross section $\sigma_0(\theta_i)$ and (2) the temporal decorrelation of the reflected signal from the water surface, τ_c. These data were collected with a repurposed radar prototype, originally developed as a demonstration unit for the terminal descent sensor (TDS) aboard the Mars Science Laboratory [*Pollard and Chen*, 2009]. This radar was deployed from a series of bridges. Section II describes the experimental configuration, sampling and sites. Section 9.2. presents observations and comparisons of $\sigma_0(\theta_i)$ from the various locations and under differing environmental conditions and for which we tabulate the corresponding surface slopes and small-scale roughness. Furthermore, the relative contribution of coherent and incoherent returns are estimated by fitting the measured voltage distributions to Rician statistics. Section 9.4. details the methodology employed to estimate the surface temporal decorrelation and then presents those results. Section 9.5. concludes this chapter and includes an introduction to future airborne acquisitions that will extend this work in support of the SWOT concept.

9.2. EXPERIMENTAL CONFIGURATION

9.2.1. Radar System Description

Designed as a testbed for the Mars Science Laboratory (MSL) landing radar, the 35.75 GHz Doppler radar was borrowed for the purpose of dedicated high spatial and temporal resolution measurements of a variety of freshwater bodies. The MSL breadboard is a versatile 1 W peak power, short-pulse radar capable of short-range operation (as close as 10 m). Key attributes of the radar are summarized in Table 9.1. We operated exclusively at the shortest pulse width of 4 ns (250 MHz band width).

Table 9.1 MSL radar breadboard parameters

Parameter	Unit	Value
Center transmit frequency	GHz	35.75
Wavelength	mm	8.4
Peak transmit power	W	1
Pulse width	ns	4
Minimum pulse pair interval	ms	0.08
Maximum pulse pair interval	ms	3.5
Burst size		100
Intraburst PRF	Hz	500
Interburst PRF	Hz	6
One-way antenna gain	dBi	33
Antenna beam width	degrees	3.0
Antenna polarization		H
Minimum range	m	10

Figure 9.1 shows the radar experiment setup. The radar was in a mobile rack, and the antenna—a horizontally polarized Gaussian optics lens antenna (beamwidth 3°)—was mounted upside down on a high-precision elevation-over-azimuth positioner at the end of a rigid boom. At typical heights above the surface, the illuminated area was O (1 m^2) and the return was beamlimited. The entire frame could be leveled and stablized in place using jack-screws on each corner of the mount. An anemometer was mounted on the frame as shown for wind speed and direction information and was logged synchronously with the radar data acquisitions. Given that for all but one of the sites the bridge height was $\approx 10\,m$ above the surface, these measurements can be considered analogous to the traditional μ_{10}, or 10 m equivalent wind traditionally quoted in ocean analysis. In the case of Ohio, the wind speeds were so low we do not expect significant deviation from the measured values at 10 m height. The radar FPGA control and processor was customized for our purposes and data was collected in one of two modes: "backscatter" mode and "correlation" mode.

In backscatter mode, the positioner was programmed to sequence through a series of incidence angles at predetermined (usually 0.1°) elevation increments. Because we were operating in a beam-filled geometry, we deliberately chose a fine elevation increment in order to be able to adequately deconvolve the antenna pattern when estimating $\sigma_0(\theta_i)$. At each angle a series of 100 pulse bursts were transmitted. Each burst was coherently averaged to reduce random measurement noise. Multiple sets of bursts were collected per elevation at an approximately 6 Hz rate, which could be further averaged incoherently. In order to sufficiently average the water dynamics, the observation time per elevation position was approximately one minute. Because of the large backscatter dynamic range experienced as a function of both incidence angle and surface conditions, we used in-line attenuators directly prior to

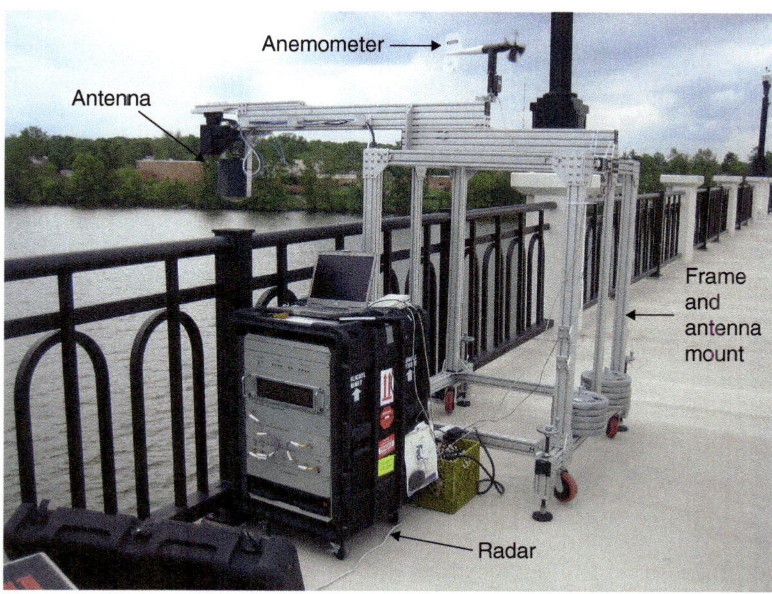

Figure 9.1 Radar experiment setup.

the receiver as necessary to avoid saturation and this value was logged. When the attenuators were changed, we would collect a region of incidence angles that overlapped with the prior configuration to verify continuity and consistency with the calibration (see Section 9.2.3.).

In temporal correlation mode a series of acquisitions were collected at a static incidence angle. The radar was capable of collecting two distinct pulse pair intervals that were inter-leaved [*Doviak and Zrnic*, 1993]. Although the products delivered by the radar were not capable of directly measuring the temporal decorrelation time, as discussed in Section 9.4., we were able to estimate the decorrelation time based on the variance of the mean of the pulse pair phase as a function of the time lag.

9.2.2. Experiment Sites and in situ Measurements

In May 2007, bridge-based Ka-band radar observations were conducted at several diverse river and reservoir locations in Ohio followed by a successive campaign in October 2008 in Redding, California. Table 9.2 provides a high-level summary of the general characteristics of the sites. To provide context, the proposed SWOT surface water science requirements [*Rodriguez*, xxxx] apply to inland water bodies greater than 1 km^2 in area and rivers greater than 100 m wide. As such, the coarse designation of "large," "medium," or "small" in Table 9.2 correspond to proposed SWOT "required," "goal," or "not required" targets, respectively. Note that to meet SWOT's proposed height accuracy requirements the normalized radar cross section, $\sigma_0(\theta_i)$ is required to be greater than 10 dB for $0.5° < \theta_i < 4.5°$.

Table 9.2 Summary of experiment locations and characteristics

Location	Site Description	Observation Height (m)
Home Road Reservoir	Reservoir	12
Ohio River	Large slow-moving river	30
Muskingum	Small fast-moving river	10
Sacramento: Diestelhorst	Medium river/laminar flow	10
Sacramento: Sundial	Medium river/turbulent flow	10
Sacramento: Stress Ribbon	Small river (deep narrow reach)	11

All of the locations exhibited distinctly different flow and water surface characteristics. In particular, it was notable that at some locations the water surface roughness was directly attributable to the local winds (e.g., Ohio River), while others had surface roughness irregardless of the wind, which was turbulence induced (e.g., Sacramento River at the Sundial Bridge). Since the area of illumination is generally very small (≈1 m^2 for all except the Ohio river where the bridge was higher), care was taken to choose locations that avoided obvious surface signatures due to flow around bridge pylons or other localized effects. At most sites data was collected both looking along the direction of flow, and at various angles (i.e., 30°, 45°, 60°) to the direction of flow. Also, at most sites we would move our location along the bridge and also the side of the bridge we were observing. Generally, however, for the data analysis shown here, the results presented seemed insensitive to the specific location on the bridge and the observation direction with respect to the river flow. It may well be that the relative insensitivity can be at least partially a result of the overall low-wind conditions observed. Indeed during one data-take on the Diestelhorst Bridge the wind speed rose to be consistent and > 3 m/s blowing directly upstream. Significant roughening was visible and long waves developed surface. In this situation one might expect an angular sensitivity to the backscatter. Unfortunately, due to the nature of the experiment logistics, specifically the scan and reconfiguration times, we were unable to test this before the wind died down once more.

9.2.3. Radar Calibration

Given the original application for which this Ka-band (35.75 GHz) radar was developed, the design does not currently include a calibration loop. Absolute calibration was therefore performed first externally by using a trihedral corner reflector of known radar cross section, from which the radar calibration constant can be derived. The receiver was also independently calibrated by inputing a 35.75 GHz sine-wave tone from a radio-frequency (RF) signal generator to the antenna port of the radar, with power levels covering its full dynamic range to characterize its linearity. The absolute accuracy of the RF source was verified with a high-accuracy power meter, resulting in an end-to-end calibration of the receiver chain to within 0.2 dB. The antenna gain and antenna patterns were measured by the antenna manufacturer and deemed accurate to within 0.3 dB. The transmitted waveform was independently measured with a Ka-band power detector and a 20 Gsamples/s digital scope; the output peak power is known to be 30 dBm (1 W) to within 1 dB. From these measurements, a second calibration constant was derived. These two independent ways of performing an absolute calibration were found to be in agreement to within 2 dB, which constitutes the maximum absolute error in the derived backscatter measurements. The overall radar stability was analyzed from several hours long observations of the corner reflector in an open environment, where the temperature of the radar electronics was allowed to vary freely (while being recorded), both during shadow and direct sunlight exposure. These observations, acquired over several days, showed an end-to-end stability of better than ± 0.3 dB(1σ). Relative calibration was also maintained by acquiring corner reflector measurements during every day of operation during the deployment.

As mentioned in the previous section, the use of in-line attenuators was necessary to avoid saturation and due to the large dynamic range experienced. The attenuator values were all verified and precisely recorded in the laboratory prior to the campaigns using a calibrated network analyzer. In the field the attenuators were inserted and removed as necessary and the Sub Miniature version A SMA connections were tightened and verified using an SMA torque wrench.

9.3. EXPERIMENTAL RESULTS FOR σ_0 AND COHERENT SCATTERING

9.3.1. Coherent and Incoherent Scattering Mechanisms

In the case of a water surface where we can neglect the presence of long-wavelength waves, as might be appropriate for many rivers and small lakes, the coherent and incoherent scattering contributions to backscattered power measurements, P_r, can be summarized by [*Tsang et al.*, 1985]

$$P_r = \frac{P_t \lambda^2}{(4\pi)^3} \frac{G^2(\theta)}{r^4} \frac{k^2 |R|^2}{\pi \cos^2 \theta} \left[|\langle I \rangle|^2 + \langle II^* \rangle \right] \quad (9.1)$$

where P_t and P_n are the transmitted and received powers, $G(\theta)$ is antenna gain for look angle θ, λ is the electromagnetic wavelength, $k = 2\pi/\lambda$ is the wave number, r is the range to the surface, and $|R|^2$ is the Fresnel reflection coefficient. The two contributions $|\langle I \rangle|^2$ and $\langle II^* \rangle$ are the coherent and incoherent contributions, respectively, for which the angular brackets indicate averaging over surface realizations. The backscattered field I is given by

$$I = \int dA \, \exp\,(2j\mathbf{k}.\mathbf{x}) \exp\,[-j2k_z h(\mathbf{x})] \quad (9.2)$$

where $\mathbf{k}$ is the component of the wave vector in the surface plane, k_z is the vertical component of the wave vector, h is the surface height, and the integral is taken over the area illuminated by the transmitted pulse at a given incidence angle. Using a conventional quasi-specular approach one can determine that $\langle I \rangle \propto A_c$, the coherent scattering area and $\langle II^* \rangle \propto A_i(\theta_i)$, the incoherent scattering area. It is notable that the coherent term in equation 9.1 depends quadratically on A_c, while the incoherent term depends linearly on A_i.

The coherent scattering area, $A_c \int dA \, \rho(x)$ is the effective scattering area for a coherent patch where $\rho(x)$ is a step window function (i.e., it has a value of 1 inside the coherent patch, and 0 outside). Note that the coherent contribution $\langle I \rangle$ decays very quickly with increasing patch roughness by the factor $\exp\left(2k_z^2 \sigma_h^2\right)$ and as such it is often neglected. However, in the next section we assess the relative contributions of coherent and incoherent scatter from our measurements.

9.3.2. Estimates of Coherent and Incoherent Components

To investigate the mixture of coherent and incoherent scattering, we examine the statistics of the return field (or, equivalently, voltage). It is well known [*Goodman*, 1985] that for a speckling random surface, the scattered field values will follow a Rayleigh distribution. On the other, for a perfectly coherent return, one expects the coherent field value to be a single phasor. For scattering from rivers, where we expect patches of flat water to be advected by the flow, and the returns to consist of mixtures of coherent and incoherent returns, we expect that the return voltage V will follow the Rice probability density function [*Goodman*, 1985]:

$$f(V \mid \upsilon, \sigma) = \frac{\upsilon}{\sigma^2} \exp\left[-\frac{V^2 - \upsilon^2}{2\sigma^2} \right] I_0\left(\frac{V\upsilon}{\sigma^2} \right) \qquad (9.3)$$

where I_0 is the modified zeroth-order Bessel function of the first kind, v can be thought of as the coherent phasor magnitude, and σ determines the expected value of the incoherent speckling component. The ratio $\mu = v/\sigma$ is then an estimate of the relative contributions from coherent and incoherent patches: when no coherent scattering is present, $\mu \to 0$, while $\mu \to \infty$ when no speckle is present.

We are not aware of an optimal estimator for μ. However, we found that a robust estimator could be obtained by computing the median (m), the 15th (p_{15}) and 85th (p_{85}) percentiles, forming the ratio $m/(p_{85} - p_{15})$ and comparing the result against tabulated values of this ratio for the Rice distributions of varying μ. We present in Figure 9.2 contrasting examples of two voltage histograms collected near-nadir ($\theta_i < 0.2°$) and their corresponding best-fit Rice distribution: purely incoherent (Rayleigh distributed) scattering from the (a) Sundial Bridge and a case with strong coherent component for a low-wind condition from the (b) Distelhorst Bridge.

Figure 9.3 shows examples of the calculated ratio V/σ for the different river conditions measured. As expected, μ decreases with incidence angle, as the incoherent

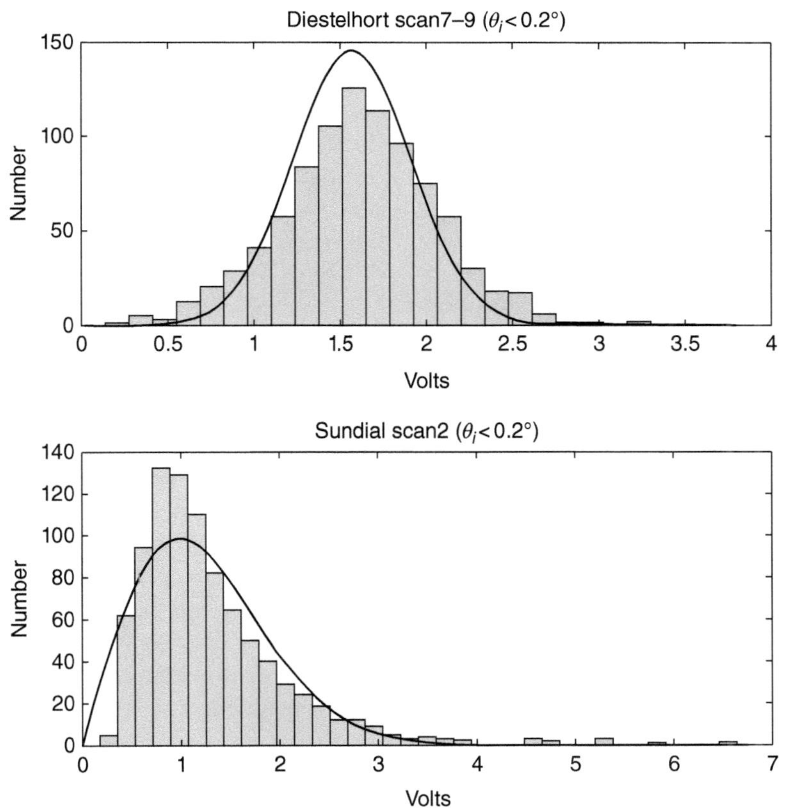

Figure 9.2 Examples for measured voltage histograms for the Sundial Bridge (*bottom*), which exhibits purely incoherent scattering; and a low-wind condition for the Diestelhorst Bridge (*top*), which exhibits significant coherent scattering. The solid line represents the Rice fit obtained by estimating the median and the ratio μ. Note the voltage axes are not on the same scale as front-end attenuation could vary between acquisitions. Rather it is the relative distribution that is important.

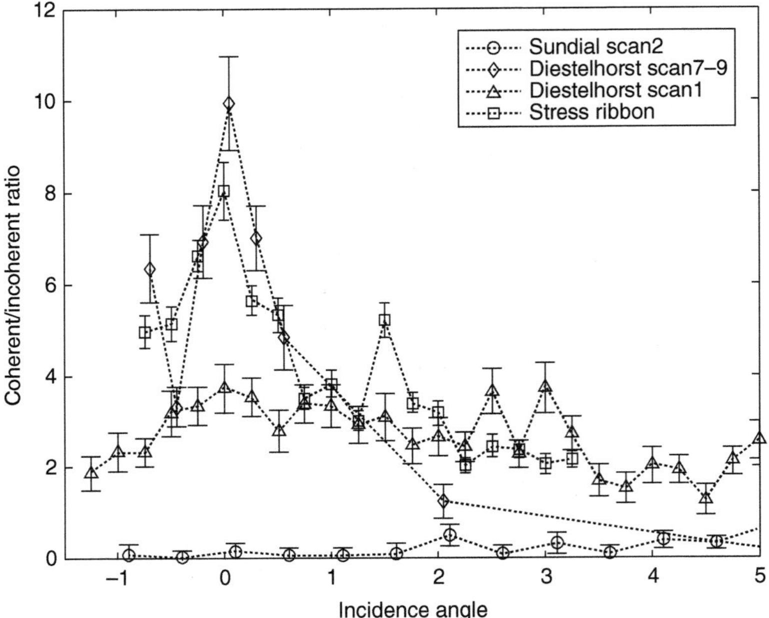

Figure 9.3 Coherent to incoherent ratio V/σ as a function of incidence angle for different river conditions. The error bars represent standard deviations of the ensemble estimate.

component starts to dominate the coherent component. However, a significant coherent component can sometimes still be measured at off-nadir angles approaching 5°. This can be explained if the coherent field is due to small flat facets: for instance, at 35.75 GHz, a coherent 5° beam width can be produced by a flat patch whose dimension is approximately 10 cm. When the scattering is observed to be broad (e.g., the Sundial Bridge), the return field statistics are consistent with incoherent (Rayleigh) returns.

9.3.3. Observations of σ_0

Measurements of the radar backscattered power, P_r, were made as a function of incidence angle, θ_i, at each location under varying conditions. For our experimental geometry we assume the following form for the radar range equation:

$$P_r\left(\theta_i\right) = \frac{P_t \lambda^2}{\left(4\pi\right)^3}$$

$$\iint \frac{G^2\left[\theta(x,y) - \theta_i, \phi(x,y)\right]\sigma_0\left[\theta(x,y)\right]}{r^4(x,y)}\, dx\, dy \quad (9.4)$$

where P_t and P_r are the transmitted and received powers, $G(\theta, \phi)$ is the gain pattern of the antenna over look angle, θ, and azimuth, ϕ, respectively, λ is the radar

wavelength, and r is the range to the water. Because we were operating in beam-limited mode, it is necessary to deconvolve the antenna pattern from the measurement in order to estimate $\sigma_0(\theta_i)$, and thereby it is also necessary to assume a functional form for $\sigma_0(\theta_i)$. For the normalized radar cross section σ_0 we use the geometric optics approximation:

$$\sigma_0 = \frac{|R|^2}{\cos^4(\theta)} \frac{\exp\left[-(\tan^2\theta/s^2)\right]}{s^2} \exp\left[-\left(2k\sigma_{hs}\right)^2\right] \quad (9.5)$$

where $|R|^2$ is the Fresnel reflection coefficient and s^2 is the surface mean-squared slope (mss). The frequency and angular dependence is introduced by the Fresnel reflection coefficient and the mss [*Freilich and Vanhoff*, 2003]. Note that to be consistent with the coherent reflection specular limit and two-scale scattering theories [*Kim and Rodriguez*, 1992; *Rodriguez and Kim*, 1992; *Tsang et al.*, 1985], we include the term $\exp[-(2k\sigma_{hs})^2]$, where k is the electromagnetic wave length and σ_{hs} is the standard deviation of small-scale height, i.e., the height contributions smaller than a few wavelengths.

Given knowledge of the measurement geometry, and the antenna patterns, we use the measurements of $P_r(\theta_i)$ to invert for the model parameters in (5): an effective Fresnel reflection coefficient, R, the mean-squared slope, s^2, and the small-scale surface roughness σ_{hs}. To illustrate the methodology, Figure 9.4 shows an example fit and the

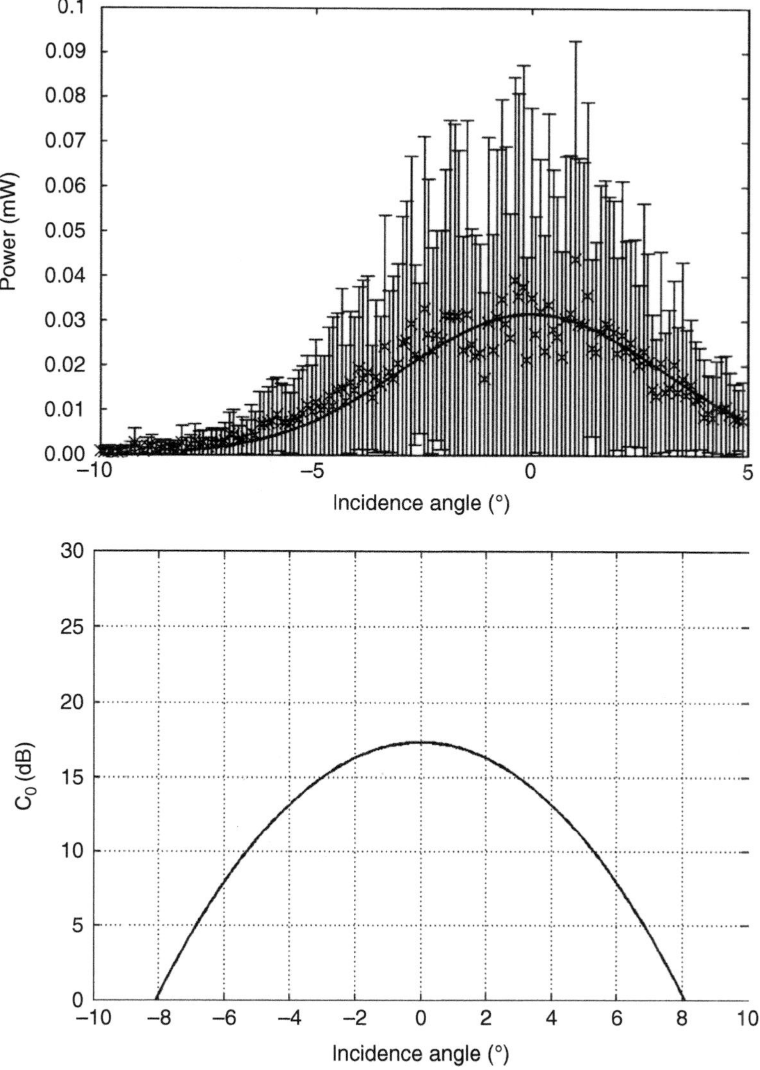

Figure 9.4 (*Top*) Example of the data fit using geometrical optics model. This case is over the Ohio River with a least squares fit resulting in a surface slope of 4.1° and small-scale roughness of 0.0. (*Bottom*) Resulting $\sigma_0(\theta_i)$ from the least squares fit to the data above.

corresponding $\sigma_0(\theta)$ for an Ohio River observation. There was a light breeze blowing throughout the acquisition, and one can see that the required 10 dB σ_0 for $-4.5° < \theta_i < 4.5°$ is readily achieved. Cases where surface conditions changed significantly during an acquisition (primarily due to changing wind conditions) were discarded from this analysis as the least-squares fit did not converge.

Figure 9.5 shows fits for two different scans (upper left and right) under different conditions observed from the Sundial Bridge over the Sacramento River. The bottom figures are the corresponding histograms of the wind speed during each scan. This particular reach had strong turbulent flow and one can see that for somewhat differing wind conditions the power rolloff is relatively invariant.

Figure 9.6 shows a summary of $\sigma_0(\theta_i)$ profiles for four locations over a variety of conditions. Figure 9.6a shows $\sigma_0(\theta_i)$ profiles from the Diestelhorst Bridge over the Sacramento River (representative of a medium river with more laminar flow). A distinct contrast is seen between the profiles of when there is a moderate wind ($u > 1$ m/s) as opposed to a low wind to still condition. Figure 9.6b shows that for the Sacramento Sundial Bridge (representative of a medium river with turbulent flow) the proposed SWOT criteria are met. Figure 9.6c shows profiles for where the Home Road Reservoir meets the proposed SWOT level across the full range of off-nadir angles. Note that for these cases there was a moderate wind consistently blowing. No low-wind data was observed. Figure 9.6d shows contrasting profiles for the Ohio River.

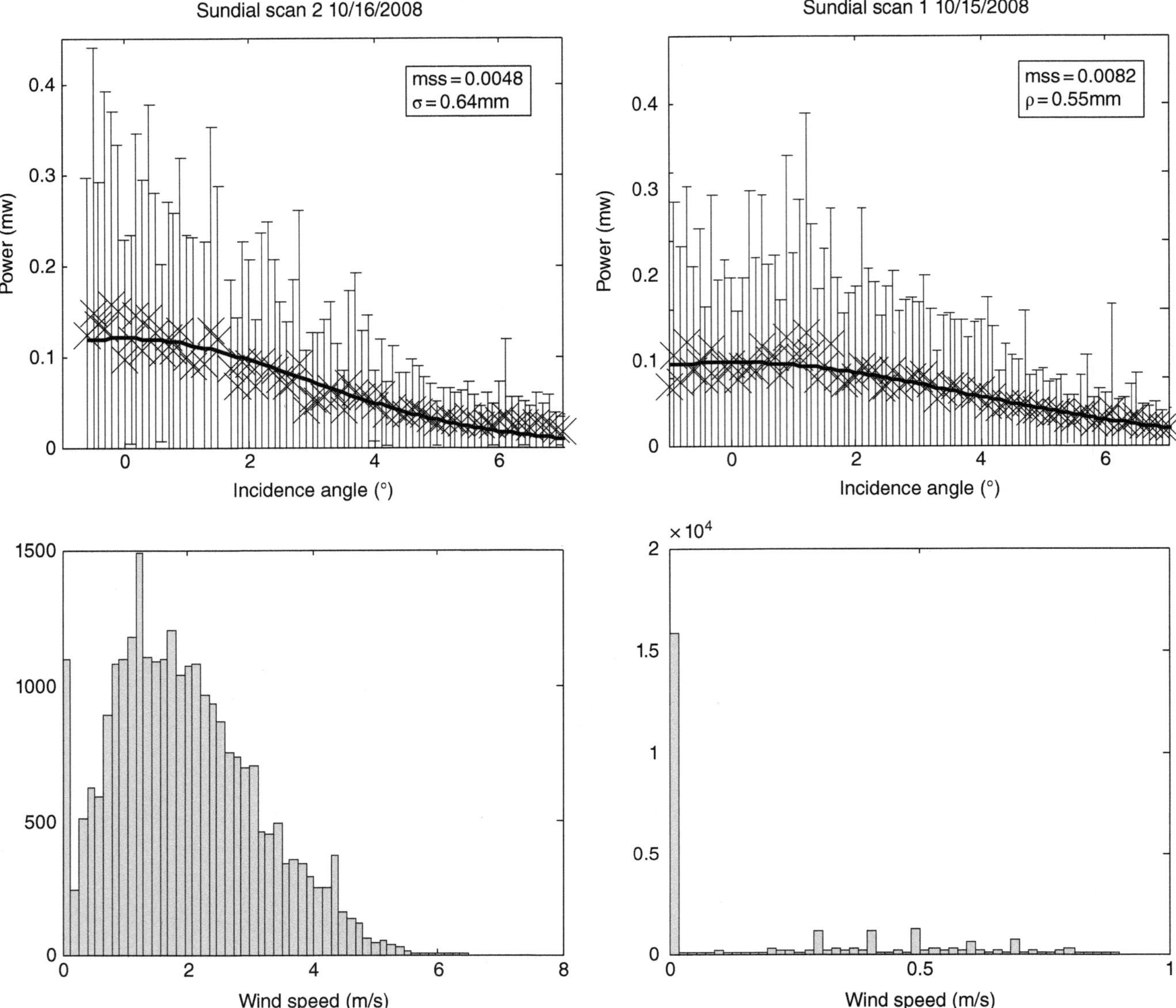

Figure 9.5 Example data fits for the Sundial Bridge (representative of turbulent flow). (*Top*) Figures show the least squares data fits to the backscattered power. (*Bottom*) Graphs show a histogram of the corresponding wind speeds during the data-take. Due to the turbulence-induced surface roughness, the backscattered power is relatively insensitive to the very different wind conditions (summarized also in Table 9.3).

The Ohio is a larger river and would represent a proposed SWOT-required river. For this case the contrasts in $\sigma_0(\theta)$ decay seemed to be primarily attributed to the wind speed, which dropped from a moderate ($u \approx 1$–$2\,$m/s) wind for the flatter σ_0 profiles (that meet requirements) to a low ($u < 0.5\,$m/s) wind for the sharply peaked $\sigma_0(\theta_i)$ characteristics. The implications of this are that SWOT may not meet proposed requirements in the outer portions of its swath instances.

Table 9.3 summarizes the corresponding slope and small-scale roughness solutions for each profile shown. Note that the range of s and σ_0 values are consistent with those observed over the ocean and one freshwater body

by *Vandemark et al.* [2004]. They are also consistent with the more recent results by *Fjortoft et al.* [2014] recognizing those observations all occured with windspeeds of $4\,$m/s in all cases as compared with our observations with almost exclusively very low wind speeds. For example, the instances of sharp rolloff of $\sigma_0(\theta_i)$ were not observed by *Fjortoft et al.* [2014]. Clearly these results are limited in as much as the locations and number of observations are limited, but they represent a significant contribution in what is currently available. Although many more scans than those shown were collected, they occured under similar conditions and at the same site the results were essentially equivalent. Profiles over the Muskingum River are

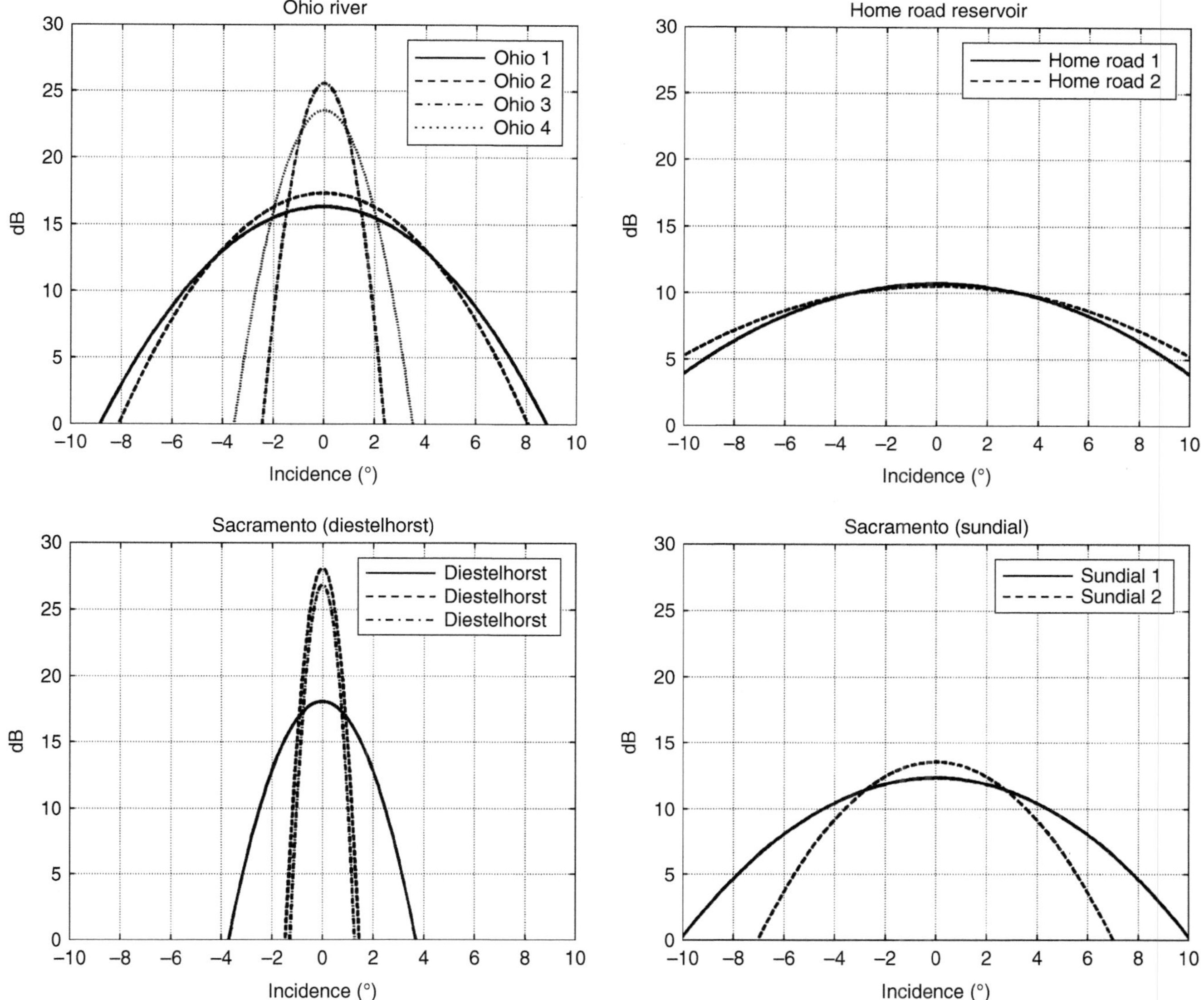

Figure 9.6 Summary of $\sigma_0(\theta_i)$ characteristics derived from four different observing locations where (a) is over the Ohio River (large river: SWOT requirement), (b) Home Road Reservoir, Ohio (open water: SWOT requirement), (c) Diestelhorst Bridge over the Sacramento (medium river laminar flow: SWOT goal), and (d) is the Sundial Bridge over the Sacramento (medium river, turbulent flow: SWOT goal). For each location and plot, the corresponding surface slope and roughness parameters are given in Table 9.3. The line sequence is solid, dashed, dash-dot, and dotted, respectively.

not shown because the near-nadir data were saturated as the backscatter was much higher than we anticipated early in these campaigns.

Of the locations observed only the Home Road Reservoir and the Ohio River would be actual required targets for SWOT. The Sacramento locations would be considered goals but not requirements. However, the variability of $\sigma_0(\theta_i)$ we observed has clear implications on meeting the proposed SWOT mission and science requirements. More observational statistics are needed beyond the scope of static bridge-based data and upcoming plans

for airborne acquisitions are introduced in Section 9.5. In the interim, however, the ground work of characterizing the range of surface variables (s^2 and σ_{hs}) that might be expected could guide sensitivity studies of SWOT.

9.4. ESTIMATION OF TEMPORAL COHERENCE TIMES

For a synthetic aperture radar system like that proposed for SWOT, the scene must remain correlated over the aperture synthesis time in order to achieve the full along-track resolution, r_a, of approximately half the antenna length (for SWOT $r_a \approx 2.5\,\text{m}$). However, if the surface decorrelates within the aperture synthesis time ($\approx 280\,\text{ms}$ for SWOT), then the resolution will be degraded and the effective resolution is given by

$$r_a = \frac{\lambda R}{2\tau_c \upsilon_p} \tag{9.6}$$

where τ_c is the $(1/e)$ surface decorrelation time and v_p is the platform velocity [*Raney*, 1980]. To illustrate this Figure 9.7 shows impact of the surface decorrelation on the effective along-track resolution assuming an orbit of approximately 890 km above the surface (the projected SWOT orbit altitude). This may impact SWOT's proposed requirement to produce a geolocated water mask at an average location of 50 m for water bodies of required size. For context, along this line we have overlaid decorrelation estimates from our field data, which we detailed subsequently. Note that the cross-track resolution varies from approximately 70 m near range to 10 m in the far range.

While the MSL radar was not able to directly measure coherence times of the water surface, it could measure the complex pulse pair product, the phase statistics of which can be used to estimate coherence times. As discussed in Section 9.2, the MSL breadboard was able to collect pulse pair products for a versatile set of temporal lags with the pulse pair phase given by the interferometric measurement:

$$\Delta\phi = \left\langle \text{Arg}\left[z(t)z^*(t+n\tau) \right] \right\rangle | N_B \tag{9.7}$$

where τ is the time interval between successive pulses and then n is the integer number of pulses where $1 \leq n \leq 7$. The averaging of $N_B = 100$ pulse pairs (a burst) was preset within the MSL firmware to reduce the impact of thermal noise on the phase measurement. Each burst would be collected within 0.2 s so it is reasonable to consider the surface conditions stationary in the broad sense within that time frame. Two different time lags could be selected and interleaved for a given burst. We selected lags of $\tau = 80\,\mu\text{s}$ and $\tau = 500\,\mu\text{s}$ to span a range of temporal lags from 80 to 3500 µs. The total burst times, T_B, for the 100 pulse pairs were 8 and 50 ms, respectively. At each individual integer setting we collected 100 bursts. The total

Table 9.3 Surface slope and small-scale roughness derived for each site

Elevation Scan	s (deg)	σ_{hs} (mm)	Wind Speed (m/s)
Ohio River 1	4.6	0.0	1.9 ± 0.6
Ohio River 2	4.1	0.0	1.9 ± 0.9
Ohio River 3	1.0	0.6	1.0 ± 0.7
Ohio River 4	1.5	0.5	0.7 ± 0.2
Home Road 1	7.9	0.3	4.2 ± 1.6
Home Road 2	8.9	0.0	4.0 ± 2.0
DiestelHorst 1	1.8	0.8	1.3 ± 1.0
DiestelHorst 2	0.6	0.8	1.0 ± 1.0
DiestelHorst 3	0.5	0.9	3.4 ± 0.3
Sundial 1	4.0	0.6	0.0 m/s with gusts to 0.9 m/s
Sundial 2	5.2	0.6	1.9 ± 1.1
Stress Ribbon	0.6	0.7	0.7 ± 0.7

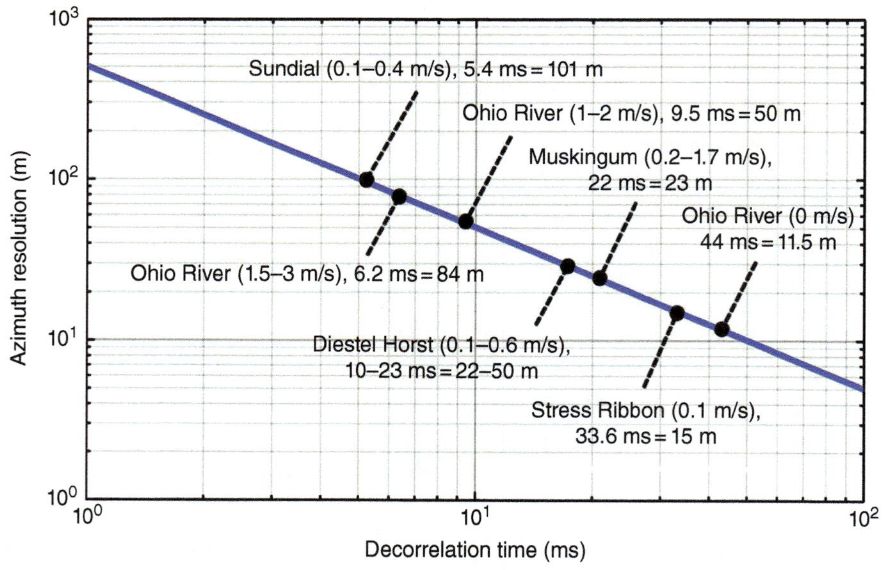

Figure 9.7 Effective along-track resolution as a function of decorrelation time for a projected SWOT orbit altitude of 890 km with observations overlaid.

acquistion time for a complete decorrelation measurement acquisition was limited by the burst rate of 6 Hz and would take approximately 11 min to complete for all possible integer lags.

If $\tau = 0$, then the probability density function (pdf) of $\Delta\phi$ is a delta function (with zero mean). However, as τ increases, the probability density function of $\Delta\phi$ broadens and the mean phase difference will shift until the phase decorrelates. In the limit where the two measurements are completely decorrelated, the phase differences are uniformly distributed between $-180°$ and $180°$ with a standard deviation of $360 / \sqrt{(12)} = 104°$ [*Chapman et al.*, 1994].

The phase standard deviation of the interferometric pulse pair measurement $\gamma = \text{std}(\Delta\phi)$, therefore, is related to the temporal decorrelation and can be expressed as

$$\gamma = \frac{1}{\sqrt{2N}} \frac{\sqrt{1-\rho^2}}{\rho} \qquad (9.8)$$

where N is the number of independent samples. We assume $\rho = e^{-(t/\tau_c)^2}$ for the surface as is often assumed for the ocean [*Raney*, 1980]. Using a least squares fit of the observations of γ as a function of lag, we estimate the decorrelation time, τ_c. Note that $N = T_B / \tau_c$ and this constraint is applied when performing the least squares fit.

Table 9.4 summarizes the derived decorrelation times from each site, along with the incidence angle of the observation and the wind speed statistics. We only reported estimates of τ_c for mean fit errors $< 8°$. In some cases if wind conditions changed, or were gusty during the data-take the estimate did not converge and these instances are ommited. The decorrelation times for the Ohio River are perhaps the most informative since there was a significant variation in the wind speed for this location for different acquisitions. In the case of the Ohio River, early in the day through afternoon was breezy resulting in $\tau_c = 6.2$ and $\tau_c = 9.5$ ms, respectively. However, into the evening the wind died to absolute calm and the decorrelation time increased to $\tau_c = 44.0$ ms. With the exception of the Home Road Reservoir, most observations occurred under light to no-wind conditions. The range of τ_c is significant and ranges from a little over 3 ms (and consistent with the ocean observations of *Chapman et al.* [1994]) up to 44.0 ms.

9.5. CONCLUSIONS AND FUTURE WORK

In this research some of the first near-nadir observations of surface water body backscatter at Ka-band was collected and analyzed in support of the proposed SWOT mission. Gradual decay of σ_0 with incidence angle is observed and the SWOT assumption of $\sigma_0 > 10$ dB is

Table 9.4 Summary of decorrelation time estimates for all sites

Site	τ_c (ms)	θ_i (deg)	Wind Speed, u (m/s)	Least Square Error (deg)
Ohio River a	6.2	1	$1.5 < u < 3.2$	2.1
Ohio River b	9.5	5	$1.1 < u < 1.9$	0.7
Ohio River c	44.0	1	0.0	0.5
Home Road a	3.5	1	$2.1 < u < 7.4$	3.7
Home Road b	4.4	4	$1.5 < u < 5.0$	5.1
Muskingum River a	18.9	5	No data	0.6
Muskingum River b	29.0	2	No data	0.3
Muskingum River c	21.7	5	$0.2 < u < 1.7$	0.4
Sacramento: Diestel-Horst a	17.2	1	0.5 ± 0.2	2.2
Sacramento: Diestel-Horst b	10.6	4	0.1 ± 0.1	1.0
Sacramento: Diestel-Horst c	2.8	12	1.3 ± 1.1	7.2
Sacramento: Diestel-Horst d	22.8	3	0.6 ± 0.8	1.0
Sacramento: Sundial a	3.6	1	0.4 ± 0.2	3.7
Sacramento: Sundial b	4.0	3	0.0	2.7
Sacramento: Sundial c	3.8	1	0.1 ± 0.1	2.4
Sacramento: Sundial d	4.0	3	0.1 ± 0.1	3.5
Sacramento: Sundial e*	5.4	1	0.0	0.5
Sacramento: Sundial f*	4.7	3	0.0	3.5
Sacramento: Sundial g*	4.9	5	0.0	0.5
Sacramento: Stress Ribbon a	33.6	1	0.1 ± 0.1	1.1
Sacramento: Stress Ribbon a	32.4	1	0.0	1.3
Sacramento: Stress Ribbon a	10.2	5	0.2 ± 0.1	2.6

Note that the sundial observations with an asterisk (e-g) were collected at a different location along the bridge (with shallower water) than the a-d observations.

confirmed when there is significant surface slope, which can occur even with slight wind roughening. However, an implication from this initial set of $\sigma_0(\theta_i)$ measurements is that, for very low wind speed conditions and little turbulence-induced roughness, there may be degraded performance at the edges of the swath or even narrowing of it. Unfortunately, a shortcoming of our results is a lack of high wind speed observations, although it is possibly relevant that the Ka-band near-nadir ocean σ_0 has been observed to fall below 10 dB for high wind speeds [*Vandemark et al.*, 2004; *Tanelli et al.*, 2006]. However, it is important to note that SWOT requirements are imposed at the 68% level. That is, surface scattering need be sufficient to meet SWOT measurement requirements over 68% of water body observations. Furthermore, and importantly, rapid temporal decorrelation would limit the azimuth resolution of SWOT, which in turn may limit the ability of KaRIN to accurately specify the spatial extent of the water body, a key variable in storage and discharge calculations. For this reason understanding the dynamics of the temporal decorrelation for different surface (current-induced turbulence) and meteorological (primarily wind) conditions is key for predicting storage change and discharge accuracies in a manner that is scalable to the global coverage projected for SWOT.

The bridge-based observations while informative are fundamentally limited by radar footprint and the portability of the ground-based configuration. In May 2009, a limited amount of Ka-band interferometric airborne data was collected by the GLISTIN airborne system [*Moller et al.*, 2011a] over inland water in support of SWOT [*Wu et al.*, 2011]. While the first of its kind and a valuable contribution, this system is not optimized for the near-term, or long-term needs of SWOT. Most notably the interferometer is constrained by and off-nadir boresite (31°) and limited to 80 MHz bandwidth (KaRIN is 200 MHz).

To address this need, the Ka-band SWOT phenomenology airborne radar (KaSPAR) has been built and is planned for flight as the primary payload for the "AirSWOT" campaigns [*Moller et al.*, 2011b]. KaSPAR is designed to mimic and characterize the SWOT geometry, scattering, and sampling (i.e., be SWOT-like) for a variety of targets and will extend the ground and airborne work by gathering more statistics from a wide variety of open water bodies (lakes, reservoirs, and varying size rivers including braided channels). These further measurements currently planned within the context of airborne campaigns will provide a far more extensive quantification of the spatial and temporal distribution of these occurrences beyond the regional cases discussed here, which will be used both to enhance our understanding of the expected ensemble average performance of SWOT at a global scale, as well as to support of the confirmation of the SWOT science requirements. Beyond this, KaSPAR will also provide data over a variety of vegetated and ice-covered regions and ultimately help maximize the science return from the proposed SWOT mission.

ACKNOWLEDGMENT

The research presented here was carried out at the Jet Propulsion Laboratory, California Institute of Technology, and Remote Sensing Solutions under contract with the National Aeronautics and Space Administration (NASA). The authors would like to thank Dr. Ernesto Rodriguez for his participation in the field campaigns and analytical guidance. We gratefully acknowledge Professor Douglas Alsdorf (OSU) and his team for support during the Ohio data collection campaign. We also thank Terry Hanson and the town of Redding, California, for the use of their bridges and their warm welcome and enthusiatic support during our Sacramento River campaign.

REFERENCES

Berry, P. A. M., J. D. Garlick, J. A. Freeman, and E. L. Mathers (2004), Blooms of σ^0 in the topex radar altimeter data, *J. Atmos. Oceanic Technol.*, *21*, 1232–1245.

Chapman, R. E., B. L. Gotwols, and R. E. Sterner (1994), On the statistics of the phase of microwave backscatter from the ocean surface, *J. Geophys. Res.*, *99*(C8), 16293–16301.

Doviak, R., and D. Zrnic (1993), *Doppler Radar and Weather Observations*, Academic Press, Orlando, Florida.

Fjortoft, R., J. M. Gaudin, J. C. Pourthie, A. Mallet, J. F. Nouvel, J. Martinot-Lagarde, H. Oriot, P. Borderies, C. Ruiz, and S. Daniel (2014), Karin on SWOT: Characteristics of near-nadir Ka-band interferometric SAR imagery, *IEEE Trans. Geosci. Remote Sens* "in press".

Freilich, M., and B. Vanhoff (2003), The relationship between winds, surface roughness, and radar backscatter at low incidence angles from TRMM precipitation radar measurements, *J. Atmos. Oceanic Technol.*, *20*, 549–562.

Fu, L.L., D. Alsdorf, E. Rodriguez, R. Morrow, N. Mognard, J. Lambin, P. Vaze, and T. Lafon (2009), The SWOT (surface water and ocean topography) mission: Spaceborne radar interferometry for oceanographic and hydrological applications, OceanObsÆ09 Community White Paper Proc. OceanObs '09.

Goodman, J. (1992), *Statistical Optics*, Wiley-Interscience, New York.

Kim, Y., and E. Rodriguez (1992), Comparison of the unified perturbation method with the 2-scale expansion, *IEEE Trans. Geosci. Remote Sens.*, *30*, 510–515.

Moller, D., S. Hensley, G. Sadowy, C. Fisher, T. Michel, M. Zawadzki, and E. Rignot (2011a), The glacier and land ice surface topography interferometer: An airborne proof-of-concept demonstration of high-precision Ka-band single-pass elevation mapping, *Trans. Geosci. Remote Sens.*, *49*, 827–842.

Moller, D., E. Rodriguez, J. Carswell and D. Esteban-Fernandez (2011b), AirSWOT - A Calibration/Validation Platform for the SWOT Mission, Proc. International Geoscience and Remote Sensing Symposium, Vancouver, Canada.

National Research Council (2007), *Earth Science and Applications from Space: National Imperatives for the Next Decade and Beyond*, The National Academies Press, Washington, DC.

Pollard, B., and C. Chen (2009), A radar terminal descent sensor for the mars science laboratory mission, Proc. Aerospace Conference, 2009 IEEE.

Raney, K. R. (1980), SAR response to partially coherent phenomena, *IEEE Trans. Antennas Propagation*, *AP-28*, 777–787.

Rodriguez, E. (custodian), Surface water and ocean topography mission (SWOT) science requirements document, available at https://swot.jpl.nasa.gov/files/swot/SWOT_Science_Requirements_Document.pdf.

Rodriguez, E., and Y. Kim (1992), A unified perturbation expansion for surface scattering, *Radio Sci.*, *27*, 79–93.

Tanelli, S., S.L. Durden, and E. Im (2006), Simultaneous measurements of Ku- and Ka-band sea surface cross sections by an airborne radar, *Geosci. Remote Sens. Lett.*, *3*(3), 359–363.

Tsang, L., J. Kong, and R. Shin (1985), Theory of Microwave Remote Sensing, Wiley-Interscience, New York.

Vandemark, D., B. Chapron, J. Sun, G. H. Crescenti, and H. C. Graber (2004), Ocean wave slope observations using radar backscatter and laser altimeters, *J. Phys. Oceanogr.*, *34*, 2825–2842.

Vincent, P., N. Steunou, E. Caubet, L. Phalippou, L. Rey, E. Thouvenot, and J. Verron (2006), Altika: A Ka-band altimetry payload and system for operational altimetry during the gmes period, *Sensors*, *6*, 208–234.

Walsh, E. J., D. C. Vandemark, C. A. Friehe, S. P. Burns, D. Khelif, R. N. Swift, and J. F. Scott (1998), Ocean wave slope observations using radar backscatter and laser altimeters, *J. Phys. Oceanogr.*, *34*, 2825–2842.

Wu, X., S. Hensley, E. Rodriquez, D. Moller, and R. Mullerschoen (2011), Near nadir Ka-band SAR interferometry: SWOT airborne experiment, in *International Geophysical and Remote Sensing Symposium Proceedings*.

Section IV: Snow

10

Snow Cover Depletion Curves and Snow Water Equivalent Reconstruction: Six Decades of Hydrologic Remote Sensing Applications

Noah P. Molotch,[1,2] Michael T. Durand,[3] Bin Guan,[4] Steven A. Margulis,[5] and Robert E. Davis[6]

10.1. INTRODUCTION

Recent works have shown that a 0.8 °C increase in global mean air temperature over the last 50 years has resulted in a decrease in snow accumulation in parts of North America and Europe [*Laternser and Schneebeli*, 2003; *Mote*, 2003]. This increase in air temperature has also caused snowmelt runoff to occur earlier over the same 50 year time period [*Bales et al.*, 2006; *Dettinger et al.*, 2004]. The distribution of snowmelt and snow water equivalent (SWE) is important for water resources because the resulting meltwater fills the reservoirs and groundwater aquifers that provide storage for subsequent irrigation, municipal use, and hydropower generation [*Dunne and Leopold*, 1978]. Current operational snowmelt runoff forecasts use empirical relationships that rely on point values of SWE measured at snow courses and snow telemetry stations (e.g., SNOwpack TELemetry (SNOTEL) stations in the United States). Operational streamflow volume forecast accuracy decreases during

extreme conditions that are not well represented in the historical record [*Pagano et al.*, 2004]. There is thus a need to develop deterministic and ensemble approaches to runoff forecasting that utilize spatially explicit land surface observations such as those from remote sensing.

Ecologically, the distribution of snow plays a vital role in the survival or demise of organisms living in nivean environments [*Halfpenny and Ozanne*, 1989]. Moreover, the timing and magnitude of snowmelt strongly influences forest health during the summer growing season [*Trujillo et al.*, 2012]. Biogeochemical cycles are dramatically influenced by snow distribution. For example, overwinter heterotrophic CO_2 efflux is sensitive to small changes in snow distribution [*Brooks et al.*, 1999]. This microbial response to snow distribution is largely dictated by the influence of snow distribution on soil temperature [*Monson et al.*, 2006]. The distribution of SWE and snowmelt affects the fate and transport of atmospheric trace elements, storing them throughout the winter accumulation season and releasing them in high concentration within the first 20% of snowmelt runoff [*Bales et al.*, 1993]. The effect of climate change on hydrological processes in seasonally snow-covered systems may affect water availability for ecosystems; the impacts to biogeochemical cycles and ecological processes are largely unstudied. These processes occur in spatially complex mosaics across the landscape. Hence, improved understanding of the processes controlling the spatiotemporal evolution of the snowpack mass and energy balance will improve the capability to assess ecosystem response to climate change.

Understanding of the physical processes controlling snowpack mass balance has improved over the last 40+ years due, in part, to advances in remote sensing and

[1]*Department of Geography, Institute of Arctic and Alpine Research, University of Colorado at Boulder, Boulder, Colorado, USA*

[2]*Jet Propulsion Laboratory, California Institute of Technology, Pasadena, California, USA*

[3]*School of Earth Sciences and Byrd Polar Research Center, Ohio State University, Columbus, Ohio, USA*

[4]*Joint Institute for Regional Earth System Science & Engineering, University of California, Los Angeles, California, USA*

[5]*Civil and Environmental Engineering, University of California, Los Angeles, California, USA*

[6]*U.S. Army Cold Regions Research and Engineering Laboratory, Hanover, New Hampshire, USA*

Remote Sensing of the Terrestrial Water Cycle, Geophysical Monograph 206. First Edition. Edited by Venkat Lakshmi.
© 2015 American Geophysical Union. Published 2015 by John Wiley & Sons, Inc.

Figure 10.1 Eastern Sierra Nevada near Lone Pine, California. Note the distinct snowline elevation. As early as the 1930s it was noted that snow persistence at lower elevation corresponded with greater snowpack water storage and greater summer streamflow.

associated integration of satellite data into distributed models [*Holko et al.*, 2011]. Concurrent with these advances, the availability of meteorological data has increased, affording physically based calculations of snowmelt rates. Recent works have combined surface energy balance observations, models, and remote sensing to develop snow depletion curves that can be used to predict both runoff and SWE. The approach uses historical relationships between rates of snow cover depletion and streamflow to predict future streamflow subsequent to a real-time snow cover observation. The technique is based on intuitive relationships between snow persistence and runoff production. If warm periods are accompanied by a rapid decline in snow extent, it is inferred that snowpack water storage is relatively low. Conversely, if prolonged warm periods are accompanied by persistent snow cover, then a deeper snowpack is implied and greater runoff is expected (Figure 10.1). Early applications of the approach used either hand-drawn maps of snow-covered area (SCA) or estimates of SCA derived from oblique or aerial photography, which were incorporated into simple regression models for predicting streamflow [*Parsons and Castle*, 1959; *Potts*, 1937].

In the 1970s the approach was extended to satellite data in which streamflow could be predicted over much larger areas than previous applications [*Martinec*, 1975; *Rango et al.*, 1977]. These early applications derived explicit depletion curves for different elevation bands whereby runoff production from different portions of a catchment could be estimated. In this regard, binary SCA data (i.e., snow presence versus snow absence) could be used to estimate the fractional SCA (i.e., fractional area with snow cover) within a given elevation band that could then be related to historical observations of streamflow. In

addition, snowmelt depletion curves were derived in which cumulative positive degree-days (i.e., the sum of air temperatures greater than 0 °C) were plotted against fractional SCA within a given elevation band. These snowmelt depletion curves were then related to the magnitude of SWE within a given elevation band. The progression of these studies spanning over 40 years is the focus of Section 10.2.

In the early 1990s improvements in spectral unmixing techniques led to the development of fractional SCA detection algorithms at the pixel scale [*Rosenthal and Dozier*, 1996]. This dramatically expanded the possibilities for depletion curve applications in that depletion curves could be developed for individual pixels rather than for larger areas such as elevation bands. As a result, the techniques developed by *Martinec* [1975] and *Rango et al.* [1977] could be used to estimate SWE for individual pixels. The approach involves the use of calorimetry in which model-based estimates of snowmelt are integrated over the period of satellite-observed snow cover to reconstruct the initial SWE before snowmelt began and to estimate proportional streamflow volumes through time. Conceptually, the approach can be articulated by considering the ablation season snowpack in the context of a simple mass balance consisting of three components: (1) the maximum SWE at the end of the accumulation season, (2) the snowmelt rate; and (3) the depletion of SCA [*Liston*, 1999]. Provided any two of these three variables are known, the mass balance can be solved. Hence the maximum SWE can be determined if the snowmelt rate can be accurately estimated from a distributed model and if satellite data are available to estimate the depletion of snow extent. This methodology, now known as the SWE reconstruction approach, has developed through several studies spanning the 1990s to the time of this writing, and is the focus of Section 10.3.

The use of depletion curves in the SWE reconstruction approach contains several limitations. The approach is retrospective in that SWE estimates are not obtainable until after snow disappearance has been observed. Thus, the approach is not applicable to the snow accumulation season, and it cannot be applied in real-time applications. Thus, several works have merged forward models and in situ observations with the SWE reconstruction approach [e.g., *Durand et al.*, 2008; *Girotto et al.*, in press; *Guan et al.*, 2013]. These recent studies are described in Section 10.4. It should be noted here that forward models such as snowpack mass balance models (e.g., SNTHERM [*Jordan*, 1991], SNOWPACK [*Lehning et al.*, 2006], Utah Energy Balance Model [*Luce et al.*, 1998], CROCUS [*Brun et al.*, 1989]) and land surface models (Variable Infiltration Capacity Model [*Liang et al.*, 1994, 1996], Simplified Simple Biosphere Model [*Xue et al.*, 1991], Biosphere-Atmosphere Transfer Scheme [*Dickinson et al.*, 1993],

Noah Land Surface Model [*Ek et al.*, 2003], and Community Land Model [*Oleson*, 2010]) can simulate the spatial distribution of snow extent and SWE. However, given that these models are not constrained by satellite observations they are not considered explicitly in this text.

The objective of this chapter is to outline applications of snow cover depletion information in hydrologic applications. The works described have improved our ability to characterize the spatial distribution of snowmelt, SWE, and associated runoff response. These works provide a foundation for future applications within the broader goal of understanding the sensitivity of snow-dominated systems to changes in climate.

10.2. HYDROLOGIC APPLICATIONS OF SNOW COVER DEPLETION OBSERVATIONS

Hydrologic applications of snow cover data prior to the 1990s were motivated, almost exclusively, by the need for improved runoff predictions. As such, and with associated data limitation, most of these earlier works use a lumped or semilumped modeling approach in which SCA and SWE are basin-wide quantities. However, understanding hydrologic response to changes in land cover and climate and associated linkages between hydrology, ecology, and biogeochemistry require a more spatially explicit understanding of the processes controlling snow distribution. Hence, more recent hydrologic applications of remotely sensed snow cover data have characterized the subbasin spatial distribution of SCA and SWE. This evolution from lumped, basin-wide representation of SCA and SWE toward pixel-specific distributions of these quantities is implicitly described below.

The use of remotely sensed snowpack information in hydrologic simulations began in the 1930s when oblique ground-based photography was used to characterize relationships between snow extent and streamflow [*Potts*, 1937]. The approach was extended by *Parsons and Castle* [1959] to aerial photography, whereby it was shown that drawings of SCA obtained from aircraft could be used to construct area depletion curves. In this application, plots of SCA versus the date of acquisition were generated for a 6 year period for the Kings River Basin, California (3885 km^2), and a 5 year period for the Kern River Basin, California (5440 mi^2) (Figure 10.2). "Discharge" depletion curves were generated by plotting SCA versus the cumulative discharge measured after the date of SCA data acquisition. Snow course data were used to determine a "snow index" representing the percent deviation from the mean historical snow course value. Forecasts of the remaining discharge, Q_i, for any given date during the melt season, i, when SCA was equal to SCA$_i$, was calculated by plotting the $Q_{i,h}$ (SCA$_i$) of the historical data against the snow index recorded that day (where $Q_{i,h}$(SCA$_i$)

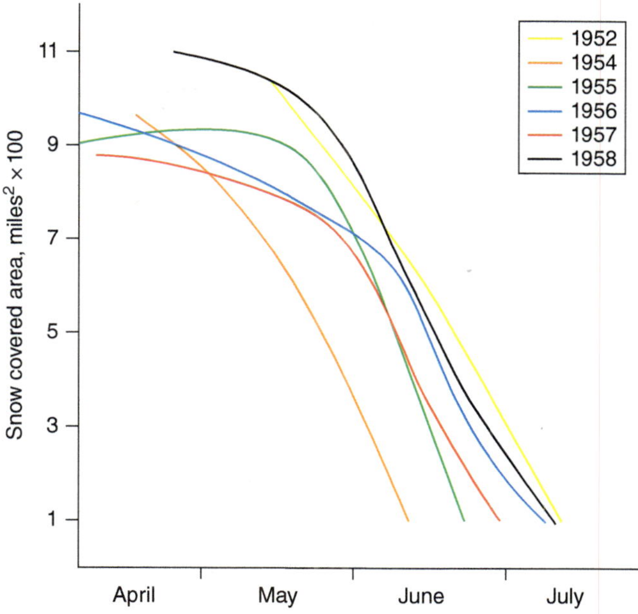

Figure 10.2 Time series of snow-covered area observed from aerial photography in the Kings River Watershed, California, April–July 1952–1958. (Adapted from *Parsons and Castle* [1959].)

is equal to the remaining discharge recorded in the historical year after SCA = SCA$_i$). Estimates of Q_i were obtained graphically by fitting a line through the plotted points. Average errors during the bulk of the melt season were within 10% of observed discharge.

A similar empirical model was developed by *Leaf* [1969] in which SCA was determined using aerial photography. These SCA measurements and observations of discharge were used to predict snowmelt runoff in the St. Louis Creek. (8 km^2), Deadhorse Creek. (2.7 km^2), and Lexen Creek. (1.2 km^2) watersheds of Colorado. Five aerial photographs were taken for each water year 1964–1968. Percent SCA (horizontal axis) was plotted versus the observed percent of total melt season discharge (vertical axis) and a line was fit to the points. Runoff forecasts were made by graphically solving for percent discharge from observations of SCA. The discharge up to the test date was divided by the percent of the total expected discharge, for the observed SCA, in order to calculate the total seasons discharge. Subtracting the observed discharge up to the test date from the total seasons discharge yielded a value for the remaining discharge.

Anderson [1974] incorporated depletion curves into a more theoretical snow accumulation/ablation model. Model inputs of temperature and precipitation were required. In this model, curves generated from plots of percent SCA and percent of maximum SWE were used to determine the SCA over which snow/atmosphere energy exchange takes place. The assumption of this method was that the depletion pattern of each basin was the same from year to year. Once the relationship was established, the percent of maximum SWE was calculated, based on an air temperature index melt rate, and used to calculate the SCA. This model was applied in the Passumpsic (1130 km²), Ammonoosuc (1023 km²), and White (1787 km²) river

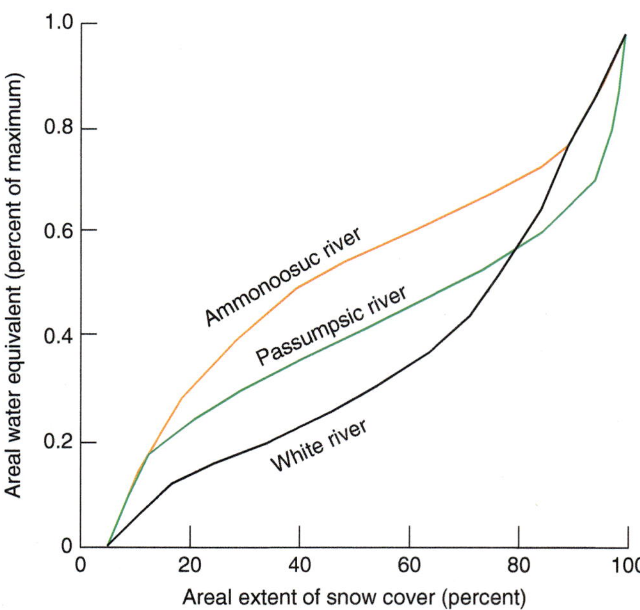

Figure 10.3 Depletion curves relating snow cover extent and snow water equivalent for three river basins in the northeastern United States. (As presented in *Anderson* [1974].)

basins in the northeastern United States (Figure 10.3). Model results showed that discharge percent bias was 0.8%, 1.3%, and −0.3% of observations for each basin, respectively.

Based on the work of *Martinec* [1975], *Rango et al.* [1977] developed a simple regression model that incorporated satellite remotely sensed SCA data. In this model the relationship between historical discharge and SCA was used to predict the discharge of the Indus (162,100 km²) and Kabul (88,600 km²) river basins of Pakistan. SCA data at 4 km resolution was obtained from the advanced Vidicon system on board various U.S. Environmental Science Services Administration (ESSA) satellites for 1967–1972 and from National Oceanic and Atmospheric Administration (NOAA) scanning radiometer data for 1973–1974. The historical time series was generated from approximately one cloud-free image per week for a 3 week period in April of each year up to 1973. The regression from these years was used to predict discharge for 1974 to within 7% and 2% of observed discharge for the Indus and Kabul rivers, respectively. The correlation coefficients of the regression equations were significant at the 95% and 99% confidence levels for the Indus and Kabul rivers, respectively. These results were consistent with earlier tests of the model performed by *Martinec* [1975]. The combined work and subsequent papers by both Martinec and Rango became the basis for the Snowmelt Runoff Model (SRM), which has been used in a variety of catchments globally and is depicted schematically in Figure 10.4.

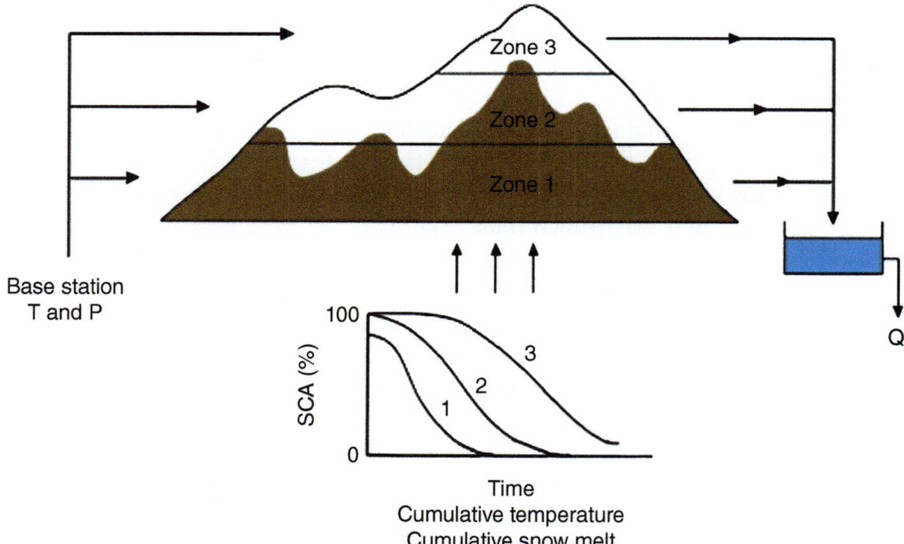

Figure 10.4 Illustration of the SRM concept using depletion curves. With cumulative snowmelt on the abscissa the integral under each curve represents the initial SWE for each elevation band. Hence, we note that the integral under the curve for zone 3 at high elevation is significantly greater than the integral under the curve for zone 1, indicating greater runoff production. (Image source: http://www.macaulay.ac.uk/hydalp/private/demonstrator_v2.0/models/srm.html.)

Remotely sensed SCA data from the Landsat Mulitspectral Scanner (MSS) was used by *Shafer et al.* [1979] to develop three regression models, applying three different independent variables to predict the dependent variable, cumulative runoff. Additionally, the Subalpine Water Balance Model (SWBM) was used in combination with Landsat MSS SCA data to predict the residual SWE for any date during the snowmelt season. In the initial regression models the first independent variable used was SCA. The second independent variable used was a snow course index and the third was a combination SCA/snow course index. Runoff predictions were made for six different drainage basins in southern Colorado ranging in size from 277 to 3756 km². In order to decrease the number of images needing processing, linear regressions between the SCA of the various basins were developed such that SCA values from neighboring basins could be used to estimate SCA values for a given basin. Subsequent analyses only required that SCA be determined for two of the six basins. Correlation coefficients of the SCA regressions of the various basins ranged from 0.81 to 0.99. Comparisons of the three regression models showed that the SCA/snow course index was best correlated with runoff. Analyses using the SWBM incorporated plots of SCA versus residual SWE to predict SWE over the basin of study. SNOTEL sites were used to see if the model was under or overpredicting SWE, and adjustments in the curves were made in order to correct any inaccuracies. The result was corrected functions of residual SWE versus SCA.

Gupta et al. [1982] also used Landsat imagery to derive SCA for use in a depletion curve technique for predicting snowmelt runoff in various subcatchments of the Beas River (12,916 km²) of the Himalaya. This work built upon the work of *Rango et al.* [1977] by defining the SCA runoff function as being logarithmic in form. Runoff forecasts were not generated, but depletion curve behavior was assessed based on the geomorphic characteristics of the subbasins. Smaller subbasins at higher elevations exhibited steeper logarithmic relationships owing to their greater snow depths. The steeper, smaller, higher elevation subcatchments exhibited larger amounts of snowmelt with corresponding decreases in SCA (i.e., curves had steeper slopes).

Depletion curves that use day-of-year as the independent variable are not able to separate the impact of melt rates and SWE on snow cover depletion rates; e.g., rapid snow cover depletion can occur because of a shallow snowpack and/or because of high melt rates. Therefore, *Rango and Martinec* [1982] developed a model in which depletion curves were generated using accumulated degree-days as the response variable; hence the response variable is inherently linked to melt rates and the depletion curve response, in theory, varies only as a function of SWE. Estimates of maximum SWE in the South Fork

Rio Grande watershed (559 km²) of southern Colorado were obtained by solving a simple water balance for SWE at the beginning of the ablation season:

$$(\mathrm{SWE} + P)C = R, \qquad (10.1)$$

where P is the cumulative precipitation during the melt season, C is the runoff coefficient, and R is the cumulative runoff. Based on observations of runoff, temperature, precipitation, and SCA, a series of depletion curves, representing different values of initial SWE, were generated. For a test year, real-time estimates of maximum SWE were obtained as follows: (1) An initial guess of SWE is made and a corresponding historical depletion curve chosen. (2) If the plot of SCA versus the temperature index does not match the chosen curve, a new curve is selected and the initial guess of SWE is updated. This approach was incorporated into the SRM in which forecasts of runoff are obtained by using the SWE estimates and predictions of the temperature index. Applications of this approach to modeling runoff were performed in the Himalaya [*Kumar et al.*, 1991], in the Rhein-Felsberg of Switzerland [*Baumgartner et al.*, 1986], in the Cordevole River Basin of the Italian Alps [*Swamy and Brivio*, 1997], and in the Rio Grande of Colorado.

Ferguson [1984] built upon the theoretical model of *Anderson* [1974] in which he developed two different snowpack mass balance equations that incorporated decreases in SWE into the calculation of SCA. The first of these equations assumes that SWE is uniform and that the snowpack melts laterally from its margins at a constant rate. Therefore, the SCA following time step i, is given by

$$\mathrm{SCA}_{i+1} = \mathrm{SCA}_i - \mathrm{SWE}/V_i, \qquad (10.2)$$

where V_i is the volume of snowmelt runoff. This equation was also used for nonuniform SWE distributions when the relationship between SWE and SCA decreases linearly and the melt rate decreases linearly at the same rate. In this case, the value for SWE is substituted by the mean SWE value. The second equation is implemented for the more realistic case when the SWE of the remaining snowpack, after some melt has occurred, is greater than the SWE over the area that melted in the previous time step. Thus as before, an inverse relationship between SWE and snow extent is assumed, but in this case the melting is assumed uniform. The equation used for this condition is

$$\mathrm{SCA}_{i+1} = \left(\mathrm{SCA}_i^2 - \mathrm{SWE}_0/V_i\mathrm{SCA}_0\right)^{1/2}, \qquad (10.3)$$

where 0 signifies initial SCA and SWE. This inverse relationship was used to calculate the initial SWE and SCA using a basin-wide water balance by constraining the product of the initial SWE and the initial SCA to be greater than the residual of the total discharge and the precipitation. The initial SCA was known and therefore the initial SWE was inferred based on the water balance constraint. Runoff was calculated as the product of the SCA, the accumulated degree-days and the melt factor.

A fourth model described by *Dunne and Leopold* [1978] was developed for the case where the above three models [i.e., equations (1)–(3)] calculated a quantity of melt for a given point that was greater than the SWE existing at that point. This fourth model predicts SCA by subtracting the melt from the depth of SWE across the snowpack. A plot of SWE (y axis) versus the percent area with SWE less than or equal to an ordinate value is used to predict the SCA. This is done by shifting the curve down the y axis as a function of the melt rate. The increased x intercept of the shifted curve indicates the new percentage of bare ground.

Buttle and McDonnell [1987] tested the four models described above in a semidistributed fashion in which the Harp Lake Four watershed (119.8 ha) was divided into three areas based on vegetation type and aspect. A fifth model was compared as well, which uses the same procedure as model 4 only it assumes that the daily melt varies inversely with the SWE over the area of interest. This is incorporated into the model by assuming that half of the SCA has a daily melt depth that is greater than the mean melt depth and that the other half is below the mean. Melt depths over areas with relatively high SWE, $M_{d\text{-high}}$, and relatively low SWE, $M_{d\text{-low}}$, are calculated as follows:

$$M_{d\text{-high}} = X - 2\text{SE}, M_{d\text{-low}} = X + 2\text{SE}, \qquad (10.4)$$

where X is equal to the mean melt depth and SE is the standard error of the melt depths. Results from these five models were compared with observations of SWE, SCA, and runoff. Snow surveys using the Meteorological Service of Canada (MSC) sampler were used to monitor snowpack conditions daily throughout the modeling period. Differences in measured daily SWE values were converted to point melt depths. The extent of bare ground was determined in the field using ground-based visual surveys. Runoff was measured at the basin outlet. Assessments of the abilities of the five different models to simulate the percentage of bare ground throughout the modeling period showed that, for areas with discontinuous snow cover, models one and two performed best and that areas with a more uniform initial snow cover were simulated best by models 4 and 5. Assumptions about variable melt rates from the edges of snow-covered areas to areas with relatively high values of SWE, in models 1, 2, and 3, caused poor model results in areas with continuous snow cover. Similar assumptions in model 5 resulted in little benefit over model 4 in these areas.

In a related work, *Ferguson* [1986] applies an exponential relationship between runoff and SCA to show that intrabasin SWE variation is exponential:

$$\text{SWE} = b(A - a)^c \qquad (0 \leq a \leq A), \qquad (10.5)$$

where b and c are basin constants, A is the initial SCA, and a is the area depleted from the previous time step. The variable c increases with increasing spatial variability in the snow cover. Equation (10.5) was then integrated over a to yield the mean basin SWE. Subsequently, the runoff (in depth of water) was calculated from the mean SWE, percent SCA, and the melt rate. Unlike the work done by *Buttle and McDonnell* [1987], this model was not tested with ground observations of SWE.

10.3. SWE RECONSTRUCTION

Liston [1999] outlines a procedure similar, in theory, to the spatially lumped approach of *Ferguson* [1986] and others but moves toward spatially distributed applications. In this research the mass balance of the snowpack for a given pixel is determined by knowing two of the following three variables: (1) the maximum SWE accumulation of the basin, (2) the melt rate, and (3) the depletion of SCA. The melt rate is calculated using an energy balance model that incorporates the radiation, sensible, and latent heat fluxes from observations of the forcing variables: air temperature, relative humidity, and wind speed. If the initial SWE distribution is known, the melt rate can be applied to the SWE surface, and the SCA depletion can be calculated throughout the melt season. Conversely, if the melt rate is known and integrated over the time required to expose a given area, then the initial SWE over that area is equal to the integrated value. *Cline et al.* [1998] implemented this approach at the pixel scale in the Emerald Lake watershed of the Sierra Nevada and was able to resolve the spatial distribution of SWE at an accuracy level of 6% relative to snow survey SWE. Conceptually, this approach can be thought of as a pixel-specific depletion curve method for obtaining estimates of SWE (Figure 10.5). In this regard, observations of SCA (vertical axis) are plotted against cumulative snowmelt flux (horizontal axis). The area underneath a given depletion curve is thus equivalent to the initial SWE before snowmelt began. Implementation of the approach requires estimates of fractional snow covered area and estimates of pixel-specific snowpack-atmosphere energy exchange (i.e., snowmelt).

Since the initial feasibility study of *Cline et al.* (1998), the pixel-specific depletion curve approach of SWE estimation has been applied widely at a variety of different spatial scales and has subsequently been labeled the *SWE reconstruction* approach [*Molotch*, 2009]. *Molotch et al.* [2004] applied the approach in the Tokopah basin, Sierra Nevada, using Landsat Thematic Mapper observations of SCA and Airborne Visible and Infrared Imaging

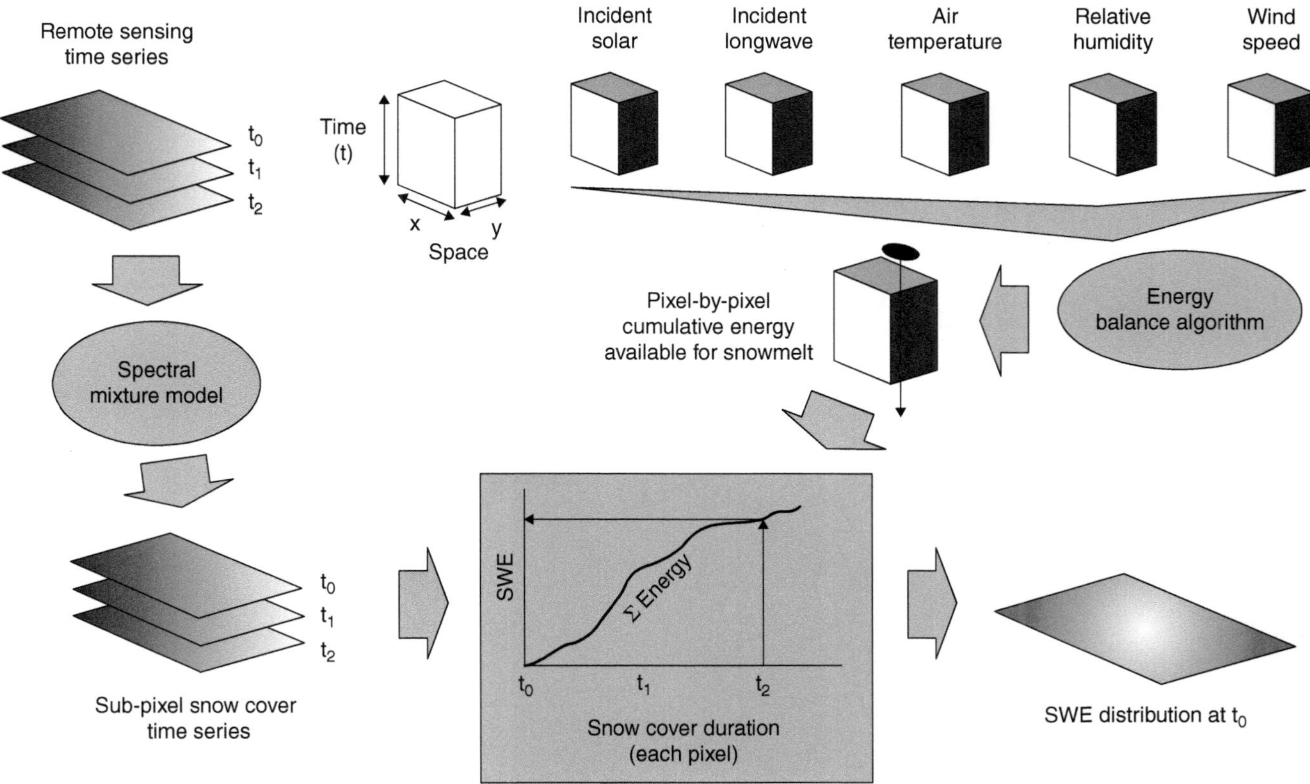

Figure 10.5 Flowcharts showing the SWE reconstruction modeling structure with three main components: (1) determination of fractional snow cover at the subpixel level using spectral unmixing and decision trees, (2) computation of the energy balance at each pixel using estimated micrometeorological data and a distributed snowmelt model, and (3) coupling the (1) and (2) to determine the initial SWE at the beginning of the model run at each pixel. (Source: *Cline et al.* [1998])

Spectroradiometer (AVIRIS) observations of snow-surface albedo. SWE estimates were then compared against distributed snow surveys to evaluate the accuracy of the SWE estimates. When running the model with albedo estimated using the USACE (1956) parameterization, SWE estimates exhibited a high bias 36% greater than observed SWE from the snow surveys. Incorporating the AVIRIS albedo data into the SWE reconstruction model improved model accuracy to within 2% of observed SWE with spatial signatures of model bias significantly reduced. Moreover, incorporation of the AVIRIS data into the reconstruction model improved correlation of modeled snowmelt and streamflow from an R^2 of 0.59–0.76 [*Molotch et al.*, 2004]. *Molotch and Bales* [2006] expanded upon this work by comparing the SWE reconstruction when using the USACE (1956), AVIRIS, and the albedo parameterization of the Biosphere Atmosphere Transfer Scheme [BATS; *Dickinson*, 1993]. Through these comparisons they showed that SWE reconstructions performed using the BATS albedo parameterization were more accurate than the simulations using the USACE parameterization but remained less accurate than the simulations using the AVIRIS albedo parameterizations (Figure 10.6).

One particular challenge of the SWE reconstruction approach is the need to characterize the distribution of snowmelt forcings (i.e., incident solar and longwave radiation, wind speed, temperature, humidity, and vapor pressure) in a spatially explicit manner over topographically complex landscapes. Hence, the aforementioned studies were limited to relatively small catchments with areas less than 20 km². The first study to extend the approach to larger scales, more relevant for water resource management, was *Molotch* [2009]. In this study, Landsat Enhanced Thematic Mapper (ETM+) SCA data were used to reconstruct SWE for the 2001 and 2002 snowmelt seasons in the 3019 km² Rio Grande Headwaters of Colorado. Forcings were derived from a combination of atmospheric reanalysis products (i.e., the North American Land Data Assimilation System [*Cosgrove et al.*, 2003]) and in situ observations of air temperature from SNOTEL stations. The results showed promise as SWE was reconstructed to within 24% of observed SWE as measured from detailed snow surveys at six different 4 km² sites where intensive snow surveys were performed.

In addition to the requirement of spatially explicit snowmelt forcings, the SWE reconstruction technique

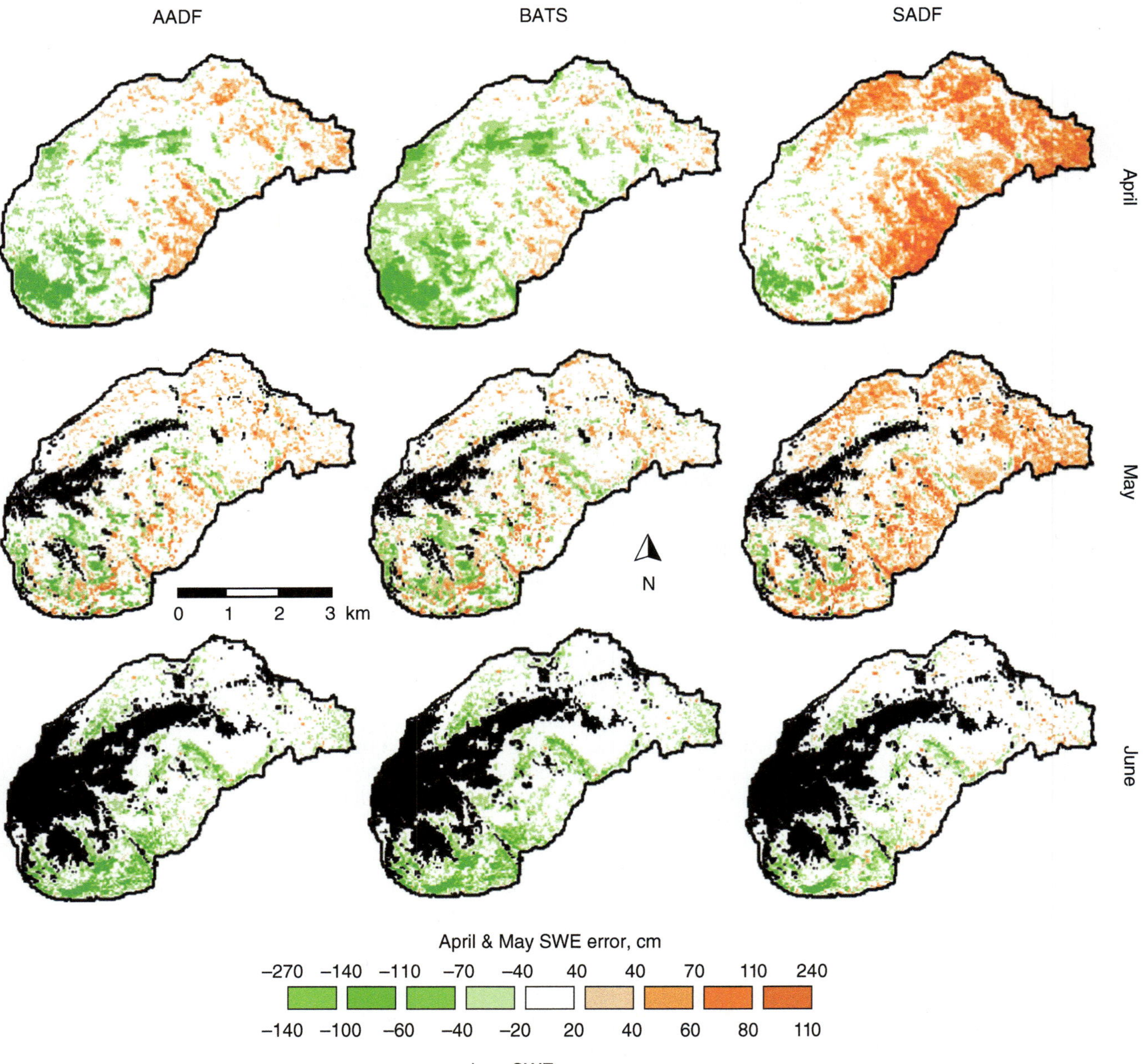

Figure 10.6 Error in reconstructed SWE using AVIRIS albedo, the BATS albedo parameterization, and the USACE albedo parameterization. White areas are those within one standard deviation of the mean observed SWE. Green and orange areas correspond to model underestimates and overestimates, respectively. (Source: *Molotch and Bales* [2006].)

also requires adequate estimates of SCA as the technique effectively relies on depletion curves constructed from cumulative snowmelt flux and fractional SCA. Given that there are a variety of polar orbiting satellites observing the land surface in visible and infrared portions of the electromagnetic spectrum, there are several choices of SCA data sources. Thus, *Molotch and Margulis* [2008] set out to compare the performance of the SWE reconstruction approach using Landsat ETM+, the Moderate Resolution Imaging Spectroradiometer (MODIS), and the Advanced Very High Resolution Radiometer (AVHRR) data as input to the model. This comparison effectively evaluated how differences in spatial, temporal, and spectral resolution influence the accuracy of the SWE reconstruction approach. The results indicated significantly different results with Landsat ETM+, MODIS, and AVHRR simulations (Figure 10.7) producing SWE estimates within 24%, 50%, and 77% of observed SWE,

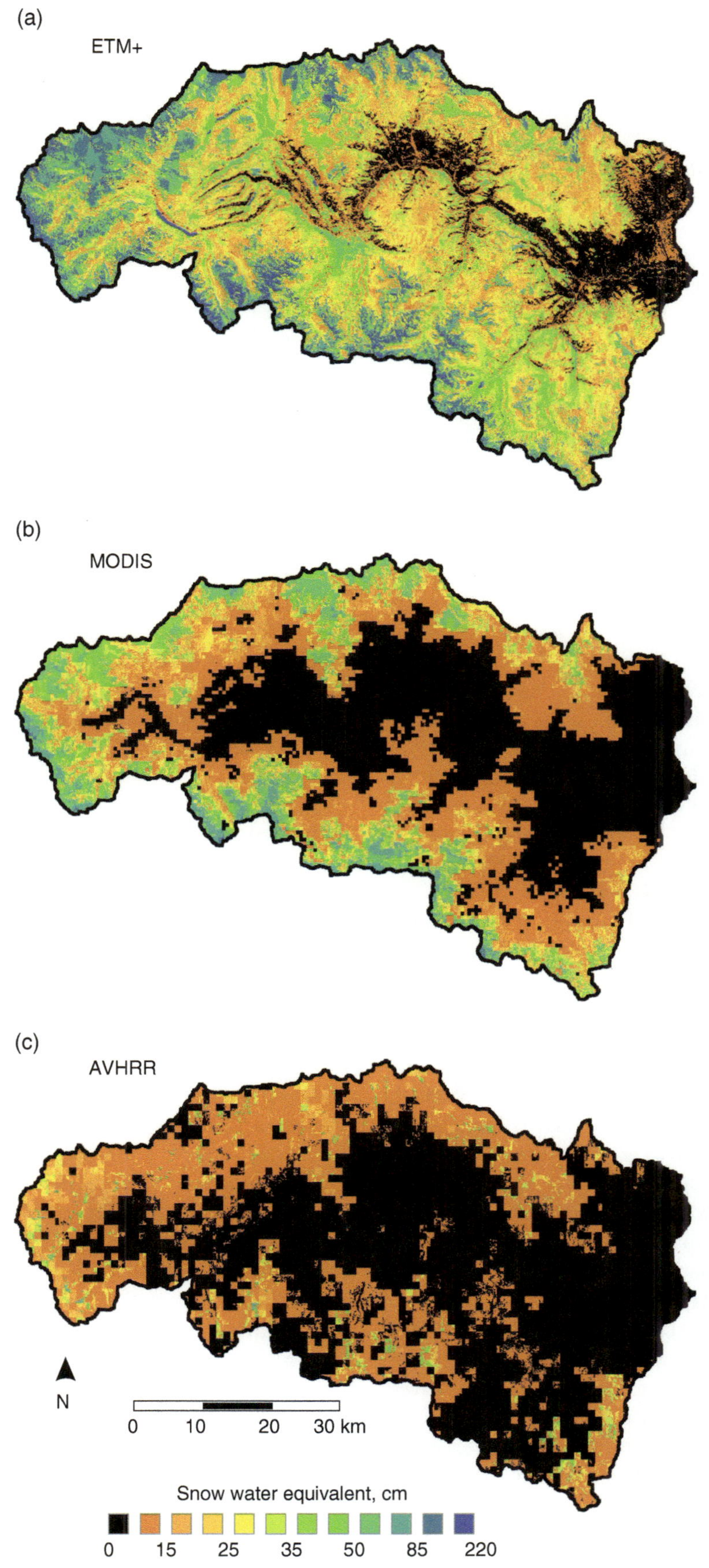

Figure 10.7 Spatial distribution of SWE in the Rio Grande headwaters on 24 April 2001 using (a) ETM + (b) MODIS and (c) AVHRR resolved at 100 m resolution. (Source: *Molotch and Margulis* [2008].)

respectively. The implications of this comparison were that accuracy in SCA estimates was more important than the temporal frequency of observations. This finding was reinforced by *Molotch et al.* [2010] whereby it was shown that SWE reconstructions using the MODIS snow-covered area and grain size algorithm [MODSCAG; *Rittger et al.*, 2013] could yield accuracy levels equivalent to Landsat ETM + simulations and with far greater accuracy than using the standard MODIS product [MOD10A1.V5; *Hall et al.*, 2001].

10.4. BLENDING SWE RECONSTRUCTIONS, FORWARD MODELS, AND SURFACE OBSERVATIONS

Given the aforementioned advances in SCA detection with MODIS, SWE reconstructions over large scales are now feasible. To illustrate this capability *Guan et al.* [2013] performed SWE reconstructions over the 60,000 km^2 Sierra Nevada of California. In this application snow-melt forcings were derived from the North American Land Data Assimilation System at 1/8 degree resolution and were downscaled using a variety of terrain-related parameterizations to 500 m resolution. In addition, this work included the first illustration of a blending procedure for merging the SWE reconstruction SWE estimates with in situ observations. This blending procedure involved calculating the residual between reconstructed SWE and measured SWE at 104 snow pillow sites. The residuals were then interpolated using inverse distance weighting to derive a spatially continuous surface of model residuals. The model residual surface was then added to the original reconstructed SWE surface to derive the blended SWE product, which was then masked by the MODSCAG SCA estimates to exclude snow cover from snow-free locations. The performance of both the reconstruction and the blended SWE estimates were determined using 17 extensive snow surveys at 6 locations with spatial sampling and with the operational snow sensor networks. The reconstructed and blended SWE estimates were also compared against NOAA's operational Snow Data Assimilation System (SNODAS). Mean bias of the methods compared to the snow surveys were −0.193 m (reconstruction), 0.001 m (blended), and −0.181 m (SNODAS). Corresponding root-mean-square errors (RMSE) were 0.252, 0.205, and 0.254 m, respectively. This work illustrated that the reconstruction approach could yield estimates of SWE with accuracy levels equivalent to SNODAS. This was an important result given that the reconstruction does not use any ground-based snow information while SNODAS heavily relies on ground-based snow observations. Hence, this work highlights the strong potential for applying the SWE reconstruction approach in mountain ranges where in situ observations are sparse or nonexistent. This work also showed that in locations with a significant in situ network, a blended product can be developed with accuracy levels above the reconstruction and SNODAS.

Blending the SWE reconstruction with surface observations has been suggested by other works as well. *Raleigh and Lundquist* [2012] used 154 snow pillow sites in California, Oregon, and Washington to compare estimates of precipitation, peak SWE, and daily SWE derived from a forward model [i.e., SNOW-17; *Anderson* E. A., 1976], a temperature-index-based SWE reconstruction model, and a blending of the SWE reconstruction and forward model. The blending procedure used in this work was simply an averaging of the precipitation and SWE estimates of the two different techniques. In terms of peak SWE the blended model exhibited the greatest accuracy, while the accuracy of the forward and reconstruction models were not statistically different. The mean absolute error (MAE) of the SWE estimates during the accumulation season were 9.6%, 13.9%, and 13.8% for the combined, reconstruction, and forward models, respectively; note the forward and reconstruction model MAE values were not statistically different. During the snow ablation season, the MAE of the combined, reconstruction, and forward models was 14.5%, 16.0%, and 31.1% of the observed peak SWE. These results indicated that a combined approach exhibited significant promise. Moreover, the significant improvement of the reconstruction model over the forward model during the ablation season is intuitive given that the reconstruction model is based on the estimated snow depletion curve.

A limitation of the aforementioned blending procedures is the inherent uncertainties in the observations, forcings, and parameters within either the forward or reconstruction modeling approaches. Data assimilation (DA) provides a means to weight these sources of uncertainties and derive ensemble-based estimates of model state variables (e.g., SWE). The approach has been applied to an SWE reconstruction model by *Durand et al.* [2008] in which the aforementioned SWE reconstruction approach is applied in a Bayesian context in which a forward land surface model is used to derive prior estimates of SCA and SWE and a DA scheme is used to update the snow cover depletion curve [*Andreadis and Lettenmaier*, 2006; *Clark et al.*, 2006; *Kolberg and Gottschalk*, 2006] to derive posterior estimates of SWE. Conceptually, the approach builds upon earlier works by deriving a set of spatial weighting functions that can be used to disaggregate coarse-scale precipitation data over complex mountainous terrain [Figure 10.8]. The method has been applied synthetically using North American Land Data Assimilation System (NLDAS) forcings and the simplified biosphere model [SSiB; *Xue et al.*, 1991] in one of the NASA Cold Lands Processes Experiment sites in Colorado [*Durand et al.*, 2008]. Results indicated that error in the peak SWE accumulation of the forward model (−0.082 m) was significantly reduced (0.0052 m) via the assimilation of the synthetic SCA data and associated updates to the precipitation disaggregation weights. The approach has since been applied using real data in the Tokopah Basin, Sierra Nevada, California [*Girotto et al.*, in press]. This work set out to compare the Bayesian SWE reconstruction technique and the aforementioned SWE reconstruction approach, which they label the deterministic SWE reconstruction. SWE was simulated for water years 1997 and 1999 with MAE values (as a percentage of observed SWE) of 7.1% and 7.2% for the Bayesian versus deterministic approaches, respectively. A series of robustness experiments were then conducted, most notably an experiment in which the number of available SCA observations was decreased. The results indicated that the Bayesian approach was much less sensitive to the number of SCA observations compared with the deterministic approach. For the Bayesian simulations using only 4 SCA observations, SWE estimates differed from the nominal simulations using 10 SCA observations by only 6% and 13% in water years 1997 and 1999, respectively. Conversely, simulations using the deterministic approach differed by 10% and 44% for the two respective years. These results highlight the utility of the Bayesian approach for SWE estimation and indicate the deterministic approach could be improved by using an analytical depletion curve instead of a temporal interpolation of SCA observations.

A major strength of the Bayesian approach is the explicit accounting of uncertainty in model forcings, parameters, and SCA observations. Analyses of these uncertainties have been documented within the literature [*Slater et al.*, 2009, 2013; *Molotch et al.*, 2010]. In this regard, uncertainties in detection of snow cover disappearance from remote sensing, estimates of incoming and outgoing shortwave and longwave radiation, and estimates of temperature, humidity, wind speed, and vapor pressure all contribute to uncertainty in reconstructed SWE, as well as uncertainty in model parameters and model structure. *Slater et al.* [2013] performed a highly constrained experiment under which a temperature index model was used to estimate melt rates, and snow disappearance timing was determined at the point scale from SNOTEL stations across the western United States. SWE was then reconstructed for over 10,000 station-years. Based on these reconstructions, it was determined that for 50% of station-years an error of +5 days in snow disappearance observation yields an SWE overestimation of 32% or greater and an error of −5 days in snow disappearance observation yields an SWE underestimation of 25%. *Slater et al.* [2013] then places uncertainty in temperature and radiation in the context of the errors associated with

errors in snow disappearance observation timing. In this regard, the results showed that a bias of +/±1.8 °C yields errors on par with a snow disappearance detection error of ±10 days. In addition, they show that a net radiation error equivalent to −20.38 and 31.66 W/m² yield errors equivalent to ±10 days of snow disappearance detection errors. The authors conclude that future work should iteratively solve the basin-scale water balance using forward and inverse (e.g., reconstruction) approaches within a complete hydrologic model. This conclusion is consistent with the aforementioned works that have combined SWE reconstructions with forward models.

Sturm and Wagner [2010] derived a climatological snow distribution pattern (CSDP) from repeated measurements of snow distribution. The basis for the CSDP is that snow distribution patterns are largely governed by the interaction of prevailing synoptic-scale weather patterns and the distribution of topography and vegetation. Thus, they argue that spatial patterns of snow accumulation are repetitive from year to year with the magnitude of accumulation varying interannually by the relative differences across the landscape remaining relatively constant. To derive the CSDP, they develop a training set of patterns from nine annual snow depth surveys observed in a small tundra basin in Alaska. Subsequently, they were able to use the CSDP and only a few depth measurements from years excluded in the training set, to estimate snow distribution with a near-zero bias and RMSE that ranged from 4.4 to 10.4 cm. The authors conclude that the accuracy of SWE estimates using the retrospective CSDP approach was consistently greater than using a physically based snow model. Similar to the findings reported above, the authors argue for a hybrid approach in which the CSDP is integrated with a physically based model; they illustrate this with an example simulation in which model accuracy is improved by up to 60%. While the CSDP was developed from distributed snow surveys, the authors indicate that the CSDP could be derived from aerial photography or satellite SCA observations, making the approach quite complimentary to the aforementioned SWE reconstruction methodologies.

10.5. CONCLUSIONS

Incorporation of remotely sensed snow observations into distributed models remains a pressing challenge in mountain hydrology. Observations of snow extent do not provide estimates of snow mass and are therefore not the most hydrologically relevant observations. However, snow extent was clearly observable in the earliest orbital Earth observations from *Tiros-1* (circa 1960), and therefore snow extent observations predate nearly all land surface endmembers. As a result, hydrologic applications of these observations have developed for decades. Operational

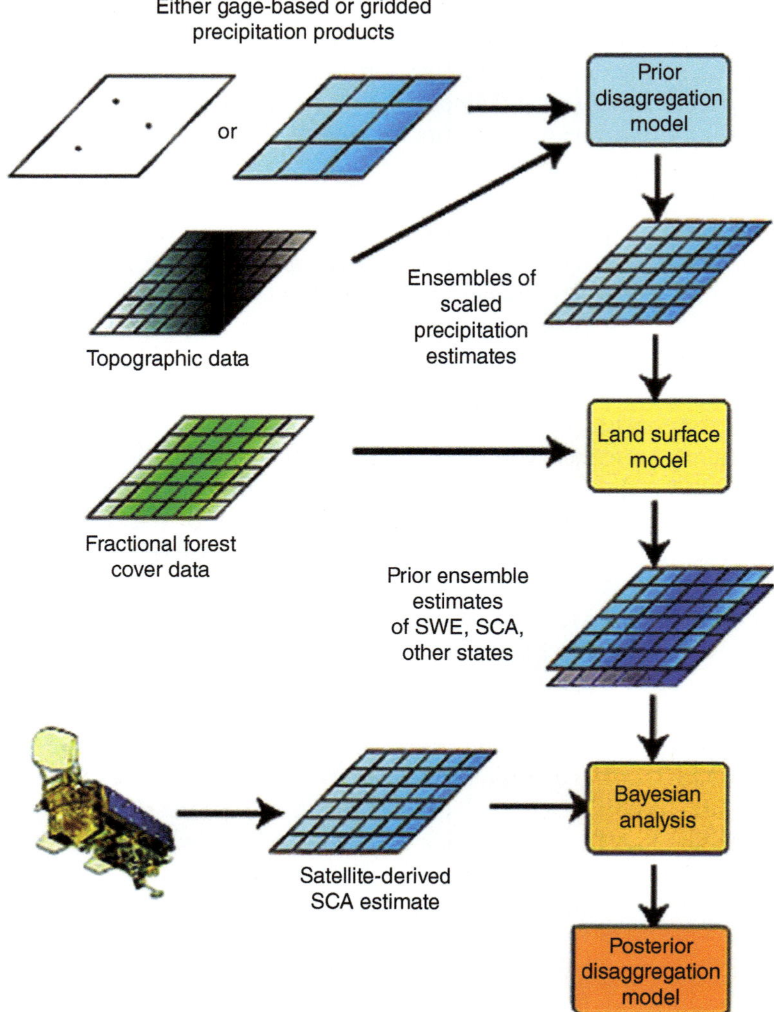

Figure 10.8 This schematic illustrates the method by which the Bayesian reconstruction combines meteorologic, topographic, vegetation, and satellite-derived snow measurements to characterize precipitation patterns. In this study, the posterior disaggregation model is used to generate posterior ensemble estimates of snow water equivalent (SWE, not shown). (Source: *Durand et al.* [2008].)

hydrologists have recognized the potential utility of snow extent observations for runoff estimation since the late 1930s, long before spaceborne measurements were imaginable [*Potts*, 1937]. Routine use of this important information in an operational context is still lacking nearly 80 years later. As a result of this long history, the literature contains a progression of hydrologic applications of snow extent observations. From the earliest applications using oblique ground-based photography and aerial photography to generate depletion curves for runoff estimation, to airborne and spaceborne observations of fractional snow covered area for the purposes of SWE reconstruction, and to data assimilation techniques that have integrated these observations within numerical models, this chapter outlines the diversity of the methods developed and applications of snow depletion observations in a hydrological context. The explicit objective in so doing is to chart a path forward as to how these techniques can be brought to bear on future operational hydrologic models, land surface models, general circulation models, and ecosystem models. The chronology of studies noted indicates that models that require high-resolution estimates of the spatial distribution of snow and/or models that require subgrid parameterizations of snow cover depletion may benefit from historical snow cover depletion observations. Countless studies focused on estimating the spatial distribution of precipitation may also benefit from inclusion of snow cover depletion information; e.g., the Parameter-Elevation Regressions on Independent Slopes Model [PRISM; *Daly et al.*, 1994]. Further developing techniques to harness snow depletion information in Earth system models may improve our ability to characterize the terrestrial water balance from catchment to global scales. Ultimately, the application of interest dictates how much effort is needed with respect to improving estimates of snowmelt runoff. In this regard, the chronology of work presented here provides a foundation for future works aimed at understanding the sensitivity of snow-dominated systems to changes in climate.

REFERENCES

Anderson, E. A. (1974), Conceptual streamflow forecasting model applied to northern New England rivers, *Proc. East. Snow Conf.*, *31*, 30–50.

Andreadis, K. M., and D. P. Lettenmaier (2006). Assimilating remotely sensed snow observations into a macroscale hydrology model, *Adv. Water Resourc.*, *29*(6), 872–886, doi:10.1016/j.advwatres.2005.08.004.

Bales, R. C., R. E. Davis, and M. W. Williams (1993), Tracer release in melting snow: Diurnal and seasonal patterns., *Hydrol. Process.*, *7*, 389–401.

Bales, R., N. Molotch, T. Painter, M. Dettinger, R. Rice, and J. Dozier (2006), Mountain hydrology of the western United States., *Water Resourc. Res.*, *42*, W08432, doi:08410.01029/02005WR004387.

Baumgartner, M. F., J. Marinec, and K. Seidel (1986), Large-area deterministic simulation of natural runoff from snowmelt basins on Landsat MSS data, *IEEE Trans. Geos. Remote Sens.*, *GE-24*(6), 1013.

Brooks, P. D., D. M. McKnight, and K. E. Bencala (1999), The relationship between soil heterotrophic activity, soil dissolved organic carbon (DOC) leachate, and catchment-scale DOC export in headwater catchments, *Water Resourc. Res.*, *35*(6), 1895–1902.

Brun, E., E. Martin, V. Simon, C. Gendre, and C. Coleou (1989), An energy and mass model of snow cover suitable for operational avalanche forecasting, *J. Glaciol.*, *35*(121), 333–342.

Buttle, J. M., and J. J. McDonnell (1987), Modelling the areal depletion of snowcover in a forested catchment, *J. Hydrol.*, *90*, 43.

Clark, M. P., A. G. Slater, A. P. Barrett, L. E. Hay, G. J. McCabe, B. Rajagopalan, and G. H. Leavesley (2006), Assimilation of snow covered area information into hydrologic and land-surface models, *Adv. Water Resourc.*, *29*, 1209–1221, doi:10.1016/j.advwatres.2005.10.001.

Cline, D. W., R. C. Bales, and J. Dozier (1998), Estimating the spatial distribution of snow in mountain basins using remote sensing and energy balance modeling, *Water Resourc. Res.*, *34*(5), 1275–1285.

Cosgrove, B. A., et al., (2003), Real-time and retrospective forcing in the North American Land Data Assimilation System (NLDAS) Project, *J. Geophys. Res.*, *108*(D22), 8842, doi:10.1029/2002JD003118.

Daly, C., R. P. Neilson, and D. L. Phillips (1994), A statistical-topographical model for mapping climatological precipitation over mountainous terrain, *J. Appl. Meteorol.*, *33*, 140–158, doi:10.1175/1520-0450(1994)033<0140:ASTMFM>2.0.CO;2.

Dettinger, M. D., D. R. Cayan, M. K. Meyer, and A. E. Jeton (2004), Simulated hydrologic responses to climate variations and change in the Merced, Carson, and American River basins, Sierra Nevada, California, 1900–2099, *Climatic Change*, *62*(1), 283–317.

Dickinson, R. E., A. Henderson-Sellers, and P. J. Kennedy (1993), Biosphere atmosphere transfer scheme (BATS) version 1e as coupled to the NCAR Community Climate Model, NCAR Tech. Note, NCAR/TN-387+STR.

Dunne, T., and L. B. Leopold (1978), Water in Environmental Planning, W.H. Freeman, New York.

Durand, M., N. P. Molotch, and S. Margulis (2008), A bayesian approach to snow water equivalent reconstruction, *J. Geophys. Res.*, *113*, doi:10.1029/2008JD009894.

Ek, M. B., K. E. Mitchell, Y. Lin, E. Rogers, P. Grunmann, V. Koren, G. Gayno, and J. D. Tarpley (2003), Implementation of Noah land surface model advances in the National Centers for Environmental Prediction operational mesoscale Eta model, *J. Geophys. Res. Atmos.*, *108*(D22). doi:10.1029/2002JD003296.

Ferguson, R. I. (1984), Magnitude and modelling of snowmelt runoff in the Cairngorm Mountains, Scotland, *Hydrol. Sci.*, *29*(1), 49.

Ferguson, R. I. (1986), Parametric modelling of daily and seasonal snowmelt using snowpack water equivalent as well as snow covered area, paper presented at Modelling Snowmelt-Induced Processes, IAHS Publication No. 155, IAHS, Wallingford, Budapest, Hungary, pp. 151–161.

Girotto, M., S. A. Margulis, and M. T. Durand (in press), Probabilistic SWE reanalysis as a generalization of deterministic SWE reconstruction techniques, *Hydrol. Process.*, *28*(12), 3875–3895, 15 June 2014. doi:10.1002/hyp.9887.

Guan, B., N. P. Molotch, D. E. Waliser, S. M. Jepsen, T. H. Painter, and J. Dozier (2013), Snow water equivalent in the Sierra Nevada: Blending snow sensor observations with snowmelt model simulations, *Water Resourc. Res.*, *49*, doi:10.1002/wrcr.20387.

Gupta, R. P., A. J. Duggal, S. N. Rao, G. Sankar, and B. B. S. Singhal (1982), Snow-cover area versus snowmelt runoff relation and its dependence on geomorphology—A study from the Beas Catchment (Himalayas, India), *J. Hydrol.*, *58*, 325.

Halfpenny, J. C., and R. D. Ozanne (1989), Winter: An Ecological Handbook, Johnson Publishing Company, Boulder, Colo.

Holko, L., L. Gorbachova, and Z Kostka, (2011), Snow hydrology in Central Europe, *Geogr. Compass*, *5*, 200–218.

Hall, D. K., G. A. Riggs, and V. V. Salomonson (2001), Algorithm theoretical basis document (ATBD) for the MODIS snow and sea ice-mapping algorithms, available at http://modis-snow-ice.gsfc.nasa.gov/atbd.html.

Jordan, R. (1991), A one-dimensional temperature model for a snow cover, Special Rep. 91-6, U.S. Army Cold Regions Res. and Eng. Lab., Hanover, N. H.

Kolberg, S. A., and L. Gottschalk (2006), Updating of snow depletion curve with remote sensing data, *Hydrol. Process.*, *20*, 2363–2380, doi:10.1002/hyp.6060.

Kumar, V. S., H. Haefner, and K. Seidel (1991), Satellite snow cover mapping and snowmelt runoff modelling in Beas basin, in *Snow Hydrology and Forests in High Alpine Areas*

(Proceedings fo the Vienna Symposium, August 1991), IAHS Publ. No. 205, pp. 101–109.

Laternser, M., and M. Schneebeli (2003), Long-term snow climate trends of the Swiss Alps (1931–99), *Int. J. Climatol.*, *23*(7), 733–750.

Leaf, C. F. (1969), Aerial photographs for operational streamflow forecasting in the Colorado Rockies, *Proc. Western Snow Conf.*, *37*, 19–28.

Lehning, M., I. Volksch, D. Gustafsson, T. A. Nguyen, M. Stahli, and M. Zappa (2006), ALPINE3D: A detailed model of mountain surface processes and its application to snow hydrology, *Hydrol. Process.*, *20*(10), 2111–2128.

Liang, X., D. P. Lettenmaier, E. F. Wood, and S. J. Burges (1994), A simple hydrologically based model of land-surface water and energy fluxes for general-circulation models, *J. Geophys. Res. Atmos.*, *99*(D7), 14,415–14,428.

Liang, X., E. F. Wood, and D. P. Lettenmaier (1996), Surface soil moisture parameterization of the VIC-2L model: Evaluation and modification, *Global Planet. Change*, *13*(1–4), 195–206.

Liston, G. E. (1999), Interrelationships among snow distribution, snowmelt, and snow cover depletion: Implications for atmospheric, hydrologic, and ecologic modeling, *J. Appl. Meteorol.*, *38*, 1474–1487.

Luce, C. H., D. G. Tarboton, and R. R. Cooley (1998), The influence of the spatial distribution of snow on basin-averaged snowmelt, *Hydrol. Process.*, *12*(10–11), 1671–1683.

Martinec, J. (1975), Snowmelt–runoff model for stream flow forecasts, *Nordic Hydrol.*, *6*, 145–154.

Molotch, N. P. (2009), Reconstructing snow water equivalent in the Rio Grande headwaters using remotely sensed snow cover data and a spatially distributed snowmelt model, *Hydrol. Process.*, *23*, doi:10.1002/hyp.7206.

Molotch, N. P., and R. C. Bales (2006), Comparison of ground-based and airborne snow-surface albedo parameterizations in an alpine watershed: Impact on snowpack mass balance, *Water Resourc. Res.*, *42*, doi:10.1029/2005WR004522.

Molotch, N. P., and S. A. Margulis (2008), Estimating the distribution of snow water equivalent using remotely sensed snow cover data and a spatially distributed snowmelt model: A multi-resolution, multi-sensor comparison, *Advances in Water Resources*, *31*(11), 1503–1514. doi:10.1016/j.advwatres.2008.07.017.

Molotch, N. P., T. H. Painter, R. C. Bales, and J. Dozier (2004), Incorporating remotely sensed snow albedo into a spatially distributed snowmelt model, *Geophys. Res. Lett.*, *31*, doi:10.1029/2003GL019063.

Molotch, N. P., S. A. Margulis, and S. M. Jepsen (2008–2010), Response to comment by AG Slater, MP Clark, and AP Barrett on "Estimating the distribution of snow water equivalent using remotely sensed snow cover data and a spatially distributed snowmelt model: A multi-resolution, multi-sensor comparison," *Adv. Water Resourc.*, *31*, 1503–1514 (2008); *32*(11), 1680–1684 (2009); *33*(2), 231–239 (2010).

Monson, R. K., D. L. Lipson, S. P. Burns, A. A. Turnipseed, A. C. Delany, M. W. Williams, and S. K. Schmidt (2006), Winter forest soil respiration controlled by climate and microbial community composition, *Nature*, *439*(7077), 711–714.

Mote, P. W. (2003), Trends in snow water equivalent in the Pacific Northwest and their climatic causes, *Geophys. Res. Lett.*, *30*(12), doi:10.1029/2003GL017258.

Oleson, K. W., et al., (2010), Technical description of version 4.0 of the Community Land Model (CLM), available at: http://www.cesm.ucar.edu/models/cesm1.0/clm/CLM4_Tech_Note.pdf.

Pagano, T., D. Garen, and S. Sorooshian (2004), Evaluation of official western US seasonal water supply outlooks, 1922–2002, *J. Hydrometeorol.*, *5*(5), 896–909.

Parsons, W. J., and G. H. Castle (1959), Aerial reconnaissance of mountain snow fields for maintaining up-to-date forecasts of snowmelt runoff during the melt period, paper presented at Western Snow Conference.

Potts, H. (1937), Snow surveys and runoff forecasting from photographs, in *Proceedings of the 5th Annual Western Interstate Snow Survey Conference, Denver, Colorado.*

Raleigh, M. S. and J. D. Lundquist (2012), Comparing and combining SWE estimates from the SNOW-17 model using PRISM and SWE reconstruction, *Water Resourc. Res.*, *48*, W01506, doi:10.1029/2011WR010542.

Rango, A., and J. Martinec (1982), Snow accumulation derived from modified depletion curves of snow coverage, in *Hydrological Aspects of Alpine and High Mountain Areas (Proceedings of the Exeter Symposium)*, IAHS Publ. No. 138, pp. 83–89.

Rango, A., V. V. Salomonson, and J. L. Foster (1977), Seasonal streamflow estimation in the Himalayan region employing meteorological satellite snow cover observations, *Water Resourc. Res.*, *13*(1), 109–112.

Rittger, K., T. H. Painter, and J. Dozier (2013), Assessment of methods for mapping snow cover from MODIS, *Adv. Water Resourc.*, *51*, 367–380, doi:10.1016/j.advwatres.2012.03.002.

Rosenthal, W., and J. Dozier (1996), Automated mapping of montane snow cover at subpixel resolution from the Landsat Thematic Mapper, *Water Resourc. Res.*, *32*(1), 115–130.

Service, R. F. (2004), As the West goes dry, *Science*, *303*(5661), 1124–1127, doi:1110.1126/science.1303.5661.1124.

Shafer, B. A., C. F. Leaf, and J. K. Marron (1979), Landsat derived snow cover as an input variable for snow melt runoff forecasting in South Central Colorado, in *Satellite Hydrology*, edited by M. Deutsch, D. Richard, and A. Rango, pp. 218–224, Am. Water Resourc. Assoc., Minneapolis, Minn.

Slater, A. G., M. P. Clark, and A. P. Barrett (2009), Comment on "Estimating the distribution of snow water equivalent using remotely sensed snow cover data and a spatially distributed snowmelt model: A multi-resolution, multi-sensor comparison" by Noah P. Molotch and Steven A. Margulis, *Adv. Water Resourc.*, *32*(11), 1567–1684, doi:10.1016/j.advwatres. 2009.09.001.

Slater, A. G., A. P. Barrett, M. P. Clark, J. D. Lundquist, and M. S. Raleigh (2013), Uncertainty in seasonal snow reconstruction: Relative impacts of model forcing and image availability, *Adv. Water Resourc.*, *55*, 165–177, doi:10.1016/j.advwatres.2012.07.006.

Sturm, M., and A. M. Wagner (2010), Using repeated patterns in snow distribution modeling: An Arctic example, *Water Resourc. Res.*, *46*(12), doi:10.1029/2010WR009434.

Swamy, A. N., and P. A. Brivio (1997), Modelling runoff using optical satellite remote sensing data in a high mountainous alpine catchment of Italy, *Hydrol. Process.*, *11*, 1475–1485.

Trujillo, E., N. P. Molotch, M. L. Goulden, A. E. Kelly, and R. C. Bales (2012), Elevation-dependent influence of snow accumulation on forest greening, *Nature Geosci.*, *5*(10), 705–709.

Xue, Y., P. J. Sellers, J. L. Kinter, and J. Shukla (1991), A simplified biosphere model for global climate studies, *J. Climate*, *4*(3), 345–364.

11

Retrieval and Validation of VIIRS Snow Cover Information for Terrestrial Water Cycle Applications

Igor Appel

11.1. INTRODUCTION

Snow cover is an important factor influencing the terrestrial water cycle and climate change. Snow cover directly affects local energy exchange between the atmosphere and land surface, significantly modifying thermodynamic fluxes and the water mass balance determining hydrological processes. Extensive seasonal and interannual changes in snow cover could be considered as one of the most important causes affecting not only local, but also regional [*Elguindi et al.*, 2005; *Ellis and Leathers*, 1999; *Zaitchik et al.*, 2007] and global [*Bamzai and Shukla*, 1999; *Cohen and Entekhabi*, 2001; *Déry et al.*, 2005b; *Yang et al.*, 2001] processes far beyond high latitudes and the areas covered by snow. Snow cover alters the atmospheric circulation through lower latitudes [*Gong et al.*, 2003] and significantly modifies soil moisture, the state of vegetation, and summer precipitation [*Delworth and Manabe*, 1998; *Shukla and Mintz*, 1982].

The information on snow cover is critically important to describe the interactions between the atmosphere and underlying surface, influencing the accuracy of land surface models and assimilation schemes determining results of hydrological simulations. However, the problem of reliable snow cover parameterization in models is far from being solved. "In a sense, snow is an inherently difficult quantity to model: … errors in snow cover quickly cascade into larger inaccuracies in albedo, soil moisture, energy exchange, and atmospheric conditions" [*Zaitchik and Rodell*, 2008]. The authors, for example, demon-

strated that the change of air temperature by several degrees could be less significant than the influence of snow cover modified by observations.

Though data assimilation for hydrological studies has a relatively short history, different approaches have been already implemented and tested [*Salamon and Feyen*, 2009; *Thirel et al.*, 2010; *Reichle et al.*, 2002]. Sophisticated schemes of physically based snow assimilation are required to establish the balance between snow state and heat and water fluxes. There are numerous works beginning from *Liston et al.* [1999] illustrating that the quality of hydrological modeling improves when the information on snow cover is assimilated.

Because of the very significant impact of snow cover, it is necessary to improve the quality of snow retrieval from remote sensing. The Visible/Infrared Imager/ Radiometer Suite (VIIRS) onboard the Suomi National Polar-Orbiting Partnership (Suomi NPP) platform launched in October 2011 is one of the currently available moderate resolution remote sensing instruments providing the spectral coverage from 412 nm to 12 μm used to generate snow cover information along with a large number of other products. On the whole, VIIRS contributes to 23 environmental data records (EDRs) and is the primary instrument for 18 EDRs. The VIIRS snow cover EDR consists of two products: the snow cover binary map (a binary product) and the snow cover fraction map (a fractional product). The binary product assesses a pixel as either covered by snow or snow free. This product classifying pixels as snow or nonsnow is named binary snow through the text of the chapter. The fractional product is defined as the fraction of the Earth's surface masked by snow. This product is named snow fraction.

IMSG at NOAA Center for Satellite Applications and Research, Washington, DC, USA

Remote Sensing of the Terrestrial Water Cycle, Geophysical Monograph 206. First Edition. Edited by Venkat Lakshmi.

175

This chapter begins from a general description of the NPP platform, the VIIRS instrument, and the requirements for the VIIRS snow products. The following sections present a consideration of the VIIRS snow algorithm and retrieval from visible and near-infrared top-of-atmosphere (TOA) reflectances. Then the focus shifts to the overall objective of the work—to quantitatively estimate the VIIRS snow information, to consider possible reasons of errors, and to outline the potential directions of snow data improvement. Optical remote sensing is used to provide the valuable snow information, first of all, snow extent and the snow cover area (SCA). These variables of interest are the center of our attention. The analysis of snow cover extent includes the consideration of errors within snow regions caused by cloud misclassification, land surface type, and the geometry of observations. The quality of SCA, the most important variable, is evaluated through a quantitative comparison of the VIIRS snow retrieval with high-resolution Landsat data in the vicinity of a snow line separating snow-covered areas from snow-free regions.

Another motivation for this work came from knowing that the quality of snow cover information provided by remote sensing can and does vary from region to region as well as from day to day depending on snow and background surface types as well as on the geometry of satellite observations and the state of the atmosphere. The variability of local snow and nonsnow reflective properties, their influence on the VIIRS snow information, and the contribution of possible algorithm parameter adjustment are also included into consideration in this chapter along with the analysis of using subpixel fractional snow cover (provided by the algorithm developed for MODIS by *Salomonson and Appel* [2006]) instead of the VIIRS binary snow classification.

At the time of this study, the VIIRS radiometric responsivity degraded very significantly in several wavelengths due to ultraviolet exposure of the primary telescope mirror [*Cao et al.*, 2014]. The quality of some sensor data records (SDRs) and VIIRS products was considered not suitable yet for reliable quantitative assessments. Therefore, specially simulated VIIRS proxy data characterized by the highest possible quality were carefully generated on the basis of reprocessing MODIS (Aqua) measurements. The set of global VIIRS proxy data for 2–9 February 2003 created and recommended for cryosphere studies is used in this chapter.

11.2. NPP PLATFORM AND VIIRS INSTRUMENT

The overall aim of the Suomi NPP mission is to provide a transition from NASA's Earth Observing System (EOS) to NOAA's Joint Polar Satellite System (JPSS). The NPP satellite launched in October 2011 carries the following instruments onboard: the Cross-track Infrared Sounder (CrIS), the Advanced Technology Microwave Sounder (ATMS), the Ozone Mapping and Profiler Suite (OMPS), the Clouds and the Earth's Radiant Energy System (CERES), and the Visible/Infrared Imager/Radiometer Suite (VIIRS). The CrIS and the ATMS use their measurements to retrieve vertical atmospheric profiles of temperature and moisture, the OMPS provides total column and the vertical profile of ozone, the CERES estimates Earth's energy budget.

VIIRS is a 22-channel scanning radiometer instrument providing observations in the wavelength range from the visible to the infrared [*Hillger et al.*, 2013] that was developed to combine the features of three previously existing sensors: the Operational Linescan System (OLS)—the operational visible/infrared scanner with a low-level light sensor (LLLS) capable of detecting visible radiation at night, the advanced very high resolution radiometer (AVHRR)—the operational visible/infrared sensor of low operational and production cost, and the moderate resolution imaging spectroradiometer (MODIS) serving as a standard of high accuracy for a wide range of satellite-based environmental measurements. The nominal VIIRS spatial resolution at nadir is 375 m for imagery (I) resolution spectral bands and 750 m for moderate (M) resolution spectral bands. The day-night band (DNB) has a 750 m spatial resolution across full scan [*Lee et al.*, 2006]. Because of an original scanning and aggregation scheme, VIIRS observations are characterized by a pixel growth factor of only two, both along a track and along a scan, giving the opportunity of getting imagery globally at 800 m resolution and 375 m at nadir. The measurements provided by VIIRS are used to retrieve 23 EDRs, being the principal source of information for 18 EDRs. Because of fine, 375 m nadir field of view of the sensor, VIIRS provides spatially detailed information at that resolution on several products including snow cover.

11.3. VIIRS SNOW PRODUCTS

Along with a large number of other land surface, ocean, atmosphere, and cryosphere products, VIIRS observations are used to routinely map and monitor the global distribution of snow cover. The generation of the snow cover products with NPP VIIRS data started in January 2012. Global daily snow cover maps are produced operationally at the finest spatial resolution of 375 m. The objective of the VIIRS snow retrieval is to achieve the performance level designed to meet the requirements stated in the NPP system specification. These specifications placing requirements on the VIIRS binary map product and the VIIRS

Table 11.1 Specification of the NPP snow cover/depth EDR for VIIRS

Paragraph	Subject	Specified Value
	a. Horizontal Cell Size (HCS)	
40.6.3-1a	1. Clear, Daytime, Nadir [VIIRS Guarantee]	0.8 km
40.6.3-1b	2. Clear, Daytime, Edge of Swath [VIIRS Guarantee]	1.6 km
	b. Horizontal Reporting Interval	
40.6.3-3a	1. Clear, Daytime [VIIRS Guarantee]	HCS
40.6.3-4	c. Snow Depth Ranges [VIIRS Guarantee]	Snow/No Snow
40.6.3-5	d. Horizontal Coverage [VIIRS Guarantee]	Land
40.6.3-7	f. Measurement Range [VIIRS & CMIS Guarantee]	0%–100% of HCS
	g. Measurement Uncertainty	
40.6.3-8	1. Clear, Daytime [VIIRS Guarantee]	10% of HCS (Snow/No Snow)
	h. Mapping Uncertainty, 3 Sigma	
40.6.3-10	1. Clear [VIIRS Guarantee]	1.5 km
40.6.3-12	i. Max Local Average Revisit Time	24 h Daytime Only
	j. Binary HCS	
40.6.3-14a	1. Clear, Day, Nadir [VIIRS Guarantee]	0.4 km
40.6.3-14b	2. Clear, Day, Edge Of Swath [VIIRS Guarantee]	0.8 km
40.6.3-16	l. Long Term Stability (C) [VIIRS Guarantee]	10%
40.6.3-17	m. Latency	NPP – 140 min.
40.6.3-18	n. Binary Map- Measurement Range [VIIRS Guarantee]	Snow/No Snow
40.6.3-19	o. Binary Map- Probability of Correct Typing [VIIRS Guarantee]	90%
40.6.3-21	p. Clear, Daytime Measurement Uncertainty Degradation If Solar Zenith Angle 70 to 85 deg [VIIRS Degradation]	40% of HCS (Snow/No Snow)
	r. Clear, Daytime Measurement Exclusions: [VIIRS Exclusion]	
40.6.3-22a	1. Snow Fraction Measurement Exclusion Condition: Horizontal Cell Contains Forest Canopy	
40.6.3-22b	2. Binary Map Probability of Correct Typing Exclusion Condition: Snow Fraction 0.2 to 0.7 or Solar Zenith Angle > 60 deg	
40.6.3-22c	3. All Measurements If Aerosol Optical Thickness > 1.0	
40.6.3-22d	4. All Measurements If Solar Zenith Angle > 85 deg	

snow fraction product are listed in Table 11.1. The specifications apply under clear, daytime conditions only. Surface properties cannot be observed through cloud cover by a visible/infrared (VIS/IR) sensor.

An original design of the VIIRS snow algorithms was created to generate the binary snow product from four VIIRS imagery bands and the snow fraction product from nine VIIRS moderate resolution bands. The bands used to retrieve binary snow are identified in bold in Table 11.2, and the bands proposed for snow fraction are italicized.

After VIIRS data products reach a certain level of maturity, they become publicly available for users from two main data archives—NOAA and NASA. The NOAA Comprehensive Large Array-Data Stewardship System (CLASS) provides access to the Suomi NPP mission products including snow EDR made publicly available on 14 March 2013. The Suomi NPP Land Product Quality Assessment website is developed by NASA to provide access to the VIIRS data. Both NOAA and NASA systems operationally ingest EDRs produced by the NPP interface data processing segment (IDPS), provide their quality control, archive, and offer public access to VIIRS information. The Land Product Quality Assessment website, in particular, was used as a source of the simulated VIIRS data calculated by IDPS and serving as the basis for the studies described in this chapter.

11.4. SNOW COVER AREA

Snow products provided by the moderate-resolution remote sensing were quantitatively estimated and considered for a large range of applications, including land surface and hydrological models. SCA estimated as the percentage of a model cell covered by snow is a major measure of spatial inhomogeneity of snow cover—a very significant feature contributing to the variability of the atmosphere-surface interaction influencing hydrological and weather forecasts as well as climate change. It is critically important that SCA be a parameter of land surface models.

The use of the snow depletion curve (SDC) illustrating related temporal changes in SCA and the snow water equivalent (SWE) [*Anderson*, 1973] obtained from remote

Table 11.2 NPP VIIRS spectral bands

VIIRS Band	Spectral Range (µm)	Nadir Spatial Resolution (m)
DNB	0.500–0.900	750 across full scan
M1	*0.402–0.422*	*750*
M2	*0.436–0.454*	*750*
M3	*0.478–0.498*	*750*
M4	*0.545–0.565*	*750*
I1	**0.600–0.680**	**375**
M5	*0.662–0.682*	*750*
M6	0.739–0.754	750
I2	**0.846–0.885**	**375**
M7	*0.846–0.885*	*750*
M8	*1.230–1.250*	*750*
M9	1.371–1.386	750
I3	**1.580–1.640**	**375**
M10	*1.580–1.640*	*750*
M11	*2.225–2.275*	*750*
I4	3.550–3.930	375
M12	3.660–3.840	750
M13	3.973–4.128	750
M14	8.400–8.700	750
M15	10.236–11.263	750
I5	**10.500–12.400**	**375**
M16	11.538–12.488	750

or in situ observations is a powerful approach to describe local conditions, employed by many researchers (*Andreadis and Lettenmaier*, 2006; *Déry et al.*, 2005a; *Kolberg and Gottschalk*, 2006]. One of the first SCA parameterizations in a one-layer description of snow state was implemented in the National Centers for Environmental Prediction (NCEP) [*Ek, et al.*, 2003] to modify the Noah one-dimensional land surface model:

$$\mathrm{SCA} = 1 - \left[\exp\left(-\frac{\gamma_{\mathrm{SWE}}}{\mathrm{SWE}_{\mathrm{SCA}=1}} \right) - \frac{\mathrm{SWE}}{\mathrm{SWE}_{\mathrm{SCA}=1}} \exp\left(-\gamma\right) \right],$$

$$(11.1)$$

where γ is the parameter used to approximate the shape of the relationship between SWE in a model cell and the relative coverage of a cell by snow (SCA).

It is customary to use the calculations on the basis of equation 11.1 globally with a predetermined magnitude of γ and the minimum SWE corresponding to complete coverage of a cell by snow ($\mathrm{SWE}_{\mathrm{SCA}=1}$), which depends on land cover type—the lowest for bare soil and the largest in the boreal forest.

Initially, simple approaches to estimate SWE on the basis of remote sensing observation on SCA were considered reasonable for global applications. The replacement of calculated SWE by the value directly determined by observed SCA could be considered as an example of the simplest method [*Rodell and Houser*, 2004] used in cells with inconsistency (absence vs. presence) between modeled SWE and observed SCA. More complex algorithms use observed SCA to modify SWE on the basis of inverted equation 11.1 (the push algorithm in *Zaitchik and Rodell* [2008].

The relationship between SCA and SWE parameterized by the equation and is characterized by significant variability. If systematic temporal observations on changes in SCA and SWE are available, then the equation could be modified for specific conditions to make the relationship more accurate. It has been demonstrated that taking into account a specific feature of SDC varying with elevation and from region to region enhances the quality of SWE simulations [*Thirel et al.*, 2011]. The experiments with a snow model using direct measurements of snow [*Clark et al.*, 2006] confirmed the improvement of information on snow distribution along with general hydrologic characteristics.

For the purpose of snow cover assimilation, SDC became a standard tool taking into account a variety of the relationships between SCA and SWE obtained usually on the basis of combining surface observations on snow depth with satellite observations on snow extent. For local research, even high-resolution satellite observations were recently used to reconstruct SDC in 90 m cells of a land surface model [*Girotto et al.*, 2014]. However, cells in land surface models are typically much larger and more frequent observations on SCA are preferable for operational applications.

Using observations on SCA provides numerous advantages. It could be used to improve our understanding of physical processes and a mathematical description of snow evolution. The relationship between SCA and SWE depends on seasons and geographic locations. Availability of satellite observations provides valuable information to study different relationships and their dependence on influencing factors. The assimilation of SCA in land surface models substantially increases the quality of information on snow cover state including SWE [*Zaitchik and Rodell*, 2008]. There are successful attempts at using information on SCA to correct not only SWE but also other parameters of land surface models, when a locally specific SDC is well defined [*Andreadis and Lettenmaier*, 2006; *Clark et al.*, 2006]. Improved retrieval of SCA helping better estimate heat and mass fluxes could be used for assimilation to initiate model calculations as well as to assess the quality of model outputs.

Following existing practice, we analyze VIIRS SCA information for 0.05° cells of a latitude/longitude grid widely used for hydrologic and climate modeling. The aggregation of VIIRS binary snow product within 0.05° climate modeling grid (CMG) cells provides the assessment of SCA for the part of a cell not obscured by cloudiness. In processing VIIRS data, only pixels with solar zenith angles not more than 85° were taken into account. Snow data were aggregated for all pixels classified as land (including desert), even those where cloud shadow, fire, heavy aerosol, or cirrus were detected. Other pixels—inland water, seawater, and coastal—were excluded from the analysis.

11.5. VIIRS SNOW ALGORITHM AND OBSERVATIONS

Surface observations on snow are not sufficiently dense and reliable to describe the spatial changes in snow distribution varying significantly as a function of many influencing factors. On the contrary, satellite observations could be considered as data smoothly changing in space and presenting unparalleled source of information for large-scale and global applications including hydrologic and land surface models [*Andreadis and Lettenmaier*, 2006; *Rodell and Houser*, 2004; *Roy et al.*, 2010; *Su et al.*, 2008; *Zaitchik and Rodell*, 2008; *Bavera and De Michele*, 2009]. Despite numerous works, microwave remote sensing information on SWE is not accurate enough for many hydrologic applications [*Pulliainen et al.*, 2010]. The difficulties of getting detailed, reliable descriptions of snow cover features investigated and demonstrated in works [*Dong et al.*, 2005; *Foster et al.*, 2005; *Rodell et al.*, 2004; *Tait and Armstrong*, 1996] are summarized by the following statement: "both model predictions and passive microwave snow water equivalent observations contain large errors" [*Dong and Peters-Lidard*, 2010].

However, visible and near-infrared satellite observations on SCA are of a much better quality [*Parajka and Bloschl*, 2006] and could be available frequently—up to several times a day in cloud-free regions. The important point for the subject under consideration in this chapter is the fact that the VIIRS data on snow cover could be efficiently used to adjust the variables determining snow evolution. For example, the information on SCA led to very noticeable improvement in important hydrological SWE estimates for the U.S. Southwest and the U.S. high plaines [*Zaitchik and Rodell*, 2008]. VIIRS provides snow information at the resolution adequate for most hydrologic applications. The VIIRS information on binary snow is retrieved using the algorithm similar to SNOWMAP [*Hall and Riggs*, 2007], originally designed for global retrieval of snow, which explains its obvious advantages.

A central feature of the binary snow algorithm, a normalized difference snow index (NDSI), is a spectral band ratio that takes advantage of the fact that snow reflectance is high in the visible wavelengths and low in the short-wave infrared region. NDSI is calculated from imagery resolution bands:

$$NDSI = R_1 - R_3 / R_1 + R_3, \qquad (11.2)$$

where R_1 and R_3 are TOA reflectances in 0.64 μm (I_1), 1.61 μm (I_3) VIIRS bands.

The VIIRS algorithm classifies snow by NDSI and an additional reflectance threshold. Pixels with NDSI > 0.4 and 0.865 μm (I_2 band) TOA reflectance > 0.11 are classified as snow. The VIIRS algorithm uses a red band in place of the green band utilized by the MODIS algorithm to achieve the resolution of 0.375 km at nadir. Our analysis demonstrated that using the red band provides the NDSI threshold nearly identical to the green band NDSI threshold—the major indicator distinguishing snow from nonsnow [*Appel and Jensen*, 2002]. The reflectance threshold for 0.64 μm VIIRS imagery resolution bands is not used in the VIIRS algorithm differing it from SNOWMAP.

The approach described above could miss snow cover under the forest canopy and incorrectly classify such forest pixels as no snow. To improve the recognition of snow in forest pixels, the VIIRS snow algorithm also uses a normalized difference vegetation index (NDVI) to classify snow-covered forest pixels by the location in an NDSI/NDVI scatterplot [*Klein et al.*, 1998]:

$$NDVI = R_2 - R_1 / R_2 + R_1, \qquad (11.3)$$

where R_1 and R_2 are TOA reflectances in 0.64 μm (I_1), 0.865 μm (I_2) VIIRS bands.

Pixels with 0.1 < NDSI < 0.4 are classified as snow cover under a forest canopy, if NDVI is greater than a threshold value depending on the pixel NDSI. Current realization of the VIIRS binary snow algorithm does not use information about the location of forests on the Earth's surface, and the additional NDVI test is part of routine process for each pixel classified as land.

Originally, an application of the multiple end-member spectral mixture analysis (MESMA) was developed for VIIRS to retrieve the snow fraction product [*Appel and Jensen*, 2002]. The spectral mixture analysis defines subpixel proportions of spectral end-members related to mappable surface constituents. It "unmixes" the mixed pixel, determining the fractions of each spectral end-member combined to produce the mixed pixel's spectral signature. Our approach was to model the spectral signature from each pixel as a combination of two components: a modeled snow reflectance spectrum and a nonsnow reflectance spectrum. The approach is based on the assumption that the nonsnow end-member spectrum for each pixel can be obtained from the

VIIRS gridded surface albedo intermediate product (IP). The performance analysis indicated that the measurement uncertainty requirement can be achieved, except for scenes with forest canopy. In the current version of the VIIRS processing system, the MESMA was temporarily replaced by the aggregation of the binary snow within 2×2 pixel blocks. Therefore, the validation of VIIRS snow products is considered only for the binary snow product. In the future, a spectral unmixing algorithm or NDSI-based approach [e.g., *Salomonson and Appel*, 2004] could be implement to calculate snow fraction.

It has been established in various studies and considered in our work with VIIRS data that the quality of the binary snow retrieval depends on several influencing factors: first of all, vegetation [*Hall et al.*, 1998; *Simic et al.*, 1998], solar and viewing geometry, and misinterpretation of clouds, but also on topography [*Hall and Riggs*, 2007], snow properties [*Klein and Barnett*, 2003], and the state of underlying nonsnow surface.

Hydrologic applications of snow remote sensing data require quantitative understanding of the errors related to those factors, which is vital to improve the quality of snow retrieval [*Dong and Peters-Lidard*, 2010]. The quality of snow identification decreases in forested areas especially noticeably in the boreal forest. The influence of the boreal forest and missing clouds on the quality of VIIRS snow retrieval is analyzed in the chapter with the emphasis on the role of the VIIRS cloud mask (VCM). However, to estimate the impact of other factors on the quality of the VIIRS snow products, the influence of cloud misinterpretation is significantly suppressed in our further analysis by the choice and consideration of the cloud-free Landsat scenes.

11.6. TEMPERATURE SCREENING

The analysis of snow mapping [*Dong and Peters-Lidard*, 2010] based on the SNOWMAP algorithm demonstrated that the algorithm underestimates the coverage by snow in western United States in shoulder seasons—the beginning and the end of the period when land is covered by snow (Figure 11.1). It was hypothesized in the publication that the errors may result from a combination of patchy snow and land surface complexity. However, another more simple explanation of the shortcoming is given by NASA and illustrated at the example of the Sierra Nevada mountain range (the region analyzed by *Dong and Peters-Lidard* [2010] in the description of snow cover algorithm changes for collection 6 on the NASA Snow & Sea Ice Global Mapping Project website (http://modis-snow-ice.gsfc.nasa.gov/).

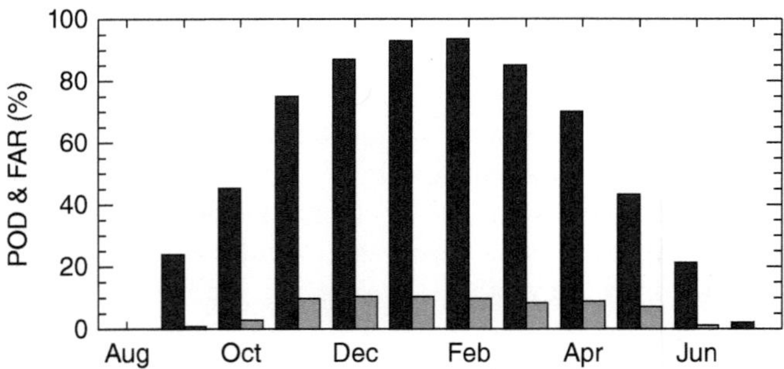

Figure 11.1 Probability of detection (black bars) and false alarm ratio (gray bars) for MODIS snow cover in each month (from February 2000 to December 2005).

The SNOWMAP algorithm initially employed a temperature screen to remove false snow cover provided by the algorithm in the warm areas where snow never exists—in low latitudes, coastal pixels, and so forth. But simultaneously with removing false snow cover, the application of the temperature screen significantly reduced the area covered by snow in spring and summer, the period of snow melting when the surface temperature of pixels with snow could be above the screen threshold under consideration. That reduction negatively impacted the quality of snow retrieval, especially in mountains. Therefore, beginning from collection 6 (C6) the surface temperature screen has been removed from MODIS data processing in NASA. It leads to the desirable extension of the period when snow observed in a seasonal cycle and to a noticeable increase in areas covered by snow. The number of snow pixels removed by the temperature screen does not change significantly from day to day (Figure 11.2), but the relative contribution of those pixels to the area of snow cover progressively increases almost to 100% (Figure 11.3) in the summer time characterized by high temperatures.

The consideration of the VIIRS retrieval in this chapter is limited to the wintertime (February), when the influence of the temperature screening is certainly negligible. But it is necessary to pay attention to the issue later to consider potential problems related to the use of temperature screening in the VIIRS data processing and its modifications.

11.7. INFLUENCE OF CLOUD MISCLASSIFICATION

On the whole, the snow distribution in the wintertime period under consideration is characterized by a relatively simple spatial structure with a clearly identifiable snow line (the boundary between snow and no snow areas) separating two zones: predominantly covered by snow north of the line and mostly snow-free areas south of the

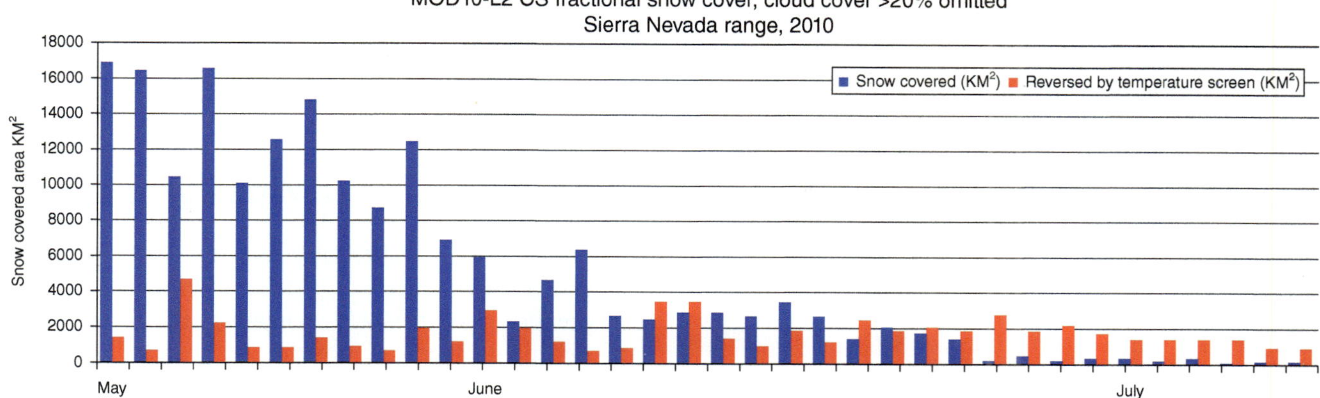

Figure 11.2 Area mapped as snow and area reversed to nonsnow by the surface temperature screen in the MOD10_L2 C5 algorithm, daily over May–July 2010.

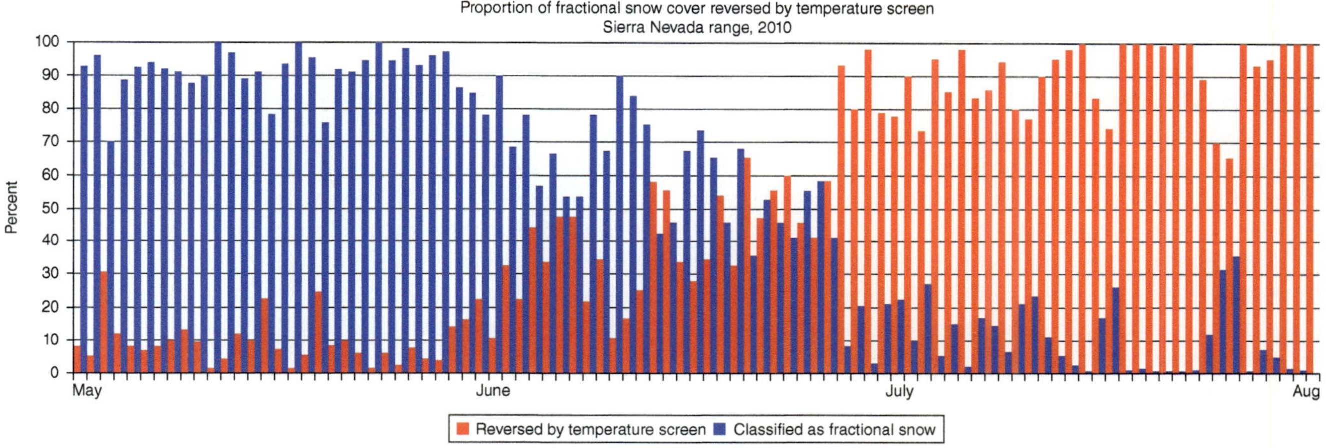

Figure 11.3 Relative area mapped as snow and reversed to nonsnow by the surface temperature screen in the MOD10_L2 C5 algorithm, daily over May–July 2010.

line in the Northern Hemisphere. VIIRS, as well as other moderate-resolution remote sensing instruments, was designed to use optical data where snow and clouds could look similar. The discrimination of snow from clouds is a complex problem that should be underlined in the works analyzing the use of satellite information to retrieve snow cover products. The application of the VIIRS snow cover algorithm to the regions completely covered by snow is used in our studies to estimate the probability of snow omission errors related to cloud misinterpretation. Siberia was chosen for the case study including one 5 min granule (Table 11.3) for each day of the period 2–9 February 2003.

The existing realization of the VIIRS snow algorithm uses VCM, assigning one of four cloud confidences to each moderate-resolution pixel: (1) confidently cloudy, (2) probably cloudy, (3) probably clear, and (4) confidently clear. The cloud confidence presents generalized characteristic of numerous individual cloud detection tests. In this study, snow products retrieved for three levels of cloud mask confidence—confidently clear, probably clear, and probably cloudy are taken into consideration. The information on cloud mask quality (poor, low, medium, and high) was ignored. The uncertainties in VCM mask influence the quality of snow products depending not only on the features of the snow retrieval algorithm but also upon the quality of cloud classification. The performance of the VIIRS snow algorithm is stratified according to the cloud mask confidence.

There are three possible options of creating cloud mask combining different confidences: (1) the "liberal" cloud mask consisting of only "confidently cloudy", (2) the "conservative" cloud mask including "confidently cloudy", "probably cloudy," and "probably clear" pixels,

Table 11.3 Probability (%) of omission errors for different cloud confidences

Day	Time	Cloud Confidence		
		Confidently Clear	Probably Clear	Probably Cloudy
02/02/2003	06:30	1	6	31
02/03/2003	05:35	1	4	15
02/04/2003	07:55	1	1	11
02/05/2003	08:40	1	1	8
02/06/2003	07:45	0	0	1
02/07/2003	05:10	1	3	5
02/08/2003	09:10	1	2	19
02/09/2003	08:15	1	1	4
Average		1	2	12

and (3) the "relaxed" cloud mask combining "confidently cloudy" and "probably cloudy" pixels. The interpretation of the VCM confidences is not quite obvious, but the pixels corresponding to probably clear and probably cloudy confidence of the VIIRS cloud mask are considered as certainly corresponding to less-than-ideal viewing conditions. The probability of omission errors increases from confidently clear pixels through probably clear to probably cloudy pixels (Table 11.3).

By definition, the likelihood of cloudiness increases from pixels classified as confidently clear to probably clear and to probably cloudy. Therefore, the frequency of applying snow retrieval to cloudy pixels also increases accordingly, which leads to the increase in omission errors. This conclusion is confirmed by the analysis of the spatial error distribution within the granules. The analysis demonstrated that the special kind of clouds often identified in false color imagery is related to omission errors in snow retrieval. The application of the VIIRS snow cover algorithm to such kind of clouds provides retrieval of nonsnow pixels.

It is not anticipated that VCM will considerably improve the identification of cloudy pixels in the nearest future. Meanwhile, because the application of the snow cover algorithm to clouds is the reason for significant errors, it seems promising to reduce the probability of the errors making snow cover retrieval only for the pixels classified by VCM as confidently clear. Such application of conservative cloud mask including confidentially cloudy, probably cloudy, and probably clear pixels improves the quality of the VIIRS snow products.

The VIIRS snow algorithm not only underestimates the snow extent but also overestimates it in other cases, retrieving false snow cover over the areas where snow never exists. The probability of these commission errors is 1%–2% over tropical areas and in contrast to omission errors does not depend of the VCM confidence. The location of the commission errors corresponds to ice clouds having a spectral signature similar to snow on the ground.

11.8. EFFECT OF OBSERVATION GEOMETRY

The geometry of VIIRS observations is also included in the stratified analysis of the VIIRS snow algorithm performance. As a first approximation, the illumination and viewing geometry of observation are parameterized by latitude and a scan angle. The analysis shows that the dependence of the probability of omission errors upon latitude is typically larger in the middle latitudes of the granules listed in Table 11.3 and decreases toward north and south (Figure 11.4). This kind of change as a function of latitude indicates a secondary role of the solar zenith angle. Figure 11.4 illustrates changes in the percentage of omission errors for different cloud confidences within narrow latitudinal ranges. The weighted average probabilities of omission errors included in Table 11.3 provide more representative information to consider total errors. In many cases the increased percentage of missing snow is observed in the regions of the boreal forest. The dependence of the snow algorithm performance upon underlying surface type is a well-based hypothesis that will be verified and assessed later in this chapter. More detailed consideration of the VIIRS snow product quality for different cloud mask confidences could be deemed necessary to explore the reasons of errors and to develop recommendations for potential improvement of the snow information.

The changes in the probability of omission errors for the varying cloud mask confidence as functions of a scan angle are considered for the same 8 granules similarly to the analysis of the latitude influence. The observations in each granule were divided in 20 equal groups of columns and the statistics of omission errors represented the scan angle influence. Figure 11.5 shows an example of the dependence of snow omission errors upon the scan angle (from west to east) for three cloud mask confidences: confidently clear, probably clear, and probably cloudy. The dependence is characterized by large changes explained by varying combinations of surface types and cloud locations. The influence of cloud misinterpretation and the underlying surface type upon the snow omission errors is larger for probably clear and probably cloudy VCM confidence and diminishes for confidently clear areas. The total omission errors matched to the estimates are included in Table 11.3.

It is important to emphasize that on the whole the omission errors for probably clear and probably cloudy

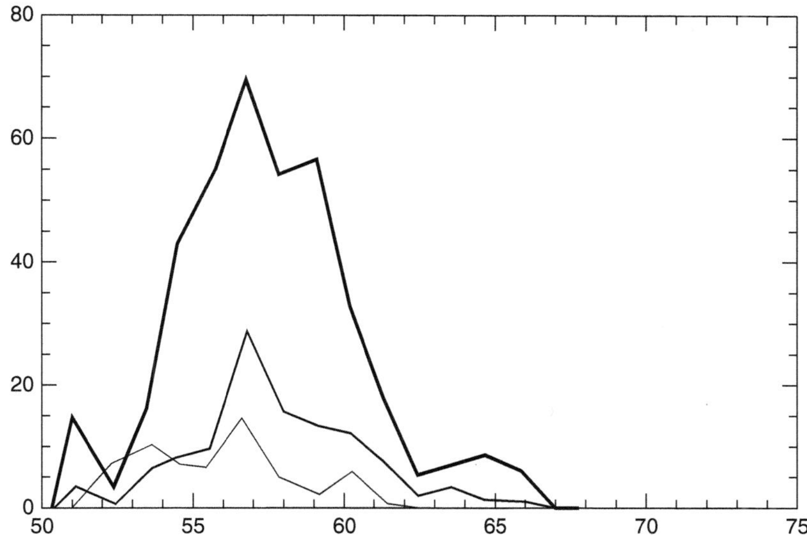

Figure 11.4 Latitudinal changes in probability (%) of missing snow for various VCM confidences—confidently clear (thin), probably clear (medium), and probably cloudy (thick)—for 5 min granule on February 2, 06:30.

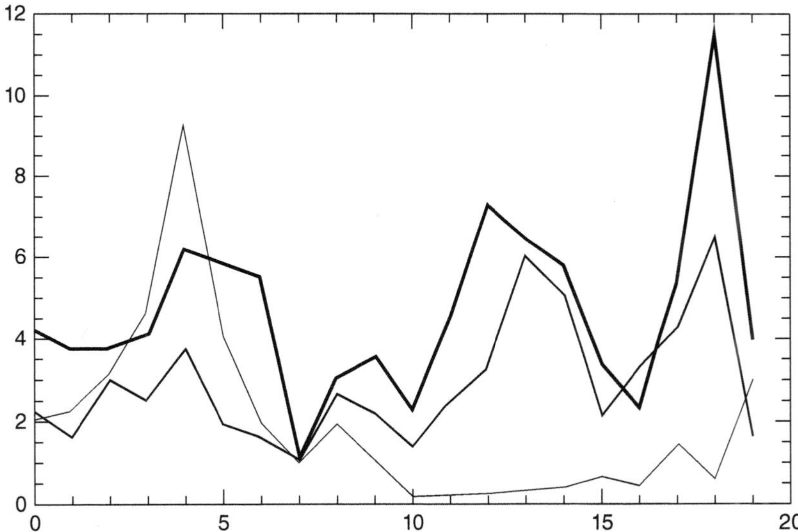

Figure 11.5 Changes in probability (%) of missing snow along scan lines from west to east for various VCM confidences—confidently clear (thin), probably clear (medium), and probably cloudy (thick) for 5 min granule on February 7, 05:10.

areas change across the scan in a similar manner, while the shape of the changes for confidently clear regions is quite different. The dissimilarity between the characteristics of omission errors is related to varying reasons for the errors. Investigating the impact of the scan angle on the frequency of omission errors, we could use the following general assumption: The smaller the influence of other factors, the more reliable assessment of the dependence under consideration could be done. Therefore the influence of the scan angle on omission errors is better illustrated in Figure 11.6 singly for confidently clear areas having weaker dependence on cloudiness and surface type. This typical case clearly demonstrates that the probability of omission (as well as commission) errors rises significantly toward ends of scans being on average larger at the western edge of swaths then at eastern edges. Such effects are caused by the influence of the bidirectional reflectance distribution function (BRDF) describing the features of non-Lambertian light reflectance by snow surface [*Appel*, 2011].

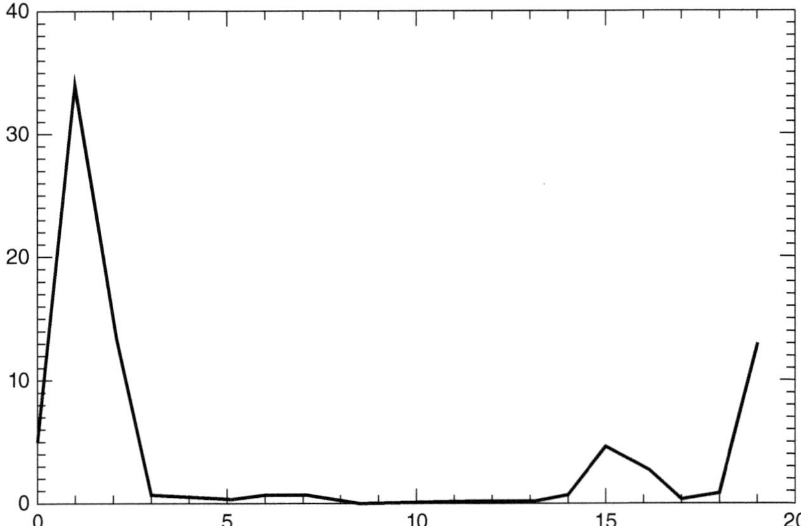

Figure 11.6 Changes in probability (%) of missing snow along scan lines from west to east for confidently clear VCM confidence for 5 min granule on February 4, 07:55.

11.9. SELECTION OF HIGH-RESOLUTION DATA

After considering the possibilities of utilizing aircraft and various high-resolution satellite data, it was concluded that the use of the Landsat Thematic Mapper observations was suitable and most effective. The Landsat band 2 (0.55 μm) and band 5 (1.64 μm) correspond to the bands used by the VIIRS snow algorithms to calculate the NDSI values serving as an indicator to detect snow. Therefore, the results of our analysis of Landsat high-resolution observations could be directly translated to VIIRS data presenting the focus of the study. The opportunity of utilizing Landsat data not only as a source of ground truth but also for investigation of reflectances, applicable to VIIRS, presents obvious advantage of choosing Landsat for investigations. An optimal approach to evaluate the quality of the VIIRS snow algorithm is based on using Landsat scenes covering a wide variety of snow conditions, first of all the areas including both snow and snow-free surfaces. Only the Landsat scenes without the identifiable influence of cloudiness in such areas are used to consider the influence of snow state and nonsnow reflective property.

The Landsat Archive of the Earth Resources Observation and Science (EROS) center, the U.S. Geological Survey (USGS), in Sioux Falls, has the most extensive set of Landsat data accessible through the Earth Explorer system, a fast geospatial search engine, providing a graphical user interface for online search, display, and order of Landsat images. Both Landsat 5 and Landsat 7 acquired the information for the period under consideration at a 30 m resolution (for most of the bands) after radiometric, geometric, and terrain corrections were applied. Landsat 5 did not have an onboard recorder and therefore supplied less systematic observations, though many ground stations acquired data locally. On the contrary, Landsat 7 had and onboard recorder, collected more information, and also used two gain regimes providing significant advantage for observations over bright snow cover surfaces.

To select Landsat 7 Enhanced Thematic Mapper Plus (ETM+) scenes with snow cover for the 8 day period under consideration, it was very helpful to use the data set prefilter, which excluded the scenes that do not meet our requirements from consideration. To reduce the number of available images to be examined for our task, it has been decided to peruse only daytime scenes with less than 10% of cloud cover. The Landsat scenes with potential snow targets were reviewed at a large resolution to analyze the details of scenes and evaluate the contamination of images by cloudiness. The selection of images with snow was simplified by the fact that snow cover is easily identifiable in most instances because of its distinctive spectral signature usually recognizable even when only three channels are used to create false color browse images.

11.10. CLASSIFICATION OF LANDSAT PIXELS

Initially, 65 Landsat scenes, including both snow and snow-free surfaces, were considered as a representative data set including a wide variety of snow conditions in quite different regions. Later, more than a dozen of the scenes with areas of open ocean (sea) water were excluded from consideration, which unfortunately removed all scenes collected in the Southern Hemisphere. The ground truth for remaining Landsat scenes was established using binary classification of 30 m Landsat pixels as snow or snow free. The classification to segregate snow-covered

Landsat pixels from snow-free pixels used the threshold of NDSI equal in accord with the SNOWMAP algorithm to 0.4. Each Landsat pixel was classified solely on the NDSI criterion. The only exception for the described binary classification was made for the Landsat pixels with negative reflectance either for Landsat band 2 or band 5. Pixels with any negative reflectance—observed mostly in mountainous areas for solar zenith angles 70° and more—were not included in further analysis.

The classification maps were made for easier interpretation using two contrasting solid colors for snow pixels (white) and for nonsnow pixels. The analysis of the Landsat binary maps included the comparison of the classification results with coregistered false color Landsat images combining three channels for exactly the same projection and resolution (Figure 11.7). The identification of snow cover in the false color images is not complicated in most instances because of a distinctive snow spectral signature usually recognizable when three channels are used to create the images. The comparison of the classification maps with pseudo color Landsat images clearly identifying snow cover demonstrates high classification quality.

The images for the cases with snow in the areas of dense vegetation were more difficult to determine snow distribution because of the large influence of vegetation concealing snow. However, it has been decided to keep the scenes with dense vegetation for future studies because such cases are important to analyze the VIIRS snow retrieval. The only scene with the certain underestimate of snow by the classification was observed in the area with a visible dusting of new thin snow not classified as snow. The underestimate of snow in the area could be explained by the low snow thickness—a very difficult and relatively rare case of snow omission.

The described comparison of the generated binary (snow/nonsnow) Landsat maps coregistered to exactly the same projection and resolution for false color Landsat images demonstrates the high quality of the classification based on a predetermined NDSI threshold. To consider the potential influence of a varying NDSI threshold, the NDSI maps presenting a whole range of NDSI changes were created. The palette shown in Figure 11.8 with different hues corresponding to every 0.1 change in NDSI was used to illustrate calculated NDSI values. The images depicting spatial distributions of NDSI values for a whole range of NDSI changes serve as a useful means to estimate the quality of the binary Landsat maps. An example of a comparison between an NDSI map and a pseudo color image is presented in Figure 11.9.

Figure 11.7 Location of Landsat scene (*top*), false color image and pixel classification (*bottom*) on February 9 (path—156, row—37).

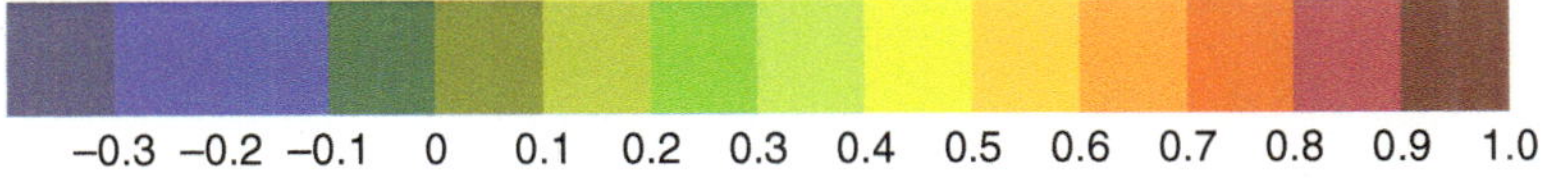

Figure 11.8 Color palette used to present spatial distribution of NDSI.

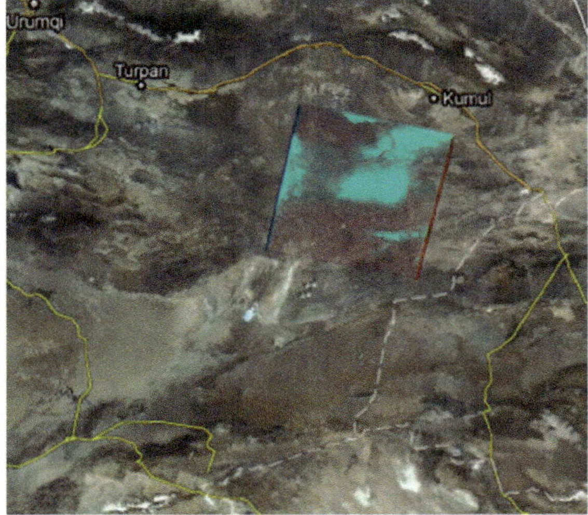

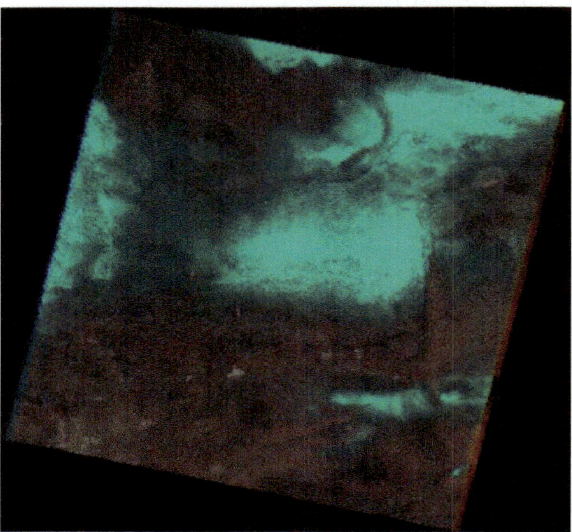

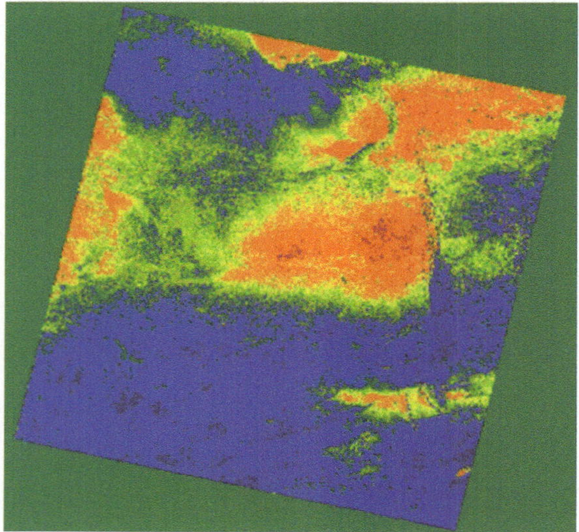

Figure 11.9 Location of Landsat scene (*top*), false color image and NDSI map (*bottom*) on February 2 (path—139, row—31).

In some parts of Landsat scenes, especially in mountainous areas, the transition from the hues corresponding to snow-free surface to the hues presenting snow is so sharp that the colors for intermediate NDSI values are not visible. In the case of such transitions even moderate changes in the NDSI threshold magnitude would not influence the binary classification of Landsat pixels. The analysis of other scenes with diffuse boundaries between snow and nonsnow are less convincing. However, as a rule, the hue matching NDSI equal to 0.5 is associated with visible snow in false color Landsat images, while lower magnitudes of NDSI correspond to snow-free surfaces. On the whole, the review of the NDSI maps confirms that the Landsat pixel classification based on the NDSI threshold equal to 0.4 is reliable and could be used to estimate the VIIRS snow product retrieval.

11.11. COREGISTRATION OF VIIRS AND LANDSAT DATA

Because the binary classification (snow or nonsnow) of Landsat pixels is characterized by high quality, the aggregation of Landsat pixels for cells of a larger size presents a reliable approximation of SCA considered a ground truth in further studies of the VIIRS information. VIIRS data include geolocation (latitude and longitude) of each pixel. To estimate the quality of the VIIRS snow information, the coordinates of all Landsat pixels need to be converted to latitude and longitude. For global studies of the VIIRS snow products, the projection containing all latitudes including equatorial areas is needed, which excludes polar projections from consideration. The latitude/longitude projection is selected to analyze the VIIRS snow algorithm performance.

Metadata for each scene of Landsat observations include the information on the Universal Transverse Mercator (UTM) zone number used for projection and UTM coordinates of one of the corner pixels in the zone. There is no real physical frame of reference for a UTM grid; the UTM northing (y) and easting (x) coordinates depend on the datum approximating the shape of the Earth. The Landsat information is related to the current datum—WGS84. Because a UTM grid is simply created by laying a square grid on the Earth datum, the calculations of UTM northing (y) and easting (x) coordinates for other pixels are made by adding predetermined 30 m of a pixel resolution to the coordinates of an image corner along both axes.

After the transformation of the UTM coordinates, the aggregation of pixels from Landsat scenes and VIIRS observations was made identically by calculating the total number of pixels and the number of snow pixels for each grid cell. Averaged aggregated SCA is found by dividing the number of snow pixels by the total number of observations in a cell. Precise coregistration of the Landsat pixels and the VIIRS binary snow product and their aggregation for identical cells of the regular geographic grid made the Landsat ground truth and the VIIRS observations completely spatially comparable. The subsets from global VIIRS data were chosen to exactly correspond to the spatial limits covered by the Landsat scenes.

11.12. PRELIMINARY COMPARISON OF LANDSAT AND VIIRS DATA

An example of SCA estimated as ground truth from classified Landsat pixels and the VIIRS SCA data is presented in Figure 11.10. Black cells in this and following figures indicate the absence of information, which means areas outside of a scene for Landsat observations and cells completely covered by cloudiness for VIIRS observations. Exactly the same color palette (Figure 11.11) corresponding to changes in SCA from 0 (dark) through lighter hues to 1 (white) simplifies the comparison of spatial changes in snow distributions for image pairs. Comparative analysis of the image pairs confirms the high accuracy of coregistration. The quality of the VIIRS snow product for the areas and days corresponding to the Landsat observations used to estimate SCA ground truth was not taken into account for the Landsat scenes selection. Therefore, several subsets of VIIRS calculations turned out to be unusable to analyze the quality of the SCA retrieval because of the absence of snow or large cloud amount in the VIIRS retrieval, though some of the cases could be used in the future to further consider the VIIRS snow data errors.

The scenes with close to 100% snow coverage both in Landsat and VIIRS data obviously illustrate a high quality of retrieval, but such cases do not provide much of additional information that could be useful to improve the VIIRS snow algorithm. However, the scenes where VIIRS retrieval provides almost complete coverage by snow while the ground truth information includes significant areas of lower SCA need special attention. Such a disagreement apparent in some parts of Figure 11.12 is related to forested areas where the currently implemented VIIRS algorithm retrieves snow even for lower NDSI, taking into account the calculations of NDVI. On the contrary, the Landsat estimates of ground truth in forested areas interpret some pixels with vegetation as nonsnow reducing SCA even in the regions covered by snow. Therefore the Landsat information on visible snow could not be compared with the VIIRS data on the snow presence in forests, and most forested areas are excluded from consideration at least at the present stage of analysis.

It has been also decided to temporally exclude from our analysis the cases with a very low percentage of snow in

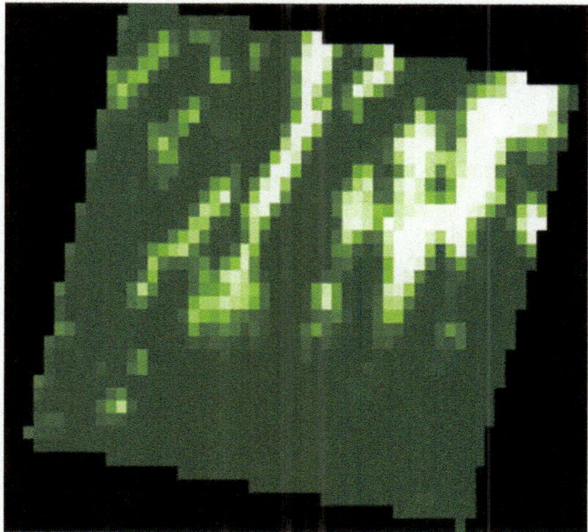

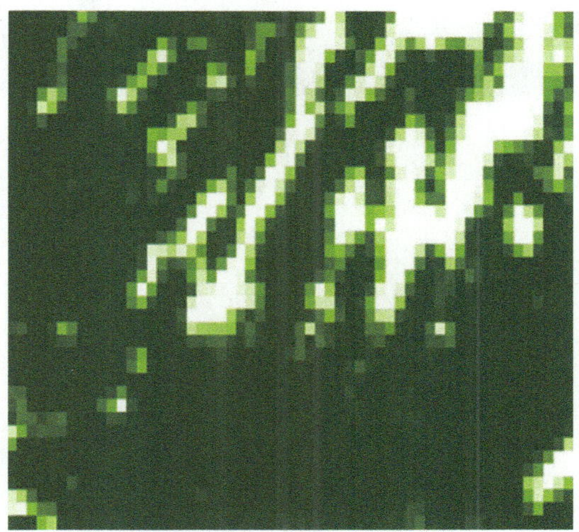

Figure 11.10 Location of Landsat scene (*top*), true and VIIRS (*bottom*) snow cover area on February 3 (path—41, row—33).

Figure 11.11 Palette used to indicate snow cover area.

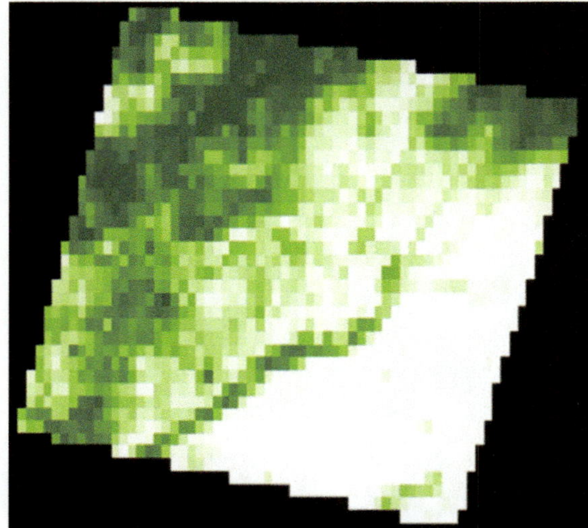
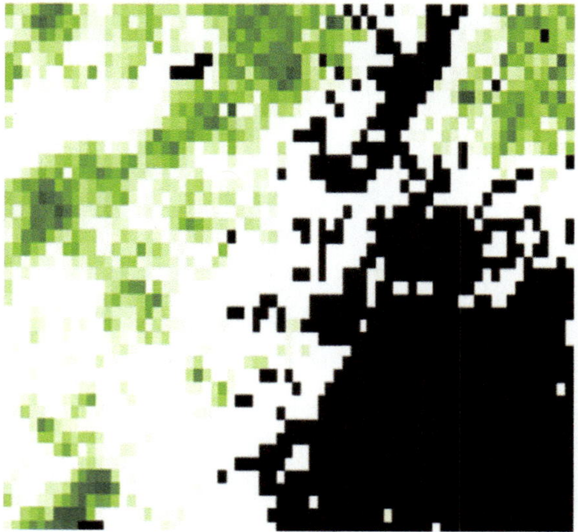

Figure 11.12 Location of Landsat scene (*top*), true and VIIRS (*bottom*) snow cover area on February 9 (path—124, row—25).

Table 11.4 Landsat scenes selected to validate VIIRS snow products

Date	Path	Row	Sun Elevation	Latitude	Longitude	Region
33	123	32	28	40 N	117 E	Eastern China
33	139	29	24	45 N	93 E	Western China
33	139	30	25	43 N	93 E	Western China
33	139	31	27	42 N	92 E	Western China
34	41	33	29	39 N	117 W	Nevada
34	41	34	30	37 N	118 W	Nevada
34	146	29	25	45 N	83 E	Western China
35	137	29	25	45 N	96 E	Mongolia
35	137	30	26	43 N	96 E	Western China
35	153	39	36	30 N	67 E	Afghanistan
36	128	30	26	43 N	110 E	Mongolia
37	30	28	24	46 N	98 W	North Dakota
39	44	27	24	47 N	119 W	Washington
39	44	31	28	42 N	121 W	Washington
40	156	35	33	36 N	64 E	Afghanistan
40	156	37	35	33 N	64 E	Afghanistan

both Landsat scenes and VIIRS retrievals. The Landsat scenes covered by more than one swath of VIIRS observation were not included in the final set for quantitative consideration either. As a result of a careful selection, only 16 scenes were considered as a representative subset characterized by good conditions of observations to make reliable quantitative validation of VIIRS SCA in the zones of the transition from snow-covered to snow-free areas. The scenes listed in Table 11.4 include a wide variety of snow conditions in quite different regions.

11.13. STATISTICAL ESTIMATES OF SCA RETRIEVAL

Quantitative evaluation of VIIRS SCA was made on the basis of using the regression of the retrieved SCA values on ground truth. The results of the regression analysis are presented in Table 11.5, including the coefficient of correlation between the VIIRS SCA and ground truth, the parameters of the linear statistical relationship between the calculated and true values (intercept and slope), as well as average true values and the VIIRS SCA for scenes. To specify the locations of Landsat granules, the names of cities, states, parks, mountains, deserts, provinces, and countries are used for the scenes. The comparisons between retrieved and true SCA show that on average there is no bias between calculated and true values. The absence of a

Table 11.5 Characteristics of VIIRS SCA regression on ground truth for landsat scenes

Date	Path	Row	Corr. Coeff.	Intercept	Slope	Mean True	Mean VIIRS	Location
33	123	32	0.84	−0.14	1.08	0.20	0.08	Beijing
33	139	29	0.89	−0.12	0.98	0.40	0.27	Altay
33	139	30	0.92	−0.14	1.10	0.61	0.53	Xinjiang 1
33	139	31	0.95	−0.03	1.00	0.20	0.17	Xinjiang 2
34	41	33	0.96	−0.01	1.06	0.20	0.22	Nevada
34	41	34	0.96	−0.01	1.11	0.07	0.07	Sierra
34	146	29	0.92	−0.05	1.09	0.68	0.68	Tian Shan
35	137	29	0.91	−0.16	1.09	0.52	0.60	Western Mongolia
35	137	30	0.89	−0.12	0.98	0.35	0.22	Gobi
35	153	39	0.74	−0.01	0.62	0.05	0.02	Pakistan
36	128	30	0.88	−0.07	1.09	0.92	0.93	Southern Mongolia
37	30	28	0.84	0.33	0.72	0.80	0.91	Dakotas
39	44	27	0.95	−0.00	1.07	0.19	0.20	Spokane
39	44	31	0.88	0.06	1.20	0.21	0.32	Oregon
40	156	35	0.95	−0.01	1.07	0.09	0.09	Northern Afghanistan
40	156	37	0.93	−0.06	1.17	0.27	0.25	Central Afghanistan

systematic deviation from true values in the retrievals of SCA indicates that improving the results of calculations is possible only when different modifications of the VIIRS algorithm are implemented for different scenes. One of potential ways to improve the VIIRS snow retrievals is to include into consideration the variability in local reflective properties of snow and nonsnow surfaces.

11.14. VARIABILITY IN LOCAL SNOW AND NONSNOW CONDITIONS

The reflectances and NDSI provided by Landsat observations illustrate fine scale variability in reflective properties of snow and nonsnow surfaces in quite different environmental conditions. The diversity of the underlying surface properties could be presented by two-dimensional histograms with x axis corresponding to visible reflectance (band 2) and y axis—to near infrared reflectance (band 5).

Figure 11.13 includes two separated groups of pixels: (a) corresponding to nonsnow (characterized by lower values of band 2 reflectance and higher band 5 reflectance) concentrated in the left part of the figure and (b) corresponding to snow (higher values of band 2 reflectance and lower band 5 reflectance) concentrated in the lower part of the figure. Figure 11.14, corresponding to the cultivated cropland, is different from Figure 11.13 because the pixels include a different percentage of both vegetation and snow. Those pixels spread from pure snow to completely covered by vegetation—from high reflectance in band 2 and low in band 5 to low reflectance in band 2 and slightly higher in band 5. On the whole, all analyzed two-dimensional histograms of the Landsat reflectances could be considered as the combinations of these three groups of pixels: nonsnow pixels, snow pixels, and pixels with different fractions of vegetation and snow. Figure 11.15 illustrates the case with all three types of pixels.

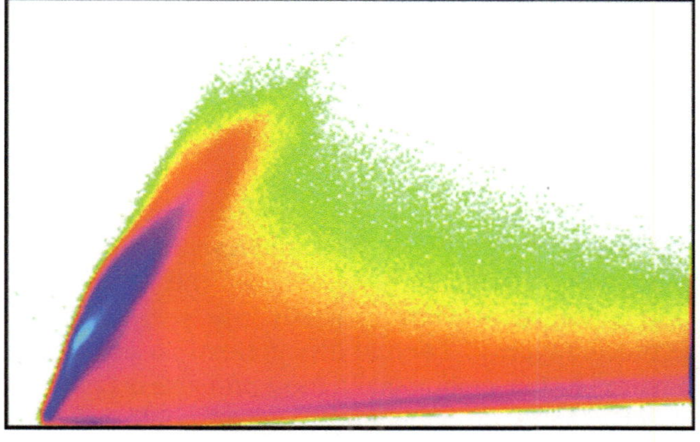

Figure 11.13 Reflectance histogram for Landsat bands 2 – x and 5 – y on February 09 (path—156, row—35).

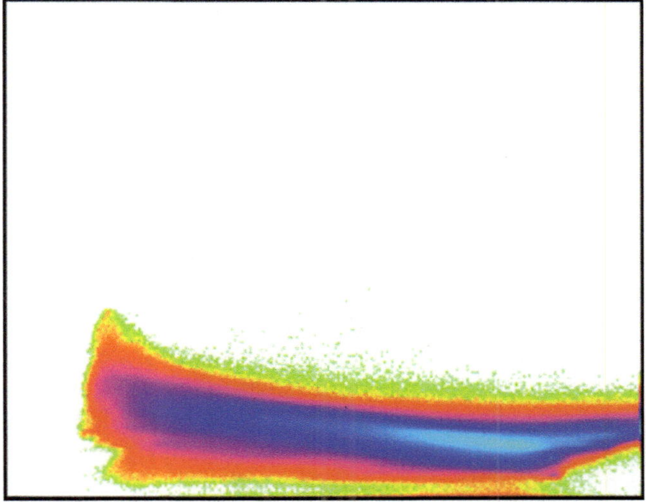

Figure 11.14 Reflectance histogram for Landsat bands 2 – x and 5 – y on February 6 (path—30, row—28).

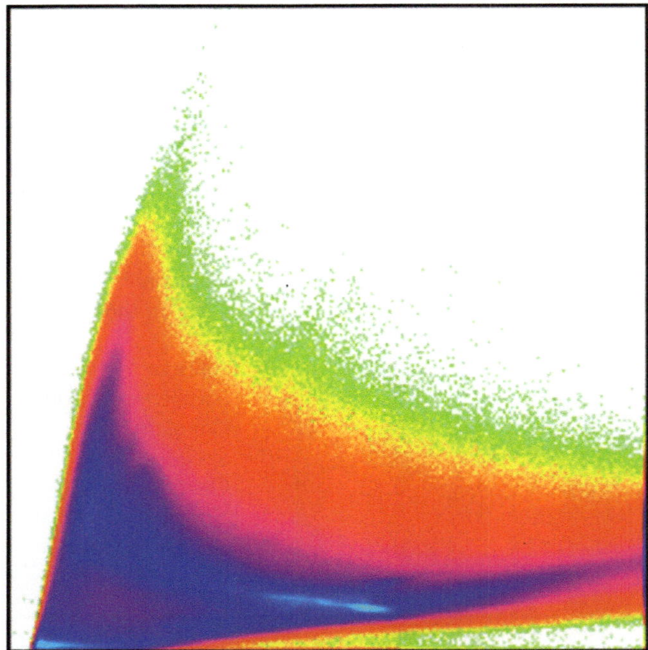

Figure 11.15 Reflectance histogram for Landsat bands 2 – *x* and 5 – *y* on February 3 (path—146, row—29).

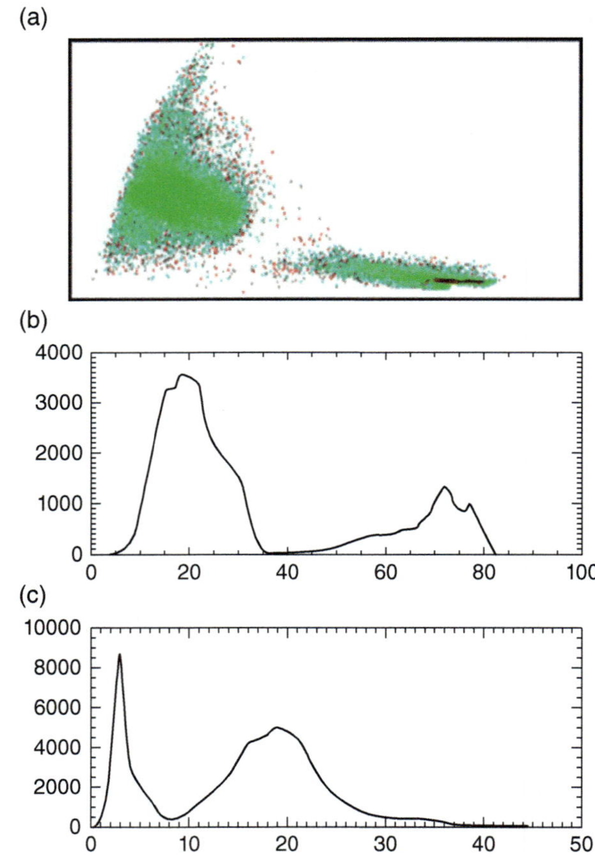

Figure 11.16 Two-dimensional histogram for VIIRS bands (a) M5 – *x* and M10 – *y* and (b) one-dimensional histograms for M5, and (c) M10 on February 2 (path—139, row—29).

Two-dimensional histograms presenting the joint probability densities for VIIRS moderate-resolution band M5 (0.64 μm) corresponding to *x* axis and band M10 (1.61 μm) corresponding to *y* axis look very similar to considered types of Landsat histograms. In a simple case (Figure 11.16a), predominant, most probable magnitudes of reflectances for snow and nonsnow could be easily determined from one-dimensional histograms for these bands (M5 and M10) presented in Figures 11.16b and 11.16c.

In a more complicated case, the histogram for VIIRS reflectances (Figure 11.17a) also remains similar to Landsat histogram (Figure 11.15) for the same scene. Such histograms are much more difficult to examine because all three groups of pixels—nonsnow pixels, snow pixels, and pixels with different concentrations of vegetation and snow—are presented. The reflectances corresponding to these groups of pixels significantly overlap each other and, for example, the one-dimensional histogram (Figure 11.17c) could not be reliably used to determine the characteristic reflectances for snow and nonsnow. However specialized, detailed analysis of VIIRS reflectances let us estimate the most probable magnitudes of snow and nonsnow reflectances, corresponding to the locations of local maximums of two-dimensional probability density functions (similar to those presented in Figures 11.16 and 11.17) for all 16 scenes under consideration (Table 11.6).

11.15. SCENE-SPECIFIC APPROACH TO SNOW RETRIEVAL

A very large variability of reflective properties characterizing different snow and underlying nonsnow states is illustrated by Figure 11.18. If such variability is taken into account, the quality of snow retrieval could be improved. One of the possible approaches to better estimate the VIIRS binary snow product for varying local conditions is to use the scene-specific NDSI thresholds. The varying NDSI thresholds are included in Table 11.7 along with the scene-specific snow and non-snow NDSI. The scene-specific NDSI thresholds correspond to the reflectances of pixels with 50% snow coverage—the reflectances calculated as average between snow and nonsnow reflectances from Table 11.6. The scene-specific snow and nonsnow NDSI are calculated using the same most probable reflectances for snow and nonsnow from the Table 11.5.

The average magnitude of the NDSI threshold for the binary classification is 0.50, which exceeds the standard value of 0.40 used in the SNOWMAP algorithm. The usual original intention of that and other snow algorithms was not to miss snow, which could explain the overestimate of snow extent provided by many existing approaches as well as by the Interactive Multisensor Snow (IMS) charts prepared at NOAA. Our evaluation of the NDSI threshold for the binary classification, showing that on average the value separating snow from nonsnow should be higher than the currently used magnitude, will cause slightly reduced SCA.

(a)

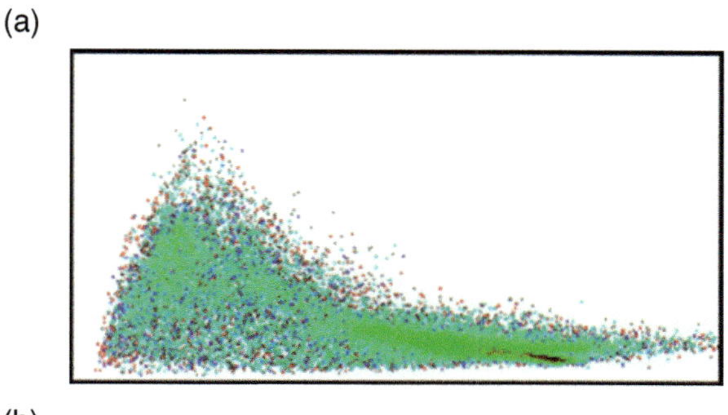

(b)

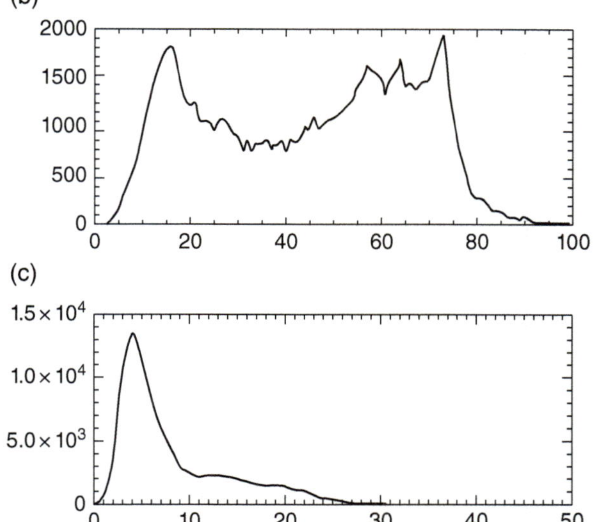

(c)

Figure 11.17 Two-dimensional histogram for VIIRS bands (a) M5 – x and M10 – y and one-dimensional histograms for (b) M5, and (c) M10 on February 3 (path—146, row—29).

Table 11.6 Most probable snow and nonsnow VIIRS reflectances for 16 scenes

VIIRS Observations		Snow Reflectance (%)		Nonsnow Reflectance (%)	
Date	Location	Band M5	Band M10	Band M5	Band M10
33	Beijing	39	7	10	11
33	Altay	72	3	15	19
33	Xinjiang 1	69	3	16	24
33	Xinjiang 2	71	5	19	32
34	Nevada	35	6	12	19
34	Sierra	35	6	14	20
34	Tian Shan	73	4	16	15
35	Western Mongolia	69	3	23	19
35	Gobi	73	3	22	20
35	Pakistan	53	5	23	50
36	Southern Mongolia	71	4	35	23
37	Dakotas	65	6	29	13
39	Spokane	55	2	8	15
39	Oregon	43	2	7	10
40	Northern Afghanistan	47	8	12	19
40	Central Afghanistan	60	7	13	18

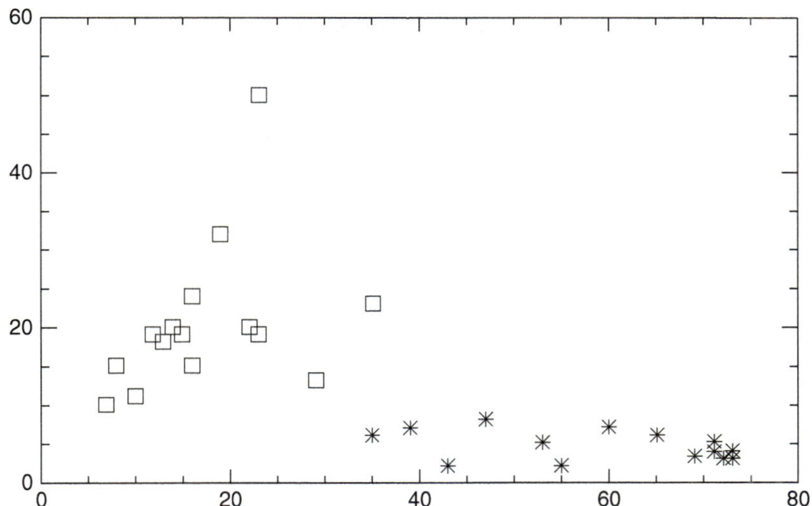

Figure 11.18 Most probable snow (asterisks) and nonsnow (squares) VIIRS reflectances (x axis — M5, y axis — M10) for 16 scenes.

Table 11.7 Magnitudes of most probable VIIRS NDSI for snow, nonsnow, and binary pixel classification threshold

VIIRS Observations		NDSI		
Date	Location	Snow	Nonsnow	50% Fraction
33	Beijing	0.70	−0.05	0.46
33	Altay	0.92	−0.12	0.59
33	Xinjiang 1	0.92	−0.20	0.52
33	Xinjiang 2	0.87	−0.25	0.42
34	Nevada	0.71	−0.23	0.31
34	Sierra	0.71	−0.18	0.31
34	Tian Shan	0.90	0.03	0.65
35	Western Mongolia	0.92	0.10	0.61
35	Gobi	0.92	0.05	0.61
35	Pakistan	0.83	−0.37	0.16
36	Southern Mongolia	0.89	0.21	0.59
37	Dakotas	0.83	0.38	0.66
39	Spokane	0.93	−0.30	0.58
39	Oregon	0.91	−0.18	0.61
40	Northern Afghanistan	0.71	−0.23	0.37
40	Central Afghanistan	0.79	−0.16	0.49
Average		0.84	−0.09	0.50

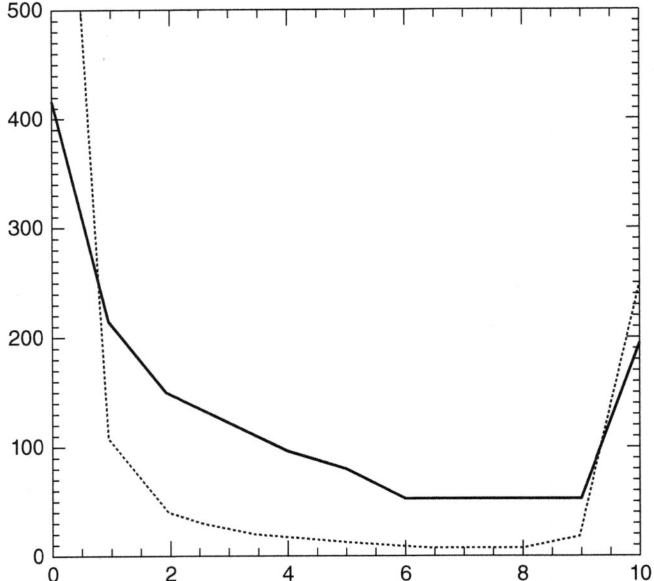

Figure 11.19 Snow cover area histograms for ground truth (solid) and for data retrieved by VIIRS algorithm (dotted) on February 4 (path—137, row—30).

It is also seen from Table 11.7 that the magnitude of the optimal NDSI threshold for binary snow/nonsnow classification varies considerably from 0.15 to 0.65, depending on local reflective properties of snow and nonsnow surfaces. Such adjustment of the parameters in snow algorithms to specific local conditions is a promising improvement leading to better quality of the VIIRS snow products. The advantages of using the scene-specific approach accounting for changes in snow and nonsnow surface reflectances within a catchment were successfully demonstrated [*Déry et al.*, 2005a] for hydrological modeling.

The motivation for tuning snow retrieval in that work came from knowing that the optimized versions of snow algorithm obtained and analyzed in *Salomonson and Appel* [2004] can and do vary from region to region as well as from day to day depending on snow and background surface types. The current research indicates how the improved SCA retrieval can be achieved by an algorithm employing scene-specific parameters characterizing local snow and nonsnow reflective properties.

11.16. SYSTEMATIC ERRORS IN BINARY SNOW PRODUCT

The VIIRS snow algorithm classifies each VIIRS pixel as snow or snow free, eliminating the real zone of transition from complete snow cover to snow-free areas from retrieval. Therefore, the algorithm even in the case of perfect distinction between snow fraction above and below 50% replaces smooth spatial changes in snow fraction by a sharp jump from 100% to 0% in neighboring pixels. Because of that algorithm feature, it is reasonable to anticipate that SCA presenting the aggregation of 0 and 1 has a tendency to overestimate the probability of extreme magnitudes of SCA close to 0% and 100%.

The analysis of SCA histograms shows that the probability of retrieving extreme magnitudes of SCA (0% and 100%) significantly and consistently exceeds the probability of the same values among ground truth data (Figure 11.19). In not one of 16 scenes under consideration, the probabilities of 0% or 100% SCA for ground truth were larger than the corresponding probabilities for the retrieved VIIRS data. The excessive probabilities of retrieving extreme SCA are associated with significantly lower probabilities of intermediate SCA (10%–90%) for all scenes and typically for a whole range of the intermediate SCA values. The comparative analysis of the histograms characterizing the probabilities of different SCA magnitudes clearly illustrates significant disadvantage of the gridded SCA that overestimates the frequency of extreme values (0% and 100%) and underestimates the appearance of intermediate values (from 10% to 90%). However, such errors are not specific for the algorithm under consideration but inherent to any binary snow retrieval.

The data in Table 11.5 also illustrate the overestimate of extreme values of SCA by the VIIRS snow retrieval. It is exceptionally rare when a linear relationship between calculated and true values is characterized by a negative intercept and by a slope exceeding 1. But such parameters of a linear regression are typical for the scenes under consideration. Analyzing the regressions, it is necessary to carefully interpret the results. Predominantly negative intercepts and slopes exceeding 1 obviously do not mean negative retrieved SCA for nonsnow parts of scenes and more than 100% for the areas completely covered by snow. A correct explanation of the calculated regressions means that the VIIRS SCA retrieval often provides the absence of snow when true snow cover area is more than zero and complete snow coverage for less than 100% ground truth coverage. These features of the VIIRS snow retrieval previously illustrated by the comparison of the histograms characterizing the probabilities of retrieved and true SCA are of prime importance. They reflect the presence of systematic errors in the VIIRS snow product: overestimate of high SCA and underestimate of small SCA.

Such systematic errors are even revealed in the regression ($y = 1.05x - 3.1$) of average for VIIRS SCA scenes on average for scenes ground truth (Figure 11.20). The parameters of the regression—unusual negative intercept and slope above 1.0—are similar to the parameters of the regression calculated for values of SCA in individual cells (Table 11.5). The overestimate of the probability of extreme SCA values close to 0% and 100% demonstrates the disadvantages of the VIIRS snow algorithm, though the coefficient of proportionality between retrieved and true SCAs averaged for scenes is equal to 0.995 when the intercept is set equal to 0.

Another quantitative way to analyze the retrieval of SCA—to estimate its stratified performance—could serve as an additional resource for improvements. The stratified performance used to assess the dependence of retrieved data quality on an affecting parameter is a very informative and effective approach not only for validation. Revealing consistent dependence of errors upon the magnitude of the influencing parameter could indicate the direction of potential algorithm modification. The influence of ground truth SCA as a stratifying parameter is considered below. It is not surprising that in the cases of significant differences between averaged VIIRS SCA and ground truth the systematic errors of the same sign are characteristic for all values of ground truth (Figure 11.21). It is reasonable to assume that such errors will be diminished when specific local conditions distinctive for individual scenes are taken into consideration by the retrieving algorithm.

Much more interesting for investigations are the cases when there is an insignificant difference between mean true and retrieved values of SCA and the error sign depends on the stratifying parameter—true SCA. All such cases are similar to a large degree. The VIIRS snow retrieval underestimates SCA for low true values (below 0.3–0.6 depending on a scene) and overestimates for higher true values (Figure 11.22).

Since such a type of error is inherent to any binary snow retrieval, we tested different approaches and reprocessed VIIRS observations using the fractional snow cover algorithm developed for MODIS by *Salomonson and Appel* [2004, 2006], who demonstrated in their works that NDSI is sensitive enough to provide the snow fraction within a pixel of moderate-resolution observations and that a fairly robust determination of the fraction of snow cover within pixels withstands to a certain degree a variability in snow and nonsnow properties. As confirmation of such conclusions, the quality of the statistical

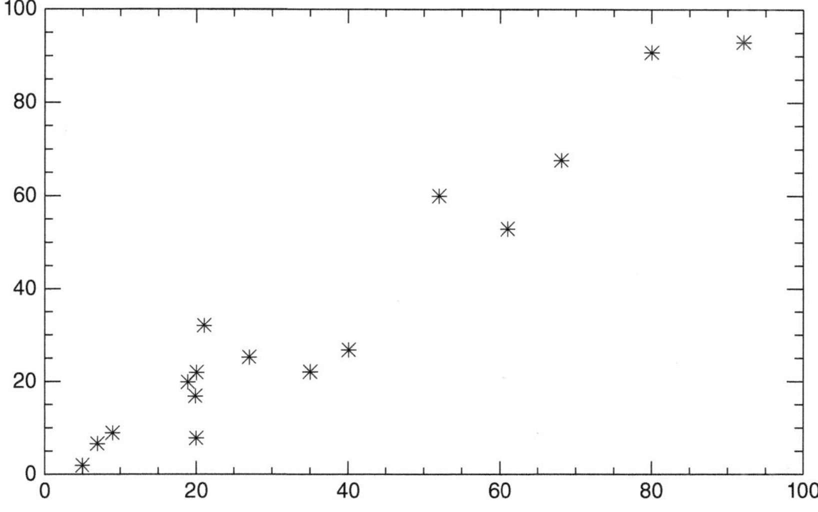

Figure 11.20 Relationship between averaged true, *x*, and retrieved, *y*, SCA for 16 scenes included in consideration.

(a)

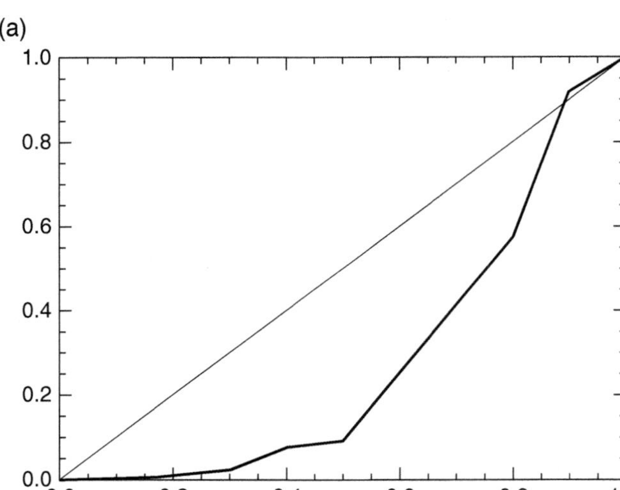

(b)

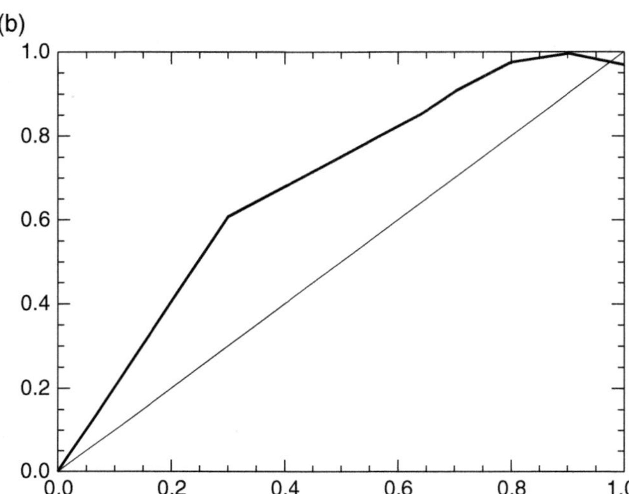

Figure 11.21 Dependence of retrieved SCA on its true value in (a) Mongolian Altay and (b) Oregon.

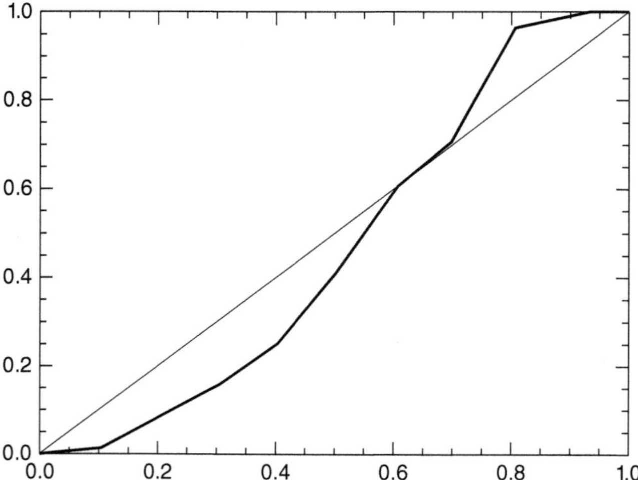

Figure 11.22 Dependence of retrieved SCA on its true value in southern Mongolia.

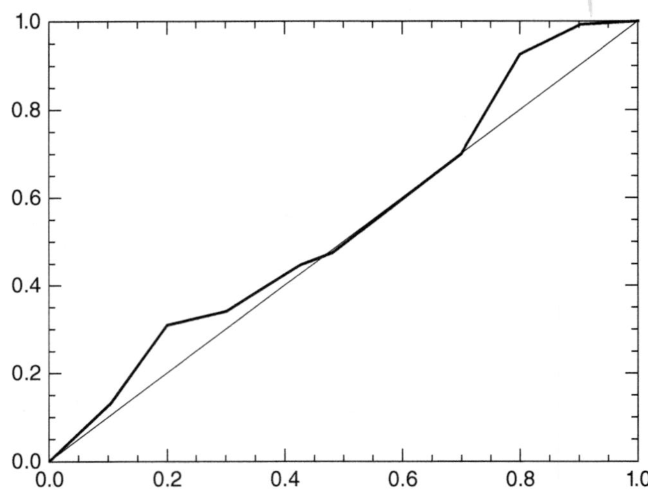

Figure 11.23 Dependence of SCA determined by MODIS fractional snow algorithm on true SCA value in southern Mongolia.

relationships between aggregated SCA and Landsat ground truth enhances in comparison with the VIIRS snow retrieval for all scenes with varying error sign, when SCA aggregated fractional snow cover calculated using the fractional snow cover algorithm [*Salomonson and Appel*, 2004, 2006].

The average increase of 2% in correlation coefficient could be considered noticeable for the correlation values above 90%. However, without tuning to local conditions the algorithm simultaneously with improvement of statistical relationships could cause a bias in data retrieval (Figure 11.23). It is important to note that the improvement of the SCA quality was provided by a standard "universal" version of the fractional snow algorithm. The adjustable version of the algorithm can and does vary from region to region and from time to time, depending on snow and background surface state as well as the conditions of observations, such as viewing and solar geometry and the state of the atmosphere.

After considering potential ways to improve the VIIRS information on SCA, it is recommended to combine allowing for the variability of snow and nonsnow reflective properties with snow fraction retrieval within a scene-specific fractional snow cover algorithm to create unbiased and consistent information on SCA distributions required for global studies, regional, and local scale hydrologic applications.

11.17. CONCLUSIONS

Because of a fine nadir field of view, and an original scanning and aggregation scheme, VIIRS observations give the opportunity of daily snow mapping at 800 m resolution globally and 375 m at nadir, which is adequate for most hydrologic applications. The information on snow cover is used to describe the interactions between the atmosphere and underlying surface influencing the accuracy of land surface models and assimilation schemes determining results of hydrological simulations. The important point for the subject under consideration in this chapter is the fact that VIIRS data on snow cover could be efficiently used to improve the quality of several variables in land surface models describing snow evolution. For example, it is a standard approach to increase the quality of SWE in models on the basis of using the relationship between SCA and SWE parameterized by SDC.

Because of the very significant impact of snow cover, it is necessary to improve not only the mathematical description of snow cover in models but the quality of snow retrieval from remote sensing as well. The VIIRS snow cover EDR consists of two products: binary snow and snow fraction. VIIRS information on binary snow is retrieved using the algorithm similar to the SNOWMAP algorithm designed for global retrieval of snow. An application of MESMA was developed for VIIRS to retrieve the snow fraction product; however, its implementation is currently postponed. Therefore, the validation of VIIRS snow retrieval in the chapter is mostly limited to binary snow product providing the valuable snow information, first of all, on snow extent and SCA, the variables determining the atmosphere-surface interaction influencing hydrological and weather forecasts as well as climate change.

The errors in snow extent related to Earth's underlying surface types and cloud misclassification are analyzed together with the VCM confidence. VCM assigns one of four cloud confidences to each pixel—(1) confidently cloudy, (2) probably cloudy, (3) probably clear, and (4) confidently clear—is responsible for the increase in the probability of snow omission errors from confidently clear pixels through probably clear to probably cloudy pixels. It is recommended to reduce the errors making snow cover retrieval only for the pixels classified by VCM as confidently clear. The VIIRS snow algorithm also retrieves false snow cover over the areas where snow never exists. The probability of these commission errors corresponding to ice clouds (with a spectral signature similar to snow on the ground) is 1%–2% over tropical areas and in contrast to omission errors does not depend of the VCM confidence.

The stratified analysis of the dependence of the VIIRS snow algorithm performance upon geometry of observations indicates a secondary role of the solar zenith angle. The influence of a scan angle on errors in ice extent is better revealed for confidently clear areas having weaker dependence on cloudiness and surface type. Typically, the results of analysis demonstrate that the probability of errors rises significantly toward ends of scans being on average larger at the western edge of swaths than at the eastern edges, which is related to the influence of BRDF describing the non-Lambertian reflectance by snow surface.

To estimate the impact of other factors on the quality of SCA, the influence of cloud misinterpretation was significantly suppressed in our further analysis by the choice of cloud-free Landsat scenes. The set of 16 thoroughly selected Landsat scenes, including a wide variety of the transitional zones from snow-covered regions to snow-free areas in quite different environmental conditions, is considered as representative data for reliable validation of SCA. The Landsat band 2 and band 5 closely correspond to the bands utilized by the VIIRS snow algorithms to calculate NDSI values used to detect snow cover and serving as an indicator of varying snow fraction. Our estimation of ground truth at a high-resolution scale was based on the hypothesis that NDSI calculated for Landsat pixels could be used to distinguish snow-covered Landsat pixels from snow-free pixels. Not only is such an approach physically based, but it is also similar to the principles currently used to retrieve the VIIRS snow information. Therefore conclusions studying high-resolution images could be directly translated to VIIRS data presenting the focus of current studies. The opportunity of utilizing Landsat data not only as a source of ground truth but also for investigation of surface reflective properties, applicable to VIIRS, presents obvious advantage of choosing Landsat as a great resource for our studies.

Good correspondence between maps of binary (snow/non–snow) Landsat pixels classification and coregistered false color Landsat images for exactly the same projection and resolution demonstrates the high quality of the classification based on a predetermined NDSI threshold. The review of NDSI images also confirms that the Landsat pixel classification is reliable and could be used to estimate the VIIRS snow products. The Landsat ground truth and the VIIRS SCA data were made completely spatially comparable after precise coregistration of the Landsat classified pixels with the VIIRS snow product and their aggregation for identical 0.05° cells of a regular latitude/longitude grid widely used for hydrologic and climate modeling. The comparisons between retrieved and true SCA demonstrate that on average there is no bias between calculated and true values.

The analysis of VIIRS SCA shows that the typical NDSI value separating snow from nonsnow is higher than the currently used threshold. It was further found that the magnitude of the optimal NDSI threshold to distinguish snow from nonsnow varies considerably—from 0.15 to 0.65 depending on local reflective properties of snow and nonsnow surfaces in the VIIRS scenes. The adjustment of the NDSI threshold to specific local conditions leads to better quality of snow products. The improved SCA retrieval can be achieved by an algorithm that takes into account the variability of snow and non-snow reflectances in space and time. The enhancement could be obtained when the "universal" parameters in the snow retrieval algorithm are made adjustable to specific local conditions taking better account of the particular non-snow-covered terrain and the snow-covered areas. The development of a scene-specific approach taking local reflective snow and nonsnow properties into account could be considered as a promising way to improve snow retrieval from VIIRS observations.

The statistical comparison of the gridded VIIRS SCA product with Landsat ground truth information is characterized by the linear relationships between true and calculated values with negative intercepts and the slopes exceeding 1. Such parameters of a linear regression, typical for the scenes under consideration, demonstrate the need to modify the current VIIRS algorithm significantly and consistently overestimating the frequency of extreme snow cover area values (0% and 100%) and underestimating the probability of intermediate values (from 10% to 90%). Experimental tests of the fractional snow cover algorithm developed for MODIS by *Salomonson and Appel* [2006] provided noticeable improvement of the VIIRS SCA estimates.

A number of enhancements to the algorithm are foreseen to improve the accuracy of snow retrievals. The optimal approach to improve moderate-resolution remote sensing information on snow cover will combine, allowing for the variability of snow and nonsnow properties with snow fraction retrieval within a scene-specific snow algorithm to create unbiased and consistent information on snow cover distribution required for global studies, regional, and local-scale hydrological applications.

REFERENCES

Anderson, E. A. (1973), National weather service river forecast system—Snow accumulation and ablation model, NOAA Tech. Memo. NWS HYDRO-17.

Andreadis, K. M., and D. P. Lettenmaier (2006), Assimilating remotely sensed snow observations into a macroscale hydrology model, *Adv. Water Resourc.*, 29(6), 872–886.

Appel, I. (2011), Improvement of the NPP VIIRS fractional snow cover EDR for creation of a climate data records using MODIS and VIIRS, 31st EARSeL Symposium, Prague, Czech Republic.

Appel, I., and K. A. Jensen (2002), Snow cover, visible/infrared imager/radiometer algorithm theoretical basis document (Version 5), SBRS Document Y2401, Raytheon Systems Company (ITSS).

Bamzai, A. S., and J. Shukla (1999), Relation between Eurasian snow cover, snow depth, and the Indian summer monsoon: An observational study, *J. Climate*, 12, 3117–3132.

Bavera, D., and C. De Michele (2009), Snow water equivalent estimation in the Mallero basin using snow gauge data and MODIS images and fieldwork validation, *Hydrol. Process.*, 23, 1961–1972. 1331–1343.

Cao, C., F. J. De Luccia, X. Xiong, R. Wolfe, and F. Weng (2014), Early on-orbit performance of the visible infrared imaging radiometer suite onboard the Suomi National Polar-Orbiting Partnership (S-NPP) satellite, *IEEE Trans. Geosci. Remote Sens.*, 52, No. 2, 1142–1156.

Clark, M. P., A. G. Slater, A. P. Barrett, L. E. Hay, G. J. McCabe, B. Rajagopalan, and G. H. Leavesley (2006), Assimilation of snow covered area information into hydrologic and land-surface models, *Adv. Water Resourc.*, 29(8), 1209–1221.

Cohen, J., and D. Entekhabi (2001), The influence of snow cover on Northern Hemisphere climate variability, *Atmos. Ocean*, 39, 35–53.

Delworth, T. L., and S. Manabe (1998), The influence of potential evaporation on variabilities of simulated soil wetness and climate, *J. Climate*, 1, 523–547.

Déry, S. J., V.V. Salomonson, M. Stieglitz, D. K. Hall, and I. Appel (2005a), An approach to using snow areal depletion curves inferred from MODIS and its application to land surface modelling in Alaska, *Hydrol. Proc*, 19, 2755–2774, doi:10.1002/hyp.5784.

Déry, S. J., J. Sheffield, and E. F. Wood (2005b), Connectivity between Eurasian snow cover extent and Canadian snow water equivalent and river discharge, *J. Geophys. Res.*, 110, D23106, doi:10.1029/2005JD006173.

Dong, J., and C. Peters-Lidard (2010), On the relationship between temperature and MODIS snow cover retrieval errors in the western US, *IEEE J. Selected Topics Earth Obser. Remote Sens.*, 3(1), 132–140, doi:10.1109/JSTARS.2009.2039698.

Dong, J., J. P. Walker, and P. R. Houser (2005), Factors affecting remotely sensed snow water equivalent uncertainty, *Remote Sens. Environ.*, 97(1), 68–82.

Ek, M. B., K. E. Mitchell, Y. Lin, E. Rogers, P. Grunmann, V. Koren, G. Gayno, and J. D. Tarpley (2003), Implementation of Noah land surface model advances in the National Centers for Environmental Prediction operational mesoscale Eta model, *J. Geophys. Res.*, 108, 8851, doi:10.1029/2002JD003296.

Elguindi, N., B. Hanson, and D. Leathers (2005), The effects of snow cover on midlatitude cyclones in the Great Plains, *J. Hydrometeorol.*, 6, 263–279.

Ellis, A. W., and D. J. Leathers (1999), Analysis of cold airmass temperature modification across the U.S. Great Plains as a consequence of snow depth and albedo, *J. Appl. Meteorol.*, 38, 696–711.

Foster J. L., C. Sun, J. P. Walker, R. E. J. Kelly, A. T. C. Chang, J. Dong, and H. Powell (2005), Quantifying the uncertainty in passive microwave snow water equivalent observations, *Remote Sens. Environ.*, 94(2), 187–203.

Girotto et al. (2014), Probabilistic SWE reanalysis as a generalization of deterministic SWE reconstruction techniques, *Hydrol. Process*, 38, 3875–3895.

Gong, G., D. Entekhabi, and J. Cohen (2003), Modeled Northern Hemisphere winter climate response to realistic Siberian snow anomalies, *J. Climate*, 16, 3917–3931.

Hall D. K. and G. A. Riggs (2007), Accuracy assessment of the MODIS snow products, *Hydrol. Process.*, *21*(12), 1534–1547.

Hall, D. K., J. L. Foster, D. L. Verbyla, A. G. Klein, and C. S. Benson (1998), Assessment of snow-cover mapping accuracy in a variety of vegetation-cover densities in central Alaska, *Remote Sens. Environ.*, *66*(2), 129–137.

Hillger, D., et al. (2013), First-Light Imagery from Suomi NPP VIIRS. *Bull. Am. Meteorol. Soc.*, *94*, 1019–1029, available at 10.1175/BAMS-D-12-00097.1.

Klein, A. G., and A. C. Barnett (2003), Validation of daily MODIS snow cover maps of the upper Rio Grande river basin for the 2000–2001 snow year, *Remote Sens. Environ.*, *86*, 162–176.

Klein, A. G., D. K. Hall, and G. A. Riggs (1998), Improving snow-cover mapping in forests through the use of a canopy reflectance model. *Hydrol. Process.*, *12*, 1723–1744.

Kolberg, S. A., and L. Gottschalk (2006), Updating of snow depletion curve with remote sensing data, *Hydrol. Process.*, *20*, 2363–2380.

Lee, T. F., S. D. Miller, F. J. Turk, C. Schueler, R. Julian, S. Deyo, P. Dills, and S. Wang (2006), The NPOESS VIIRS day/night visible sensor, *Bull. Am. Meteorol. Soc.*, *87*, 191–199.

Liston, G. E., R. A. Pielke, Sr., and E. M. Greene (1999), Improving first-order snow-related deficiencies in a regional climate model, *J. Geophys. Res.*, *104*, 19,559–19,567.

Parajka, J., and G. Bloschl (2006), Validation of MODIS snow cover images over Austria, *Hydrol. Earth Syst. Sci.*, *10*, 679–689.

Pulliainen, J., J. Lemmetyinen, A. Kontu, A. Arslan, A. Wiesmann, T. Nagler, H. Rott, M. Davidson, D. Schuettemeyer, and M. Kern (2010), Observing seasonal snow changes in the boreal forest area using active and passive microwave measurements, *IGARSS 2010*, 2375–2378.

Reichle, R. H., D. B. McLaughlin, and D. Entekhabi (2002), Hydrologic data assimilation with the Ensemble Kalman Filter, *Monthly Weather Rev.*, *130*(1), 103–114.

Rodell, M., and P. R. Houser (2004), Updating a land surface model with MODIS-derived snow cover, *J. Hydrometeorol.*, *5*, 1064–1075.

Rodell, M., P. R. Houser, and U. Jambor (2004), The global land data assimilation system, *Bull. Am. Meteorol. Soc.*, *85*(3), 381–394.

Roy, A., A. Royer, and R. Turcotte (2010), Improvement of springtime streamflow simulations in a boreal environment by incorporating snow-covered area derived from remote sensing data, *J. Hydrol.*, *390*, 35–44.

Salamon. P., and L. Feyen (2009), Assessing parameter, precipitation, and predictive uncertainty in a distributed hydrological model using sequential data assimilation with the particle filter, *J. Hydrol.*, *376*, 428–442.

Salomonson, V. V., and I. Appel (2004), Estimating fractional snow cover from MODIS using the normalized difference snow index, *Remote Sens. Environ.*, *89*, 351–360.

Salomonson, V. V., and I. Appel (2006), Development of the Aqua MODIS NDSI fractional snow cover algorithm and validation results, *IEEE Trans. Geosci. Remote Sens.*, *44*(7), 1747–1756.

Shukla, J., and Y. Mintz (1982), The influence of land-surface evapotranspiration on the earth's climate, *Science*, *214*, 1498–1501.

Simic, A., R. Fernandes, R. Brown, P. Romanov, and W. Park (1998), Validation of vegetation, MODIS, and GOES+SSM/I snow cover products overCanada based on surface snow depth observations, *Hydrol. Process.*, *18*, 1089–1104.

Su, H., Z. L. Yang, G. Y. Niu, and R. E. Dickinson (2008), Enhancing the estimation of continental-scale snow water equivalent by assimilating MODIS snow cover with the ensemble Kalman filter, *J. Geophys. Res.*, *113*, doi:10.1029/2007JD009232.

Tait, A., and R. Armstrong (1996), Evaluation of SMMR satellite-derived snow depth using ground-based measurements, *Int. J. Remote Sens.*, *17*, 657–665.

Thirel, G., E. Martin, J. F Mahfouf, S. Massart, S. Ricci, and F. Habets, (2010), A past discharges assimilation system for ensemble streamflow forecasts over France—Part 1: Description and validation of the assimilation system, *Hydrol. Earth Syst. Sci.*, *14*, 1623–1637, doi:10.5194/hess-14-1623-2010.

Thirel, G., P. Salamon, P. Burek, and M. Kalas (2011), Assimilation of MODIS snow cover area data in a distributed hydrological model, *Hydrol. Earth Syst. Sci. Discuss.*, *8*, 1329–1364, doi:10.5194/hessd-8-1329-2011.

Yang, F., A. Kumar, W. Wang, H. H. Juang, and M. Kanamitsu (2001), Snow-albedo feedback and seasonal climate variability over North America, *J. Climate*, *14*, 4245–4248.

Zaitchik, B., and M. Rodell (2008), Forward-looking assimilation of MODIS-derived snow-covered area into a land surface model, *J. Hydrometeorol.*, *10*, 130–148.

Zaitchik, B. F., J. P. Evans, and R. B. Smith (2007), Regional impact of an elevated heat source: The Zagros plateau of Iran, *J. Hydrometeorol.*, *20*, 4133–4146.

12

Seeing the Snow Through the Trees: Toward a Validated Canopy Adjustment for Satellite Snow-Covered Area

Lexi P. Coons,[1] Anne W. Nolin,[2] Kelly E. Gleason,[2] Eugene J. Mar,[2] Karl Rittger,[3] Travis R. Roth,[2] and Thomas H. Painter[3]

12.1. INTRODUCTION

12.1.1. Motivation and Research Questions

Snowpacks serve as vital reservoirs of water, supplying human and ecosystem needs. Seasonal decreases in snow-covered area (SCA) signal the initiation and duration of snowmelt runoff, so assimilation of this metric into hydrologic models provides an effective means of evaluating snow as a water resource [*Martinec and Rango*, 1995; *Clark et al.*, 2006; *McGuire et al.*, 2006; *Thirel et al.*, 2011]. Additionally, SCA is an important variable influencing climate models because the interplay between albedo and solar radiation can create either a positive or negative feedback mechanism depending on the presence of snow [*Randerson et al.*, 2006; *O'Halloran et al.*, 2012; *Jin et al.*, 2012]. Because of the influence that snow cover has on local to global hydrology and climate patterns, there is a demand for accurate and timely estimates of SCA [*Dozier*, 2011]. Satellite remote sensing has the potential to meet this demand.

Current active and passive satellite-based sensors can effectively detect snow in nonforested regions and give estimates of SCA at various spatial and temporal resolutions [*Nolin*, 2010]. Recent advances in snow remote sensing have led to improved estimates of SCA by providing subpixel, or fractional snow-covered area (*f*SCA) [*Salomonson and Appel*, 2004; *Painter et al.*, 2009]. The Moderate Resolution Imaging Spectroradiometer (MODIS) currently has an *f*SCA product; and an operational *f*SCA product, while still in the evaluation phase, will be available from a new satellite sensor, the Visible Infrared Imaging Radiometer Suite (VIIRS). This sensor, onboard NASA's recently launched Earth-observing satellite, will provide global estimates of *f*SCA as part of a mission to quantify and monitor Earth's cryosphere and its response to climate change [*Baker*, 2011]. These *f*SCA estimates apply only to the viewable fraction of snow cover in a pixel. A common canopy adjustment for *f*SCA in forests assumes that obscured, subcanopy *f*SCA resembles the viewable *f*SCA in canopy openings [*Durand et al.*, 2008; *Molotch and Margulis*, 2008; *Raleigh et al.*, 2013; *Rittger et al.*, 2013]. However, forest canopy influences snow accumulation and ablation, causing spatial variability in the distribution of snow cover [*Golding and Swanson*, 1986; *Davis et al.*, 1997; *Faria et al.*, 2000; *Storck*, 2002; *Musselman et al.*, 2008; *Varhola et al.*, 2010; *Ellis et al.*, 2011].

Despite numerous research efforts focused on the relationships between snow properties and forest characteristics [*Storck*, 2002; *Musselman et al.*, 2008; *Varhola et al.*, 2010; *Ellis et al.*, 2011], few studies have addressed how these relationships may be linked to satellite estimates of SCA [*Raleigh et al.*, 2013; *Rittger et al.*, 2013]. *Raleigh et al.* [2013] and *Rittger et al.* [2013] validated a simple forest canopy adjustment that uses fractional vegetation cover, but to date there has been no systematic assessment of snowpack spatial relationships with forest canopy structure at the scale of remote sensing pixels. Identifying forest canopy metrics that can be used to parameterize

---

[1] *USDA Forest Service, Helena National Forest, Helena, Montana, USA*

[2] *College of Earth, Ocean, and Atmospheric Sciences, Oregon State University, Corvallis, Oregaon, USA*

[3] *Jet Propulsion Laboratory, California Institute of Technology, Pasadena, California, USA*

Remote Sensing of the Terrestrial Water Cycle, Geophysical Monograph 206. First Edition. Edited by Venkat Lakshmi.

subcanopy snow at scales appropriate for remote sensing will serve as valuable input to a canopy adjustment for satellite-derived SCA.

To understand the limitations of current canopy adjustments and advance toward an improved method to account for subcanopy snow, this study aims to answer the following research questions: (1) How well do original satellite-derived fSCA and canopy-adjusted fSCA approximate subcanopy fSCA? (2) What spatiotemporal snow cover patterns explain differences between in situ snow cover and satellite-derived fSCA? (3) What canopy structure metrics can account for differences between snow cover and satellite-derived fSCA?

12.1.2. Background and Previous Work

12.1.2.1. Snow-Vegetation Interactions

Canopy snow interception and the subcanopy energy balance influence snow depth, snow water equivalent (SWE), and snowmelt rates, thereby affecting spatial and temporal patterns of snow accumulation and ablation [*Metcalfe and Buttle*, 1998; *Link and Marks*, 1999; *Storck*, 2002; *Musselman et al.*, 2008; *Varhola et al.*, 2010; *Ellis et al.*, 2011]. Maximum snow depth, SWE, and snow ablation rates are often greater in forest canopy openings because canopy characteristics, such as canopy density, tree leaf area index (LAI), and gap size, dictate snow interception [*Metcalfe and Buttle*, 1998; *Pomeroy et al.*, 2002; *Storck*, 2002; *López-Moreno and Latron*, 2008; *Varhola et al.*, 2010]. These canopy characteristics also influence the energy balance by attenuating incoming solar radiation and contributing to enhanced longwave radiation [*Pomeroy et al.*, 2009; *Ellis et al.*, 2011].

Other canopy characteristics have been linked to quantifiable differences in snow properties. For instance, tree height has been connected to significant relationships in peak SWE and maximum ablation rate, with large tree heights corresponding to lower SWE and decreased ablation rates compared to short trees [*Varhola et al.*, 2010]. Tree crown radius also affects snow properties; a notably larger decrease in snow depth has been found around individual trees with a crown radius greater than 3 m compared to trees with crown radii smaller than 2 m [*Musselman et al.*, 2008]. These forest canopy characteristics influence snow properties differently depending on climate [*Link and Marks*, 1999; *Stork*, 2002; *Sicart et al.*, 2004; *Ellis et al.*, 2011] and are important metrics to consider when predicting subcanopy snow cover, key for improving fSCA canopy adjustments.

12.1.2.2. Remote Sensing of Snow in Forests

Dense forest canopy presents a limitation to all optical remote sensing techniques for detecting snow cover. The resulting systematic underestimation of fSCA poses a challenge for hydrologic and climate models that rely on accurate estimates of SCA because 40% of the North American seasonal snow zone is forested [*Klein et al.*, 1998]. Consequently, research efforts have focused on improving estimates of subcanopy SCA in forests.

An approach to adjust for forest canopy with the current fSCA products has been established by *Molotch and Margulis* [2008]. This adjustment (fSCA$_{\text{ADJ}}$) uses the vegetation fraction (fVEG) to scale observed fSCA (fSCA$_{\text{OBS}}$) [*Molotch and Margulis*, 2008; *Raleigh et al.*, 2013; *Rittger et al.*, 2013]:

$$f\text{SCA}_{\text{ADJ}} = \frac{f\text{SCA}_{\text{OBS}}}{1 - f\text{VEG}}. \tag{12.1}$$

This approach assumes that the spatial variability of fSCA is similar in subcanopy and open locations. *Raleigh et al.* [2013] quantified the error associated with a canopy adjustment applied to fSCA from the MODIS snow-covered area and grain size (MODSCAG) algorithm [*Painter et al.*, 2009]. The study used a temporally static vegetation fraction derived from Landsat reflectance data, and results indicated that snow cover continued to be misrepresented even after application of the canopy adjustment.

12.1.3. Description of the Study Area

The central Oregon Cascades is a region where an accurate canopy adjustment is crucial since 85% of the snow zone has at least 20% forest canopy cover [*Coons*, unpublished]. This region receives 80% of its annual precipitation during winter months and at high elevations the majority of this precipitation occurs as snow [*Tague and Grant*, 2004]. Eleven study sites were selected on the west side of the Oregon Cascades representing elevations from 1100 to 1550 m in open and forested conditions. Nine of these sites are situated in the McKenzie River Basin and two are in the Middle Fork Willamette Basin (Figure 12.1). At five of the elevations, study sites were paired to include an open or low-density (LD) forest adjacent to a high-density (HD) forest. Henceforth, each site is referred to by an identifier based on its elevation and forest density category (Table 12.1).

The HD forested sites are characterized by a mature continuous canopy. The dominant tree species across the study area are Douglas fir (*Pseudotsuga menziesii*), western hemlock (*Tsuga heterophylla*), grand fir (*Abies grandis*), lodgepole pine (*Pinus contorta*), and mountain hemlock (*Tsuga mertensiana*). There are some communities of noble fir (*Abies procera*) and red alder (*Alnus rubra*) at the lower elevation sites and sporadic subalpine fir (*Abies lasiocarpa*) at the high elevation sites. The LD forested sites have sparse, patchy, or no forest canopy

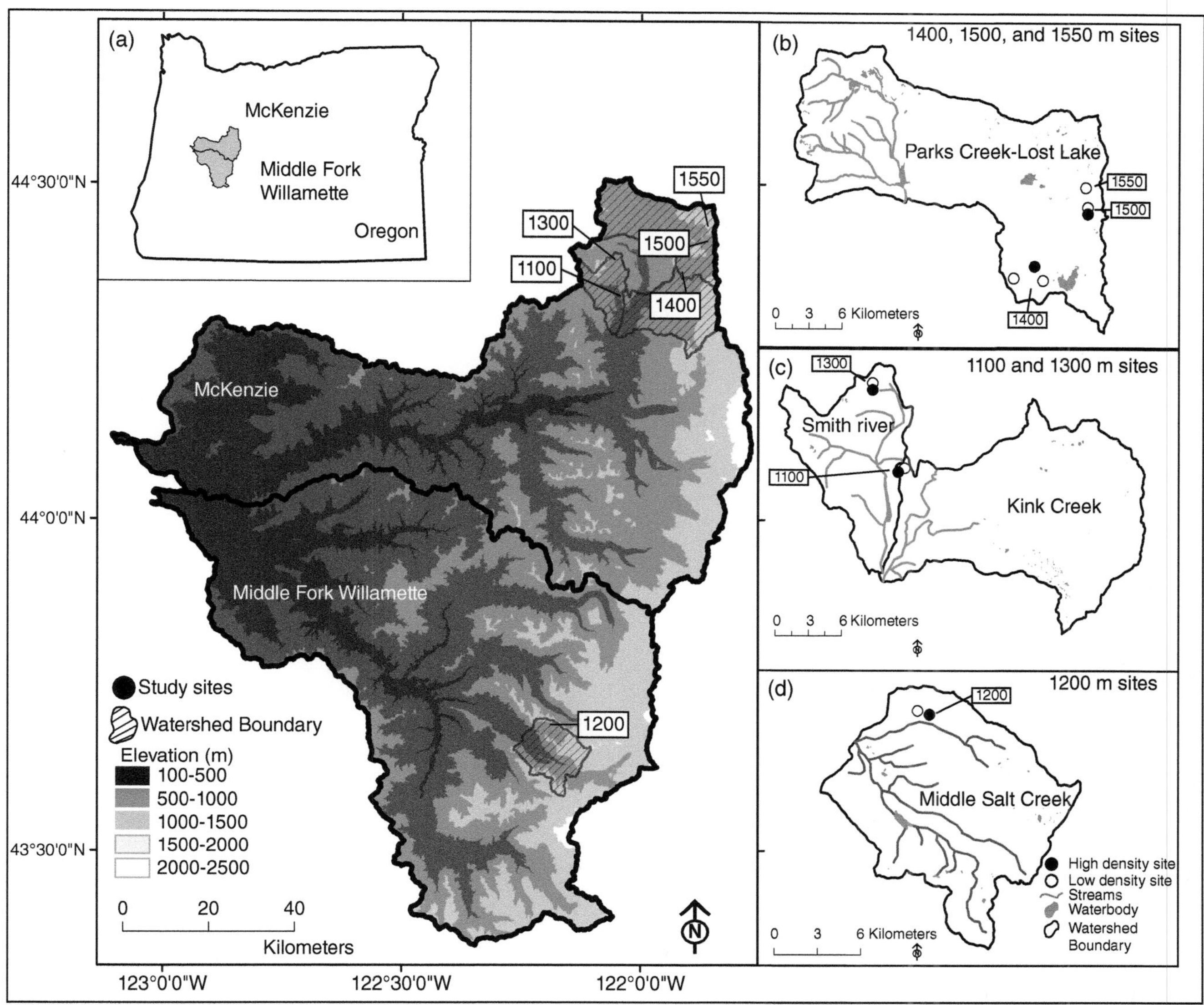

Figure 12.1 Locations of the 11 study sites in the McKenzie and Middle Fork Willamette watersheds within (a) Oregon. The locations of paired high-density (HD) and low-density (LD) sites within the indicated subbasins for (b) 1400, 1500, and 1550 m sites, (c) 1100 and 1300 m sites, and (c) 1200 m sites.

cover. They occur in open meadows or areas that have experienced canopy-removing disturbance such as timber harvest or wildfire. Figure 12.2 illustrates the relative characteristics of the HD and LD sites chosen for this study. All sites were selected in areas of low relief to exclude the effects of slope and aspect on snow accumulation and ablation.

12.2. METHODS

12.2.1. General Approach

To address the research questions presented in this study, we examined patterns in satellite fSCA, in situ snow cover, and forest canopy structure. First, original satellite-derived fSCA and canopy-adjusted fSCA (fSCA$_{ADJ}$) were both compared against in situ observations of snow cover to assess the currently employed canopy adjustment. The spatial and temporal accuracy of fSCA$_{ADJ}$ was then examined across multiple study sites, and errors were evaluated by comparing spatial and temporal patterns in forested and open sites. Errors in fSCA$_{ADJ}$ were further analyzed in comparisons with measured forest canopy characteristics.

12.2.2. Snow Observations

The distribution of snow cover across the 11 study sites was determined from in situ observations of snow cover from February to June 2012. At the 1100, 1200,

Table 12.1 Characteristics of the 11 study sites

Site ID	Elevation (m)	Slope (deg)	Aspect	Description	Tree Height (m)	Crown Radius (m)	Tree Density (trees/ha)
1100_LD	1110	5.8	SE	Open meadow	8.9	1.9	100
1100_HD	1130	9.7	E	High-density tree canopy	33.7	4.7	330
1200_LD	1200	12.6	SW	Burned forest	14.7	2.3	660
1200_HD	1200	10.5	SW	High-density tree canopy	26.3	3.6	1070
1300_LD	1320	9.4	NE	Decade-old clear cut	11.8	2.0	190
1300_HD	1330	19.4	S	High-density tree canopy	21.2	3.4	420
1400_LD	1400	2.2	SE	Low-density tree canopy	9.2	1.2	550
1400_HD	1400	4.2	NW	High-density tree canopy	9.9	1.2	1800
1500_LD	1470	1.9	SW	Burned forest, no canopy	10.4	1.1	160
1500_HD	1470	1.9	NW	High-density tree canopy	14.5	1.3	1120
1550_LD	1530	9.5	S	Burned forest, no canopy	12.8	1.5	270

Figure 12.2 Representative photographs for the (a) 1400_LD, (b) 1400_HD, (c) 1500_LD, and (d) 1500_HD study sites.

1300, 1500, and 1550 m sites, the presence or absence of snow was determined at 5 m intervals taken along a snow course passing through the site. These transects resulted in at least 50 sample locations and up to 100 sample locations for each site. Snow surveys were undertaken every 15–30 days, each survey slightly offset to avoid previously sampled locations. More intensive sampling was performed at the two 1400 m sites. At these sites snow presence/absence was recorded at 5 m intervals along 5 parallel transects through each site. The survey transects were 100 m long and were horizontally offset by 10 m, yielding a total of 100 observations per site. To investigate the temporal and spatial variability of snow cover at single locations, snow presence/absence was monitored at the same sample locations for a total of 8 surveys occurring from 15 April to 20 June 2012.

To subsequently compare with satellite fSCA, the ground observations at all sites were spatially aggregated to the 30 m resolution of a Landsat 7 Enhanced Thematic Mapper (ETM+) pixel. The sampling density yielded between 12 and 17 pixels for each site. The ground-based fractional snow-covered area (fSCA$'_{\text{GROUND}}$) was then estimated by dividing the number of locations with observed snow cover (M_{Snow}) by the total number of observation locations (N):

$$f\text{SCA}_{\text{GROUND}} = \frac{M_{\text{SNOW}}}{N}. \tag{12.2}$$

12.2.3. Forest Inventory

In the snow-free season, characteristics of the forest canopy at each site were quantified in a conventional forest inventory. At the 1400 m sites, the forest inventory occurred at the same 5 m intervals to coincide with the snow observations, while at the other sites, sample locations were selected at 50 m intervals along the snow survey transects. At these sample locations tree height, tree diameter, crown radius, and tree density were recorded. These tree properties were measured for the four trees nearest to each sample location and averaged to yield forest stand characteristics.

The forest canopy at each site was further characterized using skyward looking hemispherical photographs acquired using a Nikon Coolpix 990 digital camera equipped with an FC-E8 fisheye converter, which has a 180° field of view [*Inoue et al.*, 2004]. The hemispherical photographs were assessed with a Gap Light Analyzer 2.0 (GLA 2.0) to measure canopy closure (CC), which is the complement of the sky view fraction [*Frazer et al.*, 1999].

The resulting CC values were used to differentiate between open and subcanopy locations at 1400_LD and 1400_HD and, based on visual inspection of the hemispherical photographs and field observations, sample locations were classified into low and high CC classes. The low CC class corresponds to CC < 0.65 and were assumed to represent open locations where more snow cover would be detectable by satellite $f$sca algorithms. The high CC class includes locations with CC ≥ 0.65, approximating areas where little snow cover would be detectable. To test the relationship between snow cover in open and subcanopy locations, fSCA$_{\text{GROUND}}$ for high and low canopy classes was compared at 1400_LD and 1400_HD.

12.2.4. Landsat TM-Derived Fractional Snow-Covered Area

Cloud-free Landsat 7 ETM + images coincident with in situ snow observations were downloaded from the U.S. Geological Survey (USGS, *glovis.usgs.gov*; accessed 28 October, 2012). SWE data from snow telemetry (SNOTEL) stations near the study sites were then referenced for the dates of the acquired images to assess the possibility of snow on the forest canopy. Images acquired shortly after significant increases in SWE were avoided because snow stored in the forest canopy can lead to overestimates of ground fSCA. The fSCA was computed using the Thematic Mapper snow-covered area and grain size (TMSCAG) model, an adaptation from MODSCAG that accounts for saturation in Landsat bands 1–3 [*Painter et al.*, 2009; *Rittger et al.*, 2013]. For pixels with TMSCAG fSCA (fSCA$_{\text{TM}}$) above the detectable limit of the algorithm (0.15), equation 12.1 was applied using the TMSCAG-derived vegetation fraction (fVEG) [*Durand et al.*, 2008; *Molotch and Margulis*, 2008; *Rittger et al.*, 2013]; therefore, we assume that snow cover under forest canopy is the same as in visible areas. The fVEG in this application has the advantage that it is derived from the same sensor as fSCA while *Durand et al.* [2008], *Molotch and Margulis* [2008], and *Raleigh et al.* [2013] all use a fVEG derived from a different satellite sensor than the respective fSCA product. The fSCA$_{\text{TM}}$ and fSCA$_{\text{ADJ}}$ for all pixels overlapping the in situ snow surveys, 9–20 pixels per site, were averaged to give the overall site fSCA. These site averages were compared to the site-averaged fSCA$_{\text{GROUND}}$ as close as possible to the satellite overpass date. Because the average was taken for all pixels covering a snow survey transect, it is possible that the total fSCA for a given site may be below the detection limit of TMSCAG. For instance, if fSCA$_{\text{TM}}$ for the majority of the pixels along the transect were zero and the remaining pixels had a fSCA$_{\text{TM}}$ close to a value of 0.15, the average reported fSCA$_{\text{TM}}$ for that site would be below the detection limit of TMSCAG.

Equation 12.1 was only applied to pixels above the detection limit, and the pixels less than 0.15 were considered to be 0 for $fSCA_{TM}$ and $fSCA_{ADJ}$.

Errors in $fSCA_{ADJ}$ were determined as the difference from in situ measurements:

$$Error = fSCA_{ADJ} - fSCA_{GROUND}. \quad (12.3)$$

Negative errors, or errors of omission, represent underestimates of snow cover, and positive errors, errors of commission, correspond to overestimates of snow cover. These errors were related to site averages of CC from the hemispherical photographs and field-measured tree height, crown radius, and tree density. Error and uncertainty in the metrics $fSCA_{ADJ}$ and $fSCA_{GROUND}$ are also present due to possible sample bias, geolocation errors, and methodological uncertainties associated with TMSCAG [*Painter et al.*, 2009; *Rittger et al.*, 2013]. These uncertainties may affect the results of this error analysis, and although they were considered, no measure of such error propagation has been included.

12.3. RESULTS

12.3.1. Comparison of Satellite *f*SCA to Ground-based Snow Cover Measurements

The *Natural Resources Conservation Service* [NRCS, 2013] reported the total basin SWE for April 2012 was 111% of the long-term average in the McKenzie River Basin and 107% of the long-term average in the Middle Fork Willamette Basin. Both basins experienced a dry period during January–February and NRCS-reported SWE amounts were only about 50% of the long-term average for February. This SWE deficit was relieved by higher than average precipitation in April–May and SWE was greater than average at the end of April in both basins [*NRCS*, 2013].

There were limited cloud-free Landsat 7 scenes that coincided with the 2012 in situ snow observations. Three dates were completely cloud free over the field sites: 24 February, 14 May, and 15 June. SWE data from SNOTEL stations in the proximity (within 10 km) of the field sites indicate that snow cover was extensive across all of the field sites and on 14 May snow cover was disappearing at the low elevation sites (Figure 12.3). On 15 June the only

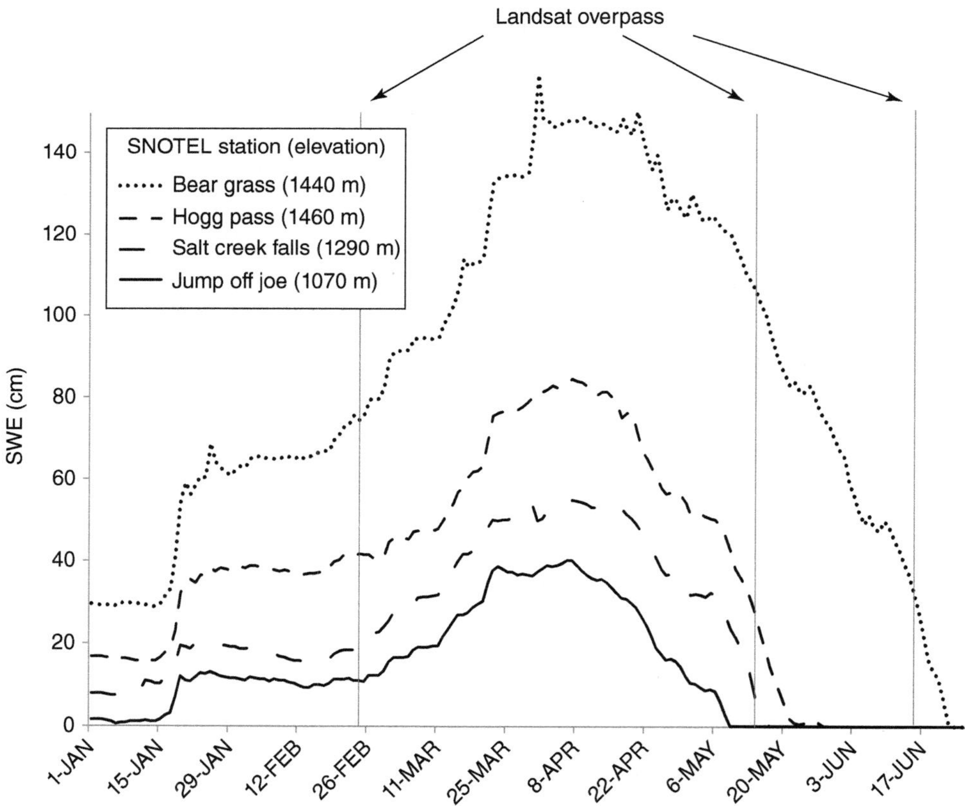

Figure 12.3 The evolution of SWE from 1 January through 20 June 2012 at SNOTEL stations in the proximity (within 10 km) of the field sites. The dates of the Landsat overpasses are indicated to show the possibility of snow in the forest canopy.

remaining snow occurred at high elevations. The SWE data also reveal the possibility of snow on the forest canopy for the 24 February overpass date. All stations indicate that about 2.5 cm of snow fell on the field sites 2 days prior to the Landsat overpass and greater amounts may have accumulated at high elevations during a storm 4 days earlier. The presence of snow in the canopy has the potential to inflate fSCA values, however, in situ surveys in early February indicate that fSCA$_{GROUND}$ was greater than 90%, so we assume errors of commission will be minimal. Snow was not present in the canopy during the 14 May and 15 June Landsat overpass.

On 24 February, application of the canopy adjustment resulted in substantial improvement over fSCA$_{TM}$ when compared to total fSCA$_{GROUND}$ at 1100_LD, 1300_LD, both 1400 m sites, and 1500_HD (Figures 12.4a, 12.4c to 12.4e). The adjustment resulted in some improvement at 1300_HD, but there continued to be an error of omission

of about 0.7 (Figure 12.4c). At the 1100_HD site, the viewable gap fraction was too low for TMSCAG to detect any snow cover, so a canopy adjustment could not be applied. Both 1200 m sites fell in a gap in satellite reflectance data that resulted from the loss of the scan line corrector (SLC) on the Landsat 7 ETM + [*NASA*, 2011]; accessed January, 2013], therefore these sites could not be included in the comparison. A canopy adjustment did not appear to be necessary for 1500_LD or 1550_LD as fSCA$_{TM}$ was comparable to fSCA$_{GROUND}$ (Figures 12.4e and 12.4f).

By the time of the satellite overpass on 14 May, the 1100 and 1200 m sites were completely snow free and unadjusted TMSCAG correctly identified no snow at these sites. There was a large error of commission for fSCA$_{ADJ}$ at 1300_HD because the adjustment resulted in an estimate of 0.7 while the ground surveys reported close to zero snow cover (Figure 12.5a). The average

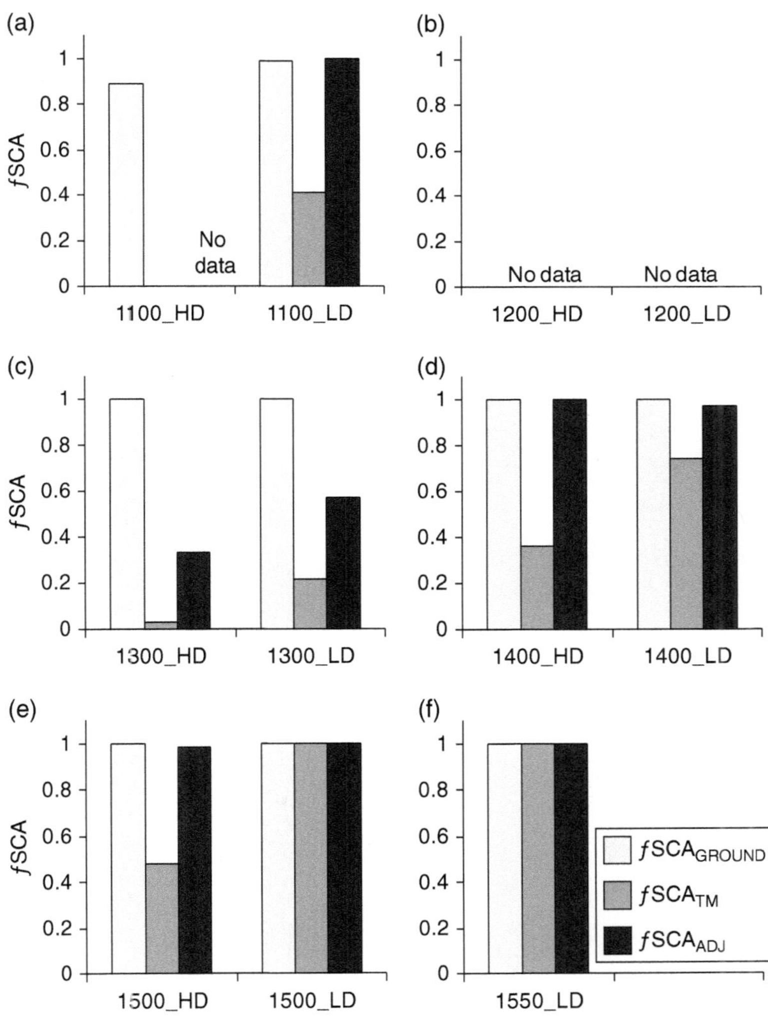

Figure 12.4 24 February site average fSCA$_{TM}$, fSCA$_{ADJ}$, and fSCA$_{GROUND}$ for elevations represented by the paired sites at (a) 1100 m, (b) 1200 m, (c) 1300 m, (d) 1400 m, (e) 1500 m, and (f) 1550 m.

$f\text{SCA}_{TM}$ for this site was below the detection limit of TMSCAG because fSCA was only slightly greater than 0.15 for pixels with observable snow cover and the fVEG was high (around 0.85). Conversely, at 1300_LD, the adjustment improved $f\text{SCA}_{TM}$ estimates by about 50%. There was also an error of commission of about 25% at the 1400_LD site and $f\text{SCA}_{TM}$ more closely approximated $f\text{SCA}_{GROUND}$ (Figure 12.5b). The 1400_HD site fell in the SLC data gap and could not be included. An adjustment may not have been necessary for 1500_LD because $f\text{SCA}_{TM}$ was within 8% of $f\text{SCA}_{GROUND}$ (Figure 12.5c). However, at 1500_HD, the canopy adjustment improved estimates of $f\text{SCA}_{GROUND}$

because $f\text{SCA}_{TM}$ was only 0.4 despite ground surveys indicating complete snow cover. The $f\text{SCA}_{TM}$ was comparable to $f\text{SCA}_{GROUND}$ for 1550_LD (Figure 12.5d), indicating that an adjustment would not be necessary for this site.

On 15 June, documented snow cover remained at only the 1400 and 1500 m sites; 1550_LD was not surveyed. Despite there being snow cover within the detectable limit of TMSCAG for 1400_HD and 1500_HD, no snow was detected by the TMSCAG algorithm (Figures 12.6a and 12.6b). Therefore, a canopy adjustment could not be applied and errors of omission up to 30% occurred in these areas.

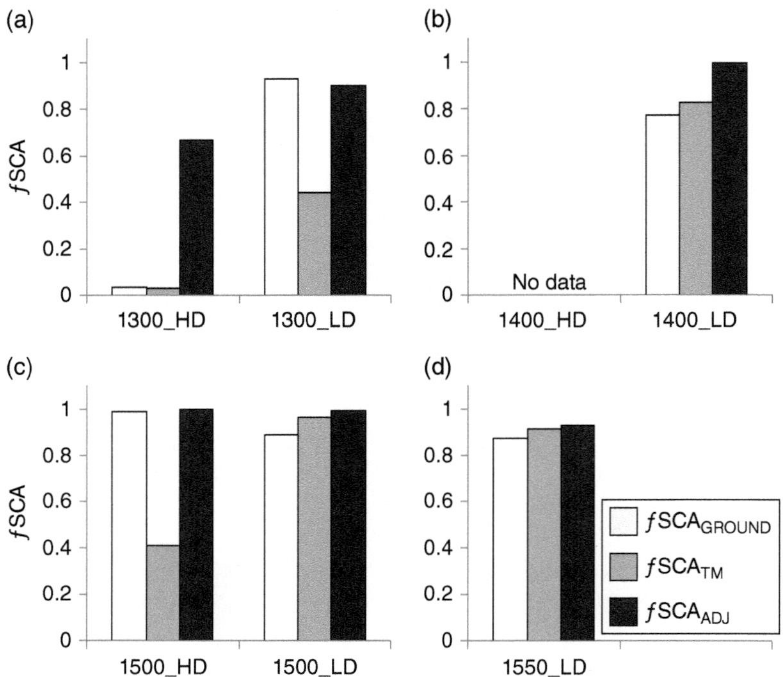

Figure 12.5 14 May differences in $f\text{SCA}_{TM}$, $f\text{SCA}_{ADJ}$, and $f\text{SCA}_{GROUND}$ for elevations represented by the paired sites at (a) 1300 m, (b) 1400 m, (c) 1500 m, and (d) 1550 m.

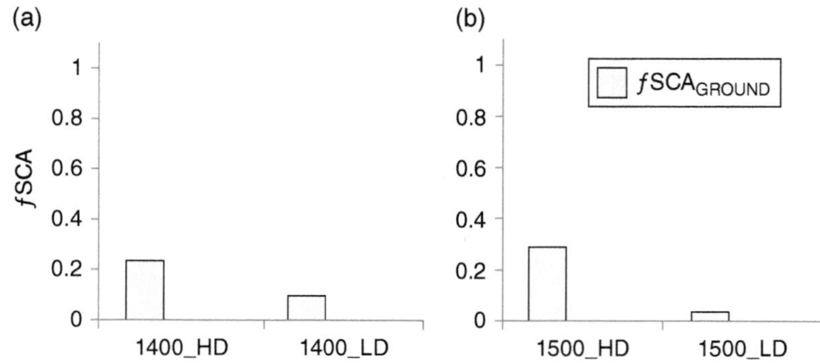

Figure 12.6 15 June $f\text{SCA}_{GROUND}$ for elevations represented by the paired sites at (a) 1400 m and (b) 1500 m. TMSCAG did not detect any fSCA for these sites.

12.3.2. Asymmetries Between Snow in Open and Subcanopy Locations

The comparison of $f\text{SCA}_{\text{ADJ}}$ to $f\text{SCA}_{\text{GROUND}}$ suggests that errors of omission occur when subcanopy snow remains undetected and errors of commission occur when the subcanopy snow fraction is less than the snow fraction in canopy gaps. To test this hypothesis and examine the snow cover variability within sites at a single elevation, the snow fraction for open (SF_o) and subcanopy (SF_{SC}) locations were compared within 1400_HD and 1400_LD. The open and subcanopy locations were classified based on the CC derived from hemispherical photographs taken at each location. The fVEG from TMSCAG is closely related to this CC (Figure 12.7). Although the significance of this relationship was not tested in this study, the observed trend suggests that fVEG from TMSCAG can serve as an effective proxy for CC and subsequently the gap fraction observable by TMSCAG. Therefore, locations with high CC can justifiably approximate the locations that have a greater potential to obscure subcanopy snow.

Both SF_o and SF_{SC} began to decrease more rapidly at 1400_LD than at 1400_HD, but different spatial and temporal patterns in SF_o and SF_{SC} were observed between the two sites. At 1400_LD, SF_{SC} and SF_o were comparable for 15 April, 4 May, and 13 May (Figure 12.8a). On 19 May, SF_o was considerably higher than SF_{SC}, with 53% of the open locations still observed to have snow cover and only 33% of the subcanopy locations retaining snow cover. This trend continued with SF_o remaining greater than SF_{SC}, which corroborates the $f\text{SCA}_{\text{ADJ}}$ error of commission on 14 May at this site. The canopy adjustment was assuming snow under the canopy, when this snow was actually the first to disappear.

In contrast to 1400_LD, the SF_o and SF_{SC} at 1400_HD were similar for the majority of the survey period, but SF_{SC} was consistently greater than SF_o (Figure 12.8b). In mid to late June, snow only remained in subcanopy locations and SF_o became zero earlier than SF_{SC}. The timing of the surveys was unable to capture the disappearance of SF_o, however, these results help to explain the inability of TMSCAG to detect snow at this site on 15 June, as the only snow remaining was in subcanopy locations with the highest CC and near the detection limit of TMSCAG at 0.15.

The comparison of SF_o and SF_{SC} was not repeated for the sites at other elevations. However, comparing the overall SF depletion for the paired high and low-density forests helps reveal if similar patterns occur at these other elevations. The SF depletion curves indicate that a longer duration of snow cover at the high-density forest, like at the 1400 m sites, is not pervasive across elevations (Figures 12.9a and 12.9b). Snow cover persisted longer in the LD forests at 1100 and 1200 m (Figures 12.9a and 12.9b), and snow lasted longer in HD forests at 1400 and 1500 m (Figures 12.9c and 12.9d). Snow survey frequency was not high enough to capture these patterns at the 1200 m paired sites.

The survey results support the differences observed between $f\text{SCA}_{\text{ADJ}}$ and $f\text{SCA}_{\text{GROUND}}$ at these sites based on the differences in melt from LD to HD. The $f\text{SCA}_{\text{GROUND}}$ was overestimated by 70% at 1300_HD on 14 May, when it is likely that snow at this site was melting under the large tree canopies and persisting in the larger canopy gaps. On 15 June, $f\text{SCA}_{\text{GROUND}}$ at 1500_HD went completely undetected by TMSCAG because similarly to the 1400_HD site, the only remaining snow was likely under high-density canopy cover and the snow fraction in that canopy was near the detection limit of TMSCAG.

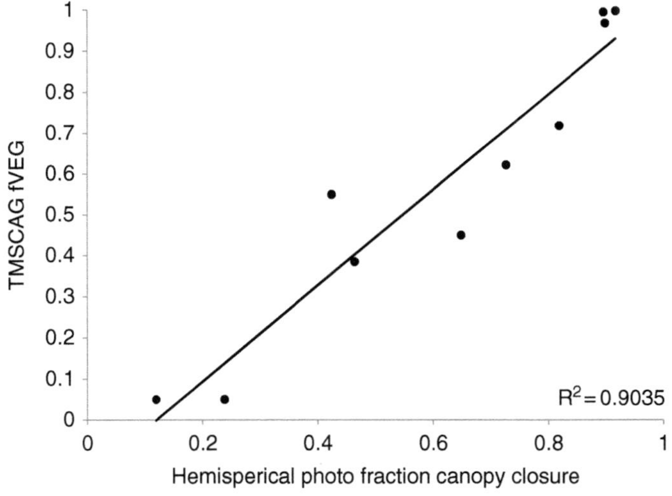

Figure 12.7 Relationship between canopy closure from hemispherical photos (CC) and fraction vegetation (fVEG) detected by TMSCAG.

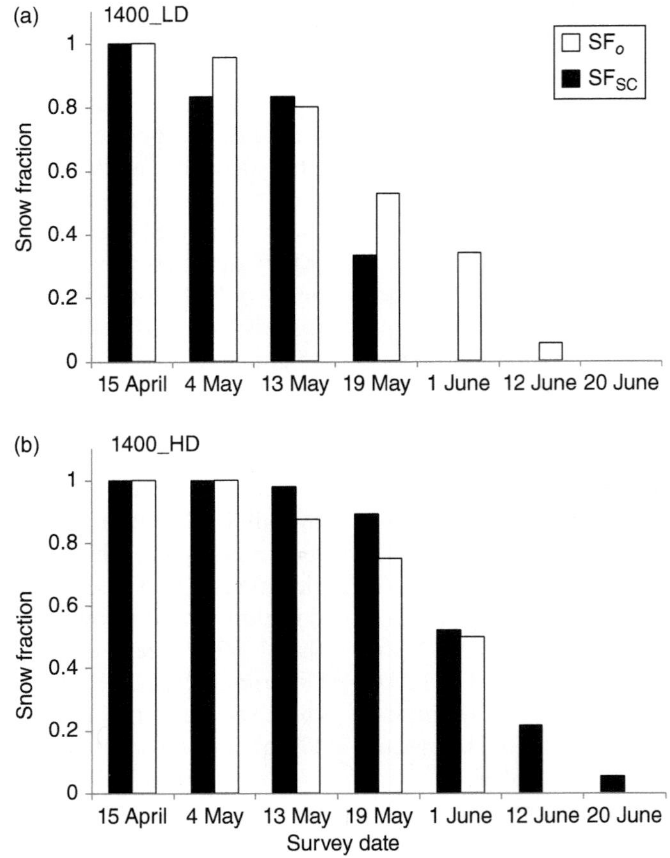

Figure 12.8 The snow fraction for (a) open (SF_o) and (b) subcanopy (SF_{SC}) observation locations in 1400_LD and 1400_HD.

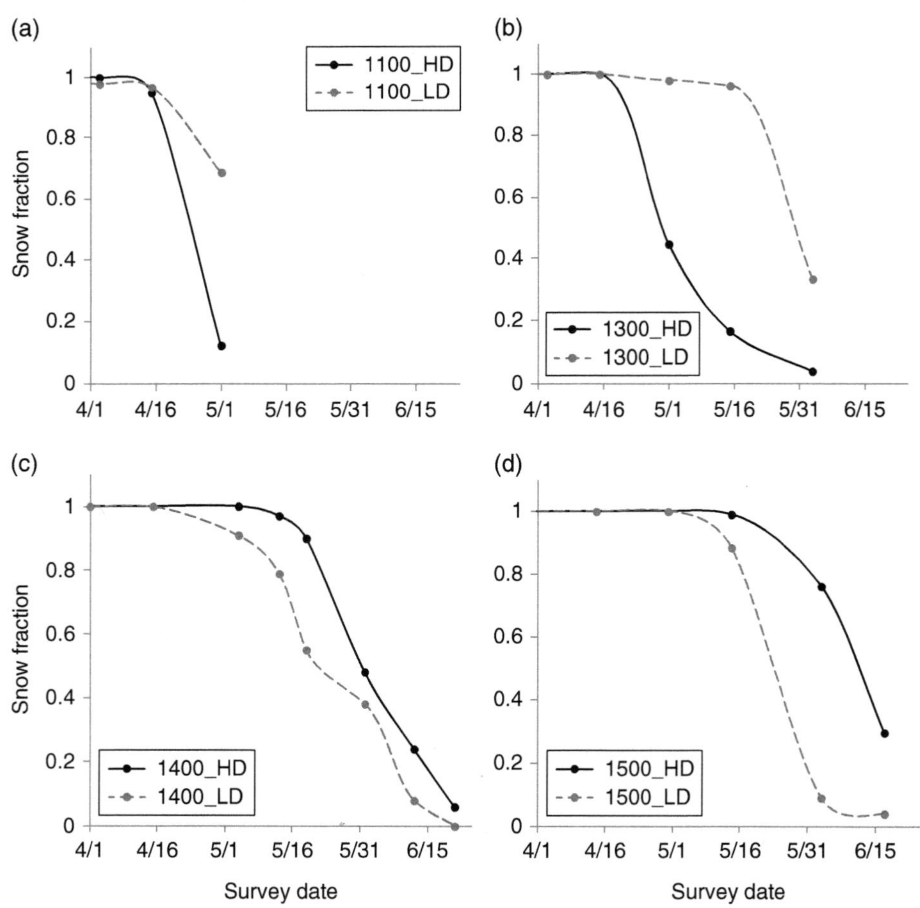

Figure 12.9 Depletion of the snow cover fraction for paired sites at (a) 1100 m, (b) 1300 m, (c) 1400 m, and (d) 1500 m elevations.

12.3.3. Forest Structure and Canopy Adjustment Error

Despite a wide range of canopy attributes at the study sites, there was a lack of a statistically significant relationship between the site-averaged canopy structure metrics of CC, tree height, canopy radius, and tree density and the average error associated with $fSCA_{ADJ}$ due to small sample sizes (maximum of 7 on 24 February) and lack of linear trend between the variables. The absence of strong relationships make it impossible to form robust inferences on the application of these canopy structure metrics to parameterize subcanopy snow and suggest that a combination of the canopy metrics may be responsible for the differences between adjusted and in situ $fSCA$.

12.4. DISCUSSION

The observed spatiotemporal snow cover patterns demonstrate that differences between $fSCA_{ADJ}$ and $fSCA_{GROUND}$ are a result of snow cover dynamics between subcanopy and open locations, not errors in the snow detection algorithm. The rapid depletion of in situ observed snow cover at 1300_HD provides evidence for the large error of commission at this site on 14 May. Enhanced melting of in situ subcanopy snow at 1400_LD also led to an error of commission on this date. Conversely, at 1400_HD, observations confirmed that the errors of omission present on 15 June occurred because late season snow was primarily retained in undetectable, subcanopy locations. A similar snow cover depletion trend between this and the 1500_HD site suggest that the same pattern is responsible for the errors of omission at 1500_HD. These errors, caused by discontinuities between observable and subcanopy snow, can be explained by physical processes linked to both elevation and canopy characteristics.

While not the explicit focus of this Chapter, the differences in snow disappearance date as functions of elevation and forest cover warrant discussion. Prior published results point to a combination of canopy interception and energy balance as responsible for these differences [*Storck*, 2002; *Musselman et al.*, 2008; *Varhola et al.*, 2010; *Ellis et al.*, 2011; *Lundquist et al.*, 2013]. For instance, the large trees at 1300_HD are likely to have increased canopy interception and, coupled with warm temperatures at this lower elevation site, may be responsible for the early disappearance of subcanopy snow relative to the nearby open site. This is qualitatively similar to the interpretations of *Lundquist et al.* [2013] who found when mean winter temperatures are warmer than $-1\,°C$, forests reduce snowpack duration by as much as 2 weeks compared with open areas. This is a result of the significant longwave contributions to the snowpack

energy balance. Our high elevation sites qualitatively resemble the colder climate sites of *Lundquist et al.* [2013] where the longwave contribution is much lower and snow persists longer in the gaps than in the forest. The relative effects of radiation balance and canopy interception on snowpacks along our elevation gradient remains to be determined. For the sites in our study, about 40–50% of snowfall is intercepted by forest canopy [*Roth and Nolin*, unpublished], comparable to values measured by *Storck* [2002] who also found that in low elevation forests, as much as 70% of intercepted snow will melt from the tree canopy before reaching the ground. The 1300 m sites are situated just above the rain-snow transition zone and are frequently subject to temperatures slightly above freezing; therefore, it is likely that a considerable portion of intercepted snow melted before it had a chance to accumulate under the forest canopy. Additionally, although sublimation is not a major contributing factor for snow loss from tree canopies in the humid, temperate climate of the Oregon Cascades [*Storck*, 2002], it may exhibit a greater influence at lower elevations [*Montesi et al.*, 2004].

The large proportion of subcanopy snow cover at 1400_HD and 1500_HD is indicative of the colder temperatures that occur at high elevations combined with the solar shading provided by the dense, continuous forest canopy [*Metcalfe and Buttle*, 1998; *Musselman et al.*, 2008]. Despite having moderate tree cover (600 trees/ha), 1400_LD did not exhibit a similar pattern. The forest at this site is characterized by clumps of trees rather than continuous forest cover. During the final 14 days of persisting snow cover at this site, the subcanopy snow fraction was between 20% and 35% lower under the clumps of trees than the snow fraction in gaps. This indicates that shading and reduced turbulent fluxes under clumps of trees did not offset the increased snow interception and enhanced longwave radiation provided by the canopy [*Musselman et al.*, 2008; *Ellis et al.*, 2011]. Additionally, the size of interclump gaps at 1400_LD may be small enough to allow sufficient forest shading along the gap margins to retard melt rates compared to a large forest opening [*Berry and Rothwell*, 1992].

Seasonal variation in the magnitude of the shortwave versus longwave radiation contributions to snowmelt [*Marks and Dozier*, 1992; *Mazurkiewicz et al.*, 2008] may also be responsible for the difference in timing of snowmelt between the LD and HD study sites at high and low elevations. Daily average temperatures at or near freezing and less total snow accumulation at the lower elevation study sites led snowmelt to occur during the cloudy, early spring conditions when the magnitude of incoming shortwave radiation was likely small relative to longwave radiation [*Marks and Dozier*, 1992; *Link and Marks*, 1999]. Low net incoming shortwave radiation made the attenuation by the forest canopy irrelevant in delaying melt, and long-wave contributions from trees and canopy may have acted to increase subcanopy ablation [*Sicart et al.*, 2004; *Pomeroy et al.*, 2009]. At the high elevation study sites, snow persisted later into the spring when cloud cover was not as frequent over the Oregon Cascades. The magnitude of incoming shortwave radiation then increased relative to longwave radiation, and the forest canopy acted to attenuate shortwave contributions, protecting subcanopy snow from rapid melt [*Link and Marks*, 1999].

The role of forest canopy in affecting the presence of snow cover may be different in other snow climate regimes [*Sturm et al.*, 1995; *Sicart et al.*, 2004, 2006]. For climates with an increased number of cloud-free days, shortwave radiation has a larger magnitude throughout the entire ablation season, and there may be an increased effect of forest canopy on ablation [*Link and Marks*, 1999]. Aspect influences forest canopy effects on both net shortwave and longwave radiation inputs to the snowpack during snowmelt [*Ellis et al.*, 2011]. Forest cover has been found to protect snow on south-facing slopes and enhance melt on north-facing slopes due to increased contributions of longwave radiation [*Ellis et al.*, 2011]. Because of these potentially confounding variables, aspect and elevation need to be included in future studies that explore the role of forest structure on the presence of subcanopy snow across different snow climates and geographic regions.

The hydrologic significance of the errors of commission and omission by $fSCA_{ADJ}$ were investigated by analyzing snow surveys of SWE and snow depth measurements at two of the study sites. A snow survey performed at 1500_HD on 15 June showed that errors of omission on this date resulted in missing on average 0.1 m SWE, which represented only 1% of the maximum annual SWE for the area covered by Landsat pixels at the 1500 m sites. Therefore, the hydrologic implications associated with errors of omission at high elevations late in the snowmelt season may be small. This is reassuring because estimating subcanopy snow when snow is not detectable in canopy gaps cannot be addressed with optical remote sensing techniques. Accounting for these errors of omission would require integration of surface monitoring and physical models with remote sensing data. However, the 70% over prediction of snow cover at 1300_HD may have greater hydrologic implications. A snow survey conducted through both 1300 m sites on 15 May indicated that the errors of omission on this date could lead to the assumption that 0.5 m SWE was remaining across these pixels when there was actually no snow. This overestimation of subcanopy snow does have the potential to be reduced with the incorporation of forest canopy characteristics to parameterize subcanopy snow. Although the results presented herein failed to reveal a relationship that can be used to parameterize subcanopy snow with a single canopy metric, further examination of a combination of variables has the possibility to reduce excessive errors of commission in satellite $fSCA$.

Field observations suggest that canopy interception and solar shading of snow are key processes that affect the observed differences between snow in subcanopy and open locations. Although interception and shading are moderated by forest canopy, a relationship could not be established to parameterize subcanopy snow with canopy closure, tree height, crown radius, or tree density. The error in $fSCA_{ADJ}$ did not exhibit a distinct pattern with any one of these metrics, but analysis combining the multiple variables could be fruitful but was not tested.

Several limitations prohibited an adequate comparison of $fSCA_{ADJ}$ error and canopy structure. Foremost was an incomplete time series of TMSCAG because of frequent cloud cover, which prevented us from capturing the onset of increased error in $fSCA_{ADJ}$. There was also a relatively small range of canopy characteristics exhibited by the study sites. For instance the study did not include any sites with small trees that may not necessitate a canopy adjustment. It is possible that this scale of analyses, forest structure averaged across an entire stand, was too large to capture the detailed canopy structure that has been tied to snow ablation under the forest canopy through moderation of solar beam transmittance [*Musselman et al.*, 2012, 2013].

The propagation of errors involved in determining $fSCA_{ADJ}$ and $fSCA_{GROUND}$ may have also contributed to the inability to develop a significant relationship. Sample bias may have been introduced with the selection of study sites and individual observation locations for both the in situ snow measurements and forest inventory, and measurement error may have occurred during manual vegetation surveys. Finally, the analyses were founded on the assumption that the in situ snow observations and forest inventory measurements adequately captured the spatial variability needed to define inferential relationships.

12.5. CONCLUSIONS

The current canopy adjustment applied to $fSCA$ from TMSCAG performs well when snow cover is homogenously distributed in open and forested areas. Errors that do occur during such times are limited to dense forests where the viewable gap fraction is near the detection limit of TMSCAG. Once snow cover displays spatial variability in the gaps, the snow cover obscured by the canopy begins to differ from that in the open. Such differences decrease the accuracy of $fSCA_{ADJ}$ and demonstrate the need to develop a more consistently accurate canopy adjustment. Our preliminary results show that the hydrologic implications of the errors may be small for high elevation snowpack but are more substantial for lower elevations where the area of snow cover is large and the cumulative error can be substantial.

While several limitations in this study undermined the ability to quantify relationships between forest structure and subcanopy snow, they provide direction for future improvements. Progress toward an effective canopy adjustment will require a continuous record of spatially coincident satellite-derived and in situ snow cover for varying forest densities. Miniature (iButton) temperature sensors have been presented as a method to track the spatial variability of snow cover [*Lundquist and Lott*, 2008; *Raleigh et al.*, 2013]. Strategic deployment of these sensors in a representative range of forest types, across multiple elevations, may provide the means to separate these confounding variables. Combined with detailed forest canopy structure metrics, these temperature sensors would offer key spatially distributed *f*SCA information at the plot scale. Canopy structure examined with snow cover data should include the metrics analyzed in this study and additional bulk canopy metrics such as effective leaf area index (LAI) and sky view fraction. These metrics can be used to estimate solar transmissivity through a forest canopy and have been linked to snow ablation rates under forests [*Musselman et al.*, 2012, 2013]. Though not available for this study, airborne and satellite LiDAR [*Hilker et al.*, 2012; *Sun et al.*, 2008] have the potential to provide cross-scale analysis of vegetation structure that can be linked to estimates from aerial photos and the Multiangle Imaging SpectroRadiometer (MISR). These *f*SCA and vegetation data products will also provide the ability to characterize other canopy structure metrics that may influence snow cover patterns such as tree clump size and spacing.

A hybrid remote sensing-modeling approach may ultimately be a successful combination for snow mapping. Prior to the snow season, canopy structure could be mapped using remote sensing techniques and subsequently used to inform a physically based snow model. During the snow season, remote sensing of snow cover can update the model and also be augmented by model output that can describe not only snow-vegetation interactions but subcanopy snow physical properties. Models, such as *Liston and Elder*'s [2006] spatially distributed snow evolution model, SnowModel, evaluate snow-vegetation interactions and can provide insight into the snow accumulation and ablation processes that may have been responsible for the errors of commission and omission in this study. Studies including physical models, continuous in situ snow cover measurements, vegetation observations, and remote sensing would need to be extended to other locations through coordinated campaigns in both maritime and continental snow climates, including Alaska's boreal forests and the Rocky Mountain West. Incorporating all the tools at hand can help elucidate the spatiotemporal snow cover patterns necessary for developing a universal canopy adjustment.

Quantifying forest-snow cover interactions is important not only for optical remote sensing but for passive microwave and radar remote sensing of snow cover as well. Forest canopy affects the microwave emissivity and radar backscatter to such an extent that accurate snow characterization in forested regions is not currently possible. While our initial goal is an accurate canopy adjustment for optical remote sensing of *f*SCA, this is a step toward overall improvements in snow remote sensing for all sensor types. This will include proper parameterization of subcanopy snow by forest canopy metrics, followed by the scaling of forest structure parameters to moderate resolutions, and inclusion of these metrics to account for areas where the current canopy adjustment is misrepresenting subcanopy snow. We anticipate that an approach using satellite observations of snow cover and forest canopy combined with process-based snow modeling will provide us with the cross-scale spatial and temporal snow characterization needed for hydrologic and climatologic applications.

ACKNOWLEDGMENTS

This research was supported by subcontracts from the NASA/Jet Propulsion Laboratory for MISR Science Team and NPP VIIRS activities; and from the National Science Foundation Water Sustainability and Climate Award, EAR-1039192. The authors would like to thank the anonymous reviewers whose contributions helped improve the quality of this manuscript.

REFERENCES

Baker, N. (2011), Joint Polar Satellite System (JPSS) VIIRS Snow Cover Algorithm Theoretical Basis Document (ATBD), pp. 1–52, Northrop Grumman Aerospace Syst., Redondo Beach, Calif.

Berry, G. J., and R. L. Rothwell (1992), Snow ablation in small forest openings in Southwest Alberta, *Can. J. Forest Res.*, *22*(9), 1326–1331.

Clark, M. P., A. G. Slater, A. P. Barrett, L. E. Hay, G. J. McCabe, B. Rajagopalan, and G. H. Leavesley (2006), Assimilation of snow covered area information into hydrologic and land-surface models, *Ad. Water Resourc.*, *29*(8), 1209–1221, doi:10.1016/j.advwatres.2005.10.001.

Davis, R. E., J. P. Hardy, W. Ni, C. Woodcock, J. C. McKenzie, R. Jordan, and X. Li (1997), Variation of snow cover ablation in the boreal forest: A sensitivity study on the effects of conifer canopy, *J. Geophys. Res.*, *102*(D24), 29,389–29,395, doi:10.1029/97jd01335.

Dozier, J. (2011), Mountain hydrology, snow color, and the fourth paradigm, *Eos Trans. Am. Geophys. Union*, *92*(43), 373–374, doi:10.1029/2011eo430001.

Durand, M., N. P. Molotch, and S. A. Margulis (2008), Merging complementary remote sensing datasets in the context of snow water equivalent reconstruction, *Remote Sens. Environ.*, *112*(3), 1212–1255, doi:10.1016/j.rse.2007.08.010.

Ellis, C. R., J. W. Pomeroy, R. L. H. Essery, and T. E. Link (2011), Effects of needleleaf forest cover on radiation and snowmelt dynamics in the Canadian Rocky Mountains, *Can. J. Forest Res.*, *41*(3), 608–620, doi:10.1139/X10-227.

Faria, D. A., J. W. Pomeroy, and R. L. H. Essery (2000), Effect of covariance between ablation and snow water equivalent on depletion of snow-covered area in a forest, *Hydrol. Process.*, *14*(15), 2683–2695, doi:10.1002/1099-1085(20001030)14:15 < 2683::aid-hyp86 > 3.0.co;2-n.

Frazer, G. W., C. D. Canham, and K. P. Lertzman (1999), Gap Light Analyzer (GLA), Version 2.0: Imaging software to extract canopy structure and gap light transmission indices from true-colour fisheye photographs, users manual and program documentation, Simon Fraser Univ., Burnaby British Columbia, Inst. of Ecosyst. Studies, Milibrook, N.Y.

Golding, D. L., and R. H. Swanson (1986), Snow distribution patterns in clearings and adjacent forest, *Water Resourc. Res.*, *22*(13), 1931–1940, doi:10.1029/WR022i013p01931.

Hilker, T., N. C. Coops, G. J. Newnham, M. van Leeuwen, M. A. Wulder, J. Stewart, and D. S. Culvenor (2012), Comparison of terrestrial and airborne LiDAR in describing stand structure of a thinned lodgepole pine forest, *J. Forestry*, *110*(2), 97–104, doi:10.5849/jof.11-003.

Inoue, A., K. Yamamoto, N. Mizoue, and Y. Kawahara (2004), Calibrating view angle and lens distortion of the Nikon fisheye converter FC-E8, *J. Forest Res.*, *9*(2), 177–181, doi:10.1007/s10310-003-0073-8.

Jin, Y., J. T. Randerson, M. L. Goulden, and S. J. Goetz (2012), Post-fire changes in net shortwave radiation along a latitudinal gradient in boreal North America, *Geophys. Res. Lett.*, *39*(13), L13403, doi:10.1029/2012GL051790.

Klein, A. G., D. K. Hall, and G. A. Riggs (1998), Improving snow cover mapping in forests through the use of a canopy reflectance model, *Hydrol. Process.*, *12*(10–11), 1723–1744, doi:10.1002/(sici)1099-1085(199808/09)12:10/11 < 1723::aid-hyp691 > 3.0.co;2-2.

Link, T., and D. Marks (1999), Distributed simulation of snow-cover mass- and energy-balance in the boreal forest, *Hydrol. Process.*, *13*(14–15), 2439–2452, doi:10.1002/(sici)1099-1085(199910)13:14/15 < 2439::aid-hyp866 > 3.0.co;2-1.

Liston, G. E., and K. Elder (2006), A distributed snow-evolution modeling system (SnowModel), *J. Hydrometeorol.*, *7*(6), 1259–1276, doi:10.1175/JHM548.1.

López-Moreno, J. I., and J. Latron (2008), Influence of canopy density on snow distribution in a temperate mountain range, *Hydrol. Process.*, *22*(1), 117–126, doi:10.1002/hyp.6572.

Lundquist, J. D., and F. Lott (2008), Using inexpensive temperature sensors to monitor the duration and heterogeneity of snow-covered areas, *Water Resourc. Res.*, *44*(W00D16), 1–6, doi:10.1029/2008wr007035.

Lundquist, J. D., S. E. Dickerson-Lange, J. Lutz, and N. C. Cristea (2013), Lower forest density enhances snow retention in regions with warmer winters: A global framework developed from plot-scale observations and modeling, *Water Resourc. Res.*, *49*, 1–15, doi:10.1002/wrcr.20504.

Marks, D., and J. Dozier (1992), Climate and energy exchange at the snow surface in the alpine region of the Sierra-Nevada 2. Snow cover energy balance, *Water Resourc. Res.*, *28*(11), 3043–3054, doi:10.1029/92WR01483.

Martinec, J., and A. Rango (1995), Seasonal runoff forecasts for hydropower based on remote sensing, in *Proceedings of the 63rd Annual Western Snow Conference, Sparks, Nevada*, pp. 10–20. http://www.westernsnowconference.org/node/430

Mazurkiewicz, A. B., D. G. Callery, and J. J. McDonnell (2008), Assessing the controls of the snow energy balance and water available for runoff in a rain-on-snow environment, *J. Hydrol.*, *354*(1–4), 1–14, doi:10.1016/j.jhydrol.2007.12.027.

McGuire, M., A. Wood, A. Hamlet, and D. Lettenmaier (2006), Use of satellite data for streamflow and reservoir storage forecasts in the Snake River Basin, *J. Water Resourc. Plan. Manag.*, *132*(2), 97–110, doi:10.1061/(asce)07339496(2006)132:2(97).

Metcalfe, R. A., and J. M. Buttle, (1998), A statistical model of spatially distributed snowmelt rates in a boreal forest basin, *Hydrol. Process.*, *12*(10–11), 1701–1722, doi:10.1002/(sici)1099-1085(199808/09)12:10/11 < 1701::aid-hyp690 > 3.0.co;2-d.

Molotch, N. P., and S. A. Margulis (2008), Estimating the distribution of snow water equivalent using remotely sensed snow cover data and a spatially distributed snowmelt model: A multi-resolution, multi-sensor comparison, *Adv. Water Resourc.*, *31*(11), 1503–1514, doi:10.1016/j.advwatres.2008.07.017.

Montesi, J., K. Elder, R. A. Schmidt, and R. E. Davis, (2004), Sublimation of intercepted snow within a subalpine forest canopy at two elevations, *J. Hydrometeorol.*, *5*(5), 763–773, doi:10.1175/1525-7541(2004)005 < 0763:soiswa > 2.0.co;2.

Musselman, K. N., N. P. Molotch, and P. D. Brooks (2008), Effects of vegetation on snow accumulation and ablation in a mid-latitude sub-alpine forest, *Hydrol. Process.*, *22*(15), 2767–2776, doi:10.1002/hyp.7050.

Musselman, K. N., N. P. Molotch, S. A. Margulis, P. B. Kirchner, and R. C. Bales, (2012), Influence of canopy structure and direct beam solar irradiance on snowmelt rates in a mixed conifer forest, *Agric. Forest Meteorol.*, *161*(August 2013), 46–56, doi:10.1016/j.agrformet.2012.03.011.

Musselman, K. N., S. A. Margulis, and N. P. Molotch (2013), Estimation of solar direct beam transmittance of conifer canopies from airborne LiDAR, *Remote Sens. Environ.*, *136*(September 2013), 402–415, doi:10.1016/j.rse.2013.05.021.

National Aeronautics and Space Administration (NASA) (2011), *Landsat 7 Science Data Users Handbook*, NASA, Greenbelt, Md.

Natural Resource Conservation Service Snowpack Reports, available at http://www.wcc.nrcs.usda.gov/cgibin/snow_rpt.pl?state=oregon, accessed May 2013.

Nolin, A. W. (2010), Recent advances in remote sensing of seasonal snow, *J. Glaciol.*, *56*(200), 1141–1150, doi:10.3189/002214311796406077.

O'Halloran, T. L., B. E. Law, M. L. Goulden, Z. Wang, J. G. Barr, C. Schaaf, M. Brown, J. D. Fuentes, M. Göckede, and A. Black (2012), Radiative forcing of natural forest disturbances, *Global Change Biol.*, *18*(2), 555–565.

Painter, T. H., K. Rittger, C. McKenzie, P. Slaughter, R. E. Davis, and J. Dozier (2009), Retrieval of subpixel snow covered area, grain size, and albedo from MODIS, *Remote Sens. Environ.*, *113*(4), 868–879, doi:10.1016/j.rse.2009.01.001.

Pomeroy, J. W., D. M. Gray, N. R. Hedstrom, and J. R. Janowicz (2002), Prediction of seasonal snow accumulation in cold climate forests, *Hydrol. Process.*, *16*(18), 3543–3558, doi:10.1002/hyp.1228.

Pomeroy, J. W., D. Marks, T. Link, C. Ellis, J. Hardy, A. Rowlands, and R. Granger (2009), The impact of coniferous forest temperature on incoming longwave radiation to melting snow, *Hydrol. Process.*, *23*(17), 2513–2525, doi:10.1002/hyp.7325.

Raleigh, M. S., K. Rittger, C. E. Moore, B. Henn, J. A. Lutz, and J. D. Lundquist (2013), Ground-based testing of MODIS fractional snow cover in subalpine meadows and forests of the Sierra Nevada, *Remote Sen. Environ.*, *128*, 44–57, doi:10.1016/j.rse.2012.09.016.

Randerson, J. T., et al. (2006), The impact of boreal forest fire on climate warming, *Science*, *314*(5802), 1130–1132.

Rittger, K., T. H. Painter, J. Dozier (2013), Assessment of methods for mapping snow cover from MODIS, *Adv. Water Resourc.*, *51*, 367–380, doi:10.1016/j.advwatres.2012.03.002.

Salomonson, V. V., and I. Appel (2004), Estimating fractional snow cover from MODIS using the normalized difference snow index, *Remote Sens. Environ.*, *89*(3), 351–360, doi:10.1016/j.rse.2003.10.016.

Sicart, J. E., J. W. Pomeroy, R. L. H. Essery, J. Hardy, T. Link, and D. Marks (2004), A sensitivity study of daytime net radiation during snowmelt to forest canopy and atmospheric conditions, *J. Hydrometeorol.*, *5*(5), 774–784, doi:10.1175/1525-7541(2004)005 < 0774:ASSODN > 2.0.CO;2.

Sicart, J. E., J. W. Pomeroy, R. L. H. Essery, and D. Bewley (2006), Incoming longwave radiation to melting snow: Observations, sensitivity, and estimation in northern environments, *Hydrol. Process.*, *20*(17), 3697–3708, doi:10.1002/hyp.6383.

Storck, P. (2002), Measurement of snow interception and canopy effects on snow accumulation and melt in a mountainous maritime climate, Oregon, United States, *Water Resourc. Res.*, *38*(11), doi:10.1029/2002wr001281.

Sturm, M., J. Holmgren, and G. E. Liston (1995), A seasonal snow cover classification system for local to global applications, *J. Climate*, *8*(5), 1261–1283, doi:10.1175/1520-0442(1995)008 < 1261:ASSCCS > 2.0.CO;2.

Sun, G., K. J. Ranson, D. S. Kimes, J. B. Blair, and K. Kovacs (2008), Forest vertical structure from GLAS: An evaluation using LVIS and SRTM data, *Remote Sens. Environ.*, *112*(2008), 107–117, doi:10.1016/j.rse.2006.09.036.

Tague, C., and G. E. Grant (2004), A geological framework for interpreting the low-flow regimes of Cascade streams, Willamette River Basin, Oregon, *Water Resourc. Res.*, *40*(4), doi:10.1029/2003wr002629.

Thirel, G., P. Salamon, P. Burek, and M. Kalas (2011), Assimilation of MODIS snow cover area data in a distributed hydrologic model, *Hydrol. Earth Syst. Sci. Disc.*, *8*, 1329–1364, doi:10.5194/hessd-9-1329-2011.

U.S. Geological Survey (USGS) (2013), Landsat ETM + SLC-off-Path: 29 Row: 30 for Scene: LE70290302013073EDC00, USGS Earth Resources Observation and Science Center (EROS), Souix Falls, S. Dak.

Varhola, A., N. C. Coops, C. W. Bater, P. Teti, S. Boon, and M. Weiler (2010), The influence of ground- and lidar-derived forest structure metrics on snow accumulation and ablation in disturbed forests, *Can. J. Forest Res.*, *40*(4), 822–826, doi:10.1139/X10-008.

13

Passive Microwave Remote Sensing of Snowmelt and Melt-Refreeze Using Diurnal Amplitude Variations

Kathryn Alese Semmens,[1,2] Joan Ramage,[2] Jeremy D. Apgar,[3] Katrina E. Bennett,[4] Glen E. Liston,[5] and Elias Deeb[6]

13.1. BACKGROUND

In many Arctic and sub-Arctic watersheds, the shoulder seasons of spring and fall are important and dynamic periods exhibiting large diurnal variations that have implications for ecology, hydrology, biology, and human systems. The spring season is characterized by snowmelt onset as above-freezing, positive air temperatures become common, while the fall season is characterized by freeze-up as temperatures start to dip below freezing. During the spring transition time the snowpack melts and refreezes, soil thaws, and runoff increases. River ice breaks up, and with melt-refreeze and thaw along the banks, transportation for many remote communities may be hampered. At the end of this transition, green-up occurs, discharge peaks, and the likelihood of flooding increases. The snowpack melts away completely, allowing vegetation to flourish with the available sunlight and water. Boreal ecosystems respond rapidly to these positive temperatures and greater available moisture with

increases in soil respiration and photosynthetic activity [*Kimball et al.*, 2004]. As such, the timing and length of the spring transition period is of consequence, and the ability to measure the timing and dynamics over wide spatial areas provides important information on ecosystem functioning and change, as well as carbon dioxide flux from spring thaw [*Bartsch et al.*, 2007]. Monitoring the diurnal amplitude variation [DAV; difference between daily minimum and maximum brightness temperature (T_b)] of the snow surface via passive microwave remote sensing provides a metric of this transition. As the surface melts, the passive microwave brightness temperature signal increases sharply in contrast to the refreeze signal at night, thus the difference from day to night is high. Tracking the diurnal changes provides information on the timing, extent, and length of melt-refreeze in both glacial and nonglacial settings.

Understanding and monitoring the DAV sheds light on the dynamics of this transition period and provides a proxy for water release timing in the terrestrial hydrologic cycle. Without remote sensing, this property is difficult to continuously observe over large spatial areas and in inaccessible, remote regions. Previous studies have shown that the end of the high DAV period relates to the rising limb of the hydrograph (freshet) with a lag of about zero to five days [*Ramage et al.*, 2006], though this relationship is expected to be variable, dependent upon basin size, topography, amount of runoff versus infiltration, and climate. For example, the snowmelt flood peak for Matanuska Glacier, Alaska, was found to be highly correlated to the accumulated area snowmelt timing (derived from T_b and DAV thresholds) so that the peak could be predicted within a five day margin [*Kopczynski et al.*, 2008].

[1]*USDA-Agricultural Research Service, Hydrology and Remote Sensing Lab, Beltsville, Maryland, USA*

[2]*Earth and Environmental Sciences Department, Lehigh University, Bethlehem, Pennsylvania, USA*

[3]*New York–New Jersey Trail Conference, Mahwah, New Jersey, USA*

[4]*International Arctic Research Center and Water and Environmental Research Center, University of Alaska, Fairbanks, Alaska, USA*

[5]*Cooperative Institute for Research in the Atmosphere, Colorado State University, Fort Collins, Colorado, USA*

[6]*U.S. Army Cold Regions Research and Engineering Laboratory, New Hanover, New Hampshire, USA*

Remote Sensing of the Terrestrial Water Cycle, Geophysical Monograph 206. First Edition. Edited by Venkat Lakshmi.
© 2015 American Geophysical Union. Published 2015 by John Wiley & Sons, Inc.

215

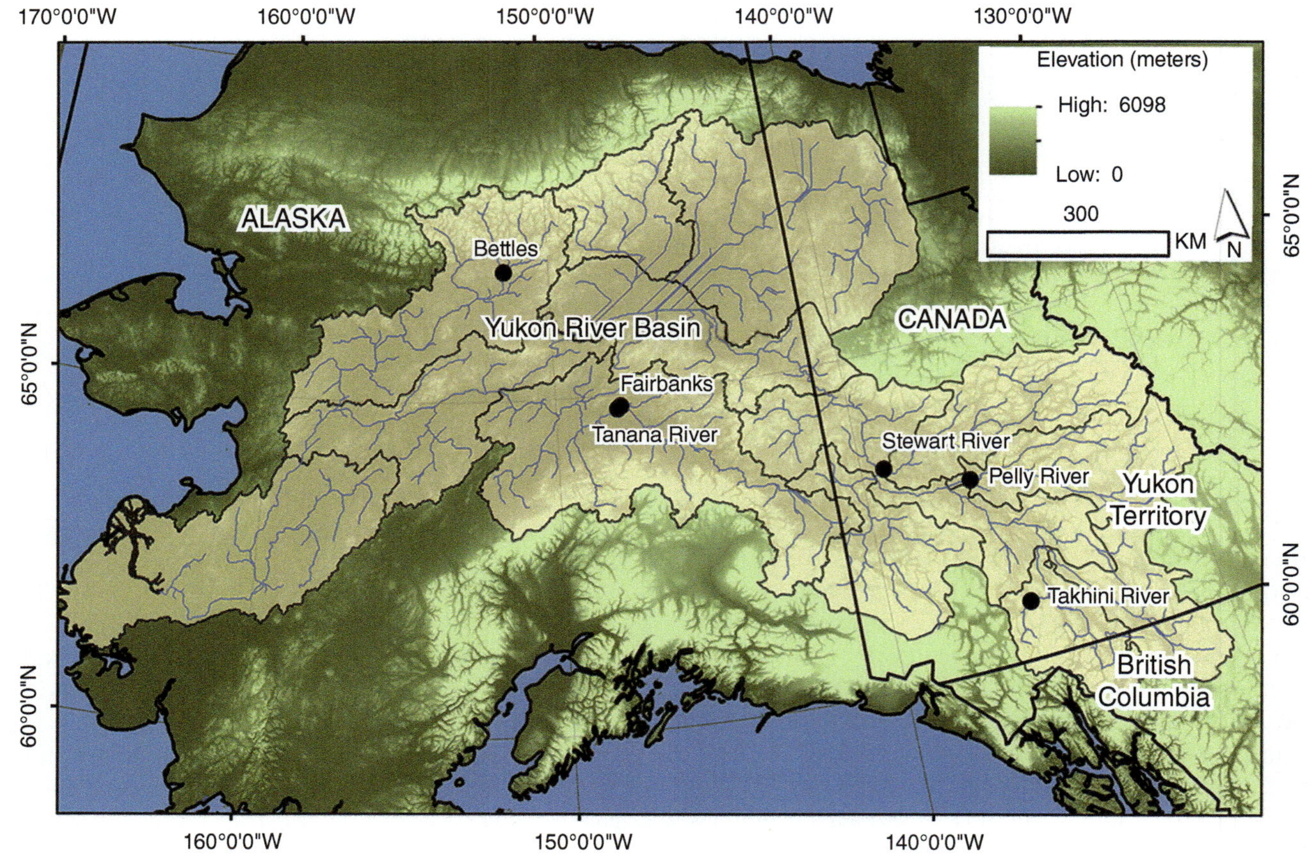

Figure 13.1 Map of the Yukon River Basin with locations of sites referenced in chapter. Bettles is the site for Figure 13.3, Fairbanks and Tanana River at Fairbanks are the sites shown in Figure 13.5, Takhini River near Whitehorse is the site for Figure 13.7, and Stewart River at Mouth and Pelly River at Pelly Crossing are shown in Figure 13.8. Bonanza Creek LTER data in Figure 13.5 are from around Fairbanks.

This chapter provides evidence for the utility of measuring and monitoring DAV in terrestrial hydrological systems via passive microwave remote sensing techniques and presents a brief overview of the development of passive microwave-derived melt, focusing in particular on DAV. Specifically, illustrations of the significance of the DAV measure in relation to other hydrological and ecological processes are provided. These examples are mostly based in the Yukon River Basin (Figure 13.1), a large (853,300 km²) sub-Arctic basin covering most of central Alaska and parts of the Yukon Territory in Canada [*Brabets et al.*, 2000], and also include experience in mid-latitude Sierra Nevada and Rocky Mountains.

13.2. DATA

In this chapter, passive microwave data from the Special Sensor Microwave/Imager (SSM/I) and the Advanced Microwave Scanning Radiometer–Earth Observing System (AMSR-E) are used in conjunction with ground-based observations, modeling results, and optical remote sensing (from Moderate Resolution Imaging Spectroradiometer, or MODIS) to illustrate the DAV technique and its potential application (Table 13.1). AMSR-E data (2003–2011) are in the form of level 2A Global Swath Spatially Resampled Brightness Temperatures V001, supplied by the National Snow and Ice Data Center [*Ashcroft and Wentz*, 2006] gridded to the Equal Area Scalable Earth-Grid (EASE-Grid) [*Brodzik and Knowles*, 2002] with nominally 12.5 × 12.5 km² resolution; 36.5 GHz has 14 × 8 km² resolution. SSM/I data are also provided in EASE-Grid format and are used to construct a longer time series of DAV as the data are available from 1987 through 2012. SSM/I data are in the form of level 3 EASE-Grid brightness temperatures gridded data for Northern Hemisphere projection with a resolution of 37 × 28 km² remapped to EASE-Grid 25 × 25 km² [*Armstrong et al.*, 1994].

13.3. MELT DETECTION METHODOLOGY

The presence of liquid water within a snowpack increases its emissivity, thus increasing brightness temperature (T_b) because T_b is a function of the surface temperature (T_s) and

Table 13.1 Data used to illustrate and validate diurnal amplitude variation technique

Data	Date	Spatial Res.	Data Sources
AMSR-E L3 daily SWE	2003–2011	25 km[a]	NSIDC
AMSR-E L2A brightness temps (36.5 GHz V_{pol})	2003–2011	12.5 km[a], 25 km[a]	NSIDC
SSM/I brightness temps (37 GHz V_{pol})	1988–2012	25 km[a]	NSIDC
Hydrometric data: Tahkini River near Whitehorse, Tanana River at Fairbanks, Stewart River at Mouth, and Pelly River at Pelly Crossing	1988–2012	Ground based	Environment Canada/United States Geological Survey
Meteorological data (air temperature and precipitation): Bettles, Alaska; Fairbanks, Alaska	1988–2012	Ground based	Global Historical Climatology Network
Digital elevation model	2000	90 m	Environment Yukon Geomatics
SnowModel (SWE depth, 2 m air temperature, precipitation)	1979–2009	10 km	*Liston and Hiemstra* [2011]
MODIS Daily Fractional Snow Cover Extent (MOD10A1 version 5)	2000–2012	500 m	*Hall et al.,* 2006
Field data (snow depth, relative volumetric water content) Swamp Angel Study Plot, Senator Beck Basin, Silvertown, Colorado	2003–2012	Ground-based	Center for Snow and Avalanche Studies (*CSAS*, 2012)

[a]Data are in EASE-grid resolution

emissivity (ε) of a material ($T_b \sim \varepsilon T_s$). As a result, passive microwave sensors can easily detect melting snow because there is a significant difference in the signal between wet (which emits close to that of a blackbody) and dry snow [*Chang et al.,* 1976; *Ulaby et al.,* 1986]. Since higher frequency wavelengths are sensitive to shallow depths of the snowpack compared to more deeply penetrating lower frequencies, the ~37 GHz vertically polarized wavelength is typically used for detecting surface melt [*Ramage et al.,* 2006]. Many different approaches to detecting melt have been developed to exploit the characteristic melt signal from passive microwave brightness temperatures; these include a single channel threshold method [*Mote et al.,* 1993], a horizontally polarized gradient ratio [*Steffen et al.,* 1993], channel differencing [*Takala and Pullianinen,* 2008], a normalized difference between the 19 GHz horizontally polarized and 37 GHz vertically polarized brightness temperatures [Cross-Polarized Gradient Ratio or XPGR, *Abdalati and Steffen,* 1995, 1997], and a single channel and diurnal amplitude variation threshold method [e.g., *Ramage and Isacks,* 2003; *Tedesco et al.,* 2009]. Microwave emission models have also been used to simulate threshold brightness temperature values [*Mote and Anderson,* 1995]. The melt detection approach has also been used for melt onset determination on Arctic sea ice [*Drobot and Anderson,* 2001].

13.4. DIURNAL AMPLITUDE VARIATION METHODOLOGY

Melt detection is the basis for determining diurnal amplitude variations indicating the surface wetness of a snowpack. To determine DAV, the running difference between daily minimum and maximum values are calculated, an approach first proposed by *Ramage and Isacks*

[2002, 2003] using SSM/I data. This calculation needs high temporal resolution, which is provided by the passive microwave sensors SSM/I and AMSR-E, so that both day and night are represented. AMSR-E collects daily observations for most areas, with the exact number depending on the latitude of the observed area (closer to the equator, most days per month have two overpasses, but some have one or none). EASE-Grid data provide T_b twice daily per grid cell, using interpolation to fill gaps, which enables the diurnal calculation. Multiple AMSR-E observations that are less than 2.5 h apart are averaged to create only two daily observations for the purposes of calculating DAV [*Apgar et al.,* 2007]. SSM/I data are also provided in a format with two observations per day.

After calculating the DAV for a given pixel, a threshold-based method is used to determine snowmelt onset and the end of melt-refreeze, or high DAV. The DAV threshold used for AMSR-E is ±18 K [*Apgar et al.,* 2007] and for SSM/I is ±10 K [*Ramage and Isacks,* 2002; *Ramage et al.,* 2006]. The melt onset date (MOD) is defined as the first day where both T_b and DAV thresholds are met for three out of five consecutive days [equation 13.1] and the end of high DAV (EHD) is defined as the last day when both T_b and DAV thresholds are exceeded [equation 13.2], where d is day, n is number of occurrences, and i is pixel:

$$\mathrm{MOD} = \min(d) \begin{bmatrix} (\text{where}(T_{b_{i,(d:d+4)}} > T_{b_{\text{thresh}}} \\ + \mathrm{DAV}_{i,(d:d+4)} > \mathrm{DAV}_{\text{thresh}}) + n \geq 3) \end{bmatrix}$$

$$(13.1)$$

$$\mathrm{EHD} = \max(d) \begin{bmatrix} (\text{where}(T_{b_{i,(d:d+4)}} > T_{b_{\text{thresh}}} \\ + \mathrm{DAV}_{i,(d:d+4)} > \mathrm{DAV}_{\text{thresh}}) + n \geq 3) \end{bmatrix}$$

$$(13.2)$$

Ramage et al. [2007] illustrated comparable melt onset, melt-refreeze, and DAV results between SSM/I and AMSR-E. While similar distributions were found for both sensors, the melt-refreeze durations were slightly longer for SSM/I, possibly due to different overpass times, which resulted in the detection of earlier melt onset.

A new algorithm, dynamic DAV (D-DAV), was developed by *Tedesco et al.* [2009], which determines an individual threshold value for each year and pixel, thus accounting for heterogeneous terrain and the occurrences of high DAV values during dry snow periods that occur in some pixels [*Tedesco et al.*, 2009]. The D-DAV approach computes the threshold as the January–February average DAV + 10 K, and the T_b threshold is determined from the bimodal distribution modeling of the T_b histogram [*Tedesco et al.*, 2009].

More recently, *Li et al.* [2012] used a Gaussian inverse distance weight (GIDW) to interpolate L2A AMSR-E 37 GHz T_b to the point scale to calculate a time series of DAV for the Kern River Basin in the Sierra Nevadas. They show DAV correspond to the temporal evolution of snow water equivalent (SWE) over the season and provide evidence that using L2A data and the GIDW technique is more accurate compared to EASE-Grid data because of the L2A's higher resolution providing more sensitivity (50% greater range in DAV compared to EASE-Grid) to melt timing dynamics across the basin. Further, they used thresholds for their GIDW derived T_b and DAV (255 and 20 K, respectively) to determine melt onset with twice the accuracy (and correlation coefficient of 0.94) of the EASE-Grid data [*Li et al.*, 2012].

Tedesco [2007] also used DAV in a glacial setting to monitor melt on the Greenland ice sheet using the 19.35 or 37 GHz SSM/I channels, enhancing previous methods by utilizing multiple frequencies for the detection of melt at different depths. While the 19.35 GHz was less sensitive to wetness than 37 GHz and required higher snow wetness values, the DAV approach produced results that matched well with reanalysis data [*Tedesco*, 2007]. Also studying a glacial system, *Monahan and Ramage* [2010] used brightness temperature and DAV thresholds to determine melt onset and melt-refreeze cycles on the Southern Patagonia Ice Field.

A diurnal difference technique was also developed for active microwave satellite data, specifically, backscatter from QuikSCAT [*Bartsch et al.*, 2007]. The average backscatter of morning and evening passes are differenced and significant thaw-refreeze cycles determined when the signal is higher than the background noise [*Kidd et al.*, 2004; *Bartsch et al.*, 2007]. *Nghiem et al.* [2001, 2005] applied a similar methodology with the K_u-band scatterometer (QuikSCAT) to map snowmelt regions on the Greenland ice field utilizing the diurnal signature (difference between morning and evening backscatter) in their melt detection algorithm.

Measuring the diurnal variation in all these contexts is useful for the information it conveys on the temporal differences of surface wetness. Since high DAV indicates a dynamic surface that has a high contrast between the day and night or between ascending and descending passes, a low DAV indicates a static surface that is always frozen, always wet, or always dry. The fluctuations in DAV differ based on the type of land surface that is being measured. For instance, a snow-covered land pixel shows a distinct melt-refreeze period in the spring when the snow surface melts and refreezes before completely melting out (Figure 13.2). In contrast, some glacier pixels show high DAV for a larger part of the year due to consistent melting and refreezing during the spring, summer, and fall months. For the glacier pixel shown in Figure 13.2, there is only low DAV during the cold, frozen winter months.

DAV reflects variability in T_b, which is dominated by changing snow conditions in early spring with the diurnal melting and refreezing which changes surface wetness characteristics and accelerates snow grain growth [*Li et al.*, 2012]. However, DAV may be affected by soil wetness conditions (when snowpack is gone), snow grain size and stratigraphy, vegetation and forest cover, topography differences within a given pixel, and temperature variations, as well as noise related to mechanical effects and atmospheric contributions. While these contributions provide some uncertainty in the results of this technique, the noise and uncertainty is minimal compared to the

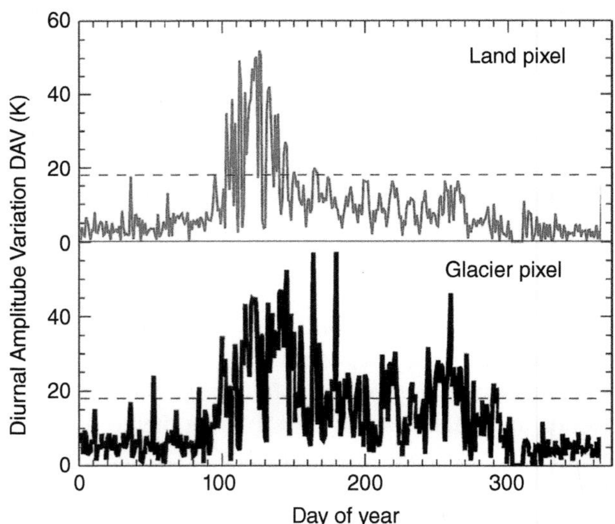

Figure 13.2 Comparison of land and glacial diurnal amplitude variation derived from AMSR-E brightness temperatures. Dashed horizontal line is the threshold of 18 K. Glacier pixels have high DAV during the majority of the melt season as the surface melts and refreezes throughout the spring and summer, whereas the land pixel has a distinct melt-refreeze period of high DAV in the spring.

robust signal of melt-refreeze in the spring. That said, improvements to DAV results from using a dynamic pixel-derived threshold, as seen in *Tedesco et al.* [2009] and from using a finer T_b processing technique and grid as seen in *Li et al.* [2012]. These improvements point to the variability in DAV that may occur due to differing pixel characteristics from elevation and vegetation, which can be better accounted for with higher resolution and dynamic thresholds. For a general understanding and large-scale basin study, the static DAV derived from EASE-Grid data is sufficient.

13.5. DAV VALIDATION AND INTERPRETATION

Validation of the interpretation and the relationship of DAV with other significant natural processes can be illustrated with a comparison of T_b and DAV time series to other data records. Figure 13.3 compares the 2003–2009 time series of the passive microwave AMSR-E brightness temperatures and DAV with ground-based station data on snow depth and air temperature, modeling data (SWE depth, precipitation, and air temperature from SnowModel; *Liston and Hiemstra*, 2011), and percent of snow cover extent (SCE) derived from the Moderate Resolution Imaging Spectroradiometer (MODIS, see Table 13.1 for more information) for observations with less than 40% cloud cover [*Bennett et al.*, in prep.]. The MODIS swaths were downloaded, reprojected into an Alaska Albers Equal Area projection, and merged into statewide GeoTIFFs. Grid cells within a 10 m buffer zone around the Bettles GHCN Global Historical Climatology Network climate station site were selected and averaged to generate a spatial mean of the SCE on each day from 1 March 2003 to 31 July 2009 (Figure 13.3, triangles).

The sharp rise in the passive microwave T_b and high DAV indicates melt onset while the termination of the high DAV and large fluctuation in T_b indicates the end of the melt-refreeze period after which the snow depth goes to zero. Both ground-based and modeling data support the end of high DAV as coinciding with depletion of the snowpack, as shown by the contemporaneous changes highlighted in the gray boxes in Figure 13.3. Air temperatures also corroborate the melt onset and refreeze timing. Results from the MODIS percent snow cover extent (March–July) provides additional evidence that the snow cover starts to disappear at melt onset and is gone by the end of the melt–refreeze/high DAV period (Figure 13.3).

Most of the previous work utilizing the DAV approach has focused on high latitude sub-Arctic environments because these areas typically have a well-defined transition from mostly frozen and dry to fluctuating between frozen and melting. However, preliminary exploration of the technique in midlatitude areas suggests a potential spatial expansion of its use in these regions. For example,

Figure 13.4 shows 2010 DAV compared with field observations at a site in Colorado (Swamp Angel Study Plot in the Senator Beck Basin; *CSAS*, 2012), providing further evidence of the importance and timing of the high DAV period in a midlatitude, alpine environment. Snow depth decreases with melt onset starting with the initial occurrence of high DAV (Figure 13.4). The end of melt-refreeze (high DAV) coincides with snow depletion. Relative water content at the base of the snowpack increases soon after melt onset as well.

Additionally, *Li et al.* [2012] illustrate the use of the DAV method in a midlatitude area, the Kern River Basin in southern Sierra Nevada, California. Snow tends to reflect a more maritime characteristic in California compared to the continental-type snow characteristic of high latitude sub-Arctic areas as well as the snow in the earlier Colorado example. *Li et al.* [2012] clearly show DAV signal sensitivity to melt timing differences between years and its utility for determining melt onset and end of melt dates, thus extending the applicability of the DAV technique to more snow types (maritime and continental).

13.6. MELT-REFREEZE RELATIONSHIP TO OTHER PROCESSES

During the period of melt-refreeze the surface of the snowpack melts slowly until its minimum temperatures are high enough to have continuous melt until depletion or snow off. The passive microwave 37 GHz channel provides surface melt observations. This wavelength does not penetrate deeply into the snowpack and therefore does not give information on melt processes at depth in the snowpack. After the high DAV/melt-refreeze period ceases, meltwater infiltrates or progresses through the basin as subsurface or overland flow, resulting in increasing discharge. For snow-dominated basins the snowmelt runoff peak is usually the highest discharge of the year and occurs several days to weeks after the end of high DAV. For glacier dominated basins, high discharge occurs relatively later compared to nival regimes and displays seasonality with more fluctuation during the summer due to the continued availability of meltwater as the surface melts and refreezes [*Jansson et al.*, 2003; *Hodgkins*, 2009]. Most glacial basins still mimic nival regimes albeit with later, larger, and longer freshets [*Fleming,* 2005]. Diurnal variations in discharge over the summer melt season result from short-term storage of meltwater in snow and subglacial channels [*Jansson et al.*, 2003]. The highest diurnal variations, streamflow, and air temperature occur during the middle of the melt season, typically around July and August [*Fleming*, 2005].

Following high DAV, green-up occurs when vegetation has enough sun and water available to flourish. Figure 13.5 shows the relationship of the timing of melt onset, snow

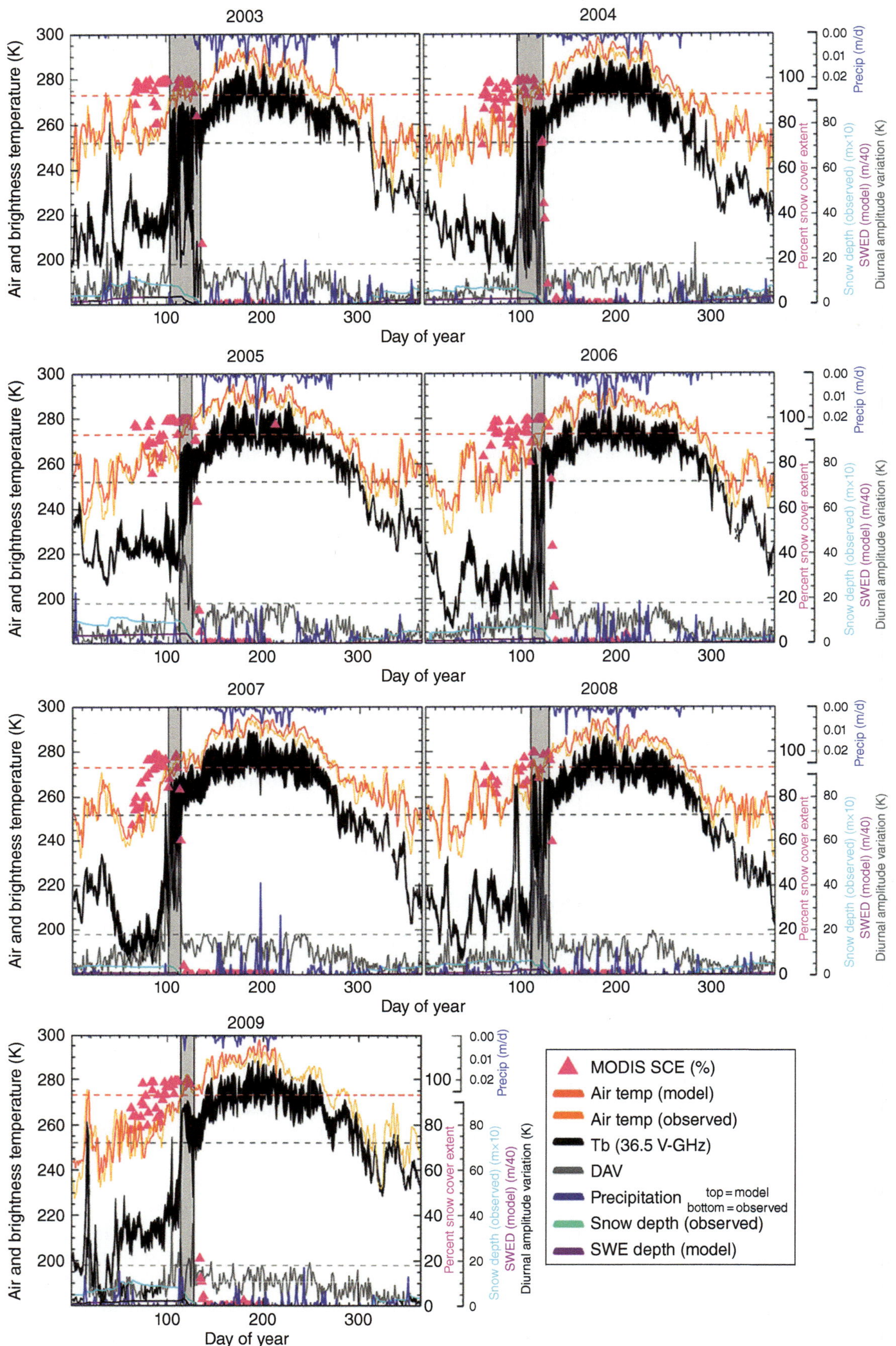

Figure 13.3 (a) [Color Version] Time series of AMSR-E brightness temperatures (T_b), DAV, precipitation, snow depth, air temperature, and snow cover extent for Bettles, Alaska (66.9133°N, 151.5225°W). Air temperature (orange), snow depth (cyan), precipitation (bottom blue) are from the Global Historical Climatology Network data at Bettles. Air temperature (red), SWE depth (purple), and precipitation (top blue) are outputs from SnowModel. Dashed lines are thresholds: black is for T_b (252 K), gray is for DAV (18 K), and red is for air temperature (273.15 K). MODIS percent snow cover extent is shown with pink triangles. Gray shaded rectangle illustrates the melt-refreeze period characterized by high DAV.

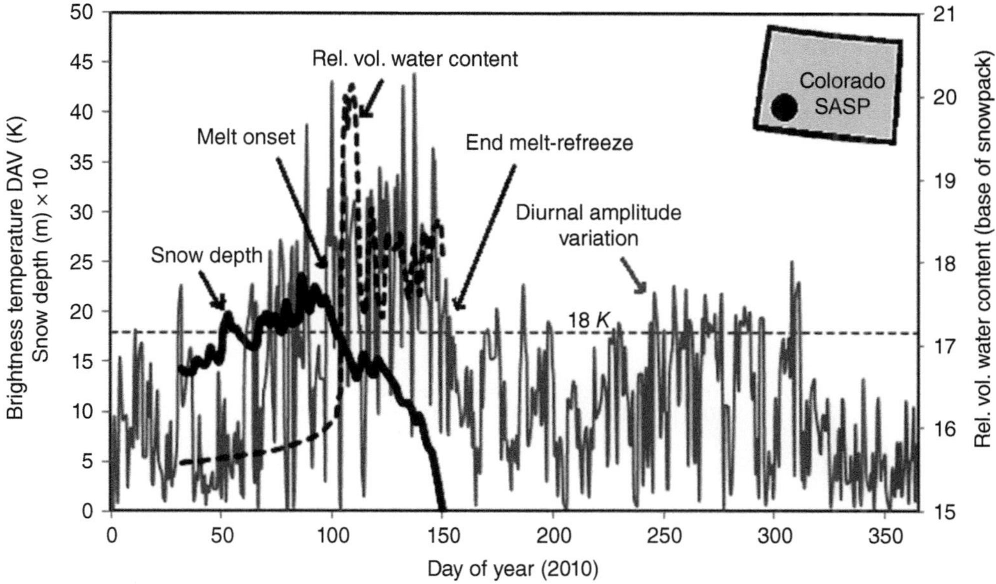

Figure 13.4 Observations of snow depth (m × 10) and relative volumetric water content at the base of the snowpack from Swamp Angel Study Plot (SASP) (37.906889°N,107.726278°W, 3371 m), Senator Beck Basin near Silverton, Colorado (CSAS, 2012) compared to AMSR-E and DAV.

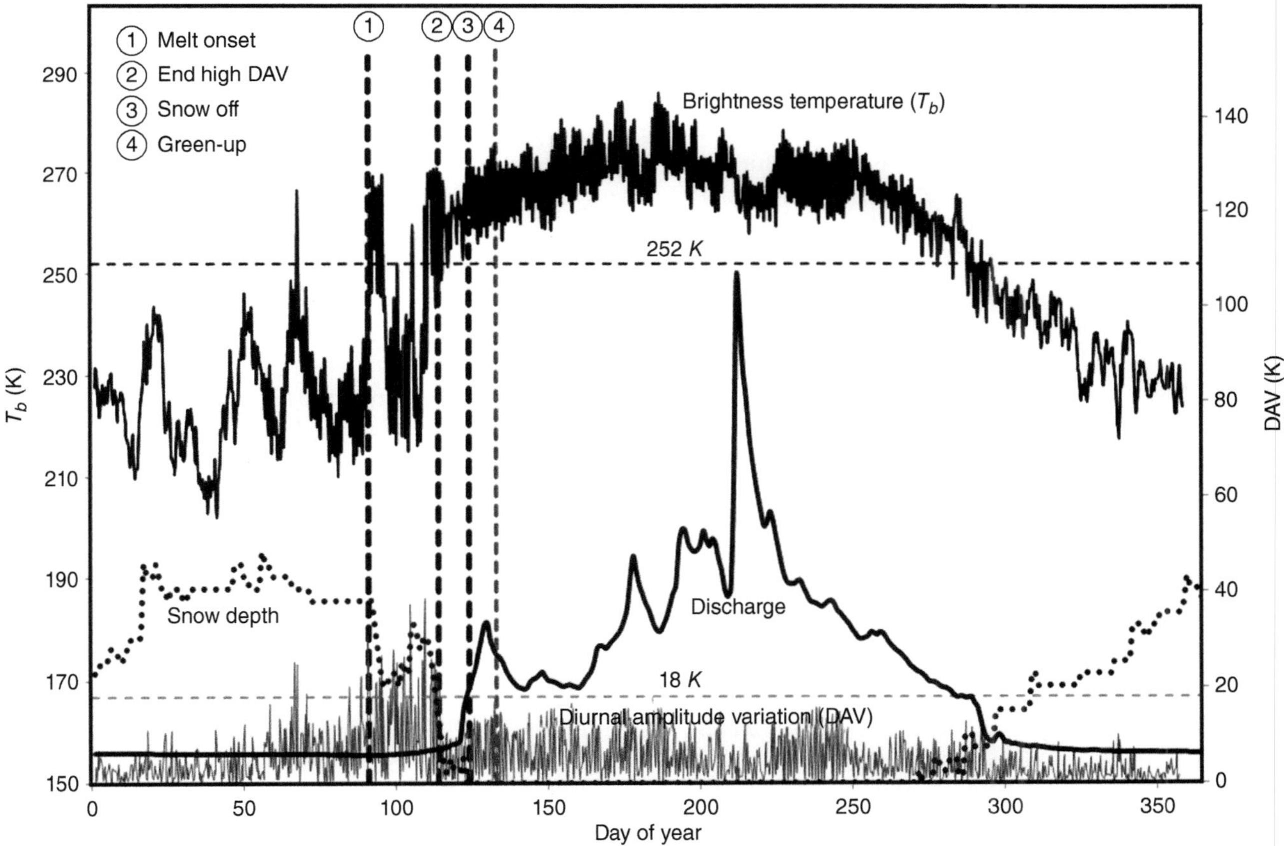

Figure 13.5 Brightness temperatures (T_b) and diurnal amplitude variation (DAV) from AMSR-E 36.5V-GHz for 2008 from the pixel in which Fairbanks is located, within the Tanana River subbasin. Discharge data is from Tanana River at Fairbanks USGS 15485500 National Water Information System. Green-up data is from Bonanza Creek Long Term Ecological Research database.

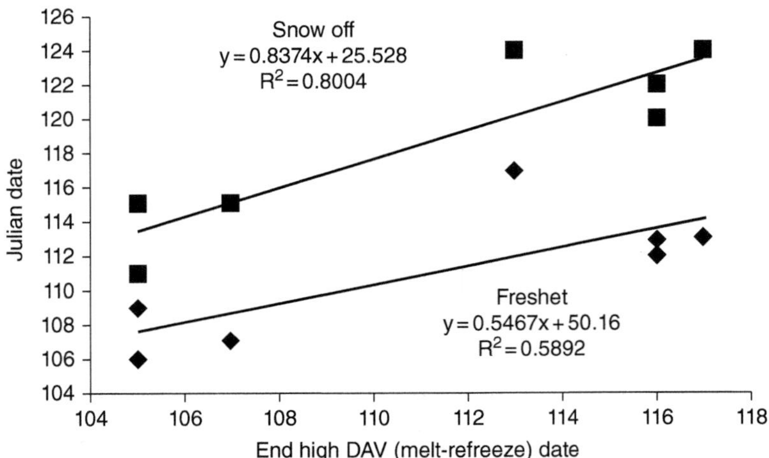

Figure 13.6 Positive linear relationship between end of melt-refreeze (high DAV) and snow off date and freshet timing based on years 2003–2010 for the Fairbanks pixel from Figure 13.5.

depth, end of high DAV or melt-refreeze, date of snow off/disappearance, and the subsequent green-up and discharge rise. T_b and DAV thresholds—snowpack is melting when $T_b > 252\,\mathrm{K}$ and $\mathrm{DAV} > 18\,\mathrm{K}$ [*Apgar et al.*, 2007]—determine dates of melt onset and end of melt-refreeze/high DAV, defined where thresholds are met for more than three of five consecutive days. The end of high DAV is followed shortly by snow off (10 days later), freshet (6 days later), snowmelt runoff peak (16 days later), and green-up (19 days later) (Figure 13.5). While these relationships are only illustrated for one year and pixel (Fairbanks, Alaska), the correlation of the end of high DAV, snow off, and freshet is consistent across years (Figure 13.6) and is typical of other terrestrial, snow-dominated areas investigated in the Yukon River Basin [*Ramage et al.*, 2007].

In addition to the temporal dynamics of the melt-refreeze period already demonstrated, the spatial progression of melt through a small basin (Takhini River Basin, Yukon Territory, Canada, 60.841 N, 135.182 W) is shown in Figure 13.7. Melt starts for parts of the basin at day 120 and spreads through the higher elevations until the entire basin reaches saturation (at the end of high DAV) when the snowpack is isothermal at day 170 and melts completely. Discharge sharply increases as the majority of the basin reaches saturation.

13.7. DAV AND SNOWMELT RUNOFF MODELING

The link between melt timing (onset and end of high DAV) and discharge can be taken a step further and applied to snowmelt runoff modeling. A simple, flux-based model called SWEHydro [adapted from *Yan et al.*, 2009] routes meltwater through a basin based on the timing of melt onset, the end of melt-refreeze, and basin terrain characteristics. The model allows for discharge modeling in areas with sparse meteorological data, relying only on remotely sensed snow and snowmelt observations [*Yan et al.*, 2009; *Ramage and Semmens*, 2012].

Specific inputs to the SWEHydro model include melt onset date, end of high DAV date, flow path distance from pixel to gauge, area of pixel, and SWE. SWEHydro takes SWE for each pixel, applies a melt rate to get a meltwater equivalent and a flow timing rate to determine how long it takes the meltwater to reach the gauge some distance away. All accumulated meltwater for each day from all pixels at the gauge (added to base flow) is summed. Different snowmelt and flow timing rates are applied for the period before and after the end of high DAV/melt-refreeze. During the melt-refreeze period, flow timing ranges from 2.4 to 4 days/10 km and after the snow is consistently melting, flow rates increase to 0.3–1.5 days/10 km. SWEHydro has been applied in various sized basins in the Yukon River Basin with reasonable results [see *Ramage and Semmens*, 2012].

Figure 13.8 illustrates the various inputs into the SWEHydro model, as well as modeled output streamflow for the Pelly River Basin. Figure 13.8e also shows how sensitive the results are to the accuracy of the input SWE. AMSR-E-derived SWE was multiplied by a factor of 0.5, 1.0, and 1.5 to determine the role of SWE in affecting hydrograph output. Results are only shown for the Pelly River Basin for a year of typical/average snow accumulation (2008), high snow relative to average (2009), and low snow relative to average (2010). Scaling the input SWE based on whether the year is an atypical high or low snow year improves the modeling results. Note that because precipitation is not included in the model, just the first part of the hydrograph (approximately days 1–180) should be considered, as the model is only meant to model the snowmelt runoff period.

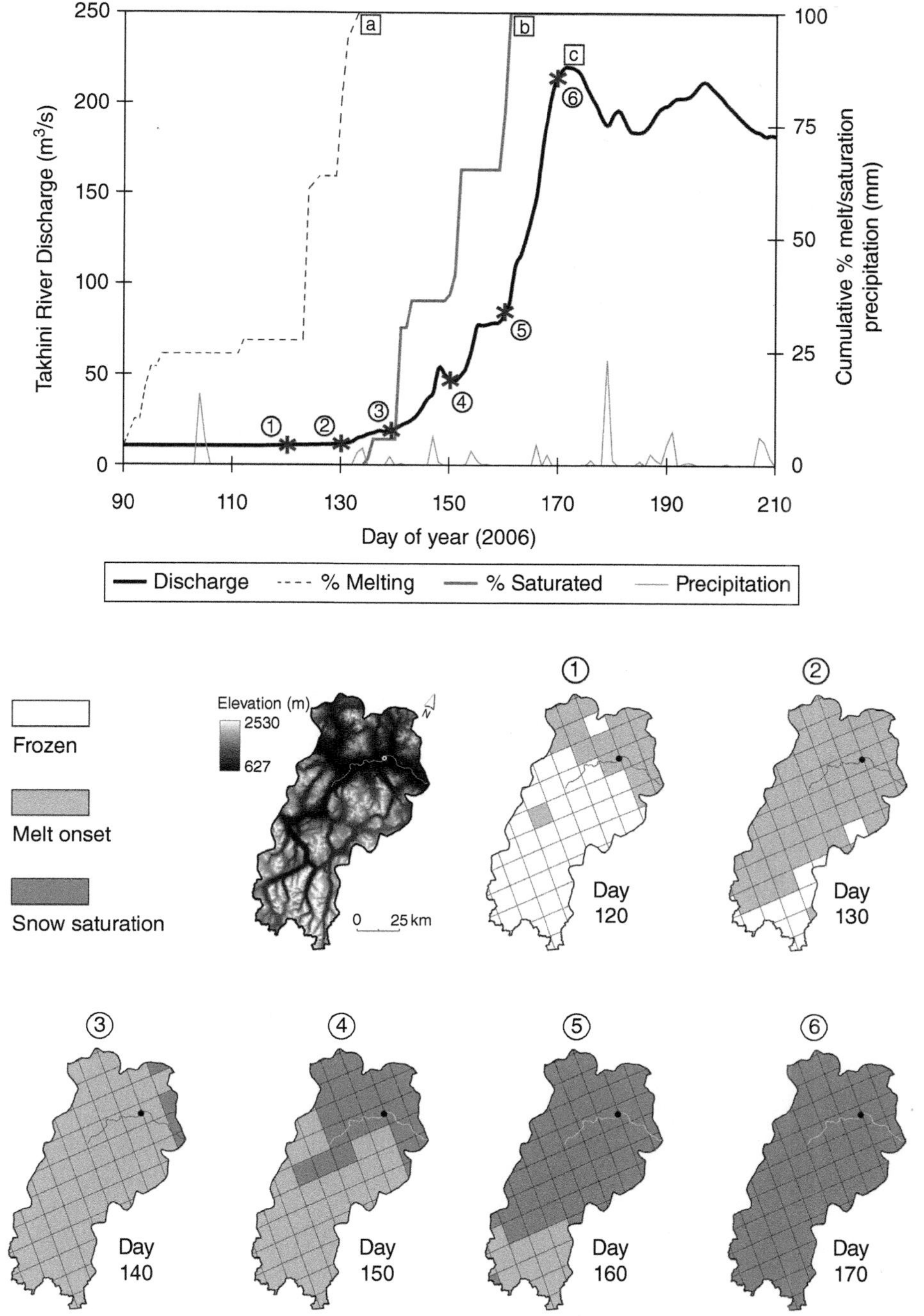

Figure 13.7 Melt progression through the Takhini River basin shows melt timing over the spatial domain of the basin from mostly frozen (day 120) through melt onset to snow saturation/end of high DAV (day 170). The top panel depicts the cumulative percentage of the basin that is (a) melting and (b) saturated at a given day of the year, as well as the (c) discharge, which reaches a peak after saturation is reached for the whole basin. Numbers along the discharge curve correspond to the numbered plots in the lower panel showing the spatial progression of melt. Discharge is from the gauge Takhini River near Whitehorse, YT (60.85°N, 135.73°W, drainage area 6990 km²). (Adapted from *Apgar* [2007].)

13.8. LIMITATIONS

Given the need for multiple observations per day that are representative of day and night conditions, the DAV approach is limited by the availability of adequate data as well as the coarse resolution of the passive microwave (25 km for SSM/I and 12.5 km for AMSR-E). If overpass times for a certain area do not enable a day-versus-night calculation, DAV may be artificially low as it does not capture the range of actual melt conditions, a situation

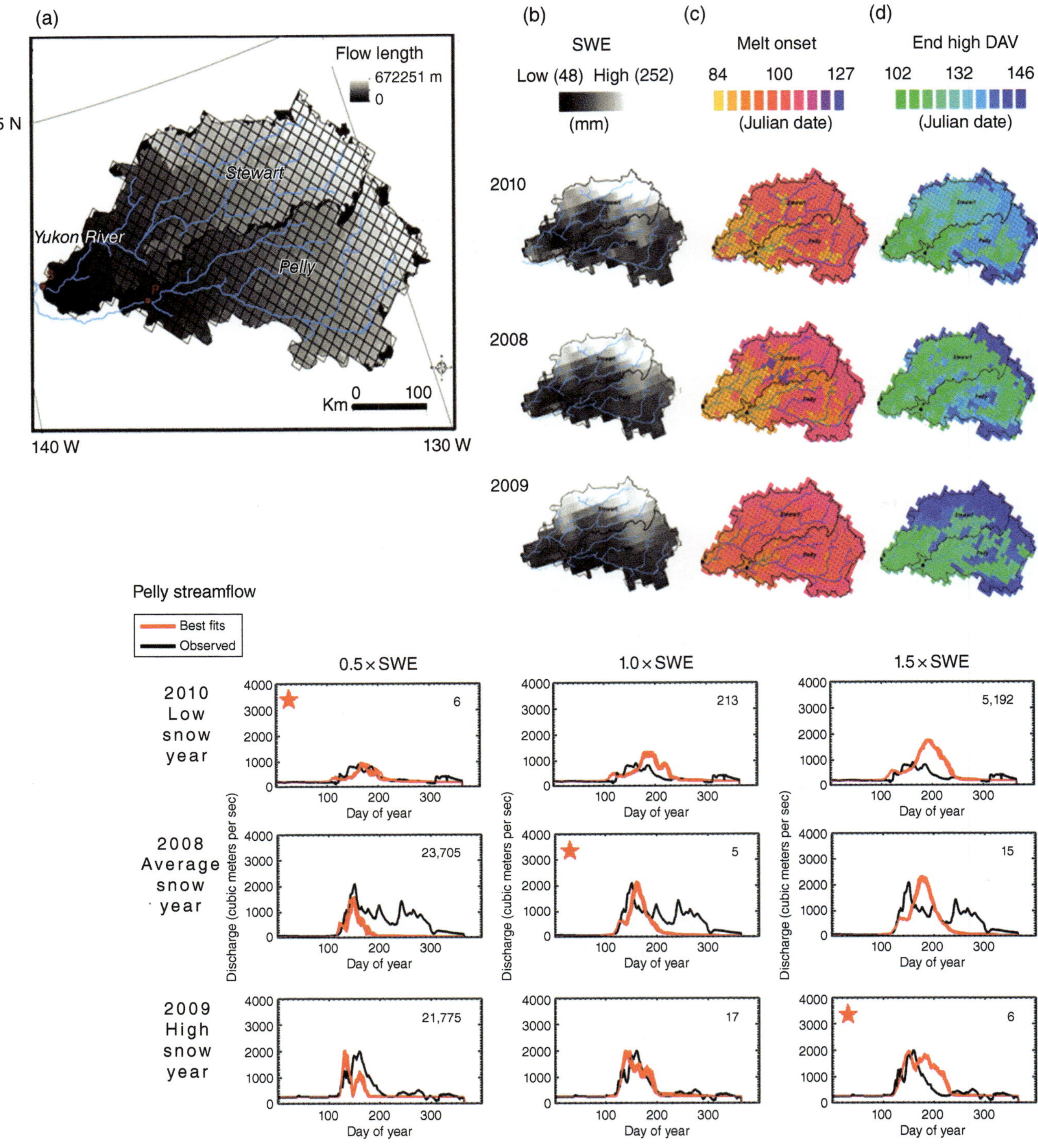

Figure 13.8 Using remotely sensed snowmelt timing to model streamflow. (a) Flow length map showing the distance (m) that meltwater flows from the pixel centroid to the channel and ultimately to the mouth with lighter colors indicating a longer path length. (b) SWE from AMSR-E level 3 data, (c) melt onset date from AMSR-E T_b and DAV thresholds, (d) end of high DAV (melt-refreeze period) date from T_b and DAV thresholds, (e) hydrograph sensitivity to variations in SWE. Lines depict observed streamflow (thick black line) and the modeled 10 best fit streamflows (thin red lines) with different SWE inputs for high, low, and average snow years. The number in the upper right of each plot is the difference in volume (cubic meters per second) estimated by the model between days 50 and 180 compared to the actual volume from the gauge data. Lower values indicate the model has less error and is a closer match to actuality. The best fitting simulation of the three SWE inputs is demarcated with a star. (Modified from *Ramage and Semmens* [2012].)

which may occur for very high polar areas such as the Russian High Arctic. For most sub-Arctic basins this will not be an issue. For instance, in the Yukon River Basin (Alaska/Canada) for which there are illustrative examples in this chapter, the overpass times for AMSR-E are generally 3:30 and 13:30 PST and for SSM/I around 8:30 and 18:30 PST. Further, data gaps in satellite observations affect the ability to calculate DAV and requires significant data interpolation in order to be used, such as with the earlier (1978–1987) Scanning Multi-channel Microwave Radiometer (SMMR). In addition, it can be difficult to differentiate between wet ground and wet snow cover, thus in areas where snow cover extent is variable the results should be compared with auxiliary information such as MODIS snow cover extent to determine whether the signal is from snow or soil.

13.9. CONCLUSIONS

Detecting wet snow and determining melt onset and melt-refreeze timing is well established with past (AMSR-E) and current (SSM/I) passive microwave sensors through the computation of diurnal amplitude variations (DAV). The spring melt-refreeze period, a time of high DAV, is linked to significant spring events (freshet, snowmelt runoff, green-up, carbon release, and fire potential) in sub-Arctic snowmelt-dominated basins, and snowmelt timing in midlatitude, western United States is increasingly important. The importance of the melt-refreeze cycle for sub-Arctic and midlatitude regions cannot be understated, having consequences for discharge and related ecological processes. For instance, the remote sensed melt timing is useful for monitoring melt progression through a basin as well as modeling streamflow in areas lacking hydrometeorological instrumentation.

Measuring changes to the critical spring melt-refreeze transition period over large, remote areas lacking comprehensive instrumentation is a pressing need for studying terrestrial hydrological systems—one that can be met using remotely sensed data enabling the calculation of DAV. An advantage to the DAV approach is its independence from drift or absolute measurement error because it is the relative difference between day and night observations. As such, there is a need for continued satellite missions to support computation of DAV, which requires high temporal resolution to maintain multiple observations per day. Without this information it is difficult to understand and assess the interconnections of related processes (physical, biological, and ecological), which often interact on short time scales. Establishing this understanding at small time scales is important for understanding a changing climate and its potential consequences.

ACRONYMS

T_b	brightness temperature (Kelvin, K)
DAV	diurnal amplitude variation (Kelvin, K)
MOD	melt onset date (Julian day of the year)
EHD	end of high DAV/end of melt-refreeze (Julian day of the year)
SWE	snow water equivalent (meters or millimeters)
SCE	snow cover extent (percent)
SSM/I	Special Sensor Microwave/Imager Satellite
AMSR-E	Advanced Microwave Scanning Radiometer–Earth Observing System
MODIS	Moderate Resolution Imaging Spectroradiometer
EASE-Grid	Equal Area Scalable Earth-Grid

ACKNOWLEDGMENTS

We appreciate data and model output from multiple sources: AMSR-E and SSM/I data were provided by the National Snow and Ice Data Center. Meteorological data were provided by NOAA National Climate Data Center and USDA SCAN and SNOTEL programs. Hydrological data were provided by the USGS National Water Information System and Environment Yukon. We would like to acknowledge financial support from Lehigh University and NASA Headquarters under the NASA Earth and Space Science Fellowship—Grant NNX10AP14H. One of us (KAS) received an AGU Travel Grant to attend the AGU Chapman Conference on Remote Sensing of the Terrestrial Water Cycle.

REFERENCES

Abdalati, W., and K. Steffen (1995), Passive microwave-derived snow melt regions on the Greenland ice sheet, *Geophys. Res. Lett.*, *22*(7), 787–790.

Abdalati, W., and K. Steffen (1997), Snowmelt on the Greenland ice sheet as derived from passive microwave satellite data, *J. Climate*, *10*, 165–175.

Apgar, J. D. (2007), Understanding the timing and variation of snowmelt in subarctic heterogeneous terrain using passive microwave AMSR-E observations, Master's thesis, Lehigh University, Bethlehem, Penna.

Apgar, J. D., J. M. Ramage, R. A. McKenney, and P. Maltais (2007), AMSR-E algorithm for snowmelt onset detection in sub-arctic heterogeneous terrain, *Hydrol. Process.*, *21*(12), 1587–1596.

Armstrong, R. L., K. W. Knowles, M. J. Brodzik, and M. A. Hardman (1994), DMSP SSM/I Pathfinder Daily EASE-Grid Brightness Temperatures [1988–2010], Nat. Snow and Ice Data Center, Boulder, Colo.

Ashcroft, P., and F. Wentz (2006), AMSR-E/aqua L2A global swath spatially-resampled brightness temperatures V002, [2003–2010], National Snow and Ice Data Center, Boulder, Colo., digital media, updated daily.

Bartsch, A., R. A. Kidd, W. Wagner, and Z. Bartalis (2007), Temporal and spatial variability of the beginning and end of daily spring freeze/thaw cycles derived from scaterrometer data, *Remote Sens. Environ.*, *106*, 360–374.

Bennett, K. E., J. E. Cherry, C.A. Hiemstra, L. Hinzman, and M. Leonawicz. In prep. Estimating, verifying and predicting snow cover depletion in Boreal watersheds of Interior Alaska from remote sensing, in situ measurements, and statistical modeling.

Brabets, T. P., B. Wang, and R. H. Meade (2000), Environmental and hydrologic overview of the Yukon River Basin, Alaska and Canada, USGS Water Resourc. Investig. Rep. 99–4204, Anchorage, Alaska. Available at http://ak.water.usgs.gov/Publications/pdf.reps/wrir99.4204.pdf

Brodzik, M. J., and K. W. Knowles (2002), EASE-Grid: A versatile set of equal-area projections and grids, in *Discrete Global Grids*, edited by M. Goodchild, Nat. Center for Geographic Inform. and Analysis, Santa Barbara, Calif.

Center for Snow and Avalanche Studies (CSAS) (2012), Archival data from Senator Beck Basin, available at http://snowstudies.org/data1.html.

Chang, A. T. C., P. Gloersen, T. Schmugge, T. T. Wilheit, and H. J. Zwally (1976), Microwave emission from snow and glacier ice, *J. Glaciol. 16*, 23–39.

Drobot, S., and M. Anderson (2001), An improved method for determining melting onset dates over Arctic sea ice using scanning multichannel microwave radiometer and special sensor microwave/imager data, *J. Geophys. Res. 106*, 24,033–24,049.

Fleming, S. W. (2005), Comparative analysis of glacial and nival streamflow regimes with implications for lotic habitat quantity and fish species richness, *River Res. Appl. 21*, 363–379.

Hall, D. K., G. A. Riggs, and V. V. Salomonson (2006). MODIS/Terra snow cover daily L3 global 500 m grid V005, 2000–2010, Boulder, Colorado USA: Nat. Snow and Ice Data Center, Boulder, Colo, digital media, updated daily.

Hodgkins, G. A. (2009), Streamflow changes in Alaska between the cool phase (1947–1976) and the warm phase (1977–2006) of the Pacific Decadal Oscillation: The influence of glaciers, *Water Resourc. Res. 45*, W06502, doi:10.1029/2008WR007575.

Jansson, P., R. Hock, and T. Schneider (2003), The concept of glacier storage: A review, *J. Hydrol.*, *282*, 116–129.

Kidd, R., K. Scipal, Z. Bartalis, and W. Wagner (2004), A diurnal difference indicator for freeze-thaw monitoring from Ku band scatterometer applied within the Siberia II project, Geoscience and Remote Sensing Symposium, 2004, IGARSS '04, *Proc. IEEE Int.*, *3*, 1671–1674.

Kimball, J. S., K. C. McDonald, S. Frolking, and S. W. Running (2004), Radar remote sensing of the spring thaw transition across a boreal landscape, *Remote Sens. Environ, 89*(2), 163–175.

Kopczynski, S., J. Ramage, D. Lawson, S. Goetz, E. Evenson, J. Denner, and G. Larson (2008), Passive microwave (SSM/I) satellite predictions of valley glacier hydrology, Matanuska Glacier, Alaska, *Geophys. Res. Lett. 35*, L16502, doi:10.1029/2008GL034615.

Li, D., M. Durand, and S. A. Margulis (2012), Potential for hydrologic characterization of deep mountain snowpack via passive microwave remote sensing in the Kern River basin, Sierra Nevada, USA, *Remote Sens. Environ.*, *125*, 34–48.

Liston, G. E., and C. A. Hiemstra (2011), The changing cryosphere: Pan-arctic snow trends (1979–2009), *J. Climate*, *24*: 5691–5712.

Monahan, P., and J. Ramage (2010), AMSR-E melt patterns on the Southern Patagonia Icefield, *J. Glaciol.*, *56*(198), 699–708.

Mote, T. L., and M. R. Anderson (1995), Variations in snowpack melt on the Greenland ice sheet based on passive microwave-measurements, *J. Glaciol.*, *41*, 51–60.

Mote, T. L., M. R. Anderson, K. C. Kuivinen, and C. M. Rowe (1993), Passive microwave-derived spatial and temporal variations of summer melt on Greenland ice sheet, *Ann. Glaciol.*, *17*, 233–238.

Nghiem, S. V., K. Steffen, R. Kwok, and W. Y. Tsai (2001), Detection of snowmelt regions on the Greenland ice sheet using diurnal backscatter change, *J. Glaciol.*, *47*(159), 539–547.

Nghiem, S. V., Steffen, K. Neumann, G. and R. Huff (2005), Mapping of ice layer extent and snow accumulation in the percolation zone of the Greenland ice sheet, *J. Geophys. Res.*, *110* (F02017), doi:10.1029/2004JF000234.

Ramage, J. M., and B. L. Isacks (2002), Determination of melt-onset and refreeze timing on southeast Alaskan Icefields using SSM/I diurnal amplitude variations, *Ann. Glaciol.*, *34*, 391–398.

Ramage, J. M., and B. L. Isacks (2003), Interannual variations of snowmelt and refreeze timing on southeast-Alaska icefields, U.S.A, *J. Glaciol.*, *49*(164), 102–116.

Ramage, J. M., and K. A. Semmens (2012), Reconstructing snowmelt runoff in the Yukon River basin using the SWEHydro model and AMSR-E observations, *Hydrol. Process.*, *26*, 2563–2572.

Ramage, J. M., R. A. McKenney, B. Thorson, P. Maltais, and S. E. Kopczynski (2006), Relationship between passive microwave-derived snowmelt and surface-measured discharge, Wheaton River, Yukon Territory, Canada, *Hydrol. Process.*, *20*(4), 689–704.

Ramage, J. M., J. D. Apgar, R. A. McKenney, and W. Hanna (2007), Spatial variability of snowmelt timing from AMSR-E and SSM/I passive microwave sensors, Pelly River, Yukon Territory, Canada, *Hydrol. Process.*, *21*, 1548–1560.

Steffen, K., W. Abdalati, and J. Stroeve (1993), Climate sensitivity studies of the Greenland ice sheet using satellite AVHRR, SMMR, SSM/I, and in situ data, *Meteor. Atmos. Phys.*, *51*, 239–258.

Takala, M., and J. Pulliainen (2008), Detection of snow melt using different algorithms in global scale, in *Microwave Radiometry and Remote Sensing of the Environment*, Inst. of Electr. and Eletron. Eng., New York. MICRORAD, 2008. DOI:10.1109/MICRAD.2008.4579496.

Tedesco, M. (2007), Snowmelt detection over the Greenland ice sheet from SSM/I brightness temperature daily variations, *Geophys. Res. Lett.*, *34*, L02504.

Tedesco, M., M. Brodzik, R. Armstrong, M. Savoie, and J. Ramage (2009), Pan arctic terrestrial snowmelt trends from spaceborne passive microwave data and correlation with the AO, *Geophys. Res. Lett.*, *36*, L21402.

Ulaby, F. T., R. K. Moore, and A. K. Fung (1986), *Microwave Remote Sensing: Active and Passive*, Vol. *III: From Theory to Applications,* Artech House, Dedham, Mass.

Yan, F., J. Ramage, and R. McKenney (2009), Modeling of high-latitude spring freshet from AMSR-E passive microwave observations, *Water Resourc. Res.*, *45*, W11408.

14

Changes in Snowpacks of Canadian Prairies for 1979–2004 Detected from Snow Water Equivalent Data of SMMR and SSM/I Passive Microwave and Related Climatic Factors

Thian Yew Gan,[1,2] **Roger G. Barry,**[1] **and Adam K. Gobena**[2]

14.1. INTRODUCTION

Recent studies on detecting changes in snowpacks have been mainly based on snow cover [e.g., *Brown and Robinson,* 2011; *Dyer and Mote,* 2007; *Hantel and Hirtl-Wielke,* in press; *Frei and Robinson,* 1999; *Brown and Goodison,* 1996] or snow depths rather than on snow water equivalent (SWE), the amount of water a snowpack would give after complete melting, which essentially is the snow depth multiplied by the average snow density. SWE is more useful than snow cover or snow depth for the management and planning of snow-dominated water resources. As a snowpack ages, or as snowfall continues, which increases its overburden load, it undergoes compaction such that ice crystals settle and metamorphose. Due to compaction, snow depth will decrease while snow density will increase, but the SWE of a snowpack should theoretically remain unchanged unless there is new snowfall or the snowpack is reduced by sublimation (direct transition from ice to water vapor) or melting. Conversely, new snowfall (sublimation) will cause a thicker (thinner) snow cover or increased (decreased) SWE, even though the snow cover area (SCA) could remain unchanged or could change marginally. In an open, sparse vegetation landscape, such as the Canadian Prairies (CP), wind activity can also be a prominent control on the distribution of SWE. Even though SCA is useful for water resources management, SCA itself cannot provide us

information about the amount of water stored in a snowpack.

Seasonal snow mass variations at mid to high latitudes are the largest signals in the changes of terrestrial freshwater storage each year [*Barry and Gan,* 2011; *Niu et al.,* 2007]. However, we cannot rely on snow gauges or ground-based, snow course measurements to accurately estimate the amount of SWE at the regional scale since snowpack generally exhibits substantial spatial variability. Other than being point measurements, it is well known that snow gauges, even when mounted with shields such as the Nipher shield, suffer from undercatch problems, especially under windy conditions. For example, the catch ratios of the Wyoming Fence to World Meteorological Organization–Double Fence Inter-Comparison Reference (WMO-DFIR) gauges were 89% and 87% at Regina (Canada) and Valdai (Russia), respectively. *Yang et al.* [2000] found that the mean catch of snowfall for the U.S. 8-inch gauge at Valdai was 44%. For the Tretyakov and Hellmann gauges, the mean catch of snowfall was 63%–65% and 43%–50%, respectively, at the northern test sites of the WMO experiment.

Another major problem concerning snow gauge data is the lack of representative information on the snowpack density. From analyzing 848 stations across Canada that were reporting daily snowfall and daily precipitation from October 2004 to February 2005, *Cox* [2005] found that the histogram of the frequency of snowfall events by snow depth/SWE ratio is dominated by a spike at the 10 : 1 ratio, a bias caused by the 10:1 approximation being used not just in forecasting but in place of actual snowpack measurements. Recognizing the inadequacy of this 10 : 1 ratio, for climate stations only equipped with a snow ruler,

[1] *National Snow and Ice Data Center (NSIDC), University of Colorado at Boulder, Boulder, Colorado, USA*

[2] *Department of Civil & Environmental Engineering, University of Alberta, Edmonton, Canada*

Remote Sensing of the Terrestrial Water Cycle, Geophysical Monograph 206. First Edition. Edited by Venkat Lakshmi.
© 2015 American Geophysical Union. Published 2015 by John Wiley & Sons, Inc.

Mekis and Hopkinson [2004] proposed an alternative for more accurately estimating the SWE at a station based on a factor called the snow water equivalent adjustment factor (SWEAF), which can range from 0.6 to 1.8. SWEAF generally increases with latitude. The province of British Columbia of Canada tends to have SWEAF less than 1.

In view of the aforementioned problems associated with snow gauge data that are point measurements, spatially distributed SWE data have been retrieved from the brightness temperature (T_B) in (Kelvins) of passive microwave remote sensing platforms such as the Scanning Multichannel Microwave Radiometer (SMMR) from 25 October 1978, to 20 August 1987, and the Special Sensor Microwave/Imager (SSM/I) since 7 September 1987. The SMMR sensor flew on NASA's *Nimbus 7* while SSM/I are mounted on the Defense Meteorological Satellite Program (DMSP) satellites of the United States Since May, 2002, T_B's retrieved from the Advanced Microwave Scanning Radiometer-EOS (AMSR-E) sensor aboard the *Aqua* satellite have also been used, and data from this sensor can estimate SWE higher than that of SSM/I [*Kelly*, 2009]. The frequencies (resolution) of SSM/I are 6.6 GHz (150 km), 19 GHz (25 km), 22 GHz (25 km), 37 GHz (25km) to 85 GHz (12.5 km), while that of AMSR-E are 6.9 GHz (50 km) to 89 GHz (5 km). These T_B values are either horizontally (H) or vertically (V) polarized.

There is a conflation between the frequency specifications of the SMMR, SSM/I, and AMSR-E gridded products, and the actual spatial measurements controlled by the instantaneous field of view (IFOV) of the respective sensors. However, the National Snow and Ice Data Center (NSIDC) resamples all the T_B's to the 25 km Equal Area Scalable Earth Grid (EASE-Grid). A good summary of these instruments can be found in *Kelly et al.* [2003]. Field measurements at large scales are logistically difficult and expensive, but there have been some intensive field studies conducted to validate remotely sensed SWE with observation such as the Cold Land Processes Experiment (CLPX) [*Cline et al.*, 2003], the SnowSTAR 2002 Transect [*Shi et al.*, 2009] and others [e.g., *Langlois et al.*, 2010; *Yueh et al.*, 2009; *Derksen et al.*, 2003; *Mote et al.*, 2003]. Using measurements from snow course transects in the former Soviet Union, *Armstrong and Brodzik* [2002] found a general tendency for nearly all of the retrieval algorithms e.g., the horizontal-based polarization algorithm of *Chang et al.* [1987] and the vertical-based counterpart of *Goodison* [1989] of passive-microwave data to underestimate SWE, especially as the forest-cover density begins to exceed 30%–40%. *Takata et al.* [2011] assessed NSIDC's global monthly SWE data (a sample size of 11,730) over Russia and the former Soviet Union with the independent INTAS-SCCONE (International Association for the promotion of co-operation with scientists from the New Independent States of the former Soviet Union —

Snow Cover Changes Over Northern Eurasia) snow course observations from the period of 1979–2000 for March only. On the basis of the snow course observations, they found a 12 mm bias in the NSIDC's data, which means a bias of 10% or less if the mean SWE is 120 mm or higher.

To analyze regional-scale SWE trends, it will not be feasible to rely on limited observed SWE measurements such as snow course data but more on passive microwave SWE data, even though such SWE products are subjected to uncertainties caused by spatiotemporal variations in snowpack properties, to different diurnal and seasonal metamorphism processes that will cause variations in T_B [*Mätzler*, 1987; *Tait and Armstrong*, 1996; *Pulliainen*, 2006; *Pardé et al.*, 2007; *Durand et al.*, 2008; *Brown et al.*, 2010], the presence of vegetation cover, the fractional IFOV presence of lakes (frozen or unfrozen) or wet snow, and the degree of heterogeneity caused by complex mountain terrains. At regional scale, reasonably accurate SWE can generally be estimated from passive microwave data [*Tait*, 1998; *Pulliainen and Hallikainen*, 2001; *Singh and Gan*, 2000; *Gan et al.*, 2009], as long as the regions where SWE retrieved from T_B data are not dominated by thick vegetation, which could complicate the microwave emissions of the snowpack, albeit SWE retrieved from the T_B data are adjusted for the effect of forest cover [see equation 14.3] [*Chang et al.*, 1996; *Derksen et al.*, 2003]. *Dong et al.* [2005] and *Sun et al.* [2004] have characterized the errors of SWE data, and *Foster et al.* [2005] had attempted to improve their algorithm with these errors in mind. In view of this limitation of passive microwave data, we have limited the data analysis to the CP where the vegetation is grassland in southern CP and boreal forest in northern CP [*Derksen et al.*, 2005] because of its semiarid climate, and so the SWE data retrieved from T_B data for the CP should be more accurate than regions dominated by dense vegetation, such as British Columbia of Canada or the United States. Based on 26 years of SWE retrieved from T_B data of SMMR and SSM/I for the CP, and the nonparametric, Mann-Kendall's approach to minimize the possible effects of uncertainties expected from remotely sensed SWE data, we analyzed the SWE trends of CP.

According to *Lemke et al.* [2007], the observed Northern Hemisphere snow cover since the 1920s had remained quite steady until around the 1980s, for since then a significant decrease in snow cover was observed. Various researchers detected substantial decreases in snow cover extent and snow depth over North America (NA) [*Dyer and Mote*, 2007] and the Northern Hemisphere [*Brown and Robinson*, 2011; *Brown and Mote*, 2009; *Frei and Robinson*, 1999]. *Dyer and Mote* [2007] demonstrated positive trends in the frequency of snow ablation events during March but a significant negative trend in May, which again indicates an earlier onset of ablation. They attributed the March trend to higher fluxes of sensible heat into the snowpack caused by warmer air, an increase in the frequency of dry moderate air masses over central Canada, and a decrease in dry polar air masses. On the basis of observed 1 April SWE, *Pierce et al.* [2008] found that about half of the SWE/*P* (*P* = water year-to-date precipitation) reductions observed in the western United States from 1950 to 1999 are the result of climate changes forced by anthropogenic greenhouse gases, ozone, and aerosols. With about 26 years of continuous SWE data from SMMR (late 1978 to 1987) and SSM/I (summer of 1987 until 2004), it is feasible to perform a trend analysis of SWE over the CP for the winter and spring seasons to detect possible changes.

From the perspective of climatic change, relative to the 1906–1970 surface temperature of North America, both observed and 58 cases of GCMs' modeled data show drastic increase in the surface temperature of NA in recent decades [*Forster et al.*, 2007]. At higher elevations, where it is still sufficiently cold for snowfall to occur, higher temperatures may increase the amount of snowfall because of the nonlinear increase in saturation vapor pressure with temperature. However, in lower elevations, a significant increase in the surface temperature probably means more rainfall at the expense of snowfall, and hence less snow on the ground. With this statement of problems in mind, the research objectives are given in Section 14.2, description of data in Section 14.3, Kendall's trend analysis in Section 14.4, discussions of results in Section 14.5, influence of climate anomalies in Section 14.6, SWE-climate data relationships in Section 14.7, and conclusions in Section 14.8.

14.2. RESEARCH OBJECTIVES

With the background information given in Section 14.1, this study has the following objectives:

1. To analyze monthly monotonic trends for the November–April SWE over the CP using the 1979–2004 SWE data retrieved from the passive microwave, SMMR/SSMI brightness temperature data.

2. To compute trend magnitudes, trend homogeneity, and the spatial distributions of trends and variability in SWE across the CP from 1979 to 2004.

3. To perform principal component analysis (PCA) on SWE and to correlate PCAs of SWE with climate anomalies.

4. To relate SWE to major climate variables, and from the results, identify possible cause(s) for the detected changes of the snowpacks of CP.

14.3. SWE RETRIEVED FROM SMMR AND SSM/I PASSIVE MICROWAVE DATA

On the basis of volumetric scattering, which is the dominant loss mechanism for microwave radiation greater than 15 GHz incident on a snowpack, it is possible to empirically relate the brightness temperature (T_B) of a certain frequency and polarization (H or V) to the SWE of the snowpacks [e.g., *Armstrong et al.*, 2001; *Gan et al.*, 2009]. The NSIDC at the University of Colorado, Boulder, provides such passive microwave T_B data at the EASE-Grid format. The 1979–1987 SMMR SWE data of NSIDC were retrieved from equation 14.1 [*Chang et al.*, 1987]:

$$\text{SWE}(\text{mm}) = 4.77(T_{B,18\text{H}} - T_{B,37\text{H}}), \qquad (14.1)$$

where $T_{B,18\text{H}}$ and $T_{B,37\text{H}}$ are the horizontally polarized T_B at 18 and 37 GHz, respectively. The 1987–2004 SSM/I SWE data of NSIDC were retrieved from equation 14.2 [*Armstrong and Brodzik*, 2001]:

$$\text{SWE}(\text{mm}) = 4.77(T_{B,19\text{H}} - T_{B,37\text{H}} - 5), \qquad (14.2)$$

which is slightly different from that of equation 14.1 because of sensitivity differences between SMMR and SSM/I sensors in detecting shallow snow. It should be noted that the F13 satellite carrying the SSM/I sensor had failed in 2007 [*P. Gibbons* of NSIDC, personal communication], and beside a slight difference in frequencies between the two sensors, there are other inhomogeneities identified between them [*Jezek et al.*, 1993], and such data are typically limited to retrieve SWE from snowpack up to 0.5 m deep [*Foster et al.*, 2005]. The daily SWE is adjusted for the surface forest cover using MODIS (Moderate Resolution Imaging Spectroradiometer) land cover data of Boston University, BU-MODIS [*NSIDC*, 2005] so that

$$\text{SWE}(\text{mm}) = \frac{\text{SWE}}{1 - \text{Forest\%}}. \qquad (14.3)$$

Any pixels with forest percentages higher than 50% are set to the 50% threshold, so that the forest correction by equation 14.3 is limited to a maximum factor of 2. *Derksen et al.* [2005] suggested that forest stem volume and canopy closure may impact microwave emission more than forest cover. Further, to ensure snowpacks are detectable by passive microwave data, SWE less than 7.5 mm is considered unreliable and set to zero [*Armstrong and Brodzik*, 2002]. From northern to southern CP, between January, February, and March, the distribution of SWE is fairly uniform with a mean SWE of about 50 mm, but by April, the mean SWE drops to below 40 mm because of melting. Given that SMMR and SSM/I data are at 25 × 25 km resolution, we expect the microscale spatial variability of snowpack to be mostly averaged out, and so basically we can only analyze meso- to regional-scale variability of snowpack from these data. Furthermore, some studies, such as that of *Clifford* [2010], found errors in the SWE record from passive microwave observations suggesting systematic biases related to grain size, not SWE, e.g., in northern Siberia, the approach of *Chang et al.* [1987], because thin snowpack and strong thermal gradient promoted large crystals that enhanced scattering, grossly overestimated SWE. Similar effect has been found in the shallow snowpack of the high plateau regions of Wyoming, which contains large amounts of depth hoar, with grain sizes up to 5 mm, which yield the coldest T_B observed across the entire basin [*Josberger et al.*, 1996].

As explained in Section 14.1, SWE retrieved from passive microwave data of 25 km resolution are subjected to errors, such as a possible mixture of deep snow on north-facing slopes but almost snow free on south-facing slopes, particularly in mountainous areas with large topographic variability, or mixed microwave emissions from trees, snow canopy, and ground surface in forested areas. Pixels of melting snow or wet snowpack could return low or zero SWE values [*Mätzler*, 1994]. *Singh and Gan* [2000] employed screening procedures to eliminate potentially erroneous data from SWE retrieved via multivariate regression algorithms for the Red River Basin.

14.4. NONPARAMETRIC, MANN-KENDALL'S TEST AND TREND MAGNITUDE (β)

There are various parametric methods available to detect trends, e.g., the linear least squares [e.g., *Dyer and Mote*, 2007]. However, the nonparametric, Mann-Kendall test [*Kendall*, 1975; *Mann*, 1945] for testing trends in seasonal time series should be more robust because it can handle nonnormality, censoring (data reported as values "less than"), missing values, and seasonality [*Hirsch et al.*, 1982]. To safeguard against nonrepresentative results, only data sets with few missing values were tested. According to *Berryman et al.* [1988], who gave a comprehensive review on the practical limits of nonparametric tests for monotonic trends, among 12 types of such tests, Kendall and Spearman's tests have the highest asymptotic efficiency of 0.98. Kendall's statistic S_k for season k (where $k = 1, 2, \ldots . q$) is given as

$$S_k = \sum_{j=1}^{n-1} \sum_{i=j+1}^{n} \mathrm{sgn}(X_{jk} - X_{ik}), \qquad (14.4)$$

where n is the sample size, S_k tends to be Gaussian normal asymptotically, and

$(X_{jk} - X_{ik})$	$\mathrm{sgn}(X_{jk} - X_{ik})$
>0	+1
=0 (tie)	0
<0	−1

Using a two-sided test at a significant level α, the trend is statistically significant at α if the absolute value of the standard normal variate, $|Z_k| > Z_{\alpha/2}$, where Z_k is given as

$$Z_k = \begin{cases} \dfrac{S_k - 1}{\sigma_k} & \text{if } S_k > 0 \\[2mm] 0 \quad \text{(tie)} & \text{if } S_k = 0 \\[2mm] \dfrac{S_k + 1}{\sigma_k} & \text{if } S_k < 0 \end{cases} \qquad (14.5)$$

and σ_k is the standard deviation of S_k. Details of Kendall's test are given in *Gan* [1995] and *Hirsch et al.* [1982]. Since S_k is not based on the magnitudes of individual SWE, it is expected to be more robust than say, the simple linear regression. In addition, results of the nonparametric approach should not be affected by SWE retrieved from passive microwave data of two different sensors, SMMR [equation 14.1] before 1987 and SSM/I [equation 14.2] after 1987. There have been past studies conducted on the T_B inhomogeneity between SMMR and SSM/I (*Jezek et al.*, 1993; *Derksen et al.*, 2003). To circumvent this problem of inhomogeneity between SMMR and SSM/I data, we did the trend analysis of SWE data for CP involving both SMMR (data before 1987) and SSM/I data (data between 1987 and 2007) using the nonparametric approach of Mann-Kendall, which does not depend on the underlying probability distribution of the SWE data, because it simply counts the recurrence frequency. Furthermore, equation 14.4 shows that S_k is not based on the magnitudes of individual SWE but on the number of times SWE has been increasing minus the number of times SWE has been decreasing as we move, for a particular month, of one year to the next until the study period (26 years) is completed. On the other hand, because the metamorphism of the snowpack could change from year to year, which means that grain size and other snowpack properties could also change from year to year, and since changes to T_B also depend on snowpack properties (not just the SWE), the trend estimated by the nonparametric algorithm may still (or may not) be biased.

14.5. DISCUSSIONS OF RESULTS

The SCA over NA has been shown to increase during autumn and early winter (November–January) but decrease over early spring [*Frei and Robinson*, 1999; *Dyer and Mote*, 2006], which indicates an earlier onset of the spring snowmelt. In terms of SWE, both statistically significant decreasing and increasing trends detected at $\alpha/2 = 0.05$, $\alpha/2 = 0.025$, and $\alpha/2 = 0.01$ for the CP are shown in Figure 14.1. Other than November, up to 45% of detected decreasing trends in SWE (February) for 1979–2004 are statistically significant at $\alpha/2 = 0.05$, which is about three times the

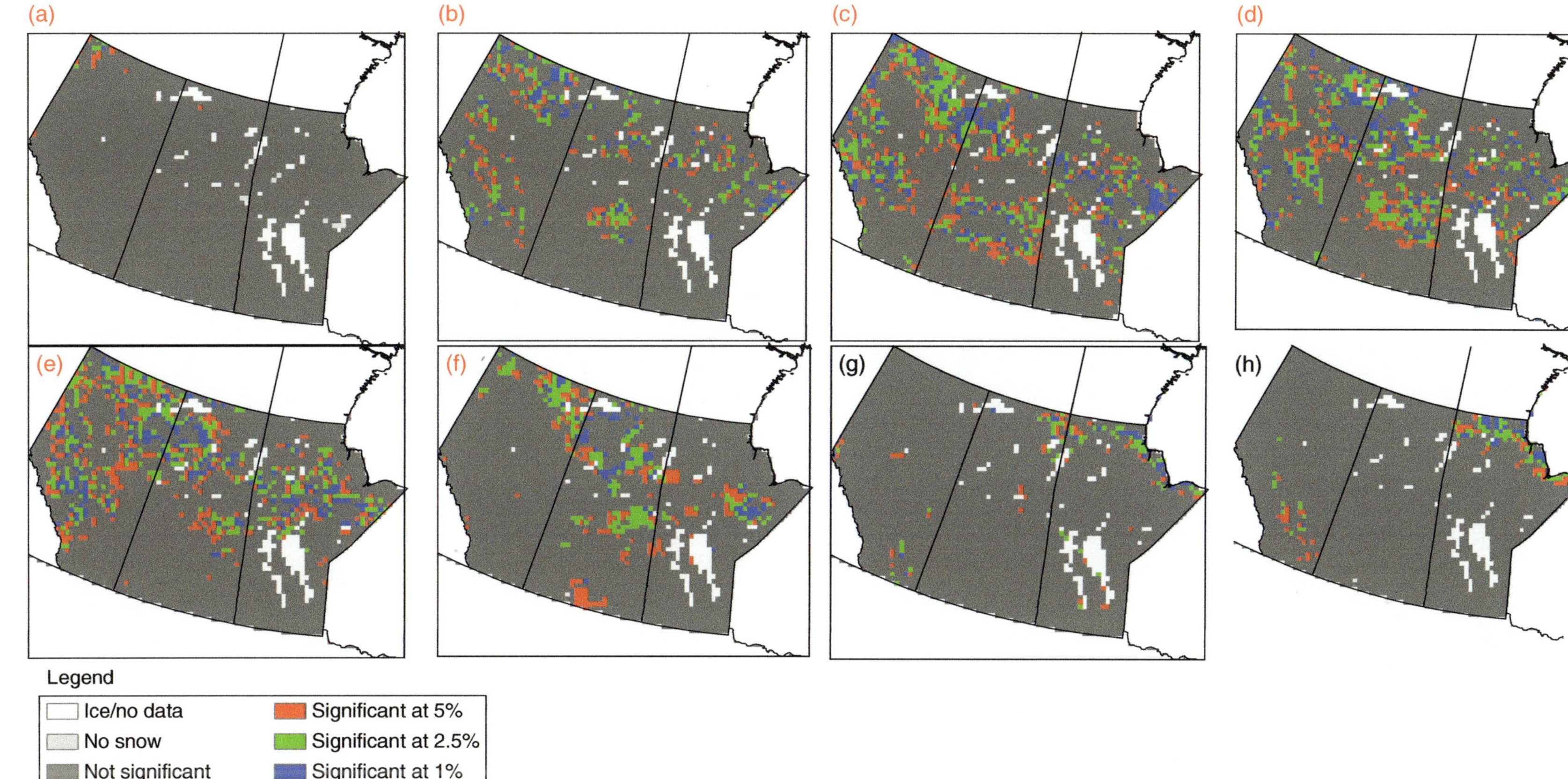

Figure 14.1 Spatial distributions of both **decreasing** [(a) November, (b) December, (c) January, (d) February, (e) March, and (f) April] and **increasing** [(g) March and (h) April] trends in SWE, the snow cover extend (light color) in Canadian Prairies (CP) for detected trends that are statistically significant at $\alpha/2=0.05$ (red), $\alpha/2=0.025$ (green) and $\alpha/2 =0.01$ (blue), respectively.

detected increasing trends in SWE (see Table 14.1 and Figure 14.2). Conversely, for the CP, significant decreasing trends in SWE are more extensive in December to March than in November and April, and most extensive in January to March. The distributions of significant decreasing trends vary from month to month without any clear pattern. In the CP, only scattered increasing trends are found near northeastern Manitoba and southern Alberta.

Simple linear regressions fitted to the mean monthly SWE of CP show mostly monthly decreasing trends in SWE (except for April) ranging from about –0.03 (November) to –1.1 (March) mm/yr (Figure 14.3). In addition, a more rigorous approach to compute the trend magnitude β_k in millimeters/year was used. It was based on an estimator extended from that proposed by *Sen* [1968] and defined as

$$\beta_k = \text{median} \frac{X_{jk} - X_{ik}}{j - i} \qquad k = 1, 2, \ldots q, \quad (14.6)$$

where $1 < i < j < n$.

The overall mean trend magnitudes estimated using equation 14.6 range from about –0.02 to –1.1 mm/yr (histograms in Figure 14.4 and boxplots in Figure 14.5), or an average of 0.63 mm/yr, while the mean (median) of negative trends alone is about –1.52 (–1.32) mm/yr (Table 14.2), respectively. These values are of smaller range than that estimated from simple regressions, which range from 0.71 to –1.1 mm/yr (Figure 14.3). Given the generally low goodness-of-fit statistics associated with each monthly simple regression shown in Figure 14.3, the former values should be more representative. At an average decreasing trend magnitude of 0.63 mm/yr, it means that the overall SWE of CP has decreased by about 16 mm or possibly more, which if expressed in terms of snow depth will be about 6–8 cm (assuming an average snow density of about 200–250 g/cm³) from 1979–2004; this could have significant impacts on the spring snowmelt of the CP. The decreasing trend magnitudes of November and April are relatively modest compared to December–March, partly because of much less snow present on the ground in November and April than in the winter months for the CP. Overall, only

Table 14.1 Results of kendall's nonparametric test on statistically significant increasing [(1)–(5)] and decreasing [(6)–(10)] trends of monthly SWE for 1979–2004 at different significant levels[a]

Month	(1) $\alpha/2$ <.95	(2) $\alpha/2$ >.95	(3) $\alpha/2$ >.975	(4) $\alpha/2$ >.99	(5) $\alpha/2$ >.995	(6) $\alpha/2$ ≥.05	(7) $\alpha/2$ <.05	(8) $\alpha/2$ <.025	(9) $\alpha/2$ <.01	(10) $\alpha/2$ <.005	Total Pixels Analyzed
Jan.	535	77	71	45	82	1099	275	333	224	274	3015
Feb.	391	101	105	66	117	819	257	336	236	456	2883
Mar.	460	95	104	54	129	1067	303	319	218	240	2989
Apr.	550	95	103	64	72	787	186	185	88	120	2252
Oct.	7	2	5	0	1	26	9	11	1	0	62
Nov.	157	16	7	3	5	113	6	5	1	0	313
Dec.	366	41	51	28	10	634	156	211	122	290	1908

[a]Significant decreasing trends are about three or more times that of statistically significant increasing trends.

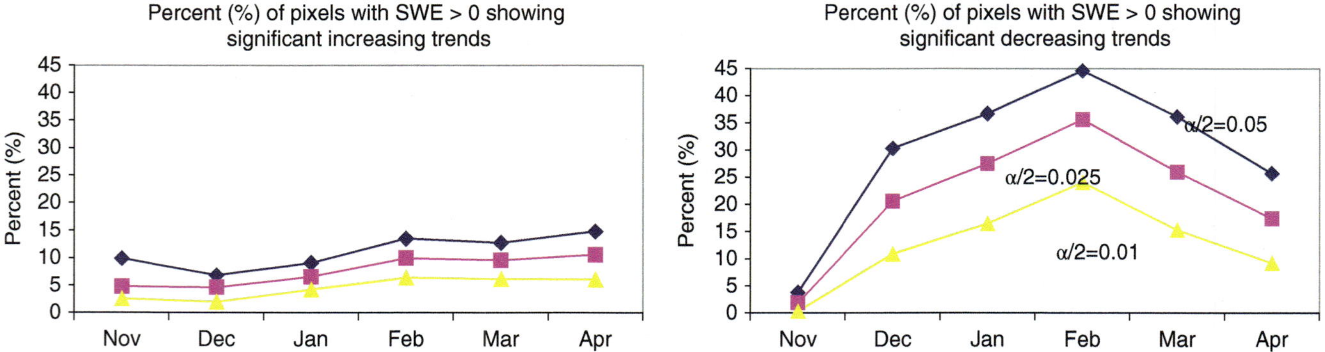

Figure 14.2 Pixels covered with snow (SWE > 0) show more frequent (in percentage, %) statistically significant decreasing trends at two-sided significant levels, $\alpha/2 = 0.05$, $\alpha/2 = 0.025$, and $\alpha/2 = 0.01$, estimated based on the nonparametric Kendall statistics (K-S) of Canadian Prairies for 1979–2004, assuming K-S to be Gaussian normally distributed.

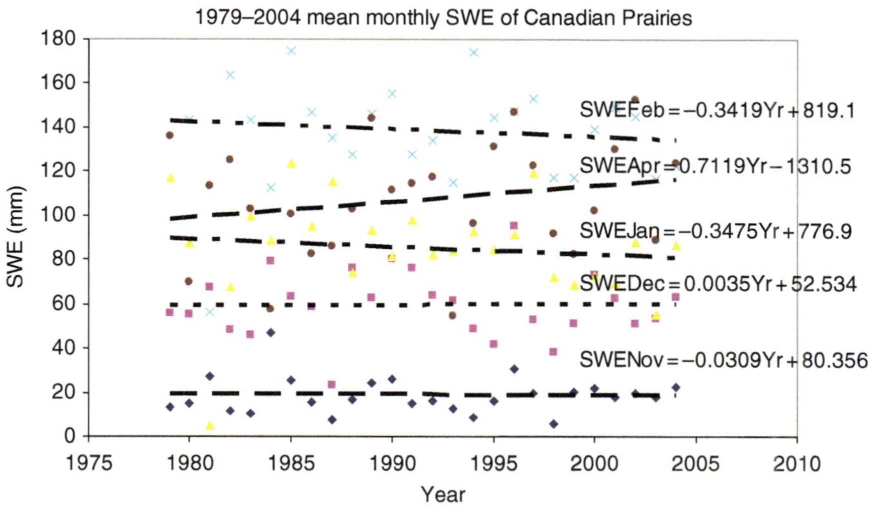

Figure 14.3 Simple linear regressions fitted to the mean monthly SWE of Canadian Prairies, showing mostly monthly overall decreasing trends (except for April), of trend magnitudes ranging from about +0.7 to −1.1 mm/yr between 1979 and 2004.

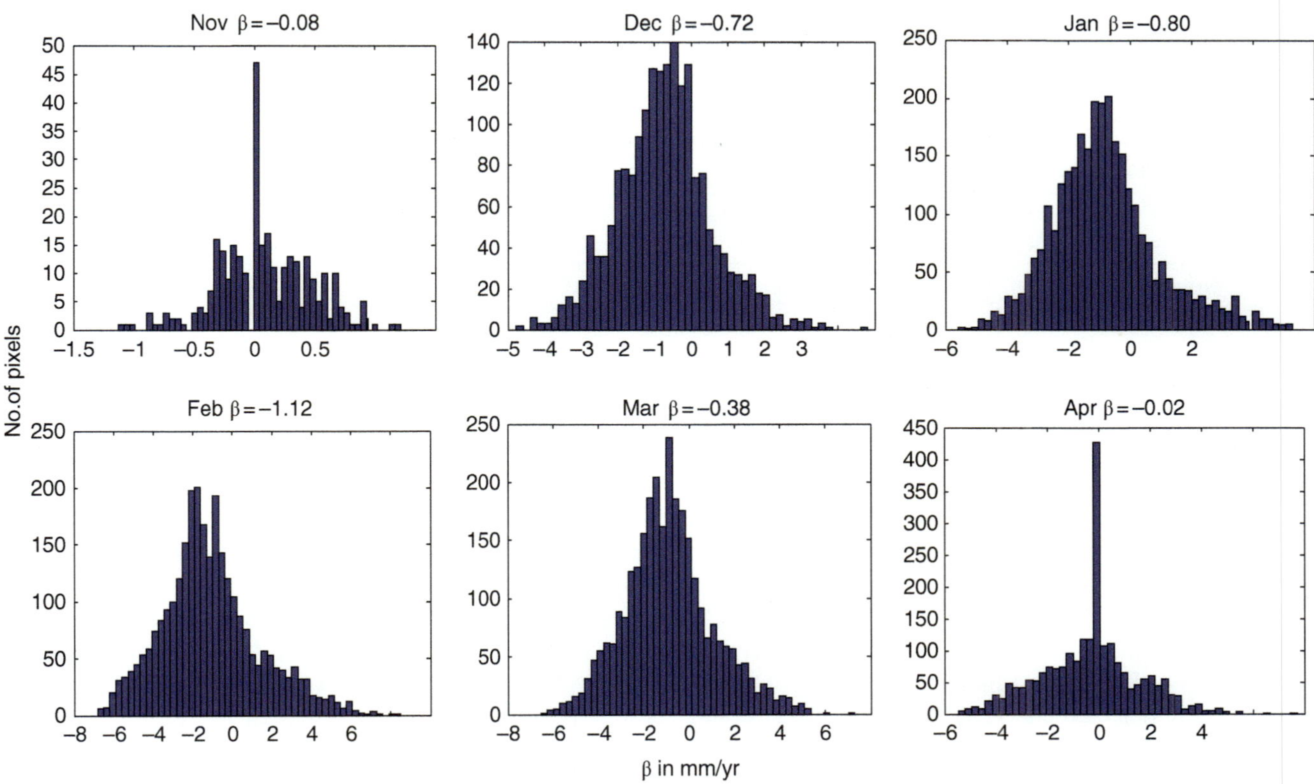

Figure 14.4 Histograms and mean of monthly trend magnitudes (β in mm/yr) of monthly SWE of SMMR and SSM/I passive microwave data of Canadian Prairies from 1979 to 2004 for November–April showing more negative than positive trends.

scattered increasing trends are detected in the CP, but in isolated cases there had been outliers where increasing trends in SWE are as high as 8 mm/yr as shown by the boxplots of β (Figure 14.5), which could be partly attributed to the effects of climatic anomalies discussed in Section 14.6.

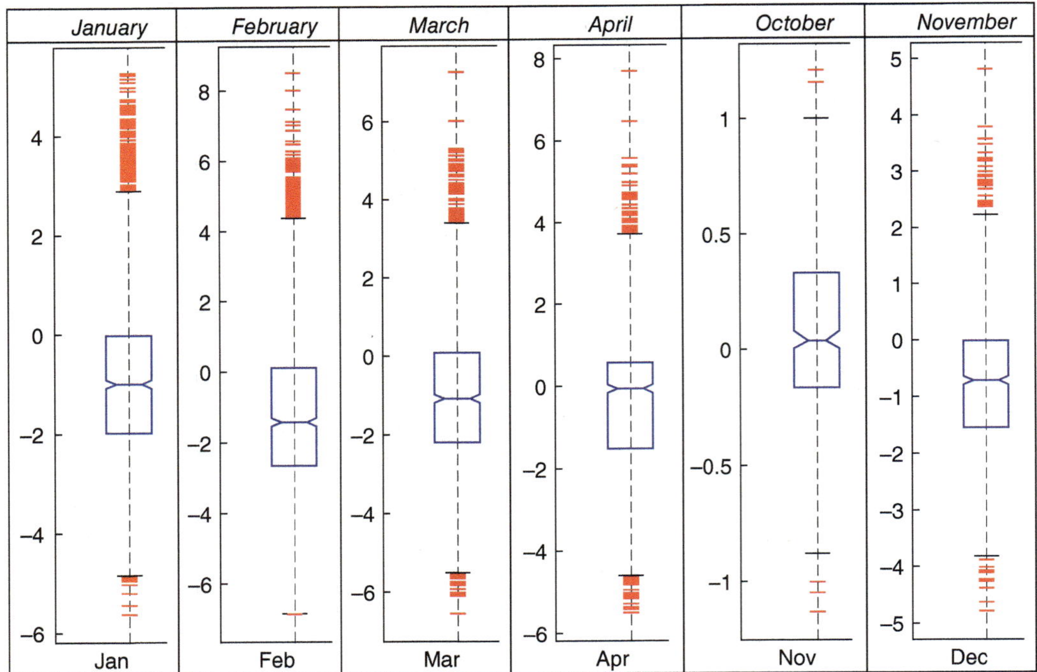

Figure 14.5 Boxplots of monthly SWE trend magnitudes β in mm/yr showing mostly negative median/mean β except for November in the Canadian Prairies, and some outliers where in some instances, β exceeds 8 mm/yr.

Table 14.2 Mean monthly SWE, means, and medians of monthly trend magnitudes (β in mm/yr) for the SWE of the Canadian Prairies between 1979 and 2004

Month	Mean Monthly SWE (mm)	Overall Mean of Negative and Positive Trends	Mean of Negative Trends Only (mm/Year)	Median of Negative Trends Only
November	32.6	0.08	−0.32	−0.27
December	47.7	−0.72	−1.29	−1.1
January	49.9	−0.80	−1.61	−1.44
February	50.8	−1.12	−2.30	−2.0
March	48.3	−0.89	−1.85	−1.62
April	37.6	−0.34	−1.73	−1.5
Average	44.5	−0.63	−1.52	−1.32

14.6. INFLUENCE OF CLIMATIC ANOMALIES ON SWE

Besides the possible impact of climate warming (Section 14.7), climatic anomalies could also contribute to part of the detected changes, especially to increasing SWE trends, even though they occurred less frequently and were more isolated than decreasing trends. The possible effects of the Pacific Decadal Oscillation (PDO), Pacific/North American (PNA) teleconnection pattern, and the El Nino–Southern Oscillation (ENSO) are examined using the PDO, PNA, Southern Oscillation Index (SOI), and Nino3 indices (Tables 14.3–14.5) since these three climate anomalies have been shown to affect the precipitation and stream flow of western Canada [e.g., *Gan et al.*, 2007; *Gobena and Gan*, 2006]. The PDO is represented by the leading principal component (PC1) of monthly sea surface temperature (SST) anomalies in the North Pacific [*Mantua and Hare*, 2001]; Nino3 is a time series of monthly SST anomalies averaged over the window (5°N–5°S,150°W–90°W) in the equatorial Pacific Ocean; SOI is a time series of the normalized monthly differences in sea level pressure (SLP) at Tahiti (≈ 150°W, 18°S) and at Darwin (≈ 130°E, 13°S); and the Pacific/North American (PNA) pattern represents a quadripole of 700 mb geopotential height anomalies, with opposite anomalies centered over the Aleutian low and western Canada, and between the Hawaiian Islands and southeastern United States [*Wallace and Gutzler*, 1981].

In order to reduce the dimensionality and capture any dominant signals in the SWE data, a PCA was conducted on the pixels with SWE > 0. Scree plots of the January–April 1979–2004 SWE (Figure 14.6) show that in most cases about 65% of the total variance is explained by the first four principal

Table 14.3 Pearson's correlations (ρ_p) between PC1 to PC4 and corresponding monthly PDO (columns 1–6), one-season lag, seasonal PDO of OND, NDJ, DJF, and JFM[a,b]

ρ_p	Monthly PDO						OND				NDJ	DJF		JFM
Col	1	2	3	4	5	6	7	8	9	10	11	12	13	14
Mon	Nov	Dec	Jan	Feb	Mar	Apr	Jan	Feb	Mar	Apr	Mar	Mar	Apr	Apr
PC1	−0.24	0.09	**0.39**	**−0.40**	**0.47**	−0.3	0.19	−0.27	0.30	−0.24	0.32	**0.39**	**−0.34**	**−0.40**
PC2	−0.11	0.28	0.12	0.35	−0.15	0.09	0.23	0.30	−0.31	**0.34**	−0.32	−0.31	0.27	0.16
PC3	−0.19	−0.22	0.17	−0.27	−0.10	0.31	−0.19	0.21	−0.35	0.22	−0.33	−0.24	0.29	0.34
PC4	−0.26	−0.20	0.25	−0.17	0.02	−0.15	**−0.37**	0.14	−0.08	0.18	0.12	0.07	−0.04	−0.09

[a]The ρ_p between PCs of October, November, and December and PDO-JAS (averaged of July–August–September) are not shown because no statistically significant case is detected.
[b]Bolded numbers mean statistically significant at $\alpha/2 = 0.05$.

Table 14.4 Spearman's rank correlation (ρ_s) between PC1, PC2, and PC3 and the corresponding monthly PDO (cols. 1 to 6), one-season lag, seasonal PDO of OND, NDJ, DJF, and JFM

ρ_s	Monthly PDO						OND				NDJ	DJF		JFM
Col	1	2	3	4	5	6	7	8	9	10	11	12	13	14
Mon	Nov	Dec	Jan	Feb	Mar	Apr	Jan	Feb	Mar	Apr	Mar	Mar	Apr	Apr
PC1	**−0.39**	**−0.40**	0.27	−0.12	0.27	−0.23	0.22	−0.08	0.17	−0.16	0.19	0.19	−0.25	−0.26
PC2	−0.12	0.31	0.21	0.32	−0.20	0.07	0.19	0.34	−0.30	0.30	−0.34	−0.26	0.16	0.07
PC3	−0.21	0.10	0.13	0.24	−0.03	0.35	0.22	0.18	−0.32[#]	0.31	−0.30	−0.15	0.31	0.38
PC4	−0.28	**0.49**	0.21	−0.15	0.03	−0.18	0.17	0.11	0.12	0.23	0.12	0.04	0.05	−0.03

Table 14.5 Pearson's correlations ($\boldsymbol{\rho}_p$) and Spearman's rank correlation ($\boldsymbol{\rho}_s$) between PC1, PC2, PC3, and PC4 with the corresponding monthly PNA, Nino3, and SOI, showing comparable influence of PNA and PDO over the SWE of Canadian Prairies in all PCs

		Monthly PNA						Monthly Nino3			Monthly SOI		
	Month	Nov	Dec	Jan	Feb	Mar	Apr	Nov	Mar	Apr	Nov	Mar	Apr
$\boldsymbol{\rho}_p$	PC1	0.15	−0.06	−0.15	0.00	0.35	−0.20	−0.25	0.12	−0.21	0.19	−0.19	−0.01
	PC2	−0.18	−0.22	0.29	−0.23	−0.25	−0.22	−0.16	0.08	0.02	0.33	0.06	−0.02
	PC3	0.20	0.31	0.01	0.15	0.13	**−0.44**	−0.21	0.03	0.25	0.04	−0.14	**−0.42**
	PC4	**0.40**	−0.12	0.04	**0.47**	0.33	−0.07	0.19	−0.01	0.04	−0.01	0.09	−0.31
$\boldsymbol{\rho}_s$	PC1	0.31	0.01	−0.10	0.14	**−0.40**	−0.27	**−0.49**	−0.21	−0.16	0.38	−0.23	−0.09
	PC2	−0.20	−0.22	0.23	−0.15	0.01	−0.16	−0.11	−0.21	0.14	0.18	0.18	−0.01
	PC3	0.20	0.30	−0.05	0.11	0.08	**−0.41**	−0.18	−0.19	0.11	−0.05	−0.10	**−0.46**
	PC4	**0.42**	−0.01	0.12	**0.39**	0.01	−0.07	0.15	−0.01	0.05	−0.12	0.04	−0.22

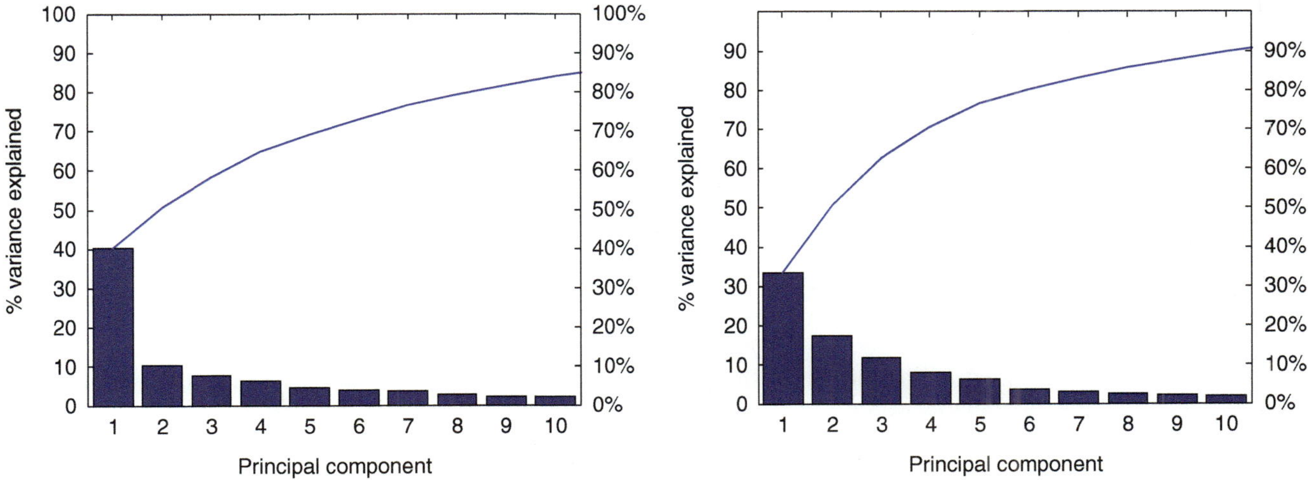

Figure 14.6 Scree plots of the January–April 1979–2004 SWE (October–December not shown), which show that the first four PCs generally explaining more than 65% of the total variance.

components (PCs). The PCA for November is based on 73 pixels only, compared to 2337 pixels for March, mainly because in November only a small part of the CP are covered with snowpack detectable by the passive microwave sensors.

Pearson correlation coefficients (ρ_p) of Table 14.3 show several statistically significant ρ_p between monthly PC1, PC2, PC3, and PC4 of CP's SWE (see Figure 14.1 for the window of data analyzed) and monthly (columns 1–6) or seasonal (OND to JFM) PDO (columns 8–14). No significant ρ_p was found between PDO-JAS and SWE of November–December, which means that winter to early spring SWE of CP are more affected by PDO than the autumn SWE, but their relationships can vary widely from negligible, e.g., ρ_p of PC1-PDO for December is 0.09, to fairly significant, e.g., ρ_p of PC1-PDO for March is 0.47. The characteristics between PC1 and PDO for March are quite similar to each other (Figures 14.7f and 14.7h), such that the March PC1 score and PDO are positive until about 1988, after which the values are mostly negative.

Table 14.4 also shows several statistically significant Spearman's rank correlation ρ_s, but they are of different months than statistically significant ρ_p shown in Table 14.3. Apparently, PDO exerts some influence on the overall snowpack SWE of CP during January–March. However, the effect is not consistent, nonextensive, and is likely quite unpredictable.

PNA also exert some influence on the SWE of CP, with significant ρ_s and ρ_p in November and February (PC4), March (PC1), and April (PC3), while ENSO represented by Nino3 and SOI exhibits smaller influence over the snowpacks of CP (only significant ρ_s in November and significant ρ_s and ρ_p in April; see Table 14.5), even though *Hsieh and Tang* [2001] found ENSO to have affected the snowpack of British Columbia, Canada, and the impact of ENSO on the precipitation of western Canada/NA has been established [*Gan et al.*, 2007; *Cayan et al.*, 1999; *Shabbar et al.*, 1997; *Ropelewski et al.*, 1986], and Nino3 and SOI are found to be significantly related to PDO and PNA in some months (Table 14.6). Monthly PDO is most significantly related to the corresponding monthly PNA, except for December and March. PDO is more consistently related to Nino3 and SOI than PNA.

Sobolowski and Frei [2007] found significant Spearman's ranked correlation between preceding seasonal (JAS or OND) ENSO, NAO, and PNA (PDO) indices with the PC1 (PC3) of CP winter SWE (JFM) of window (35°N–55°N)(130°W–50°W) over the 1980–1997 period. The PC1 relationship is located mainly in the south central to western regions of Canada, while the PC3 relationship stretches from the midwest to Atlantic Canada, which means that these relationships are strongest at large spatial scales. They suggested that these relationships could vary over time, but they did not have enough SWE data to explore how such relationships could change with time.

To gain a better insight into the influence of ENSO (in terms of Nino3 and SOI), PDO, and PNA on the snowpacks of CP, the first four principal components (PC1–PC4) of monthly SWE are compared with these climate anomalies (Figure 14.7). The influence of PDO (see Figure 14.7(h) and Figure 14.8 is approximately represented by PC1 for November, January, and March when PC1 is mainly positive until about 1988 and is mainly negative thereafter, and it likely describes SWE variations of low frequency. It seems that PDO is more correlated to PC1 until 1988, and after that PDO's influence became fuzzy. In contrast, the PC1 of December, February, and April are mostly negative until about 1988 and are mainly positive thereafter. PC2–PC4 exhibit interannual fluctuations between positive and negative scores and the characteristics vary from month to month. Given PDO is of interdecadal time scale, and with only 26 years of data analyzed, our understanding of PDO's influence on the snowpack of CP is somewhat limited. The relationships between the PCs and PNA are relatively scattered and mostly not statistically significant.

14.7. SWE-TEMPERATURE AND SWE-PRECIPITATION RELATIONSHIPS

Hantel and Hirtl-Wielke [2007] showed that in the Alpine region snow cover duration could decrease by a maximum of 4 weeks/°C of warming in winter at an elevation of 700 m above mean sea level. However, above the 700 m level, the reduction rate of snow cover duration will fall rapidly. Higher air temperatures in the last three decades in CP could be a major contributor to the fairly widespread reduction in its SWE, since sensible, latent heat and net incoming solar fluxes contribute to melting or sublimation, especially when air temperature >0°C. Similarly, possible changes to precipitation (e.g., snowfall) could be partly responsible for the detected changes in SWE. For example, annual precipitation for latitudes north of 50°N had been observed to increase by about 4% in the last five decades, though this increase varied spatially and temporally [*Groisman et al.*, 2005]. In lower latitude regions, not much change in the annual precipitation had been observed but less snowfall and more rainfall had been observed in some places, e.g., the lower Missouri River Basin [*Berger et al.*, 2002] and in New England [*Huntington et al.*, 2004].

To find possible climatic factors leading to a general reduction in the snowpacks of CP, the next step is to correlate changes of SWE to gridded precipitation and air temperature data to explore the possible relationships between climate (temperature and precipitation) and SWE. Trend analysis of both the gridded, 2-m air temperature

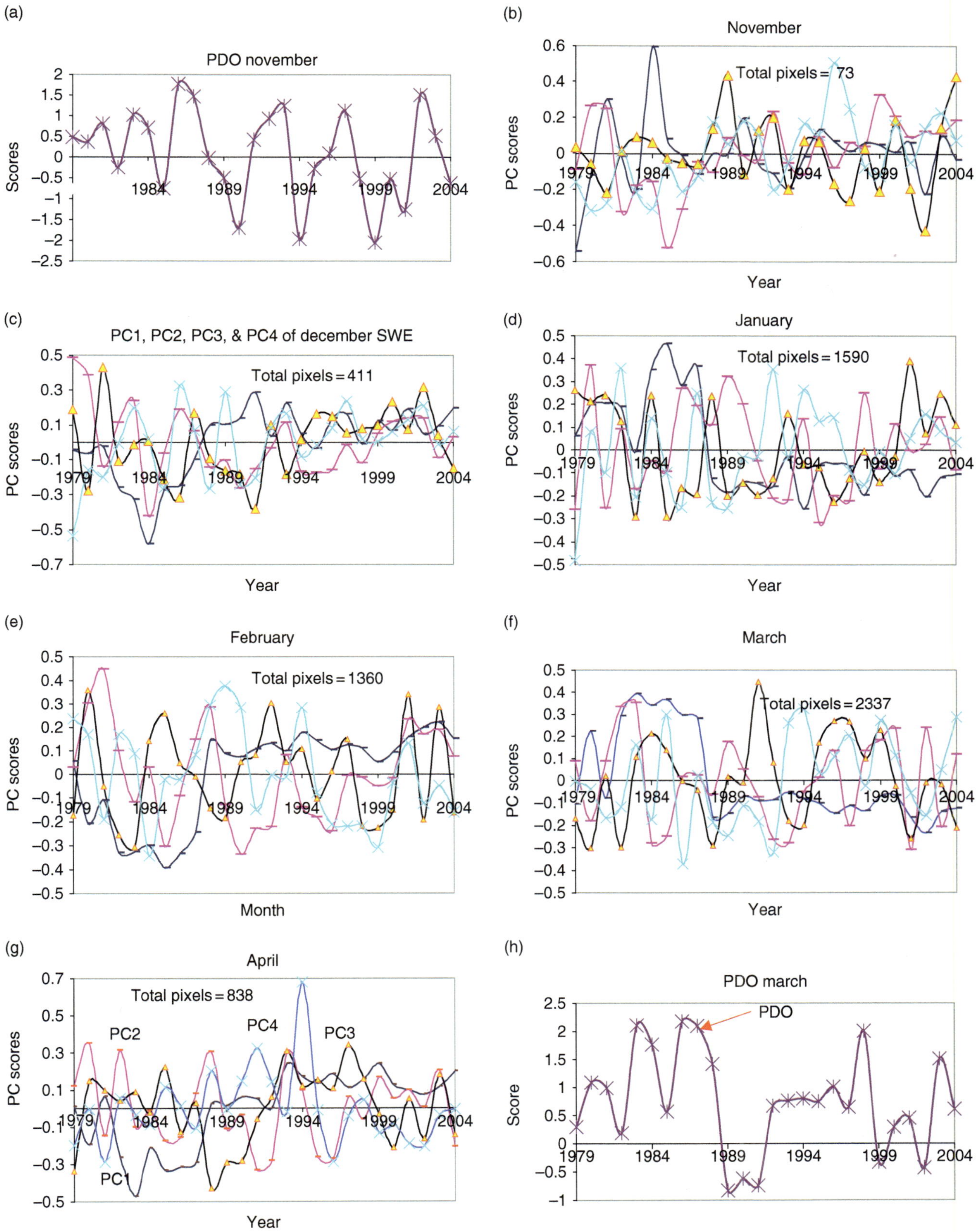

Figure 14.7 (b)–(g) Shows the PC1, PC2, and PC3 of monthly SWE of 1979–2004 for October, November, December, January, February, March, and April, respectively, while (a) and (h) shows the February PDO. The respective number of pixels selected to perform the PCA are as listed in the figure.

Table 14.6 Monthly Pearson's correlation coefficients between PDO, PNA, Nino3, and SOI

		PDO	PNA	Nino3	SOI		PDO	PNA	Nino3	SOI
PDO	Oct	1	**−0.44**	**0.45**	−0.46	Jan	1	**−0.39**	0.28	−0.22
PNA			1	0.08	−0.03				**−0.44**	**0.56**
PDO	Nov	1	−0.29	0.37	−0.26	Feb	1	−0.31	0.36	**−0.45**
PNA			1	0.05	−0.15			1	**−0.46**	0.34
PDO	Dec	1	−0.20	0.25	−0.16	Mar	1	−0.07	**0.54**	−0.43
PNA				−0.29	0.37			1	−0.30	0.13
PDO						Apr	1	−0.40	0.33	**0.45**
PNA								1	−0.09	0.44

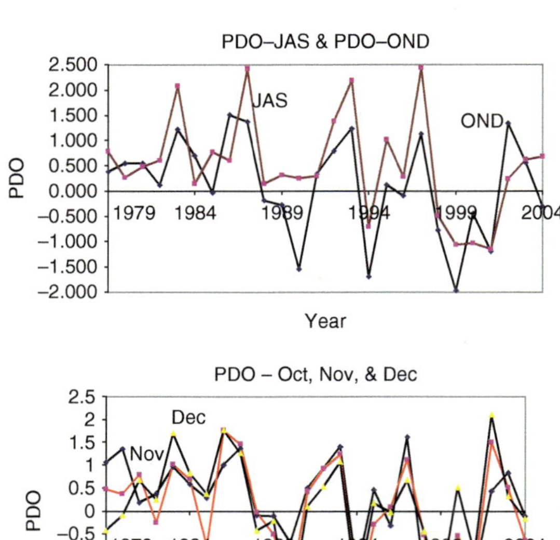

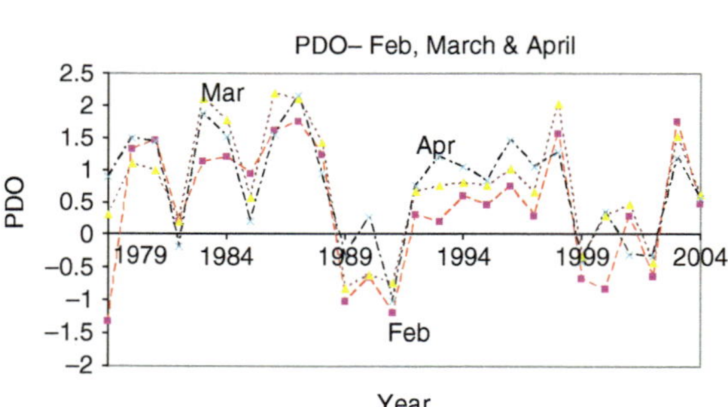

Figure 14.8 Plots of seasonal PDO (JAS and OND) and monthly PDO showing PDO to be primarily of the positive phase prior to 1988, and after which PDO fluctuates widely between positive and negative phases.

data of the University of Delaware (figure not shown) and that of the North American Regional Reanalysis (NARR) showed hardly any statistically significant increasing temperature trends (Figure 14.9a), in contrast to fairly extensive statistically significant, decreasing SWE trends based on passive microwave data (Figure 14.1) in the CP. The decreasing precipitation trends detected

are also limited to Alberta in February. However, extensive significant negative correlations are detected between SWE and temperature in the CP, especially for March and for the northern parts of the CP, and there are similarities between the distribution of negative correlation areas and that of decreasing trends detected from the SWE data, even though there are some discrepancies between them (Figures 14.1c–14.1f). For January, the patterns of negative correlation areas resemble that of decreasing SWE trends except the negative correlation areas are not statistically significant. On the basis of NARR and University of Delaware 2 m surface air temperature data, warmer air temperature in the CP between 1979 and 2004, though not statistically significant, might have contributed to significant decreasing SWE trends detected from passive microwave data.

There is limited agreement between areas of detected decreasing precipitation trends based on gridded data of the University of Delaware (figure not shown) and NARR (Figure 14.9b) and decreasing SWE trends that are statistically significant. Furthermore, unlike the extensive negative correlation patterns detected between SWE and temperature data that are statistically significant (especially for February to April), the correlations detected between SWE and precipitation (blue for positive and red for negative correlations) in the CP are mostly not statistically significant, except for limited areas in Alberta and Manitoba for March (Figure 14.9d). Therefore, it seems that extensive decreasing SWE trends detected are more likely caused by warming than by decreasing precipitation, which played a relatively minor role compared to temperature, given that there is less agreement between the areas of significant precipitation-SWE correlation and areas of decreasing SWE trends that are significant.

Even though climatic anomalies such as PDO could contribute to part of the detected changes, it seems that extensive decreasing trends in SWE detected in the CP are caused more by increasing temperatures than by decreasing precipitation, for rising temperatures have generally resulted in rain rather than snow especially in areas and seasons where the average air temperature is

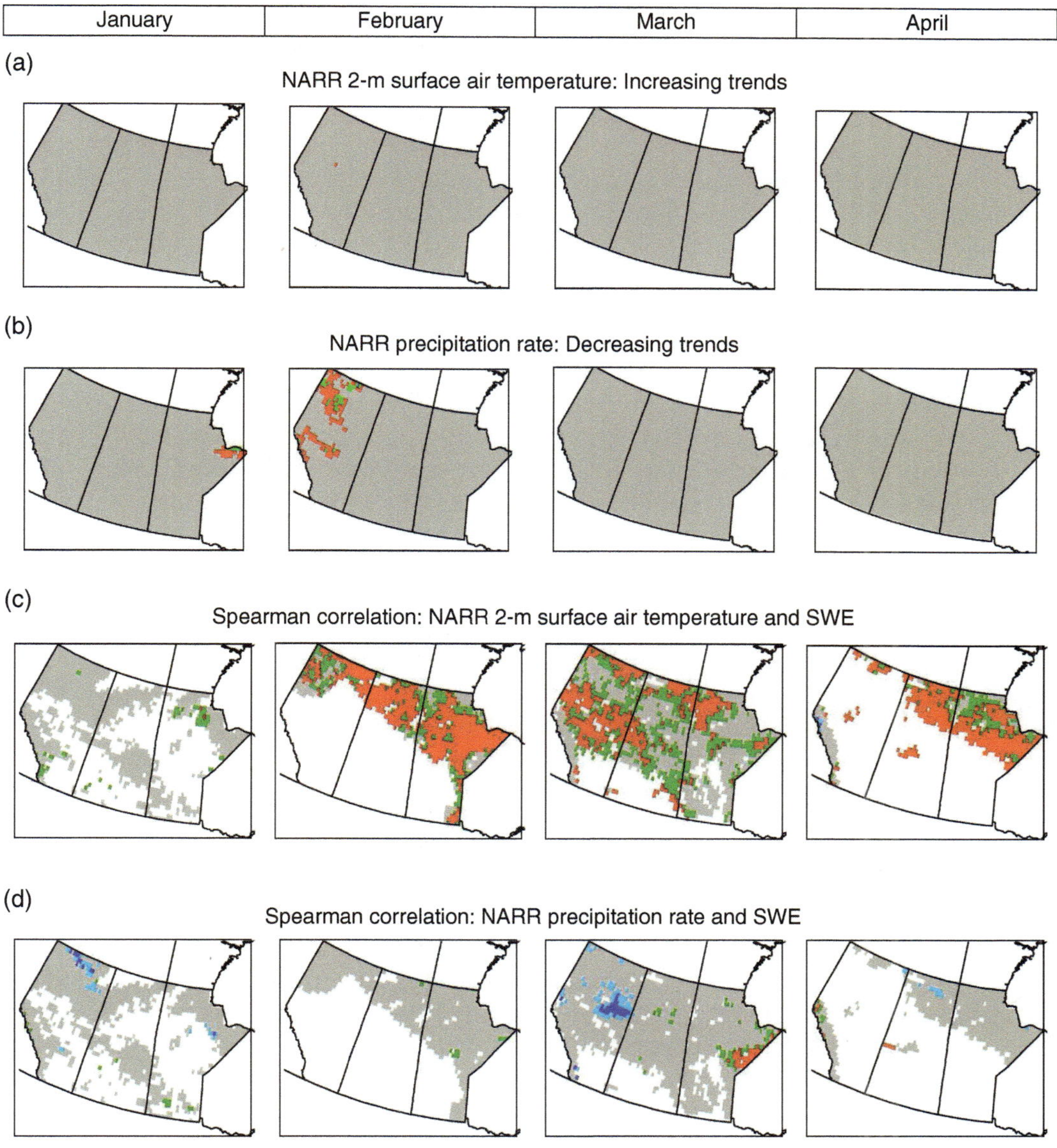

Figure 14.9 Monthly (a) NARR surface temperature increasing trends and (b) NARR precipitation decreasing trends based on a two-tailed significance level of 5% (red), 2.5% (green), and 1% (blue); (c) NARR surface temperature-SWE correlation and (d) NARR precipitation-SWE correlation, where color shadings are for 5% significance level (green for negative and light blue for positive correlations) and 1% significance level (red for negative and dark blue for positive correlations). For (c) and (d) the gray colors are grids for which correlations are not significant at the 5% level, whereas the white areas are grids for which correlations were not computed since at least one SWE value is missing out of the 26 years of record.

close to 0°C, even though warmer temperature could also lead to more snowfall in areas where air temperature is predominantly below 0°C. Through three spring indicators—lilacs, honeysuckles, and stream flow—*Cayan et al.* [2001] found earlier onset of the spring season by up to 3 weeks in western North America since the 1970s. By simulating the snow energy balance to climatic changes projected by nine regional climate models to the end of the 21st century in the Pyrenees, *Moreno et al.* [2008]

concluded that the most significant changes to future snowpack processes are related to temperature. *Tedesco* [2007] concluded that record melting of snow in southern Greenland (which is marginally further north than the northern CP) in 2007 was caused by warmer surface temperature. The aforementioned observations, and the global-scale decline of snow and ice cover since 1980 and increasing rate of decline during the past decade, as noted by *Lemke et al.* [2007] in their IPCC 2007 report, likely support our conclusion that warmer temperature caused the detected decline in SWE of CP.

14.8. SUMMARY AND CONCLUSIONS

Since the late 1980s, both the snow water equivalent (SWE) retrieved from 1979–2004 SMMR and SSM/I passive microwave data for the Canadian Prairies (CP) show negative anomalies. On the basis of the nonparametric Kendall test applied to the SWE data of SMMR and SSM/I for CP, statistically significant decreasing trends have been detected. Up to 45% (February) of the detected decreasing trends in SWE for 1979–2004 are statistically significant at $\alpha/2 = 0.05$, which is about three times more frequent than detected increasing trends in SWE. Significant decreasing trends in SWE are more extensive in January–March than in November and April.

The overall mean trend magnitudes are about −0.6 mm/yr, which for 1979–2004 would mean an overall reduction of snow depth of about 6–8 cm (assuming a snowpack density of 0.20–0.25 g/cm³), which can have significant impacts on the CP relying on spring snowmelt for water supply. For November, January, and March, the PC1 of CP's SWE are found to resemble the Pacific Decadal Oscillation (PDO), marginally correlated to the Pacific North American (PNA) pattern and to El Nino–Southern Oscillation (ENSO). The PDO and the SWE PC1 scores for these few months show similar variations, such that the PC1 score and PDO are positive until about 1988, after which the values are mostly negative.

Lastly, even though temperature data itself did not exhibit significant increasing trends, from the extensive, significant negative correlations detected between SWE and gridded temperature data of NARR and the University of Delaware for the CP, and the lack of extensive significant correlations between SWE and gridded precipitation data in the CP, and related findings such as global-scale decline of snow cover and warming temperature trends, longer rainfall seasons, etc., it seems that the extensive decreasing trends in SWE detected in the CP are more attributed to increasing temperatures than to decreasing precipitation, even though climate anomalies could also play a minor role on part of the detected trends.

14.9. RECOMMENDATIONS FOR FUTURE RESEARCH

Four possible studies that will contribute to passive microwave snow research are listed below:

1. *Derksen et al.* [2005] found good agreements between in situ snow measurements and SWE derived from the Meteorological Service Canada (MSC) land cover sensitive algorithms of the form, $SWE = K_1 + K_2(T_{B,37V} - T_{B,19V})$, applied to vertically polarized SSM/I data of 37 and 19 GHz for northern boreal forests of northern Manitoba. The values of K_1 and K_2 depend on the land use, such as open, coniferous, deciduous, and sparse forests. They found the retrieved SWE pattern agrees with SWE reanalysis data set of *Brown et al.* [2003] and SWE simulated by the Canadian Regional Climate Model [*MacKay et al.*, 2003]. However, they also found consistently lower SWE retrieved from SSM/I data compared to in situ measurements across denser boreal forest to the south and open tundra to the north, and the absolute SWE differences between them increase with larger mean snow depths. They also found a wide range of SWE values measured on the ground in each SSM/I pixel but noted a high percentage of SSM/I SWE agrees within ±20 mm of the median in situ measurements, even though such observations are somewhat limited by the large-scale mismatch between ground measurements and large SSM/I pixels (625 km² in area). It will be interesting to do a similar trend analysis using the SWE data of MSC, given that the SWE data of NSIDC are based on horizontally polarized, while that of MSC are based on vertically polarized, passive microwave data and different retrieving algorithms, and possibly the latest database of *Foster et al.* [2011].

2. Using AMSR-E and radiosonde data sets over two widely separated regions in the continental United States, *Wang and Tedesco* [2007] found that even under a clear sky, the atmospheric absorption on passive microwave measurements near 19 and 37 GHz could account for as much as 25–50% to the estimation of SWE. Conversely, an AMSR-E estimated SWE of 10 cm could become about 14 cm when measured at the ground level under the same atmospheric condition. It will be interesting to examine atmospheric contributions at regional scale to passive microwave T_B.

3. The simple equation 14.3 that likely does not provide proper assessment of vegetation contribution to the T_B signal can be improved by estimating the impact of stem volumes of coniferous and deciduous forests on T_B signal [*Karam et al.*, 1992; *Mätzler*, 1994; *Paloscia*, 1995; *Ferrazzoli and Guerriero*, 1996; *Kruopis et al.*, 1999; *Roy et al.*, 2004; *Pampaloni*, 2004]. A possible equation to account for the contribution of vegetation to T_B signal is

$$T_{B-FOV,p} = F\tau_{veg,p}T_{B-Snow,p}\left(1 - \frac{T_{veg}}{T_{Snow}}\right) - F\tau_{veg,p}^2\left(1 - \frac{T_{B-Snow,p}}{T_{snow}}\right)T_{veg}$$
$$+ (1-F)T_{B-Snow,p} + FT_{veg},$$

$$(14.7)$$

where F is the fraction of stem volume, $\tau_{veg,p}$ is the dielectric constant of vegetation at frequency p.

4. Lastly, effects of snowpack geophysical properties such as grain size and correlation length on snow

microwave emissions in terms of T_B at 19 and 37 GHz for different classes of snow (Prairie, tundra, taiga, alpine, and maritime) can be simulated using electromagnetic models such as the Helsinki University of Technology (HUT) snow emission model, the microwave emission model of layered snowpacks (MEMLS), a dense-medium radiative-transfer theory model (DMRT), and a strong fluctuation theory (SFT) model. Such experiments can help us to understand interactions between electromagnetic waves and snow media and can benefit remote sensing of snow [*Tedesco and Kim*, 2006].

ACKNOWLEDGMENTS

The first author was funded by a visiting fellowship of CIRES (Cooperative Institute of Research in Environmental Science). The SMMR and SSM/I SWE data were downloaded from the website of National Snow & Ice Data Center (NSIDC) *http://nsidc.org/data*, NARR data were downloaded from *http://www.cdc.noaa.gov/data/gridded/data.narr.html*, NOAA's Climate Diagnostics Center website; and University of Delaware data were downloaded from http://climate.geog.udel.edu/~climate/html_pages/archive.html. The third author was partly funded by the Canadian Foundation of Climate and Atmospheric Science (CFCAS).

REFERENCES

Armstrong R., and M. J. Brodzik (2002), Hemispheric-scale comparison and evaluation of passive microwave snow algorithms, *Ann. Glaciol.*, *34*, 38–44.

Armstrong R., M. J. Brodzik, and M. H. Savoie (2001), Recent Northern Hemisphere snow extent: A comparison of data derived from visible and microwave sensors, *Geophys. Res. Lett.*, *23*(19), 3673–3676.

Barry, R., and T. Y. Gan (2011), *Global Cryosphere*, Past, Present and Future, 472 pages, Cambridge University Press, Cambridge, ISBN: 9780521769815.

Berger, C. L., A. R. Lupo, P. Browning, M. Bodner, M. D. Chambers, and C. C. Rayburn, (2002), A climatology of northwest Missouri snowfall events: Long-term trends and interannual variability, *Phys. Geogr.*, *23*, 427−448.

Berryman, D., B. Bobee, D. Cluis, and J. Haemmerli (1988), Nonparametric tests for trend detection in water quality time series, *Water Resourc. Bull.*, *24*(3), 545–556.

Brown, R., and D. Robinson (2011), Northern Hemisphere spring snow cover variability and change over 1922–2010 including an assessment of uncertainty, *Cryosphere*, 5, 219–229, doi:10.5194/tc-5-219-2011.

Brown, R., C. Derksen, and L. Wang (2010), A multi-data set analysis of variability and change in Arctic spring snow cover extent, 1967–2008, *J. Geophys. Res.*, *115*, D16111, doi:10.1029/2010JD013975.

Brown, R., and P. Mote (2009). The response of Northern Hemisphere snow cover to a changing climate, *J. Climate*, 22, 2124–2145.

Brown, R. D. (2000), Northern Hemisphere snow cover variability and change, 1915–97, *J. Climate*, *13*, 2339–2355.

Brown, R. D., and Goodison, B. E. (1996), Interannual variability in reconstructed Canadian snow cover, 1915–1992, *J. Climate*, 9(6), 1299–1318.

Brown, R., B. Brasnett, and D. Robinson (2003), Gridded North American monthly snow depth and snow water equivalent for GCM evaluation, *Atmos.-Ocean*, *41*(1), 1–14.

Cayan, D. R., K. T. Redmond and L. G. Riddle (1999), ENSO and hydrologic extremes in the Western United States, *J. Climate*, *12*, 2881–2893.

Cayan, D. R., S. A. Kammerdiener, M. D. Dettinger, J. M. Caprio, and D. H. Peterson (2001), Changes in the onset of spring in the western United States, *Bull. Am. Meteorol. Soc.*, *82*, 399–415.

Chang, A. T. C., J. L. Foster, and D. K. Hall (1987), Nimbus-7 SMMR derived global snow cover parameters, *Ann. Glaciol.*, *9*, 39–44.

Chang, A., J. L. Foster, and D. K. Hall (1996), Effects of forest on the snow parameters derived from microwave measurements during the boreal, *Hydrol. Proc.*, *10*, 1565–1574.

Cline, D., et al. (2009), NASA Cold Land Processes Experiment (CLPX 2002/03): Airborne remote sensing, *J. Hydrometeorol.*, *10*, 338–346.

Cox, J. (2005), The snow/snow water equivalent ratio and its predictability across Canada, MSc thesis, McGill Univ., Montreal, Canada.

Clifford, D. (2010), Global estimates of snow water equivalent from passive microwave instruments: History, challenges and future developments, *Int. J. Remote Sens.*, *31*(14), 3707–3726.

Derksen, C., A. Walker, and B. Goodison (2003), A comparison of 18 winter seasons of in situ and passive microwave derived snow water equivalent estimates in Western Canada, *Remote Sens. Environ.*, *88*(3), 271–282.

Derksen, C., A. Walker, and B. Goodison (2005), Evaluation of passive microwave snow water equivalent retrievals across the boreal forest/tundra transition of western Canada, *Remote Sens. Environ.*, *96*, 315–327, doi:10.1016/j.rse.2005.02.014.

Dong, J., J. P. Walker, and P. Houser (2005), Factors affecting remotely sensed snow water equivalent uncertainty, *Remote Sens. Environ.*, *97*(1), 68–82.

Durand, M., E. J. Kim, and S. A. Margulis (2008), Quantifying uncertainty in modeling snow microwave radiance for a mountain snowpack at the Point-Scale, including stratigraphic effects, *IEEE Trans. Geosci. Remote Sens.*, *46*(6), 1753–1767.

Dyer, J. L., and T. L. Mote (2006), Spatial variability and trends in observed snow depth over North America, *Geophy. Res. Lett.*, *33*, L16503, doi:10.1029/2006GL027258.

Dyer, J. L., and T. L. Mote (2007), Trends in snow ablation over North America, *Int. J. Climatol.*, *27*, 739–748, doi:10.1002/joc.1426.

Ferrazzoli, P., and L. Guerriero (1996), Passive microwave remote sensing of forests: A model investigation," *IEEE Trans. Geosci. Remote Sens.*, *34*(2), 433–443.

Forster, P., et al. (2007), Changes in atmospheric constituents and in radiative forcing, in *Climate Change 2007: The Physical Science Basis*, Contribution of Working Group I to the Fourth Assessment Report of the Intergovernmental Panel on Climate Change, edited by S. Solomon, D. Qin, M. Manning, Z. Chen, M. Marquis, K. B. Averyt, M. Tignor and H. L. Miller, Cambridge Univ. Press, Cambridge, United Kingdom.

Foster, J. L., C. Sun, J. P. Walker, R. E. J. Kelly, A. T. C. Chang, J. Dong, and H. Powell (2005), Quantifying the uncertainty in passive microwave snow water equivalent observations, *Remote Sens. Environ.*, *94*, 187–203.

Foster, J. L., et al. (2011), A blended global snow product using visible, passive microwave and scatterometer satellite data, *Int. J. Remote Sens.*, *32*(5), 1371–1395.

Frei, A., and D. A. Robinson (1999), Northern hemisphere snow extent: Regional variability 1972–1994, *Int. J. Climatol.*, *19*, 1535–1560.

Gan, T. Y., A. Gobena, and Q. Wang (2007), Precipitation of Western Canada—Wavelet, scaling, and multifractal analysis and teleconnection to large-scale climate anomalies, *J. Geophys. Res. Atmos.*, *112*, D10110, doi:10.1029/2006JD007157.

Gan, T. Y., O. Kalinga, and P. R. Singh (2009), Comparison of snow water equivalent retrieved from SSM/I passive microwave data using artificial neural network, projection pursuit & nonlinear regressions, *Remote Sens. Environ.*, *25*(21), 4593–4615, doi:10.1016/j.rse.2009.01.004.

Gobena, A. K., and T. Y. Gan (2006), Low-frequency variability in southwestern Canadian streamflow: Links to large-scale climate anomalies, *Int. J. Climatol.*, *26*, 1843–1869, doi:10.1002/joc.1336.

Goodison, B. E. (1989), Determination of areal SWE on the Canadian Prairies using passive microwave satellite data, in *International Geoscience and Remote Sensing Symposium (IGARSS), Proceedings*, vol. 3, pp. 1243–1246, IEEE, Vancouver.

Groisman, P. Ya., R. W. Knight, D. R. Easterling, T. R. Karl, G. C. Hegerl and V. N. Razuvaev (2005), Trends in intense precipitation in the climate record, *J. Climate.*, *18*, 1326–1350.

Hantel, M., and L.-M. Hirtl-Wielke (2007), Sensitivity of Alpine snow cover to European temperature, *Int. J. Climatol.*, *27*, 1265–1275, doi:10.1002/joc.1472.

Hirsch, R. M., J. R. Slack, and R. A. Smith (1982), Techniques of trend analysis for monthly water quality data, *Water Resourc. Res.*, *18*(1), 107–121.

Hsieh, W. W., and B. Tang (2001), Interannual variability of accumulated snow in the Columbia basin, British Columbia, *Water Resourc. Res.*, *37*, 1753–1760.

Huntington, T. G., G. A. Hodgkins, B. D. Keim, and R. W. Dudley (2004), Changes in the proportion of precipitation occurring as snow in New England (1949–2000), *J. Climate.*, *17*(13), 2626–2636.

Jezek, K., C. Merry, and D. Cavalieri (1993), Comparison of SMMR and SSM/I passive microwave data collected over Antarctica, *Ann. Glaciol.*, *17*, 131–136.

Josberger, E. G., P. Gloersen, A. Chang, and A. Rango (1996), The effects of snowpack grain size on satellite passive microwave observations from the Upper Colorado River Basin, *JGR-Oceans*, *101*(C3), 6679–6688.

Karam, M. A., A. K. Fung, R. H. Lang, and N. S. Chauhan (1992), A microwave scattering model for layered vegetation, *IEEE Trans. Geosci. Remote Sens.*, *30*(4), 767–784.

Kelly, R. (2009), The AMSR-E snow depth algorithm: Description and initial results, *Japanese J. Remote Sens.*, *29*(1), 307–317.

Kelly, R. E. J., A. T. C Chang, L. Tsang, and J. L. Foster (2003), Development of a prototype AMSR-E global snow area and snow volume algorithm, *IEEE Trans. Geosci. Remote Sens.*, *41*(2), 230–242.

Kendall, M. G. (1975), *Rank Correlation Methods*, Charles Griffin, London.

Kruopis, J., A. Praks, N. Arslan, H. Alasalmi, J. Koskinen, and M. Hallikainen (1999), Passive microwave measurements of snow-covered forest areas in EMAC'95, *IEEE Trans. Geosci. Remote Sens.*, *37* (6), 2699–2705.

Langlois, A., A. Royer, and K. Goïta (2010), Analysis of simulated and spaceborne passive microwave brightness temperatures using in situ measurements of snow and vegetation properties, *Can. J. Remote Sens.*, *36* (1), S135–S148.

Lemke, P., et al. (2007), Observations: Changes in snow, ice and frozen ground, In: *Climate Change 2007: The Physical Science Basis*, Contribution of Working Group I to the Fourth Assessment Report of the Intergovernmental Panel on Climate Change, edited by S. Solomon, D. Qin, M. Manning, Z. Chen, M. Marquis, K. B. Averyt, M. Tignor, and H. L. Miller, Cambridge Univ. Press, Cambridge, United Kingdom.

López-Moreno, J. I., S. Goyette, M. Beniston, and B. Alvera (2008), Sensitivity of the snow energy balance to climatic changes: Prediction of snowpack in the Pyrenees in the 21st century, *Climate Res.*, *36*, 203–217, doi:10.3354/cr00747.

MacKay, M., F. Seglenieks, D. Verseghy, E. Soulis, K. Snelgrove, and A. Walker (2003), Modeling Mackenzie Basin surface water balance during CAGES with the Canadian Regional Climate Model, *J. Hydrometeorol.*, *4*(4), 748–767.

Mann, H. B. (1945), Non-parametric tests against trend, *Econometrica*, *13*, 245–259.

Mantua, N. J., and S. R., Hare (2001), The Pacific Decadal Oscillation, *J. Oceanogr.*, *58*, 35–44.

Mätzler, C. (1987), Applications of the interaction of microwaves with the natural snow cover, *Remote Sens. Rev.*, *2*, 259–387.

Mätzler, C. (1994), Passive microwave signatures of landscapes, *Meteorol. Atmos. Phys.*, *54*, 241–260.

Mekis, È., and R. Hopkinson, (2004). Derivation of an improved snow water equivalent adjustment factor map for application on snowfall ruler measurements in Canada. Proceedings, *14th Conference on Climatology*, Seattle, WA, January 12–15. np.

Mote, T. L., A. J. Grundstein, and D. A. Robinson (2003), A comparison of modeled, remotely sensed and measured snow water equivalent in the northern Great Plains, *Water Resource. Res.*, *39*(8), 1209, doi:10.1029/2002WR001782.

National Snow and Ice Data Center (NSIDC) (2005), EASE-Grid version of BU-MODIS land cover characteristic, National Snow and Ice Data Center, Digital Media, available at ftp://sidads.colorado.edu/pub/EASE/.Boulder.

Niu, G.-Y., K.-W. Seo, Z.-L. Yang, C. Wilson, H. Su, J. L. Chen, and M. Rodell (2007), Retrieving snow mass from GRACE terrestrial water storage change with a land surface model, *Geophys. Res. Lett.*, *34*, L15704, doi:10.1029/2007GL030413.

Pardé, M., K., Goïta, and A. Royer (2007), Inversion of a passive microwave snow emission model for water equivalent estimation using airborne and satellite data, *Remote Sens. Environ.*, *111*(2–3), 346–356.

Pierce, D. W., et al. (2008), Attribution of declining Western U.S. snowpack to human effects, *J. Climate*, doi:10.1175/2008JCLI2405.1.

Paloscia, S. (1995), Microwave emission from vegetation, Passive Microwave Remote Sensing of Land–Atmosphere Interactions, edited by B. J. Choudhury, Y. H. Kerr, E. G. Njoku, and P. Pampaloni, VSP BV, Netherlands.

Pulliainen, J. (2006), Mapping of snow water equivalent and snow depth in boreal and sub-Arctic zones by assimilating spaceborne microwave radiometer data and ground-based observations, *Remote Sens. Environ.*, *101*, 257−269.

Pulliainen, J., and M. Hallikainen (2001), Retrieval of regional snow water equivalent from space-borne passive microwave observations, *Remote Sens. Environ.*, *75*, 76–85.

Ropelewski, C. F., and M. S. Halpert (1986), North American precipitation and temperature patterns associated with the El Niño-Southern Oscillation (ENSO), *Monthly Weather Rev.*, *114*, 2352–2362.

Roy, V. K. Goïta, A. Royer, A. E. Walker, and B. E. Goodison (2004), Snow water equivalent retrieval in a Canadian boreal environment from microwave measurements using the HUT snow emission model, *IEEE Trans. Geosciences Remote Sensing*, *42*(9), 1850–1859.

Sen, P. K. (1968), Estimates of the regression coefficient based on Kendall's tau, *J. Am. Stat. Assoc.*, *63*, 1379–1389.

Shabbar, A., B. Bonsal, and M. Khandekar (1997), Canadian precipitation patterns associated with the Southern Oscillation. *J. Climate*, *10*, 3016–3027.

Shi, X., M. Sturm, G. E. Liston, R. E. Jordan, and D. P. Lettenmaier (2009), SnowSTAR2002 transect reconstruction using a multilayered energy and mass balance snow model, *J. Hydrometeorol.*, *10*, 1151–1167, doi:10.1175/2009JHM1098.1.

Singh, P. S., and T. Y. Gan (2000), Retrieval of snow water equivalent using passive microwave brightness temperature data, *Remote Sens. Environ.*, *74*(2), 275–286.

Sobolowski, S., and A. Frei (2007), Lagged relationships between North American snow mass and atmospheric teleconnection indices, *Int. J. Climatol.*, *27*, 221–231, doi:10.1002/joc.1395.

Sun, C., J. P. Walker, and P. R. Houser (2004), A methodology for snow data assimilation in a land surface model, *J. Geophys. Res.*, *109*, D08108. doi:10.1029/2003JD003765.

Tait, A. (1998), Estimation of snow water equivalent using passive microwave radiation data, *Remote Sens. Environ.*, *64*, 286–291.

Tait, A., and R. Armstrong (1996), Evaluation of SMMR satellite-derived snowdepth using ground-based measurements, *Int. J. Remote Sens.*, *17*(4), 657–665.

Takala, M., et al. (2011), Estimating northern hemisphere snow water equivalent for climate research through assimilation of space-borne radiometer data and ground-based measurements, *Remote Sens. Environ.*, doi:10.1016/j.rse.2011.08.014

Tedesco, M. (2007), A new record in 2007 for melting in Greenland, *Eos*, *88*(39), September 27.

Tedesco, M., and E. J. Kim (2006), Intercomparison of electromagnetic models for passive microwave remote sensing of snow, *IEEE Trans. Geosci. Remote Sens.*, *44*(10), 2654–2666.

Wallace, J. M., and D. S. Gutzler (1981), Teleconnections in the geopotential height field during the Northern Hemisphere winter, *Monthly Weather Rev.*, *109*, 784–812.

Wang, J. R., and M. Tedesco (2007), Identification of atmospheric influences on the estimation of snow water equivalent from AMSR-E measurements, *Remote Sens. Environ.*, *111*(2–3), 398–408.

Yang, D., D.L. Kane, L. D. Hinzman, B. E. Goodison, J. R. Metcalfe, Y. T. Louie, G. H. Leavesley, D. G. Emerson, and C. L. Hanson (2000), An evaluation of the Wyoming Gauge system for snow measurement, *Water Resourc. Res.*, *36*(9), 2665–2677.

Yueh, S., S. Dinardo, A. Akgiray, R. West, D. Cline, and K. Elder (2009), Airborne Ku-band polarimetric radar remote sensing of terrestrial snow cover, *IEEE Trans. Geosci. Remote Sens.*, *47*(10), 3347–3364.

Section V: Soil Moisture

15

Some Issues in Validating Satellite-Based Soil Moisture Retrievals from SMAP with in Situ Observations

Thomas J. Jackson, Michael Cosh, and Wade Crow

15.1. INTRODUCTION

The Soil Moisture Active Passive (SMAP) satellite is scheduled for launch in the fall of 2014 [*Entekhabi et al.*, 2010]. SMAP incorporates a combined L-band radiometer radar to provide global products every 2–3 days. It complements other satellite-based systems, such as the Advanced Microwave Scanning Radiometer series (AMSR) [*Njoku et al.*, 2003], the Advanced Scatterometer (ASCAT) [*Wagner et al.*, 2013], and the Soil Moisture Ocean Salinity (SMOS) [*Kerr et al.*, 2012], by providing a surface soil moisture product with a 36 km gridding using only the radiometer. But, in addition, it will also provide a 3 km radar and a 9 km combined product.

As the first of NASA's decadal survey missions, efforts are being made to implement both best practices and innovations, which include calibration and validation of the remote sensing products (Cal/Val). This chapter reviews some of the best practices as related to soil moisture validation using in situ network observations that have been incorporated.

There are four primary reasons why calibration and validation are necessary for a successful satellite mission: mission requirements, quality assurance, data integration, and science.

From the point of view of the satellite project and science team, validation is usually one of the requirements of the funding agency that is used to assess the mission success. These requirements define the mission goals and design. In the case of the SMAP mission, the requirement is an accuracy of 0.04 m^3/m^3 at a spatial resolution of at least 10 km.

In addition to mission requirements criteria, validation provides the users with quality assurance, which in theory should lead to greater confidence and acceptance, resulting in more to widespread use of the mission products. This in turn provides support for the mission and its successors.

Validation also provides a common basis from which a single climate record can be compiled by integrating multiple satellite products of different origin. For soil moisture, this is a new problem that the community is beginning to address [*Wagner et al.*, 2012]. It will be necessary to resolve the differences in soil moisture products from AMSR, ASCAT, SMOS, Aquarius [*Bindlish et al.*, 2014], and SMAP.

Finally, from the scientific perspective and considering microwave remote sensing in particular, there will always be some uncertainty in the products due to ambiguities in defining contributing area and depth for the various satellite mission sensors and frequencies. Validation can help us resolve these issues and improve algorithms through a careful evaluation of algorithm performance and anomalies.

The purpose of this chapter is to provide an overview of some of the issues that were addressed in the development of the SMAP Calibration/Validation (Cal/Val) Plan. The full plan is available at http://smap.jpl.nasa.gov/. In the following sections, some sources of available guidance on the design of a validation program are discussed. How this translates to soil moisture will be considered. In situ observations play a major role in the validation of satellite-based soil moisture and several aspects of using these data resources will be discussed.

USDA-ARS Hydrology and Remote Sensing Laboratory, Beltsville, Maryland, USA

Remote Sensing of the Terrestrial Water Cycle, Geophysical Monograph 206. First Edition. Edited by Venkat Lakshmi.
© 2015 American Geophysical Union. Published 2015 by John Wiley & Sons, Inc.

Finally, the implementation of these ideas into the SMAP Cal/Val plan will be described.

15.2. GUIDANCE ON CAL/VAL PLANNING

The first step in establishing a Cal/Val program is the review of the guidance available and precedents established by related satellite projects. A study by *Baret et al.* [2009] attempted to summarize the international programs concerned with terrestrial Earth observation. Looking at these from a soil moisture perspective, the most relevant is the Global Earth Observing System of Systems (GEOSS).

Within GEOSS, the Committee on Earth Observing Satellites (CEOS) and its Working Group on Calibration and Validation (WGCV) are the primary resource for Cal/Val design (http://www.ceos.org/index.php?option= com_content&view=category&layout=blog&id=75& Itemid=116). The purpose of the WGCV is to ensure long-term confidence in the accuracy and quality of Earth observation data and products. The WGCV has several subgroups, one of which is land product validation (LPV) (http://lpvs.gsfc.nasa.gov/). The objective of the LPV subgroup is to foster quantitative validation of higher level global land products derived from remotely sensed data, in a traceable way, and to relay results so they are relevant to users. The LPV has six focus groups; one of these is soil moisture.

The WGCV provides the basic definitions of calibration and validation.

• Calibration is the process of quantitatively defining the system response to known, controlled signal inputs.

• Validation is the process of assessing by independent means the quality of the data products derived from the system outputs.

The LPV provides guidance on structuring a Cal/Val program that considers the issues that are specifically associated with land products using four stages:

• *Stage 1:* Product accuracy is assessed from a small (typically < 30) set of locations and time periods by comparison with in situ or other suitable reference data.

• *Stage 2:* Product accuracy is estimated over a significant set of locations and time periods by comparison with reference in situ or other suitable reference data. Spatial and temporal consistency of the product and with similar products have been evaluated over globally representative locations and time periods. Results are published in the peer-reviewed literature.

• *Stage 3:* Uncertainties in the product and its associated structure are well quantified from comparison with reference in situ or other suitable reference data. Uncertainties are characterized in a statistically robust way over multiple locations and time periods representing global conditions. Spatial and temporal consistency

of the product and with similar products have been evaluated over globally representative locations and periods. Results are published in the peer-reviewed literature.

• *Stage 4:* Validation results for stage 3 are systematically updated when new product versions are released and as the time series expands.

A satellite program should attempt to get through at least stage 2 during its Cal/Val phase with a continuing commitment to stage 4.

Other GEOSS subgroups that require soil moisture include the Global Climate Observation System (GCOS) and Global Terrestrial Observation System (GTOS). GCOS has established a Quality Assurance and Earth Observations program (QA4EO) and the GTOS has essential climate variables (ECV) measurement standards (http://www.wmo.int/pages/prog/gcos/index.php?name= EssentialClimateVariables).

The ECVs are those variables that play key roles in the international infrastructure. Being recognized as an ECV is valuable to data providers; however, it comes with the burden of supporting the product with uncertainty and validation studies.

The ECVs are derived from the need for systematic and sustained observations to assess change (environmental or climate) of the three parts of the Earth system, the least satisfied of which is the land component.

A list of land ECVs (soil moisture and other variables) was established and a report that assessed each in terms of importance/urgency, readiness, supporting documents, and existing expertise. This assessment was then used to prioritize supporting activities. Soil moisture was in the top tier priority. For soil moisture there were two very clear messages that are relevant here:

• There are no standards or protocols.

• Because of the size of satellite products, upscaling of points is a big issue for validation.

The other program mentioned above is QA4EO. The focus of this is defining the uncertainty of data and harmonizing diverse recourses. This is very much concerned with reference standards and applies to both in situ and satellite observations.

15.3. SOIL MOISTURE VALIDATION CHALLENGES

Of principle concern in this section is why validation is needed and what are the expectations of validation. Here the focus is on soil moisture and begins with a discussion of some of the challenges we face in the implementation and specifically with in situ observations.

An overarching issue for characterizing soil moisture with point observations is multiscale variability. To start this discussion we begin by considering only bare soils and factors that influence variability. Vegetation will be

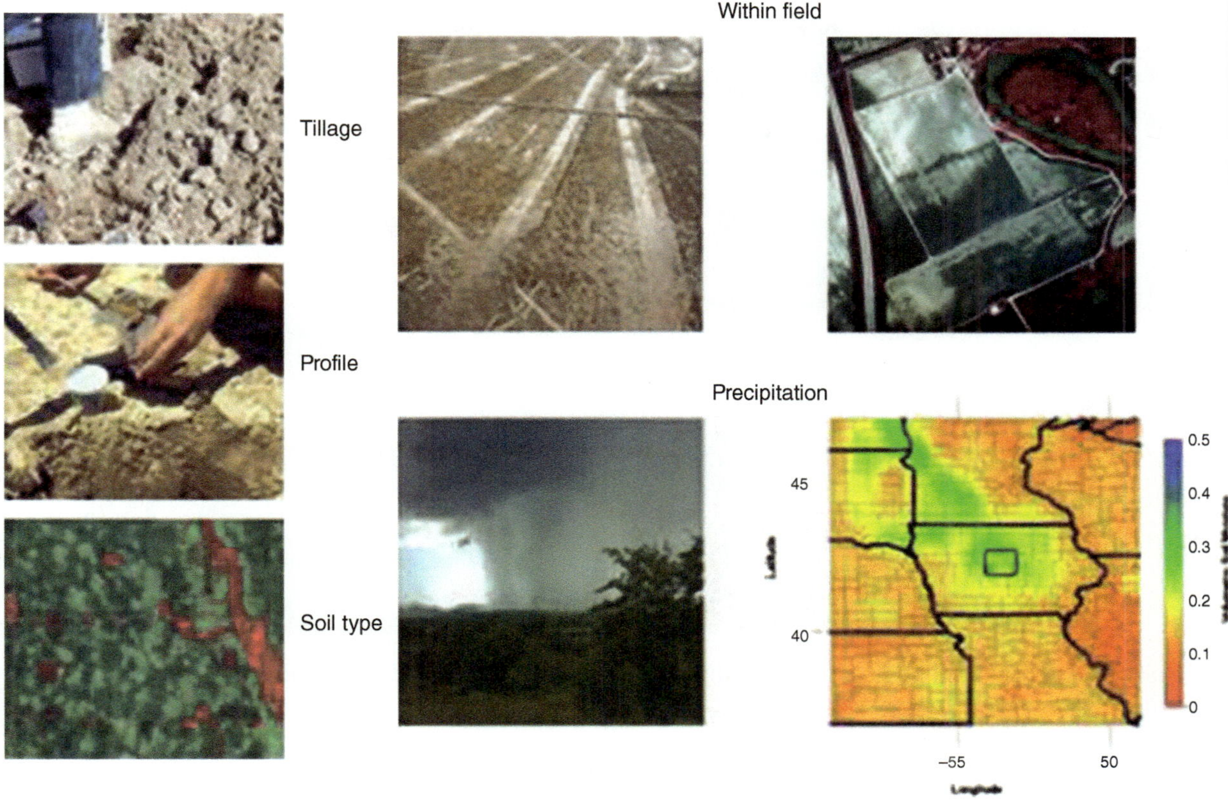

Figure 15.1 Factors contributing to multiscale variability of soil moisture.

another source of variation. At any in situ observing point we will need to consider very small scale effects associated with soil type and tillage (if applied) and the distribution of these vertically in the soil profile. At the field or management unit scale, these effects are compounded by spatial structures resulting from field operations and topography. Precipitation from both localized and synoptic events can further influence variability. The significance of these variations is a derivative of the relatively large footprint size of a satellite sensor (historically 25–50 km). Figure 15.1 illustrates what some of these sources of variability look like. However, even if the satellite sensor has a high resolution for soil moisture of say 1 km, relating this to a single point is still difficult.

Some aspects of this mismatch of scales are illustrated in Figure 15.2. The disparity of spatial scales of points and satellite footprints is four orders of magnitude. In addition, oftentimes the position of the footprint relative to the point cannot be controlled and may change on a day-to-day basis. Thus, establishing a scaling function can prove to be a significant challenge.

The discussion above suggests that it may be very difficult to validate a satellite soil moisture product using an in situ observation. However, we know from both theory and field campaigns that if we have enough sample points available, it is possible to characterize a footprint. In addition, there are tools that can be used to scale a point to a footprint under some conditions [*Crow et al.*, 2012].

Another major issue involving in situ data is that the available resources involve different types of ground-based sensors and standards. Each of these methods involves different techniques, volumes, depths, calibrations, and latency issues. If we are going to use all of the available resources, we have to figure out how to harmonize them.

We also face the challenge for soil moisture validation of dealing with multiple satellites that have different characteristics in terms of measurement technique, contributing depth, spatial resolutions, and uncertainty.

Finally, various data users and providers (not just remote sensing) have used different units of measurement.

15.4. THE SMAP CAL/VAL PLAN

As noted earlier, the objective of this presentation is to describe the basis for the SMAP soil moisture Cal/Val plan. SMAP provides a variety of level 1 (instrument), level 2/3 (science), and level 4 (value-added) products. Here the discussion focuses on L2/L3 soil moisture, which is tied most directly to the in situ soil moisture observations. More details on SMAP can be found at http://smap.jpl.nasa.gov/.

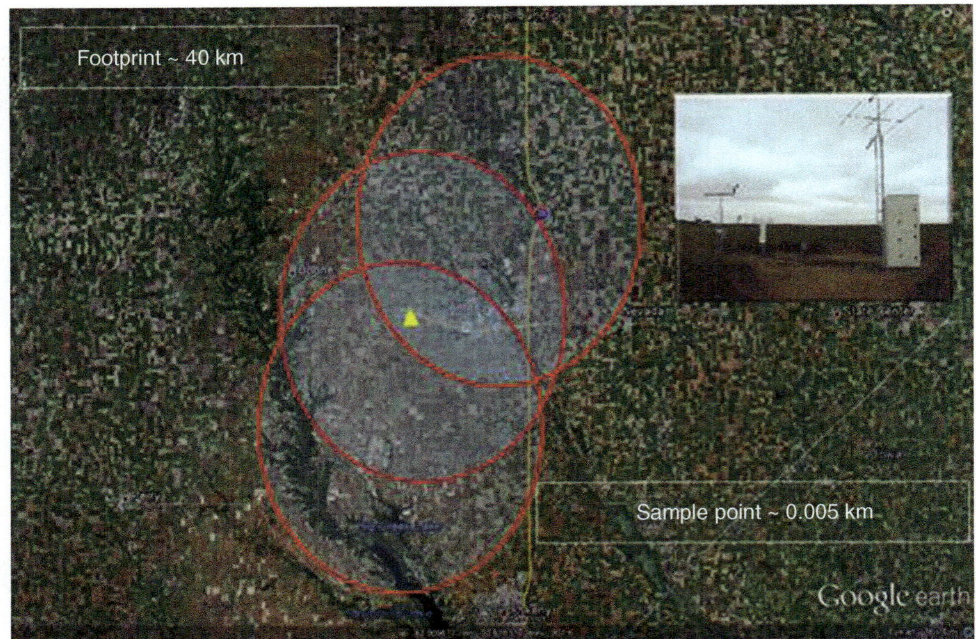

Figure 15.2 Illustration of the disparity of scales of in situ and satellite observations of soil moisture for a a typical low-frequency radiometer.

The starting point for the SMAP Cal/Val plan was the set of mission requirements, which specifically state that the soil moisture product will have an accuracy of 0.04 m³/m³ for the 0–5 cm soil moisture at a spatial resolution of 10 km. This product is referred to as L2_SM_AP and is based on the combined active and passive algorithm. The actual resolution and gridding of the product is 9 km. In addition to the AP, SMAP provides two other soil moisture products: L2_SM_P (passive only on a 36 km grid) and L2_SM_A (active only on a 3 km grid). Providing Cal/Val of all three products presents significant problems in terms of replicate sampling and upscaling. Mission requirements also included conducting a Cal/Val program.

In constructing the Cal/Val program, activities were separated into pre- and postlaunch phases. The prelaunch phase is mostly concerned with algorithm development and infrastructure; whereas postlaunch is focused on validation and algorithm improvement. Although validation will be a continuing activity throughout the mission life, a major constraint on the postlaunch validation is that the assessments must be completed within 15 months of launch (Cal/Val phase). This means that the entire Cal/Val process and resources must be ready to go. There is no time for resolving issues with the methodologies and data during this period.

SMAP will adopt the CEOS validation stages as described in *Baret et al.* [2009] and described earlier.

In attempting to address the requirements, the validation stages, and the challenges of validating a satellite soil moisture product, we have developed a plan that will utilize five different but complementary methodologies (Table 15.1). The core validation sites (CVS) are intended to support the stage 1, and all the others, except field campaigns, will support stage 2 and beyond.

Early in the science development of SMAP, primarily through a Cal/Val working group, each of these methodologies was scrutinized to identify issues that we might address during the prelaunch phase that would improve validation. These are listed in Table 15.1. For each issue actions were then defined and implemented. Those related to in situ observations are described below in more detail.

As noted in Table 15.1, core validation sites and sparse networks share the issues of sensor calibration and limited number of sites. Sensor calibration also involves comparability. There are many different types of sensors available that are based upon different physical principles. Although many of these have a default calibration, all can benefit from site calibration using thermogravimetric methods. In addition to the basic physics exploited, the designs differ. Some are installed horizontal at a depth and others vertically. The volume measured can also vary.

15.4.1. In Situ Sensor Calibration and Comparability

In order to effectively utilize these diverse resources for SMAP validation, it was suggested that all validation sites conduct a site-based calibration (thermogravimetric) to the 0–5 cm layer. The benefit to SMAP is that all data sets would be comparable to its surface soil moisture product.

Table 15.1 SMAP validation methodologies

Methodology	Role	Constraints	Resolution
Core validation sites	Accurate estimates of products at matching scales for a limited set of conditions	• In situ sensor calibration • Limited number of sites	• In situ testbed • Cal/Val partners
Sparse networks	One point in the grid cell for a wide range of conditions	• In situ sensor calibration • Upscaling • Limited number of sites	• In situ testbed • Scaling methods • Cal/Val partners
Satellite products	Estimates over a very wide range of conditions at matching scales	• Validation • Comparability • Continuity	• Validation studies • Distribution matching
Model products	Estimates over a very wide range of conditions at matching scales	• Validation • Comparability • Continuity	• Validation studies • Distribution matching
Field campaigns	Detailed estimates for a very limited set of conditions	• Resources • Schedule conflicts	• Simulators • Partnerships

Note: Applies to all L2–L4 products.

Recognizing that the responsibility for providing this calibration/comparability might fall upon SMAP, NASA supported the establishment of the SMAP Marena Oklahoma In Situ Sensor Testbed (SMAP-MOISST) (http://ars.usda.gov/Research/docs.htm?docid=20251). Although many such sites would be desirable, budgetary concerns were restrictive, so a single representative site was established, taking into account as many critical factors as possible. These included long-term access to a low vegetation landscape in a region of interest for soil moisture studies. The topographic variability also had to be relatively low and soil type should be homogenous across the approximate 700 m site dimension. This dimension was dictated by the inclusion of the COSMOS (Cosmic Ray Soil Moisture Observing System), which has a sensing diameter of approximately 700 m [*Zreda et al.*, 2012]. Setting up the site required some additional trade-offs between replications and resources. So, the team established four replicate base stations within the domain. Each of these provides measurements at standard depths and included many of the most widely used sensors as well as several experimental approaches. These included the aforementioned COSMOS, global positioning system reflectometry [*Larson et al.*, 2008], and distributed temperature sensing systems [*Selker et al.*, 2006]. Calibration and scaling of the sensors and site are addressed with a series of intensive sampling campaigns that include thermogravimetric sampling. This work was initiated in 2010.

15.4.2. Cal/Val Partner Sites

SMAP will produce global products. Validation should attempt to be robust and incorporate diverse land cover and climate conditions, geographic distribution, and be statistically significant. Providing this information is a major issue facing SMAP, as well as other satellite-based soil moisture programs, because there are a limited number of core validation sites available. As described earlier, these sites provide accurate estimates of SMAP products at matching scales. We have found these to be very useful in our previous efforts at satellite-based soil moisture validation [*Jackson et al.*, 2010; *Jackson et al.*, 2012].

Recognizing this issue, one of the Cal/Val program actions was to increase the number of these sites and to ensure that SMAP will get the data in a timely manner. The approach used was to have NASA issue an announcement of opportunity for participation in the project as a Cal/Val partner. These agreements would be no exchange of funds and allowed international participation. In exchange for providing in situ data to SMAP, the Cal/Val partners would receive early access to SMAP data. The call provided guidance and minimum requirements for CVS, which included in situ observing sites that provide well-characterized estimates of an L2–L4 product at a matching spatial scale (i.e., 0–5 cm surface soil moisture) as well as the following:

• Calibration of the in situ sensors
• Upscaling strategy provided
• Data provided in a timely manner
• Long-term commitment by the sponsor/host
• An area that is homogeneous or has a uniform mixture of land covers at the product scale
• Represents an extensive or important biome and complements the overall set of sites

The response to the call was good for surface soil moisture and included sites that could address all three spatial resolutions of the soil moisture products. Geographic coverage distribution was also good. In addition to the CVS, several networks responded that did not meet the criteria for being a CVS. These were mostly sparse operational networks. Although they have upscaling issues, these data will have a role in stage 2 validation of SMAP (Table 15.1). These networks have several positive features: they are usually operational, low latency, and well documented. As a result, a second category of Cal/Val partners was created for contributing validation sites.

Following this solicitation, the project opened up the opportunity for additional participants. Several valuable additional resources have been engaged that were not a part of the initial solicitation process.

15.4.3. Upscaling

Regardless of what advances may be made in the spatial resolution of satellite-based soil moisture mapping, there will always be a disparity of scales between the satellite footprint and what we can measure on the ground (~4 orders of magnitude for SMAP). This is clearly illustrated by Figures 15.1 and 15.2.

The challenges to upscaling will depend upon whether or not there are replicate samples available within a product grid or if observations are available only at a single point. SMAP initiated an assessment of upscaling methodologies that is reported in *Crow et al.* [2012]. That review presents the protocols that can be employed. With an adequate number of samples, upscaling can be as simple as averaging. However, more sophisticated methods can be used. For sparse networks (single points), it is much more difficult to conduct a traditional validation assessment. Approaches that use intensive field campaigns and/or temporary networks can be employed [*Cosh et al.*, 2013]. In addition, upscaling errors associated with the direct comparison of sparse network with footprint-scale retrievals, can be estimated—and corrected for—using statistical techniques such as triple colocation analysis.

15.5. SUMMARY

SMAP, as well as other soil moisture missions, have specific requirements for validation that include accuracy. One of the most important methodologies available for validating satellite soil moisture is data from in situ observing networks. The technology of in situ soil moisture remote sensing has also advanced over the same period that satellite sensing has progressed. As a result of more reliable sensors and improved data acquisition systems, the number of in situ soil moisture networks available has increased. Unfortunately, these networks have evolved without national or international standardization, which presents challenges to the validation of satellite-based soil moisture remote sensing, which requires the integration of numerous networks. In addition to the issues of a limited number of sites and their standardization, the validation of satellite-based soil moisture products faces the challenge of resolving the disparate scales of the sensor footprints and the in situ sensor. The actions that SMAP has taken to ameliorate issues with calibration, number of sites, and upscaling should lead to a robust validation in a timely manner.

REFERENCES

Baret, F., J. Nightingale, S. Garrigues, C. Justice, and J. Nickeson (2009), Report on the CEOS land product validation subgroup meeting, *Earth Observ.*, *21*(6), 26–30.

Bindlish, R., T. J. Jackson, M. Cosh, T. Zhao, and P. O'Neill (2014), Global soil moisture from the Aquarius/SAC-D satellite: description and initial assessment, *IEEE Geosc. and Remote Sens. Lett.*, in press.

Cosh, M. H., T. J. Jackson, C. Smith, B. Toth, and A. Berg (2013), Validating the BERMS in situ soil water content data record with a large scale temporary network, *Vadose Zone J.*, *12*, doi:10.2136/vzj2012.0151.

Crow, W. T., A. A. Berg, M. H. Cosh, A. Loew, B. P. Mohanty, R. Panciera, P. de Rosnay, D. Ryu, and J. P. Walker (2012), Up-scaling sparse ground-based soil moisture observations for the validation of coarse-resolution satellite soil moisture products, *Rev. Geophys. 50*(2), doi:10.10229/2011RG000372.

Entekhabi, D., et al. (2010), The soil moisture active and passive (SMAP) mission, *Proc. IEEE*, *98*, 704–716.

Jackson, T. J., M. H. Cosh, R. Bindlish, P. J. Starks, D. D. Bosch, M. S. Seyfried, D. C. Goodrich, and M. S. Moran (2010), Validation of advanced microwave scanning radiometer soil moisture products, *IEEE Trans. Geosc. and Remote Sens.*, *48*, 4256–4272.

Jackson, T. J., R. Bindlish, M. H. Cosh, T. Zhao, P. J. Starks, D. D. Bosch, M. S. Moran, M. S. Seyfried, Y. Kerr, and D. Leroux (2012), SMOS validation of soil moisture and ocean salinity (SMOS) soil moisture over watershed networks in the *U.S.*, *IEEE Trans. Geosci. Remote Sens.*, *50*, 1530–1543.

Kerr, Y. H., et al. (2012), The SMOS soil moisture retrieval algorithm, *IEEE Trans. Geos. Remote Sens.*, *50*, 1384–1403.

Larson, K. M., E. E. Small, E. Gutmann, A. Bilich, J. Braun, and V. Zavorotny (2008), Use of GPS receivers as a soil moisture network for water cycle studies, *Geophysi. Res. Lett.*, *35*, L24405.

Njoku, E., T. Jackson, V. Lakshmi, T. Chan, and S. Nghiem (2003), Soil moisture retrieval from AMSR-E, *IEEE Trans. Geosci. Remote Sens.*, *41*, 215–229.

Selker, J. S., L. Thevanez, H. Huwald, A. Mallet, W. Luxemburg, N. van de Giesen, M. Stejskal, J. Zeman, M. Westhoff, and M. B. Parlange (2006), Distributed fiber-optic temperature sensing for hydrologic systems, *Water Resourc. Res.*, *42*, W12202, doi:10.1029/2006WR005326.

Wagner, W., W. Dorigo, R. de Jeu, D. Fernandez, J. Benveniste, E. Haas, and M. Ertl (2012), Fusion of active and passive microwave observations to create an Essential Climate Variable data record for soil moisture, ISPRS Annals of the Photogrammetry, Remote Sensing and Spatial Information Sciences, Volume I-7, XXII ISPRS Congress, 25 August–01 September 2012, Melbourne, Australia. pp. 315–321.

Wagner, W., et al. (2013), The ASCAT soil moisture product: A review of its specifications, validation results, and emerging applications, *Meteorol. Zeits.*, *22*(1), 5–33.

Zreda, M., W. J. Shuttleworth, X. Zeng, C. Zweck, D. Desilets, T. Franz, and R. Rosolem (2012), COSMOS: The Cosmic-Ray Soil Moisture Observing System, *Hydrol. Earth Sys. Sci.*, *16*, 4079–4099.

16

Soil Moisture Retrieval from Microwave (RADARSAT-2) and Optical Remote Sensing (MODIS) Data Using Artificial Intelligence Techniques

Nasreen Jahan and Thian Yew Gan

16.1. INTRODUCTION

Even though soil moisture only represents a small fraction of the global water budget, it is nonetheless an important hydrologic variable in many water-related applications such as hydrological modeling, crop growth modeling, and streamflow forecasting. Accurate and timely measurements of soil moisture are also essential for effective irrigation management, crop selection, and plant stress determination. However, the accurate estimation of soil moisture at regional or larger scale is difficult because soil moisture varies highly over space and time, and ground measurements are often time consuming and expensive. Past studies [e.g., *Biftu and Gan*, 1999; *Sokol et al.*, 2004; *Said et al.*, 2008; *Baghdadi et al.*, 2006] have shown that airborne and satellite active microwave sensors can be used to retrieve soil moisture of bare soil or areas with sparse vegetation by developing empirical relationships between soil moisture and microwave backscattering. *Moran et al.* (2004) reported that a soil moisture retrieval model primarily based on microwave remote sensing such as synthetic aperture radar (SAR) sensors is an efficient approach for obtaining spatially distributed soil moisture. The advantages of radar (radio detection and ranging) remote sensing are that its active microwave radiation can penetrate cloud cover and also it can operate both day and night.

Department of Civil and Environmental Engineering, University of Alberta, Edmonton, Canada

16.2. SOIL MOISTURE RETRIEVAL MODELS

An estimation of soil moisture from radar data is generally done by either theoretical [*Fung et al.*, 1992] or empirical [*Oh et al.*, 1992; *Dubois et al.*, 1995], or semiempirical models [*Chen et al.*, 1995; *Oh*, 2004]. The integral equation model (IEM) of *Fung et al.* [1992] is one of the most widely used theoretical models in inversion procedures of SAR images for retrieving soil moisture. The inversion procedure of IEM requires three roughness parameters: the standard deviation of surface height, σ; the surface correlation length, L; and the shape of surface autocorrelation function, ACF. Unfortunately, these parameters are difficult to measure in the field [*Baghdadi et al.*, 2008]. Therefore, different inversion algorithms have been proposed to retrieve soil moisture based on IEM [*Thoma et al.*, 2006; *Altese et al.*, 1996].

In December 2007, the Canadian Space Agency (CSA) launched RADARSAT-2, the follow-on satellite to RADARSAT-1. RADARSAT-2 is the first commercial spaceborne SAR satellite that produces fully polarimetric data sets (HH, HV, VV, and VH). The first letter in the terms HH, HV, VV, VH indicates the polarization of the transmitted signal (H denotes horizontally polarized radiation and V denotes vertically polarized radiation) and the second letter indicates the polarization of the received signal. It acquires images at incidence angles ranging from 20° to 60° and a resolution varying from 3 to 100 m. The sensor's quad polarization mode enables it to capture the comprehensive characteristics of the scattering field of a surface. The horizontally polarized mode provides useful information about the underlying soil condition such as soil moisture [*Soria-Ruiz et al.*, 2007; *McNairn and Brisco*,

Remote Sensing of the Terrestrial Water Cycle, Geophysical Monograph 206. First Edition. Edited by Venkat Lakshmi.
© 2015 American Geophysical Union. Published 2015 by John Wiley & Sons, Inc.

2004]. On the other hand the VV-polarized backscatter is useful in determining vegetation growth stage, height, type, and leaf water content [*Martin et al.*, 1989] while HV (or VH) backscatter can provide information about crop conditions such as crop yield [*McNairn et al.*, 2004] and biomass of crop [*Ferrazzoli et al.*, 1997]. In very recent years few studies have utilized these polarimetric data from the RADARSAT-2 to retrieve soil moisture [*Gherboudj et al.*, 2010; *Hajnsek et al.*, 2009; *Merzouki et al.*, 2011; *Pierdicca et al.*, 2010; *Pasolli et al.*, 2011]. *Pasolli et al.* [2011] found that HV backscatter from RADARSAT-2 was able to disentangle the vegetation effect on the radar signal and use of polarimetric data improved the soil moisture estimation capability in an alpine area of Italy. *Merzouki et al.* [2010] estimated soil moisture from RADARSAT-2 data by using the Dubois model and a lookup table (LUT) approach applied to the Oh model. *Gherboudj et al.* [2011] used RADARSAT-2 linear polarization data and ground measurements (acquired over the mature crops) in the semiempirical water-cloud model to retrieve the soil moisture. They estimated soil surface roughness, crop height, vegetation water content, and soil moisture with a relative error of 19%, 10%, 25.5%, and 32%. *Merzouki et al.* [2011] compared the modeled backscatter, from the Dubois, Oh, and IEM models against the RADARSAT-2 measured backscatter and found some bias and discrepancies. They emphasized that significant study remains to be done to improve the accuracy of the radar backscatter models.

16.3. RESEARCH OBJECTIVES

The objective of this study is to examine the application of three artificial intelligence (AI) techniques, namely artificial neural network (ANN), adaptive neuro-fuzzy interference system (ANFIS), and the support vector machine (SVM) to retrieve soil moisture over vegetated soil and compare the results with observation and IEM-based inverse model. In recent years, AI techniques have become increasingly popular in hydrology and water resources studies because of their excellent capability of analyzing nonlinear relationships among hydrologic variables. ANN is one of the most popular AI techniques. Some past studies [*Said et al.*, 2008] have used ANN in retrieving soil moisture. Besides radar backscatter, those models [*Said et al.*, 2008; *Ramirez-Beltran et al.*, 2008] have used additional predictors (e.g., vegetation variables such as moisture of stems and leaves, density of stems, plant water content, terrain elevation, precipitation, soil moisture, etc.) obtained through field measurements.

The ANFIS provides the benefits of neural network and fuzzy logic in a single framework and demonstrated its great ability of functional mapping between input and output and generalization in past studies. Examples of ANFIS application are available in rainfall-runoff modeling [*Talei and Chua*, 2012], river flow prediction [*Pramanik and Panda*, 2009], groundwater modeling [*Daliakopoulos et al.*, 2005], etc. However, application of ANFIS in modeling soil moisture from radar data is rare.

The SVM is based on statistical learning theory and can be used to predict a variable through the use of a trained model that utilizes past data. This learning strategy was developed by Vapnik and coworkers in the early 1990s for classification problems and later *Vapnik* [1995] extended SVM for regression problems. Although, SVMs have successfully been used for pattern recognition and regression in bioinformatics and AI, there are also a few applications of SVM in hydrology. *Lin et al.* [2009] applied SVM to forecast hourly typhoon rainfall in the Fei-Tsui Reservoir watershed in northern Taiwan while *Yang et al.* [2006b] has applied SVM for modeling continental scale evapotranspiration. Among the three AI methods, several examples of ANN application in soil moisture retrieval from radar data are available; however, there are only a few applications of SVM and ANFIS in modeling soil moisture from radar data. Moreover most of the past studies were done for bare soil and used single polarized data; therefore, we expect that this study of soil moisture retrieval under vegetated conditions will be very useful in mapping soil moisture with more operational advantage in support of agricultural monitoring.

The specific objectives of this study include the following:

1. To examine the sensitivity of HH, VV, HV, and VH radar backscatter from RADARSAT-2 to soil moisture.

2. To assess the accuracy of soil moisture retrieved from the quad-polarization data using an inversion procedure based on IEM and three artificial intelligence techniques—ANN, ANFIS, and SVM.

3. To test usefulness of optical satellite data (surface temperature and vegetation index) as supplementary predictors in the AI techniques.

16.4. DATA AND STUDY SITES

16.4.1. Study Site

The study site is located in the Paddle River Basin (53°52′N, 115°32′W) of Alberta, Canada (Figure 16.1). The basin consists of about 50% mixed forest, 21% coniferous forest, 15% agricultural land, 11% pasture land with short grass, 2% water body, and 1% impervious land [*Biftu and Gan*, 1999]. The predominant vegetation of this basin is deciduous and aspen forest [*Alberta Energy and Natural Resources*, 1977]. The slope of the basin is 3%–5% and the annual runoff coefficient is of about 0.28. The major soil type of this basin is characterized as orthic gray luvisol belonging to the Hubalta series, nonsaline, and moderately fine-textured glacial till. The

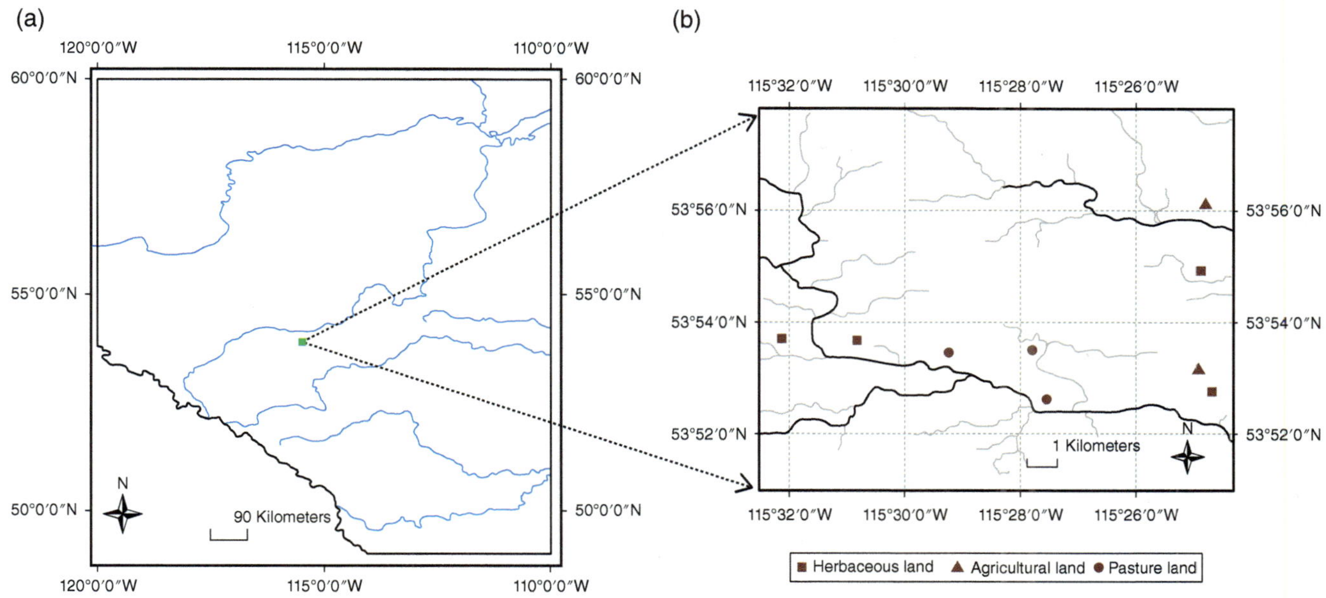

Figure 16.1 (a) Location of the study area within the province of Alberta, Canada and (b) location of individual study sites within the Paddle River Basin.

Table 16.1 Description of 10 RADARSAT2 images acquired over the Paddle River Basin at descending orbit

Acquisition Date	Scene Center		Beam Mode	Angle of Incidence
	Longitude	Latitude		
29 May 2009	115°34′04″ W	53°52′33″ N	FQ8	27.83
22 June 2009	115°34′04″ W	53°52′28″ N	FQ8	27.83
16 July 2009	115°33′40″ W	53°52′30″ N	FQ8	27.83
17 May 2010	115°32′38″ W	53°53′33″ N	FQ4	23.20
24 May 2010	115°33′32″ W	53°53′14″ N	FQ8	27.83
10 June 2010	115°33′13″ W	53°53′29″ N	FQ4	23.20
4 July 2010	115°32′42″ W	53°53′31″ N	FQ4	23.20
19 May 2011	115°33′39″ W	53°53′47″ N	FQ8	27.83
5 June 2011	115°33′12″ W	53°53′54″ N	FQ4	23.20
12 June 2011	115°33′52″ W	53°53′40″ N	FQ8	27.83

typical soil texture is clay loam at a depth of 0–10 cm [*Biftu and Gan*, 1999]. The mean temperature of the basin is from − 15.5 °C in January to 15.6 °C in July, and the mean annual precipitation is about 500 mm. The field sites chosen for collecting soil moisture data are comprised of agricultural land, pasture land (with short grass), and herbaceous land (with grass, weed, short bush). In the study period of 2009–2011, data were collected during the early growing period (Table 16.1).

16.4.2. Data

16.4.2.1. Soil Moisture Data

Soil moisture data from 9 sites (4 agricultural, 3 pasture land, 2 herbaceous land) (Figure 16.2) were collected in selected 3 days of 2009, 4 days of 2010, and 3 days of 2011 (Table 16.1) when the RADASAT-2 satellite passed over the study site. These dates were chosen because on these dates SAR images were acquired with low incidence angle, which is desirable for soil moisture retrieval since at low incidence angle the effect of surface roughness and vegetation on radar backscatter is minimal. The sites were chosen at intersite distances of 1–10 km from each other. On each day, soil moisture data were collected within 2–3 h of RADARSAT's image acquisition. At each site, data was collected from 25 sampling points and at an integrated depth of 0–5 cm. The sampling points were located at approximately 50 m apart and covered a plot of about 200 m × 200 m in area. On those 10 days a total of 2250 soil samples were collected from the 9 sites.

Figure 16.2 Photographs of three land cover types used in this study: (a) herbaceous land, (b) agricultural land, and (c) pasture land.

The soil samples were analyzed in the soil laboratory at the University of Alberta to obtain gravimetric soil moisture and then converted to volumetric soil moisture.

16.4.2.2. Radar Data

Ten RADARSAT-2 images were provided by the Canadian Centre for Remote Sensing (CCRS) under the Scientific Operational Applications Research – Education (SOAR-E) initiative, which is a joint program between the MacDonald Dettwiler and Associates Ltd.– Geospatial Services Inc. (MDA-GSI) and the government of Canada, represented by the Canadian Space Agency (CSA) and CCRS. The images were obtained in the form of SAR georeferenced extra fine resolution product (Path Image Plus, SGX) format. The images were obtained in descending mode and low incidence angles. Details of the images are given in Table 16.1.

Each scene was subjected to radiometric calibration, speckle filtering, and georeferencing. Radiometric calibration was done following the approach of *Shepherd* [2000] using the LUT provided with the data. In this study speckle was filtered using the 7×7 gamma map filter [*Lopes et al.*, 1993], which minimizes the loss of texture information important for mapping forested and agricultural area. Next the images were geometrically corrected by coregistering to a Landsat TM image with respect to many ground control points uniformly distributed across the whole image and based on a first-order polynomial transformation. The average root-mean-squared error (RMSE) was found to be less than two pixels in both X and Y directions for all RADARSAT-2 images. The original image with a pixel spacing of $3.125 \times 3.125\,\text{m}$ was then resampled to an image of $25 \times 25\,\text{m}$ resolution using the nearest neighbor method.

Lastly, the backscatter values of the study sites were extracted from the processed images.

16.4.2.3. Landsat and MODIS Data

The Landsat images of the study site dates were acquired from the U.S. Geological Survey (USGS) (http://landsat.usgs.gov/). However, no Landsat image of the study area was available for the same dates when RADARSAT images were acquired because Landsat images are available at 16 day intervals. Moreover, all these available images suffer from partial to full cloud covers over the study area, which hindered the use of these Landsat images. For example, we could not get any cloud-free Landsat images for May, 2010. Therefore, instead of Landsat images, we used images of the Moderate Resolution Imaging Spectroradiometer (MODIS) sensor, which views the entire Earth surface every 1–2 days. For each of the selected dates, MODIS images were available. However, in some days, due to the cloud cover effects, data of some portions of the study area were not available, and for those effected portions we used the spectral data of images acquired the next day. MODIS surface reflectance products and land surface temperature (LST) data were obtained from the USGS' Land Processes Distributed Active Archive Centre. Reflectance products are atmospherically corrected and available at 250 m resolution. On the other hand, MODIS LST, is available at 1 km resolution and retrieved using the split window algorithm and the thermal infrared bands of MODIS [*Wan and Dozier*, 1996].

16.4.2.4. Digital Elevation Data

The digital elevation model (DEM) data of the Paddle River Basin (PRB) was collected through the GeoBase portal (www.geobase.ca.) of the Canadian Council on Geomatics (CCOG). The resolution of the data set is 0.75 arc seconds. From the elevation data, topographic variables such as slope and aspect were calculated using ArcGIS. A detail description of the DEM data quality is available in http://www.geobase.ca/geobase/en/data/cded/description.html.

16.4.2.5. Soil Properties Data

The soil properties of the study sites were obtained from the Agricultural Region of Alberta Soil Inventory Database (AGRASID), which is a database describing the soil properties and landscapes within the agricultural region of Alberta. Since the beginning of soil mapping in Alberta in 1920, more than 70 reconnaissance soil surveys have been conducted by different organizations, such as the Alberta Research Council, Agriculture and Agri-Food Canada, and the University of Alberta. Details of the soil properties are available in the website of Alberta Agriculture (http://www1.agric.gov.ab.ca/). For each site we extracted the information of percent of sand, silt, and clay and the water holding capacity of the soil.

16.5. SOIL MOISTURE RETRIEVAL ALGORITHMS

16.5.1. Input Variable Selection

To select the input variables in the AI-based soil moisture retrieval models, we used the following procedures:

1. *Selection of Preliminary Candidate Input Variables Based on Physical Basis* The selection of appropriate model input variables is crucial in data-driven prediction models since it provides the basic driving factors behind the system to be modeled. A number of variables with physical basis were chosen as candidate input variables to the soil moisture retrieval algorithm. The preliminary candidate variables selected were radar backscatter, land surface temperature (LST), normalized difference vegetation index (NDVI), leaf area index (LAI), soil information (percent of sand, silt, and clay and water holding capacity), terrain information (elevation, slope, and aspect), incidence angle, co-polarization ratio, and cross-polarization ratio.

HH backscatter can penetrate the vegetation and provide useful information about soil moisture. As radar backscatter contains signals from both vegetation and soil moisture, we have also chosen VV backscatter as a candidate variable since it is useful to distinguish crop types and vertical structures of vegetation [*McNairn and Brisco*, 2004]. Studies also showed that VV backscatter is related to the leaf water content [*Martin et al.*, 1989], which should be related to the status of available soil moisture. On the other hand, HV (or VH) backscatter can provide information regarding crop conditions such as the productivity [*McNairn et al.*, 2004] and biomass of crops [*Ferrazzoli et al.*, 1997]. We also examined the possibility of using cross- and co-polarization ratios in retrieving soil moisture given that some past studies found them to be useful information in soil moisture retrieval [*Oh*, 2004].

The NDVI is the reflectance difference between the visible red (R) and near-infrared (NIR) bands, over their sum [equation 16.1]:

$$NDVI = \frac{NIR - R}{NIR + R}. \qquad (16.1)$$

NDVI and LAI are indices related to vegetation types, growth, and density. Studies have shown that NDVI of vegetation is related to the spatial and temporal patterns of precipitation given that precipitation is the primary source of soil moisture [*Jahan and Gan*, 2011]. NDVI and LAI are also related to vegetation properties such as surface roughness, which affect the radar backscatter [*Gupta et al.*, 2002; *Makkeasorn et al.*, 2006]. Some recent studies used NDVI as a predictor to simulate soil moisture [*Yang et al.*, 2006a; *Makkeasorn et al.*, 2006]. Studies using optical data to simulate soil moisture have also reported that LST and NDVI can be used together to model the crop stress and soil moisture condition effectively because LST is controlled by soil moisture [*Carlson*, 2007; *Patel et al.*, 2009]. In general the higher the soil moisture, the cooler should be the land surface. Therefore, in addition to radar backscatter, we have also considered NDVI, LAI, and LST as candidate input variables in the preliminary analysis.

We have also examined the potential of using soil properties (percent of sand, silt, and clay and water holding capacity), terrain information (elevation, slope, and aspect), and incidence angles (ϑ) as candidate input variables to the AI techniques. After a rainfall event, soil properties are part of the controlling factors behind the soil infiltration process and the water holding capacity of soil [*Makkeasorn et al.*, 2006]. In addition, terrain features such as elevation, slope, and aspect may also affect the radar backscatter [*Makkeasorn et al.*, 2006].

2. Selecting Key Variables from the Candidate Input Variables Using Regression and Eigenvalue Analysis Jolliffe [1973] discussed eight methods for selecting variables and referred to the process as *discarding variables*. These eight methods were based on three basic approaches: correlation analysis, clustering of variables, and principal components [*Rencher*, 2002]. In the second step, we performed correlation analysis and principal component method based on eigenvalue analysis to select the key variables through excluding the less useful variables.

At first we performed correlation analysis between each candidate input variables and soil moisture. Results shows that HH, VV, and HV radar backscatter, LST, and NDVI are significantly correlated with soil moisture at 5% significance level. Among these variables, correlation analysis showed that co-polarization and cross-polarization ratios are not well correlated with soil moisture. Further, we also found minimal correlation between soil moisture and LAI, elevation, and percent of sand, silt, and clay.

In screening the input variables, we then conducted eigenvalue analysis to discard less important variables following the procedure described in *Rencher* [2002]. According to *Rencher* [2002], we first computed the eigenvalues, eigenvectors, and principal components of all candidate input variables. Next, components with high eigenvalues that account for a specified percentage of the total variance, e.g., 90%, were retained. Then, input variables with large coefficients (in absolute value) in the selected principal components were retained, if the

Table 16.2 Eigenvalues of principal components

Principal Component (PC)	Eigenvalue	% of Variance Explained	Cumulative % of Variance Explained
PC1	2.86	31.82	31.82
PC2	1.86	20.65	52.46
PC3	1.13	12.59	65.05
PC4	0.99	10.98	76.03
PC5	0.76	8.47	84.50
PC6	0.58	6.48	90.98
PC7	0.42	4.65	95.63
PC8	0.30	3.34	98.98
PC9	0.09	1.02	100.00

Table 16.3 Eigenvectors of first six principal components

Variable	Eigenvectors of First Six Components (V)					
	V1	V2	V3	V4	V5	V6
σ°_{VH}	0.83	−0.23	0.38	0.26	−0.24	−0.09
σ°_{VV}	0.96	−0.21	0.08	0.21	−0.12	0.35
σ°_{HH}	1.00	−0.20	−0.03	0.09	−0.20	0.14
LST	−0.46	0.68	0.47	0.19	−0.83	1.00
NDVI	0.24	0.44	1.00	−0.23	1.00	0.18
Slope	0.26	1.00	0.34	−0.35	−0.48	−0.75
Aspect	−0.25	−0.86	0.68	−0.13	−0.75	−0.53
Water holding capacity	−0.48	−0.93	0.54	0.14	0.26	0.14
Angle	−0.25	0.46	0.16	1.00	0.18	−0.57

variables have not been previously selected. This process was repeated, by re-computing the new principal components after a candidate variable is retained or deleted. Through this iterative process we found some variables, namely, co-polarization and cross-polarization ratios, LAI, elevation, percent of sand, silt, and clay to be less useful. This finding is similar to the finding from the correlation analysis. Consequently, the final set of retained variables includes radar backscatter, LST, NDVI, soil information (water holding capacity), terrain information (slope and aspect), and incidence angle.

The results of eigenvalue analysis (final iteration) are given in Tables 16.2 and 16.3. The eigenvectors are scaled so that the largest value in each component is 1. Given the first six principal components explain about 90% of the total variance (Table 16.2), we decided to keep these six components. This amount of variance should be sufficient for most descriptive purposes [*Rencher*, 2002]. Table 16.3 shows the contribution of the input variables in different components. It is noticed that the largest coefficients in component 1 correspond to HH, VV, and HV radar backscatters, respectively. This result indicates that radar backscatters have notable influence in component 1, which explains about 31% of the total variance. In component 2, slope, aspect, and water holding capacity have the largest coefficients. On the other hand NDVI and angle are the most influencing variables in component 3 and 4, respectively, and LST is the dominating variable in both components 5 and 6. In summary, on the basis of the eigenvalue and correlation analysis we selected radar backscatter, LST, NDVI, soil information (water holding capacity), and terrain information (slope and aspect and incidence angle) as input variables as these variables have been found to be the dominating variables in the first six principal components and well correlated with soil moisture, and thus they should possess sufficient information as input to the AI model to retrieve the soil moisture.

We also used the variable inflation factor (VIF) as a diagnostic tool to check the multicollinearity between the retained predictor variables to make sure that there is no redundancy between the retained variables. Usually, VIF values greater than 10 suggest the presence of multicollinearity [*Chatterjee et al.*, 2000]. In our study, the computed VIF for each variable has a value that is less than 10.

3. *Determining Optimum Number of Variables* To determine the optimum number of input variables, we started training the network with a minimal number of input variables (i.e., only radar backscatter) from the retained variables, and we gradually increased the number of input variables and trained the corresponding network. This procedure of adding and changing the combination of predictors enable us to determine which combination of predictors work out as the most useful sets of variables in retrieving the soil moisture. The performance of each set of input variables is evaluated in terms of RMSE. We also used the Akaike information criterion (AIC), which has been widely used for determining the optimum set of input variables [*Panchal et al.*, 2010; *May et al.*, 2011], to determine the optimal set of input variables by finding an optimal trade-off between the number of input variables and the modeling accuracy.

16.5.2. Inversion Procedure to Retrieve Soil Moisture Using the Integral Equation Model

In this study we used an IEM-based inversion procedure to retrieve soil moisture. IEM is a theoretical backscatter model developed for randomly rough dielectric surface and the complete version of IEM is theoretically valid for all scales of roughness and a wide range of wavelength. However, due to the complexity of a complete version, an approximate version of IEM is more practical [*Altese et al.*, 1996; *Biftu and Gan*, 1999]. Details of IEM is given in *Fung* [1992].

The IEM calculates the radar backscatter($\sigma°$) on the basis of the radar frequency, polarization, angle of incidence, dielectric constant, the root mean-squared (rms) surface height (σ), surface correlation length (L), and autocorrelation function (ACF). This model can be inverted to estimate the dielectric constant (ε), which is related to the soil moisture, once the other parameters are known. However, an accurate estimation of the surface roughness is still challenging, especially at large basins [*Baghdadi et al.*, 2006]. Recently, *Baghdadi et al.* [2008] reported that the roughness variables estimated from field measurements are very sensitive to the profile length. *Oh and Kay* [1998] found that the accuracy of the rms surface height and correlation length estimated for a surface depend on the length and the horizontal resolution of the roughness profiles. To overcome the problems associated with the roughness variables, some of the recent studies have proposed notable improvements to the initial version of the IEM model. *Baghdadi et al.* [2006] proposed an empirical approach to replace the measured correlation length by a model parameter that considers both the true correlation length and the imperfections of the IEM. This calibration parameter determined through calibration is dependent on the roughness, incident angle, polarization, and wavelength. Consequently their model-simulated backscatters agreed closely with the observed counterparts. *Wu and Chen* [2004] also proposed some other improvements to the IEM model and compared the model results with both numerical simulations and laboratory measurements. Since these new models have not been tested extensively in other areas, we have decided to test the original version of IEM [*Fung*, 1992] and followed the inversion procedure of *Biftu and Gan* [1999].

In summary, according to IEM, $\sigma° = f$ (frequency, polarization, $\vartheta, \varepsilon, \sigma, L$, correlation function). We obtain the backscatter coefficient and the radar configuration (frequency, polarization, ϑ) from the radar image, and we assume the correlation function as exponential because previous studies [*Biftu and Gan*, 1999; *Chen et al.*, 1995] showed that the exponential function is less sensitive to the roughness variables than the Gaussian function. So the remaining unknowns in the IEM are roughness variables (σ and L) and ε. Now from the measured soil moisture (θ) of 2009 and 2010, ε can be computed by inverting the empirical equation [equation 16.2] of *Topp and Davis* [1985]:

$$\theta = (-530 + 292\varepsilon - 5.5\varepsilon^2 + 0.043\varepsilon^3) \times 10 \quad (16.2)$$

Then using the ε obtained from equation 16.2 and radar backscatter from RADARSAT-2 images, the surface parameters (σ and L) of IEM will be optimized using the global optimization algorithm, the shuffled-complex evolution [*Duan et al.*, 1993]. Once the σ and L are optimized from the calibration (2009–2010) data,

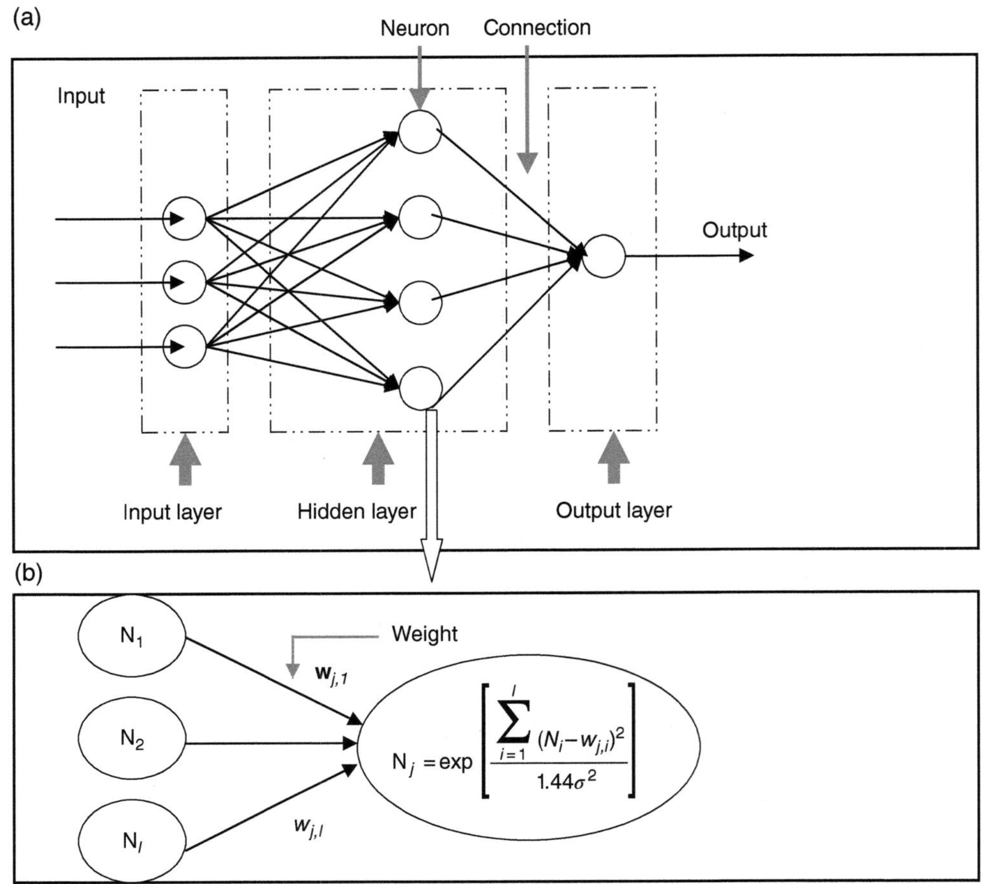

Figure 16.3 Architecture of a typical radial basis neural network.

then dielectric constants [hence soil moisture using equation 16.2] will be computed for each image of the validation data set (2011) using these optimized surface variables (σ, L) and the radar data (radar configuration and backscatter) through IEM inversion.

16.5.3. Artificial Neural Network

The ANN is developed on the basis of the neuron cell structure of the biological nervous system and can fit almost any type of nonlinear input-output relationships [*Hsu et al.*, 1995; *Phil*, 2000]. ANN has been found to be a robust tool for modeling many nonlinear hydrological processes because they are able to capture the nonlinear relationships of such processes without requiring the users to understand the physics of the process [*Nor et al.*, 2007]. In that regards, using an ANN to retrieve soil moisture is an appealing idea. In this study, the radial basis function (RBF) network is used to retrieve soil moisture from radar data. RBF networks are widely used for nonlinear modeling, pattern recognition, and modeling of complex and chaotic dynamical systems [*Nor et al.*, 2007; *Islam et al.*, 2005].

Shown in Figure 16.3, RBF neural networks (RBFNN) consist of one input and one hidden and one output layers [*Broomhead and Lowe*, 1988]. The input layer takes the input to the network while the hidden layer processes the input by certain nonlinear activation functions and sends the computed value to the output layer. The activation function (also known as transfer function) allows nonlinearity in the network [*Zealand*, 1997]. The output layer computes the final response of the network. Each layer of a RBF consists of a certain number of neurons and connections. The main controlling parameters of any ANN are the strengths of the connections between neurons represented by weights and biases. RBF is a supervised, feed-forward neural network that uses a linear transfer function for the output layer and a nonlinear transfer function, normally Gaussian, for the hidden layer. The value of a neuron in the hidden layer is computed as

$$N_j = \exp\left(-\frac{\sum_{i=1}^{I} (N_i - w_{j,i})^2}{1.44\sigma_i^2} \right), \tag{16.3}$$

Table 16.4 Input cases for radial basis function neural network (RBFNN) and results

Input Case	Input Variables	Calibration Results		Validation Results	
		R	RMSE(%)	R	RMSE(%)
A-1	Radar backscatter (σ°_{HH}, σ°_{VV}, σ°_{HV})	0.70	4.51	0.73	4.21
A-2	Radar backscatter, LST, NDVI	0.76	4.22	0.73	3.90
A-3	Radar backscatter, LST, NDVI, angle, soil properties	0.80	3.84	0.77	3.55
A-4	Radar backscatter, LST, NDVI, angle, terrain properties (slope and aspect)	0.82	3.61	0.76	3.53
A-5	Radar backscatter, soil properties, angle, terrain properties	0.75	3.77	0.70	4.32
A-6	LST, NDVI, angle, soil properties, and terrain properties	0.68	4.65	0.41	5.85
A-7	Radar backscatter, LST, NDVI, angle, soil properties, and terrain properties	0.83	3.56	0.76	3.52

where N_i is the value of the ith (where $i = 1, 2,\ldots, I$) unit in the input layer, N_j is the value of the jth (where $j = 1, 2,\ldots, J$) unit in the hidden layer, $w_{j,i}$ are weights connecting the jth unit of the hidden layer to the ith unit of input layer, and σ_1 is an adjustable parameter called width or spread. The optimum value of σ_1 is chosen by trial-and-error approach [*Phil*, 2000]. The number of neurons in the hidden layer is kept the same as the number of input patterns.

The output of the hidden layer (N_j) is passed on as input to the output layer. The neurons of the output layer do not have any activation function but a bias b_k is added. The value of an output neuron is computed as

$$O_k = \sum_{j=1}^{I} (N_j w_{k,j}) + b_k, \tag{16.4}$$

where O_k is the value of the kth ($k = 1,2,\ldots,K$) neuron of the output layer, and $w_{k,j}$ is the weight connecting the jth neuron in the hidden layer to the kth neuron of the output layer. A more detailed description of the radial basis function network is available in *Phil*, [2000]. In this study, different networks were tested by changing the combination of input variables (radar backscatter, NDVI, LST, soil properties, terrain information, and incidence angles). For each case of input (Table 16.4), we tested different σ for the RBFNN and the best network is presented here. In this study RBFNN modeling was done using Matlab.

16.5.4. Adaptive Neuro-Fuzzy Interference System

A fuzzy interference system (FIS) is a fuzzy-logic-based data-driven modeling approach that uses IF–THEN rules and logical operators (AND, OR, NOT, etc.) to determine the qualitative relationships between input-output. An FIS consists of four basic steps: (1) For each input and variable, a set of membership functions must be defined. A membership function is a mathematical function that defines the degree of membership of an element in a fuzzy set. Membership functions can be triangular, Gaussian, sigmoidal, etc. (2) A set of IF–THEN rules relating the membership functions of each input variable to the output should be defined. As, for example, one rule would be: IF the backscatter (antecedents) is low (represented by a membership function), THEN the soil moisture (consequent) is low (represented by a membership function). (3) All rules are mathematically evaluated through implication and the results are combined using aggregation. (4) The output of the aggregation process is a fuzzy set for each output variable, and defuzzification determines a single output value from the set. The adaptive neuro-fuzzy interference system (ANFIS) refers to the integration of the FIS and feed-forward neural networks in a single framework. The FIS performs the nonlinear mapping from its input space to the output space through IF-THEN fuzzy rules while the back-propagation neural network (or a combination of both back-propagation and least squares method) optimizes the fuzzy membership parameters. ANFIS was first introduced by *Jang* [1993]. In recent years a number of studies have employed [*Talei and Chua*, 2012; *Pramanik and Panda*, 2009; *Daliakopoulos et al.*, 2005] ANFIS in hydrologic modeling because of its promising ability of tuning the parameters by minimizing the global error of the model.

The current study has used Takagi-Sugeno FIS [*Takagi and Sugeno*, 1985], where the consequent part of the FIS is constructed by taking a weighted linear combination of crisp input rather than a fuzzy set. In the first-order

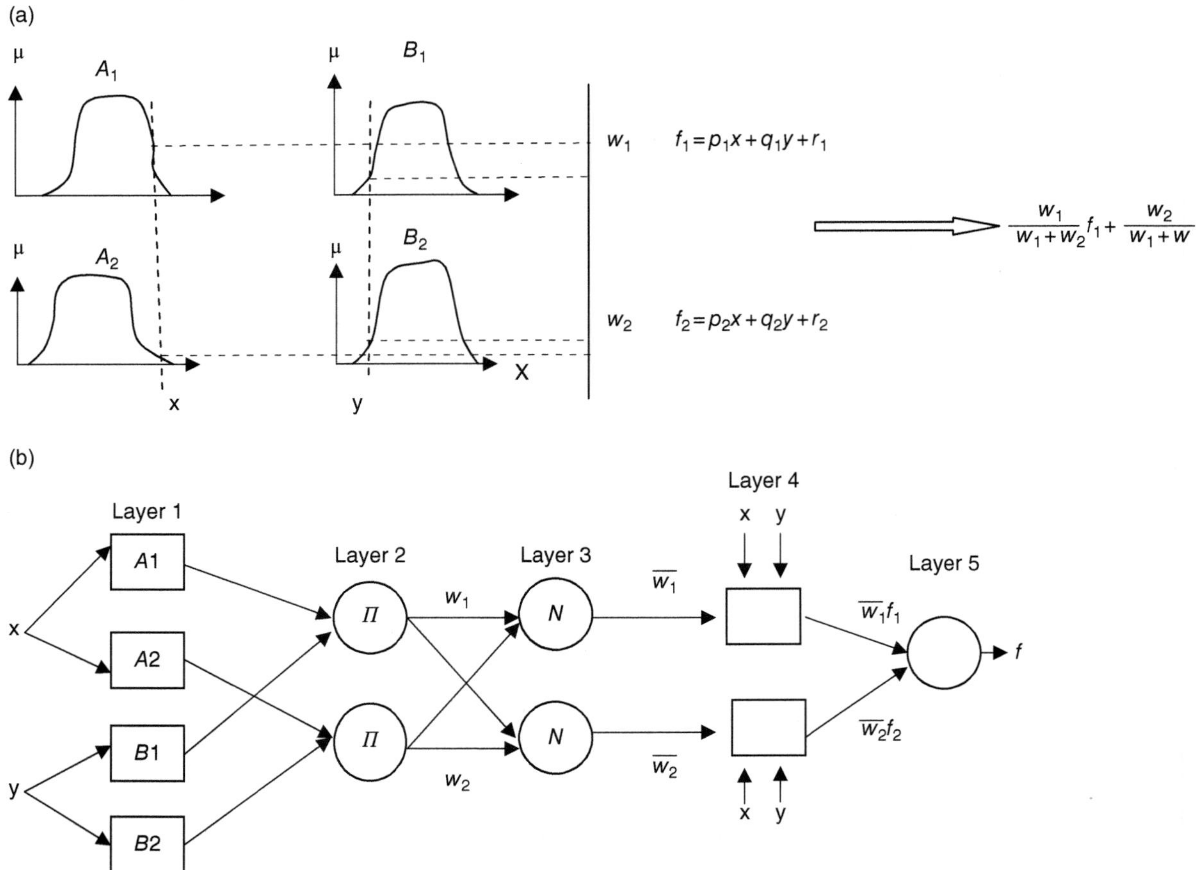

Figure 16.4 (a) Two inputs of first-order Sugeno fuzzy model with two rules and (b) equivalent ANFIS structure.

Takagi-Sugeno FIS, a typical fuzzy IF-THEN rule, with two inputs x and y, can be expressed as

$$\text{Rule 1}: \text{IF } x \text{ is } A_1 \text{ and } y \text{ is } B_1, \text{THEN } f_1 = p_1 x + q_1 y + r_1, \quad (16.5)$$

$$\text{Rule 2}: \text{IF } x \text{ is } A_2 \text{ and } y \text{ is } B_2, \text{THEN } f_2 = p_2 x + q_2 y + r_2, \quad (16.6)$$

where A_1, A_2 and B_1, B_2 are membership values of input variables x and y, respectively; p_1, q_1, r_1 and p_2, q_2, r_2 are the parameters of the output functions f_1 and f_2. Figure 16.4a illustrates a basic Takagi-Sugeno FIS, which produces an output function f from input variables x and y by applying a generalized bell-shaped member function. The corresponding ANFIS structure, consisting of five neural network layers, is shown in Figure 16.4b. The description of five layers is given below.

Layer 1 (Fuzzification Layer): The first layer generates the fuzzy membership values for each input variable using membership function. The output of the first layer is

$$O_{1,i} = \mu_{A_i}(x), \quad \text{for } i = 1, 2,$$
$$O_{1,i} = \mu_{B_{i-2}}(y), \quad \text{for } i = 3, 4, \quad (16.7)$$

where x, y are the crisp input to the node i, A_i and B_{i-2} are the linguistic variable, such as low, medium, or high, characterized by appropriate membership functions $\propto_{A_i}$ and $\mu_{B_{i-2}}$. Any continuous and piecewise differentiable function can be used as a member function, such as Gaussian, generalized bell-shaped, trapezoidal-shaped, or triangular-shaped function. In this study, we used the generalized bell-shaped member function [equation 16.8], which is a commonly used membership function [*Jang*, 1993; *Pramanik and Panda*, 2009]:

$$\mu_{A_i} = \frac{1}{1 + \left[(x - c_i / a_i)^2\right]^{b_i}}, \quad (16.8)$$

where $\{a_i, b_i, c_i\}$ is the premise parameter set. With the change of these parameters, the bell-shaped function varies accordingly and thus produces various forms of membership functions on linguistic label A_i with a maximum value of 1 and minimum value of 0.

Layer 2 (Rule Layer): In this layer, each node (labeled as π) calculates the firing strength of each fuzzy rule by multiplying input signals from the respective fuzzification nodes and sends the product out:

$$O_{2,i} = w_i = \mu_{A_i}(x) \times \mu_{B_i}(y), \qquad i=1,2, \qquad (16.9)$$

where w_i is the firing strength of the ith rule. In this layer, each rule is compared with fuzzy values, and when the antecedent part of the rule is true, then the consequent part becomes active and this is called rule firing.

Layer 3 (Implication Layer): In this layer, each node (labeled as N) normalizes the firing strength by taking the ratio of the ith rule's firing strength to the sum of all rules' firing strengths:

$$O_{3,i} = \overline{w}_i = \frac{w_i}{\sum w_i}, \qquad i=1,2. \qquad (16.10)$$

Layer 4 (Defuzzification Layer): In this layer, the nodes calculate the output of the ith fuzzy rule by multiplying the normalized firing strength and consequent function. *Jang* [1993] used the linear combination of all inputs as the consequent function:

$$O_{4,i} = \overline{w}_i f_i = \overline{w}_i (p_1 x + q_1 y + r_1), \qquad (16.11)$$

where $\overline{w}_i$ is the output of the third layer, (p_1, q_1, r_1) is the parameter set called the consequent parameter, which needs to be adjusted.

Layer 5 (Aggregation Layer): This single-node layer calculates the overall output as the summation of all incoming signals from all the defuzzification nodes:

$$O_{5,i} = \sum \overline{w}_i f_i = \frac{\sum w_i f_i}{\sum w_i}. \qquad (16.12)$$

The procedure of adjusting the parameters or weights of a neural network is called the learning rules. The basic learning rule of an adaptive neural network for optimizing the premise and the consequent parameters or weights is the back-propagation algorithm, which is based on the gradient descent rule. However, the gradient descent method is generally slow and is likely to be trapped in local minima. Therefore, *Jang et al.* [1993] developed a faster learning algorithm called the hybrid method, which combines the gradient descent method and the least squares estimate (LSE) to optimize the parameters. In this study we used the hybrid learning rules to tune all parameters.

16.5.5. Support Vector Machine

The support vector machine (SVM) is a statistical machine learning technique that transforms nonlinear regression into linear regression by mapping the original low-dimensional input space to a higher dimensional feature space by a nonlinear mapping function ϕ and then performs linear regression in the feature space. SVM is based upon the structural risk minimization (SRM) theory where both the empirical error (e.g., mean-squared error) and the model complexity should be minimized simultaneously. The use of the SRM principle enhances the general capability of SVMs.

The main objective of SVM regression is to determine the functional dependency between independent variables $\{x_1, x_2, \ldots, x_n\}$ and dependent variables $\{y_1, y_2, \ldots, y_n\}$. In other words, SVM finds a function $f(x)$ that yields the output $\hat{y}$, which is the best estimate of the actual output y with a small error tolerance of ε. In the other words, we do not care about errors as long as they are less than ε, but any deviation larger than this will not be accepted [*Vapnik*, 1995]. This is done through the use of an ε-insensitive loss function (described later). First, the input vector x is mapped onto a higher dimensional feature space by a using a nonlinear function $\phi(x)$. Then the linear regression is performed in the feature space and can be expressed as

$$\hat{y} = f(x) = \langle w \cdot \phi(x) \rangle + b, \qquad (16.13)$$

where w is the weight vector, b is the bias, and $\langle w \cdot \phi(x) \rangle$ is the dot product between w and $\phi(x)$. On the basis of the structural risk minimization principle, w and b are estimated by minimizing the following structural risk function:

$$\text{Minimize } \frac{1}{2}\|w\|^2 + C\sum_{i=1}^{n} L_\varepsilon(y, x, f(x)), \qquad (16.14)$$

where $L_\varepsilon(y, x, f(x))$ is the Vapnik's ε-insensitive loss function [*Vapnik*, 1995], defined as

$$L_\varepsilon(y, x, f(x)) = \begin{cases} 0 & \text{for}\, |y - f(x) \leq \varepsilon| \\ |y - f(x)| - \varepsilon & \text{otherwise,} \end{cases}$$

$$(16.15)$$

where $\|w\|$ is the Euclidean norm of the weight vector. Minimizing $\|w\|^2$ corresponds to minimizing the model complexity. The parameter C determines the trade-off between the model complexity and the training error [*Smola and Scholkopf*, 1998]. A large value of C means that the model complexity part will be negligible during optimization while a small value of C means that the training error (also called empirical error) has less influence in the optimization formulation. If $C=1$, then it represents that both the model complexity is as important as the empirical error. The loss function of equations 16.14 and 16.15 are represented by two variables (called slack variables) when data cannot be estimated by the function under the precise ε. By introducing two

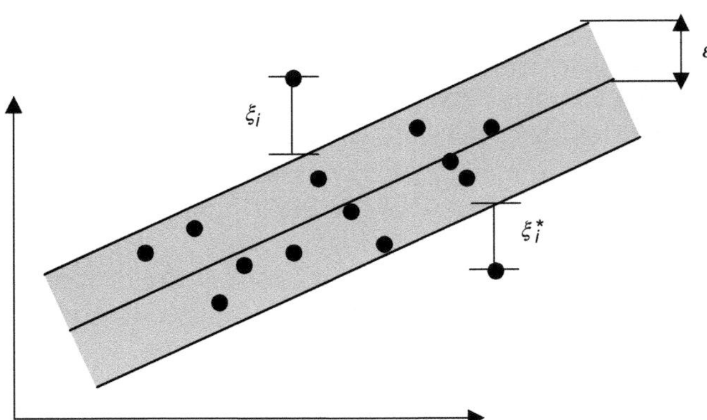

Figure 16.5 Details of ε band with slack variables ξ_i and ξ_i^* and data points (black circles).

slack variables, the optimization problem of equation 16.14 can be expressed as

$$\text{Minimize } \frac{1}{2}\|w\|^2 + C\sum_{i=1}^{n}(\xi_i + \xi_i^*)$$

$$\text{Subject to } y_i - (\langle w \cdot \phi(x_i)\rangle + b) \le \varepsilon + \xi_i \qquad (16.16)$$
$$(\langle w \cdot \phi(x_i)\rangle + b) - y_i \le \varepsilon + \xi_i^*$$
$$\xi_i, \xi_i^* \ge 0, \qquad i = 1,2,3,\ldots,n,$$

where ξ_i and ξ_i^* are slack variables, represents the lower and upper training errors, respectively, and determine the degree to which sample points are penalized if the error is larger than ε (Figure 16.5). The optimization problem of equation 16.16 can be solved by dual formulation. The key idea of dual formulation is to construct a Lagrange function using both the objective function and corresponding constraints by introducing a dual set of Lagrange multipliers, α and α^* [*Smola and Scholkopf*, 1998; *Mangasarian*, 1969]. The partial derivatives of this Lagrange function with respect to w, b, ξ_i and ξ_i^* should be zero to satisfy the optimizing condition. Substituting the partial derivative output into equation 16.16 transforms this optimization equation into the following form:

$$\text{Minimize } \varepsilon \sum_{i=1}^{n}(\alpha_i^* + \alpha_i) - \sum_{i=1}^{n} y_i(\alpha_i^* - \alpha_i)$$
$$+ \frac{1}{2}\sum_{i,j=1}^{n}(\alpha_i^* - \alpha_i)(\alpha_j^* - \alpha_j)\langle \phi(x_i) \cdot \phi(x_j)\rangle$$

$$\text{Subject to } \sum_{i=1}^{n}(\alpha_i^* - \alpha_i) = 0, \qquad 0 \le \alpha_i, \alpha_i^* \le C, i = 1,2,\ldots,n.$$

$$(16.17)$$

Then solving equation 16.17 with constraints determines the Lagrange multipliers (α and α^*), and the regression equation 16.13 can be rewritten as

$$y = f(x) = \sum_{i=1}^{n}(\alpha_i^* - \alpha_i)\langle \phi(x_i) \cdot \phi(x)\rangle + b$$
$$= \sum_{i=1}^{n}(\alpha_i^* - \alpha_i)K(x_i, x) + b, \qquad (16.18)$$

where $K(x_i, x) = \langle \varphi(x_i) \cdot \phi(x)\rangle$ is the kernel function that determines the nonlinear dependence between the two input variables x and x_i. It should be mentioned that it is not necessary to know the analytical form of the nonlinear transformation function, which is difficult to find. In equation 16.17, only the dot product $K(x_i, x) = \langle \phi(x_i) \cdot \phi(x)\rangle$ is necessary for the optimization, and we can generalize the dot product to other functions. The kernel function used in this study is the RBF, which has the following form:

$$K(x_i, x) = \langle \phi(x_i) \cdot \phi(x)\rangle = \exp\left(-\frac{1}{2\sigma^2}\|x_i - x\|^2\right). \qquad (16.19)$$

There are different other kernel functions such as linear, polynomial, and sigmoid. In this study the RBF kernel is used because it requires only one parameter, which makes the computation process easier [*Yang et al.*, 2006b]. The steps involved in SVM modeling are as follows: (1) selecting a suitable kernel function and kernel parameter (kernel width, σ), (2) specifying the ε for the width of an insensitive error band, and (3) specifying the parameter, C for the cost of error. More details about SVM are available in *Vapnik* [1995, 1998]. The SVM modeling was done using the SVM software WEKA [*Hall et al.*, 2009], developed at the University of Waikato, NewZealand.

16.6. DISCUSSIONS OF RESULTS

16.6.1. Sensitivity of Radar Backscatter to Soil Moisture

In this study we first employed linear regression (LR) to assess the sensitivity of soil moisture to radar backscatter. The RADARSAT backscatter of 9 study sites, each averaged over an area of $200\,\text{m} \times 200\,\text{m}$ (averaged over 64 pixels of 25 m resolution) were regressed against their corresponding mean soil moisture (averaged from 25 samples per site collected over an area of $200\,\text{m} \times 200\,\text{m}$ each day). Figure 16.6 shows the results obtained from the regression. The linear regression plots shows that HH, VV, and HV radar backscatters are linearly correlated with the soil moisture. The correlation coefficients (R) between radar backscatters (HH or HV or VV) and soil moisture are found to be similar (ranging from 0.62

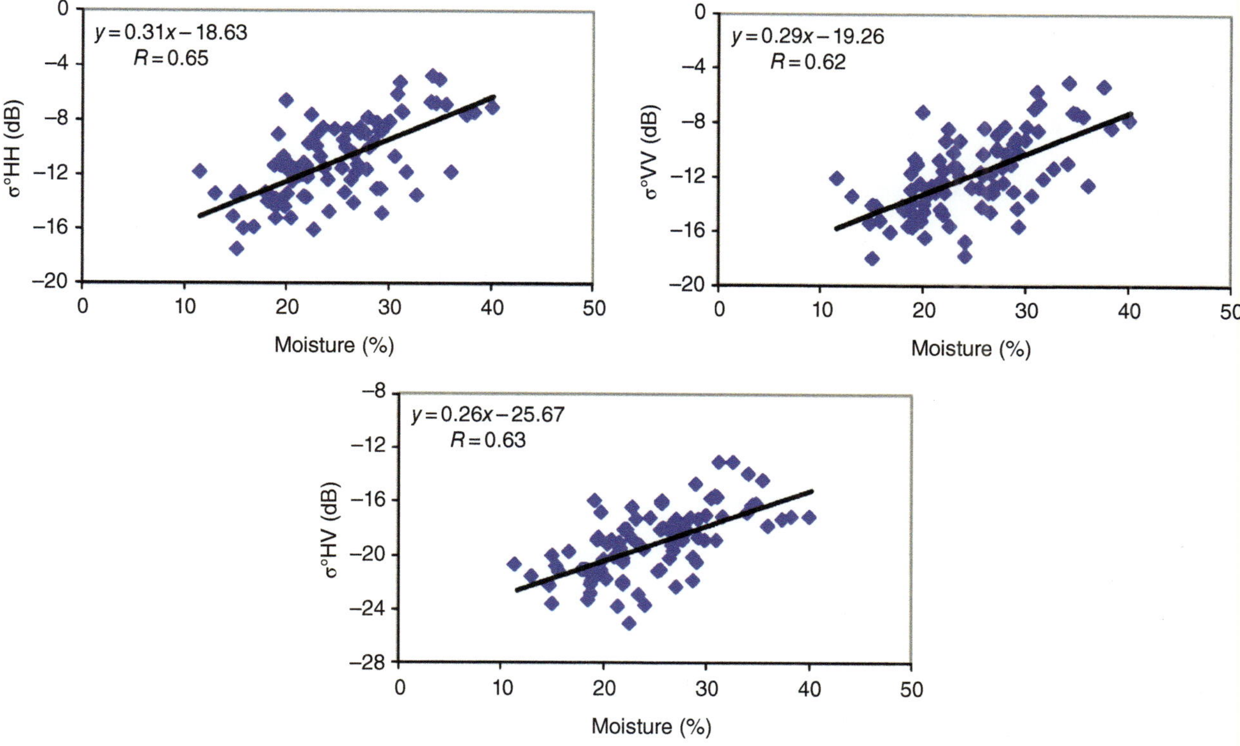

Figure 16.6 Sensitivity of radar backscatter to volumetric soil moisture with respect to the combined data of 2009–2011.

to 0.65). Our result is consistent with the results reported in some previous studies using multipolarized data [*Sokol et al.*, 2004; *Baghdadi and Zribi*, 2006]. *Sokol et al.* [2004] found that the correlation coefficients between soil moisture and HH, VV, and HV radar backscatter of the spaceborne imaging radar (SIR-C) for bare fields were similar and varied within a narrow range.

16.6.2. Integral Equation Model

Under this study we collected satellite data and soil moisture data on 10 selected days. Among these 10 days, all the data acquired on the first 2 years: 3 days of 2009 and 4 days of 2010 (total 7 days) from all the 9 study sites has been used for calibrating the IEM and the AI models. On the other hand all the data collected in the last year (2011) from all the study sites has been used for validating the calibrated models. Data of 2009 and 2010 has been used as training data set as it contains a wide range of moisture conditions (11.53% to 40.19%), which is desired for training the AI methods.

We also performed some preliminary investigations by using all the data of 2010 and 2011 from all the sites as the training data set and all the data from 2009 as the validation data set (not shown in this study). But in this case we got lower RMSE in the validation stage and therefore we did not report it here.

Figure 16.7 shows the IEM results in which the correlation coefficients obtained are 0.70 and 0.67 and RMSEs are 9.74 and 7.25 (percent of soil moisture) during the calibration (2009 and 2010) and validation periods (2011), respectively. Even though IEM is a theoretical model, the results of IEM are poor because of possible difficulties and limitations in applying IEM. First, IEM was originally developed for bare soil, and so the effect of vegetation was not explicitly incorporated in this model [*Bindlish and Barros*, 2001]. Therefore applying IEM to a vegetated landscape may incur some errors or uncertainties to the results [*Thoma et al.*, 2006]. Second, in this study, instead of using measured surface roughness, we have calibrated the surface roughness parameters of IEM using the known soil moisture of 2009–2010 and corresponding radar backscatter and configuration to avoid the limitations of field measurement of surface roughness reported in past studies [*Baghdadi et al.*, 2008; *Thoma et al.*, 2006; *Oh and Kay*, 1998]. A range of surface roughness values, taken from the literature for agricultural, pasture, and herbaceous land uses were used in the optimization algorithm [*Biftu and Gan*, 1999; *Jackson et al.*, 1997]. Later those calibrated surface roughness values were used to retrieve soil moisture for the validation stage of 2011, assuming these calibrated values as time invariant. Although examples [*Biftu and Gan*, 1999; *Thoma et al.*, 2006] of using calibrated roughness

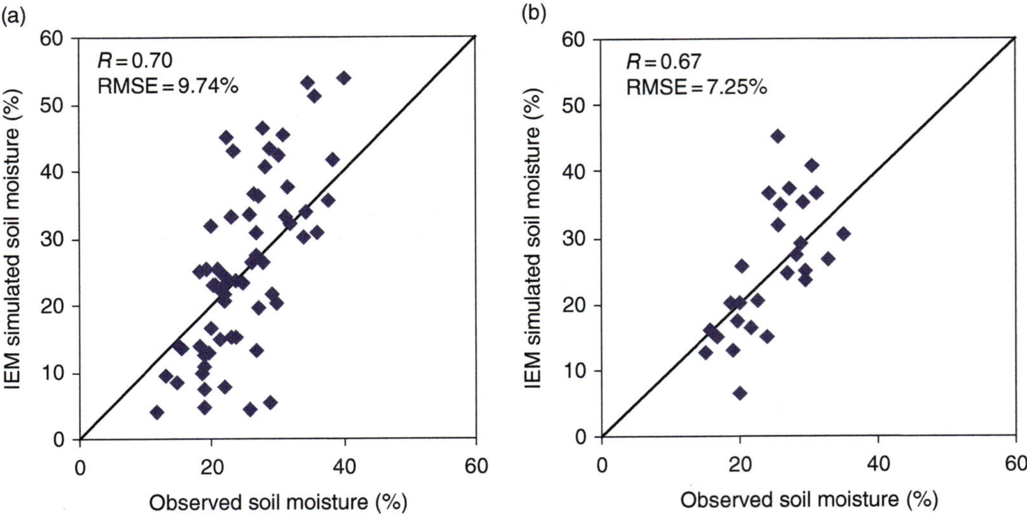

Figure 16.7 Regression between observed and IEM-simulated volumetric soil moisture during (a) the calibration and (b) the validation stage.

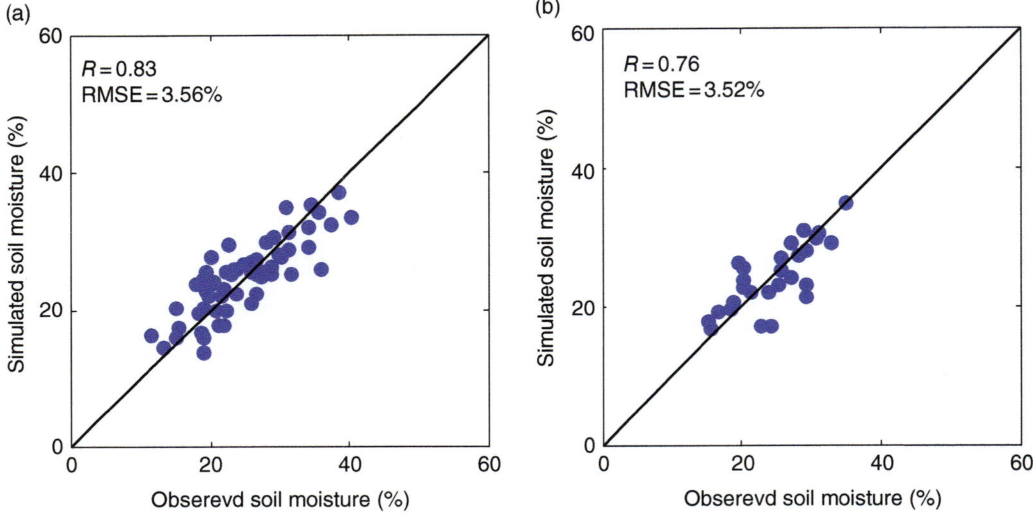

Figure 16.8 Observed soil moisture versus radial basis function neural network (RBFNN) simulated soil moisture where radar backscatter, incidence angle, LST, NDVI, and soil and terrain properties were used as predictors.

parameter as time invariant is available in the literature, it could possibly lead to some errors. We tried to minimize this problem by choosing spring and early summer of each year for our study when the vegetation was at the early growth stage. In all years, field measurements and corresponding image acquisitions were done during the end of May to the middle of July (approximately 6 weeks). Over this short period it may be reasonable to assume similar roughness for the pasture and herbaceous lands as reported by *Leconte et al.* [2004] for these vegetation types. On the hand, for agricultural fields, variation of surface roughness may be small during the early growing season because of the combined effect of roughness increase due to vegetation growth and roughness decrease due to weathering and rainfall erosion

[*Leconte et al.*, 2004]. Third, the presence of residual speckle, uncertainty in the field measured soil moistures, and row effect in agricultural field also may produce some errors in the simulated soil moisture.

16.6.3. Radial Basis Function Neural Network

Results from the RBFNN are presented in Table 16.4 and Figure 16.8. For each input case we tested different network by varying the width parameter σ and different significant combination of input variables are shown in Table 16.4. The performances of the networks are presented in terms of R and RMSE. Results show that when radar backscatters were used as the only predictor, then the result obtained is similar to that of the linear regressions, which gave $R = 0.67$ (0.68), 0.65 (0.64), and 0.66 (0.64) while using HH, VV, and HV input, respectively, during the calibration period of 2009–2010 (validation period of 2011). On the other hand, the worst result was obtained when LST, NDVI, incident angles, and soil and terrain variables were used without the radar backscatter (input case A-6, Table 16.4). This indicates that radar backscatters are essential to estimate near surface soil moisture. Results from cases A-3 and A-4 were comparable where radar backscatter, LST, NDVI, and angles along with either soil or terrain properties were used as predictors. It is noteworthy that the results of case A-5 (all variables except LST and NDVI were used as predictors) were inferior to the results of the input case A-7 where all variables (including LST and NDVI) were used. Better results obtained for the case A-7 than for the case A-5 probably indicates the importance of LST and NDVI in soil moisture retrieval. The R of case A-7 is similar to cases A-4 and A5. However, the RMSE (RMSE = 3.56 and 3.52% during the calibration and the validation stages, respectively) of case A-7 are less than the other cases because these input variables together probably provide valuable information about soil and vegetation surface conditions, which control the soil moisture status of a vegetated surface. For each of these input cases we computed AIC to determine the optimal set of input variable. Given a set of candidate models for the data, the preferred model is the one with the minimum AIC value. Hence AIC not only rewards goodness of fit but also includes a penalty for increasing number of input variables. This penalty discourages overfitting. In this study we found case A-7 has the lowest AIC (AIC = 173) followed by case A-4 (AIC = 178) and A-3 (AIC = 182).

16.6.4. Adaptive Neuro-Fuzzy Interference System

The input cases used in RBFN have also been used here for the ANFIS model. Data of 2009 and 2010 were used to train the ANFIS network while data of 2011 was used to validate the trained model, as before. For the construction of a Takagi-Sugeno model, a different type of membership function can be used. If expert knowledge is available, then the fuzzy sets of the antecedent variables can be defined through their membership function (MF), using prior knowledge and experience. However, it is extremely difficult to define the function, and an inappropriate function may lead to results that are not repeatable by others or for other data sets. However, if expert knowledge is not available, data-driven identification techniques, such as fuzzy clustering algorithms or grid partitioning, can be applied. Such an approach has the major advantage of being repeatable by different researchers and is, therefore, used in their studies. In this study, grid partitioning has been used.

The number of membership functions chosen should be a trade-off between the model complexity and the model accuracy. Since we have no a priori knowledge on the number of membership functions to be used, therefore in order to remain linguistically interpretable, the maximum number of membership functions assigned to each input of the ANFIS was varied between two and three. The input data are scaled so that they lie between 0 and 1, since the MFs of the ANFIS takes values between 0 and 1. The values of R and RMSE are used here to determine the performance of the model, as before. As the variables are independent, therefore AND was used as the logical link between variables. The type of membership function used for all the models was of generalized bell (gbell) type, which is a direct generalization of Cauchy distribution used in the probability theory with three parameters, as shown in equation 16.8. Due to its smoothness and concise expression, it is popularly used in many applications to specify the fuzzy sets [*Jang et al.*, 1993].

Results from the ANFIS are presented in Figure 16.9. Among all the input cases, A-7 gave the best result for ANFIS in terms of RMSE and AIC. A comparison between the RBFNN and ANFIS results shows that for the input cases A-7, ANFIS produced slightly worse results than the RBFNN in terms of RMSE, although R was higher for the former case. During the calibration stage, the correlation between actual and ANFIS-simulated soil moisture was 0.85 and RMSE was 3.70% while during the validation stage the correlation was 0.81 and RMSE was 3.53%. However, for the other input cases results from ANFIS were either similar or slightly better than the results from RBFNN.

16.6.5. Support Vector Machine

We used the data of 2009–2010 for calibrating the SVM model, which was then validated with the data of 2011, same as before. All the input variables were scaled to the range of -1 to 1 following standard SVM techniques to eliminate the influence of variables with different magnitudes. The following procedure was followed during model calibration: First, we selected the widely used RBF kernel, which requires only one parameter, σ [*Yang et al.*, 2006b]. Second, optimal values of C (cost of errors), ε (width of insensitive error band), and σ (kernel parameter) were searched using a grid search method [*Chang and Lin*, 2005]. The SVM parameters that produced the lowest cross-validation errors were selected. At the beginning, we conducted a coarse grid search for C (2^{-2}, 2^{-1}, 2^0, ..., 2^4), ε (2^{-5}, 2^{-4}, 2^{-3}, ..., 2^{-1}), and σ (2^{-5}, $2^{-4.5}$, 2^{-4}, ..., 2^4) to identify the values of C, ε, and σ that produce the lowest mean RMSE during the cross validation. We then used a gradually finer grid search until the variance of the RMSE was smaller than 0.01.

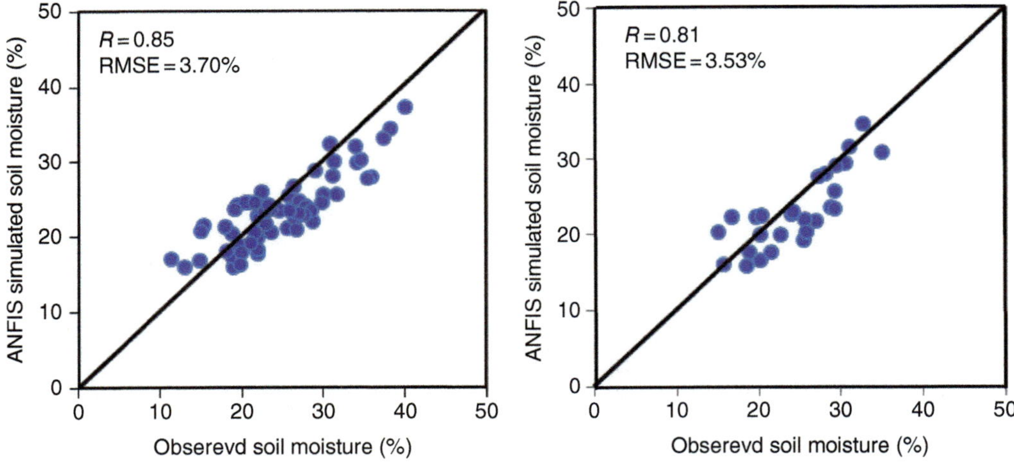

Figure 16.9 Observed soil moisture versus ANFIS-simulated soil moisture where radar backscatter, incidence angle, LST, NDVI, and soil and terrain properties were used as predictors.

Table 16.5 Input cases for support vector machine and results

Input Case	Input Variables	Calibration results		Validation results	
		R	RMSE (%)	R	RMSE (%)
A-1	Radar backscatter (σ^o_{HH}, σ^o_{VV}, σ^o_{HV})	0.70	4.49	0.77	4.10
A-2	Radar backscatter, LST and NDVI	0.77	4.13	0.76	3.55
A-3	Radar backscatter, LST, NDVI, angle, soil properties	0.85	3.35	0.78	3.51
A-4	Radar backscatter, LST, NDVI, angle, terrain properties (slope and aspect)	0.82	3.57	0.76	3.63
A-5	Radar backscatter, soil properties, angle, terrain properties	0.79	3.77	0.70	4.28
A-6	LST, NDVI, angle, soil properties, and terrain properties	0.76	4.16	0.49	4.40
A-7	Radar backscatter, LST, NDVI, angle, soil properties, and terrain properties	0.88	3.23	0.84	3.11

We explored seven cases of input data, the same as RBFNN, to determine the usefulness of different input variables in soil moisture retrieval. For each input case, we followed the above procedure to train the SVM model. The trained model was later used to simulate the soil moisture of 2011. Results of different input cases are given in Table 16.5. During the training stage, the lowest AIC and RMSE were obtained for case A-7. The performance of SVM was compared with respective RBFNN and ANFIS input cases in terms of RMSE and R. The results (Table 16.5) show that SVM performs slightly better than the RBFNN and ANFIS in all input cases, except for cases A-4 and A-5 where the results of SVM were comparable to RBFNN. Among the seven input cases of SVM, the worst result was obtained when radar backscatter was the only input. Including other variables such as LST and NDVI, soil and terrain properties significantly improved the results. Poor result was also obtained

when backscatter was omitted from the input data set (RMSE = 4.16% and 4.40% during the calibration and validation stages, respectively). The SVM model driven by all the predictors produced the best overall result (RMSE = 3.23% and 3.11% for the calibration and validation stage, respectively) among all of the retrieval algorithms and the input cases tested (Figure 16.10).

Recent studies have also reported that the superiority of SVM over ANN in different fields of hydrology [*Kalra and Ahmad*, 2009; *Asefa et al.*, 2005; *Yang et al.*, 2006b; *Lin et al.*, 2009]. According to the statistical learning theory, SVMs possess better generalization ability than ANNs and the optimization algorithm for SVM is more robust than that of ANN [*Lin et al.*, 2009]. Therefore in general a more reliable model can be obtained by using SVM rather than ANN. A drawback of RBFNN is that it produces a network with as many hidden neurons as there are input vectors. Because of this, RBFNN may

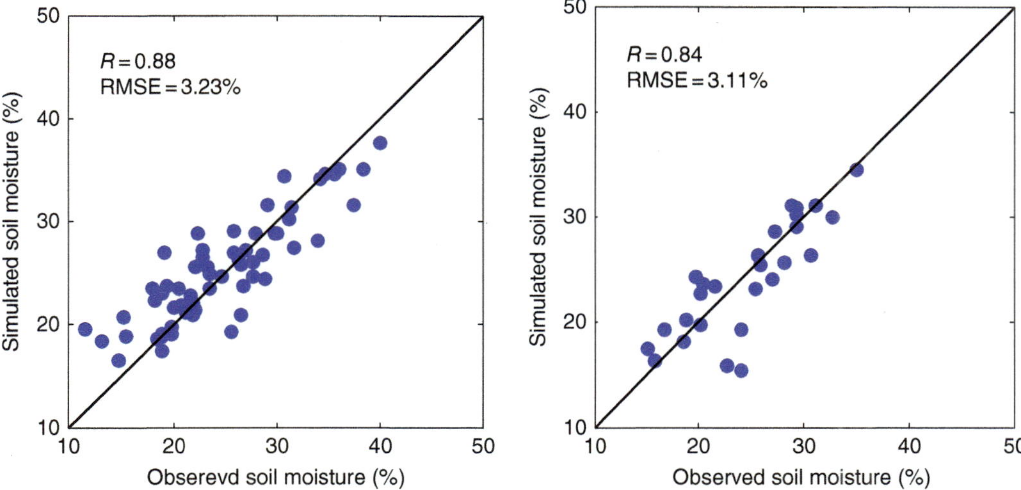

Figure 16.10 Observed soil moisture versus SVM simulated soil moisture where radar backscatter, incidence angle, LST, NDVI, and soil and terrain properties were used as predictors.

sometimes produce less accurate results when many hidden neurons are needed to accurately define a network. Probably because of these advantages, SVM gave better results than the RBFNN. However, there are also some drawbacks of SVM. The time taken to tune a SVM model could be too long for some applications of large databases [*Oyang et al.*, 2005].

Although we attempted to incorporate different information of soil and vegetation in the AI techniques, however, there are numerous other factors, such as difference in tillage condition and soil compaction, that control soil moisture but were not included as predictors since these information are seldom available. Therefore, some differences between the actual and the simulated soil moisture were noticed. The presence of residual speckle, uncertainty in the field-measured soil moistures and uncertainties in the other input data may also cause the discrepancies between actual and simulated soil moisture. Given the data-driven nature of machine learning techniques (i.e., ANN, ANFIS, or SVM), it is generally beneficial to have long data series to train the network. But in this study, we have only 10 days of data (90 observations), which may not be sufficient enough to adequately train the AI techniques. However, examples of using short data set for training machine learning techniques are also available in the literature. For example, *Alcazar et al.* [2008] used 46 cases to train a neural network to estimate environmental flows. *Kagoda et al.* [2010] suggest that the amount of data necessary to train an ANN or SVM depends largely on the complexity of the relationship between the input and output data series, which is difficult to know a priori. Therefore, it is difficult to determine the optimum amount of input for successful training of any machine learning technique.

16.6.6. Results at the Watershed Scale

At field scale, the observed soil moisture of each field was computed by averaging all 25 samples (at approximately 50 m spacing over a plot size of 200 m × 200 m) collected, while the soil moisture retrieved from RADARSAT-2-SAR data was based on the average backscatters of all pixels covering that same area. So the average soil moisture obtained from 25 samples may not be exactly the same as the actual soil moisture over that 200 × 200 m plot. Another possible error was that the row direction effect of agricultural fields, which can significantly affect radar backscatters, was ignored. Two fields with the same crop type and conditions but of different row directions may produce different backscatters. In case of HH or VV polarizations, the radar backscatters should be notably higher for fields with row directions perpendicular to the radar look direction, than for backscatters obtained from fields with row directions parallel to the radar look direction, if other conditions are the same [*McNairn and Brisco*, 2004; *Leconte et al.*, 2004]. The increase of radar backscatters due to the row direction effect is generally within a few several decibels but can be as high as 10 decibels [*McNairn and Brisco*, 2004]. Therefore, ignoring the row effect might induce some error in retrieving soil moisture from agricultural fields.

With reference to field measurements, soil moisture retrieved from RADARSAT-2-SAR data by the IEM, ANN, ANFIS, and SVM models achieved correlation coefficients (*R*) of 0.69, 0.81, 0.84, and 0.85, respectively, at the field scale when the total dataset (2009–2011) was considered. To assess the performance of the models at the watershed scale, we averaged the soil moisture data of all nine sites, collected on a particular day. Figure 16.11

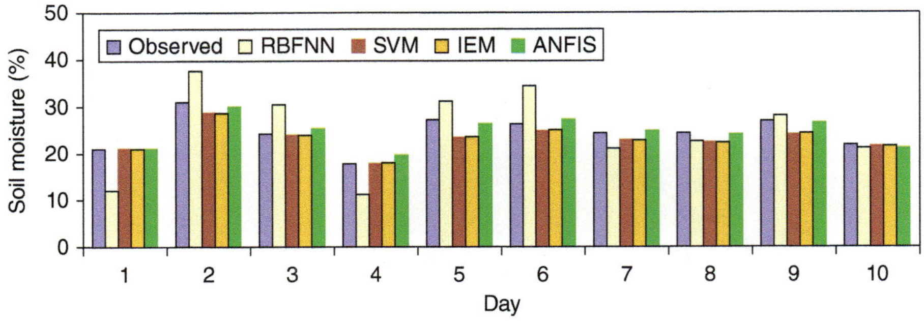

Figure 16.11 Observed soil moisture versus IEM, RBFNN, ANFIS, and SVM-simulated soil moisture for the 10 dates of 2009–2011 at the watershed scale. Each bar represents average soil moisture over the 9 study sites for a particular day.

shows the results obtained from the IEM model and the best RBFNN, ANFIS, and SVM models, for retrieving soil moisture from radar backscatter data at the watershed scale. The results indicate an improvement at the watershed scale (RMSE of 5.56%, 1.18%, 1.91%, and 0.98% of soil moisture, for the IEM, RBFNN, ANFIS, and SVM, respectively) over the field scale (RMSE of 9.06%, 3.55%, 3.65%, and 3.19%, for the IEM, RBFNN, ANFIS, and SVM, respectively) for the combined data of 2009, 2010, and 2011. Results for the watershed scale were better than the field scale because at a larger spatial scale, the errors caused by heterogeneities at field scale partially canceled out each other so the overall errors were reduced. A number of studies have also reported more accurate soil moisture retrieval results obtained at the watershed than at the field scales [*Leconte et al.*, 2004; *Thoma et al.*, 2006; *Kelly et al.*, 2003]. Since the grid resolutions used to model basin hydrology in a distributed manner are mostly larger than the field size (200 m × 200 m) used in this study, we can argue that the AI techniques applied for the soil moisture retrieval from the RADARSAT-2-SAR data have the potential to provide valuable soil moisture information at higher spatial resolution than soil moisture obtained from passive microwave sensors such as the Advanced Microwave Sounding Radiometer (AMSR)-E [*Njoku et al.*, 2003], Scanning Multichannel Microwave radiometer (SMMR) or soil moisture and ocean salinity (SMOS) [*Kerr at al.*, 2001], which can only provide soil moisture information at low spatial resolutions (25 km for AMSER-E and SMMR and 40 km for SMOS).

16.7. SUMMARY AND CONCLUSIONS

This study investigated the potential for retrieving soil moisture from the newly available quad-polarization backscatter data set from RADARSAT-2. Ten RADARSAT-2 images were obtained in 2009–2011 along with simultaneous field measurement of soil moisture at nine selected sites of the Paddle River Basin, Canada. Data of 2009–2010 was used to calibrate the IEM and three AI techniques (RBFNN, ANFIS, and SVM algorithms) and data of 2011 were used to validate the calibrated algorithms. In the IEM algorithm, calibrated surface roughness parameters were used instead of measured surface roughness variables to avoid the uncertainty usually induced from the profilometer used to measure surface roughness at fields. In this study roughness variables were calibrated from the radar images and soil moisture data of 2009–2010, and later these calibrated values were applied to retrieve the soil moisture from 2011's data. Results from this study can be summarized as follows:

1. All HH, VV, and HV radar backscatter were significantly correlated with soil moisture (R ranging from 0.62 to 0.65).

2. This study demonstrates a promising prospective of AI techniques in soil moisture retrieval from satellite data alone without requiring any surface roughness measurements, which limits the utility of IEM in regional scale application.

3. The AI techniques performed better than the IEM. The assumption of calibrated surface roughness as a time-invariant variable could be one of the limitations of applying IEM in this study.

4. In general, SVM was slightly better than RBFNN or ANFIS because of the better generalization capability of SVM over RBFNN or ANFIS. Among all the methods and input cases tested, the best result was obtained with SVM (RMSE of 3.23% and 3.11% for the calibration and validation stage, respectively) when radar backscatter, LST, NDVI, incidence angles, and soil and terrain properties were used as input variables.

5. The models showed better performances at the watershed scale (RMSE of 5.56%, 1.18%, 1.91%, and 0.98%) over the field scale (RMSE of 9.06%, 3.55%, 3.65%, and 3.19%) for the IEM and case A-7 of RBFNN, ANFIS, and SVM, respectively, and for the combined

data of 2009 to 2011, which demonstrates the potential of accurate soil moisture retrieval at larger spatial scale.

6. This study demonstrates the potential of microwave and optical data fusion in mapping soil moisture under vegetated condition with more operational advantage in support of agricultural monitoring.

REFERENCES

Alberta Energy and Natural Resources (AENR) (1977), A study of watershed management options for controlling flood runoff in the headwaters of the Paddle River Basin, Rept. 44, Edmonton, Alberta, Canada, pp. 1–39.

Alcazar, J., A. Palau, and C. Vega-Garcia (2008), A neural net model for environmental flow estimation at the Ebro River Basin, Spain, *J. Hydrol.*, *349*(1–2), 44–55.

Altese, E., O. Bolognani, and M. Mancini (1996), Retrieving soil moisture over bare soil from ERS1 synthetic aperture radar data: Sensitivity analysis based on a theoretical surface scattering model and field data, *Water Resourc. Res.*, *32*, 653–661.

Asefa, T., M. Kemblowski, U. Lall, and G. Urroz (2005), Support vector machines for nonlinear state space reconstruction: Application to the Great Salt Lake time series, *Water Resourc. Res.*,*41*,W12422,doi:10.1029/2004WR003785.

Baghdadi, N., and M. Zribi (2006), Evaluation of radar backscatter models IEM, Oh and Dubois using experimental observations, *Int. J. Remote Sens.*, *27*(18), 3831–3852.

Baghdadi, N., N. Holah, and M. Zribi (2006), Calibration of the integral equation model for SAR data in C-band and HH and VV polarizations, *Int. J. Remote Sens.*, *27*(3–4), 805–816.

Baghdadi, N., et al. (2008), Operational performance of current synthetic aperture radar sensors in mapping soil surface characteristics: Application to hydrological and erosion modelling, *Hydrol. Process.*, *22*(1), 9 – 20.

Biftu, G. F., and T. Y. Gan (1999), Retrieving near-surface soil moisture from Radarsat SAR data, *Water Resourc. Res.*, *35*(5), 1569–1579.

Bindlish, R., and A. P. Barros (2001), Parameterization of vegetation backscatter in radar-based soil moisture estimation, *Remote Sens. Environ.*, *76*, 130–137.

Broomhead, D. S., and D. Lowe (1988), Multivariate functional interpolation and adaptive networks, *Complex Syst.*, *2*, 302–355.

Carlson, T. N. (2007), Review: An overview of the "triangle method" for estimating surface evapotranspiration and soil moisture from satellite imagery, *Sensors*, *7*, 1612–1629.

Chang, C. C., and C. J. Lin (2005), LIBSVM: A Library for Support Vector Machines,Version 2.8. Software available at http://www.csie.ntu.edu.tw/~cjlin/libsvm/

Chatterjee, S., A. S. Hadi, and B. Price (2000), *Regression Analysis by Example*, 3rd ed., Wiley, New York.

Chen, K. S., S. K. Yen, and W. P. Huang (1995), A simple model for retrieving bare soil moisture from radar scattering coefficients, *Remote Sens. Environ.*, *54*, 121–126.

Daliakopoulos, I., P. Coulibaly, and I. Tsanis (2005), Groundwater level forecasting using artificial neural networks, *J. Hydrol.*, *309*(1–4), 229–240.

Duan, Q., V. K. Gupta, and S. Sorooshian (1993), A Shuffled complex evolution approach for effective and efficient global minimization, *J. Optimiz. Theory Appl.*, *76*(3), 501–521.

Dubois, P., J. Van Zyl, and T. Engman (1995), Measuring soil moisture with imaging radars, *IEEE Trans. Geosci. Remote Sens.*, *33*, pp. 915–926.

Ferrazzoli, P., S. Paloscia, P. Pampaloni, G. Schiavon, S. Sigismondi, and D. Solimini (1997), The potential of multifrequency polarimetric SAR in assessing agricultural and arboreous biomass, *IEEE Trans. Geosci. Remote Sens.*, *35*(1), 5–17.

Fung, A. K., Z. Li, and K. S. Chen (1992), Backscattering from a randomly rough dielectric surface, *IEEE Trans. Geosci. Remote Sens.*, *30*, 356–369.

Gherboudj, I., R. Magagi, A. A. Berg, and B. Toth (2011), Soil moisture retrieval over agricultural fields from multipolarized and muti-angular Radarsat-2 SAR data, *Remote Sens. Environ.*, *115*, 33–43.

Gupta, R. K., T. S. Prasad, and D. Vijayan (2002), Estimation of roughness length and sensible heat flux from WiFS and NOAA AVHRR data, *Adv. Space Rese.*, *29*(1), 33–38.

Hajnsek, I., T. Jagdhuber, H. Schcon, and K. P. Papathanassiou (2009), Potential of estimating soil moisture under vegetation cover by means of PolSAR, *IEEE Trans. Geosci. Remote Sens.*, *47*(2), 442–454.

Hall, M., E. Frank, G. Holmes, B. Pfahringer, P. Reutemann, and I. H. Witten. (2009), The WEKA data mining software: An update, *SIGKDD Explor.*, *11*(1), 10–18.

Hsu, K., H. V. Gupta, and S. Sorooshian (1995), Artificial neural networking modeling of the rainfall-runoff Process, *Water Resourc. Res*, *3*(10), 2517–2530.

Islam, R. (2005), Improved quantitative estimation of rainfall by radar, M.S. Thesis, Univ. of Manitoba, Canada.

Jackson, T. J., H. McNairn, M. A. Weltz, B. Brisco, and R. J. Brown (1997), First order surface roughness of active microwave observations for estimating soil moisture, *IEEE Trans. Geosci. Remote Sens.*, *35*(4), 1065–1069.

Jahan, N., and T. Y. Gan (2011), Modeling vegetation-climate relationship in a central mixed wood forest of Alberta using normalized difference and enhanced vegetation indices, *Int. J. Remote Sens.*, *32*(2), 313–335.

Jang, J.-S.R. (1993), ANFIS: Adaptive-network-based fuzzy inference system, *IEEE Trans. Syst. Man Cybernet.*, *23*(3), 665–685.

Jolliffe, I. T. (1973), Discarding variables in a principal component analysis, II: Real data, *Appl. Statist.*, *22*, 21–31.

Kagoda, P. A., J. Ndiritu, C. Ntuli, and B. Mwaka (2010), Application of radial basis function neural network to short term stream flow forecasting, *Phys. Chem. Earth*, *35*, 571–581.

Kalra, A. and S. Ahmad (2009), Using oceanic–atmospheric oscillations for long lead time streamflow forecasting, *Water Resourc. Res.*, *45*, W03413, doi:10,1029/2008WR006855.

Kelly, R. E. J., T. J. A. David, and P. M. Atkinson (2003), Explaining temporal and spatial variation in soil moisture in a bare field using SAR imagery, *Int. J. Remote Sens.*, *24*(15), 3059–3074.

Kerr, Y. H., P., Waldteufel, J. P. Wigneron, J. M. Martinuzzi, J. Font, and M. Berger (2001), Soil moisture retrieval from space:

The soil moisture and ocean salinity (SMOS) mission, *IEEE Trans. Geosci. Remote Sens.*, *39*, 1729 – 1735.

Leconte, R., F. Brissette, M. Galarneau, and J. Rousselle (2004), Mapping near surface soil moisture with Radarsat-1 aperture radar data, *Water Resourc. Res.*, *40*, W01515, doi:10.1029/2003WR002312.

Lin, G. F, G. R. Chen, M. C. Wu, and Y. C. Chou (2009), Effective forecasting of hourly typhoon rainfall using support vector machines, *Water Resourc. Res.*, *45*, W08440.

Lopes, A., E., Nezry, R. Touzi, and H. Laur (1993), Structure detection and statiscal adaptive speckle filtering in SAR images, *Int. J. Remote Sens.*, *14*, 1735–1758.

Makkeasorn, A., B. Chang, S. Beaman, C. Wyatt, and C. Slater (2006), Soil moisture estimation in a semi-arid watershed using RADARSAT-1 satellite imagery and genetic programming, *Water Resourc. Res.*, *42*, W09401.

Mangasarian, O. L. (1969), *Nonlinear Programming*, McGraw-Hill, New York.

Martin, R. D., G. Asrar, and E. T. Kanemasu (1989), C-band scatterometer measuremnts of a tall grass prairie, *Remote Sens. Environ.*, *29*, 281–292.

May, R., G. Dandy, and H. Maier (2011), Review of input variable selection methods for artificial neural networks in artificial neural networks, in *Methodological Advances and Biomedical Applications*, edited by K. Suzuki, InTech, Rijeka, Croatia, doi:10.5772/16004.

McNairn, H., and B. Brisco (2004), The application of C band polarimetric SAR for agriculture: A review, *Can. J. Remote Sens.*, *30*(3), 525–542.

McNairn, H., K. Hochheim, and N. Rabe (2004), Applying polarimetric radar imagery for mapping the productivity of wheat crops, *Can. J. Remote Sens.*, *30*(3), 517–524.

Merzouki, A., H. McNairn, and A. Pacheco (2010), Evaluation of the Dubois, Oh, and IEM radar backscatter models over agricultural fields using C-band RADARSAT-2 SAR image data, *Cana. J. Remote Sens.*, *36*(Suppl. 2), S274–S286.

Merzouki, A., H. McNairn, and A. Pacheco (2011), Mapping soil moisture using RADARSAT-2 data and local autocorrelation statistics, *IEEE J. Select. Topics Appl. Earth Observ. Remote Sens.*, *4*(1), 128–137.

Moran, M. S., C. D. Peters-Lidard, J. M. Watts, and S. McElroy (2004), Estimating soil moisture at the watershed scale with satellite-based radar and land surface models, *Can. J. Remote Sens.*, *30*, 1 – 22.

Njoku, E. G., T. J. Jackson, V. Lakshmi, T. K. Chan, and S. V. Nghiem (2003), Soil moisture retrieval from AMSR-E, *IEEE Trans. Geosci. Remote Sens.*, *41*(2), 215–229.

Nor, N. I. A., S. Harun, and A. H. M. Kassim (2007), Radial basis function modeling of hourly streamflow hydrograph, *J Hydrol. Eng.*, *112*, 113–123.

Oh, Y. (2004), Quantitative retrieval of soil moisture content and surface roughness from multipolarized radar observations of bare soil surfaces, *IEEE Trans. Geosci. Remote Sens.*, *42*, 596–601.

Oh, Y., and Y. Kay (1998), Condition for precise measurement of soil surface roughness, *IEEE Trans. Geosci. Remote Sens.*, *36*, 691–695.

Oh, Y., K. Sarabandi, and F. T. Ulaby (1992), An empirical model and inversion technique for radar scattering form bare soil surfaces, *IEEE Trans. Geosci. Remote Sens.*, *30*, 370–381.

Oyang, Y. J., S. C. Hwang, Y. Y. Ou, C. Y. Chen, and Z. W. Chen (2005), Data classification with radial basis function networks based on a novel kernel density estimation algorithm, *IEEE Trans. Geosci. Remote Sens.*, *16*(1), *225–236*.

Panchal. G., A. Ganatra, Y. P. Kosta, and D. Panchal (2010), Searching most efficient neural network architecture using Akaike's information criterion (AIC), *Int. J. Computer Appl.*, *(5)*, 41–44.

Pasolli, L., et al., (2011), Estimation of soil moisture in an alpine catchment with RADARSAT2 images, *Appl. Environ. Soil Sci.*, *2011*, Art. ID 175473, 1–12.

Patel, N. R., R. Anapashsha, S. Kumar, S. K. Saha, and V. K. Dadhwal (2009), Assessing potential of MODIS derived temperature/vegetation condition index (TVDI) to infer soil moisture status, *Int. J. Remote Sens.*, *30*(1), 23–39.

Phil, P. (2000), *Neural Networks*, 2nd ed., Grass Root Series, MacMillan, Hampshire, England.

Pierdicca, N., L. Pulvirenti, and C. Bignami (2010), Soil moisture estimation over vegetated terrains using multitemporal remote sensing data, *Remote Sens. Environ.*, *114*(2), 440–448.

Pramanik, N., and R. K. Panda (2009), Application of neural network and adaptive neuro-fuzzy inference systems for river flow prediction, *Hydrol. Sci. J.*, *54*(2), 247–260.

Ramirez-Beltran, N. D., J. M. Castro, E. Harmsen, and R. Vaquez (2008), Stochastic transfer function model and neural networks to estimate soil moisture, *J. Am. Water Resourc. Associ.*, *44*(4), 847–865.

Rencher, A. C. (2002), *Methods of Multivariate Analysis*, Wiley, Hoboken, NJ.

Said, S., U. C. Kothyari, and M. K. Arora (2008), ANN based soil moisture retrieval over bare and vegetated surface areas using ERS-2 SAR data, *J. Hydrol. Eng.*, *13*(6), 461–471.

Shepherd, N. (2000), Extraction of beta-nought and sigma-nought from RADARSAT CDPF Products, Doc. AS97-5001, Can. Space Agency, Ontario, Canada.

Smola, A. J. and B. Scholkopf (1998), A tutorial on support vector regression, Tech. Rep. Ser. NC2-TR-1998-030, NeuroCOLT2, University of London, U.K.

Sokol, J., H. Mcnairn, and T. J. Pultz (2004), Case studies demonstrating the hydrological applications of C band multipolarized and polarimetric SAR, *Can. J. Remote Sens.*, *30*(3), 470–483.

Soria-Ruiz, J., H. McNairn, Y. Fernandez-Ordonez, and J. Bugden-Storie (2007), Corn monitoring and crop yield using optical and Radarsat-2 images, in *Proceedings of Geoscience and Remote Sensing Symposium*, edited by Pei-Gee Peter Ho, Barcelona, Spain, pp. 3655–3658.

Takagi, T. and M. Sugeno, (1985), Fuzzy identification of systems and its applications to modeling and control, *IEEE Trans. Syst., Man Cybernet.*, *15*(1), 116–132.

Talei, A., and L. H. C. Chua (2012), Influence of lag time on event-based rainfall-runoff modeling using the data driven approach, *J. of Hydrol.*, *438*, 223–233.

Thoma, D. P., M. S. Moran, R. Bryant, M. Rahman, C. D. Holifield-Collins, S. Skirvin, E. E. Sano, and K. Slocum, (2006), Comparison of four models to determine surface soil moisture from C-band radar imagery in a sparsely vegetated semiarid landscape, *Water Resourc. Res.*, *42*, W01418, doi:10.1029/2004WR003905.

Topp, G. C., and J. L. Davis, (1985), Measurements of soil water content using time domain reflectometry: A field evaluation, *Soil Sci. Soci. of Am. J.*, *49*, 19–24.

Vapnik, V. N. (1995), *The nature of statistical learning theory*, Springer, New York.

Vapnik, V. N. (1998), *Statistical Learning Theory*, Wiley, New York.

Wan, Z., and J. Dozier, (1996), A generalized split-window algorithm for retrieving land-surface temperature from space, *IEEE Trans. Geosci Remote Sens.*, *34*, 892–905, doi:10.1109/36.508406.

Wu, T. D., and K. S. Chen, (2004), A reappraisal of the validity of the IEM model for backscattering from rough surfaces, *IEEE Trans. Geosci. Remote Sens.*, *42*, 743–753.

Yang, H., J. Shi, Z. Li, and H. Guo, (2006a), Temporal and spatial soil moisture change pattern detection in an agricultural area using multi-temporal Radarsat Scan SAR data, *Int. J. Remote Sens.*, *27*(19), 4199–4212.

Yang, F., M. A. White, A. R. Michaelis, K. Ichii, H. Hashimoto, P. Votava, A. Zhu, and R. R. Nemani, (2006b), Prediction of continental-scale evapotranspiration by combining MODIS and Ameriflux data through support vector machine, *IEEE Trans. Geosci. Remote Sens.*, *44*(11), 3452–3461.

Zealand, C. M. (1997), Short term stream flow forecasting using artificial neural networking, M.Sc. Thesis, Univ. of Manitoba, Canada.

17

AMSR-E Soil Moisture Disaggregation Using MODIS and NLDAS Data

Bin Fang and Venkat Lakshmi

17.1. INTRODUCTION

The land surface is spatially heterogeneous—soils, vegetation, and topography and, as a result, the hydrological responses are represented by soil moisture, surface temperature, and evapotranspiration and are also spatially variable. A combination of hydrological models, ground observations have been used for capturing hydrological variability in both space and time [*Jackson and Schmugge*, 1989; *Schmugge and Jackson*, 1994 *Lakshmi et al.*, 2011]. Remote sensing has a rich heritage of soil moisture mapping [*Schmugge et al.*, 2002; *Njoku et al.*, 2003]. Passive microwave remote sensing is able to obtain soil moisture in near real time by simple retrieval algorithms [*Njoku and Entekhabi*, 1996; *Njoku and Li*, 1999], especially with the high sensitivity of L-band to soil moisture. However, due to the limitation of satellite wavelength of L-band and antenna diameter, the spatial resolution of L-band microwave remote sensing imagery is restricted to tens of kilometer scales [*Schmugge et al.*, 1974; *Schmugge and Jackson*, 1994].

The Soil Moisture and Ocean Salinity (SMOS) and Soil Moisture Passive/Active (SMAP) are the first two satellites dedicated for soil moisture monitoring using L-band radiometers. The SMOS mission satellite, launched by European Space Agency (ESA) in 2009, provides soil moisture observations of 40 km on average at L-band [*Kerr et al.*, 2012]. The NASA SMAP mission satellite is designed and to be launched in 2014 and will provide the microwave radiometer for soil moisture product at 36 km and radar soil moisture at 3 km [*Entekhabi et al.*,

2010]. The Advanced Microwave Scanning Radiometer (AMSR-2) onboard Global Change Observation Mission (GCOM)-W1 measures microwave energy from land surface and provides high accurate soil moisture product of two spatial resolutions 0.1°/0.25° (10/25 km), with six frequency bands between 7 and 89 GHz and revisit times of 1–2 days [*Imaoka et al.*, 2010; http://www.jaxa.jp/pro jects/sat/gcom_w/index_e.html]. The coarse soil moisture can be disaggregated by combining radiometer and radar data and/or land surface model outputs, based on the relationships among vegetation cover, soil moisture, and surface temperature [*Lakshmi et al.*, 1997; *Narayan et al.*, 2004, 2006; *Carlson*, 2007; *Narayan and Lakshmi*, 2008; *Das et al.*, 2010; *Piles et al.*, 2011; *Merlin et al.*, 2013; *Fang et al.*, 2013]. In order to improve the soil moisture downscaling algorithms derived from the relationships among soil moisture, land surface temperature variation, and vegetation, there are several issues that should be considered. Soil property data sets are seldom available at fine spatial scales, and detailed vegetation information can be time consuming to generate at appropriate resolutions. Consequently, the land-atmosphere interactions related variables, such as evaporation, can be used to improve the soil moisture downscaling algorithms.

Previous studies have tried to estimate soil evaporation using both mechanical and physical methods [*Mahfouf and Noilhan*, 1991; *Chanzy and Bruckler*, 1993; *Yamanaka et al.*, 1998; *Brunsell et al.*, 2001]. Other studies have suggested methods to retrieve soil evaporation by using remote sensed soil moisture [*Nishida et al.*, 2003; *Zhang et al.*, 2003]. The surface skin temperature has been used to estimate soil evaporation [*Nishida et al.*, 2003; *Kustas et al.*, 1990]; *Merlin et al.* [2010a] applied soil efficiency

Department of Earth and Ocean Sciences, University of South Carolina, Columbia South Carolina, USA

Remote Sensing of the Terrestrial Water Cycle, Geophysical Monograph 206. First Edition. Edited by Venkat Lakshmi.

models to SMOS data for retrieving and then downscaling soil moisture products derived from microwave radiometry.

The mechanical-based methods work generally better than the physical methods since soil evaporation can be expressed using simple formulations. Two types of mechanical methods for estimating soil evaporation have been developed. The first method utilizes the resistance of vapor diffusion r_{ss}, which uses soil surface temperature to estimate soil evaporation [*Monteith*, 1981; *Camillo and Gurney*, 1986; *Passerat de Silans*, 1989; *Kondo et al.*, 1992; *Sellers et al.*, 1992; *Daamen and Simmonds*, 1996; *Merlin et al.*, 2010b]. However, this method is only applicable when water flow is determined by vapor transport diffusion. The second method uses soil evaporation efficiency to estimate soil evaporation [*Merlin et al.*, 2010a].

In this study we aim to (1) disaggregate the North American Land Data Assimilation System (NLDAS)-derived soil temperature by the relationships among soil temperature (0–10 cm layer), land surface temperature, and fractional vegetation cover; and (2) derive the soil evaporation efficiency variable of *Merlin et al.* [2010a] using the disaggregated soil temperature as well as calculate the soil moisture from two soil evaporation models [*Noilhan and Planton*, 1989; *Lee and Pielke*, 1992] to disaggregate AMSR-E soil moisture.

17.2. DATA

17.2.1. Little Washita River Watershed Micronet

Little Washita River Watershed Micronet is a valuable resource for soil moisture validation. The Little Washita River Watershed is located in southwestern Oklahoma, which is composed of 20 Micronet stations within a 625 km^2 region [*Basara et al.*, 2009; *McPherson et al.*, 2007; *Cosh et al.*, 2004] and measures soil moisture at the near surface, providing an accurate estimate of the 0–5 cm soil layer. The locations of Micronet stations are shown in Figure 17.1.

For the length of this study, there were nine stations of Micronet soil moisture observations consistently available to provide the large-scale surface estimate. The nearest network average available at 1:30 P.M. of Little Washita local time was used for comparison. The reliable stations were geolocated for comparison with the disaggregated as well as AMSR-E and NLDAS soil moisture data. Figure 17.2 shows the averaged and standard deviation values of the soil moisture from the nine Micronet stations in Little Washita of the 4 months in this study. It is observed that the averaged soil moisture has a similar trend as the standard deviation of soil moisture.

17.2.2. NLDAS Data

The NLDAS (http://ldas.gsfc.nasa.gov/nldas/) phase 2 hourly forcing data derived from NOAA's (National Oceanic and Atmospheric Administration) Noah land surface model, which was developed for the NCEP (National Centers for Environmental Prediction) mesoscale eta model and soil hydraulic properties data set were used. NLDAS provides real-time land surface model output data at 1/8° (12.5 km) spatial resolution for variables such as surface temperature, solar radiation, vegetation indices, and soil wetness. In this study, three variables were used to retrieve 1 km resolution soil moisture. These included skin surface temperature, which refers to the averaged top surface temperature of vegetation, bare soil, and snow, soil temperature at 0–10 cm depth, and field capacity. The data was processed at Oklahoma local time 1:30 P.M. for matching up with *Aqua* satellite overpass. The NLDAS data set has been extensively validated [*Lohmann et al.*, 2004; *Robock et al.*, 2003; *Schaake et al.*, 2004].

17.2.3. MODIS Data

The Moderate Resolution Imaging Spectroradiometer (MODIS) onboard *Aqua* satellite has 36 spectral bands covering visible and infrared bands and provides 44 products. In this study, MODIS products, including daytime (local time 1:30 P.M.) daily Land Surface Temperature (LST) (MYD11A1) and biweekly normalized difference vegetation index (NDVI) (MYD13A2) of the most consecutive cloud-free days in May 2004 and June–August 2005 were selected. For acquiring daily NDVI, the biweekly NDVI imageries of these months were collected and fitted using sinusoidal regression among each MODIS pixel to interpolate for the daily NDVI.

17.2.4. AMSR-E Data

Soil moisture is retrieved from L-band microwave observations using the AMSR onboard the *Aqua* satellite (AMSR-E) using the single-channel algorithm (SCA) [*Jackson*, 1993; *Jackson et al.*, 1999, 2010; *Njoku et al.*, 2003]. The AMSR-E soil moisture is produced at 1/4° resolution and covers the years from 2002 to 2011. Previous research for AMSR-E soil moisture validation has shown that the error of retrieved soil moisture is less than 0.1 m^3/m^3 in low vegetation areas [*Njoku and Li*, 1999; *Njoku and Chan*, 2006; *Mladenova et al.*, 2011].

In this study, the footprint based AMSR-E level-2 soil moisture retrieval was geolocated and gridded at 1/4° spatial resolution by averaging all the soil moisture values within each AMSR-E pixel. The variables used in this study, including MODIS 1 km LST, 1 km NDVI, NLDAS 1/8° soil temperature, and soil moisture and field capacity,

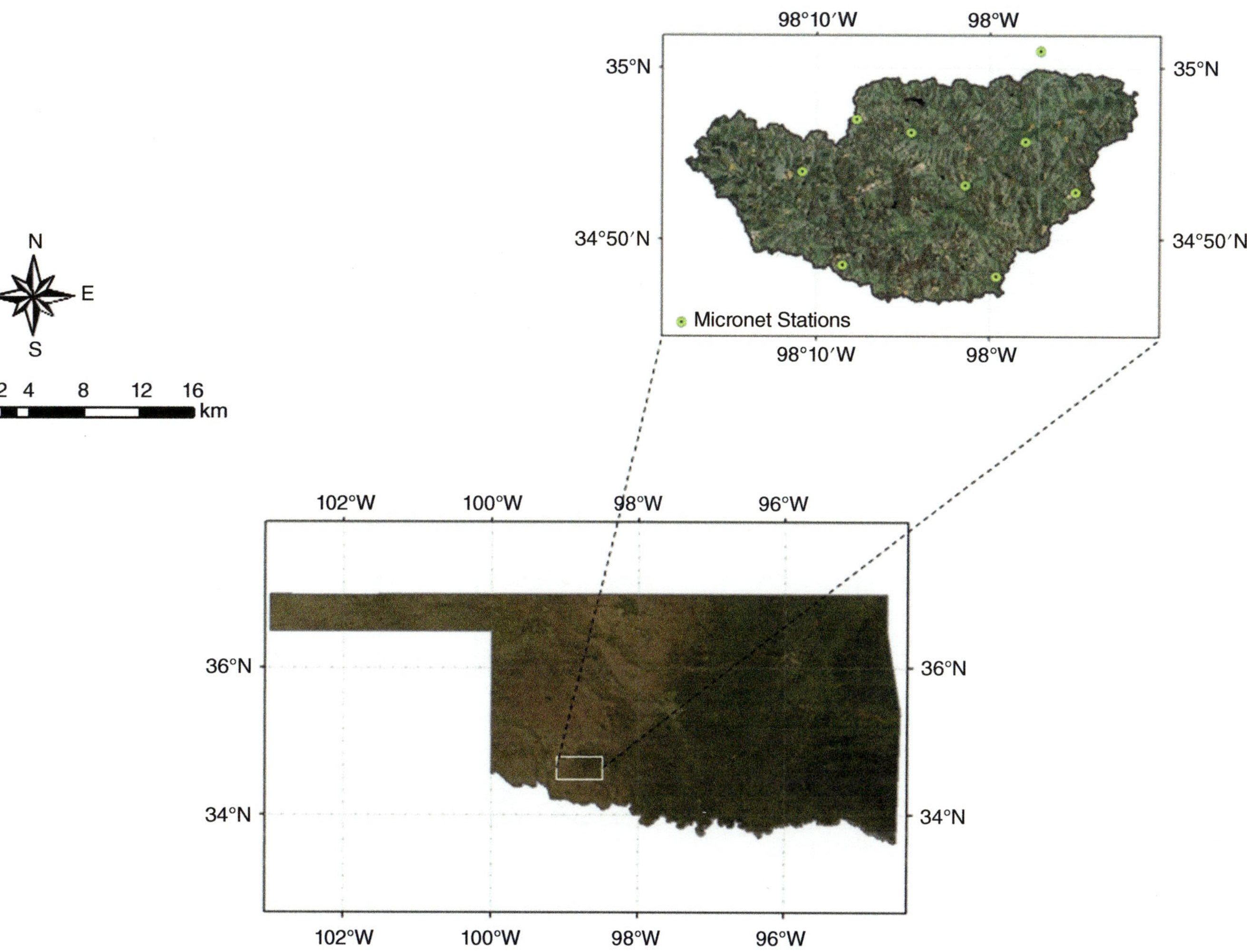

Figure 17.1 Maps of Oklahoma and Little Washita Watershed boundary. The Micronet soil moisture stations are shown in green dots.

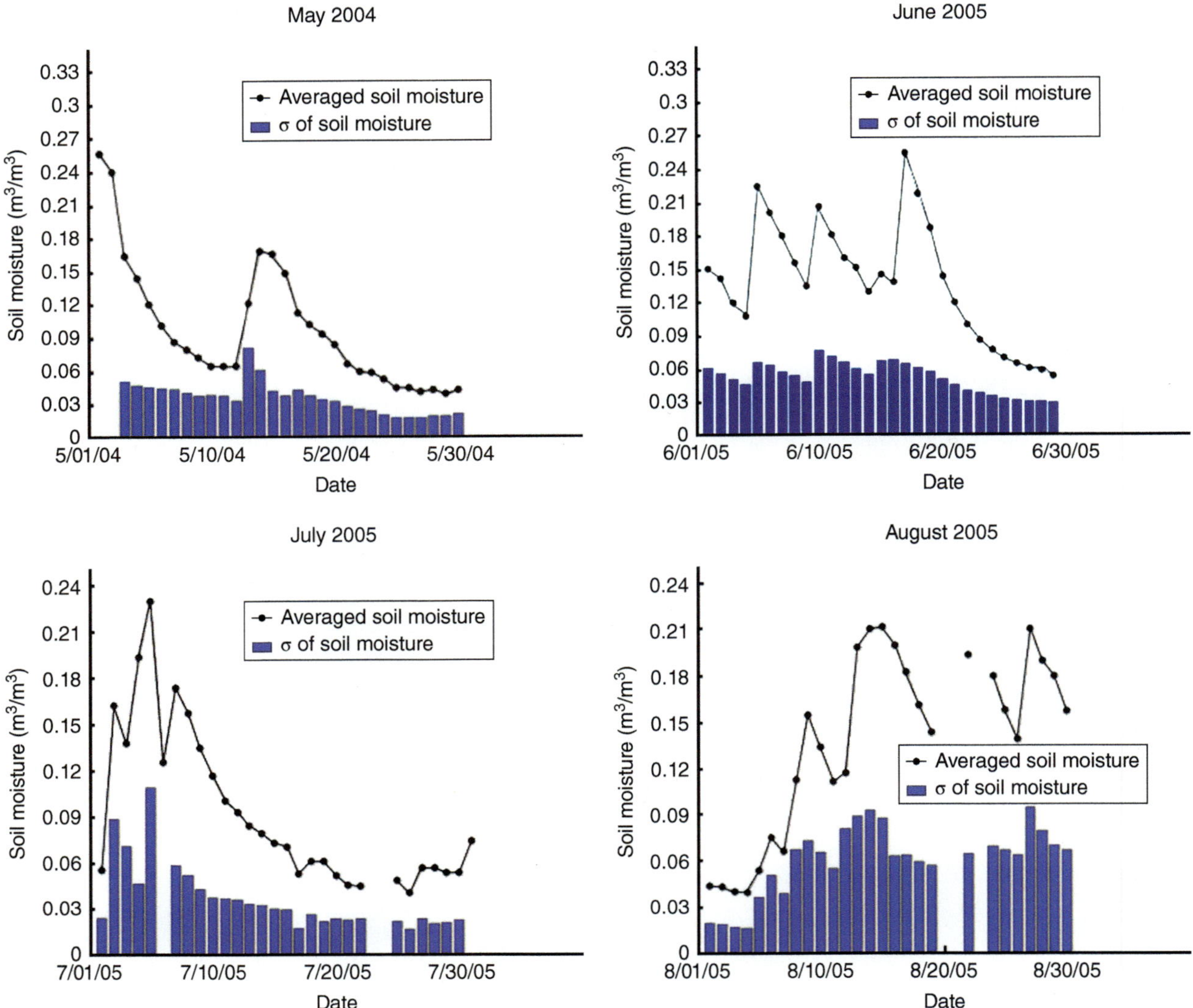

Figure 17.2 Averaged and standard deviation values of soil moisture observations from the Micronet in May 2004, June 2005, July 2005, and August 2005. The gaps represent the days of missing soil moisture measurements.

AMSR-E 1/4° soil moisture for 22 July 2005 are shown for the Little Washita Watershed in Figure 17.3.

17.3. METHODOLOGY

Several assumptions were made in this study: (a) The land surface temperature is a linear combination of soil temperature and vegetation temperature. These are proportional to fractional photosynthetic vegetation cover [*Gutman and Ignatov*, 1998; *Merlin et al.*, 2010b]. (b) The top layer soil moisture is a function of soil evaporation efficiency and field capacity. Previous experiments have shown similar nonlinear relationships under different soil depths and soil types, and the models K89 and LP92 are applicable for the soil moisture disaggregation model building [Komatsu, 2003; *Lee and Pielke*, 1992; *Noilhan and Planton*, 1989]. (c) We assume that the field capacity

among each NLDAS scale pixel is homogeneous and variation at the 1 km scale is not accounted for in this study.

The methodology will be presented in three sections: (a) The method to disaggregate NLDAS soil temperature T_{soil} to 1 km by using MODIS LST and NDVI; (b) the 1 km resolution θ will be calculated from the two soil evaporation efficiency models K03 and LP92 using the disaggregated T_{soil}; and (c) the disaggregated 1 km θ will be used to correct AMSR-E soil moisture retrievals.

17.3.1. Soil Temperature Disaggregation

The land surface temperature from satellite T_{rad} can be expressed as a linear combination of soil temperature T_{soil} and vegetation temperature T_{veg} as

$$T_{\text{rad}} = \left(1 - f_{\text{G98}}\right)T_{\text{soil}} + f_{\text{G98}}T_{\text{veg}}, \qquad (17.1)$$

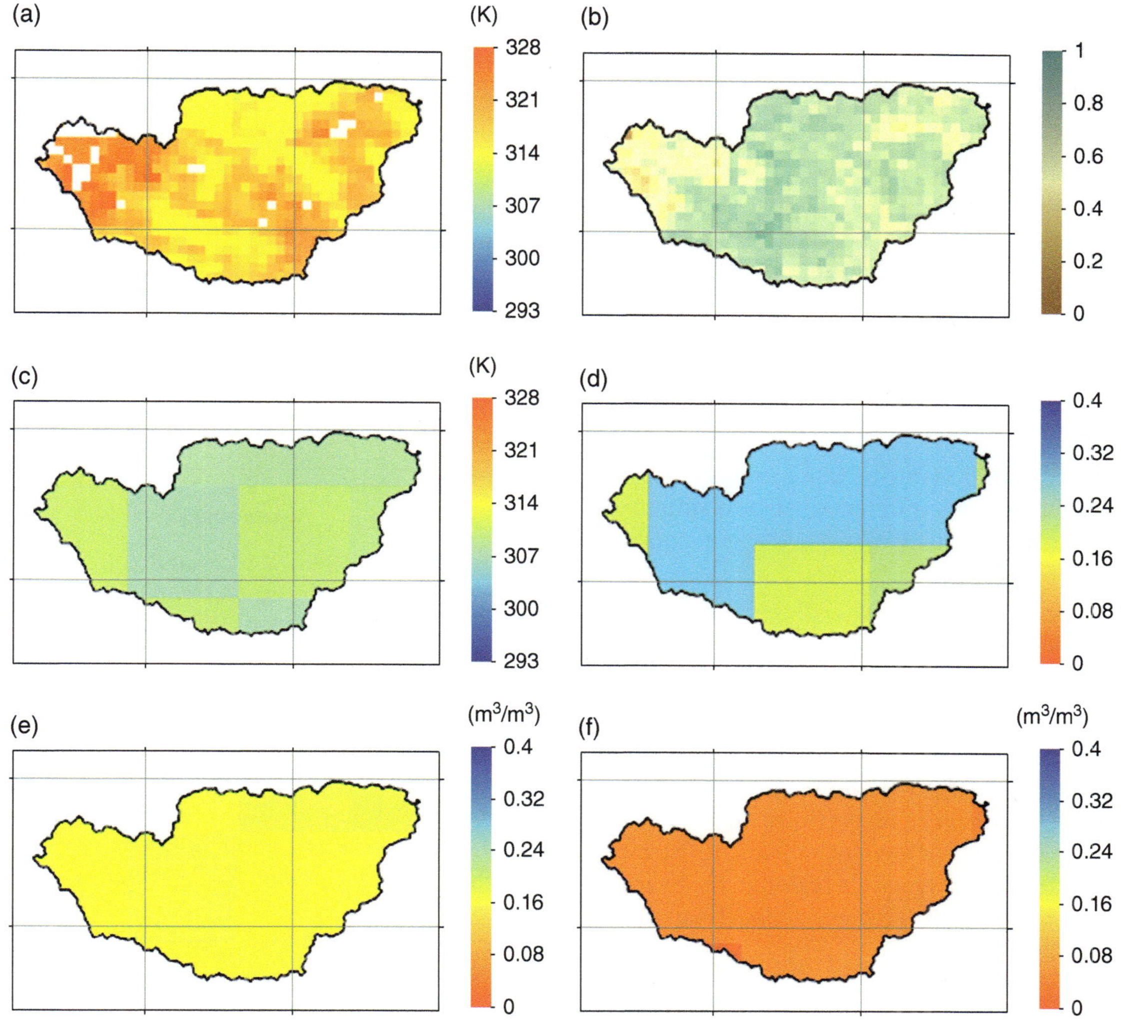

Figure 17.3 Maps of variables used in building the soil moisture downscaling algorithm over Little Washita Watershed region from 22 July 2005: (a) daytime 1 km resolution MODIS land surface temperature, (b) 1 km resolution MODIS NDVI, (c) NLDAS 1/8° resolution soil temperature at 0–10 cm depth, (d) NLDAS-derived field capacity, (e) NLDAS 1/8° resolution soil moisture at 0–10 cm depth, and (f) AMSR-E 1/4° resolution soil moisture.

where f_{G98} is the fractional vegetation cover using the *Gutman and Ignatov* [1998] method calculated using MODIS-derived NDVI as

$$f_{G98} = \frac{\text{NDVI} - \text{NDVI}_S}{\text{NDVI}_v - \text{NDVI}_S} \qquad (17.2)$$

where NDVI_v and NDVI_S are NDVI values corresponding to fully vegetated and bare soil, respectively. In this study we denoted the minimum NDVI value of imagery as NDVI_S and the maximum NDVI value of imagery as NDVI_v.

High spatial resolution vegetation temperature is usually difficult to obtain. Therefore, and LST disaggregation method using fractional vegetation cover and surface temperature at high resolution [*Merlin et al.*, 2010] was applied to NLDAS soil temperature in this study.

Bivariate regression analysis among three parameters T_{rad}, T_{soil}, and f_{G98} is expressed as

$$T_{soil} = a \times f_{G98} + b \times T_{rad} + c, \qquad (17.3)$$

where a, b, and c are the fitting coefficients.

The variation of T_{soil} of the MODIS pixels among each NLDAS pixel can be adjusted by the variations of f_{G98} and T_{rad} as

$$\Delta T_{soil} = \frac{\partial T_{soil}}{\partial f_{G98}} \Delta f_{G98} + \frac{\partial T_{soil}}{\partial f_{rad}} \Delta T_{rad}, \qquad (17.4)$$

where ΔT_{soil}, Δf_{G98}, and ΔT_{rad} are the differences of T_{soil}, f_{G98}, and T_{rad} between MODIS and disaggregated NLDAS pixels, $\partial T_{soil} / \partial f_{G98}$ and $\partial T_{soil} / \partial T_{rad}$ are the partial derivatives of T_{soil} to f_{G98} and T_{rad}.

This regression model was derived at 1/8° NLDAS scale and applied to f_{G98} and T_{rad} at 1 km to calculate the total derivative ΔT_{soil}. Then ΔT_{soil} was added up to each MODIS pixel to obtain corrected 1 km T_{soil}.

17.3.2. Soil Evaporation Efficiency Model

Soil evaporation efficiency β can be defined as a ratio of soil evaporation rate between soil surface and water surface. The standard equation of soil evaporation efficiency [*Komatsu*, 2003] is expressed as

$$E = \beta \rho_a \Delta q / r_a, \qquad (17.5)$$

where E is the evaporation rate of water from soil surface, Δq is the specific humidity difference between air and soil surface, ρ_a is the mass density of air, and r_a is evaporative resistance.

The soil evaporation efficiency β at 1 km is estimated using disaggregated soil temperature T_{1km} by

$$\beta = \frac{T_{max} - T_{1km}}{T_{max} - T_{min}}, \qquad (17.6)$$

where T_{max} and T_{min} are the maximum and minimum soil temperatures of the imagery, respectively.

Practically, these parameters are difficult to be measured and collected on a large-scale region. Consequently, simplified soil evaporation efficiency model was developed as [*Komatsu*, 2003]

$$\beta = 1 - \exp\left(-\theta_{mm}/\theta_c\right), \qquad (17.7)$$

where θ_{mm} is experimental soil moisture data at 1–3 mm and θ_c is a semiempirical parameter field capacity.

The θ_c is determined by humidity, soil type, and wind speed, defined as

$$\theta_c = \theta_{c0}\left(1 + \frac{r_{ah}^{ref}}{r_{ah}}\right), \qquad (17.8)$$

where θ_{c0} is a soil-dependent parameter ranging between 0.01 and 0.04 vol/vol, r_{ah}^{ref} is the reference aerodynamic resistance, while r_{ah} is the aerodynamic resistance [*Komatsu*, 2003].

However, this model is only applicable for the soil layer depth of several millimeters. In recent years several semiempirical soil evaporation efficiency models have been developed for a soil layer in the centimeter scale, which are consistent with the sensing depth of NLDAS as follows:
(a) NP89 Method [*Noilhan and Planton*, 1989]:

$$\beta_{mod,NP89} = 0.5 - 0.5\,\cos\left(\pi\theta / \theta_{C,NP89}\right), \qquad (17.9)$$

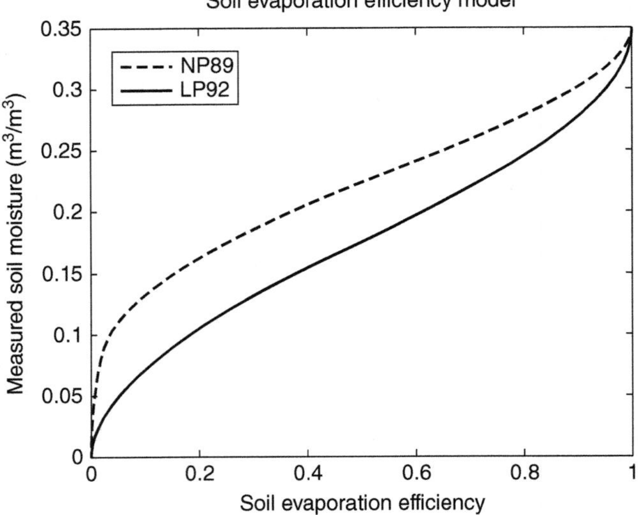

Figure 17.4 Correlation of measured soil evaporation efficiency and soil moisture (θ_C=0.35). NP89 is the model proposed by *Noilhan and Planton* [1989], while LP92 is the model proposed by *Lee and Pielke* [1992].

where $\theta_{c,NP89}$ is the soil moisture field capacity of the NP89 experiment and θ is the measured soil moisture. This model was derived from the comparison of relative humidity, soil moisture, and field capacity.
(b) LP92 method [*Lee and Pielke*, 1992]:

$$\beta_{mod,LP92} = \left[0.5 - 0.5\,\cos\left(\pi\theta / \theta_{C,LP92}\right)\right]^2, \qquad (17.10)$$

where $\theta_{C,LP92}$ is the soil moisture field capacity of the LP92 experiment and θ is measured soil moisture. This model was derived from several numerical atmospheric models and is particularly suitable for sand or silt loam soil.

Both of the soil evaporation efficiency models correspond to various soil depth and types, including agricultural soil, sand, and cornstarch. The β-θ curves under different soil types demonstrate similar patterns. Figure 17.4 shows the β-θ curves of these two models when θ_C equals 0.35.

To determine the field capacity θ_C, NLDAS provides gridded θ_C at 1/8° covering the whole of Oklahoma. The NLDAS θ_C was resampled to 1 km scale by comparing the coordinates of NLDAS and MODIS pixels to match up with 1 km soil temperature pixels.

17.3.3. AMSR-E Soil Moisture Correction

The AMSR-E soil moisture was disaggregated using the calculated 1 km daily soil moisture θ from models NP89 and LP92. Among each 1/4° AMSR-E pixel, the difference between AMSR-E soil moisture Θ and averaged 1 km disaggregated soil moisture θ within this AMSR-R pixel boundary was added to each 1 km pixel. These cloud-contaminated or no-data pixels were discarded.

The AMSR-E soil moisture correction equation is given by

$$\theta_c(i, j) = \theta(i, j) + \left[\Theta - \frac{1}{N}\Sigma_{i,j}\theta(i, j)\right] \quad (17.11)$$

where $\theta_c(i, j)$ is 1 km corrected soil moisture, $\theta(i, j)$ is soil evaporation efficiency model output 1 km soil moisture, Θ is AMSR-E descending orbit soil moisture, and N is the number of 1 km pixels among each AMSR-E pixel.

17.4. RESULTS AND DISCUSSIONS

17.4.1. Correlation Analysis

Figure 17.5a shows the bivariate correlation between the MODIS daily 12.5 km land surface temperature and fractional vegetation cover as well as aggregated 12.5 km NLDAS soil temperature, of 4 days from 20 May 2004, 26 June 2005, 22 July 2005, and 9 August 2005 covering Oklahoma, while Figure 17.5b displays the relationship between surface soil temperature and fractional vegetation cover, and Figure 17.5c shows the relationship between surface soil temperature and surface temperature. All the scatterplots show good correlation between the three variables and inverse relationships of surface soil temperature-surface temperature and surface-temperature-fractional

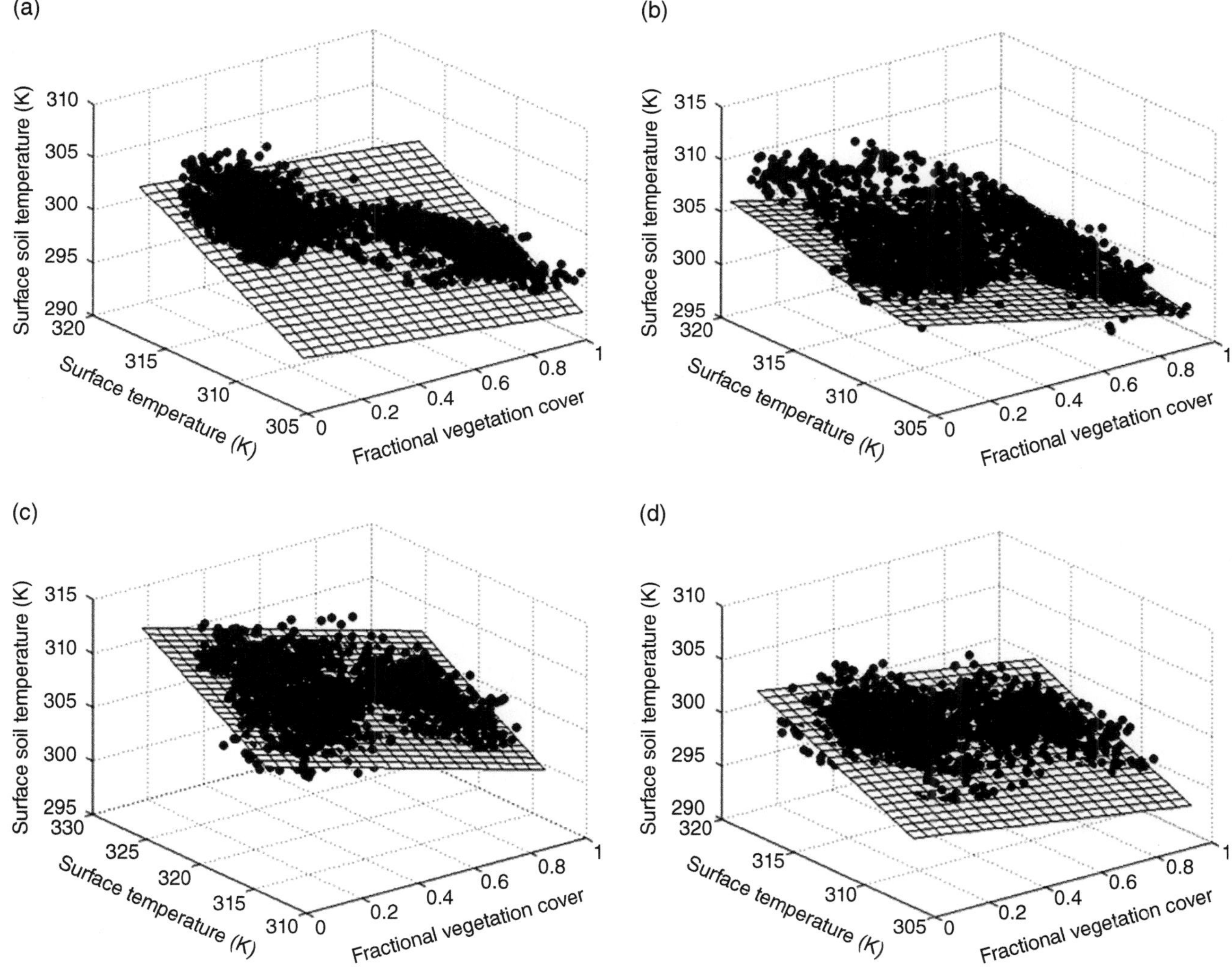

Figure 17.5a Bivariate correlation analysis among surface soil temperature (0–10 cm), surface temperature and fractional vegetation cover from (a) 20 May 2004, (b) 26 June 2005, (c) 22 July 2005 and (d) 9 August 2005, at 12.5 km spatial resolution.

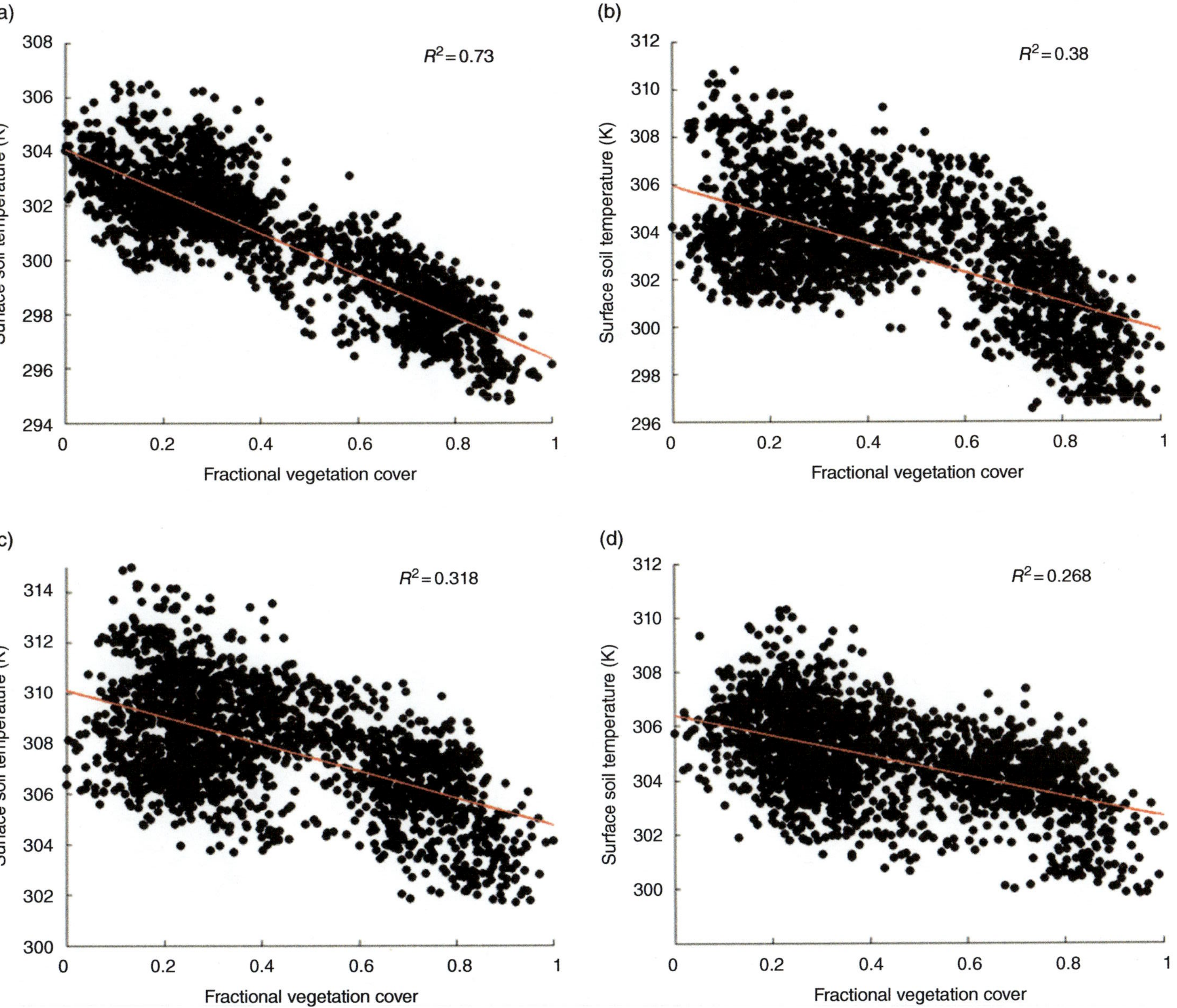

Figure 17.5b Correlation analysis between surface soil temperature (0–10 cm) and fractional vegetation cover from (a) 20 May 2004, (b) 26 June 2005, 22, (c) July 2005, and (d) 9 August 2005, at 12.5 km spatial resolution.

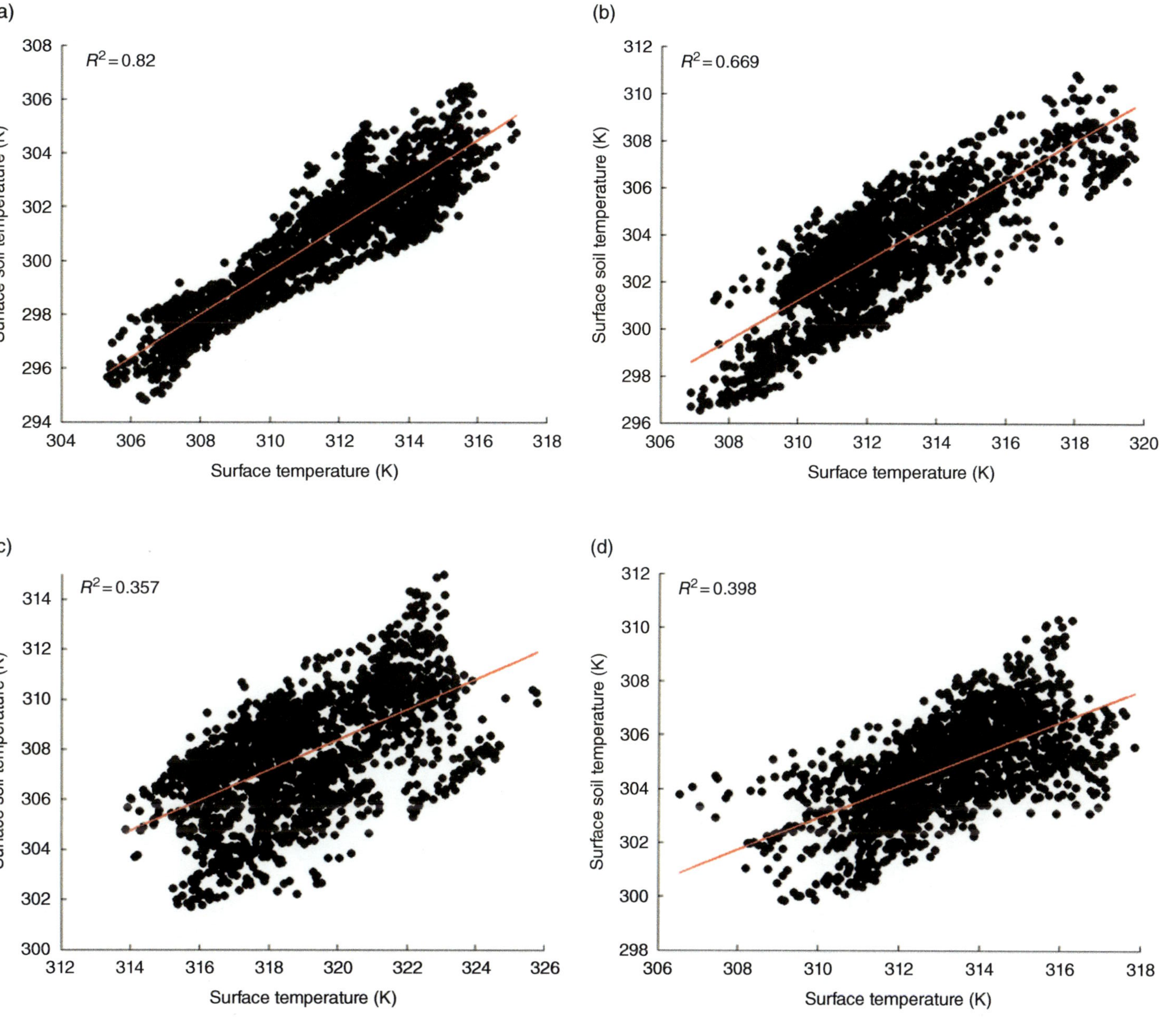

Figure 17.5c Correlation analysis between surface soil temperature (0–10 cm) and surface temperature from (a) 20 May 2004, (b) 26 June 2005, (c) 22 July 2005, and (d) 9 August 2005, at 12.5 km spatial resolution.

vegetation cover. The R^2 of $f_{G98} - T_{soil}$ and $f_{G98} - T_{rad}$ for 20 May 2004 are 0.73 and 0.82, for 26 June 2005 are 0.38 and 0.669, for 22 July 2005 are 0.318 and 0.357, and for 9 August 2005, are 0.268 and 0.398, respectively. The bivariate correlation coefficients for the other days in the study period show similar values.

17.4.2. Soil Variables Mapping

In Figure 17.6a–b soil variables, including soil evaporation efficiency, disaggregated 1 km soil temperature, and NLDAS soil moisture of 4 days with low cloud cover were chosen to examine soil moisture dry-down patterns. The western and northeast parts of the watershed correspond to high soil temperature and low soil evaporation efficiency, while the center part of the basin corresponds to lower soil temperature and higher soil evaporation efficiency. The regional variations in temperature responses

to seasonal change are better displayed in the 1 km than AMSR-E soil moisture.

Corresponding to these days, the soil moisture at different resolutions, 1 km disaggregated soil moisture output from NP89 and LP92 models, AMSR-E and NLDAS were mapped (Figure 17.7a–b). Compared with the coarser spatial resolution soil moisture from AMSR-E and NLDAS, which contain only 3–7 pixels, the disaggregated results demonstrate soil moisture spatial distribution pattern with far more detail and have around 625 pixels of 1 km. However, the difference in soil moisture ranges among the three scales must be noted. NLDAS soil moisture values are higher than the disaggregated and AMSR-E soil moisture. A major reason is the mismatch of soil sensing depth between microwave remote sensing imagery and land surface model output. The microwave satellite sensor can only penetrate soil layer of 0–2 cm depth, matching the sensing depths of

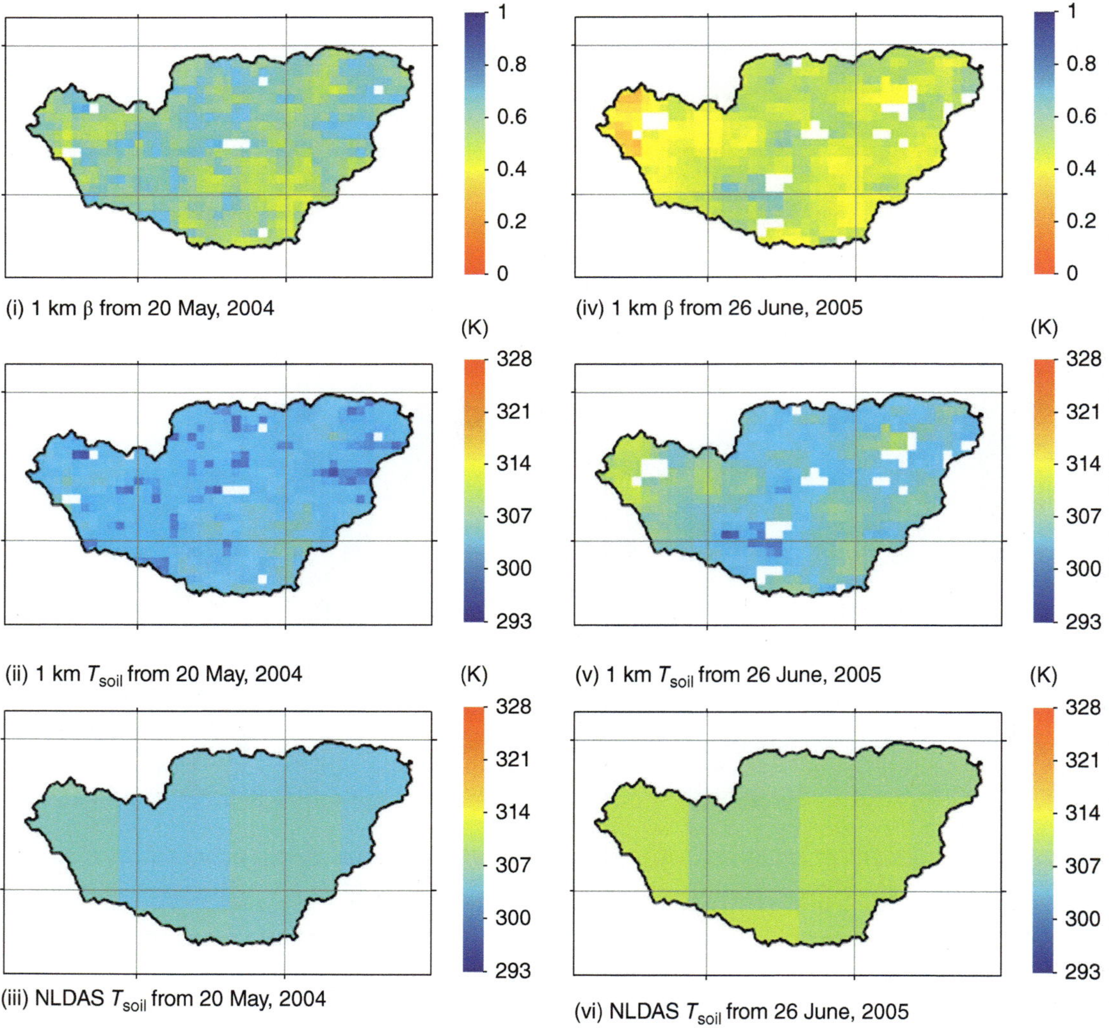

Figure 17.6a Maps of 1 km soil evaporation efficiency, 1 km disaggregated soil temperature, and NLDAS soil temperature in Little Washita Watershed region from 20 May 2004 and 26 June 2005.

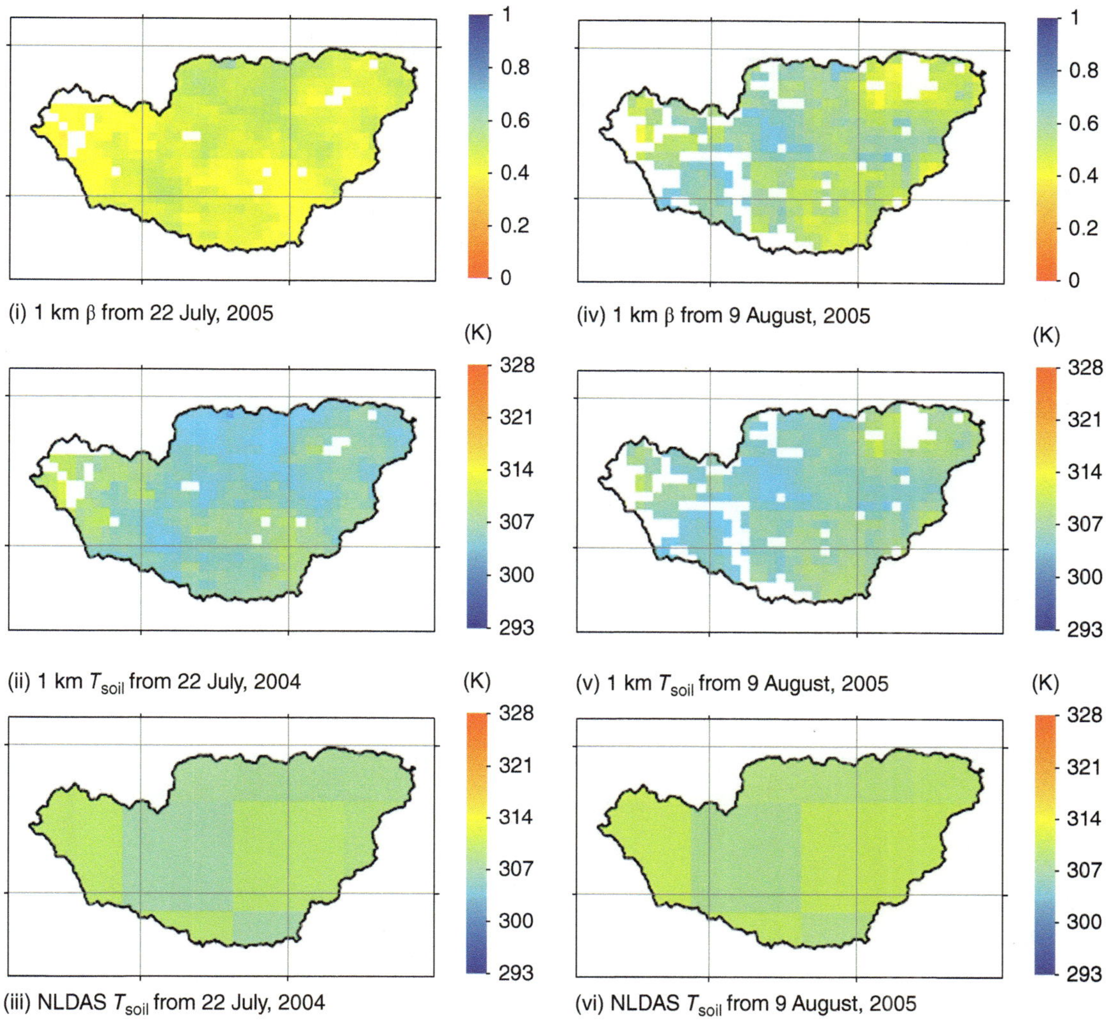

Figure 17.6b Maps of 1 km soil evaporation efficiency, 1 km disaggregated soil temperature, and NLDAS soil temperature in Little Washita Watershed region from 22 July 2005 and 9 August 2005.

the two experimental soil evaporation efficiency models NP89 and LP92, while the NLDAS soil moisture is at 0–10 cm, recording the soil water content corresponding to a deeper layer. Moreover, an unnatural transition of soil moisture distribution pattern in the southeast portion of the watershed could be observed, due to the coarse resolution of NLDAS-derived field capacity. It also can be observed that the soil moisture of 22 July 2005 shows the least spatial variation compared to the other days, while 26 June 2005 shows the most variability. If comparing the days shown in Figures 17.6 and 17.7 with soil moisture dry-down curves in Figure 17.2, soil moisture of 20 May 2004, 26 June 2005, and 9 August 2005, was showing an apparent drying trend, while the soil moisture of 22 July 2005 was temporally invariant. For 26 June 2005, the soil moisture curve showed a very rapid drying trend.

17.4.3. Validation with Micronet data

Tables 17.1–17.4 display the statistics of 1 km disaggregated soil moisture derived from NP89 and LP92, AMSR-E and NLDAS-validated with Little Washita Micronet observations and include slope, root-mean-squared error (RMSE), unbiased RMSE, and spatial standard deviation. Table 17.5 shows the overall averaged statistical variables of each month. We set the threshold of the valid number of point pair as 5 and the days with a lesser number of point pairs were dropped. The bar plots of the statistical values of Table 17.1–17.5 are also shown in Figure 17.8a–e.

The statistical variables are listed as below.

$$m = \frac{\Delta\hat{\theta}}{\Delta\theta}, \tag{17.12}$$

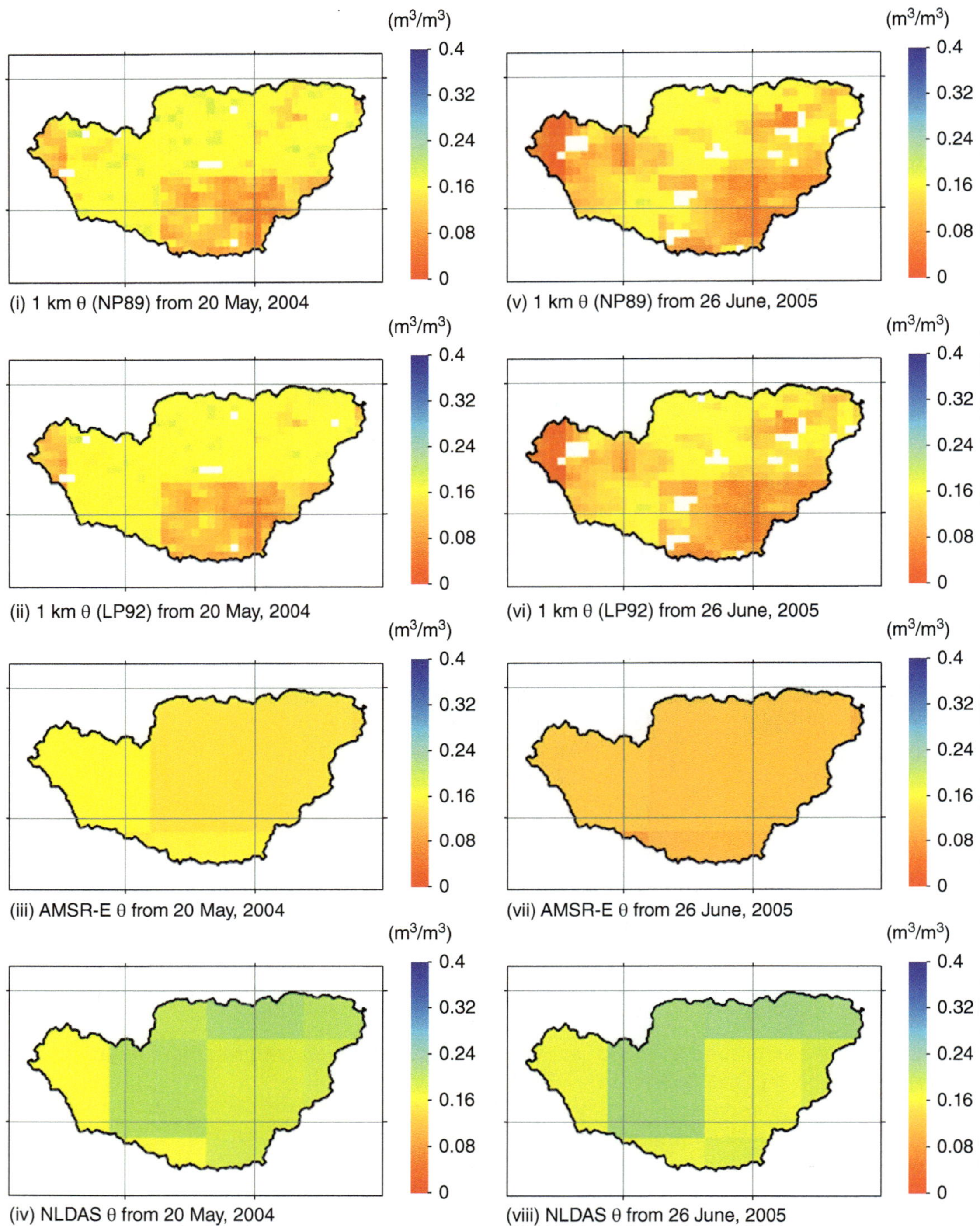

Figure 17.7a Maps of 1 km disaggregated soil moisture derived from NP89 and LP92 models, AMSR-E, and NLDAS soil moisture in Little Washita Watershed region from 20 May 2004 and 26 June 2005.

$$\text{RMSE} = \sqrt{\frac{\Sigma_{i=1}^{n}\left(\hat{\theta}_i - \theta_i\right)^2}{n}}, \quad (17.13)$$

$$\text{RMSE*} = \sqrt{\frac{\Sigma_{i=1}^{n}(\hat{\theta}_i^{\,*} - \theta_i)^2}{n}}, \quad (17.14)$$

$$\sigma = \sqrt{\frac{\Sigma_{i=1}^{n}\left(\hat{\theta}_i - \bar{\theta}\right)^2}{n}}, \quad (17.15)$$

where m is the slope, which is the ratio of estimated soil moisture $\hat{\theta}$ (1 km disaggregated, AMSR-E and NLDAS) to Micronet moisture observations θ (Micronet). RMSE*

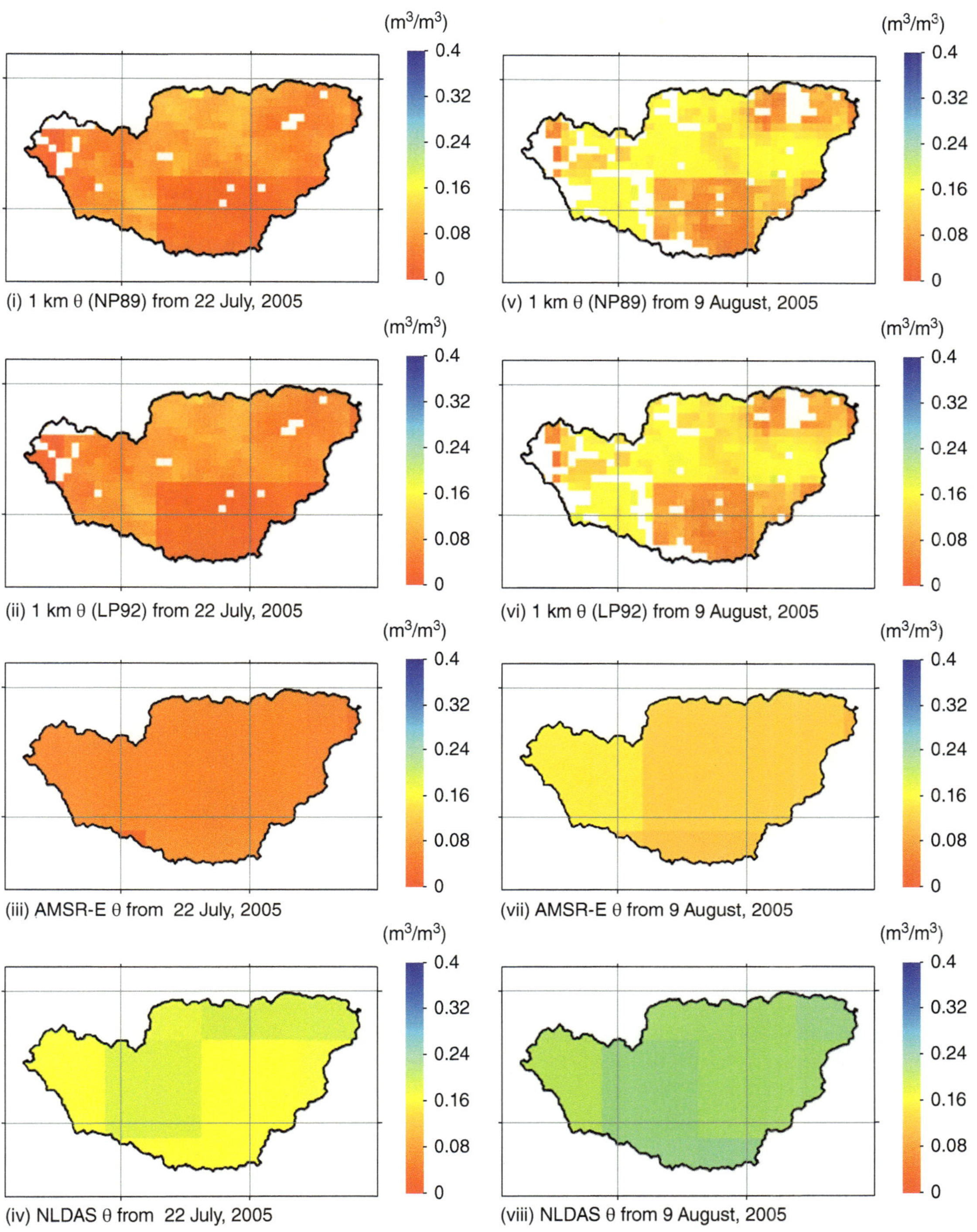

Figure 17.7b Maps of 1 km disaggregated soil moisture derived from NP89 and LP92 models, AMSR-E, and NLDAS soil moisture in Little Washita Watershed region from 22 July 2005 and 9 August 2005.

is unbiased RMSE and $\hat{\theta}_i^*$ is the predicted soil moisture derived from linear regression fitting between $\hat{\theta}$ and θ. and σ is spatial standard deviation of each soil moisture data set.

From Tables 17.1–17.5 and Figure 17.8a–e we can summarize the following: (a) The slopes of all four soil moisture sources are biased, i.e., the slopes are different from 1 when compared with the Micronet soil moisture. The slope of the disaggregated soil moisture ranges from −0.042 to 0.056 m³/m³. Specifically, for July 2005 and August 2005, slopes of NP89 and LP92 are better than NLDAS and closer to AMSR-E. (b)

Table 17.1 Soil moisture validation variables

Day	Data Set	Slope	RMSE (m³/m³)	Unbiased RMSE (m³/m³)	Spatial Standard Deviation (m³/m³)	Number of Points
4 May 2004	NP89	**0.019**	**0.045**	**0.043**	**0.013**	8
	LP92	**0.016**	**0.045**	**0.043**	**0.013**	
	AMSR-E	0.068	0.045	0.044	0.005	
	NLDAS	0.097	0.108	0.107	0.01	
	Micronet				0.043	
5 May 2004	NP89	**0.051**	**0.044**	**0.042**	**0.014**	8
	LP92	**0.05**	**0.044**	**0.042**	**0.014**	
	AMSR-E	0.098	0.039	0.039	0.006	
	NLDAS	0.106	0.123	0.123	0.01	
	Micronet				0.042	
6 May 2004	NP89	**0.056**	**0.049**	**0.041**	**0.015**	8
	LP92	**0.052**	**0.049**	**0.041**	**0.016**	
	AMSR-E	0.108	0.038	0.030	0.006	
	NLDAS	0.120	0.136	0.136	0.01	
	Micronet				0.041	
15 May 2004	NP89	**0.04**	**0.04**	**0.036**	**0.013**	5
	LP92	**0.044**	**0.041**	**0.036**	**0.014**	
	AMSR-E	−0.003	0.057	0.037	0.003	
	NLDAS	0.048	0.072	0.036	0.009	
	Micronet				0.037	
20 May 2004	NP89	**−0.069**	**0.056**	**0.03**	**0.019**	8
	LP92	**−0.129**	**0.057**	**0.03**	**0.023**	
	AMSR-E	0.05	0.037	0.03	0.002	
	NLDAS	0.15	0.111	0.026	0.009	
	Micronet				0.03	
23 May 2004	NP89	**0.049**	**0.065**	**0.022**	**0.015**	7
	LP92	**0.006**	**0.065**	**0.022**	**0.016**	
	AMSR-E	0.077	0.044	0.022	0.002	
	NLDAS	0.221	0.119	0.019	0.009	
	Micronet				0.022	
24 May 2004	NP89	**0.247**	**0.079**	**0.017**	**0.014**	7
	LP92	**0.282**	**0.077**	**0.017**	**0.015**	
	AMSR-E	−0.013	0.06	0.01	0.002	
	NLDAS	0.234	0.12	0.017	0.009	
	Micronet				0.018	

Note: Slope, RMSE, unbiased RMSE and spatial standard deviation of downscaled soil derived from NP89 and LP92 models, AMSR-E, and NLDAS comparing with Little Washita Watershed Micronet in May 2004.
The values in bold indicate 1 km disaggregated results.

The RMSE and unbiased RMSE of disaggregated soil moisture range from 0.042 to 0.069 m³/m³ and 0.031 to 0.055 m³/m³, respectively. For May 2004 and June 2005, RMSE and unbiased RMSE for disaggregated soil moisture of NP89 and LP92 are better than NLDAS and slightly worse than AMSR-E. For July 2005 and August 2005, RMSE for the two model disaggregated soil moistures are better than both AMSR-E and NLDAS. (c) The spatial standard deviation of disaggregated soil moisture ranges from 0.013 to 0.017. Comparing the spatial standard deviation of each soil moisture data set with ground observations, NP89 and LP92 methods disaggregated soil moisture are closer to Micronet soil moisture measurements than AMSR-E and NLDAS. (d) Comparing the accuracy and spatial variation between the two evaporation efficiency models, LP92 has better RMSE but a worse unbiased RMSE than NP89, while the spatial standard deviation is always better than NP89.

Figure 17.9 shows that NP89 and LP92 have much more variance than AMSR-E and NLDAS. AMSR-E and NLDAS comparison results appear tightly grouped but are obviously biased. In Figure 17.10 the correlation of all the detrended soil moisture comparisons with

Table 17.2 Soil moisture validation variables

Day	Data Set	Slope	RMSE (m³/m³)	Unbiased RMSE (m³/m³)	Spatial Standard Deviation (m³/m³)	Number of Points
2 June 2005	NP89	**0.012**	**0.052**	**0.052**	**0.016**	7
	LP92	**−0.001**	0.053	**0.052**	**0.017**	
	AMSR-E	0.088	0.052	0.041	0.008	
	NLDAS	0.001	0.104	0.052	0.01	
	Micronet				0.052	
4 June 2005	NP89	**0.156**	**0.043**	**0.036**	**0.011**	5
	LP92	**0.185**	**0.042**	**0.037**	**0.014**	
	AMSR-E	0.11	0.05	0.025	0.006	
	NLDAS	0.059	0.113	0.026	0.003	
	Micronet				0.05	
6 June 2005	NP89	**−0.044**	**0.062**	**0.057**	**0.012**	6
	LP92	**−0.055**	**0.063**	**0.055**	**0.012**	
	AMSR-E	0.013	0.063	0.059	0.011	
	NLDAS	0.013	0.078	0.058	0.007	
	Micronet				0.059	
13 June 2005	NP89	**−0.103**	**0.084**	**0.057**	**0.025**	5
	LP92	**−0.1**	**0.081**	**0.058**	**0.027**	
	AMSR-E	−0.072	0.075	0.058	0.023	
	NLDAS	0.002	0.14	0.14	0.013	
	Micronet				0.059	
15 June 2005	NP89	**−0.047**	**0.087**	**0.057**	**0.011**	6
	LP92	**−0.05**	**0.085**	**0.057**	**0.011**	
	AMSR-E	−0.028	0.075	0.059	0.009	
	NLDAS	0.017	0.16	0.061	0.009	
	Micronet				0.062	
18 June 2005	NP89	**−0.14**	**0.066**	0.024	**0.012**	5
	LP92	**−0.145**	**0.067**	0.026	**0.013**	
	AMSR-E	−0.077	0.071	0.031	0.01	
	NLDAS	−	0.075	0.055	0.01	
	Micronet				0.055	
19 June 2005	NP89	**−0.096**	**0.062**	**0.038**	**0.013**	5
	LP92	**−0.104**	**0.064**	**0.04**	**0.014**	
	AMSR-E	−0.019	0.069	0.032	0.005	
	NLDAS	−0.002	0.086	0.052	0.01	
	Micronet				0.052	
20 June 2005	NP89	**−0.081**	**0.051**	**0.039**	**0.012**	6
	LP92	**−0.103**	**0.052**	**0.039**	**0.013**	
	AMSR-E	0.008	0.049	0.045	0.003	
	NLDAS	0.003	0.113	0.046	0.01	
	Micronet				0.047	
22 June 2005	NP89	**−0.093**	**0.045**	**0.033**	**0.013**	6
	LP92	**−0.111**	**0.045**	**0.034**	**0.014**	
	AMSR-E	−0.003	0.038	0.037	0.004	
	NLDAS	0.002	0.138	0.037	0.011	
	Micronet				0.037	
24 June 2005	NP89	**0.046**	**0.047**	**0.033**	**0.014**	6
	LP92	**0.051**	**0.045**	**0.033**	**0.015**	
	AMSR-E	0.073	0.035	0.026	0.004	
	NLDAS	0.002	0.147	0.033	0.012	
	Micronet				0.033	
26 June 2005	NP89	**−0.061**	**0.052**	**0.03**	**0.016**	5
	LP92	**−0.079**	**0.051**	**0.03**	**0.018**	
	AMSR-E	0.067	0.037	0.023	0.003	
	NLDAS	0.004	0.143	0.03	0.013	
	Micronet				0.03	

(Continued)

Table 17.2 (Continued)

Day	Data Set	Slope	RMSE (m³/m³)	Unbiased RMSE (m³/m³)	Spatial Standard Deviation (m³/m³)	Number of Points
27 June 2005	NP89	**−0.064**	**0.056**	**0.026**	**0.011**	6
	LP92	**−0.072**	**0.054**	**0.027**	**0.012**	
	AMSR-E	0.024	0.039	0.016	0.001	
	NLDAS	0.005	0.144	0.028	0.012	
	Micronet				0.028	
29 June 2005	NP89	**0.036**	**0.06**	**0.026**	**0.011**	6
	LP92	**0.044**	**0.058**	**0.027**	**0.011**	
	AMSR-E	0.072	0.044	0.015	0.002	
	NLDAS	0.003	0.14	0.027	0.012	
	Micronet				0.027	

Note: Slope, RMSE, unbiased RMSE, and spatial standard deviation of downscaled soil derived from NP89 and LP92 models, AMSR-E, and NLDAS comparing with Little Washita Watershed Micronet in June 2005.
The values in bold indicate 1 km disaggregated results.

Table 17.3 Soil moisture validation variables

Day	Data Set	Slope	RMSE (m³/m³)	Unbiased RMSE (m³/m³)	Spatial Standard Deviation (m³/m³)	Number of Points
3 July 2005	NP89	**0.035**	**0.064**	**0.038**	**0.012**	6
	LP92	**0.035**	**0.064**	**0.049**	**0.012**	
	AMSR-E	0.021	0.074	0.059	0.004	
	NLDAS	0.056	0.099	0.048	0.005	
	Micronet				0.065	
8 July 2005	NP89	**0.002**	**0.054**	**0.048**	**0.012**	7
	LP92	**−0.007**	**0.055**	**0.048**	**0.013**	
	AMSR-E	0.042	0.066	0.037	0.004	
	NLDAS	0.057	0.086	0.041	0.007	
	Micronet				0.048	
10 July 2005	NP89	**0.258**	**0.027**	**0.024**	**0.016**	6
	LP92	**0.331**	**0.026**	**0.024**	0.02	
	AMSR-E	0.078	0.035	0.026	0.004	
	NLDAS	0.076	0.11	0.027	0.006	
	Micronet				0.031	
11 July 2005	NP89	**0.035**	**0.035**	**0.035**	**0.014**	5
	LP92	**0.011**	**0.036**	**0.036**	**0.014**	
	AMSR-E	0.048	0.039	0.039	0.004	
	NLDAS	0.037	0.115	0.035	0.008	
	Micronet				0.036	
12 July 2005	NP89	**0.075**	**0.033**	**0.031**	**0.012**	7
	LP92	**0.085**	**0.032**	**0.031**	**0.013**	
	AMSR-E	0.073	0.034	0.033	0.005	
	NLDAS	0.048	0.114	0.031	0.008	
	Micronet				0.033	
20 July 2005	NP89	**0.085**	**0.048**	**0.02**	**0.013**	7
	LP92	**0.072**	**0.046**	**0.02**	**0.013**	
	AMSR-E	0.165	0.031	0.014	0.005	
	NLDAS	−0.125	0.125	0.02	0.009	
	Micronet				0.021	
22 July 2005	NP89	**−0.143**	**0.038**	**0.019**	**0.012**	6
	LP92	**−0.184**	**0.037**	**0.019**	**0.013**	
	AMSR-E	0.108	0.022	0.014	0.003	
	NLDAS	−0.153	0.123	0.019	0.009	
	Micronet				0.021	

Note: Slope, RMSE, unbiased RMSE, and spatial standard deviation of downscaled soil derived from NP89 and LP92 models, AMSR-E, and NLDAS comparing with Little Washita Watershed Micronet in July 2005.
The values in bold indicate 1 km disaggregated results.

Table 17.4 Soil moisture validation variables

Day	Data Set	Slope	RMSE (m³/m³)	Unbiased RMSE (m³/m³)	Spatial Standard Deviation (m³/m³)	Number of Points
2 August 2005	NP89	**0.321**	**0.038**	**0.016**	**0.02**	6
	LP92	**0.336**	**0.036**	**0.017**	**0.022**	
	AMSR-E	0.322	0.019	0.018	0.008	
	NLDAS	−0.081	0.129	0.017	0.006	
	Micronet				0.018	
9 August 2005	NP89	**−0.013**	**0.086**	**0.074**	**0.014**	5
	LP92	**−0.031**	**0.088**	**0.073**	**0.015**	
	AMSR-E	0.025	0.097	0.068	0.006	
	NLDAS	0.005	0.092	0.074	0.005	
	Micronet				0.074	
18 August 2005	NP89	**0.027**	**0.053**	**0.053**	**0.01**	6
	LP92	**0.009**	**0.055**	**0.054**	**0.01**	
	AMSR-E	0.07	0.053	0.044	0.006	
	NLDAS	0.007	0.091	0.055	0.009	
	Micronet				0.055	
25 August 2005	NP89	**−0.077**	**0.081**	**0.057**	**0.02**	6
	LP92	**−0.069**	**0.081**	**0.058**	**0.022**	
	AMSR-E	−0.113	0.099	0.054	0.016	
	NLDAS	−0.009	0.104	0.06	0.01	
	Micronet				0.061	
29 August 2005	NP89	**0.01**	**0.076**	**0.064**	**0.015**	6
	LP92	**0.015**	**0.077**	**0.064**	**0.016**	
	AMSR-E	−0.006	0.094	0.064	0.006	
	NLDAS	−0.031	0.097	0.061	0.01	
	Micronet				0.064	
30 August 2005	NP89	**0.015**	**0.077**	**0.061**	**0.017**	6
	LP92	**0.022**	**0.078**	**0.061**	**0.018**	
	AMSR-E	−0.002	0.095	0.061	0.007	
	NLDAS	−0.037	0.103	0.057	0.011	
	Micronet				0.061	

Note: Slope, RMSE, unbiased RMSE, and spatial standard deviation of downscaled soil derived from NP89 and LP92 models, AMSR-E, and NLDAS comparing with Little Washita Watershed Micronet in August 2005.
The values in bold indicate 1 km disaggregated results.

Table 17.5 Overall soil moisture validation variables of the 4 months

Day	Data Set	Slope	RMSE (m³/m³)	Unbiased RMSE (m³/m³)	Spatial Standard Deviation (m³/m³)	Number of Points
May 2004	NP89	**0.056**	**0.054**	**0.033**	**0.015**	51
	LP92	**0.046**	**0.054**	**0.033**	**0.016**	
	AMSR-E	0.055	0.046	0.03	0.004	
	NLDAS	0.14	0.113	0.066	0.009	
	Micronet				0.033	
June 2005	NP89	**−0.037**	**0.059**	**0.039**	**0.014**	74
	LP92	**−0.042**	**0.058**	**0.04**	**0.015**	
	AMSR-E	0.02	0.054	0.036	0.007	
	NLDAS	0.009	0.122	0.05	0.01	
	Micronet				0.046	
July 2005	NP89	**0.05**	**0.043**	**0.031**	**0.013**	44
	LP92	**0.049**	**0.042**	**0.032**	**0.014**	
	AMSR-E	0.076	0.043	0.032	0.004	
	NLDAS	−0.001	0.11	0.032	0.007	
	Micronet				0.036	
August 2005	NP89	**0.047**	**0.069**	**0.054**	**0.016**	35
	LP92	**0.047**	**0.069**	**0.055**	**0.017**	
	AMSR-E	0.049	0.076	0.052	0.008	
	NLDAS	−0.025	0.103	0.054	0.008	
	Micronet				0.056	

Note: Slope, RMSE, unbiased RMSE, and spatial standard deviation of downscaled soil derived from NP89 and LP92 models, AMSR-E, and NLDAS comparing with Little Washita Watershed Micronet.
The values in bold indicate 1 km disaggregated results.

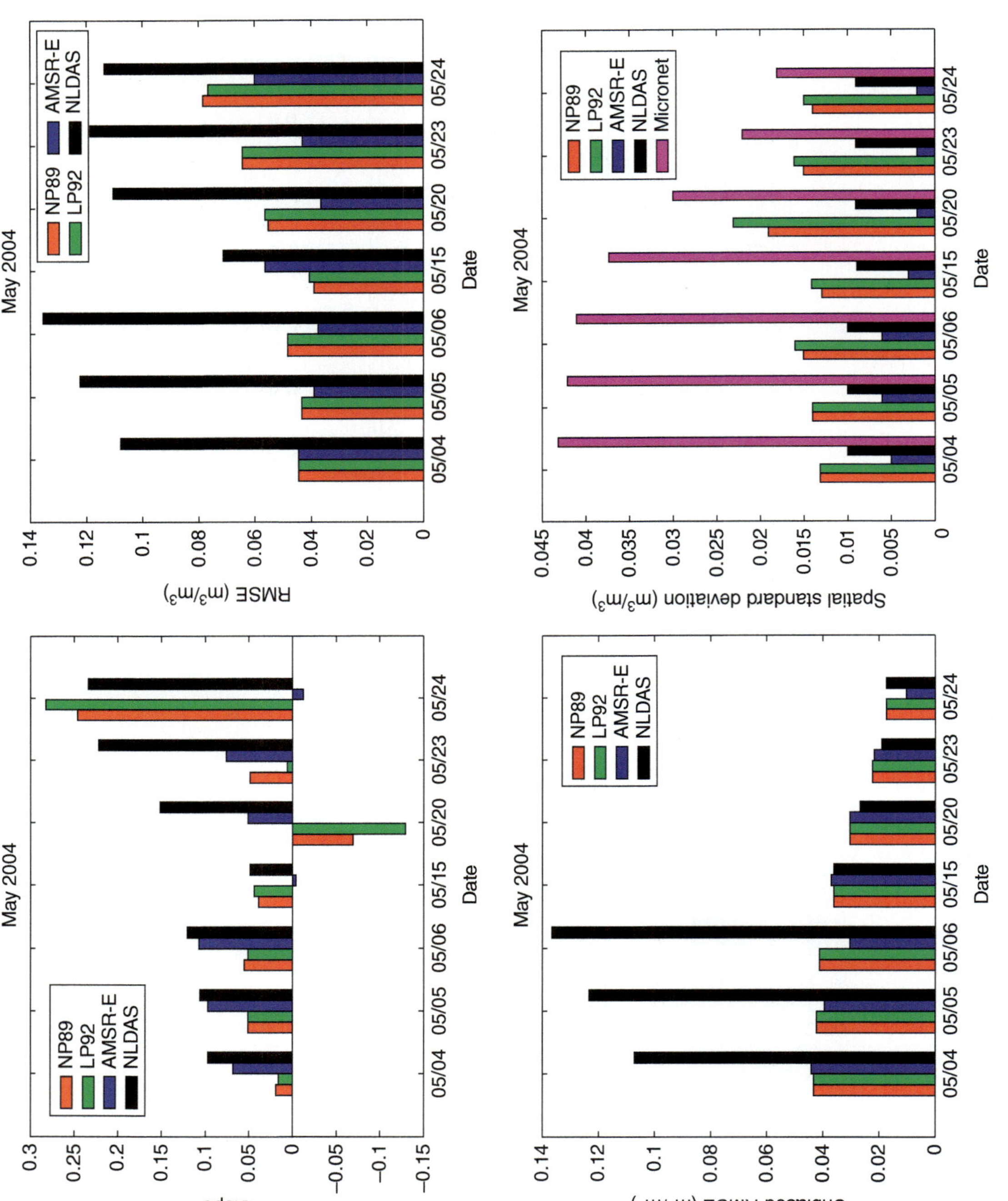

Figure 17.8a Bar plots of soil moisture validation variables: slope, RMSE, unbiased RMSE, and spatial standard deviation of downscaled soil derived from NP89 and LP92 models, AMSR-E, and NLDAS comparing with Little Washita Watershed Micronet in May 2004.

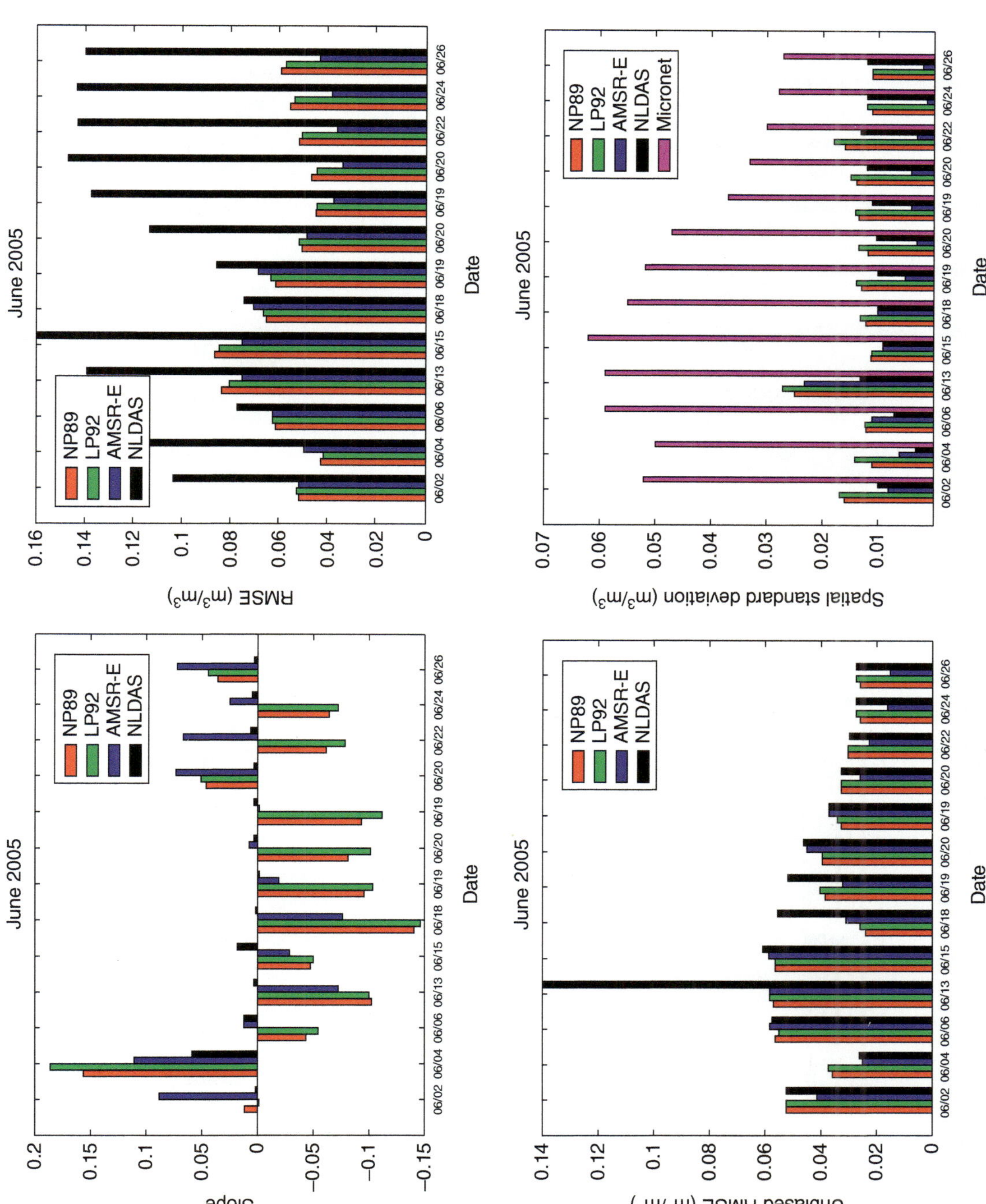

Figure 17.8b Bar plots of soil moisture validation variables: slope, RMSE, unbiased RMSE, and spatial standard deviation of downscaled soil derived from NP89 and LP92 models, AMSR-E, and NLDAS comparing with Little Washita Watershed Micronet in June 2005.

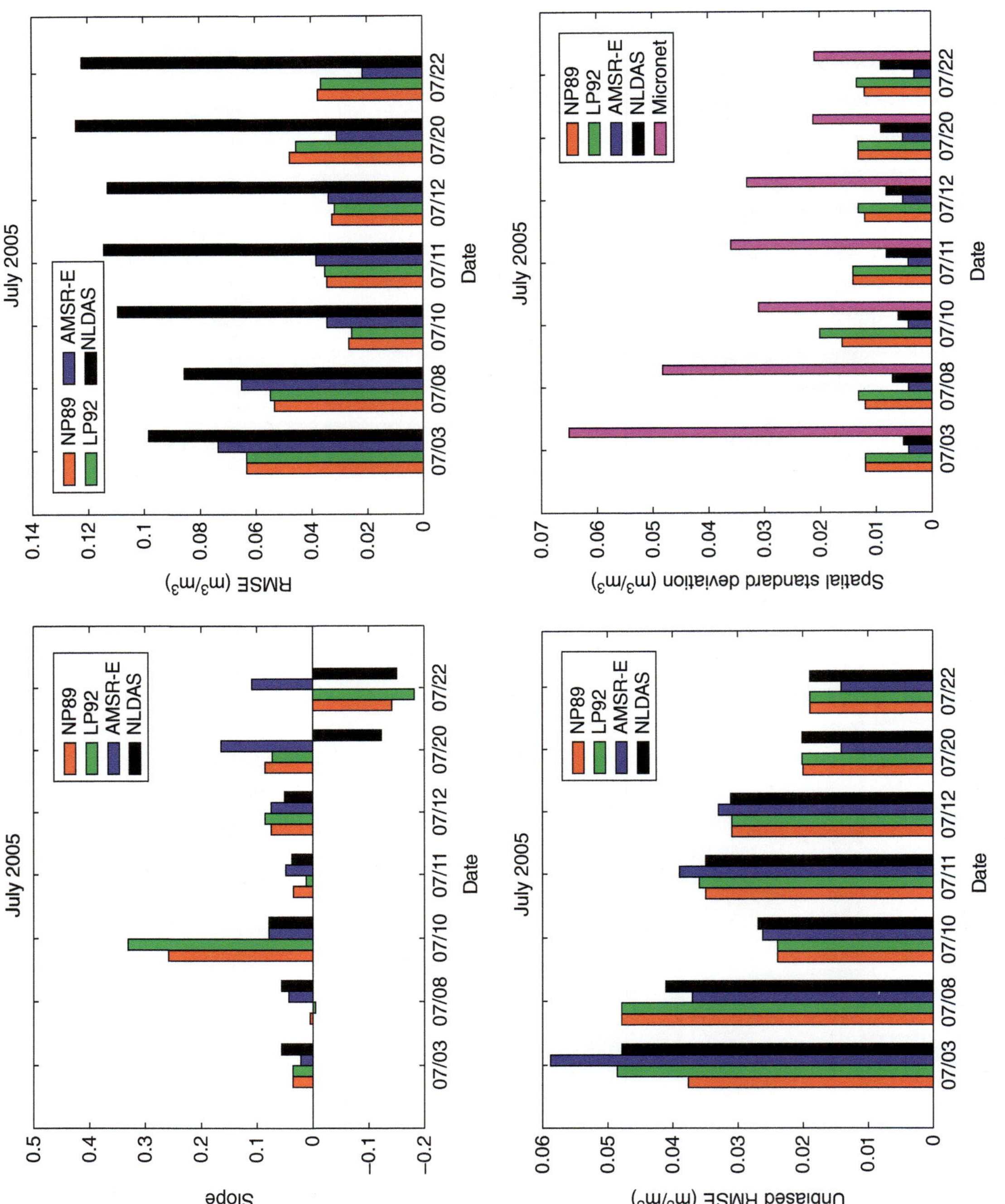

Figure 17.8c Bar plots of soil moisture validation variables: slope, RMSE, unbiased RMSE, and spatial standard deviation of downscaled soil derived from NP89 and LP92 models, AMSR-E, and NLDAS comparing with Little Washita Watershed Micronet in July 2005.

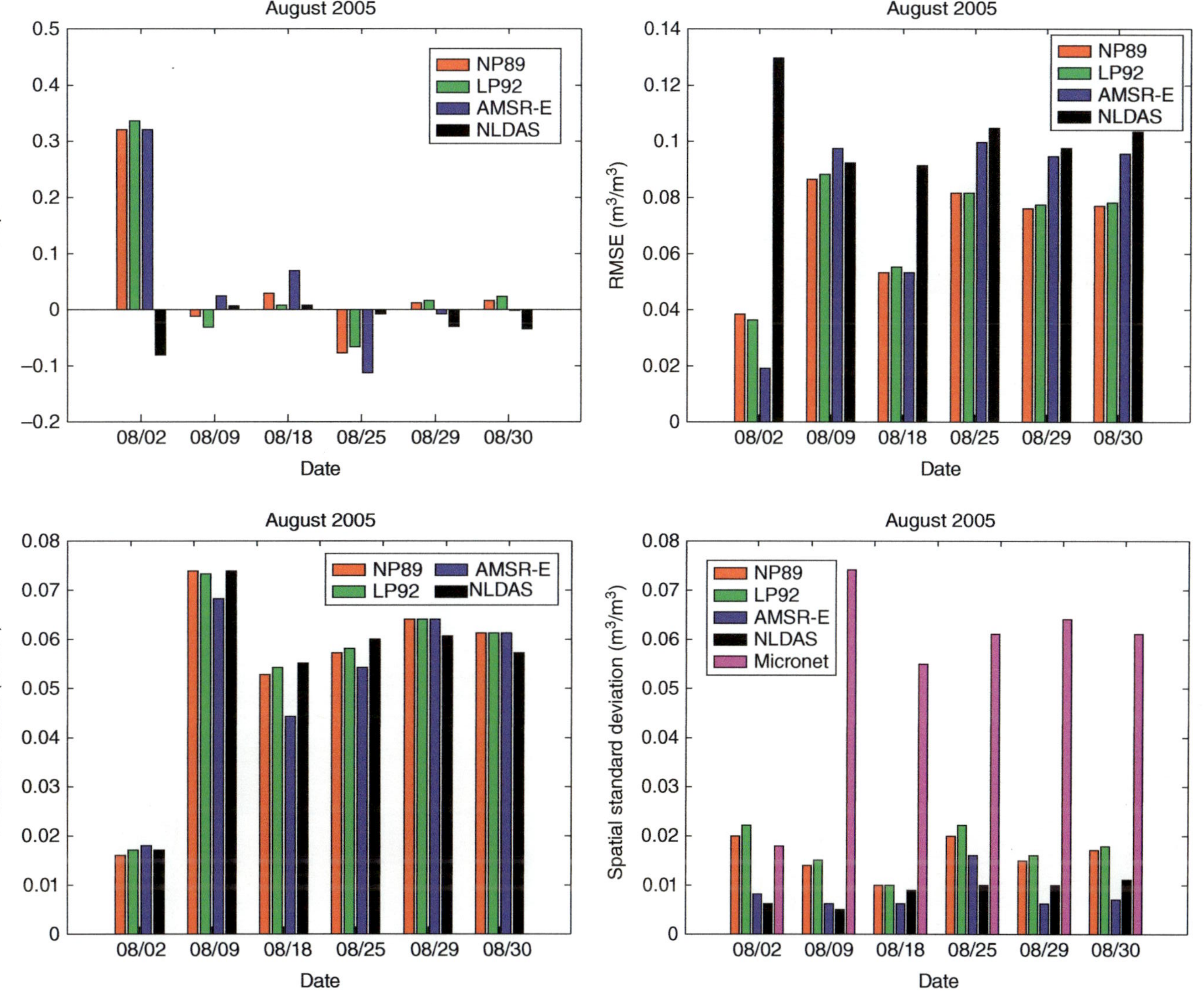

Figure 17.8d Bar plots of soil moisture validation variables: slope, RMSE, unbiased RMSE, and spatial standard deviation of downscaled soil derived from NP89 and LP92 models, AMSR-E, and NLDAS comparing with Little Washita Watershed Micronet in August 2005.

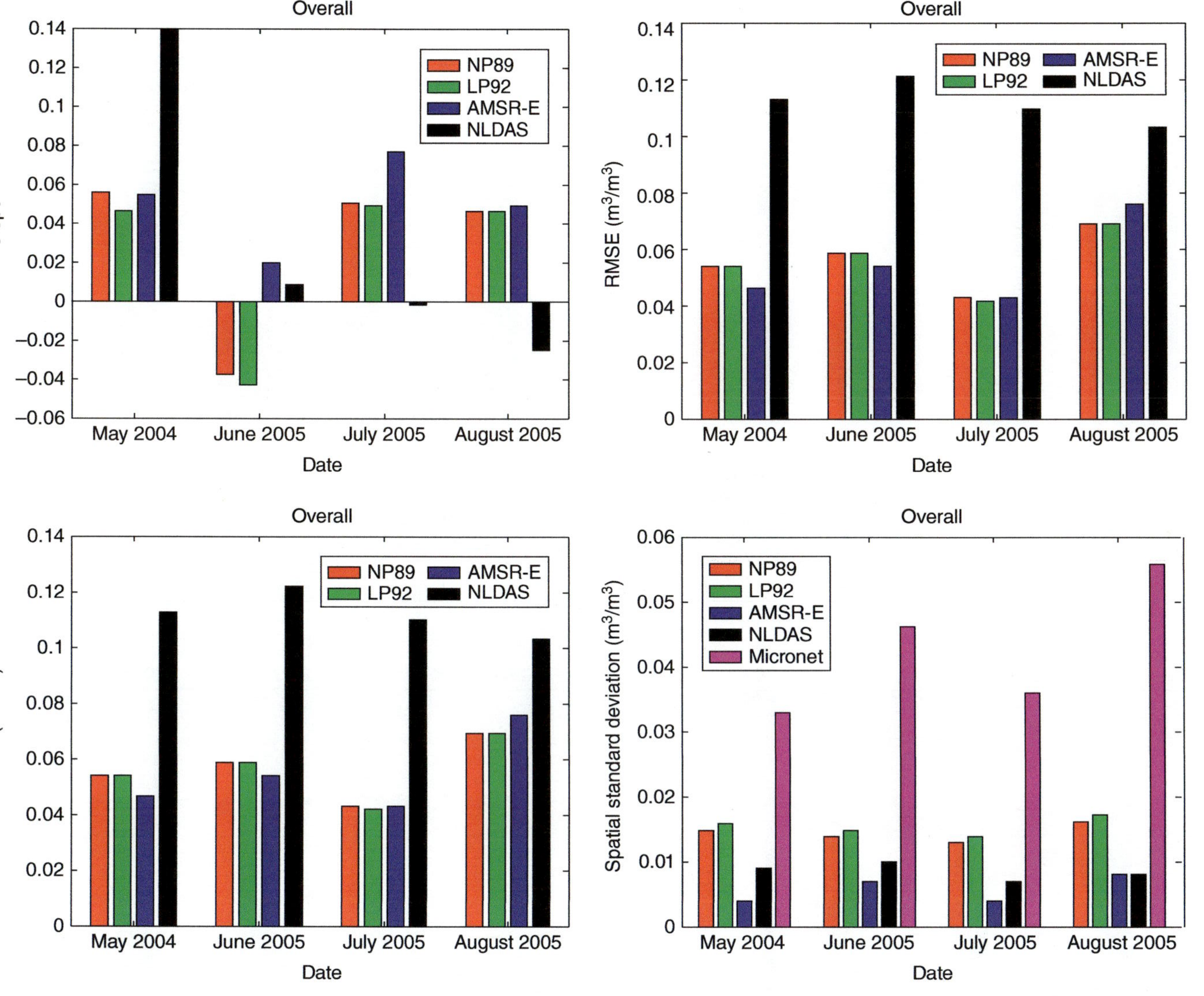

Figure 17.8e Bar plots of overall soil moisture validation variables: slope, RMSE, unbiased RMSE, and spatial standard deviation of downscaled soil derived from NP89 and LP92 models, AMSR-E, and NLDAS comparing with Little Washita Watershed Micronet.

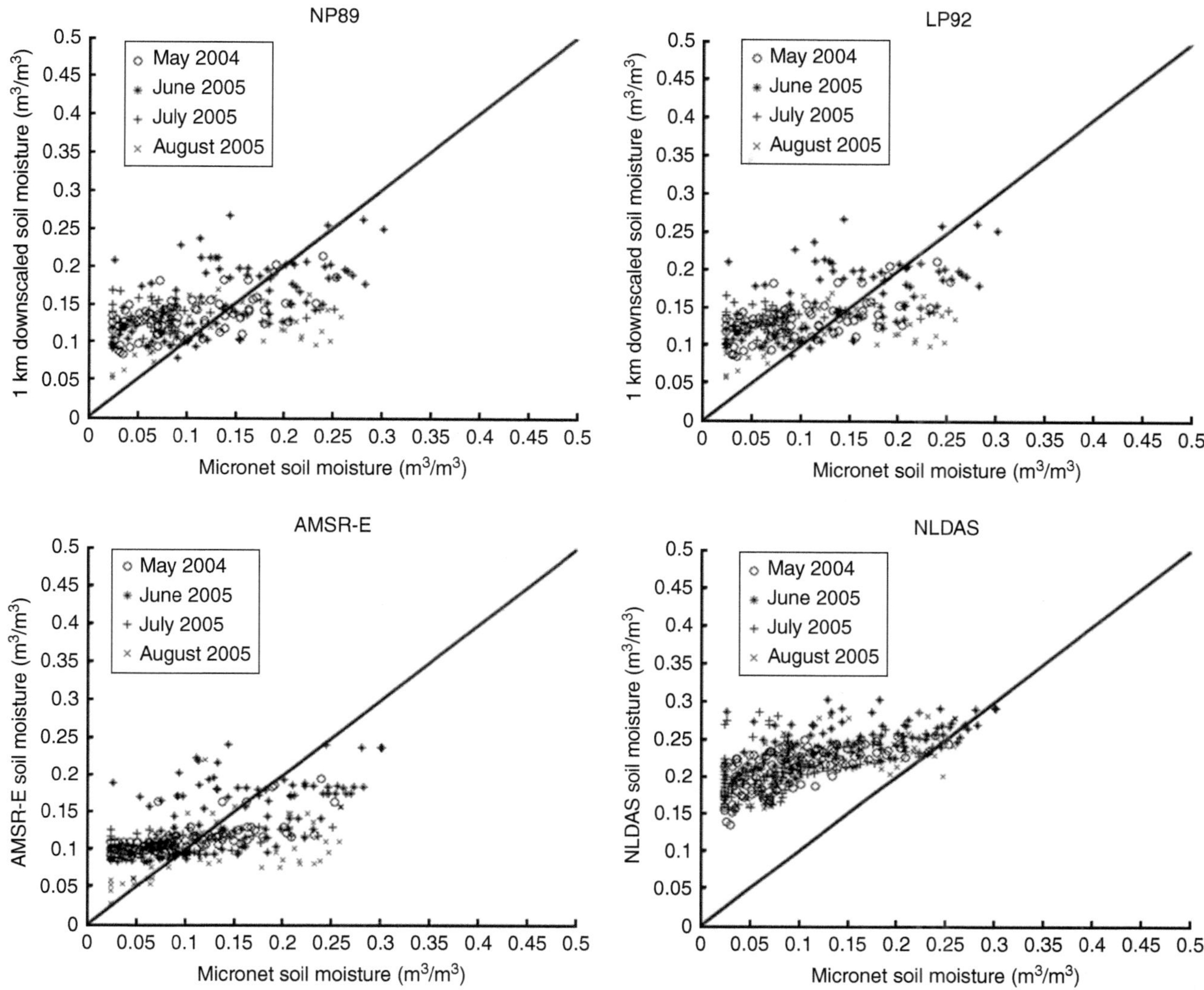

Figure 17.9 Overall scatterplots of 1 km disaggregated soil moisture derived from NP89 and LP92 models, AMSR-E, and NLDAS soil moisture comparing with Little Washita Micronet soil moisture observations in May 2004, June 2005, July 2005, and August 2005.

Micronet are greatly enhanced (as compared to the raw data in Figure 17.9). It should be pointed out that the NLDAS forcing data has been scaled to ground observations [*Luo et al.*, 2003], so the scatterplot of its correlation with Micronet soil moisture in Figures 17.9 and 17.10 tends to be more convergent than the disaggregated and AMSR-E soil moisture, but biased wet. This explains the reason the unbiased RMSE of NLDAS is significantly smaller and closer to NP89 and NP92. Figure 17.11 compares the spatial standard deviation between Micronet and the four soil moisture data sets. It can be noted that all the soil moistures are biased low compared to the Little Washita Micronet. However, the comparisons of NP89 and LP92 spatial standard deviation are definitely better than AMSR-E and NLDAS as they are closer to the corresponding Micronet spatial standard deviation.

Two factors significantly influence soil moisture validation: sensing depth and scale. (a) The mismatch of the sensing depth influences the soil temperature disaggregation. The 0–10 cm soil layer NLDAS soil temperature was downscaled and corrected using MODIS LST and NDVI, as well as AMSR-E soil moisture with sensing depth of a few of millimeters. Another sensing depth inconsistency comes from soil moisture validation. Micronet soil moisture observations are at 0–5 cm depth, while microwave sensor retrieved AMSR-E soil moistures corresponds to less than 1 cm depth and the disaggregated 1 km soil temperature is at 0–10 cm. (b) The mismatch of soil moisture spatial scales also needs to be discussed. The Little Washita Micronet records soil moisture observations at a point scale, while the disaggregated, remotely sensed and land surface model output soil moisture corresponds to the kilometer scale. The bias of soil moisture comparison as well as the spatial

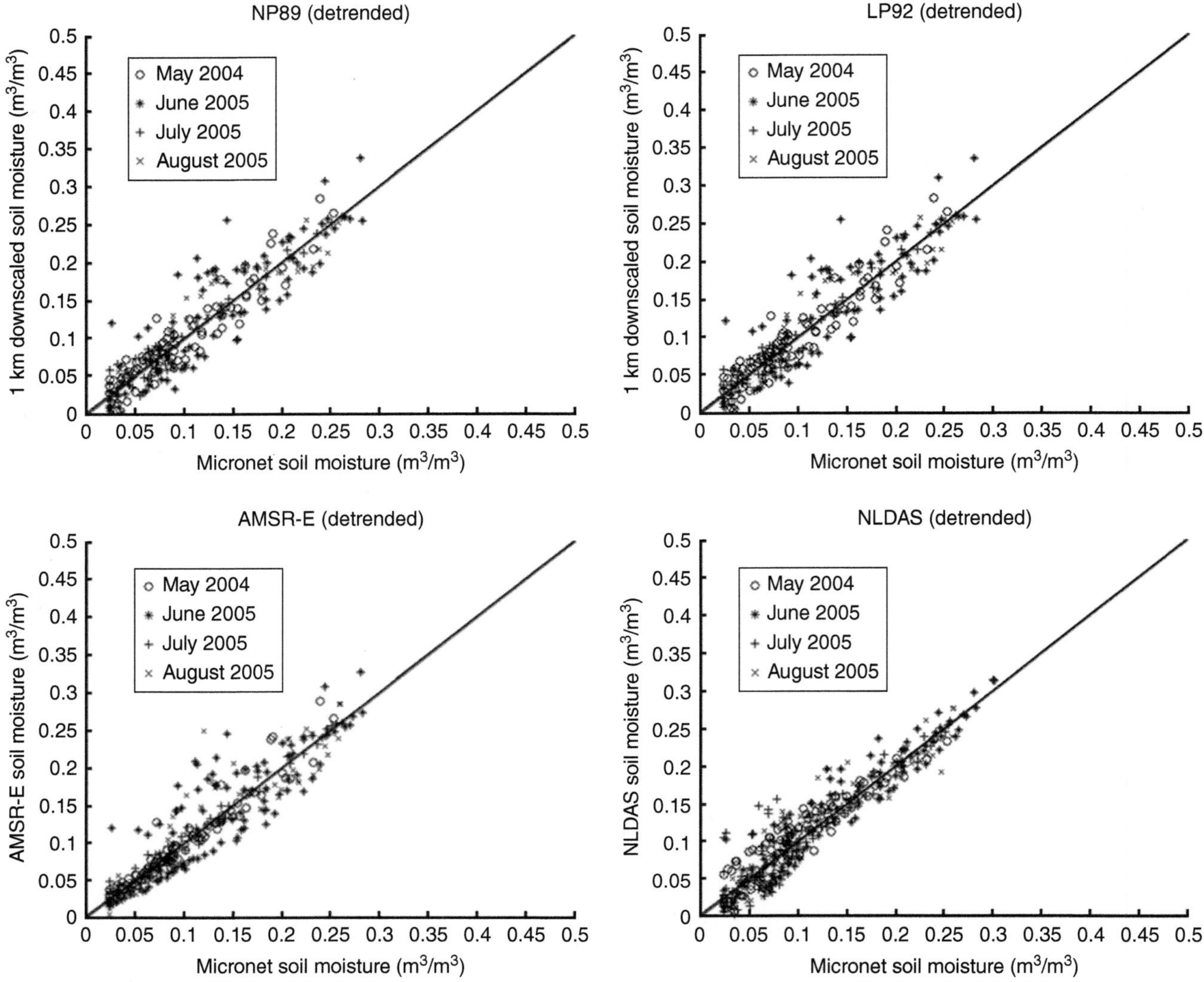

Figure 17.10 Overall scatterplots of detrended 1 km disaggregated soil moisture derived from NP89 and LP92 models, detrended AMSR-E, and NLDAS soil moisture comparing with Little Washita Micronet soil moisture observations in May 2004, June 2005, July 2005, and August 2005.

standard deviation comparison between Micronet observations and estimated soil moistures can be noted (see Figures 17.9 and 17.11).

The validation results prove that the soil evaporation efficiency disaggregated soil moisture is a compromise between accuracy and spatial variation. It retains the accuracy of satellite soil moisture but also significantly improves the variability from the other sources.

17.5. CONCLUSIONS

In this study the AMSR-E soil moisture was disaggregated through three steps. First, the 1/8° spatial resolution NLDAS soil temperature of top soil layer was disaggregated to 1 km by using a bivariate regression analysis among top layer soil temperature, land surface temperature, and fractional vegetation cover, and the total derivative of land surface temperature and fractional vegetation cover to soil temperature was used for correcting the disaggregated soil temperature. Second, the 1 km soil temperature was then used for estimating soil evaporation efficiency and retrieving 1 km soil moisture from two experimental soil evaporation efficiency models. Third, in order to eliminate the inconsistency among different sensors, including AMSR-E and MODIS as well as land surface models output, the disaggregated 1 km soil moisture was bias corrected using AMSR-E soil moisture. Table 17.6 compares this study to some soil moisture disaggregation studies. It can be concluded that the RMSE and unbiased RMSE obtained in this study are as good as most of the previous studies and better than some of them, such as *Piles et al.* [2011] and *Fang*

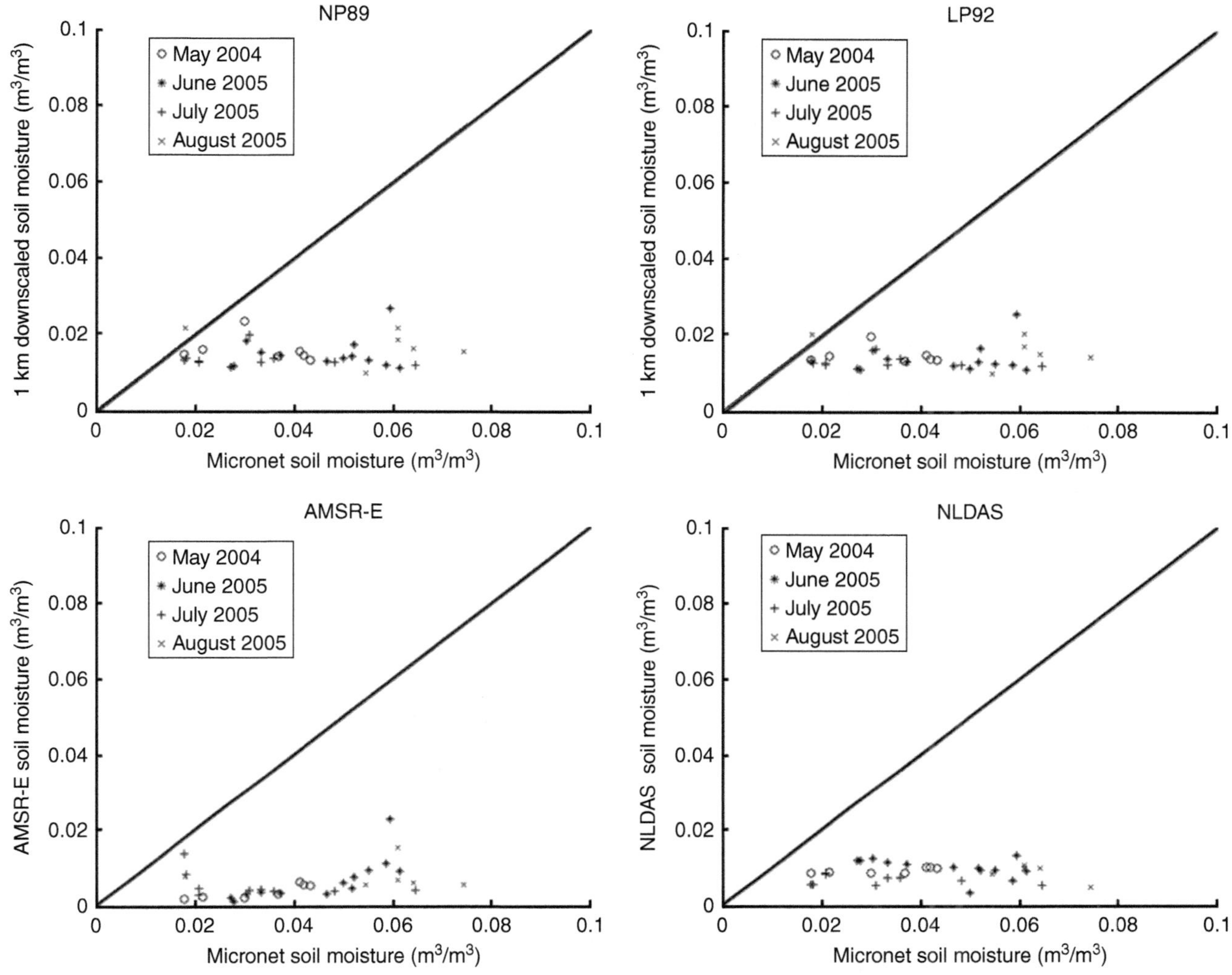

Figure 17.11 Overall scatterplots of spatial standard deviation of 1 km disaggregated soil moisture derived from NP89 and LP92 models, AMSR-E, and NLDAS soil moisture comparing with Little Washita Micronet soil moisture observations in May 2004, June 2005, July 2005, and August 2005.

et al., [2013]. An obvious improvement of this study is that it estimates instantaneous soil moisture rather than using averaged value of two instant times as daily average from *Fang et al.*, [2013]. Another advantage is that this study improved the soil moisture to 1 km scale as well as a daily temporal coverage, and this has not been achieved in most other studies.

The study results demonstrate that (1) strong correlations among top layer soil temperature, land surface temperature, and fractional vegetation cover and soil temperature variations were discovered and can be used to disaggregate soil temperature to a 1 km scale. (2) The disaggregated 1 km soil moisture maps definitely captures the spatial distribution pattern and variation in greater detail than AMSR-E and NLDAS. (3) NLDAS soil moisture was found apparently wet biased [*Mo et al.*, 2012], mostly

due to the mismatch of the soil layer depth. (4) Validation using the Little Washita River Watershed Micronet soil moisture shows that the NP89 and LP92 1 km soil moistures sustain accuracy as well as improve spatial variation. Although the slope of disaggregated soil moisture of 2 months are better than either AMSR-E or NLDAS, the spatial standard deviation of disaggregated soil moisture output from NP89 and LP92 is always better than AMSR-E and NLDAS, while the RMSE is better than NLDAS and is close to AMSR-E.

However this study has various limitations that need to be addressed. First, our study is based on a couple of assumptions; so mismatches and uncertainties exist during the soil temperature and soil moisture disaggregation. (a) The data set for this algorithm comes from the sources with different sensing depths. MODIS and

Table 17.6 Studies of disaggregating microwave soil moisture using fine-scale remote sensing imagery and model data

Author	Methodology	Time and Region	Result
Narayan and Lakshmi [2008]	Fused AMSR-E and TMI soil moisture with TRMM-PR backscatter data	Entire year of 2003, Little Washita Micronet	The RMSE ranges 0.01–0.36 comparing with AMSR-E validation of 0.102–0.239
Merlin et al. [2008]	Disaggregated LPRM soil moisture using MODIS data by soil evaporation efficiency model	NAFE 2006 (October–November) in Murrumbidgee, Australia	RMSE ranges 1.4%–1.8%
Merlin et al. [2010a]	Disaggregated LPRM soil moisture using MODIS data by combined soil evaporation efficiency model	NAFE 2006 (October–November) in Murrumbidgee, Australia	Average slope is 0.94 and the most accuracy with an error of 0.012
Das et al. [2011]	Merged L-band SMAP soil moisture and radar data using a linear relationship model	PALS SMEX02 in Iowa, USA, 2002	The RMSE ranges 0.015–0.02 cm³/cm³
Piles et al. [2011]	Disaggregated SMOS data using MODIS and multiple regression model	Jan – Feb 2010, Murrumbidgee Australia	R^2 ranges 0.14–0.21 and RMSE ranges 0.09–0.17 m³/m³
Kim and Hogue [2012]	Improved AMSR-E soil moisture using MODIS products	SMEX 04 in San Pedro River Basin, 2004	Spatial correlation ranges of 0.08–0.34
Fang et al. [2013]	Disaggregated AMSR-E soil moisture using MODIS and NLDAS data	Oklahoma Mesonet and Little Washita Micronet, 2004–2005	Overall spatial slope is 0.208, RMSE is 0.059, unbiased RMSE is 0.027 m³/m³ and spatial standard deviation is 0.023 m³/m³ for Micronet
Fang et al. [2013]	Disaggregated AMSR-E soil moisture using MODIS and NLDAS data by soil evaporation efficiency model	Little Washita Micronet, 2004–2005	RMSE ranges 0.042–0.069 m³/m³, while unbiased RMSE ranges 0.031–0.055 m³/m³ and spatial standard deviation ranges 0.013–0.017
Merlin [2012]	Disaggregated SMOS soil moisture based on physical and theoretical scale change (DisPATCh) method	AACES experiments in Southeastern Australia, 2010	R^2 ranges 0.7–0.85 during summer AACES
Merlin [2013]	Disaggregated SMOS soil moisture using the DisPATCh method as well as calibration procedure	Apr–Oct 2011 campaign in Catalunya, Spain	R^2 ranges 0.2–0.5

AMSR-E respond to soil depths of a couple of millimeters, while NLDAS data is at 0–10 cm and Micronet observation is at 0–5 cm. (b) Vegetation temperature at fine spatial scales is difficult to obtain, the bivariate regression analysis was used, and we also presumed straightforward linear relationship between top layer soil temperature (0–10 cm), land surface temperature, and fractional vegetation cover. (c) Soil evaporation efficiency was estimated by an empirical equation using disaggregated soil temperature. (d) Field capacity data used in this study is at 1/8° and the spatial variation between 1/8° and 1 km has been ignored. (e) The inconsistencies of imagery scales among MODIS, AMSR-E, and NLDAS would cause mismatch of pixels between different spatial resolutions. These issues will be addressed in ent spatial resolutions. These issues will be addressed in future work that will focus on solving inconsistencies and uncertainties on scale and sensing depth, as well as consider using other newer sources of remotely sensed soil moisture, such as SMOS and SMAP.

REFERENCES

Basara, J. B., B. G. Illston, C. A. Fiebrich, R. A. McPherson, J. P. Bostic, P. Browder, and K. Kesler (2009), An Overview of the Oklahoma City Micronet. In Eighth Symposium on the Urban Environment.

Brunsell, N. A., and R. R. Gillies (2001), The effect of emissivity on evaporation, IAHS Pub. 276–280.

Camillo, P. J., and R. J. Gurney (1986), A resistance parameter for bare-soil evaporation models, *Soil Sci.*, *141*(2), 95–105.

Carlson, T. (2007), An overview of the "triangle method" for estimating surface evapotranspiration and soil moisture from satellite imagery, *Sensors*, *7*, 1612–1629.

Chanzy, A., and L. Bruckler (1993), Significance of soil surface moisture with respect to daily bare soil evaporation, *Water Resourc. Res.*, *29*(4), 1113–1125.

Cosh M. H., T.J. Jackson, R. Bindlish, and J.H. Prueger (2004), Watershed scale temporal and spatial stability of soil moisture and its role in validating satellite estimates, *Remote Sens. Environ.*, *92*, 427–435.

Daamen, C. C., and L. P. Simmonds (1996), Measurement of evaporation from bare soil and its estimation using surface resistance, *Water Resourc. Res.*, *32*(5), 1393–1402.

Das, N. N., D. Entekhabi, and E. G. Njoku (2011), An algorithm for merging SMAP radiometer and radar data for high-resolution soil-moisture retrieval. Geoscience and Remote Sensing, *IEEE Transactions on*, *49*(5), 1504–1512.

Entekhabi, D., et al. (2010), The soil moisture active passive (SMAP) mission, *Proc. IEEE*, *98*(5), 704–716.

Fang, B., V. Lakshmi, R. Bindlish, T. J. Jackson, M. Cosh, and J. Basara (2013), Passive microwave soil moisture downscaling using vegetation index and skin surface temperature, *Vadose Zone J.*, *12*(3).

Gutman, G., and A. Ignatov (1998), The derivation of the green vegetation fraction from NOAA/AVHRR data for use in numerical weather prediction models, *Int. J Remote Sens.*, *19*(8), 1533–1543.

Imaoka, K., M. Kachi, H. Fujii, H. Murakami, M. Hori, A. Ono, and H. Shimoda (2010), Global Change Observation Mission (GCOM) for monitoring carbon, water cycles, and climate change, *Proc. IEEE*, *98*(5), 717–734.

Jackson, T. J. (1993), III. Measuring surface soil moisture using passive microwave remote sensing. *Hydrol. Process. 7*(2), 139–152.

Jackson, T. J., and T. J. Schmugge (1989), Passive microwave remote sensing system for soil moisture: Some supporting research, *IEEE Trans. Geosci. Remote Sens.*, *27*(2), 225–235.

Jackson, T. J., D. M. Le Vine, A. Y. Hsu, A. Oldak, P. J. Starks, C. T. Swift, J. D. Isham, and M. Haken (1999), Soil moisture mapping at regional scales using microwave radiometry: The Southern Great Plains Hydrology Experiment. *IEEE Trans. Geosci. Remote Sens.*, *37*(5), 2136–2151.

Jackson, T. J., A. J. Gasiewski, A. Oldak, M. Klein, E. G. Njoku, A. Yevgrafov, S. Christiani, and R. Bindlish (2002), Soil moisture retrieval using the C-band polarimetric scanning radiometer during the Southern Great Plains 1999 Experiment, *IEEE Trans. Geosci. Remote Sens.*, *40*(10), 2151–2161.

Jackson, T. J., M. H. Cosh, R. Bindlish, P. J. Starks, D. D. Bosch, M. Seyfried, D. C. Goodrich, M. S. Moran, and J. Du (2010), Validation of advanced microwave scanning radiometer soil moisture products, *IEEE Transs. Geosci. Remote Sens.*, *48*(12), 4256–4272.

Kerr, Y. H., P. Waldteufel, P. Richaume, J. P. Wigneron, P. Ferrazzoli, A. Mahmoodi, and S. Delwart (2012), The SMOS soil moisture retrieval algorithm, *Geoscience and Remote Sensing, IEEE Transactions on, 50*(5), 1384–1403.

Kim, J., and T. S. Hogue (2012), Improving spatial soil moisture representation through integration of AMSR-E and MODIS products, *IEEE Trans. Geosci. Remote Sens.*, *50*(2), 446–460.

Komatsu, T. S. (2003), Toward a robust phenomenological expression of evaporation efficiency for unsaturated soil surfaces, *J. Appl. Meteorol.*, *42*(9), 1330–1334.

Kondo, J., N. Saigusa, and T. Sato (1992), A model and experimental study of evaporation from bare-soil surfaces, *J. Appl. Meteorol.*, *31*, 304–312.

Kustas, W. P., and C. S. Daughtry (1990), Estimation of the soil heat flux/net radiation ratio from spectral data, *Agric Forest Meteorol.*, *49*(3), 205–223.

Lakshmi, V., et al. (1997), Evaluation of special sensor microwave/Imager satellite data for regional soil moisture estimation over the Red River Basin, *J. Appl. Meteorol.*, *36*(10), 1309–1328.

Lakshmi, V., S. Hong, E. E. Small, and F. Chen (2011), The influence of the land surface on hydrometeorology and ecology: New advances from modeling and satellite remote sensing, *Hydrol. Res.*, *42*(2–3) 95–112.

Lee, T. J., and R. A. Pielke (1992), Estimating the soil surface specific humidity. *J. Appl. Meteorol.*, *31*, 480–484.

Lohmann, D., et al. (2004), Streamflow and water balance intercomparisons of four land surface models in the North American Land Data Assimilation System project, *J. Geophys. Res. Atmos.*, *109*(D7).

Luo, L., et al. (2003), Validation of the North American Land Data Assimilation System (NLDAS) retrospective forcing over the southern Great Plains, *J. Geophys. Res.*, *108*(D22), 8843.

Mahfouf, J. F., and J. Noilhan (1991), Comparative study of various formulations of evaporations from bare soil using in situ data, *J. Appl. Meteorol.*, *30*(9), 1354–1365.

McPherson, R. A., C. A. Fiebrich, K. C. Crawford, J. R. Kilby, D. L. Grimsley, J. E. Martinez, and A. J. Sutherland (2007), Statewide monitoring of the mesoscale environment: A technical update on the Oklahoma Mesonet, *J. Atmos. Ocean. Technol.*, *24*(3), 301–321.

Merlin, O., J. P. Walker, A. Chehbouni, and Y. Kerr (2008), Towards deterministic downscaling of SMOS soil moisture using MODIS derived soil evaporative efficiency, *Remote Sens. Environ.*, *112*(10), 3935–3946.

Merlin, O., A. Al Bitar, J. P. Walker, and Y. Kerr (2010a), An improved algorithm for disaggregating microwave-derived soil moisture based on red, near-infrared and thermal-infrared data, *Remote Sens. Environ.*, *114*(10), 2305–2316.

Merlin, O., B. Duchemin, O. Hagolle, F. Jacob, B. Coudert, G. Chehbouni, and Y. Kerr (2010b), Disaggregation of MODIS surface temperature over an agricultural area using a time series of Formosat-2 images, *Remote Sens. Environ.*, *114*(11), 2500–2512.

Merlin, O., A. Al Bitar, V. Rivalland, P. Béziat, E. Ceschia, and G. Dedieu (2011), An analytical model of evaporation efficiency for unsaturated soil surfaces with an arbitrary thickness, *J. Appl. Meteorol. Climatol.*, *50*(2), 457–471.

Merlin, O., C. Rudiger, A. Al Bitar, P. Richaume, J. P. Walker, and Y. H. Kerr (2012), Disaggregation of SMOS soil moisture in Southeastern Australia, *IEEE Trans. Geosci. Remote Sens. 50*(5), 1556–1571.

Merlin, O., M. J. Escorihuela, M. A. Mayoral, O. Hagolle, A. Al Bitar, and Y. Kerr (2013), Self-calibrated evaporation-based disaggregation of SMOS soil moisture: An evaluation study at 3 km and 100 m resolution in Catalunya, Spain, *Remote Sens. Environ.*, *130*, 25–38.

Mladenova, I., et al. (2011), Validation of AMSR-E soil moisture using L band radiometer data from the National Airborne Field Experiment (NAFE 2006), *Remote Sens. Environ.*, *115*(8), 2096–2103.

Mo, K. C., L. C. Chen, S Shukla, T. J. Bohn, and D. P. Lettenmaier (2012), Uncertainties in North American land data assimilation systems over the contiguous United States, *J. Hydrometeorol.*, *13*(3), 996–1009.

Monteith, J. L. (1981), Evaporation and surface temperature, *Q. J. R. Meteorol. Soc.*, *107*(451), 1–27.

Narayan, U., V. Lakshmi and E.G. Njoku, (2004), Retrieval of soil moisture from passive and active L/S band sensor (PALS) observations during the Soil Moisture Experiment in 2002 (SMEX02). Remote Sensing of Environment, *92*(4) 483–496.

Narayan, U., and V. Lakshmi (2008), High resolution change detection using TMI-PR and AMSR-E soil moisture data, *Water Resourc. Res.*, *44*(6).

Narayan, U., V. Lakshmi, and T. Jackson (2006), A simple algorithm for spatial disaggregation of radiometer derived soil moisture using higher resolution radar observations, *IEEE Trans. Geosci. Remote Sens.*, *44*(6), 1545–1554.

National Aeronautics and Space Administration (NASA) (2011), National Space Science Data Center, NSSDC ID: 1978-041A, available at http://nssdc.gsfc.nasa.gov/nmc/spacecraftDisplay.do?id=1978-041A.

Nishida, K., R. R. Nemani, S. W. Running, and J. M. Glassy (2003), An operational remote sensing algorithm of land surface evaporation, *J. Geophys. Res. Atmo.* (1984–2012), *108*(D9).

Njoku, E. G., and S. K. Chan (2006), Vegetation and surface roughness effects on AMSR-E land observations, *Remote Sens. Environ. 100*(2), 190–199.

Njoku, E. G., and D. Entekhabi (1996), Passive microwave remote sensing of soil moisture. *J. Hydrol. 184*(1), 101–129.

Njoku, E. G., and L. Li (1999), Retrieval of land surface parameters using passive microwave measurements at 6–18 GHz, *IEEE Trans. Geosci. Remote Sens.*, *37*(1), 79–93.

Njoku, E. G., T. J. Jackson, V. Lakshmi, T. K. Chan, and S. V. Nghiem (2003), Soil moisture retrieval from AMSR-E, *IEEE Trans. Geosci. Remote Sens. 41*(2), 215–229.

Noilhan, J., and S. Planton (1989), A simple parameterization of land surface processes for meteorological models, *Monthly Weather Rev.*, *117*(3), 536–549.

North American Land Data Assimilation Systems (NLDAS) (2011), NLDAS-2 Data, available at http://ldas.gsfc.nasa.gov/.

Passerat de Silans, A., L. Bruckler, J. L. Thony, and M. Vauclin (1986), Numerical modeling of coupled heat and water flows during drying in a stratified bare soil—Comparison with field observations, *J. Hydrol.*, *105*(1), 109–138.

Piles, M., A. Camps, M. Vall-Llossera, I. Corbella, R. Panciera, C. Rudiger, Y. H. Kerr, and J. Walker (2011), Downscaling SMOS-derived soil moisture using MODIS visible/infrared data, *IEEE Trans. Geosci. Remote Sens.*, *49*(9), 3156–3166.

Robock, A., et al. (2003), Evaluation of the North American Land Data Assimilation System over the southern Great Plains during the warm season, *J. Geophys. Res. Atmosp.* (1984–2012), *108*(D22).

Schaake, J. C., et al. (2004), An intercomparison of soil moisture fields in the North American Land Data Assimilation System (NLDAS), *J. Geophys. Res. Atmos.* (1984–2012), *109*(D1).

Schmugge, T., and T. J. Jackson (1994), Mapping surface soil moisture with microwave radiometers, *Meteorol. Atmos. Phys.*, *54*(1–4), 213–223.

Schmugge, T., P. Gloersen, T. Wilheit, and F. Geiger (1974), Remote sensing of soil moisture with microwave radiometers, *J. Geophys. Res.*, *79*(2), 317–323.

Schmugge, T. J., W. P. Kustas, J. C. Ritchie, T. J. Jackson, and A. Rango (2002), Remote sensing in hydrology, *Adv. Water Resourc.*, *25*(8), 1367–1385.

Sellers, P. J., M. D. Heiser, and F. G. Hall (1992), Relations between surface conductance and spectral vegetation indices at intermediate ($100\,m^2$ to $15\,km^2$) length scales, *J. Geophys. Res. Atmos.* (1984–2012), *97*(D17), 19033–19059.

Yamanaka, T., A. Takeda, and J. Shimada (1998), Evaporation beneath the soil surface: Some observational evidence and numerical experiments, *Hydrol. process.*, *12*(13–14), 2193–2203.

Zhang, R., X. Sun, Z. Zhu, H. Su, and X. Tang (2003), A remote sensing model for monitoring soil evaporation based on differential thermal inertia and its validation, *Sci. China Ser. D Earth Sci.*, *46*(4), 342–355.

18

Assessing Near-Surface Soil Moisture Assimilation Impacts on Modeled Root-Zone Moisture for an Australian Agricultural Landscape

R. C. Pipunic,[1] D. Ryu,[1] and J. P. Walker[2]

18.1. INTRODUCTION

Soil moisture content is an important component of the hydrologic cycle, particularly over vegetation root-zone depths where its variation is linked to the relative fractions of evaporative and sensible heat flux (L_E and H) feedbacks to the lower atmosphere, surface runoff, and groundwater recharge [*Brutsaert*, 2005]. Quantifying these processes across catchments using land surface models (LSMs), therefore, depends on soil moisture state prediction. Improved characterization of root-zone soil moisture quantities has the potential to contribute toward better predictions for a range of hydrological processes — information that will ultimately benefit agricultural and land-use management decisions (e.g., better irrigation scheduling), numerical weather prediction (NWP; e.g., through improved L_E and H feedbacks), and emergency management (e.g., improved flood prediction). While an imperfect model structure means that improving certain model variables will not necessarily lead to improvements in predictions of all other model variables [*Drusch*, 2007], improved root-zone soil moisture can translate to improvement in predictions of other water-balance-related quantities [*Pipunic et al.*, 2013]. Therefore, the ability to routinely improve root-zone moisture prediction is an important aim, and the impact on other hydrologic variables of interest may contribute to a better understanding of model structural inaccuracies.

Inherent LSM uncertainty, resulting from errors in input data (meteorological forcing and parameter information on soil and vegetation properties) and model structural inaccuracies, is the impetus for data assimilation techniques such as the ensemble Kalman filter [EnKF: *Evensen*, 1994], where observed information is used to sequentially update/correct LSM states through time, based on both modeled and observed error statistics. For routine constraint of root-zone soil moisture prediction across catchments, assimilating relevant remotely sensed data is ideal given their broad spatial coverage at regular repeat intervals.

Brightness temperature observations from passive microwave remote sensors have proven particularly suitable for deriving spatial estimates of soil moisture [*Kerr et al.*, 2010; *Njoku et al.*, 2003]. However, these estimates have major limitations, including coarse spatial resolution (>10 km) and shallow sensing depth, which varies depending on a sensor's spectral frequency and the near-surface moisture conditions but is typically within the top few centimeters of soil at most. Therefore, the impact from assimilating such data products must be able to adequately translate to the model's deeper layers in order to improve root-zone estimates. A number of studies using synthetic data or in situ field data have shown near-surface moisture assimilation can improve root-zone predictions [e.g., *Pipunic et al.*, 2013; *Kumar et al.*, 2009; *Pipunic et al.*, 2008; *Walker et al.*, 2001; *Entekhabi et al.*, 1994], with some modest improvements to deeper soil moisture from assimilating remotely sensed near-surface moisture shown by *Reichle et al.* [2007]. More recent work demonstrating the value of assimilating remotely sensed near-surface moisture for

[1]Department of Infrastructure Engineering, University of Melbourne, Parkville, Victoria, Australia

[2]Department of Civil Engineering, Monash University, Clayton, Victoria, Australia

---

Remote Sensing of the Terrestrial Water Cycle, Geophysical Monograph 206. First Edition. Edited by Venkat Lakshmi.
© 2015 American Geophysical Union. Published 2015 by John Wiley & Sons, Inc.

deeper LSM moisture state predictions includes *Draper et al.* [2012] and *Liu et al.* [2011].

Remotely sensed data assimilation in a spatially distributed modeling framework generally involves greater uncertainty than is the case with more controlled scenarios using synthetic or one-dimensional field-based data. This is partly due to varying spatial scales of different data sets used for model input in addition to their measurement error, while the uncertainty in assimilated remotely sensed data may also be high [*Reichle et al.*, 2007]. Furthermore, the coverage of well-calibrated independent in situ data is relatively sparse for many landscapes, posing a problem for validation on a global scale. Consequently, the availability of such data should be taken advantage of to support research that can contribute to a clearer understanding of the benefits and limitations of remotely sensed near-surface soil moisture assimilation in terms of optimizing deeper root-zone moisture predictions.

This study utilizes in situ soil moisture profile data from the OzNet monitoring network in southeastern Australia [*Smith et al.*, 2012] to validate LSM assimilation results. The impact from assimilating a near-surface soil moisture data product — derived from the Advanced Microwave Scanning Radiometer for the Earth Observing System (AMSR-E) observations of the top ~1–2 cm of soil — on deeper soil moisture profile predictions from the Community Atmosphere Biosphere Land Exchange model [CABLE; *Kowalczyk et al.*, 2006] is examined. CABLE is the land surface component coupled with the Met Office unified model as part of ACCESS 1.3 (Australian Community Climate and Earth-System Simulator) used for climate simulations [*Kowalczyk et al.*, 2013] and is also planned for use in Australia's NWP in the near future [*Law et al.*, 2012]. It has yet to be rigorously tested with the assimilation of remotely sensed data for improving moisture prediction. A $100 \times 100 \, \text{km}^2$ agricultural landscape in the Yanco region of New South Wales, Australia, is the focus of the experiment, where predicted moisture states were validated against 0–5 cm, 0–30 cm, and 0–60 cm in situ moisture data from 12 OzNet sites. The aim was to show if assimilating the near-surface moisture product demonstrated potential as a dependable way to improve CABLE root-zone moisture prediction for this environment. Knowledge gained here from assimilating AMSR-E data is also assumed relevant to use of data from the successor sensor AMSR-2 [*Imaoka et al.*, 2010].

The EnKF algorithm was used to update the prognostic soil moisture and temperature states for each of CABLE's six soil layers. Two specific procedures were followed here in implementing the EnKF, which can potentially remedy some common issues, but which do not appear to have been thoroughly tested in real-world applications. First, we perturbed model parameter values for key soil hydraulic properties determined from field-sampled soil [*McKenzie et al.*, 2000], where perturbation was guided by the associated error information supplied in the form of 5th and 95th percentile values. Perturbing time-invariant parameters ensures the LSM state ensemble spread is maintained and avoids potential problems with ensemble collapse that will render the filter ineffective. By using parameters with error estimates determined from analysis of real field samples, this approach for generating ensembles and maintaining their spread has a sound basis when compared to directly perturbing state predictions at each time step or applying covariance inflation [e.g., *Anderson and Anderson*, 1999] prior to state updating, using values estimated through trial and error. Second, the bias correction scheme of *Ryu et al.* [2009] was implemented to remove state ensemble biases caused by nonlinear model physics. This is the first implementation of this technique in a real remotely sensed data assimilation application.

18.2. DATA SETS AND EXPERIMENTAL SETUP

The $100 \times 100 \, \text{km}^2$ study area for this assimilation experiment incorporates 12 OzNet soil moisture monitoring sites in the Yanco region within the Murrumbidgee catchment [*Smith et al.*, 2012]. This mostly agricultural region is dominated by crops/pasture and located in the southeast of Australia within the Murray-Darling Basin. CABLE simulations were run at 5 km spatial resolution over the area while the AMSR-E soil moisture product used for assimilation [*Owe et al.*, 2008] was provided at 25 km resolution (Figure 18.1).

18.2.1. Model Specifications

The CABLE model version 2.0 was used for this research. This model calculates water and energy exchanges at the land surface for both soil and vegetation surfaces, with detailed descriptions of the model physics provided by *Kowalczyk et al.* [2006]. The soil scheme consists of a six layer soil profile over 4.60 m with layer thicknesses of: 2.2, 5.8, 15.4, 40.9, 108.5, and 287.2 cm, respectively, from top to bottom. Vertical water movement between layers is based on the Richards equation with the relationships of *Clapp and Hornberger* [1978] for hydraulic conductivity (K) and air entry potential (ψ_{aep}). CABLE can only be assigned with the one set of soil parameter values which apply to all six layers; therefore, depth varying soil hydraulic properties are not represented.

Calculations of L_E are done separately for soil and vegetation canopy surfaces, and thus the leaf area index

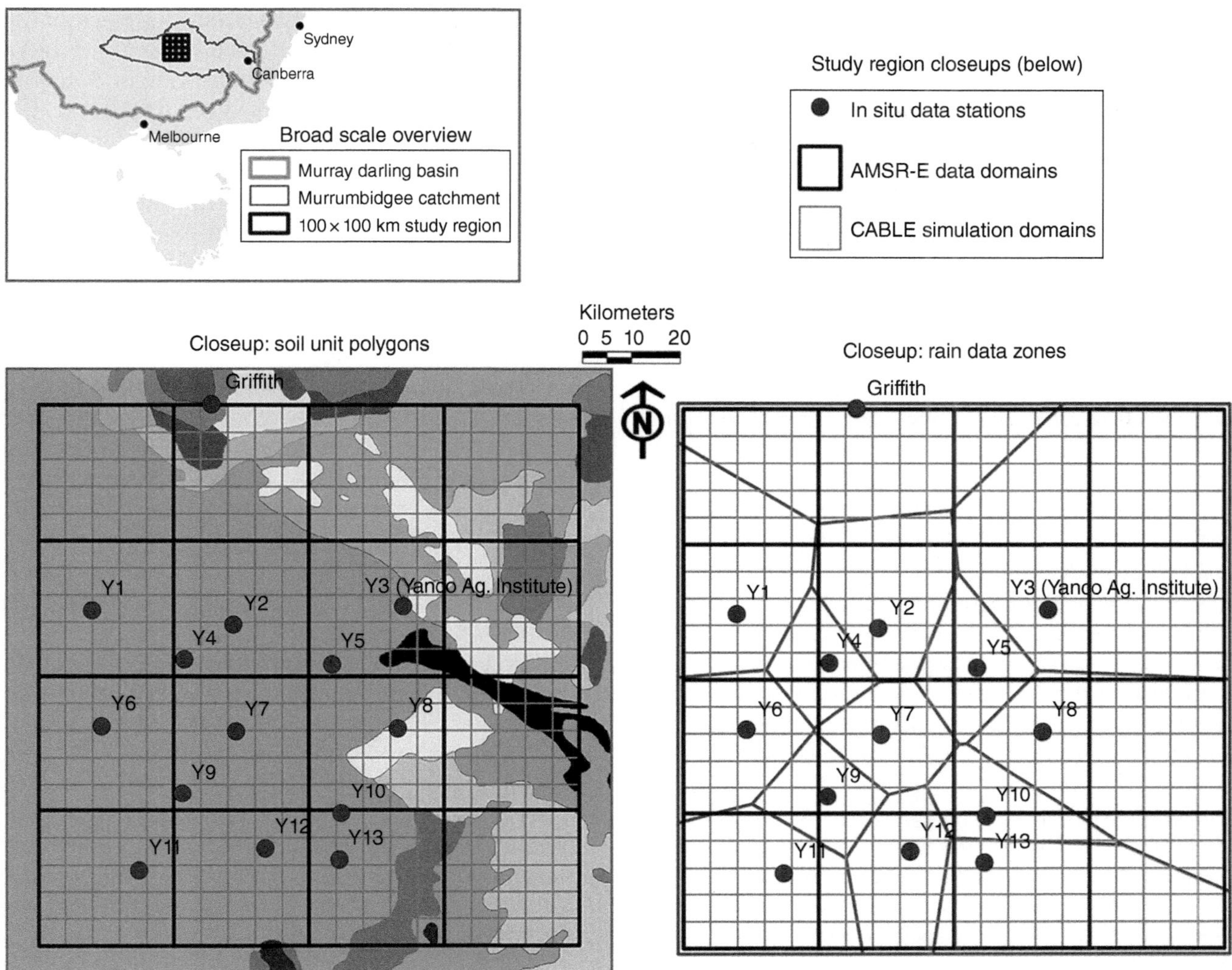

Figure 18.1 An overview of the study region location in the top left and a closeup of it in the two bottom panels. The closeups show: 25 km AMSR-E soil moisture product pixels (black squares); 5 km CABLE simulation pixels (gray squares); and locations of OzNet in situ soil moisture and rainfall stations (Y1–Y13) with Griffith meteorological station (dots). The closeup on the left shows polygons representing the mapped soil units used for prescribing model soil parameters to simulation pixels, and the closeup on the right shows the zones corresponding to in situ stations (dark gray lines), which were used to assign rainfall to simulation pixels. All other meteorological forcing from Y3 and Griffith stations were assigned to simulation pixels based on nearest proximity to either station.

(LAI) is an important model parameter for defining the fraction of bare soil to vegetation surface for energy balance calculations. A Penman-Monteith-based calculation is used for the canopy L_E, which incorporates a term linked to stomatal conductance and hence a water availability term, which consists of the available moisture content in the soil layers weighted by the relative fraction of roots in each. Vegetation root fractions for each soil layer are user-specified parameters, which are constant through the simulation time. The soil component of L_E is calculated based on the potential evaporation weighted by moisture availability in the top soil layer.

18.2.2. Model Inputs

Meteorologic forcing data determines the CABLE time step interval and simulation period, and the essential variables include short- and long-wave incoming radiation, air temperature, rainfall, wind speed, and specific humidity. Referring to Figure 18.1, forcing data is available for the Griffith and Yanco Agricultural Institute (station Y3) sites, while rainfall data is also available for each of the soil moisture stations (Y1–Y13). All of these data were used in this study and are provided on a 30 min time series, which is the integration time step for the simulations performed.

Some of the key soil parameters relevant to CABLE include wilting point (θ_{wilt}), field capacity (θ_{fc}), bulk density (ρ_{soil}), and hydraulic conductivity at saturation (K_s). Values for these used in this study are from *McKenzie et al.* [2000] and are based on field sample analysis associated with mapped spatial soil units from the Atlas of Australian Soils [*Northcote et al.*, 1960–1968; see Figure 18.1] and soil type interpretations from *Northcote* [1979]. Each mapped soil unit is associated with A-horizon and B-horizon property data for at least one Dominant Principle Profile Form and possibly one or more Subdominant Principle Profile Forms. In this study the A-horizon data was used since the focus is on a generally low vegetation region including large areas of grassland with vegetation rooting depth assumed to be predominantly shallow (~0.5–1.0 m). For each spatial soil unit only the Dominant Principle Profile Form data was used.

McKenzie et al. [2000] relied on the work of *Williams et al.* [1992] for determining parameters θ_{fc} (defined as soil moisture content at 0.1 bar pressure head) and θ_{wilt} (defined as soil moisture content at 15 bar pressure head), where *Williams et al.* [1992] demonstrated pedotransfer functions for predicting the *Campbell* [1974] water retention curve using basic soil characteristic information obtainable from field samples including ρ_{soil}, soil texture, and soil structure classifications (for which they provide details). *McKenzie et al.* [2000] references a range of work conducted by CSIRO Land and Water and some published data sets as providing the basis for the K_s estimates they provide.

In summary, the key soil parameter values from *McKenzie et al.* [2000] associated with soil units in Figure 18.1 are ρ_{soil}, θ_{wilt}, θ_{fc}, and K_s. For all of these except K_s, the 5th, 50th, and 95th percentile values were included, and the 50th percentile values used for the spin-up and open loop (OL). Other key soil parameters for CABLE — θ_{sat} (soil moisture content at saturation), ψ_{aep}, and the b parameter from *Campbell* [1974] equations — were calculated using the data from *McKenzie et al.* [2000]. The θ_{sat} calculation was done using *Williams et al.* [1992]

$$\theta_{sat} = \left[1 - (\rho_{soil} / 2650)\right] \times 0.93, \qquad (18.1)$$

where 2650 kg/m³ is a standard value for density of mineral solids. Values for ψ_{aep} and b were generated from θ_{fc}, θ_{wilt}, and θ_{sat} using [*Campbell*, 1974]

$$\psi = \psi_{aep} (\theta / \theta_{sat})^{-b}, \qquad (18.2)$$

where ψ is the pressure head (m) for a soil with moisture content θ, and θ_{fc} in this data set corresponds to $\psi = 1$ m and θ_{wilt} to $\psi = 150$ m. Hence there is a value for θ_{sat} and two unique values of ψ corresponding to θ_{fc} and θ_{wilt},

which enables the use of two simultaneous equations to solve for the two unknowns ψ_{aep} and b. For the key vegetation parameter LAI, the remotely sensed Moderate Resolution Imaging Spectroradiometer (MODIS) data product MYD15A2 (*Aqua*) LAI/fPAR (fraction of photosynthetically active radiation) was used (courtesy of NASA Land Processes Distributed Active Archive Center; LP DAAC), which is an 8 day composite of 1 km spatial resolution data.

Generating spatially distributed inputs from the forcing and sourced parameter data for running CABLE over the study region involved assigning the data to the 5 km simulation pixels. For point scale data, the study region was subdivided using Theissen polygons and values assigned to pixels based on the point site in nearest proximity. Thus the region was split into two for assigning all measured meteorological forcing variables (except rainfall) from the Griffith and Y3 station sites, while 14 subdivided zones were used to assign rainfall data from all soil moisture stations (Y1–Y13) including the Griffith site (see Figure 18.1).

The 1 km scale MODIS-based LAI data were spatially averaged within each 5 km simulation pixel domain. Key soil parameter data based on the work of *McKenzie et al.* [2000], and associated with the soil units in Figure 18.1, were assigned to pixels according to the dominant fraction of the soil unit polygons within each pixel. All CABLE vegetation parameters other than LAI, and soil parameters other than those previously discussed, were assigned the default global values provided with the CABLE code.

18.2.3. Assimilation Data

The AMSR-E near-surface soil moisture data used for this study was the version 04 level 3A product, on a ~25 × 25 km resolution grid, derived using the Land Parameter Retrieval Model (LPRM) developed jointly by the Vrije Universiteit Amsterdam and NASA [VUA-NASA; *Owe et al.*, 2008]. *Crow et al.* [2010] found that this algorithm produced good-quality soil moisture data relative to other algorithms in a global evaluation. The data represents moisture in only the top ~1–2 cm of soil, and its availability corresponds to two repeat satellite overpasses per day at most for a geographic region — the descending overpass (~01:30–02:00 local time) and ascending overpass (~13:30–14:00 local time). Previous work [*Su et al.*, 2013; *Draper et al.*, 2009b] indicates that data from the descending (local morning) overpass provides superior moisture estimates for a similar region in the Murrumbidgee catchment to that focused on in this study. Hence assimilation of AMSR-E near-surface soil moisture was performed only once per day (subject to data availability) for 1:30 A.M. local time.

18.2.4. Assimilation Algorithm

The EnKF after *Evensen* [1994] is one variant of filters based on the original Kalman filter [*Kalman*, 1960] and was used for assimilation in this study. The general form of the Kalman filter can be represented as

$$\mathbf{X}_k^a = \mathbf{X}_k^f + \mathbf{K}(\mathbf{Z}_k - \mathbf{Z}_k^f), \qquad (18.3)$$

where the model state vector $\mathbf{X}$ consists of 12 state values in this experiment—soil moisture and soil temperature for the six CABLE soil layers—and the observation, $\mathbf{Z}$, is the AMSR-E soil moisture value. Subscript k refers to the assimilation time step, superscript f denotes model predicted values, and superscript a denotes analyzed (updated) values. Thus the innovation here $(\mathbf{Z}_k - \mathbf{Z}_k^f)$ represents the observation-based AMSR-E moisture minus the CABLE-predicted soil moisture for the top 2.2 cm thick soil layer.

The Kalman gain ($\mathbf{K}$) depends on the relative uncertainties between model predictions and observed data, and weights the innovations to determine the degree to which predicted states $\mathbf{X}_k^f$ are adjusted. Therefore, defining model and observation error is crucially important and affects the filter performance, yet is very challenging especially for complex nonlinear models [*Crow and Reichle*, 2008]. The EnKF is a Monte Carlo approach to Kalman filtering based on generating ensembles of simulation predictions about a mean (or "true") value, and using the ensemble spreads for different predictions to calculate the error covariances required for $\mathbf{K}$.

Factors contributing to LSM prediction error include uncertainty in model structure, input forcing, and parameter data. *Richter et al.* [2004] demonstrated the sensitivity of LSM water balance prediction to soil parameters, with the implication that soil parameter uncertainty can play a major part in overall water balance prediction error, particularly for soil moisture. The most comprehensive set of quantitative uncertainty information available was for soil hydraulic properties estimated from field data analysis. Hence, we endeavored to define model error based on observed uncertainty information from data analysis of key soil hydraulic parameters that directly affect modeled soil moisture dynamics through the Richards equation. As such, CABLE ensemble predictions were generated using soil parameter ensembles for parameters that had an associated uncertainty range—the procedure is discussed in the following section.

Since soil parameters are time invariant, the same parameter ensembles are applied to every simulation time step. Consequently, the model prediction ensembles were able to maintain a constant spread for the EnKF, thus

avoiding the potential for ensemble collapse and filter divergence [*Whitaker and Hamill*, 2002]. These problems can occur for implementations where underestimates of model uncertainty (via ensemble spread) persist over time, and are further reduced with successive updates until ensembles collapse toward a single member. At this point the EnKF is ineffective since model predictions are weighted as highly certain, with their trajectory diverging from observed information [*Slater and Clark*, 2006]. If perturbing only forcing inputs such as rainfall [e.g., *Turner et al.*, 2008], time steps with no rain are typically associated with no rainfall error (hence no perturbation), which may result in an underestimated ensemble spread for state predictions. Consequently, applying state perturbation or covariance inflation [*Anderson and Anderson*, 1999] to ensure the ensemble spread correctly represents the overall model error relies largely on trial and error. The soil parameter perturbation applied in this work at least has some basis in uncertainty estimates from field data with no tuning required to maintain the ensemble spread.

Another feature of the EnKF application used for this work is the procedure to minimize unintended bias in model state ensembles during assimilation, as demonstrated by *Ryu et al.* [2009]. Essentially, for a model perturbed with zero-mean Gaussian noise as is typical for EnKF implementation, nonlinear model processes can result in biased ensemble state predictions that can lead to degraded impacts on deeper soil states. Potential problems with biased ensembles from parameter perturbation is also described in earlier work by *De Lannoy et al.* [2006]. The implemented bias correction procedure of *Ryu et al.* [2009] is discussed more in the following section.

18.3. METHODOLOGY

Eight years of data were compiled for the study region (Figure 18.1)—from 1 July 1 2002 to 30 June 2010—including meteorological forcing and parameters (with time varying LAI) at the 5 km scale, and AMSR-E moisture data at the 25 km scale for the 1:30 A.M. local time overpass. Following a 10 year spinup to provide initial soil moisture and temperature state values for all six CABLE soil layers, a single unperturbed OL simulation was run with the full 8 years of compiled input data. The first 4 years of this—1 July 2002 to 30 June 2006— were used for examining model AMSR-E bias and determining factors for correction. The following 4 years—1 July 2006 to 30 June 2010—was the main experiment period where assimilation was performed after bias correction.

There were three main components to the experimental work. First was determining and removing the systematic bias between AMSR-E moisture data and CABLE moisture predictions for its top soil layer. Setting up the parameter error perturbation along with the online bias correction from *Ryu et al.* [2009] was then carried out for EnKF implementation. Finally, the assimilation was performed and the resulting soil moisture predictions assessed against in situ data.

18.3.1. Observed-Modeled Bias Removal

The statistical basis for data assimilation is to correct for random model errors. Therefore, systematic biases between modeled and observed states need to be accounted for and removed prior to assimilation. Rescaling remotely sensed soil moisture data to match the climatology of the model predicted near-surface moisture is somewhat standard practice [e.g., *Draper et al.*, 2009a; *Drusch et al.*, 2005].

A rescaling approach was applied to match the mean and standard deviation of the AMSR-E data series to that of the CABLE predicted series for the top 0–2.2 cm soil layer. A limitation to removing long-term model observation climatological differences is the relatively short record of remotely sensed data (~10 years for AMSR-E). This makes it difficult to know if the rescaling based on a certain period is representative of future differences, which has potential implications for rescaling new data as it becomes available in real-world modeling applications. To test this, the rescaling relationship between AMSR-E moisture data and CABLE was determined for an initial 4 year period, which was then applied to rescale data over the following 4 year period for use in the data assimilation experiment.

Rescaling the AMSR-E moisture series for the period 1 July 2002 to 30 June 2006 removed much of the bias relative to CABLE. Using the same mean and standard deviation as a benchmark, the rescaling was applied for the assimilation period of 1 July 2006 to 30 June 2010 for which much of the bias was also removed (Figure 18.2), and the resulting 4 year rescaled AMSR-E data series was used in the assimilation.

18.3.2. EnKF Implementation

Uncertainty in both the observed data and the model predictions forms the basis for data assimilation but remain difficult to define. Major sources of error for LSMs such as CABLE are model physics, where complex physical interactions are represented with generalized relationships; inaccurate parameters related to the water balance due to a general lack of quality and quantity (including spatial detail) of such data, errors in key meteorological forcing inputs, and, in the shorter term, uncertain initial state variables. For the EnKF to work optimally, the total model error from all of these sources needs to be adequately represented by the ensemble spread, which at present is extremely challenging.

The strategy here was perturbing key CABLE soil parameter inputs for which uncertainty information was available, resulting in a consistent ensemble spread for soil state predictions reflecting the impact of this uncertainty. State values for all six CABLE soil layers were also perturbed for the initial time step of the 1 July 2006 to 30 June 2010 assimilation period. An ensemble of random zero-mean Gaussian variates, together with the standard deviation (σ) of available uncertainty estimates, were used

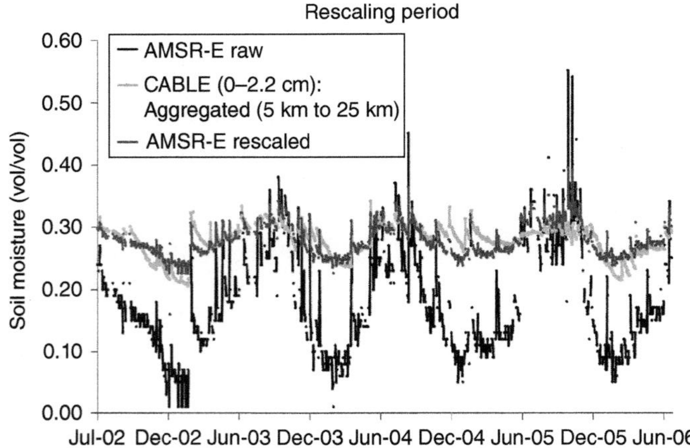

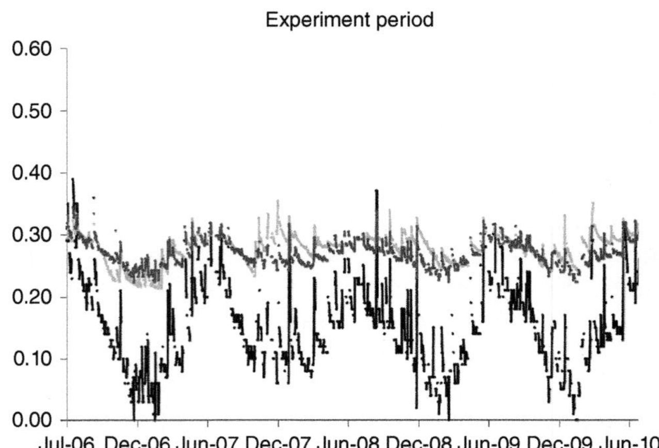

Figure 18.2 Time series plots showing bias removal between AMSR-E (local morning overpass) and CABLE soil moisture for a sample AMSR-E domain in the study region. The mean and standard deviation of CABLE predictions over the rescaling period in the left-hand plot was used to rescale the AMSR-E series. The plot on the right indicates this was adequate for removing much of the bias over the experiment period, and the resulting AMSR-E data were used for assimilation here.

to generate ensembles for the different inputs within the 95% confidence interval ($\pm 1.96\sigma$). The ensemble size used here was 30 members. Perturbation of soil parameters was performed once and the resulting parameter ensembles used throughout the whole experiment period. Care was taken to ensure ensemble member values for particular parameters were sensible in relation to corresponding members for other parameters (e.g., each perturbed θ_{fc} value, ranked from minimum to maximum, is matched with a perturbed θ_{wilt} value that is in the corresponding position of minimum to maximum rank ordered θ_{wilt} values, to avoid physically unrealistic combinations of these two parameters).

With no knowledge of the true initial state values across the study region, uncertainty estimates were made for their ensemble generation with: $\sigma = 0.04$ vol/vol for the six soil moisture states and $\sigma = 4\,°C$ for the six soil temperature states. The 5th and 95th percentile values provided with θ_{wilt}, θ_{fc}, and ρ_{soil} soil parameters sourced directly from *McKenzie et al.* [2000] were used for estimating σ values, from which ensembles were generated for each of the parameters with 95% confidence interval.

Corresponding ensemble member values for θ_{fc} and θ_{wilt} were sorted in ascending order to ensure a reasonable available water range between the two, with an assumption of some correlation between the two with changes in soil texture. Ensemble members for ρ_{soil} were used to calculate an ensemble for θ_{sat} using equation 18.1, and θ_{sat} ensemble member values were sorted in ascending order to correspond with ascending values of θ_{fc} and θ_{wilt} members. The ensembles for all these parameters were then used to calculate ensembles for ψ_{aep} and b using equation 18.2. Calculating the ensemble values for these (and θ_{sat}) ensured consistency in the relationships between corresponding ensemble members for all related parameter values.

Values for K_s are available with the *McKenzie et al.* [2000] soils map data and are classified according to a corresponding $\log_{10}$ linear scale with increments of 0.5 per class, with uncertainty information given only in terms of $\pm n$ classes. The strategy here involved generating ensembles for inputs within a 95% confidence interval, therefore $\pm n$ was treated as the bounds for this interval, and the $\log_{10}$ linear scale value for K_s associated with each soil unit was treated as the ensemble mean value (used in spinup and OL). Ensemble generation was done here with the $\log_{10}$ linear scale data assuming a Gaussian distribution, and the resulting distribution of values was transformed back to provide ensembles for K_s distributed on a $\log_{10}$ scale with units of millimeters/hour.

Observational error applied to the AMSR-E (VUA-NASA) moisture product for the assimilation was $\sigma = 0.06$ vol/vol [*Parinussa et al.*, 2011]. While validation of this product by *Draper et al.* [2009b] over the Murrumbidgee catchment (where this experiment was performed) resulted in an error estimate of ~0.02–0.04 vol/vol (after rescaling the product to minimize bias relative to in situ validation data), the more conservative estimate of 0.06 vol/vol was used.

As previously discussed, a problem with the EnKF is that even when applying Gaussian perturbations for error representation, predicted state ensemble means are often biased during the assimilation due to nonlinear model processes. The approach of *Ryu et al.* [2009] used to address this involved running an additional ensemble member with no model perturbation (the OL) in parallel with the other 30 members resulting from perturbations. Prior to calculating model error covariances for **K** [equation 18.3] at each EnKF state update time step, the ensemble mean of the perturbed predictions is corrected (by correcting each perturbed ensemble member accordingly) to remove bias relative to the single unperturbed member. The bias-corrected ensemble of predictions is then used as per normal for covariance calculations, and the resulting mean of the updated state ensemble is used as the initial state value for the unperturbed (OL) member for the next prediction period between observations.

Ryu et al. [2009] describe specific soil moisture biases introduced by ensemble end members near the bounds of the model soil moisture range, defined in this study by θ_{wilt} and θ_{sat}, which are the lower and the upper bounds, respectively. In many models (including CABLE) any members less than θ_{wilt} will be reset to θ_{wilt} and any members greater than θ_{sat} are reset to θ_{sat}. Therefore, as ensemble means approach these bounds for either near dry or near saturated conditions, the normal distribution of the ensembles implies that a number of members will exceed and therefore be reset, leading to a concentration of members with boundary values. This skews the ensemble distributions and biases the means away from the boundaries. To correct for this bias, the *Ryu et al.* [2009] approach needs to adjust the skewed members toward the boundaries, which has the potential to excessively reduce the ensemble spread and lead to filter divergence. However, it was found that this problem can be avoided by perturbing model soil parameters, which ensured an adequate spread is maintained, at least in terms of uncertainty introduced by the key soil parameters.

Finally, given the spatial disparity between model simulation (5 km) and AMSR-E product (25 km) resolutions (Figure 18.1), the average of predictions for all 5 km simulation pixels within each AMSR-E footprint domain were first calculated, and the up-scaled values used to calculate innovations with the corresponding ASMR-E data [bracketed term in equation 18.3]. The resulting update terms (innovation weighted by **K**) were then applied back to the states for each individual 5 km simulation domain. This approach ensured the spatial variability of moisture between 5 km simulation pixels was maintained.

18.3.3. Assessing the Assimilation

The experimental simulation was performed for the 4 year period spanning from 1 July 2006 to 30 June 2010. The AMSR-E soil moisture data for the ~01:30 A.M. local overpass time were assimilated into the top soil layer of the CABLE LSM, and the impact on updates made to its deeper root-zone moisture state predictions was assessed for 12 OzNet station sites: Y1 – Y10, Y12, and Y13 (Figure 18.1).

Assimilation outputs for the 5 km simulation pixels were compared against available in situ data from collocated OzNet stations over the 4 year period. Moisture values for the different soil layers in CABLE, and also those measured in situ at the stations (for 0–5 cm—except for Y3 which was 0–7 cm, 0–30 cm, and 30–60 cm depths) were depth averaged where necessary to enable direct comparisons for 0–5 cm, 0–30 cm and 0–60 cm depth ranges. The calculation of depth averaged values for comparisons is summarized in Table 18.1. For this predominantly pasture/crop region the root zone is defined as 0–60 cm.

To facilitate meaningful/bias-free error assessment of assimilation results, the in situ observed data series for each validation site were rescaled (by matching their series mean and standard deviation) to the OL series for the experiment period. This is consistent with the assimilation performed relative to the CABLE OL climatology (using rescaled AMSR-E data). Metrics used to compare both OL and assimilation/updated (UP) outputs with rescaled in situ data were the root-mean-squared error (RMSE) and the coefficient of determination (R^2).

The overall impact over the study region was assessed for the three soil depths, where mean differences between UP and OL in terms of both RMSE and R^2 values were calculated across the 12 sites. A permutation test (with 10,000 resamples) was applied where the statistical significance of these mean differences (indicating the significance of the overall impact from data assimilation) was tested. Therefore, the null hypothesis was that the mean differences between UP and OL across the 12 sites in terms of RMSE and R^2 values is zero. Determining the significance of any of the mean differences (therefore the significance of the overall assimilation impacts) was through p values resulting from the tests.

Normalized values for RMSE and R^2, which represent improvement or degradation from OL to UP, were also calculated and plotted to summarize the overall assimilation impacts for the three depths. The normalized RMSE is simply the ratio between the RMSE for UP and OL:

$$\mathrm{NRMSE} = \frac{\mathrm{RMSE_{UP}}}{\mathrm{RMSE_{OL}}}, \qquad (18.4)$$

while for R^2 the ratio between the unexplained variance in UP to the unexplained variance in OL was used:

$$\mathrm{NR}^2 = \frac{\left(1 - R^2{}_{\mathrm{UP}}\right)}{\left(1 - R^2{}_{\mathrm{OL}}\right)}. \qquad (18.5)$$

Values of NRMSE = 1 and NR^2 = 1 from equations 18.4 and 18.5 indicate no impact on CABLE from the assimilation in terms of changes in residual error (RMSE) or explained variance (R^2), respectively. Values for each that are <1 indicate improvement, and values for each that are >1 represent degradation from the assimilation according to the respective metrics.

Table 18.1 Calculations applied for averaging modeled and in situ soil moisture data over different depths (in cm) to enable direct comparisons between them

	In Situ Soil Moisture	CABLE Soil Moisture
0–5 cm (0–7 cm for Y3 only)	Direct measurements	$\dfrac{\left(2.2 \times \theta^M_{0-2.2}\right) + \left(2.8 \times \theta^M_{2.2-8}\right)}{5}$ $\dfrac{\left(2.2 \times \theta^M_{0-2.2}\right) + \left(4.8 \times \theta^M_{2.2-8}\right)}{7}$
0–30 cm	Direct measurements	$\dfrac{\left(2.2 \times \theta^M_{0-2.2}\right) + \left(5.8 \times \theta^M_{2.2-8}\right) + \left(15.4 \times \theta^M_{8-23.4}\right) + \left(6.6 \times \theta^M_{23.4-64.3}\right)}{30}$
0–60 cm (root zone)	$\dfrac{\left(30 \times \theta^I_{0-30}\right) + \left(30 \times \theta^I_{30-60}\right)}{60}$	$\dfrac{\left(2.2 \times \theta^M_{0-2.2}\right) + \left(5.8 \times \theta^M_{2.2-8}\right) + \left(15.4 \times \theta^M_{8-23.4}\right) + \left(36.6 \times \theta^M_{23.4-64.3}\right)}{60}$

Note: Moisture content data is represented by θ, with superscripts indicating whether it is an in situ measurement (I) or model prediction (M), and subscripts indicating the depth range of the data (i.e., for CABLE they represent model soil layer depths).

18.4. RESULTS AND DISCUSSION

A complete summary of assessments for assimilation impacts over the root zone is displayed in Table 18.2, showing R^2 and RMSE metrics for both OL and UP relative to in situ data. Mean differences between OL and UP for these metrics across the 12 validation sites are also included along with corresponding p values from the permutation tests, which tested whether the mean differences are significantly different to zero as an indication of how significant the overall assimilation impacts were.

From Table 18.2, the greatest improvements in UP compared to OL were for the 0–5 cm depth, with improved RMSE and R^2 scores for 7 out of 12 and 10 out of 12 stations respectively. The 0–5 cm depth was also the only depth range where the overall improvement from assimilation across all sites was statistically significant, but only in terms of mean improvement in R^2 for which a p value of 0.009 indicates significance at the 99% confidence level. For 0–30 cm and 0–60 cm depths, there was improvement in terms of RMSE for 7 out 12 and 6 out of 12 stations, respectively, and improvement in R^2 for 8 out of 12 and 7 out of 12 stations, respectively. None of the overall changes from the assimilation across the 12 sites for 0–30 cm and 0–60 cm depths, as per average RMSE and

R^2 differences, were statistically significant according to corresponding p values from permutation tests. These results demonstrate some difficulty in improving modeled soil moisture states with depth from the assimilation of a shallow moisture product. It also appears that the assimilation may be slightly better at improving the temporal variance of predictions as indicated by R^2 than reducing overall residual error as indicated by reduced RMSE.

Figure 18.3 provides a visual summary of overall improvement to soil moisture in the 3 depths across the 12 sites. It shows median and interquartile ranges of NRMSE and NR^2 [equations 18.4 and 18.5] for the 12 sites. All points in the plot fall below 1 for both NMRSE and NR^2, indicating some overall improvement based on the median values of these metrics. By contrast, the mean difference values for RMSE and R^2 in Table 18.2 indicate slight overall degradation for some depths. However, given the very small sample size of 12 stations, the means are likely to be skewed by one or two extreme differences for each depth (the validity of the permutation tests for means is based on the resampling of 10,000 sets of differences).

Plots in Figure 18.4 are time series results illustrating the assimilation impacts on the root-zone moisture prediction, sampled from three sites. One where there was

Table 18.2 Root-mean-squared error (RMSE) and coefficient of determination (R^2) assessments of predicted soil moisture from the 4 year assimilation experiment relative to in situ data for the 12 OzNet sites Y1–Y10, Y12, and Y13

Site	0–5 cm				0–30 cm				0–60 cm			
	RMSE		R^2		RMSE		R^2		RMSE		R^2	
	OL	UP	OL	UP	OL	UP	OL	UP	OL	UP	OL	UP
Y1	**0.021**	**0.018**	**0.34**	**0.63**	**0.022**	**0.018**	**0.19**	**0.53**	**0.021**	**0.019**	**0.14**	**0.36**
Y2	0.008	0.026	0.39	0.39	0.007	0.028	0.50	0.11	0.007	0.029	0.48	0.01
Y3	**0.017**	**0.012**	**0.36**	**0.66**	**0.018**	**0.014**	**0.25**	**0.46**	**0.017**	**0.014**	**0.25**	**0.44**
Y4	0.007	0.007	**0.54**	**0.63**	**0.011**	**0.009**	**0.25**	**0.40**	0.010	0.010	**0.29**	**0.33**
Y5	0.008	0.009	**0.49**	**0.65**	**0.012**	**0.011**	**0.08**	**0.38**	0.009	0.012	0.36	0.18
Y6	**0.014**	**0.011**	**0.16**	**0.43**	**0.018**	**0.014**	**0.00**	**0.11**	**0.018**	**0.015**	**0.01**	**0.04**
Y7	0.007	0.007	**0.49**	**0.59**	0.009	0.009	**0.26**	**0.43**	**0.009**	**0.008**	**0.21**	**0.45**
Y8	**0.007**	**0.006**	**0.48**	**0.58**	0.007	0.009	0.58	0.41	0.007	0.010	0.55	0.23
Y9	**0.022**	**0.016**	**0.12**	**0.32**	**0.024**	**0.020**	**0.01**	**0.05**	**0.024**	**0.020**	4.2E-7	0.02
Y10	0.006	0.011	0.57	0.37	0.007	0.011	0.42	0.38	0.007	0.010	0.46	0.45
Y12	**0.008**	**0.006**	**0.40**	**0.67**	**0.011**	**0.009**	**0.10**	**0.21**	**0.012**	**0.010**	2.1E-3	**0.06**
Y13	**0.008**	**0.007**	**0.33**	**0.46**	0.007	0.008	0.47	0.39	0.008	0.008	0.37	0.29
μ (mean differences)	-2.5×10^{-4}		0.14		-5.8×10^{-4}		0.06		-1.3×10^{-3}		−0.02	
p value for μ ($H_0: \mu = 0$)	0.476		0.009		0.408		0.196		0.305		0.612	

Note: Values for each metric in bold indicate an improvement from the assimilation, and the nonbold font indicates either no improvement or a degraded impact. The bottom two rows include the means of differences between OL and UP in terms of RMSE and R^2, along with corresponding p values from permutation tests for whether the mean differences significantly differ from zero.

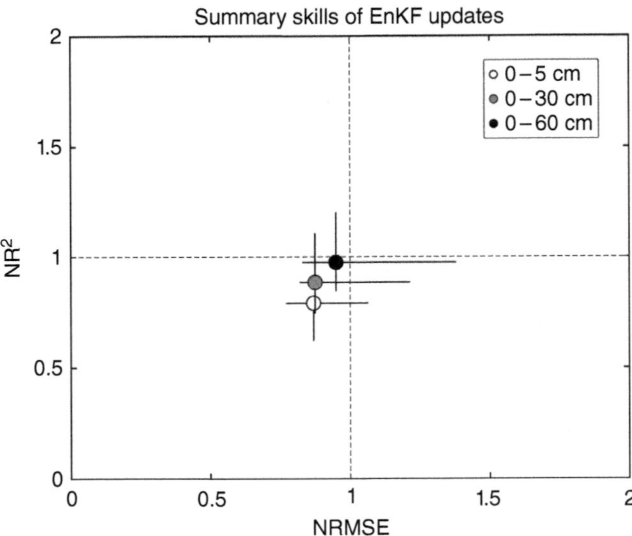

Figure 18.3 Plot showing the median NRMSE against the median NR^2 [equations 18.4 and 18.5] across the 12 validation sites for the 3 soil depths examined, along with interquartile ranges (25th and 75th percentiles). Grid lines on the plot at 1 for both NRMSE and NR^2 signify no difference between UP and OL, while values <1 signify improvement from the assimilation and values >1 signify degradation.

clear root-zone improvement according to both RMSE and R^2 (Y1), one where there was degradation (Y5), and one where there was minor improvement in either one or both metrics (Y6). The Y6 site is of particular interest as the in situ data contains information from isolated irrigation events. The root-zone time series plots shown in Figure 18.4 highlight the ability of the assimilation to translate information on AMSR-E and CABLE differences for the near-surface down to deeper root-zone states in CABLE. Another feature to note from Figure 18.4 is the ensemble extremes in each plot, which demonstrate the maintained ensemble spread over the 4 year experiment period due to the input of perturbed parameter ensembles.

For Y5, where the UP root-zone series was degraded overall from OL (Table 18.2), most of the adjustments appear to be in the right direction relative to the in situ data (except for around December 2007), but they overshoot it, which is generally the case for most other sites where the root zone has also been degraded. This may be an issue of imbalanced model-observation error representation or coupling strength between the near surface and deeper soil layers in CABLE. A more detailed investigation is necessary to better understand the degraded results.

For the Y6 series where known irrigation events are present in the in situ data (~September 2006 and 2007), there is negligible evidence of it in the modeled series.

While it is expected that AMSR-E is able to detect such events relative to dry surrounds, and for such information to be assimilated into the model, in this case the irrigation was known to be at the paddock scale only (in the order of ~100s of m^2) and hence the lack of impact. This highlights a particular limitation of broad-scale (~10s of km^2) moisture data from microwave remote sensing in terms of constraining water balances and providing useful information for agricultural decisions relevant to specific small-scale areas (~100s of meters to ~1 km). Also relevant to this issue is the large-scale discrepancy between model inputs (forcing and parameters), assimilated data products, and the point scale validation data, which adds to the difficulty of defining uncertainty and evaluating results in experiments such as this.

18.5. CONCLUSIONS

Assimilating remotely sensed near-surface soil moisture into the CABLE LSM was shown to make some improvement to deeper moisture state prediction. However, the only overall improvement across all 12 validation sites that was statistically significant (at the 99% confidence level) was the average improvement in predicted variance by 0.14 over the 0–5 cm depth. The ability for improvement was also shown to drop off with increasing depth—for the full 0–60 cm root zone, some improvement was evident for only just over half of the validation sites. While a larger sample size of validation data would enable stronger conclusions to be drawn, investigating near-surface and deeper moisture state coupling in CABLE, and the sensitivity of the near-surface and deeper state error correlations to different perturbations, may also provide insight into the cause of degraded assimilation impacts.

Perturbing key soil parameters with informed uncertainty estimates successfully maintained predicted ensemble spreads for the EnKF, without applying covariance inflation or perturbations based on trial and error. Consequently, the online bias correction that was implemented worked well, with no evidence of it leading to ensemble collapse, which may be a problem in some cases when ensemble mean moisture state values approach the extremes of the model soil moisture range. The presence of small-scale irrigation in the study region explicitly highlighted the issue of uncertainty from spatial scale discrepancies between data used for model input, assimilation, and validation, particularly for agricultural areas. Therefore, information from broad-scale remotely sensed data (~10s of km) such as AMSR-E may not always provide water balance information that is useful for decision making at the farm/paddock scale.

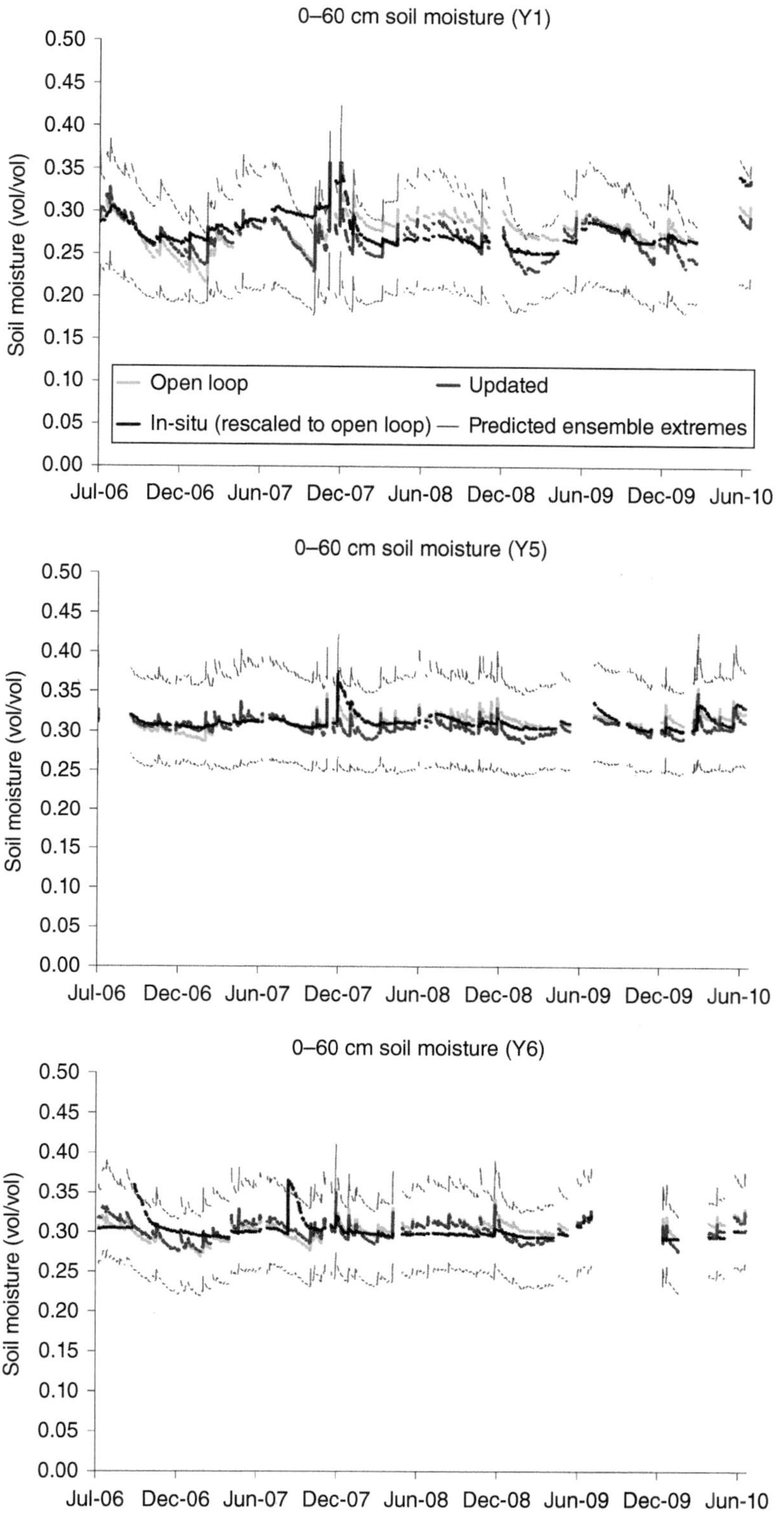

Figure 18.4 Time series plots of data sampled at 1:30 A.M. (local) time for three OzNet station sites, with a priori output from the assimilation procedure. Shown are the impacts on CABLE root-zone moisture predictions resulting from the assimilation of AMSR-E soil moisture data at previous update times, including a priori maximum and minimum ensemble end members resulting from the perturbed parameter ensemble inputs. UP for station Y1 (top) was an overall improvement on OL (see Table 18.2), for Y5 (middle) it was degraded overall, and for Y6 (bottom) there was some improvement (only in terms of RMSE). Note the impact on Y6 in situ data from irrigation in ~ September 2006 and 2007.

ACKNOWLEDGMENTS

This work was supported by the Department of Primary Industries (DPI), Victoria, Australia, and the Australian Research Council (ARC) through ARC Linkage Project LP0989441. We would also like to acknowledge the landowners in the Yanco region who allowed the OzNet stations to operate on their property. Rodger Young, field engineer at The University of Melbourne, provided valuable insights from his records and personal experience of many years work on the Yanco OzNet stations—in particular for confirming the occurrence and extent of irrigation events relevant to data sets used in this study. We also thank Lionel Siriwardena from the Department of Infrastructure of Engineering, The University of Melbourne, and Kaighin McColl from his time as research assistant at the University of Melbourne for helping to prepare the meteorological forcing data sets used in this study.

REFERENCES

Anderson, J. L., and S. L. Anderson (1999), A Monte Carlo implementation of the nonlinear filtering problem to produce ensemble assimilations and forecasts, *Monthly Weather Rev., 127*, 2741–2758.

Brutsaert, W. (2005), *Hydrology: An Introduction*, Cambridge Univ. Press, Cambridge, United Kingdom.

Campbell, G. S. (1974), A simple model for determining unsaturated conductivity from moisture retention data, *Soil Sci., 117*, 311–314.

Clapp, R. B., and G. M. Hornberger (1978), Empirical equations for some soil hydraulic properties, *Water Resourc. Res., 14*, 601–604.

Crow, W. T., and R. H. Reichle (2008), Comparison of adaptive filtering techniques for land surface data assimilation, *Water Resourc. Res., 44*, W08423.

Crow, W. T., D. G. Miralles, and M. H. Cosh (2010), A quasi-global evaluation system for satellite-based surface soil moisture retrievals, *IEEE Trans. Geosci. Remote Sens., 48*, 2516–2527.

De Lannoy, G. J. M., P. R. Houser, V. R. N. Pauwels, and N. E. C. Verhoest (2006), Assessment of model uncertainty for soil moisture through ensemble verification, *J. Geophys. Res., 111*, D10101.

Draper, C. S., J.-F. Mahfouf, and J. P. Walker (2009a), An EKF assimilation of AMSR-E soil moisture into the ISBA land surface scheme, *J. Geophys. Res., 114*, D20104.

Draper, C. S., J. P. Walker, P. Steinle, R. d. Jeu, and T. Holmes (2009b), An evaluation of AMSR-E derived soil moisture over Australia, *Remote Sens. Environ., 113*, 703–710.

Draper, C. S., R. H. Reichle, G. J. M. De Lannoy, and Q. Liu (2012), Assimilation of passive and active microwave soil moisture retrievals, *Geophys. Res. Lett., 39*, L04401.

Drusch, M. (2007), Initializing numerical weather prediction models with satellite-derived surface soil moisture: Data assimilation experiments with ECMWF's Integrated Forecast System and the TMI soil moisture data set, *J. Geophys. Res., 112*, D03102.

Drusch, M., E. F. Wood, and H. Gao (2005), Observation operators for the direct assimilation of TRMM microwave imager retrieved soil moisture, *Geophys. Res. Lett., 32*, L15403.

Entekhabi, D., H. Nakamura, and E. G. Njoku (1994), Solving the inverse problem for soil moisture and temperature profiles by sequential assimilation of multifrequency remotely sensed observations, *IEEE Trans. Geosci. Remote Sens., 32*, 438–448.

Evensen, G. (1994), Sequential data assimilation with a nonlinear quasi-geostrophic model using Monte Carlo methods to forecast error statistics, *J. Geophys. Res. Oceans, 99*, 10,143–10,162.

Imaoka, K., M. Kachi, M. Kasahara, N. Ito, K. Nakagawa, and T. Oki (2010), Instrument performance and calibration of AMSR-E and AMSR2, *Int. Arch. Photogram. Remote Sens. Spatial Inform. Sci., 38*, 13–16.

Kalman, R. E. (1960), A new approach to linear filtering and prediction problems, *Trans. ASME J. Basic Eng., 82*, 35–45.

Kerr, Y. H., et al. (2010), The SMOS mission: New tool for monitoring key elements of the global water cycle, *Proc. IEEE, 98*, 666–687.

Kowalczyk, E. A., Y. P. Wang, R. M. Law, H. L. Davies, J. L. McGregor, and G. Abramowitz, (2006), The CSIRO Atmosphere Biosphere Land Exchange (CABLE) model for use in climate models and as an offline model, Paper 013, CSIRO Marine and Atmospheric Research, Aspendale, Australia.

Kowalczyk, E. A., L. Stevens, R. M. Law, M. Dix, Y. P. Wang, I. N. Harman, K. Haynes, J. Srbinovsky, B. Pak, and T. Ziehn (2013), The land surface model component of ACCESS: Description and impact on the simulated surface climatology, *Austral. Meteorol. Oceanogr. J., 63*, 65–82.

Kumar, S. V., R. H. Reichle, R. D. Koster, W. T. Crow, and C. D. Peters-Lidard, (2009), Role of subsurface physics in the assimilation of surface soil moisture observations, *J. Hydrometeorol., 10*, 1534–1547.

Law, R. M., M. R. Raupach, G. Abramowitz, I. Dharssi, V. Haverd, A. J. Pitman, L. Renzullo, A. Van Dijk, and Y.-P. Wang, (2012), The Community Atmosphere Biosphere Land Exchange (CABLE) model roadmap for 2012–2017, Tech. Rep. 057, *The Centre for Australian Weather and Climate Research (CAWCR)*, Aspendale, Australia.

Liu, Q., R. H. Reichle, R. Bindlish, M. H. Cosh, W. T. Crow, R. de Jeu, G. J. M. De Lannoy, G. J. Huffman, and T. J. Jackson, (2011), The contributions of precipitation and soil moisture observations to the skill of soil moisture estimates in a land data assimilation system, *J. Hydrometeorol., 12*, 750–765.

McKenzie, N. J., D. W. Jacquier, L. J. Ashton, and H. P. Cresswell, (2000), Estimation of soil properties using the Atlas of Australian Soils, Tech. Rep. 11/00, CSIRO Land and Water, Canberra, Australia.

Njoku, E. G., T. J. Jackson, V. Lakshmi, T. Chan, and S. V. Nghiem (2003), Soil moisture retrieval from AMSR-E, *IEEE Trans. Geosci. Remote Sens., 41*, 215–229.

Northcote, K. H. (1979), *A Factual Key for the Recognition of Australian Soils,* 4th ed., Rellim Technical Publications, Glenside, Australia.

Northcote, K. H., et al. (1960–1968), Atlas of Australian Soils, Sheets 1–10, with explanatory data, CSIRO Australia and Melbourne Univ. Press, Melbourne, Australia.

Owe, M., R. A. M. de Jeu, and T. R. H. Holmes, (2008), Multi-sensor historical climatology of satellite-derived global land surface moisture, *J. Geophys. Res., 113,* F01002.

Parinussa, R. M., A. G. C. A. Meesters, Y. Y. Liu, W. Dorigo, W. Wagner, and R. A. M. de Jeu, (2011), Error estimates for near-real-time satellite soil moisture as derived from the land parameter retrieval model, *IEEE Geosci. Remote Sens. Lett., 8,* 779–783.

Pipunic, R. C., J. P. Walker, and A. W. Western, (2008), Assimilation of remotely sensed data for improved latent and sensible heat flux prediction: A comparative synthetic study, *Remote Sens. Environ., 112,* 1295–1305.

Pipunic, R. C., J. P. Walker, A. W. Western, and C. M. Trudinger, (2013), Assimilation of multiple data types for improved heat flux prediction: A one-dimensional field study, *Remote Sens. Environ., 136,* 315–329.

Reichle, R. H., R. D. Koster, P. Liu, S. P. P. Mahanama, E. G. Njoku, and M. Owe, (2007), Comparison and assimilation of global soil moisture retrievals from the Advanced Microwave Scanning Radiometer for the Earth Observing System (AMSR-E) and the Scanning Multichannel Microwave Radiometer (SMMR), *J. Geophys. Res., 112,* D09108.

Richter, H., A. W. Western, and F. H. S. Chiew, (2004), The effect of soil and vegetation parameters in the ECMWF land surface scheme, *J. Hydrometeorol., 5,* 1131–1146.

Ryu, D., W. T. Crow, X. Zhan, and T. J. Jackson, (2009), Correcting unintended perturbation biases in hydrologic data assimilation, *J. Hydrometeorol., 10,* 734–750.

Slater, A. G., and M. P. Clark, (2006), Snow data assimilation via an ensemble Kalman filter, *J. Hydrometeorol., 7,* 478–493.

Smith, A. B., J. P. Walker, A. W. Western, R. I. Young, K. M. Ellet, R. C. Pipunic, R. B. Grayson, L. Siriwardena, F. H. S. Chiew, and H. Richter (2012), The Murrumbidgee soil moisture monitoring network data set, *Water Resourc. Res., 48,* W07701.

Su, C.-H., D. Ryu, R. I. Young, A. W. Western, and W. Wagner (2013), Inter-comparison of microwave satellite soil moisture retrievals over the Murrumbidgee Basin, southeast Australia, *Remote Sens. Environ., 134,* 1–11.

Turner, M. R. J., J. P. Walker, and P. R. Oke (2008), Ensemble member generation for sequential data assimilation, *Remote Sens. Environ., 112,* 1421–1433.

Walker, J. P., G. R. Willgoose, and J. D. Kalma (2001), One-dimensional soil moisture profile retrieval by assimilation of near-surface observations: A comparison of retrieval algorithms, *Adv. Water Resourc., 24,* 631–650.

Whitaker, J. S., and T. M. Hamill (2002), Ensemble data assimilation without perturbed observations, *Monthly Weather Rev., 130,* 1913–1924.

Williams, J., P. J. Ross, and K. L Bristow (1992), Prediction of the Campbell water retention function from texture, structure and organic matter, in *Proceedings of the International Workshop on Indirect Methods for Estimating the Hydraulic Properties of Unsaturated Soils,* edited by M. T. Van Genuchten, F. J. Leij, and L. J. Lund pp. 427–441 Univ. Calif., Riverside, Calif.

19

Assimilation of Satellite Soil Moisture Retrievals into a Hydrologic Model for Improving River Discharge

Feyera A. Hirpa,[1] Mekonnen Gebremichael,[1] Thomas M. Hopson,[2] Rafal Wojick,[3] and Haksu Lee[4,5]

19.1. INTRODUCTION

Soil moisture, commonly defined as the amount of water in a unit volume of soil, accounts for only 0.15% of liquid freshwater on Earth [*Dingman*, 1994; *Western et al.*, 2002], but it is considered the most important parameter linking the key components of hydrological, biological, and geochemical processes [*NRC*, 2007; *Owe et al.*, 2008]. It determines the precipitation partitioning into surface runoff and infiltration and influences the partitioning of the incoming solar energy into latent and sensible heat components. Since it controls the water and energy balance between land surface and atmosphere, soil moisture plays a critical role in a variety of water and energy balance modeling such as numerical weather prediction, rainfall-runoff modeling, radiative transfer modeling, and climate and agricultural modeling [e.g., *Dunne et al.*, 1975; *Rodrıguez-Iturbe*, 2000; *Lu et al.*, 2005; *Kim and Wang*, 2007; *Koster et al.*, 2004, 2010; *Dirmeyer et al.*, 2009; *Albergel et al.*, 2010; *Bolten et al.*, 2010; *Mei and Wang*, 2011].

Soil moisture is highly variable over space and time, and its relationship with hydrologic response is highly nonlinear. Hence, obtaining reliable data over a suitable spatial and temporal scale for hydrologic applications is challenging [e.g., *Reichle et al.*, 2004]. Ground-based observations are sparse and not usually representative of the large spatial scale [e.g., *Brocca et al.*, 2011]. In fact, the mere total number of ground stations currently contributing to the global in situ soil moisture database (operated by the International Soil Moisture Network at the Vienna University of Technology in cooperation with the Global Soil Moisture Databank of Rutgers University) is less than 1500, with 57% of those stations coming from one nation, the United States (see http://ismn.geo.tuwien.ac.at/networks/). Global land surface models could produce soil moisture over larger spatial scales, but separate models often produce different soil moisture values even when identical forcings were used to run the models [*Entin et al.*, 1999], and they usually fail to reproduce the true soil moisture.

Satellite-based soil moisture retrievals have been recently shown to be useful for several hydrometeorological applications. Particularly, advanced soil moisture retrievals from passive and active microwave satellite remote sensing [e.g., *Wagner et al.*, 1999; *Njoku et al.*, 2003, *Njoku*, 2004; *Owe et al.*, 2008; *Bartalis et al.*, 2007] are now globally available at fine spatial (~25 km or less) and temporal (mostly daily) scales needed for hydrologic applications. The most commonly used sensors for soil moisture are the Advanced Microwave Scanning Radiometer for Earth Observing System (AMSR-E) onboard the *Aqua* satellite [*Njoku et al.*, 2003: *Njoku*, 2004; *Owe et al.*, 2008] and the Advanced SCATterometer (ASCAT) onboard the *MetOp* (Meteorological Operational) satellite [*Wagner et al.*, 1999; *Bartalis et al.*, 2007]. Soil moisture estimates derived from these sensors have been evaluated through

[1] *Department of Civil & Environmental Engineering, University of Connecticut, Storrs, Conneticut, USA*

[2] *Research Applications Laboratory, National Center for Atmospheric Research, Boulder, Colorado, USA*

[3] *Department of Civil & Environmental Engineering, Massachusetts Institute of Technology, Cambridge, Massachusetts, USA*

[4] *NOAA National Weather Service, Office of Climate, Weather, and Water Services, Silver Spring, Maryland, USA*

[5] *Len Technologies, Oak Hill, Virginia, USA*

Remote Sensing of the Terrestrial Water Cycle, Geophysical Monograph 206. First Edition. Edited by Venkat Lakshmi.
© 2015 American Geophysical Union. Published 2015 by John Wiley & Sons, Inc.

319

comparison with in situ or model-simulated soil moistures [e.g., *Wagner et al.*, 2007; *Draper et al.*, 2009; *Albergel et al.*, 2009, 2010; *Jackson et al.*, 2010; *Brocca et al.*, 2010, 2011; *Liu et al.*, 2011; *Hain et al.*, 2011] and reported to correspond significantly well with the reference soil moisture.

Studies have shown that the remotely sensed soil moisture observations have given promising results in enhancing river discharge simulation when assimilated into rainfall-runoff models [e.g., *Pauwels et al.*, 2001; *Parajka et al.*, 2006; *Crow and Ryu*, 2009; *Draper et al.*, 2011, 2012; *Brocca et al.*, 2012, to name a few]. The most common approach is sequential updating of the initial soil moisture of hydrologic model using the satellite observations; however, accurate initialization of the model soil moisture alone does not guarantee a significant improvement of the river discharge due to the highly nonlinear relationship between runoff and soil moisture [*Hirpa*, 2013]. Another methodology proposed by *Crow et al.* [2005], and later evaluated on the Sacramento soil moisture accounting (SAC-SMA) hydrologic model over several basins [*Crow and Ryu*, 2009], is dual correction of the soil moisture state and antecedent precipitation index (API). The dual-correction data assimilation has been proved to be more effective in enhancing river discharge when compared to soil-moisture-only updating. The *Crow and Ryu* [2009] study was carried out using model-generated synthetic soil moisture data, and hence it would be important to test the approach using real satellite soil moisture retrievals.

In the current study, a dual updating experiment of soil moisture state and total channel inflow (TCI) of the SAC-SMA model [*Burnash et al.*, 1973] using ensemble Kalman filter (EnKF) was performed. It is analogous to the *Crow and Ryu* [2009] work in that dual updating of two separate states using sequential data assimilation was carried out in both cases, but also with the following important differences. First, the current study used observed satellite soil moisture retrievals, while the previous one used synthetic soil moisture. Second, in the current work rainfall was not directly corrected, but instead the TCI (the aggregate flow component) of the model was corrected using the satellite soil moisture observations.

The objective of our work is to evaluate the added skill of river discharge due to assimilation of satellite soil moisture compared to the open-loop (without data assimilation) simulation. Ground-based river discharge measurement was used as a reference to evaluate the results of data assimilation and the open-loop simulations. The study aims at answering the following scientific questions. How do active and passive microwave satellite soil moisture retrievals improve the river discharge when assimilated into rainfall-runoff model? How does the degree of discharge improvement differ between the sur-

face and root-zone satellite soil moisture? Do the dual soil moisture state and the TCI updating of the SAC-SMA model perform better than soil-moisture-only update? The remainder of this work is organized as follows. Section 19.2 presents the study area, hydrologic model, and description of satellite soil moisture retrievals. Data matching and data assimilation methods are described in section 19.3. Results are presented in section 19.4, and finally, conclusions are summarized in section 19.5.

19.2. STUDY AREA, MODEL, AND DATA

19.2.1. Study Area

The study area is the Greens Bayou Watershed (shown in Figure 19.1a) located in Harris County, Texas. The basin has a drainage area of $178\,km^2$ with the outlet [the U.S. Geological Survey (USGS) site ID of 08076000] located at about $17\,km$ northeast of the city of Huston, Texas. It is characterized by humid subtropical climate receiving mean annual rainfall of $1300\,mm$ [*Hirpa*, 2013]. The land use and land cover extracted from the National Land Cover Data 2006 reveals that most of the basin is developed with low and medium intensity (Figure 19.1b). The basin has an estimated impervious area of 28.8% [*Kuzmin et al.*, 2008]. Only a small proportion of the area is covered with forest, which presents a good opportunity to evaluate the satellite soil moisture with minimal influence of vegetation on the microwave signal.

Flooding in the basin occurred numerous times in the last few decades, mainly brought on by tropical storms. The floods of June 2001, November 2003, and October 2006 (all from last the two decades) are among the top 10 historical flooding events ever recorded in the Greens Bayou Basin. The most extreme flood was caused by tropical storm Alison in June 2001, following more than $650\,mm$ of rainfall in 10 hours between 8 and 9 June (http://www.nws.noaa.gov/os/assessments/pdfs/allison.pdf). Greens Bayou is one of several basins under the West Gulf River Forecast Center (WGRFC) selected for data assimilation experiment [*Seo et al.*, 2009].

19.2.2. Hydrologic Model and Inputs

The Sacramento soil moisture accounting model [SAC-SMA, *Burnash et al.*, 1973] was used for the streamflow simulation. The SAC-SMA is used by the National Weather Service (NWS) for river forecasting for the United States. It is a physically based conceptual model with two vertical soil zones: the upper zone representing the short-term surface soil and interception storage and the lower

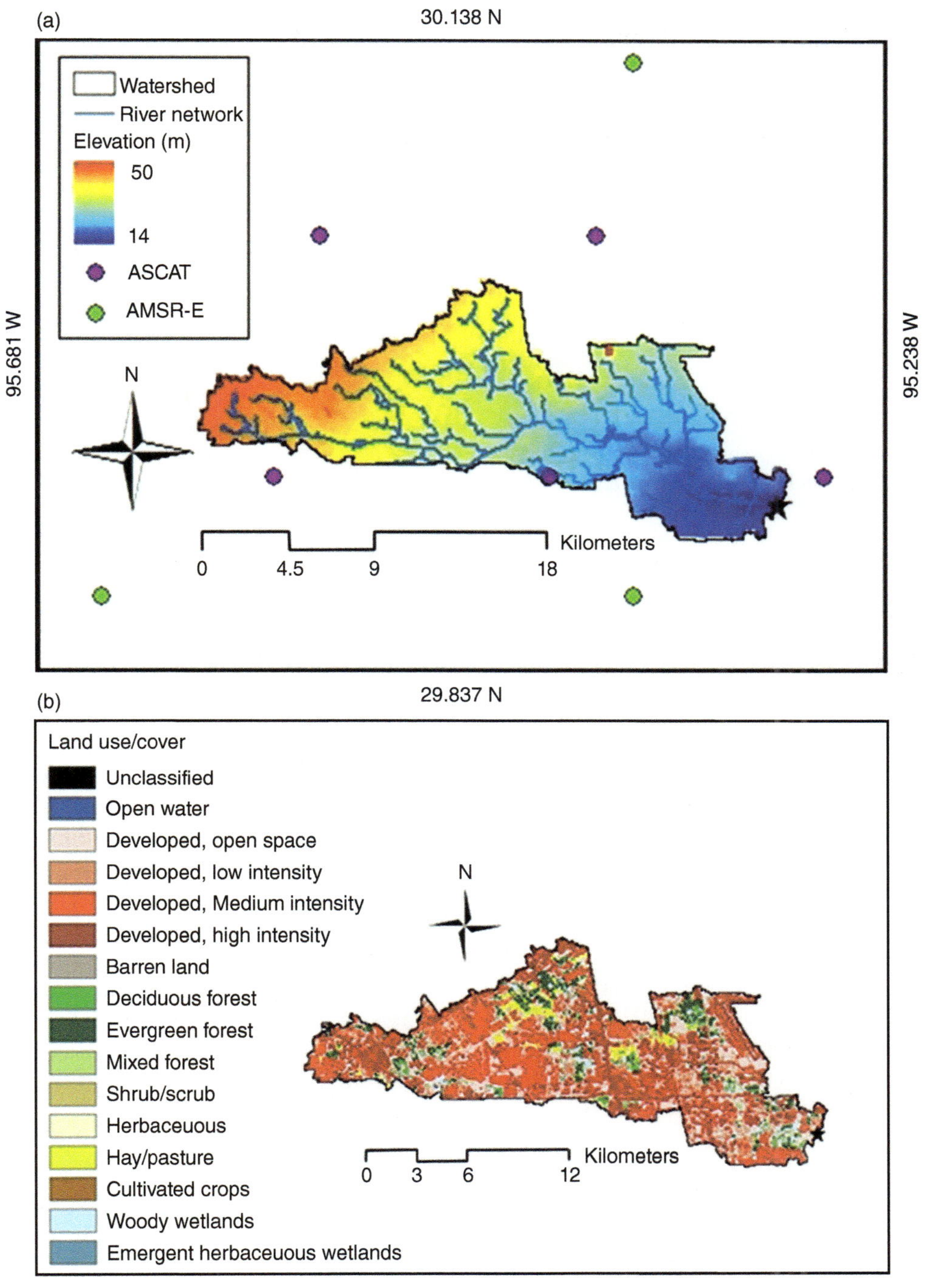

Figure 19.1 Greens Bayou Basin (area = 178 km²) located in eastern Texas. (a) center of satellite soil moisture grids for AMSR-E (three green dots) and ASCAT (five purple dots. (b) Land use/cover of the basin extracted from national land cover data (NLCD 2006) [*Fry et al.*, 2011].

zone representing the deeper soil moisture and longer groundwater storages (see Figure 19.2 for a schematic diagram). The zones have free water and tension water elements, where the free water (fast flow component) is dominantly driven by gravitation forces but may also be depleted by evapotranspiration, percolation, and horizontal flow, while the tension water (slow flow component) is driven by evapotranspiraion and diffusion.

The model has six soil moisture states and four flow components. The six soil moisture states are the upper zone tension water content (UZTWC), the upper zone free water content (UZFWC), the lower zone tension

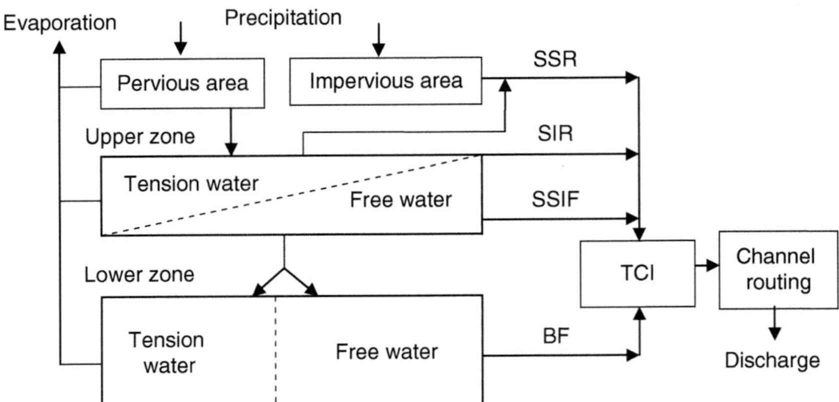

Figure 19.2 Schematic diagram of the SAC-SMA model with two soil zones (upper and lower zone) and four flow components. The flow components (SSR, SIR, SSIF, and BF) are aggregated to produce the total channel inflow (TCI). River discharge is generated after channel routing of TCI using unit hydrograph.

water content (LZTWC), the lower zone free primary content (LZFPC), the lower zone free supplemental content (LZFSC), and the total tension water content including impervious area (ADIMC). The flow components are related to the soil moisture states as follows. The surface saturation-excess runoff (SSR) takes place once the tension water capacity of the soil is filled, and thus it is a function of ADIMC. The surface infiltration-excess runoff (SIR) occurs when the rainfall rate exceeds the infiltration capacity of the upper zone (UZTWC and UZFWC). The subsurface interflow (SSIF) is a function of the UZFWC. And the base flow (BF) is generated from the free water contents of the lower zone (LZFPC and LZFSC). The aggregation of the four flow components produces the total channel inflow (TCI) as indicated in Figure 19.2.

The SAC-SMA has 16 calibrated parameters, and the empirical unit hydrograph method is used for channel flow routing. For the study basin, the model parameters and empirical unit hydrograph were estimated by *Seo et al.* [2009] using Adjoint-Based OPTimizer [AB_OPT; see also *Kuzmin et al.*, 2008; *Koren et al.*, 2008]. The calibrated unit hydrograph [*Seo et al.*, 2009] has a width of 32 with time to peak of 5h. The model is run at hourly time steps with mean areal precipitation and potential evapotranspiration as inputs, while channel inflow and actual evapotranspiration are model outputs. The rainfall input was hourly precipitation estimates produced by the multisensor precipitation estimator (MPE) from real-time radar. The spatially nonuniform bias in the hourly real-time radar measurement was corrected using information from rain gauge data [*Seo and Breidenbach*, 2002]. The climatological mean areal potential evaporation is estimated from the Penman equation [*Penman*, 1948] and then adjusted by the normalized vegetation difference index (NVDI) climatology [*Koren et al.*, 1998]. The study

period covers four calendar years from 2007 to 2010. The satellite soil moisture retrievals (presented in the following sections) are all available during this period.

19.2.3. AMSR-E Soil Moisture Data

The soil moisture estimates are based on brightness temperature observed from the AMSR-E sensor onboard the *Aqua* satellite. *Aqua* follows a polar sun-synchronous orbit crossing the equator at approximately 1:30 and 13:30 local time for descending (nighttime) and ascending (daytime) overpasses, respectively. The AMSR-E is a passive-microwave radiometer that measures brightness temperature at six different frequencies (6.9, 10.7, 18.7, 23.8, 36.5, and 89.0 GHz) for both vertical and horizontal polarizations [*Njoku et al.*, 2003]. Several different algorithms have been developed to retrieve surface soil moisture from the AMSR-E brightness temperature from microwave channels. The two commonly used retrievals are the official AMSR-E soil moisture product developed at NASA [*Njoku et al.*, 2003; *Njoku*, 2004], and the Land Parameter Retrieval Model (LPRM) AMSR-E soil moisture [*Owe et al.*, 2008] developed by the collaboration of the VU University Amsterdam and NASA.

The *Njoku et al.* [2003] retrieval method uses normalized polarization ratios from AMSR-E brightness temperature in order to minimize the effect of surface temperature on soil moisture signal. Then the soil moisture value for each grid cell is estimated using a long-term deviation from monthly mean polarization ratio. The LPRM [*Owe et al.*, 2008] soil moisture is based on a forward modeling optimization approach to solve microwave radiative transfer equation at three different frequencies (C-, X-, and K-bands). The AMRS-E based LPRM land surface soil moisture and vegetation optical depth retrievals at C- (6.9 GHz) and X- (10.7 GHz) bands, and land surface temperature at K-band (37GHz) are globally available from June 2002 to October 2011. Deeper root-zone soil moisture was derived by assimilating the surface LPRM soil moisture from descending orbit into a two-layer Palmer water balance model using one-dimensional ensemble Kalman filter [*USDA*, 2012]. The global daily 0.25° gridded LPRM root-zone soil moisture was produced at NASA for the time period from June 2002 to December 2010.

Evaluation studies carried out on the AMSR-E retrievals indicate that the LPRM AMSR-E products show better correlation when compared with in situ soil moisture [e.g., *Wagner et al.*, 2007; *Draper et al.*, 2009; *Jackson et al.*, 2010; *Brocca et al.*, 2011] compared to the *Njoku et al.* [2003] retrievals. Hence, only the LPRM AMSR-E surface soil moisture retrieval was used in this study, and for ease of terms, we refer to the LPRM AMSR-E simply as AMSR-E in the reminder of the chapter. The X-band level 3 ascending surface soil moisture (AMSR-E A-SSM), descending surface soil moisture (AMSR-E D-SSM) and, root-zone soil moisture (AMSR-E D-RZ) from 4 years (2007–2010) were assimilated into the SAC-SMA model separately.

19.2.4. ASCAT Soil Moisture Data

The Advanced SCATterometer (ASCAT) soil moisture estimates are based on radar backscatter (active microwave at C-band) from sensor onboard EUMETSAT's *MetOp* satellite launched in 2007. *MetOp* follows a sun-synchronous orbit crossing the equator at approximately 9:30 and 21:30 local time for descending and ascending overpasses, respectively. The backscatter coefficients are converted to surface soil moisture using empirical change detection method developed at Vienna University of Technology by *Wagner et al.* [1999, 2007]. The method assumes that the highest long-term backscattering coefficient values correspond to the wettest soil conditions, while the lowest backscattering coefficients correspond to the driest soil condition. Based on this assumption, the relative soil moisture data in the form of degree of saturation ranging from 0% to 100% are provided as surface soil moisture data. Detail description of the ASCAT soil moisture used in this study can also be found in *Bartalis et al.* [2007].

Soil water index (SWI) data are derived from the ASCAT surface soil moisture using a two-layer model representing upper and lower soil layers, where the soil profile for a soil depth of 1 m was estimated from the surface soil moisture using an exponential filter [*Wagner et al.*, 1999]. The exponential filter has one parameter referred to as characteristic time length (T) that controls the temporal variation of the SWI. The data is produced at eight different characteristic values ranging from 1 to

100 days, with 1 day and 100 days representing the highest and lowest temporal variations, respectively. In this study, after comparing the temporal variability of the SWI time series with that of the SAC-SMA simulations, the data corresponding to $T = 1$ was chosen for the assimilation. The SWI is denoted as root-zone soil moisture in the reminder of this chapter. The ASCAT surface and root-zone soil moisture retrievals from four 4 (2007–2010) were separately assimilated into the SAC-SMA model.

19.3. METHOD

19.3.1. Cumulative Distribution Function Mapping

There is discrepancy between the soil moisture fields from the SAC-SMA model and satellite retrievals in terms of spatial scales, temporal resolution, and the soil moisture extent. The model has spatially lumped soil moisture accumulation (mm) with values ranging from zero (completely dry) to maximum capacity (fully saturated) determined by the model parameters based on the soil properties (e.g., porosity). In contrast, the AMSR-E soil moisture provides gridded volumetric soil moisture (m^3/m^3) at mostly daily and 25km resolutions. The ASCAT product is expressed in terms of the degree of saturation ranging from 0% to 100% at 12.5km and generally daily scales for the study basin. To overcome this lack of correspondence among the data sets, first the satellite retrievals were spatially aggregated to the basin scale to match the SAC-SMA model. Then they are transformed from their native values (i.e., volumetric soil moisture or saturation percentage) into the SAC-SMA model (soil moisture accumulation, mm) using the cumulative distribution function (CDF) matching method. The CDF matching, also sometimes referred to as quantile-to-quantile mapping, has been widely used for data matching purposes in several applications [e.g., *Reichle and Koster*, 2004; *Lee and Anagnostou*, 2004; *Hopson and Webster*, 2010; *Liu et al.*, 2011; *Brocca et al.*, 2011; *Draper et al.*, 2012].

19.3.2. Data Assimilation experiment

Two separate data assimilation (DA) experiments were carried out using each satellite soil moisture estimate: (1) soil moisture updating (SMU) and (2) dual soil moisture and total channel inflow (TCI) updating (STU). The relationship between soil moisture states and river discharge (and the TCI) is highly nonlinear, especially for the upper soil layer that controls the dominant proportion of runoff for the study basin [*Hirpa*, 2013]. Hence, there is a limit to translating accurate initialization of the soil moisture into equivalently better river discharge estimation, in particular after the soil reaches saturation and the surface runoff becomes a direct result of rainfall. In this case, soil moisture state updating does not necessarily yield a significant improvement in runoff accuracy, and, therefore, additional flux (e.g., rainfall or more) correction could also be considered. *Crow and Ryu* [2009] presented such a case where they updated the combined soil moisture state and antecedent precipitation index (API), and reported that the dual-API-soil-moisture correction produced the best result when the runoff root-mean-squared error (RMSE) is considered. In the current study, the soil moisture and TCI dual assimilation experiment (STU) was carried out. The direct rainfall correction of the *Crow and Ryu* [2009] was replaced by the TCI update because the most accurate rainfall from ground observation was used as input to the SAC-SMA model.

The EnKF, [*Evenson*, 1994, 2003] was implemented into the SAC-SMA model for the data assimilation of the satellite soil moisture retrievals. In the EnKF, finite number ($N = 100$ in the present case) of ensemble members of the model state (X_t) are generated at the start of the model run, and the hydrologic model is run forward to predict the state ensemble at the next time [see equation (19.1)].

$$X_t = \left[x_t^1, x_t^2, \ldots, x_t^{N-1}, x_t^N \right] \overset{\text{model run}}{\to} X_{t+1}$$
$$= \left[x_{t+1}^1, x_{t+1}^2, \ldots, x_{t+1}^{N-1}, x_{t+1}^N \right], \tag{19.1}$$

where x_t^i is the prior estimates of ith ensemble member of the model state at time t; and N is the number of ensembles. The ensemble represents the distribution of the state, and the ensemble mean is the best state estimate ("truth"). The prediction spread is computed from the covariance of the ensemble [as in equation (19.2)]:

$$P_{t+1} = \text{cov}\left(X_{t+1} \right). \tag{19.2}$$

Whenever satellite soil moisture is available, then each state ensemble member is updated by additive value proportional to the difference (innovation) between predicted state and satellite soil moisture estimates as follows:

$$X_{t+1}^+ = X_{t+1} + K_{t+1}\left[Y_{t+1} - HX_{t+1} \right]. \tag{19.3}$$

The + superscript on the left-hand side indicates the posterior state (after the update), Y is satellite soil moisture estimate, H is linearized measurement model (identity matrix in this case), and K is Kalman gain computed as

$$K_{t+1} = P_{t+1}H^{\text{T}}\left[HP_{t+1}H^{\text{T}} + R_{t+1} \right]^{-1}, \tag{19.4}$$

where R is the observation error covariance matrix, and other terms are as described above.

In this study, the model states (X_t) are defined as the soil moisture state (vector of six elements) in the SMU experiment; and the combination of soil moisture states and the TCI (vector of size 7) for the STU experiment. The model parameters and the empirical unit hydrograph were calibrated offline and then assumed to be time invariant; hence they are not included in the dynamic state space update. The AMSR-E and ASCAT retrievals have been reported to have comparable error estimates [e.g., see *Draper et al.*, 2012]. Following the *Draper et al.*, [2012] work, the same error covariance (R) was assumed for both soil moisture retrievals.

The following is the summary of the data assimilation experiments. For the soil moisture updating (SMU), the CDF-mapped satellite soil moisture retrievals were separately assimilated into the SAC-SMA model for updating the six soil moisture states. The five data sets from satellite retrievals (three from AMSR-E and two from ASCAT) were independently assimilated using the ensemble Kalman filter. For the dual assimilation (STU), however, in parallel with the soil moisture states, the TCI was also modified using the assimilation of the satellite-based data sets. The STU could be more effective in improving the discharge simulation due to the dual correction of the TCI (flux) and the soil moisture (state) at the assimilation time.

19.4. RESULTS

All five satellite soil moisture estimates were mapped to the SAC-SMA model space, accumulations (mm), over the study period of 2007–2010 using the CDF matching method. Figure 19.3 presents the time series plot of the upper zone and the lower zone for a selected year 2009 (similar pattern was found for other years too). All remotely sensed soil moisture retrievals have been mapped to each of the six soil moisture states and the TCI, but for brevity of presentation only selected plots are shown in Figure 19.3. The upper zone (Figure 19.3a) from the SAC-SMA model and the mapped surface soil moistures (SSM) from the ASCAT, the descending and ascending AMSR-E correspond reasonably well for the year, except for the winter months (January and February). The upper layer moisture patterns are fairly similar for spring and summer months and the remotely sensed moisture could replicate the general model dynamics. However, there was less agreement for the winter months between surface soil moisture from the model and remotely sensing. The transform from the AMSR-E retrievals overestimate the soil moisture compared to the SAC-SMA model, while the ASCAT transform showed more temporal variations. In the lower zone (Figure 19.3b), both the satellite-based root-zone (RZ) soil moisture estimates detect the driest moisture during summer season, and the overall moisture dynamics was in good agreement with the SAC-SMA model.

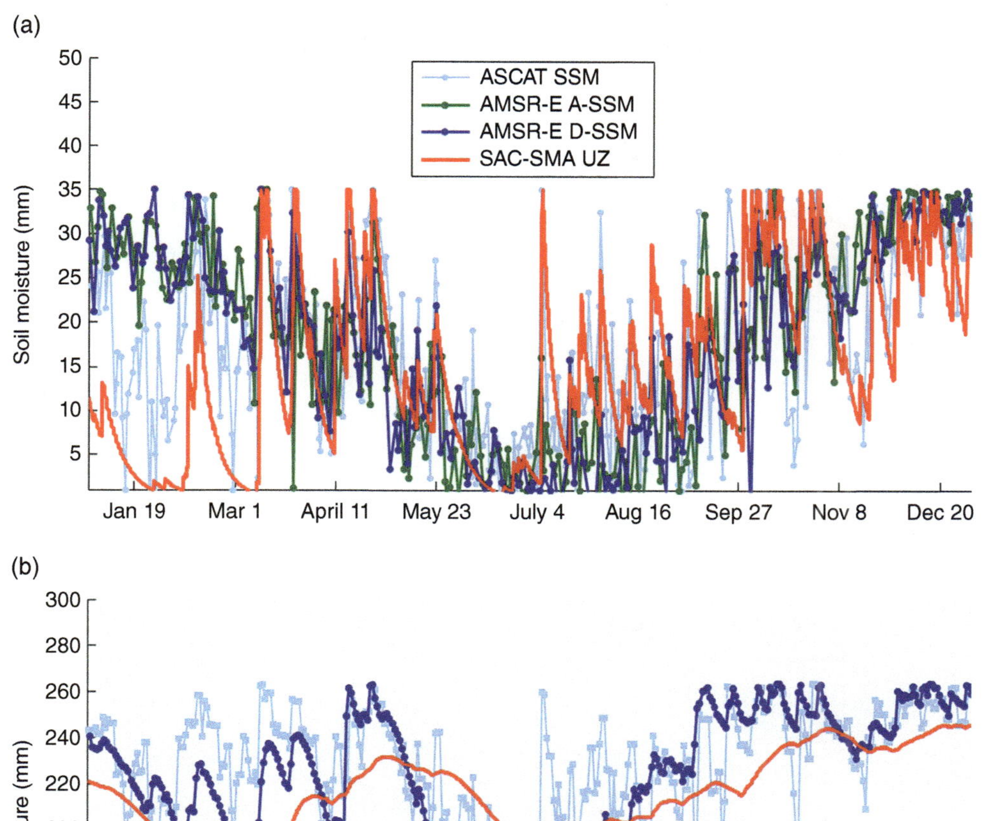

Figure 19.3 The AMSR-E, ASCAT, and SAC-SMA simulated soil moisture for (a) upper and (b) lower soil zones as defined by the SAC-SMA model for 2009. The satellite soil moisture retrievals are mapped to SAC-SMA model space using CDF matching approach. The surface soil moisture (SSM) for the upper zone and the root-zone (RZ) soil moisture for the lower zone are shown.

19.4.1. AMSR-E Assimilation

The AMSR-E surface soil moisture from descending orbit (AMSR-E D-SSM), the AMSR-E surface soil moisture from ascending orbit (AMSR-E A-SSM), and the AMSR-E root zone from descending orbit (AMSR-E D-RZ) were separately assimilated to the SAC-SMA model over a period of 4 years (2007–2010) using EnKF. Each of these retrievals were mapped to the six model soil moisture states and the TCI using the CDF matching before assimilation experiment was carried out. Figure 19.4 compares the RMSE of river discharge, computed with reference to in situ discharge, from the model without data assimilation (open loop), soil moisture

update (SMU) and dual soil moisture and TCI update (STU) using the three AMSR-E retrievals. The discharge RMSE for soil-moisture-only update was lower than that of the open-loop model, and the dual-assimilation discharge RMSE was the lowest compared to open loop and SMU for all three AMSR-E satellite retrievals. This indicates that (1) assimilation of AMSR-E surface and root-zone soil moisture into SAC-SMA hydrologic model improves the river discharge simulation, and (2) dual soil moisture and TCI update (STU) for SC-SMA model is better for discharge simulation compared to soil moisture state update only (SMU). This finding is consistent with the results of the synthetic experiment of *Crow and Ryu* [2009] where they also showed that the dual soil moisture/

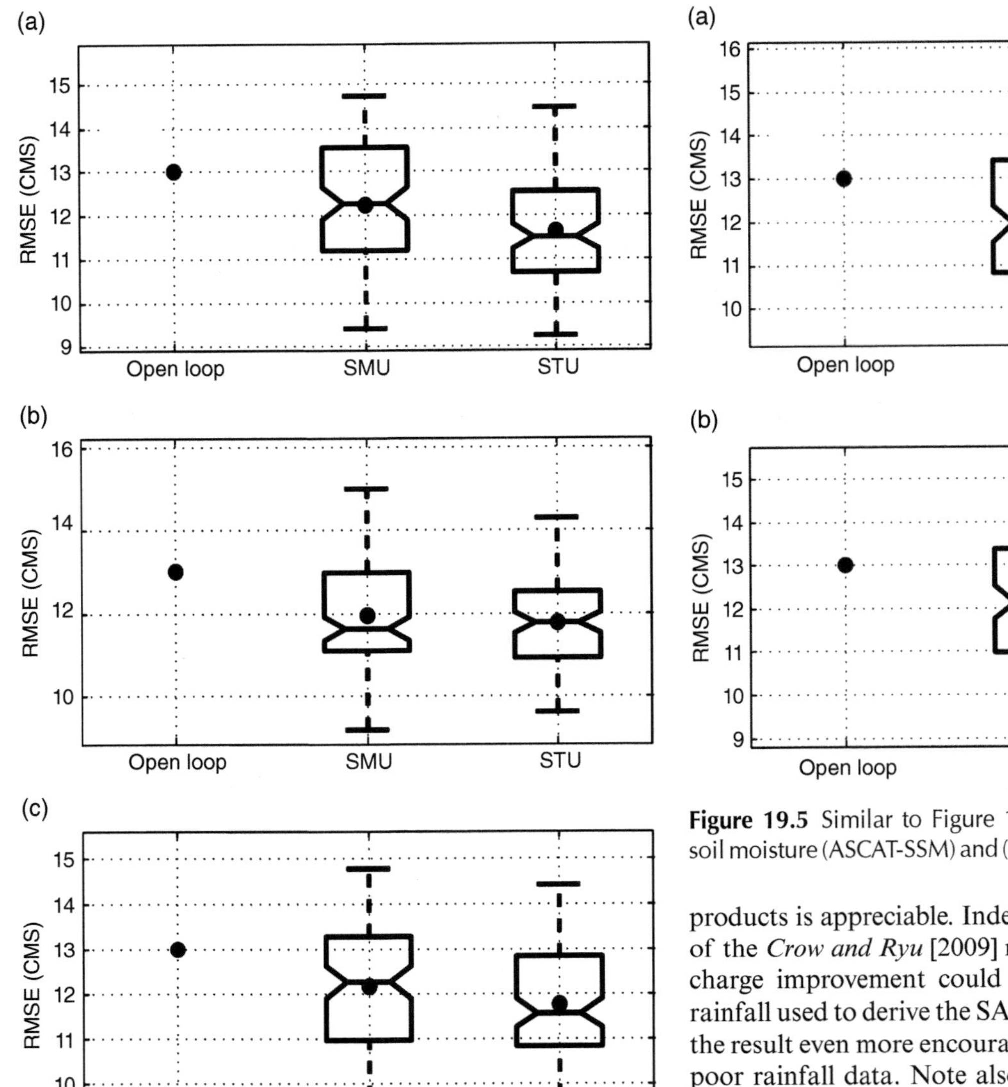

Figure 19.4 River discharge improvement from assimilation of AMSR-E soil moisture for the soil moisture updating (SMU) and soil moisture and TCI updating (STU). The RMSE are shown for (a) AMSR-E surface soil moisture from descending orbit (AMSR-E D-SSM), (b) AMSR-E surface soil moisture from ascending orbit (AMSR-E A-SSM), and (c) AMSR-E root-zone soil moisture from descending orbit (AMSR-E D-RZ). The shaded circles denote the ensemble mean for each case. The RMSE 25th, 50th and 75th percentiled are indicated in the box plots.

rainfall correction produced better discharge compared to open-loop and soil-moisture-only update. Given that the accuracy of open-loop discharge was boosted by the highly accurate ground-based rainfall input in our study, the slender but consistent RMSE enhancement (just under 10%) over the open loop after using the AMSR-E

Figure 19.5 Similar to Figure 19.4, except (a) ASCAT surface soil moisture (ASCAT-SSM) and (b) ASCAT root zone (ASCAT-RZ).

products is appreciable. Indeed, the synthetic experiment of the *Crow and Ryu* [2009] revealed that only minor discharge improvement could be achieved when accurate rainfall used to derive the SAC-SMA model, which makes the result even more encouraging especially for areas with poor rainfall data. Note also that the SAC-SMA model here is run at hourly time scale while the satellite soil moisture is currently only available at most on a daily basis, and we believe that more discharge improvement could be achieved with more frequent moisture data.

19.4.2. ASCAT Assimilation

The ASCAT data assimilation was carried out for surface (ASCAT-SSM) and root-zone (ASCAT-RZ) soil moisture using the EnKF method. Both surface and root-zone ASCAT were independently mapped to the six model soil moisture states, and the TCI using the CDF matching before assimilation experiment was performed. The discharge RMSE results for open loop, SMU, and STU are shown in Figure 19.5. Similar to the AMSR-E above, the discharge RMSE improved with ASCAT assimilation into the SAC-SMA model, and the best result was obtained with dual soil moisture and TCI updating (STU) for both surface and root zone. These results were consistent with the AMSR-E findings,

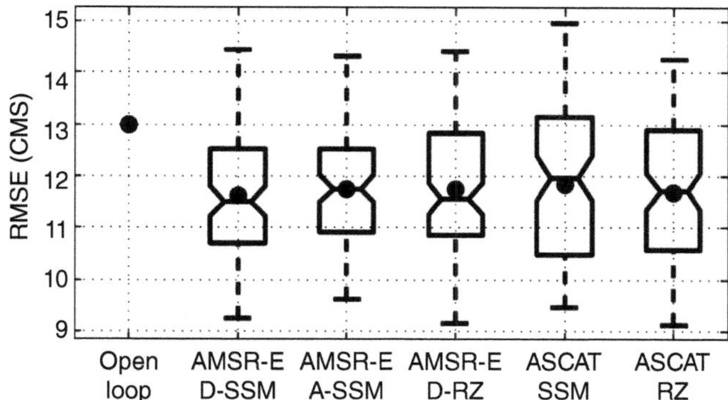

Figure 19.6 Comparison of the discharge RMSE after the soil moisture state and TCI updates using different satellite soil moisture retrievals. Other details are similar to Figure 19.4.

regardless of the surface or root-zone retrievals. Unlike the previous study by *Brocca et al.* [2012], there was no appreciable difference in discharge enhancement found between the surface and root-zone ASCAT assimilation. The previous study reported that the root-zone assimilation produced better correspondence with in situ discharge compared to the surface soil moisture over basins in Italy. It is not entirely clear as to why different results were obtained, but we suspect that it could be due to a mismatch in optimal characteristic time length used in the exponential filter to create the root-zone soil moisture from the surface. Regardless, the ASCAT soil moisture retrievals (SSM and RZ) clearly improved the river discharge and showed encouraging results for future active microwave soil moisture missions with higher spatial and temporal resolutions.

19.4.3. Comparison of the STU

In this section, the discharge skills of the dual soil moisture and TCI update (STU) for each of the five satellite soil moisture retrievals were compared. Figure 19.6 shows the discharge RMSE computed from assimilation of each soil moisture data set along with the open loop. The mean of the ensemble RMSE for AMSR-E D-SSM is slightly lower than all others and ASCAT-SSM has marginally higher RMSE. However, the RMSE difference among the satellite assimilated was inconsiderable compared to the improvement from the open loop. Furthermore, the spread of the RMSE was found to be wider for ASCAT-SSM, possibly owing to the more temporal variability observed in the ASCAT time series (compared in Figure 19.3). The close similarity of the AMSR-E and ASCAT was in agreement with the reported findings of the *Draper et al.* [2012] assimilation into the NASA catchment land surface model. The previous study recommended that joint assimilation of the AMSR-E and ASCAT into the NASA catchment land surface

model provides better accuracy and converge for soil moisture, but whether the combined assimilation also results in improved discharge remains to be confirmed. Overall, the value of soil moisture data assimilation for river discharge simulation was demonstrated and sheds positive light on future soil moisture missions with higher spatial and temporal resolutions, such as Soil Moisture Active Passive [SMAP; *Entekhabi et al.*, 2010].

19.5. CONCLUSIONS

Surface and root-zone satellite soil moisture retrievals from passive and active microwave sensors were separately assimilated to the SAC-SMA hydrologic model, and their contribution to river discharge improvement was evaluated. The passive microwave sensor is the AMSR-E radiometer onboard the *Aqua* satellite, and the active microwave sensor is the ASCAT onboard the *MetOp* satellite. A CDF matching approach was applied to remove the mismatch between the remotely sensed and model soil moisture and to convert the satellite retrievals to the model space (accumulation in mm). Two separate data assimilation experiments were carried out and compared: soil moisture state updating and dual soil moisture and total channel inflow correction.

The major conclusions are summarized as follows:
• All the AMSR-E and ASCAT data sets considered improved the river discharge compared to the open loop (without data assimilation).
• The best result was obtained when dual soil moisture and TCI (STU) update was carried out, compared to soil-moisture-only state update (SMU).
• There was no major difference between surface and root-zone soil moisture retrievals in the degree of the discharge enhancement. This is the case for both AMSR-E and ASCAT data sets.
• Both passive and active microwave retrievals produced similar discharge results.

ACKNOWLEDGMENTS

This research is supported by the National Aeronautics and Space Administration (NASA) Earth and Space Science Fellowship (NESSF) awarded to Feyera A. Hirpa. He is grateful for the support.

REFERENCES

Albergel, C., C. Rüdiger, D. Carrer, J. C. Calvet, N. Fritz, V. Naeimi, Z. Bartalis, and S. Hasenauer (2009), An evaluation of ASCAT surface soil moisture products with in-situ observations in southwestern France, *Hydrol. Earth Syst. Sci., 13*, 115–124.

Albergel, C., et al. (2010), Cross-evaluation of modelled and remotely sensed surface soil moisture with in situ data in southwestern France, *Hydrol. Earth Syst. Sci.*, *14*, 2177–2191.

Bartalis, Z., W. Wagner, V. Naeimi, S. Hasenauer, K. Scipal, H. Bonekamp, J. Figa, and C. Anderson (2007), Initial soil moisture retrievals from the METOP-A Advanced Scatterometer (ASCAT), *Geophys. Res. Lett.*, *34*, L20401, doi:10.1029/2007GL031088.

Bolten, J. D., W. T. Crow, X. Zhan, T. J. Jackson, and C. A. Reynolds (2010), Evaluating the utility of remotely sensed soil moisture retrievals for operational agricultural drought monitoring, *IEEE J. S. A. Earth Obs. Remote Sens.*, *3*(1), 57–66.

Brocca, L., F. Melone, T. Moramarco, W. Wagner, V. Naeimi, Z. Bartalis, and S. Hasenauer (2010), Improving runoff prediction through the assimilation of the ASCAT soil moisture product, *Hydrol. Earth. Syst. Sci.*, *14*, 1881–1893, doi:10.5194/hess-14-1881-2010.

Brocca, L., S. et al. (2011), Soil moisture estimation through ASCAT and AMSR-E sensors: An intercomparison and validation study across Europe, *Remote Sens. Environ.*, *115*, 3390–3408, doi:10.1016/j.rse.2011.08.003.

Brocca, L., T. Moramarco, F. Melone, W. Wagner, S. Hasenauer, and S. Hahn (2012), Assimilation of surface and root zone ASCAT soil moisture products into rainfall-runoff modeling, *IEEE Trans. Geosci. Remote Sens.*, *50*, 2534–2555, doi:10.1109/TGRS.2011.2177468.

Burnash, R. J. C, R. L. Ferral, and R. A. McGuire (1973), A generalized streamflow simulation system: Conceptual modeling for digital computers, Tech. Rep., U.S. Natl. Weather Serv. and Calif. Dept. of Water Resour., Sacramento, Calif.

Crow, W. T., and D. Ryu (2009), A new data assimilation approach for improving runoff prediction using remotely-sensed soil moisture retrievals, *Hydrol. Earth Syst. Sci.*, *13*, 1–16, doi:10.5194/hess-13-1-2009.

Crow, W. T., R. Bindlish, and T. J. Jackson (2005), The added value of space borne passive microwave soil moisture retrievals for forecasting rainfall-runoff ratio partitioning, *Geophys. Res. Lett.*, *32*, L18401, doi:10.1029/2005GL023543.

Dingman, S. L. (1994), *Physical Hydrology*, Macmillan, New York.

Dirmeyer, P. A., C. A. Schlosser, and K. L. Brubaker (2009), Precipitation, recycling and land memory: An integrated analysis, *J. Hydrometeorol*, *10*, 278–288, doi:10.1175/2008JHM1016.1.

Draper, C. S., J. P Walker, P. Steinle, R. A. M. De Jeu, and T. R. H. Holmes (2009), An evaluation of AMSR-E derived soil moisture over Australia, *Remote. Sens. Environ.*, *113* (4), 703–710, doi:10.1016/j.rse.2008.11.011.

Draper, C. S., J. F. Mahfouf, J. C. Calvet, E. Martin, and W. Wagner (2011), Assimilation of ASCAT near-surface soil moisture into the SIM hydrological model over France, *Hydrol. Earth Syst. Sci.*, *15*, 3829–3841, doi:10.5194/hess-15-3829-2011.

Draper, C. S., R. H. Reichle, G. J. M. De Lannoy, and Q. Liu (2012), Assimilation of passive and active microwave soil moisture retrievals, *Geophys. Res. Lett.*, *39*, L04401, doi:10.1029/2011GL050655.

Dunne, T., T. R. Moore, and C. H. Taylor (1975), Recognition and prediction of runoff-producing zones in humid regions, *Hydrol. Sci. Bull.*, *20*, 305–327.

Entekhabi, D., et al. (2010), The Soil Moisture Active and Passive (SMAP) mission, *Proc. IEEE*, *98*, 704–716.

Entin, J. K., A. Robock, K. Y. Vinnikov, V. Zabelin, S. Liu, A. Namkhai, and T. Adyasuren (1999), Evaluation of Global Soil Wetness Project soil moisture simulations, *J. Meteorol. Soc. Jpn.*, *77*, 183–198.

Evensen, G. (2003), The ensemble Klaman filter: Theoretical formulation and practical implementation, *Ocean dy.*, *53*, 343–367.

Evensen, G. (1994), Sequential data assimilation with a nonlinear quasi-geographic model using Monte Carlo methods to forecast error statistics, *J. Geophys. Res.*, *99*, 10,143–10,162.

Fry, J., G. Xian, S. Jin, J. Dewitz, C. Homer, L. Yang, C. Barnes, N. Herold and J. Wickham (2011), Completion of the 2006 National Land Cover Database for the Conterminous United States, PE&RS, *77*(9), 858–864.

Hain, C. R., W. T. Crow, J. R. Mecikalski, M. C. Anderson, and T. Holmes (2011), An intercomparison of available soil moisture estimates from thermal infrared and passive microwave remote sensing and land surface modeling, *J. Geophys. Res. Let.*, *116*, D15107, doi:10.1029/2011JD015633.

Hirpa, F. A, (2013), Hydrologic Data Assimilation for Operational Streamflow Forecasting, *Doctoral Dissertation*, University of Connecticut.

Hopson, T. M, and P. J., Webster (2010), A 1–10-day ensemble forecasting scheme for the major river basins of Bangladesh: Forecasting severe floods of 2003–07, *J. Hydrometeorol.*, *11*, 618–641, doi:10.1175/2009JHM1006.1.

Jackson, T. J., M. H. Cosh, R. Bindlish, P. J. Starks, D. D. Bosch, M. Seyfried, D. C. Goodrich, M. S. Moran, and J. Du (2010), Validation of advanced microwave scanning radiometer soil moisture products, *IEEE Trans. Geosci. Remote Sens.*, *48*(12), 4256–4272.

Kim, Y., and G. Wang (2007), Impact of initial soil moisture anomalies on subsequent precipitation over NorthAmerica in the coupled land–atmosphere model CAM3–CLM3, *J. Hydrometeor.*, *8*, 513–533.

Koren, V., J. Schaake, Q. Duan, M. Smith, and S. Cong (1998), ET upgrades to NWSRFC-Project plan, HRL Internal Rep., Hydrologic Laboratory, Office of hydrologic Development, Silver Spring, Md.

Koren, V, F. Moreda, and M. Smith (2008), Use of soil moisture observations to improve parameter consistency in watershed calibration, *Phys. Chem. Earth. PartsA/B/C,* *33*(17–18), 1068–1080.

Koster, R. D. et al. (2004), Regions of strong coupling between soil moisture and precipitation, *Science*, *305*, 1138–1140.

Koster, R. D., S. P. P. Mahanama, B. Livneh, D. P. Lettenmaier, and R. H. Reichle (2010), Skill in streamflow forecasts derived from large-scale estimates of soil moisture and snow, *Nat. Geosci. 3*, 613–616, doi:10.1038/NGEO944.

Kuzmin, V., D. J. Seo, and V. Koren (2008), Fast and efficient optimization of hydrologic model parameters using a priori estimates and stepwise line search, *J. Hydrol.*, *353*(1–2), 109–128.

Lee, K. H., and E. N. Anagnostou (2004), A combined passive/active microwave remote sensing approach for surface variable retrieval using Tropical Rainfall Measuring Mission observations, *Remote Sens. Environ.*, *92*, 112–125, doi:10.1016/j.rse.2004.05.003.

Liu, Y. Y., R. M. Parinussa, W. A. Dorigo, R. A. M. De Jeu, W. Wagner, A. I. J. M. van Dijk, M. F. McCabe, and J. P. Evans (2011), Developing an improved soil moisture dataset by blending passive and active microwave satellite-based retrievals, *Hydrol. Earth Syst. Sci.*, *15*, 425–436, doi:10.5194/hess-15-425-2011.

Lu, H., T. Koike, H. Fujii, N. Hirose, and K. Tamagawa (2005), A radiative transfer model and an algorithm for soil moisture including very dry conditions, *Proc. IEEE Int. Geosci. Remote Sens. Symp. 2*, 1131–1134.

Mei, R., and G. Wang (2011), Impact of sea surface temperature and soil moisture on summer precipitation in the United States based on observational data, *J. Hydrometeor.*, *12*, 1086–1099.

National Research Council (NRC) (2007), Earth science and applications from space: National imperatives for the next decade and beyond, Committee on Earth Science and Applications from Space: A community assessment and strategy for the future, Natl. Acad. Press, Washington, D.C.

Njoku, E. (2004), AMSR-E/Aqua daily L3 surface soil moisture, V001, Natl. snow and Ice Data Center, Boulder, Colo.

Njoku, E. G., T. J. Jackson, V. Lakshmi, T. K. Chan, and S. V. Nghiem (2003), Soil moisture retrieval from AMSR-E, *IEEE Trans. Geosci. Remote*, *41*, 215–229, doi:10.1109/TGRS.2002.808243.

Owe, M., R. De Jeu, and T. Holmes (2008), Multisensor historical climatology of satellite-derived global land surface moisture, *J. Geophys. Res.*, *113*, F01002, doi:10.1029/2007JF000769.

Parajka, J., V. Naeimi, G. Bloschl, W. Wagner, R. Merz, and K. Scipal (2006), Assimilating scatterometer soil moisture data into conceptual hydrologic models at the regional scale, *Hydrol. Earth Sys. Sci.*, *10*, 353–368, doi:10.5194/hess-10-353-2006.

Pauwels, V. R., R. Hoebeb, N. E. Verhoest, and F. P. De Troch (2001), The importance of the spatial patterns of the remotely sensed soil moisture in the improvement of discharge prediction for small scale basins through data assimilation, *J. Hydrol.*, *251*, 88–102.

Reichle, R. H., and R. D. Koster (2004), Bias reduction in short records of satellite soil moisture, *Geophys. Res. Lett.*, *31*, L19501, doi:10.1029/2004GL020938.

Reichle, R. H., R. D. Koster, J. Dong, and A. A. Berg (2004), Global soil moisture from satellite observations, land surface models, and ground data: Implications for data assimilation, *J. Hydrometeorol.*, *5*, 430–442.

Rodrıguez-Iturbe, I. (2000), Ecohydrology: A hydrologic perspective of climate-soil-vegetation dynamics, *Water Resour. Res.*, *36*, 3–9.

Seo, D. J., and J. P. Breidenbach (2002), Real-time correction of spatially non uniform bias in radar rainfall data using rain gauge measurements, *J. Hydrometeor.*, *3*, 93-111.

Seo, D. J., L., Cajina, R. Corby, and T. Howieson (2009), Automatic state updating for operational streamflow forecasting via variational data assimilation, *J. Hydrol.*, *367*, 255–275.

U.S. Department of Agriculture (USDA) (2012), USDA Agricultural Research Center, LPRM/AMSR-E/aqua daily L3 descending and 2-layer Palmer water balance model root zone soil moisture, Goddard Earth Sciences Data and Information Services Center (GES DISC), Greenbelt, Md.

Wagner, W., G. Lemoine, and H. Rott (1999), A method for estimating soil moisture from ERS scatterometer and soil data, *Remote Sens. Environ.*, *70*(2) 191–207.

Wagner, W., V. Naeimi, K. Scipal, R. A. M. De Jeu, and J. Martinez-Fernandez (2007), Soil moisture from operational meteorological satellites, *Hydrol. J.*, *15*, 121–131.

Western, A. W., R. B. Grayson, and G. Bloschl (2002), Scaling of soil moisture: A hydrologic perspective, *Annu. Rev. Earth Planet. Sci.*, *30*, 149–180, doi:10.1146/annurev.earth.30.091201.140434.

20

NASA Giovanni: A Tool for Visualizing, Analyzing, and Intercomparing Soil Moisture Data

William Teng,[1,2] Hualan Rui,[1,2] Bruce Vollmer,[1] Richard de Jeu,[3] Fan Fang,[1,2] Guang-Dih Lei,[1,2] and Robert Parinussa[3]

20.1. INTRODUCTION

There are many existing satellite soil moisture algorithms and their derived data products, but there is no simple way for a user to intercompare the products or analyze them together with other related data (e.g., precipitation). An environment that facilitates such intercomparison and analysis would be useful for validation of satellite soil moisture retrievals against in situ data and for determining the relationships among different soil moisture products. The latter relationships are particularly important for applications users, for whom the continuity of soil moisture data, from whatever source, is critical. A recent example was provided by the sudden demise of the EOS *Aqua* AMSR-E (Advanced Microwave Scanning Radiometer-Earth Observing System) sensor and the end of its related soil moisture data production, as well as the end of other soil moisture products that had been based on the AMSR-E data.

For more than 16 years, NASA Giovanni (Geospatial Interactive Online Visualization ANd aNalysis Infrastructure) has provided users worldwide with a simple but powerful way to explore (visualize and analyze) the large variety and vast amounts of data [*Acker and Leptoukh*, 2007; *Berrick et al.*, 2009]. These data are from NASA Earth observation sensors, including those on the EOS platforms of *Terra*, *Aqua*, and *Aura*, as well as smaller platforms like TRMM (Tropical Rainfall Measuring Mission); from model data sets, such as GLDAS and NLDAS

(Global and North American Land Data Assimilation System, respectively); and from various, separately funded studies and projects, such as MAIRS (Monsoon Asia Integrated Regional Study). Data available via Giovanni are grouped into "portals." For example, there are 4 such portals of data from the MAIRS project and 16 portals of data from TRMM, 5 of which are for intercomparison.

As part of two NASA-funded projects, with team members who are end users from the National Oceanic and Atmospheric Administration (NOAA) National Weather Service (NWS) and the U.S. Department of Agriculture (USDA) World Agricultural Outlook Board (WAOB), a number of level 3 (gridded) soil moisture products have been incorporated into a beta prototype Giovanni portal (Intercomparison of Soil Moisture Products). The purpose of this effort, beyond satisfying the requirements of the projects, is to create an environment, as part of the NASA Giovanni family of portals, that facilitates intercomparisons and other uses of soil moisture products. [There are other apparently similar systems for exploration of remote sensing data, e.g., LAS (Live Access Server) of the NASA Jet Propulsion Laboratory (JPL) PODAAC (Physical Oceanography Distributed Active Archive Center), but a comparison of Giovanni with other systems is beyond the scope of this chapter. A compilation of such systems in the NASA data community can be found at https://earthdata.nasa.gov/data/data-tools/data-visualization-analysis-tools.]

The following sections of this chapter (1) provide some additional details on Giovanni, (2) describe the data products currently available in the Giovanni Intercomparison of Soil Moisture Products portal, (3) illustrate the portal's user interface and currently available services, and (4) give two examples of how the portal might be

[1]*NASA Goddard Space Flight Center, Goddard Earth Sciences Data and Information Services Center (GES DISC), Greenbelt, Maryland, USA*

[2]*ADNET Systems, Inc., Rockville, Maryland, USA*

[3]*Vrije Universiteit, Amsterdam, The Netherlands*

Remote Sensing of the Terrestrial Water Cycle, Geophysical Monograph 206. First Edition. Edited by Venkat Lakshmi.
© 2015 American Geophysical Union. Published 2015 by John Wiley & Sons, Inc.

used for and facilitate hydrologic applications. Evaluation and intercomparison of soil moisture products and with other data sources are left to the prospective users of the soil moisture portal.

20.2. NASA GIOVANNI

Giovanni is a NASA data exploration system that provides a simple and intuitive way to visualize, analyze, and access vast amounts of Earth sciences remote sensing data, without having to download the data, to be familiar with various and often complicated data formats, and to import into other systems or write custom code for basic display and analysis (http://disc.sci.gsfc.nasa.gov/giovanni/overview/index.html). Developed at the NASA Goddard Earth Sciences Data and Information Services Center (GES DISC), with inputs from users both at the Goddard Space Flight Center (GFSC) and from around the world, Giovanni obviates much of the data "prework" that users would otherwise need to undertake, thus resulting in weeks to months of time saved. It allows users to rapidly explore data, so that, e.g., spatial-temporal variability, anomalous conditions, and patterns of interest can be directly and quickly observed and analyzed online. Giovanni currently provides 20 some services, such as Lat-Lon Map (Time-averaged), Latitude-Time Hovmöller Plot, Correlation Plot, and Time Series (Area-averaged), different subgroups of which are offered by the various portals. Users of Giovanni have included researchers, modelers, applications users, policy- and decision-makers, those in the education community, and proposal writers. Giovanni has contributed to many users' science research efforts and applications. One metric of this contribution is the number of peer-reviewed publications of research that used Giovanni in some way. Since 2004, there have been 963 such publications (as of 30 June 2014).

Giovanni is an evolving system. The earliest portal, before the term "Giovanni" was coined, is the highly popular TOVAS (TRMM Online Visualization and Analysis System) [*Liu et al.*, 2013]. To take advantage of advancing technology and reduce costs in maintaining and extending Giovanni, in response to user-driven and system interoperability requirements, a radical engineering of the underlying infrastructure was undertaken, resulting in "Giovanni, Version 3" (or G3). The Giovanni Intercomparison of Soil Moisture Products portal is in the G3 family of portals. There is an ongoing effort to develop the next generation of "Giovanni, Version 4" (or G4), designed to be faster, more interactive, and easier to learn.

20.3. SOIL MOISTURE AND RELATED DATA PRODUCTS IN NASA GIOVANNI

Table 20.1 lists those soil moisture and related data products that are currently included in the beta prototype Giovanni Intercomparison of Soil Moisture Products portal. This suite of products will be modified or extended,

Table 20.1 Level 3 (gridded) soil moisture and related products currently available in the beta prototype Giovanni portal, "Inter-comparison of Soil Moisture Products."

Soil Moisture (SM) and Related Products	Spatial Resolution	Spatial Coverage	Temporal Resolution	Temporal Coverage	Reference
EOS[a] *Aqua* AMSR-E[b] (SM)	1/4 deg	Global	Daily	6/2002–10/2011	*Njoku et al.* [2003]
LSMEM[c]-TMI[d] (SM)	1/8 deg	U.S. up to 40°N	Daily	1/1998–12/2004	*Gao et al.* [2006]
LPRM[e]-AMSR-E (SM)	1/4 deg	Global	Daily	6/2002–10/2011	*Owe et al.* [2008] *de Jeu et al.* [2008]
LPRM-TMI (SM)	1/4 deg	Global 40°N–40°S	Daily	1/1998–present	*Owe et al.* [2008] *de Jeu et al.* [2008]
LPRM-WindSat (SM)	1/4 deg	Global	Daily	2/2003–7/2011	*Parinussa et al.* [2012]
LPRM-RZSM[f] (AMSR-E)	1/4 deg	Global	Daily	6/2002–12/2010	*Bolten et al.* [2010]
TMPA[g] precipitation	1/4 deg	Global 50°N–50°S	Daily	1/1998–present	*Huffman et al.* [2007]
EOS *Aqua* AIRS[h] surface air temperature	1 deg	Global	Daily	8/2002–present	*Aumann et al.* [2003]

[a]Earth Observing System.
[b]Advanced Microwave Scanning Radiometer – Earth Observing System.
[c]Land Surface Microwave Emission Model.
[d]TRMM Microwave Imager.
[e]Land Parameter Retrieval Model.
[f]Root-zone soil moisture.
[g]TRMM Multisatellite Precipitation Analysis.
[h]Atmospheric Infrared Sounder.

based on user feedback and availability of resources. The current six level-3 (gridded) soil moisture products were used or produced as parts of the two previously mentioned projects with NOAA NWS and USDA WAOB.

The LSMEM-TMI (Land Surface Microwave Emission Model-TRMM Microwave Imager) product currently in the Giovanni portal is an early limited version, covering 1998–2004 [*Gao et al.*, 2006]. As part of the NASA Making Earth System Data Records for Use in Research Environments (MEaSUREs) Program, there is an ongoing effort, led by Princeton University, to produce Earth System Data Records (ESDRs) for the global terrestrial water cycle [*MacCracken et al.*, 2013]. One of these ESDRs is LSMEM-based soil moisture, using the AMSR-E brightness temperature as input. All of these water cycle ESDRs will be delivered to the GES DISC, at which time the current LSMEM-TMI product in the Giovanni Soil Moisture portal will be replaced by the newer LSMEM-AMSR-E ESDR.

The Land Parameter Retrieval Model (LPRM) was originally developed by NASA GSFC and Vrije Universiteit Amsterdam (VUA) [*Owe et al.*, 2008; *de Jeu et al.*, 2008]. Subsequent development of the LPRM algorithm has been carried out by VUA, e.g., error estimates [*Parinussa et al.*, 2011], with operational production of the LPRM-AMSR-E data done by the GES DISC. More recent work has focused on replacing the lost AMSR-E/*Aqua* data stream, by applying LPRM to the TRMM Microwave Imager (TMI) and WindSat brightness temperatures, to generate LPRM-TMI and LPRM-WindSat data, respectively. For the latter, see *Parinussa et al.* [2012] and *Li et al.* [2010]. Current temporal coverage of LPRM-WindSat is limited to that of the input data available from the Naval Research Laboratory (NRL). There is an ongoing effort to fill in the missing time periods. The LPRM-TMI, among all the currently "live" (i.e., forward processed) retrieved soil moisture products, has the longest temporal coverage (1998–present), although it is relatively limited by the TRMM spatial coverage of 40°N – 40°S.

The LPRM-RZSM product was produced as part of a separate NASA-funded project, working with the USDA Foreign Agricultural Service (FAS) to improve its soil moisture inputs and, thus, its crop yield forecasts [*Bolten et al.*, 2010]. Input to the LPRM-RZSM assimilated product is the LPRM-AMSR-E surface soil moisture.

There are other existing and upcoming soil moisture products that could potentially be added to this Giovanni Soil Moisture portal (resources permitting). These include the ongoing missions of (1) the European Space Agency (ESA): Soil Moisture and Ocean Salinity (SMOS), launched in 2009 [*Kerr et al.*, 2001, 2010, 2012] and (2) the Japan Aerospace Exploration Agency (JAXA): Advanced Microwave Scanning Radiometer 2 (AMSR2) aboard the Global Change Observation Mission 1st-Water "SHIZUKU" (GCOM-W1), launched in 2012 [*Kachi et al.*, 2008; *Oki et al.*, 2010] (LPRM-AMSR2 data will soon go into production at the GES DISC, with incorporation into the Giovanni Soil Moisture portal to follow thereafter.); the upcoming NASA mission: Soil Moisture Active Passive (SMAP), scheduled for launch in 2015 [*Entekhabi et al.*, 2010]; as well as model outputs from NLDAS [*Mitchell et al.*, 2005] and GLDAS [*Rodell et al.*, 2004].

An initial effort has been made by GES DISC and VUA to test the feasibility of incorporating into the Giovanni portal the soil moisture data from the ESA Water Cycle Multimission Observation Strategy (WACMOS) project [*Dorigo et al.*, 2011; *Liu et al.*, 2011]. Although the effort was successful, further work awaits additional resources.

20.4. NASA GIOVANNI INTERCOMPARISON OF SOIL MOISTURE PORTAL

Figure 20.1 shows the user interface of the Giovanni Intercomparison of Soil Moisture portal. The interface was designed to be simple and intuitive. User selections (e.g., those outlined in red) are spatial region, parameter(s) of interest, temporal range, and Giovanni service (e.g., Lat-Lon Map, Time-averaged). Once the resulting plot is made, additional user preferences can be specified (e.g., color bar, min-max values). Figure 20.2 shows the resulting soil moisture area maps from the user selections shown in Figure 20.1.

In addition to Lat-Lon Map, Time-averaged, there are currently three other available services (or visualization types) in the Giovanni Soil Moisture portal: (1) Time series, (2) Scatter plot, Time-averaged, and (3) Animation. Other Giovanni services will be added as appropriate and needed, based on user feedback and resources available. Figure 20.3 shows a scatter plot between LPRM-AMSR-E and LPRM-TMI soil moisture for an area near Austin, Texas. For the same general area, Figure 20.4 shows three time series plotted together, one of precipitation and two of soil moisture, which can be useful, e.g., to intercompare the two soil moisture products, in relationship to the precipitation product.

Once the user is satisfied with the plot results, several data format options are available for downloading from all Giovanni portals, including HDF (Hierarchical Data Format), NetCDF (Network Common Data Format), ASCII (American Standard Code for Information Interchange), and KMZ (Keyhole Markup Language-Zipped) (Figure 20.5). Individual input files and generated output plot files (both data and image) can be downloaded.

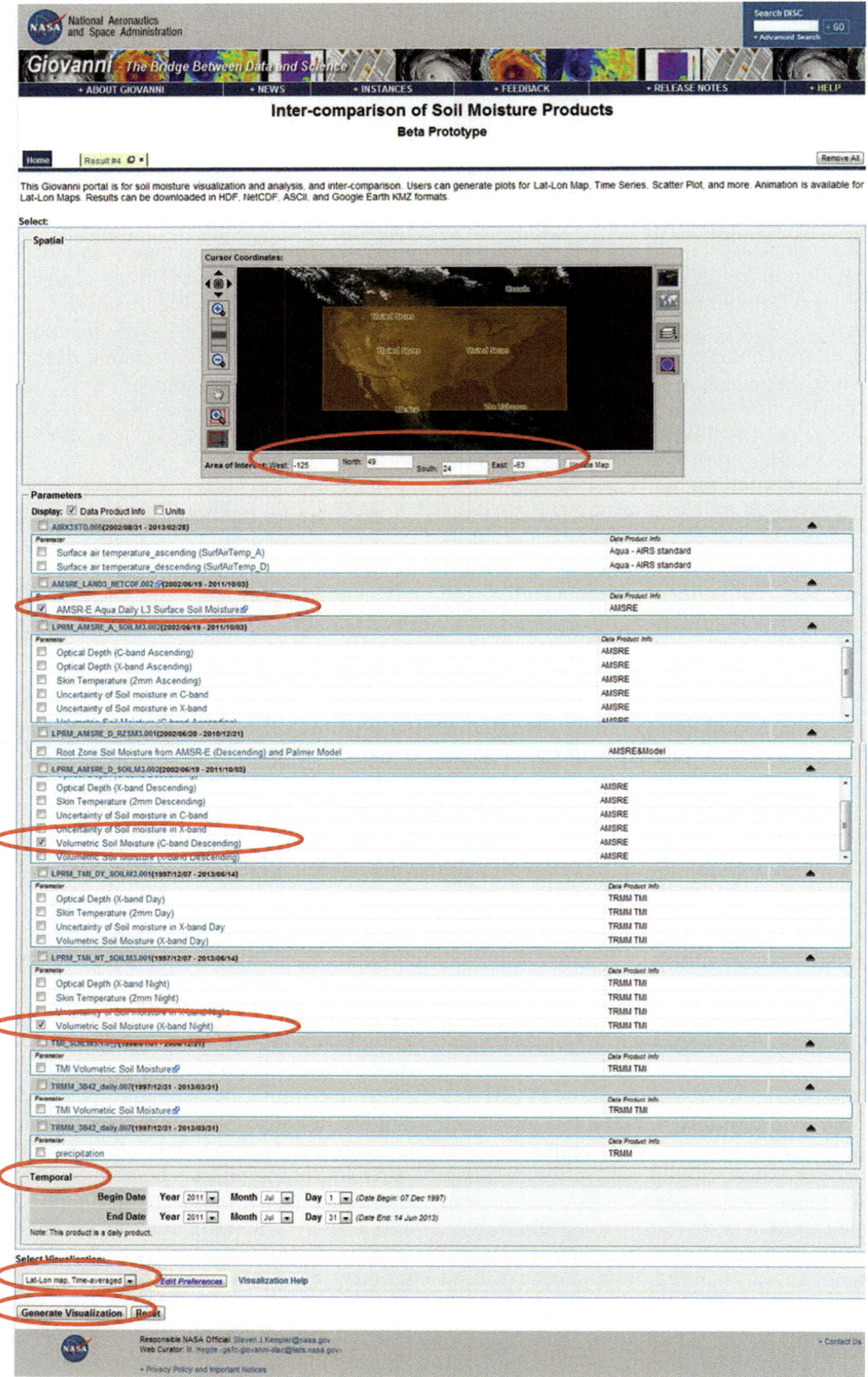

Figure 20.1 User interface of the Giovanni Intercomparison of Soil Moisture portal. Red ovals are example user selections (i.e., spatial: 125°W–63°W, 24°N–49°N (U.S.); parameter: soil moisture (three products); temporal: 1–31 July 2011; visualization: Lat-Lon map, Time-averaged) that result in the area maps shown in Figure 20.2.

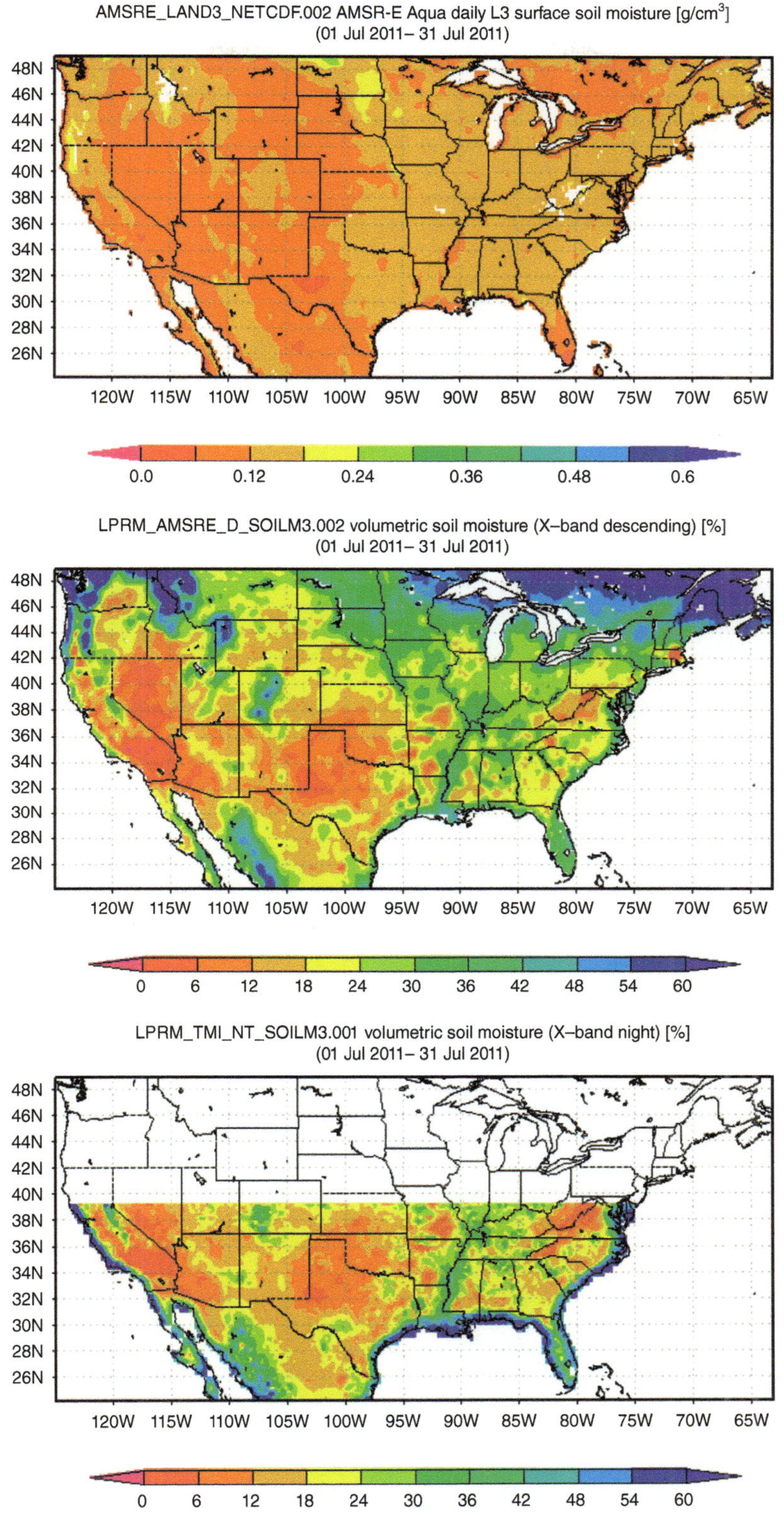

Figure 20.2 Giovanni Intercomparison of Soil Moisture portal outputs: Area maps (Lat-Lon map, Time-averaged) of soil moisture from (top to bottom) EOS *Aqua* AMSR-E, LPRM-AMSR-E, and LPRM-TMI (see Table 20.1), for the United States and 1–31 July 2011.

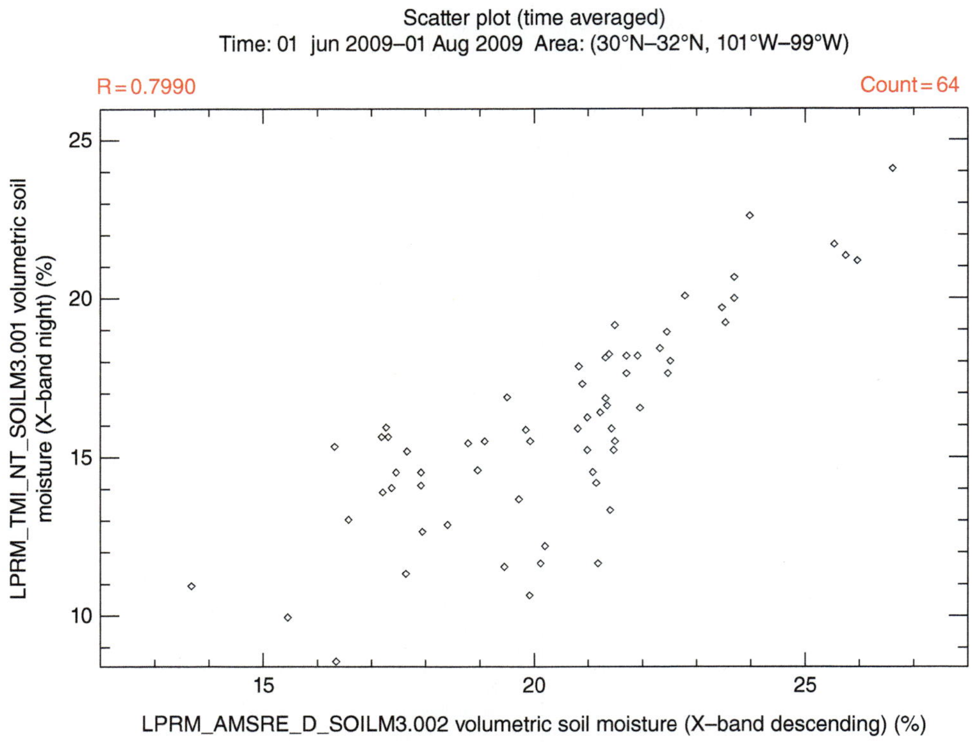

Figure 20.3 Scatter plot between LPRM-AMSR-E and LPRM-TMI soil moisture, for an area near Austin, Texas, and 1 June to 1 August 2009.

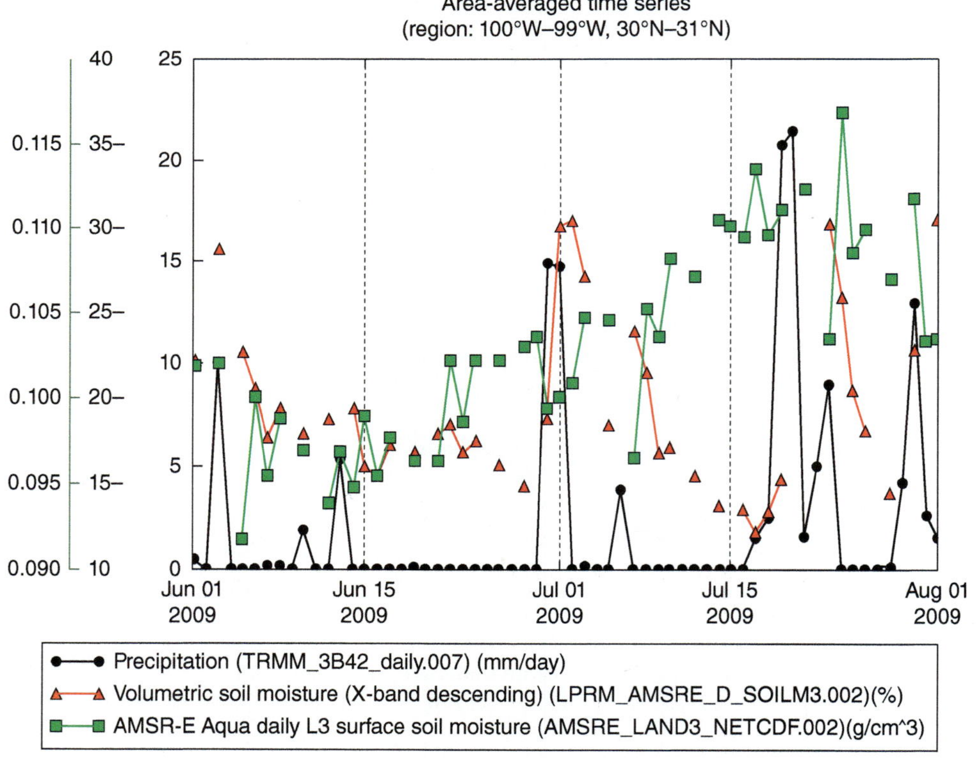

Figure 20.4 Time series, area-averaged, of TRMM precipitation and LPRM-AMSR-E and EOS *Aqua* AMSR-E soil moisture, for an area near Austin, Texas, and 1 June to 1 August 2009.

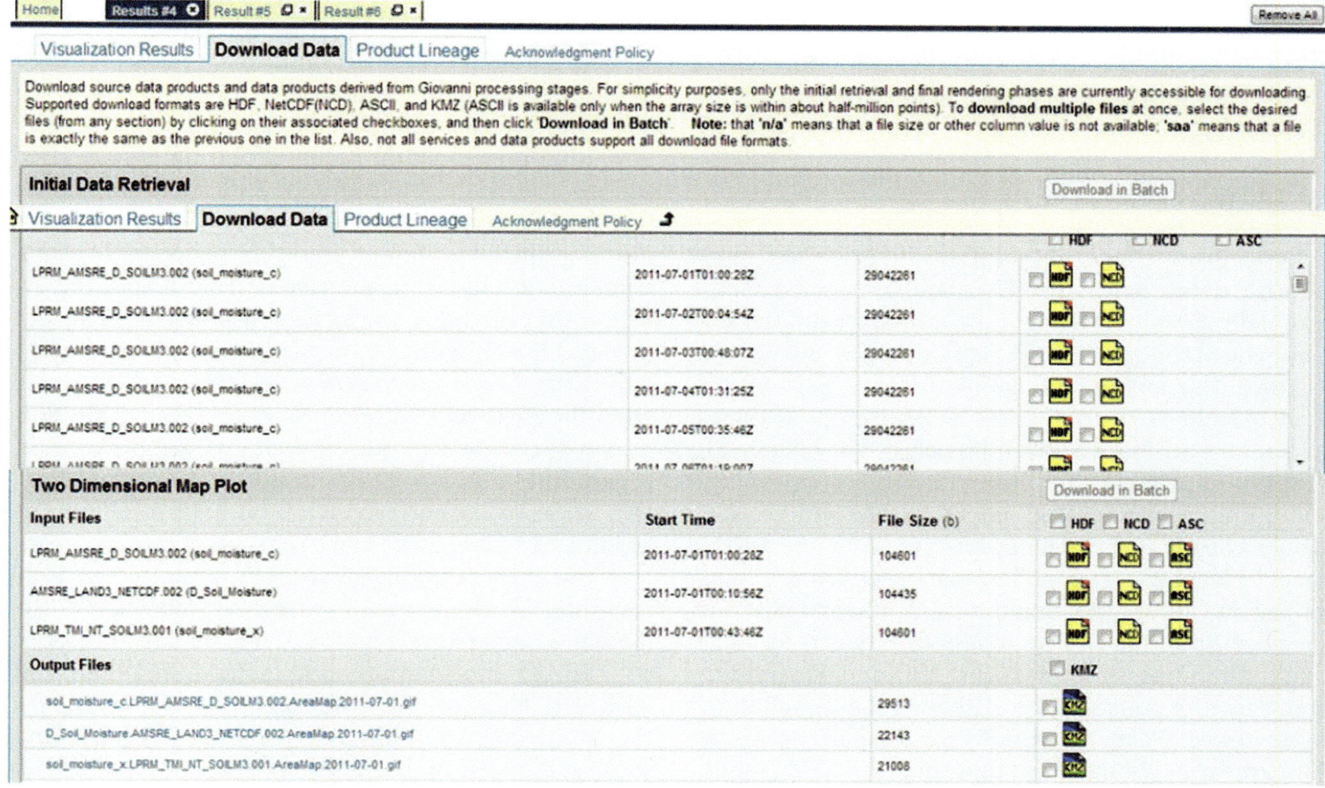

Figure 20.5 "Download Data" tab in Giovanni portals provides data format options for downloading individual input files and generated output plot files (both data and image).

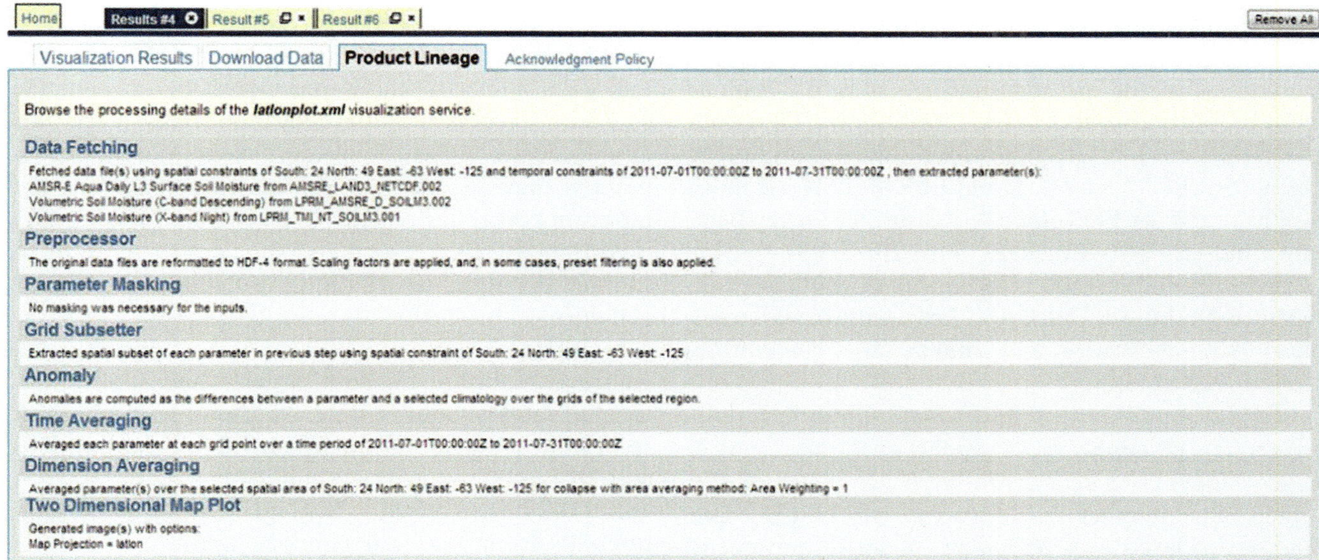

Figure 20.6 "Product Lineage" tab in Giovanni portals provides information on the lineage (or provenance) of the Giovanni-generated output files, describing all the intermediate steps taken to result in a particular output plot.

Information on the lineage (or provenance) of the Giovanni-generated output files is also available, describing all the intermediate steps taken to result in a particular output plot (Figure 20.6).

20.5. APPLICATION OF GIOVANNI INTERCOMPARISON OF SOIL MOISTURE PORTAL

20.5.1. Texas Drought of 2011

Drought and excessive heat in 2011 created major adverse effects across the U.S. Southern Great Plains, especially Texas and Oklahoma, as shown by the U.S. Drought Monitor (USDM) map of August 30, 2011, in contrast with a more "average" year map of September 1, 2009 (Figure 20.7). Total direct losses to crops, livestock, and timber in these areas approached $10 billion. The USDM maps are based on a convergence of multiple sources of evidence. How might surface soil moisture data retrieved from satellites compare with the USDM maps? Figure 20.8 shows Giovanni-generated area maps of soil moisture, vegetation optical depth, and skin temperature (from the LPRM-AMSR-E product) for Texas and surrounding areas for August 2011 and, for comparison, those for August 2009. Although the overall differences in spatial patterns between the 2 years are also seen, they are not as contrasted as those in the USDM. Specific features or areas of interest can be quickly identified from these Giovanni outputs, for more detailed analyses.

Land surface models comprise another source of data for comparison with satellite-retrieved data. Figure 20.9 shows the NLDAS Noah model output for 0–100 cm soil moisture content for a 0.125° grid box over Austin, Texas, for the entire time period of available coverage (so far) and for the year 2011. The extraordinary nature of the 2011 drought in Texas is particularly evident in the context of the historical record (some 34 years so far). In comparison, Figure 20.10 shows Giovanni-generated area-averaged time series, centered on Austin, Texas, for LPRM-AMSR-E and LPRM-TMI soil moisture data and for EOS *Aqua* AIRS surface air temperature and LPRM-AMSR-E and LPRM-TMI skin temperature. The trends of both sets of plots are consistent with those of NLDAS Noah, with overall minimum (soil moisture) and maximum (temperature) around August 2011.

20.5.2. Tropical Storm Lee

During September 2011, tropical storm (TS) Lee, due to its large size and slow forward movement, dumped heavy rainfall over southern Louisiana, Mississippi, Alabama, and the Florida panhandle. The area map of the NLDAS total precipitation, over the period 2–9 September is shown in Figure 20.11a, with TS Lee's track clearly visible. Figure 20.11b shows the corresponding area map of the LPRM-AMSR-E soil moisture, averaged over the same time period, resulting in the storm track that is not as distinct. A storm's effect on soil moisture typically has a time lag relative to that on precipitation. Figure 20.11c shows the same soil moisture data, but averaged with a 2 day time lag (i.e., 4–11 September). The storm track is somewhat more distinct. Finally, a shorter averaging period after a rain event should show a more distinct soil moisture response. Figure 20.11d shows the same soil moisture data, but averaged for only the 2 days after an intense rain event during the storm over Alabama, Georgia, and Tennessee. Now, the effect of TS Lee can be clearly seen. This location is the middle of the three boxed areas shown in Figure 20.11a. Figure 20.12 shows, for these three areas, time series of the corresponding NLDAS total precipitation and NLDAS Mosaic 0–10 cm soil moisture content and their relationship to each other, in terms of responses to the storm. Figure 20.13 shows, for the middle boxed area of Figure 20.11a, similar time series of TRMM precipitation and LPRM-AMSR-E soil moisture. There are far fewer data points available for these satellite data, but the precipitation peak and slight soil moisture time lag are similar to those shown by the NLDAS data.

20.6. SUMMARY

The NASA Giovanni Intercomparison of Soil Moisture Products portal is versatile and extensible, with many possible uses, for research and applications (e.g., natural disasters, agriculture). It is currently being used by the USDA WAOB, as part of one of the projects that have supported the portal's development. For the education community, the Giovanni family of portals has been used in classrooms, workshops, and other venues. For example, the current beta prototype of the soil moisture portal has already been used in one of VUA's classroom exercises. The portal should also be useful to support upcoming missions, e.g., for prelaunch SMAP activities.

Maintaining continuity of data and minimizing data gaps are important for both research and applications users of satellite mission data. As discussed in an earlier section, to mitigate the loss of the EOS *Aqua* AMSR-E and keep the data gap to a minimum, the GES DISC and VUA have applied LPRM to brightness temperature data from TMI and WindSat, to produce LPRM-TMI and LPRM-WindSat products, respectively (both level 2 or swath and level 3 or gridded). (NB: "Continuity"

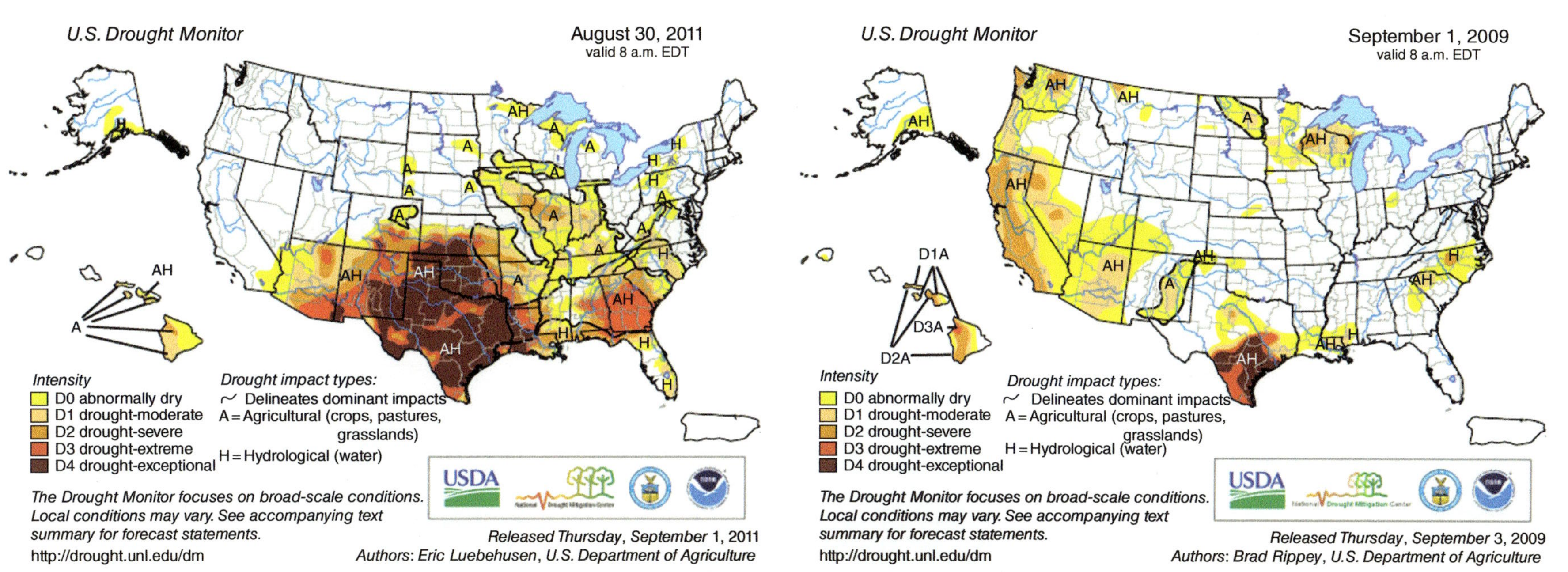

Figure 20.7 U.S. Drought Monitor map of 30 August 2011, showing most of Texas in "exceptional drought," compared with a similar map for 1 September 2009, a more "average" year.

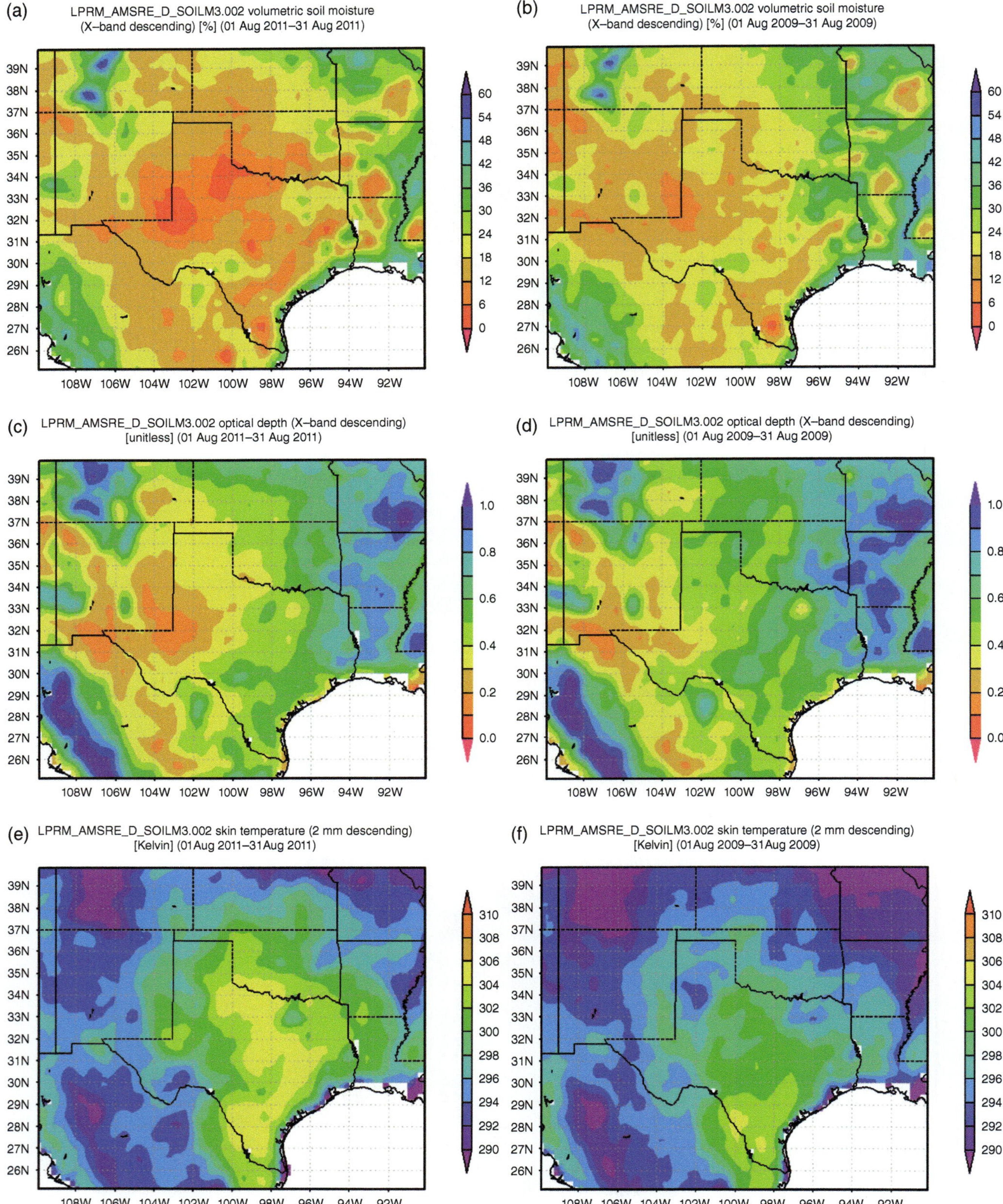

Figure 20.8 Giovanni-generated area maps of (a,b) soil moisture, (c,d) vegetation optical depth, and (e,f) skin temperature from the LPRM-AMSR-E product, for Texas and surrounding areas for August 2011 and August 2009.

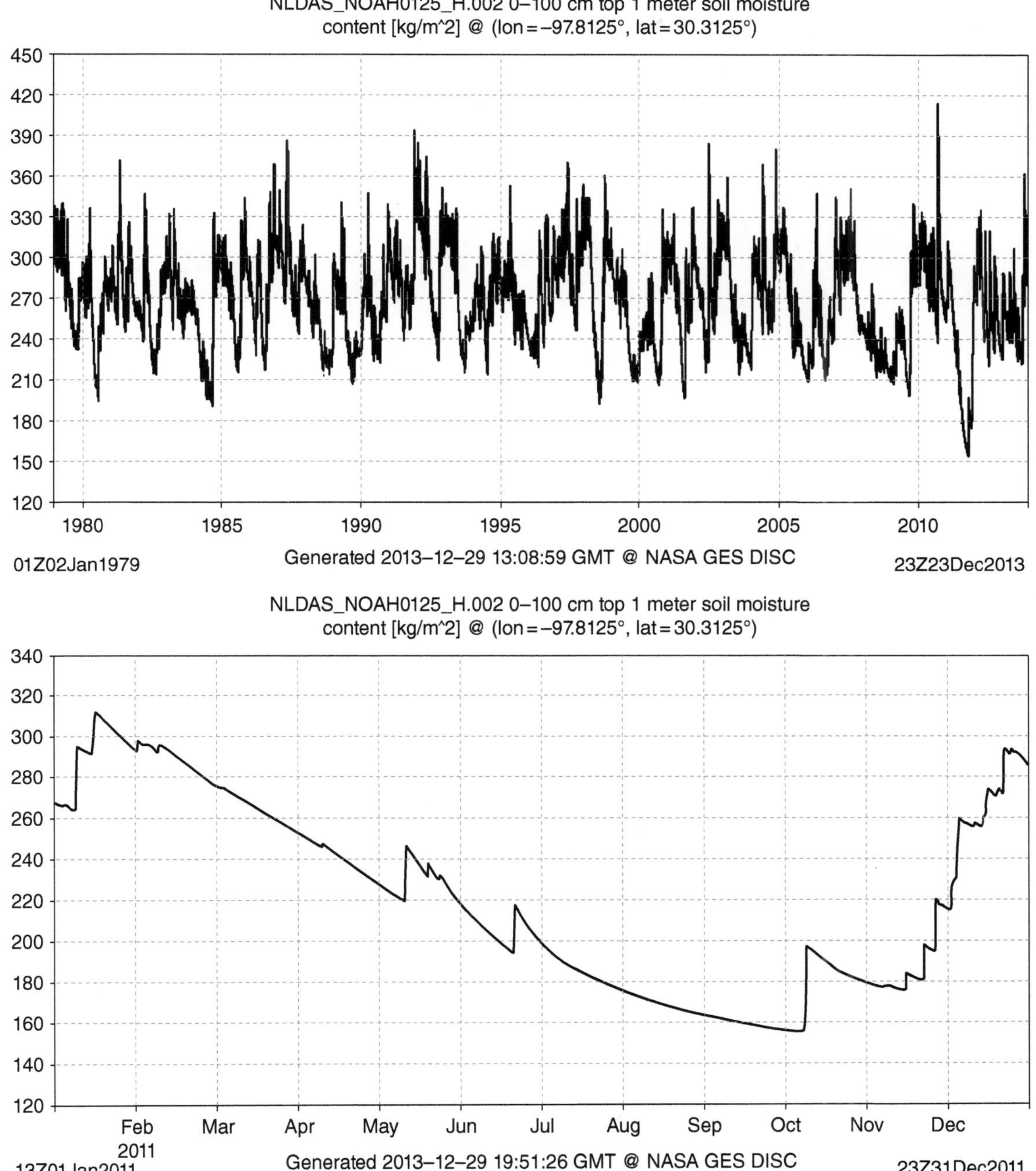

Figure 20.9 NLDAS Noah model output for 0–100 cm soil moisture content for a 0.125° grid box over Austin, Texas, for the entire time period of available coverage (so far) (top) and for the year 2011 (bottom).

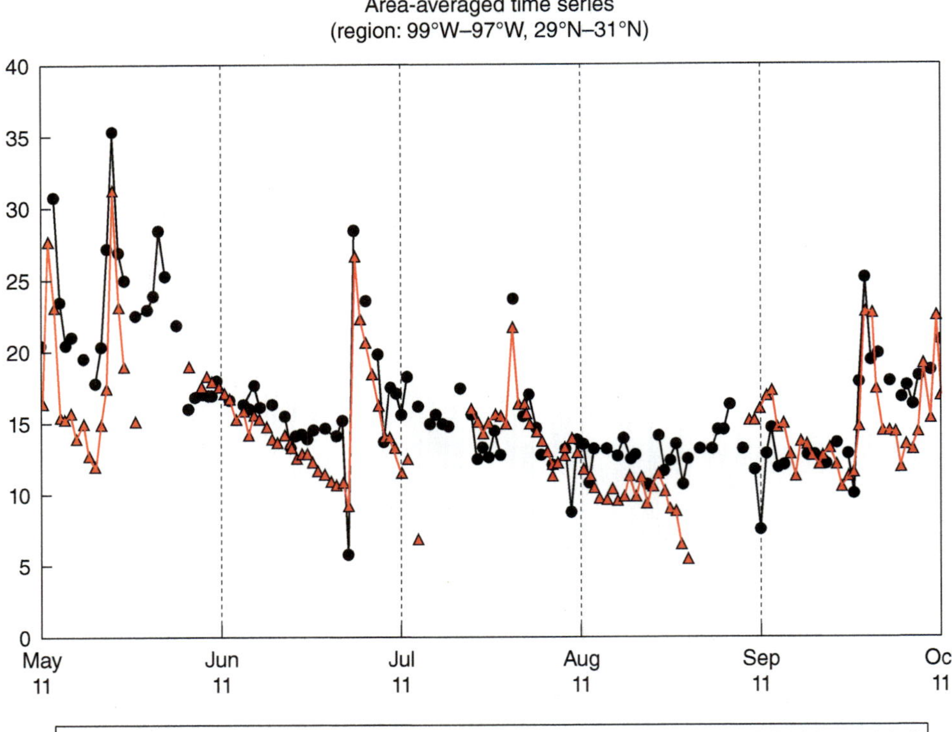

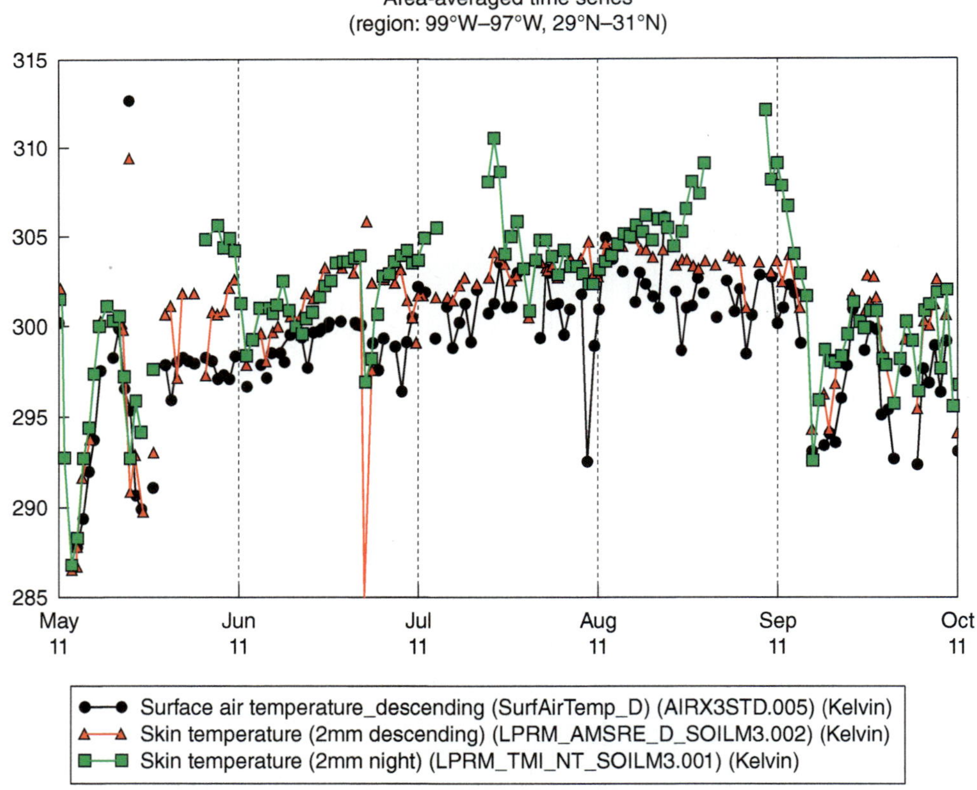

Figure 20.10 Giovanni-generated time series, area-averaged, centered on Austin, Texas, for soil moisture from LPRM-AMSR-E and LPRM-TMI (top) and surface temperature from EOS Aqua AIRS (surface air temperature) and LPRM-AMSR-E and LPRM-TMI (skin temperature) (bottom), covering the time period from 1 May to 1 October 2011.

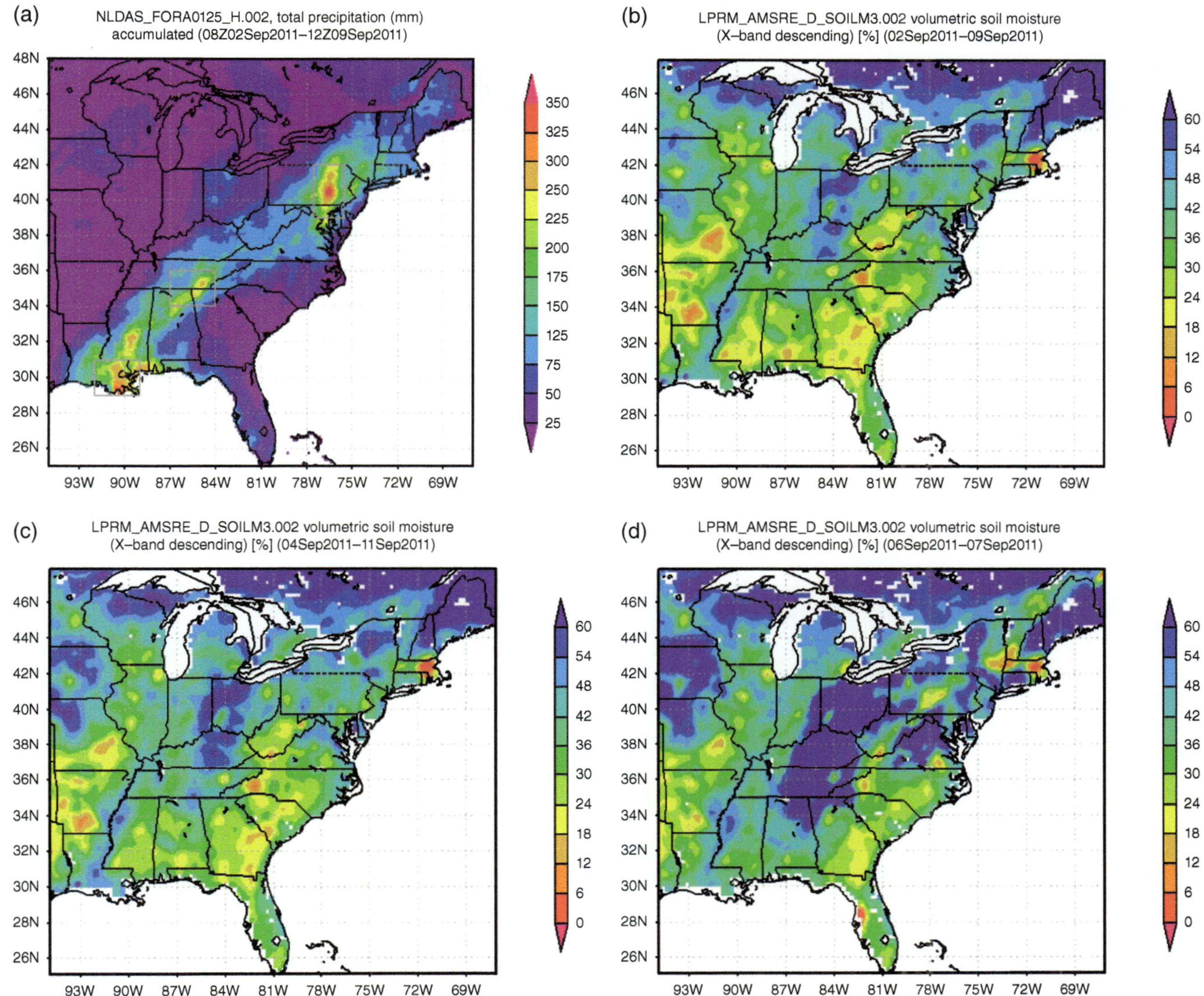

Figure 20.11 The track and effect of Tropical Storm Lee, 2–9 September 2011, shown by Giovanni-generated area maps of (a) NLDAS total precipitation; (b) LPRM-AMSR-E averaged soil moisture, over the same time period; (c) averaged soil moisture, with a 2-day time lag (i.e., 4–11 September); and (d) averaged soil moisture, for the 2 days after an intense rain event over Alabama, Georgia, and Tennessee.

as used here simply refers to that of temporal coverage provided by Giovanni of some soil moisture data; it does not refer to that of a long time series, i.e., an Earth Science Data Record, or ESDR, such as those being produced as part of the ESA WACMOS project.) As previously noted, there is an ongoing effort to extend the coverage of LPRM-WindSat, and the LPRM-AMSR2 data will soon be in production.

In all these developments, user inputs and feedback are critical (on services, data, and user interface). The Giovanni Intercomparison of Soil Moisture Products portal, along with all other portals in the Giovanni family, are created, maintained, and enhanced for the many users of NASA data worldwide. To contribute, please check out http://gdata1.sci.gsfc.nasa.gov/daac-bin/G3/gui.cgi?instance_id=soilmoisture_daily. The new version of Giovanni (G4) will feature more interactive data exploration, with order-of-magnitude performance improvement, browser-based interactive plotting, and new services (e.g., seasonal data visualization). An early release of G4 can be accessed at http://giovanni.gsfc.nasa.gov/giovanni/.

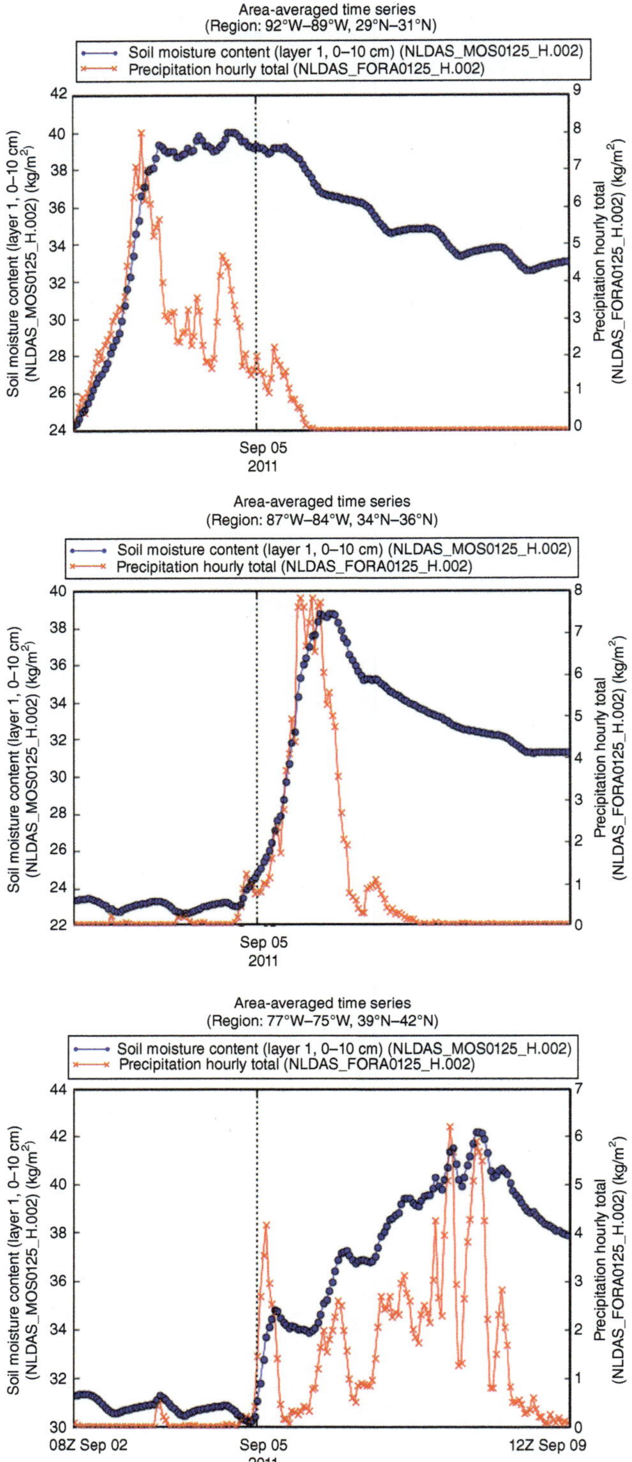

Figure 20.12 Time series of NLDAS total precipitation and NLDAS Mosaic 0–10 cm soil moisture content for the three boxed areas shown in Figure 20.11a and for the same time period of 2–9 September 2011. Plots were generated by NASA Giovanni NLDAS hourly portal. Geographical locations for the top, middle, and bottom plots are, respectively, parts of Louisiana/Mississippi, parts of Alabama/Georgia/Tennessee, and parts of Pennsylvania/Maryland.

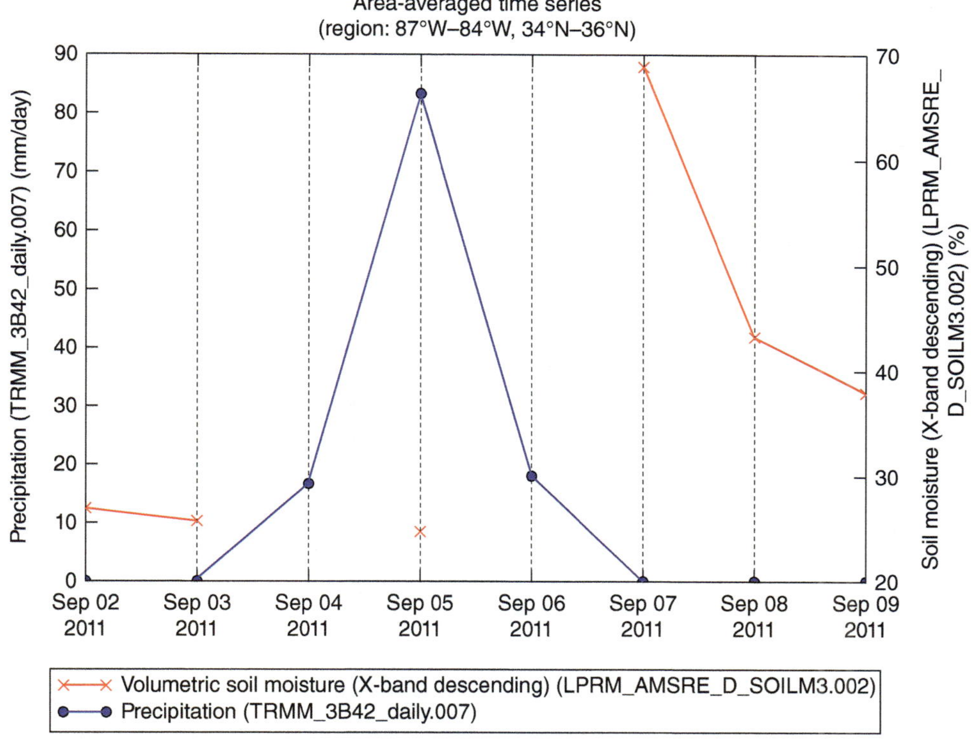

Figure 20.13 Time series of TRMM precipitation and LPRM-AMSR-E soil moisture, for the same middle boxed area in Figure 20.11a, showing a similar precipitation peak and slight soil moisture time lag as those shown by the NLDAS data (compare middle plot of Figure 20.12).

ACKNOWLEDGMENT

Development of the Giovanni Intercomparison of Soil Moisture Products portal is partially supported by NASA NNH05ZDA001N-DECISION, NASA NNH08ZDA001N-DECISIONS, NASA GES DISC, and Vrije Universiteit Amsterdam.

REFERENCES

Acker, J. and G. Leptoukh (2007) Online analysis enhances use of NASA Earth science data, *EOS, Trans. Am. Geophys. Union*, *88*(2), 14–17, doi: 10.1029/2007EO020003.

Aumann, H. H., et al. (2003), AIRS/AMSU/HSB on the Aqua mission: Design, science objectives, data products, and processing systems, *IEEE Trans. Geosci. Remote Sens.*, *41*, 253–264, doi:10.1109/TGRS.2002.808356.

Berrick, S., G. Leptoukh, J. Farley, and H. Rui (2009), Giovanni: A Web services workflow-based data visualization and analysis system, *IEEE Trans. Geosci. Remote Sens.*, *47*(1), 106–113, doi:10.1109/TGRS.2008.2003183.

Bolten, J. D., W. T. Crow, X. Zhan, T. J. Jackson, and C. A. Reynolds (2010), Evaluating the utility of remotely sensed soil moisture retrievals for operational agricultural drought monitoring, *IEEE J. Select. Topics Appl. Earth Observ. Remote Sens.*, *3*(1), 57–66, doi:10.1109/JSTARS.2009.2037163.

de Jeu, R. A. M., W. Wagner, T. R. H. Holmes, A. J. Dolman, N. C. van de Giesen, and J. Friesen (2008), Global soil moisture patterns observed by space borne microwave radiometers and scatterometers, *Surv. Geophys.*, *29*(4–5), 399–420, doi:10.1007/s10712-008-9044-0.

Dorigo, W., R. de Jeu, Y. Liu, R. Parinussa, W. Wagner, B. Su, and D. Fernandez-Prieto (2011), The WACMOS multi-mission soil moisture product: 30 years of soil moisture in support of climate change studies, *Geophys. Res. Abstr.*, *13*, EGU2011-10916, EGU General Assembly.

Entekhabi, D., et al. (2010), The Soil Moisture Active Passive (SMAP) Mission, *Proc. IEEE*, *98*(5), 704–716, doi:10.1109/JPROC.2010.2043918.

Gao, H., E. F. Wood, T. J. Jackson, M. Drusch, and R. Bindlish (2006), Using TRMM/TMI to retrieve soil moisture over the southern United States from 1998–2002, *J. Hydrometeorol.*, *7*, 23–38, doi:10.1175/JHM473.1.

Huffman, G. J., R. F. Adler, D. T. Bolvin, G. Gu, E. J. Nelkin, K. P. Bowman, Y. Hong, E. F. Stocker, and D. B. Wolff (2007), The TRMM Multisatellite Precipitation Analysis (TMPA): Quasi-global, multi-year, combined-sensor precipitation estimates at fine scales. *J. Hydrometeorol.*, *8*, 38–55, doi:10.1175/JHM560.1.

Kachi, M., K. Imaokaa, H. Fujiia, A. Shibatab, M. Kasaharaa, Y. Iidaa, N. Itoa, K. Nakagawaa, and H. Shimoda (2008), Status of GCOM-W1/AMSR2 development and science activities, *Proc. SPIE, 7106*, 71060P-1–71060P-8, doi:10.1117/12.801228.

Kerr, Y. H., P. Waldteufel, J.-P. Wigneron, J.-M. Martinuzzi, J. Font, and M. Berger (2001), Soil moisture retrieval from space: The Soil Moisture and Ocean Salinity (SMOS) Mission, *IEEE Trans. Geosci. Remote Sens., 39*(8), 1729–1735, doi:10.1109/36.942551.

Kerr, Y. H., et al. (2010), The SMOS Mission: New tool for monitoring key elements of the global water cycle, *Proc. IEEE, 98*(5), 666–687, doi:10.1109/JPROC.2010.2043032.

Kerr, Y. H., et al. (2012), The SMOS soil moisture retrieval algorithm, *IEEE Trans. Geosci. Remote Sens., 50*(5), 1384–1403, doi:10.1109/TGRS.2012.2184548.

Li, L., P. W. Gaiser, B.-C. Gao, R. M. Bevilacqua, T. J. Jackson, E.G. Njoku, C. Rüdiger, J.-C. Calvet, and R. Bindlish (2010), WindSat global soil moisture retrieval and validation, *IEEE Trans. Geosci. Remote Sens., 48*(5), 2224–2241, doi:10.1109/TGRS.2009.2037749.

Liu, Y. Y., R. M. Parinussa, W. A. Dorigo, R. A. M. de Jeu, W. Wagner, A. I. J. M. van Dijk, M. F. McCabe, and J. P. Evans (2011), Developing an improved soil moisture dataset by blending passive and active microwave satellite-based retrievals, *Hydrol. Earth Sys. Sci., 15*, 425–436, doi:10.5194/hess-15-425-2011.

Liu, Z., D. Ostrenga, W. Teng, and S. Kempler (2014), Developing online visualization and analysis services for NASA satellite-derived global precipitation products during the Big Geospatial Data era, in *Big Data: Techniques and Technologies in Geoinformatics (H.A. Karimi, ed.)*, Taylor & Francis, Boca Raton, pp. 91–116.

MacCracken, R., et al. (2013), The global terrestrial water cycle Earth System Data Record, paper presented at the AGU Fall Meet., San Francisco, Calif., December 9–13.

Mitchell, K. E., et al. (2004), The multi-institution North American Land Data Assimilation System (NLDAS): Utilizing multiple GCIP products and partners in a continental distributed hydrological modeling system, *J. Geophys. Res., 109*, D07S90, doi:10.1029/2003JD003823.

Njoku, E. G., T. J. Jackson, V. Lakshmi, T. K. Chan, and S. V. Nghiem (2003), Soil moisture retrieval from AMSR-E, *IEEE Trans. Geosci. Remote Sens., 41*(2), 215–229, doi:10.1109/TGRS.2002.808243.

Oki, T., K. Imaoka, and M. Kach (2010), AMSR instruments on GCOM-W1/2: Concepts and applications, *IEEE Int. Geosci. Remote Sens. Symp.*, 1363–1366, doi:10.1109/IGARSS.2010.5650001.

Owe, M., R. de Jeu, and T. Holmes (2008), Multisensor historical climatology of satellite-derived global land surface moisture, *J. Geophys. Res., 113*, F01002, doi:10.1029/2007JF000769.

Parinussa, R. M., A. G. C. A. Meesters, Y. Y. Liu, W. Dorigo, W. Wagner, and R. A. M. de Jeu (2011), Error estimates for near-real-time satellite soil moisture as derived from the Land Parameter Retrieval Model, *IEEE Geosci. Remote Sens. Lett., 8*(4), 779–783, doi:10.1109/LGRS.2011.2114872.

Parinussa, R. M., T. R. H. Holmes, and R. A. M. de Jeu (2012), Soil moisture retrievals from the WindSat spaceborne polarimetric microwave radiometer, *IEEE Trans. Geosci. Remote Sens., 50*(7), 2383–2694, doi:10.1109/TGRS.2011.2174643.

Rodell, M., et al. (2004), The Global Land Data Assimilation System, *Bull. Am. Meteorol. Soc., 85*, 381–394.

Section VI: Groundwater

21

Monitoring Aquifer Depletion from Space: Case Studies from the Saharan and Arabian Aquifers

Mohamed Sultan,[1] Mohamed Ahmed,[1,2] John Wahr,[3] Eugene Yan,[4] and Mustafa Kemal Emil[1]

21.1. INTRODUCTION

In arid and semiarid regions of the world, the demand for freshwater resources is increasing due to increasing populations and the scarcity of freshwater supplies. Examples of these regions include Middle East countries where freshwater resource scarcity contributes to political instability, disputes, and conflicts [*Amery*, 1997]. Many of these countries are blessed by extensive freshwater supplies in fossil aquifers of regional extent that were recharged during the previous wet climatic periods. Given the paucity of freshwater resources in these countries, these regional aquifers are extremely important for the sustenance of their general population. However, challenges are involved with the utilization of these resources, the most significant of which is unsustainable overexploitation.

There is ample evidence to suggest that these aquifers were mostly recharged in previous wet climatic periods, tens to hundreds of thousands years ago, yet these fossil aquifers still receive modest local recharge in interleaving dry periods such as those prevailing at present [*Abouelmagd et al.*, 2012; *Sultan et al.*, 2011, 2013]. Despite the modest recharge of these regional aquifers in

Middle Eastern countries, they have been utilized for long periods in a sustainable manner. Until recently, many of these systems were at, or near, steady state conditions, where water loss from the system by natural discharge was largely compensated for by the modest local recharge of these aquifers in areas of relatively high precipitation. In the Saharan deserts, nomads built their communities in and around natural depressions, where the groundwater table intersects topographic relief and natural discharge occurs [i.e., *Fetter*, 2001]. In these depressions and during dry periods, the nomads adopted sustainable practices and only utilized as much water as nature provided them. With increasing demands on freshwater supplies came aggressive extraction practices and losses to natural and artificial discharge that far exceeded the modest recharge, which only occurs locally. Under such conditions, these systems are better described as being in transient rather than steady state conditions. Of prime importance is the ability to evaluate the status of such systems. Are they receiving modern recharge? Are they being used in a sustainable manner, or are they being depleted? If the latter, how long could they be used and what would the optimum extraction rates be? Answering these questions has traditionally been accomplished through the construction and calibration of groundwater flow models; this requires the collection of extensive subsurface field data, including temporal water levels, hydraulic parameters, and well records. Such data are difficult to obtain for many of the aquifers, including the aquifers examined in this study, given their locations in less developed parts of the world, the inaccessibility of many of these regions, and the general absence of local funding to support the collection of data sets.

[1] *Department of Geosciences, Western Michigan University, Kalamazoo, Michigan, USA*

[2] *Department of Geology, Faculty of Science, Suez Canal University, Ismailia, Egypt*

[3] *Department of Physics, University of Colorado at Boulder, Boulder, Colorado, USA*

[4] *Environmental Science Davison, Argonne National Laboratory, Argonne, Illinois, USA*

Remote Sensing of the Terrestrial Water Cycle, Geophysical Monograph 206. First Edition. Edited by Venkat Lakshmi.
© 2015 American Geophysical Union. Published 2015 by John Wiley & Sons, Inc.

The launch of the Gravity Recovery and Climate Experiment (GRACE) mission and the acquisition of monthly mass solutions over the past 12 years are now providing reliable and cost-effective alternatives for investigating the temporal mass variations of large aquifer systems worldwide. In this chapter, we utilize GRACE data in conjunction with other readily available remote sensing data sets, as well as field and geochemical data, to investigate the spatial and temporal mass variations over two of the largest fossil aquifer systems, the Nubian Sandstone Aquifer System (NSAS) in northeast Saharan Africa, and the Arabian Peninsula Aquifer System (APAS) in the Arabian Peninsula (Figure 21.1), and to investigate the factors (i.e., natural and/or anthropogenic) that control these variations.

21.2. GRACE MISSION AND MONITORING FOSSIL AQUIFERS

The GRACE satellite mission was launched in March of 2002 to map the temporal variations in the Earth's global gravity field on a monthly basis at scales of a few hundred kilometers and greater [*Tapley et al.*, 2004a, 2004b]. After removing the atmospheric and the oceanic contributions, the largest time-variable gravity signals are expected to come from changes in the terrestrial water

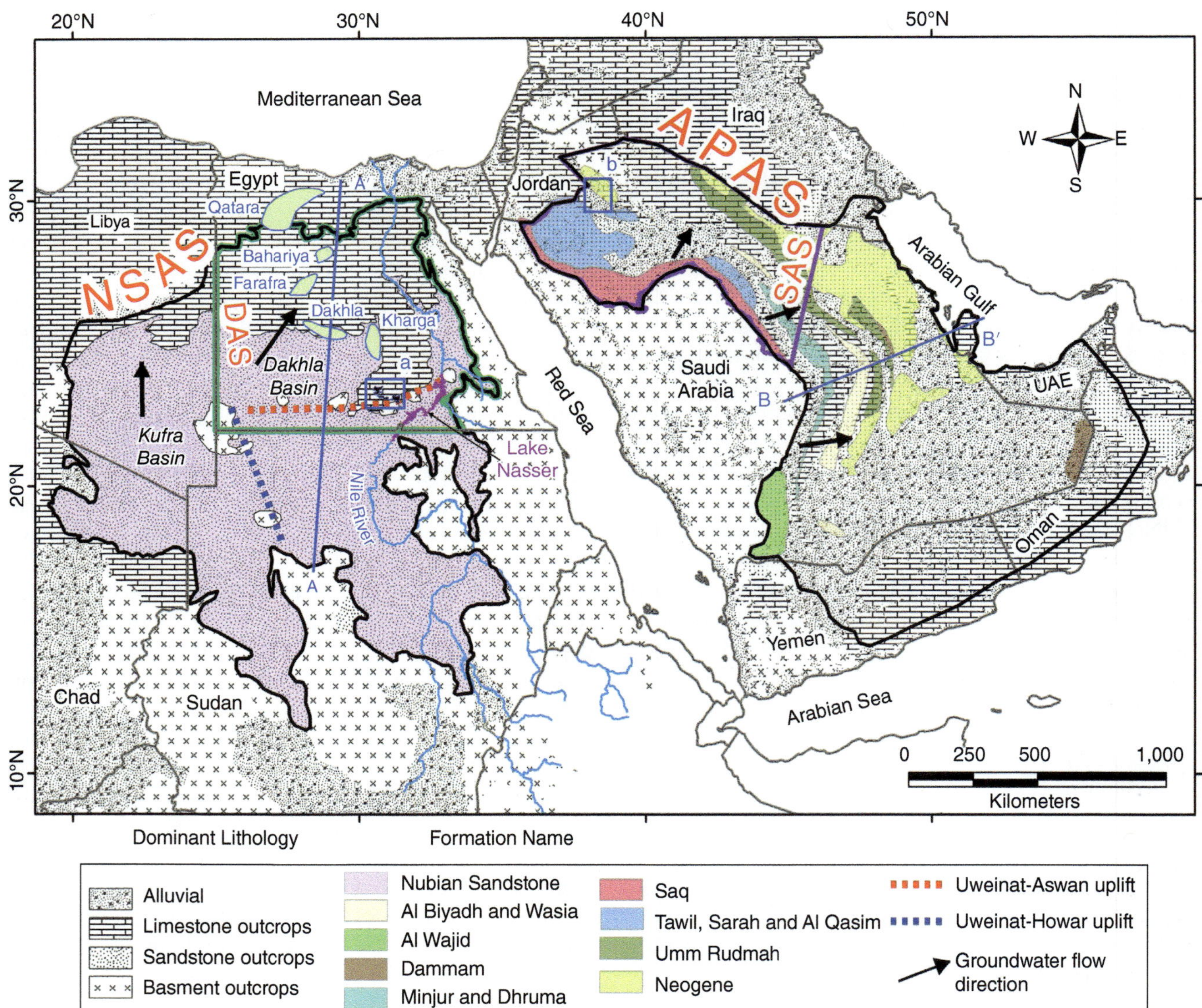

Figure 21.1 Simplified geologic map showing the distribution of the APAS and the NSAS outcrops in the Arabian and Nubian Shield, respectively. Also shown are the outlines of NSAS, APAS, Dakhla Aquifer System (DAS), and Saq Aquifer System (SAS), the Kufra and Dakhla basins, the oases, groundwater flow directions (thick black arrows), basement uplifts, Lake Nasser, Tushka lakes (Box a), Nile River, Wadi AlSharhan (Box b), and the location of traverses A–A' and B–B', for which schematic cross sections are shown in Figure 21.2.

storage (TWS) [*Wahr et al.*, 1998]. This refers to the total, vertically integrated, water content in an area regardless of the reservoir (surface water, groundwater, soil moisture and permafrost, snow and ice, and wet biomass) in which it resides. GRACE has been successfully used to spatially and temporally monitor: (1) elements of hydrologic cycles on subbasin scales in Africa [*Ahmed et al.*, 2011], (2) floods [*Reager and Famiglietti*, 2009] and droughts [*Yirdaw et al.*, 2008], (3) melting of glaciers and ice sheets [*Chen et al.*, 2006, 2009; *Velicogna and Wahr*, 2005, 2006], and (4) aquifer depletion rates [*Gonçalvès et al.*, 2013; *Rodell et al.*, 2009; *Scanlon et al.*, 2012; *Tiwari et al.*, 2009; *Voss et al.*, 2013].

During wet climatic conditions, such as those that prevailed during recharge periods of the NSAS and APAS, soil moisture increases, fluvial systems, and drainage networks are developed, underlying aquifers are recharged, rising groundwater tables discharge in lowlands and depressions, and interactions between surface runoff and groundwater flow systems intensify. In contrast, during the dry climatic conditions similar to present conditions over the study areas, there is less soil moisture, surface drainage patterns dry up and may get buried under encroaching aeolian deposits, aquifer recharge is reduced and localized, groundwater tables are lowered, and groundwater discharge in lowlands decreases.

Obviously, GRACE-derived TWS data could not be used to address the impacts of climatic changes on the NSAS and the APAS aquifers over the timescales described above (tens to hundreds of thousands of years), but this data could possibly be used to investigate the impacts of the more recent climatic changes (e.g., global warming and/or multiyear variability [*Chen et al.*, 2009; *Jacob et al.*, 2012; *Velicogna*, 2009]). Global warming influences sea and land surface temperatures, which in turn affect precipitation rates and patterns. On land, it increases evaporation, dries land surfaces, intensifies droughts, and prolongs drought duration; over oceans, it increases the water-holding capacity of air and intensifies storms and precipitation over the coastal plains and mountainous source areas along storm trajectories [*Trenberth*, 2011]. A fraction of this added precipitation could ultimately lead to enhanced infiltration and recharge of the fossil aquifers if the increased precipitation was not offset by enhanced evapotranspiration.

Anthropogenic practices could significantly affect TWS as well. For example, TWS in arid areas could be reduced by the excessive exploitation of fossil aquifers. Dams, on the other hand, impound surface water, induce infiltration, and increase recharge from the lakes to groundwater, and hence increase the TWS [*Ahmed et al.*, 2014]. In this chapter, we monitor trends in GRACE-derived TWS data over the NSAS and the APAS over a period of 10 years (January 2003–September 2012) and investigate the nature of the factors (e.g., climatic and/or human pressure related) controlling these variations.

21.3. GEOLOGY AND HYDROGEOLOGY OF THE STUDY AREA

The study areas—the NSAS in northeast Africa and the APAS in the Arabian Peninsula—have similar geologic and hydrogeologic settings. The Arabian and Nubian Shields in northeast Africa and in the Arabian Peninsula were formed by accretion of a complex of island arcs and interleaving oceanic basins that were later accreted against the African continent 550–950 Ma [*Kroner et al.*, 1987]. The Arabian and Nubian Shields remained contiguous until about 25 Ma [*Bohannon*, 1986], in the late Oligocene to early Miocene period, as the Red Sea rift propagated northward and separated the Arabian plate from the once-contiguous Arabian-Nubian plate [*Bosworth et al.*, 2005].

Rifting was associated with uplift; the shoulders of the rift along the Red Sea and the Gulf of Suez and along the Dead Sea transform fault were elevated by as much as 4 km [*Garfunkel and Bartov*, 1977], exposing the underlying crystalline rock and the overlying thick (up to 2.5 km) [*Garfunkel and Bartov*, 1977] sedimentary successions to extensive erosion. The uplift and erosion devastated the thick sedimentary successions overlying the rising Red Sea Hills; however, it brought to the surface deeply buried sedimentary sequences, providing ample opportunities to recharge the now-exposed sequences at the Red Sea foothills. These outcrops flank the Red Sea Hills and dip to the east on the Arabian side and to the west on the African side (Figures 21.1 and 21.2).

The NSAS comprises two major aquifer units, the Nubian Aquifer System (NAS) and the Post-Nubian Aquifer System (PNAS) (Figure 21.2a). The NAS covers large territories (area: $2 \times 10^6 \text{km}^2$) in four countries: Sudan (17%), Libya (34%), Chad (11%), and Egypt (38%). It consists mainly of sandstone and interleaving confining layers of Cambrian to Cenomanian ages. The PNAS extends over the northern sections of eastern Libya and the northern parts of the Western Desert of Egypt. The NAS and PNAS are separated by low-permeability Upper Cretaceous to Lower Tertiary units but are hydraulically connected via structures and by the frequent absence of low-permeability units due to non-deposition. These stratigraphic relationships are demonstrated in Figure 21.2a, a generalized schematic cross section along a south-north trending transect (location in Figure 21.1). Within the NAS, the water-bearing strata reaches thicknesses of up to 3.5 km, with the majority of these waters in Egypt (41.5 vol. %), and Libya (41.5 vol. %), and less of it in Chad (12.8 vol. %) and Sudan (9 vol. %) [*CEDARE*, 2001].

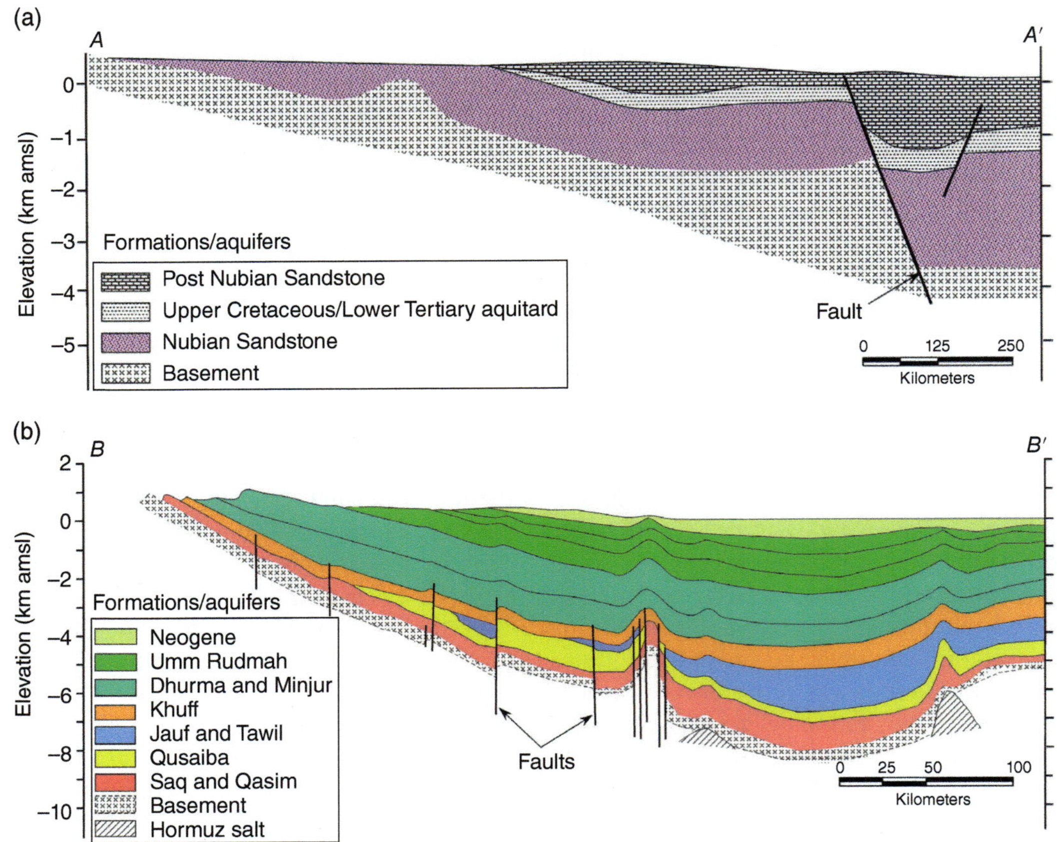

Figure 21.2 Schematic cross sections: (a) south to north schematic cross section across the NSAS along line A–A′ shown in Figure 21.1 modified from *CEDARE* [2001](b) SW to NE schematic cross section across the APAS along line B–B′ shown in Figure 21.1, modified from *Konert et al.* [2001] and *Alsharhan and Nairn* [1994].

The NAS comprises two interconnected major basins: the Kufra Basin in Libya, Sudan, and Chad and the Dakhla Basin in Egypt [*CEDARE*, 2001] (Figure 21.1). Water in the Kufra Basin flows from the south to the north, and that in the Dakhla Basin flows from south-southwest to north-northeast (average gradient ~0.5 m/km) and discharges naturally at oases in the Western Desert of Egypt [*Thorweihe and Heinl*, 2002] (Figure 21.1). In both basins, the aquifer host rock is pre-Mesozoic sandstone with minor intercalations of shale and clay of deltaic and marine origin [*Hesse et al.*, 1987; *Thorweihe*, 1982]. The sandstone is exposed in northeast Chad, southeast Libya, northern Sudan, and southwest Egypt. North of the 25th parallel in Egypt, the NAS is confined beneath thick marine shale of Upper Cretaceous age (Figure 21.1). North of the 29th parallel, northward groundwater flow is limited by saline water. Where the aquifer is confined, it is divided into four thick (tens to hundreds of meters thick) horizons of sandstone layers that are separated by relatively thin (meters to tens of meters thick) shale layers.

The APAS (area: 1.8×10^6 km²) extends across the entire kingdom of Saudi Arabia; in the northern parts of the Arabian Peninsula, it extends into Jordan and Iraq; in the southern parts of the Arabian Peninsula it extends to southern Saudi Arabia, Oman, Yemen, and the United Arab Emirates (Figure 21.1). Thick sequences of sedimentary formations ranging in age from Cambrian to recent time rest unconformably over the crystalline basement of the Red Sea Hills; they dip gently to the east and thicken in the same direction, reaching thicknesses of up to 5 km in the vicinity of the Arabian Gulf. These stratigraphic relationships are demonstrated in Figure 21.2b, a generalized schematic cross section along a southwest-northeast trending transect (location in Figure 21.1). Groundwater in the APAS flows from the west to the east (Figure 21.1) and discharges naturally in the Arabian Gulf and in inland sabkhas proximal to the Arabian Gulf [*Sultan et al.*, 2008].

Groundwater in the APAS is hosted primarily in sandstone, limestone, and dolomite formations separated by interleaving confining shale units. The APAS contains several major aquifers (Figure 21.2b) that curve in an arc along the edge of the Arabian Shield from northwest to southeast. These aquifers are grouped into three main groups: (1) Paleozoic sandstone (e.g., Saq, Al Wajid,

Al Qasim, Tawil, and Sarah) and limestone (e.g., Khuff) aquifers; (2) Mesozoic sandstone (e.g., Minjur, Dhruma, Biyadh, and Wasia) and limestone and dolomite (e.g., Umm Radhum and Dammam) aquifers; and (3) Cenozoic aquifers that include the Neogene deposits [*Al Alawi and Abdulrazzak*, 1994; *Ministry for Higher Education*, 2000]. The APAS also contains secondary aquifers such as Jauf, Skaka, Khuff, and Jilh [*Ministry for Higher Education*, 2000]. In general, the various aquifers of the APAS are unconfined in recharge areas (outcrop locations at the foothills of the Red Sea mountainous areas) and become confined eastward, away from outcrop locations.

There is an agreement among researchers that the NSAS and the APAS were largely recharged in previous wet climatic periods. This suggestion is supported in the NSAS by the old ages of the NSAS groundwater samples and by the progressive increase in ages of groundwater along the flow direction from the recharge areas in the southwest (SW) to the northeast (NE) in the Western Desert of Egypt. Carbon-14 (^{14}C), chlorine-36 (^{36}Cl), and krypron-81 (^{81}Kr) ages for these samples ranged from 30,000 years in East Uweinat (SW of Western Desert) to a million years in the Bahariya oasis in the northern parts of the Western Desert (Figure 21.1) [*Patterson et al.*, 2005; *Sturchio et al.*, 2004]. Two models were advanced to explain the mechanics of recharge in the previous wet climatic periods: (1) large-scale flow from mountainous recharge areas in Chad [*Ball*, 1927; *Sanford*, 1935], and (2) autochthonous recharge throughout the present desert area [*Abouelmagd et al.*, 2012; *Sultan et al.*, 1997; *Thorweihe*, 1982].

Similarly, groundwater in the APAS originated largely from paleo-precipitation over the Red Sea Hills during moist intervals of the Late Pleistocene that recharged the aquifers cropping out at the foothills of the Red Sea mountains [*Sultan et al.*, 2008]. This suggestion is supported by (1) the progressive decrease in hydraulic head and increase in groundwater salinity from west to east, (2) the presence of major east-west trending drainage networks that channeled precipitation from the Red Sea Hills toward recharge areas at the foothills of the mountains [*Sultan et al.*, 2008], and (3) the radiocarbon ages of groundwater samples from a number of these aquifers (Saq: 22,000–28,000 ka; Biyadh and Wasia: 8000–16,000 ka; Umm Radhuma: 10,000–28,000 ka) [*Alsharhan*, 2003; *Beaumont*, 1977; *Otkun*, 1971]. These apparent radiocarbon ages are probably mixed ages of relatively young (8000–10,000 ka) waters that were precipitated during the last pluvial ages with waters that were recharged during earlier pluvials. Because the older water does not contain ^{14}C, the mix will yield relatively young ^{14}C ages, but will yield much older ages (hundreds of thousands up to millions of years) if dated with ^{36}Cl, ^{81}Kr, or helium [*Heinl and Brinkmann*, 1989], as is the case

with the NSAS groundwater [*Patterson et al.*, 2005; *Sturchio et al.*, 2004].

Recent work has also shown that, locally, these aquifers are receiving additional modern recharge in areas where relatively high precipitation occurs. Examples of such areas include the recharge areas at the foothills of mountainous source areas for the NSAS in Sudan, Chad, and in Sinai [*Sultan et al.*, 2011, 2013], and at the foothills of Red Sea Hills in the Jazan area for the APAS [*Sultan et al.*, 2008].

21.4. METHODOLOGY

In this section we briefly describe the data used in this study and explain in detail the processing steps used to identify temporal and spatial variations over the NSAS and the APAS in monthly (1) GRACE gravity field solutions, (2) precipitation extracted from Tropical Rainfall Measuring Mission (TRMM) data, and (3) soil moisture extracted from the European Space Agency (ESA) Essential Climate Variable (ECV) project. The spatial and temporal analysis of these data together with other relevant data sets (geologic, geochemical, remote sensing, and hydrologic models outputs) provides insights into the nature (i.e., climatic or anthropogenic) of the factors controlling the observed GRACE mass variations over the investigated aquifer systems.

21.4.1. Data Description

21.4.1.1. GRACE Data
GRACE gravity field solutions (RL05) that span the period from January 2003 through September 2012 from the GRACE database provided by the University of Texas Center of Space Research (UT-CSR) (available at ftp://podaac.jpl.nasa.gov/allData/grace/L2/CSR/RL05) were analyzed over the course of this study. The GRACE solutions are represented in terms of fully normalized spherical harmonic decompositions up to degree 60. Over the study area, all GRACE-derived products were interpreted as reflecting changes in TWS [*Ahmed et al.*, 2011], given (1) the slow rates of the mass variations in the underlying solid earth, (2) the absence of large earthquakes and glacial isostatic adjustment in the region, and (3) the small to negligible contributions related to mass fluctuations from the adjacent ocean [*Wahr et al.*, 1998] combined with the fact that model-based corrections were applied by UT-CSR to remove time-variable oceanic and atmospheric gravity signal from the raw GRACE measurements [*Tapley et al.*, 2004a, 2004b]. Moreover, the ocean general circulation model used to remove oceanic contributions includes the Red Sea. Therefore, assuming the model is accurate enough in this region, there should be little contamination of our results by the Red Sea.

21.4.1.2. Surface Water Data

We examined the temporal variations in the height of the surface water bodies within the NSAS in Egypt, namely Lake Nasser (area: 5250 km²) and Tushka lakes (area: 1300 km²) (Figure 21.1). There were no significant surface water bodies within the APAS area. Lake Nasser surface water levels were extracted from two main sources: (1) the Hydroweb database (available at http://www-apache.legos.obs-mip.fr/en/soa/hydrologie/hydroweb) at Laboratoire d'Etudes en Geophysique et Oceanographie Spatiales (LEGOS/GOHS) [*Cretaux et al.*, 2011], and (2) the U.S. Department of Agriculture's Foreign Agricultural Service (USDA-FAS) Global Reservoir and Lake Monitor (GRLM) database (available at http://www.pecad.fas.usda.gov/cropexplorer/global_reservoir). Both sources are primarily based on satellite altimetry data (e.g., Topography Experiment/Poseidon [Topex/Poseidon], European Remote Sensing [ERS-2], Geosat Follow-On [GFO], Jason-1, and Environmental Satellite [ENVISAT]). The first source provides Lake Nasser absolute height data from September 1992 through October 2011, and the second provides height anomalies from September 1992 through March 2013. We merged the data from both sources to generate a monthly time series of Lake Nasser height over the examined period (January 2003–September 2012).

Since Biblical times, the floods of the Nile River have been noted to follow cyclic patterns. In 1993 Lake Nasser started to rise; it reached its maximum water level of 182 m above mean sea level (amsl) and spilled over into the Tushka spillway, filling the first Tushka depression by late 1998. The lake covered an area of 400 km², with a lake stage of 172 m amsl. By 2000, the excess water from Lake Nasser spread farther west, forming three additional lakes across a lowland area of 750 km², with lake stages decreasing from east to west (162 m amsl at the second lake to 147 m at the fourth lake). By 2002, a fifth lake to the northwest developed, with a lake stage of 144 m amsl. All five newly developed Tushka lakes covered a total area of 1300 km², which is more than 20% of the area of Lake Nasser at maximum capacity. Starting in 2003, successive low Nile flows resulted in a general decrease in the Lake Nasser surface water levels and the flow into the Tushka lakes ceased. The reduction of flow into the Tushka lakes coupled with excessive evaporation lowered the Tushka lake levels, diminished their areas, reduced their reservoir capacity, and led to the progressive demise of the lakes over time.

21.4.1.3. Rainfall and Soil Moisture Data

Rainfall data was extracted from TRMM (version: 3B42.007A) data spanning the same GRACE period (January 2003–September 2012). TRMM is a joint mission between the National Space Development Agency (NASDA) of Japan and the National Aeronautics and Space Administration (NASA) of the United States, launched in 1997 as part of the Earth Observing System (EOS) [*Huffman et al.*, 2007]. TRMM provides global (50°N–50°S) data on rainfall using microwave and visible-infrared sensors. Instantaneous rainfall estimates are obtained every 3 h with a 0.25° × 0.25° footprint and continuous coverage from 1998 to the present. The primary rainfall instruments of TRMM are the Precipitation Radar (PR), TRMM Microwave Imager (TMI), and the Visible and Infrared Scanner (VIRS) [*Kummerow et al.*, 1998].

The soil moisture data (January 2003–December 2010) was extracted from the European Space Agency–Essential Climate Variable (ESA-ECV) project data. The ECV soil moisture data is a merged product of both active and passive soil moisture products. The active sensors include the scatterometer measurements from ERS-1&2, Meteorological Operational (METOP) Advanced Scatterometer (ASCAT), and Active Microwave Instrument–Windscat (AMI-WS), while the passive sensors include the radiometer measurements from Scanning Multichannel Microwave Radiometer (SMMR), Special Sensor Microwave Imager (SSM/I), TMI, and Advanced Microwave Scanning Radiometer–Earth Observing System (AMSR-E) [*Liu et al.*, 2012; *Wagner et al.*, 2012]. The homogenized and merged product presents surface (top 0.5–2 cm) soil moisture measurements with a global coverage, spatial resolution of 0.25°, and temporal resolution of one day.

21.4.2. Data Processing and Analysis

21.4.2.1. GRACE Data

The processing of GRACE data involved the removal of the temporal (January 2003–September 2012) mean from GRACE monthly solutions. To spatially identify the nature (i.e., increasing or decreasing) of the regional trends in GRACE-derived TWS over the selected aquifer systems, we converted the monthly GRACE gravity field solutions to grids (0.5° × 0.5°) of equivalent water thickness using a Gaussian smoothing function with a radius of 350 km [*Wahr et al.*, 1998]; then we simultaneously fit a trend and seasonal terms (i.e., annual and semiannual components) at each grid point. The Gaussian smoothing function with a 350 km radius was selected after trying several Gaussian radii because it revealed regional trends free from the linear, north-south oriented features (i.e., stripes) [*Swenson and Wahr*, 2006]. The smaller the radius of the smoothing function, the more visible the stripes become. The solution for the trend is shown in Figure 21.3.

Examination of Figure 21.3 reveals that two areas show significant TWS depletions: (1) the northeastern

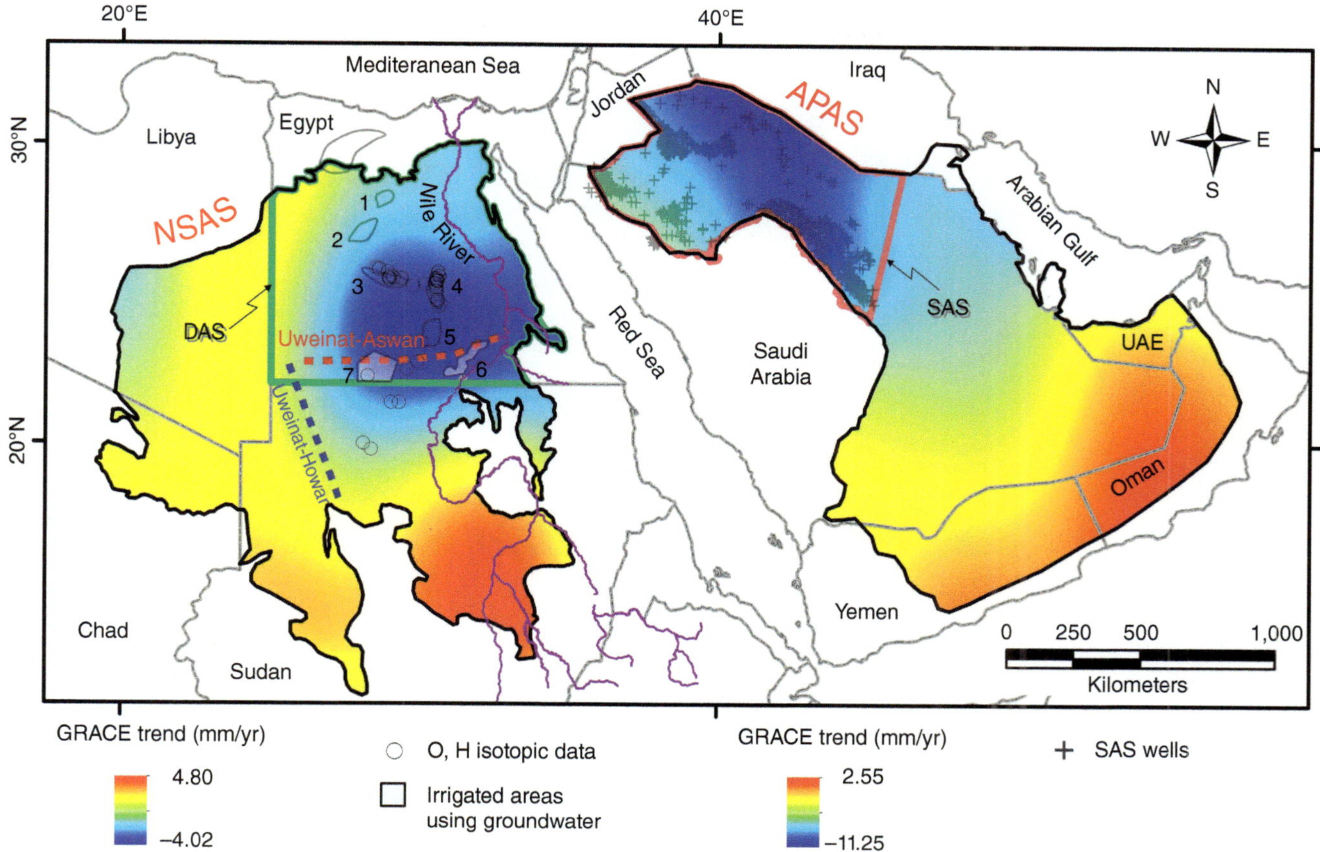

Figure 21.3 Linear trend (mm/yr) generated from smoothed (350 km; Gaussian) GRACE monthly solutions that span the period from January 2003 to September 2012 over the NSAS and the APAS. Also shown are the locations of irrigated areas within DAS (1: Baharya; 2: Farafra; 3: Dakhla; 4: Kharga; 5: Baris; 6: west of Nile River; 7: East Uweinat), basement uplifts, wells tapping the SAS, and groundwater samples located north and south of the Uweinat uplift that were analyzed for their O and H isotopic compositions.

part of the NSAS in Egypt, which correlates with the distribution of the Dakhla Aquifer System (DAS; area: 0.66×10^6 km²; Figure 21.3) in the NSAS; and (2) the northern parts of the APAS, which correlates with the distribution of the Saq Aquifer System (SAS; area: 0.46×10^6 km²; Figure 21.3) within the APAS. We adopt the term "DAS" to refer to the NAS and the overlying PNAS within the Dakhla Basin and the term "SAS" to refer to the Saq and the overlying Post-Saq Aquifer Systems in northern Saudi Arabia.

Following the identification of the regional trends, detailed studies were then conducted to quantify the temporal TWS variations over DAS and SAS. We first generated the equivalent water thickness grids using a Gaussian smoothing function with a radius of 200 km, given the relatively larger area of both DAS and SAS, and summed the results for all grid points lying within an individual aquifer. Because GRACE solutions are affected by the truncation of the spherical harmonics at degree ≤ 60 and by applying smoothing functions [*Landerer and Swenson*,

2012], we then adopted standard procedures to quantify these contributions and to rescale the data in ways that minimize these effects. We first generated the average sensitivity kernel functions (Figure 21.4) for our DAS and SAS averages using the methods described in *Jacob et al.* [2012]. Ideally, a sensitivity function should be 1.0 inside an aquifer, and 0 outside. However, this can never be perfectly realized in practice. In our case, Figure 21.4 shows large sensitivities in the centers of DAS and SAS and lower sensitivities along their edges. In addition, the functions are nonzero outside the boundaries of the DAS and SAS, which can lead to "leakage" from adjacent mass signals. To correct for these imperfect sensitivity functions and to recover relatively unbiased TWS estimates for DAS and SAS, we determined and applied scaling factors using methods described, for example, by *Velicogna and Wahr* [2006]. Briefly, we chose a synthetic mass distribution across the region, and computed the "true" DAS and SAS signals predicted by that mass distribution. We then converted that mass distribution into Stokes coefficients up to

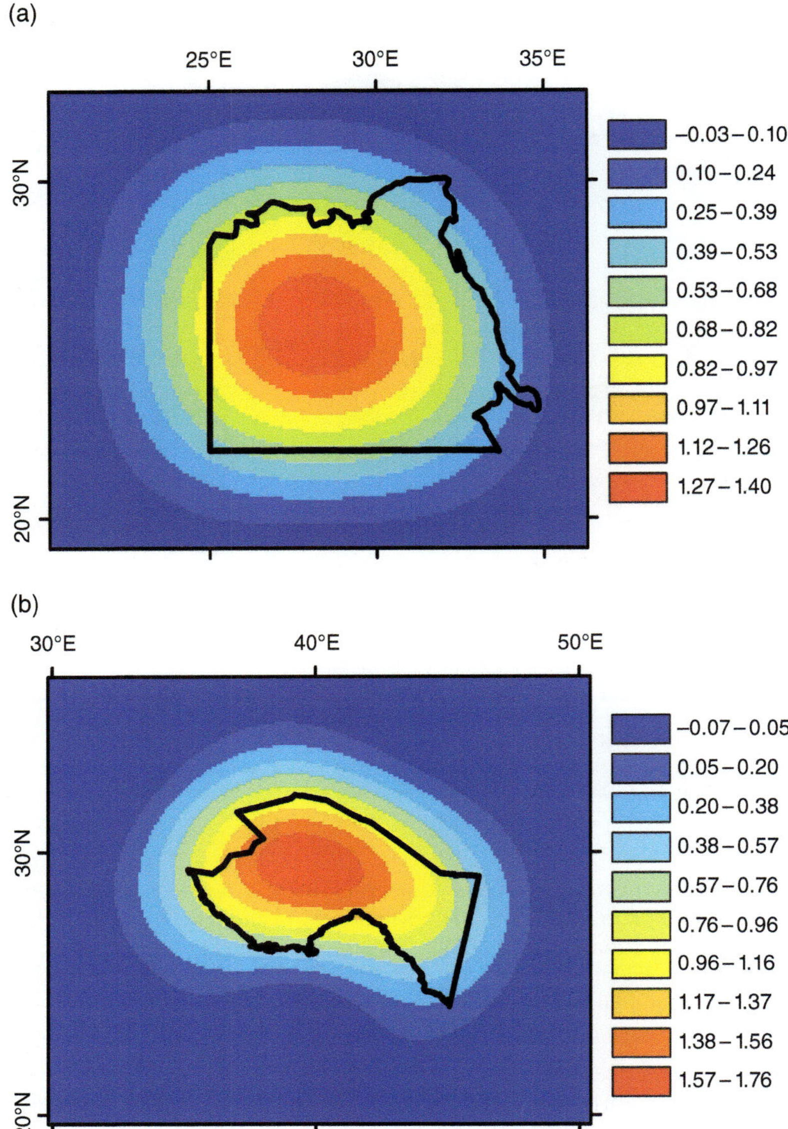

Figure 21.4 GRACE averaging sensitivity kernel function. The unscaled, dimensionless averaging function was used to estimate GRACE TWS over (a) the DAS in Egypt and (b) the SAS in Saudi Arabia.

degree 60, and estimated the apparent DAS and SAS mass distributions from those Stokes coefficients by applying the same analysis we used for the real GRACE data. The ratio of the true aquifer mass signal to the mass signal estimated using the GRACE approach is the scale factor for that aquifer; we multiply the actual GRACE solutions by that scale factor to obtain our final scaled TWS estimate. We have done this for two synthetic mass distributions: one where the mass anomaly pattern is set equal to the trend results computed for 350 km Gaussian averaging (Figure 21.3; but extended to cover the entire globe), and the other where the water storage is set to be 1.0 inside an aquifer and 0 outside. We use the average of the resulting two scale factors as our preferred scale factor, and we assume that the uncertainty in that scale factor is ±(differ-

ence between the two scale factors). Note that this choice for the uncertainty is equivalent to assuming the total width of the uncertainty interval to be twice the difference between the scale factors. Our results for the scaling factors are 1.8 ± 0.4 for DAS and 1.6 ± 0.4 for SAS. Finally, for the generated scaled TWS time series for DAS and SAS (Figure 21.5), we fitted a trend and seasonal terms and calculated the associated measurement errors using the approach described in *Tiwari et al.* [2009]. The final TWS trend uncertainties were calculated by adding, in quadrature, the contributions from measurement errors to the effects of errors in the scaling factors. The annual GRACE-derived TWS declines for the DAS in Egypt and the SAS in Saudi Arabia are estimated at -4.22 ± 1.50 mm and -13.00 ± 4.00 mm, respectively (Figure 21.5).

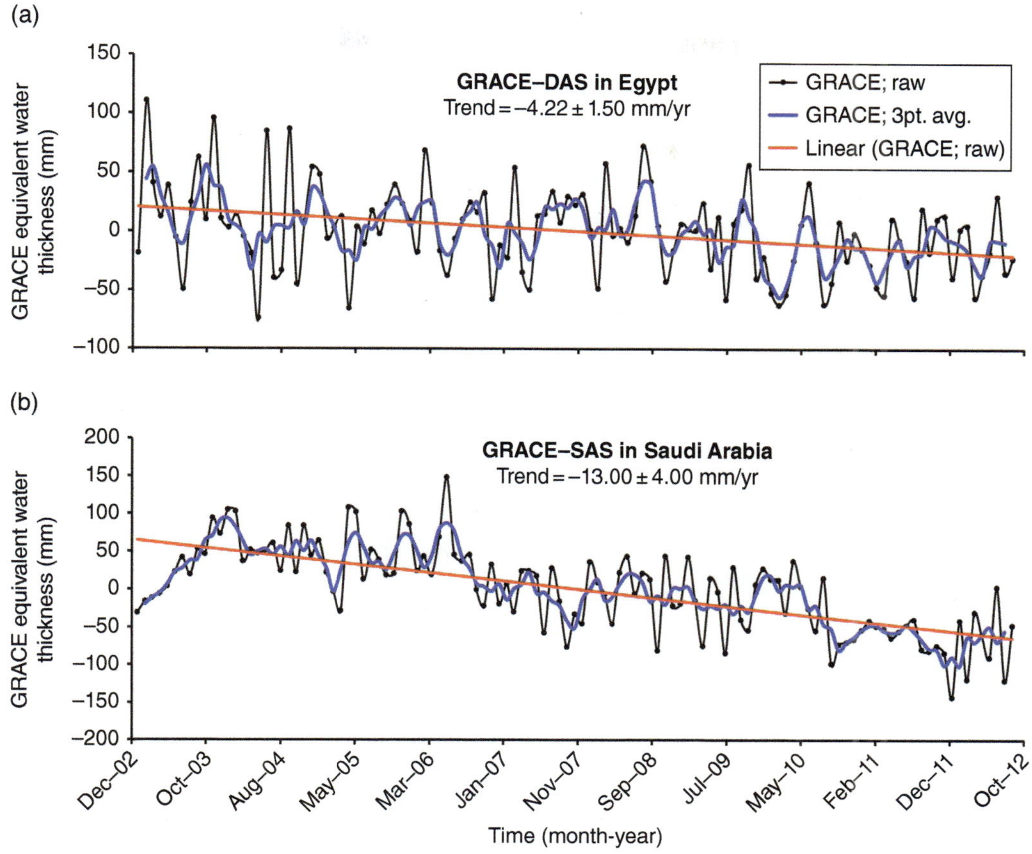

Figure 21.5 Time series for scaled GRACE monthly (January 2003–September 2012) solutions over (a) the DAS in the NSAS and (b) the SAS in the APAS.

21.4.2.2. Surface Water Data

The following procedures were applied to estimate the contributions of surface water variations in both Lake Nasser and the Tushka lakes to the GRACE-derived TWS over DAS. Following the generation of the Lake Nasser height time series, we generated the lake monthly height anomalies (with respect to the temporal mean). We then converted these height anomalies into monthly Stokes coefficients up to degree 60. Equivalent water thickness grids were then generated using a Gaussian smoothing function with a radius of 200 km. The results for all grid points lying within the DAS were then summed to generate the time series. The trend was then calculated as in the case of GRACE data. The Lake Nasser trend was estimated to be –0.45 mm/yr.

A surface water model was constructed and calibrated by adjusting routing parameters against lake stages identified for 1998–2002 to investigate the long-term hydrologic impacts of the Tushka lakes [*Sultan et al.*, 2013; *Yan et al.*, 2003]. Outputs of these models indicated that the total volume of water naturally diverted from Lake Nasser to the five Tushka lakes from late 1998 to early 2002 is about $2.74 \times 10^{10}\,m^3$, of which

$1.06 \times 10^{10}\,m^3$ were lost to evaporation, leaving behind an estimated volume of $1.68 \times 10^{10}\,m^3$. Examination of temporal Landsat Thematic Mapper (TM) and Enhanced Thematic Mapper (ETM) satellite imagery acquired for the time period from 2003 to the present reveals a progressive decline in the Tushka lakes' areas to the extent that three of the lakes have completely dried by year 2012 and the size of the remaining two lakes was drastically reduced (Figure 21.6). The volume of storage in the Tushka lakes changes from $1.68 \times 10^{10}\,m^3$ in 2002 to $8.88 \times 10^9\,m^3$ in 2012. Over the investigated period (January 2003–September 2012), about 53% of the water in the Tushka lakes was lost to evaporation, leaving only $7.92 \times 10^9\,m^3$ in storage. We used the same approach as in the case of Lake Nasser to calculate the contribution of the Tushka lakes to the GRACE-derived TWS. The Tushka lakes trend was estimated to be –0.68 mm/yr.

It is difficult to obtain meaningful estimates for the accuracy of the Nasser and Tushka lakes corrections. We note, however, that when we view a Gaussian-averaged map of the trends across the entire region after these corrections have been made, there is no apparent signal either over these lakes or over the Red Sea. Moreover, the

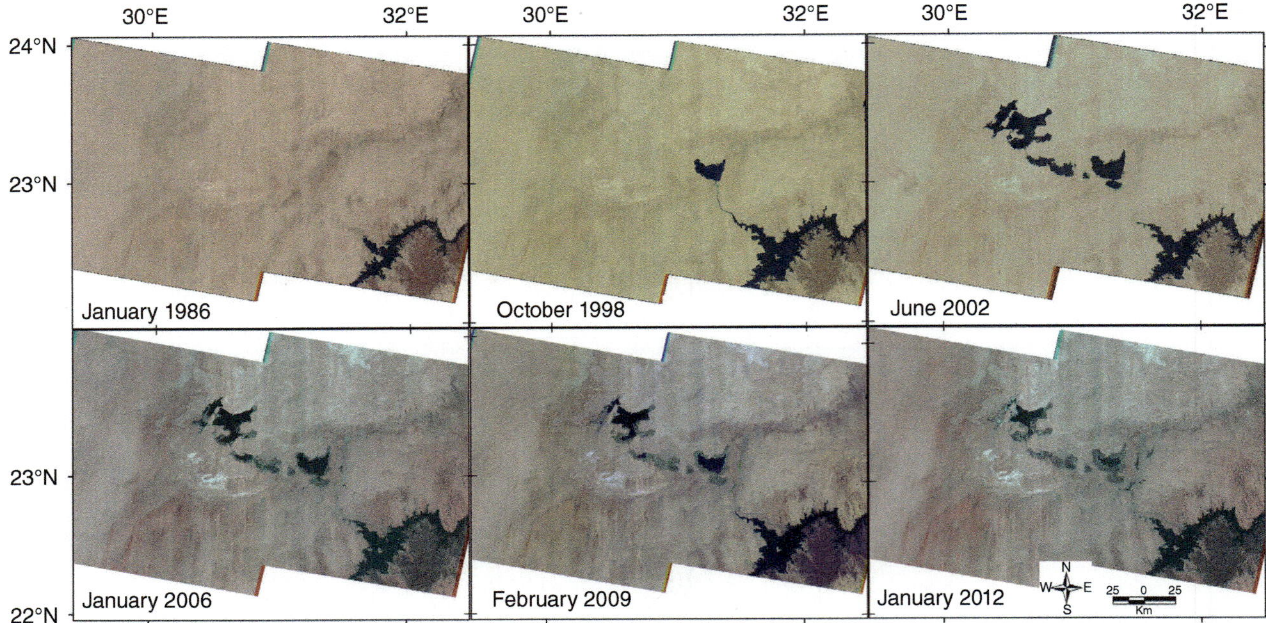

Figure 21.6 Temporal satellite imagery spanning the time period from January 1986 to January 2012, showing variations in number and size of the Tushka lakes. The images in the top row were extracted from the Landsat TM, while those in the bottom row were extracted from the Landsat ETM.

Red Sea lies far enough from the DAS and SAS that it would not have been likely to contaminate our solutions even if there had been a significant signal there. The implication is that whatever errors exist in the corrections for the lakes and for the Red Sea, they are far smaller than the total TWS uncertainties of our results as given above.

21.4.2.3. Rainfall and Soil Moisture Data

Rainfall and soil moisture data were used to investigate the climatic controls on the temporal TWS variations observed in GRACE data. The precipitation rates were processed as follows: (1) total monthly rainfall images were generated from the 3 hourly precipitation data; (2) an average annual precipitation (AAP) image was extracted (Figure 21.7); (3) a trend image was then generated from the monthly precipitation as in the case of GRACE data (Figure 21.8); and (4) the rainfall time series for individual aquifers were generated by averaging the results for all grid points lying within these aquifers (Figure 21.9). Similar procedures were applied to the soil moisture data, where the daily (January 2003–December 2010) soil moisture data was converted to average monthly measurements that were then spatially averaged over both SAS and DAS systems to generate the soil moisture time series (Figure 21.10). The uncertainty associated with the rainfall and soil moisture trends was calculated using the approach described in *Tiwari et al.* [2009].

21.5. DISCUSSION AND FINDINGS

Figure 21.5a shows the scaled GRACE-derived TWS time series over the DAS in Egypt and the SAS in Saudi Arabia. The depletions (-4.22 ± 1.50 mm/yr) over the DAS are related to variations in water volume at all depths, including variations in groundwater, soil moisture, and surface water (i.e., Lake Nasser and Tushka lakes). The contributions of soil moisture over the DAS are negligible, given the rare rainfall events throughout the investigated period and during the preceding two decades, as described below. If we removed the combined Lake Nasser (-0.45 mm/yr) and Tushka lakes (-0.68 mm/yr) contributions from the TWS trend, the remainder (-3.08 ± 1.5 mm/yr) would represent variations in TWS that are largely related to temporal variations in groundwater storage. This corresponds to a loss by extraction and natural discharge amounting to $-2.04 \pm 0.99 \times 10^9$ m³/yr of the DAS groundwater across the examined area.

Similarly, Figure 21.5b shows the scaled GRACE-derived TWS time series over the SAS in Saudi Arabia; a depletion rate of -13.00 ± 4.00 mm/yr is observed over the investigated period. These depletions are related to variations in the groundwater and soil moisture volumes. There are no contributions from surface water bodies that are absent from the area. The reported GRACE trend corresponds to a decline of $-6.11 \pm 1.83 \times 10^9$ m³/yr in the combined groundwater and soil moisture volumes.

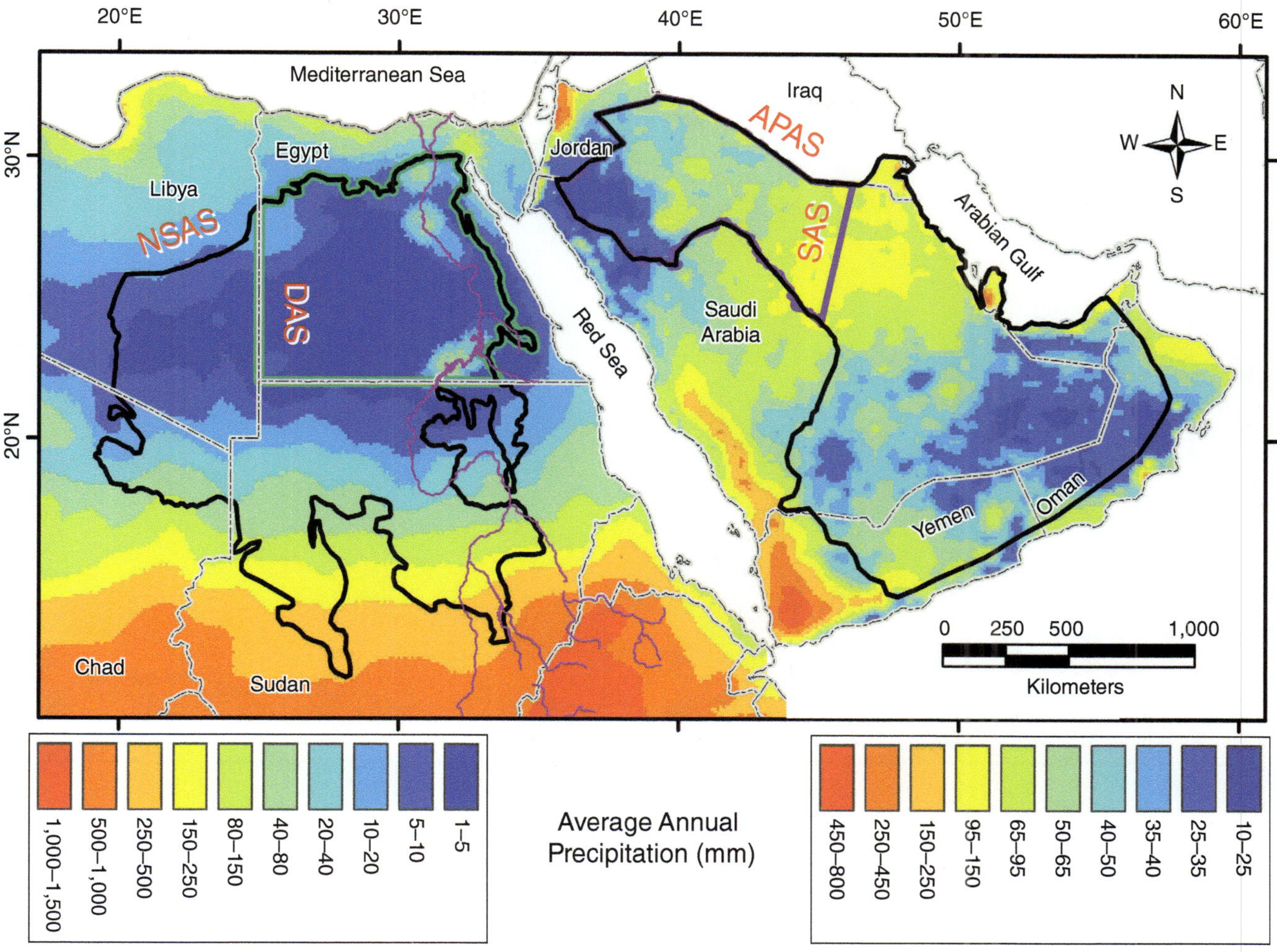

Figure 21.7 Average annual (January 2003–September 2012) precipitation over the NSAS and the APAS extracted from TRMM data.

The question of whether the observed declines in TWS over the DAS and the SAS are related to a decline in precipitation over these areas was addressed by examining (1) images that display AAP (Figure 21.7), trends in precipitation (Figure 21.8), and precipitation time series (Figure 21.9), all generated from TRMM data; and (2) soil moisture time series (Figure 21.10). Examination of these figures over the DAS and the SAS areas show negligible to low precipitation (AAP: DAS: 7 mm; SAS: 60 mm) over the investigated period, and near-steady trends in precipitation (DAS: -0.02 ± 0.01 mm/yr; SAS: -0.38 ± 0.18 mm/yr) and slightly increasing trends in soil moisture (DAS: 0.87 ± 0.15 mm/yr; SAS: 0.94 ± 0.44 mm/yr), suggesting that the observed decline in TWS over the DAS and the SAS throughout the investigated time period is not likely to be caused by a climate-change-related decrease in precipitation and recharge. If the decline was related to drier conditions throughout the investigated period compared to earlier periods, highly reduced AAP rates would be expected for the investigated period compared to those for the preceding period. This is apparently not the case; the AAP for the investigated period approximates the AAP for the preceding 10 years (1992–2001) (DAS: 9 mm; SAS: 78 mm). The AAPs for the preceding years were extracted from the Climate Prediction Center (CPC) Merged Analysis of Precipitation (CMAP) data.

The spatial correlation between the distribution of the areas showing declining TWS centered over DAS and groundwater-irrigated areas in the oases (Baris, Kharga, Dakhla, Farafra, Baharyia) and in the reclaimed lands west of Lake Nasser and in the East Uweinat) (Figure 21.3) suggests that the observed TWS depletions are largely related to increased groundwater extraction that is not compensated for by recharge and/or groundwater flow.

Exploitation of groundwater from the DAS for the development of agricultural communities in the Western Desert of Egypt has been on the rise since the 1960s. As a result, the groundwater table in the Western Desert oases declined by some 60 m and the artesian wells and springs

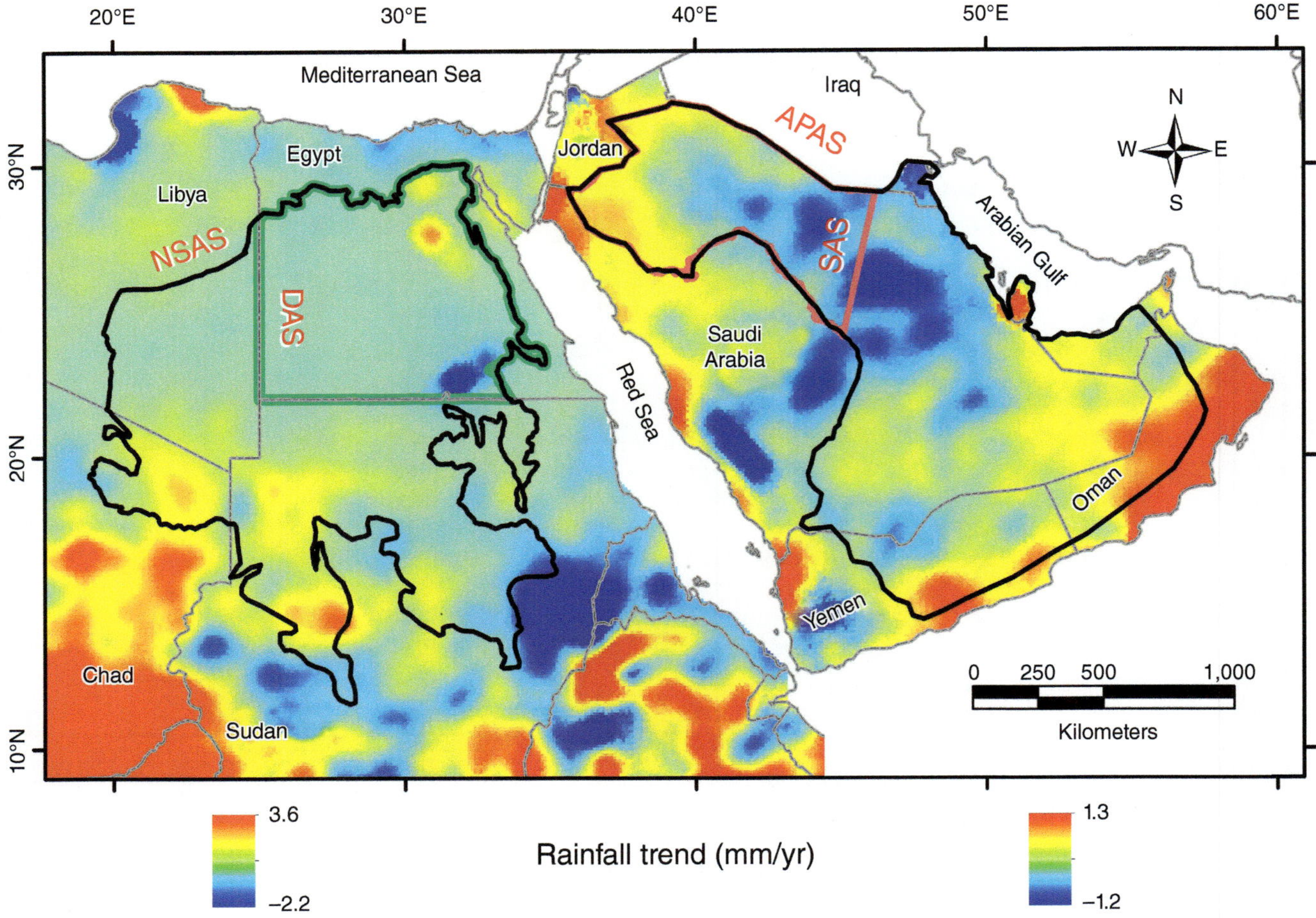

Figure 21.8 Linear trend (mm/yr) image generated from monthly (January 2003–September 2012) TRMM data over NSAS and APAS and surroundings.

have been replaced by deep wells [*Bakhbakhi*, 2006]. Additional large irrigation schemes are underway. One of these projects is the East Uweinat project (Figure 21.3); the lands reclaimed by this project amounted to 1200 ha in 1992 and 4200 ha in 2003, and the target is to reclaim a total of 75,000 ha by 2022, all of which will be irrigated using groundwater from the DAS [*FAO*, 2005; *Salem and Pallas*, 2002; *Salem*, 2007]. Another major project that is underway is the New Valley project, which utilizes groundwater from the DAS and excess surface water channeled from Lake Nasser to reclaim lands along a path subparallel to, and to the west of the Nile Valley. Upon completion, $540 \times 10^6 \, m^3/yr$ of groundwater will be extracted from the DAS [*Al-Eryani et al.*, 2006].

Similarly, extensive programs were initiated for groundwater extraction from the SAS to develop agricultural communities in northern Arabia. Temporal satellite images (Landsat TM and ETM) over one of these areas, Wadi AlSarhan (Figure 21.1, Box b), show the progressive increase in agricultural lands through the years (Figure 21.11; 1987, 1991, 2000, and 2012). As is the case

in the Western Desert of Egypt, groundwater abstraction remained insignificant ($\sim 100 \times 10^6 \, m^3/yr$) until the 1960s, the years that signaled the onset of projects aimed at the development of agricultural communities in the area. Since that time, there has been a progressive increase in groundwater extraction in 1984: $2 \times 10^9 \, m^3$; in 2005: $8.7 \times 10^9 \, m^3$) [*BRGM*, 2008], most of which came from the Saq Formation. Extraction from this source was estimated at $1.4 \times 10^9 \, m^3$ in 1984 and rose to $5.7 \times 10^9 \, m^3$ in 2005; the extracted amounts for each of these two years constituted approximately 65% of the total extracted volume of water from the SAS [*BRGM*, 2008]. In the next section, we investigate aquifer depletion rates in relation to relevant geochemical, geologic, and field data and outputs from hydrologic models.

21.5.1. DAS in Egypt

Our estimates for the annual groundwater decline in the DAS should be considered conservative, given that we ignored recharge from the Tushka lakes and Lake Nasser.

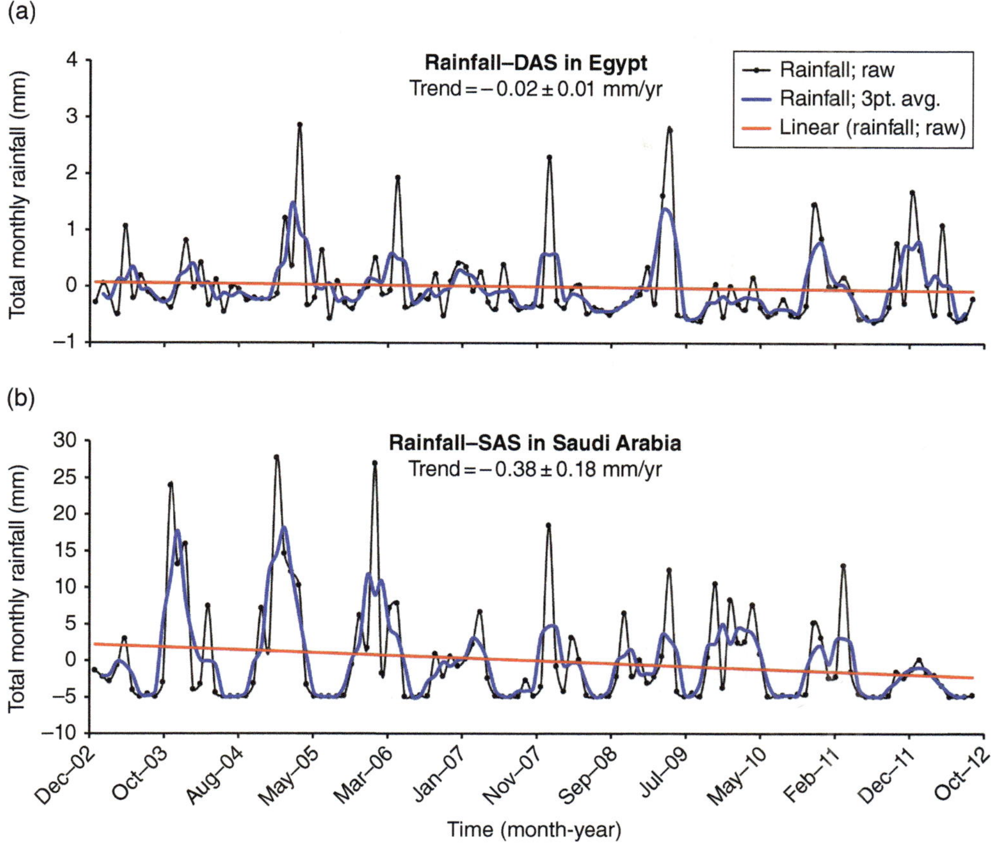

Figure 21.9 Time series for TRMM monthly (January 2003–September 2012) data over (a) the DAS in NSAS and (b) the SAS in APAS.

Accounting for these factors would increase the estimated groundwater losses. Our estimated combined groundwater losses to natural discharge and extraction are approximately twice the measured and modeled estimates for extraction (0.506×10^9 m³/yr) [*Bakhbakhi*, 2006] and natural discharge combined. The latter is of the same order of magnitude as the extraction [*Heinl and Brinkmann*, 1989]. This discrepancy could be explained by the fact that an estimated 0.7×10^9 m³/yr of groundwater discharge could have gone undetected in the Nile River and Gulf of Suez [*Sultan et al.*, 2007]. The progressive increase in groundwater extraction in the DAS is projected to bring the losses to 2.8×10^9 m³/yr by 2070 [*Heinl and Brinkmann*, 1989].

A question that needs to be addressed is why the NSAS-related TWS depletions are observed over the Dakhla Basin in the Western Desert (i.e., DAS) but not over the Kufra Basin in Libya, despite the fact that the reported annual losses to groundwater extraction in Libya (0.83×10^9 m³) exceed the extraction in Egypt (0.506×10^9 m³) [*Bakhbakhi*, 2006]. As described earlier, there are two major NAS basins, the Kufra and the Dakhla; precipitation over the Kufra and Dakhla outcrops in both countries is negligible to absent

(Figure 21.7), and thus these aquifers are not recharged locally. However, they are presumably replenished by groundwater flow from recharge areas in the south (i.e., Chad and Sudan). For the Dakhla Basin, replenishment by groundwater flow from recharge areas in Sudan, where substantial precipitation occurs over the NAS outcrops (Figures 21.1 and 21.7), is hindered by a major east-west trending structural uplift, the Uweinat-Aswan uplift (Figure 21.1 and 21.3), that restricts the south to north groundwater flow. That is not the case with the Kufra basin, where groundwater moves freely from the south to the north. This hypothesis is supported by the large differences in the isotopic compositions of groundwater on either side of the uplift (north of uplift: 37 samples, average isotopic composition δ^{18}O: $-10.7‰ \pm 0.9‰$; south of uplift: 7 samples, average isotopic composition δ^{18}O: $-8.6‰ \pm 1.4‰$; Figure 21.3) [*Ahmed et al.*, 2014; *Sultan et al.*, 2013]. The depletions observed over the DAS are absent from the remaining regions of the NSAS (Figure 21.3), suggesting that they are probably at nearly steady state conditions.

We estimate that the DAS could be mined at present depletion rates ($-2.04 \pm 0.99 \times 10^9$ m³/yr) for a period approaching 2500 years. This estimate is based on modeled

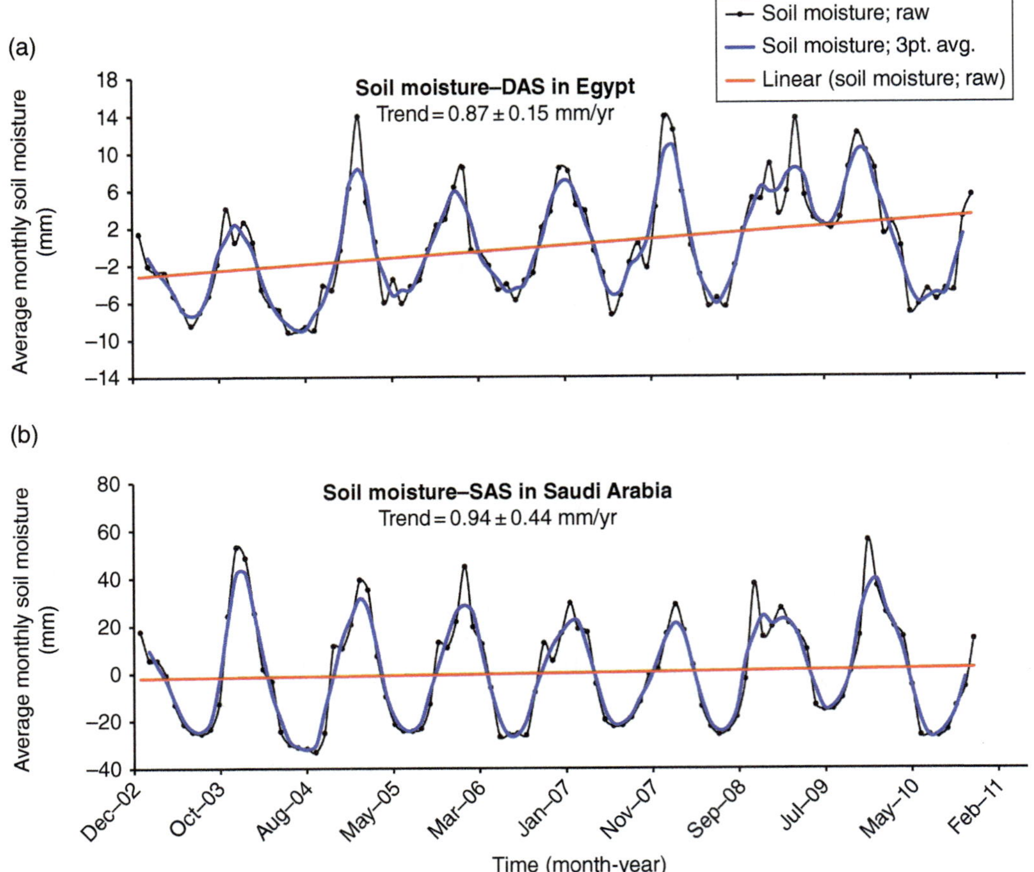

Figure 21.10 Time series for soil moisture monthly (January 2003–December 2010) data over (a) the DAS in NSAS and (b) the SAS in APAS.

recoverable groundwater volumes $(5180 \times 10^9\,\text{m}^3)$ [*Bakhbakhi*, 2006] that assume a maximum water decline of 100 m in the unconfined aquifer areas and 200 m in the confined aquifer. If we were to adopt the projected rates of artificial extraction in 2070 $(2.8 \times 10^9\,\text{m}^3/\text{yr})$ [*Heinl and Brinkmann*, 1989] and the present natural discharge rates $(0.506 \times 10^9\,\text{m}^3/\text{yr})$, the DAS could be mined for some 1550 years; it could be mined for approximately 350 years if the extraction rates double every 50 years. Between years 1979 and 2003 extraction increased from $0.47 \times 10^9\,\text{m}^3/\text{yr}$ [*Amer et al.*, 1979] to $1.1 \times 10^9\,\text{m}^3/\text{yr}$ [*Ebraheem et al.*, 2003].

21.5.2. SAS in Saudi Arabia

Over the SAS system, our GRACE-derived depletions $(-6.11 \pm 1.83 \times 10^9\,\text{m}^3/\text{yr})$ are consistent with the reported extraction $(-8.73 \times 10^9\,\text{m}^3)$ less recharge (3 mm; $1.38 \times 10^9\,\text{m}^3$) for the SAS in 2005 [*BRGM*, 2008]. As is the case with the NSAS and DAS, the depletions observed over the SAS are absent from the remaining regions of the APAS, as evidenced by the large number of wells

tapping the SAS (Figure 21.3), which suggests that the remaining regions of the APAS are probably at nearly steady state conditions.

We attribute the TWS depletions over the SAS and their absence over the remaining regions of the APAS to (1) exploitation of the Saq aquifer, which is heavy within the SAS region but absent elsewhere in the APAS; and (2) limited compensation for the heavy extraction by recharge from precipitation. The Saq aquifer is heavily utilized because of its large dimensions (area: $0.46 \times 10^6\,\text{km}^2$; thickness: 400–1200 m), and high storage capacity (storage: $258,400 \times 10^6\,\text{m}^3$; available: $103,600 \times 10^6\,\text{m}^3$) [*FAO*, 2009] compared to all other aquifers in the Kingdom of Saudi Arabia. The Saq Formation and the remaining SAS aquifers are fed primarily by runoff from precipitation over the adjoining northern and central sections of the Red Sea Hills, which receive modest precipitation (~60 mm/yr) (Figure 21.7). The remaining regions of the APAS to the south of the SAS have not been exploited like the SAS and are being readily recharged by the relatively higher amounts of runoff from the added precipitation

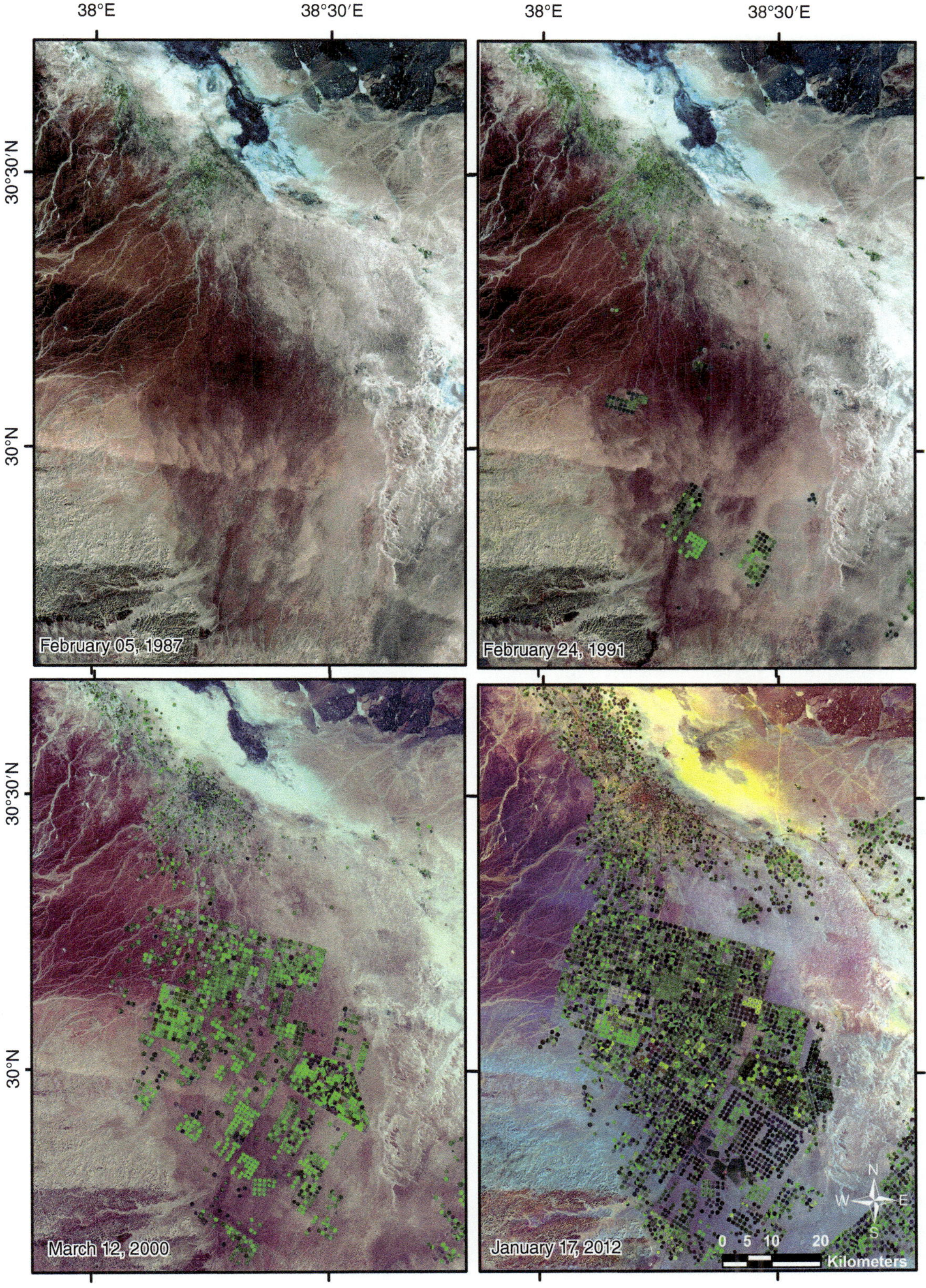

Figure 21.11 Temporal satellite imagery acquired in 1987, 1991, 2000, and 2012, showing a progressive increase in agricultural lands in the Wadi AlSarhan area in northeast Saudi Arabia (location in Figure 21.1). The agricultural fields in the images are about 1 km across and use a center-pivot irrigation system. On these images, the new vegetation appears bright green, dry vegetation or uncultivated fields appear rust colored, and dry and barren surfaces (i.e., desert) are pink and yellow. The 1987, 1991, and 2000 scenes are Landsat TM images, while the 2012 scene is a Landsat ETM image.

(~400 mm/yr) over their source areas in the Red Sea Hills. The total proven reserves for all fresh aquifers in the kingdom range from $900 \times 10^9 \, \text{m}^3$ [*FAO*, 2009] to $2185 \times 10^9 \, \text{m}^3$ [*Marget et al.*, 2006], and the available reserves range from $430 \times 10^9 \, \text{m}^3$ to $1000 \times 10^9 \, \text{m}^3$ [*FAO*, 2009]. Thus, we estimate that SAS could be mined for 70–160 years at the present GRACE-derived extraction and depletion rates ($-6.11 \pm 1.83 \times 10^9 \, \text{m}^3/\text{yr}$).

21.6. SUMMARY AND IMPLICATIONS

Many of the countries in arid and semiarid parts of the world are mining aquifers such as the NSAS and the APAS that were largely recharged in previous wet climatic periods tens to hundreds of thousands of years ago; recharge occurs through infiltration in the unconfined parts of the aquifers. During these periods, the gradients are high and large-scale flow occurs under steady state conditions. During dry conditions such as those prevalent in the present, precipitation is reduced and localized and recharge is diminished. During these periods, the reduced precipitation can no longer maintain the equilibrium conditions since the gradients are reduced with time. We describe our systems during these periods as being in near-steady conditions. With excessive extraction, the systems quickly move toward becoming highly unsteady.

Examination of the GRACE-derived TWS trend over the selected aquifer systems reveals two areas of significant depletions, the first correlated with the distribution of the DAS in the NSAS, and the second with the distribution of the SAS in the northern sections of the APAS. Scaled TWS time series were extracted for the DAS in Egypt and for the SAS in Saudi Arabia. From these, the annual combined groundwater and soil moisture depletion rates were extracted for the DAS ($2.04 \pm 0.99 \times 10^9 \, \text{m}^3/\text{yr}$) and for the SAS ($6.11 \pm 1.83 \times 10^9 \, \text{m}^3/\text{yr}$). Climatic changes are not responsible for the observed TWS depletions; instead, anthropogenic factors are, namely groundwater extraction, which is not compensated for by modern recharge and/or replenishment by groundwater flow. In case of the DAS, compensation by groundwater flow from the south is hindered by the presence of the Uweinat-Aswan basement uplift.

We estimate that the DAS in Egypt, if mined at present extraction rates and under the existing GRACE-derived depletion conditions, could last for a period approaching 2500 years. If, instead, we were to adopt the projected rates of artificial extraction in 2070 and present natural discharge rates, it will be consumed in 1550 years, and it will be consumed in an even shorter time period (350 years) if the extraction rates continue to double every 50 years. Given the range of reported available reserves ($430 \times 10^9 \, \text{m}^3$ to $1000 \times 10^9 \, \text{m}^3$), we estimate that

SAS could be mined for 70–160 years at the present extraction rates and under the existing GRACE-derived depletion conditions.

We interpret the observed depletions over the DAS and the SAS and their absence across the remaining regions of the NSAS and the APAS to indicate that both aquifers are largely at near-steady conditions, yet the DAS and SAS are not. They are both witnessing transient conditions. Finally, we suggest that the methodologies advocated here could be readily applied for assessment and optimum management of a large suite of fossil aquifers worldwide.

ACKNOWLEDGMENTS

Funding was provided by the National Aeronautics and Space Administration, Earth Science Division grants NNX08AJ85G and NNX12AJ94G awarded to Western Michigan University; and, at the University of Colorado, by NASA GRACE funding, and by NASA's "Making Earth Science Data Records for Use in Research Environments (MEaSUREs) Program."

REFERENCES

Abouelmagd, A., M. Sultan, A. Milewski, A. Kehew, N. Sturchio, F. Soliman, R. V. Krishnamurthy, and E. Cutrim (2012), Toward a better understanding of palaeoclimatic regimes that recharged the fossil aquifers in North Africa: Inferences from stable isotope and remote sensing data, *Palaeogeogr. Palaeoclimatol. Palaeoecol.*, *329–330*, 137–149.

Ahmed, M., M. Sultan, J. Wahr, E. Yan, A. Milewski, W. Sauck, R. Becker, and B. Welton (2011), Integration of GRACE (Gravity Recovery and Climate Experiment) data with traditional data sets for a better understanding of the time-dependent water partitioning in African watersheds, *Geology*, *39*, 479–482, doi: 410.1130/G31812.31811.

Ahmed, M., M. Sultan, J. Wahr, and E. Yan (2014), The use of GRACE data to monitor natural and anthropogenic induced variations in water availability across Africa, *Earth-Science Reviews*, *136*, 289–300.

Al Alawi, J., and M. Abdulrazzak (1994), Water in the Arabian Peninsula: Problems and perspectives, in *Water in the Arab World; Perspectives and Prognoses*, edited by P. Rogers and P. Lydon, pp. 171–202, Harvard Univ. Press, Cambridge, Mass.

Al-Eryani, M., B. Appelgren, and S. Foster (2006), Social and economic dimensions of non-renewable resources, in *Nonrenewable Groundwater Resources: A Guidebook on Socially Sustainable Management for Water-Policy Makers*, edited by S. Foster and D. P. Loucks, pp. 25–34, United Nations Educational, Scientific and Cultural Org., Paris.

Alsharhan, A. S. (2003), Petroleum geology and potential hydrocarbon plays in the Gulf of Suez rift basin, *Egypt*, *AAPG Bull.*, *87*, 143–180.

Alsharhan, A. S., and A. E. M. Nairn (1994), The Late Permian carbonates (Khuff Formation) in the western Arabian Gulf: Its hydrocarbon parameters and paleogeographical aspects, *Carbonates and Evaporites*, *9*(2), 132–142.

Amer, A., S. Nour, and M., Mishriki (1979), A finite element model of the Nubian Aquifer System in Egypt, Groundwater Seminar, Egyptian Ministry of Land Reclamation, Cairo, Egypt.

Amery, H. (1997), Water security as a factor in Arab-Israeli wars and emerging peace, *Studies in Conflict and Terrorism*, *20*, 95–104.

Bakhbakhi, M. (2006), Nubian Sandstone Aquifer System, in *Non-Renewable Groundwater Resources: A Guidebook on Socially Sustainable Management for Water-Policy Makers*, edited by S. Foster and D. P. Loucks, pp. 75–81, United Nations Educational, Scientific and Cultural Org. Paris.

Ball, J. (1927), Problems of the Libyan Desert, *Geogr. J.*, *70*, 21–38, 105–128, 209–224.

Beaumont, P. (1977), Water and development in Saudi Arabia, *Geogr. J.*, *143*(1), 42–60.

Bohannon, R. G. (1986), Tectonic configuration of the Western Arabian Continental-Margin, Southern Red-Sea, *Tectonics*, *5*(4), 477–499.

Bosworth, W., P. Huchon, and K. McClay (2005), The Red Sea and Gulf of Aden Basins, *Journal of African Earth Sciences*, *43*(1–3), 334–378.

Bureau de Recherches Géologiques et Minières (BRGM) (2008), Investigations for updating the groundwater mathematical model(s) of the Saq and overlying aquifers, Ministry of Water and Electricity—Kingdom of Saudi Arabia, pp. 1–144.

Center for Environment and Development for the Arab Region and Europe (CEDARE) (2001), *Regional Strategy for the Utilization of the Nubian Sandstone Aquifer System— Hydrogeology*, Vol. II, Center for Environ. Devel. for the Arab Region and Europe, Cairo, Egypt.

Chen, J. L., C. R. Wilson, and B. D. Tapley (2006), Satellite gravity measurements Confirm accelerated melting of Greenland Ice Sheet, *Science*, *313*(5795), 1958–1960.

Chen, J. L., C. R. Wilson, D. Blankenship, and B. D. Tapley (2009), Accelerated Antarctic ice loss from satellite gravity measurements, *Nat. Geosci.*, *2*, 859–862.

Cretaux, J., et al. (2011), SOLS: A lake database to monitor in the near real time water level and storage variations from remote sensing data, *Adv. Space Res.*, *47*, 1497–1507.

Ebraheem, A., H. Garamoon, S. Raid, P. Wycisk, and A. Seif Al Nasr (2003), Numerical modeling of groundwater resource management options in the East Oweinat area, SW Egypt, *Hydrol. Mediterranean Semiarid Reg.*, *278*, 15–23.

Fetter, C. W. (2001), *Applied Hydrogeology*, Prentice-Hall, Upper Saddle River N. J.

Food and Agriculture Organization of the United Nations (FAO) (2005), Aquastat Country Profile Egypt, available at http://www.fao.org/ag/agl/aglw/aquastat/countries/egypt/index.stm

Food and Agriculture Organization of the United Nations (FAO) (2009), Groundwater management in Saudi Arabia, Draft Synthesis Rep. FAO, Rome.

Garfunkel, Z., and Y. Bartov (1977), The tectonics of the Suez rift, *Geol. Surv. Israel Bull.*, 71.

Gonçalvès, J., J. Petersen, P. Deschamps, B. Hamelin, and O. Baba-Sy (2013), Quantifying the modern recharge of the "fossil" Sahara aquifers, *Geophys. Res. Lett.*, *40*, 2673–2678.

Heinl, M., and P. J. Brinkmann (1989), A groundwater model of the Nubian aquifer system, *Hydrol. Sci. J.*, *34*(4), 425–447.

Hesse, K. H., A. Hissese, O. Kheir, E. Schnacker, M. Schneider, and U. Thorweihe (1987), Hydrogeological investigations in the Nubian Aquifer system, Eastern Sahara, paper presented at Research in Egypt and Sudan, Dietrich Reimer, Berlin. pp. 397–464

Huffman, G., R. Adler, D. Bolvin, G. Gu, E. Nelkin, K. Bowman, Y. Hong, E. Stocker, and D. Wolff (2007), The TRMM multi-satellite precipitation analysis: Quasi-global, multi-year, combined-sensor precipitation estimates at fine scale, *J. Hydrometeorol.*, *8*, 38–55.

Jacob, T., J. Wahr, W. Pfeffer, and S. Swenson (2012), Recent contributions of glaciers and ice caps to sea level rise, *Nature*, *482*, 514–518.

Konert, G., A. M. Afifi, S. A. Al-Hajri, and H. J. Droste (2001), Paleozoic stratigraphy and hydrocarbon habitat of the Arabian plate, *GeoArabia*, *6*(3), 407–442.

Kroner, A., R. Greiling, T. Reischman, I. M. Hussein, R. J. Stern, S. Durr, J. Kruger, and M. Zimmer (1987), Pan-African crustal evolution in the Nubian segment of Northest Africa, *Proterozoic Lithospher. Evol.*, *17*, 235–257.

Kummerow, C., W. Barnes, T. Kozu, J. Shiue, and J. Simpson (1998), The tropical rainfall measuring mission (TRMM) sensor package, *J. Atmos. Oceanic Technol.*, *15*, 809–817.

Landerer, F. W., and S. Swenson (2012), Accuracy of scaled GRACE terrestrial water storage estimates, *Water Resourc. Res.*, *48*, W04531.

Liu, Y. Y., W. A. Dorigo, R. M. Parinussa, R. A. M. De Jeu, W. Wangner, M. F. McCabe, J. P. Evans, and A. I. J. M. Van Dijk (2012), Trend-preserving blending of passive and active microwave soil moisture retrievals, *Remote Sen. Environ.*, *123*, 280–297.

Marget, J., R. Foster, and A. Droubi (2006), Concept and importance of non-renewable groundwater resources, in *A Guidebook on Socially- Sustainable Management for Water-Policy Makers*, edited by S. Foster and D. P. Loucks, UNESCO, Paris, pp. 13–24.

Ministry for Higher Education, S. A. (2000), *Atlas of Saudi Arabia*, Saudi Arabian Ministry for Higher Education Riyadh, Kingdom of Saudi Arabia.

Otkun, G. (1971), Paleozoic sandstone aquifers in Saudi Arabia, paper presented at International Association of Hydrogeologists: Tokyo Congress.

Patterson, L. J., et al. (2005), Cosmogenic, radiogenic, and stable isotopic constraints on groundwater residence time in the Nubian Aquifer, Western Desert of Egypt, *Geochem. Geophys. Geosyst.*, *6*(1), 1–19.

Reager, J. T., and J. S. Famiglietti (2009), Global terrestrial water storage capacity and flood potential using GRACE, *Geophys. Res. Lett.*, *36*, L23402.

Rodell, M., I. Velicogna, and J. Famiglietti (2009), Satellite-based estimates of groundwater depletion in India, *Nature*, *460*, 999–1002.

Salem, O. M. (2007), Management of shared groundwater basins in Libya, *Afr. Water J.*, *1*(1), 109–120.

Salem, O. M., and P. Pallas (2002), The Nubian Sandstone Aquifer System, paper presented at Managing Shared Aquifer Resources in Africa, ISARM-AFRICA, Tripoli, Libya, 2–4 June 2002, pp. 19–21

Sanford, K. S. (1935), Sources of water in north-western Sudan, *Geogr. J.*, *84*, 421–431.

Scanlon, B. R., L. Longuevergne, and D. Long (2012), Ground referencing GRACE satellite estimates of groundwater storage changes in the California Central Valley, USA, *Water Resourc. Res.*, *48*, W04520.

Sturchio, N. C., et al. (2004), One million year old groundwater in the Sahara revealed by krypton-81 and chlorine-36, *Geophys. Res. Lett.*, *31*, L05503.

Sultan, M., N. Sturchio, F. A. Hassan, M. A. R. Hamdan, A. M. Mahmood, Z. El Alfy, and T. Stein (1997), Precipitation source inferred from stable isotopic composition of Pleistocene groundwater and carbonate deposits in the Western Desert of Egypt, *Quatern. Res.*, *48*, 29–37.

Sultan, M., E. Yan, N. Sturchio, A. Wagdy, K. Abdel Gelil, R. Becker, M. Manocha, and A. Milewski (2007), Natural discharge: A key to sustainable utilization of fossil groundwater, *J. Hydrol.*, *335*, 25–36.

Sultan, M., N. Sturchio, S. Al Sefry, A. Milewski, R. Becker, I. Nasr, and Z. Sagintayev (2008), Geochemical, isotopic, and remote sensing constraints on the origin and evolution of the Rub Al Khali aquifer system, Arabian Peninsula, *J. Hydrol.*, *356*, 70–83.

Sultan, M., et al. (2011), Modern recharge to the Nubian Aquifer, Sinai Peninsula: Geochemical, geophysical, and modeling constraints, *J. Hydrol.*, *403*, 14–24.

Sultan, M., M. Ahmed, N. Sturchio, Y. Eugene, A. Milewski, R. Becker, J. Wahr, D. Becker, and K. Chouinard (2013), Assessment of the vulnerabilities of the Nubian Sandstone Fossil Aquifer, North Africa, in *Climate Vulnerability: Understanding and Addressing Threats to Essential Resources*, edited by R. A. Pielke, pp. 311–333, Elsevier, Academic, Oxford.

Swenson, S., and J. Wahr (2006), Post-processing removal of correlated errors in GRACE data, *Geophys. Res. Lett.*, *33*, L08402.

Tapley, B. D., S. Bettadpur, J. C. Ries, P. F. Thompson, and M. M. Watkins (2004a), GRACE measurements of mass variability in the Earth System, *Science*, *305*, 503–505, doi:10.1126/science.1099192.

Tapley, B. D., S. Bettadpur, M. Watkins, and C. Reigber (2004b), The Gravity Recovery and Climate Experiment: Mission overview and early results, *Geophys. Res. Lett.*, *31*, L09607.

Thorweihe, U. (1982), *Hydrogeologie des Dakhla Beckens (Agypten)*, Berliner Geowiss. Abh. 1–58.

Thorweihe, U., and M. Heinl (2002), Groundwater resources of the Nubian Aquifer System, Aquifers of Major Basins-non-renewable Water Resources, Modified synthesis; Observatiore du Sahara et du Sahel (OSS), Tunis, 24p.

Tiwari, V. M., J. Wahr, and S. Swenson (2009), Dwindling groundwater resources in northern India, from satellite gravity observations, *Geophys. Res. Lett.*, *36*, L18401.

Trenberth, K. E. (2011), Changes in precipitation with climate change, *Climate Res.*, *47*, 23–138, doi:110.3354/cr00953.

Velicogna, I. (2009), Increasing rates of ice mass loss from the Greenland and Antarctic ice sheets revealed by GRACE, *Geophys. Res. Lett.*, *36*, L19503.

Velicogna, I., and J. Wahr (2005), Ice mass balance in Greenland from GRACE, *J. Geophys. Res.*, *32*, L18505, doi:10.1029/2005GL023955.

Velicogna, I., and J. Wahr (2006), Measurements of time-variable gravity show mass loss in Antarctica, *Science*, *311*(5768), 1754–1756.

Voss, K., J. Famiglietti, M. Lo, C. Linage, M. Rodell, and S. Swenson (2013), Groundwater depletion in the Middle East from GRACE with implications for transboundary water management in the Tigris-Euphrates-Western Iran region, *Water Resourc. Res.*, *49*, 1–11.

Wagner, W., W. Dorigo, R. De Jeu, D. Fernandez, J. Benveniste, E. Haas, and M. Ertl (2012), Fusion of active and passive microwave observations to create an essential climate variable data record on soil moisture, paper presented at XXII ISPRS Congress, Melbourne, Australia.

Wahr, J., M. Molenaar, and F. Bryan (1998), Time variability of the Earth's gravity field: Hydrological and oceanic effects and their possible detection using GRACE, *J. Geophys. Res.*, *103*(B12), 30,205–30,229, doi:10.1029/98JB02844.

Yan, E., R. Becker, M. Sultan, and E. Ballerstein (2003), Development of the Tushka lakes in the southwestern desert of Egypt, *Geol. Soc. Am. Annu. Meet. Abstr. Progr.*, *35*(7), 315.

Yirdaw, S. Z., K. R. Snelgrove, and C. O. Agboma (2008), GRACE satellite observations of terrestrial moisture changes for drought characterization in the Canadian Prairie, *J. Hydrol.*, *356*(1–2), 84–92.

22

Dominant Patterns of Water Storage Changes in the Nile Basin During 2003–2013

J. L. Awange,[1] E. Forootan,[2] K. Fleming,[3] and G. Odhiambo[4]

22.1. INTRODUCTION

Arguably the world's longest river, the Nile traverses some 6500 km from south to north through Tanzania, Uganda, Kenya, Rwanda, Burundi, Democratic Republic of Congo (DRC), Eritrea, Ethiopia, South Sudan, Sudan, and Egypt [*Sahin*, 1985; *Sutcliffe and Parks*, 1999]. The Nile basin, with an area of 3,400,000 km², supports the livelihoods of over 200 million people [e.g., *Awange*, 2012].

Yet, even though its basin is huge in area, the volume of the Nile's runoff is relatively low [e.g., *Sutcliffe and Parks*, 1999], thus making its meager waters very precious, particularly to the downstream countries such as Sudan and Egypt, which are almost completely dependent upon the Nile for their very survival. One significant factor that has been reported to threaten the river's water resources involves direct anthropogenic influences [see, e.g., *Hamouda et al.*, 2009]. This is largely due to increased human population, which puts increased pressure on domestic water needs, the supply of hydroelectric power, and increased agricultural activities, all of which are coupled to the need to sustain economic growth. Moreover, not only are the demands on the basin's water increasing, but the available water supplies appear to be decreasing, with environmental degradation of the upper Blue Nile catchment having increased throughout the 1980s [*Whittington and McClelland*, 1992]. For instance, Ethiopia, with its extreme hydrological variability and seasonality, is currently constructing hydroelectric dams within the Blue Nile subbasin [*Berhane et al.*, 2013]. Moreover, there is the threat that emanates from natural factors such as the changing climate, which has been the subject of numerous studies [e.g., *Yates and Strzepek*, 1998, and the references therein]. Combined and individually, direct human influences (population growth, unsustainable water usage, and development) and climatic change (droughts, floods, and desertification, both natural and anthropogenic) threaten the Nile's ability to supply crucially needed water to the people of the basin.

Therefore, for the proper management of the Nile basin's water resources, accurate monitoring of its spatial and temporal changes is required to allow informed policy decisions. Most studies dealing with the Nile basin to date, however, have dealt mainly with modeling the impacts of climate change [e.g., *Conway*, 2005; *Beyene et al.*, 2010; *Baldassarre et al.*, 2011], with very little being reported on how to monitor the spatial and temporal variations in the basin's overall stored water (surface, groundwater, and soil moisture) in a holistic manner. The reason for such few studies [e.g., *Senay et al.*, 2009; *Becker et al.*, 2010; *Bonsor et al.*, 2010; *Crétaux et al.*, 2011; *Ahmed et al.*, 2011; *Sultan et al.*, 2013; *Boy et al.*, 2012] has been partly attributed to its large size, as well as the

[1]*Western Australian Centre for Geodesy and The Institute for Geoscience Research, Curtin University, Perth, Australia*

[2]*Institute of Geodesy and Geoinformation, Bonn University, Bonn, Germany*

[3]*Physics of Earthquakes and Volcanoes Centre Potsdam, GFZ German Research Centre for Geosciences, Potsdam, Germany*

[4]*Department of Geography and Urban Planning, UAE University, Al Ain, United Arab Emeritus*

Remote Sensing of the Terrestrial Water Cycle, Geophysical Monograph 206. First Edition. Edited by Venkat Lakshmi.

lack of appropriate monitoring techniques that could cover such a vast spatial extent. For instance, the hydrological water balance involves the flow of surface water, the movement of deeper groundwater, and the coupling of the land, ocean, and atmosphere through evaporation and precipitation. Monitoring these components requires an accuracy and completeness of geographical data coverage that challenges conventional ground-based measurement capabilities.

The acquisition of continuous measurements of land surface properties is important for the analyses of environmental change over any scale, with satellite-derived information becoming increasingly important when considering regional to global extents. For example, time series of satellite images have an important role in the monitoring of regional and global land cover change [e.g., *Sulla-Menashe et al.*, 2011]. Satellite sensors are well suited to large-scale monitoring tasks because they provide repeatable measurements at spatial scales that are appropriate for capturing the effects of changes in key properties of an ecosystem. Likewise, time series analysis is useful for identifying, modeling, and forecasting the dynamics (trends, cycles, periodicities) of variables observed sequentially over time. One example of satellite monitoring is the Gravity Recovery and Climate Experiment (GRACE), which was designed to monitor climate-related mass redistribution within the Earth system using measurements of changes in the terrestrial gravity field. After correcting for tidal, ocean, and atmospheric variations, the temporal variation in the gravity field is largely caused by changes in the distribution of water [e.g., *Tapley et al.*, 2004a, 2004b; *Wahr et al.*, 1998; *Rodell and Famiglietti*, 2001; *Awange et al.*, 2008, 2013; *Awange*, 2012; *Awange and Kiema*, 2013; *Forootan et al.*, 2012]. Currently, river basins of the order of 400,000 km^2 in area can be successfully studied using the GRACE products [*Swenson et al.*, 2003]. Given that the Nile basin has an area of more that 3,400,000 km^2, GRACE products can potentially be used to analyze changes in its water storage. Another satellite mission of interest is the Tropical Rainfall Measuring Mission (TRMM), which is observing precipitation over the Earth between latitudes 50°N and 50°S, hence providing data for regions poorly served by ground-based rain gauges, which includes much of the Nile basin. Similarly, among modern satellite observation-based products used to monitor terrestrial ecosystems, the normalized digital vegetation index (NDVI) is of primary importance thanks to its capability of providing synoptic information about the status of vegetation over wide areas with high acquisition frequency [*Millington et al.*, 1994].

The main task of this work is to apply space-based satellite approaches to remotely sense the changes in the Nile basin's water storage over the last decade. To achieve this, use is made of GRACE products whose results are compared to precipitation, as observed by TRMM over the same period, to see whether changes in the Nile basin's waters are due to climatic change and/or human effects. This is achieved by constructing total water storage (TWS) time series for the Nile basin from the GRACE gravity data for the period between 2003 and 2013. The computation is done while taking care of the leakage caused by the basin's shape. GRACE spatiotemporal TWS changes reveal the general pattern of water changes within the basin, including all the tributaries in its catchments, lakes, soil moisture, and groundwater. TRMM products have previously been applied to the region, e.g., by *Owor et al.*, [2009] and *Nicholson et al.* [2003], and have been used to study the impact of precipitation variations on TWS changes within the basin. Furthermore, Moderate Resolution Imaging Spectroradiometer (MODIS) 500 m 16 day maximum value NDVI composites (MOD13A1) were used to evaluate surface vegetation conditions, which are then related to changes in stored water within the basin [e.g., *Omute et al.*, 2012]. Note that this study did not make use of satellite altimetry products, and the interested reader is referred to studies by *Becker et al.* [2010] and *Crétaux et al.* [2011] for the use of such information.

In the next section, we briefly describe the Nile basin. The methodology followed (data used in the study and the analysis methods) is outlined in section 22.3. Section 22.4 presents and discusses the results. Finally, section 22.5 summarizes the major findings.

22.2. THE NILE BASIN

The Nile basin (Figure 22.1) extends over a wide latitude range (from ~4°S to ~32°N) [e.g., *Sutcliffe and Parks*, 1999] and has two major tributaries, the White Nile and the Blue Nile. The White Nile rises in the Great Lakes region of Eastern Africa (i.e., its furthest tributary is the River Kagera, which flows into Lake Victoria; see, e.g., *Awange and Ong'ang'a* [2006] and flows northward through Uganda and southern Sudan.

From Lake Victoria, the waters are discharged to Lake Kyoga, which also receives water from its surrounding 75,000 km^2 catchment before flowing on to Lake Albert. In addition to the waters received from Lake Kyoga, Lake Albert is supplied by its upstream Semiliki basin and the Lake Edward subbasin (Figure 22.1). Together, lakes Edward, Albert, and George form the western edge of the Nile basin, comprising an area of 48,000 km^2, of which 7800 km^2 is open water [*Yates and Strzepek*, 1998; *Sutcliffe and Parks*, 1999]. In the Sudd swamp region, where a great deal of the water of the White Nile is lost via evaporation, the Nile benefits from water being supplied by two

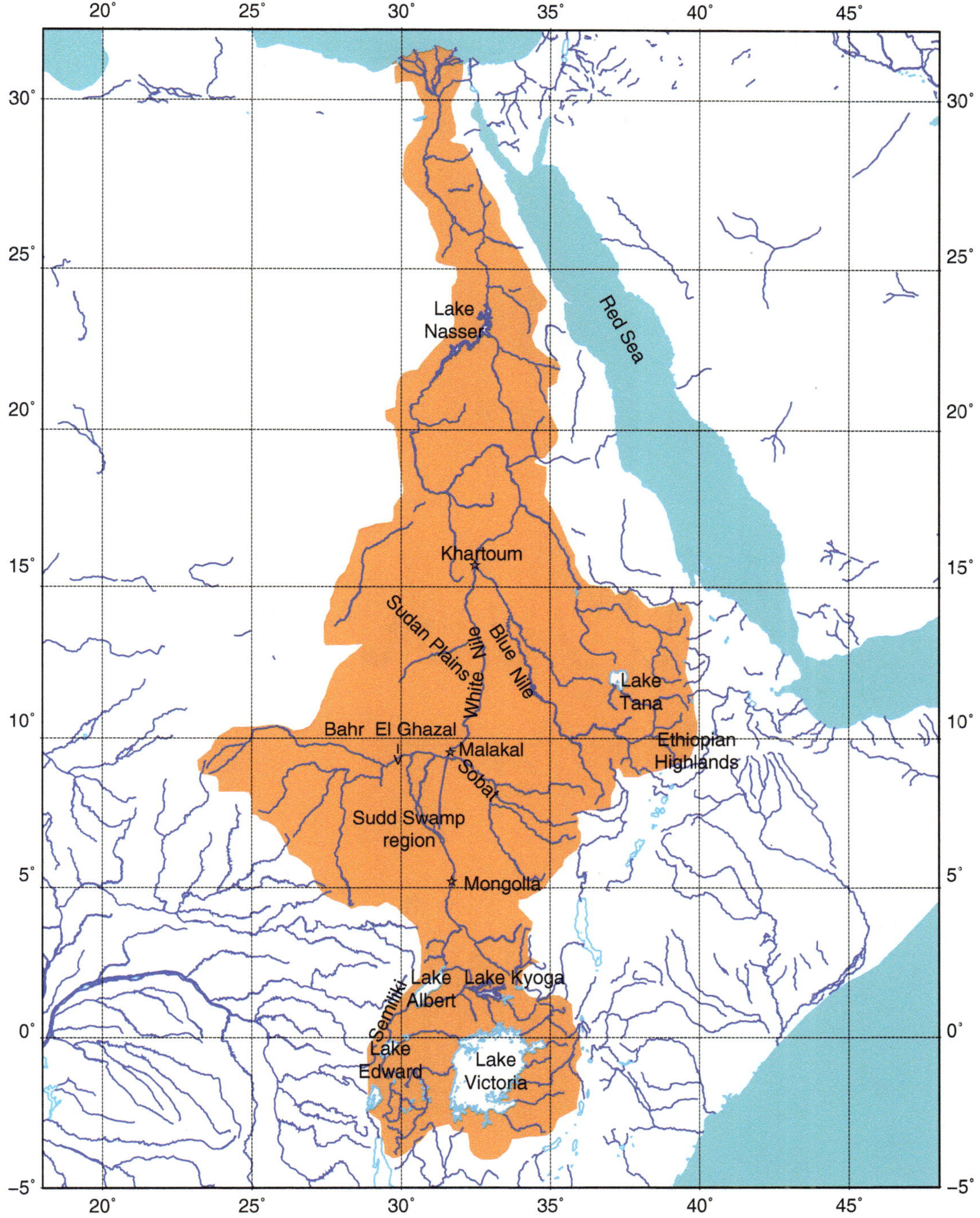

Figure 22.1 The Nile basin (dark shaded region) with the major features.

other basins, the Bahr-El-Ghazal (500,000 km²) to the west and the Sobat (150,000 km²) to the east, before exiting at Malakal. Rainfall within the White Nile generally occurs over two seasons, March–May (MAM) and October–December (OND) [e.g., *Awange et al.*, 2007; *Omondi et al.*, 2012, 2013b].

The Blue Nile starts at Lake Tana in the Ethiopian Highlands, draining major parts of its western highland, and then flows into Sudan from the southeast before meeting the White Nile at Khartoum in Sudan, where it provides most of the water. The rainfall within the Ethiopian Highlands is confined to the single season of June–September (JJAS) [e.g., *Sutcliffe and Parks*, 1999; *Omondi et al.*, 2012, 2013b]. From Khartoum, the Nile passes through Egypt and ends its journey by flowing into the Mediterranean Sea via the Nile Delta.

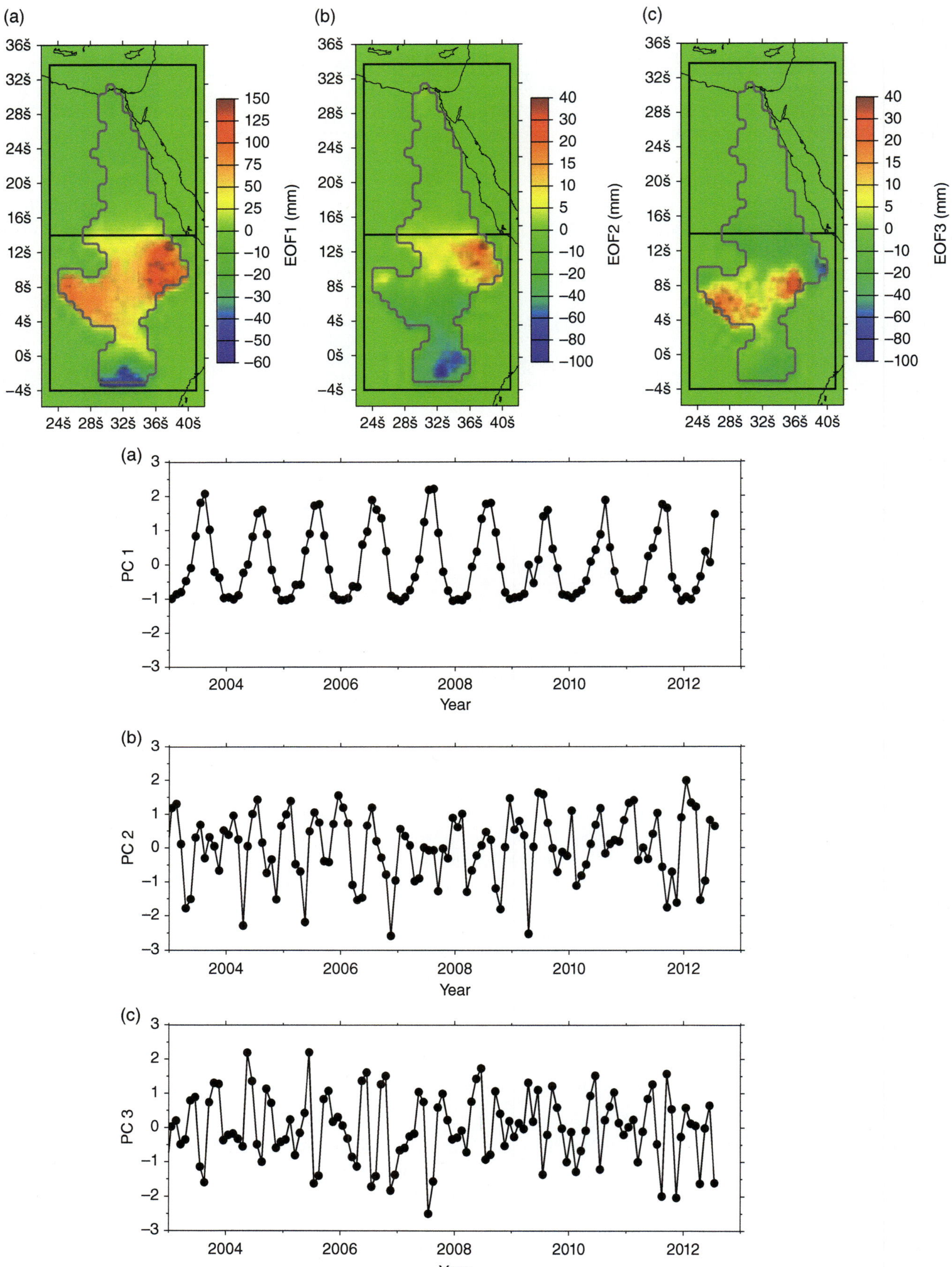

Figure 22.2 PCA decomposition of the TRMM-derived rainfall variations over the Nile basin. EOFs of rainfall changes over the Nile basin are shown on top and arranged with respect to the variance they represent from left to right. The corresponding temporal variations (PCs) are shown on the bottom. PC1 and EOF1 present 69% of the rainfall variance, PC2 and EOF2 present 7% of the variance, and 5% of the variance is the contribution of PC3 and EOF3.

For the purpose of this study, the Nile basin in Figure 22.1 is divided into two sectors, divided at 15°N (see, e.g., Figure 22.2). The reason for this subdivision is that the Nile basin falls within equatorial and tropical sectors where rainfall seasons are both bimodal (March–April–May and October–November–December) and unimodal (June–July–August) and hence the southern and northern sectors, respectively.

22.3. DATA SETS AND METHODOLOGY

22.3.1. Gravity Recovery and Climate Experiment

The Gravity Recovery and Climate Experiment (GRACE) is a U.S. (National Aeronautics and Space Administration, NASA) and German (Deutsche Zentrum für Luft- und Raumfahrt, DLR) space mission that has been providing products that describe the temporal variation in the terrestrial gravity field arising from mass movements within the Earth's system. GRACE was launched on 17 March, 2002, and consists of two almost identical spacecraft in the same orbit separated by ~220 km, having an initial altitude of ~500 km and an inclination of 89.5°. The GRACE satellites operate in what is termed as satellite-to-satellite tracking in the low-low mode. By this is meant that the satellites measure the separation between them using a K-band microwave ranging system, while at the same time employing GNSS (Global Navigation Satellite System) to determine their precise location [*Rummel et al.*, 2002; *Tapley et al.*, 2004b]. Satellite laser-ranging reflectors (SLR) are also used as an independent check [e.g., *Tapley et al.*, 2004a] while the effects of nongravitational forces acting on the satellite are removed using measurements from onboard accelerometers [*Tapley et al.*, 2004b].

The main product of the GRACE mission is its time-variable gravity fields in the form of normalized spherical harmonic solutions, of which the latest release five (RL05) monthly solutions, provided by the German Research Centre for Geosciences (GFZ) [*Dahle et al.*, 2012], covering 2003–2013, are used here. GRACE level 02 products still contain correlated errors among the higher order spherical harmonics, which is known as the striping pattern [*Kusche*, 2007]. In order to remove these stripes, we applied the de-correlation filter denoted by DDK2 of *Kusche et al.* [2009] to the GFZ-RL05 solutions. The filtered solutions can also be downloaded from http://icgem.gfz-potsdam.de/ICGEM/TimeSeries.html.

For computing monthly TWS change over the Nile basin, the following items are considered:

1. Residual gravity field solutions with respect to the temporal average of 2003–2013 were computed.

2. The residual coefficients were then convolved with a basin function, considering the basin boundary shown in Figure 22.2 [see, e.g., *Fenoglio-Marc et al.*, 2006, 2012]. The selected basin boundary has a convex shape that itself reduces the leakage error, as outlined in *Longuevergne et al.* [2010].

3. To account for the leakage, we used a simple approach that assumes a uniform mass distribution with the value of one inside the Nile basin and no mass outside the basin ($S_1 = 1$, is a uniform mass in the basin). Then, we transformed the uniform mass into spherical harmonics. The obtained coefficients are filtered with the same DDK2 filter as applied to the GRACE products. Afterward, the total surface mass of the basin was calculated from the filtered coefficients (S_2, synthesized uniform mass in the basin). The ratio of S_1/S_2 reflects the effect of the truncation of the spherical harmonics as well as signal attenuation due to filtering. A more detailed discussion of the leakage problem can be found, for example, in *Klees et al.* [2007]. An alternative statistical approach for reducing the leakage is addressed in *Forootan et al.* [2014].

The derived ratio is multiplied by the coefficients derived in item 2, and the results were transformed into $0.5° \times 0.5°$ TWS maps within the Nile basin, following *Wahr et al.* [1998]. An average uncertainty of ~15 mm was computed based on the formal error of the GFZ-RL05 products.

22.3.2. Tropical Rainfall Measuring Mission

The TRMM was launched in 1997 and is a joint mission of NASA and the Japan Aerospsace Exploration Agency (JAXA). TRMM orbits at an altitude of about 400 km with an inclination of 35° and orbital period of around 91 min, thus completing 16 revolutions per day. TRMM was designed to monitor and study tropical rainfall over the latitude range ±50° over inaccessible areas such as the oceans and poorly sampled terrain. The primary onboard instruments are the TRMM microwave imager (TMI), the precipitation radar (PR), and the Visible and Infrared Radiometer System (VIRS) [*Kummerow et al.*, 1998, 2000]. TRMM products have been employed in a number of studies of precipitation in Africa, where they have been found to be adequate when compared with ground truth observations [e.g., *Nicholson et al.*, 2003].

The product employed in this work is referred to as the TRMM and Other Data Precipitation Data Set (denoted as 3B43). TRMM-3B43 provides monthly rainfall (average hourly rate) between latitudes 50°N/50°S over a $0.25° \times 0.25°$ grid. Over time, the products 3B43 are updated as the processing techniques and methods for integrating the different data sets are improved. In this work, we use the latest version, number 7, which has been found in the case of Australia to be a significant improvement over the previous version 6 [*Fleming and Awange*, 2013], which itself showed a high correlation with in situ derived products [*Fleming et al.*, 2011], owing to such changes as the use of additional satellites and a superior means of incorporating rain gauge information from the Global Precipitation Climatological Centre [*Huffmann and Bolvin*, 2012; *Fleming and Awange*, 2013]. The use of TRMM satellite data for this study is informed by the works of *Dinku et al.* [2007, 2008, 2010], and *Beyene and Meissner* [2010] on the validation of satellite product in the region, which includes nearly half of Ethiopia (i.e., the Blue Nile portion). In particular, *Dinku et al.* [2010] reported good performance of TRMM-3B43 rainfall product. We computed correlations between time series of TRMM version 7 and its former version 6 products over the Nile basin and obtained a correlation of greater than 0.7 for more than 75% of the time series, thus justifying its use for studying the precipitation pattern over the Nile basin.

22.3.3. NDVI Data

Vegetation monitoring by remotely sensed data has been usually carried out using vegetative indices, which are mathematical transformations designed to assess the spectral contributions of green plants to multispectral observations [*Li et al.*, 2004]. These indices are mainly derived from reflectance data from discrete red (R) and near-infrared (NIR) bands. Although there are several vegetation indices, one of the most widely used is the normalized difference vegetation index. NDVI values range from +1.0 to −1.0. It operates by contrasting intense chlorophyll pigment absorption in the red against the high reflectance of leaf mesophyll in the near infrared, hence NDVI = [NIR − R]/[NIR + R]. NDVI is, therefore, indicative of plant photosynthetic activity and has been found to be highly related to the green leaf area index and the fraction of photosynthetically active radiation absorbed by vegetation [*Veroustraete et al.*, 2002]. Areas of barren rock, sand, or snow usually show very low NDVI values (e.g. 0.1 or less). Sparse vegetation such as shrubs and grasslands or crops may result in moderate NDVI values (N 0.2–0.5). High NDVI values (N 0.6–0.9) correspond to dense vegetation such as that found in tropical forests or crops at their peak growth stage. Because of these properties, NDVI can be utilized as an indicator of vegetation stress, particularly due to water shortage, which may result from reduced precipitation [*Millington et al.*, 1994]. This is the case whenever water availability is the main limiting factor for vegetation processes and therefore controls leaf pigment content and integrity. NDVI values of vegetation usually offer an efficient and objective means for evaluating phonological characteristics and have long been used to monitor vegetation conditions and changes in vegetation cover [*Woodcock et al.*, 2001; *Omute et al.*, 2012].

In this study, time series analysis of the NDVI data for the Nile basin was carried out using MODIS data from NASA, in the form of 16 day composite MODIS surface reflectance products (MOD13A1) for the years 2000–2009, provided by the NASA Distributed Active Archive Center in HDF-EOS format (http://reverb.echo.nasa.gov/reverb). These data were derived from the Earth-Observing System (EOS) *Terra* MODIS surface reflectances, which have been corrected for molecular scattering, ozone absorption, and aerosols. The index is produced globally over land at 16 day compositing intervals and enables consistent spatial and temporal comparisons of vegetation condition.

The MODIS re-projection tool (http://edc.usgs.gov/programs/sddm/modisdist/) was used to merge the 20 tiles that cover the Nile basin. The data were then subject to the Nile basin study boundary, re-projected from a sinusoidal to Universal Transverse Mercator (UTM) projection using a nearest-neighbor re-sampling routine, and entered into a 250 m × 250 m grid cell multilayer image stack. Separate data stacks were developed for both the original NDVI data and Quality Assessment Science Data Sets (QASDS). Details documenting the MODIS NDVI compositing process and QASDS can be found at NASAs MODIS website https://lpdaac.usgs.gov/products/modis_products_table/mod13a3. The NDVI data stack was first filtered to eliminate anomalous high (hikes) and low (drops) values. These were effectively eliminated by removing data values that suddenly decreased or increased and then immediately returned to near the previous NDVI value. The threshold for the removal of pseudo hikes and drops was set at ± 0.15% [see, e.g., *Knight et al.*, 2006] to achieve the best setting (determined qualitatively) to eliminate most anomalous points, while not inadvertently removing good data points, resulting in smoother temporal proles. It should be pointed out that 0.15% is not by any means a standard value but is somewhat subjective and is mainly based on the fact that sudden spikes and drops in NDVI would, in most cases, refer to errors in the data as one would not expect that to occur even in cases where conditions change, i.e., the phenological effect takes time. What this simply means is that it takes a while (several weeks) before noticeable changes in vegetative cover occurs as a result of weather (mainly precipitation) changes. So, sudden spikes (drops and hikes) in NDVI would simply indicate error in the data, which must be corrected.

22.4. RESULTS AND DISCUSSIONS

22.4.1. Analysis of TRMM Data

The statistical method of principal component analysis (PCA) is applied to spatiotemporal rainfall grids in order to identify and extract their dominant variability

[Preisendorfer, 1988]. PCA, here, is applied through an eigenvalue decomposition of the autocovariance matrix of the data into sets (modes) of empirical orthogonal functions (EOFs) and principal components (PCs) corresponding to the spatial and temporal variability, respectively. Details of the PCA approach, when it is applied to time series of TWS and rainfall maps, can be found in, e.g., *Forootan and Kusche* [2012, 2013] and *Forootan et al.* [2012]. All the PCA-derived orthogonal components shown here passed *North et al.*'s [1982] significance test.

From the results of PCA applied to the TRMM products, we will only show the first three components. A point that should be made immediately is that the TRMM data over the region has been already subjected to a trend analysis in *Omondi et al.* [2013b], who found no significant trend over the period of study and also over the longer period of 1998–2012. This study, therefore, performed no trend assessment but refers an interested reader to the study above. Examining the PCA results, we see from the EOFs (Figure 22.2) that there is very little signal north of latitude 15°N. Considering the first mode, representing 69% of the total variability, which corresponds to the annual signal (EOF1 and PC1 in Figure 22.2), we note the difference in sign between the eastern (Ethiopia) and westernmost parts when compared to the southern extreme of the basin in the vicinity of Lake Victoria (see EOF1 in Figure 22.2). This is similar to the second mode, representing 7% of the total variability (EOF2 and PC2 in Figure 22.2), which in turn corresponds to the semiannual cycles. We also note the peak of PC1 to be around the middle of the year, corresponding to the rainy season in the Ethiopian Highlands. The third mode representing 3% of the total variability appears to show variability in precipitation to be restricted to an area corresponding to a band extending from the Sudd Swamp region to the Ethiopian Highlands, with a change in sign from west to east (EOF3 and PC3 in Figure 22.2), again with a variability approaching semiannual. Correlation between PC2 and PC3 in Figure 22.2 shows the maximum lag of 2–3 months between the peak of rainfall over Lake Victoria (see EOF2 and PC2 in Figure 22.2) and the Ethiopian Highlands (see EOF3 and PC3 in Figure 22.2). From the PCA results, there seems to be very little change in the rate of precipitation over the Nile during the period of study.

In order to better illustrate the interannual variability of rainfall over the basin, we removed the dominant annual and semiannual cycles from the original rainfall time series using a least squares adjustment procedure. The residuals were then subjected to the PCA decomposition, which are shown in Figure 22.3. The temporal components were subjected to the least squares spectral analysis [LSSA; *Vaniček*, 1969] as described, e.g., in *Sharifi et al.* [2013]. The reported dominant cycles passed

a Fisher significance test (for details on the Fisher test, see, e.g., *Sharifi et al.*, [2013]. PC1 in Figure 22.3 contains two main dominant cyclic components with the period of 1.66 years and 4 months. Its corresponding spatial pattern (EOF1 in Figure 22.3) indicates that these cycles are dominant over the southern part of the basin. The remaining amplitude of the cyclic component with the period of 4 months is captured by PC2 and EOF2 of Figure 22.3. The EOF2 shows the dipole structure of the rainfall variability over the Lake Victoria basin and parts of South Sudan and Ethiopia. The computed spectrum of PC3 in Figure 22.3 contained several peaks that barely passes the performed Fisher significance test and are therefore not interpreted here.

22.4.2. GRACE

The PCA method was also applied to the GRACE-TWS maps of the Nile basin, with the three dominant EOFs of the GRACE-TWS and their corresponding temporal components (PCs) shown in Figure 22.4. Similar to the rainfall results, the components are arranged with respect to the variance they represent. EOF1 representing 79% of the total variability shows a spatial north-south dipole structure, which as PC1 indicates, corresponds to the annual TWS changes of the basin. EOF2 representing 9% of the total variability is concentrated over the Lake Victoria basin, with PC2 showing an average decline of −22 mm/yr between 2002 to the middle of 2006 and a positive trend of 13 mm/yr between 2007 and 2013. Mode 3 (EOF3 and PC3), representing 5% of the total variability, shows the northern (including Egypt) and western (including South Sudan) parts of the basin are exhibiting decreasing mass between 2003 and 2012, with an average linear rate of −3 mm/yr. In contrast, the Ethiopian Highlands in the east are exhibiting a mass gain, with an average linear rate of 2.8 mm/yr. Note that the patterns derived from PCA represent the dominant variability of TWS signal over the region. Hence the reason why it is preferred here over a simple computation of the trend and seasonal components. The first two dominant modes of GRACE-TWS, shown in Figure 22.4, indicate the seasonal transport of water from the headwater regions (e.g., South Sudan) and the reservoirs of the lower Nile basin, e.g., Lake Victoria.

Similar to the rainfall data, the dominant annual and semiannual cycles were removed from TWS and the residuals were decomposed using the PCA (shown in Figure 22.5). Using the LSSA, PC1 in Figure 22.5 contains a dominant seasonal cycle with the period of ~3 − 4 months. A long-term trend, which is dominant over the Lake Victoria basin is also noticed. PC2 contains a dominant cycle of 1.42 years period and several cycles with interannual periods. Justification of the reliability of the

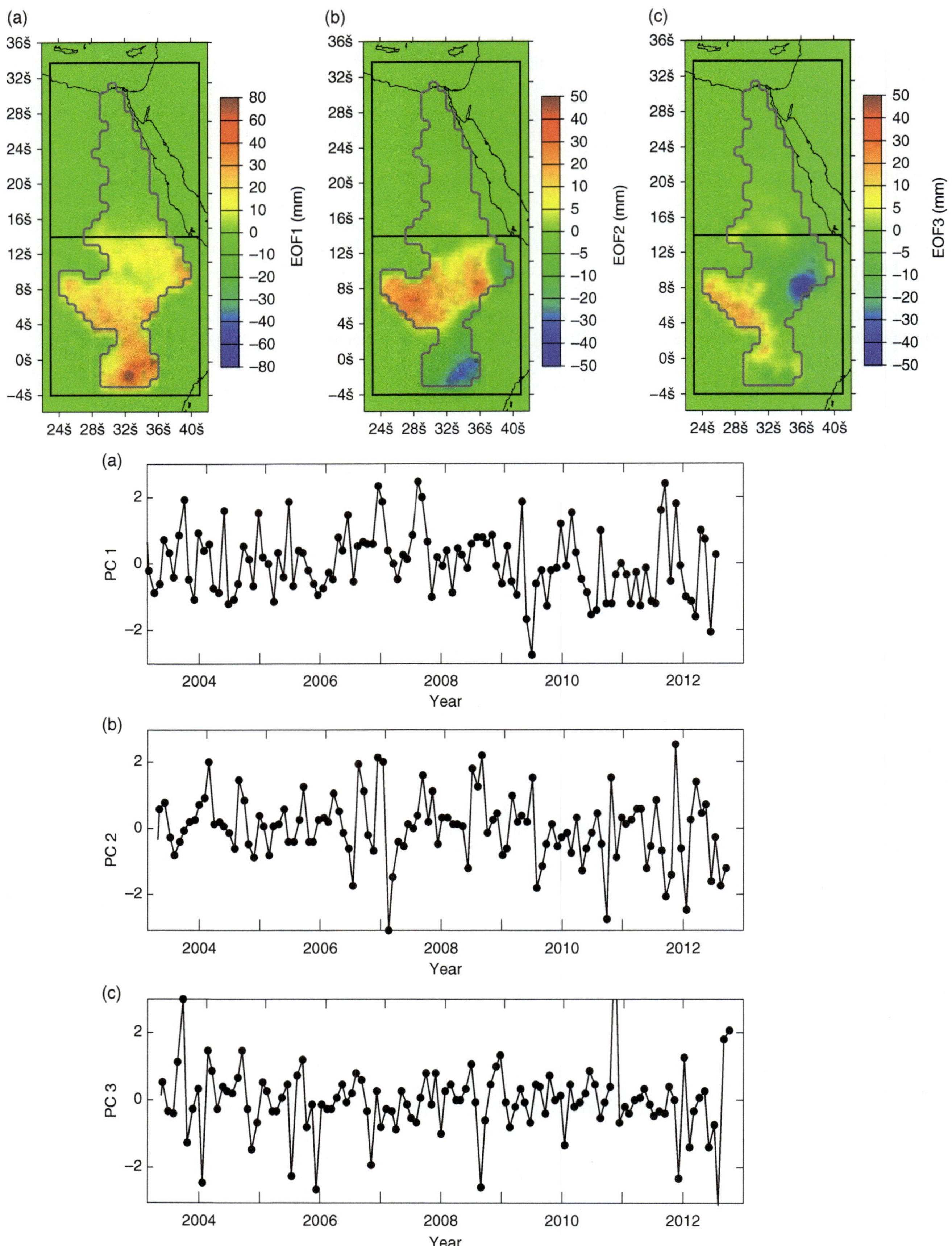

Figure 22.3 PCA decomposition of the TRMM-derived rainfall variations over the Nile basin after removing the dominant annual and semiannual cycles. EOFs and PCs are arranged similar to Figure 22.2.

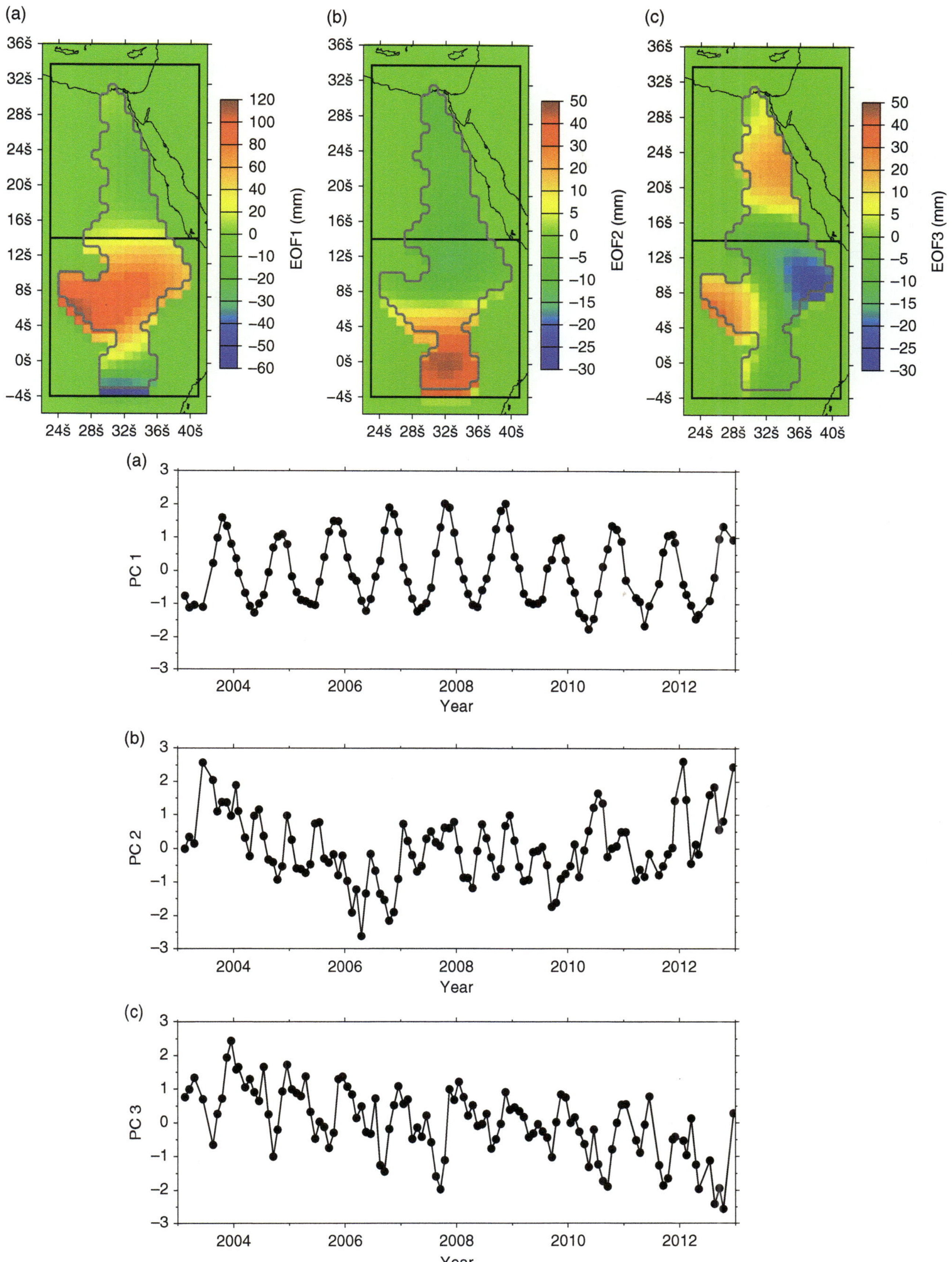

Figure 22.4 Overview of the PCA decomposition of TWS changes from the GRACE products over the Nile basin. EOFs represent anomaly maps of the TWS variations. PCs are their unitless normalized temporal evolution. PC1 and EOF1 present 79% of the TWS variance, PC2 and EOF2 present 9% of the variance, and 5% of the variance is the contribution of PC3 and EOF3.

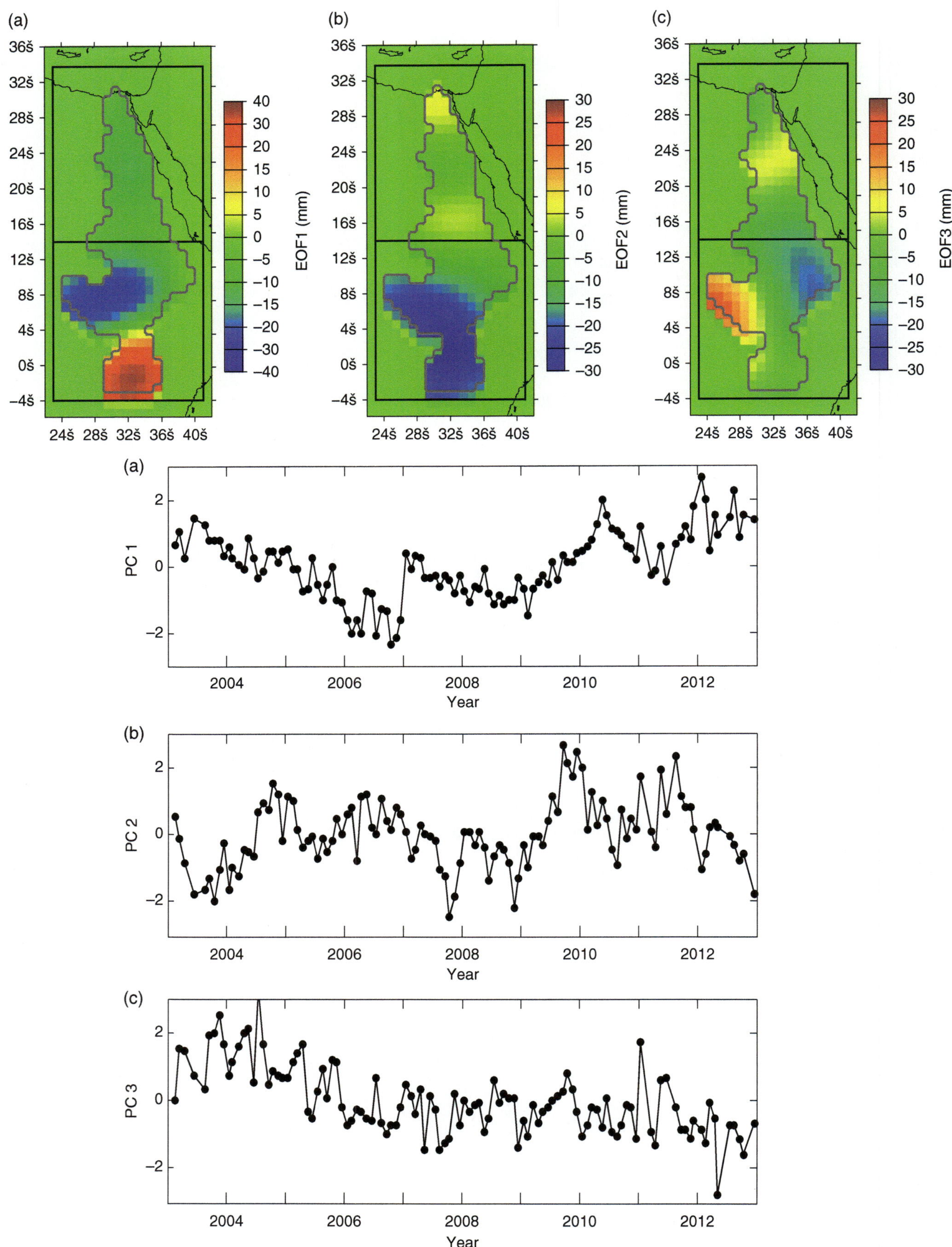

Figure 22.5 PCA decomposition of TWS variations over the Nile basin after removing the dominant annual and semiannual cycles. EOFs and PCs are arranged similar to Figure 22.4.

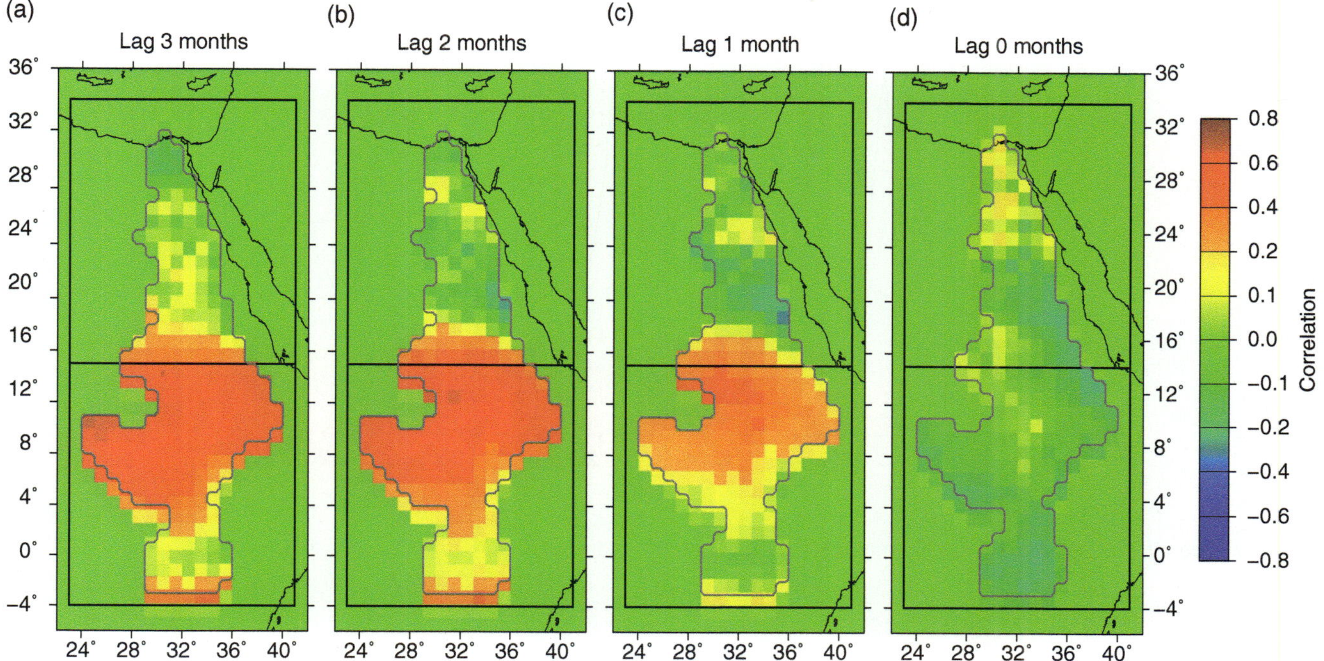

Figure 22.6 Correlations between the GRACE and TRMM time series for different time lags over the Nile basin. The figure shows a lag of 2 and 3 months is the possible optimum correlation between rainfall and TWS changes of the region.

computed cycles in the second mode (PC2 and EOF2) and several seasonal cycles in the third mode (PC3 and EOF3) requires longer data sets.

To assess the relationships between rainfall and TWS changes derived from GRACE within the Nile basin, we first reconstructed the main variability of TWS and rainfall using the three PCA modes shown above. Then, correlations between the TWS and rainfall time series, with a lag of 0, 1, 2, and 3 months at 95% level of confidence were computed. The results (Figure 22.6) show the maximum correlation arises for the lag of 2 months. Correlations derived from a 3 month lag are similar to that derived from a 2 month lag. Note that before computing those correlations, TRMM grids were interpolated to 0.5° × 0.5° grids, to be comparable with the GRACE grids.

22.4.3. Normalized Digital Vegetation Index

The NDVI trend analysis results are presented in Figure 22.7. This analysis reveals a cyclic seasonal trend with peak values recorded during the rainfall seasons (Figure 22.7). A downward trend is also noticed, albeit with some irregularity, with the NDVI monthly mean values for 2004 and 2006 being particularly lower than the rest of the years, coinciding with drought years [*Rojasa et al.*, 2011; *Sergio et al.,* 2012] when rainfall was significantly lower within the Nile basin. In these years, there was no month with NDVI values equal to or greater than 0.6, indicating that vegetative cover was more sparse. In other words, the years referred to were characterized by minimal/low precipitation, hence the low vegetative cover/growth as shown by the low NDVI values.

The downward trend in the NDVI values coincides with the decreasing rainfall trend observed by the ground stations within the southern half of the Nile basin such as Bukoba. These findings corroborate those of *Omute et al.*, [2012], who found that for Lake Victoria, weather was a major factor governing the water level trends. A further study by *Rasmus and Kjeld* [2011] on rain use efficiency (the ratio of vegetation productivity to annual precipitation) revealed similar patterns of NDVI trends with rainfall. However, the NDVI analysis in our study covers only part of the period for which the rainfall analysis was carried out. The analysis reveals that climate has substantial control on NDVI through annual rainfall [see, e.g., *Omute et al.*, [2012]. That said, much more detailed statistical analysis involving other factors, such as land use, overgrazing, soil type, fire, and floods, which might also influence vegetation at local scales, should be used to separate the rainfall impacts on NDVI from other factors. While high positive correlations between NDVI and rainfall portrays "normal" land cover performance [e.g., *Agulara et al.*, 2012], low correlation might indicate areas impacted upon by human influence [*Li et al.*, 2004].

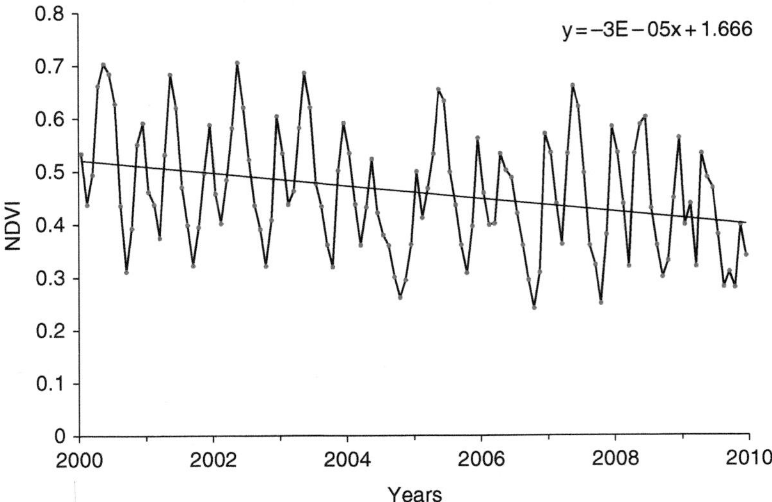

Figure 22.7 Nile basin's monthly mean NDVI between the years 2000 and 2009.

22.5. CONCLUSIONS

This study has highlighted the following features of the state of the Nile basin:

1. Trend (not shown here) and PCA do not show a significant overall decrease in rainfall magnitude over the Nile basin despite the high variability of precipitation, e.g., low rainfall during the "short rains" of 2005 or the massive rainfall during the El- Niño of 2007 and 2012.

2. PCA results indicate a decrease in TWS in the northern part of the basin at a rate of 3 mm/y, while the southern part experienced a decrease at a rate of 22 mm/y between 2002 and 2006, and an increase thereafter from 2007 to 2013 at a rate of 13 mm/y. Over the whole period of study (2003–2013), the Ethiopian Highlands gained stored water at a rate of 2.8 mm/y. A damp in the magnitude of the annual signals are found in 2005 and 2010 form the PCA data of TRMM rainfall and GRACE-TWS maps over the basin (e.g., PC1 in Figures 22.2 and 22.4), which indicates the importance of the rainfall magnitude in the total water storage changes over the basin.

3. A lag of 1–2 months is derived between rainfall and GRACE-TWS time series. This might represent a large-scale relationship between accumulated rainfall patterns and storage changes over the basin. From Figure 22.6, this relationship is stronger over the southern part of the basin.

4. NDVI results show a downward trend in vegetation cover for the years characterized by low precipitation.

Considering the fact that rainfall conditions over the last decade have stayed relatively constant over the basin, it can be concluded that the 2002–2006 decline in the stored water within the southern part of the Nile basin, home to the Lake Victoria basin, is attributed to anthropogenic factors consistent with previous findings of *Awange et al.* [2008] and *Swenson and Wahr* [2009]. The subsequent upsurge in the water levels from 2007 to the present could be due to the ENSO rainfalls; see, e.g., *Omondi et al.* 2012, 2013a, 2013b].

ACKNOWLEDGMENTS

E. Forootan is grateful for the financial support by the German Research Foundation (DFG) under the project BAYES-G. The authors are grateful to the data used in this study. This work is a TiGeR publication no. 486.

REFERENCES

Aguilara, C., J. C. Zinnert, M. J. Polo, and D. R. Young (2012), NDVI as an indicator for changes in water availability to woody vegetation, *Ecol. Indicators*, *23*, 290–300, doi:10.1016/j.ecolind.2012.04.008.

Ahmed, M., M. Sultan, J. Wahr, E. Yan, A. Milewski, W. Sauck, R. Becker, and B. Welton (2011), Integration of GRACE (Gravity Recovery and Climate Experiment) data with traditional data sets for a better understanding of the time-dependent water partitioning in African watersheds, *Geology*, doi:10.1130/G31812.1.

Awange, J. L. (2012), *Environmental Monitoring Using GNSS*, Springer-Verlag, Berlin.

Awange, J. L., and J. B. K. Kiema (2012), *Environmental Geoinformatics: Monitoring and Management*, Springer-Verlag, Berlin.

Awange, J. L., and O. Ong'ang'a (2006), *Lake Victoria: Ecology Resource and Environment*, Springer-Verlag, Berlin.

Awange, J. L., J. Aluoch, L. Ogallo, M. Omulo, and P. Omondi (2007), An assessment of frequency and severity of drought in the Lake Victoria region (Kenya) and its impact on food security, *Climate Res.*, *33*, 135–142.

Awange J. L., M. Sharifi, G. Ogonda, J. Wickert, E. Grafarend, and M. Omulo (2008), The Falling Lake Victoria Water Levels: GRACE, TRIMM and CHAMP satellite analysis of the lake basin, *Water Resourc. Manag.*, *22*, 775–796, doi:10.1007/s11269-007-9191-y.

Awange, J., E. Forootan, J. Kusche, J. K. B. Kiema, P. Omondi, B. Heck, K. Fleming, S. Ohanya, and R. M. Gonçalves (2013), Understanding the decline of water storage across the Ramser-lake Naivasha using satellite-based methods, *Adv. Water Resourc.*, *60*, 7–23, doi:10.1016/j.advwatres.2013.07.002.

Baldassarre, G. D., et al. (2011), Future hydrology and climate in the River Nile basin: A review, *Hydrol. Sci. J.*, *56*(2), doi:10.1080/02626667.2011.557378.

Becker, M., W. LLovel, A. Cazenave, A. Güntner, and J-F. Crétaux (2010), Recent hydrological behavior of the East African great lakes region inferred from GRACE, satellite altimetry and rainfall observations, *C. R. Geosci.*, *342*, 223–233.

Berhane, G., M. Kristine, A. Nawal, and W. Kristine (2013), Water leakage investigation of micro-dam reservoirs in Mesozoic sedimentary sequences in Northern Ethiopia, *J. Afr. Earth Sci.*, *79*, 98–110, doi:10.1016/j.jafrearsci.2012.10.004.

Beyene, E. G., and B. Meissner (2010), Spatio-temporal analyses of correlation between NOAA satellite RFE and weather stations' rainfall record in Ethiopia, *Int. J. Appl. Earth Observ. Geoinform.*, *125*, 569–575.

Beyene, T., D. P. Lettenmaier, and P. Kabat (2010), Hydrologic impacts of climate change on the Nile River Basin: Implications of the 2007 IPCC scenarios, *Climatic Change*, *100*, 433–461, doi:10.1007/s10584-009-9693-0.

Bonsor, H. C., M. M. Mansour, A. M. MacDonald, A. G. Hughes, R. G. Hipkin and T. Bedada (2010), Interpretation of GRACE data of the Nile basin using a groundwater recharge model, *Hydrol. Earth Syst. Sci. Discuss.*, *7*, 4501–4533, doi:10.5194/hessd-7-4501-2010.

Boy, J. P., J. Hinderer, and C. De Linage (2012), Retrieval of large-scale hydrological signals in Africa from GRACE time-variable gravity fields, *Pure Appl. Geophys.*, *169*(8), 1373–1390.

Conway, D. (2005), From headwater tributaries to international river: Observing and adapting to climate variability and change in the Nile basin, *Global Environ. Change*, *15*, 99–114, doi:10.1016/j.gloenvcha.2005.01.003.

Conway, D., C. Mould, and B. Woldeamlak (2004), Over one century of rainfall and temperature observations in Addis Ababa, Ethiopia, *Int. J. Climatol.*, *24*, 77–91.

Crétaux J. F., et al. (2011), SOLS: A lake database to monitor in the near real-time water level and storage variations from remote sensing data, *Adv. Space Res.*, *47*, 1497–1507, doi:10.1016/j.asr.2011.01.004.

Dahle, C., F. Flechtner, C. Gruber, R. König, G. Michalak, and K.-H. Neumayer (2012), GFZ GRACE Level-2 processing standards document for Level-2 product release 0005, Sci. Tech. Rep. STR12/02-Data, GFZ, Potsdam, Germany.

Dinku, T., P. Ceccato, E. Grover-Kopec, M. Lemma, S. J. Connor, and C. F. Ropelewski (2007), Validation of satellite rainfall products over East Africa's complex topography, *Int. J. Remote Sens.*, *28*(7), 1503–1526, doi:10.1080/01431160600954688.

Dinku, T., S. Chidzambwa, P. Ceccato, S. J. Connor, and C. F. Ropelewski (2008), Validation of high-resolution satellite rainfall products over complex terrain, *Int. J. Remote Sens.*, *29*(14), 4097–4110, doi:10.1080/01431160701772526.

Dinku, T., P. Ceccato, K. Cressman, and S. J. Connor (2010), Evaluating detection skills of satellite rainfall estimates over desert locust recession regions, *Int. J. Appl. Meteorol. Climatol.*, *49*, 1322–1332.

Fenoglio-Marc, L., J. Kusche, and M. Becker (2006), Mass variation in the Mediterranean Sea from GRACE and its validation by altimetry, steric and hydrologic fields, *Geophys. Res. Lett.*, *33*(19), doi:10.1029/2006GL026851.

Fenoglio-Marc, L., R. Rietbroek, S. Grayek, M. Becker, J. Kusche, and E. Stanev (2012), Water mass variation in the Mediterranean and Black Sea, *J. Geodyn.*, *59–60*, 168–182, doi:org/10.1016/j.jog.2012.04.001.

Fleming, K., and J. L. Awange (2013), Comparing the version 7 TRMM 3B43 monthly precipitation product with the TRMM 3B43 version 6/6A and BoM datasets for Australia, *Austral. Meteorol. Oceanogr. J. 63*, 421–426

Forootan, E., and J. Kusche (2012), Separation of global time-variable gravity signals into maximally independent components, *J. Geodesy*, *86*(7), 477–497, doi:10.1007/s00190-011-0532-5.

Forootan, E., and J. Kusche (2013), Separation of deterministic signals, using independent component analysis (ICA), *Stud. Geophys. Geod.*, *57*, 17–26, doi:10.1007/s11200-012-0718-1.

Forootan, E., J. L. Awange, J. Kusche, B. Heck, and A. Eicker (2012), Independent patterns of water mass anomalies over Australia from satellite data and models, *Remote Sens. Environ.*, *124*, 427–443, doi:0.1016/j.rse.2012.05.023.

Forootan, E., R. Rietbroek, J. Kusche, M. A. Sharifi, J. Awange, M. Schmidt, P. Omondi, and J. S. Famiglietti (2014), Separation of large scale water storage patterns over Iran using GRACE, altimetry and hydrological data, *J. Remote Sens. Environ.*, *140*, 580–595, doi:10.1016/j.rse.2013.09.025.

Hamouda, M. A., M. N. Nour El-Din, and F. I. Moursy (2009), Vulnerability assessment of water resources systems in the Eastern Nile basin, *Water Resourc. Manag.*, *23*, 2697–2725. doi:10.1007/s11269-009-9404-7.

Huffmann, G., and D. Bolvin (2012), TRMM and other data precipitation data set documentation, available at ftp://precip.gsfc.nasa.gov/pub/trmmdoc/3B42_3B43_doc.pdf, accessed 25 September 2012.

Klees, R., E. A. Zapreeva, H. C. Winsemius, and H. H. G. Savenije (2007), The bias in GRACE estimates of continental water storage variations, *Hydrol. Earth Syst. Sci. Discuss.*, *11*, 1227–1241.

Knight, J. F., R. L. Lunetta, J. Ediriwickrema, and S. Khorram (2006), Regional scale land-cover characterization using MODIS-NDVI 250 m multi-temporal imagery: A phenology based approach, *GISci. Remote Sens.*, *43*(1), 1–23.

Kummerow, C., W. Barnes, T. Kozu, J. Shiue, and J. Simpson (1998), The Tropical Rainfall Measuring Mission (TRMM) sensor package, *J. Atmos. Oceanic Technol.*, *15*(3), 809–817, doi:10.1175/1520-0426(1998)015<0809:TTRMMT>2.0.CO;2.

Kummerow, C., et al. (2000), The status of the Tropical Rainfall Measuring Mission (TRMM) after two years in orbit, *J. Appl. Meteorol., 39*(12), 1965–1982, doi:10.1175/1520-0450(2001)040<1965:TSOTTR>2.0.CO;2.

Kusche, J. (2007), Approximate decorrelation and non-isotropic smoothing of time-variable GRACE-type gravity field models, *J. Geodesy, 81*, 733–749. doi:10.1007/s00190-007-0143-3.

Kusche, J., R. Schmidt, S. Petrovic, and R. Rietbroek (2009), Decorrelated GRACE time-variable gravity solutions by GFZ, and their validation using a hydrological model, *J. Geodesy, 83*, 903–913. doi:10.1007/s00190-009-0308-3.

Li, J., J. Lewis, G. Rowland, L. Tappan, and L. Tieszen (2004), Evaluation of land performance in Senegal using multi-temporal NDVI and rainfall series, *J. Arid Environ., 59*(3), 463–480.

Longuevergne, L., B. R. Scanlon, and C. R. Wilson (2010), GRACE hydrological estimates for small basins: Evaluating processing approaches on the High Plains Aquifer, USA, *Water Resourc. Res.,46*, W11517, doi:10.1029/2009WR008564.

Millington, A. C., J. Wllens, J. J. Settle, and R. J. Saull (1994), Explaining and monitoring land cover dynamics using multi-temporal analysis of NOAA-AVHRR imagery, in *Environmental Remote Sensing from Regional to Global Scales*, edited by G. M. Foody and P. J. Curran, pp. 223–232, Wiley, Chichester, West Sussex, England.

Nicholson, S., et al. (2003), Validation of TRMM and other rainfall estimates with a high-density gauge dataset for West Africa: Part II: Validation of TRMM rainfall products, *J. Appl. Meteorol., 42*(10), 1355–1368, doi:10.1175/1520-0450(2003)042<1355:VOT.

North, G. R., T. L. Bell, R. F. Cahalan, and F. J. Moeng (1982), Sampling errors in the estimation of empirical orthogonal functions, *Monthly Weather Rev., 110*, 699–706, doi:10.1175/1520-0493(1982)110<0699:SEITEO>2.0.CO;2

Omondi, P., J. Awange, L. A. Ogallo, R. A. Okoola, and E. Forootan (2012), Decadal rainfall variability modes in observed rainfall records over East Africa and their relations to historical sea surface temperature changes, *J. Hydrol., 464–465*, 140–156, doi:10.1016/j.jhydrol.2012.07.003.

Omondi, P., et al. (2013a), Changes in temperature and precipitation extremes over the Greater Horn of Africa region from 1961 to 2010, *Int. J. Climatol.*, doi:10.1002/joc.3763.

Omondi, P., J. Awange, L. A. Ogallo, J. Ininda, and E. Forootan (2013b), The influence of low frequency sea surface temperature modes on delineated decadal rainfall zones in Eastern Africa region, *Adv. Water Resourc.*, doi:10.1016/j.advwatres.2013.01.001.

Omute, P., R., Corner and J. L. Awange (2012), The use of NDVI and its derivatives for monitoring Lake Victoria's water level and drought conditions, *Water Resourc. Manag., 26*, 1591–1613, doi:10.1007/s11269-011-9974-z.

Owor, M., R. G. Taylor, C. Tindimugaya, and D. Mwesigwa (2009), Rainfall intensity and groundwater recharge: Empirical evidence from the Upper Nile basin, *Environ. Res. Lett., 4*, 035,009, doi:10.1088/1748-9326/4/3/035009.

Preisendorfer, R. (1988), *Principal Component Analysis in Meteorology and Oceanography*, Elsevier, Amsterdam.

Rasmus, F., and R. Kjeld (2011), Analysis of trends in the Sahelian rain-use efficiency using GIMMS NDVI, RFE and GPCP rainfall data, *Remote Sens. Environ. 115*(2), 438–451.

Rodell, M., and J. S. Famiglietti (2001), An analysis of terrestrial water storage variations in illinois with implications for the gravity recovery and climate experiment (GRACE), *Water Resourc. Res., 37*, 1327–1340.

Rojasa, O., A. Vrieling, and F. Rembold (2011), Assessing drought probability for agricultural areas in Africa with coarse resolution remote sensing imagery, *Remote Sens. Environ., 115*(2), 343–352.

Rummel, R., G. Balmino, J. Johannessen, P. Visser, and P. Woodworth, (2002), Dedicated gravity field missions—Principles and aims, *J. Geodyn. 33*, 3–20.

Sahin, M. (1985), *Hydrology of the Nile Basin*, Developments in Water Science No. 21, Elsevier, N. Y.

Senay, G. B., K. Asante and G. Artan (2009), Water balance dynamics in the Nile basin, *Hydrol. Process., 23*, 3675–3681, doi:10.1002/hyp.7364.

Sergio, M. V-S, et al. (2012), Challenges for drought mitigation in Africa: The potential use of geospatial data and drought information systems, *Appl. Geogr., 34*, 471–486.

Sharifi, M. A., E. Forootan, M. Nikkhoo, J. Awange, and M. Najafi-Alamdari (2013), A point-wise least squares spectral analysis (LSSA) of the Caspian Sea level fluctuations, using TOPEX/Poseidon and Jason-1 observations, *Ad. Space Res., 51*(5), 858–873. http://dx.doi.org/10.1016/j.asr.2012.10.001.

Sulla-Menashe, D., M. A. Friedl, O. N. Krankina, A. Baccini, C. E. Woodcock, A. Sibley, G. Sun, V. Kharuk, and V. Elsakov (2011), Hierarchical mapping of Northern Eurasian land cover using MODIS data, *Remote Sens. Environ., 115*, 392–403.

Sultan, M., M. Ahmed, N. Sturchio, Y. Eugene, A. Milewski, R. Becker, J. Wahr, D. Becker, and K. Chouinard (2013), Assessment of the vulnerabilities of the Nubian sandstone fossil aquifer, North Africa, in (R. A. Pielke, Editor), Climate Vulnerability: Understanding and addressing threats to essential resources. Elsevier, 311-333 pp.

Sutcliffe, J. V., and Y. P. Parks (1999), An outline of the Nile basin, (H. Salz and Z.E. Kundzewicz, Editors) *The Hydrology of the Nile*, Spec. Pub. 5, IAHS, 1999.

Swenson, S., and J. Wahr (2009), Monitoring the water balance of Lake Victoria, East Africa, from space, *J. Hydrol. 370*(1–4), 163–176, doi:10.1016/j.jhydrol.2009.03.008.

Swenson, S., J. Wahr, and P. Milly (2003), Estimated accuracies of regional water storage variations inferred from the Gravity Recovery and Climate Experiment (GRACE), *Water Resourc. Res., 39*(8), 1223, doi:10.1029/2002WR001808.

Tapley, B., S. Bettadpur, J. Ries, P. Thompson, and M. Watkins (2004a), GRACE measurements of mass variability in the Earth system, *Science, 305*, 503–505, doi:10.1126/science.1099192.

Tapley, B., S. Bettadpur, M. Watkins, and C. Reigber (2004b), The gravity recovery and climate experiment: Mission overview and early results, *Geophys. Res. Lett., 31*(L09607), doi:10.1029/2004GL019920.

Vanicék, P. (1969), Approximate spectral analysis by least-squares fit: Successive spectral analysis, *Astrophys. Space Sci.*, *4*, 387–391, doi.:10.1007/BF00651344.

Veroustraete, F., H. Sabbe, and H. Eerens (2002), Estimation of carbon mass fluxes over Europe using the C-Fix model and Eruflux data, *Remote Sens. Environ.*, *83*, 376–399.

Wahr, J., M. Molenaar, and F. Bryan (1998), Time variability of the Earth's gravity field: Hydrological and oceanic effects and their possible detection using GRACE, *J. Geophys. Res.*, *103*(B12), 30,205–30,229, doi:10.1029/98JB02844.

Whittington, D., and E. McClelland (1992), Opportunities for regional and international cooperation in the Nile basin, *Water Int.*, *17*(3), 144–154, doi:10.1080/02508069208686134.

Woodcock, C. E., S. A. Macomber, M. Pax-Lenney, and W. B. Cohen (2001), Monitoring large areas for forest change using Landsat: Generalization across space, time and Landsat sensors, *Remote Sen. Environ.*, *78*, 194–203.

Yates, D. N., and K. M. Strzepek (1998), Modelling the Nile basin under climate change, *J. Hydrol. Eng.*, *3*(2), 98–108, doi:10.1061/(ASCE)1084-0699(1998)3:2(98).

Use of Multifrequency Synthetic Aperture Radar (SAR) to Support Regional-Scale Groundwater Potential Maps

Gregory S. Babonis[1] and Matthew W. Becker[2]

23.1. INTRODUCTION

Understanding the role of groundwater transport is of vital interest in global water management for both resource regulation and quality, yet little is known about the influence of subsurface water cycling at the subcontinental or continental scale [*Oki and Kanae*, 2006]. One indicator of shallow groundwater flow is surface water, which can be considered as an outcropping of groundwater and directly reflects how subsurface heads drive water from higher to lower elevations. Closed wetlands and seepage lakes, which are fed primarily by groundwater, make ideal proxies for surface water measurements. If the elevations of surface water sources can be measured on regional scales, then numerical models of groundwater flow can be used to predict large-scale groundwater storage and transport in cases where surface water is a surface expression of groundwater. However, field collection of accurate elevation data can be costly and, in some cases, logistically prohibitive. Remote sensing techniques can provide a solution to regional-scale data collection problems.

One challenge to using remote sensing in regions of shallow water table is that vegetation is typically dense. Open water surfaces or peat flats that reflect groundwater elevations may be obscured by vegetation canopy. The purpose of this work is to evaluate the use of unsupervised or minimally supervised classifications of radar images for the identification of open wetlands that may be used to derive groundwater elevations in suitable environments.

Active radar reflection is highly dependent upon wavelength. For example, the P-band (68 cm wavelength) of the Airborne Synthetic Aperture Radar (AirSAR) penetrates canopy cover and scatters primarily from the ground, the C-band (5.7 cm wavelength) scatters primarily from the top of the canopy, and the L-band (25 cm wavelength) scatters primarily from tree branches within the canopy. As a consequence, wetlands might be classified as canopied or noncanopied by ratioing radar bands. We investigate the potential for this approach in this chapter using AirSAR as an example radar data set. AirSAR flew between 1988 and 2004 as part of NASA's Earth Science Enterprise. Although AirSAR does not provide global coverage, it does offer opportunity for research in polarimetric radar imagery. In addition, similar technology may become increasingly available with the development of uninhabited aerial vehicle platforms such as UAVSAR (http://uavsar.jpl.nasa.gov/).

By removing wetlands with significant canopy from the data set, elevation data can be extracted from wetland locations using synthetic aperture radar or other companion data sets. These elevation data can then be used to measure groundwater elevation over vast areas in an efficient manner. Although other suitable land classification data may be available in the United States, there are vast boreal forests in the Northern Hemisphere with poor or no land cover data, but reasonably good topographic data from, for example, the Shuttle Radar Topography Mission (SRTM) [*Rodriguez et al.*, 2004]. For the purposes of verifying our approach, we use a highly verified

---

[1] *Department of Geology, State University of New York at Buffalo, Buffalo, New York, USA*

[2] *Department of Geology, California State University at Long Beach, Long Beach, California, USA*

Remote Sensing of the Terrestrial Water Cycle, Geophysical Monograph 206. First Edition. Edited by Venkat Lakshmi.

land cover data set with which to compare our AirSAR-based classification of wetland cover.

23.1.1. Previous Work

Using different radar frequencies to resolve details from above and below vegetated environments is a long-standing remote sensing technique and one of the strong advantages of active microwave radar versus optical platforms [*Jensen*, 2007]. *Krohn et al.* [1983] employed the L-band (25 cm), single-polarization, SAR system on the satellite SEASAT in one of the first studies of the response of lowland vegetation types in temperate forests to radar backscatter. The sensitivity of radar backscatter to the presence of moisture in vegetated regions was explored by *Place* [1985], who used SEASAT to map forested wetlands.

From the 1980s to the present, there has been extensive research into the structure of wetlands, forests, and other vegetative cover involving C- and L-band radar detection of subcanopy flooding, seasonal hydrology patterns, and wetland classification [*Baghdadi et al.*, 2001; *Hess et al.*, 1990; *Lang and Kasischke*, 2008; *Mougin et al.*, 1999; *Sokol et al.*, 2004; *Townsend*, 2002]. *Neeff et al.* [2005] employed the difference between P-band (68 cm) and X-band (3 cm) wavelengths to determine elevations of the forest floor and canopy crown scattering centers in order to estimate tree stand information for ecological concerns. Similarly, *Woodhouse et al.* [2006] investigated the accuracy of tree height measurements and associated scattering phenomena by employing X-band preferential scattering from tree canopy crowns.

Fredrick et al. [2007] employed surface water elevations from SRTM and ground truthing to calibrate regional-scale groundwater models of the study area also considered here (Figure 23.1). The critical assumption in this work, and in ours, is that lakes and wetlands in this region dominated by glacial outwash are highly connected to groundwater. The study demonstrated that remote sensing techniques are useful for developing regional-scale groundwater models using elevation data from frozen lakes. In addition, it was noted that open wetland elevation data had similar influence on parameter estimation as lake elevation data. Fredrick et al. concluded that in regions where lake elevation data are sparse, open wetland elevations could be used to calibrate regional-scale groundwater models. A review of the role of SAR in hydrologic applications can be found in *Becker* [2006].

23.1.2. Study Area

The study site is a 12×25 km swath of the Northern Highland Lakes Region (NHLR) of Vilas County Wisconsin (Figure 23.1). Much of the topography of the NHLR is the result of glacial activity during the Wisconsin Glaciation, 10,000–23,000 years ago [*Dott and Attig*, 2004]. The study area lies in a pitted outwash plain just south of a terminal moraine and contains numerous kettle lakes created by ice calving into deformable proglacial and subglacial fluvial sediment deposits from the Laurentide Ice Sheet. The modern terrain is characterized by 30–50 m of sandy glacial outwash overlying pre-Cambrian igneous bedrock [*Hunt et al.*, 2006]. As a result, aquifers tend to be unconfined and have a strong hydraulic connection to surface waters. Many kettle lakes are classified as seepage lakes, which are fed and drained almost entirely by groundwater.

Vilas County comprises the majority of the NHLR. The land cover in Vilas County is 70% forested, 17.2% wetlands, and 16% surface water [Vilas County Land, Air, and Water Conservation Department 2000]. It contains approximately 1355 separate wetlands, ranging from emerging/meadow and open to densely forested and vegetated peat bogs. Many of these wetlands, 31.3%, have been classified as seepage wetlands [*Piurek*, 2007]. A seepage wetland is a wetland not connected to surface hydrology by visible inflow or outflow. These wetlands develop within ancestral seepage lakes. Their strong hydraulic connection to groundwater makes them ideal for use as elevation head inputs in regional-scale groundwater flow models.

23.2. METHODOLOGY

This investigation consisted of field work to collect ground-truth data, a computational component to process and classify the AirSAR image, the extraction of elevation data from SRTM Digital Elevation Models (DEMs), and the creation of a groundwater potential map. The data sources used for the study are listed in Table 23.1.

23.2.1. Classification Scheme

The classification scheme was designed to be comparable to an existing and highly verified land cover data set known as WISCLAND. The Wisconsin Initiative for Statewide Cooperation on Landscape Analysis and Data (WISCLAND) land cover classification was produced in 1998 by the State of Wisconsin (Figure 23.1). It is based upon the synthesis of Landsat Thematic Mapper (TM) satellite imagery from 1991 to 1993, collected between May and October to account for spectral changes in land cover over a growing season. The Landsat images were corrected by the U.S. Geological Survey (USGS) Earth Resources Observation Systems

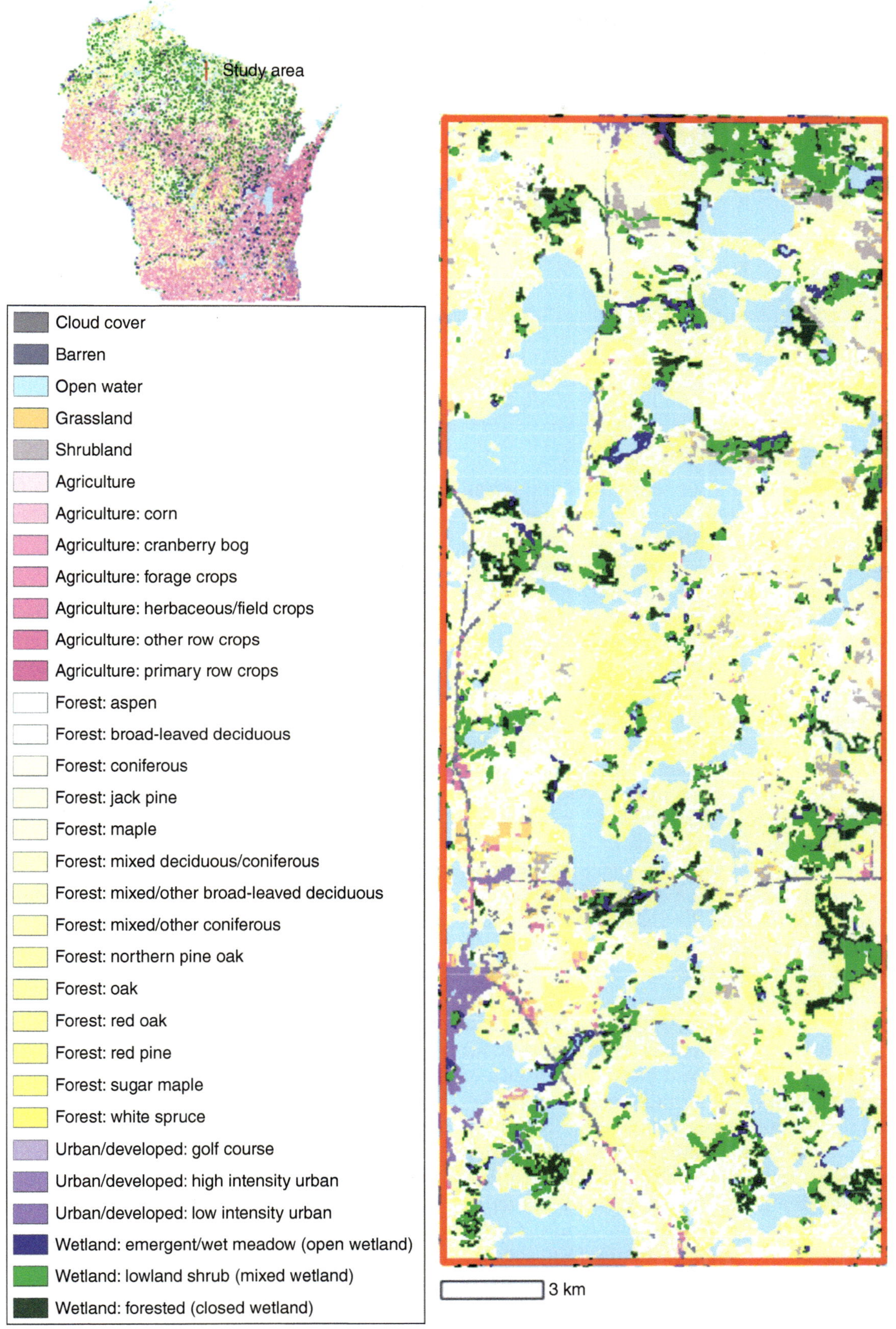

Figure 23.1 WISCLAND land cover data set used for identification of potential wetland study areas [WiDNR, 1998].

Table 23.1 Data sets used in this project, their primary purpose, amounts of each type of data, resolution, and date published or collected

Data Set	Purpose	Amount	Resolution/Scale	Date
WISCLAND land cover data	Layer of open, closed, and mixed wetlands	1 classified raster image	120 m	1998
GPS points	Ground truth	79 location points	10 m horizontal accuracy	26–28 October, 2007
Field photographs	Ground truth	645	2048×1536 and 2576×1932 pixels	26–28 October, 2007
AirSAR image	Wetland identification and classification	3	5 m (40 MHz) 10 m (20 MHz)	20 March, 2002
Double quadrangle maps	Wetland and road locations	5 raster images	1:24,000	1981–1982
TM				
NED	Elevation Data	2 Scenes	30 m	1 February, 1999
SRTM	Elevation Data	2 Scenes	30 m	11 February, 2000

(EROS) Data Center to an error not exceeding one pixel, or approximately 30 m. Errors less than one pixel represent georegistration or geolocation errors. Additionally, each wetland type classified by WISCLAND was ground truthed through a combined process of stratified random sampling from USGS 1:24,000-scale quadrangle maps, ground observations made at randomly selected field sites, and comparison to archived wetland locations in the Wisconsin Wetland Inventory, which contains the location of approximately 5.3 million acres of wetlands throughout the state [*WiDNR*, 1998]. Land cover maps of this caliber are rare, especially in remote areas because of the amount of ground truthing required to produce the necessary training sets. The intention of this work is to investigate the use of AirSAR as an alternative method of wetland identification. As WISCLAND is among the best data sets of its kind, AirSAR results that either meet or exceed the accuracy of WISCLAND are assumed to be appropriate for use in areas with less accurate existing land cover maps.

Three wetland classes (Figure 23.2) were used to represent the different vegetation cover geometries of the study area: (1) open—a vegetated understory is open to incident electromagnetic (EM) radiation; (2) closed—dense deciduous or coniferous canopy cover dominate; and (3) mixed—some canopy cover exists, but vegetated understory is also exposed, even if sparse. This classification system reflects the different types of scattering mechanisms influencing EM radiation return. Closed wetlands scatter EM radiation primarily from the tree canopy, open wetlands primarily from the understory. A mixed wetland scatters EM radiation both from the canopy and understory. The relative contribution of this scattered radiation is partially a function of the vegetation morphology (e.g., conifer/deciduous, leaf-on/off) and density (thick versus sparse). Every return signal is assumed to be a mixture of direct scatter from both open vegetation and canopy tops, volume scattering within the canopy, and trunk-ground interactions, including double bounce. These classes correspond to the three WISCLAND wetland types: emergent/wet meadow, lowland shrub, and forested. These WISCLAND classifications are based primarily on vegetation type. However, during field work, we confirmed that emergent/wet meadow corresponded to open wetlands, lowland shrub to mixed wetlands, and forested to closed wetlands.

23.2.2. Fieldwork and Regions of Interest Creation

Ground truth fieldwork was performed over a 3 day period from 26–28 October 2007. These dates were chosen so that deciduous trees would be leaf free as was the case during the AirSAR acquisition date (20 March, 2002). While this means that the data sets being compared are not from the same years (WISCLAND, 1998; AirSAR, 2002; fieldwork, 2007), we verified that none of the scenes were covered in deep snowfall or were under extreme drought or flooding conditions. Because the regional hydrology is controlled by damming of interconnected lakes (locally referred to as "flowages"), water elevations change less than a meter from year to year. Thus the extent and morphology of wetlands is unusually stable in the study area.

A total of 31 wetlands were visited, photographed, and marked with the global positiming system (GPS). A total of 79 GPS points and 645 photographs were collected. GPS measurements were collected as close to the center of the wetland as possibly, preferably in clear-sky

Figure 23.2 Typical wetlands in the NHLR. Two open wetlands: (a) Hermanson Springs and (b) Fallison Lake. Two forested wetlands: (c) Crystal Bog, surrounded by forested wetland and (d) Allequash Lake. Two mixed wetlands: (e) Mud Creek area and (f) north of Trout Springs.

locations (although this was not possible in forested wetlands), and with repeat measurements for accuracy. A tape measure was used to measure from the GPS point to notable features, such as water bodies, wetland edges, changes in wetland type, or photograph sites. Each site was classified as an open, closed, or mixed

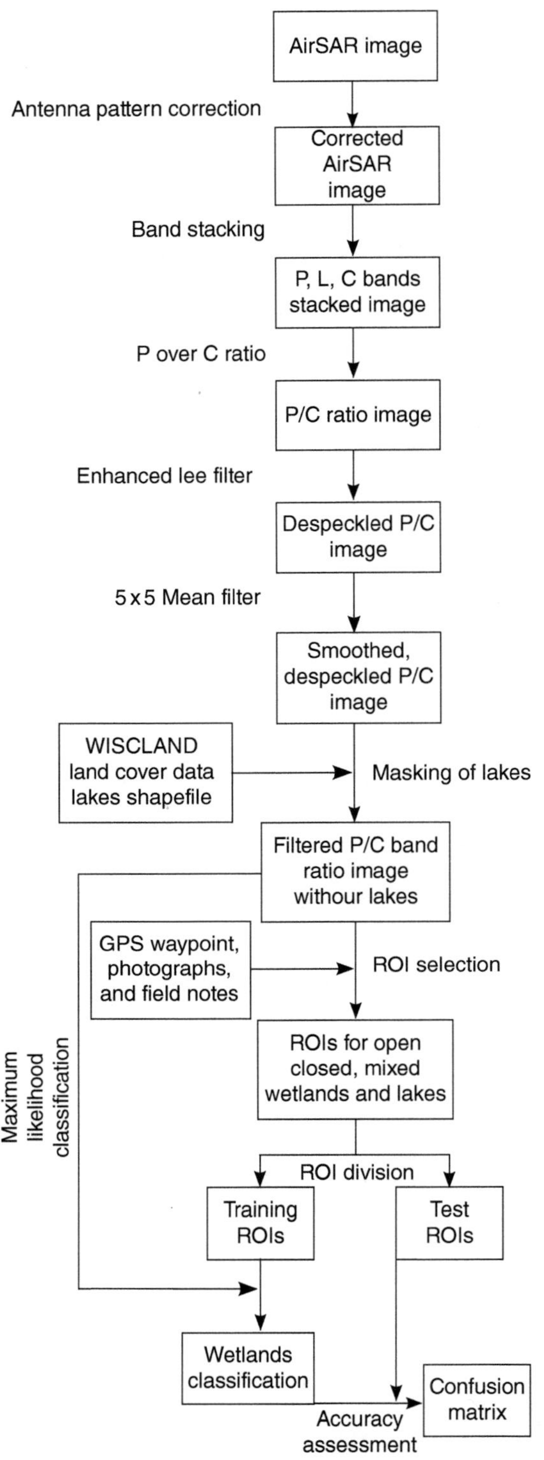

Figure 23.3 Work flow diagram of the image processing steps used to classify the AirSAR image into open, closed, and mixed wetlands.

wetland (Figure 23.3). Predominant vegetation within a typical open wetland was sedge grasses, leatherleaf, or other short-stalked hydrophytic plant types. In most cases, open wetlands either had water present at the surface or closely bordered a body of water. Forested wetlands were composed of coniferous black spruce and pine, and deciduous birch, maple, alder, and tamarack. In some forested wetlands a densely vegetated understory of shrubs was present. Water was not visible at the surface. Mixed wetlands exhibited a combination of vegetation from open and forested wetlands. In some cases, surface water was present at mixed wetlands, as well as thick peat and leatherleaf understories. Many wetlands consisted of several areas with differing classifications (open, closed, or mixed), and these areas were carefully documented.

The GPS locations, photographs, and notes collected during the field study were used to map representative areas for each wetland visited. From these mapped wetland areas, we created three wetland (open, closed, and mixed) regions of interest (ROIs) for training and test sets for AirSAR classification. We discounted wetlands that were located along the edges of the AirSAR image where some antenna pattern effects made them poor ROIs, and we only used mapped wetland areas that were at least 50 pixels in area to assure representative samples. Twenty-one open wetland ROIs contained an average of 79 pixels, 6 closed wetland ROIs contained an average of 51 pixels, and 10 mixed wetland ROIs contained an average of 66 pixels. An additional ROI was also created to exclude lakes within the region. The ROIs were split into training sets for the classification and test sets for validation.

23.2.3. AirSAR Processing

A work flow diagram of all the processing steps undertaken to classify the AirSAR image into open, closed, and mixed wetland can be seen in Figure 23.4. AirSAR data were converted to GeoTIFF format and processed using the ENVI ® image analysis software package. After performing a standard antenna pattern correction, the C-, L-, and P-band images were layer stacked into one composite image so that band math operations could be performed. An enhanced Lee filter was employed in order to reduce random noise. This filter is a standard deviation based filter that reduces speckle in radar imagery while preserving texture information [*Lopes et al.*, 1990]. A 5 × 5 moving-average filter was also applied to the resulting de-speckled image to smooth the image. Because each pixel corresponds to 5 m, smoothing over the 25 m kernel ensured that each GPS point—and its 10 m radius of potential error— would be associated with an appropriate value from the AirSAR image, and that small inhomogeneities in the mapped wetland areas (for example, a small stand of trees within an open wetland) would not result in an overly granular image. This was important for the later

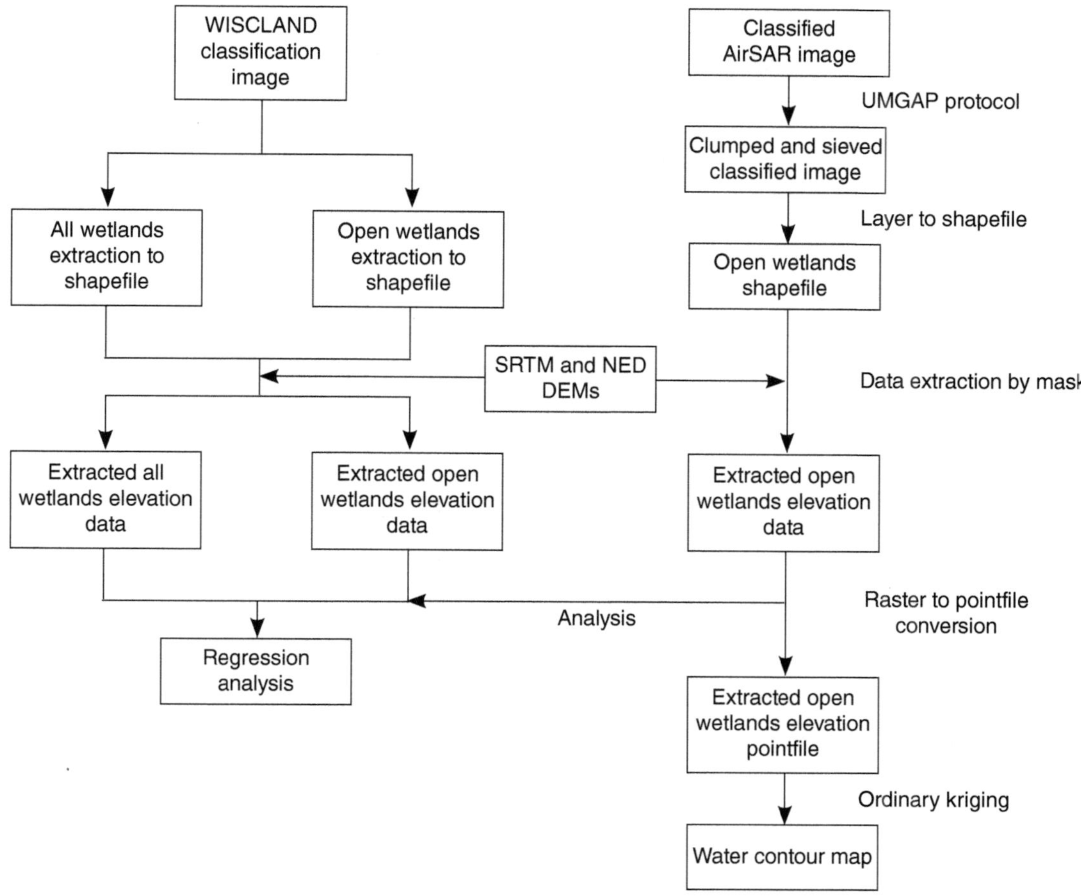

Figure 23.4 Work flow diagram of postclassification steps to extract elevation data from DEMs and create groundwater contours from those data.

training and test ROIs used for classification and accuracy assessment.

A P/C-band ratio image was created. The reasoning for ratioing was twofold: First, the ratio of two different wavelengths of radar reduces the effect of any common factors, such as topography, upon the data. Second, this ratio accentuates the effect of variation in vegetative cover, due to the dominance of different scattering centers in these bands (tree trunks and ground for P-band and canopy crown for C-band).

For regions where canopy cover dominates, strong P- and C-band backscatter returns are expected, and result in pixel values close to 1 in the ratioed image. For regions where there is little vegetative covering, P-band returns are diminished while C-band returns energy from smaller vegetation, making the pixel values for those locations close to zero. The remaining backscatter values can be attributed to a distribution of vegetation and open spaces. Thus, the backscatter returns can be classified into three regions corresponding to three different wetland cover geometries: open, closed, and mixed.

23.2.4. AirSAR Classification

The AirSAR scene was initially classified in an unsupervised fashion utilizing the ISODATA algorithm [*Jensen*, 2007] into a maximum of four separate classes: unclassified, open, mixed, and closed. Further classification efforts involved dividing the ROIs into two separate sets: one for a supervised classification training set and one for an accuracy assessment test set. They were separated so that each wetland classification type (closed, mixed, and open) was represented by roughly the same number of pixels in the training and test sets.

A maximum likelihood supervised classification was performed using the statistics of ROI classes for segmenting the filtered, P/C-band ratio image. Maximum likelihood classification is a Bayesian statistical tool in which the pixels in each user-defined class sample are assumed to be normally distributed. The training set is used to estimate the variance and covariance of each of the classes, and characterizes each class by a mean vector and covariance matrix. No probability threshold was used, forcing all pixels to be classified. Because the maximum likelihood classification algorithm classifies the entire input scene, and we only collected ground-truth data about wetland types and no other land cover classes (i.e., farmland, urban areas, roads, etc.), a reclassified WISCLAND land cover data set was used to cut out only the wetland areas from the classification. Our accuracy analysis is based upon the wetland classification performance, not upon the ability to map the extent of wetlands.

23.2.5. Accuracy Analysis

The classifications were compared to the reserved ROI test set and to the WISCLAND land cover data set (Figure 23.1). The classified AirSAR image was subjected to the same postprocessing steps employed in the creation of the WISCLAND data set. Confusion matrices, or error matrices, were used to assess the accuracy of the classification. Accuracy is analyzed in terms of user's and producer's accuracy. The producer's accuracy is calculated from the proportion of correctly classified pixels to the total number of pixels belonging to a particular ground-truth class. The difference between the producer's accuracy and a perfect result is the error of omission. User's accuracy measures the probability that each pixel was correctly assigned to a corresponding class during the classification process. The difference between the user's accuracy and a perfect result is the error of commission.

Cohen's kappa was also used to measure the agreement between the AirSAR classifications and the ground truth. It is often considered a more robust measure of agreement, as it considers agreement due to chance. According to the Cohen kappa evaluation system, a kappa value of 0.61–0.80 corresponds to a substantial agreement.

A comparison of the separability of the three different classes of wetlands used in this study was performed. The separability was determined using the Jeffries-Matusita (J-M) distance, which depends upon the difference between the probability functions of different classes [*Jensen*, 2007]. A value ≥ 1.9 indicates good separability; < 1.9 indicates poor separability.

23.2.6. Elevation Extraction

The postclassification processes necessary to translate wetland elevation data from SRTM DEMs to groundwater contours is diagrammed in Figure 23.4. The AirSAR-supervised classification was processed according to the Upper Midwest Gap Analysis Protocol (UMGAP) [*Lillesand*, 1998], which was used in the original creation of WISCLAND, and included the application of a clumping and sieving algorithm, operating on an 18 × 18 pixel window (90 m × 90 m) to remove any pixels that had less than 8 adjacent pixels of the same class.

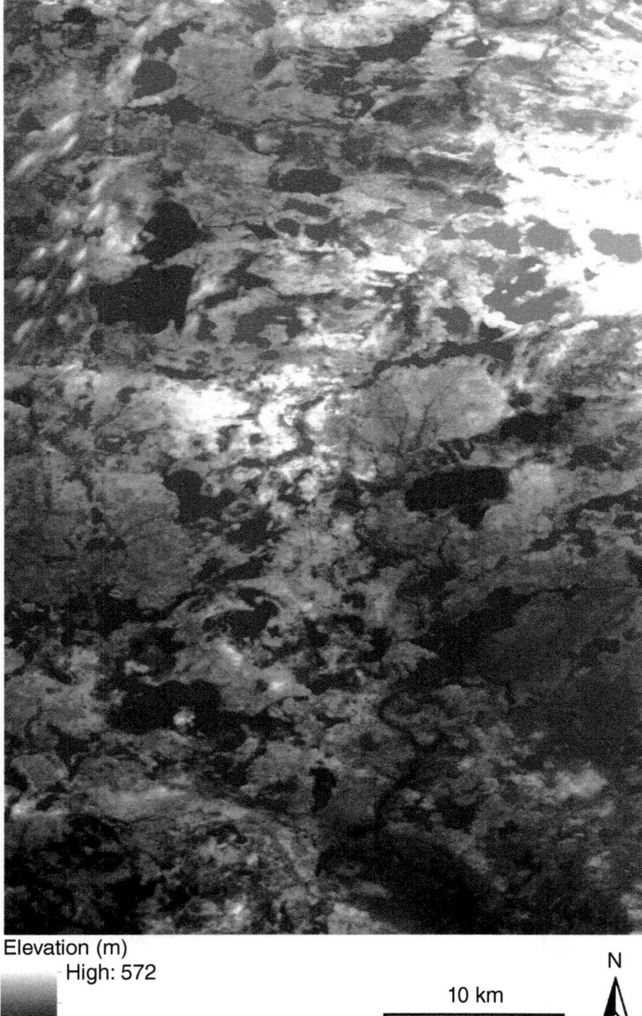

Figure 23.5 SRTM DEM for the study area. Elevation is given in meters above sea level.

The final step was to assess whether AirSAR would provide accurate elevation data for regional-scale groundwater models as compared to elevations extracted using WISCLAND. The open wetland classes were extracted from the classified AirSAR image and the WISCLAND land cover data set and used to extract elevation data from a National Elevation Dataset (NED) and SRTM DEM (Figure 23.5).

23.3. RESULTS

23.3.1. AirSAR Processing and Classification

The results of the P/C-band ratioing and the enhanced Lee and mean filtering can be seen in Figure 23.6. The enhanced Lee filter reduced random noise from radar

(a)　　　　　　　　　　(b)

Figure 23.6 (a) Antenna-pattern-corrected, P/C-band ratio image. (b) The same image with the enhanced Lee and mean filters applied. The two images have been histogram matched to one another other. The difference in brightness may come from image file compression when being exported.

backscatter, and the mean filter smoothed the image over a 25 m kernel. The enhanced Lee filter was applied to the P-band and C-band images both before and after ratioing. This was done to determine if filtering after ratioing would remove information rather than noise from the image. Subsequent classifications on the images with enhanced Lee filtering performed before ratioing, and after, showed a 6% increase in classification accuracy by the images filtered after ratioing.

The unsupervised and maximum likelihood classifications are compared to the WISCLAND classification in Figure 23.7. Both the unsupervised and maximum likelihood AirSAR classifications have much better spatial resolution than the Landsat-based WISCLAND classification. Overall, the three classifications seem fairly similar in terms of spatial extent and predominant wetland composition (Figure 23.8).

23.3.2. Accuracy Assessment

A confusion matrix comparing the AirSAR wetland classifications with the ground-truth test ROIs is shown in Table 23.2. The overall wetland accuracy for the Vilas County region was 78% in the WISCLAND classification, whereas the AirSAR unsupervised and supervised classifications achieved 75% and 84% overall wetland accuracies, respectively. More detailed user's accuracy assessments are not available for each WISCLAND county. A comparison of the user's accuracy (error of commission) of our unsupervised AirSAR classification and the *statewide* WISCLAND classifications (Table 23.2) shows that both the open and closed wetland classifications have better user's accuracy than WISCLAND. However, the unsupervised classification was less capable of separating mixed wetlands from the other two classes, with a user's accuracy of about 50%. The Cohen's kappa coefficient of 0.58 represents a 58%, or poor, agreement between the ground-truth and the subsequent classification.

For the supervised maximum likelihood classification, both open and closed wetlands show good agreement with the ground-truth (>85%) for both user's and producer's accuracy. The producer's accuracy (error of omission) for closed wetlands is very high, 99.5%. This means that the number of pixels incorrectly excluded from the closed wetland class is 1 pixel out of 202. A comparison of the user's accuracy (error of commission) of our maximum likelihood AirSAR classification and the *statewide* WISCLAND classifications (Table 23.2) shows that the AirSAR classification was more accurate than WISCLAND for open (88%–86%) and closed (85%–72%) wetlands and slightly less accurate than WISCLAND for mixed wetlands (74%–79%). The Cohen's kappa coefficient of 0.72 represents a 72%, or substantial, agreement between the ground truth and the subsequent classification.

Jeffries-Matusita distances between each of the AirSAR wetland classes were computed to determine the separability. The J-M distance between open and closed wetlands (1.89) and between open and mixed wetlands (1.89) shows good separability. However, the J-M distance between closed and mixed wetlands (0.57) shows poor separability.

23.3.3. Elevation Extraction

To evaluate the effect of wetland classification on extraction of elevation from wetlands, we use the NED as a reference. The NED DEM elevations are considered to be more representative of ground elevations (rather than canopy top elevations) because of the extensive ground truthing incorporated in their production [*Gesch*, 2007].

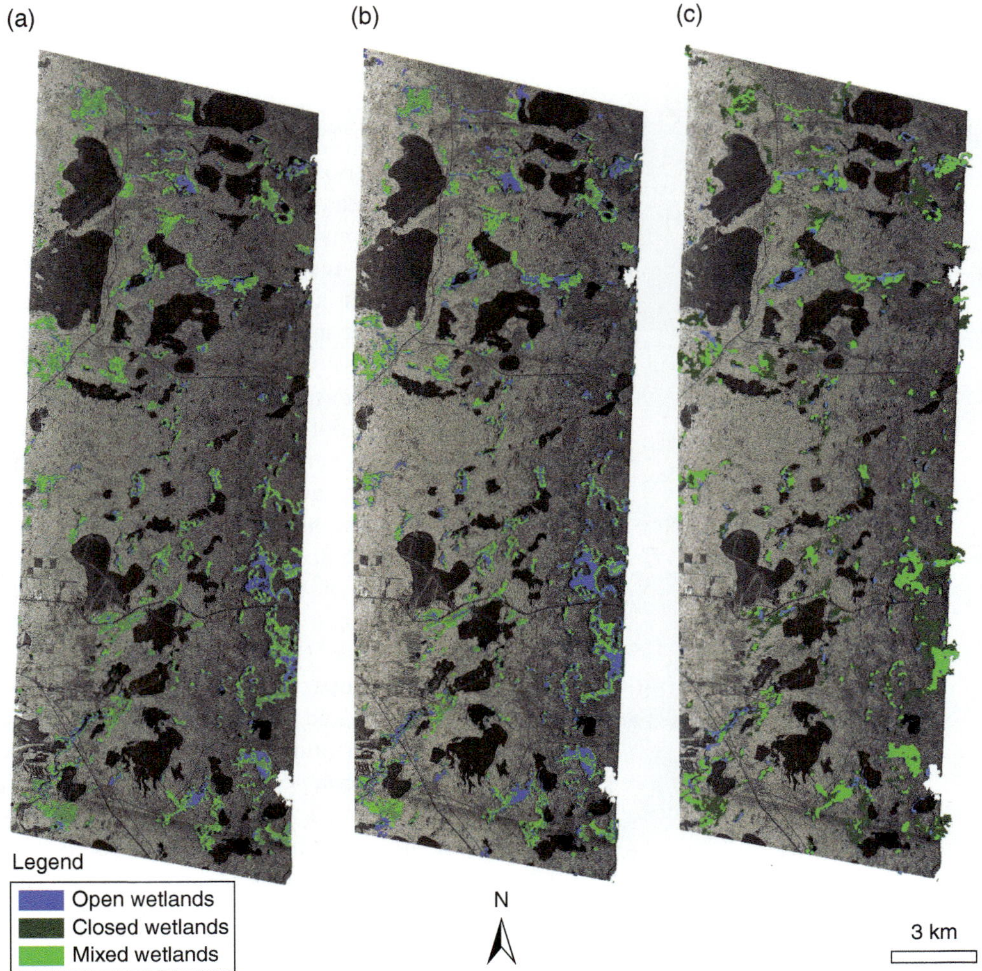

Figure 23.7 (a) The cut-out AirSAR unsupervised classification compared to (b) the cut-out AirSAR maximum likelihood classification.

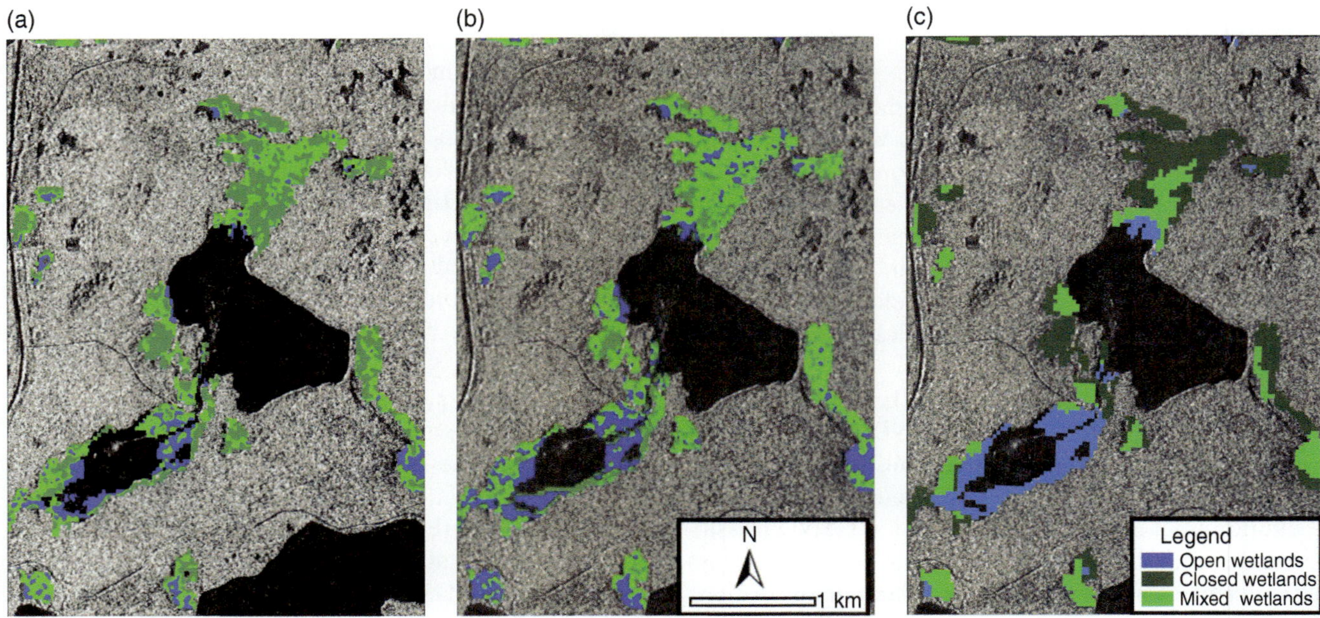

Figure 23.8 Closeup of Allequash Lake region wetlands. (a) The maximum likelihood classification has smaller and more numerous wetland types than (b) the WISCLAND classification, but the overall character of the wetland region is similar.

Table 23.2 Confusion matrix generated by comparing the test ROIs to the AirSAR maximum likelihood supervised and unsupervised classifications

Class	AirSAR Supervised Classification		WISCLAND
	Error of Omission (Producer's Accuracy)	Error of Commission (User's Accuracy)	Error of Commission (User's Accuracy)
Open wetland	12.6% (87.4%)	12.4% (87.6%)	14.0% (86.0%)
Closed wetland	00.5% (99.5%)	14.8% (85.2%)	27.8% (72.3%)
Mixed wetland	33.5% (66.5%)	25.9% (74.1%)	20.7% (79.3%)
Overall accuracy	*84%*		*78%*
Kappa coefficient	*0.72*		
	AirSAR Unsupervised Classification		
Open wetland	33.6% (66.5%)	1.4% (98.6%)	
Closed wetland	0.0% (100.0%)	22.0% (78.0%)	
Mixed wetland	12.4% (87.6%)	49.1% (50.9%)	
Overall accuracy	*75%*		
Kappa coefficient	*0.58*		

Note: Error of commission (user's accuracy) and overall accuracy for WISCLAND is shown for comparison.

Although there are clearly residual errors in the NED elevations when compared pixel by pixel, it does provide some indication of the relative elevation error before and after removing closed and mixed wetlands from the analysis.

For consistency, all pixels in the AirSAR image outside of wetlands were masked using the WISCLAND classification. Then, elevation residuals were calculated as the difference between SRTM and NED elevations on a 30 m grid. The mean residual for all wetlands was found to be 2.1 m (3.3 m standard deviation). After removal of closed and mixed wetlands as classified by AirSAR, the mean residual was found to be 1.0 m (3.2 m standard deviation). Thus, the SRTM elevations exceeded the NED elevations pre- and postclassification, presumably due to the presence of canopy, but this effect was reduced by removal of the closed wetlands.

23.4. DISCUSSION

Our AirSAR-supervised classification of wetlands outperformed (84% accuracy) the Landsat-based WISCLAND wetland classification (78% accuracy) for Vilas County. The unsupervised classification did not perform as well, but its overall accuracy (75%) approached that of WISCLAND. While the accuracy for each class is not available for *each* WISCLAND county, user's accuracy for the entire state of Wisconsin reveals that this improvement is likely due to an increase (13%) in the accuracy of closed wetland supervised classification using AirSAR. This improvement was anticipated because of AirSAR's ability to penetrate canopy and better discern closed wetlands obscured by tree cover. The

AirSAR-supervised classification showed a slight (2%) improvement compared to WISCLAND for the classification of open wetlands, but more substantial (5%) deterioration over WISCLAND for the classification of mixed wetlands. The poor performance of the AirSAR-supervised classification of mixed wetlands may be influenced by the difficulty in defining and identifying mixed wetlands in an objective fashion during ground truthing. The major discrepancies between the AirSAR and WISCLAND classification are between open and mixed and closed and mixed wetlands. Very few areas were classified as open in one image and closed in another. The AirSAR-supervised classification resulted in a Cohen's kappa value of 0.72 when compared to the ground truth. According to the Cohen kappa evaluation system, a Kappa value of 0.61–0.80 corresponds to a substantial agreement. The Cohen kappa value is considered a more robust sign of agreement because it takes into account that the agreement could have occurred by chance.

Both WISCLAND– and AirSAR-supervised classifications identified the predominant type of any individual wetland. However, the AirSAR classification provided more details within a wetland body than did the WISCLAND classification. This is due to the greater spatial resolution of AirSAR over WISCLAND; AirSAR has a spatial resolution of 5 m and classified wetlands 25 m or larger. WISCLAND is based on Landsat, which has a 30 m spatial resolution, but the classification technique used only identified objects greater than 142 m in size. Even with its greater spatial resolution, however, the AirSAR classification was less accurate than WISCLAND for identifying mixed wetland types, according to our ground-truth data. This indicates that certain vegetation types may be indicative of mixed wetlands that are better identified by the visible-infrared spectrum than radar.

23.5. CONCLUSIONS

The use of P/C, band radar ratios was able to discriminate between open (uncanopied) and closed (canopied), and mixed (partially canopied) wetlands in our study area. This was true in both supervised and unsupervised classification. P-band, being the longest wavelength employed in this classification, penetrates deepest into vegetated canopy and scatters primarily from trunk and ground interactions but not from short vegetation. C-band, being the shortest wavelength, primarily scatters from the canopy crown or short vegetation. For open areas, where P-band returns are weak and C-band returns are moderate, the P-band/C-band ratio is low, making open wetlands appear darker than they would in either a P-band or C-band image. For areas of dense canopy, both the P-band and C-band returns are high, making the closed wetlands appear brighter.

Based upon our ground-truth data, the supervised AirSAR classification of wetlands produced by this study is more accurate than the supervised Landsat-based WISCLAND classification of Wisconsin land cover types (84% and 78%, respectively). While AirSAR was most accurate in classifying open wetlands, the largest improvements over WISCLAND were in classifying closed wetlands, where radar had the advantage of being able to penetrate forested canopy. Overall, our classification method performs about as well as the best Landsat land use/land cover data set available. It would seem, however, that vegetation type is a slightly better approach to identifying mixed wetlands given the better performance of Landsat over radar data.

When all wetlands (as masked by WISCLAND) were used to extract SRTM water/wetland surface elevations, the residual between SRTM and NED elevations had a mean of 2.1 m (standard deviation 3.3 m). After removing all but the open wetlands as classified by AirSAR, the residual between SRTM and NED elevations had a mean of 1.0 m (standard deviation 3.2 m). Thus, using the radar band ratio reduced the estimated SRTM elevation error in half. SRTM elevations are based upon C-band returns that do not penetrate canopy, which explains the improvement. A comparison of mean residual values of open wetlands shows that there is almost no difference between using the open wetland elevations from WISCLAND versus those classified by AirSAR. Therefore, AirSAR P/C band ratio images represent a reliable basis for estimating elevations representative of water levels, as evaluated by comparison with NED.

AirSAR P/C-band ratio images perform as well as WISCLAND land use map in both classifying wetland types and extracting ground elevations. Even the unsupervised classification performed reasonably well, with only a 3% loss of accuracy over WISCLAND. AirSAR data are available only over specific regions. However, the same approach should be applicable to some existing or future global radar data sets, provided the spatial accuracy is sufficient. In regions where land cover data are not available, therefore, radar band ratios may present a useful option for removing the influence of canopy from measurement of water surface elevations in wetlands. In northern latitudes, water surface elevations reflect long-term trends in terrestrial water storage. In turn, water storage is highly coupled to carbon storage. The use of radar imaging of wetland water elevations has potential as a tool for monitoring carbon and water cycles in critical ecosystems such as northern boreal forests.

ACKNOWLEDGMENTS

This work was funded through the National Research Council and the Goddard Space Flight Center (#39498), NASA New Investigator Program (NAG5-10608), and the NSF Information Technology Research Program (IIS-0426557). We are grateful for the support of Trout Lake Station and North Temperate Lakes LTER Project. The comments of the two anonymous reviewers are gratefully acknowledged.

REFERENCES

Baghdadi, N., M. Bernier, R. Gauthier, R. and I. Neeson (2001), Evaluation of C-band SAR data for wetlands mapping, *Int. J. Remote Sens.*, *22*(1), 71–88.

Becker, M. W. (2006), Potential for satellite remote sensing of ground water, *Ground Water*, *44*(2), 306–318.

Dott, Jr., R. H., and J. W. Attig (2004), *Roadside Geology of Wisconsin*, Mountain Press Pub. Co., Missoula, Mont.

Fredrick, K. C., M. W. Becker, L. S. Matott, A. Daw, K. Bandilla, and D. M. Flewelling, (2007), Development of a numerical groundwater flow model using SRTM elevations, *Hydrogeol. J.*, *15*, 171–181.

Gesch, D. B. (2007), The National Elevation Dataset, in *Digital Elevation Model Technologies and Applications—The DEM User's Manual*, edited by D. Maune, pp. 99–118, Am. Soci. for Photogramm. and Remote Sens., Bethesda, Md.

Hess, L.L., J.M. Melack, and D. S. Simonett (1990), Radar detection of flooding beneath the forest canopy: A review, *Int. J. Remote Sens.*, *11*(7), 1313–1325.

Hunt, R. J., M. Strand, and J. F. Walker (2006), Measuring groundwater-surface water interaction and its effect on wetland stream benthic productivity, Trout Lake watershed, northern Wisconsin, USA, *J. Hydrol.*, *320*, 370–384, doi:10.1016/j.jhydrol.2005.07.029.

Jensen, J. R. (2007), *Remote Sensing of the Environment: An Earth Resource Perspective*, Prentice Hall Series in Geographic Infromation Science, Pearson Prentice Hall, Upper Saddle River, N.J.

Krohn, M.D., N.M. Milton, and D.B. Segal (1983), SEASAT synthetic aperture radar (SAR) response to lowland vegetation types in eastern Maryland and Virginia, *J. Geophys. Res. Oceans Atmos.*, *88*(NC3), 1937–1952.

Lang, M. W., and E. S. Kasischke (2008), Using C-band synthetic aperture radar data to monitor forested wetland hydrology in Maryland's coastal plain, USA, *IEEE Trans. Geosci. Remote Sens.*, *46*(2), 535–546.

Lillesand, T., J. Chipman, D. Nagel, H. Reese, M. Bobo, and R. Goldman (1998), Upper Midwest Gap analysis program image processing protocol, report prepared for the U.S. Geological Survey, Enviornmental Management Technical Center, Onalaska, Wis., June 1998.

Lopes, A., R. Touzi, and E. Nezry (1990), Adaptive speckle filters and scene homogeneity, *IEEE Trans. Geosci. Remote Sens.*, *28*(6), 992–1000.

Mougin, E.,C. Proisy,G. Marty, F. Fromard, H. Puig, J. L. Betoulle, and J. P. Rudant (1999), Multifrequency and multipolarization radar backscattering from mangrove forests, *IEEE Trans. Geosci. Remote Sens.*, *37*(1), 94–102.

Neeff, T., L. V. Dutra, J. R. dos Santos, C. D. Freitas, and L. S. Araujo (2005), Tropical forest measurement by interferometric height modeling and P-band radar backscatter, *Forest Sci.*, *51*(6), 585–594.

Oki, T., and S. Kanae (2006), Global hydrological cycles and world water resources, *Science*, *313*(5790), 1068–1072.

Piurek, R. B. (2007), Classification of wetland ground-water interactions using remote sensing, SUNY at Buffalo, Buffalo, N.Y.

Place, J. L. (1985), Mapping of forested wetlands: Use of SEASAT radar images to complement conventional sources, *Professional Geogr.*, *37*(4), 463–469.

Rodriguez, E., C. S. Morris, J. E. Beltz, E. C. Chapin, J. M. Martin, and W. Daffer (2004), An assessment of the SRTM topographic products, final report to NIMA, Jet Propulsion Lab., Calif. Inst. of Technol.

Sokol, J., H. NcNairn, and T. J. Pultz (2004), Case studies demonstrating the hydrological applications of C-band multipolarized and polarimetric SAR, *Can. J. Remote Sens.*, *30*(3), 470–483.

Townsend, P. A. (2002), Relationships between forest structure and the detection of flood inundation in forested wetlands using C-band SAR, *Int. J. Remote Sens.*, *23*(3), 443–460.

Vilas County Land, Air, and Water Conservation Department, August (2000), Vilas County Land and Water Resource Management Plan. Vilas Co. Land, Air, and Water ConservationDepartment, http://www.co.vilas.wi.us.

Wisconsin Department of Natural Resources (WiDNR) (1998), *User's Guide to WISCLAND Land Cover Data*, available at http://dnr.wi.gov/maps/gis/datalandcover.html#metadata.

Woodhouse, I. H., Izzawati, E. D. Wallington, and Turner D. (2006), Edge effects on tree height retrieval using X-band interferometry, *IEEE Geosci. Remote Sens. Lett.*, *3*(3), 344–348.

24

Monitoring Subsidence Associated with Groundwater Dynamics in the Central Valley of California Using Interferometric Radar

Tom G. Farr and Zhen Liu

24.1. INTRODUCTION

California's Central Valley produces one quarter of the nation's food, much of it irrigated with groundwater. In the northern Central Valley, surface water is more plentiful than groundwater, but in the more arid southern part, groundwater accounts for more than 50% of the annual water needs. The Central Valley is, in effect, California's largest reservoir [*Faunt*, 2009]. Groundwater levels in the Central Valley undergo seasonal fluctuations with longer term changes caused by dry years. In the southern part of the Central Valley, including the San Joaquin basin, a long-term decline in water levels indicates overdraft of the groundwater resource. Climate change models generally predict longer and more intense dry periods in the future, leading to concern among water resource managers that groundwater overdrafts will become worse.

In order to make better use of groundwater, water resource managers need to know the extent of groundwater depletion, both spatially and volumetrically, and to be able to monitor it over long periods. Water wells provide one solution, but owing to funding limitations, a lack of wells, and the difficulty of mandating government monitoring of private wells, less direct methods are needed. One less direct indicator of groundwater depletion that has been recognized for many years is ground surface subsidence [see *Galloway et al.*, 1999, for a review]. Therefore, a technique for mapping and monitoring subsidence and rebound should provide an important indicator of groundwater state and dynamics for water resource managers.

Interferometric synthetic aperture radar (InSAR) is a technique whereby surface change occurring between two radar imaging passes may be measured and mapped to high precision. The ability to map surface deformation of a few millimeters over large areas at spatial resolutions of a few 10's of meters has opened up new possibilities for remote monitoring of groundwater resources as well as many other areas of geology and geophysics [e.g., *Smith*, 2002; *Madsen and Zebker*, 1998]. Several groups have made studies of the effects of groundwater withdrawal and recharge on InSAR measurements of deformation of the Earth's surface [e.g., *Reeves et al.*, 2011; *Calderhead et al.*, 2011; *Lu and Danskin*, 2001; *Amelung et al.*, 1999]. Preliminary evaluations of InSAR applications to groundwater monitoring have been made over the last few years in Los Angeles [*Bawden et al.*, 2001], the Antelope Valley [*Galloway et al.*, 1998], Las Vegas [*Hoffmann et al.*, 2001; *Bell et al.*, 2008], the Santa Clara Valley [*Sneed et al.*, 2003], the Coachella Valley [*Sneed and Brandt*, 2007], and the southern Central Valley [*Farr*, 2009, 2010, 2011]. The Arizona Department of Water Resources (ADWR) has implemented a comprehensive InSAR program for monitoring land subsidence in Arizona [*ADWR*, 2012].

While measuring and understanding subsidence as a function of groundwater dynamics will greatly improve management of that important resource, the effect of subsidence on infrastructure can also benefit from using InSAR to map and monitor its extent. Roads can be broken by fissures, pipelines have been exhumed, and the slope of the land can be altered, changing drainage

Jet Propulsion Laboratory, California Institute of Technology, Pasadena, California, USA

Remote Sensing of the Terrestrial Water Cycle, Geophysical Monograph 206. First Edition. Edited by Venkat Lakshmi.

patterns. This last effect has proven to be a significant problem on the California Aqueduct, where over 1 m of freeboard has been added since 1995 in order to preserve flow. Areas of low relief that have subsided are also subject to flooding.

This chapter presents results of a continuing study of the application of InSAR to monitoring of land subsidence as a function of groundwater dynamics. We define, for the first time, the full extent as well as the evolution from 2007 to 2011 of a large subsidence bowl in the southern San Joaquin Valley and present the results in several different formats geared to different audiences. Further development, including subsurface geologic information, may allow more quantitative estimates of groundwater change based on InSAR subsidence histories.

24.2. BACKGROUND: GROUNDWATER

The aquifer system of the southern Central Valley has both unconfined and confined parts caused by alternating layers of coarse and fine-grained sediments. Water in the coarse-grained, unconfined or water table aquifers may be extracted or recharged easily and causes only minor "elastic" compaction reflected as seasonal subsidence and rebound of water levels and the land surface (Figure 24.1). Most water wells exploit the deeper confined aquifers, and withdrawal of water from them causes drainage of the fine-grained confining layers called aquitards. Much more water is available in the aquitards. These, however, drain more slowly and compact both elastically as well as inelastically. In general, if water levels are not drawn too low, when pumping ceases, water recharges the aquitards and

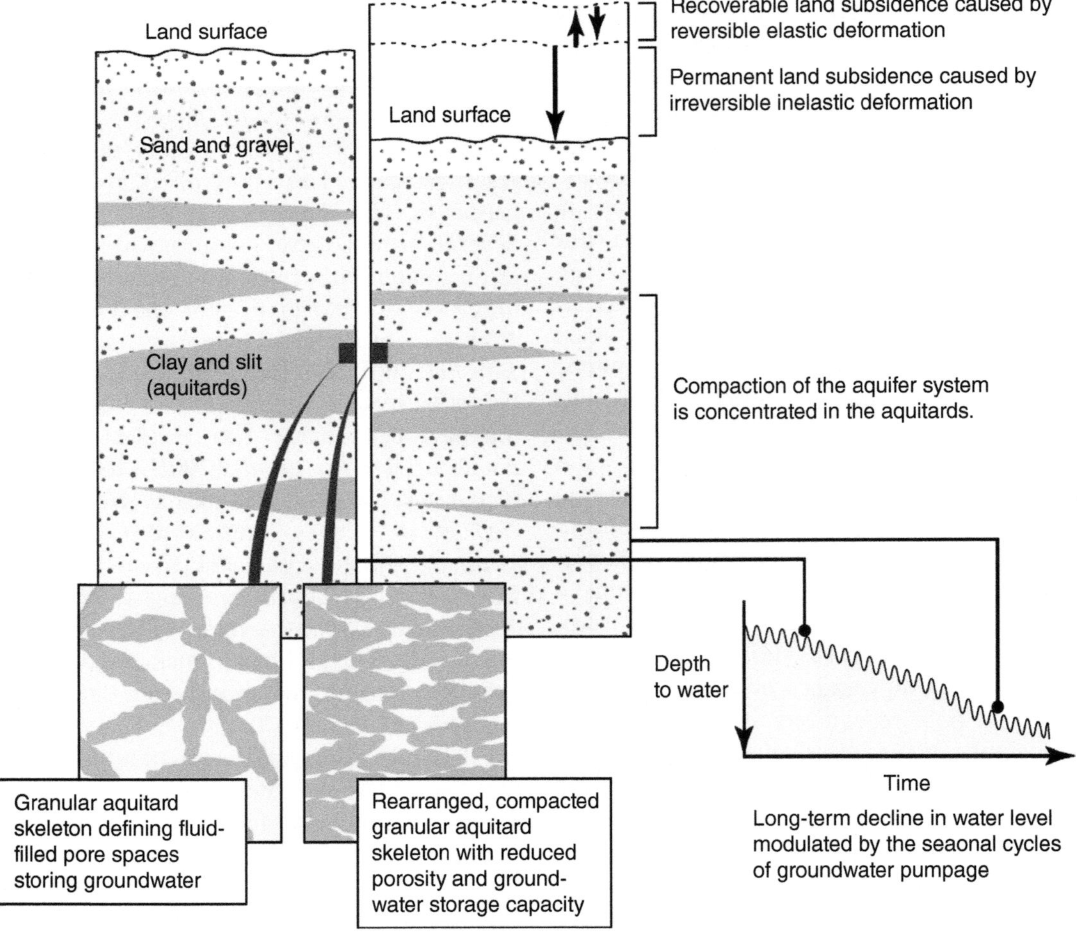

Figure 24.1 When water is pumped from an aquifer system, the fine-grained aquitards compact, causing water levels to decline and the land surface to subside. Seasonal recharge reverses the subsidence as an elastic response. However, if the water levels are drawn too low, long-term irreversible compaction occurs leading to permanent loss of groundwater storage capacity. The plot at lower right is a simplification, assuming relatively consistent pumping (decreases in water level) and recharge (increases). However, wet and dry years cause large variations in both [from *Galloway et al.*, 1999].

their structure expands. However, if water levels are drawn too low, then an irreversible compaction of the fine-grained aquitards occurs and water cannot recharge the layers, causing permanent loss of groundwater storage capacity and permanent subsidence. Much larger and longer term subsidence results from aquitard compaction [Figure 24.1; see *Galloway et al.*, 1999, pp. 8–13; *Bertholdi et al.*, 1991, for reviews].

24.3. BACKGROUND: InSAR

Ground surface deformation such as subsidence and rebound can be measured and mapped with repeat passes of a radar sensor [see *Madsen and Zebker*, 1998, for a review]. Most applications have used satellite radar systems, although airborne systems are also available. The technique works by acquiring images from the same viewing geometry at two different times between which a change has occurred. The phases of the returns from the two acquisitions are subtracted to create a phase difference map that can be processed to create a map of changes in distance along the line of sight of the satellite precise to fractions of the radar wavelength. Typical wavelengths are 3–25 cm. There are, however, some noise factors that must be considered: orbital error, atmospheric noise, static topography error, and decorrelation error. The orbital errors and static topography errors are largely handled in the InSAR data processing. Atmospheric water vapor and other variations in Earth's troposphere can introduce phase delay artifacts that can mimic surface changes. This is usually dealt with by analyzing many interferometric pairs and averaging (stacking) if the ground deformation rate is assumed to be steady. For time-varying deformation signals, more sophisticated spatiotemporal filtering can be applied during InSAR time series analysis to mitigate the effects of atmosphere noise. The exploitation of many interferograms and spatiotemporal filtering can effectively remove the dominant part of the atmospheric phase artifacts, leaving their influence on the estimated ground movement less significant [e.g., *Galloway et al.*, 1998; *Lanari et al.*, 2004; *Ozawa and Ueda*, 2011; *Chaussard et al.*, 2013]. Another problem, especially acute in agricultural areas such as the Central Valley, is small-scale surface changes near the scale of the radar wavelength. Crops blowing in the wind or fields plowed between acquisitions can spoil the phase coherence between the two radar images and cause loss of information. This effect can be ameliorated by using a longer wavelength and/or selecting interferometric pairs that have small orbital baseline separations and temporal differences.

After many pairs of radar images over an area have been processed into interferograms, they can be further analyzed to create a time series of surface deformation.

This is done by an InSAR time series inversion algorithm called the small baseline subset (SBAS) method [e.g., *Berardino et al.*, 2002; *Sansosti et al.*, 2010]. SBAS makes use of interferometric pairs that have small spatial (perpendicular) baselines and short temporal separations. The time series is constructed pixel by pixel and requires no assumptions on the continuity or stability of phase in time. Proper choice of the reference location and reference time are needed in order to tie together the relative InSAR measurements into a consistent time series. The reference location (or pixel) is usually chosen at a place that is stable compared to the deformation of interest based on in situ geodetic measurements or a priori information, while the reference time can be set at either the beginning or middle of the time series, depending on the time history and user preference.

The InSAR time series analysis produces a history of line-of-sight (LOS) surface displacements similar to Global Positioning System (GPS) time series observations but with much higher spatial resolution. Unlike stacking, which has no temporal resolution, InSAR time series recover both long-term mean velocities and time-varying components, while at the same time isolating atmospheric (as well as surface deformation) into the respective SAR image epochs [e.g., *Lanari et al.*, 2004]. This technique has been applied successfully for imaging non-steady-state deformation at volcanoes [*Lundgren et al.*, 2004], aquifer dynamics [*Lanari et al.*, 2004], deforming plate boundaries [*Lundgren et al.*, 2009], and more recently, episodic volcanic unrest in Long Valley Caldera [*Liu et al.*, 2011]. After the initial time series inversion, temporal and spatial filtering can be applied to further suppress atmospheric noise and smooth the deformation time series. Since atmospheric noise is spatially correlated but temporally uncorrelated, its net effect on the InSAR means the LOS deformation rate is negligible. The estimated measurement accuracy for InSAR displacement time series is on the order of ~ subcentimeter to ~ centimeter depending on the InSAR acquisitions and noise levels [e.g., *Galloway et al.*, 1998; *Ozawa and Ueda*, 2011; *Chaussard et al.*, 2013].

24.4. DATA

24.4.1. InSAR

The Jet Propulsion Laboratory (JPL), along with several other institutions, is a member of the WinSAR consortium (http://winsar.unavco.org). As such, we have been acquiring data on California from several imaging radar satellites since 1991. The main goal of WinSAR has been monitoring of the San Andreas and other active faults, but all available passes of the satellites

Table 24.1 Past, present, and future InSAR-capable satellites and nominal characteristics

Acronym	Dates	Resolution (m)	Swath (km)	Incidence Angles	Minimum Revisit (days)	Band[a]/pol	Agency[b]
ERS 1,2	1991–2010	25	100	25°	35	CVV	ESA
Envisat	2002–2010	25	100	15–45°	35	CVV, CHH	ESA
PALSAR[a]	2006–2011	10–100	40–350	10–60°	46	L-quad	JAXA
Radarsat 1	1995–2013	10–100	45–500	20–49°	24	CHH	CCRS
Radarsat 2	2008–	3–100	25–500	10–60°	24	C-quad	CCRS
TerraSAR-X	2007–	1–16	5–100	15–60°	11	X-quad	DLR
Cosmo-Skymed	2007–	1–100	10–200	20–60°	1	X-quad	ASI
PALSAR-2	2014	3–100	50–350	30–45°	14	L-quad	JAXA
Sentinel-1	2014	5–40	80–250	20–45°	14	C-dual	ESA
NASA SAR	2020?	35	350	15–60°?	8?	L-quad	NASA

[a]Wavelengths: X ~ 3 cm, C ~ 5 cm, L ~ 25 cm.
[b]ESA=European Space Agency, JAXA=Japan Aerospace Exploration Agency, CCRS=Canada Center for Remote Sensing, DLR=German Space Agency, ASI=Italian Space Agency.
[c]Phased-array L-band synthetic aperture radar.

covering California have been archived. As none of these radar satellites are operated by NASA, coverage is not always optimized, but preliminary work has identified numerous scenes that could be used to monitor ground deformation in the Central Valley [*Farr*, 2009, 2010, 2011]. The imaging radars that have been used (Table 24.1) have varying wavelength (X-band, 3 cm; C-band, 5.5 cm; L-band, 23.5 cm), polarization (VV, HH, HV), incidence angle (ranging from 10° to 60°), azimuth (ascending and descending orbits), and minimum revisit periods (11–46 days). These different imaging parameters present a challenge, but we have shown that it is possible to merge them into a long time series [*Farr*, 2009, 2010, 2011].

The WinSAR archive contains SAR coverage of the Central Valley back to 1991 and up to 2011, when the last satellite under the WinSAR free data agreements ceased functioning. Several satellites still collect data, but the costs are generally too high to continue the series. Future satellites, including Europe's Sentinel and Japan's PALSAR-2, may provide new free or low-cost data suitable for long-term monitoring.

24.4.2. Global Positioning System

Compared to the InSAR line-of-sight measurements, point-based in situ continuous GPS (CGPS) measurements provide three-dimensional (3D) displacement components that have dense temporal sampling (daily or higher), are in general highly accurate in horizontal but less accurate in vertical, and limited when resolving small to regional-scale deformation processes because of coarse station distribution and spacing. Continuous GPS sites in the Central Valley consist mainly of sites from the Earthscope Plate Boundary Observatory (PBO) network

(http://pbo.unavco.org/) (Figure 24.2). Available sites are widely spaced and limited in capturing subsidence area as shown in Figure 24.2. Most sites have deformation measurements since 2006.

These measurements provide cross validation to InSAR results and enable us to tie relative InSAR measurements to actual deformation. The CGPS solutions used in this study are from the NASA Making Earth Science Data Records for Use in Research Environments (MEaSUREs) Earth Science Data Records (ESDRs) system, downloadable from ftp://garner.ucsd.edu/pub/timeseries/measures/. The combined solution is obtained by combining SOPAC (Scripps Orbit and Permanent Array Center) GLOBK and JPL GIPSY solutions using the Quasi-Observation Combination Analysis (QOCA) software (http://gipsy.jpl.nasa.gov/qoca) [*Dong et al.*, 1998]. GPS time series analysis has been performed on the combined solutions to estimate seasonal variations, earthquakes and hardware-related offsets, linear rates, and to remove outliers ("cleaned") and common-mode errors ("filtered"). More details about the GPS time series analysis can be found in *Liu et al.* [2010, 2011]. The cleaned and filtered time series solutions are then compared with InSAR results. For the CGPS sites in the Central Valley (Figure 24.2), mean position uncertainties of the time series are ~2 mm for the horizontal and ~3 mm for the vertical. Note that the CGPS time series provided by MEaSUREs are in ITRF2008 reference frame, while InSAR LOS time series is with regard to the selected reference pixel. To ensure that the GPS measurements are comparable to InSAR, we differentiate GPS time series with regard to the GPS site located at the same location as the reference pixel used in the InSAR time series analysis. This removes the frame difference between the two data sets.

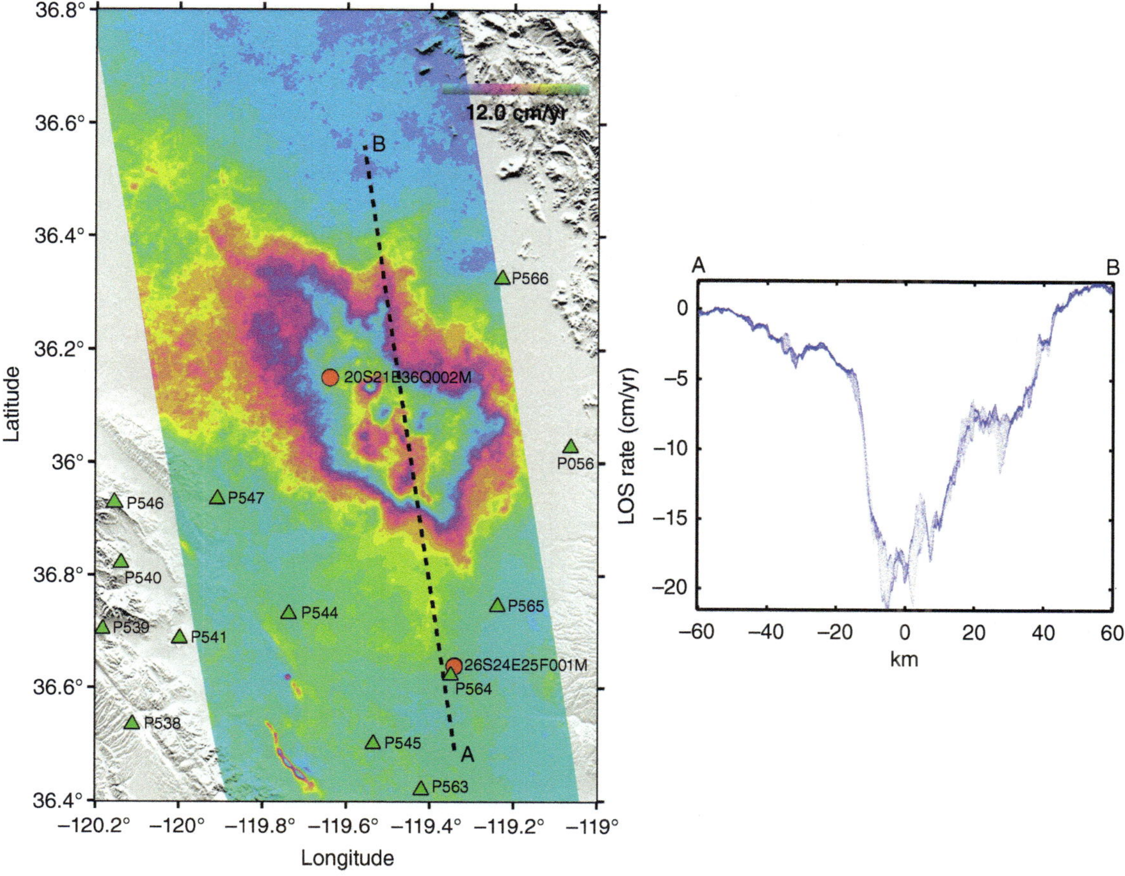

Figure 24.2 Mean line-of-sight velocity of L-band ALOS PALSAR interferometry in the southern Central Valley between June 2007 and December 2010 (left), and subsidence rate along the chosen profile (right). Also overlaid are continuous GPS sites (green triangles) and two wells (red dots). GPS sites are from Earthscope Plate Boundary Observatory (PBO) network. Comparisons between derived LOS time series at the locations of the two wells (20S21E36Q002M, 26S24E25F001M) and GPS site P564 are shown in Figure 24.3. The differential GPS time series relative to the GPS site P725 (Lon. -119.7456, Lat. 37.0889) is used in the comparison.

24.4.3. Wells

Thousands of water wells have been drilled to various depths throughout the Central Valley. Many have been monitored at one time or another, with water heights reported on the California Department of Water Resources (CADWR) online system: http://www.water.ca.gov/waterdatalibrary/. Recently the California Statewide Groundwater Elevation Monitoring Project (CASGEM; http://www.water.ca.gov/groundwater/casgem/) has taken up the challenge of compiling well data. While water well data are available as far back as the early 1900s, measurements are often sporadic with large gaps in temporal coverage. Very few wells have been monitored more often than once per quarter for more than a few years.

Height of water in wells is one of the few spatially extensive data sets available with which to monitor the state of the groundwater. While confined and unconfined aquifers may have different water heights or hydraulic heads, most wells are perforated at multiple depths, making it difficult to separate the two. In addition, these wells permit groundwater to migrate between aquifers, diminishing the difference between them. Currently, the California Department of Water Resources contours water height measurements on an annual basis to estimate changes in groundwater storage [*CADWR*, 2003, 2013; *McGuire*, 2013]. We have extracted water height data for numerous wells in the Southern Central Valley (SCV) with measurements in the period of our InSAR observations.

While water well monitoring provides the majority of information about the state of groundwater and aquifers in the Central Valley, a few specially equipped wells have been set up to monitor compaction of the aquifers directly. These extensometers have been installed and operated by the U.S. Geological Survey. New extensometers are currently being deployed and in the future will help assess InSAR measurements of subsidence (http://ca.water.usgs.gov/projects/central-valley/land-subsidence-monitoring-network.html).

24.5. RESULTS

Our early results for the Central Valley used C-band (5.5 cm) satellite systems [*Farr*, 2009, 2010, 2011]. The relatively short wavelength was affected by vegetation and plowing, but enough acquisitions were available to create time series that detected subsidence features.

One of the challenges of using satellites to produce long-term records is that the lifetime of a satellite is usually measured in years, while long-term records are considered to be at least a few decades long. However, various radars have been in orbit since 1991, so that if their records can be merged, a longer record may be obtained. Because InSAR measures deformation in the line of sight to the satellite, both incidence angle and azimuth differences come into play. One simplification is that if we assume changes in the land surface are due mainly to vertical movement, then the deformation observed by any orbiting radar can be projected into the vertical direction. The variation of horizontal velocities due to tectonic motion across the Central Valley are less than 5 mm/yr [*Shen et al.*, 2012], and the vertical tectonic deformation rate is on the order of ~1–2 mm/yr [e.g., *Hammond et al.*, 2012].

Another challenge to the merging of multiple InSAR satellite data sets is that it may be difficult to use the same reference pixel and reference time if the lifetimes of the satellites do not overlap or the ground footprints of the satellites are not the same. GPS stations may help define the stable reference area or pixels that are shared by multiple satellite measurements. These can be used along with the long, dense time series to tie different satellite measurements together in space and time. This is currently a work in progress.

Large-scale tectonic plate movements may also bias subsidence observations made from different satellites, but the network of GPS stations in the Central Valley (Figure 24.2) can provide control points with which the InSAR data can be calibrated. That the subsidence rates we have seen in the SCV are much faster than tectonic rates suggests that the large-scale slower tectonic motion has a negligible effect. However, for future measurements of slower subsidence and interpretation using satellite geodesy, influence of tectonic movements has to be taken into account.

The longer wavelength of L-band SAR sensors (~25 cm) can significantly improve temporal coherence, enabling time series analysis even with limited SAR acquisitions as compared to C-band satellite systems. As an example of this, we processed ~17 SAR scenes acquired by the Japanese Space Agency's L-band ALOS PALSAR between 21 June 2007 and 30 December 2010 from ascending track 218, frame 690–730, which spans the southern portion of the Central Valley. We used the JPL/Caltech ROI_PAC software in the processing: major steps include topography phase correction based on a global digital elevation model (e.g., Shuttle Radar Topography Mission), baseline reestimation for orbital error correction, phase unwrapping, filtering, and geocoding. We selected InSAR pairs with perpendicular baselines and temporal separations that satisfy a priori criteria to obtain better coherence. Then a variant of the SBAS InSAR time series inversion algorithm was applied to these selected interferograms to solve for the LOS displacement time series and mean LOS deformation rate at each pixel. Figure 24.2 shows the mean LOS deformation rate and subsidence rate along the profile traversing through a region that has undergone rapid subsidence.

The resultant InSAR LOS displacement time series for the southern Central Valley (Figure 24.3) are in good agreement with GPS time series. Comparisons of InSAR time series with water level change at selected wells are also consistent, suggesting ground deformation from satellite remote sensing can serve as a useful proxy indicator for subsurface groundwater change with time.

Figure 24.3 shows that water heights in two wells are generally correlated with subsidence history measured by InSAR and GPS. Figure 24.3a shows measurements at a shallow well within the deep subsidence "bowl": Water levels show a seasonal cycle with an amplitude of only a few feet. The water level also shows a long-term decline for most years until the wetter winter of 2010–2011. The associated InSAR time series follows the long-term decline with little indication of a seasonal signal or a rebound in 2010–2011 probably due to the fact that it is sensitive to compaction of much deeper aquitards. The lack of a strong rebound in the wetter years indicates that the time constant for equilibration of the deep aquitards is not short; thick aquitards such as the Corcoran Clay that underlies the area may take many years to rebound after water levels increase. Several nearby deeper wells also show little seasonal signal, but a much larger long-term water level decline (20–50 ft).

Figure 24.3b shows the water level and surface change for the selected location south of the subsidence bowl. The well was drilled to 812 ft and screened from 300 ft to the bottom. However, it is outside the area of the Corcoran Clay [*Page*, 1986], so the aquifers it samples are likely not completely confined. Water levels show a strong seasonal component with an amplitude of ~60 ft and a long-term decrease of ~100 ft from ~2007 to 2010. The wetter period of 2010–2011 sees an uptick in the water level. The InSAR and GPS measurements track the seasonal and longer term water levels with an amplitude of about 5 cm and long-term subsidence of about 12 cm. The rebound of water levels in 2010–2011 is only weakly reflected in the InSAR and GPS measurements, possibly due to longer time constants of aquitards in the area.

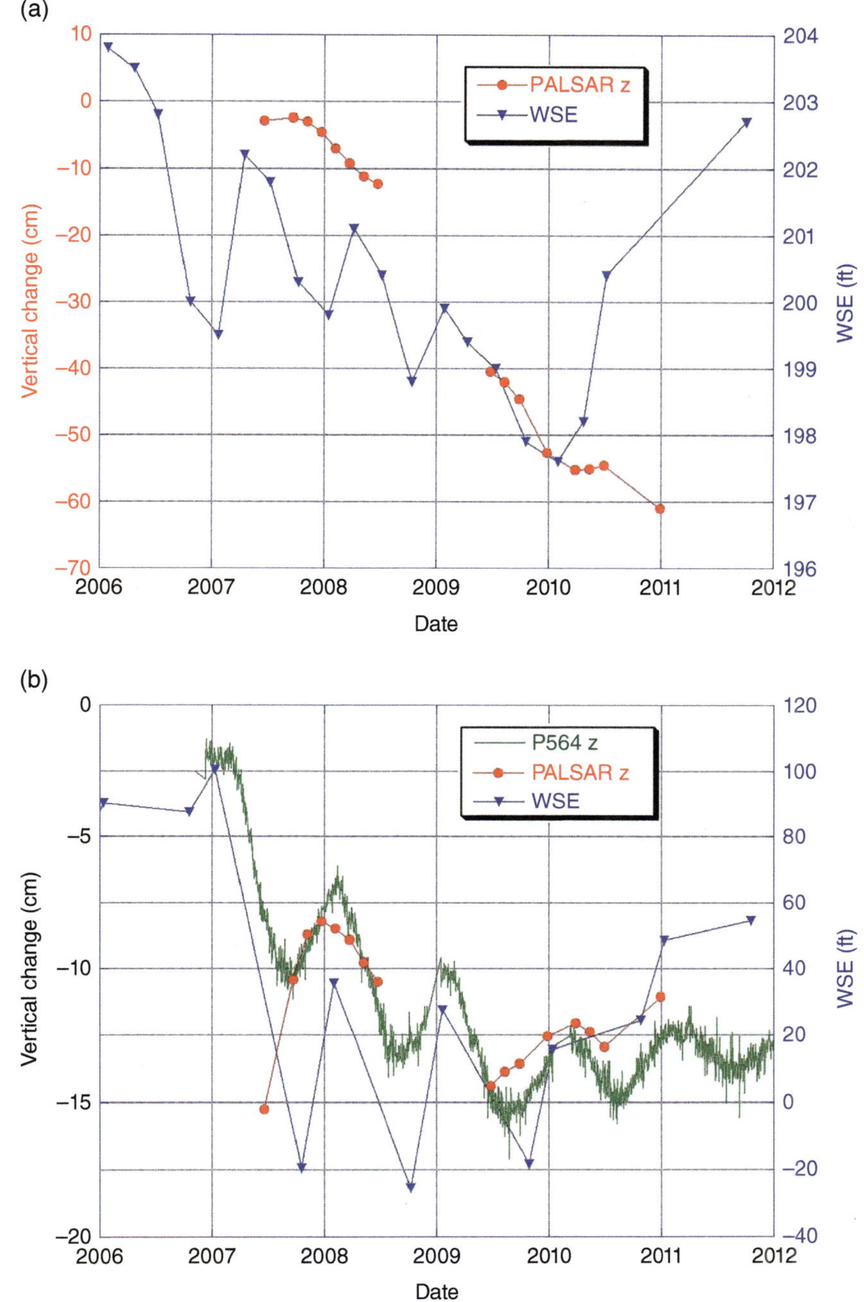

Figure 24.3 Comparisons between InSAR time series, water surface elevations (WSE) in wells, and continuous GPS measurements of vertical deformation two sites. The locations of the wells and GPS station are shown in Figure 24.2. The InSAR measurements have been projected to the vertical direction. Well data from http://www.water.ca.gav/waterdatalibrary/. (a) Measurements at well 20S21E36Q002M show a seasonal signal with a long-term drop in water and land elevation. A recovery of water level in 2010 and 2011 is not reflected in the InSAR measurements. (b) Measurements at well 26S24E25F001M and GPS site P564. A seasonal signal can be seen in all three measurements with less long-term deformation. The PALSAR and GPS vertical changes refer to the left-hand vertical axis. The PALSAR time series was offset to align with GPS because of its different reference time from that of the GPS data.

It is clear from these two examples that land surface movements correlate with groundwater-level fluctuations in the Central Valley. Work is continuing on integrating knowledge of the subsurface geohydrology with other observations and models of groundwater state to obtain a more quantitative understanding of groundwater storage changes. In the meantime, we can produce qualitative and semiquantitative information products to aid water resource managers.

24.6. APPLICATIONS TO GROUNDWATER MONITORING

24.6.1. Animation of Surface Deformation

Once a time series has been calculated, with smoothing, interpolation, reprojection, and a lot of exaggeration, a series of movie frames can be produced showing how the land surface has deformed through time. This was done for the Los Angeles area several years ago [*Lanari et al.*, 2004; http://photojournal.jpl.nasa.gov/catalog/PIA13202]. We have also produced a preliminary version for the Central Valley that incorporates the above PALSAR data (http://photojournal.jpl.nasa.gov/catalog/PIA16293). This type of product has the advantage of being immediately recognizable to nonscientists, either the public or managers unfamiliar with these types of measurements. The animations depict semiquantitatively the seasonal and longer term changes in surface heights with millimeter precision. Thus, one can follow the recharge of aquifers in winter followed by their drawdown in summer. Longer term subsidence caused by overpumping shows up as pits that deepen and persist for years. In particular, for the Central Valley animation, the long-term formation of the subsidence bowl masks the smaller seasonal oscillations.

24.6.2. Maps of Subsidence "Hot Spots"

Of more use to water resource managers are maps that show problem areas. A primary concern is the detection of "critical areas of overdraft" [*CADWR*, 2003], meaning areas where groundwater is being pumped at an unsustainable rate, causing subsidence and eventual compaction of the aquifer system. At some point, this compaction becomes irreversible, leading to permanent loss in storage capacity. The Arizona Department of Water Resources already uses InSAR data to produce maps of hot spots [*ADWR*, 2012], but this is not done on a routine basis anywhere else. Using time series maps, areas of long-term subsidence can be detected easily and amounts of subsidence spanning decades measured.

24.6.3. WebGIS Map

In order to make the time series information available in a more quantitative manner, a WebGIS interface similar to that produced by the Italian Research Center and available at http://webgis.irea.cnr.it can be produced. Using this web interface, a user can click on any pixel (typically a few 10's of meters in size) and obtain the time series of surface deformation for that pixel. Thus, plots may be made of seasonal and longer term subsidence and compared with GPS measurements or water height in nearby wells such as those shown in Figure 24.3. This provides better spatial and temporal resolution. This interface should be of interest to hydrologists studying subsidence.

24.6.4. Map of Groundwater Storage Changes

A factor often estimated by water resource managers is the change in groundwater storage from year to year [*CADWR*, 2003]. Currently, this is done by contouring water level measurements in wells on an annual basis and subtracting the contour maps [*CADWR*, 2003, 2013; *McGuire*, 2013]. It may be possible to estimate storage changes to first order from subsidence maps by assuming a constant specific yield for the aquifer system and subtracting subsidence contour maps from year to year. This will require extensive evaluation and testing as the subsurface heterogeneity may be too great to allow such extrapolation.

24.7. CONCLUSIONS AND FUTURE PROMISE

We hope this is just the start of a long-term monitoring program for California's Central Valley groundwater. Canada's RADARSAT-2 continues to operate and collect data over the Central Valley, and we hope to make use of it as well as potentially the several X-band satellites that currently operate. Continuing the time series of InSAR observations into future decades will help follow groundwater dynamics, especially in areas with long time constants of aquitard drainage and recharge.

Other remote sensing techniques are currently being used to observe other facets of the hydrology of California. Gravity changes have long been recognized as a way to measure mass changes underground related to groundwater movement [e.g., *Pool and Eychaner*, 1995]. Gravity maps from space produced by GRACE are now showing depletion of groundwater on regional scales [*Famiglietti et al.*, 2011]. While the spatial resolution of these maps is a few hundred kilometers, they measure directly groundwater loss from unconfined aquifers, which may be combined with InSAR measurements of groundwater movement within confined aquifers to create a fuller picture of groundwater dynamics.

Snow and surface water are of primary importance in the hydrologic cycle of California, and new techniques are today being applied to their mapping and monitoring. The Airborne Snow Observatory is flying in the Sierra Nevada to measure the depth and water content of snow (http://aso.jpl.nasa.gov/) and the Airborne Surface Water Ocean Topography experiment is flying over rivers and lakes to measure their height and flow (http://swot.jpl.nasa.gov/airswot/). Future spaceborne versions of these technologies will soon provide better information for water resource managers.

ACKNOWLEDGMENTS

The Alaska Satellite Facility (http://www.asf.alaska.edu/) archives and distributes the WinSAR data used in this study. The WinSAR consortium (http://winsar.unavco.org/) had the foresight to negotiate with international data providers for SAR coverage of California and provide an invaluable archive for future research. The Canadian, European, and Japanese space agencies contributed the data over the years to WinSAR, which were used for this study. The chapter was greatly improved thanks to two anonymous reviewers. This work was carried out at the Jet Propulsion Laboratory, California Institute of Technology, with funding from the Terrestrial Hydrology and Earth Surface & Interior Programs NASA.

REFERENCES

Amelung, F., D.L. Galloway, J.W. Bell, H.A. Zebker, and R.J. Laczniak (1999), Sensing the ups and downs of Las Vegas: InSAR reveals structural control of land subsidence and aquifer-system deformation, *Geology*, 27, 483–486.

Arizona Department of Water Resources (ADWR) (2012), InSAR Program, available at http://www.azwater.gov/AzDWR/Hydrology/Geophysics/documents/ADWRInSARProgramFactSheet.pdf; http://www.azwater.gov/AzDWR/Hydrology/Geophysics/InSAR.htm.

Bawden, G.W., W. Thatcher, R.S. Stein, K.W. Hudnut, and G. Peltzer (2001), Tectonic contraction across Los Angeles after removal of groundwater pumping effects, *Nature, 412*, 812–813.

Bell, J.W., F. Amelung, A. Ferretti, M. Bianchi, and F. Novali (2008), Permanent scatterer InSAR reveals seasonal and long-term aquifer-system response to groundwater pumping and artificial recharge, *Water Resourc. Res., 44*, doi:10.1029/2007WR006152.

Berardino, P., G. Fornaro, R. Lanari, and E. Sansosti (2002), A new algorithm for surface deformation monitoring based on small baseline differential SAR interferograms, *IEEE Trans. Geosci. Remote Sens., 40*, doi:10.1109/TGRS.2002.803792.

Bertoldi, G.L., R.H. Johnston, and K.D. Evenson (1991), Ground water in the Central Valley, California—A summary report, U.S. Geological Survey Professional Paper 1401-A.

Calderhead, A.I., R. Therrien, A. Rivera, R. Martel, and J. Garfias (2011), Simulating pumping-induced regional land subsidence with the use InSAR and field data in the Toluca Valley, Mexico, *Adv. Water Resourc., 34*, 83–97, doi:10.1016/j.advwatres.2010.09.017.

California Department of Water Resources (CADWR) (2003), California's groundwater, Bull. 118, Update 2003.

California Department of Water Resources (CADWR) (2013), Calculating annual change in groundwater storage using groundwater level data, draft appendix E of California Groundwater Update 2013, available at http://www.waterplan.water.ca.gov/docs/meeting_materials/caucus/2013.05.29/Task_4_TM_MainDraft_20130524.pdf.

Chaussard, E. F. Amelung, H. Abidin, and S. H. Hong (2013), Sinking cities in Indonesia: ALOS PALSAR detects rapid subsidence due to groundwater and gas extraction, *Remote Sens. Environ., 128*, 150–161, doi:10.1016/j.rse.2012.10.015.

Dong, D., T. A. Herring, and R. W. King (1998), Estimating regional deformation from a combination of space and terrestrial geodetic data, *J. Geodesy, 72*, 200–214.

Famiglietti, J.S., M. Lo, S.L. Ho, J. Bethune, K.J. Anderson, T.H. Syed, S.C. Swenson, C.R. de Linage, and M. Rodell (2011), Satellites measure recent rates of groundwater depletion in California's Central Valley, *Geophys. Res. Lett., 38*, L03403, doi:10.1029/2010GL046442.

Farr, T.G. (2009), Groundwater monitoring with InSAR, Abstr., paper presented at the Water Information Management Systems Workshop, San Diego, Calif., available at http://www.westgov.org/wswc/09wims.html.

Farr, T.G. (2010), Groundwater monitoring with InSAR, Abstr., paper presented at the IGCP 565 Workshop 3: Separating Hydrological and Tectonic Signals in Geodetic Observations, Reno, Nev., available at http://www.igcp565.org/workshops/Reno_2010/.

Farr, T.G. (2011), Remote monitoring of groundwater with orbital radar, Abstr., paper presented at the Groundwater Resources Assoc. Ann. Mtg., Sacramento, Calif., available at http://www.grac.org/am2011.asp.

Faunt, C.C. (Ed.) (2009), Groundwater availability of the Central Valley aquifer, California, USGS Prof. Paper 1766.

Galloway, D.L., K.W. Hudnut, S.E. Ingebritsen, S.P. Phillips, G. Peltzer, F. Rogez, and P.A. Rosen (1998), Detection of aquifer system compaction and land subsidence using interferometric synthetic aperture radar, Antelope Valley, Mojave Desert, California, *Water Resourc. Res., 34*, 2573–2585.

Galloway, D.L., D.R. Jones, and S.E. Ingebritsen (1999), Land subsidence in the United States, U.S. Geol. Surv. Circular 1182.

Hammond, W.C., G. Blewitt, Z.H. Li, H.P. Plag and C. Kreemer (2012), Contemporary uplift of the Sierra Nevada, western United States, from GPS and InSAR measurements, *Geology, 40*(7), 667–670, doi:10.1130/G32968.

Hoffmann, J., H.A. Zebker, D.L. Galloway, and F. Amelung (2001), Seasonal subsidence and rebound in Las Vegas Valley, Nevada, observed by synthetic aperture radar interferometry, *Water Resourc. Res., 37*, 1551–1566.

Lanari, R., P. Lundgren, M. Manzo, and F. Casu (2004), Satellite radar interferometry time series analysis of surface deformation for Los Angeles, California, *Geophys. Res. Lett., 31*, doi:10.1029/2004GL021294.

Liu, Z., S. Owen, D. Dong, P. Lundgren, F. Webb, E. Hetland, and M. Simons (2010), Estimation of interplate coupling in the Nankai trough, Japan using GPS data from 1996 to 2006, *Geophys. J. Int., 181*, 1313–1328, doi:10.1111/j.1365-246X.2010.04600.x.

Liu, Z., D. Dong, and P. Lundgren (2011), Constraints on time-dependent volcanic dynamics at Long Valley Caldera from 1996 to 2009 using InSAR and geodetic measurements, *Geophys. J. Int., 187*, 1283–1300, doi:10.1111/j.1365-246X.2011.05214.x.

Lu, Z., and W. R. Danskin (2001), InSAR analysis of natural recharge to define structure of a ground-water basin, San Bernardino, California, *Geophys. Res. Lett., 28*, 2661–2664.

Lundgren, P., F. Casu, M. Manzo, A. Pepe, P. Berardino, E. Sansosti, and R. Lanari (2004), Gravity and magma induced spreading of Mount Etna volcano revealed by satellite radar interferometry, *Geophys. Res. Lett.*, *31*, L04602.

Madsen, S. N., and H. A. Zebker (1998), Imaging radar interferometry, in Principles and Applications of Imaging Radar, Manual of Remote Sensing, vol. *2*, edited by F. M. Henderson and A. J. Lewis, pp. 359–380, Wiley, N.Y.

McGuire, V. L. (2013), Water-level and storage changes in the High Plains Aquifer, predevelopment to 2011 and 2009–11, USGS Sci. Inv. Rept. 2012–5291.

Ozawa, T., and H. Ueda (2011), Advanced interferometric synthetic aperture radar (InSAR) time series analysis using interferograms of multiple-orbit tracks: A case study on Miyake-Jima, *J. Geophys. Res.*, *116*, B12407, doi:10.1029/2011JB008489.

Page, R. W. (1986), Geology of the fresh ground-water basin of the Central Valley, California, with texture maps and sections, USGS Prof. Paper 1401-C.

Pool, D. R., and J. H. Eychaner (1995), Measurements of aquifer-storage change and specific yield using gravity surveys, *Ground Water*, *33*, 425–432.

Reeves, J. A., R. Knight, H. A. Zebker, W. A. Schreuder, P. S. Agram, and T. R. Lauknes (2011), High quality InSAR data linked to seasonal change in hydraulic head for an agricultural area in the San Luis Valley, Colorado, *Water Resourc. Res.*, *47*, doi:10.1029/2010WR010312.

Sansosti, E., F. Casu, M. Manzo, and R. Lanari (2010), Spaceborne radar interferometry techniques for the generation of deformation time series: An advanced tool for Earth's surface displacement analysis, *Geophys. Res. Lett.*, *37*, L20305, doi:10.1029/2010GL044379.

Shen, Z. K., M. Wang, and Z. Yue (2012), Unified Western US Crustal Motion Map, SCEC Annual Meet. http://www.scec.org/meetings/2012am/SCECProceedingsXXII_2012.pdf

Smith, L. C. (2002), Emerging applications of Interferometric Synthetic Aperture Radar (InSAR) in geomorphology and hydrology, *Ann. Assoc. Am. Geog.*, *92*, 385–398.

Sneed, M., and J. T. Brandt (2007), Detection and measurement of land subsidence using global positioning system surveying and Interferometric Synthetic Aperture Radar, Coachella Valley, California, 1996–2005, USGS. Sci. Investig. Rep. 2007–5251.

Sneed, M., S. V. Stork, and R. J. Laczniak (2003), Aquifer-system characterization using InSAR, in *US Geological Survey Subsidence Interest Group Conference Proceedings, 2001*, USGS Open-File Rept. 03-308. K. R. Prince and D. L. Galloway, eds.

Section VII: Data and Modeling

25

NLDAS Views of North American 2011 Extreme Events

Hualan Rui,[1,2] **Bill Teng,**[1,2] **Bruce Vollmer,**[1,2] **David Mocko,**[1,3] **and Guang-Dih Lei**[1,2]

25.1. INTRODUCTION

Year 2011 was one of the most extreme years of natural disasters in recent history. Over the course of the year, weather-related extreme events, such as floods, heat waves, blizzards, tornadoes, and wildfires, caused tremendous loss of human life and properties. Numerous studies and research projects have focused on acquiring observational and modeling data and revealing linkages between the intensity and frequency of extreme events, the global water and energy cycle, and global climate change. However, drawing definitive conclusions is still a challenge due to observational data inadequacy and scarcity, particularly at the land surface and subsurface.

The North American Land Data Assimilation System (NLDAS) data [*Mitchell et al.*, 2004; *Xia et al.*, 2012a, 2013a; http://ldas.gsfc.nasa.gov/nldas/; http://www. emc.ncep.noaa.gov/mmb/nldas/) has high spatial and temporal resolutions (0.125° × 0.125°, hourly since January 1979) and multiple water- and energy-related variables, including precipitation, soil moisture, snow cover/amount, runoff, evapotranspiration, latent heat, etc. NLDAS is an excellent data source for supporting meteorological and hydrological research [e.g., *Lee et al.*, 2010; *Mo et al.*, 2011; *Anderson et al.*, 2013, and the NLDAS Drought Monitor http://www.emc.ncep.noaa.gov/mmb/nldas/drought/].

This chapter illustrates the breadth of NLDAS Phase 2 (NLDAS-2) data by showing examples of descriptions for extreme events of 2011 in North America, including Hurricane Irene, Tropical Storm Lee, the July heat wave, and the February blizzard. For these, snowfall and snow water equivalent are used for illustrating the winter blizzard event; precipitation, 0–10 cm soil moisture, total runoff, and surface wind fields are shown for describing the two east coast tropical storm events; and 2 m above-ground air temperature data is selected for demonstrating the heat wave event. The NLDAS-2 streamflow data from the National Centers for Environmental Prediction (NCEP) (ftp://nomad6.ncep.noaa.gov/pub/raid2/wd20yx/nldas/ Streamflow/) is also shown for describing the effect of the heavy rainfall from the Tropical Storm Lee on rivers.

In addition, this chapter also briefly introduces the recently released NLDAS monthly means and monthly climatology data sets, and provides a description of the major characteristics of NLDAS data and data accessing methods.

25.2. 2011 EXTREME EVENTS IN THE UNITED STATES

Based on a National Oceanic and Atmospheric Administration (NOAA) announcement issued on 19 January 2012, two additional severe weather events during 2011 reached the $1 billion damage threshold, which raised 2011's billion-dollar disasters count from 12 to 14 events. And, therefore, the year 2011 has been classified as a year of climate extremes in the United States (http:// www.noaanews.noaa.gov/stories2012/20120119_global_ stats.html). The list of the 14 billion-dollar disasters (Table 25.1) shows that extreme weather events occurred

[1]*NASA Goddard Space Flight Center, Goddard Earth Sciences Data and Information Services Center (GES DISC), Greenbelt, Maryland, USA*

[2]*ADNET Systems, Inc., Rockville, Maryland, USA*

[3]*Science Applications International Corp., Greenbelt, Maryland, USA*

Remote Sensing of the Terrestrial Water Cycle, Geophysical Monograph 206. First Edition. Edited by Venkat Lakshmi.
© 2015 American Geophysical Union. Published 2015 by John Wiley & Sons, Inc.

Table 25.1 List of the 14 billion-dollar disasters 2011 in United States

Extreme Event	Date
Groundhog Day blizzard	29 January–3 February 2011
Midwest/Southeast tornadoes	4–5 April 2011
Southeast/Midwest tornadoes	8–11 April 2011
Midwest/Southeast tornadoes	14–16 April 2011
Southeast/Ohio Valley/Midwest tornadoes	25–28 April 2011
Midwest/Southeast tornadoes	22–27 May 2011
Midwest/Southeast tornadoes and severe weather	18–22 June 2011
Southern Plains/Southwest drought and heat wave	Spring–Fall 2011
Mississippi River flooding	Spring–Summer 2011
Rockies and Midwest severe weather	10–14 July 2011
Upper Midwest flooding	Summer 2011
Hurricane Irene	20–29 August 2011
Texas, New Mexico, Arizona wildfires	Spring–Fall 2011
Tropical Storm Lee	Early September 2011

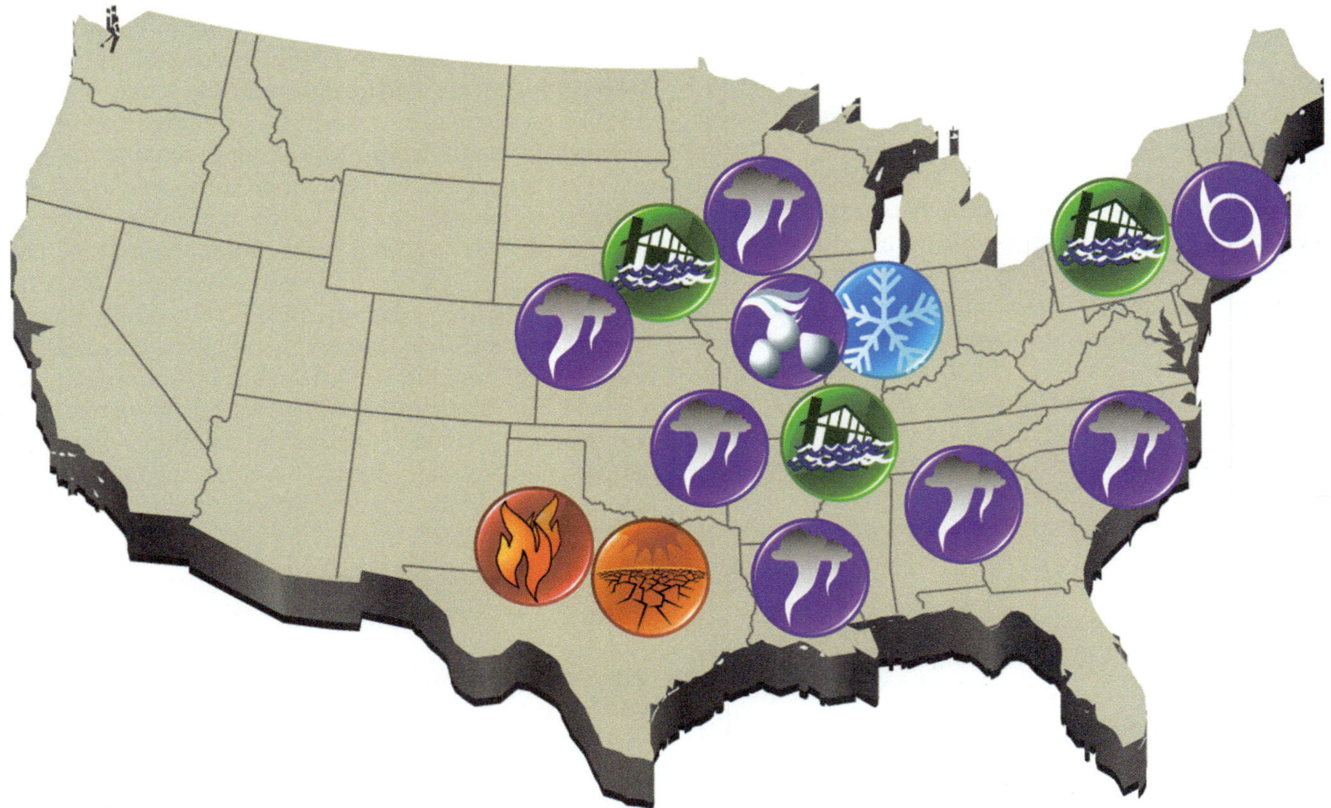

Figure 25.1 Spatial distribution of billion-dollar disasters 2011 in United States (Courtesy: NOAA http://www. noaanews.noaa.gov/stories2012/20120119_global_stats.html).

in all seasons of 2011. However, the spatial distribution of the billion-dollar disasters (Figure 25.1) shows that all of them occurred in the eastern half of the United States. The one thing that these events all have in common is they are related to extremes in water and/or energy, particularly near the land surface, where humans live and work. Most of them occurred in large river basins.

Information provided by NLDAS data on land surface properties and soil and water characteristics could play a unique role in exploring and understanding the extreme events and addressing underlining science questions.

The bar chart in Figure 25.2a shows interannual variations of the number of billon-dollar weather/climate disasters in 1980–2011. The number of disaster events and

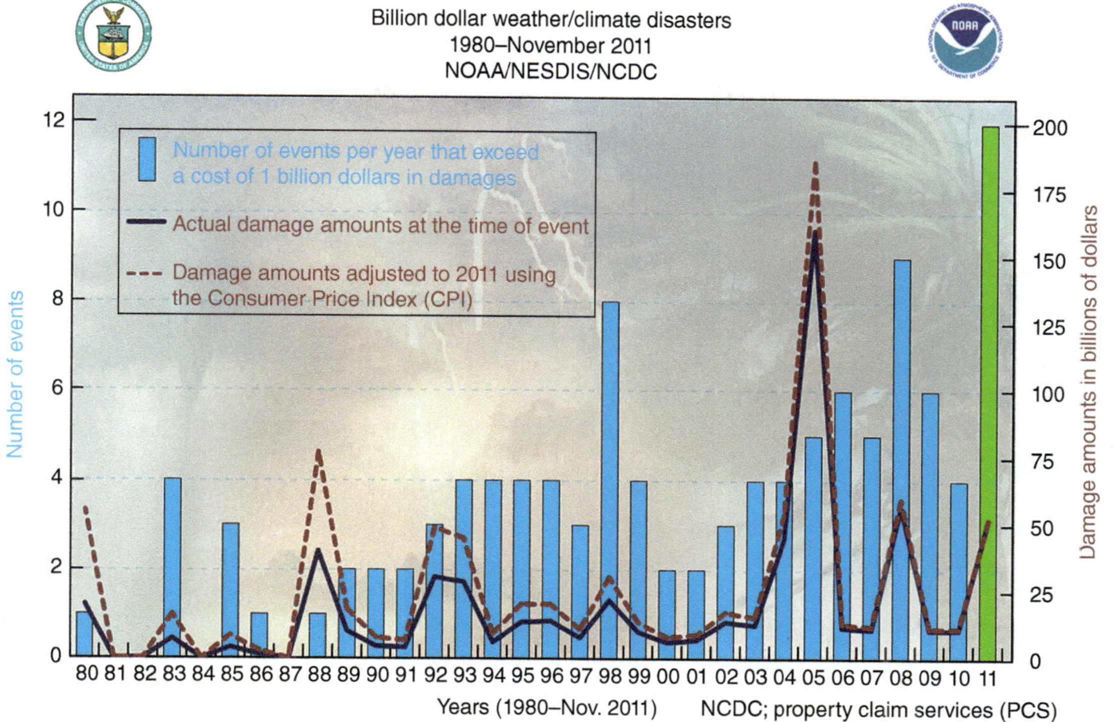

Figure 25.2a Billion-dollar weather/climate disasters (1980–2011) (Courtesy: NOAA, http://www.noaanews. noaa.gov/stories2011/20111207_novusstats.html).

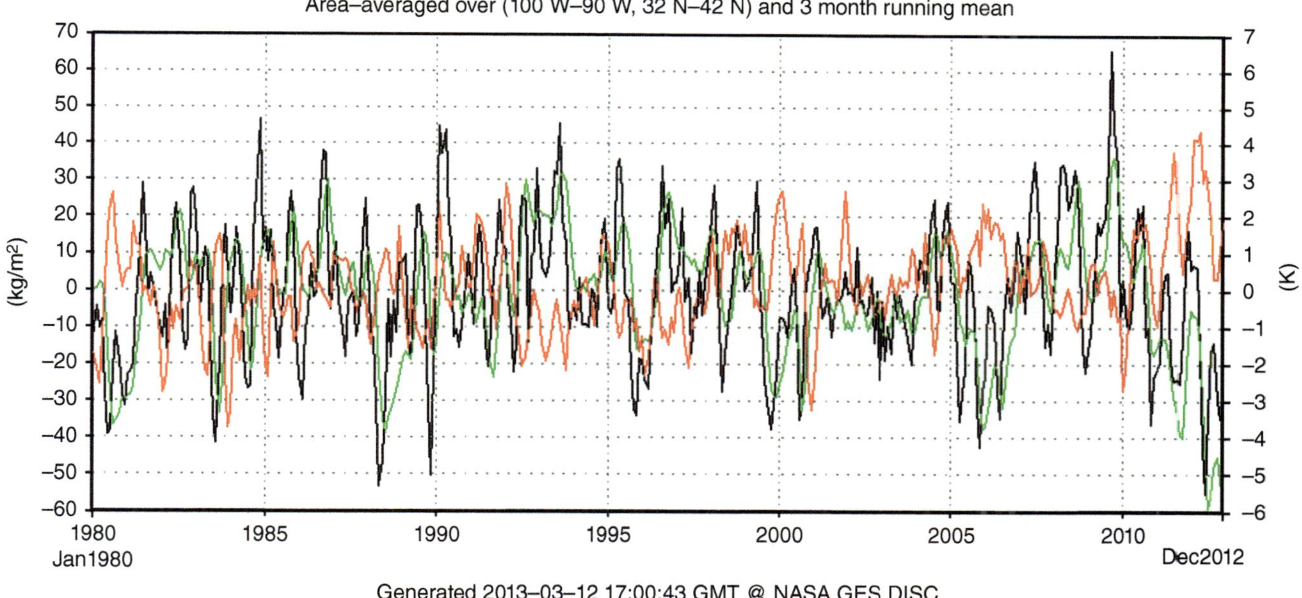

Figure 25.2b Area-averaged time series of anomaly precipitation, temperature, and soil moisture content. The precipitation anomalies (black line) are the monthly accumulated precipitation from NLDAS primary forcing, with the climatology removed. The temperature anomalies (red line) are the monthly 2 m above ground temperature from NLDAS-2 primary forcing, with the climatology removed. The soil moisture anomalies (green line) are the monthly top 1 m soil moisture content from NLDAS-2 Noah model, with the climatology removed. The climatology is a 30 year (1980–2009) monthly climatology. The region for the area-averaged is the central United States (100°W–90°W, 32°N–42°N); 3 month running mean was applied to each of the time series.

associated economic losses show pronounced interannual variations with an increasing trend, particularly during the latest 10 years, although the upward trend is partially due to the total economic increase and other factors. It is important to pinpoint global weather/climate anomalies and local environmental conditions in addressing extreme events and variations, and in mitigating impacts of the disasters. In these efforts, NLDAS, with a set of water- and energy-related variables and the long temporal coverage (1979–present), is capable for facilitating the interannual variability study. Phase 2 of the NLDAS project, detailed in [*Xia et al.* 2012a, 2013a], goes from January 1979 to the present day, typically updated 3.5 days behind the current time. As an example, shown in Figure 25.2b is the central United States area-averaged time series from NLDAS-2 data. Three variables from 1979 to 2012 are selected for demonstration: (black line) the interannual variations and month-to-month changes of precipitation; (red line) 2 m above ground temperature; and (green line) top 1 m soil moisture from one of the NLDAS LSMs. It is clear that larger precipitation anomalies correspond well to the anomaly peaks of soil moisture, and they are generally negatively correlated to the anomaly peaks of surface temperature. For this central U.S. region, periods with low precipitation thus tend

to have less soil moisture and higher surface temperatures. It is also interesting to notice that some of the largest anomalies in all three variables occur after 2010.

We have selected four of these 2011 extreme events to demonstrate the capability of the NLDAS system in capturing and describing their structures and variations. NLDAS data will also be used to view the detailed evolutions of these extreme events and to analyze relationships between different variables for consistency and coherency.

25.3. GROUNDHOG DAY BLIZZARD

The first billion-dollar disaster of 2011 is the Groundhog Day blizzard, a large winter storm that impacted several central, eastern, and northeastern states, with total losses greater than $1.8 billion. At least 36 deaths are attributed to this sprawling storm.

The two-and-half-day (12Z 30 January to 23Z 2 February 2011) accumulated snowfall (Figure 25.3a) seen by the NLDAS-2 Noah Land Surface Model (LSM) shows a large amount of snowfall from Oklahoma, Missouri, Illinois, Indiana, Wisconsin, to New York, Massachusetts, and other northeastern states, with the heaviest snowfall in Wisconsin along the eastern shores of

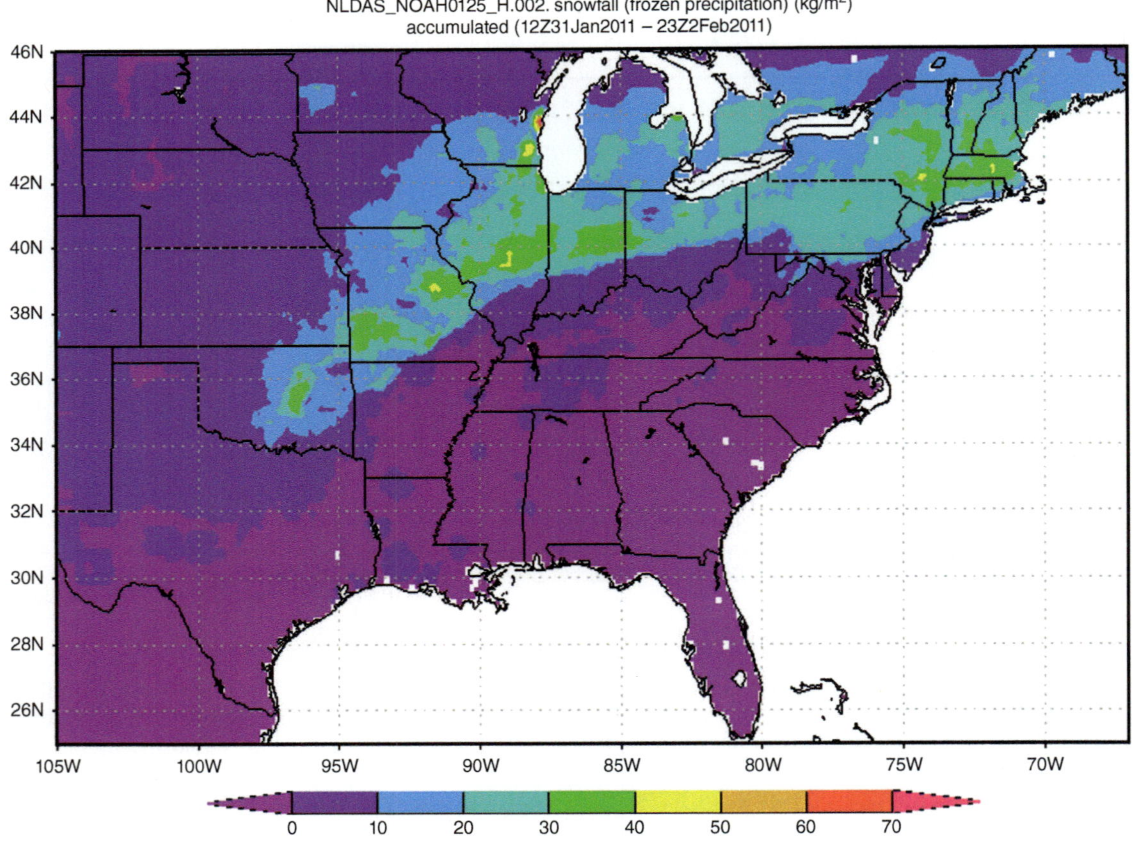

Figure 25.3a Accumulated snowfall from NLDAS-2 Noah model, accumulated between 12Z 31 January 2011 and 23Z 2 February 2011. "Z" refers to coordinate universal time (UTC), also known as Greenwich mean time.

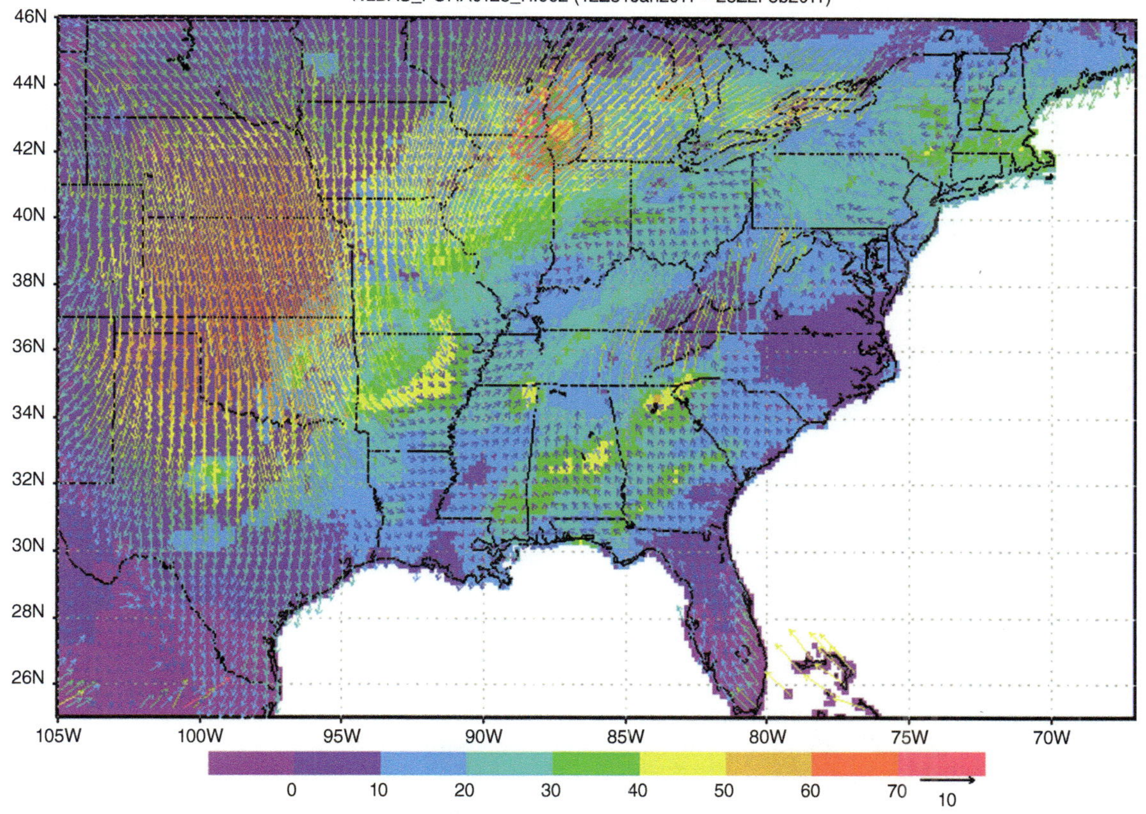

Figure 25.3b NLDAS-2 10 m above ground wind (averaged between 12Z 31 January and 23Z 2 February 2011) overlaying on the NLDAS-2 precipitation (liquid and frozen rain, accumulated between the same temporal range).

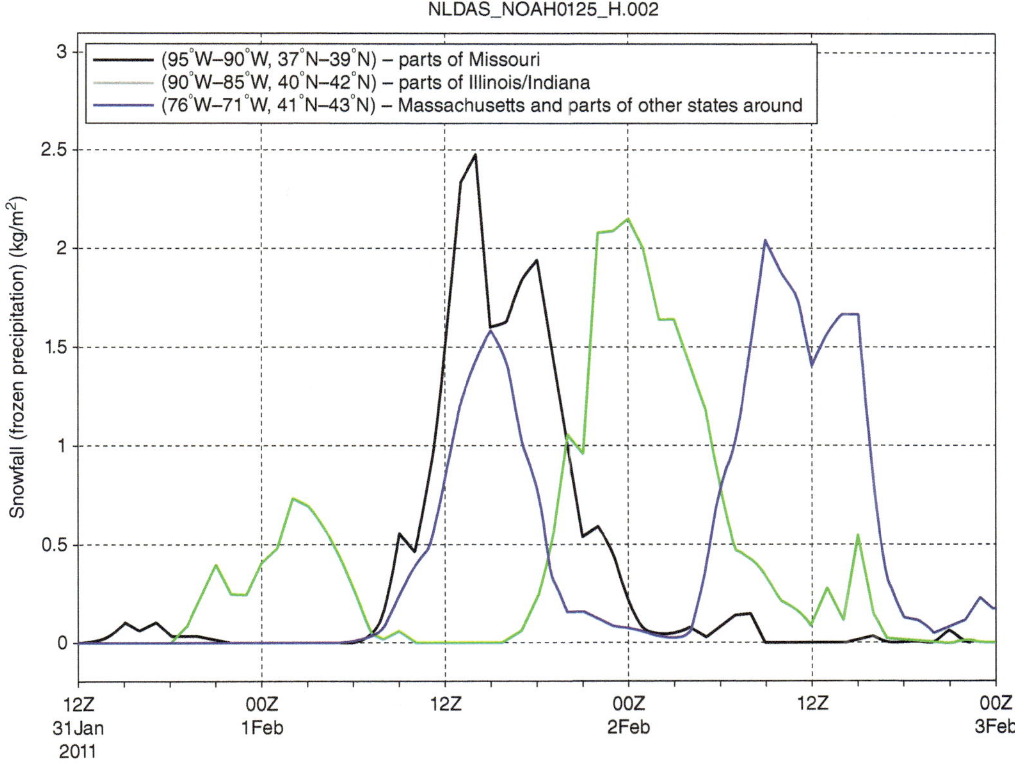

Figure 25.3c Time series of the snowfall from NLDAS-2 Noah model, area-averaged over the parts of Missouri (black line: 95°W–90°W, 37°N–39°N), the parts of Illinois/Indiana (green line: 90°W–80°W, 40°N–42°N), and Massachusetts and the parts of other northern states around (blue line: 76°W–71°W, 41°N–43°N).

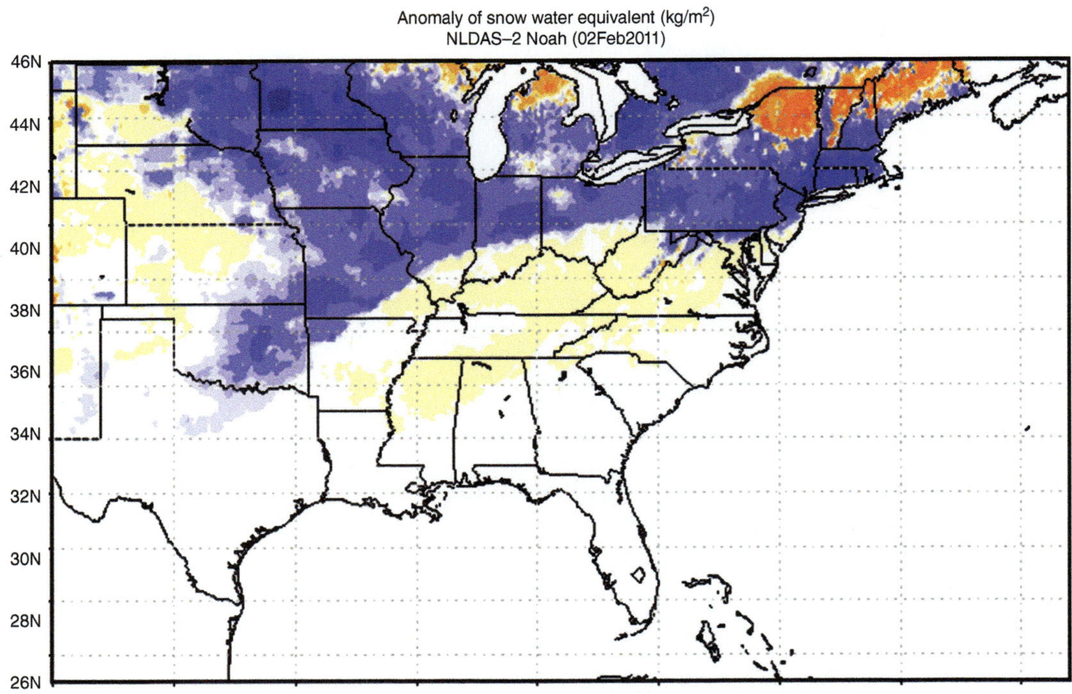

Figure 25.3d Anomaly of snow water equivalent (SWE) from NLDAS-2 Noah model for 2 February 2011, with respect to 28 year (1980–2007) daily climatology. The NLDAS-2 Noah SWE daily climatology data are provided by Dr. Youlong Xia Environmental Modeling Center (EMC), National Centers for Environmental Prediction (NCEP), National Oceanic and Atmospheric Administration (NOAA).

the Great Lakes. Very strong winds were associated with this storm for an extended period of time and many locations had blizzard conditions. The two-and-half-day average wind (Figure 25.3b), overlaying on the accumulated precipitation (snowfall + rainfall), shows the heavy winds along the storm track. The accumulated precipitation map (Figure 25.3b background image) indicates that the storm brought not only heavy snow and blizzard conditions to northern states but also heavy rainfall to southern states.

With the hourly NLDAS data, the snowfall amount can be viewed hour by hour. The time series of the NLDAS-2 Noah snowfall (Figure 25.3c) shows that Missouri (black line) had a snowfall peak during the middle of the night on February 1 to an area-average snowfall rate of about 2 mm/h (liquid equivalent). Illinois and Indiana (green line) had two snowfall peaks, as did Massachusetts and the other surrounding northeastern states, as the storm progressed from west to east. Note that these rates are in units of mass of liquid; a rough calculation is a ratio of 10:1, resulting in snowfall accumulations as high as 2 cm/h.

The high-resolution NLDAS data is clearly able to capture and describe the details in evolutions and structures of this unusual snow blizzard, and information provided here is very much consistent with in situ and remote sensing observations during the events, as well as results from various model simulations and assimilations.

The daily anomaly of snow water equivalent (SWE) from the NLDAS-2 Noah model for 2 February 2011 (Figure 25.3d) shows that most areas along the blizzard path had the SWE anomaly greater than 50 kg/m². SWE is one of the most important cold season process variables, and it quantifies the amount of frozen moisture storage and will in turn determine the amount and timing of runoff during subsequent spring melt [*Pan et al.*, 2003; *Sheffield et al.*, 2003].

NLDAS contains many other winter weather-related variables, such as snow depth, snow cover, snowmelt, snow phase-change heat flux, 2-m above ground temperature, soil temperatures, and averaged surface albedo.

25.4. HEAT WAVE ON THE GREAT PLAINS

During the spring to fall seasons of 2011, drought and excessive heat had major impacts across Texas, Oklahoma, New Mexico, Arizona, southern Kansas, and western Louisiana. The total direct economic losses have approached $10 billion.

Looking at Figure 25.2b again, the central United States had positive anomalies of the 2 m above ground temperature (red line) during 2010–2012, with magnitudes exceeding 3 K. This region suffered extreme drought, with negative anomaly of the top 1 m soil moisture exceeding 50 kg/m² (green line). The significant low and high peaks of soil moisture anomalies are inversely correspondent to the peaks of the temperature anomalies.

A close examination of drought anomalies in July 2011, using the monthly averaged 2 m above ground temperature from NLDAS-2 primary forcing, as shown in Figure 25.4a, reveals that the heat wave is centered in Texas and Oklahoma with temperatures above 305° K (90°F).

Figure 25.4b shows the time series of the average daily NLDAS-2 primary forcing 2 m above ground temperatures for the Oklahoma region from 100 to 94.5°W and 34 to 37°N. It indicates a persistent high temperature throughout the entire month with very little relief, and the highest daily average temperature reached an astonishing 309 K (96.5°F). The monthly average 2 m above ground temperature, shown as the red line in the figure, was 307.4 K (93.6°F), is comparable to 88.9°F, the averaged statewide temperature for all of Oklahoma in July 2011 as reported by the National Climatic Data Center using weather station data.

The diurnal cycle time series over a slightly larger region centered over Oklahoma is plotted in Figure 25.4c for July 2011 using the hourly 2 m above ground air temperature from NLDAS-2 primary forcing. There were more than 20 days in July 2011 with area-averaged daily maximum temperature above 100°F (310.93 K, red line), and both daily maximum and minimum temperatures were trending upward during July. To examine spatial coverage of instantaneous temperature maxima, plotted in Figure 25.4d is the hourly 2 m above ground temperature at 21Z 28 July 2011 (one of the peaks shown in Figure 25.4c). Temperatures above 312.5 K (103°F) extended from large areas of Texas, the entire state of Oklahoma, the majority of Kansas, to more than the northern half of Missouri, and beyond.

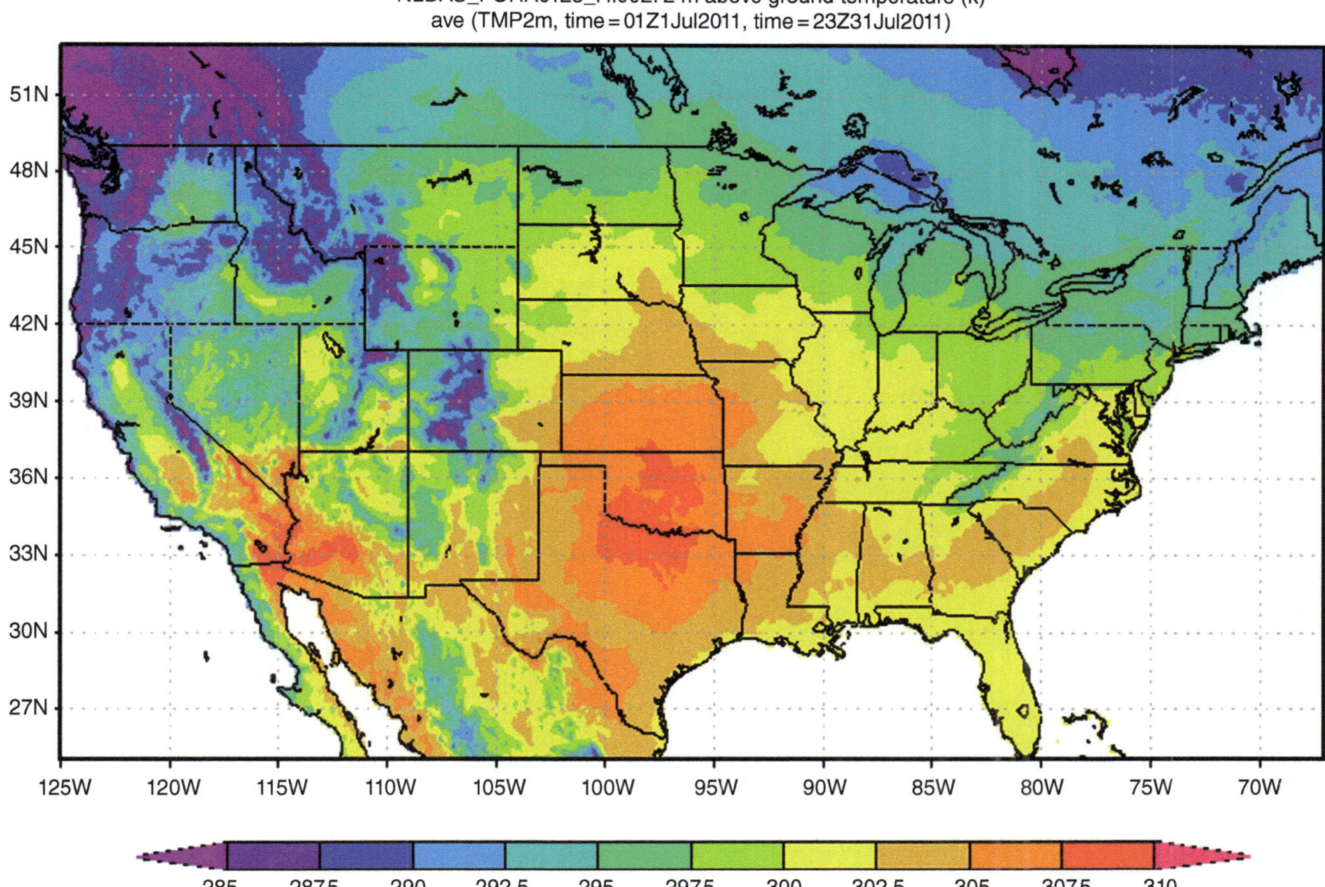

Figure 25.4a Monthly averaged 2 m above ground temperature for NLDAS-2 primary forcing data for July 2011.

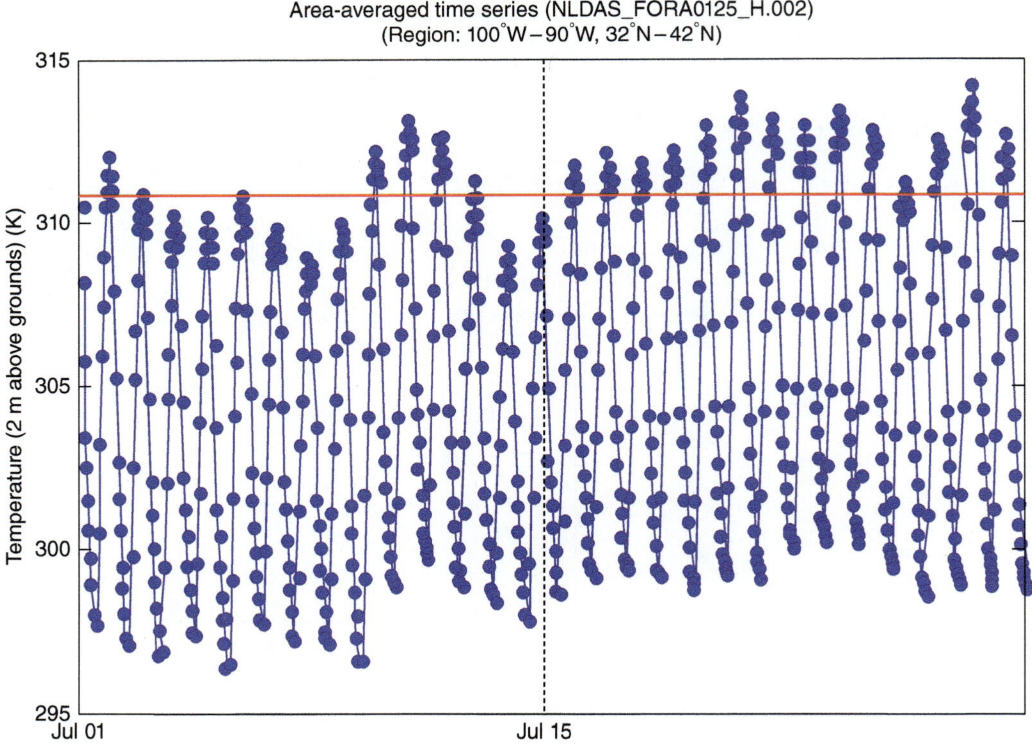

Figure 25.4b Daily average 2 m above ground temperatures of NLDAS-2 primary forcing over most of Oklahoma, July 2011, in degrees Kelvin. The red line indicates the average temperature of the data points displayed in the plot.

Figure 25.4c Time series of 2 m above ground air temperatures from hourly NLDAS-2 primary forcing data for the central United States, July 2011. The red line indicates 100°F (310.93 K).

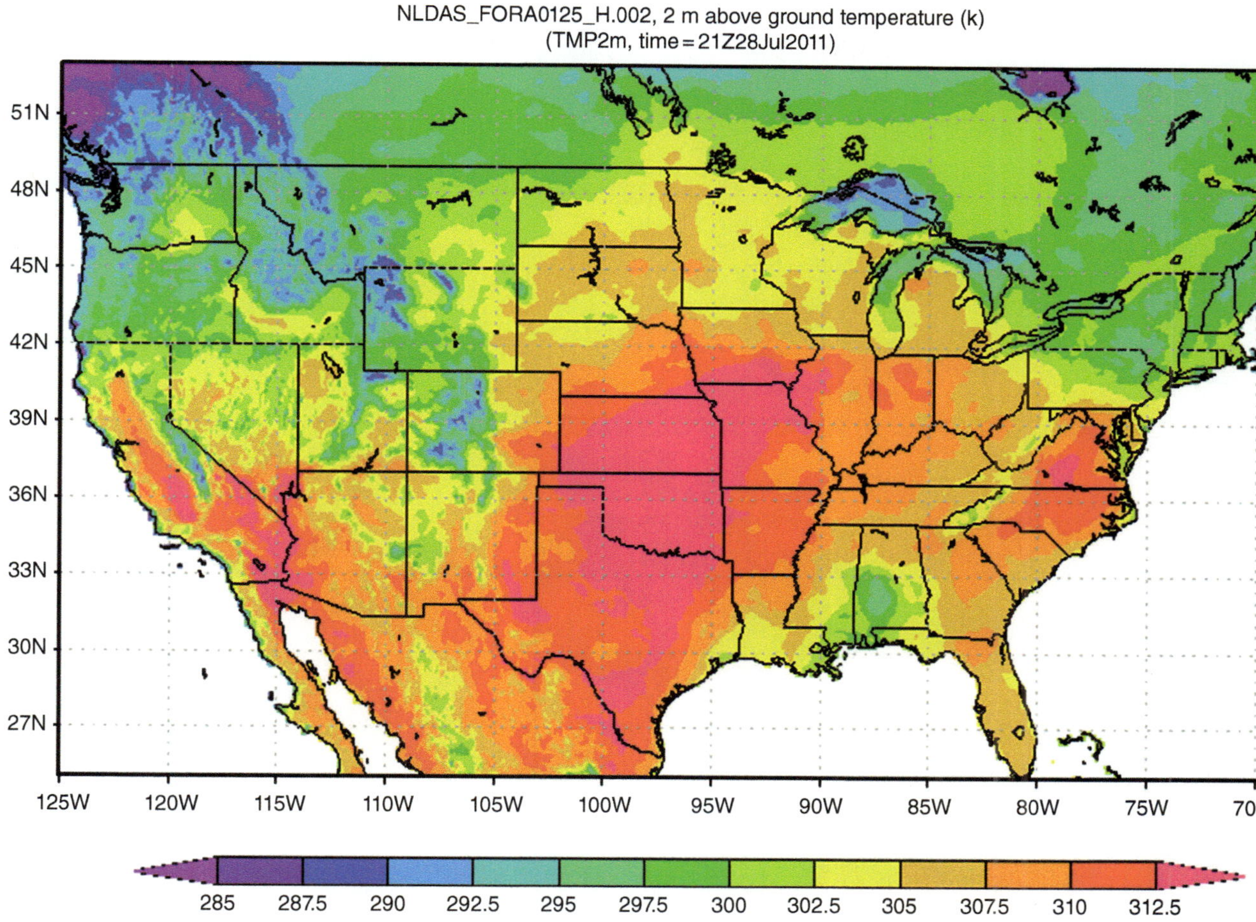

NLDAS_FORA0125_H.002, 2 m above ground temperature (k)
(TMP2m, time = 21Z28Jul2011)

Figure 25.4d NLDAS-2 primary forcing hourly 2 m above ground temperature at 21Z 28 July 2011.

25.5. HURRICANE IRENE

Hurricane Irene, as it moved northward along the east coast of the United States from the Caribbean Sea through the New England states, brought tremendous rainfall and wind and caused more than \$7.3 billion in damages. The passage of Hurricane Irene killed at least 55 people and left millions of people in the dark due to power outages. Fallen trees caused power outages that persisted for more than a week. Flood damage was particularly extensive in the state of Vermont, with parts of several nearby coastal states declared disaster areas.

An animation of hourly wind overlay on precipitation can be viewed with the Giovanni NLDAS hourly portal (http://gdata1.sci.gsfc.nasa.gov/daac-bin/G3/gui.cgi?instance_id=NLDAS0125_H, http://disc.sci.gsfc.nasa.gov/gesNews/images/hurricane_nldas_rainfall_animation) to illustrate status and evolutions of Hurricane Irene in greater detail. A few images from the animation (Figure 25.5a) capture the major characteristics of Irene, as seen from satellites images (http://www.nasa.gov/mission_pages/hurricanes/archives/2011/h2011_Irene.html). Irene made the first landfall in eastern North Carolina early on 27 August and then moved northward along the mid-Atlantic Coast, where the well-defined center and strong winds remained for more than 10 h. While it continued moving northward to Virginia and Maryland, Irene was slowly weakening. After merging with an approaching cold front and getting energized, Irene made landfall in southeastern New Jersey on 28 August and then passed near New York City a few hours later, dropping torrential rainfall that caused widespread flooding. The time series of hourly precipitation (Figure 25.5b) from NLDAS-2 primary forcing, averaged over (86°W–67°W, 25°N–53°N) shows three rainfall peaks at 03Z 27 August 2011 for North Carolina, 19Z 27 August 2011 for Virginia, and 13Z 28 August 2011 for New York. The heaviest rainfall was observed at 13Z 28 August 2011 in New York (Figure 25.5c, left). The NLDAS-2 Noah soil moisture of 14Z 28 August 2011 (Figure 25.5c, right) shows high soil moisture content centered on New York at the same time.

The total runoff from NLDAS-2 Noah model [*Xia et al.*, 2012a, 2012b] shows large positive daily anomaly on 28 August 2011 (Figure 25.5d) and the positive anomaly region overlays with the region of the high rain and soil moisture.

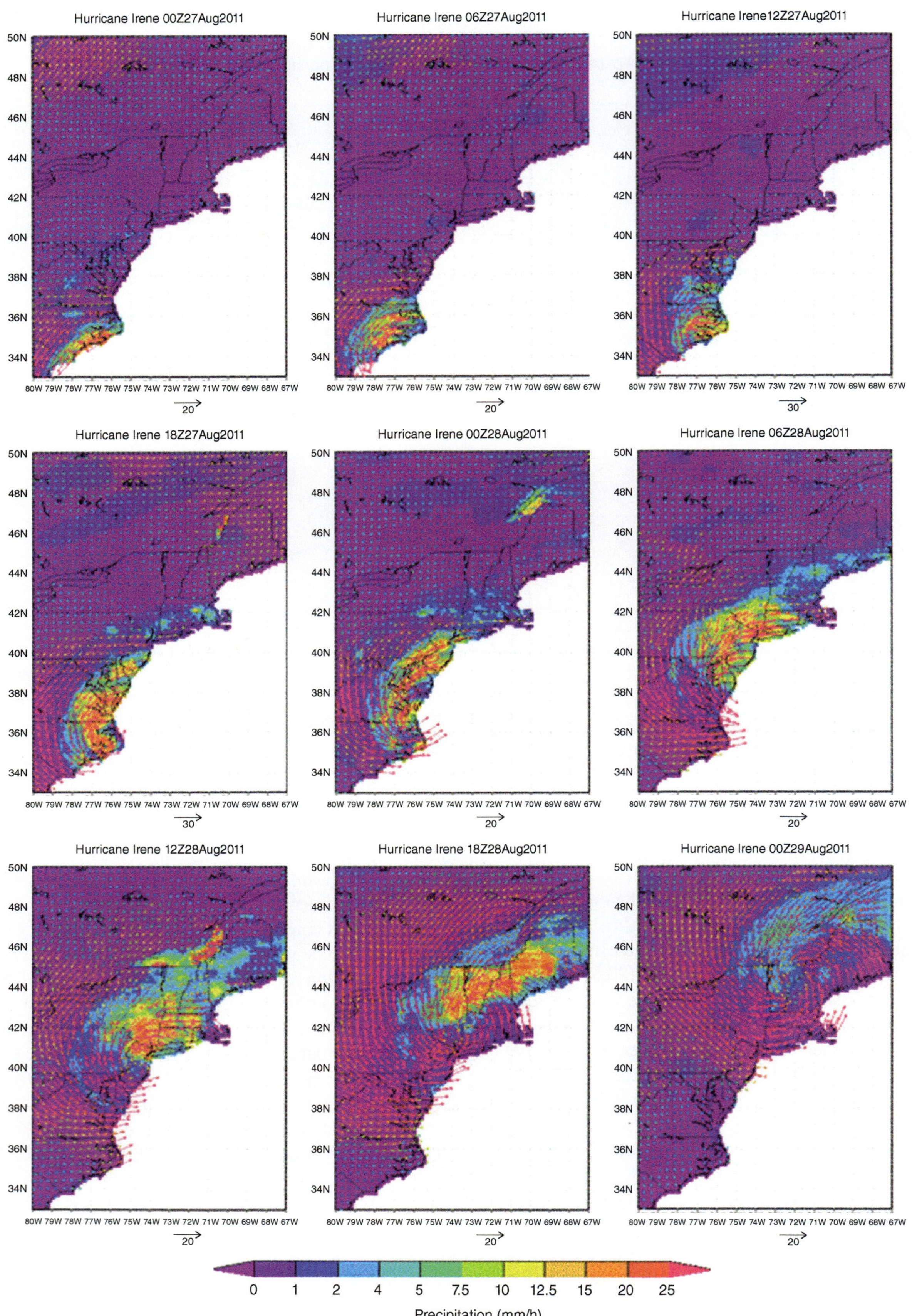

Figure 25.5a NLDAS-2 primary forcing hourly wind overlaid on NLDAS-2 hourly precipitation every 6 h from 00Z 27 August 2011 to 00Z 29 August 2011.

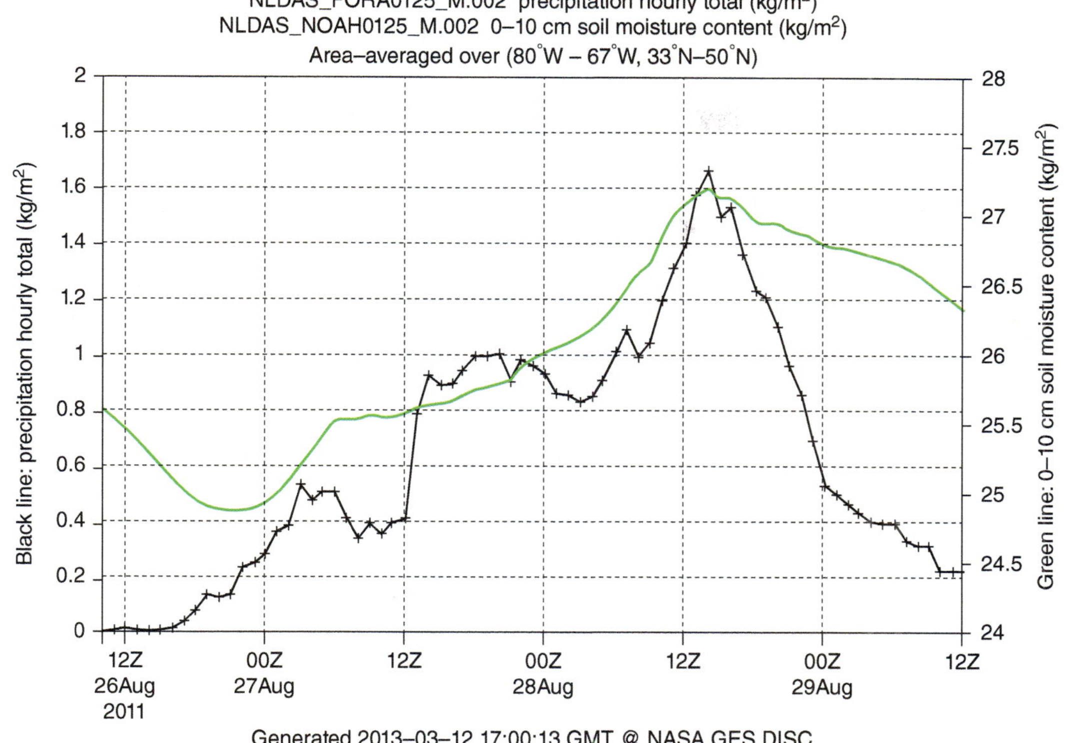

Figure 25.5b Time series of hourly precipitation (black line, from NLDAS-2 primary forcing) and 0–10 cm soil moisture content (green line, from NLDAS-2 Noah model), averaged over 80°W–67°W, 33°N–53°N.

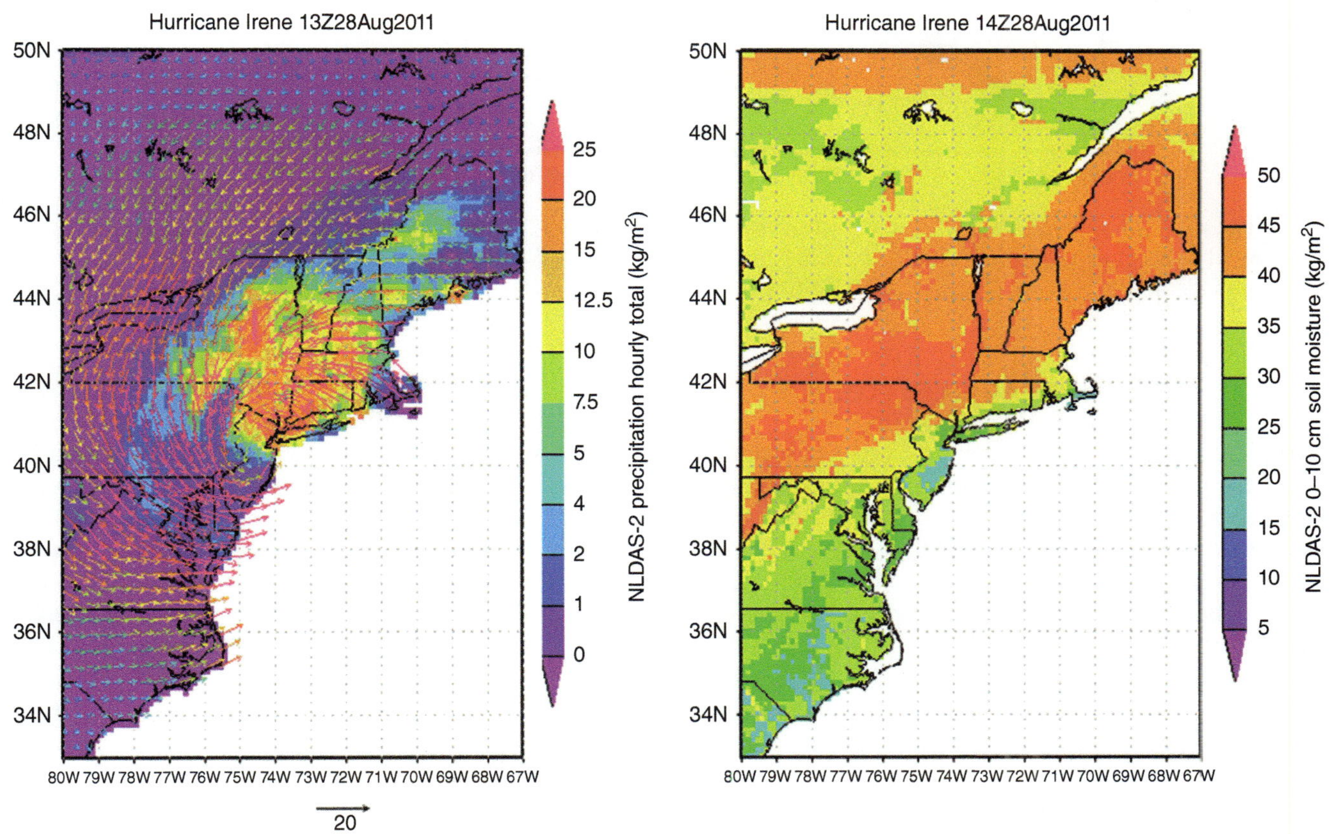

Figure 25.5c Hourly precipitation (left, from NLDAS-2 primary forcing) and 0–10 cm soil moisture (right, from NLDAS-2 Noah model), at 14Z 28 August 2011.

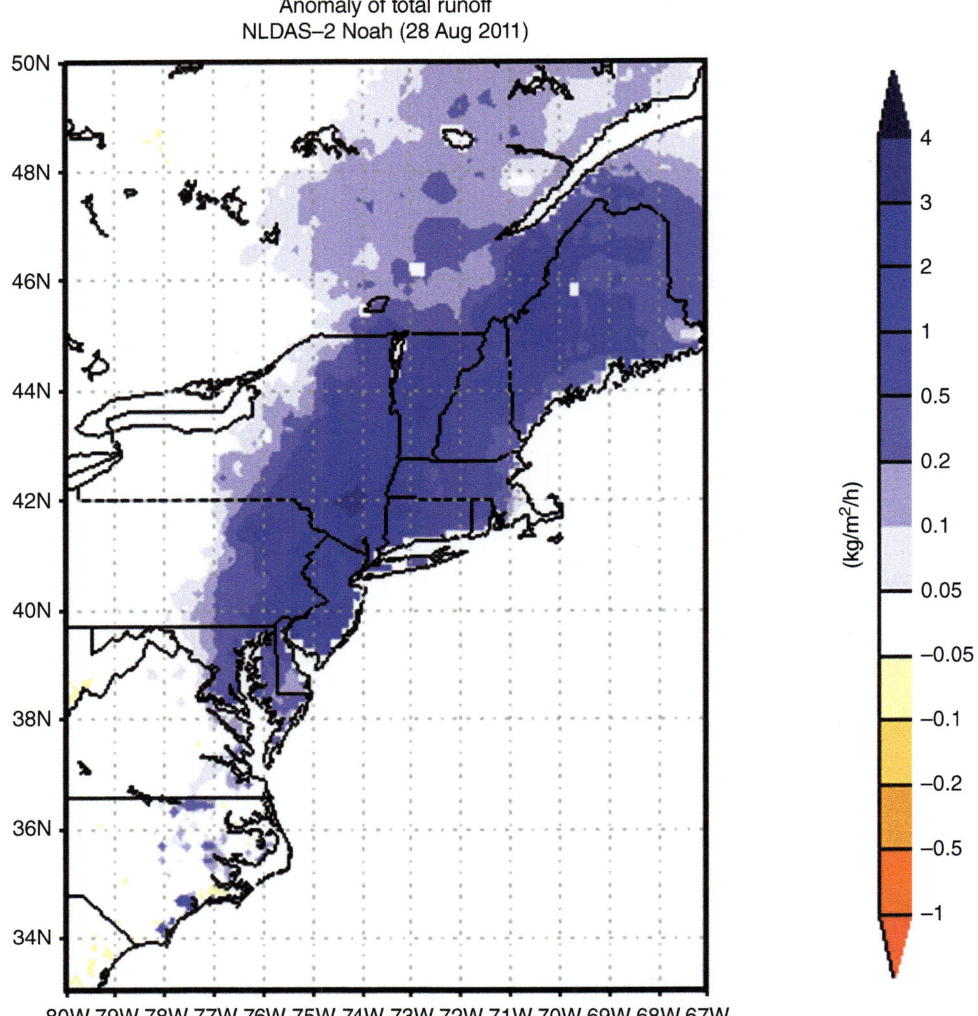

Figure 25.5d Anomaly of total runoff from NLDAS-2 Noah model for 28 August 2011 (averaged over 00Z–23Z), with respect to 28 year (1980–2007) daily climatology. The NLDAS-2 Noah total runoff daily climatology data are provided by Dr. Youlong Xia, EMC/ NCEP/NOAA.

25.6. TROPICAL STORM LEE

Tropical Storm Lee poured huge amounts of water on top of the already saturated Northeast and again inundated many inland cities, causing at least 21 deaths and more than $1.0 billion in damages. During early September 2011, as the east coast, middle Atlantic, and New England states were still recovering from record flooding caused by Hurricane Irene, Tropical Storm (TS) Lee compounded previous damages by dumping additional rainfalls, as well as strong winds to the Northeast states and cities. Examples of these data views are shown below.

As an animation of hourly precipitation (http://disc.sci.gsfc.nasa.gov/hydrology/gesNews/nldas_views_ts_lee) from the NLDAS-2 primary forcing showed from 2–4 September Tropical Storm Lee produced heavy rains that saturated Louisiana, stalled off the coast, continuously expanded and intensified, and then slowly weakened and moved northeastward, spreading rain to Mississippi and Alabama. The animation also shows that, as Lee dissipated and moved northeastward, the rain system interacted with subtropical and middle-latitude systems and brought heavy rain to Virginia, Maryland, Pennsylvania, and New York. NLDAS-2 accumulated precipitation between 08Z 2 September and 12Z 9 September (Figure 25.6a) shows three heavy rain centers, one over Louisiana and Mississippi, another over Alabama/ Georgia/Tennessee, and another over Pennsylvania, with accumulated rainfall exceeding 10 inches (254 mm).

The area-averaged time series of hourly precipitation (Figure 25.6b) for the three heavy rain regions depicted

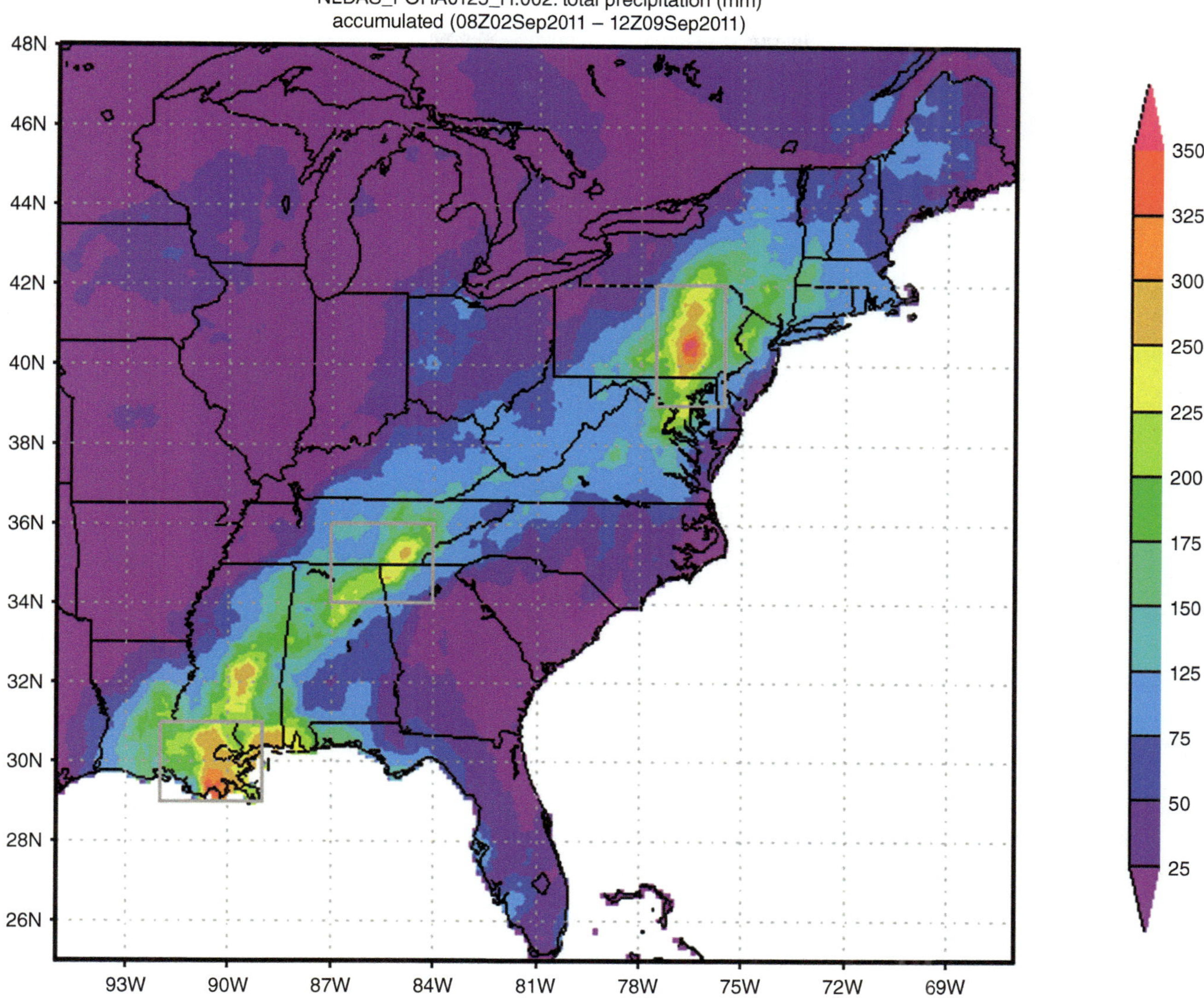

Figure 25.6a Accumulated precipitation between 08Z 2 September and 12Z 9 September 2011 (NLDAS-2 primary forcing) shows three heavy rain centers, with accumulated rainfall exceeding 10 inches (254 mm).

by the boxes in Figure 25.6a shows temporal variations of rainfall and clearly marked the start and end of the heaviest precipitation. The time series of precipitation and soil moisture overlay (Figure 25.6c) shows the soil moisture peaks well correlated with the rainfall peaks. The persistence of high soil moisture content following the heavy rain period during Tropical Storm Lee has contributed to flash flooding in many areas.

The anomaly of total runoff from NLDAS-2 Noah model for 5 September 2011 (Figure 25.6d) shows larger positive anomaly over Louisiana, Mississippi, and Tennessee, where Tropical Storm Lee dropped heavy rain during 2–6 September 2011. The anomaly of routed streamflow from the NLDAS-2 Noah model for 8 September 2011 also shows large positive anomaly over the path of Tropical Storm Lee, with the largest anomaly over Pennsylvania, coinciding with the double peak rainfall over Pennsylvania in Figure 25.6b. Runoff and streamflow are very important water cycle variables. They are affected by weather and directly impact on the water quality and living creatures in the streams. The runoff and streamflow data from NLDAS [*Xia et al.*, 2012b], along with precipitation, soil moisture, evaporation, snow water (SWE), have been used for monitoring drought/flood over United States (http://www.emc.ncep.noaa.gov/mmb/nldas/drought/Stream/).

25.7. NLDAS DATA

NLDAS is a collaboration project among several groups (NOAA/NCEP/EMC, NASA/GSFC, Princeton University, University of Washington, NOAA/OHD,

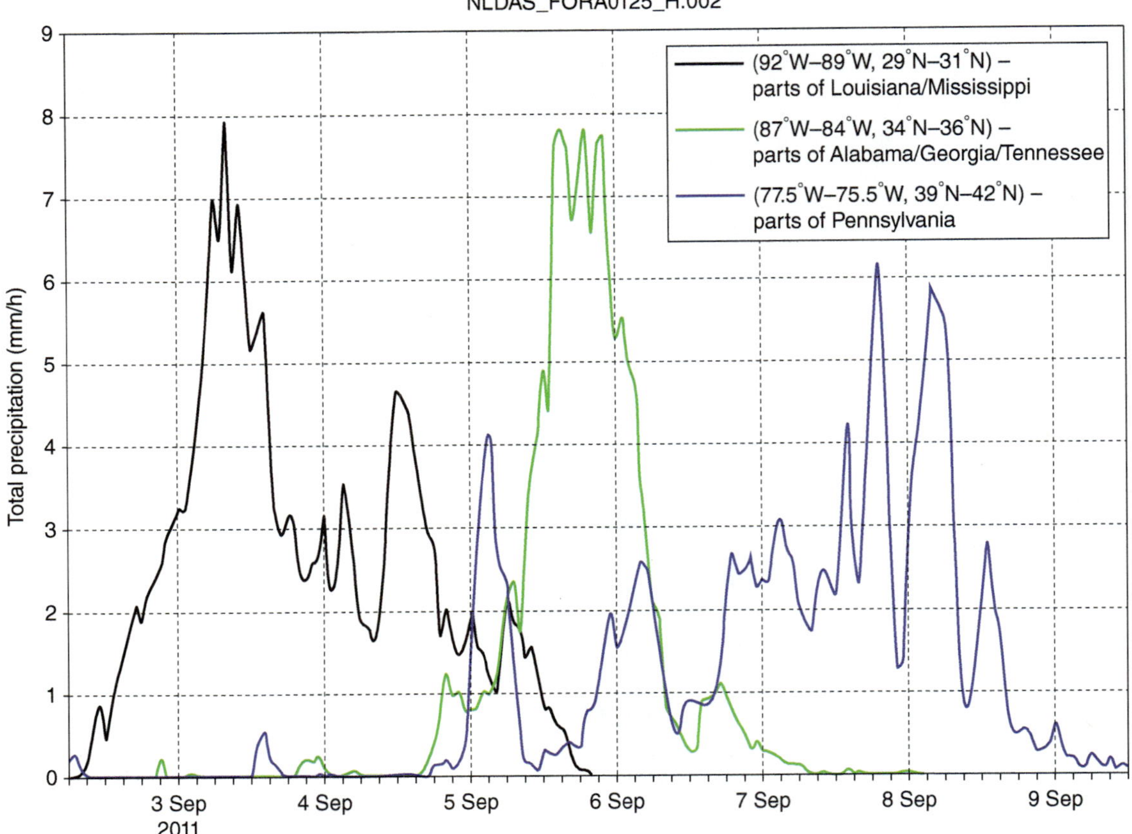

Figure 25.6b Area-averaged time series of hourly precipitation (NLDAS-2 primary forcing) for the three heavy rain regions depicted by the boxes in Figure 25.6a shows clearly when the heaviest rain started and ended.

and NOAA/NCEP/CPC) and is a core project of NOAA/MAPP. To date, NLDAS, with satellite- and ground-based observational forcing data, has produced more than 34 years (1979 to present) of quality-controlled, spatially and temporally consistent, land surface model data [*Xia et al.*, 2012a, 2013b; and *Luo et al.*, 2002]. The original NLDAS-2 data generated by NOAA/NCEP/EMC, both retrospective and real time, are accessible via ftp from http://www.emc.ncep.noaa.gov/mmb/nldas/.

To further facilitate analysis of water and energy budgets and trends, the NASA GSFC Hydrological Sciences Laboratory (HSL) has generated NLDAS monthly and monthly climatology data sets. NLDAS monthly data are generated from the NLDAS hourly data, consisting of monthly accumulations for precipitation, runoff, evaporation, and snowmelt, and monthly averages for other variables. NLDAS monthly climatology data are generated from the NLDAS monthly data as an 11 year (1997–2007) monthly average for NLDAS Phase 1 (NLDAS-1) data, and a 30 year (1980–2009)

monthly average for NLDAS Phase 2 (NLDAS-2) data. More information about NLDAS is provided in NLDAS README documents at http://disc.sci.gsfc.nasa.gov/hydrology/documentation and the NLDAS websites. Table 25.2 lists all the NLDAS data sets currently available at the NASA GES DISC. The NLDAS-2 data sets from Sacramento (SAC) model will also become accessible from the Hydrology Data Holdings portal in the near future.

The NLDAS data sets (listed in Table 25.2) currently available from the NASA GES DISC can be accessed, along with the README documents, from the Hydrology Data Holdings Portal, http://disc.sci.gsfc.nasa.gov/hydrology/data-holdings, through the following access methods:

• **Mirador Search and Download:** http://mirador.gsfc.nasa.gov/

• **Simple Subset Wizard (SSW):** http://disc.gsfc.nasa.gov/SSW/

• **Direct ftp:** ftp://hydro1.sci.gsfc.nasa.gov/data/s4pa/NLDAS/

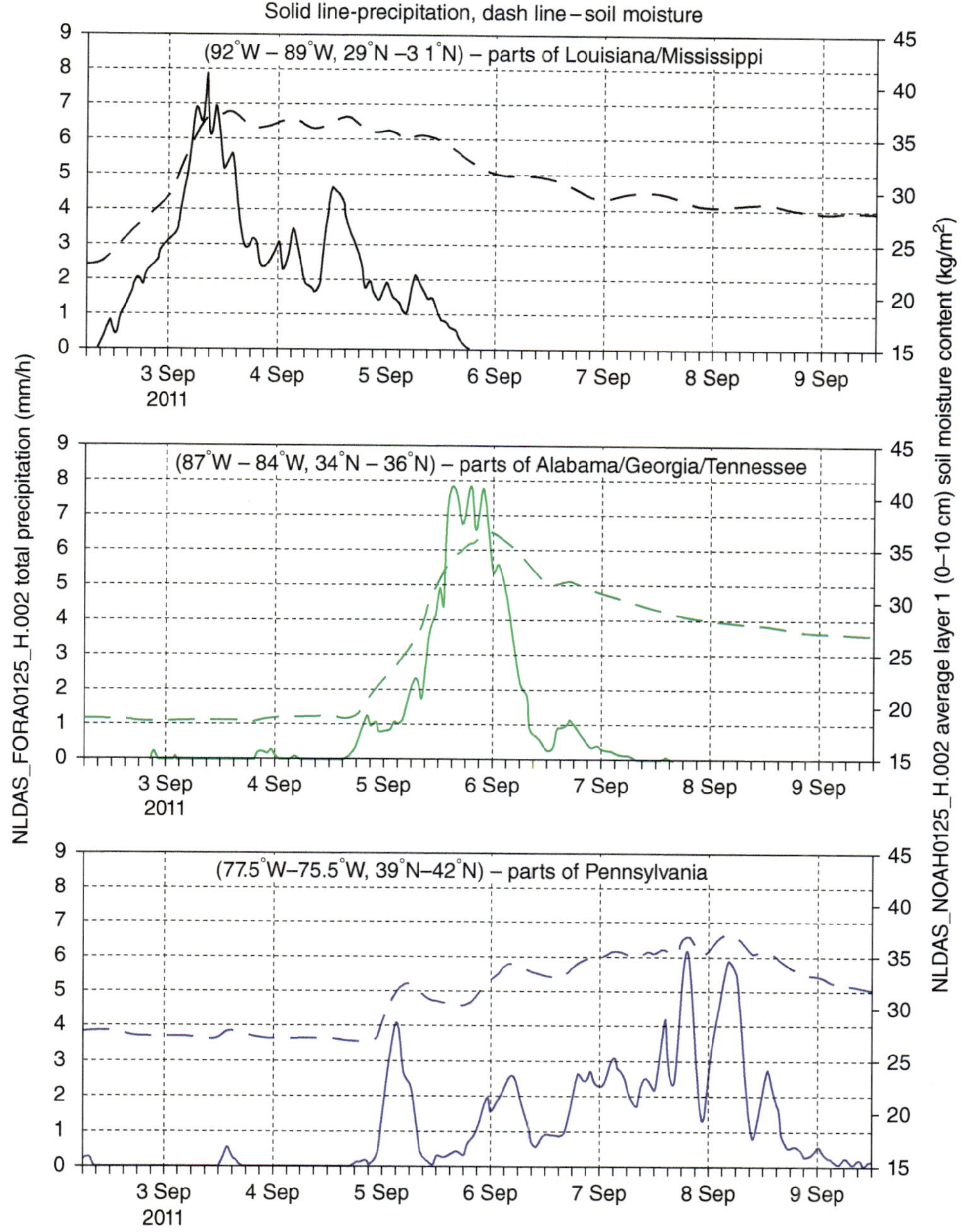

Figure 25.6c Time series of precipitation and 0–10 cm soil moisture, from NLDAS-2 primary forcing and Noah model, respectively, averaged over the three heavy rain regions depicted by the boxes in Figure 25.6a.

• **GrADS Data Server (GDS):** http://hydro1.sci.gsfc.nasa.gov/dods/

• **Giovanni Visualization and Analysis:** Giovanni is a Web-based application developed by the GES DISC that provides a simple and intuitive way to visualize, analyze, and access vast amounts of Earth science remote sensing data without having to download the data. NLDAS data sets can be accessed through the following Giovanni portals:

○ NLDAS Hourly Portal for hourly data visualization and analysis: http://gdata1.sci.gsfc.nasa.gov/daac-bin/G3/gui.cgi?instance_id=NLDAS0125_H

○ NLDAS Monthly Portal for monthly, and monthly climatology and anomaly visualization and analysis: http://gdata1.sci.gsfc.nasa.gov/daac-bin/G3/gui.cgi?instance_id=NLDAS0125_M

○ NLDAS Monthly Climatology Portal for monthly climatology visualization and analysis: http://gdata1.

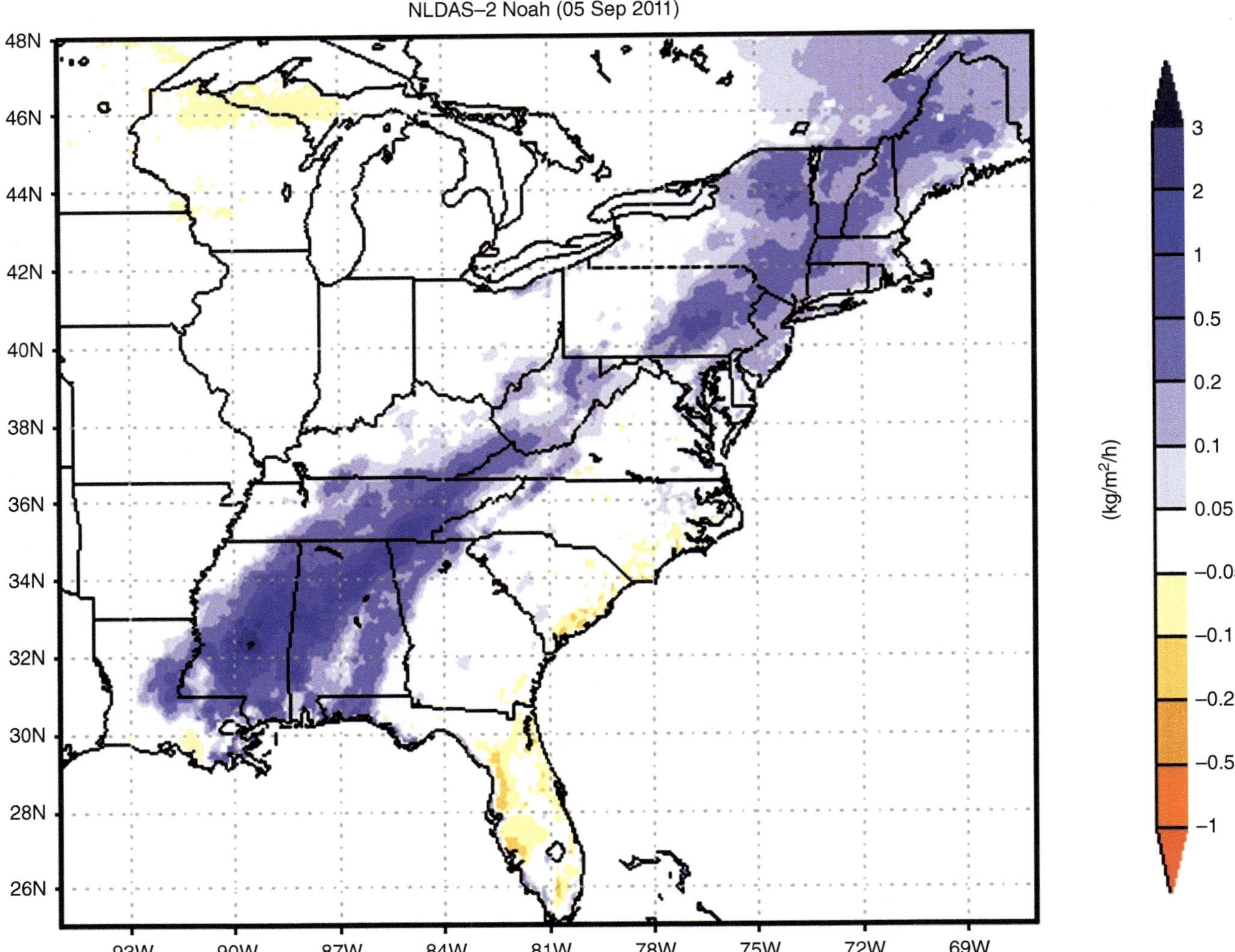

Figure 25.6d Anomaly of total runoff (no infiltrating) from NLDAS-2 Noah model for 5 September 2011 (averaged over 00Z–23Z), with respect to 28 year (1980–2007) daily climatology. The NLDAS-2 total runoff daily climatology data are provided by Dr. Youlong Xia, EMC/NCEP/NOAA.

sci.gsfc.nasa.gov/daac-bin/G3/gui.cgi?instance_id=
NLDAS0125_MClim

25.8. SUMMARY

Four of the 2011 billion-dollar weather/climate disasters are illustrated by using NLDAS-2 primary forcing and Noah model data. The high-resolution data is clearly able to capture and describe the details in evolutions and structures of these extreme weather events, such as winter blizzard, hurricane, heat wave, and drought [*Ek et al.,* 2011; *Mo et al.,* 2011]. The data shows very well the major characteristics of these extreme events, spatially and temporally. The information provided by NLDAS-2 data is very much consistent with in situ and remote sensing observations during the events. NLDAS-2 data is an excellent data source for case studies of extreme events.

To date, NLDAS has generated more than 34 (1979–present) years of data. These quality-controlled, spatially and temporally consistent, terrestrial hydrological data could play an important role in characterizing the spatial and temporal variability of water and energy cycles and, thereby, improve our understanding of the land-surface-atmosphere interaction and the impact of land-surface processes on climate extremes.

NLDAS data are accessible from the Hydrology Data Holdings portal at the NASA GES DISC. Giovanni NLDAS portals provide a simple and intuitive way to visualize, analyze, and inter-compare NLDAS data without having to download the data.

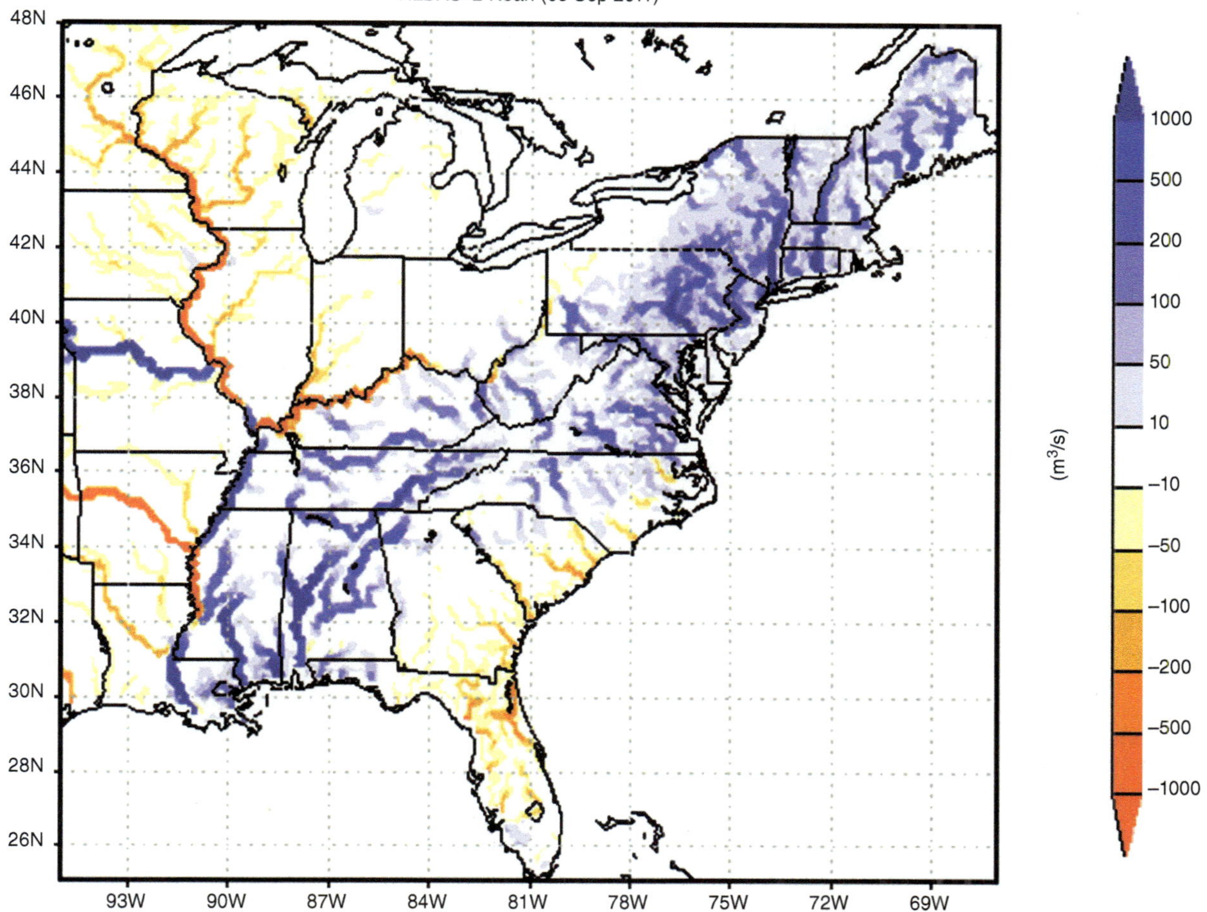

Figure 25.6e Anomaly of routed stream flow from NLDAS-2 Noah model for 5 September 2011 (averaged over 00Z–23Z), with respect to 28 year (1980–2007) daily climatology. The NLDAS-2 hourly streamflow and daily climatology data are provided by Dr. Youlong Xia, EMC/NCEP/NOAA.

Table 25.2 NLDAS data set available at NASA GES DISC

Data Set Short Name	Description	Temporal Coverage
Hourly		
NLDAS_FOR0125_H.001	NLDAS-1 hourly forcing	00Z01 August 1996–23Z31 Dec 2007
NLDAS_FORA0125_H.002	NLDAS-2 hourly primary forcing	13Z01 January 1979–present
NLDAS_FORB0125_H.002	NLDAS-2 hourly secondary forcing	13Z01 January 1979–present
NLDAS_MOS0125_H.002	NLDAS-2 hourly Mosaic model	00Z02 January 1979–present
NLDAS_NOAH0125_H.002	NLDAS-2 hourly Noah model	01Z02 January 1979–present
NLDAS_VIC0125_H.002	NLDAS-2 hourly VIC model	01Z02 January 1979–present
Monthly		
NLDAS_FOR0125_M.001	NLDAS-1 monthly forcing	August 1996–December 2007
NLDAS_FORA0125_M.002	NLDAS-2 monthly primary forcing	January 1979–present
NLDAS_FORB0125_M.002	NLDAS-2 monthly secondary forcing	January 1979–present
NLDAS_MOS0125_M.002	NLDAS-2 monthly Mosaic model	January 1979–present
NLDAS_NOAH0125_M.002	NLDAS-2 monthly Noah model	January 1979–present
NLDAS_VIC0125_M.002	NLDAS-2 monthly VIC model	January 1979–present
Monthly Climatology		
NLDAS_FOR0125_MC.001	NLDAS-1 monthly climatology forcing	January 1997–December 2007
NLDAS_FORA0125_MC.002	NLDAS-2 monthly climatology primary forcing	January 1980–December 2009
NLDAS_FORB0125_MC.002	NLDAS-2 monthly climatology secondary forcing	January 1980–December 2009
NLDAS_MOS0125_MC.002	NLDAS-2 monthly climatology Mosaic model	January 1980–December 2009
NLDAS_NOAH0125_MC.002	NLDAS-2 monthly climatology Noah model	January 1980–December 2009
NLDAS_VIC0125_MC.002	NLDAS-2 monthly climatology VIC model	January 1980–December 2009

REFERENCES

Anderson, M., et al. (2013), An intercomparison of drought indicators based on thermal remote sensing and NLDAS-2 simulations with U.S. drought monitor classifications, *J. Hydrometeorol.*, *14*, 1035–1056, doi:10.1175/JHM-D-12-0140.1.

Ek, M. B., et al. (2011), North American Land Data Assimilation System Phase 2 (NLDAS-2): Its development and application to support US drought monitoring and prediction of National Integrated Drought Information System (NIDIS), *GEWEX News*, *21*, 7–9.

Lee, S., et al. (2010), Assessing the hydrologic performance of the EPA's nonpoint source water quality assessment decision support tool using North American Land Data Assimilation System (NLDAS) products, *J. Hydrol.*, *387*(3), 212–220, doi:10.1016/j.jhydrol.2010.04.009.

Luo, L., et al. (2002), Validation of the North American Land Data Assimilation System (NLDAS) retrospective forcing over the Southern Great Plains, *J. Geophys. Res.*, *108*(D22), 8843, doi:10.1029/2002JD003246, 2003.

Mitchell, K. et al. (2004), The multi-institution North American Land Data Assimilation System (NLDAS): Utilizing multiple GCIP products and partners in a continental distributed hydrological modeling system, *J. Geophys. Res.*, *109*, D07S90, doi:10.1029/2003JD003823.

Mo, K. C., L. N. Long, Y. Xia, S. K. Yang, J. E. Schemm, and M. Ek (2011), Drought indices based on the climate forecast system reanalysis and ensemble NLDAS, *J. Hydrometeorol.*, *12*, 181–205, doi:10.1175/2010JHM1310.1.

Pan, M., et al. (2003), Snow process modeling in the North American Land Data Assimilation System (NLDAS): 2. Evaluation of model simulated snow water equivalent, *J. Geophys. Res.*, *108*(D22), 8850, doi:10.1029/2003JD003994.

Sheffield, J., et al. (2003), Snow process modeling in the North American Land Data Assimilation System (NLDAS): 1. Evaluation of model-simulated snow cover extent, *J. Geophys. Res.*, *108*(D22), 8849, doi:10.1029/2002JD003274.

Xia, Y., et al. (2012a), Continental-scale water and energy flux analysis and validation for the North American Land Data Assimilation System project phase 2 (NLDAS-2): 1. Intercomparison and application of model products, *J. Geophys. Res.*, doi:10.1029/2011JD016048.

Xia, Y., et al. (2012b), Continental-scale water and energy flux analysis and validation for North American Land Data Assimilation System project phase 2 (NLDAS-2): 2. Validation of model-simulated streamflow, *J. Geophys. Res.*, *117*, D03110, doi:10.1029/2011JD016051.

Xia, Y., et al. (2013a), Validation of Noah-simulated soil temperature in the North American land data assimilation system phase 2, *J. Appl. Meteorol. Climatol.*, *52*, 455–471, doi:10.1175/JAMC-D-12-033.1.

Xia, Y., B. A. Cosgrove, M. B. Ek, J. Sheffield, L. Luo, E. Wood, K. Mo, and the NLDAS Team (2013b), Overview of the North American Land Data Assimilation System (NLDAS), in *Land Data Observation, Modeling and Assimilation*, edited by S. Liang et al., pp. 337–375, World Scientific, Singapore, doi:10.1142/9789814472616_0011.

26

Growth Studies of *Mytilus californianus* Using Satellite Surface Temperatures and Chlorophyll Data for Coastal Oregon

Jessica R. Price and Venkat Lakshmi

26.1. INTRODUCTION

The advancement of remote sensing technology has led to better understanding of the spatial and temporal variation in many physical and biological parameters, such as temperature [*Wan et al.*, 2010], salinity [*Gabarro et al.*, 2004], soil moisture [*Njoku et al.*, 2003], vegetation cover [Gillies and Carlson, 1995], and community composition [*Gould*, 2000]. Remotely sensed surface temperatures have been used to better understand land cover changes [*Lambin and Ehrlich*, 1997], detection of infectious disease [*Lobitz et al.*, 1999], and predicting the distribution of species [*Nagendra*, 2001]. In addition, remotely sensed chlorophyll-*a* concentration has been used to identify areas of upwelling, algal blooms, productivity, and food availability [*Antoine et al.*, 1996]. Within marine rocky intertidal ecosystems, temperature and food availability influence species abundance, physiological performance, and distribution of mussel species [*Harley et al.*, 2006; *Helmuth et al.*, 2006a, 2006b; *Miezkowska et al.*, 2005, 2006]. However, since the temperature that mussel species experience is different from the air temperature due to physical and biological characteristics (size, color, gaping, etc.), it is difficult to accurately predict the thermal stresses they experience. Current methods to determine the temperature mussel species experience range from in situ field observations, temperature loggers, temperature models, and using other temperature variables (air and water) [*Fitzhenry et al.*, 2003; *Lima and Wethey*, 2009]. Methods to determine food availability (chlorophyll-*a* concentration used as a proxy) for mussel species are mostly done at specific study sites using water sampling [*Menge et al.*, 2008; *Page and Ricard*, 1990]. This implies that estimation of temperature and food availability across large spatial scales and long temporal scales is not a trivial task given the spatial heterogeneity. However, this estimation is an essential step in determination of the impact of changing climate on vulnerable ecosystems such as the marine rocky intertidal system. This research takes a novel approach to understanding the spatial variability of mussel body growth using remotely sensed surface temperatures and chlorophyll-*a* concentration.

The purpose of this study is to investigate whether remotely sensed data (surface temperature and chlorophyll-*a* concentration) can be used to understand the spatial variation in the growth of the rocky intertidal mussel species, *Mytilus californianus*. *Menge et al.* [2008] showed that in situ water temperature and chlorophyll-*a* concentration explained 44.5% of the variance in mussel growth for sites along the Oregon coast. We use a similar approach but include more sites and exclusively use satellite data. We focus on using remotely sensed data because it is widely available (both spatially and temporally), easy to use, and potentially very valuable in addressing the following ecological questions [*Miller et al.*, 2005]:

1. Do the patterns in remotely sensed surface temperature chlorophyll-*a* concentration and mussel growth vary with site location?

2. Is there a relationship between remotely sensed variables and mussel body growth?

In this study we used multiple sites to better understand the spatial variability of remotely sensed surface temperatures, remotely sensed chlorophyll-*a* concentration, and mussel body growth.

Department of Earth and Ocean Sciences, University of South Carolina, Columbia, South Carolina, USA

Remote Sensing of the Terrestrial Water Cycle, Geophysical Monograph 206. First Edition. Edited by Venkat Lakshmi.
© 2015 American Geophysical Union. Published 2015 by John Wiley & Sons, Inc.

26.2. DATA AND METHODS

26.2.1. Study System and Sites

Marine rocky intertidal ecosystems are distributed globally, but, depending on the geographical location and local characteristics, intertidal ecosystems can vary drastically. The distribution of organisms within these intertidal zones is directly influenced by physical and biological factors [*Dayton*, 1971]. Local weather and tide conditions along with substrate characteristics can impact the upper and lower ranges of species. Biological factors such as competition and predation also alter species distribution and performance. The present study focuses on the rocky intertidal ecosystems along the Oregon coastline. Each specific study site used has different characteristics that influence species but are similar enough to support the same species.

The sites used in this study were chosen based on the mussel growth monitoring data that was available from the Partnership for Interdisciplinary Studies of Coastal Oceans (PISCO) (Table 26.1). There are eight sites along the Oregon coast that have 7–8 years of mussel growth monitoring data. These study sites were Cape Meares (CMRX00), Boiler Bay (BBYX00), Fogarty Creek (FCKX00), Seal Rock (SRKX00), Yachats Beach (YBHX00), Gull Haven (GHVX00), Cape Arago (CARX00), and Cape Blanco (CBLX00). The coordinates for the mussel growth study sites can be seen in Table 26.2. Each PISCO study site was defined as areas with four corner coordinates that make up the northwestern, northeastern, southwestern, and southeastern corners. We took the four corner coordinates and used them to locate the pixels that could be used for surface temperature and chlorophyll-*a* concentration. The closest land surface pixel that contained the intertidal zone was used to collect what we refer to as the intertidal surface temperature (IST). Additionally, a pixel located 1 km inland from the study site was used to collect land surface temperature (LST) because it did not fall in the intertidal zone. Six ocean pixels (~12 km × ~8 km) were chosen to collect the sea surface temperature and chlorophyll-*a* concentration. All the study sites are shown in Figure 26.1.

26.2.2. Study Species

The present study focuses on the sessile invertebrate, the California mussel (*Mytilus californianus*), because it is considered a dominant species in the rocky intertidal ecosystems of the North Pacific [*Paine*, 1966; *Dayton*, 1971]. It lives within the mid-intertidal zone, which is flooded and exposed daily by mixed semidiurnal tides. *Mytilus californianus* makes large "mussel beds" that provide structure to the intertidal zone. Furthermore,

Table 26.1 Information about all datasets used in study

Type of Data	Source	Frequency (units)	Years	Website address
Terra SST	NASA Jet Propulsion Laboratory (JPL)	Daily °C	2000–2011	http://podaac.jpl.nasa.gov/dataaccess
Terra LST	NASA Land Processes Distribution Active Archive Center (LP DAAC)	Daily °C	2000–2011	https://lpdaac.usgs.gov/get_data
Terra Chlorophyll-*a*	NASA Ocean Color	Monthly (mg/m^{-3})	2000–2011	http://oceandata.sci.gsfc.nasa.gov
Mussel growth	Partnership for Interdisciplinary Studies of Coastal Ocean (PISCO)	Annual (mm)	2000–2009	http://data.piscoweb.org

Table 26.2 Description of geographic region for each study site used starting from the northernmost study site and ending with the southernmost study site

Study Site		Bounding Coordinates			
		Northern	Southern	Western	Eastern
Cape Meares, OR	CMRX00	45.47224	45.47035	−123.97260	−123.97206
Boiler Bay, OR	BBYX00	44.83171	44.83033	−124.05929	−124.06078
Fogarty Creek, OR	FCKX00	44.83833	44.83696	−124.06001	−124.05847
Seal Rock, OR	SRKX00	44.49979	44.49933	−124.08455	−124.08428
Yachats Beach, OR	YBHX00	44.31928	44.31841	−124.10932	−124.10840
Strawberry Hill, OR	SHLX00	44.25009	44.24923	−124.1144	−124.11543
Gull Haven, OR	GHVX00	44.20675	44.20368	−124.11651	−124.11704
Cape Arago, OR	CARX00	43.30778	43.30164	−124.39991	−124.40232
Cape Blanco, OR	CBLX00	42.84025	42.83969	−124.56580	−124.56512

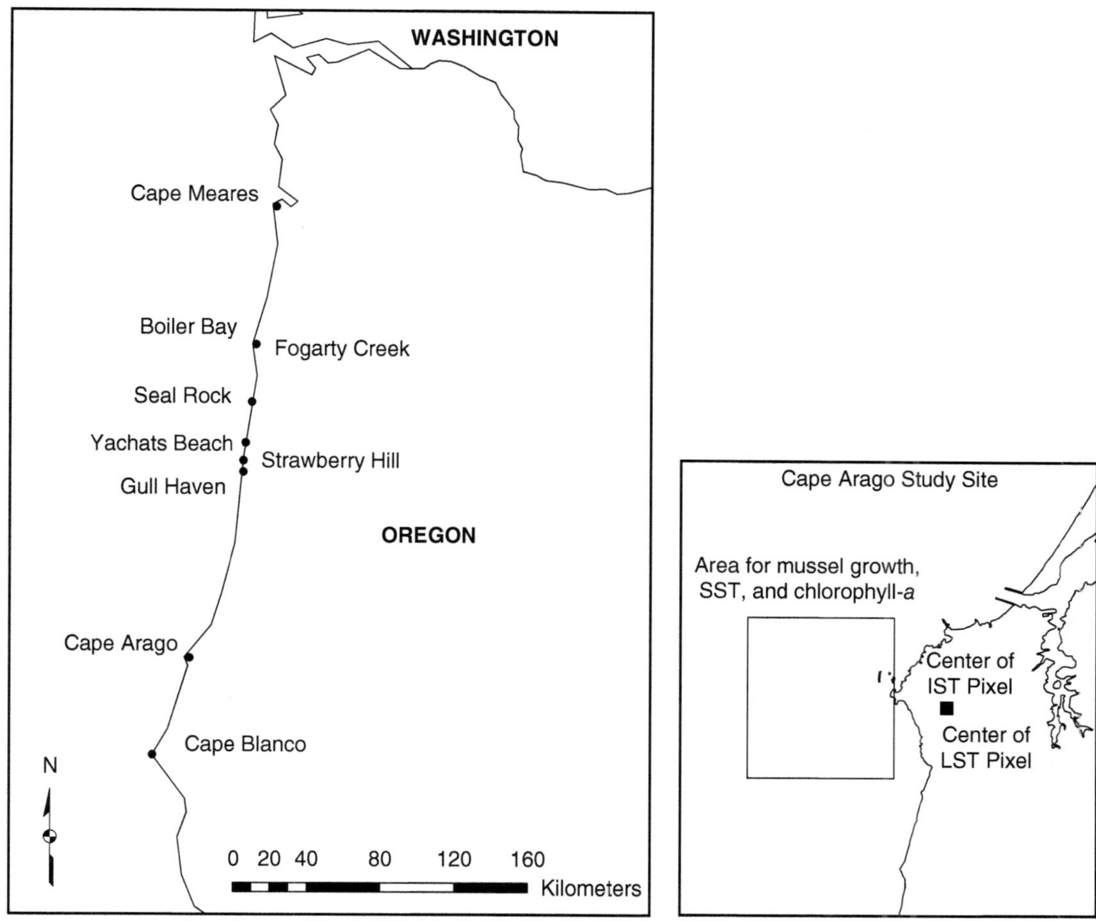

Figure 26.1 Map of the Oregon coastline with the eight study sites where the solid squares represent the center of the land surface temperature pixel, the solid circle represent the center of the intertidal surface temperature pixel, and the open square represent the area where mussel growth, sea surface temperature, and chlorophyll-*a* concentration was collected.

they are prey for a variety of species including, sea stars (*Piaster ochraceus*), various bird species, and marine vertebrates.

26.2.3. Mussel Growth

Mussel growth was determined by using the mussel growth monitoring data from PISCO. The protocol for mussel growth monitoring from PISCO is as follows (http://data.piscoweb.org):

All mussels that are 40–50 mm in height are collected from a common source, Bob Creek, Oregon. The mussels were then notched at the posterior end of the valves once using a wood file. The mussels were kept in a water table at Hatfield Marine Science Center until deployment. The mussels were then transplanted to the intertidal zone and were placed in the mid-zone-exposed location and held down with vexar plastic mesh for approximately 6 weeks. At 2-week intervals the mesh was domed to help the mussels orient themselves in a more natural posture and allow them

to attach stronger. After 6 weeks the mesh was removed. The mussels are left on the shore for approximately 12 months. After this period, the mussels were collected and stored frozen in the lab until they could be processed. In the lab, the initial size was recorded as the height from the anterior end to the notch scar, and growth was measured from the notched scar to the edge of the posterior end. The total height, length, and the width were also measured.

The initial length was subtracted from the final length to get the total growth (length) for each individual measured. Since mussel collection did not always happen exactly one year after deployment, the final growth was divided by the collection frequency to get a comparable measurement. For example, if the initial length of a mussel was 35.8 mm and the final length was 61.3 mm after 1.06 years, then the final growth would be 25.5 mm and the comparable growth would be 24.057 mm. This was done to get a measurement that can be compared with other collections and at different sites since collections varied for location and year. Once this was done for each mussel at each site, the initial growth, final growth, total growth, and comparable growth were each averaged so that all replicates in the same zone with the same deploy date, collect date, deploy site, and source site were averaged together. For example, all of the mussels that had the deploy date as 18 April 2000, collect date as 10 May 2000, deploy site as Seal Rock (SRKX00), source site as Bob Creek (BCKX00), and were located within the low intertidal zone were averaged by replicate, and then the replicates were averaged together to get annual averages. Then the size-specific proportional growth rate [*Menge et al.*, 2008] was calculated for each study site:

$$\text{Size-specific proportional growth rate} = \text{shell growth} \times \text{initial length}^{-1} \times \text{day}^{-1} \times 1000.$$

26.2.4. Satellite Data

Surface temperature and chlorophyll-*a* concentration was obtained from NASA's sun-synchronous Moderate Resolution Imaging Spectroradiometer (MODIS) aboard the Earth-Observing System (EOS) (Table 26.1). Sea surface temperature (SST), land surface temperature (LST), intertidal surface temperature (IST), and chlorophyll-*a* concentration from the *Terra* daytime overpass (10:30 A.M. descending overpass) were used in this study. Land surface temperature is at a spatial resolution of 1 km, sea surface temperature is at 4 km, and chlorophyll-*a* concentration is at a 4 km spatial resolution.

26.2.4.1. Surface Temperature

Daily *Terra* SST was obtained from the Physical Oceanography Distribution Active Archive Center (PO. DAAC) website (http://podaac.jpl.nasa.gov/) via the ftp protocol. Land surface temperatures from MODIS *Terra* satellites were obtained from NASA Land Processes Distribution Active Archive Center (LP DAAC) website (https://lpdaac.usgs.gov/get_data). *Terra* surface temperatures were available from February 2000 to December 2011. Daily surface temperatures were used for the Oregon study sites to compute the monthly and annual averages used. Days without temperature data were not used.

26.2.4.2. Chlorophyll-a

Monthly chlorophyll-*a* concentration data from MODIS *Terra* satellite was obtained from the Ocean Color website (http://oceandata.sci.gsfc.nasa.gov/). Monthly *Terra* chlorophyll-*a* concentration data were available from February 2000 to October 2011. Monthly chlorophyll-*a* concentration data were used to compute the annual averages and anomalies used for the study

locations. The monthly chlorophyll-*a* concentrations for September 2010 for Fogarty Creek and Boiler Bay had abnormally high chlorophyll-*a* concentrations (80.781 mg/m^3). These chlorophyll-*a* concentrations were still used in the analysis. There were few monthly chlorophyll-*a* concentration observations (eight) for Strawberry Hill over the entire study period. Temperature and chlorophyll-*a* concentration anomalies were calculated in the following manner:

$$\text{Annual anomaly} = \text{Annual average} - \text{Average of entire study period for the variable measured.}$$

Positive values mean that the year had a higher average and negative values mean that year had lower values.

26.3. RESULTS

26.3.1. Monthly Trends of Land, Intertidal, and Sea Surface Temperatures and Chlorophyll-*a* Concentration

The highest average LST (Figure 2.26a) and IST (Figure 26.2b) occurred during the months of June through August. Seal Rock, Yachats Beach, and Boiler Bay experienced the highest average LST during the months of June through August (Figure 26.2a). Cape Blanco, the southernmost study site, had the highest average LST during the months of January, February, March, November, and December (Figure 26.2a). Similar to LST, Seal Rock (SRKX00) had the highest average IST compared to the other study sites (Figure 26.2b). The highest average SST occurred during the months of June through September at Cape Meares, the northernmost study site. The lowest average SST also occurred at the Cape Meares site during the month of February. Fogarty Creek, Gull Haven, Seal Rock, Yachats Beach, and Boiler Bay had similar seasonal patterns during the summer months with the month of July being lower than the month of June (Figure 26.2c). Cape Blanco and Cape Arago, the two southernmost study sites, had the lowest average SSTs and had similar trends during the summer months (Figure 26.2c).

There is no clear pattern in average chlorophyll-*a* concentration by month at any study site (Figure 26.3). There appears to be higher chlorophyll-*a* concentrations from May to September with large variability. Cape Meares, the northernmost study site, has the lowest average concentration with very little variability within any given month (Figure 26.3). Fogarty Creek and Boiler Bay have large standard errors for September because of the abnormally high chlorophyll-*a* concentrations as mentioned previously.

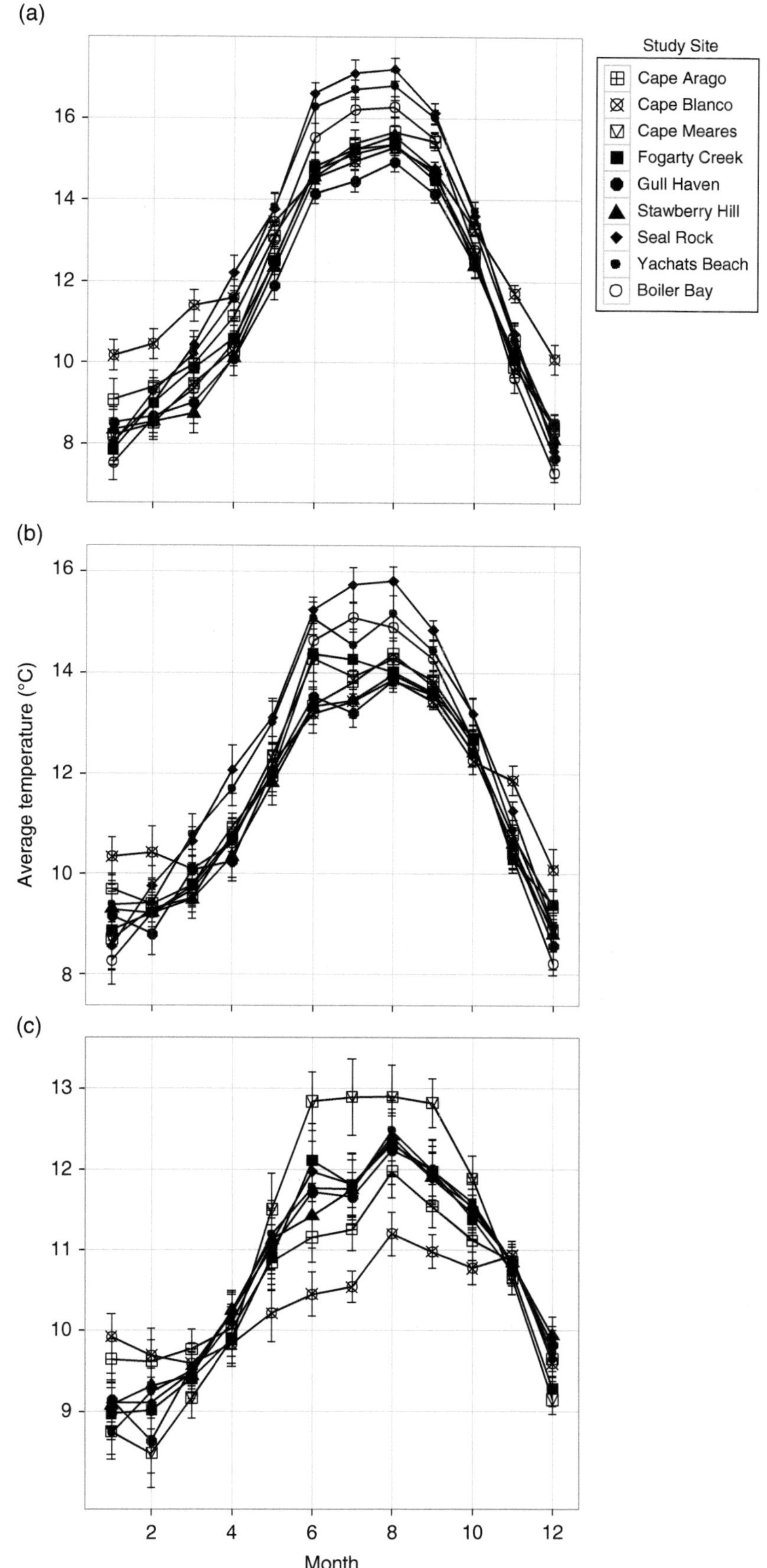

Figure 26.2 (a) Average land surface temperature, (b) intertidal surface temperature, and (c) sea surface temperature for each calendar month (January–December) over the entire study period with standard error bars shown for each month.

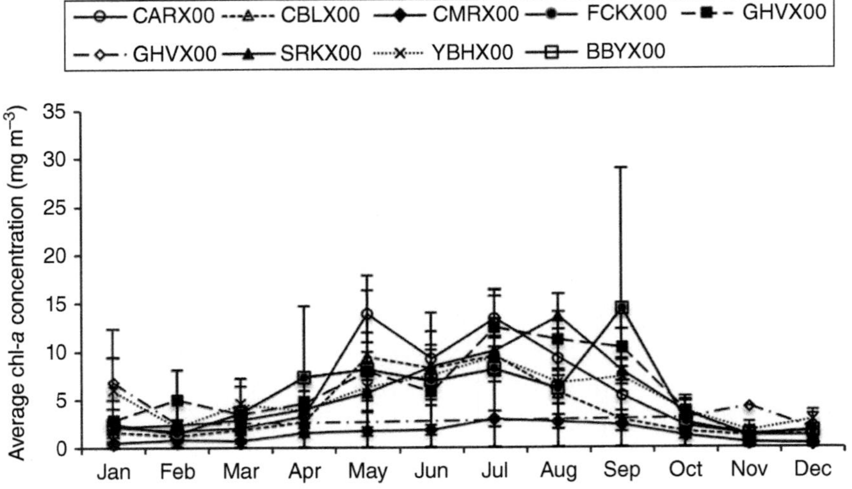

Figure 26.3 Average chlorophyll-*a* concentration for each calendar month (January–December) over the 12 year period with standard error bars shown for each month. The data for Strawberry Hill is not used in this figure because it had very few monthly averages. The chlorophyll-*a* concentration during September at Fogarty Creek and Boiler Bay show high variability.

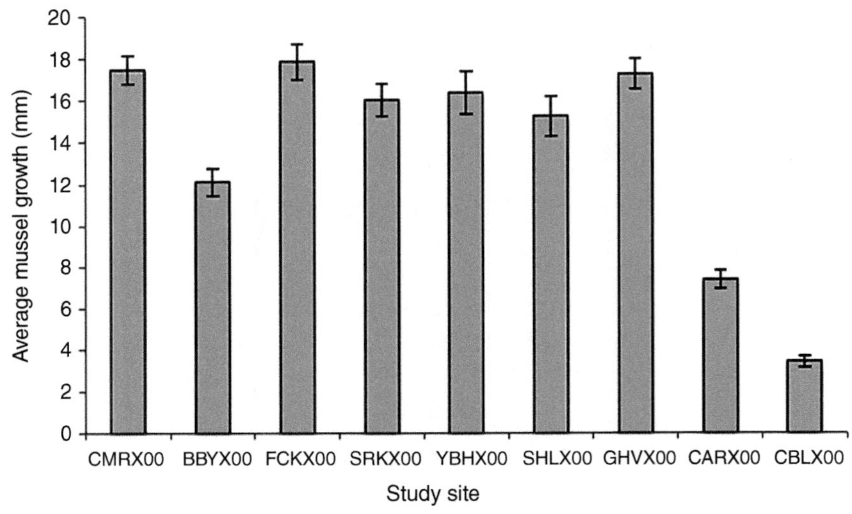

Figure 26.4 Mean mussel growth for each study site over the entire study period. Sites are listed from left to right from north to south, Cape Meares (CMRX00), Boiler Bay (BBYX00), Fogarty Creek (FCKX00), Seal Rock (SRKX00), Yachats Beach (YBHX00), Gull Haven (GHVX00), Cape Arago (CARX00), and Cape Blanco (CBLX00). Standard error bars with the same letters are not different $p > 0.05$, bars with different letters are different at $p < 0.05$.

26.3.2. Twelve Year Average of Land, Intertidal, and Sea Surface Temperature and Chlorophyll-*a* Concentration

Cape Arago and Cape Blanco, the two southernmost study sites, have significantly lower mussel growth than Cape Meares, Fogarty Creek, Yachats Beach, and Gull Haven (Figure 26.4). Seal Rock, Yachats Beach, and Cape Blanco had higher average LST than other sites [χ^2 = 270.90, degrees of freedom (df) = 8, *p*-value < 0.05, Friedman's chi square, Figure 26.5]. Seal Rock and Yachats Beach had significantly higher average IST than any other study site (χ^2 = 174.85, df = 8, *p* value < 0.05, Friedman's chi square, Figure 26.5). Cape Meares, the northernmost study site, had significantly higher average SST than Boiler Bay, Fogarty Creek, and Cape Blanco (χ^2 = 270.90, df = 8, *p* value < 0.05, Friedman's chi square, Figure 26.5). Cape Meares (CMRX00) had the lowest average chlorophyll-*a* concentration (Figure 26.6).

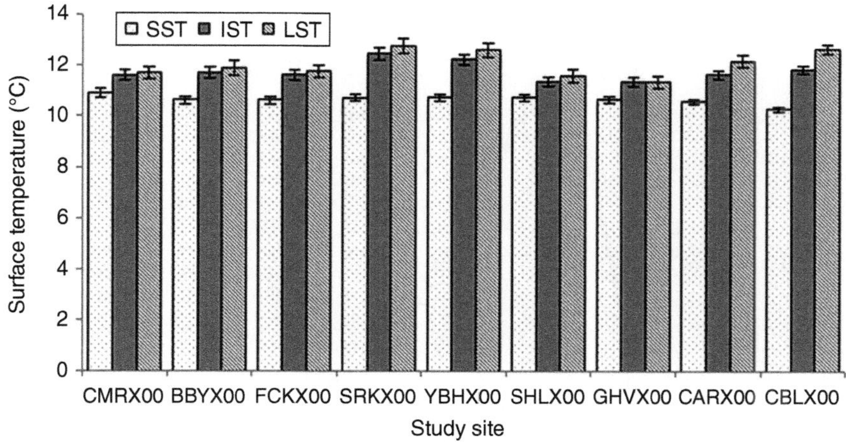

Figure 26.5 Average sea surface temperature, intertidal surface temperature, and land surface temperature at nine sites along the Oregon coast. Sites listed from left to right from north to south were Cape Meares (CMRX00), Boiler Bay (BBYX00), Fogarty Creek (FCKX00), Seal Rock (SRKX00), Yachats Beach (YBHX00), Strawberry Hill (SHLX00), Gull Haven (GHVX00), Cape Arago (CARX00), and Cape Blanco (CBLX00). Standard error bars are shown.

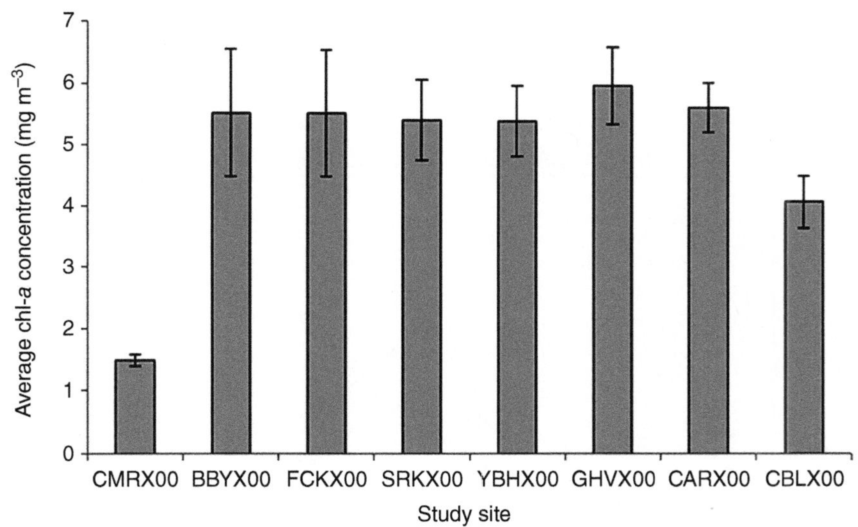

Figure 26.6 Average chlorophyll-*a* at eight sites along the Oregon coast. Sites listed from left to right from north to south were Cape Meares (CMRX00), Boiler Bay (BBYX00), Fogarty Creek (FCKX00), Seal Rock (SRKX00), Yachats Beach (YBHX00), Gull Haven (GHVX00), Cape Arago (CARX00), and Cape Blanco (CBLX00). Standard error bars are shown for each study site.

26.3.3. Anomaly Analysis

The SST anomalies for 2003 through 2005 are high (≥ +0.5°C) and low (≤ −0.5°C) during 2008 and 2009 (Figure 26.7). All of the study sites had similar SST anomalies trends (Figure 26.7). All study sites had similar patterns in size-specific growth rate (Figure 26.8). Cape Blanco and Cape Arago, the two southernmost study sites, had the lowest size-specific growth rate. Fogarty Creek had the highest size-specific growth rate. The highest size-specific growth rates occurred during 2004–2005

for all the study sites. Fogarty Creek, Yachats Beach, and Seal Rock had high size-specific growth rates from 2008 to 2009 (Figure 26.8).

26.3.4. Correlation Analysis

Sea surface temperature was not significantly correlated to chlorophyll-*a* concentration except at Cape Meares (Table 26.3). Interestingly, chlorophyll-*a* concentration was correlated to IST at all but two study site, Yachats Beach and Strawberry Hill (Table 26.3). The

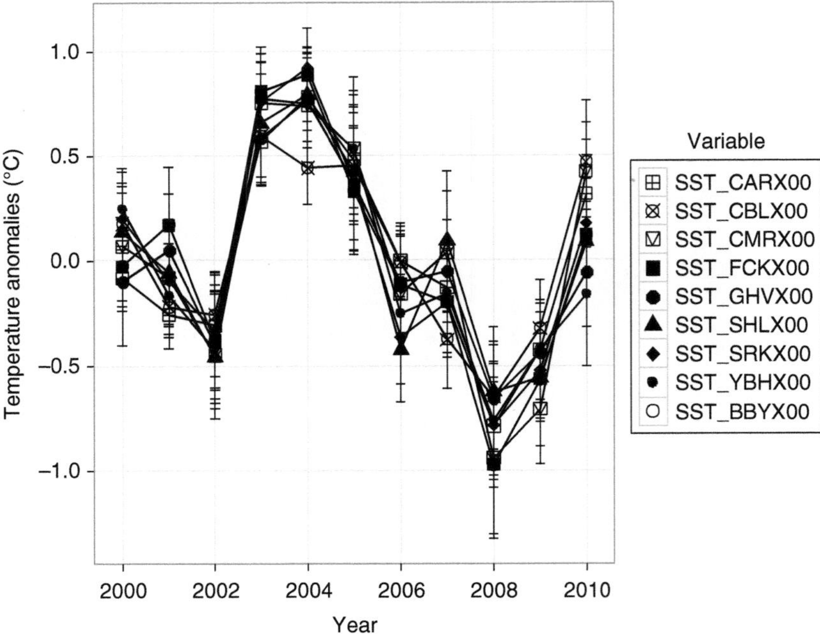

Figure 26.7 Average sea surface temperature anomalies by year for each study site with standard error bars shown for each year.

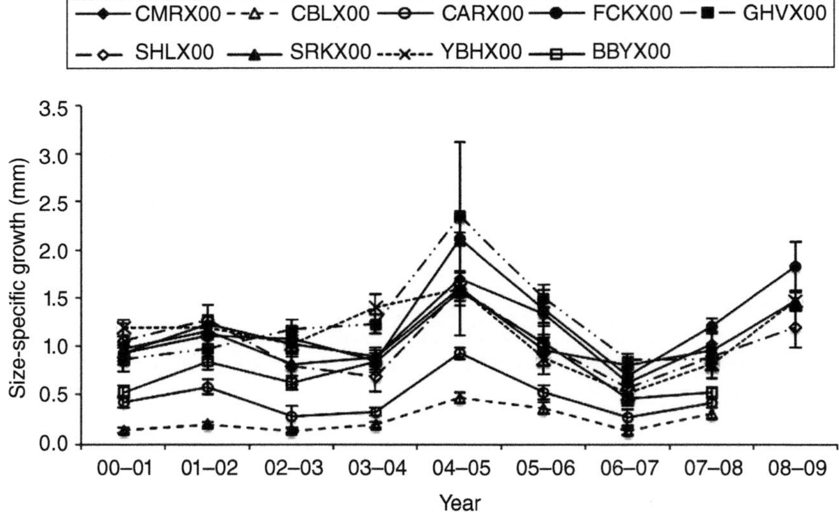

Figure 26.8 Size-specific growth rate for each year of study period from 2000 to 2009 at each study site with standard error bars shown.

strongest correlation between mean monthly IST and chlorophyll-*a* concentration was at Cape Meares ($r = 0.468$, p value < 0.001, df $= 137$, Pearson's r correlation, Table 26.3). Chlorophyll-*a* concentration was significantly correlated to LST at all study sites except Strawberry Hill (Table 26.3). The strongest correlation between LST and chlorophyll-*a* concentration was at Cape Meares ($r = 0.540$, p value < 0.001, df $= 137$, Pearson's r correlation, Table 26.3). Mean annual surface

temperature anomalies and mean annual chlorophyll-*a* concentration anomalies were not significantly correlated (p value > 0.05).

26.3.5. Linear Regressions

Remotely sensed surface temperatures (SST, IST, and LST) and chlorophyll-*a* (Chl-*a*) concentration were not good predictor variables for mussel growth at Cape Meares

Table 26.3 Pearson *r* correlation test between mean monthly *Terra* surface temperature and mean monthly *Terra* chlorophyll-*a* concentration for all study sites

Temperature Type	Study Site	df	Pearson *r*
SST	CMRX00	121	0.346[c]
	BBYX00	91	0.177
	FCKX00	91	0.177
	SRKX00	118	0.175
	YBHX00	113	−0.063
	SHLX00	6	−0.045
	GHVX00	119	0.053
	CARX00	115	0.125
	CBLX00	120	0.048
IST	CMRX00	137	0.468[c]
	BBYX00	102	0.263[b]
	FCKX00	102	0.238[a]
	SRKX00	134	0.355[c]
	YBHX00	126	0.171
	SHLX00	6	0.182
	GHVX00	133	0.252[b]
	CARX00	128	0.298[c]
	CBLX00	134	0.329[c]
LST	CMRX00	137	0.540[c]
	BBYX00	102	0.277[b]
	FCKX00	102	0.265[b]
	SRKX00	134	0.412[c]
	YBHX00	126	0.225[a]
	SHLX00	6	0.213
	GHVX00	133	0.324[c]
	CARX00	128	0.367[c]
	CBLX00	135	0.403[c]

Note: Significant levels: [a]0.05, [b]0.01, and [c]0.001.

(p value > 0.05). Mean annual SST was a good predictor variable for size-specific mussel growth at Boiler Bay ($F = 9.623$, p value = 0.022, $r^2 = 0.552$, $n = 7$). However, no other variable (IST, LST, or Chl-*a*) was a good predictor of mussel growth. Neither surface temperatures nor chlorophyll-*a* concentration were good predictors of mussel growth at Cape Blanco, Fogarty Creek, Strawberry Hill, Seal Rock, or Yachats Beach. Mean annual SST was a good predictor of size-specific mussel growth at Gull Haven ($F = 7.329$, p value = 0.035, $r^2 = 0.475$, $n = 7$). At Cape Arago, mean annual chlorophyll-*a* concentration was a good predictor of size-specific mussel growth ($F = 9.029$, p value = 0.024, $r^2 = 0.532$, $n = 7$). No single variable was a good predictor of size-specific mussel growth at Cape Blanco. Approximately 24% of variation in size-specific mussel growth was explained by mean annual SST and LST when all study sites were used (Figure 26.9).

There was a significant linear relationship between size-specific mussel growth and SST anomalies when all sites were combined ($F = 12.44$, p value < 0.001, $r^2 = 0.132$, df = 74).

26.4. CONCLUSIONS AND DISCUSSION

The purpose of this study was to investigate the potential of using satellite surface temperatures and chlorophyll-*a* concentration to better understand the spatial variability of the body growth of *Mytilus californianus*. While satellite data cannot replace in situ observations, they may prove helpful in understanding the spatial variability of certain physical parameters in areas that do not have in situ observations. However, according to the results in this present study, satellite surface temperature and chlorophyll-*a* concentration are unable to capture local-scale spatial variability in mussel growth across multiple study sites. This could be explained by (1) coarse resolution of the data, (2) annual averages rather than seasonal averages were used in this study, (3) variation in upwelling along the Oregon coast, or (4) spatial resolution of the satellite pixels used were not representative of the study sites. Based on the results of Menge et al. [2008] and other previous studies (Table 26.4), we should see some significant relationship among temperature, chlorophyll-*a* concentration, and mussel body growth at our study sites. *Menge et al.* [2008] found that 44.5% of the variation in mussel growth along the Oregon coast was explained by water temperature and food (chlorophyll-*a*). The only study site that had a significant relationship between chlorophyll-*a* concentration and mussel growth was Cape Arago. Interestingly, 53% of the variation seen in size-specific mussel growth was explained by mean annual chlorophyll-*a* concentration at only the Cape Arago site. This suggests that the resolution of the remotely sensed chlorophyll-*a* concentration data is too coarse to accurately measure the food availability that the mussels experience in the intertidal sites. Only two other study sites had significant relationships with the remotely sensed variables. Mean annual SST explained 55% of the variation seen in the size-specific mussel growth at Boiler Bay and 47% at Gull Haven.

We can conclude from our results that satellite data are unable to capture the local-scale spatial variability in mussel growth. Mechanistically, these results make sense because remotely sensed satellite chlorophyll-*a* concentration and sea surface temperature are measured at a coarse scale away from the coast and the mussel growth behavior is dominated by the micro-scale climate surrounding the organisms. Chlorophyll-*a* concentration varies dramatically away from the coast and is unlikely able to reflect the fine-scale variability that intertidal organisms experience. Further studies are needed before the full value, or lack thereof, of satellite coastal oceanographic information can be realized.

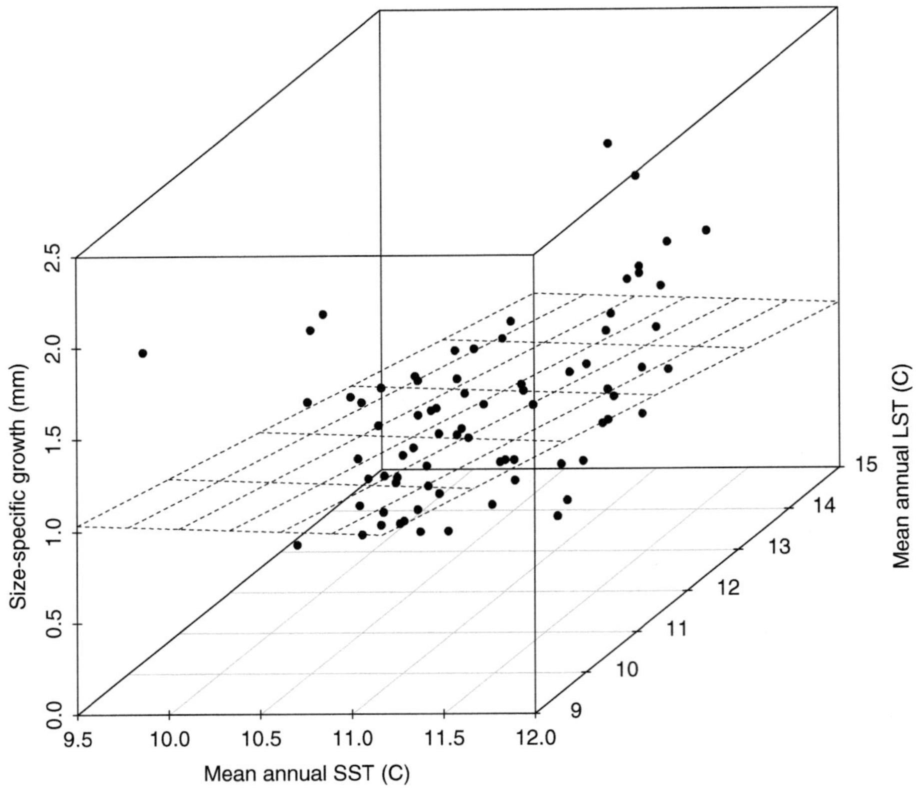

Figure 26.9 Influence of mean annual temperature (SST and LST) on the size-specific growth rate of *M. californianus* using a multiple regression test. All of the study sites were included in this regression analysis.

Table 26.4 Relevant findings from previous studies

Study	Type of Study & Data	Study Species	Study Location	Relevant Findings
Page and Ricard [1990]	Field study	*Mytilus edulis*	California	Water temperature not correlated to mussel growth
	All in situ data (water samples for chl-*a*, water temperature, mussel shell growth)			Food availability correlated to mussel growth
Blanchette et al. [2006]	Field and remotely sensed data	Macrophytes and invertebrates	California	Species abundance correlated to SST
	SST from AVHRR			Mussel abundance, growth, and recruitment were correlated with SST
	In situ data (water samples for nutrient and chl-*a*, algal tissue, species abundance)			No spatial pattern in chl-*a*
Blanchette et al. [2007]	Field study mussel size, abundance, and distribution	*M. californianus*	California	Wave exposure and intertidal temperature was correlated to mussel growth, not chl-*a*
Menge et al. [2008]	Field study All in situ data	*M. californianus*	Oregon	Growth and reproduction were positively correlated with chl-*a*
Schneider et al. [2010]	Laboratory experiment	*M. galloprovincialis*	Mussels collected from Washington	Food availability did not affect survival
	Thermal treatments			Higher thermal treatments caused higher levels of mortality
	Food availability	*M. trossulus*	Experiment done at USC	Higher thermal treatments and decreased food availability had strongest effect on species

ACKNOWLEDGMENTS

This study was supported by NASA funding—NNX11AP77G Program Manager Woody Turner. We used the data collected by the Partnership for Interdisciplinary Studies of Coastal Oceans, a long-term ecological consortium funded primarily by the Gordon and Betty Moore Foundation and David and Lucile Packard Foundation.

REFERENCES

Antoine, D., J.-M. André, and A. Morel (1996), Oceanic primary production: 2. Estimation at global scale from satellite (Coastal Zone Color Scanner) chlorophyll, *Global Biogeochem. Cycles, 10*, 57–69.

Blanchette, C. A., B. R. Broitman, and S. D. Gaines (2006), Intertidal community structure and oceanographic patterns around Santa Cruz Island, CA, *Marine Biol., 149*, 689–701.

Blanchette, C. A., B. Helmuth, and S. D. Gaines (2007), Spatial patterns of growth in the mussel, *Mytilus californianus*, across a major oceanographic and biogeographic boundary at Point Conception, California, USA, *J. Exp. Marine Biol. Ecol., 34*, 126–148.

Dayton, P. K. (1971), Competition, disturbance, and community organization: The provision and subsequent utilization of space in a rocky intertidal community, *Ecol. Monogr., 41*, 351–389.

Fitzhenry, T., P. M. Haplin, and B. Helmuth (2003), Testing the effects of wave exposure, site, and behavior on intertidal mussel boy temperatures: Applications and limits of temperature logger design, *Marine Biol., 145*, 339–349.

Gabarró, C., J. Font, A. Camps, M. Vall-llossera, and A. Julià (2004), A new empirical model of sea surface microwave emissivity for salinity remote sensing, *Geophys. Res. Lett., 31*, L01309, doi:10.1029/2003GL018964.

Gould, W. (2000), Remote sensing of vegetation, plant species richness, and regional biodiversity hotspots, *Ecol. Appl., 10*, 1861–1870.

Harley, C. D., A. R. Hughes, K. M. Hultgren, B. G. Miner, C. J. B. Sorte, C. S. Thornber, L. F. Rodriguez, L. Tomanek, and S. L. Williams (2006), The impacts of climate change in coastal marine systems, *Ecol. Lett., 9*, 228–241.

Helmuth, B., N. Mieszkowska, P. Moore, and S. J. Hawkins (2006a), Living on the edge of two changing worlds: Forecasting the responses of rocky intertidal ecosystems to climate change, *Annu. Rev. Ecol. Evol. Syst., 37*, 373–404.

Helmuth, B., B. R. Broitman, C. A. Blanchette, S. Gilman, P. Halpin, C. D. G. Harley, M. J. G. O'Donnell, E. Hofmann, B. Menge, and D. Strickland (2006b), Mosaic patterns of thermal stress in the rocky intertidal zone: Implications for climate change, *Ecol. Monogr., 76*, 461–479.

Lambin, E. F., and D. Ehrlich (1997), Land-cover in sub-Saharan Africa (1982–1991): Application of a change index based on remotely sensed surface temperature and vegetation indices at a continental scale, *Remote Sens. Environ, 61*, 181–200.

Lima, F. P., and D. S. Wethey (2009), Robolimpets: Measuring intertidal body temperatures using biomimetic loggers, *Limnol. Oceanogr. Methods, 7*, 347–353.

Lobitz, B., L. Beck, A. Huq, B. Woods, G. Fuchs, A. S. G. Faruque, and R. Colwell (1999), Climate and infectious disease: Use of remote sensing for detection of Vibrio cholera by indirect measurement, *Natl. Acad. Sci., 97*, 1438–1443.

Menge, B. A., et al. (2003), Coastal oceanography sets the pace of rocky intertidal community dynamics, *Proc. Natl. Acad. Sci., 100*, 12,229–12,234.

Menge, B. A., F. Chan, and J. Lubchenco (2008), Response of a rocky intertidal ecosystem engineer and community dominant to climate change, *Ecol. Lett., 11*, 151–162.

Mieszkowska, N., et al. (2005), Marine biodiversity and climate change: Assessing and predicting the influence of climatic change using intertidal rocky shore biota, Marine Biological Association of The United Kingdom. Occasional Publications, *20*, 1–53.

Mieszkowska, N., et al. (2006), Changes in the range of some common rocky shore species in Britain—A response to climate change? *Hydrobiology, 555*, 241–251.

Miller, R. L., C. E. Del Castillo, and B. A. Mekee (2005), Remote Sensing of Coastal Aquatic Environments, pp. 101–127, Springer, the Netherlands.

Nagendra, H. (2001), Using remote sensing to assess biodiversity, *Int. J. Remote Sens., 22*, 2377–2400.

Njoku, E. G., T. J. Jackson, V. Lakshmi, T. K. Chan, and S. V. Nghiem (2003), Soil moisture retrieval from AMSR-E, *IEEE Trans. Geosci. Remote Sens., 41*, 215–229.

Page, H. M., and Y. O. Ricard (1990), Food availability as a limiting factor to *Mytilus edulis* growth in California coastal waters, *Fish. Bull. 88*, 677–686.

Paine, R. T. (1966), Food web complexity and species diversity, *Am. Nat., 100*, 65–75.

Schneider, K. R., L. E. Van Thiel, and B. Helmuth (2010), Interactive effects of food availability and aerial body temperature on the survival of two intertidal *Mytilus* species, *J. Therm. Biol., 35*, 161–166.

Wan, Z., Y. Zhang, Q. Zhang, and Z. L. Li (2010), Quality assessment and validation of the MODIS global land surface temperature, *Int. J. Remote Sens., 25*, 261–274.

27

Impact of Assimilating Spaceborne Microwave Signals for Improving Hydrological Prediction in Ungauged Basins

Yu Zhang,[1,2,5] Yang Hong,[1–3,5] Jonathan J. Gourley,[4] Xuguang Wang,[5,6] G. Robert Brakenridge,[7] Tom De Groeve,[8] and Humberto Vergara[1,2]

27.1. INTRODUCTION

Insufficient ground gauge observations have been historical barriers in hydrological predictions. Over the globe, especially in Africa, it is much more common for a given basin to be only sparsely or not monitored at all by in situ observation networks. However, recent advancement in satellite remote sensing technology bears the promising potential to overcome the limited spatial coverage of in situ observation networks, thus providing the potential for hydrological predictions by being creatively used as the forcing (e.g., satellite precipitation estimation), calibration basis (e.g., passive microwave streamflow signal), and sources for assimilation (e.g., satellite-detected soil moisture estimation and passive microwave streamflow signals). This forecast system based entirely on remote sensing information thus enhances the reliability of streamflow prediction in poorly gauged basins and makes streamflow prediction possible even in ungauged basins.

Considering hydrological modeling in those basins with limited ground surface observation networks, a great deal of success has been achieved through the recent availability of remote sensing precipitation data [e.g., *Hong et al.*, 2004; *Huffman et al.*, 2007; *Joyce et al.*, 2004; *Sorooshian et al.*, 2000; *Turk and Miller*, 2005]. Besides utilizing the remote sensing precipitation data as forcing, remote sensing soil moisture data can also facilitate hydrological prediction by data assimilation approaches [e.g., *Brocca et al.*, 2010, 2012; *Crow and Ryu*, 2009; *Crow et al.*, 2005; *Gao et al.*, 2007; *Matgen et al.*, 2012; *Pauwels et al.*, 2002], which is promising for those basins with sparsely or even without in situ soil moisture observations. As a traditional way, the hydrological prediction accuracy is commonly improved by calibrating hydrologic models and through assimilating in situ soil moisture observations and gauge-based streamflow measurements into hydrological models [e.g., *Aubert et al.*, 2003; *Clark et al.*, 2008; *Pauwels and De Lannoy*, 2006]. The use of streamflow estimates from remote sensing methods is a new area being explored, also for model calibration and data assimilation. Recently, the Global Flood Detection System (GFDS, http://www.gdacs.org/flooddetection/), began using a passive microwave sensor, Advanced Microwave Scanning Radiometer–Earth Observing System (AMSR-E), together with the Tropical Rainfall Measurement Mission (TRMM) microwave imager (TMI), to measure surface brightness temperatures, which can be used creatively to infer streamflow and thus

[1]*School of Civil Engineering and Environmental Science, University of Oklahoma, Norman, Oklahoma, USA*

[2]*Advanced Radar Research Center, University of Oklahoma, Norman, Oklahoma, USA*

[3]*Water Technology for Emerging Region (WaTER) Center, University of Oklahoma, Norman, Oklahoma, USA*

[4]*NOAA/National Severe Storms Laboratory, Norman, Oklahoma, USA*

[5]*Center for Analysis and Prediction of Storms, University of Oklahoma, Norman, Oklahoma, USA*

[6]*School of Meteorology, University of Oklahoma, Norman, Oklahoma, USA*

[7]*CSDMS, INSTAAR, University of Colorado, Boulder, Colorado, USA*

[8]*Joint Research Centre of the European Commission, Ispra, Italy*

Remote Sensing of the Terrestrial Water Cycle, Geophysical Monograph 206. First Edition. Edited by Venkat Lakshmi.
© 2015 American Geophysical Union. Published 2015 by John Wiley & Sons, Inc.

show the potential to monitor floods over the globe [*Brakenridge et al.*, 2007]. While prior studies have evaluated the potential application of the AMSR-E sensor for discharge estimation and flood detection [*Salvia et al.*, 2011; *Temimi et al.*, 2007, 2011], they all required in situ streamflow information. In addition to AMSR-E, very recently, MOderate Resolution Imaging Spectroradiometer (MODIS) has also been applied by *Tarpanelli et al.* [2013] to estimate the streamflows in medium-sized basins adopting the same methodology developed by *Brakenridge et al.* [2007] at daily scale; good results show the potential of applying this for smaller basins thanks to the higher spatial resolution of MODIS data (250 m) compared to the spatial resolution of AMSR-E at 25 km; likewise the discharge estimation from AMSR-E, using MODIS also requires in situ discharge observations.

In this study, the passive microwave streamflow signals from AMSR-E are utilized directly, without in situ streamflow observations, in a hydrologic model to calibrate the hydrological model based on the approach in *Khan et al.* [2012]. Then, the frequency (exceedance probability) of the remote sensing streamflow signals is assimilated into the hydrological model through the data assimilation approach that was applied in *Zhang et al.* [2013] in order to demonstrate probabilistic flood prediction for an African basin. Though conducted in the same research region – Cubango River basin with the same forcing data (refer to section 27.2.2 data sources for detailed information) and a similar data assimilation approach (refer to section 27.2.4 for detailed information), this study is conducted in the frequency domain and is independent from the in situ streamflow observation. The application of in situ steamflow observation in this study is set up as the benchmarks to evaluate the hydrological performance of the remote sensing streamflow signals, which demonstrates an innovative way for improving the hydrological prediction in ungauged basins with outcomes as probabilistic-based predictions. In contrast, [*Zhang et al.*, 2013] conducted the hydrological prediction in the actual streamflow domain for the same research basin. It first converted the remote sensing streamflow signals into streamflow based on the in situ streamflow observation by adopting the algorithm that was developed by *Brakenridge et al.* [2007]. Then, the remote sensing signal converted streamflow was applied to calibrate and update the model, thus providing hydrological predictions in actual streamflow domain (i.e., in units of cubic meters/second).

Section 27.2 describes the study basin, the data applied, and the algorithm. Then section 27.3 discusses the results of both calibration and data assimilation in frequency domain conducted by the streamflow signals (experiment 3) compared to the results obtained by the in situ streamflow in both actual domain (experiment 1) and frequency domain (experiment 2). Finally, the conclusion is drawn in section 27.4.

27.2. STUDY BASIN, DATA SOURCES, AND METHODOLOGY

27.2.1. Study Basin

The Okavango River, which runs for about 1100 km from central Angola and flows through Namibia and Botswana, is the fourth longest river in southern Africa (Figure 27.1). The Okavango catchment is approximately 413,000 km^2; it originates in the headwaters of central Angola, then the Cubango and Cuito tributaries meet to form the Cubango-Okavango River near the border of Angola and Namibia and flow into the Okavango Delta in Botswana. The upper stream region belongs in a subtropical climate zone with annual precipitation around 1300 mm while the downstream region, which contains the Kalahari Desert, belongs to the semiarid climate zone with annual precipitation around 450 mm [*Hughes et al.*, 2006; *Milzow et al.*, 2009a]. The headwater region, which is the northern part of the basin, is mainly covered by the ferralsols soil with a lower hydraulic conductivity. The headwater region also has a high forest cover and contributes significantly to the river runoff [*Hughes et al.*, 2006]. The rest of the basin is dominated by arenosals soil (www.sharing-water.net), which is very porous with high hydraulic conductivity, so that water drains rapidly, leaving little moisture for plants. As mentioned by *Hughes et al.* [2006], around 95% of inflow is lost in the atmosphere due to high potential evapotranspiration rate and only a small portion contributes to groundwater.

Several studies in the Okavango River Basin have investigated the hydrological response under climate change [*Andersson et al.*, 2006; *Hughes et al.*, 2006, 2011; *McCarthy et al.*, 2003; *Milzow et al.*, 2009b]. Since the Okavango River Basin is one of the most important economic and water resources in southern Africa, additional studies have been solicited to assist in the decision making for water management in this basin. The main tributary of Okavango River—the Cubango River, which is mainly located in Angola, is selected as the study basin. It accounts for a majority of the available water resources in the Okavango River. The Rundu Gauge station is the outlet of the Cubango River; at Rundu Gauge, both gauge-based streamflow and the remote sensing discharge estimates (i.e., the AMSR-E and TMI streamflow signals) are available.

27.2.2. Data Sources

This study develops an advanced exceedance probability-based, flood prediction framework, which is based entirely on satellite remote sensing data without a requirement of conventional in situ hydrological measurements. The in situ streamflow observation, with daily temporal

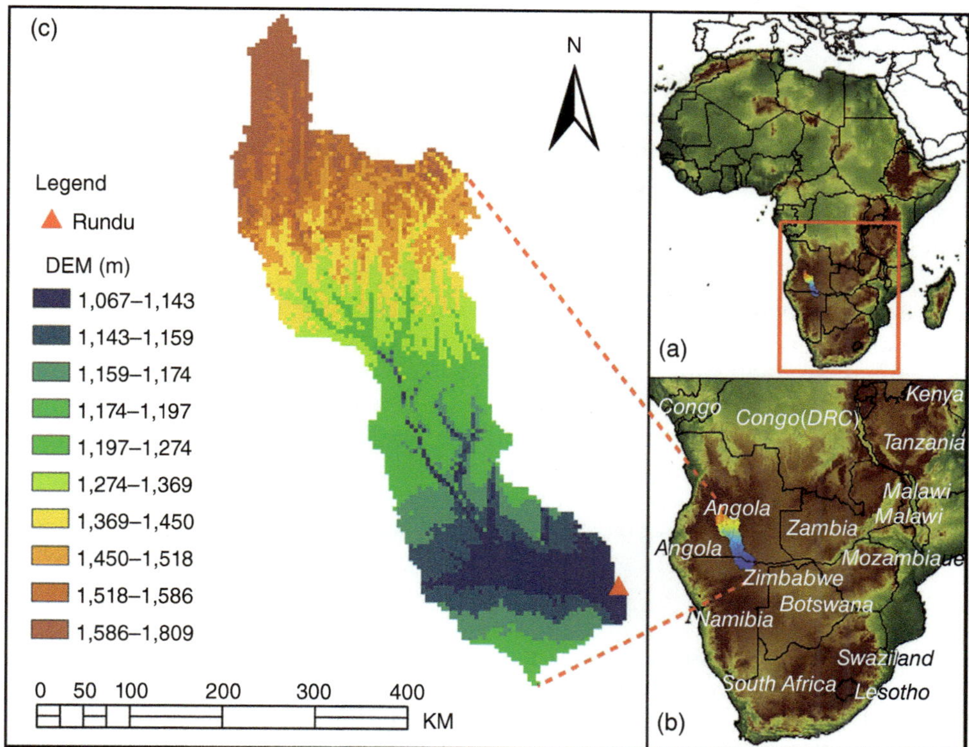

Figure 27.1 Research region—Cubango River basin.

resolution, is only used in this study to evaluate the exceedance probability-based hydrological prediction algorithm. The following proposed data sets were applied in this study:

27.2.2.1. TRMM Real-Time (RT) Satellite Precipitation Estimates

The TRMM satellite precipitation estimates are taken as forcing data into hydrological modeling in this study since the Okavango River Basin is poorly gauged [*Milzow et al.*, 2011]. TRMM multi-satellite precipitation analysis (TMPA) provides two standard 3B42-level products: the near-real-time 3B42 RT, which uses the TRMM combined instrument data set to calibrate the data and the post-real-time research product 3B42 V7 (level 7), which adjusts the rainfall accumulation by gauge analysis [*Huffman et al.*, 2007]. Both 3B42 RT and 3B42 V7 products are quasi-global with coverage from 50°N to 50°S latitude. In this study, the TRMM 3B42 RT with a spatial resolution of 0.25° (approximate to 25 km in the tropical area) and temporal resolution of three hourly is processed into daily accumulations as well as basin averages and applied as the forcing data to drive the hydrological model.

27.2.2.2. Famine Early Warning System (FEWS) Network Potential Evapotranspiration

Potential evapotranspiration (PET) comes from the FEWS NET (http://igskmncnwb015.cr.usgs.gov/Global/)

with a temporal resolution of monthly and spatial resolution of 0.25° and is likewise processed into daily and basin averages as additional forcing to the model.

27.2.2.3. The Passive Microwave Streamflow Signal from TRMM and Aqua

The GFDS uses near-real-time, satellite-based, remote sensing data to monitor floods over the globe on a daily scale. In this system, a passive microwave sensor, AMSR-E, together with the TRMM TMI sensor is used to measure the brightness temperature at 36.5 GHz (AMSR-E, 1 June 2002–4 October 2011) or 37 GHZ (TRMM, 1 January 1998 to the present) at river measurement sites. Microwave brightness at this frequency is sensitive to changes in surface water extent and thus flooding [*Brakenridge et al.*, 2007, 2012]. A "wet" pixel (usually centered over a river reach and adjacent floodplain; pixel dimensions are ~10 km at low and middle latitudes) is selected to measure the brightness temperature of the measurement (M) area while the "driest" (brightest) pixel within a bracketing 7 × 7 pixel grid measures a calibrating brightness temperature (C). The ratio of the measurement and calibration brightness temperature is referred as the streamflow signal is retrieved for all measurement sites [equation. 27.1].

$$M/C \text{ Ratio} = Tb_m/Tb_c. \tag{27.1}$$

A major benefit of the passive microwave sensor approach using AMSR-E (now AMSR-2 aboard GCOM-W, http://mirs.nesdis.noaa.gov/amsr2.php) and TRMM is that it is not restricted severely by cloud cover and thereby provides the frequent data needed for monitoring dynamic river discharge changes. For further detailed information regarding the GFDS streamflow signals, please refer to *Brakenridge et al.* [2007, 2012] and *Kugler and Groeve*, [2007].

27.2.2.4. Ground-Based Streamflow Observation

Besides the passive microwave streamflow signal data at Rundu for both calibration and assimilation (will be specified in section 27.2.5), ground-based streamflow observation at Rundu, Namibia, was used to evaluate the performance of the proposed "exceedance probability based flood-prediction framework" [*Khan et al.*, 2012] in an upstream catchment—Cubango of around 95,000 km^2.

27.2.3. Model

In this study, a simplified and lumped version of the CREST (**C**oupled **R**outing and **E**xcess **ST**orage [*Wang et al.*, 2011]) was applied, together with the satellite data and the EnSRF (ensemble square root filter) data assimilation approach, to provide exceedance probability-based hydrological predictions over the Cubango basin. The model structure is shown in Figure 27.2. After precipitation passes the canopy layer, the excess precipitation that reaches the soil surface is P_{soil}, which is divided into excess rainfall R and infiltration water I through the variable infiltration curve (VIC; [*Liang et al.*, 1994]). After that, the excess rainfall R is further separated into overland excess rainfall R_o and interface excess rainfall, and this procedure is governed by K, which is closely related to the saturated soil hydraulic conductivity (the interface excess rainfall $= K \dfrac{R}{P_{\text{soil}}}$ when $P_{\text{soil}} > k$, the interface excess rainfall $= R$ when $P_{\text{soil}} \leq k$). Next, the interface excess rainfall is evapotranspired through three soil layers and then the interface excess rainfall reduces to R_I The overland excess rainfall R_o flows through three overland flow linear reservoirs while the interface excess rainfall R_I flows through one interflow reservoir; those two procedures are governed by the overland reservoir discharge multiplier LEAKO ($R_{o,\text{out}} = \text{LEAKO*} R_o$) and the interflow reservoir discharge multiplier LEAKI ($R_{I,\text{out}} = \text{LEAKI*} R_I$), respectively, and then form the total runoff as Q ($Q = R_{o,\text{out}} + R_{I,\text{out}}$).

27.2.4. EnSRF

A sequential data assimilation technique, the EnSRF is applied to assimilate passive microwave streamflow signals into CREST. Unlike the traditional EnKF, which requires perturbing both forcing data and observations, the EnSRF only perturbs the forcing data and the ensemble mean is updated by the observation. *Whitaker and Hamill* [2002] demonstrated that there is no additional computational cost by EnSRF relative to EnKF, and

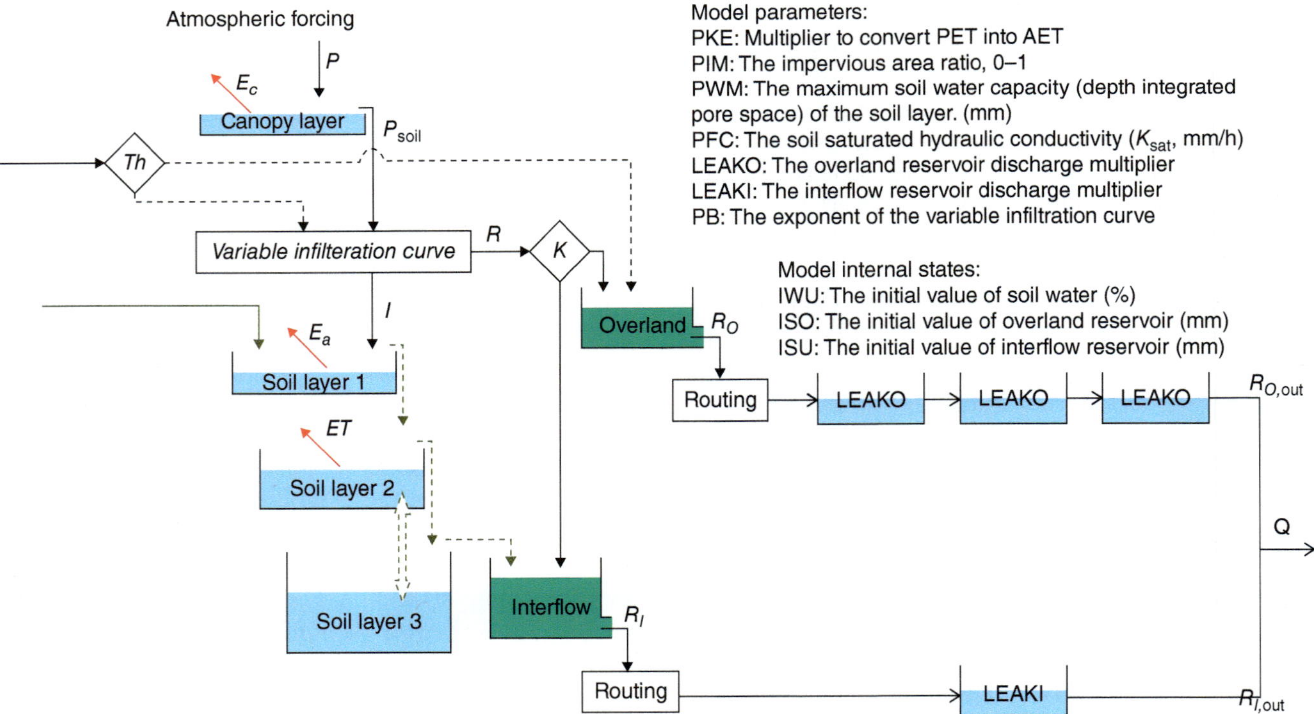

Figure 27.2 Structure of CREST model.

EnSRF performs more accurately than EnKF for the same ensemble size. But it still remains a research topic to compare the accuracy and efficiency of different sequential data assimilation approaches (e.g., EnKF, EnSRF). The major equations of EnSRF are listed below:

$$X^{\mathrm{a}} = X^{\mathrm{b}} + K(y - H(X^{\mathrm{b}})), \qquad (27.2)$$

where

$X^{(a)}$ = updated estimate of the analyzed state ($n \times 1$ dimension and n is the number of ensembles)

$X^{(b)}$ = background model forecast, which is also referred to the first guess in data assimilation ($n \times 1$ dimension)

y = observation ($p \times 1$ dimension and p is the number of observations), which is the streamflow measurements in this study

H = observation operator that converts the states in the model into observation space ($p \times n$ dimension)

$\hat{K}$ = traditional Kalman gain

Let's denote the ensemble X^b as

$$X^{\mathrm{b}} = x_1^{\mathrm{b}}, x_2^{\mathrm{b}}, \ldots, x_n^{\mathrm{b}}, \qquad (27.3)$$

where we ignore time index and the subscript represents the ensemble member. The ensemble mean is then defined as

$$\overline{X^{\mathrm{b}}} = \frac{1}{n}\sum_{i=1}^{n} x_i^{b}. \qquad (27.4)$$

The perturbation from the mean for the ith member is

$$x_i'^{\mathrm{b}} = x_i^{\mathrm{b}} - \overline{x^{\mathrm{b}}}. \qquad (27.5)$$

Then X'^{b} is defined as a matrix formed from the ensemble of perturbations:

$$X'^{\mathrm{b}} = (x_1'^{\mathrm{b}}, x_2'^{\mathrm{b}}, \ldots, x_n'^{\mathrm{b}}). \qquad (27.6)$$

An estimation of background error covariance is defined as

$$\hat{P}^{\mathrm{b}} = \frac{1}{n-1} X'^{\mathrm{b}} (X'^{\mathrm{b}})^{\mathrm{T}}. \qquad (27.7)$$

However, in practice, we do not calculate $\hat{P}^{\mathrm{b}}$, but rather calculate $\hat{P}^{\mathrm{b}}H^{\mathrm{T}}$, and $H\hat{P}^{\mathrm{b}}H^{\mathrm{T}}$ are evaluated by the following equations:

$$\hat{P}^{\mathrm{b}}H^{\mathrm{T}} = \frac{1}{m-1}\sum_{i=1}^{m}\left(X_i^{\mathrm{b}} - \overline{X}^{\mathrm{b}}\right)\left\{H\left[X_i^{\mathrm{b}} - \overline{H(X^{\mathrm{b}})}\right]\right\}^{\mathrm{T}}, \qquad (27.8)$$

$$H\hat{P}^{\mathrm{b}}H^{\mathrm{T}} = \frac{1}{m-1}\sum_{i=1}^{m}\left[H\left(X_i^{\mathrm{b}}\right) - H\left(\overline{X}^{\mathrm{b}}\right)\right]$$
$$\left\{H\left[X_i^{\mathrm{b}} - \overline{H(X^{\mathrm{b}})}\right]\right\}^{\mathrm{T}}. \qquad (27.9)$$

Here, m is the ensemble size. Then the traditional Kalman gain $\hat{K}$ can be calculated by equation 27.10:

$$\hat{K} = \hat{P}^{\mathrm{b}}H^{\mathrm{T}}\left(H\hat{P}^{\mathrm{b}}H^{\mathrm{T}} + R\right)^{-1}, \qquad (27.10)$$

where R is the observation error covariance with a dimension of $p \times p$. In EnSRF, the reduced Kalman gain $\widetilde{K}$ is used to update the deviation from the ensemble mean as estimated by the following equation:

$$\widetilde{K} = \left(1 + \sqrt{\frac{R}{H\hat{P}^{\mathrm{b}}H^{\mathrm{T}} + R}}\right)^{-1}\widetilde{K}. \qquad (27.11)$$

The ensemble mean can be updated by

$$\overline{X}_i^{\mathrm{a}} = \overline{X}_i^{\mathrm{b}} + \hat{K}\left[y - H\left(\overline{X}_i^{\mathrm{b}}\right)\right]. \qquad (27.12)$$

The perturbation (deviation of ensemble mean) can be updated by

$$X_i'^{\mathrm{a}} = X_i'^{\mathrm{b}} + \hat{K}H\left(X_i'^{\mathrm{b}}\right). \qquad (27.13)$$

The final analysis follows as

$$X_i^{\mathrm{a}} = \overline{X}_i^{\mathrm{a}} + X_i'^{\mathrm{a}}. \qquad (27.14)$$

As mentioned above, when the EnSRF is applied, the forcing data (which is the precipitation in this study) needs to be perturbed. Precipitation perturbations in this study are defined as

$$P_i = P + \varepsilon_i, \qquad (27.15)$$

where ε_i is a random noise factor drawn from a Gaussian distribution:

$$\varepsilon_i \sim N(0, r). \qquad (27.16)$$

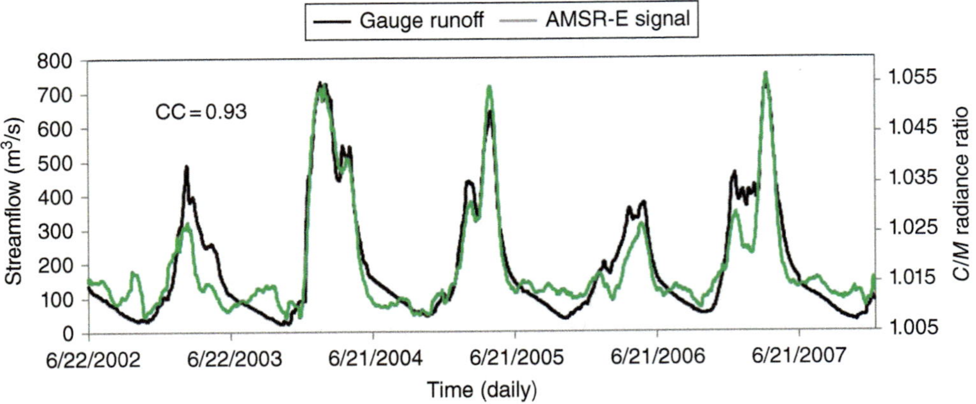

Figure 27.3 Time series of gauge streamflow observation plotted against primary *y* axis and *C/M* Radiance ratio plotted against secondary *y* axis.

Table 27.1 List of experiments design

Experiment		Calibration Data Source	Data Assimilated into Model	Calibration Objective Function
1		Gauge streamflow	Gauge streamflow	Min(RMSE)
2		Gauge streamflow frequency	Gauge streamflow frequency	Max(CC)
3	(a) Before threshold applied	AMSR-E signal frequency	AMSR-E signal frequency	Max(CC)
	(b) After threshold applied			

At each time step, an independent rainfall error is generated by Gaussian distribution [refer to equations (27.15)–(27.16)] and added to the original basin average precipitation.

27.2.5. Experimental Design

The *C/M* radiance ratio, which is the reciprocal of *M/C* ratio signal [equation 27.1], is correlated at a significant level with observed streamflow especially during the peak flow periods, as shown in Figure 27.3. Based on the high correlation coefficient between the gauge-based streamflow and the *C/M* radiance ratio, an innovative calibration method—the flood frequency approach—was proposed by *Khan et al.* [2012], which first requires the conversion of model-simulated streamflow into exceedance probability and then takes "max(CC)" (CC refers to correlation coefficient) as the objective function to conduct the automatic hydrological calibration via the shuffled complex evolution algorithm from the University of Arizona [SCE-UA; *Duan et al.*, 1994]. The flood frequency approach utilizes the period of recorded observations to compute the frequency or exceedance probability. This approach essentially normalizes the streamflow observations from absolute units (m³/s) to dimensionless values in the frequency domain. The same approach can

be applied to any time series data (i.e., passive microwave streamflow signal) as long as there is a sufficiently long record to represent climatological conditions and the signal is temporally correlated to streamflow.

As shown by Table 27.1, experiment 1, which was conducted in absolute streamflow units (m³/s), is the traditional gauged-based approach to model calibration and data assimilation. It sets the reference to be compared to the frequency-based in situ and remote sensing approaches in experiments 2 and 3. In experiment 2, streamflow observations from the Rundu Gauge are used to automatically calibrate the model parameters as in experiment 1, but using the exceedance probability approach described in *Khan et al.* [2012]; and then the gauge streamflow frequency was assimilated into CREST model via EnSRF. Experiment 3, which represents the advanced exceedance probability-based streamflow prediction framework, is designed similarly to experiment 2, but the exceedance probability of the observed streamflow is replaced with the frequency of the AMSR-E signals. Experiment 3 is thus based entirely on remote sensing data and applies generally to ungauged basins. Results from the experiments with no data assimilation are referred to as "open loop," while the components that employ the EnSRF are referred to as "assimilation." Results from all experiments are evaluated using gauge-observed streamflow at the Rundu station.

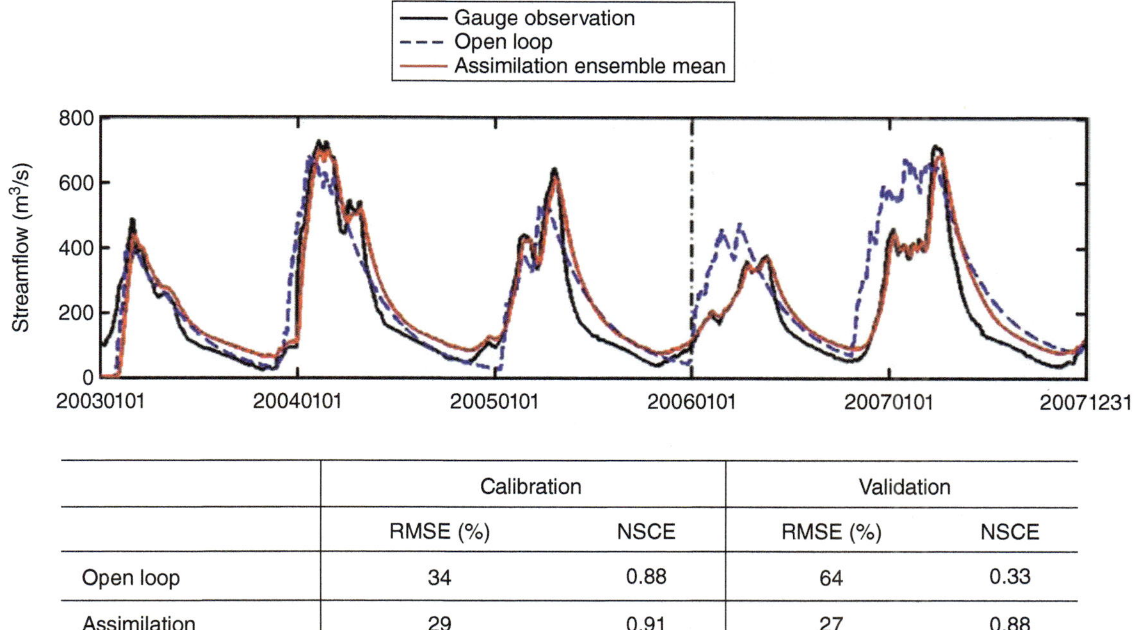

	Calibration		Validation	
	RMSE (%)	NSCE	RMSE (%)	NSCE
Open loop	34	0.88	64	0.33
Assimilation	29	0.91	27	0.88

Figure 27.4 Impact of assimilating gauge streamflow into CREST in experiment 1. The statistic indices in the table are calculated against the observed streamflow. Note that to the left side of the dark dashed line is the Calibration period from 2003 to 2005; to the right side of the dark dashed line is the validation period from 2006 to 2007; the same for Figure 27.5 and 25.6.

The modeling performance for both open loop and assimilation was evaluated by the statistic indices normalized root mean-squared error (RMSE) and Nash-Sutcliffe coefficient of efficiency (NSCE):

$$\text{RMSE}\,(\%) = \frac{\sqrt{\Sigma\left(x_i - y_i\right)^2 / n}}{\overline{x}} \times 100, \quad (27.17)$$

$$\text{NSCE} = 1 - \frac{\Sigma\left(x_i - y_i\right)^2}{\Sigma\left(x_i - \overline{x}\right)^2}, \quad (27.18)$$

where x_i is the observed streamflow and y_i is the simulated streamflow.

27.3. RESULTS AND DISCUSSION

Zhang et al. [2013] have conducted a sensitivity analysis in order to better understand the spread of precipitation [r in equation 27.16], ensemble size [n in equation 27.3] and observation error [R in equation 27.10] and their impact to data assimilation efficiency. In this study, the same spread of precipitation (50%, as the precipitation are the same for those three experiments) and ensemble size (20) are applied to all of those three experiments. For the observation error R, it is usually assumed from the "actual" observation error based on experience. In

experiments 1 and 2, the observation error is assumed as 8% according to the report [*Sauer and Meyer*, 1992] from the U.S. Geological Survey (USGS), which indicates 8% represents the common streamflow observation error; as the AMSR-E streamflow signals shows, overestimation during low flows (Figure 27.3), a larger observation error of 10% is assumed in experiment 3.

Experiment 1 is the reference experiment; the model was calibrated by gauge-based streamflow observations for the period 2003–2005 with a computed RMSE of 34% and NSCE of 0.88. Then, the model was validated for the period 2006–2007 in which the RMSE shot up to 64% and the NSCE dropped to 0.33. In order to enhance the hydrological performance, the gauge streamflow observation was assimilated into the well-calibrated lumped CREST model via EnSRF at a daily time step. After assimilation, the modeling performance was improved significantly during both calibration and validation periods. (Note: The statistical evaluation excludes the first half-year due to the bad first guesses at the beginning for each experiment.) The two simulations illustrated in Figure 27.4 serve as the stream gauge-based reference for the open-loop and assimilation experiments focused on the use of the gauge streamflow and the microwave streamflow signals in frequency domain hereafter.

In experiment 2, the sources of data for model calibration are the same, but the simulated and observed

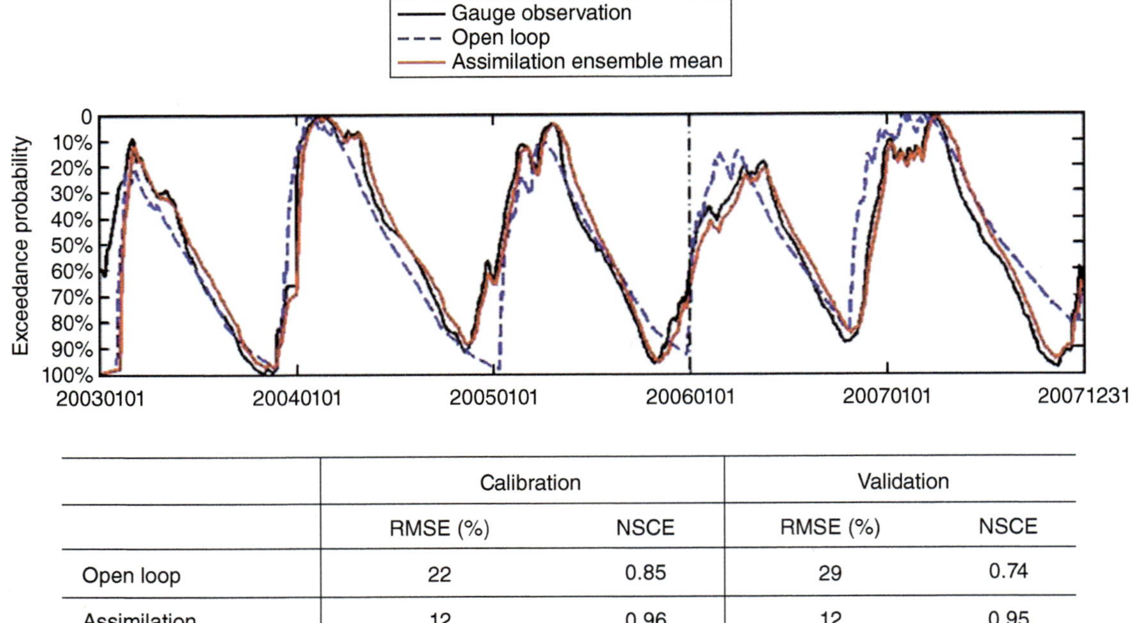

	Calibration		Validation	
	RMSE (%)	NSCE	RMSE (%)	NSCE
Open loop	22	0.85	29	0.74
Assimilation	12	0.96	12	0.95

Figure 27.5 Impact of assimilating gauge streamflow frequency into CREST in Experiment 2. The statistic indices in the tables are calculated against the exceedance probability of observed streamflow.

streamflow data have been converted to the frequency domain and expressed as a exceedance probability (Figure 27.5). This conversion upgraded the skill of the open loop simulation compared to the one in experiment 1 in terms of RMSE; it has fallen from 34% to 22%, though with a slight decrease of NSCE from 0.88 to 0.85 during the calibration period. For the validation period, the modeling skill improved significantly relative to experiment 1; RMSE decreased from 64% to 29% while NSCE increased from 0.33 to 0.74. For further improvement, the assimilation simulation, which employed the EnSRF by assimilating the gauge streamflow data in the frequency domain, resulted in a better overall skill compared to the assimilation run in experiment 1.

Figure 27.3 shows the time series of the passive microwave C/M radiance ratio (green line), which is used as the streamflow proxy for automatically estimating the model parameters. The C/M radiance ratio matches well with the gauge streamflow observations during the high-flow period, but shows noise during the low-flow period because of the insensitivity of the AMSR-E and TMI sensors to low flows. In experiment 3(a), the sources of data for model calibration are the C/M radiance ratios, and the C/M radiance ratios have been converted into the frequency domain and also expressed as the exceedance probability (Figure 27.6a). The application of C/M radiance ratio frequency degraded the skill of the open-loop simulation compared to the ones in both experiments 1 and 2 during the calibration period, but enhanced the open-loop simulation during the validation period with NSCE increasing from 0.33 (experiment 1)

and 0.74 (experiment 2) to 0.81. However, after assimilation, the streamflow signal indicates a small peak near November 2003 that was not observed by the stream gauge (Figure 27.6a). This error was not reflected in the open-loop simulation; however, by assimilating the C/M radiance ratio with noise into the model during the low flows, errors during low flows result. The performance of the simulations was poor for low flows but remarkable for high flows. This latter feature prompted us to devise experiment 3(b), the same as the assimilation component of experiment 3(a), but the radiance ratio data are assimilated only if the exceedance probability is < 30%. In other words, the C/M radiance ratio data are trusted only during high-flow conditions. After application of this subjectively chosen threshold, the red curve in Figure 27.3b illustrates very similar performance during high flows as in experiment 3(a) (red curve in Figure 27.6a), but the prior problems during low flows have been alleviated. The RMSE (26% during calibration period and 23% during validation period) is even better than the reference simulations in experiment 1 that assimilated gauge streamflow (in absolute units) but worse than the simulation in experiment 2. The NSCE of 0.79 and 0.84 during calibration and validation periods, respectively, is only a slight reduction from the reference values in both experiments 1 and 2. Nonetheless, this reduction is quite modest considering experiment 3(b) is based entirely on remote sensing data thus can provide probability-based streamflow prediction in ungauged basins.

In addition to the improvement of hydrological prediction by assimilation of remote sensing streamflow signal

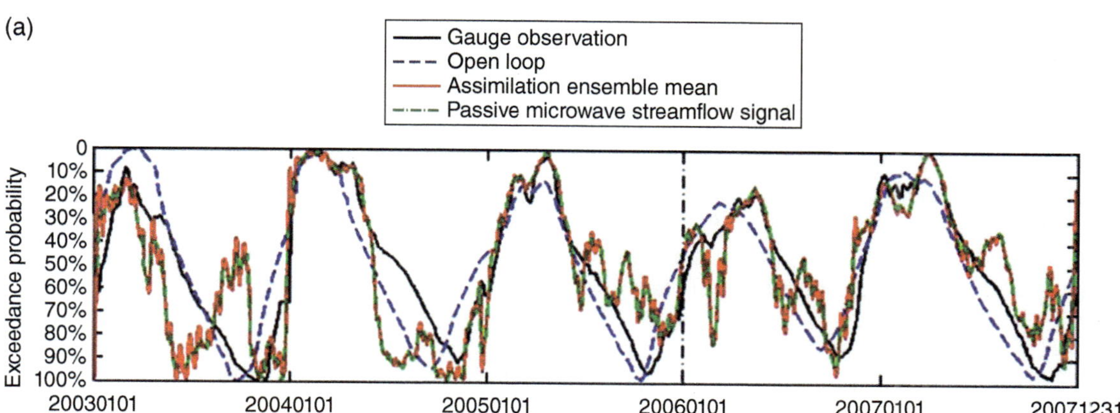

	Calibration		Validation	
	RMSE (%)	NSCE	RMSE (%)	NSCE
Open loop	27	0.77	27	0.81
Assimilation	36	0.61	31	0.69

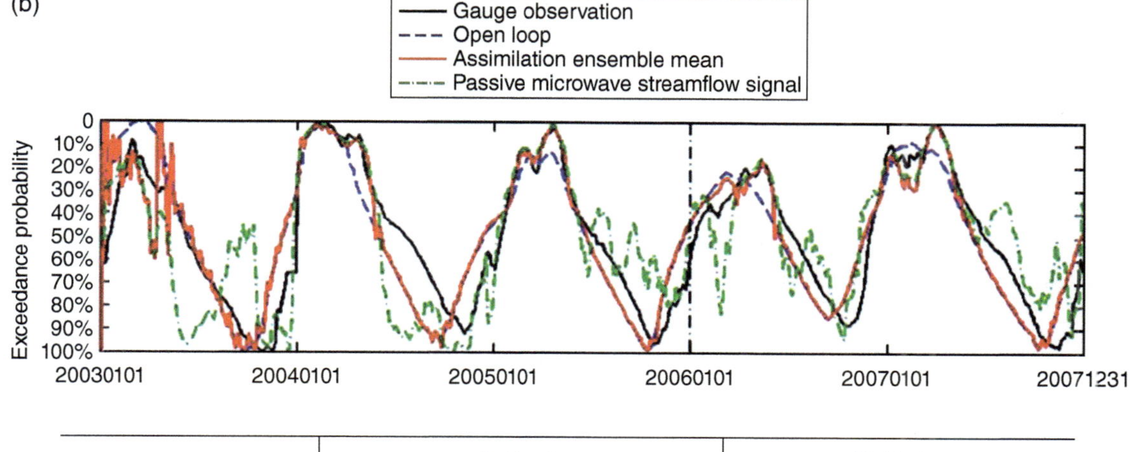

	Calibration		Validation	
	RMSE (%)	NSCE	RMSE (%)	NSCE
Open loop	27	0.77	25	0.81
Assimilation	26	0.79	23	0.84

Figure 27.6 Impact of assimilating Passive Microwave signal frequency into CREST in experiment 3: before threshold and (b) after threshold. The statistic indices in the table are calculted against the exceedance probability of observed streamflow.

frequencies, it is also interesting to explore the parameter values and ranges between those that were calibrated in the actual streamflow and frequency domains. Concerning parameter ranges, upper and lower bounds were set on all parameter values based on physical constraints and on past experience with those parameters that are more intangible. We found out that there are slight differences with the parameters PKE (multiplier to convert PET), PIM (the impervious area ratio), LEAKI (the interflow reservoir discharge multiplier), and PB (the exponent of the variable infiltration curve). However, there are large differences with the parameters PWM (the maximum soil water capacity), PFC (the soil saturated hydraulic conductivity) and LEAKO (the overland reservoir discharge multiplier). From our experience, those three parameters with the largest differences are the most sensitive parameters in this model, which control the peak volume and timing. In addition, when the parameter set calibrated from the frequency domain was applied into the actual streamflow domain, simulated streamflow showed strong overestimation compared to gauge observations. Nevertheless, the simulated streamflow and gauge observations were still well correlated, which indicates that the consistent overestimation in the actual domain does not impact its hydrological performance in the frequency domain.

Overall, the lumped CREST coupled with state estimation through an EnSRF approach can effectively improve flood prediction using remote sensing data alone in the Cubango River basin. A limitation, as mentioned by *Khan et al.* [2012] is that the use of AMSR-E signals for streamflow estimation is limited to medium- and large-scale basins. Moreover, the signal was found to be uncorrelated with observed streamflow during low-flow periods. These constraints must be considered when using the GFDS streamflow signals to infer streamflow for hydrologic model calibration and state estimation.

27.4. CONCLUSION

The application of remote sensing data, alone, to force, calibrate, and update a hydrologic model is a major contribution of this study. More generally, the approach developed and benchmarked herein can have great potential for predicting floods for the vast number of river basins throughout the world that are poorly gauged or even ungauged. In the Cubango River basin, data from an in situ streamflow gauge was used for model calibration and data assimilation in a traditional manner, providing a benchmark for evaluating the use of the passive microwave sensor-derived streamflow signals as a proxy for streamflow. Then, the passive microwave streamflow signals were converted into exceedance probability, i.e., in the frequency domain, to be applied similarly as the traditional approach for calibration and assimilation.

The major outcomes from this study are summarized as follows:

• In the absence of data assimilation (i.e., open loop), model performance was limited due to the inherent deficiencies of the model structure but was more likely dominated by bias in the rainfall forcing from the TRMM 3B42 RT algorithm.

• The implementation of the EnSRF in all experiments resulted in a significant improvement over the open loop simulations except experiment 3(a).

• When the GFDS streamflow signals converted to the frequency domain were substituted as the streamflow proxy for the open loop simulation in experiment 3(a), there was a significant reduction in model skill compared to using gauged streamflow in both actual and frequency domains during the calibration period, but there was a significant enhancement during the validation period. However, the assimilation of

the GFDS signals during the calibration period degraded the RMSE to 36% (from 27% for open loop) and the NSCE to 0.61 (from 0.77 for open loop), which was worse than the values in the reference experiments 1 and 2. This characteristic was found to be a result of poor sensitivity of the GFDS signal during low-flow periods.

• The final experiment 3(b) assimilated the AMSR-E signal only if the exceedance probability was < 30%; i.e., during high-flow periods. The application of this threshold resulted in model skill that was comparable to what was obtained in the reference experiment 1 but slightly worse than in experiment 2. Nonetheless, this reduction is quite modest considering experiment 3(b) is based entirely on remote sensing data, and this approach can be applied to those ungauged basin over the globe.

Given the real-time availability of satellite-based precipitation and AMSR-E and TMI-like passive microwave streamflow signal information, we argue that this work contributes to the decadal initiative of prediction in ungauged basins. Moreover, this study presents a potential paradigm shift in the use of streamflow exceedance probabilities, different from traditional methods reliant on in situ streamflow observation for calibration, and toward new techniques and new types of observations. These observations and new methods are particularly imperative for the vast sparsely gauged or ungauged basins around the world. More promisingly, assimilation of remote sensing information for improving hydrological prediction can be increasingly appreciated and supported by the current TRMM and anticipated GPM (Global Precipitation Mission, to be launched in early 2014), together with the future SMAP (Soil Moisture Active and Passive, to be launched in 2014). Both missions are anticipated to provide better precipitation and surface wetness estimates in terms of coverage, accuracy, and resolutions, which promises to further improve flood predictions in combination with the proposed framework in this study.

REFERENCES

Andersson, L., J. Wilk, M. C. Todd, D. A. Hughes, A. Earle, D. Kniveton, R. Layberry, and H. H. Savenije (2006), Impact of climate change and development scenarios on flow patterns in the Okavango River, *J. Hydrol.*, *331*(1), 43–57.

Aubert, D., C. Loumagne, and L. Oudin (2003), Sequential assimilation of soil moisture and streamflow data in a conceptual rainfall–runoff model, *J. Hydrol.*, *280*(1), 145–161.

Brakenridge, G. R., S. V. Nghiem, E. Anderson, and R. Mic (2007), Orbital microwave measurement of river discharge and ice status, *Water Resourc. Res.*, *43*(4), W04405.

Brakenridge, G. R., S. Cohen, A. J. Kettner, T. De Groeve, S. V. Nghiem, J. P. M. Syvitski, and B. M. Fekete (2012), Calibration of satellite measurements of river discharge using a global hydrology model, *J. Hydrol.*, *475*(0), 123–136.

Brocca, L., F. Melone, T. Moramarco, W. Wagner, V. Naeimi, Z. Bartalis, and S. Hasenauer (2010), Improving runoff prediction through the assimilation of the ASCAT soil moisture product, *Hydrol. Earth Syst. Sci.*, *14*(10), 1881–1893.

Brocca, L., T. Moramarco, F. Melone, W. Wagner, S. Hasenauer, and S. Hahn (2012), Assimilation of surface- and root-zone ASCAT soil moisture products into rainfall-runoff modeling, *IEEE Trans. Geosci. Remote Sens., 50*(7), 2542–2555.

Clark, M. P., D. E. Rupp, R. A. Woods, X. Zheng, R. P. Ibbitt, A. G. Slater, J. Schmidt, and M. J. Uddstrom (2008), Hydrological data assimilation with the ensemble Kalman filter: Use of streamflow observations to update states in a distributed hydrological model, *Adv. Water Resourc.*, *31*(10), 1309–1324.

Crow, W., and D. Ryu (2009), A new data assimilation approach for improving runoff prediction using remotely-sensed soil moisture retrievals, *Hydrol. Earth Syst. Sci.*, *13*(1), 1.

Crow, W., R. Bindlish, and T. Jackson (2005), The added value of spaceborne passive microwave soil moisture retrievals for forecasting rainfall-runoff partitioning, *Geophys. Res. Lett.*, *32*(18).

Duan, Q., S. Sorooshian, and V. K. Gupta (1994), Optimal use of the SCE-UA global optimization method for calibrating watershed models, *J. Hydrol.*, *158*(3–4), 265–284.

Gao, H., E. F. Wood, M. Drusch, and M. F. McCabe (2007), Copula-derived observation operators for assimilating TMI and AMSR-E retrieved soil moisture into land surface models, *J. Hydrometeorol. 8*(3), 413–429.

Hong, Y., K. L. Hsu, S. Sorooshian, and X. G. Gao (2004), Precipitation estimation from remotely sensed imagery using an artificial neural network cloud classification system, *J. Appl. Meteorol.*, *43*(12), 1834–1852.

Huffman, G. J., D. T. Bolvin, E. J. Nelkin, D. B. Wolff, R. F. Adler, G. Gu, Y. Hong, K. P. Bowman, and E. F. Stocker (2007), The TRMM Multisatellite Precipitation Analysis (TMPA): Quasi-global, multiyear, combined-sensor precipitation estimates at fine scales, *J. Hydrometeorol.*, *8*(1), 38–55.

Hughes, D. A., L. Andersson, J. Wilk, and H. H. Savenije (2006), Regional calibration of the Pitman model for the Okavango River, *J. Hydrol.*, *331*(1), 30–42.

Hughes, D., D. Kingston, and M. Todd (2011), Uncertainty in water resources availability in the Okavango River basin as a result of climate change, *Hydrol. Earth Syst. Sci.*, *15*(3), 931–941.

Joyce, R. J., J. E. Janowiak, P. A. Arkin, and P. P. Xie (2004), CMORPH: A method that produces global precipitation estimates from passive microwave and infrared data at high spatial and temporal resolution, *J. Hydrometeorol.*, *5*(3), 487–503.

Khan, S. I., H. Yang, H. J. Vergara, J. J. Gourley, G. R. Brakenridge, T. De Groeve, Z. L. Flamig, F. Policelli, and Y. Bin (2012), Microwave satellite data for hydrologic modeling in ungauged basins, *Geosci. Remote Sens. Lett., IEEE*, *9*(4), 663–667.

Kugler, Z., and T. D. Groeve (2007), The global flood detection system, JRC Sci. Tech. Rep., EUR 23303 EN.

Liang, X., D. P. Lettenmaier, E. F. Wood, and S. J. Burges (1994), A simple hydrologically based model of land surface water and energy fluxes for general circulation models, *J. Geophys. Res. Atmos.*, *99*(D7), 14,415–14,428.

Matgen, P., F. Fenicia, S. Heitz, D. Plaza, R. De Keyser, V. Pauwels, W. Wagner, and H. Savenije (2012), Can ASCAT-derived soil wetness indices reduce predictive uncertainty in well-gauged areas? A comparison with *in situ* observed soil moisture in an assimilation application, Adv. *Water Resourc*, *44*, 49–65.

McCarthy, J. M., T. Gumbricht, T. McCarthy, P. Frost, K. Wessels, and F. Seidel (2003), Flooding patterns of the Okavango wetland in Botswana between 1972 and 2000, *AMBIO J. Human Environ.*, *32*(7), 453–457.

Milzow, C., L. Kgotlhang, P. Bauer-Gottwein, P. Meier, and W. Kinzelbach (2009a), Regional review: The hydrology of the Okavango Delta, Botswana—Processes, data and modelling, *Hydrogeol. J.*, *17*(6), 1297–1328.

Milzow, C., L. Kgotlhang, W. Kinzelbach, P. Meier, and P. Bauer-Gottwein (2009b), The role of remote sensing in hydrological modelling of the Okavango Delta, Botswana, *J. Environ. Manag.*, *90*(7), 2252–2260.

Milzow, C., P. E. Krogh, and P. Bauer-Gottwein (2011), Combining satellite radar altimetry, SAR surface soil moisture and GRACE total storage changes for hydrological model calibration in a large poorly gauged catchment, *Hydrol. Earth Sys. Sci.*, *15*(6), 1729–1743.

Pauwels, V. R., and G. J. De Lannoy (2006), Improvement of modeled soil wetness conditions and turbulent fluxes through the assimilation of observed discharge, *J. Hydrometeorol.*, *7*(3), 458–477.

Pauwels, V., R. Hoeben, N. E. Verhoest, F. P. De Troch, and P. A. Troch (2002), Improvement of TOPLATS-based discharge predictions through assimilation of ERS-based remotely sensed soil moisture values, *Hydrol. Process.*, *16*(5), 995–1013.

Salvia, M., F. Grings, P. Ferrazzoli, V. Barraza, V. Douna, P. Perna, C. Bruscantini, and H. Karszenbaum (2011), Estimating flooded area and mean water level using active and passive microwaves: The example of Parana River Delta floodplain, *Hydrol. Earth Syst. Sci.*, *15*(8), 2679–2692.

Sauer, V. B., and R. W. Meyer (1992), Determination of error in individual discharge measurements, USGS Open-File Rep. 92–144. http://pubs.usgs.gov/of/1992/ofr92-144/pdf/ofr92-144.pdf.

Sorooshian, S., K.-L. Hsu, G. Xiaogang, H. V. Gupta, B. Imam, and D. Braithwaite (2000), Evaluation of PERSIANN system satellite-based estimates of tropical rainfall, *Bull. Am. Meteorol. Soc.*, *81*(9), 2035–2046.

Tarpanelli, A., L. Brocca, T. Lacava, F. Melone, T. Moramarco, M. Faruolo, N. Pergola, and V. Tramutoli (2013), Toward the estimation of river discharge variations using MODIS data in ungauged basins, *Remote Sens. Environ.*, *136*, 47–55.

Temimi, M., R. Leconte, F. Brissette, and N. Chaouch (2007), Flood and soil wetness monitoring over the Mackenzie River Basin using AMSR-E 37GHz brightness temperature, *J. Hydrol.*, *333*(2–4), 317–328.

Temimi, M., T. Lacava, T. Lakhankar, V. Tramutoli, H. Ghedira, R. Ata, and R. Khanbilvardi (2011), A multi-temporal analysis of AMSR-E data for flood and discharge monitoring during the 2008 flood in Iowa, *Hydrol. Process.*, *25*(16), 2623–2634.

Turk, F. J., and S. D. Miller (2005), Toward improved characterization of remotely sensed precipitation regimes with MODIS/AMSR-E blended data techniques, *IEEE Trans. Geosci. Remote Sens.*, *43*(5), 1059–1069.

Wang, J., et al. (2011), The coupled routing and excess storage (CREST) distributed hydrological model, *Hydrol. Sci. J.*, *56*(1), 84–98.

Whitaker, J. S., and T. M. Hamill (2002), Ensemble data assimilation without perturbed observations, *Monthly Weather Rev.*, *130*(7), 1913–1924.

Zhang, Y., Y. Hong, X. Wang, J. J. Gourley, J. Gao, H. J. Vergara, and B. Yong (2013), Assimilation of passive microwave streamflow signals for improving flood forecasting: A first study in Cubango River Basin, Africa, *IEEE J. Select. Topics Appl. Earth Observ. Remote Sens.*, *PP*(99), 1–16.

<h1 style="text-align:center">28</h1>

Application of High-Resolution Images from Unmanned Aircraft Systems for Watershed and Rangeland Science

A. Rango,[1] E. R. Vivoni,[2,3] C. A. Anderson,[3] N. A. Pierini,[3] A. Schreiner-McGraw,[2]
S. Saripalli,[2] A. Slaughter,[1] and A. S. Laliberte[4]

28.1. INTRODUCTION

Much information has been in the news in recent years about the use of unmanned aircraft systems (UAS) in border security, antiterrorist operations, and military reconnaissance. Civil applications of UAS are on the rise as well, especially for assisting law enforcement personnel in their normal duties. Although most civil applications of UAS are derived from the military establishment, the civil applications are growing at a rapid pace. Here, we present a summary and illustrations of civil applications of UAS for watershed and rangeland science at the Jornada Experimental Range, New Mexico.

28.2. HISTORY OF UAS DEVELOPMENT FOR CIVIL APPLICATIONS

As recently as 2004, only about 2% of the 2400 unmanned aircraft systems (UAS) were operating in the civil market [*Newcome*, 2004]. The first UAS to take photography for aerial mapping purposes was the Radioplane in 1955. Many civilian applications have the need for an aerial map base, and the aerial photos can be used as a baseline that can be used for change detection purposes. Regular land cover assessment allows detection of

changes that are important for land management agencies such as the Bureau of Land Management (BLM) and the Natural Resources Conservation Service (NRCS) as well as similar state agencies. Additional changes in infrastructure such as dams and irrigation canals can also be detected, especially when UAS digital photography is employed with decimeter resolution. Agencies in charge of infrastructure seldom get to use temporal images of these structures, so they can be evaluated if safety concerns might be an issue.

Forest applications of UAS have become of widespread interest and include forest fire monitoring and recovery and forest condition assessment [*Horcher and Visser*, 2004; *Ambrosia et al.*, 2003; *Zajkowski*, 2003]. Forests in the Mediterranean region have been mapped with UAS [*Dunford et al.*, 2009]. Precision agriculture is another possible application with UAS. Hyperspatial digital photography has been conducted in Hawaii to predict coffee bean ripeness [*Herwitz et al.*, 2003] and in California to map crop vigor in vineyards [*Johnson et al.*, 2003]. Another contribution of UAS to precision agriculture is the use of miniature helicopters for targeted spraying of crops [*Schrage et al.*, 1999]. Multiple sensors such as visible, thermal, and multispectral were also used on an unmanned helicopter for precision viticulture applications [*Turner and Lucieer*, 2011].

Land located in the western United States is usually dominated by nonurbanized and noncultivated areas and are often called wildlands. These wildlands are also prime areas for UAS flights because of lack of population, larger distances between airports, and low precipitation amounts leading to fewer cultivated areas. Additionally, because of these characteristics, these areas generally

[1] *USDA-ARS Jornada Experimental Range, Las Cruces, New Mexico, USA*

[2] *School of Earth and Space Exploration, Arizona State University, Tempe, Arizona, USA*

[3] *School of Sustainable Engineering and Built Environment, Arizona State University, Tempe, Arizona, USA*

[4] *Earthmetrics, Brownsville, Oregon, USA*

Remote Sensing of the Terrestrial Water Cycle, Geophysical Monograph 206. First Edition. Edited by Venkat Lakshmi.
© 2015 American Geophysical Union. Published 2015 by John Wiley & Sons, Inc.

consist of undeveloped drainage basins and rangelands. So, the most logical air space in which to fly UAS is also one of the areas of the country most likely to benefit from the developing technology of UAS. In fact two important applications of UAS, watershed and rangeland studies, require many of the same input variables for hydrologic modeling and rangeland health calculations. These applications require location information about types of vegetation, percent cover of vegetation versus bare soil, and susceptibility to erosion (soil slopes and types). Although the outputs from hydrological and rangeland health calculations are put to different uses, the products potentially provided by UAS for these two applications are related, i.e., the same land areas are used for hydrologic modeling and rangeland health assessment.

Both watershed and rangeland studies require digital elevation model (DEM) data that can be derived from the UAS stereo aerial photography. Both types of studies require the classification of vegetation type, bare soil, and the relative positioning of these two types of land cover. Hydrologic modeling requires the identification of different types of vegetation and their relationship to stream channels and contributing areas. Rangeland health assessment requires measurement of bare soil gaps between vegetation patches.

The resolution necessary for these applications needs to be on the order of 5–10 cm as determined from tests with rangeland and watershed scientists. Because the image area from the digital camera is small due to low-altitude flights at 215 m, both orthorectification and mosaicking were necessary to overcome the inherent instability of a small UAS platform [*Laliberte et al.*, 2011]. In the following, we summarize and illustrate some of our efforts in using UAS for watershed and rangeland studies.

28.3. LOCATION OF STUDY AREAS

28.3.1. Jornada Experimental Range

The Jornada Range Reserve was created when President Taft signed an Executive Order on 4 May 1912. This experimental area was created from withdrawn public domain land north of Las Cruces, New Mexico, and totals 783 km². The ranch headquarters is 29 km north of New Mexico State University and Las Cruces (Figure 28.1a). Data on the grazing intensity was begun in 1912, and the collection of climate data was started in 1915. When created in 1912, the Jornada Range Reserve was established within the U.S. Department of Agriculture (USDA) Bureau of Plant Industry. In 1915, the Jornada Range Reserve was transferred to the U.S. Forest Service. In 1927 the Jornada Range Reserve was

renamed the Jornada Experimental Range (JER). In 1954, JER was transferred to the Agricultural Research Service (ARS). Since that time, JER has received designation as a Biosphere Reserve in UNESCO's Man and the Biosphere Program and as an Ecological Reserve by the Institute of Ecology. JER was also one of the original sites of the National Science Foundation's Long Term Ecological Research (LTER) program. Most recently, JER became a site in the National Ecological Observation Network (NEON).

Although formal JER measurements were begun in 1912, earlier baseline vegetation data were acquired using General Land Office records of the JER area taken in 1858. As an example of the possibility of using these data in the Jornada basin, *Buffington and Herbel* [1965] reconstructed vegetation changes in the area between 1858 and 1963. The period of record of vegetation change has now been extended to 140 years, using ground surveys and remote sensing data [*Gibbens et al.*, 2005]. The remote sensing database at JER includes over 6000 aerial photos from 1936 to the present. In addition, we have developed a UAS program that acquires hyperspatial imagery. We now have more than 90,000 images taken by UAS, at approximately 6 cm resolution, and about 200 image mosaics. Supplementing this imagery we have numerous satellite scenes of JER that include Landsat, ASTER, SPOT, IKONOS, QuickBird, Worldview, and GeoEye.

As mentioned previously, ground-based measurements have been made in both the LTER and NEON programs. Other national ground-based networks represented at JER include the NRCS Soil Climate Analysis Network (SCAN), the National Oceanic and Atmospheric Administration (NOAA) Climate Reference Network (CRN), and the COsmic-ray Soil Moisture Observing System (COSMOS) sites.

The JER has perfected UAS flight procedures to acquire hyperspatial images. In the United States, there are two major types of airspace in which to fly. The Federal Aviation Administration (FAA) regulates flights in the National Airspace System (NAS), and various military bases regulate flights in Restricted Military Airspace (RMA). Regulations differ between these two types of airspace. Fortunately at JER, approximately 50% is in NAS and 50% is in RMA (Figure 28.1b). As a result, we have learned to fly UAS efficiently in either airspace and how to acquire the necessary permissions.

28.3.2. Use of UAS at the Jornada Experimental Range

The UAS have flown at the JER numerous times in both the NAS and RMA. Both fixed-wing and rotary-wing UAS have been employed. The fixed-wing aircraft

(a)

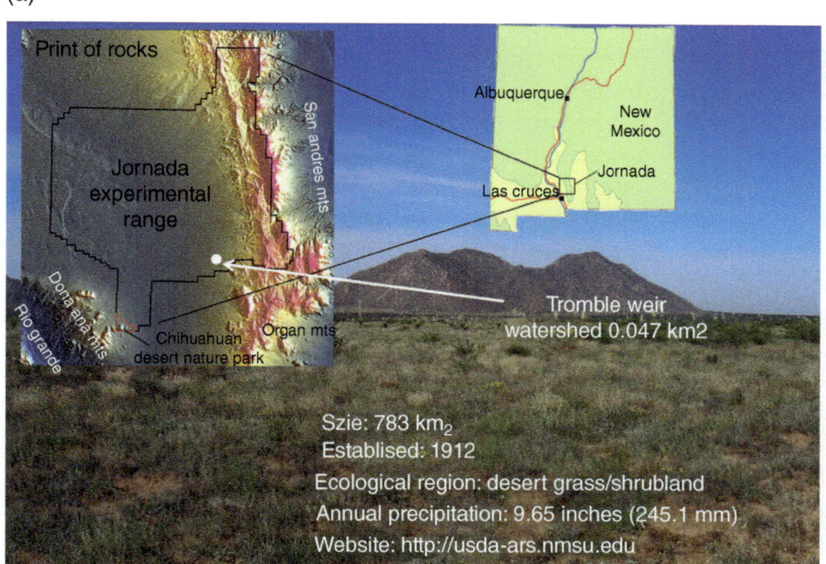

(b)

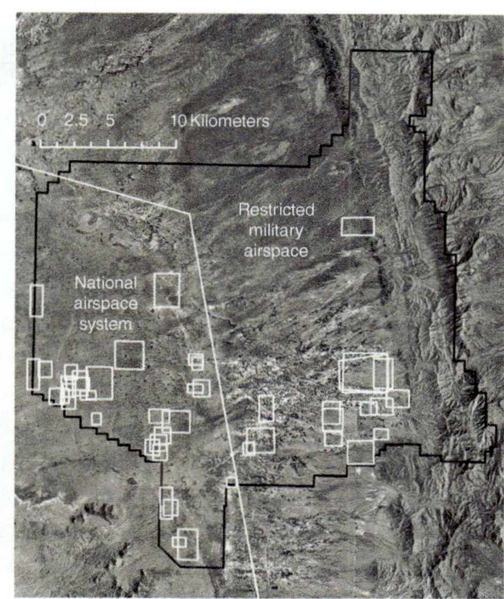

(c)

(d)

Figure 28.1 (a) Location of study site within the Jornada Experimental Range (JER) and the state of New Mexico. (b) National and restricted airspace boundaries over JER, with white rectangles representing the areas cf JER where UAS data has been collected. (c) Photograph of fixed-wing Bat-3 UAS. (d) Photograph of Rotomotion SR 30 VTOL helicopter UAS.

that have been used were two different models, both made by the MLB Co. in Mountain View, California. Our first system was the Bat-3 that has a 1.8 m wingspan and weighs 10 kg with a payload of 1.4 kg. It has a flight duration of 2–5 h (Figure 28.1c). The sensors it carries are a Canon SD900 digital camera, a Tetracam 6-band multispectral camera, and a daylight video camera.

The second system is the Bat-4 with a wingspan of 4 m, weight of 45 kg, with a payload of 14 kg. Whereas the Bat-3 takes off from a catapult, the Bat-4 takes off on wheels from a runway. At the moment, there is room for additional sensors on the Bat-4. It currently carries a Canon EOS 5D Mark II 21mp still camera, the Tetracam 6 band multispectral camera, and a TASE 150 FCB-EX 1020 color video camera. Flight duration is up to 6 h. Both the Bat-3 and Bat-4 can provide overlapping imagery that can be used for stereo analysis for DEMs.

Two different helicopters have been tested at Jornada. The first was the Yamaha R-Max Research Helicopter and the second one was the Rotomotion SR 30 VTOL (Figure 28.1d). In general the fixed-wing aircraft are a more stable data collection platform. The rotary-wing aircraft are more maneuverable and possess the capability to provide a variable horizontal resolution by vertical precision flying over known targets at different altitudes.

28.4. FLYING IN THE FAA-NAS

To fly in the NAS, an application for a Certificate of Authorization (COA) must be made to the FAA. The UAS to be flown must be owned by a public government entity. Since a decision by FAA on the COA application can take 3–4 months, this time period must be built into the preliminary planning for UAS flights. Once the COA is approved, the UAS flight team must abide by the details of the COA. Jornada UAS have flown under FAA-COAs in New Mexico, Idaho, and Arizona. The same Jornada UAS have flown in RMA in New Mexico and Alaska.

External pilots, who stay within line-of-sight distance of the airplane and can take over flying the airplane through radio control in an emergency, and internal pilots, who are the computer controllers of the airplane flights, must either be licensed private pilots or at a minimum take all the training, testing, and medical exams as licensed pilots. FAA will decide if a COA application will require that external or internal pilots possess a manned pilots license in order to be the pilot in command (PIC) of a UAS mission. Additionally, for the UAS mission, the PIC is responsible for filing a Notice to Airmen 48 hours before the UAS mission just as if they were flying a manned aircraft. During the UAS mission, observers must be posted around the target UAS flight area. These observers remain in contact with UAS pilots (both external and internal) and need the same requirements as the external and internal pilots. Both pre and postflight briefings will be conducted by the PIC. If incidents with the UAS occur, they will be reported to FAA. Reliability of the UAS must be checked before and after flights, and logs must be kept to assure that records can be provided to FAA upon request. The flight team members must maintain their UAS flight proficiency by flying the UAS every 90 days at a minimum.

Because the current regulations for flying in FAA-NAS prohibits sending UAS out of line-of-sight distance from the external pilot, the maintenance of line-of-sight from the external pilot to the UAS is a critical factor to both the FAA and a UAS flight team. In our experiments, we have determined that this line-of-sight distance between the external pilot and the UAS is approximately 1.1 km [*Rango and Laliberte*, 2010], depending upon environmental conditions and size of the UAS (in this case Bat-3, shown in Figure 28.1c, has a wing span of 1.8 m). In order to keep the UAS in line-of-sight and to avoid numerous launches and landings, the Jornada flight team has developed an approach that transports the external pilot and any necessary observers or flight crew members by surface vehicle along the path of the UAS [*Rango et al.*, submitted]. Because the UAS and the surface vehicle traveling over rough terrain can go at similar speeds, we have devised a long-distance, continuous pilot move method. This is an essential point for operational UAS missions.

The UAS is launched and placed into a circle hold pattern near the ground control station or home location. The external pilot and observer are picked up by the transport vehicle and driven a short distance (0.4 km) beyond the location of the initial circle hold. A flight team member in the transport vehicle calls by radio to the internal pilot to send the UAS to the next circle hold, and the transport vehicle travels along the road at approximately the same speed as the UAS while keeping within the required line-of-sight distance. By keeping the external pilot slightly ahead of the UAS, both the external pilot and an accompanying observer can see the UAS while keeping their "backs to the wind." In order to keep the UAS away from the sun in the external pilot's field of view, the UAS flight path is planned to be north of the transport vehicle route (in the Northern Hemisphere). Intermediate circle holds are placed along the flight path. When the UAS is in a circle hold, the internal pilot can send it to the next point as soon as the external pilot requests a move. This continues until the pilot reaches the image acquisition area when the UAS is put into a paint pattern to acquire photography. On completion of image acquisition, the UAS is sent to the final circle hold in the continuous move flight pattern. Flight data is downloaded and the transport vehicle moves into position for return to the home location.

28.5. APPLICATIONS TO WATERSHED AND RANGELAND SCIENCE AT THE JORNADA TROMBLE WEIR SITE

Because 6 cm resolution data from the Bat-3 UAS were not previously available, the new 1 m digital elevation model provides much improved fluvial geomorphic characteristics for the small Tromble Weir watershed, which are summarized in Table 28.1 and illustrated in Figures 28.2b–28.2d. Prior investigators using a Global Positioning System (GPS) unit conducted a walking survey of the basin and recorded GPS coordinates of what they determined to be the watershed boundary. The UAS-derived DEM was used to determine the basin boundary automatically using geographical information system (GIS) software. The walking ground survey yielded a drainage area of the Tromble Weir watershed of 56,988 m² (Figure 28.2a). The UAS basin boundary is shown in Figure 28.2b and the drainage area is 46,734 m². Additionally, the shape of the drainage basin and stream channels is much more realistic using the UAS approach as confirmed by visual inspection of the drainage network in the field. There is a 22% difference in the drainage areas between the traditional and UAS methods. Additionally, the drainage density is different, 0.013/m for the ground survey approach and 0.032/m for the UAS DEM approach which indicates a more detailed drainage

Table 28.1 Characteristics of the watershed terrain and hydrography derived from the UAS-based digital elevation model at 1 m resolution

Area (m²)	Elevation (m²)			Slope (deg)			Aspect (% of watershed area)			Main Channel Slope (m/m)	Main Channel Length (m)	Drainage Density (1/m)	T_c (Min)
	Min	Mean	Max	Min	Mean	Max	North	South	West				
46,734	1450.5	1458.3	1467.5	0	3.93	45	39.2	39.1	20.3	0.039	286	0.03	5.3

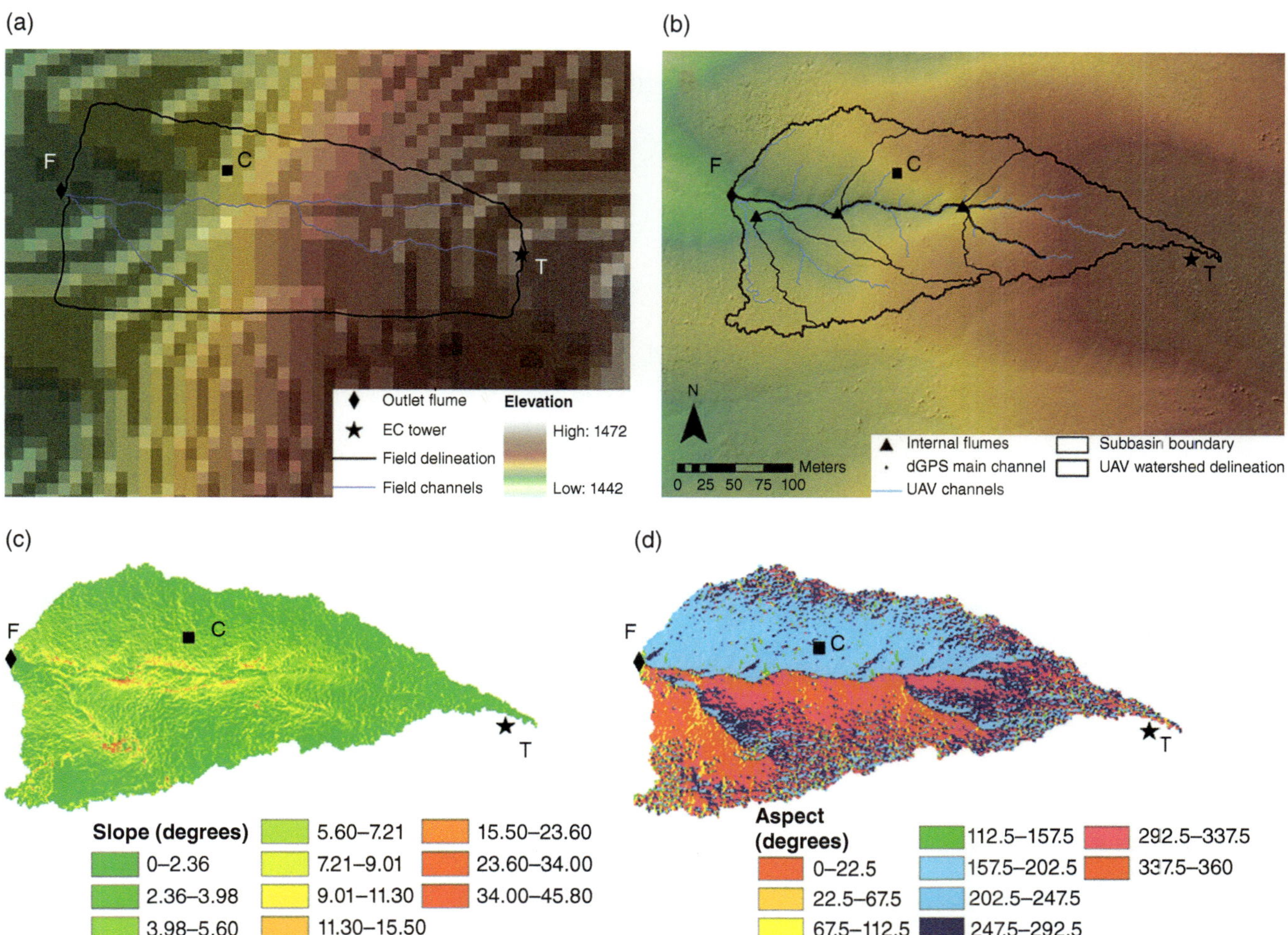

Figure 28.2 Terrain and hydrologic analyses from UAS-derived digital elevation model (DEM). (a) 10 m DEM obtained from U.S. Geological Survey topographic map along with a field-derived watershed boundary and channel network prior to the acquisition of UAS products. (b) UAS-derived 1-m DEM and the resulting watershed boundary, subbasin boundaries, and channel network. A dGPS survey was conducted to verify the main channel. The location of the EC tower (T), COSMOS probe (C), and outlet flume (F) are shown as well as three internal flumes. (c) Watershed slope (degrees) obtained from the UAS DEM at 1 m resolution. (d) Watershed aspect (degrees) obtained from the UAS DEM at 1 m resolution.

network, as was also confirmed in the field. The 1 m DEM was used to calculate the time of concentration (T_c) for the Tromble Weir watershed using the Kirpich equation [Kirpich, 1940], resulting in 5.3 min, which matches fairly well with estimates based on field observations of rainfall and runoff [*Templeton et al.*, in review]. This indicates that the fast watershed response can be estimated from UAS products. Because the individual UAS images are 213 m × 160 m, the single frames need to be mosaicked to provide complete coverage of a drainage basin. The Jornada UAS team has developed their own mosaicking method [*Laliberte et al.*, 2011]. Mosaics are produced each time the Tromble Weir watershed is flown on a quarterly basis.

Classification of the vegetation in the Tromble Weir watershed yielded some interesting results that were not initially expected. First, it was expected that the dominant vegetation type would either be creosote or mesquite shrubs, but it turned out that the shrub mariola was dominant within the watershed, as shown in Figure 28.3 and Table 28.2. The UAS vegetation classification was used to estimate that 34% of the basin was covered by vegetation and 66% of the area by bare soil, consistent with line-intercept vegetation transects reported by *Templeton et al.* (in review). Table 28.2 shows the percent cover by vegetation type for the entire watershed at the outlet flume, for the footprint of the eddy covariance tower [*Baldocchi*, 2003] and for the COSMOS probe footprint [*Zreda et al.*, 2012], as illustrated in Figure 28.4. Each sensor has a different footprint within the landscape, which contribute to the aggregation of information measured by the device (e.g., runoff upstream of the flume, evapotranspiration

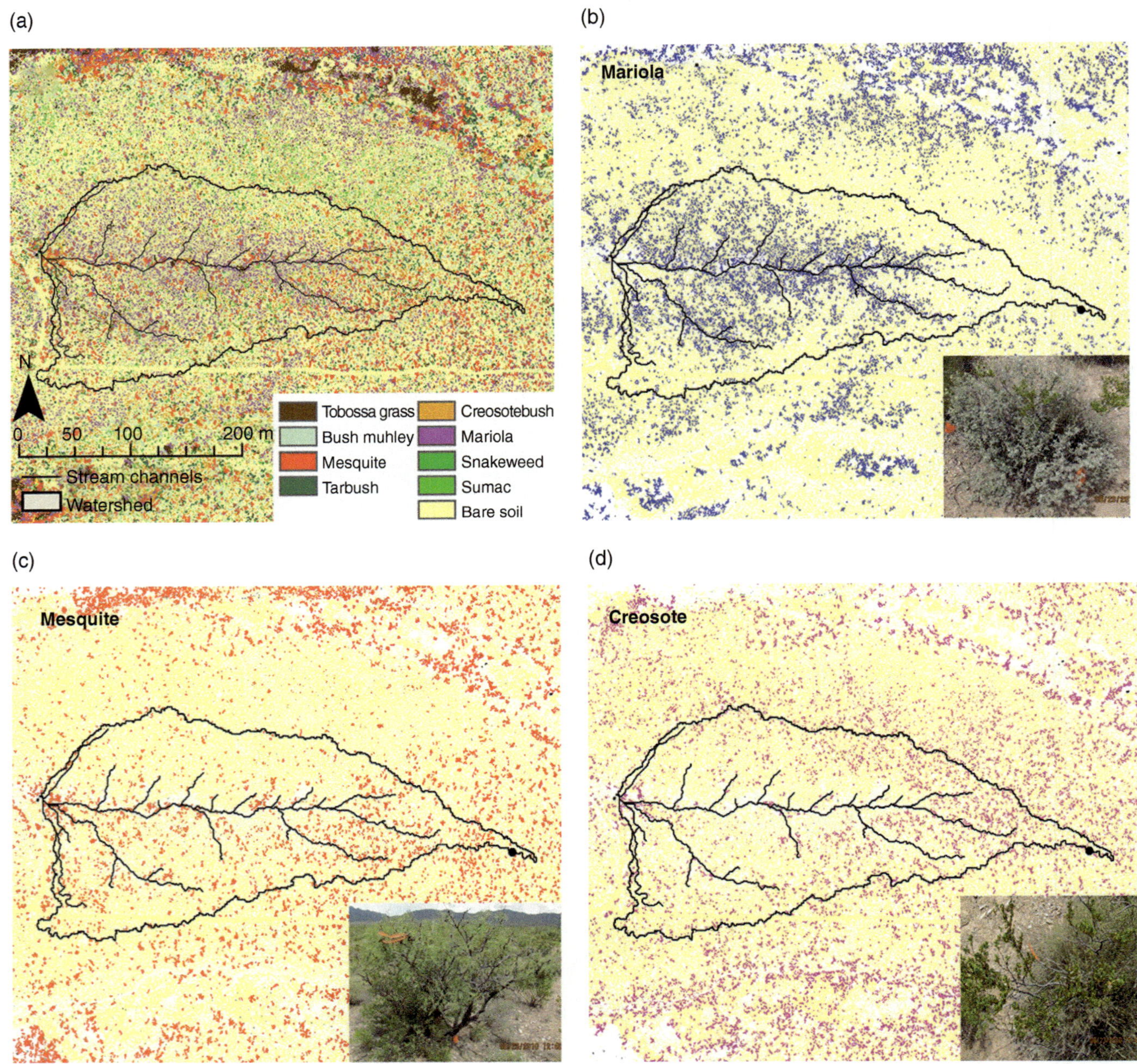

Figure 28.3 (a) Species-specific vegetation classification map for the region near the study watershed derived from the fixed-wing UAS imagery. The spatial distributions of individual species are shown for (b) *Parthenium incanum* (mariola), (c) *Prosopis glandulosa* (mesquite), and (d) *Larrea tridentata* (creosote bush). Insets show a field photograph of each shrub type.

Table 28.2 Land cover percentages for three regions derived from the UAS-based vegetation classification

Classification	Watershed Coverage (%) Area = 46,700 m²	EC Footprint Coverage at 50% Contribution (%) Area = 17,500 m²	COSMOS Footprint Coverage at 50% Contribution (%) Area = 34,600 m²
Bare soil	65.95	67.14	67.00
Parenthium incanum (mariola)	11.94	3.84	6.20
Prosopis glandulosa (mesquite)	6.47	8.92	8.30
Larrea tridentata (creosote bush)	5.82	7.79	7.21
Muhlenbergia porteri (bush muhly)	2.89	4.62	4.17
Flourensia cernua (tar bush)	2.48	4.85	4.57
Gurierrezia sarothrae (snakeweed)	1.82	1.46	1.34
Pleuraphis mutica (tobosa grass)	1.4	0.53	0.69
Rhus sp. (sumac)	1.15	0.52	0.55
Sporobolus airoides (alkali sacaton)	0.04	0.1	0.00
No data	0.02	0.23	0.00

Note: See Figure 28.4 for a map of the regions.

upwind of the eddy covariance (EC) tower, and soil moisture near the COSMOS station). Because these regions are relatively close together, they would traditionally be assumed to have similar vegetation cover. However, it is apparent from the UAS data that this is not the case with variations in the percentage of each vegetation type and in the dominance of individual species. By knowing the land cover in a particular sensor footprint, it is possible to improve the interpretation of the sensor data collected in each footprint, rather than assuming that the entire watershed or area of interest is homogeneous. When the vegetation types were superimposed over hillslopes and soil types, it was discovered that creosote and mesquite tend to grow on relatively flat deep soils. Mariola on the other hand tends to grow on steeply sloping hillslopes with shallow soils. Such information is important for detailed hydrologic and rangeland models that consider landscape characteristics. The use of UAS remote sensing on such small basins is required for land cover management because satellites such as Landsat, QuickBird, and Worldview cannot provide the hyperspatial resolution necessary for decision making.

By using UAS to monitor the normalized difference vegetation index (NDVI) several times a year, we can observe changes in vegetation growth. The acquisition of images in different seasons, such as during the wet monsoon in contrast to the dry spring at the Jornada allows us to capture differences in vegetation greenness. Figure 28.5 provides an example of a vegetation index change due to heavy monsoon rainfall at the Jornada. The UAS approach with 6 cm resolution allows very small basins or study areas to be monitored throughout the year.

The stereo photographic measurements obtained by the UAS can be used to develop a DEM as mentioned earlier, but they can also be used to obtain a Canopy Height Model (CHM). Such a model has been used to obtain shrub heights at Jornada. As shown in Figure 28.6, the approach has been evaluated using shrub heights obtained from field measurements and from the image-derived CHM. These results provide promise that UAS data can be used to estimate shrub canopy height and to monitor their changes with time.

The 6 cm resolution images from UAS overflights provide new hyperspatial data important for hydrology and rangeland science. For many applications it appears that satellite and manned aircraft photography are not sufficient despite resolution in the 40 cm to 1 m range. The 5 – 6 cm resolution opens up new possibilities for data input for hydrologic modeling and rangeland health calculations. Other forms of field data acquired through line-point-intercept surveys have a better resolution than the UAS imagery, but they are often more costly and many times not enough trained personnel are available. In a comparison of UAS and ground approaches [*Laliberte et al.*, 2010], the UAS measurements were found to save time and expense when more than eight 50 m × 50 m plots are measured. When changes with time need to be assessed, the UAS, with up-to-date GPS systems, can guarantee returning to the same exact location to duplicate earlier measurements.

28.6. CONCLUSION

The UAS provide a new way to acquire hyperspatial data with a resolution of 6 cm that has not been available in the past. This hyperspatial data can be used to obtain detailed 1 m DEMs, mosaics of entire watersheds, detailed vegetation classification of bare soils and vegetation type, and input to models in watershed and rangeland science. Because of improved UAS guidance systems, UAS can be used to repetitively cover the same areas for change detection. UAS can provide frequent and affordable aerial coverage of study areas with high resolution data to fill in gaps in ground observation networks and between satellite overpasses.

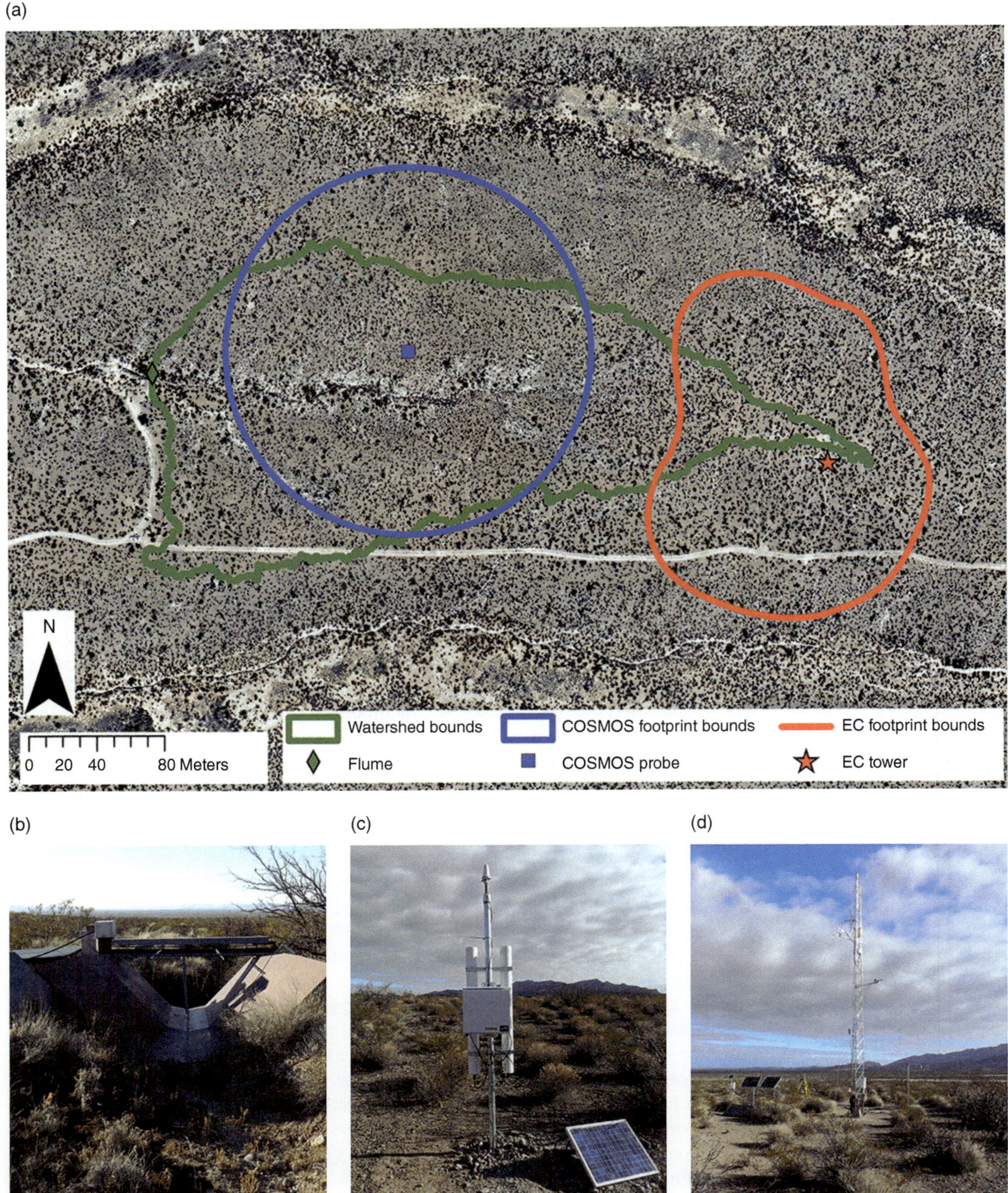

Figure 28.4 (a) Instrument locations and associated contributing areas relevant to their measurements overlaid on UAS imagery. The outlined footprints of the COSMOS and EC tower describe the 50% contributing areas. The land cover within each contributing area is listed in Table 28.1. Photographs of the instruments featured: (b) watershed outlet flume, (c) COSMOS probe, and (d) EC tower.

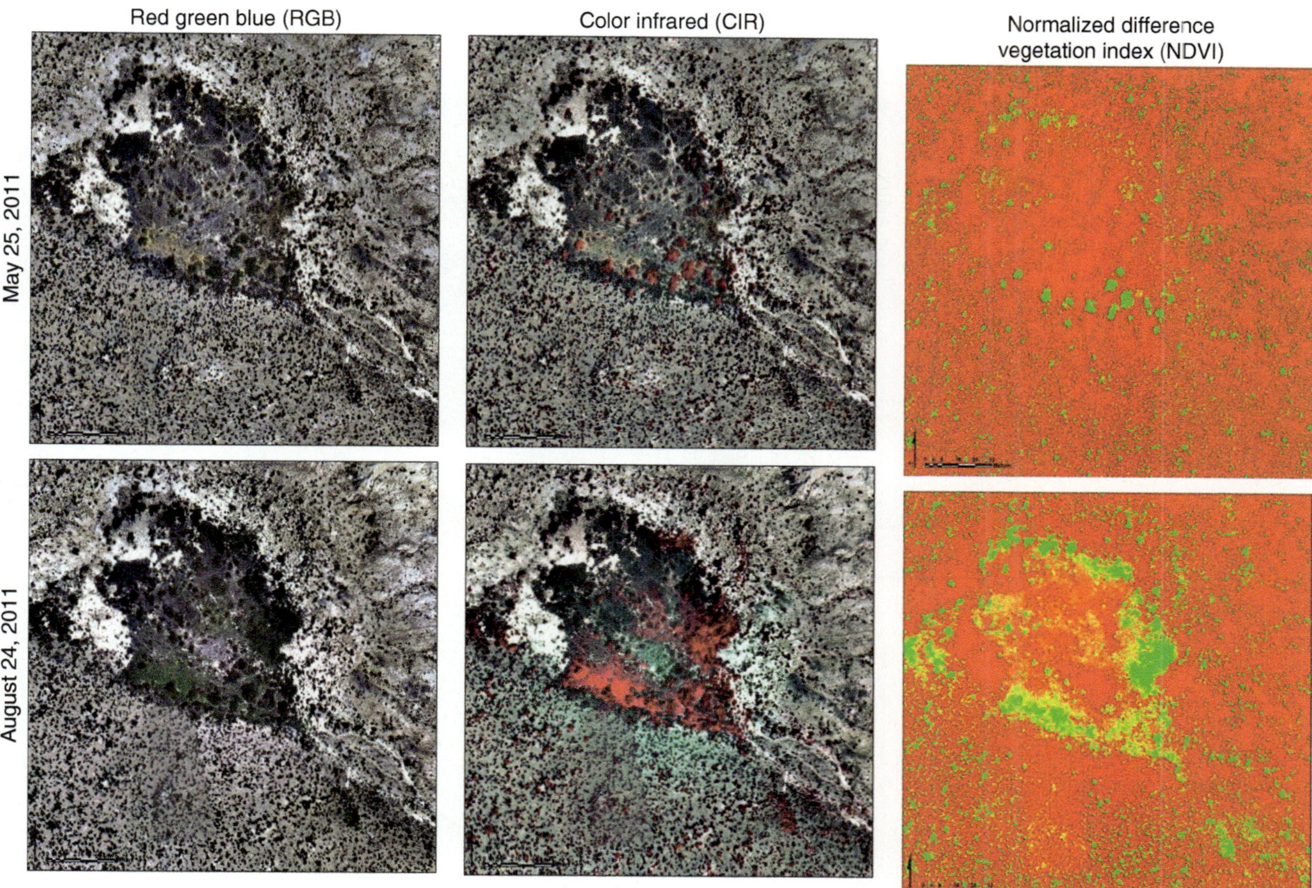

Figure 28.5a The re-green-blue (RGB) images present a view of a widening section or bead of a runoff channel flowing southeast to northwest. shrubs and grasses retard the runoff and benefit from increased soil moisture.

Figure 28.5b The color infrared (CIR) image shown the reflected near-infrared band (825–925 nm). In this image infrared reflectance is presented in red and visible data is in green and blue. Infrared wavelengths are strongly reflected by chlorophyll in living plant material so growing plants show up as red in a CIR image.

Figure 28.5c NDVI is a wicely used vegetation index. In this image green is a strong vegetation index response (0.4–1.0), yellow is moderate (0.2–0.4), and red is low to no vegetation (0.0–0.2). In the May image we see that only the shrubs show active chlorophyll. By late August the channel bed has filled in with grasses and annuals in response to the August rains.

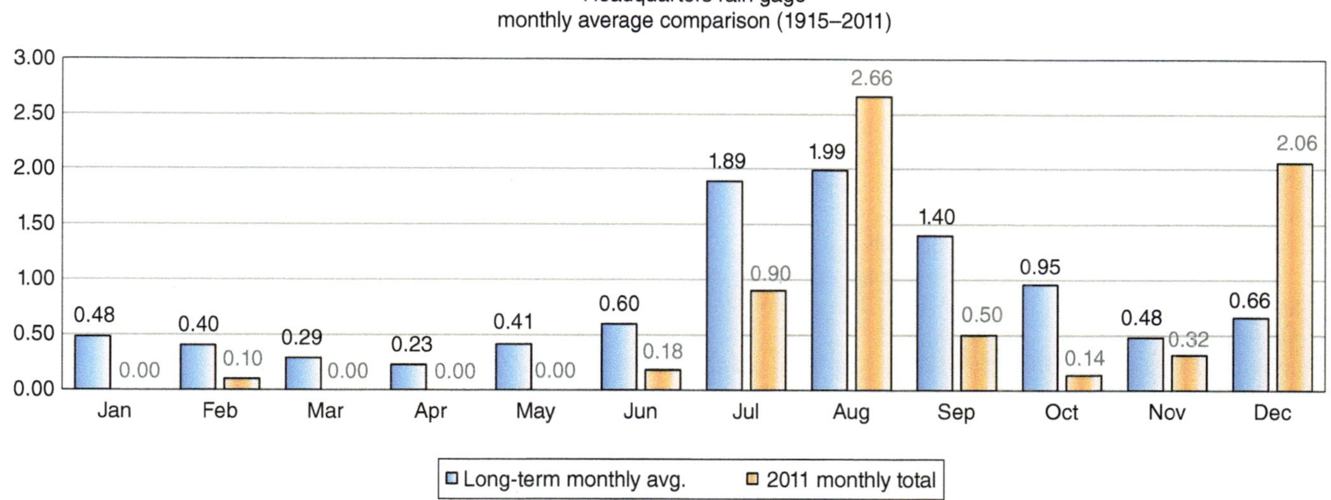

Figure 28.5d The upper series of images in Figures 28.5a–c are derived from data acquired on 25 May 2011. As shown in the rain gauge chart (d), May of 2011 was the third consecutive month without rain and followed 5 months with only 0.10″ of rain. The lower series of images are derived from data acquired on 24 August 2011. As shown in the rain gauge chart, August of 2011 has above average rainfall at 2.66″ in addition to an inch of rain in the preceding 2 months. The image data were acquired toward the end of August and shows the result of the recent rainfall (1″ = 25.4 mm).

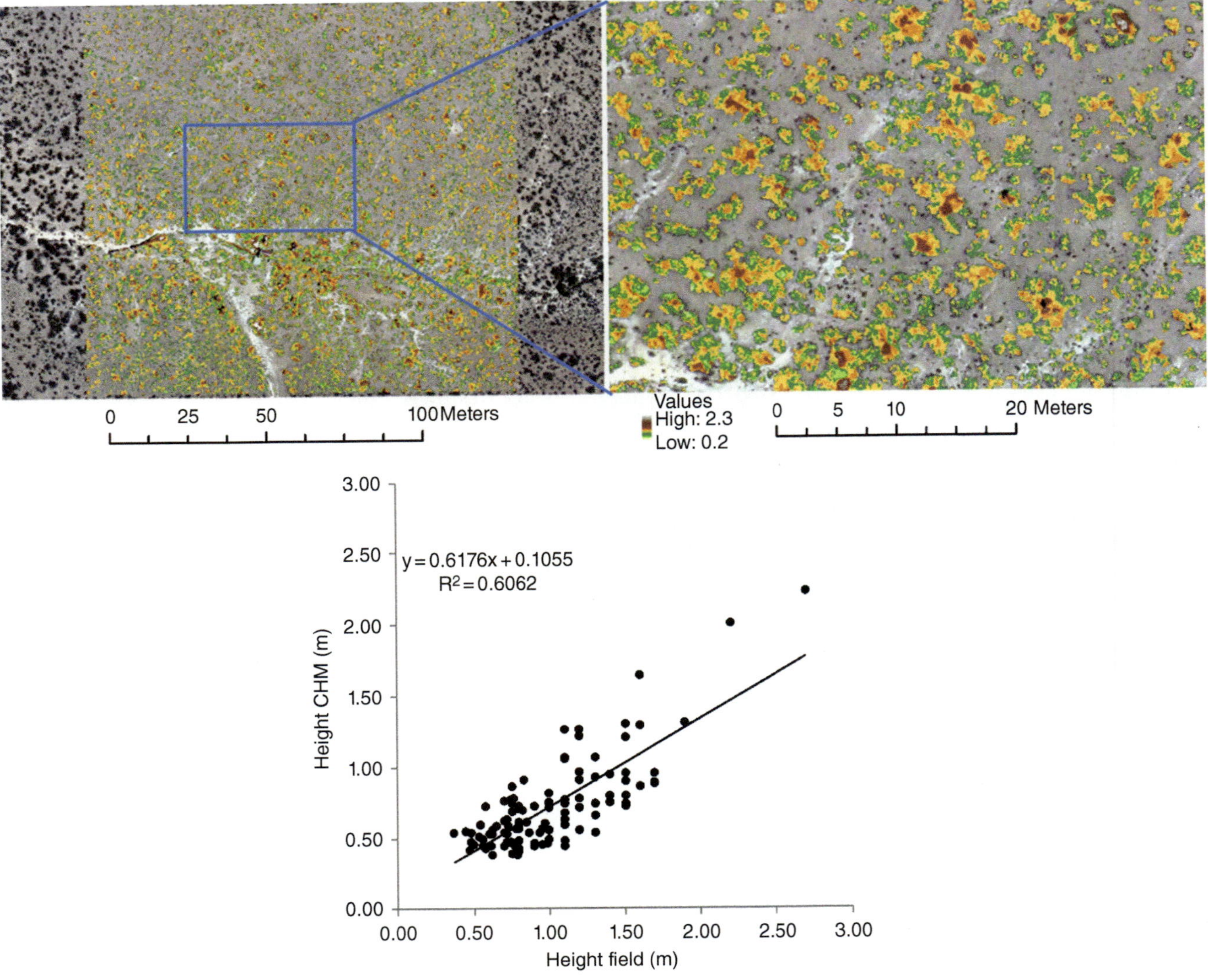

Figure 28.6 Comparison of shrub heights obtained from field measurements and from UAS image-derived CHM ($n = 106$) at the Tromble Weir watershed.

REFERENCES

Ambrosia, V. G., S. S. Wegener, D. V. Sullivan, S. W. Buechel, S. E. Dunagan, J. A. Brass, J. Stoneburnn, and S. M. Schoening (2003), Demonstrating UAV-acquired real-time thermal data over fires, *Photogram. Eng. Remote Sens.*, *69*, 391–402.

Baldocchi, D. D. (2003), Assessing the eddy covariance technique for evaluating carbon dioxide exchange rates of ecosystems: Past, present and future, Global Change Biol., *9*(4), 479–492.

Buffington, L. C., and C. H. Herbel (1965), Vegetation changes on a semidesert grassland range from 1858 to 1963, *Ecol. Monogr.*, *35*, 139–164.

Dunford, R., K. Michel, M. Gagnage, H. Piegay, and M. L. Tremelo (2009), Potentials and constraints of unmanned aerial vehicle technology for the characterization of Mediterranean riparian forest, *Int. J. Remote Sens.*, *30*, 4915–4935.

Gibbens, R. P., R. P. McNeely, K. M. Havstad, R. F. Beck, and B. Nolen (2005), Vegetation changes in the Jornada Basin from 1858 to 1998, *J. Arid Environ.*, *61*(4), 651–668.

Herwitz, S. R., et al. (2003), Coffee field ripeness detection using high resolution imaging systems of a solar-powered UAV, in *Proc., 30th Symp. Rem. Sens. Environ.*, TS-12.4, Honolulu, Hawaii.

Horcher, A., and R. J. M. Visser (2004), Unmanned aerial vehicles: Application for natural resource management, Proc. Council of Forest Eng., (COFE) Conference Proceedings, April 27–30, 2004. Machines and People, the Interface, Hot Springs, AR.

Johnson, L. F., S. Herwitz, S. Dunagan, B. Lobitz, D. Sullivan, and R. Slye (2003), Collection of ultra high spatial and spectral resolution image data over California vineyards with a small UAV, in *Proc. 30th Symp. Rem. Sen. Environ.*, TS-12.4, Honolulu, Hawaii.

Kirpich, Z. P. (1940), Time of concentration in small agricultural watersheds, *Civil Eng.*, *10*(6), 362.

Laliberte, A. S., J. E. Herrick, A. Rango, and C. Winters (2010), Acquisition, orthorectification, and object-based classification of unmanned aerial vehicle (UAV) imagery for rangeland monitoring, *Photogram. Eng. Remote Sens.*, 76(6), 661–672.

Laliberte, A. S., M. A. Goforth, C. M. Steele, and A. Rango (2011), Multispectral remote sensing from unmanned aircraft: Image processing workflows and applications for rangeland environments, *Remote Sens.*, 3(11), 2529–2551.

Newcome, L. R. (2004), Unmanned aviation: A brief history of unmanned aerial vehicles, Am. Inst. of Aeronaut. Astronaut., Reston, Va.

Rango, A., and A. S. Laliberte, (2010), Impact of flight regulations on effective use of unmanned aircraft systems for natural resources applications, *J. App. Remote Sens.*, 4, 043, 539, doi:10.117/1.3474649.

Rango, A., A. Laliberte, C. Maxwell, A. Slaughter, and C. Winters (submitted), An UAS approach to monitor and secure rangeland resources under current FAA regulations, *J. Unmanned Vehicle Sys.*

Schrage, D. P., Y. K. Yillikci, S. Lui, J. V. R. Prasad, and S. V. Hanaguad (1999), Instrumentation of the Yamaha R-50/RMax helicopter testbeds for airloads identification and follow-on research, in *Proc. 25th European Rotocraft Forum*, P4:1–13. September 14–16, 1999. Rome, Italy.

Templeton, R. C., E. R. Vivoni, L. A. Méndez-Barroso, N. A. Pierini, C. A. Anderson, A. Rango, A. S. Laliberte, and R. L. Scott (2014), High-resolution characterization of a semiarid watershed: Implications on evapotranspiration estimates, *J. Hydrol. 509*, 306–319.

Turner, D., and A. Lucieer (2011), Development of an Unmanned Aerial Vehicle (UAV) for hyper resolution vineyard mapping based on visible, multispectral, and thermal imagery, in *Proc. of the 34th Int. Symp. on Rem. Sens. of Environment*, Sydney, Australia, p. 4. April 10–15, 2011.

Zajkowski, T. J. (2003), Unmanned aerial vehicles: Remote sensing technology for the USDA Forest Service. Project Rep. RSAC-1507-RPT1, Remote Sens. Appl. Center, Salt Lake City, Utah.

Zreda, M., W. J. Shuttleworth, X. Zeng, C. Zweck, D. Desilets, T. E. Franz, R. Rosolem, and P. A. Ferre (2012), COSMOS: The COsmic-ray Soil Moisture Observing System, *Hydrol. Earth Syst. Sci.*, 16(11), 4079–4099.

29

Simulation of Water Balance Components in a Watershed Located in Central Drainage Basin of Iran

Ammar Rafiei Emam,[1,2] Martin Kappas,[1] and Karim C. Abbaspour[3]

29.1. INTRODUCTION

Water scarcity is one of the main problems in arid and semiarid regions. In Iran, as a semiarid country, water resources have experienced raising pressures due to increasing demand and recurrent droughts. Considering climate change as projected by the Intergovernmental Panel on Climate Change (IPCC) [IPCC, 2013], Iran faces severe water shortages in the next decades. Furthermore, the freshwater of the Raza-Ghahavand watershed is exploited by Iran in ever increasing demand for sanitation, drinking, manufacturing, and agriculture. Successful management of any resources requires accurate knowledge of the resource available, the uses to which it may be put, and the competing demands for the resource. Therefore, estimation of water balance is critical for water management and development planning in watersheds. Understanding the water balance in the basin is necessary to achieve sustainable water management.

The water balance is defined as "the balance between incoming water from precipitation and snowmelt and outgoing water by evapotranspiration, groundwater recharge and stream flow" [Dunne, 1978]. Figure 29.1 shows the water balance concept.

The main objective of this research is to assess the temporal and spatial variation of water resources in the semiarid

Razan-Ghahavand watershed based on hydrologic modeling. The second objective is to calibrate, validate, uncertainty, and sensitivity analysis of Soil and Water Assessment Tools (SWAT) hydrologic model of Razan-Ghavand basin based on river discharge. Finally the calibrated hydrologic model is used for climate change and land use change assessment.

29.2. MATERIAL AND METHODS

29.2.1. Study Area

The study area is a watershed called Razan-Ghahavand with a surface area about $3100\,km^2$ located in a central drainage basin of Iran (Figure 29.2). The difference between minimum and maximum elevation is $1265\,m$ with a maximum altitude of $2842\,m$ and a minimum altitude of $1577\,m$. The Razan-Ghahavand watershed can be seen as a representative case study in comparison to other watersheds in the central drainage basin of Iran. The watershed is not located at the headwater of the basin. The main river, called Gharehchay, enters the watershed from an eastward direction and exits the watershed in a westward direction. Simultaneously, two branches (Sirab Khomigan and Zehtaran) of reach drain water enter the main river from the north. The climate in this area is semiarid with an average annual precipitation about $290\,mm$ and a mean annual temperature about $11\,°C$. Most of the area is allocated by rangelands with different animal stocking capacity. Arable land (e.g., wheat, alfalfa, etc.) is located in the center of basin and covers 30% of the watershed. Groundwater recharge is the main source of irrigation in this part of the watershed. According to the soil taxonomic classification [Soil Survey Staff, 2010],

[1] Department of Cartography, GIS and Remote Sensing, University of Goettingen, Goettingen, Germany

[2] Department of Desert Management, Yazd University, Yazd, Iran

[3] Eawag, Swiss Federal Institute of Aquatic Science and Technology, Dübendorf, Switzerland

---

Remote Sensing of the Terrestrial Water Cycle, Geophysical Monograph 206. First Edition. Edited by Venkat Lakshmi.

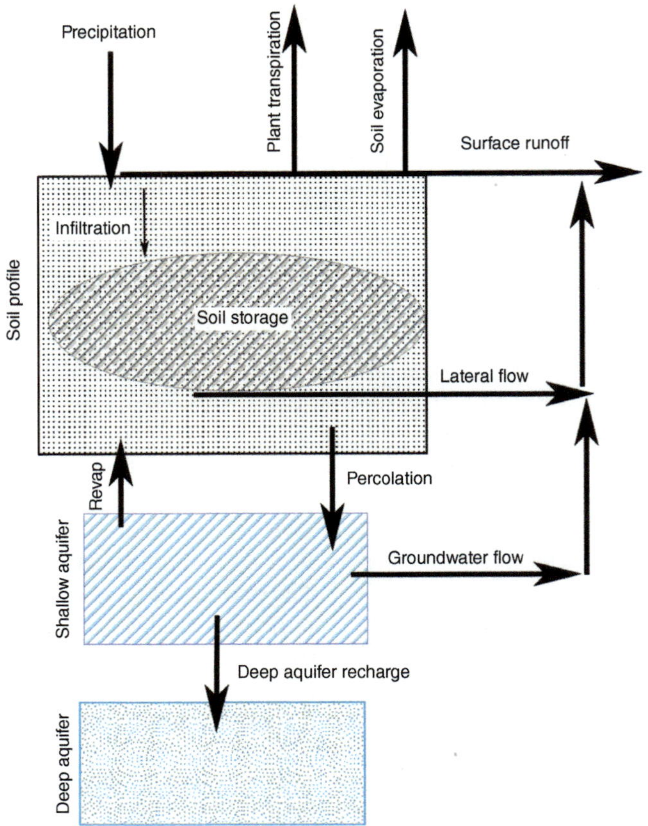

Figure 29.1 Water balance concept.

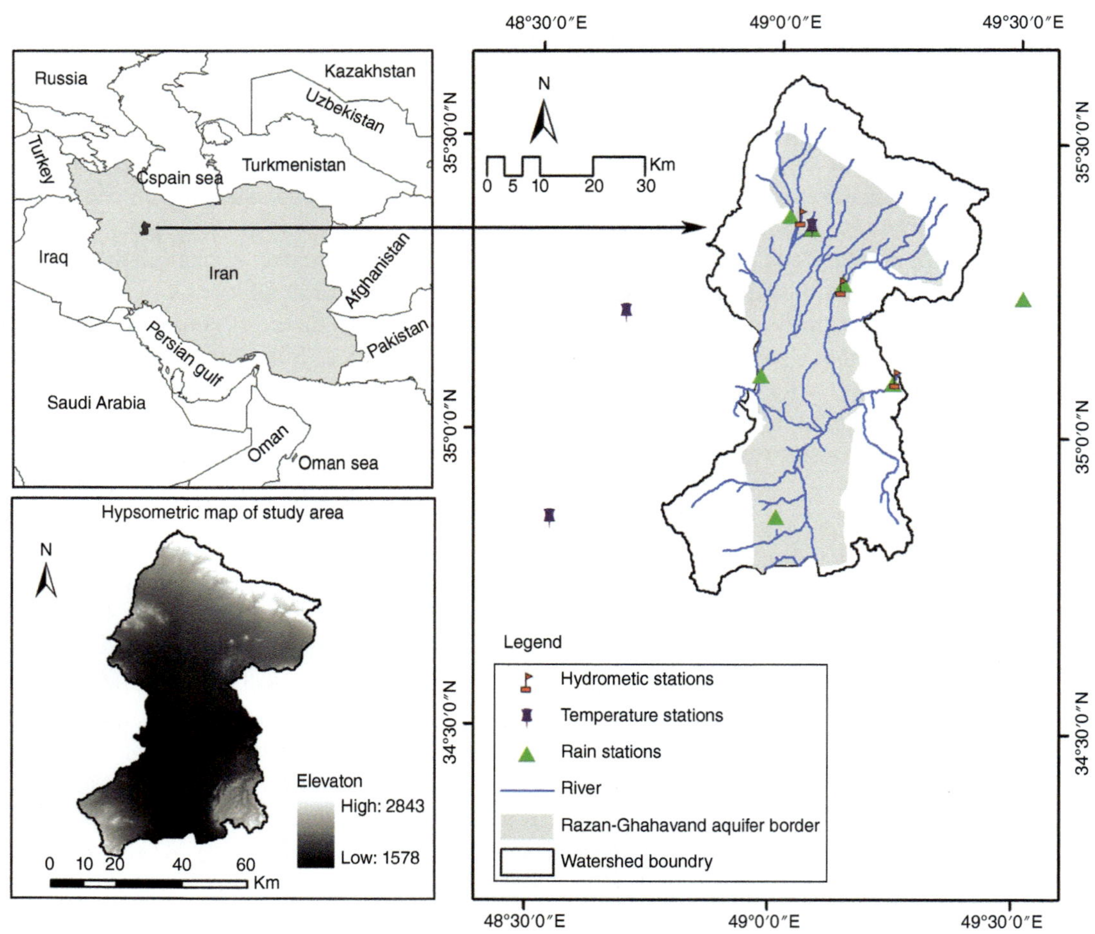

Figure 29.2 Razan-Ghahavand study area.

Table 29.1 Overview of weather data availability

No.	Type	Name	Range of Data Availability	Time Scale
1	Rain gauge station	Sarab-Khomigan	1976–2008	Daily
2	Rain gauge	Zehtaran	1982–2008	Daily
3	Rain gauge	Khanabad	1983–2008	Daily
4	Rain gauge	Famenin	1991–2007	Daily
5	Rain gauge	Omarabad	1984–2008	Daily
6	Rain gauge	Ghahavand	1982–2008	Daily
7	Rain gauge	Damagh	1990–2008	Daily
8	Synoptic station	Hamedan	1979–2008	Daily
9	Synoptic station	Nozheh	1976–2008	Daily
10	Climatology station	Dargazin	1976–2008	Daily

Aridisols and Entisols with typic Haplocalcids and lithic Xerorthents are the most frequent soils in the watershed. The soil moisture regimes inside the area vary from weak aridic to xeric, where the soil temperature regime of the area is mesic. Figure 29.2 shows the rain gauges and the synoptic and hydrometric stations in the basin. Table 29.1 shows the characteristics of weather stations. Characteristics of soil properties are shown in Table 29.2.

29.2.2. Land Use Mapping

Land use is one of the most important layers in hydrologic models and for the creation of hydrologic response units (HRU). To model surface runoff with the Soil Conservation Service (SCS) curve number (CN) method, land use information is directly needed. Remotely sensed data from Landsat TM (year 2009; source: http://earth explorer.usgs.gov/) are used to derive land use/land cover (LULC) information. Preprocessing is performed with auxiliary data including topographic map information at a scale of 1:25.000 and ground control points (GCPs) collected by Global Positioning System (GPS)-based field studies.

After preprocessing of the remote sensing data such as georeferencing with GCPs and fieldwork to check the various land cover/use in the landscape, a supervised land use classification is created. Figure 29.3 shows photos of typical areas with their dominant land use/cover.

Seven LULC classes are defined for supervised classification of the remote sensing data: Agriculture, dry farming, rangeland in good, moderate, or poor condition, bare lands, and urban areas. Supervised classification using different algorithms is used in order to create the LU map with highest accuracy. The maximum likelihood classification (MLC) is preferred, which produces a minimum error in classification under the assumption that in each class the spectral data are normally distributed. Generally, more spectral confusion is found between various classes: Usually there are similar classes with various spectral and different classes with the same spectral content.

The capability of MLC in comparison to other algorithms is reported by *Mengistu and Salami* [2007], *Reis* [2008], and *Diallo et al.* [2009].

The LU accuracy assessment will be thoroughly implemented by means of the error matrix and kappa statistics. This is commonly accepted as a standard approach in remote sensing. Error matrices as cross tabulation of the mapped class versus the reference class is used to assess the classification accuracy. Randomly selected points are chosen as a reference class for the accuracy assessment. Overall accuracy, user's and producer's accuracies, as well as the kappa statistics are then derived from the error matrices [*Congalton*, 2004].

29.2.3. Hydrologic Model

In this study the physical-based hydrologic model SWAT [*Arnold et al.*, 1998] is used. SWAT can be applied in watersheds that vary from small surface areas to bigger surface areas. Many studies show the capability of this model in hydrological assessments [*Schmalz et al.*, 2008; *Rostamian et al.*, 2008; *Schmalz & Fohrer*, 2009; *Srinivasan et al.*, 2010; *Betrie et al.*, 2011]. SWAT uses daily data as an input and delivers daily, monthly, or yearly output. The local water balance represented in SWAT is based on four storage volumes: snow, soil profile (0–2 m), shallow aquifer (2–20 m), and deep aquifer (>20 m). The core for hydrological modeling builds the water balance equation:

$$\mathrm{SW}_t = \mathrm{SW}_0 + \sum_{i=1}^{t}\left(R_{\mathrm{day}} - Q_{\mathrm{surf}} - E_a - W_{\mathrm{seep}} - Q_{\mathrm{gw}}\right),$$

where SW_t is the final soil water content, SW_0 is the initial soil water content, R_{day} is the amount of daily precipitation, Q_{surf} is the amount of surface runoff, W_{seep} is the amount of percolation and bypass flow exiting the soil profile bottom daily, E_a is the amount of evapotranspiration on day, and Q_{gw} is the amount of return flow on day.

Table 29.2 Soil characteristics in different land types[a]

Id	Soil Class	HYDGRP	Soil Depth (mm)	Texture[b]	Layer 1			Layer 2			Layer 3			Layer 4			Layer 5		
					Depth	SOL_BD	SOL_K	Depth	SOL_BD	SOL_K	Depth	SOL_BD	SOL_K	Depth	SOL_BD	SOL_K	Depth	SOL_BD	SOL_K
1	I	B	100	SaL	0-100	1.45	9.5												
2	II	B	500	SiL-L	0-15	1.37	10.7	15-50	1.43	10.9									
3	III	A	1300	L-SiCL-SiCL_SiC	0-19	1.34	5.9	19-39	1.3	4.2	39-84	1.27	2.9	84-130	1.23	2.8			
4	IV	A	1100	L-SaCL-CL-L	0-10	1.42	7.3	10-26	1.38	4.1	26-46	1.32	2.7	46-110	1.35	5.2			
5	V	B	1000	SiCL-SiCL-L	0-13	1.28	4.7	13-40	1.27	3.1	40-100	1.38	5.7						
6	VI	C	1200	SiCL-SiC-SiC-	0-18	1.27	5.2	18-68	1.22	2.9	68-120	1.21	2.8						
7	VII	C	1000	SiCL-SiC-SiC-SiC-	0-13	1.26	4.2	13-36	1.23	3	36-66	1.21	2.7	66-100	1.24	3.1			
8	VIII	D	1450	L-CL-C-SiCL-C	0-11	1.35	5.1	11-30	1.31	2.9	30-60	1.24	2.1	60-117	1.27	3.3	117-145	1.2	2.2
9	IX	C	840	CL-CL-CL-CL-	0-10	1.34	4.9	10-25	1.29	2.6	25-54	1.29	2.5	54-84	1.31	2.4			
10	X	D	1500	CL-C-C-C-C	0-10	1.36	4	10-30	1.29	1.7	30-60	1.25	2.1	60-120	1.28	1.6	120-150	1.23	2.2
11	XI	C	1200	SiCL-SiC-C-SiL	0-10	1.27	5.3	10-22	1.19	2.8	22-77	1.18	2.9	77-120	1.26	2.7			
12	XII	D	1420	SaL-C-C-SiC	0-20	1.47	10.7	20-34	1.24	1.8	34-83	1.21	2.1	83-142	1.26	3.2			
13	XIII	B	650	SiCL-C-CL-	0-15	1.31	5.9	15-37	1.23	2.4	37-65	1.32	2.6						

[a]HYDGRP: soil hydrologic group; Depth: depth of soil layer (cm); SOL_BD: soil bulk density (g/cm^3); SOL_K: saturated hydraulic conductivity (mm/h).
[b]Texture: Sa: Sandy; L:Loamy; Si: Silty, C: Clay (e.g., SiCL: silty clay loam).

Figure 29.3 Typical land use types inside the watershed: (a) bare land, (b) rangeland good condition in north watershed, (c) rangeland poor condition in south watershed, and (d) agriculture (e.g., alfalfa).

The water balance is calculated based on hydrological response units (HRUs), which are created by integration of land use, soil type, and slope class in terms of homogenous landscape units. All hydrological assessments are calculated on these HRUs. In this research various water balance components such as surface runoff, evapotranspiration, and percolation are estimated.

Surface runoff is simulated by modification of the SCS curve number (CN) method from daily rainfall and based on land use, soil hydrological group, and antecedent soil moisture [*USDA Soil Conservation* Service, 1972]. The SCS curve number, which is a watershed-specific coefficient, represents the runoff potential of particular land cover and soil. The SCS-CN method is based on the following relationship between rainfall and runoff [*USDA-SCS*, 1972]:

$$Q = \frac{(R - 0.2S)^2}{R + 0.8S}, \quad R > 0.2S,$$
$$Q = 0, \quad\quad\quad\quad R <= 0.2S,$$

where Q is the daily surface runoff (mm), R is the daily rainfall (mm), and S is the retention parameter. The retention parameter (S) is related to curve number, and this CN was calculated according to soil and land use condition by the following equation:

$$S = 254\left(\frac{100}{\text{CN}} - 1\right).$$

The value of CN ranges from 0 to 100. A table of initial CN is developed by SCS as a function of the soil type, land use, and antecedent moisture condition (AMC). The soil condition was classified in four different categories, ranked A–D according to the potential of soils on runoff production. The difference of soil groups on surface runoff was discussed by *Melese & Shih* [2002]. Soil class A mostly consist of deep soil with well-drained sands and gravels, high infiltration, low runoff potential, and high water transmission rates (greater than 0.30 in./h). Class B consists mostly of moderately deep to deep soil, moderately well to well-drained soils with moderately fine to moderately course textures, and the rate of water transmission is moderate (0.15–0.30 in./h). Soil class C has moderately fine to fine texture, with low infiltration and slow water transmission rate (0.15–0.30 in./h). Class D consists of clay soils with very slow infiltration rates and high runoff potential. They have a very low water transmission rate (0.00–0.05 in./h).

The curve number is defined in three antecedent moisture conditions by the USDA. AMC-I is the lower limit of moisture representing a dry condition, AMC-II is the average moisture condition, and AMC-III is the upper limit of moisture representing a wet condition. Usually AMC-II is used for CN value estimation. In this study the AMC-II class of soil moisture is applied to each land use pattern.

Potential evapotranspiration (PET) is estimated using the Hargreaves method, which requires daily precipitation and minimum and maximum temperature [*Hargreaves et al.,* 1985]. After simulation of PET, actual evapotranspiration (AET) was estimated based on *Ritchie* [1972] methodology, according to leaf area index (LAI) simulated by crop growth component inside SWAT. The Hargeaves equation is

$$\lambda E_0 = 0.0023 H_0 \left(T_{mx} - T_{mn} \right)^{0.5} \left(\overline{T}_{av} + 17.8 \right),$$

where λ is the latent heat of vaporization MJ/Kg, E_0 is the potential evapotranspiration (mm/d), H_0 is the extraterrestrial radiation (MJ/m^2 d), T_{mx} is the maximum air temperature for a given day (°C), T_{mn} is the minimum air temperature for a given day (°C), $\overline{T}_{av}$ is the mean air temperature for a given date (°C).

Percolation is derived from a storage routing technique combined with a crack-flow model to predict flow through each soil layer in the profile [*Neitsch et al.,* 2009]. The crack-flow model allows percolation of infiltrated rainfall. The storage routing technique is based on the following equation:

$$W_i = SW_{oi} \left(1 - \exp \left[\frac{-\Delta t}{TT_{perc}} \right] \right)$$

where W_i is water percolation (mm), SW_{oi} is drainable water (mm), Δt is the length of time step (h), TT_{perc} and is the travel time for percolation through the layer (h).

29.2.4. Calibration and Sensitivity Analysis

The SUFI-2 algorithm [*Abbaspour et al.,* 2007] is used for calibration, validation, and uncertainty analysis. In this algorithm all uncertainties such as uncertainty in

parameter, conceptual model, and input data are depicted into the parameter ranges, and the process tries to capture most of the measured data within the 95% prediction uncertainty (95PPU) band. To calibrate the model, monthly river discharge at three hydrometric stations is used. Furthermore, the model is calibrated from 2001 to 2008, including 2 years as a warmup period and validated for 1997–2000. Figure 29.1 shows the position of hydrometric stations in the basin used for the calibration and sensitivity analysis.

The uncertainty of the output is quantified by 95PPU, which is calculated at the 2.5% and 97.5% levels of cumulative distribution of an output variable. To assess the goodness of calibration and uncertainty performance two indices are used: the P factor and R factor. P factor is the percentage of measured data bracketed by the 95PPU, and the R factor is 95PPU divided by the standard deviation of measured data. P factors around one, bracket most of data within the band, and R factors near zero, narrowest band, are ideal. Nash-Sutcliffe model efficiency (NS) and coefficient of determination (R^2) were used to assess the SWAT model. The equation of NS is given as follows:

$$NS = 1 - \frac{\sum_{i-1}^{n} (Q_{obs} - Q_{sim})^2}{\sum_{i=1}^{n} [Q_{obs} - \mathrm{mean}(Q_{obs})]^2},$$

where Q_{obs} and Q_{sim} are the measured and simulated data, respectively, and n is the total number of data records. NS varies between minus infinity to 1. Usually when NS is more than 0.5, the accuracy of the model is very good, and negative values show the model is not suitable [*Zhi et al.,* 2009]. The coefficient of determination, R^2, is calculated as follows:

$$R^2 = \frac{\left[\sum (Q_{obs,i} - \overline{Q_{obs}})(Q_{sim,i} - \overline{Q_{sim}}) \right]^2}{\sum (Q_{obs,i} - \overline{Q_{obs}})^2 \sum (Q_{sim,i} - \overline{Q_{sim}})^2},$$

where $\overline{Q_{obs}}$ and $\overline{Q_{sim}}$ are the mean measured and simulated data, respectively.

A weighted version of Nash-Sutcliffe coefficient (*g*) was used as an objective function to compare the monthly measured and simulated discharges in multiple discharge stations:

$$g = \sum_{i}^{n} w_i \, NS_i.$$

Many studies show the capability of SUFI-2 for calibration and uncertainty analysis [*Abbaspour et al.,* 2007; *Schuol et al.,* 2008; *Faramarzi et al.* 2009; *Yang et al.,* 2008].

29.3. RESULTS & DISCUSSION

29.3.1. Hydrological Response Unit

The integration of land use, soil type, and slope layers derive 831 HRUs spread over 138 subbasins. The following details of these single information layers (land use, soil, and slope information) for hydrologic modeling are discussed.

Land use is obtained by remotely sensed data (Landsat ETM) and field studies (GPS-based land use mapping, soil samples, and soil mapping). In case of land use mapping, the result shows an overall accuracy for the supervised classification of 71.4% and kappa coefficient of 0.66, which represents a reliable classification. Figure 29.4 shows the results of remotely sensed analysis to produce a land use map, and Table 29.3 presents the statistical results of each land use class.

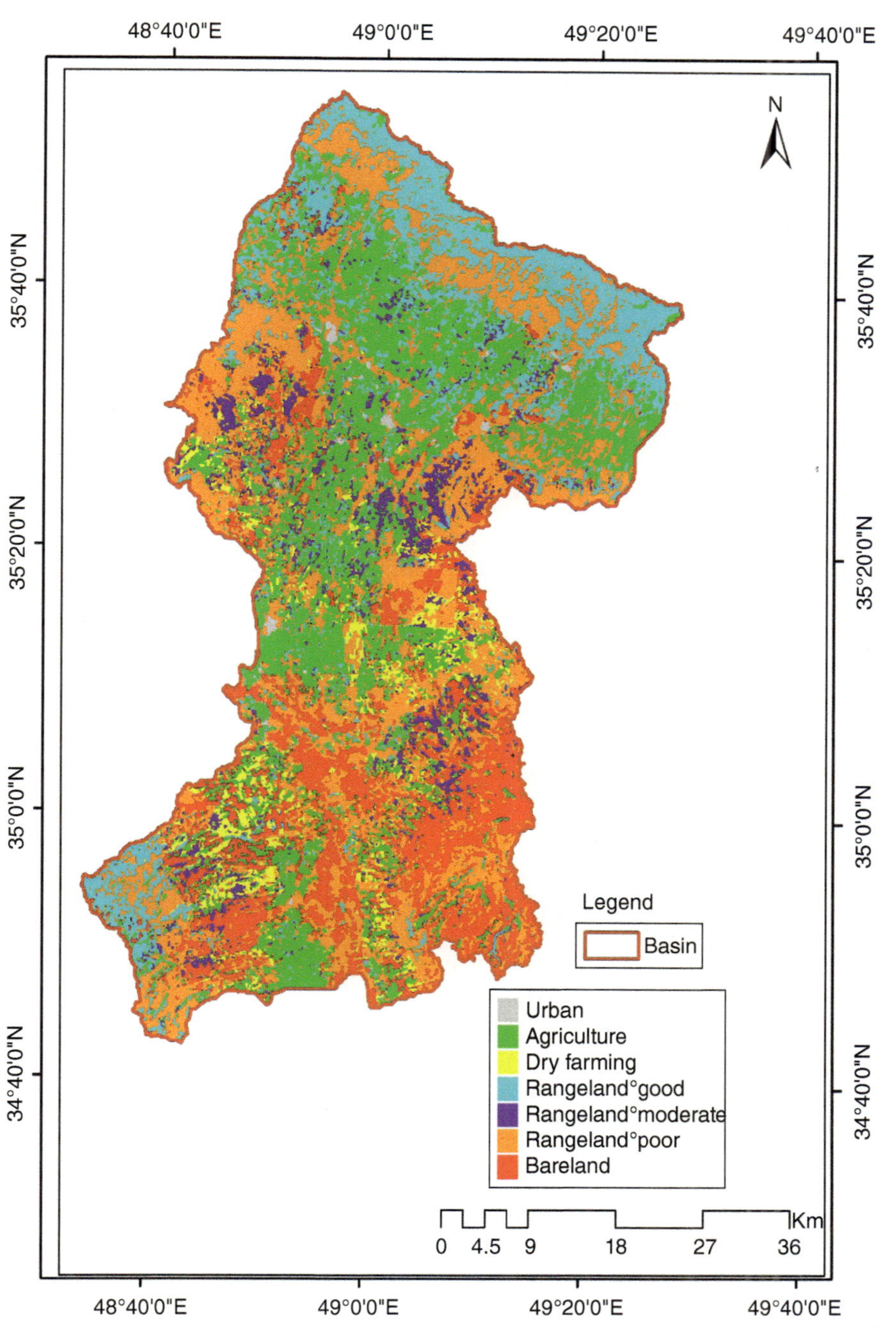

Figure 29.4 Land use map.

Table 29.3 Statistical results of land use classes

	Agg.	Land Use*	Area (%)	Area (km²)
1	URBN	Urban	0.6	19
2	AGRR	Agriculture	26	814
3	HAY	Dry farming	5.0	156
4	PAST	Rangeland_G	14.6	456
5	RNGB	Rangeland_M	5.7	177
6	SWRN	Rangeland_P	30.8	964
7	BRLD	bareland_erosion	17.3	542
			100.0	3128

*G:Good, M:Moderate, P:Poor

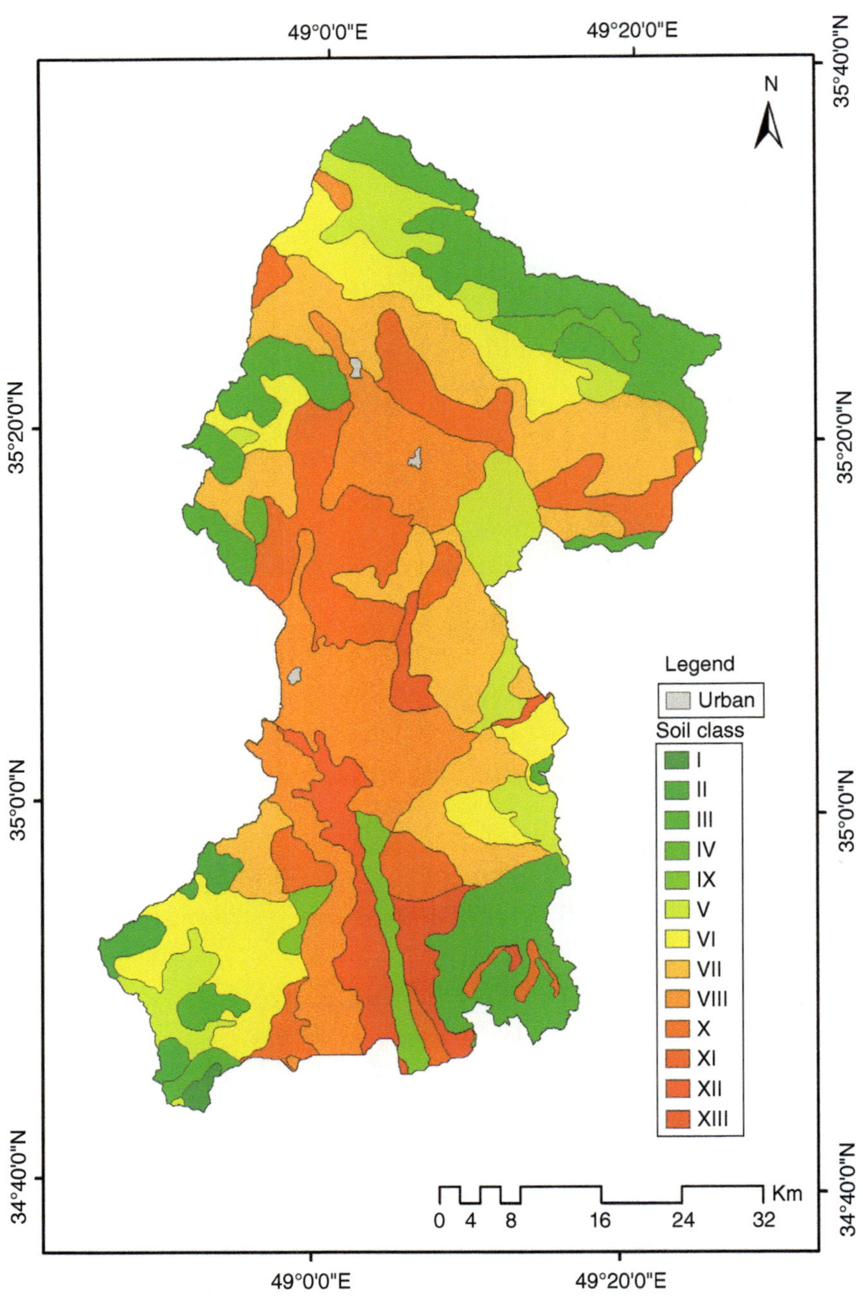

Figure 29.5 Soil map.

Table 29.4 Information of soil map

Soil Class	Soil Depth (mm)	No. of Layers
I	100	1
II	500	2
III	1300	4
IV	1100	4
V	1000	3
VI	1200	3
VII	1000	4
VIII	1450	5
IX	840	4
X	1500	5
XI	1200	4
XII	1420	4
XIII	650	3
URBAN	—	—

The second information layer is build up by soil mapping (deriving a soil map). The soil map is generated using soil profile information in each land type up to 5 layers. The soil map includes 13 soil types (Figure 29.5 and Table 29.4).

The last information layer needed for hydrologic modeling is contents information about slope angle distribution (Figure 29.6). It is created by the Digital Elevation Model (DEM) analysis (DEM resolution: 15 m scale) inside ArcGIS (Vers. 9.3, Spatial Data Analyst) with an output of 3 classes. Slope values less than 2% are predominant in the basin (Table 29.5). These land parcels (slope < 2%) are mostly farmlands in the center of the watershed. The south of the watershed is marked by lowlands with degraded soils.

29.3.2. Model Calibration

The Razan-Ghahavand watershed is calibrated using the parameters in Table 29.6. The parameters of Table 29.6 include their initial and final ranges, which are used in the calibration process. The first results of the SWAT model did not present the streamflow correctly. The sensitivity analysis shows that some parameters are very sensitive against streamflow. Table 29.7 shows the most sensitive parameters based on their location in the watershed by assigning the subbasin numbers to the parameters. The result from the t statistic provides a measure of sensitivity (larger in absolute values and more sensitive), and p values determine the significance of the sensitivity. A value closer to zero has more significance. According to sensitivity analysis, snow, soil, groundwater, and infiltration parameters are the most sensitive parameters in the watershed. The most sensitive parameter (curve number) is located in the subbasin with the

hydrometric stations Sirab Khomigan (No. 41) and Zehtaran (No. 14) as their outlets. These stations are located in high elevation in comparison to Omarabad station (No.71) hydrometric station. Therefore CN2 is the most sensitive parameter at these stations. Second parameter after CN is snow parameter (minimum melt rate for snow during the year). Snow melt is the most sensitive parameter due to the mountainous character of Razan-Ghahavand basin, where much of the streamflow is controlled by snowmelt.

Figure 29.7 shows the results of calibration and validation for three hydrometric stations. The calibration results for Q (m³/s) are shown for measured/observed data, the 95% prediction uncertainty (95PPU), and the best fit of optimization. The gray curves represent the 95PPUs of the discharge simulation together with observed discharges (blue line) and best discharge simulation (red line) at three hydrometric stations (Q_14, Q_41, and Q_71). The three hydrometric stations are located at different rivers (Zehtaran, Sirabe-Khomigan, and Gharehchay) and represent a large area of the catchment. The general trend of the discharge both for calibration and validation simulations for all stations look pretty good. The model simulated the variation in time of peaks quit well, but the uncertainty interval at peaks in some stations (for instance, in March 2005 at the Q_14 station) is extremely large. In Table 29.8 the statistical results of this analysis are summarized, which show that the overall performance of the model is good. The R factor at the outlet of the watershed is less than 1 and shows a good calibration result. The p factor is small and revealed that the actual uncertainty is likely larger. Moreover the NS and R^2 for all three stations are bigger than 0.5, which is acceptable and shows good calibration results.

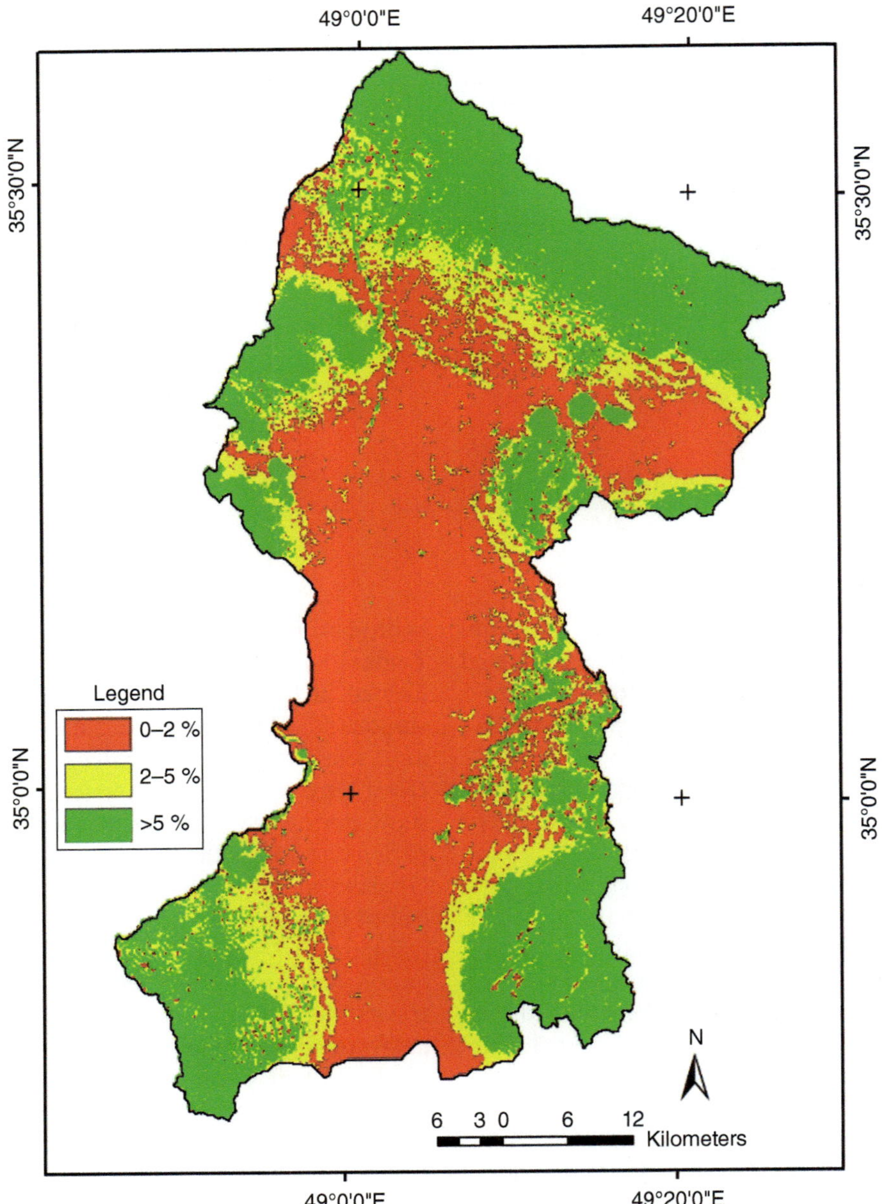

Figure 29.6 Slope map.

Table 29.5 Information of slope map

	Class	Area (%)
1	0–2%	48.89
2	2–5%	20.87
3	> 5%	30.24

29.3.3. Quantification of Components

For the extraction of water components the average monthly 95PPUs of water components and average annual components (2003–2008) were simulated. As pointed out in Figure 29.8, precipitation in April is highest with an amount around 50 mm and precipitation in September is lowest (0.2 mm).

Figure 29.8a–29.8c shows the average monthly values of actual evapotranspiration (AET), surface runoff (SURQ) and soil water (SW) content. The trend of 95PPUs for actual evapotranspiration is similar to precipitation, and the uncertainty values are not high. In April the rate of AET is high, due to highest precipitation and also because of more transpiration by crops due to growth and development of vegetation (both in arable and rangelands). In September AET is low, due to the end of the growing season of crops and decreasing precipitation. On the other

Table 29.6 Parameters for river discharge calibration and their initial and final ranges[a]

Parameter Name[2]	Definition	Initial Range		Final Range	
r__CN2.mgt	SCS runoff curve number for moisture condition II	−0.5	0.5	−0.13	0.14
v__GW_DELAY.gw	Groundwater delay time (days)	0.0	500	302	401
v__ALPHA_BF.gw	Base flow alpha factor (days)	0	1	0.56	0.68
v__REVAPMN.gw	Threshold water in shallow aquifer	0	500	0.22	0.30
v__GW_REVAP.gw	Revap coefficient	0.02	0.2	0.18	0.20
v__RCHRG_DP.gw	Aquifer percolation coefficient	0	1	0.90	0.96
v__GWQMN.gw	Threshold water level in shallow aq. for baseflow	0	5000	3900	4050
r__SOL_AWC().sol	Available water capacity factor	−0.5	0.5	0.24	0.49
r__SOL_K().sol	Saturated hydraulic conductivity	−0.95	0.95	−0.90	−0.86
r__SOL_BD().sol	Soil bulk density	−0.5	0.5	−0.34	−0.29
v__SOL_ALB().sol	Moist soil albedo	0	0.25	0.14	0.16
v__EPCO.hru	Plant uptake compensation factor	0.01	1	0.36	0.42
v__ESCO.hru	Soil evaporation compensation factor	0.01	1	0.01	0.42
v__SLSUBBSN.hru	Average slope length (m)	10	150	134	150
v__OV_N.hru	Manning's n value for overland flow	0	0.8	0.46	0.80
r__CH_N2.rte	Manning's n value for main channel	0	0.3	0.04	0.11
v__CH_K2.rte	Effective hydraulic conductivity in main channel alluvium	0	150	92	100
v__ALPHA_BNK.rte	Base flow alpha factor for bank storage (days)	0	1	0.22	0.48
r__HRU_SLP.hru	Average slope steepness	−0.95	0.95	0.19	0.68
v__SFTMP.bsn	Snowfall temperature	−5	5	−4.7	4
v__SMTMP.bsn	Snowmelt base temperature (°C)	−5	5	3.3	5
v__SMFMX.bsn	Melt factor for snow on 21 Jun	0	10	1.8	5
v__SMFMN.bsn	Melt factor for snow on 21 Dec	0	10	0.004	5.5
v__TIMP.bsn	Snowpack temperature lag factor	0.01	1	0.05	0.70
v__SURLAG.bsn	Surface runoff lag coefficient	1	24	3.5	20

[a]The final ranges are based on the parameters in the outlet station (Omarabad station).
[b]r__: refers to rational changes; v__: refers to substitute changes.

Table 29.7 Most sensitive parameters including their t value and p value

Parameter Name	t Value	p Value
r__CN2.mgt________23,26,27,28,29,32,33,34,37,38,41	−44.67	0.00
r__CN2.mgt________1,2,3,4,5,6,12,13,14	−15.06	0.00
v__SMFMN.bsn	−11.31	0.00
v__ALPHA_BNK.rte________1,2,3,4,5,6,12,13,14	−9.39	0.00
v__ALPHA_BNK.rte________23,26,27,28,29,32,33,34,37,38,41	−9.27	0.00
r__SOL_AWC().sol________23,26,27,28,29,32,33,34,37,38,41	7.3	0.00
v__SMFMX.bsn	−4.93	0.00
v__TIMP.bsn	−4.89	0.00
v__CH_K2.rte________23,26,27,28,29,32,33,34,37,38,41	2.47	0.01
V__GW_DELAY.gw________1,2,3,4,5,6,12,13,14	2.34	0.02
v__GWQMN.gw________23,26,27,28,29,32,33,34,37,38,41	2.02	0.04
r__SOL_AWC().sol________1,2,3,4,5,6,12,13,14	1.97	0.05
v__GW_REVAP.gw________23,26,27,28,29,32,33,34,37,38,41	1.78	0.07
r__CH_N2.rte________1,2,3,4,5,6,12,13,14	−1.66	0.10
v__SMTMP.bsn	1.72	0.09
V__GW_DELAY.gw________7-11,15-22,24,25,30,31,35,36,39,40,42-138	1.64	0.10

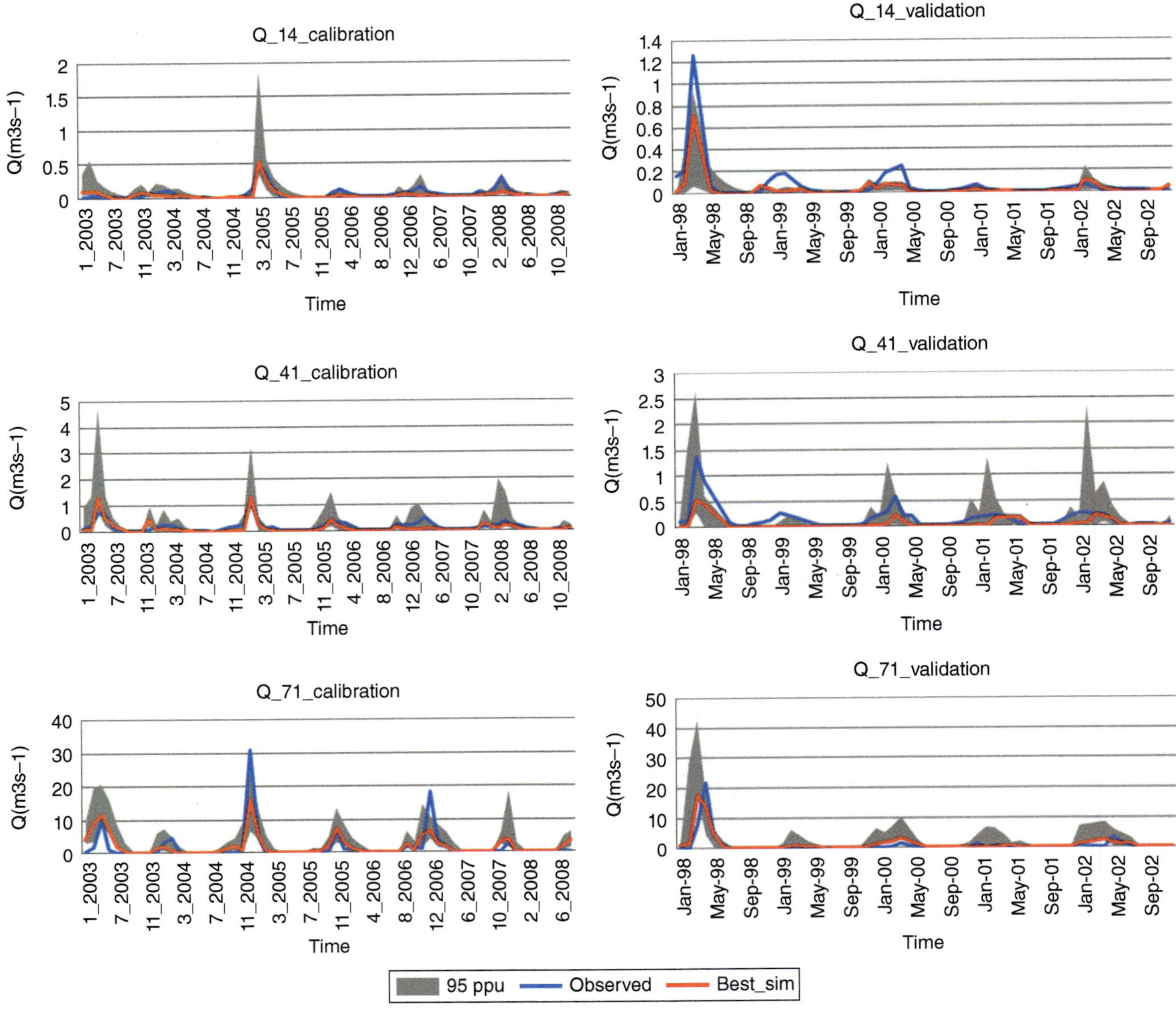

Figure 29.7 Calibration and validation of river discharge.

Table 29.8 Statistical result of calibration and validation

	Calibration (2003–2008)				Validation (1998–2002)			
Station	*P* Factor	*R* Factor	R^2	NS	*P* Factor	*R* Factor	R^2	NS
14	0.66	1.41	0.68	0.63	0.44	0.36	0.91	0.72
41	0.60	1.99	0.70	0.53	0.51	1.33	0.78	0.42
71	0.30	0.79	0.62	0.60	0.45	1.02	0.69	0.66

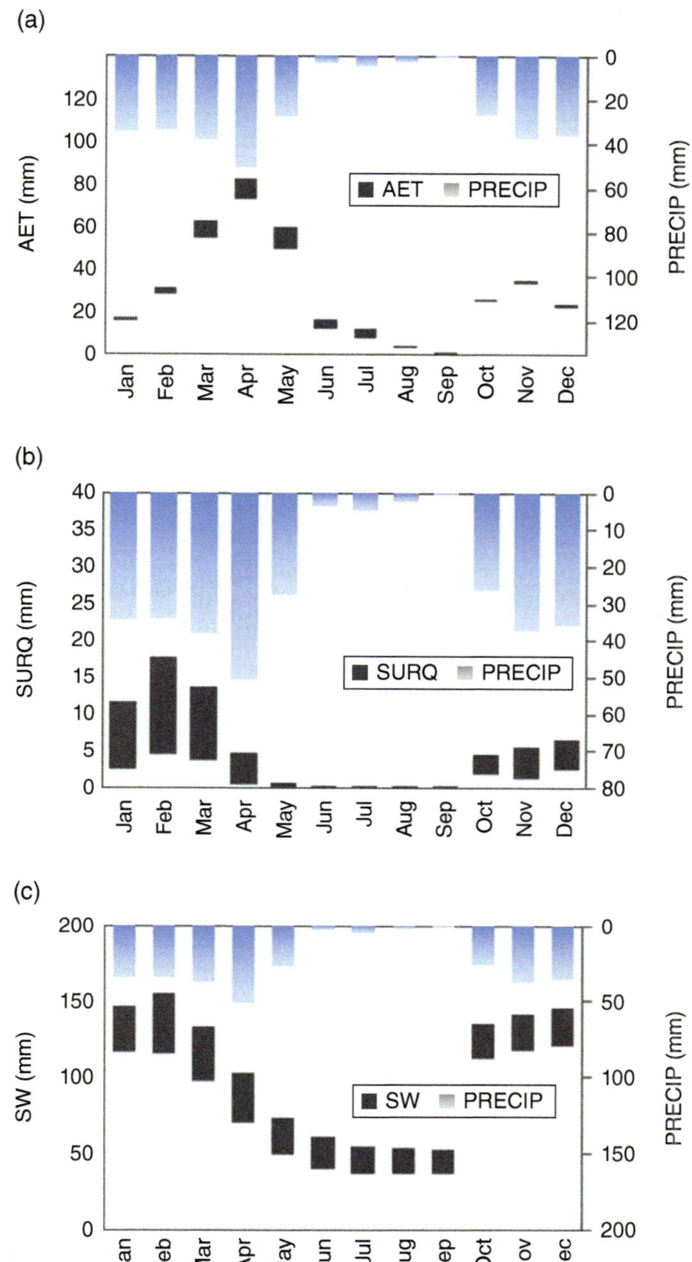

Figure 29.8 a–c: Average monthly 95PPU (2003–2008) of actual evapotranspiration (AET), surface runoff (SURQ), and soil water content (SW) against with precipitation (PRECIP).

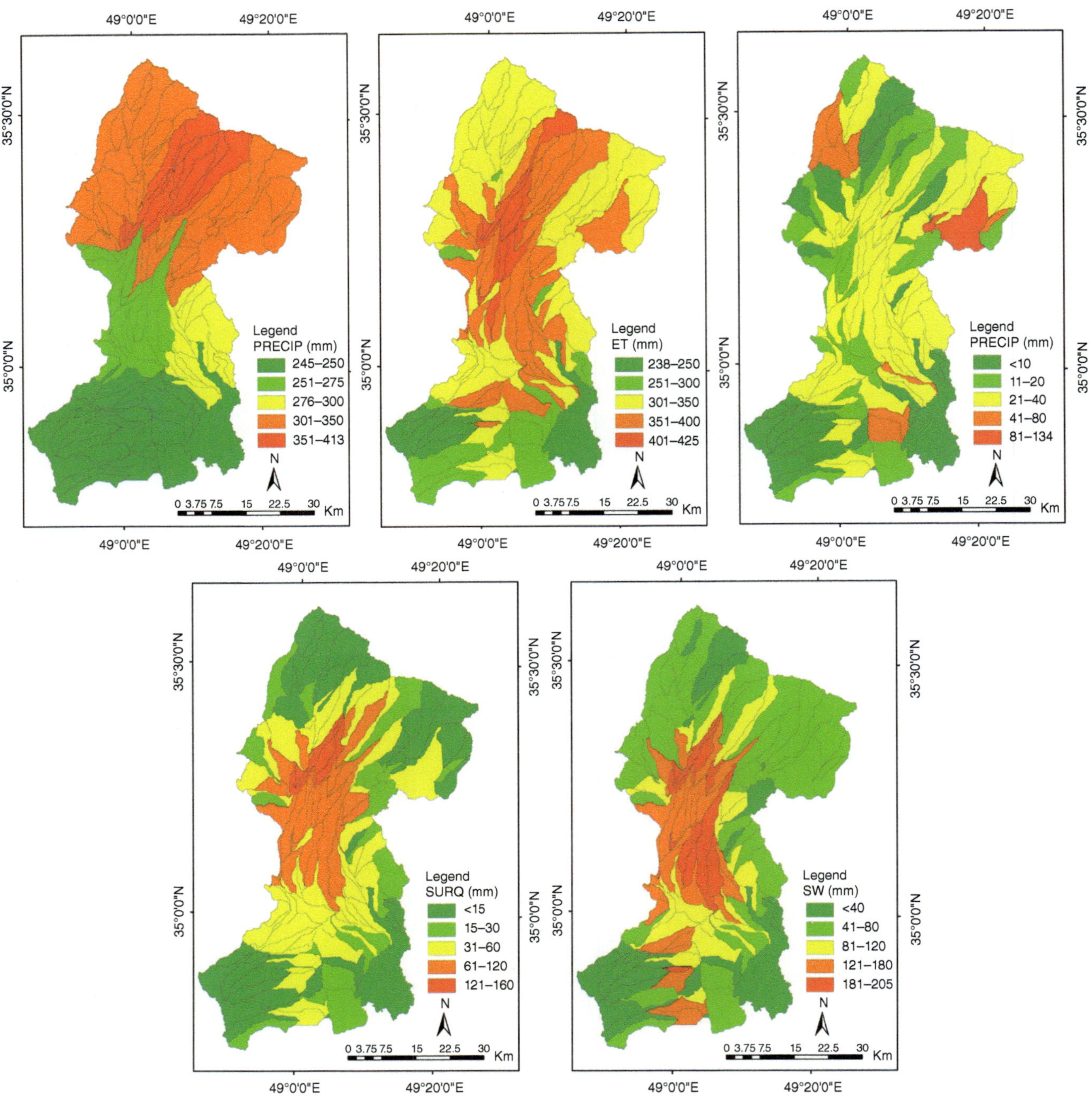

Figure 29.9 Average spatial distribution of yearly average of precipitation (PRECIP), actual evapotranspiration (AET), percolation (PERC), surface runoff (SURQ), and soil water content (SW) for the years 2003–2008.

hand surface runoff is highest in January, February, and March. Soil water content during the summer is lowest and during the winter is highest. Surface runoff in summer is lowest due to decreasing precipitation in this period of time. Moreover, soil water content in February is highest because of low evapotranspiration and pretty high precipitation.

The assessment of spatial distribution of evapotranspiration (ET) shows that the annual average values of ET vary from 238 mm in the south to 425 mm in the north of the watershed. The range of precipitation varies from 245 mm in the south to 413 mm in the north. The yearly average of surface runoff varies from 1 to 160 mm in various subbasins at which soil water varies from 13 to 206 mm in the whole watershed. The highest average percolation is 134 mm in the north east of the watershed. In Figure 29.9 the spatial distribution of these components together with their spatial pattern in the watershed are presented.

29.4. CONCLUSION

Based on DEM analysis inside GIS, land use classification and soil information the applicability of the SWAT model in this central drainage basin of Iran is tested. A GIS-based model for the Razan-Ghahavand watershed is now available and can be developed step by step into the future (e.g., integration of new and more precise base data). The river discharge data from 1997 to 2008 are used for model calibration and validation and the extraction of water balance components. The general trend of the discharge simulations (both for calibration and validation) looks pretty good for all stations, but the uncertainty interval at single stations is quite large. Overall, the research results show that water balance components like ET, surface runoff and percolation can be modeled with a hydrological model like SWAT, and the model output produces robust information for watershed management. This information can be used for adjusted planning and management of the watershed to meet the requirements for better water harvesting, finding initial suitable places for artificial aquifer recharge, and future land rehabilitation by adapted rangeland management.

REFERENCES

Abbaspour, K. C. (2007), *User Manual for SWAT-CUP, SWAT Calibration and Uncertainty Analysis Programs*, Eawag: Swiss Fed. Inst. of Aquat. Sci. and Technol., Dübendorf, Switzerland.

Abbaspour, K. C., J. Yang, I. Maximov, R. Siber, K. Bogner, J. Mieleitner, J. Zobrist, and R. Srinivasan (2007), Modelling hydrology and water quality in the pre-alpine/alpine Thur watershed using SWAT, *J. Hydrol.*, *333*, 413–430, doi:10.1016/j.jhydrol.2006.09.014.

Arnold, J. G., R. Srinivasan, R. S. Muttiah, and J. R. Williams (1998), Large area hydrologic modeling and assessment–Part 1: Model development, *J. Am. Water Resourc. Assoc.*, *34*, 73–89.

Betrie, G. D., Y. A. Mohamed, A. van Griensven, and R. Srinivasan (2011), Sediment management modelling in the Blue Nile Basin using SWAT model, *Hydrol. Earth Syst. Sci.*, *15*, 807–818, doi:10.5194/hess-15-807-2011.

Congalton, R. G. (2004), *Remote Sensing and GIS Accuracy Assessment*, CRC Press, Boca Raton, Fla.

Diallo, Y., G. Hu, and X. Wen (2009), Applications of remote sensing in land use/land cover change detection in Puer and Simao Counties, Yunnan Province, *J. Am. Sci.*, *5*, 157–166.

Dunne, T. (1978), *Water in Environmental Planning*, W.H. Freeman, New York.

Faramarzi, M., K. C. Abbaspour, R. Schulin, and H. Yang (2009), Modelling blue and green water resources availability in Iran, *Hydrol. Process*, *23*, 486–501.

Hargreaves, G. L., G. H. Hargreaves, and J. P. Riley (1985), Agricultural benefits for Senegal River Basin, *J. Irrigation Drainage Eng.*, *111*, 113–124.

Intergovernmental Panel on Climate Change (IPCC) (2013), Climate change 2013: The physical science basis, in *Working Group I to the Fifth Assessment Report of the Intergovernmental Panel on Climate Change*, edited by T. F. Stocker et al., Cambridge Univ. Press, Cambridge, United Kingdom and New York.

Melese, A. M., and S. F. Shih (2002), Spatially distributed storm runoff depth estimation using LANDSAT images and GIS, *Computer Electron. Agric.*, *37*, 173–183.

Mengistu, D. A., and A. T. Salami (2007), Application of remote sensing and GIS inland use/land cover mapping and change detection in a part of south western Nigeria, *Afr. J. Environ. Sci. Technol.*, *1*, 99–109.

Neitsch, S. L., J. G. Arnold, J. R. Kiniry, and J. R. Williams (2009), Soil and water assessment tools, Theoretical documentation, ARS, Tex.

Reis, S. (2008), Analyzing land use/land cover changes using remote sensing and GIS in Rize, North-East Turkey, *Sensors*, *8*, 6188–6202.

Ritchie, J. T. (1972), A model for predicting evaporation from a row crop with incomplete cover, *Water Resourc. Res.*, *8*, 1204–1213.

Rostamian, R., A. Jaleh, M. Afyuni, S. F. Mousavi, M. Heidarpour, A. Jalalian, and K. C. Abbapour (2008), Application of SWAT model for estimation runoff and sediment in two mountainous basins in central Iran, *Hydrol. Sci. J. 53*(5), 977–988.

Schmalz, B., and N. Fohrer (2009), Comparing model sensitivities of different landscapes using the ecohydrological SWAT model, *Adv. Geosci. 21*, 91–98.

Schmalz, B., F. Tavares, and N. Fohrer (2008), Modelling hydrological processes in mesoscale lowland river basins with SWAT—Capabilities and challenges, *Hydrol. Sci. J.*, *53*(5), 989–1000

Schuol, J., K. C. Abbaspour, R. Srinivasan, and H. Yang (2008), Estimation of freshwater availability in the West African sub-continent using the SWAT hydrologic model, *J. Hydrol.*, *352*, 30–49.

Srinivasan, R., X. Zhang, and J. Arnold (2010), SWAT ungauged: Hydrological budget and crop yield predictions in the Upper Mississippi River Basin, *Trans. ASABE*, *53*(5), 1533–1546.

Soil Survey Staff (2010), Keys to Soil Taxonomy, 11th ed., USDA Natural Resources Conservation Service, Washington, D.C.

U.S. Department of Agriculture Soil Conservation Service (USDA SCS) (1972), *SCS National Engineering Handbook*, Sect. 4, Hydrology, USDA, Washington, D.C.

Yang, J., P. Reichert, K. C. Abbaspour, J. Xia, and H. Yang (2008), Comparing uncertainty analysis techniques for a SWAT application to the Chaohe Basin in China, *J. Hydrol.*, doi:10.1016/j.jhydrol.2008.05.012.

Zhi, L., L. Wen-Zhao, Z. Xun-Chang, and Z. Fen-li (2009), Impact of land use change and climate variability on hydrology in an agricultural catchment on the Loess Plateau of China, *J. Hydrol.*, *377*, 35–42, doi:10.1016/j.jhydrol.2009.08.007.

30

Estimating Water Use Efficiency in Bioenergy Ecosystems Using a Process-Based Model

Zhangcai Qin[1] and Qianlai Zhuang[1,2]

30.1. INTRODUCTION

Bioenergy, made available from materials derived from biological sources, has been widely considered as one of the major renewable and sustainable energy sources to enhance energy security and mitigate climate change [*Beringer et al.*, 2011; *Fargione et al.*, 2010]. Food grain is currently the most popularly used biomass feedstock for biofuel production, e.g., maize (*Zea Mays* L.) grain for bioethanol and soybeans [*Glycine max* (L.) Merr.] for biodiesel. However, traditional biofuels have many unintended consequences concerning feedstock availability, food security, environmental sustainability, and societal welfare. During the last two decades, maize grain production increased about 65%. The major contribution is from grain yield, with about 1.96% increase annually [*FAOSTAT,* 2012]. But most of maize grain in the United States is used for human consumption, livestock feed, or other purposes. Only about 30% of grain harvested was used for ethanol production in 2009 [*USDA*, 2010]. The slowly growing maize production may not be able to support the rapidly increasing biofuel demand. Food-based feedstocks alone may limit further biofuel expansion, for instance, to reach the 2022 U.S. biofuel target [*U.S. Congress,* 2007]. In addition, conventional food-based biofuel development can be a threat to food security, due to competitive consumption of cropland, water, and nutrient resources that could otherwise be used for food production [*Fargione et al.*, 2010]. Indirect land use impacts on ecosystem services, such as monoculture and deforestation, also limit further expansion of conventional biofuel development [*Fargione et al.*, 2010; *Searchinger et al.*, 2008].

Second-generation biofuel (or advanced biofuel), produced from various types of biomass, is increasingly recognized as another option for bioenergy development. Among many tested species, switchgrass (*Panicum virgatum* L.) and *Miscanthus* (e.g., *Miscanthus giganteus*) were often studied for its characteristics, adaptation, and environmental impacts in the United States [*Fargione et al.*, 2010; *Heaton et al.*, 2008; *McIsaac et al.*, 2010]. Switchgrass is a perennial, warm-season cellulosic crop native to North America. It is widely distributed over the United States, especially the prairies of the Midwest [*Wright and Turhollow*, 2010]. Switchgrass was originally used for soil conservation, forage production, and other purposes. It is more recently used as biomass crop for biofuel production and electricity and heat production [*McLaughlin and Adams Kszos*, 2005; *Meyer et al.*, 2010]. *Miscanthus* is a genus of several species of perennial grasses mostly native to subtropical and tropical regions of Asia. It was used as biofuel crop in Europe since the 1980s and then introduced to the United States recently [*Heaton et al.*, 2008; *Stewart et al.*, 2009]. Several commonly shared characteristics make switchgrass and *Miscanthus* favorite choices as biofuel feedstock resources. First of all, they can produce abundant biomass, with much higher production than maize or soybeans. It normally ranges from 5 to 15 Mg dry matter (DM)/ha^{-1} with maximum production of over 20 Mg DM/ha for switchgrass, and about 20–30 Mg DM/ha with maximum of 60 Mg DM/ha for *Miscanthus* [*Heaton et al.*, 2008;

[1] *Department of Earth, Atmospheric and Planetary Sciences, Purdue University, West Lafayette, Indiana, USA*

[2] *Department of Agronomy, Purdue University, West Lafayette, Indiana, USA*

Remote Sensing of the Terrestrial Water Cycle, Geophysical Monograph 206. First Edition. Edited by Venkat Lakshmi.

Wright and Turhollow, 2010]. In addition, these cellulosic crops have high input use efficiency. They are perennial rhizomatous plants, capable of cycling nutrients seasonally between the above- and below-ground vegetation, and thus minimizing fertilizer application. It was reported that switchgrass and especially *Miscanthus* require no or very small amounts of nitrogen (N) fertilizer, if any, while maize growth may double or even triple the N demand [*Fargione et al.*, 2010; *Lewandowski et al.*, 2003]. Switchgrass and *Miscanthus* are C4 plants and therefore normally more photosynthetically efficient than C3 plants (e.g., soybeans) [*Heaton et al.*, 2004; *Skinner et al.*, 2012]. Also, the agronomic management for these cellulosic crops is relatively less sophisticated and more energy saving than food crops. Planting is required only once, and fertilization or tillage is less frequently needed [*Skinner et al.*, 2012]. Compared with maize, these crops may potentially release less life-cycle greenhouse gases due to less N fertilizer application [*Hillier et al.*, 2009; *Skinner et al.*, 2012].

However, conversion of food crops to cellulosic crops for biofuel production will likely alter carbon (C), N, and more importantly water dynamics, which may eventually impact the water use efficiency in terms of C production per unit of water loss. Earlier studies from either field observations [*Heaton et al.*, 2008; *Wright and Turhollow*, 2010] or model simulations [*Qin et al.*, 2012] showed that cellulosic crops (e.g., *Miscanthus*) could have higher biomass productivity than food crops (e.g., maize). Also, there was evidence indicating that cellulosic crops may use more water than food crops to produce biomass. For instance, a field experiment conducted in Urbana, Illinois, indicated that evapotranspiration from *Miscanthus* was about 104 mm/yr greater than under maize-soybean [*McIsaac et al.*, 2010]. Modeling experiments also suggested that *Miscanthus* consumes more water than maize in order to support crop growth (e.g., [*VanLoocke et al.*, 2010; *Zhuang et al.*, 2013]). It is therefore important to assess water use efficiency to relate biomass production to water consumption.

In order to evaluate regional or national impacts of bioenergy expansion on carbon-water relationship, large-scale models are required to incorporate spatially explicit information of climate, soil, and vegetation [*Le et al.*, 2011; *Vanloocke et al.*, 2010]. In ecosystem models, C and N dynamics are normally simulated to assess biomass production, C exchange, and possibly greenhouse gas emissions. Hydrology components, if available in the model, can be applied to simulate water cycle at large scales. The hydrological simulation together with biomass and C simulation will further be used to estimate regional water balance and water use efficiency [*Le et al.*, 2011; *VanLoocke et al.*, 2010]. For a given region, the amount of biomass production, ecosystem C balance, and biofuel production relative to the amount of water used can be used to interpret water use efficiency at different scales and for different purposes [*VanLoocke et al.*, 2012; *Zhuang et al.*, 2013]. In this study, the main goal is to assess water balance and water use efficiency of different crop ecosystems over the conterminous United States. It is assumed that conventional grain crop, maize, and two cellulosic crops, switchgrass and *Miscanthus*, will be grown on current maize cropland as potential energy crops for biomass production. By using an extended ecosystem model, we propose: (1) to estimate spatially explicit ecosystem productivity and water balance and (2) to assess water use efficiency of ecosystem production, in terms of biomass production and ecosystem C balance with regard to ecosystem water loss.

30.2. MATERIALS AND METHODS

30.2.1. Overview

In order to produce biofuel from biomass feedstocks, conventional maize, switchgrass, and *Miscanthus* can be grown on current maize-producing areas. In the conterminous United States, maize is traditionally considered as food crop, but a considerable proportion of its grain was devoted to ethanol production since the late 2000s. Switchgrass and *Miscanthus* are widely considered as possible alternatives to maize as energy crops in the temperate regions and can be potentially grown on croplands due to their adaptability to different soil and climate environments [*Davis et al.*, 2012; *Fargione et al.*, 2010; *Heaton et al.*, 2004]. A biogeochemical model was used to simulate biomass production and corresponding water dynamics for separate crop ecosystems.

A well-documented process-based model, the Terrestrial Ecosystem Model (TEM), coupling with a hydrological model was first described with emphasis on C dynamics and hydrology. Then, TEM was calibrated and extrapolated to specific regions (i.e., maize-producing areas) to simulate grid-by-grid C fluxes and pools, water balance, and water use efficiency, using spatially referenced data describing local climate, soil, and vegetation characteristics. Finally, spatial analyses were conducted to assess national biomass production, ecosystem C exchange, water consumption, and water use efficiency.

30.2.2. Model Description and Parameterization

TEM is a global-scale ecosystem model, originally designed to estimates C and N fluxes and pool sizes in terrestrial ecosystems at a monthly time step using spatial climate and ecological data [*McGuire et al.*, 1992; *Raich et al.*, 1991]. The model has been updated and

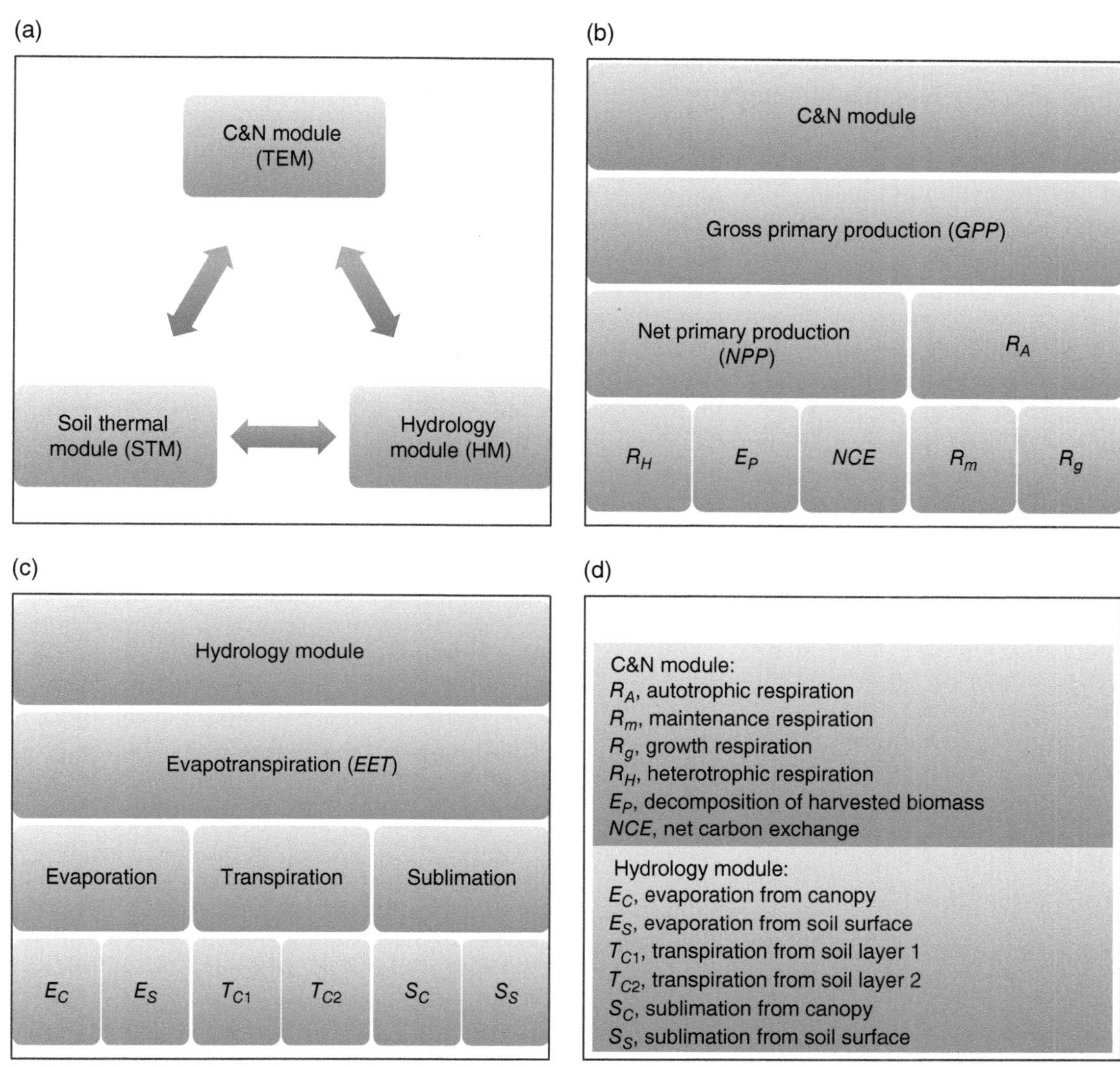

Figure 30.1 Terrestrial Ecosystem Model. (a) The TEM version used in this study includes the core C&N module (original TEM), as well as STM and HM modules. (b) Sketch of carbon allocation in TEM. (c) Sketch of EET modeling in the HM framework. (d) Variable abbreviations in the modules. Transpiration is split into T_{C1} and T_{C2} according to soil layers; sublimation would be considered if snowfall present.

further developed to integrate more up-to-date knowledge and algorithms for different simulating purposes [*Felzer et al.*, 2004; *Melillo et al.*, 2009; *Zhuang et al.*, 2003, 2010]. The TEM version used in this study includes the core module of TEM describing C and N dynamics, the soil thermal module simulating soil thermal dynamics, and a hydrological module of terrestrial ecosystems (Figure 30.1a).

In TEM, the C and N cycles are simulated by dividing the total ecosystem C and N stock into separate vegetation and soil pools, and simulating C and N dynamics using multiple flux variables. Of these fluxes, gross primary production (GPP) is the governing variable describing the rate at which the plant produces useful chemical energy. GPP can be further split into autotrophic respiration, which indicates energy loss due to plant growth and maintenance, and net primary production (NPP), which represents the rate of net "useful" energy produced by an ecosystem's producers (e.g., forest, grass) (Figures 30.1b and 30.1d). In TEM, GPP is modeled as a function of the maximum rate of C assimilation (C_{MAX}) and multivariate factors:

$$\text{GPP} = C_{MAX} \cdot \prod f(X_i), \tag{30.1}$$

where $f(X_i)$ is a scalar used to simulate impacts of physiological, biogeochemical, or environmental variable or process X_i on GPP. The X_i's in TEM include, but are not limited to, factors such as irradiance of photosynthetically active radiation, atmospheric CO_2

concentration, relative canopy conductance, air temperature, moisture, and nitrogen availability [*McGuire et al.*, 1992; *Raich et al.*, 1991]. NPP is mostly referred to as net biomass production of an ecosystem in terms of carbon fixation rate. It is modeled as the difference between GPP and autotrophic respiration. The net carbon exchange (NCE) between the terrestrial biosphere and the atmosphere is described as the remaining C flux in NPP after heterotrophic respiration (R_H) and decomposition of harvested biomass (E_P) [equation (30.2)] [*McGuire et al.*, 2001]. NCE represents the net C flux at the ecosystem scale, with a positive value showing a CO_2 sink and a negative value showing a CO_2 source. In this study, only 30% of maize stover was collected for biofuel use, the rest was returned to soil to maintain soil fertility [*Payne*, 2010]:

$$NCE = NPP - R_H - E_P. \qquad (30.2)$$

The hydrological cycle in TEM consisted of processes of precipitation (rainfall and snowfall), sublimation, evaporation, interception, throughfall, percolation, transpiration, runoff, and drainage [*Zhuang et al.*, 2002]. Evapotranspiration is a significant water loss from ecosystem and mostly used to quantify water consumption of an ecosystem. In TEM, estimated evapotranspiration (EET) is calculated as a total of evaporation (E), transpiration (T) and sublimation (S):

$$\begin{aligned} EET &= E + T + S \\ &= \left(E_C + E_S\right) + \left(T_{C1} + T_{C2}\right) + \left(S_C + S_S\right), \end{aligned} \qquad (30.3)$$

where evaporation, transpiration, and sublimation are separately subdivided into multiple components according to canopy and soil layers (Figures 30.1c and 30.1d). In the model, EET is also constrained by potential evapotranspiration (PET) based on Jensen-Haise formulation [*Jensen and Haise*, 1963] and soil moisture [*Zhuang et al.*, 2002]. In crop ecosystems described in this study, EET simulated in TEM represents the water consumption for crop growth, without further drainage considered. Further modeling details can be found in previous studies [*Zhuang et al.*, 2002, 2003].

The well-organized TEM was then parameterized for maize, switchgrass, and *Miscanthus* ecosystems. Most parameters are constant and have been defined in previous studies (e.g., [*McGuire et al.*, 1992; *Zhuang et al.*, 2003]). Some others, either soil-specific or species-specific parameters, were calibrated using observational data with respect to climate, soil conditions, ecosystem C and N pools, and fluxes. The TEM version used here has been well parameterized for maize, switchgrass, and *Miscanthus* and applied to large regions to assess C and water dynamics [*Qin et al.*, 2011, 2013; *Zhuang et al.*, 2013].

30.2.3. Model Application and Regional Analysis

By assuming that maize, switchgrass, and *Miscanthus* will be grown on the currently available maize-producing areas in the conterminous United States, TEM was applied to simulate the ecosystem C, N, and water dynamics separately for three different ecosystems. The model was forced by spatially referenced information on climate, elevation, soil, and vegetation. The grid-by-grid model outputs, including spatially explicit NPP, NCE, and EET, were used to analyze spatial and regional dynamics of biomass, C, and water.

The model input data describing climate, elevation, soil, and vegetation were organized at a 0.25° latitude $\times$ 0.25° longitude spatial resolution. Specifically, the driving climate data, including the monthly air temperature, precipitation, and cloudiness, were based on CRU (Climatic Research Unit) [*Mitchell and Jones*, 2005]. Annual atmospheric CO_2 concentrations were derived from the Mauna Loa records (http://www.esrl.noaa.gov/gmd/ccgg/trends/). The elevation data were collected from the Shuttle Radar Topography Mission (SRTM) [*Farr et al.*, 2007]. Soil data indicating soil texture of sand, silt, and clay content were based on the Food and Agriculture Organization/Civil Service Reform Committee (FAO/CSRC) digitization of the FAO/UNESCO soil map of the world (1971). Vegetation map describing national crop distribution were extracted from a global crop harvest area database [*Monfreda et al.*, 2008]. These time series data, mostly climate data, were collected from 1900 to 2000, with a time step of one month.

Model simulations were run separately for each crop over the United States at a monthly time step. For each simulation, TEM was first run to equilibrium using the data of 1900, to determine the model initial conditions. Then the model was spun-up for 150 years by repeatedly using the first 50 years' data. The transient simulations were finally run through the 1900–2000 period, and grid-level model outputs of the 1990s were collected for regional analysis. For each ecosystem (i.e., maize, switchgrass, and *Miscanthus*), biomass production (i.e., NPP), C balance (i.e., NCE), and water loss (i.e., EET) were estimated in the model simulations. Water use efficiency (WUE), generally defined as biomass production or yield gain per unit of water consumption, was also used to measure the efficacy of economic gain (e.g., NPP) or ecological gain (e.g., NCE) relative to an environmental cost of water loss [*Ito and Inatomi*, 2012; *Niu et al.*, 2011; *VanLoocke et al.*, 2012]. In this study, biomass WUE (WUE_B) and carbon WUE (WUE_C) were defined in terms of biomass production [equation (30.4)] and C balance [equation (30.5)] at the cost of unit water loss, respectively:

$$WUE_B = NPP / EET, \qquad (30.4)$$

$$WUE_C = NCE / EET. \qquad (30.5)$$

Spatial analyses were conducted to estimate spatial distribution and regional average of C and water dynamics, based on spatially explicit TEM simulations. The decadal averages of the 1990s were presented to show biomass production, C exchange, and water use efficiency.

30.3. RESULTS

30.3.1. Ecosystem Productivity

Ecosystem production, in terms of NPP and NCE, varies among different ecosystems and also differs over space due to spatial heterogeneity of climate, soil, and vegetation conditions. As reported previously, most biomass production concentrates in the intensive cropping areas in the Midwest [*Qin et al.*, 2012]. The grid-level NPP statistics (not area weighted) show that (Figures 30.2a–30.2c) maize has a relatively small spatial variation, with 500–900 g C/m^{-2} of NPP at most grids. *Miscanthus*, however, shows widely distributed NPP with most grids ranging from 1100 to 1900 g C/m^{-2}. From the perspective of national average, *Miscanthus* produces twice as much NPP as maize or switchgrass could (Table 30.1).

In TEM, NCE accounts for the net C balance at the ecosystem scale, and the flux is highly dependent on the spatially explicit environmental conditions such as soil and climate. For maize, most of the intensive cropping areas in the Midwest (except the Illinois area) act as net C sources (Figure 30.3a). Swichgrass and *Miscanthus*, however, have vast areas showing positive

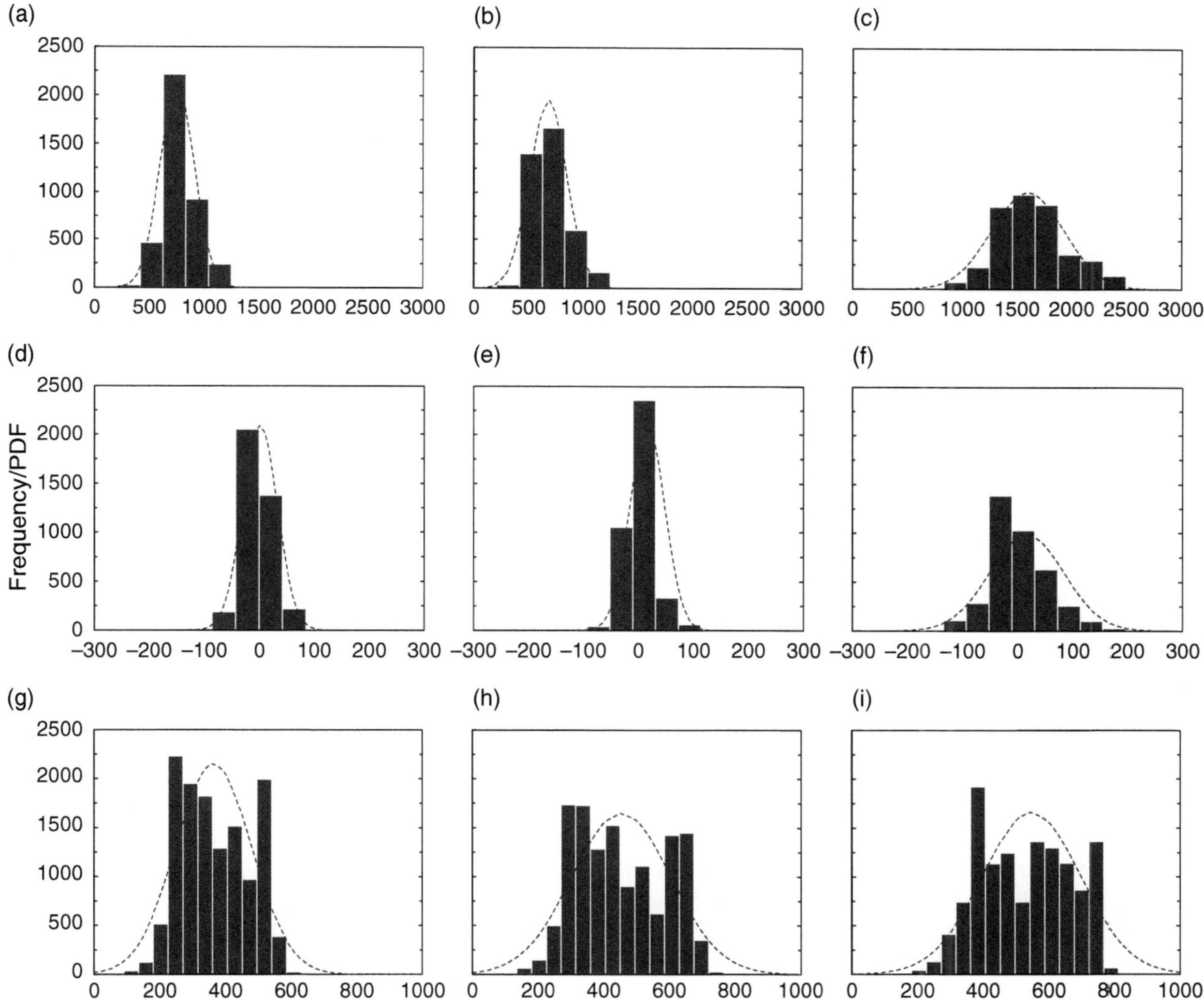

Figure 30.2 Spatial variations of the estimated NPP, NCE, and EET. Grid-level estimates were made for NPP (g C/m^{-2} yr^{-1}) of (a) maize, (b) switchgrass, and (c) *Miscanthus*; NCE (g C/m^{-2} yr^{-1}) of (d) maize, (e) switchgrass, and (f) *Miscanthus*; and EET (mm) of (g) maize, (h) switchgrass, and (i) *Miscanthus*. The NPP, NCE, and EET are presented with bars showing frequency and dashed lines indicating Gaussian distribution. Actual vegetation area of grid is not considered.

Table 30.1 Estimated national average NPP, NCE, and EET

Crop	NPP (g C/m^{-2})	NCE (g C/m^{-2})	EET (mm)
Maize	713 (66)	−1.9 (0.3)	347 (42)
Switchgrass	622 (49)	7.1 (1.0)	440 (45)
Miscanthus	1513 (122)	11.0 (1.9)	523 (51)

Note: Net primary production (NPP), net carbon exchange (NCE), and evapotranspiration (EET) are reported as corresponding decadal averages (1990s) with temporal standard deviations in parentheses. NPP details can also be found in *Qin et al.* [2012].

NCE and therefore potentially mitigate C emissions (Figures 30.3b and 30.3c). Spatially, about 90% of the cropping grids have an NCE (not area weighted) ranging from −50 to 50 g C/m^{-2} in maize and switchgrass ecosystems. Even though switchgrass has a positive mean NCE and maize has a negative value, they share similar spatial variation with a standard deviation (SD) of about 30 g C/m^{-2} (Figures 30.2d and 30.2e). *Miscanthus*, however, has a more spatially heterogeneous NCE distribution, with about 60% grids ranging from −50 to 50 g C/m^{-2} and 86% ranging from −100 to 100 g C/m^{-2} (Figure 30.2f). Considering actual cropping area, the maize ecosystem produces a national average NCE of −1.9 g C/m^{-2}. The switchgrass and *Miscanthus* ecosystems produce 9 and 12.9 g C/m^{-2} more NCE than maize, respectively, both acting as C sinks at national scales (Table 30.1).

The model results suggest that crop switching from maize to switchgrass may cause a net decrease of NPP of 91 g C/m^{-2} nationally but, meanwhile, may create a national C sink. If switched to *Miscanthus*, the ecosystem would increase both biomass production and potential C mitigation.

30.3.2. Evapotranspiration at Ecosystem Scales

Cellulosic crop ecosystems, especially a *Miscanthus* ecosystem, show a significantly higher evapotranspiration than a maize ecosystem, as simulated from TEM. As shown in our previous report [*Zhuang et al.*, 2013], EET distributes mainly along the dominant maize-producing areas in the Midwest, with especially high annual EET in the states of Illinois and Indiana. Compared with a maize ecosystem, switchgrass has an overall higher EET and *Miscanthus* has the highest EET among all. Annual EET varies dramatically over space. It was estimated that, across the majority of cropping areas, the actual water loss through EET is 200–550 mm in maize ecosystems and increases to 250–600 mm in switchgrass and 300–800 in *Miscanthus* ecosystems [*Zhuang et al.*, 2013]. Statistically, the maize

ecosystem has the lowest mean EET, as well as the smallest spatial EET variation among all three crop systems (Figure 30.2g). Switchgrass (Figure 30.2h) and *Miscanthus* (Figure 30.2i), show respective increases of one quarter and one half on the basis of maize in terms of both mean and variation.

According to estimates based on maize harvested areas, the national average EET of switchgrass and *Miscanthus* is 27% and 51% higher than that of maize, respectively (Table 30.1). If, as hypothesized here, crop switching from maize to bioenergy crops such as switchgrass or *Miscanthus* occurs, the annual EET will increase in most areas. Due to land cover change from maize to switchgrass, the EET increases about 90 mm on average, with 60–120 mm increase in most places. If crop is changed to *Miscanthus*, the EET of most locations will increase 140–210 mm, with an annual average increase of 176 mm (Table 30.1). Similar results have also been reported, mostly for the Midwest of the United States. By applying a multilayer canopy model, *Le et al.* [2011] estimated evapotranspiration under climate condition of 2005. It was reported that switchgrass and *Miscanthus* have, respectively, 118 and 208 mm higher total EET than maize. *Vanloocke et al.* [2012] simulated 30 year (1973–2002) hydrology using an ecosystem model and estimated that switchgrass EET is 25–150 mm higher and *Miscanthus* is 50–200 mm higher than maize EET. These results are comparable, and the differences are partly caused by input data, including climate and soil data, model structure, simulation year, and the study regions. It was believed that the EET differences between maize and these bioenergy crops are mostly resulted from the density and architecture of aboveground foliage [*Le et al.*, 2011].

30.3.3. Water Use Efficiency

The WUE$_B$ and WUE$_C$ were determined, respectively, as NPP and NCE produced at cost of each unit of water loss through evapotranspiration. Generally, the spatial distribution of WUE$_B$ shows that the switchgrass ecosystem has lower efficiency than maize and *Miscanthus* ecosystems (Figures 30.4a–30.4c). *Miscanthus*, in particular, has the highest WUE in most Midwest areas (Figure 30.4c). This is understandable considering that switchgrass produces the lowest NPP but with higher EET in comparison with maize. *Miscanthus* consumes even more water than switchgrass, but its biomass production more than doubles that of maize or switchgrass, and thus owns much high WUE$_B$. Unlike NPP and EET, the pattern of WUE$_B$ spatial distribution is similar among different ecosystems (Figures 30.5a–30.5c). Averaged over the whole cropping area, the national WUE$_B$ of maize, switchgrass, and

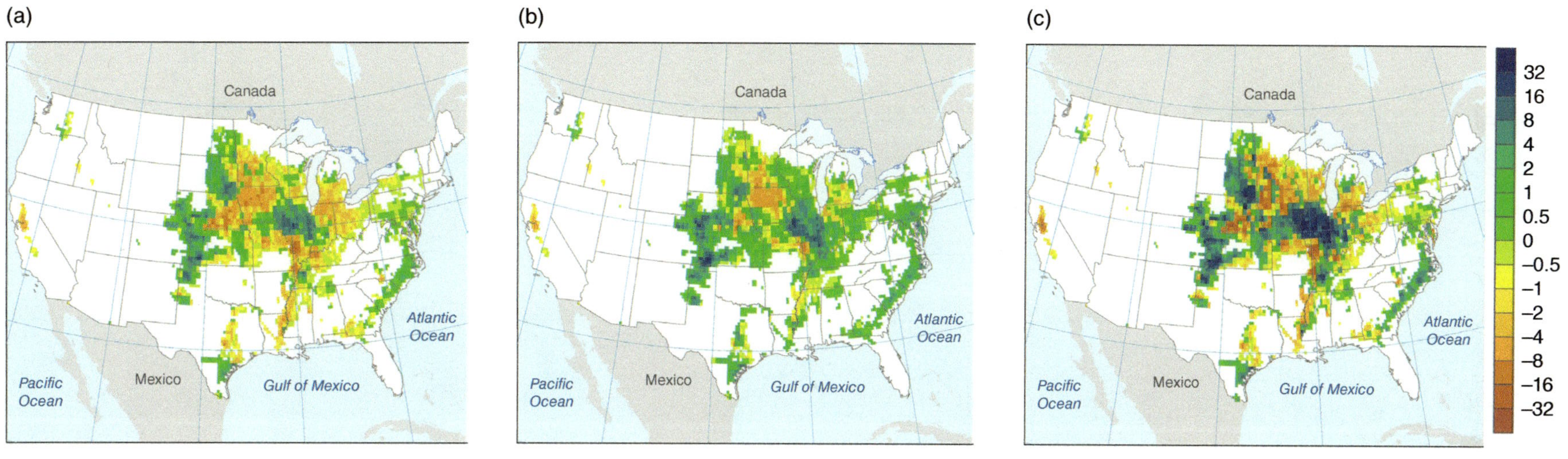

Figure 30.3 Estimated spatial NCE over the maize-producing areas in the United States. Spatial estimates were made for NCE (g C/m^{-2} yr^{-1}) of (a) maize, (b) switch-grass, and (c) *Miscanthus*. Grid-level NCE values are area weighted. Note: Spatial estimates for NPP [*Qin et al.*, 2012] and EET [*Zhuang et al.*, 2013] were reported previously.

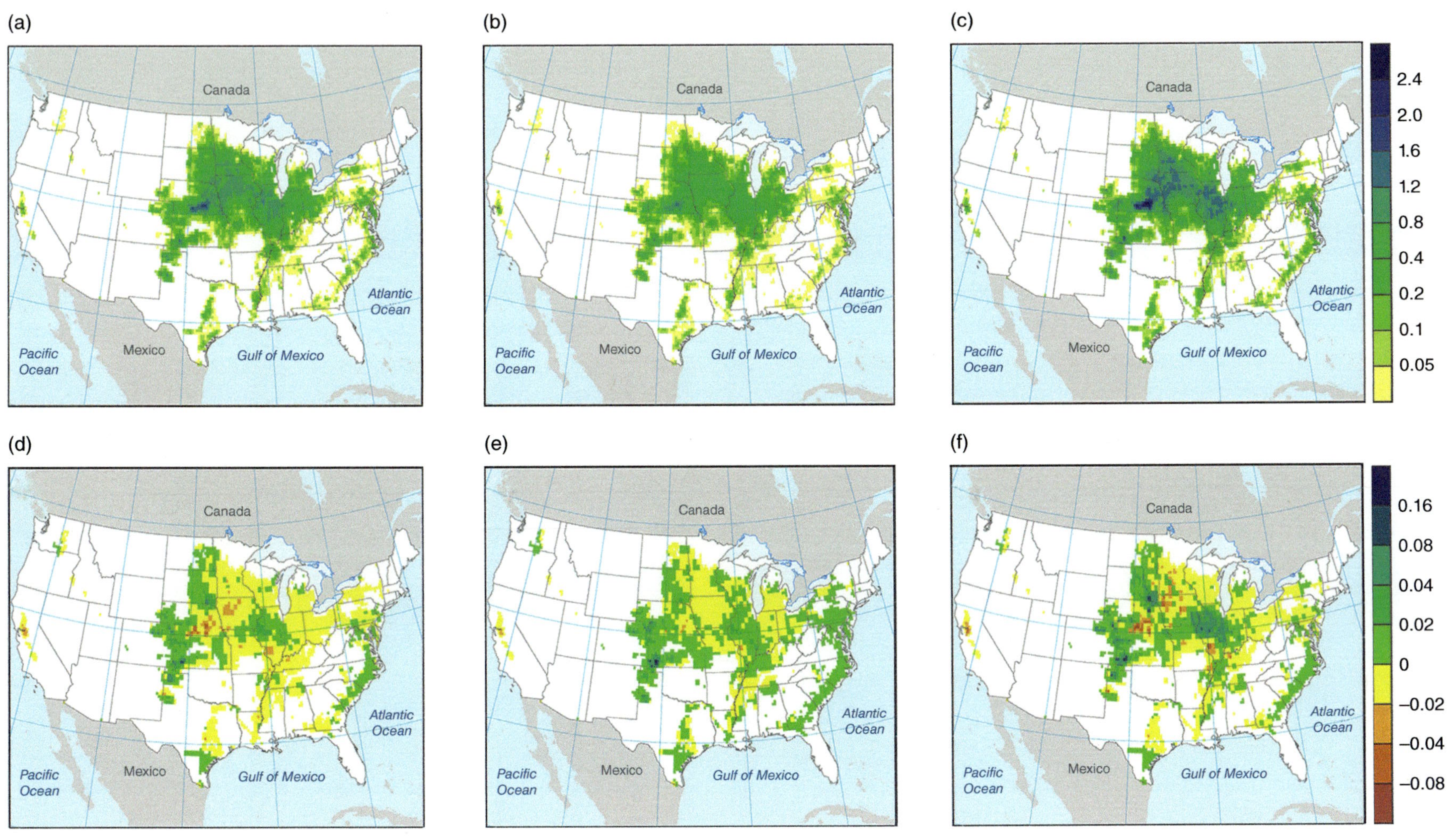

Figure 30.4 Estimated spatial WUE of NPP and NCE over the maize-producing areas in the United States. Spatial estimates were made for WUE of NPP (kg C/m^{-3}) of (a) maize, (b) switchgrass, and (c) *Miscanthus*, and WUE of NCE (kg C/m^{-3}) of (d) maize, (e) switchgrass, and (f) *Miscanthus*. Grid-level values are area weighted.

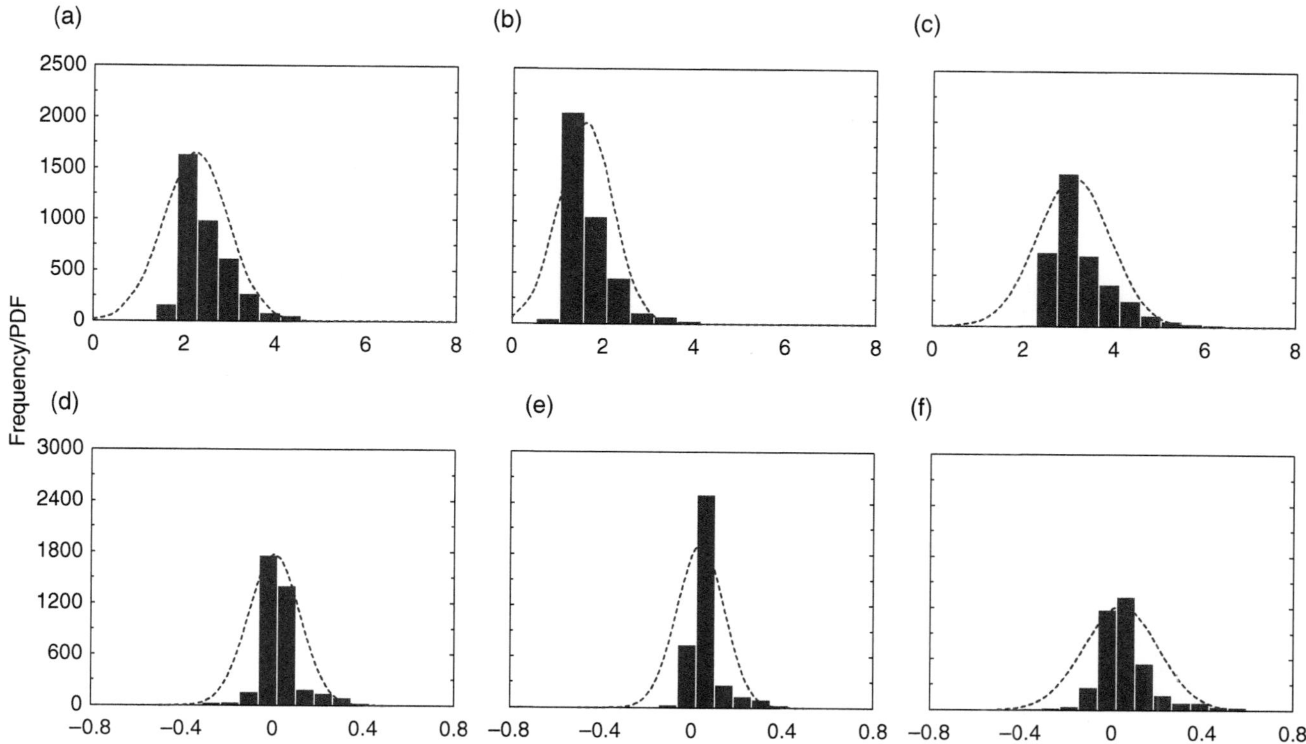

Figure 30.5 Spatial variations of estimated WUE. Grid-level estimates were made for WUE of NPP (kg C/m⁻³) of (a) maize, (b) switchgrass, and (c) *Miscanthus*, and WUE of NCE (kg C/m⁻³) of (d) maize, (e) switchgrass, and (f) *Miscanthus*. The WUE of NPP and NCE are presented with bars showing frequency and dashed lines indicating Gaussian distribution. Actual vegetation area of grid is not considered.

Miscanthus are 2.3, 1.6, and 3.1 kg C/m⁻³, respectively (Table 30.2). That is, with each unit of water loss, *Miscanthus* could produce 35% more NPP than maize and 94% more than switchgrass.

The WUE_C spatial distribution shows the size of NCE flux relative to EET and the net impact of C mitigation (Figures 30.4d–30.4f). For NCE, the negative WUE_C indicates a net C source and the positive one indicates a C sink. Apparently, the maize ecosystem has more areas acting as net C sources and the WUE_C is 0 ± 0.02 kg C/m⁻³ at most sites (Figure 30.4d). Statistics shows that over 80% of grids lie between −0.05 and 0.05 kg C/m⁻³ of WUE_C (Figure 30.5d). For switchgrass, most of the cropping area shows a small positive WUE_C (Figure 30.4e), with about 65% of grids in the range of 0–0.05 kg C/m⁻³ (Figure 30.5e). A *Miscanthus* ecosystem has much higher spatial variations than maize and switchgrass ecosystems, with about 90% of grids ranging between −0.2 and 0.3 kg C/m⁻³ (Figure 30.5f). Overall, the maize ecosystem acts as a net C source with smallest WUE_C at national scales, and cellulosic ecosystems act as a net C sink with similar WUE_C. For each cubic meter of water loss, maize produces 5.0 kg C of C emissions. Switchgrass and *Miscanthus* mitigate 18.4 and 22.5 kg C of C emissions, respectively (Table 30.2).

Table 30.2 National average water use efficiency

Crop	Water Use Efficiency	
	WUE_B (kg C m⁻³)[*]	WUE_C (g C m⁻³)[†]
Maize	2.3 (0.7)	−5.0 (0.7)
Switchgrass	1.6 (0.6)	18.4 (2.0)
Miscanthus	3.1 (0.8)	22.5 (3.4)

Note: Water use efficiency of biomass[*] and carbon exchange[†] were calculated as in equations (30.4) and (30.5), respectively. Results are reported as decadal averages (1990s) with temporal standard deviations in parentheses.

30.4. DISCUSSION

30.4.1. Input Use Efficiency as a Measure of Resource Allocation

Climate conditions, soil fertility, land availability, and water availability are several dominant environmental factors determining crop growth and biomass production. Land and water, in particular, are two major resources for producing biomass feedstock. Among the three energy crops, each will exclude others from using a certain amount of water at a given region. The goal of this study is to evaluate the ecosystem production of these crops under the same environmental conditions, but with different cropping systems. As estimated here and reported elsewhere [*Heaton et al.*, 2008; *Qin et al.*, 2012], cellulosic crops (particularly *Miscanthus*) are capable of accumulating a considerable amount of C (e.g., NPP) and using soil nutrients efficiently (e.g., nitrogen), at a given land area. They have higher land use efficiency (LUE) than many food-based crops (e.g., maize) in terms of biomass production per land area. This is mostly because cellulosic crops have a high photosynthetic productivity due to important characteristics such as high efficiency of solar radiation interception and conversion [*Heaton et al.*, 2008], large leaf area, and long canopy duration [*Dohleman and Long*, 2009; *Heaton et al.*, 2004]. Switchgrass and *Miscanthus* could also produce positive NCE, making great contributions to C mitigation (Table 30.1). Compared with annual plants (e.g., maize), these perennial plants could survive multiple years with less soil disturbance due to agricultural management such as tillage and rotation [*Heaton et al.*, 2004]. Also, the cellulosic ecosystems can sequester a large amount of C in belowground biomass and keep a relatively high level of soil carbon [*Kahle et al.*, 2001; *Lee et al.*, 2007].

It is possible that maize may outweigh cellulosic crops in terms of biomass-based WUE, due to its lower water loss during growth. Switchgrass is indeed less efficient than maize due to its lower biomass productivity and higher water use. *Miscanthus*, however, is still more productive in biomass production when using the same amount of water (Table 30.2). It was reported that, per cubic meter of water depletion, about 1.1–2.7 kg maize yield was produced globally [*Zwart and Bastiaanssen*, 2004]. That is about 1.0–2.4 kg C/m^{-3} of WUE$_B$. The results for the United States in this study fall in the upper end of this range, and comparable with other site [*Hickman et al.*, 2010] or regional estimates [*VanLoocke et al.*, 2012]. A similar estimation in the Midwest also found that *Miscanthus* has higher and switchgrass has lower WUE$_B$ than maize [*VanLoocke et al.*, 2012]. It suggests that productive *Miscanthus* compensates its high LUE for water loss and still results in a relatively high WUE$_B$. Switchgrass, however, is highly water consuming but with no comparable LUE or biomass productivity. In terms of WUE$_C$, the model experiments here and elsewhere [*VanLoocke et al.*, 2012] suggest that switchgrass and *Miscanthus* could positively affect ecosystem C sequestration while maize has a negative impact. From the perspective of resource use (mainly land and water), *Miscanthus* rather than switchgrass could be an efficient substitute to maize as biomass feedstock resource.

Other resources besides land and water should also be factored into the consideration of crop switch. For example, fertilizer application and nutrient uptake are key factors determining crop nutrient use efficiency (NUE). It was reported that cellulosic crops may require less fertilization than maize due to their high NUE [*Fargione et al.*, 2010; *Lewandowski et al.*, 2003]. This may further benefit greenhouse gas mitigation since nitrogen fertilizer contributes significantly to ecosystem nitrous oxide emissions [*Hoben et al.*, 2011]. It should also be noted that input use efficiency assesses the biomass or C productivity relative to resource input but does not necessarily consider the economic, temporal, or spatial availability of these resources. Especially in the study, water input only accounted for precipitation and did not consider possible irrigation and other agricultural practices (e.g., tillage and rotation) that may affect water available for crop growth. Further analyses regarding issues of water availability still await future study.

30.4.2. Limitations and Future Needs

Unlike crop models that focus on crop yield estimation, ecosystem models are often used to estimate biogeochemical cycles in natural or agricultural ecosystems. Even with simulated C dynamics and additional algorithms describing C allocation and crop yield formation, ecosystem models, such as TEM, are still lacking detailed information and processes on agricultural management, more often lacking supporting data, which may further impact the accuracy of biomass prediction [*Qin et al.*, 2012]. More data of agricultural practices, such as irrigation, rotation, tillage, fertilizer application, and timing of planting, will improve ecosystem model predictability. Still, caution should be used when interpreting spatial heterogeneity of these practices and corresponding spatial data [*Davis et al.*, 2012; *VanLoocke et al.*, 2010].

Miscanthus would not be grown on croplands for biofuel simply because of its high LUE and WUE. Many other factors should also be included in a life-cycle assessment for certain biofuels [*Davis et al.*, 2009]. Issues such as food security, economic viability, and ethical concerns could all affect decision making [*Fargione et al.*, 2010; *Pimentel et al.*, 2010; *Tilman et al.*, 2009]. From the environmental perspective, many issues concerning biofuel development and land use introduce large uncertainties into regional estimations of large-scale bioenergy expansion. Growing bioenergy crops, especially cellulosic crops, instead of conventional food crops, may have fundamental impacts on ambient climate (e.g., greenhouse gas, air temperature, moisture) [*Bessou et al.*, 2011; *Hallgren et al.*, 2013], soil quality (e.g., soil carbon, soil acidity) [*Cayuela et al.*, 2010; *Clifton-Brown et al.*, 2007], water quality (e.g., N and P concentration), as well as water quantity [*Behnke et al.*, 2012; *Skinner et al.*, 2012]. These impacts are important but not well studied for the newly established ecosystems, such as switchgrass and

Miscanthus. More evidence from field observations is required to improve ecosystem modeling and large-scale model extrapolation.

In our study, we only considered maize cropland for biofuel cropping. However, other types of land could also serve as potential biofuel land sources. For example, conservation reserve program land could be properly cultivated to produce biomass [*Lee et al.*, 2013]. Marginal lands, including most abandoned or degraded cropland and grassland where most traditional food crops may not survive due to poor soil or climate conditions, could be used to grow cellulosic crops with high environmental stress resistance [*Gelfand et al.*, 2013; *Varvel et al.*, 2008]. Switchgrass, under this circumstance, could be much more competitive than maize with higher LUE. But still, besides biomass production, other environmental issues including water availability, nutrient sustainability, and those understudied problems for cropland should be further investigated to uncover the potential consequences of growing cellulosic crops (e.g., switchgrass and *Miscanthus*) on marginal lands.

30.5. SUMMARY

To assess WUE of bioenergy crops (i.e., maize, switchgrass, and *Miscanthus*) grown on cropland, an ecosystem model was used to estimate regional ecosystem productivity and evapotranspiration over the conterminous United States. Compared with maize, switchgrass has relatively lower biomass productivity while *Miscanthus* has much higher productivity. Nationally, both cellulosic crops have higher net carbon exchange than maize, acting as net C sinks. Further analyses suggest that, in terms of biomass production at the cost of unit water loss, the productive *Miscanthus* compensates its high land use efficiency for water loss and results in the highest water use efficiency among three bioenergy crops. Switchgrass, however, is highly water-consuming but with no comparable biomass productivity. Its water use efficiency is the lowest among the three crops. At given water loss level, switchgrass and *Miscanthus* ecosystems sequester a similar amount of carbon, while the maize ecosystem releases carbon. More evidence from field observations is required to improve ecosystem modeling and large-scale extrapolation analysis of other environmental impacts. Further analyses on using other land sources (e.g., marginal lands) should also be conducted for future biofuel development.

ACKNOWLEDGMENT

The authors are thankful to Yaling Liu and anonymous reviewers for their valuable and constructive comments. Computing is supported by Rosen Center for Advanced Computing (RCAC) at Purdue University. This study is supported through projects funded by the NASA Land Use and Land Cover Change program (NASA-NNX09AI26G), Department of Energy (DE-FG02-08ER64599), the NSF Division of Information & Intelligent Systems (NSF-1028291), and the NSF Carbon and Water in the Earth Program (NSF-0630319).

REFERENCES

Behnke, G. D., M. B. David, and T. B. Voigt (2012), Greenhouse gas emissions, nitrate leaching, and biomass yields from production of Miscanthus × giganteus in Illinois, USA, *BioEnergy Res.*, *5*(4), 801–813.

Beringer, T. I. M., W. Lucht, and S. Schaphoff (2011), Bioenergy production potential of global biomass plantations under environmental and agricultural constraints, *GCB Bioenergy*, *3*(4), 299–312.

Bessou, C., F. Ferchaud, B. Gabrielle, and B. Mary (2011), Biofuels, greenhouse gases and climate change, *Sustain. Agric.*, *2*, 365–468.

Cayuela, M. L., O. Oenema, P. J. Kuikman, R. R. Bakker, and J. W. Van Groenigen (2010), Bioenergy by-products as soil amendments? Implications for carbon sequestration and greenhouse gas emissions, *GCB Bioenergy*, *2*(4), 201–213.

Clifton-Brown, J. C., J. Breuer, and M. B. Jones (2007), Carbon mitigation by the energy crop, Miscanthus, *Global Change Biol.*, *13*(11), 2296–2307.

Davis, S. C., K. J. Anderson-Teixeira, and E. H. DeLucia (2009), Life-cycle analysis and the ecology of biofuels, *Trends Plant Sci.*, *14*(3), 140–146.

Davis, S. C., W. J. Parton, S. J. D. Grosso, C. Keough, E. Marx, P. R. Adler, and E. H. DeLucia (2012), Impact of second-generation biofuel agriculture on greenhouse-gas emissions in the corn-growing regions of the US, *Front. Ecol. Environ.*, *10*, 69–74.

Dohleman, F. G., and S. P. Long (2009), More productive than maize in the Midwest: How does Miscanthus do it? *Plant Physiol.*, *150*(4), 2104–2115.

FAOSTAT (2012), FAOSTAT, available at http://faostat.fao.org/, accessed May 2012.

Fargione, J., R. J. Plevin, and J. D. Hill (2010), The ecological impact of biofuels, *Annu. Rev. Ecol. Evol. Systemat.*, *41*(1), 351–377.

Farr, T. G., P. A. Rosen, E. Caro, R. Crippen, R. Duren, S. Hensley, M. Kobrick, M. Paller, E. Rodriguez, and L. Roth (2007), The shuttle radar topography mission, *Rev. Geophys.*, *45*(2), RG2004.

Felzer, B., D. Kicklighter, J. Melillo, C. Wang, Q. Zhuang, and R. Prinn (2004), Effects of ozone on net primary production and carbon sequestration in the conterminous United States using a biogeochemistry model, *Tellus B*, *56*(3), 230–248.

Gelfand, I., R. Sahajpal, X. Zhang, R. C. Izaurralde, K. L. Gross, and G. P. Robertson (2013), Sustainable bioenergy production from marginal lands in the US Midwest, *Nature*, *493*, 514–517.

Hallgren, W., C. A. Schlosser, D. Kicklighter, and A. Sokolov (2013), Climate impacts of a large-scale biofuels expansion, *Geophys. Res. Lett*, *40*, 1624–1630.

Heaton, E., T. Voigt, and S. P. Long (2004), A quantitative review comparing the yields of two candidate C4 perennial biomass crops in relation to nitrogen, temperature and water, *Biomass Bioenergy*, *27*(1), 21–30.

Heaton, E. A., F. G. Dohleman, and S. P. Long (2008), Meeting US biofuel goals with less land: The potential of Miscanthus, *Global Change Biol.*, *14*(9), 2000–2014.

Hickman, G. C., A. Vanloocke, F. G. Dohleman, and C. J. Bernacchi (2010), A comparison of canopy evapotranspiration for maize and two perennial grasses identified as potential bioenergy crops, *GCB Bioenergy*, *2*(4), 157–168.

Hillier, J., C. Whittaker, G. Dailey, M. Aylott, E. Casella, G. M. Richter, A. Riche, R. Murphy, G. Taylor, and P. Smith (2009), Greenhouse gas emissions from four bioenergy crops in England and Wales: Integrating spatial estimates of yield and soil carbon balance in life cycle analyses, *GCB Bioenergy*, *1*(4), 267–281.

Hoben, J. P., R. J. Gehl, N. Millar, P. R. Grace, and G. P. Robertson (2011), Nonlinear nitrous oxide (N_2O) response to nitrogen fertilizer in on-farm corn crops of the US Midwest, *Global Change Biol.*, *17*(2), 1140–1152.

Ito, A., and M. Inatomi (2012), Water-use efficiency of the terrestrial biosphere: A model analysis focusing on interactions between the global carbon and water cycles, *J. Hydrometeorol.*, *13*(2), 681–694.

Jensen, M. E., and H. R. Haise (1963), Estimating evapotranspiration from solar radiation, *Proc. Am. Soc. Civil Eng., J. Irrig. Drainage Div.*, *89*, 15–41.

Kahle, P., S. Beuch, B. Boelcke, P. Leinweber, and H. R. Schulten (2001), Cropping of Miscanthus in Central Europe: Biomass production and influence on nutrients and soil organic matter, *Eur. J. Agron.*, *15*(3), 171–184.

Le, P. V. V., P. Kumar, and D. T. Drewry (2011), Implications for the hydrologic cycle under climate change due to the expansion of bioenergy crops in the Midwestern United States, *Proc. Natl. Acad. Sci.*, *108*(37), 15,085–15,090.

Lee, D. K., V. N. Owens, and J. J. Doolittle (2007), Switchgrass and soil carbon sequestration response to ammonium nitrate, manure, and harvest frequency on conservation reserve program land, *Agron. J.*, *99*(2), 462–468.

Lee, D., E. Aberle, C. Chen, J. Egenolf, K. Harmoney, G. Kakani, R. L. Kallenbach, and J. C. Castro (2013), Nitrogen and harvest management of Conservation Reserve Program (CRP) grassland for sustainable biomass feedstock production, *GCB Bioenergy*, *5*(1), 6–15.

Lewandowski, I., J. M. O. Scurlock, E. Lindvall, and M. Christou (2003), The development and current status of perennial rhizomatous grasses as energy crops in the US and Europe, *Biomass Bioenergy*, *25*(4), 335–361.

McGuire, A. D., J. M. Melillo, L. A. Joyce, D. W. Kicklighter, A. L. Grace, B. Moore Iii, and C. J. Vorosmarty (1992), Interactions between carbon and nitrogen dynamics in estimating net primary productivity for potential vegetation in North America, *Global Biogeochem. Cycles*, *6*(2), 101–124.

McGuire, A. D., et al. (2001), Carbon balance of the terrestrial biosphere in the twentieth century: Analyses of CO_2, climate and land use effects with four process-based ecosystem models, *Global Biogeochem. Cycles*, *15*(1), 183–206.

McIsaac, G. F., M. B. David, and C. A. Mitchell (2010), *Miscanthus* and switchgrass production in Central Illinois: Impacts on hydrology and inorganic nitrogen leaching, *J. Environ. quality*, *39*(5), 1790–1799.

McLaughlin, S. B., and L. Adams Kszos (2005), Development of switchgrass (Panicum virgatum) as a bioenergy feedstock in the United States, *Biomass Bioenergy*, *28*(6), 515–535.

Melillo, J. M., J. M. Reilly, D. W. Kicklighter, A. C. Gurgel, T. W. Cronin, S. Paltsev, B. S. Felzer, X. Wang, A. P. Sokolov, and C. A. Schlosser (2009), Indirect emissions from biofuels: How important? *Science*, *326*(5958), 1397.

Meyer, M. H., J. Paul, and N. O. Anderson (2010), Competive ability of invasive Miscanthus biotypes with aggressive switchgrass, *Biol. Invas.*, *12*(11), 3809–3816.

Mitchell, T. D., and P. D. Jones (2005), An improved method of constructing a database of monthly climate observations and associated high-resolution grids, *Int. J. Climatol.*, *25*(6), 693–712.

Monfreda, C., N. Ramankutty, and J. A. Foley (2008), Farming the planet: 2. Geographic distribution of crop areas, yields, physiological types, and net primary production in the year 2000, *Global Biogeochem. Cycles*, *22*(1), 1–19.

Niu, S., X. Xing, Z. Zhang, J. Xia, X. Zhou, B. Song, L. Li, and S. Wan (2011), Water-use efficiency in response to climate change: From leaf to ecosystem in a temperate steppe, *Global Change Biol.*, *17*(2), 1073–1082.

Payne, W. A. (2010), Are biofuels antithetic to long-term sustainability of soil and water resources? in *Advances in Agronomy*, vol. 105, edited by D. L. Sparks, pp. 1–46, Elsevier Academic, San Diego.

Pimentel, D., A. Marklein, M. A. Toth, M. N. Karpoff, G. S. Paul, R. McCormack, J. Kyriazis, and T. Krueger (2010), Environmental and economic costs of biofuels, *Human Ecol.*, *37*(1), 349–369.

Qin, Z., Q. Zhuang, X. Zhu, X. Cai, and X. Zhang (2011), Carbon consequences and agricultural implications of growing biofuel crops on marginal agricultural lands in China, *Environ. Sci. Technol.*, *45*(24), 10,765–10,772.

Qin, Z., Q. Zhuang, and M. Chen (2012), Impacts of land use change due to biofuel crops on carbon balance, bioenergy production, and agricultural yield, in the conterminous United States, *GCB Bioenergy*, *4*(3), 277–288.

Raich, J. W., E. B. Rastetter, J. M. Melillo, D. W. Kicklighter, P. A. Steudler, B. J. Peterson, A. L. Grace, B. Moore Iii, and C. J. Vorosmarty (1991), Potential net primary productivity in South America: Application of a global model, *Ecol. Appl.*, *1*(4), 399–429.

Searchinger, T., R. Heimlich, R. A. Houghton, F. Dong, A. Elobeid, J. Fabiosa, S. Tokgoz, D. Hayes, and T. H. Yu (2008), Use of US croplands for biofuels increases greenhouse gases through emissions from land-use change, *Science*, *319*(5867), 1238.

Skinner, R., W. Zegada-Lizarazu, and J. Schmidt (2012), Environmental impacts of switchgrass management for bioenergy production, *in Switchgrass: A Valuable Biomass Crop for Energy* (ed. A Monti). Springer-Verlag, London.

Stewart, J., Y. O. Toma, F. G. FernáNdez, A. Y. A. Nishiwaki, T. Yamada, and G. Bollero (2009), The ecology and agronomy of Miscanthus sinensis, a species important to bioenergy crop development, in its native range in Japan: A review, *GCB Bioenergy*, *1*(2), 126–153.

Tilman, D., R. Socolow, J. A. Foley, J. Hill, E. Larson, L. Lynd, S. Pacala, J. Reilly, T. Searchinger, and C. Somerville (2009), Beneficial biofuels—The food, energy, and environment trilemma, *Science*, *325*(5938), 270–271.

U.S. Congress (2007), The Energy Independence and Security Act of 2007 (H.R. 6), available at http://energy.senate.gov/public/index.cfm?FuseAction=IssueItems.Detail&IssueItem_ID=f10ca3dd-fabd-4900-aa9d-c19de47df2da&Month=12&Year=2007, accessed May 2011.

U.S. Department of Agriculture (USDA) (2010), USDA agricultural projections to 2019, Rep. OCE-2010-1, USDA, Washington, D.C.

Vanloocke, A., C. J. Bernacchi, and T. E. Twine (2010), The impacts of Miscanthus × giganteus production on the Midwest US hydrologic cycle, *GCB Bioenergy*, *2*(4), 180–191.

VanLoocke, A., T. E. Twine, M. Zeri, and C. J. Bernacchi (2012), A regional comparison of water use efficiency for miscanthus, switchgrass and maize, *Agric. Forest Meteorol.*, *164*, 82–95.

Varvel, G. E., K. P. Vogel, R. B. Mitchell, R. Follett, and J. Kimble (2008), Comparison of corn and switchgrass on marginal soils for bioenergy, *Biomass Bioenergy*, *32*(1), 18–21.

Wright, L., and A. Turhollow (2010), Switchgrass selection as a "model" bioenergy crop: A history of the process, *Biomass Bioenergy*, *34*(6), 851–868.

Zhuang, Q., A. McGuire, K. O'neill, J. Harden, V. Romanovsky, and J. Yarie (2002), Modeling the soil thermal and carbon dynamics of a fire chronosequence in interior Alaska, *J. Geophys. Res.*, *107*, 8147.

Zhuang, Q., A. McGuire, J. Melillo, J. Clein, R. Dargaville, D. Kicklighter, R. Myneni, J. Dong, V. Romanovsky, and J. Harden (2003), Carbon cycling in extratropical terrestrial ecosystems of the Northern Hemisphere during the 20th century: A modeling analysis of the influences of soil thermal dynamics, *Tellus B*, *55*(3), 751–776.

Zhuang, Q., J. He, Y. Lu, L. Ji, J. Xiao, and T. Luo (2010), Carbon dynamics of terrestrial ecosystems on the Tibetan Plateau during the 20th century: An analysis with a process-based biogeochemical model, *Global Ecol. Biogeogr.*, *19*(5), 649–662.

Zhuang, Q., Z. Qin, and M. Chen (2013), Biofuel, land and water: Maize, switchgrass or Miscanthus? *Environ. Res. Lett.*, *8*(1), 015,020.

Zwart, S. J., and W. G. M. Bastiaanssen (2004), Review of measured crop water productivity values for irrigated wheat, rice, cotton and maize, *Agric. Water Manag.*, *69*(2), 115–133.

31

Watershed Reanalysis of Water and Carbon Cycle Models at a Critical Zone Observatory

Xuan Yu,[1] Christopher Duffy,[1] Jason Kaye,[2] Wade Crow,[3] Gopal Bhatt,[1] and Yuning Shi[4]

31.1. INTRODUCTION

Terrestrial water and nutrient cycling are generally described as interacting hydrological and biogeochemical processes with various assumptions about the degree of coupling. Biogeochemical processes are generally assumed to be relatively slow in comparison to rainfall-runoff dynamics. From a hydrologic point of view, the assumption is typically that biophysical properties of vegetation are fixed or simply respond to the seasonal changes in the energy-water conditions without long-term changes. This assumption may provide adequate short-term hydrological model performance, however, longer term vegetation and soil biophysical contributions to hydrological processes are not resolved with this approach [*Brolsma and Bierkens*, 2007; *Miller et al.*, 2010; *Smucker and Hopmans*, 2007].

Stand-level process-based ecosystem models explicitly calculate carbon and water fluxes at time steps varying from hourly to monthly at a point, emphasizing one-dimensional (1D) transport and redistribution of water, energy, and nutrients [*Hanson et al.*, 2004]. The carbon and nitrogen fluxes are determined by ecophysiological characteristic of the plant function type (PFT). In this formulation, the water cycle is fundamentally one dimensional and mainly driven by the meteorological data with limited descriptions of landscape gravitational effects from surface or subsurface lateral flow.

Comparative studies suggest that different models have varied degrees of strength, and the improvements might be made toward synthesis of a model that simulates dynamics of the coupled water and carbon cycle [*Castellano et al.*, 2012; *Ichii et al.*, 2010; *Morales et al.*, 2005; *Siqueira et al.*, 2006]. Coupling hydrological and biogeochemical models is one of the major steps toward understanding the multiscale interactions between water and carbon cycles [*Morales et al.*, 2005]. *Wolf et al.* [2008] linked an ecosystem model with a hydrological model to test the sensitivity of the ecosystem model to hydrology and temperature. The results demonstrated that soil moisture and soil temperature are the most sensitive driving factors of carbon fluxes, particularly soil carbon emissions. The vegetation simulation was improved in the hydrologic transport model SWAT (Soil Water Assessment Tool) by implementing a field-scale plant model [*Kiniry et al.*, 2008]. An increasing number of studies have shown that only a complete representation of laterally connected ecologic and hydrologic processes could comprehensively improve the understanding of the dynamics of the ecohydrosystem [*Yi et al.*, 2009].

Recently, catchment-scale research networks are beginning to provide the experimental test beds necessary to evaluate the impact of model assumptions and their relative impact on changes in ecosystem services [*Banwart et al.*, 2011]. *Duffy et al.* [2011] argue that with the emergence of a real-time and spatially distributed embedded sensor

[1]Department of Civil & Environmental Engineering, The Pennsylvania State University, University Park, Pennsylvania, USA

[2]Department of Crop & Soil Sciences, The Pennsylvania State University, University Park, Pennsylvania, USA

[3]USDA-ARS, Hydrology and Remote Sensing Laboratory, Beltsville, Maryland, USA

[4]Earth and Environmental Systems Institute, The Pennsylvania State University, University Park, Pennsylvania, USA

Remote Sensing of the Terrestrial Water Cycle, Geophysical Monograph 206. First Edition. Edited by Venkat Lakshmi.
© 2015 American Geophysical Union. Published 2015 by John Wiley & Sons, Inc.

network, a high-resolution hydrological modeling is needed to integrate a wide range of space and time observation data. At the same time, the meteorological community is providing land-data assimilation products that give high resolution as well as atmospheric forcing on a continental scale at hourly intervals North American Land Data Assimilation System phase 2 (NLDAS-2; *Xia et al.*, 2012a]. Another example is the Airborne Microwave Observatory of Subcanopy and Subsurface (AirMOSS) project, which is attempting to significantly reduce the current uncertainty in the net ecosystem exchange (NEE) estimates of water and carbon over North America. AirMOSS will meet this challenge by providing high-resolution regional observations of root-zone soil moisture (RZSM) over the major North American biomes [*Chapin et al.*, 2012]. RZSM is defined here as the total moisture in the soil column from surface to the root zone. To measure RZSM, AirMOSS will build an ultra-high-frequency (UHF) synthetic aperture radar (SAR) that has the capability to penetrate through substantial vegetation canopies and subsurface to retrieve information from the subsurface depths as large as 1.2 m depending on the soil moisture content.

The overall goal of this chapter is to compare the physics-based watershed model Penn State Integrated Hydrologic Model (PIHM) and an ecophysiological model BioGeochemical Cycles (Biome-BGC) to gain insight into the strengths and weakness of each in the context of a new watershed sensor and reanalysis data set. To bring in correspondence with the intensive regional observation, this chapter focuses on model evaluation and assessment of coupling with primary objectives to (a) demonstrate how hydrological and biogeochemical models could resolve the multiple sources of high-resolution regional observation, (b) determine how coupled models could help with the interpretation of interaction between water and carbon cycles, and (c) how these, in turn, influence the evolution and design of process-based coupling in the future generations of integrated environmental models.

31.2. SITE DESCRIPTION AND FIELD MEASUREMENTS

31.2.1. Site Description

The study took place in one of the National Critical Zone Observatory (CZO) program sites, Susquehanna Shale Hills CZO (SSHCZO), an 8 hectare experimental hill-slope catchment, located in the Ridge-and-Valley Appalachians in central Pennsylvania (Figure 31.1). The SSHCZO has been the focus of several interdisciplinary studies, including biochemistry, ecology, geomorphology, meteorology, hydrology, and pedology [*Brantley et al.*, 2007], which provide a unique and well-documented watershed to benefit cross-disciplinary science [*Anderson et al.*, 2008]. The emerging observing system has improved the understanding of catchment-scale water cycle processes [*Thomas et al.*, 2013; *Jin,* 2011]. Nearly 30 years ago, Shale Hills was the subject of an intensive series of irrigation experiments conducted by the Forest

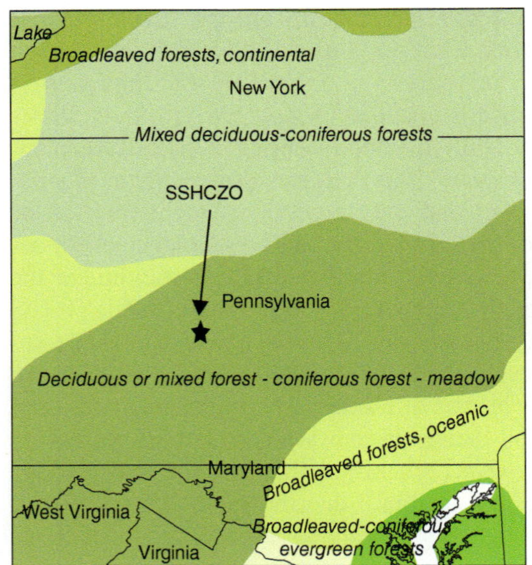

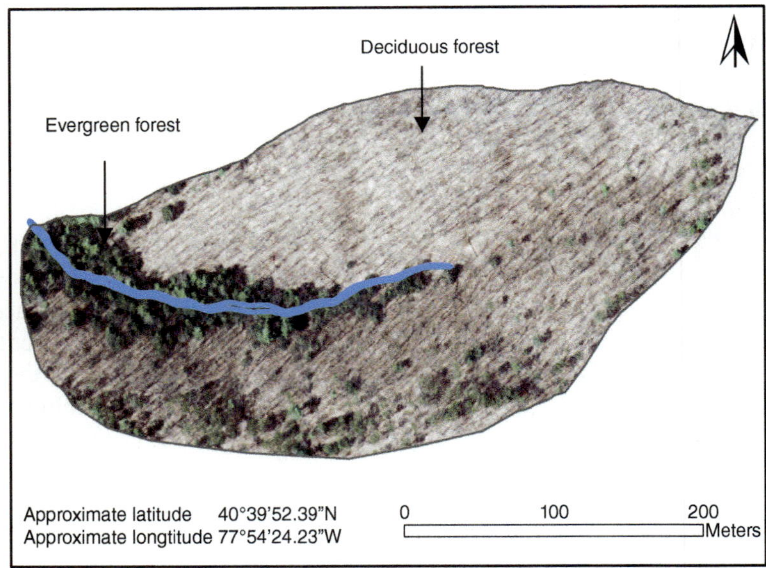

Figure 31.1 Location of study site SSHCZO. Figure (left) shows the position of SSHCZO in central Pennsylvania, where major land cover type is deciduous or mixed forest. Figure (right) shows the watershed boundary of SSHCZO and the LiDAR image.

Hydrology group at the Pennsylvania State University in 1974 [*Lynch*, 1976]. The experiment provided insight into the physical mechanisms involved in the runoff and streamflow generation at SSHCZO and revealed the effects of antecedent soil moisture on the runoff peak and timing. The hydrological processes involving the irrigation experiment were reproduced by a numerical study involving application of a fully coupled physics-based model [*Qu and Duffy*, 2007]. Spatial hydropedologic heterogeneity of SSHCZO was studied by year-round soil moisture monitoring across the watershed [*Lin*, 2006].

The physiography of the SSHCZO can be described as an erosion-cut, deep V-shaped valley with thin residual soils overlying the low-permeability Rose Hill Shale Formation [*Lynch*, 1976]. The 8 hectare watershed is entirely forested with an ephemeral first-order stream discharging into Shaver's Creek (185 km^2), which flows into the Juniata River, the second largest tributary of the Susquehanna River.

31.2.2. Landscape

A 1m digital elevation model (DEM) derived from aircraft LiDAR (light detection and ranging) [*Guo*, 2010] was used to delineate the watershed, with elevation varying from 258m at the outlet to 309m at the highest point. To prepare the simulation of PIHM, the watershed was spatially decomposed into 535 triangles, with 21 channel segments using the PIHMgis tool [*Bhatt et al.*, 2014; *Bhatt*, 2012] (Figure 31.2). Notably, the swales were explicitly represented in the model through the high-resolution domain decomposition (Figure 31.2). In addition, the average channel gradient in the model was 4.3%, which is consistent with previous field observation study [*Lynch*, 1976].

31.2.3. Soil

The residual soils of SSHCZO were derived from the thin-bedded acid gray Rose Hill Shale [*Lynch*, 1976]. The Soil Survey Geographic (SSURGO) database show two major soil types (Berks-Weikert association and Ernest silt loam) and smaller portions of the other soil classifications (Berks shaly silt loam and Berks-Weikert shaly silt loam). *Lin* [2006] identified five soil series (Weikert, Berks, Rushtown, Blairton, and Ernest series) in the catchment and based on the samples collected from field soil pits and laboratory characterization. The soil classification maps from the two different sources are overlapped in Figure 31.3. The SSURGO map provides major soil class information for the entire watershed, which could not distinguish between upland and riparian regions, whereas the survey soil classification shows improved spatial distribution, which is correlated with the landscape.

Figure 31.2 Three-dimensimal representation of the landscape in PIHM. Watershed is decomposed spatially into triangular meshes to efficiently represent the topography of the landscape in the model. Note that the swales are clearly identifiable in the triangular mesh.

31.2.4. Land Cover

The SSHCZO is a highly forested catchment with dense canopy vegetation. The National Land Cover Dataset (NLCD) 2001 data [*U.S. Geological Survey*, 2001: http://www.mrlc.gov/nlcd.php] shows three major types of forest cover. Field studies mapped tree geospatial data, including the geographic coordinate of each stand, species name, tree height, diameter at the breast height, and crown class at SSHCZO [*Meinzer et al.*, 2013]. Based on the spatial information of trees, a finer resolution land cover data product was generated according to the tree species and density of the trees. The improved land cover data sets were classified into nine classes, which preserve the spatial distribution of trees (Figure 31.4).

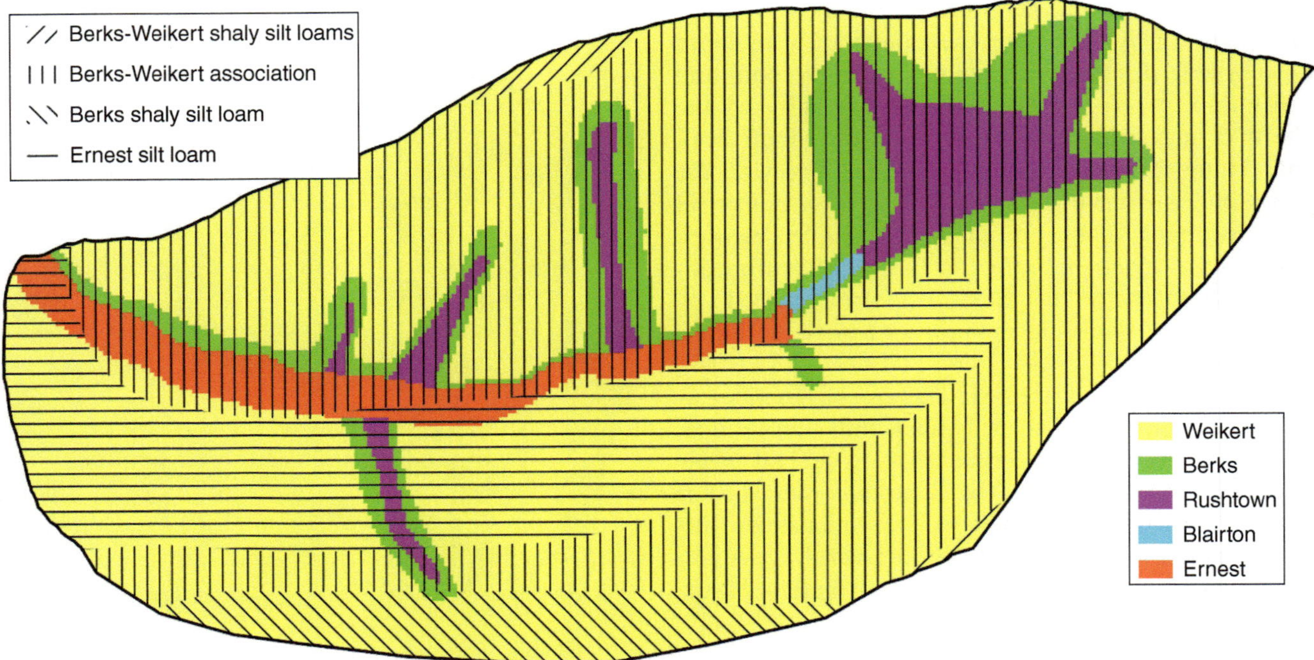

Figure 31.3 Refined spatial classifications of soil using national soil data product and field-scale soil survey. The SSURGO map (in shading) provides major soil class information that is unable to distinguish between upland and riparian regions. The soil classification based on field-scale survey and laboratory analysis (in color) provides an improved spatial distribution, which is correlated with the landscape.

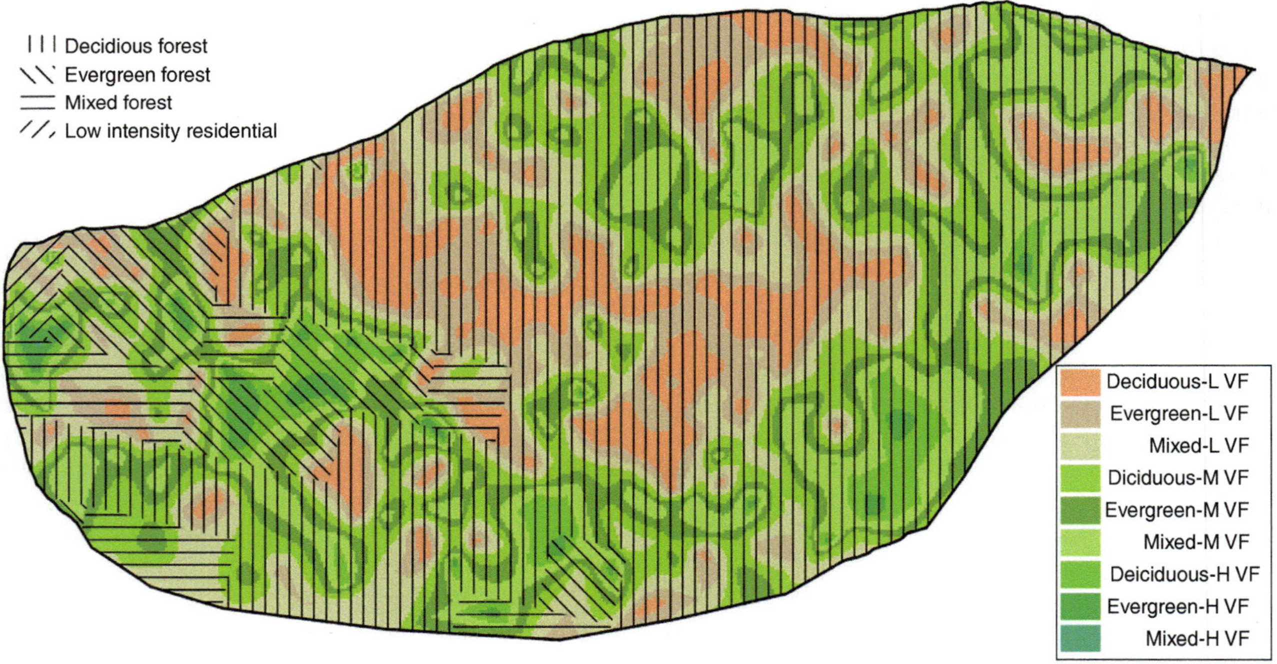

Figure 31.4 Spatial layers of (a) National Land Cover Data (in shading) and (b) a high-resolution land cover classification derived using tree survey data (in color). The watershed was classified into nine classes, which preserve the spatial distribution of trees.

31.2.5. Hydrometeorological Observations

The climate in central Pennsylvania represents a humid continental climate. The summers are humid and winters are severely cold. Extreme temperatures have been recorded above 100°F and below −20°F. Precipitation is relatively well distributed year round. Usually, Atlantic tropical cyclones bring summer rainfall and winter precipitation occurs in the form of snow.

The real-time hydrologic monitoring sensor-network (RTHnet) has been set up on the top of a ridge of the watershed to record precipitation, air temperature, and relative humidity every 10 min since 2006. The historic observations are available at State College or University Park Airport from Federal Aviation Administration (FAA) or Citizen Weather Observer Program (CWOP) [*Pennsylvania State Climatologist*, 2013]. Other meteorological variables from 1979 to 2012 were extracted from NLDAS-2 meteorological forcing data set available at 1 h temporal resolution.

The Cosmic-ray Soil Moisture Observing System (COSMOS) initiated one site within SSHCZO recently, and the data are available since 2011 [*Zreda et al.*, 2012]. The COSMOS volumetric soil moisture reflects the aggregate watershed condition.

31.2.6. Carbon Observations

Developing a better understanding of the carbon cycle is one of the key goals of Critical Zone (CZ) research [*Brantley et al.*, 2007]. The observations relating to the carbon cycle include distribution of total soil carbon, soil CO_2 concentrations, dissolved organic carbon (DOC), and soil microorganisms. Comprehensive soil water samples and laboratory analyses have been reported to understand DOC export and soil organic carbon (SOC) storage [*Andrews et al.*, 2011]. The significant spatial heterogeneity of SOC storage suggests spatially distributed modeling of the carbon cycle could be more realistic and meaningful.

The eddy covariance flux tower on the top ridge of the watershed started recording the latent heat flux and net ecosystem exchange of carbon dioxide in 2009.

31.3. PARAMETERIZING THE MODELS

We applied PIHM for the hydrological modeling. PIHM is a physics-based fully coupled hydrologic model. It simulates interception, throughfall, infiltration, recharge, evapotranspiration, overland flow, subsurface flow, and channel routing in a fully coupled scheme for multiscale distributed hydrological simulations [*Qu and Duffy*, 2007]. The resolution of triangular mesh allows the user to vary according to the geomorphological or hydrological characteristics of the watershed, and the triangles can be constrained by point observations [e.g., streamflow, groundwater level, soil moisture, leaf area index (LAI)], and the watershed boundary conditions [*Kumar et al.*, 2009]. The model resolves hydrological processes for land surface energy, overland flow, channel routing, and subsurface flow, governed by partial differential equations (PDEs). The system is discretized on a triangular mesh and projected prism from canopy to bedrock [*Qu and Duffy*, 2007]. The model also includes canopy interception, evapotranspiration (ET), snow accumulation, snowmelt, infiltration, and recharge within the fully coupled system. PIHM uses a semidiscrete finite-volume formulation for solving the system of coupled PDEs, resulting in a system of ordinary differential equations (ODEs) representing all processes within the prismatic control volume. The local system is assembling throughout the entire model domain, and the global ODE system is solved using the implicit scheme of CVODE C-language Variable-coefficients ODE solver [*Cohen and Hindmarsh*, 1996].

Hydrologic reanalysis of SSHCZO was developed through long-term (32 years) continuous modeling of the watershed using PIHM. Model inputs are listed in Table 31.1. The model parameters were calibrated against observed discharge data in 2009 and validated against streamflow data from the 1974 forestry experiment of Lynch described by *Yu et al.* [2013]. One important feature of the SSHCZO is the impact of tropical storms that contribute important late summer and early fall moisture to the watershed. These episodic events often have the impact of ending the summer drought, increasing soil moisture, raising the shallow water table, and initiating streamflow again in this ephemeral watershed. This is an important element of the overall upland catchment response, and Figure 31.5 illustrates some of the important storms in SSHOCZO.

Table 31.1 Model inputs for PIHM and Biome-BGC

	PIHM	Biome-BGC
Meteorological variables	NLDAS II/CWOP/RTHnet/Irrigation records in 1974 [*Pennsylvania State Climatologist*, 2013]	CWOP/RTHnet [*Pennsylvania State Climatologist*, 2013]
Vegetation parameters	NLDAS Monthly Vegetation Parameters [*NLDAS vegetation parameters*, 2011]	Plant functional types [*White et al.*, 2000]
Topographic parameters	LiDAR [*Guo*, 2010]	LiDAR [*Guo*, 2010]
Soil parameters	Field survey [*Lin*, 2006]	Field survey [*Lin*, 2006]

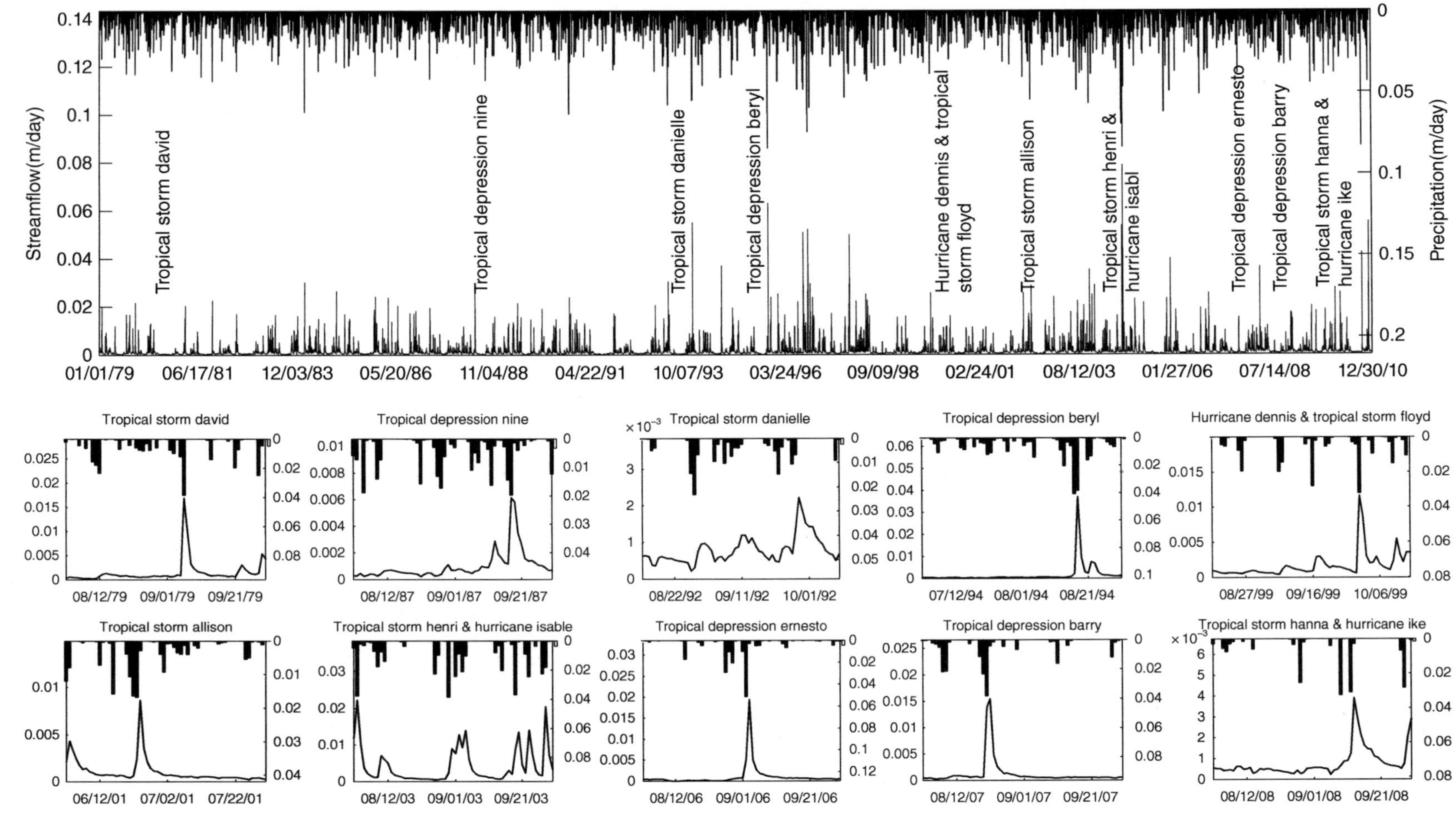

Figure 31.5 Hydrological reanalysis of the watershed shows daily streamflow (on the bottom axis) response simulated using PIHM, forced with hourly precipitation (on the top axis) and meteorological data over 32 year (1979–2010) period. Figures showed variability in streamflow response to different hurricanes emphasizing the role of antecedent moisture in the generation of peak flow.

Table 31.2 Tree survey in SSHCZO

Common Name	Scientific Name	Species	Stem Number	Total Basal Area (m²)
Sugar maple	*Acer saccharum*	DBF	122	8.40
Pignut hickory	*Carya glabra*	DBF	122	7.75
Mockernut hickory	*Carya tormentosa*	DBF	178	13.31
Eastern white pine	*Pinus strobus*	ENF	59	7.02
Virginia pine	*Pinus virginiana*	ENF	96	7.39
White oak	*Quercus alba*	DBF	282	26.77
Chestnut oak	*Quercus prinus*	DBF	558	47.50
Red oak	*Quercus rubra*	DBF	287	39.25
Eastern hemlock	*Tsuga canadensis*	ENF	164	15.76
Other			249	25.08

Note: Data was retrieved from http://criticalzone.org/shale-hills/.
DBF: deciduous broadleaf forest; ENF: evergreen needleleaf forest.

Biome-BGC is an ecosystem process model that estimates storage and flux of carbon, nitrogen, and water. Reasons for the selection of Biome-BGC are (1) meteorological inputs are similar to that of PIHM, (2) the model parameters for different types of vegetation are available and ready to use [*White et al.*, 2000], and (3) the moderate complexity of the model structure, which estimates same fluxes and storage of energy and water as PIHM. Parameterizing the forest land cover for SSHOCZO implements National Land Cover Classification and a forest survey result [*Naithani et al.*, 2013]. Chestnut oak (*Quercus prinus*) is the dominant tree species, although there is a wide diversity with some 25 species (Table 31.2) in the inventory [*Meinzer et al.*, 2013]. Biome-BGC [*Thornton et al.*, 2002] was used for the ecosystem modeling of the parameters for deciduous broad leaf forest (DBF) as shown in (Table 31.3) [*White et al.*, 2000]. We ran a spin-up to bring the model into equilibrium using the meteorological data from 1904 to 2012. Some of the parameters were adjusted to match the annual mean of the Moderate Resolution Imaging Spectroradiometer (MODIS) products, such as gross primary product (GPP), net primary product (NPP), evapotranspiration (ET), and the high-resolution observed net ecosystem exchange (NEE) from the flux tower (Figure 31.6). The simulated NEE were correctly reproduced during the spring and autumn months (Table 31.4). MODIS products were retrieved at 1 × 1 km spatial grid centering at 40.66455°N, 77.70673°W.

31.4. RESULTS

31.4.1. Water Cycle

The volume-integrated behavior of simulated water budgets using the two models was similar as shown in Figure 31.7. The ET routine used in PIHM and Biome-BGC is also comparable as both models use Penman-Monteith to calculate bare soil evaporation, canopy-intercepted water evaporation, and the leaf transpiration. However, the differences were noted in the time step used and how stomatal conductance of the canopy is modeled. Biome-BGC has the mechanism to scale maximum stomatal conductance for the simulation of the drivers of stomatal closure, which is not available in the PIHM formulation.

The simulated hydrological cycle at a daily time scale demonstrated significant differences in ET, soil moisture, and discharge (Figure 31.8) due to the fundamental differences in the representation of hydrological processes (Table 31.5). PIHM preserves spatial variability present in the landscape and other geospatial properties of the watershed and then couples the equations representing various hydrologic processes to calculate the movements of surface and subsurface water at each model grid. Biome-BGC conceptualizes the whole watershed as a 1D bucket model. Soil moisture and streamflow response to precipitation simulated are near linear in Biome-BGC. It is easy to notice that the antecedent soil moisture effects were rarely interpreted in Biome-BGC.

The PIHM-simulated spatial pattern in soil moisture and groundwater was able to capture the wet valley floor and dry ridge top (Figure 31.9). During September 1987, tropical depression Nina brought 13 cm of rain to some parts of Pennsylvania. The PIHM simulation showed that before the tropical season, the significant groundwater storage was confined to the swales and the near stream area, which is a typical behavior in the dry season. After the storm, the watershed was saturated, and the flooding event resulted in an increase of groundwater storage and in spreading of lateral flow across most places of the watershed.

The physics-based distributed hydrologic modeling resolved lateral redistribution of soil moisture and captured the hydrological function of landscape, vegetation, and soil properties. We feel the distributed water budget is

Table 31.3 Ecophysiological parameters at SSHCZO

Parameter	Original	Improved	Unit
Transfer growth period as fraction of growing season	0.2		—
Litter fall as fraction of growing season	0.2		—
Annual leaf and fine root turnover fraction	1		year^{-1}
Annual live wood turnover fraction	0.7	0.56	year^{-1}
Annual whole-plant mortality fraction	0.005	0.003	year^{-1}
Annual fire mortality fraction	0.0025	0.00001	year^{-1}
New fine root C : new leaf C	1	0.3	—
New stem C : new leaf C	1.57	2.32	—
New live wood C : new total wood C	0.1	0.27	—
New coarse root C : new stem C	0.23	0.1	—
Current growth proportion	0.5		—
C : N of leaves	35		kg C/kg N
C : N of leaf litter	41.7	48.7	kg C/kg N
C : N of fine roots	42		kg C/kg N
C : N of live wood	50		—
C : N of dead wood	442	742	—
Leaf litter labile proportion	0.39		—
Leaf litter cellulose proportion	0.36		—
Leaf litter lignin proportion	0.25		—
Fine root labile proportion	0.3		—
Fine root cellulose proportion	0.45		—
Fine root lignin proportion	0.25		—
Dead wood cellulose proportion	0.76		—
Dead wood lignin proportion	0.24		—
Canopy water interception coefficient	0.041	0.0011	LAI^{-1} d^{-1}
Canopy light extinction coefficient	0.7	0.4	—
All-sided to projected leaf area ratio	2		—
Canopy average specific leaf area	19.9		m^2/kg C
Ratio of shaded SLA: sunlit SLA	2		—
Fraction of leaf N in Rubisco	0.08	0.18	—
Maximum stomatal conductance	0.005		m/s
Cuticular conductance	0.00001		m/s
Boundary layer conductance	0.01		m/s
Leaf water potential: start of conductance reduction	−0.6		MPa
Leaf water potential: complete conductance reduction	−2.3		MPa
VPD: start of conductance reduction	930	969	Pa
VPD: complete conductance reduction	4100	3700	Pa

Note: Parameters were chosen as dominated specie *Quercus prinus*.

a crucial element of the carbon cycle estimation because of the sensitive infiltration, recharge, and lateral flow to high-frequency meteorological events [*Medvigy et al.*, 2010]. Coupling a distributed hydrological model and terrestrial ecosystem model suggested dramatic spatial differences in net ecosystem productivity within a watershed [*Peng et al.*, 2013]. It is also important to mention that lateral redistribution of shallow groundwater supports soil moisture in the lower parts of the watershed during summer months by capillary rise. This would be difficult to capture in Biome-BGC, but it is fundamental to the PIHM model. The detailed watershed heterogeneity was found to play a critical role in identifying DOC export and SOC storage. Explicit groundwater table and lateral

groundwater flow is important to understand DOC delivery in a watershed [*Mei et al.*, 2012].

Hydrologic sensitivity of vegetation dynamics has been widely observed through tree harvesting and growth comparisons in paired catchment studies. Here we tested the hydrologic sensitivity of the model based on the responses simulated to the changes in LAI. PIHM simulations were forced using default seasonal LAI curves from NLCD and then compared with Biome-BGC-simulated LAI to simulate hydrologic response for the year 1974. The result suggested that phenology could lead to major errors in simulated peak flow of the hydrologic model (Figure 31.10; Table 31.6). The errors were mainly due to the model parameters that were calibrated for one year (2009) and

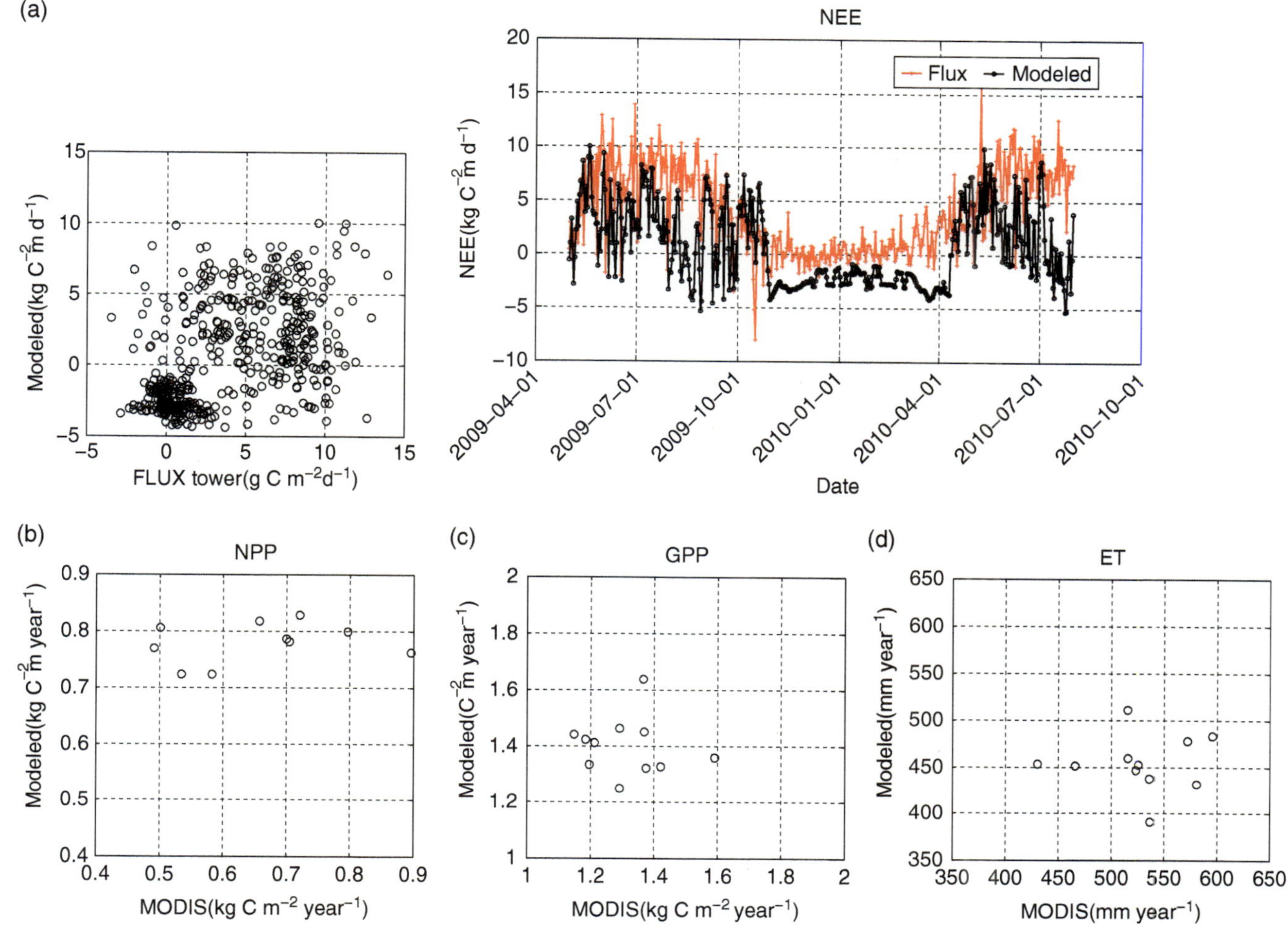

Figure 31.6 Model validation of biogeochemical processes: (a) modeled NEE compared with flux tower observed NEE and modeled (b) NPP, (c) GPP, and (d) ET compared with MODIS products.

Table 31.4 Model performance of Biome-BGC

	Pearson's Correlation Coefficient	Error (%)
NEE	0.56	−17.6
NPP	0.14	23.3
GPP	0.60	6.8
ET	0.45	−13.8

were then used for the entire reanalysis period. Thus dynamic LAI in the PIHM model provided a better match between simulated and observed peak flow, for long-term hydrological simulation, suggesting the importance of dynamic vegetation for interannual and longer time scale hydrologic changes in the watershed.

31.4.2. Carbon Cycle

The Biome-BGC-simulated GPP, NEE, and maximum LAI suggested that vegetation dynamics varied dramatically during the past 109 years (Figure 31.11). Figures 31.11a, 31.11b and 31.11d show that the temperature is well correlated with GPP and NEE. Variation in maximum LAI is more complicated, which is also influenced by the availability of soil moisture in the growing season.

The simulated GPP replicate the seasonal carbon flux in the watershed as shown in Figure 31.12. The NEE was underestimated during the dormant season (Figure 31.6), as well as GPP. This is probably due to the simple spatial perspective of Biome-BGC. Furthermore, the model simulation considers only one type of vegetation.

Another important factor in resolving the watershed carbon dynamics is the DOC delivery. Biome-BGC simplified the DOC leaching processes using a one-dimensional assumption. Field observations show that the DOC export varies significantly during different seasons of a year, which is influenced by both discharge and temperature [*Andrews et al.*, 2011]. PIHM simulates both base flow and surface runoff in a coupled spatially distributed manner across the entire watershed (Figure 31.13), which has the potential to help the carbon module to capture the different behavior of stormflow and base flow [*Buffam et al.*, 2001].

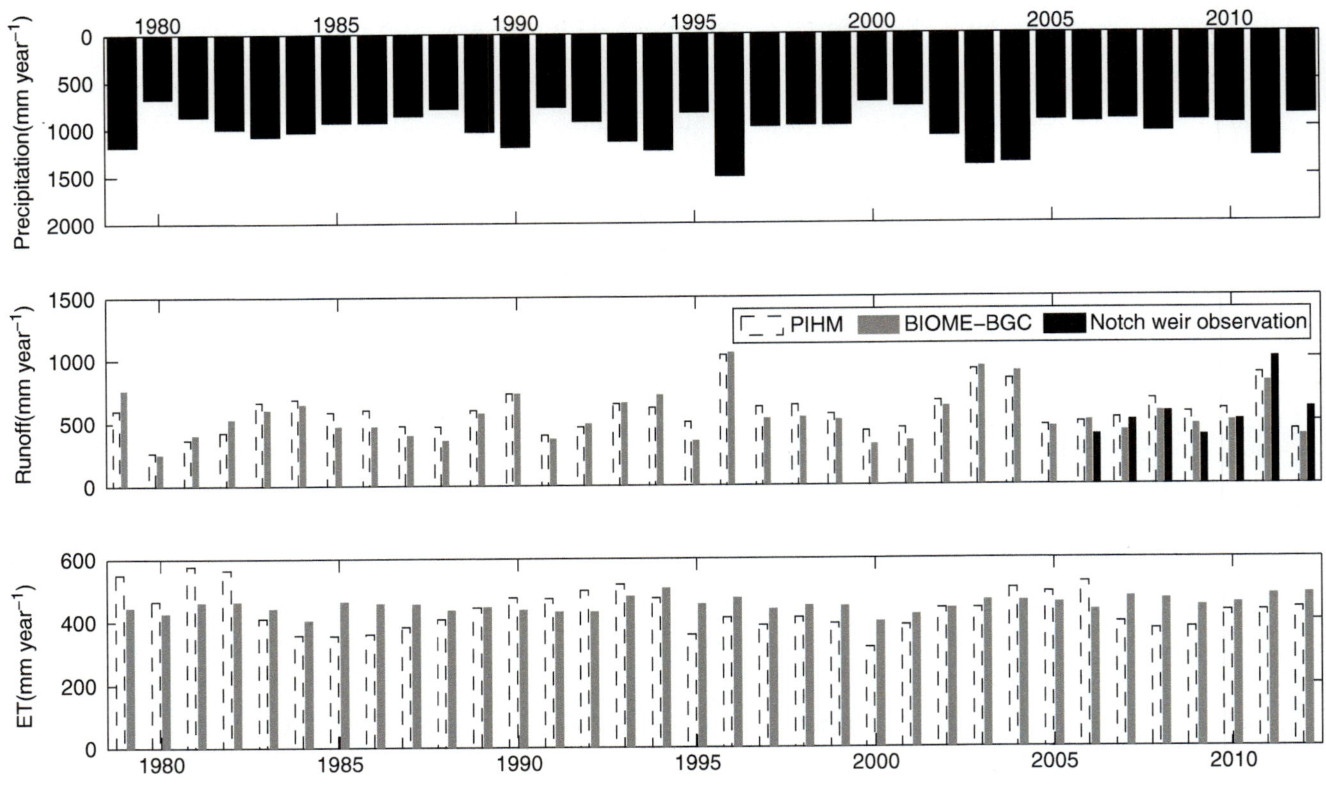

Figure 31.7 Annual water budget of SSHCZO.

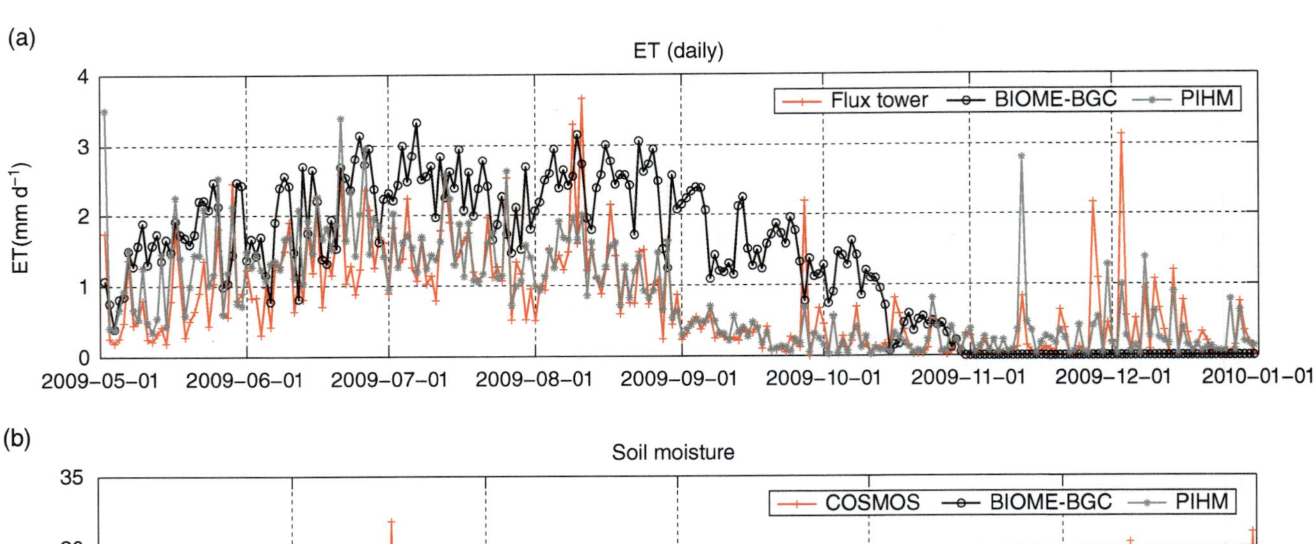

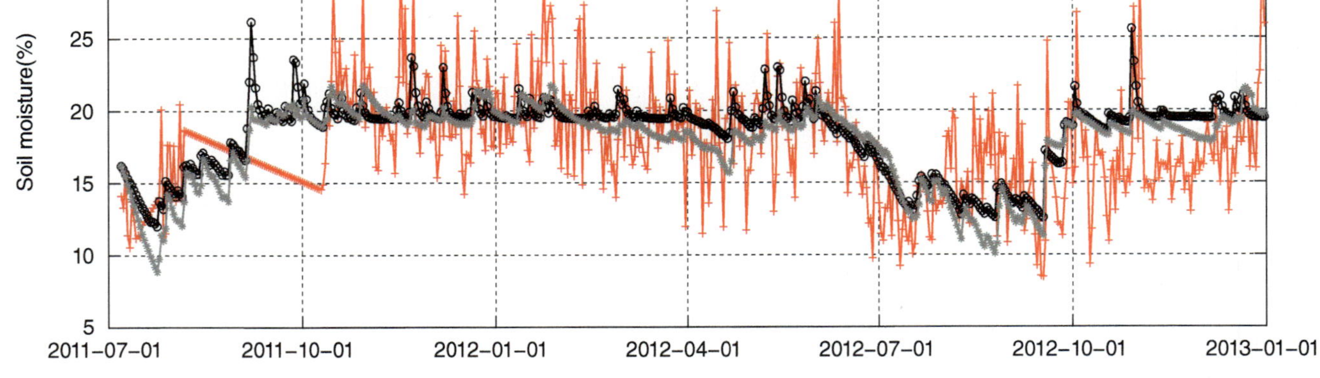

Figure 31.8 Assessment of daily water cycle of SSHCZO demonstrates significant differences in estimates of (a) evapotranspiration (b) soil moisture, and (c) discharge from the PIHM and Biome-BGC.

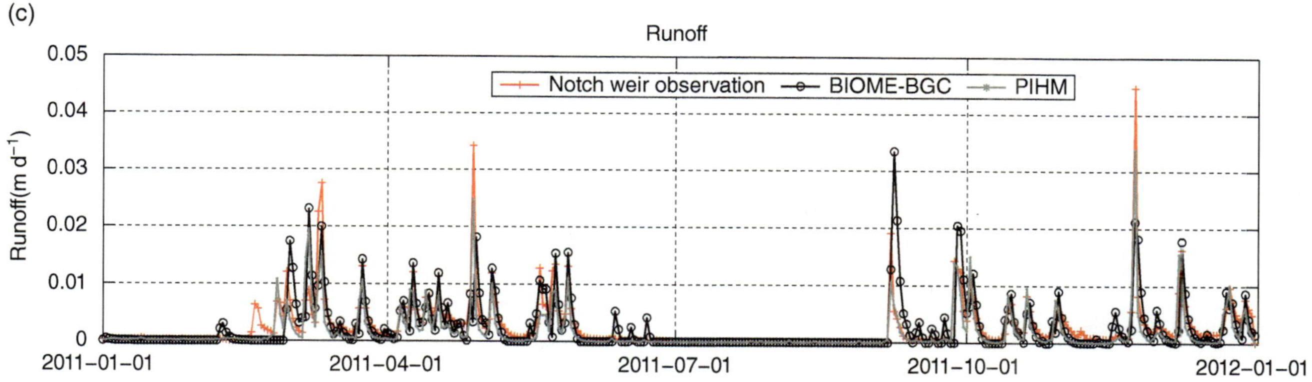

Figure 31.8 (*Continued*)

Table 31.5 Evaluate PIHM and Biome-BGC at the simulation of daily water cycle

	PIHM	Biome-BGC
ET	0.67	0.47
Soil moisture	0.82	0.60
Runoff	0.90	0.67

Note: Pearson's correlation coefficient was used to evaluate the simulated daily results.

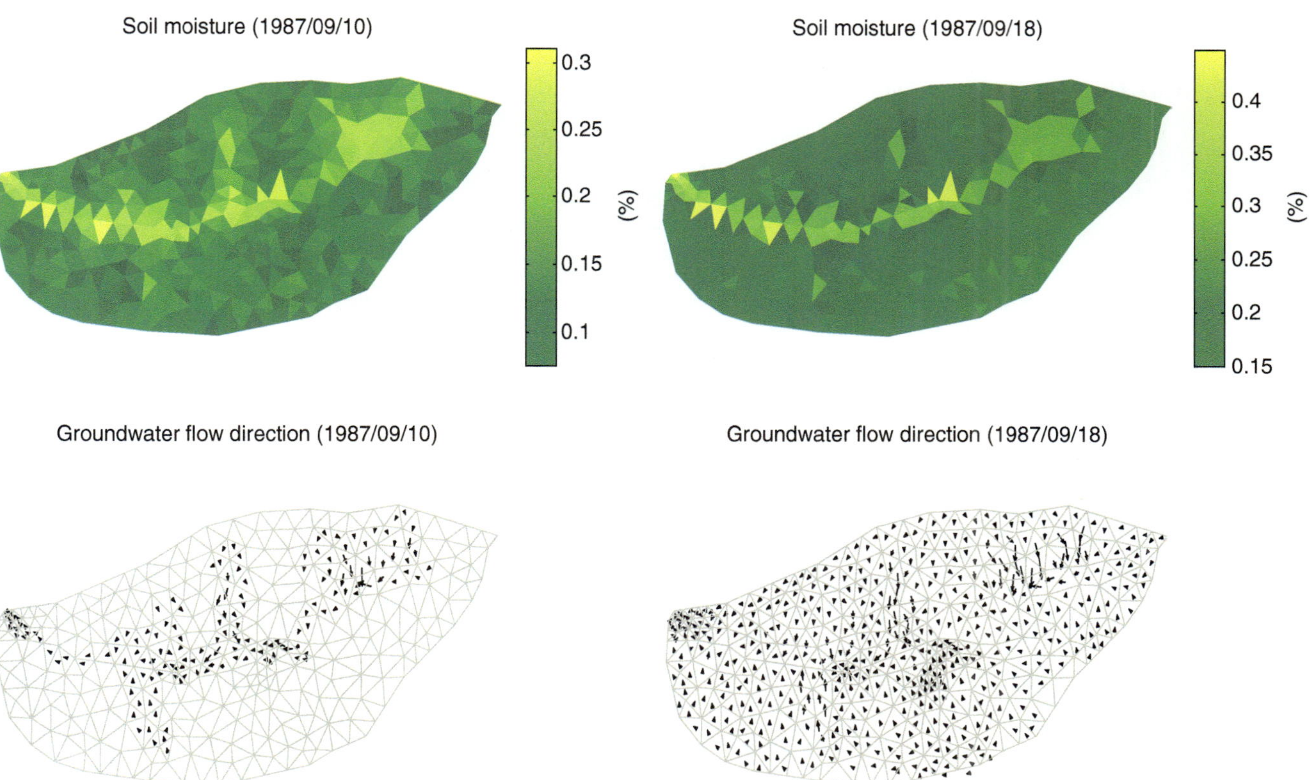

Figure 31.9 Spatial moisture distribution and groundwater flow direction before and after a precipitation event. Arrow indicates the simulated groundwater flow; length of the arrow suggests the relative volume of lateral groundwater flow.

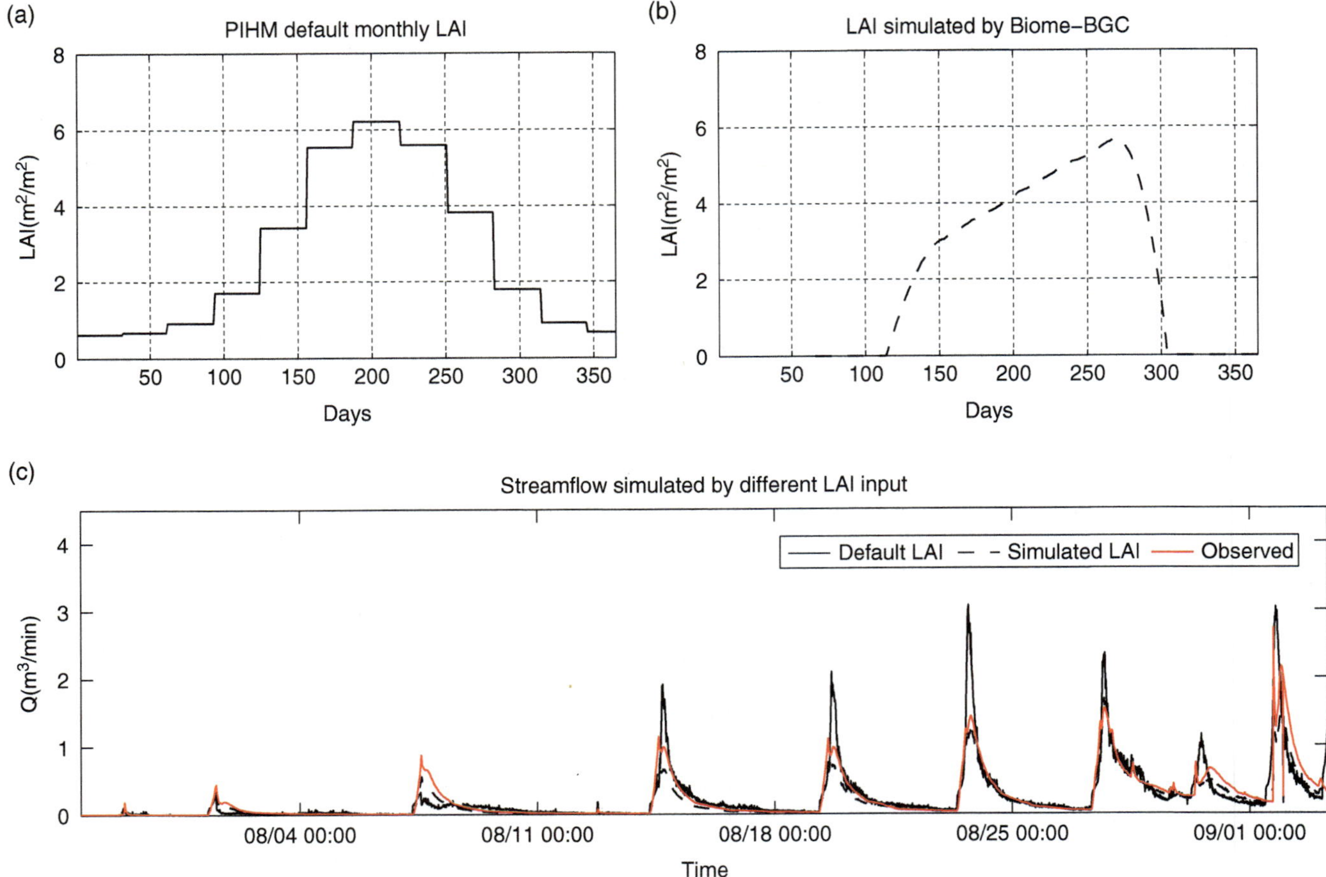

Figure 31.10 Streamflow sensitivity to LAI. Two types of LAI (a and b) input resulted in difference in (c) streamflow simulation.

Table 31.6 Streamflow sensitivity for the 1974 experiment at Shale Hills

Event	Duration	Model Result from Default LAI	Model Result from Simulated LAI
1	1–7 Aug	0.84	0.96
2	7–14 Aug	0.87	0.98
3	14–19 Aug	0.89	0.94
4	19–23 Aug	0.92	0.94
5	23–27 Aug	0.94	0.98
6	27–31 Aug	0.80	0.96
Total	1–31 Aug	0.87	0.97

Note: Pearson's correlation coefficient between simulated and observed streamflow was used to evaluate the streamflow sensitivity.

31.5. DISCUSSION

31.5.1. Critical Processes in Physics-Based Water and Carbon Cycle Modeling

One of the critical processes of the coupled water and carbon cycle modeling is the vegetation dynamics. The plant growth involving the competition for light, nutrients, and water actively interfere with the water cycle by taking away the precipitated water back to the atmosphere. A model intercomparison study [*Morales et al.*, 2005] showed that models incorporating dynamic vegetation processes performed clearly well in the estimation of evapotranspiration. *Kiniry et al.* [2008] argued that plant modeling could increase the simulation accuracy of water fluxes and water quality in SWAT-type models. In this

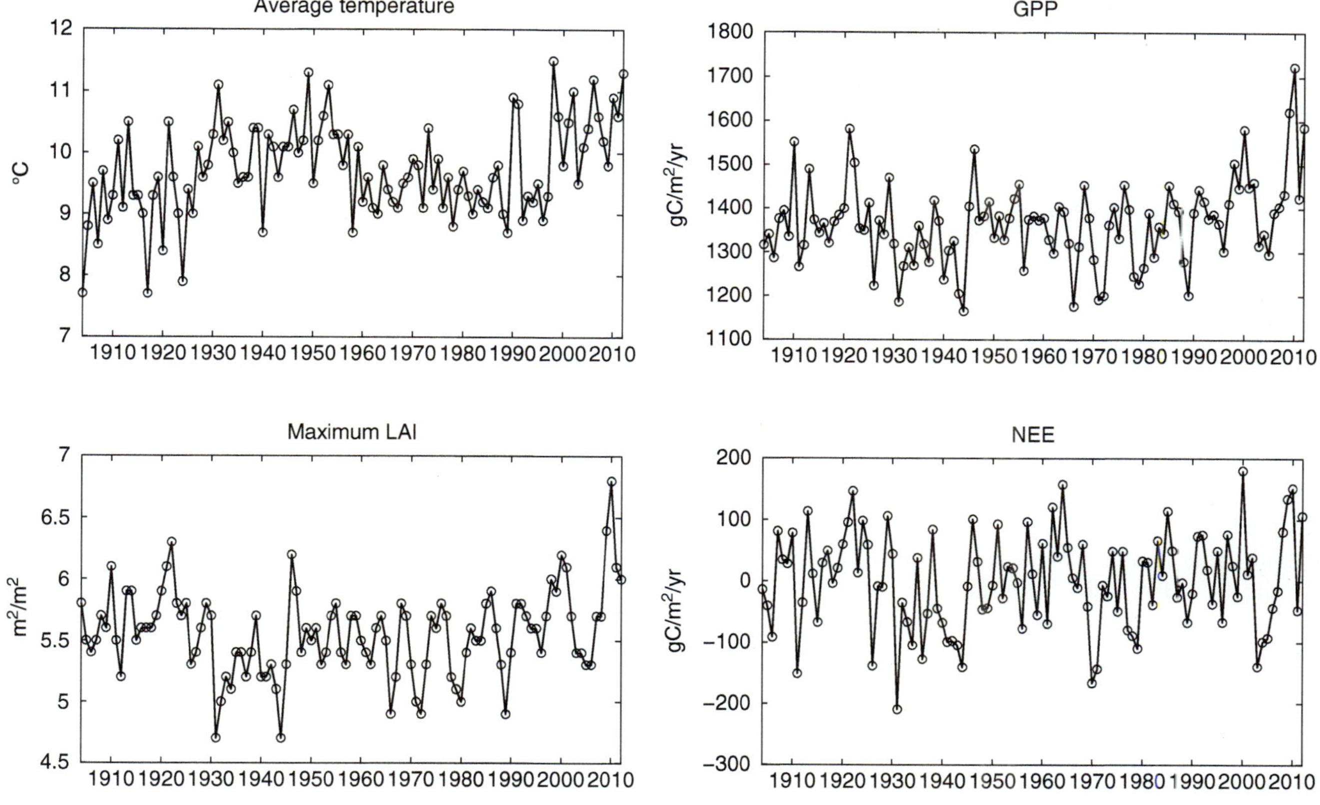

Figure 31.11 Annual carbon budget of SSHCZO simulated using Biome-BGC.

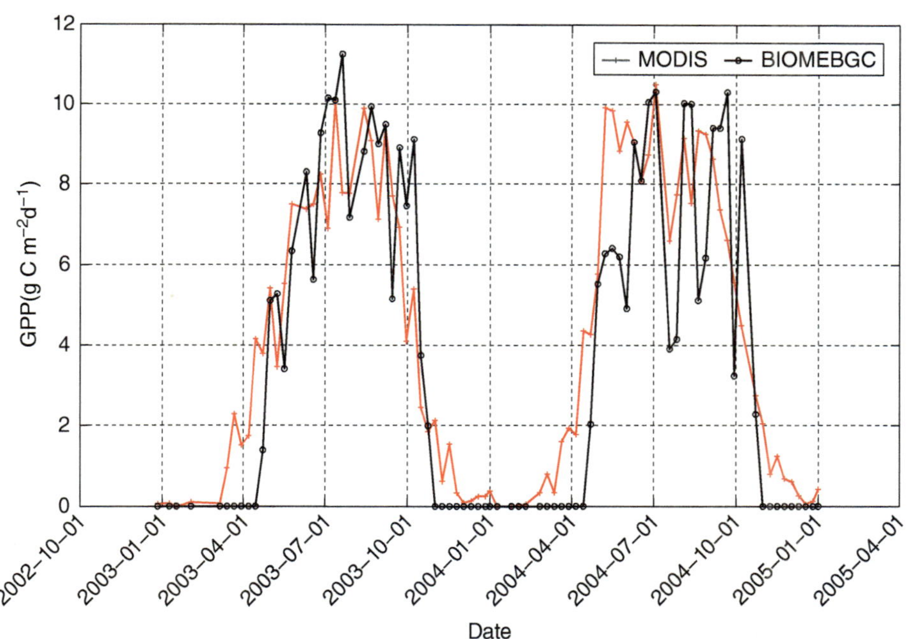

Figure 31.12 Assessment of daily carbon cycle of SSHCZO. The Pearson's correlation coefficient is 0.78.

study, we tested Biome-BGC in the modeling of growth dynamics of *Quercus prinus*. The improvement was well explained by the simulated peak flow during the experiment in 1974.

Another critical processes of the coupled water and carbon modeling is the spatial redistribution of water, carbon, and nutrient. Most current ecohydrological models couple the water and carbon cycle at spatially

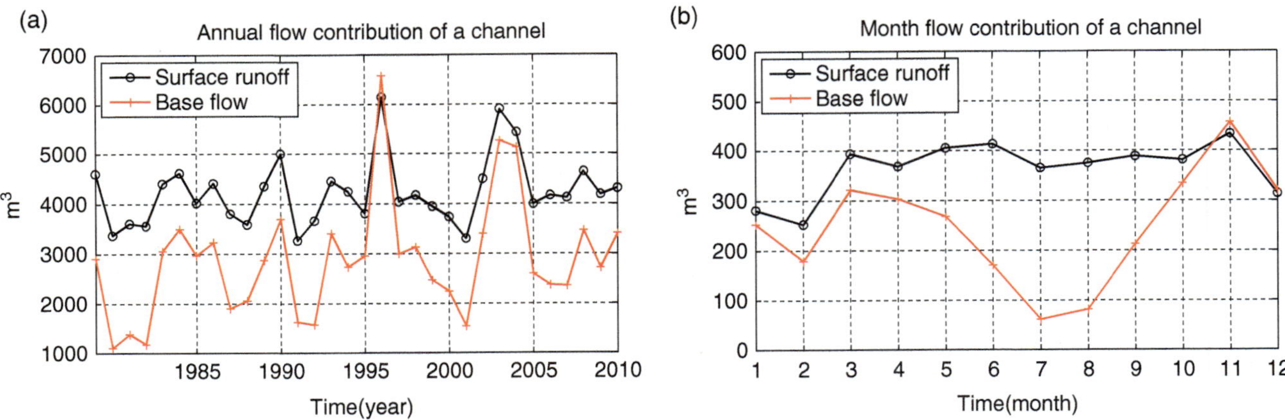

Figure 31.13 Comparison between base flow and surface runoff.

distributed computational model grids, which could explain the spatial heterogeneity of water and carbon dynamics. However, without the incorporation of impacts originating from lateral surface and subsurface flow, nutrient redistribution within the watershed is not well simulated [*Hwang et al.*, 2012]. The method selected in this study involves comparing a physics-based distributed hydrologic model, PIHM, with a terrestrial ecosystem model. PIHM preserves the spatial connectivity and simulates lateral surface and subsurface water fluxes. At each computation grid the module of biogeochemical coupled with hydrologic processes simulates the vertical nutrient fluxes. With lateral water fluxes between each grid, the lateral nutrient redistribution can be explicitly modeled. Additionally, local environmental parameters determine the vegetation dynamics that would potentially improve the vegetation water consumption modeling in hydrological cycle.

31.5.2. Importance of Hydrological Reanalysis on Models and Data

The concept of "reanalysis" can be traced back to atmospheric studies at the end of the last century. *Trenberth* [1995] suggested that the reanalysis is the repeating activity of incorporating the observations from multiple sources together. Land data reanalysis products (e.g., NLDAS-2) mainly focus on the surface hydrologic variables, such as soil moisture, runoff, and evapotranspiration using different land surface models (LSMs). However, the accuracies of LSM runoff predictions are far from satisfactory [*Lohmann et al.*, 2004; *Xia et al.*, 2012b]. Hydrologic modeling has been a useful tool in understanding the water movement and storage in different mediums. Therefore, revisiting a watershed is frequently reported in the literature as a result of improved observation data and modeling strategy. For example, *Loague and Freeze* [1985] compared a unit hydrograph model, a

quasi-physically based model, and a regression model in simulation of events from the well-known R-5 catchment, located near Chickasha, Oklahoma [*Loague and Freeze*, 1985]. The results suggested that the application of the physically based model could be improved with better characterization of spatial hydraulic properties. Therefore, spatial infiltration information was gathered in the following study [*Loague and Gander*, 1990] improving the performance of the physically based model [*Loague*, 1990]. Furthermore, better simulation for R-5 events were achieved after the development of the physics-based model with improved representation of processes [*VanderKwaak and Loague*, 2001]. Similarly, the SSHCZO has been revisited frequently along with the data and model improvement. The first research was a regression model relating antecedent soil moisture and storm flow, antecedent flow, and antecedent soil moisture [*Lynch*, 1976]. Due to the importance of the antecedent soil moisture conditions in hydrologic processes analysis, an irrigation system was designed and installed at SSHCZO for correlation analysis between antecedent soil moisture and quick flow and storm flow. However, due to limited irrigation experiments, the research did not include the extreme events when antecedent soil moisture reaches maximum or minimum. Application of PIHM at SSHCZO [*Qu and Duffy*, 2007] explained the antecedent soil moisture effects on storm hydrographs corresponding to a series of irrigation experiments. As improved data sets became available and model data processing toolkits emerged [*Bhatt et al.*, 2014], Shale Hills watershed was revisited [*Li*, 2010; *Yu et al.*, 2013]. *Shi et al.* [2013] used the MODIS LAI to drive the model and improved the discharge prediction by coupling PIHM with an advanced land surface scheme. This study showed that dynamic modeled vegetation LAI could improve the streamflow prediction and is necessary to the long-term ecohydrologic simulation. With the advances in various observation data sets, the reanalysis of coupled water-carbon cycle would be repeated to improve the simulation of spatial variability in water and nutrient fluxes.

Watershed reanalysis is an objective, quantitative method of synthesizing all sources of information. *Mirus and Loague* [2013] quantitatively characterized the runoff processes based on the previously simulated and observed catchment hydrologic responses. The simulation scenarios demonstrated the type of surface/subsurface water interaction. *Beven* [2007] concluded the role of data and model in a learning process: "Data will be required to characterize research domains, to drive model predictions, to evaluate the results of model predictions and constrain predictive uncertainty, to reject some models previously considered feasible, and to monitor changes in system response; the role of models is to extrapolate data, in both time and space." Through the watershed reanalysis of model and data we expect to test our understanding of the terrestrial ecosystem cycle (including atmospheric, hydrological, biogeochemical, and energy balances), to develop the scenarios and model projections under changing environment, and to assess the impacts on human and ecosystem services risk and vulnerability, water access, food production, food security, and sustainable development [*Duffy et al.*, 2011].

31.6. SUMMARY AND FUTURE WORK

In this study, the long-term hydrologic and biogeochemical reanalysis and comparison of a headwater catchment was demonstrated. The reanalysis products were simulated from a physics-based spatially distributed hydrologic model, PIHM, and a terrestrial ecosystem model, Biome-BGC. The processes and output of the two models demonstrated favorable potential of the water and nutrient cycle coupling. First, spatial and temporal variability in lateral water movement and storage simulated with PIHM could help understand the nutrient redistribution within the watershed. Second, the vegetation dynamics in Biome-BGC could improve the plant water use calculation in PIHM, improving the peak-flow simulation and related hydrologic predictions.

The AIRMOSS [*Chapin et al.*, 2012] has started regional soil moisture and carbon flux observation across North America (http://airmoss.jpl.nasa.gov/). Development of a coupled PIHM-BGC model and its application at the Harvard Forest (http://www.pihm.psu.edu/applications/harvard_forest.html) and Tonzi Ranch (http://www.pihm.psu.edu/applications/willow_creek.html) has been proposed. Future research would investigate an integrated water and carbon simulation tool PIHM-BGC to match AirMOSS data with the model and to understand regional near-surface carbon and water flux.

ACKNOWLEDGMENTS

This research was funded by grants from the National Science Foundation, EAR 0725019 Shale Hills-Susquehanna Critical Zone Observatory. Biome-BGC version 4.1.2 was provided by Peter Thornton at the National Center for Atmospheric Research (NCAR), and by the Numerical Terradynamic Simulation Group (NTSG) at the University of Montana. NCAR is sponsored by the National Science Foundation.

REFERENCES

Anderson, S. P., R. C. Bales, and C. J. Duffy (2008), Critical zone observatories: Building a network to advance interdisciplinary study of Earth surface processes, *Mineral. Mag.*, *72*(1), 7.

Andrews, D. M., H. Lin, Q. Zhu, L. Jin, and S. L. Brantley (2011), Hot spots and hot moments of dissolved organic carbon export and soil organic carbon storage in the Shale Hills catchment, *Vadose Zone J.*, *10*(3), 943–954.

Beven, K. (2007), Towards integrated environmental models of everywhere: Uncertainty, data and modelling as a learning process, *Hydrol. Earth Syst. Sci.*, *11*, 460–467, doi:10.5194/hess-11-460-2007.

Bhatt, G. (2012), A distributed hydrologic modeling system: Framework for discovery and management of water resources, Ph.D. Dissertation, Pennsylvania State University.

Bhatt, G., M. Kumar, and C. J. Duffy (2014), A tightly coupled GIS and distributed hydrologic modeling framework, Environmental Modelling & Software, *62*, 70–84.

Brantley, S. L., M. B. Goldhaber, and K. V. Ragnarsdottir (2007), Crossing disciplines and scales to understand the Critical Zone, *Elements*, *3*(5), 307.

Buffam, I., J. N. Galloway, L. K. Blum, and K. J. McGlathery (2001), A stormflow/baseflow comparison of dissolved organic matter concentrations and bioavailability in an Appalachian stream, *Biogeochemistry*, *53*(3), 269–306.

Castellano, M. J., J. P. Kaye, H. Lin, and J. P. Schmidt (2012), Linking Carbon saturation concepts to nitrogen saturation and retention, *Ecosystems*, *15*(2), 175–187.

Chapin, E., et al. (2012), AirMOSS: An Airborne P-band SAR to Measure Root-Zone Soil Moisture. 2012 IEEE Radar Conference (Radar).

Cohen, S. D., and A. C. Hindmarsh (1996), CVODE, a stiff/nonstiff ODE solver in C, *Computers Phys.*, *10*(2), 138–143.

Duffy, C., L. Leonard, G. Bhatt, X. Yu, and L. Giles (2011), Watershed reanalysis: Towards a national strategy for model-data integration, in *e-Science Workshops (eScienceW), 2011 IEEE Seventh International Conference On*, pp. 61–65, IEEE, doi:10.1109/eScienceW.2011.32.

Guo, Q. (2010), CZO dataset: Shale Hills—Digital elevation model (DEM), GIS/map data, land cover, LiDAR, soil survey (2010), doi:10.5069/G9VM496T.

Hanson, P. J., J. S. Amthor, S. D. Wullschleger, K. B. Wilson, R. F. Grant, A. Hartley, D. Hui, J. Hunt, D. W. Johnson, and

J. S. Kimball (2004), Oak forest carbon and water simulations: Model intercomparisons and evaluations against independent data, *Ecol. Monogr.*, *74*(3), 443–489.

Hwang, T., L. E. Band, J. M. Vose, and C. Tague (2012), Ecosystem processes at the watershed scale: Hydrologic vegetation gradient as an indicator for lateral hydrologic connectivity of headwater catchments, *Water Resourc. Res.*, *48*(6), W06514.

Ichii, K., T. Suzuki, T. Kato, A. Ito, T. Hajima, M. Ueyama, T. Sasai, R. Hirata, N. Saigusa, and Y. Ohtani (2010), Multimodel analysis of terrestrial carbon cycles in Japan: Limitations and implications of model calibration using eddy flux observations, *Biogeosciences*, *7*(7), 2061–2080.

Jin, L., D. M. Andrews, G. H. Holmes, H. Lin, and S. L. Brantley (2011), Opening the "black box": Water chemistry reveals hydrological controls on weathering in the Susquehanna Shale Hills Critical Zone Observatory, *Vadose Zone J.*, *10*(3), 928–942.

Kiniry, J. R., J. D. MacDonald, R. K. Armen, B. Watson, G. Putz, and E. P. Ellie (2008), Plant growth simulation for landscape-scale hydrological modelling, *Hydrol. Sci. J.*, *53*(5), 1030–1042.

Kumar, M., G. Bhatt, and C. Duffy (2009), An efficient domain decomposition framework for accurate representation of geodata in distributed hydrologic models, *Int. J. Geogr. Inform. Sci.*, *23*, 1569–1596.

Li, W. (2010), Implementing the Shale Hills watershed model in application of PIHM, Master's Dissertation, Pennsylvania State University.

Lin, H. (2006), Temporal stability of soil moisture spatial pattern and subsurface preferential flow pathways in the Shale Hills Catchment, *Vadose Zone J.*, *5*(1), 317.

Loague, K. (1990), R-5 revisited 2. Reevaluation of a quasi-physically based rainfall-runoff model with supplemental information, *Water Resourc. Res.*, *26*(5), 973–987.

Loague, K. M., and R. A. Freeze (1985), A comparison of rainfall-runoff modeling techniques on small upland catchments, *Water Resourc. Res.*, *21*(2), 229–248.

Loague, K., and G. A. Gander (1990), R-5 revisited: 1. Spatial variability of infiltration on a small rangeland catchment, *Water Resourc. Res.*, *26*(5), 957–971.

Lohmann, D., et al. (2004), Streamflow and water balance intercomparisons of four land surface models in the North American Land Data Assimilation System project, *J. Geophys. Res.*, *109*, D07S91.

Lynch, J. A. (1976), Effects of antecedent soil moisture on storm hydrographs, Ph.D. Thesis, Pennsylvania State University.

Medvigy, D., S. C. Wofsy, J. W. Munger, and P. R. Moorcroft (2010), Responses of terrestrial ecosystems and carbon budgets to current and future environmental variability, *Proc. Nat. Acad. Sci.*, *107*(18), 8275–8280.

Mei, Y., G. M. Hornberger, L. A. Kaplan, J. D. Newbold, and A. K. Aufdenkampe (2012), Estimation of dissolved organic carbon contribution from hillslope soils to a headwater stream, *Water Resourc. Res.*, *48*(9), W09514.

Meinzer, F. C., D. R. Woodruff, D. M. Eissenstat, H. S. Lin, T. S. Adams, and K. A. McCulloh (2013), Above- and belowground controls on water use by trees of different wood types in an eastern US deciduous forest, *Tree Physiol.*, *33*(4), 345–356.

Miller, G. R., X. Chen, Y. Rubin, S. Ma, and D. D. Baldocchi (2010), Groundwater uptake by woody vegetation in a semi-arid oak savanna, *Water Resourc. Res.*, *46*(10).

Mirus, B. B., and K. Loague (2013), How runoff begins (and ends): Characterizing hydrologic response at the catchment scale, *Water Resourc. Res.*, *49*(5), 2987–3006.

Morales, P., M. T. Sykes, I. C. Prentice, P. Smith, B. Smith, H. Bugmann, B. Zierl, P. Friedlingstein, N. Viovy, and S. Sabaté (2005), Comparing and evaluating process based ecosystem model predictions of carbon and water fluxes in major European forest biomes, *Global Change Biol.*, *11*(12), 2211–2233.

Naithani, K. J., D. C. Baldwin, K. P. Gaines, H. Lin, and D. M. Eissenstat (2013), Spatial distribution of tree species governs the spatio-temporal interaction of leaf area index and soil moisture across a forested landscape, *PLOS ONE*, *8*(3), e58704.

NLCD, (2001), National Land Cover Database, Multi-Resolution Land Characteristics Consortium. Retrieved from http://www.mrlc.gov/.

NLDAS vegetation parameters, (2011), available at: http://ldas.gsfc.nasa.gov/nldas/NLDASmapveg.php, accessed 12 March 2011.

Peng, H., Y. Jia, Y. Qiu, C. Niu, and X. Ding (2013), Assessing climate change impacts on the ecohydrology of the Jinghe River basin in the Loess Plateau, China, *Hydrol. Sci. J.*, *58*(3), 651–670.

Pennsylvania State Climatologist (2013), available at http://www.climate.psu.edu/data/networkdescription.php, accessed 20 March 2013.

Qu, Y., and C. J. Duffy (2007), A semidiscrete finite volume formulation for multiprocess watershed simulation, *Water Resourc. Res.*, *43*, W08419.

Shi, Y., K. J. Davis, C. J. Duffy, and X. Yu (2013), Development of a coupled land surface hydrologic model and evaluation at a critical zone observatory, *J. Hydrometeorol.*, doi:10.1175/JHM-D-12-0145.1.

Siqueira, M. B., G. G. Katul, D. A. Sampson, P. C. Stoy, J. Y. Juang, H. R. McCarthy, and R. Oren (2006), Multiscale model intercomparisons of CO_2 and H_2O exchange rates in a maturing southeastern US pine forest, *Global Change Biol.*, *12*(7), 1189–1207.

Smucker, A. J., and J. W. Hopmans (2007), Preface: Soil biophysical contributions to hydrological processes in the vadose zone, *Vadose Zone J.*, *6*(2), 267–268.

Thomas, E., H. Lin, C. Duffy, P. Sullivan, G. H. Holmes, L. Jin, and S. L. Brantley (2013), Spatiotemporal patterns of water stable isotope compositions at the Shale Hills Critical Zone: Linkages to subsurface hydrologic processes, *Vadose Zone J.*, doi:10.2136/vzj2013.01.0029.

Thornton, P. E., B. E. Law, H. L. Gholz, K. L. Clark, E. Falge, D. S. Ellsworth, A. H. Goldstein, R. K. Monson, D. Hollinger, and M. Falk (2002), Modeling and measuring the effects of disturbance history and climate on carbon and water budgets in evergreen needleleaf forests, *Agric. Forest meteorol.*, *113*(1), 185–222.

Trenberth, K. E. (1995), Atmospheric circulation climate changes, *Climatic Change*, *31*(2), 427–453.

VanderKwaak, J. E., and K. Loague (2001), Hydrologic-response simulations for the R-5 catchment with a comprehensive physics-based model, *Water Resourc. Res.*, *37*(4), 999–1013.

White, M. A., P. E. Thornton, S. W. Running, and R. R. Nemani (2000), Parameterization and sensitivity analysis of the BIOME-BGC terrestrial ecosystem model: Net primary production controls, *Earth Interactions*, *4*(3), 1–85.

Wolf, A., E. Blyth, R. Harding, D. Jacob, E. Keup-Thiel, H. Goettel, and T. Callaghan (2008), Sensitivity of an ecosystem model to hydrology and temperature, *Climatic Change*, *87*(1), 75–89.

Xia, Y., et al. (2012a), Continental-scale water and energy flux analysis and validation for North American Land Data Assimilation System project phase 2 (NLDAS-2): 2. Validation of model-simulated streamflow, *J. Geophys. Res.*, *117*, D03110.

Xia, Y., et al. (2012b), Continental-scale water and energy flux analysis and validation for the North American Land Data Assimilation System project phase 2 (NLDAS-2): 1. Intercomparison and application of model products, *J. Geophys. Res.*, *117*, D03109.

Yi, S., A. D. McGuire, J. Harden, E. Kasischke, K. Manies, L. Hinzman, A. Liljedahl, J. Randerson, H. Liu, and V. Romanovsky (2009), Interactions between soil thermal and hydrological dynamics in the response of Alaska ecosystems to fire disturbance, *J. Geophys. Res*, *114*, G02015.

Yu, X., G. Bhatt, C. J. Duffy, and Y. Shi (2013). Parameterization for distributed watershed modeling using national data and evolutionary algorithm, *Computers Geosci.*, *58*, 80–90, doi:10.1016/j.cageo.2013.04.025.

Zreda, M., W. J. Shuttleworth, X. Zeng, C. Zweck, D. Desilets, T. Franz, R. Rosolem, and T. P. A. Ferre (2012), COSMOS: The COsmic-ray soil moisture observing system, *Hydrol. Earth Syst. Sci.*, *9*, 4079–4099.

32

Challenges for Observing and Modeling the Global Water Cycle

Kevin E. Trenberth

32.1. INTRODUCTION

In this chapter an outline is given of issues related to observing and garnering information about the global water cycle. The challenges in developing an adequate climate observing system, especially for the hydrological cycle-related variables, are outlined in section 32.2. Section 32.3 provides a brief summary of some recent evaluations of the state of knowledge about the global aspects of the water cycle, and section 32.4 outlines the Global Energy and Water Exchanges (GEWEX) programmatic plans to address these issues over the next decade.

32.2. OBSERVATIONS AND INFORMATION

Observations of planet Earth and all of the climate components and forcings are increasingly needed for planning and decisions related to climate sciences in the broadest sense. Moreover, the fact that the climate is changing adds a new dimension because the climate of the past is no longer a good guide to the future. For water, there are increasing satellite sensor capabilities that are improving analyses of all aspects of the water cycle and creating new information, but they create challenges for constructing a reliable climate record owing to the changes in observing methods. New sensors expected in the next few years create new opportunities. However, synthesizing the multivariate information requires sophisticated assimilation systems and very good models.

National Center for Atmospheric Research (NCAR), Boulder, Colorado, USA

Figure 32.1 presents a schematic view of the changing observing system [*Trenberth et al.*, 2013]. A century ago the main observations were from the surface, although kites were used to provide limited vertical information. Various balloon-borne instruments played a major role over the twentieth century along with aircraft observations. Substantial advances were made in the International Geophysical Year 1957–1958 and satellite imagery began in the 1960s. It was not until the 1970s though that satellite sounding capabilities came into being, and the Global Weather Experiment in 1979 provided the incentive for truly global syntheses of meteorological data.

However, the operational lifetime of satellites is on the order 5 years or so. Prior to the 1990s many were shorter lived than this, and it is only in the late 1990s that satellites and their instruments have lasted longer than the planned lifetime. Nonetheless, sensors degrade over time, outgassing contaminants cloud the window to space, and calibration can be more and more difficult to achieve reliably. For climate purposes, calibration is essential. Most instruments lack absolute calibration, and therefore the climate record is only reliably achieved if there is a period of overlap between old and new instruments. Continuity has proven to be essential.

The main satellites used for climate and water are polar orbiters or precessing satellites. These are complemented by the geostationary satellites that provide more continuous coverage but with lower resolution.

New observing systems and data processing systems are wonderful but potentially cause havoc for climate as they destroy the continuity unless properly managed, hopefully with an overlap of old and new systems. Nearly all satellite data sets contain large spurious variability associated with changing instruments and satellites,

Remote Sensing of the Terrestrial Water Cycle, Geophysical Monograph 206. First Edition. Edited by Venkat Lakshmi.
© 2015 American Geophysical Union. Published 2015 by John Wiley & Sons, Inc.

511

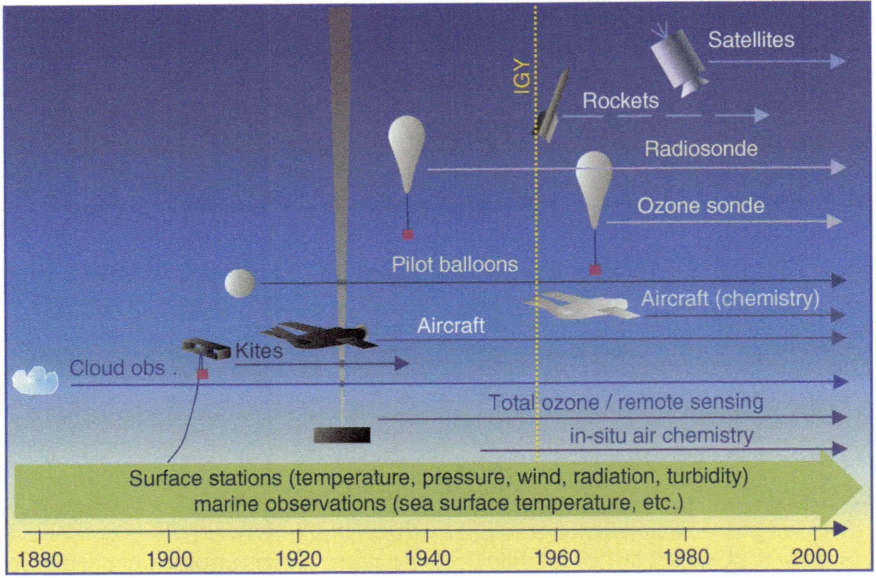

Figure 32.1 Changes in the mix and increasing diversity of observations over time create challenges for a consistent climate record. (From *Trenberth et al.* [2013].)

orbit decay and drift, calibration drift, and changing methods of analysis. Only two data sets (SSM/I* water vapor and MSU satellite temperatures) were used in the Intergovernmental Panel on Climate Change (IPCC) fourth assessment report (AR4) [*Trenberth et al.*, 2007b].

In the 1990s, it was a struggle to obtain a single time series of a variable. This has changed, and there is now a proliferation and multiple data sets for most variables, all purporting to be the "one." All differ, often substantially. Many were produced for a particular purpose and may serve that purpose well, but it has often left users confused. At NCAR, we have recognized this situation and begun developing the Climate Data Guide (http:// climatedataguide.ucar.edu/), which provides the key insights needed to select the data that best align with particular goals, including critiques of data sets by experts from the research community. The characteristics, strengths, and limitations of data sets are discussed.

For water-related variables, remote sensing from space has been a boon but also a trap for the unwary. The SSM/I instruments have provided varying amounts and quality of data with identifiable discontinuities in mid-1987, January 1992, 1999, 2000, and 2007. The major transition from TOVS* to ATOVS* that began in November 1998 (NOAA 12 to NOAA 15) and was mostly

complete by March 2001 (NOAA 14 to NOAA 16) caused major discontinuities in temperature and water vapor channels and thus in many products. The introduction of AIRS in late 2002 and Global Positioning System (GPS) radio occultation (RO) in 2002 also brought notable improvements in water-vapor-related information in the atmosphere. The RO observations received a major boost in April of 2006 with the COSMIC* array.

Adequate analysis, processing, metadata, archival, access, and management of the resulting data and the data products have created further challenges in spite of the new computational tools. Moreover, volumes of data continue to grow and the challenge is to distill information out of the increasing data.

One method of synthesis of multivariate data is via reanalysis using a comprehensive four dimension data assimilation system. Reanalysis is an activity to reprocess past observations in a fixed, state-of-the-art assimilation system based on data assimilation in numerical models. Most reanalysis activities have been for the atmosphere, but some exist for the ocean, sea ice, and land variables. Freezing the analysis system removes the spurious variations that otherwise appear in the operational analyses, but the effects of observing system changes are still apparent. An early attempt to balance regional energy and water budgets using reanalyses [*Roads et al.*, 2002] as part of a GEWEX project illustrated the state of the art at that time.

An evaluation has been made of 8 fairly current atmospheric reanalyses with respect to their global hydrological cycles [*Trenberth et al.*, 2011; see also *Lorenz and Kunstmann*, 2012]. The reanalyses include those from NASA/Goddard, MERRA (Modern Era Retrospective-Analysis for Research

*SSM/I: Special Sensor Microwave Imager; TOVS: TIROS Operational Vertical Sounder; TIROS: Television and Infrared Observation Satellite; ATOVS: Advanced TOVS; COSMIC: Constellation Observing System for Meteorology, Ionosphere, and Climate. MSU: Microwave Sounder Unit; AIRS: Atmospheric InfraRed Sounder; TOA: Top of Atmosphere.

and Applications); European Centre for Medium Range Weather Forecasts (ECMWF), ERA-40 and ERA-Interim (ERA-I); from NOAA, NCEP (National Centers for Environmental Prediction); the Climate Forecast System (CFS) reanalysis CFSR, and R1 and R2 earlier generation reanalyses; the Japanese Reanalysis (JRA), and the C20R reanalysis using only sea level pressure.

The assessment of the hydrological cycles reveals substantial shortcomings and sensitivity to the observing system and its changes over time. Using model-based precipitation P and evaporation E, the time and area average E-P for the oceans, P-E for land, and the vertically integrated atmospheric moisture transport from ocean to land should all be identical but are not close in most reanalyses, and often differ significantly from observational estimates of the surface return flow based on net river discharge into the oceans. Most reanalysis models, the exception being MERRA, have too intense water cycling (P and E) over the ocean. There is a tendency for moisture to be quickly rained out in models, with the lifetime too short and the recycling too large in models. The large-scale moisture budget divergences are more stable in time and similar across reanalyses than model-based estimates of E-P. Trenberth et al. [2011] also evaluated the global energy flows in the same reanalyses and noted that the global mean E and P differ, sometimes by as much as 14% (in ERA-40). In CFSR and ERA-I the difference is 5%. (See Figure 32.2.)

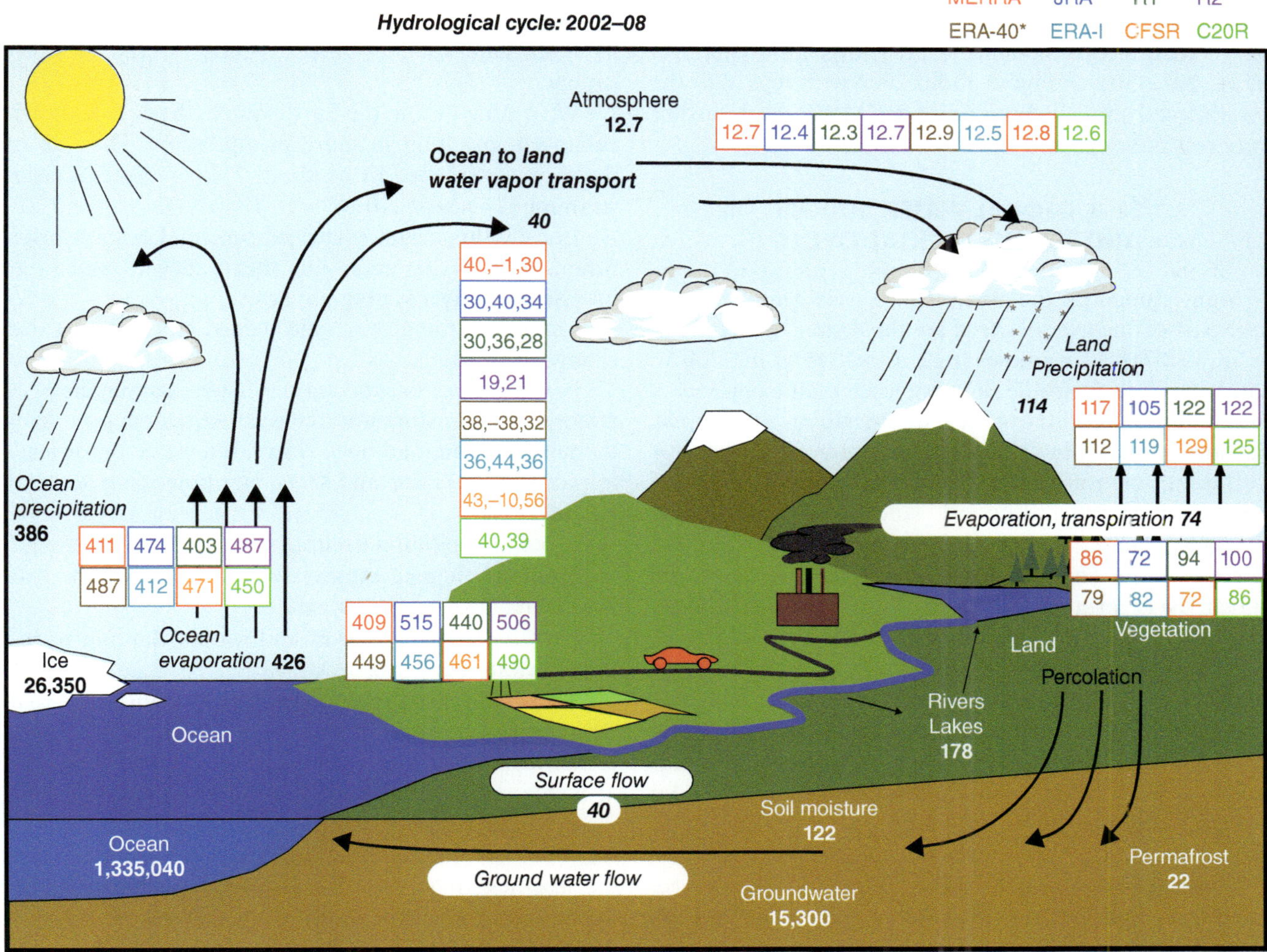

Figure 32.2 Background figure shows the estimates of the observed hydrological cycle adjusted from *Trenberth et al.* [2007a] to apply to the 2002 – 2008 period, with units in 1000 km³ for storage, and 1000 km³/yr for exchanges. Superposed are values from the eight reanalyses for 2002–2008, color coded as given at top right. The exception is for ERA-40, which is for the 1990s. For the water vapor transport from ocean to land, the three estimates given for each are (i) the actual transport estimated from the moisture budget (based on analyzed winds and moisture), (ii) the E-P from the ocean, and (iii) P-E from the land, which should be identical. (From *Trenberth et al.* [2011].)

Even if the assimilating model has a balanced energy budget, when Sea Surface Temperatures (SSTs) are specified, there is an infinite heat and moisture source or sink. Moreover, there is no feedback on the SSTs from surface fluxes. The assimilation of observations creates an analysis increment that violates conservation but that can provide useful information about the observations and the assimilating model. The result is potentially large energy imbalances at Top-of-Atmosphere (TOA) and at surface, and also moisture (E vs. P); see *Trenberth and Fasullo* [2013b]. However, the TOA and surface balances can be strong diagnostics of model bias problems. Hence the latest reanalyses still have multiple problems and must be evaluated and used with caution. The ERA-I reanalysis is regarded as the most stable and closest to reality, although it suffers from several deficiencies, most notably in too much evaporation and spurious moisture sources over land [*Albergel et al.*, 2012]. Its vintage is about 2004 with regard to the operational models running at ECMWF, and further progress has occurred since.

32.3. GLOBAL WATER BUDGET AND HYDROLOGICAL CYCLE

Many studies on the global water cycle deal with only one part of the overall cycle; see the review by *Trenberth et al.* [2007a]. Few studies [e.g., *Trenberth et al.*, 2007a, 2011; *Oki and Kanae*, 2006] have attempted to provide a synthesized, quantitative view of the global water cycle. Greatest difficulties arise from lack of reliable data for surface evapotranspiration, oceanic precipitation, and terrestrial runoff, although land precipitation and changes in water storage also contain uncertainties; see *Trenberth and Fasullo* [2013b]. Regional closure of the water cycle over many large river basins has been attempted by *Vinukollu et al.* [2011], *Sahoo et al.* [2011], and *Pan et al.* [2012]; but, unless adjusted, they do not adequately close the water budget, and the imbalances highlight the outstanding observational and modeling limitations. *Syed et al.* [2009, 2010] deduce the continental discharge as a residual. *Trenberth and Fasullo* [2013a] were able to close the water budget for North America to within about 10%.

Challenges also exist for the changes in the water cycle over time. It is possible that systematic errors, such as from the absence of data from some areas, may be ignored, as they may not influence overall changes over time. However, as the climate changes partly from human activities, the water cycle is also changing [*Trenberth*, 2011]. Moreover, demand for water continues to increase owing to growing population, enhanced agricultural and industrial development, and other human activities such as transformation of landscape and construction of dams and reservoirs, so that very little of the land surface remains in a natural state. This affects the disposition of water when it hits the ground: how much runs off and how much finds its way to rivers or infiltrates into the soil and percolates to depths to replenish the underground water reservoirs.

As noted by *Trenberth and Asrar* [2013], the main impacts of a warmer climate on global water cycle include the following:

• With warming, higher atmospheric temperatures increase the water holding capacity of the atmosphere by about 7% per degree Celsius [e.g., *Trenberth et al.*, 2003].

• Over the ocean where there is ample water supply, the relative humidity remains about the same, and hence the observed moisture goes up at about this rate: an increase in total column water vapor of about 4% since the 1970s [*Trenberth et al.*, 2007b].

• Over land the response depends on the moisture supply.

• With more heat in the Earth system the evaporation is enhanced, resulting in more precipitation. The rate of increase is estimated to be about 2% per degree Celsius warming [*Trenberth*, 2011].

• Locally this means increased potential evapotranspiration, and in dry areas this means drying and more intense and longer lasting droughts.

• Larger warming over land versus the ocean further changes monsoons.

• Precipitation occurs mainly from convergence of atmospheric moisture into the weather system producing the precipitation, and hence increased water vapor leads to more intense rains and snow, and potentially to more intense storms.

• More precipitation occurs as rain rather than snow.

• However, higher temperatures in mid-winter over continents favor higher snowfalls.

• Snowpack melts quicker and sooner, leading to less snowpack in the spring.

• These conditions lead to earlier runoff and changes in peak streamflow.

Hence there is a risk of more extremes, such as floods and droughts.

The pattern of observed changes to date indicates wetter conditions in higher latitudes across Eurasia, east of the Rockies in North America, and in Argentina, but drier conditions across much of the tropics and subtropics [*IPCC*, 2007; *Dai et al.*, 2009; *Trenberth*, 2011; *Dai*, 2011], and this pattern is referred to as "The rich get richer and the poor get poorer" syndrome (the wet areas get wetter while the arid areas get drier). However, a major part of this patten also seems to be associated with a poleward expansion of the subtropical dry zones [*Scheff and Frierson*, 2012]. This pattern is projected to continue into the future [*IPCC*, 2007].

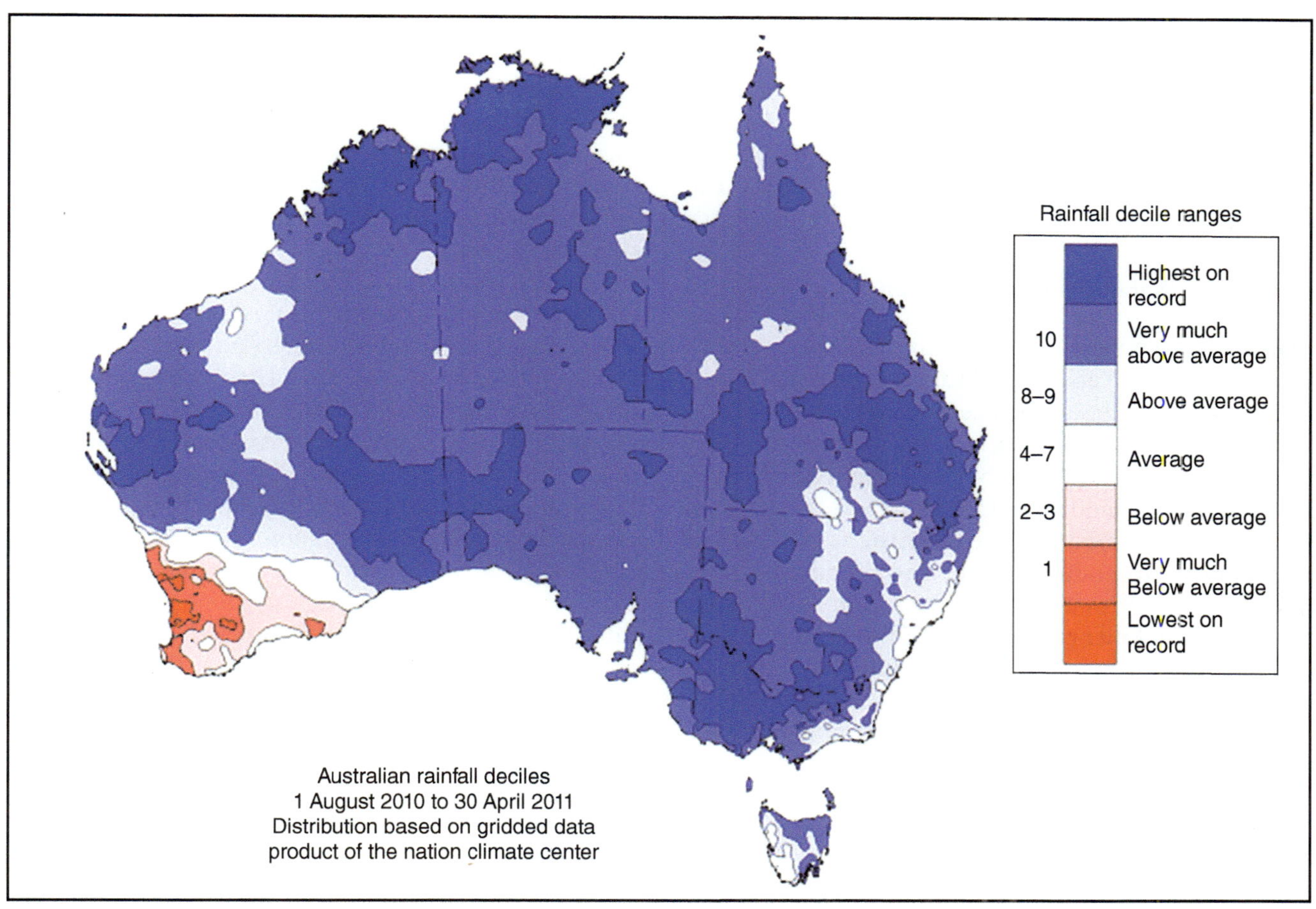

Figure 32.3 Australian rainfall from 1 August 2010 to 30 April 2011, expressed in terms of decile ranges (key at right). These were the wettest months in this wet spell. (From Bureau of Meteorology, downloaded at http://www. bom.gov.au/jsp/awap/rain/index.jsp 18 October, 2011.)

Over land there is a strong negative correlation between precipitation and temperature throughout the tropics and over continents in summer, but a positive correlation in the extratropics in winter [*Trenberth and Shea*, 2005]. The latter arises from the baroclinic storms that advect warm moist air ahead of and into the storm, combined with the ability of warmer air to hold much more moisture. The former arises from the nature of the atmospheric circulation interactions with land. In cyclonic conditions increased cloud and rain provide more soil moisture and thus partitions the decreased surface energy more into latent energy (higher evaporation) instead of sensible heat (lower temperatures). Anticyclonic conditions favor sunshine (more available energy), less rain and soil moisture, and the larger surface energy raises temperatures instead of evaporating moisture. The result is more likely either hot and dry or cool and wet conditions, but not the alternatives.

On global land, there is large variability in precipitation from year to year and decade to decade associated especially with El Niño-Southern Oscillation (ENSO), but there has been an increase overall in land precipitation and the two wettest years are 2010 and 2011. In particular, major flooding in Pakistan, Australia, and Colombia was associated with record high SSTs in the second half of 2010 into 2011 [*Trenberth*, 2012] and led to a dramatic drop in sea level of about 5mm [*Boening et al.*, 2012]. About 2.5mm sea level equivalent ended up in Australia (Figure 32.3) in Lake Eyre because the interior of Australia does not have a riverine system. The prospects for more intense precipitation but longer dry spells lead to the increased risk of flooding and drought, which pose major challenges for the society at large and those who have to manage water resources for food, fiber, and energy production and human consumption and leisure.

32.4. GRAND CHALLENGE ON WATER

The GEWEX Science Steering Group (SSG) has identified four GEWEX Science Questions (GSQs). These emerged from in-depth discussions and subsequent circulation to all GEWEX panel members for commentary.

Three of these GSQs deal with water and are discussed further here, and two of those have been combined into a more general water resource grand challenge for the World Climate Research Programme (WCRP), as detailed in section 32.4.1. The third is given in section 32.4.2. The material here is adapted from *GEWEX* [2013] and *Trenberth and Asrar* [2013].

32.4.1. First Water Resource Question

How can we better understand and predict precipitation variability and changes, and how do changes in land surface and hydrology influence past and future changes in water availability and security?

The vast majority of water comes from precipitation — either directly or indirectly through runoff from distant locations. From a climate perspective, it is therefore an imperative to understand the natural variability of precipitation, as well as its susceptibility to change from external forcings. Within GEWEX, the Global Precipitation Climatology Project (GPCP) [*Huffman et al.*, 2009] has been a focus of improving estimates of precipitation. Because of its inherently intermittent nature, precipitation is difficult to determine reliably with a few instantaneous observations of rates such as from available satellites. Improved observations and analysis products related to precipitation and the entire hydrological cycle, and their use in evaluating and improving weather, climate, and hydrological models are important and tractable over the next 5–10 years. However, it has been even more difficult to reliably simulate precipitation. *Gleckler et al.* [2008] evaluated global model intercomparison projects to assess model errors and how they were improving over time. Using a synthesized metric of signal and assessing the root mean squared errors in several variables, they found that while the noise-to-signal ratio had improved to less than 20% for temperature and about 45% for sea level pressure, it remained above 70% for precipitation. In the tropics order 1 errors occur with the spurious movement of the Inter-Tropical Convergence Zone across the equator with the seasons, for instance.

The specific questions that will be addressed over the next 5–10 years include:

• How well can precipitation be described by various observing systems, and what basic measurement deficiencies and model assumptions determine the uncertainty estimates at various space and time scales?

• How do changes in climate affect the characteristics (distribution, amount, intensity, frequency, duration, type) of precipitation — with particular emphasis on extremes of droughts and floods?

• How do models become better and how much confidence do we have in global and regional climate predictions of precipitation?

• How do changes in the land surface and hydrology influence past and future changes in water availability and security?

• How do changes in climate affect terrestrial ecosystems, hydrological processes, water resources, and water quality, especially water temperature?

• How can new observations lead to improvements in water management?

• How can better climate models contribute to improvements in water management?

The detailed approach to addressing these questions is outlined in *Trenberth and Asrar* [2013] and in *GEWEX* [2013]. There are real prospects for improved data sets of precipitation, soil moisture, evapotranspiration, and related variables such as water storage and sea surface salinity in the next 5–10 years. Key areas of development include the Global Precipitation Mission, which will bring greater accuracy and better sampling of precipitation; CloudSat and EarthCARE missions, which advance understanding of the role that clouds and aerosols play in the climate system; Soil Moisture and Ocean Salinity (SMOS), Aquarius, and the future Soil Moisture Active Passive (SMAP) data, which provide new information on soil moisture and ocean salinity; and Gravity Recovery and Climate Experiment (GRACE), which can be used to provide water storage change estimates. In addition in situ observations from flux towers on land and buoys and Argo floats for the ocean will help close the water and energy budgets. These will help close the water budget over land and provide improved information for products related to water availability and quality for decision makers and for initializing climate predictions from seasons to years in advance.

Hence the improvements will come from ongoing and planned satellite missions as well as greater use of in situ observations; their evaluation and analysis to document means, variability, patterns, extremes, and probability density functions; their use to confront models in new ways and to improve our understanding of atmospheric and land surface processes that in turn improve simulations of precipitation; and new techniques of data assimilation and forecasts that can lead to improved predictions of the hydrological cycle across scales, from catchments to regions to the entire globe, including hydrogeological aspects of groundwater recharge. New atmospheric reanalyses that include coupling to the surface will be carried out and made available. In particular, attention is needed on the use of realistic land surface complexity with all anthropogenic effects taken into account, instead of a fictitious natural environment. This encompasses all aspects of global change, including water management, land use and land cover change, and urbanization. The ecosystem response to climate variability and responsive vegetation to such changes must be included, as must cryospheric changes such as dynamics of permafrost and thawing and changes in mountain glaciers.

The focus on these scientific questions should lead to improved understanding and prediction of precipitation and water variability, enhance the evaluation of the vulnerability of water systems, especially to extremes, which are vital for considerations of water security and can be used to increase resilience through good management and governance.

32.4.2. Second Water Resource Question

How does a warming world affect climate extremes, especially droughts, floods, and heat waves, and how do land area processes, in particular, contribute?

A warming world is expected to alter the occurrence and magnitude of extremes such as droughts, heavy rainfalls and floods, as well as the geographic distribution of rain and snow [*IPCC*, 2007; *Trenberth*, 2011, 2012]. Such changes are related to an acceleration of the hydrologic cycle and circulation changes and include the direct impact of warmer conditions on atmospheric water vapor amounts, rainfall intensity, and snow-to-rain occurrence. How well are models able to handle extremes and how can we improve their capability? New improved and updated data sets at high frequency (e.g., hourly) are needed to properly characterize many of these facets of Earth's climate and to allow for assessment against comparable model data sets. New research activities are needed to promote analyses quantifying which changes are consistent with our expectations and how we can best contribute to improving their prediction in a future climate. Confronting models with new observationally based products will lead to new metrics of performance and highlight shortcomings and developmental needs that will focus field experiments, process studies, numerical experimentation, and model development. New applications should be developed for improved tracking and warning systems, and assessing changes in risk of drought, floods, river flow, storms, coastal sea level surges, and ocean waves.

The specific questions [*GEWEX*, 2013] that will be addressed over the next 5–10 years include:

• What are the short-term, mid-term, and strategic requirements for the existing observing systems and data sets, and which observations are needed to accurately quantify trends in the intensity and frequency of extremes on different space/time scales?

• How can models be improved in their simulation and predictions or projections of the magnitude and frequency of extremes?

• How can the phenomena responsible for extremes be better simulated in models?

• How can we promote development of applications for improved tracking and warning systems arising from extremes?

There is major concern that the occurrence, character, and intensity of extremes will change in the future, as the climate changes due to human activities, and this will have enormous consequences for society and the environment. Yet addressing changing extremes satisfactorily is a daunting task, and it will be difficult to keep up with society's expectations. As noted above, huge improvements in near-global spatial and temporal coverage for precipitation, soil moisture, and other hydrological variables provide opportunities for new data sets, products, improved models, and model applications, making it an opportune time to fully address extremes.

The climate system does not neatly package such extremes. Extremes may be highly localized in time and in space. Drought in one region frequently means heavy precipitation not that far away. The worst extremes are generally compound events, which often are consequences of a chain of events that may be related at the global scale despite their regional implications. Flooding may be accentuated due to saturated soils from previous storms and/or from snowmelt. Furthermore, coastal flooding may involve storm surge effects, local precipitation, and remote snowmelt signals.

Because of its importance, there are many efforts focusing at least in part on extremes. One focus is on drought, although there is certainly interest in other hydrometeorological extremes and related issues, such as statistical analyses. Another issue addresses tropical and extratropical cyclones and associated marine storms as well as extreme sea level variability and change that is connected to storm surges. GEWEX, with its focus on the water cycle and on land surface processes, with strong observational capabilities from global to local and with numerous links with society, is a natural home for addressing many types of extremes. The main GEWEX focal point is to increase efforts on hydrometeorological extremes including drought, heat waves, cold outbreaks, floods, storms, and heavy precipitation events, including hazardous winter snowfalls and hail.

Prospects for advancements are excellent because of new observations, research, modeling and prediction activities already underway and planned. Key areas of development include: (i) utilization of the new global and regional data sets outlined above and from improved data assessment to better characterize extremes on different spatial scales and promote evaluations of model results; (ii) ensure continued development of the Global Drought Information System; (iii) facilitate a number of intercomparison projects aimed at comparison of characteristics of extremes in different data sets (in situ, reanalyses, and satellites), and revealed by different models; (iv) initiate a parallel activity centered on capabilities of statistical methodologies to deal with the complexity of extremes, including their clustering in space and time, and with sparse and regionally unevenly distributed data; (v) initiate multimethod activities and encourage documentation and data inventory centered on a few megaextreme events; examine cold season extremes such as snowstorms, rain-on-snow episodes, freezing precipitation, and prolonged cold weather events.

Drought has devastating consequences whenever and wherever it occurs. Water resources can be strained and adverse effects occur in agriculture. Heat waves are often but not always linked with drought. Health effects can be profound. Prolonged cold weather episodes are a critical feature of mid- and subpolar latitudes in winter. They are disruptive and costly. Isolated extreme rainfalls, as well as continuous periods of heavy and moderate precipitation, occur everywhere with numerous impacts including flooding, devastation of ecosystems, and havoc in urban regions. Storms in different parts of the world are the means by which precipitation, often linked with strong winds, occur, and changes in their paths, intensity, and frequency have enormous consequences, sometimes devastating. A recent example is superstorm Sandy on the east coast of the United States in October 2012. Warming conditions imply that regions accustomed to receiving snow should experience more rain, and changing times of runoff and peak streamflow, with large consequences for ecosystems, hydrologic risks, and water resources. These examples highlight the importance of progress in the area of climate extremes, both in terms of their observations and analysis, and in terms of improved modeling and prediction.

32.5. CONCLUSIONS

The successful implementation of WCRP grand challenges and associated science questions described here depend significantly on GEWEX imperatives: observations and data sets, their analysis, process studies, model development and exploitation, applications, technology transfer to operationalize results, and research capacity development and training of the next generation of scientists. They involve all of the GEWEX panels and will benefit greatly from strong interactions with other WCRP projects and other sister global change research programs such as the International Geosphere-Biosphere Programme (IGBP), International Human Dimensions (IHDP), etc.

Closure of the observed regional and global water budget over the past decade has progressed significantly but remains a major challenge. Thus it continues to be a science imperative for the research community to better observe and understand all aspects of the water cycle in order to improve models that can predict reliably its future variability and change as a major source

of information for decision makers for water resources, food production, and management of risks associated with extreme events. Many potential products could be invaluable to water resource managers on several time horizons, extending well beyond the one week weather scale to seasonal, interannual, and decadal predictions, and climate change projections.

ACKNOWLEDGMENTS

The research of Trenberth is partially sponsored by NASA under grant NNX09AH89G. GEWEX information and publications are available at http://www.gewex.org/

REFERENCES

Albergel, C., P. De Rosnay, G. Balsamo, L. Isaksen, and J. Muñoz-Sabater (2012), Soil moisture analyses at ECMWF: Evaluation using global ground-based in situ observations, *J. Hydrometereol.*, *13*, 1442–1460.

Boening, C., J. K. Willis, F. W. Landerer, R. S. Nerem, and J. Fasullo (2012), The 2011 La Niña: So strong, the oceans fell, *Geophys. Res. Lett.*, *39*, L19602,doi:10.1029/2012GL053055.

Dai, A. (2011), Drought under global warming: A review, *Wiley Interdisciplinary Rev. Climate Change*, doi:10.1002/wcc.81.

Dai, A., T. Qian, K. E. Trenberth, and J. D. Milliman (2009), Changes in continental freshwater discharge from 1949–2004. *J. Climate*, *22*, 2773–2791.

GEWEX (2013), *GEWEX Plans for 2013 and Beyond: GEWEX Science Questions*, GEWEX Doc. Ser. 2012-2, available at http://www.gewex.org/pdfs/GEWEX_Science_Questions_final.pdf.

Gleckler, P. J., K. E. Taylor, and C. Doutriaux (2008), Performance metrics for climate models, *J. Geophys. Res.*, *113*, D06104, doi:10.1029/2007JD008972.

Huffman, G. J., R. F. Adler, D. T. Bolvin, and G. Gu (2009), Improving the global precipitation record: GPCP version 2.1, *Geophys. Res. Lett.*, *36*, L17808, doi:10.1029/2009GL040000.

Intergovernmental Panel on Climate Change (IPCC) (2007), *Climate Change 2007. The Physical Science Basis*, edited by S. Solomon et al. Cambridge Univ. Press, New York, p. 996.

Lorenz, C., and H. Kunstmann (2012), The hydrological cycle in three state-of-the-art reanalyses: Intercomparison and performance analysis, *J. Hydrometereg.*, *13*, 1397–1420.

Oki, T., and S. Kanae (2006), Global hydrological cycles and world water resources, *Science*, *313*, 1068–1072.

Pan, M., A. K. Sahoo, T. J. Troy, R. K. Vinukollu, J. Sheffield, and E. F. Wood (2012), Multisource estimation of long-term terrestrial water budget for major global river basins, *J. Climate*, *25*, 3191–3206.

Roads, J. O., M. Kanamitsu, and R. Stewart (2002), CSE water and energy budgets in the NCEP-DOE Reanalysis II, *J. Hydrometeorol*, *3*, 227–248.

Sahoo, A. K., M. Pan, T. J. Troy,R. K. Vinukollu, J. Sheffield, and E. F. Wood (2011), Reconciling the global terrestrial water budget using satellite remote sensing, *Remote Sens. Environ.*, *115*, 1850–1865.

Scheff, J., and D. M. W. Frierson (2012), Robust future precipitation declines in CMIP5 largely reflect the poleward expansion of model subtropical dry zones, *Geophys. Res. Lett.*, *39*, L18704, doi:10.1029/2012GL052910.

Syed, T. H., J. S. Famiglietti, and D. P. Chambers (2009), GRACE-based estimates of terrestrial freshwater discharge from basin to continental scales, *J. Hydrometeoral*, *10*, 22–40.

Syed, T. H., J. S. Famiglietti, D. P. Chambers, J. K. Willis, and K. Hilburn (2010), Satellite-based global-ocean mass balance estimates of interannual variability and emerging trends in continental freshwater discharge, *Proc. Natl. Acad. Sci.*, *42*, 17,916–17,921, doi:10.1073/pnas.1003292107.

Trenberth, K. E. (2011), Changes in precipitation with climate change, *Climate Res.*, *47*, 123–138, doi:10.3354/cr00953.

Trenberth, K. E. (2012), Framing the way to relate climate extremes to climate change, *Climatic Change*, *115*, 283–290, doi:10.1007/s10584-012-0441-5.

Trenberth, K. E., and G. Asrar (2013), Challenges and opportunities in water cycle research: WCRP contributions, *Surv. Geophys.*, *35*, 515–532. doi:10.1007/s10712-012-9214-y.

Trenberth, K. E., and J. T. Fasullo (2013a), North American water and energy cycles, *Geophys. Res. Lett.*, *40*, 365–369, doi:10.1029/2012GL054084.

Trenberth, K. E., and J. T. Fasullo (2013b), Regional energy and water cycles: Transports from ocean to land, *J. Climate*, *26*, 7837–7851. doi:10.1175/JCLI-D-00008.1.

Trenberth, K. E., and D. J. Shea (2005), Relationships between precipitation and surface temperature, *Geophys. Res. Lett. 32*, L14703, doi:10.1029/2005GL022760.

Trenberth, K. E., A. Dai, R. M. Rasmussen, and D. B. Parsons (2003), The changing character of precipitation. *Bull. Am. Meteorol. Soc.*, *84*, 1205–1217.

Trenberth, K. E., L. Smith, T. Qian, A. Dai, and J. Fasullo (2007a), Estimates of the global water budget and its annual cycle using observational and model data, *J. Hydrometeorol*, *8*, 758–769.

Trenberth, K. E., et al. (2007b), Observations: Surface and atmospheric climate change, in *Climate Change 2007: The Physical Science Basis*, edited by S. Solomon et al., pp. 235–336 Cambridge Univ. Press, New York.

Trenberth, K. E., J. T. Fasullo, and J. Mackaro (2011), Atmospheric moisture transports from ocean to land and global energy flows in reanalyses, *J. Clim.*, *24*, 4907–4924, doi:10.1175/2011JCLI4171.1.

Trenberth, K. E., R. A. Anthes, A. Belward, O. Brown, E. Haberman, T. R. Karl, S. Running, B. Ryan, M. Tanner, and B. Wielicki (2013), Challenges of a sustained climate observing system, in *Climate Science for Serving Society: Research, Modelling and Prediction Priorities*, edited by G. A. Asrar and J. W. Hurrell, pp. 13–50, Springer, 484 pp, Dordrecht.

Vinukollu R. K., R. Meynadier, J. Sheffield, and E. F. Wood (2011), Multi-model, multi-sensor estimates of global evapotranspiration: Climatology, uncertainties and trends, *Hydrol. Process.*, *25*, 3993–4010.

33

Integrated Assessment System Using Process-Based Eco-Hydrology Model for Adaptation Strategy and Effective Water Resources Management

Tadanobu Nakayama

33.1. INTRODUCTION

Human activity has dramatically changed ecosystem dynamics in East Asia (Figure 33.1). The change from rural or wilderness to urban or agricultural uses has greatly affected their surrounding ecosystems. Furthermore, environmental pollution is recently becoming intertwined with degradations of water environment, imbalance of hydrologic cycle, thermal environment, and contamination. Because nearly 80% of the world's population is now exposed to high levels of threat to water security from rising demand due to population growth, urbanization, and industrialization [*Gleick and Palaniappan*, 2010], we need a cumulative diagnosing threat framework to offer a tool for prioritizing policy and management responses to the crisis and to assure global water security for both humans and ecosystem biodiversity [*Jackson et al.*, 2001; *Baron et al.*, 2002; *Vorosmarty et al.*, 2010]. To facilitate sustainable development, it is necessary to quantify the mechanisms of ecosystem change through integrated approach, and to carry out adaptation strategy to develop sustainably and recover actively the local natural environments that have been damaged in the past. A numerical model is one of very powerful tools in this process of river–basin–wide initiative approach. Though stationarity assumption—the idea that natural systems fluctuate within an unchanging envelope of variability—has long been composed by human disturbances in river basins, some researches started to point out that the stationarity should no longer serve as a central, default assumption in water

resource risk assessment and planning and that we need alternative nonstationarity approaches to predict global patterns of streamflow and to ensure sustainable water availability under a global climate change [*Milly et al.*, 2005, 2008]. Global depletion of groundwater resources is also becoming serious in addition to surface water in many countries, including the High Plains of the central United States, the North China Plain (NCP), northern India, the Sahel, South Africa, and Australia, though most people are often unfamiliar with its dynamic nature view as important storage and reservoir of water resources [*Alley et al.*, 2002; *Milly et al.*, 2008; *Wada et al.*, 2010]. Because countries should not go to the worst-case situation, such as war over water, and have to solve their water shortages through trade and international agreements [*Barnably*, 2009], it has a possibility that water-market-based policy instruments would help to attain the targets in the most cost-effective manner. Recently, there have been some interesting studies of virtual (water saved in one region by the importation of goods produced from water used in another region) and real water transfer and their relationship to actual water demand, water scarcity, and water conflicts [*Yang and Zehnder*, 2001; *Hoekstra and Hung*, 2005; *Ma et al.*, 2006]. Although it is generally recognized that in the short term its globalization may prevent malnourishment, famine, and conflicts, we have to keep in mind that in the long term the globalization of water resources might reduce the societal resilience with respect to water limitations in that it leaves fewer options available to cope with exceptional droughts and crop failure [*D'Odorico et al.*, 2010].

In China, hydroclimate is diverse between north and south and its extreme is becoming seriously remarkable mainly due to global climate change and anthropogenic

Center for Global Environmental Research, National Institute for Environmental Studies (NIES), Ibaraki, Japan

Remote Sensing of the Terrestrial Water Cycle, Geophysical Monograph 206. First Edition. Edited by Venkat Lakshmi.
© 2015 American Geophysical Union. Published 2015 by John Wiley & Sons, Inc.

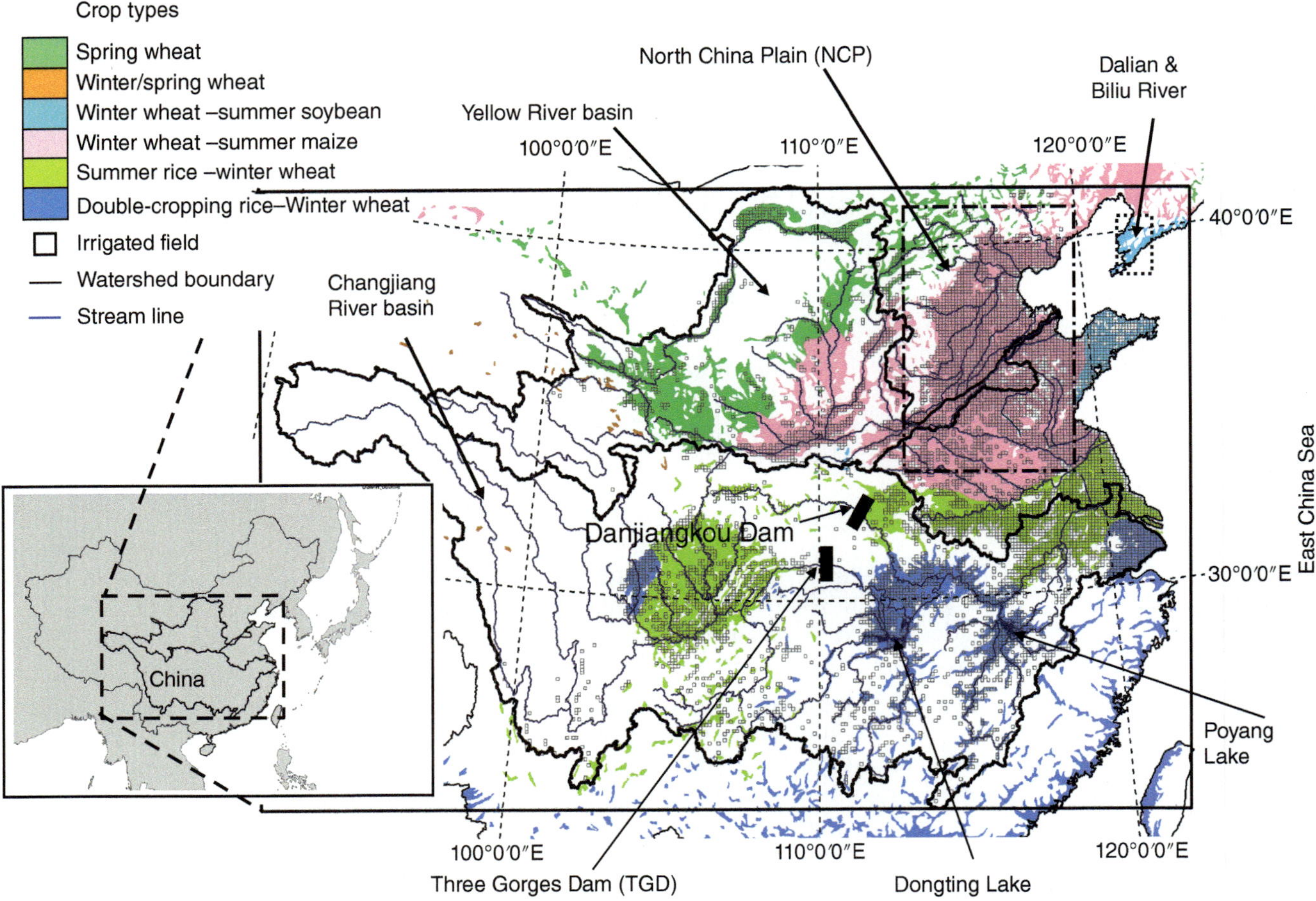

Figure 33.1 Crop types in the agricultural areas of the Changjiang and Yellow river basins in China [*Chinese Academy of Sciences*, 1988]. Irrigation areas are also overshaded in the figure [*Liu*, 1996]. Black squire line is the border of the simulation area with grid of 300 × 200 blocks in the Albers (WGS, 1984) coordinates. Black dash-dotted line is the border of the North China Plain (NCP), which includes the downstream area of the Yellow River. Black dotted line is the border of Dalian and Biliu river region.

effect (Figure 33.1) [*Nakayama*, 2011a, 2011b; *Nakayama and Watanabe*, 2008b). In the Changjiang River (Yangtze) basin in the humid south, deforestation and land reclamation induce serious soil erosion and increase floods, in particular, shrinkage of flood storage ability around lakes is serious [*Zhao et al.*, 2005; *Nakayama and Watanabe*, 2008b; *Nakayama and Shankman*, 2013a]. Although the recently completed Three Gorges Dam (TGD) will provide flood control and other benefits, changes in the aquatic environment of the estuary and the East China Sea might be brought about by changes in pollutant loads caused by the deposition of large amounts of sediment carried from the upper region in the dam and the artificial control of discharge volume [*Yang et al.*, 2006]. Some recent researches have pointed out that the impact of the TGD on flood occurrence in Changjiang downstream is also increasing problems against the original justifications for building the dam [*Shankman and Liang*, 2003; *Zhao et al.*, 2005; *Nakayama and Watanabe*, 2008b;

Nakayama and Shankman, 2013a, 2013b). On the other hand, the Yellow River (Huanghe) in the semiarid north is well known for high sand content, frequent floods, unique channel characteristics in the lower reach (the river bed is higher than the land outside the banks), and the limited water resources. In particular, the NCP, located downstream of Yellow River with a semiarid climate (Figure 33.1), has suffered serious water deficit problems due to the fact that it is one of the most important grain cropping areas in China [*Brown and Halweil*, 1998; *Shimada*, 2000; *Chen et al.*, 2003; *Nakayama et al.*, 2006]. Furthermore, it includes some vast metropolitan areas such as Beijing and Tianjin. This water stress has intensified water use conflicts between the upstream and downstream areas and also between agriculture and municipal/industrial sectors there. In particular, this region has changed from a water-rich one in the 1950s to a water-poor one currently, accompanied by various ecosystem degradations such as the dry-out of the Yellow

River, the near closure of the Hai River, and groundwater degradation in the NCP [*Liu and Xia*, 2004; *Nakayama*, 2011a, 2011b; *Nakayama et al.*, 2006]. This has led to the occurrence of large depression cones and land subsidence in urban areas, and the expansion of the area of saline-alkaline land due to seawater intrusion in the coastal zone [*Brown and Halweil*, 1998; *Liu and Xia*, 2004; *Nakayama*, 2011a; *Nakayama et al.*, 2006]. Under these serious situations, some researchers stated that China's environmental pressure already exceeds its carrying capacity of this densely populated land [*Varis and Vakkilainen*, 2001]. For example, Dalian as an industrial city, located at the southern tip of the Liaodong Peninsula (Figure 33.1), has inherently little freshwater supply. Most of the city's water necessary for its fast economic growth is drawn at a great distance from there, the Biliu Reservoir at the middle of the river. Since the completion of the dam in 1984, most of the water has been piped to Dalian, and as a result, the catchment has changed from water-rich to water-poor, as evidenced by drying-out of the downstream river, groundwater fall, and seawater intrusion, all seen in other regions of northern China [*Ren et al.*, 2002; *Nakayama*, 2011b; *Nakayama et al.*, 2006, 2010]. The author's previous studies [*Nakayama*, 2012c; *Nakayama and Shankman*, 2013b] evaluated general characteristics of the ecohydrological process in the entire Changjiang and Yellow river basins in the extension of separated basins to evaluate optimum amount of transferred water, socioeconomic, and environmental consequences there after constructing the basic structure of the model with different functions of representative crops (wheat, maize, soybean, and rice). To compensate for the export of virtual water from the highly irrigated regions and to reduce water scarcity in the north, China is accelerating huge a project for driving water from Changjiang to the Yellow and Hai rivers (South-to-North Water Transfer Project; SNWTP) [*Rich*, 1983; *Yang and Zehnder*, 2001; *Li et al.*, 2007]. The chief objective of this project is to remedy the essential mismatch of China's land and water resources. Though it is estimated that TGD and SNWTP would generally remove these imbalances as much as possible [*Yang and Zehnder*, 2001; *Li et al.*, 2007], there is urgent need to clarify the relations of these two extremes of drought and flood affected by human activity, develop a process-based model, and present an integrated approach for appropriate water resource management to achieve sustainable development on a regional scale [Nakayama, 2012c; *Nakayama and Shankman*, 2013b].

The objective of this study is to present an advanced method of combination between numerical model, satellite image, and statistical analysis in order to clarify complex and unavoidable relationships between water resource and economic growth, which has greatly affected

ecosystem degradation and its serious burden on the environment. In particular, the author's various studies about different land covers such as natural, irrigated, and urban areas in China [*Nakayama*, 2011a, 2011b, 2012c; *Nakayama and Shankman*, 2013a, 2013b; *Nakayama and Watanabe*, 2008b; *Nakayama et al.*, 2010] are reviewed and summarized in order to synthesize the impact of anthropogenic activity on ecohydrological changes on a catchment or basin scale for adaptation strategy and effective water resources management in the future. The process-based model presented in this study includes complex subsystems to simulate hydrological change resulting from human intervention such as drying-up of the rivers and groundwater degradation and to evaluate the linkage between urban development and sustainable water resource management. The model also includes different growth processes of representative crops and predicts hydrologic change after TGD and SNWTP to estimate whether dilemmas between water stress, crop productivity, and ecosystem degradation would diminish in the basins. Effective method of satellite image analysis in agricultural fields is also described to estimate uneven crop yield and its relation to water availability in surface water and groundwater. This integrated approach will have some roles not only to re-consider this complex process from the viewpoint of hydrologic and biogeochemical cycles, but also to clarify how the substantial pressures of the complex nature of the water problems can be overcome by effective trans-boundary solutions to support decision making on sustainable development.

33.2. METHODS

33.2.1. Development into Coupled Human and Natural Systems

A process-based NICE (National Integrated Catchment-based Eco-Hydrology) model includes surface unsaturated-saturated water processes and assimilates land surface hydrothermal processes describing phenology with satellite data (Figure 33.2) [*Nakayama*, 2008a, 2008b, 2009, 2010, 2011a, 2011b, 2012a, 2012b, 2012c, 2013; *Nakayama and Fujita*, 2010; *Nakayama and Hashimoto*, 2011; *Nakayama and Shankman*, 2013a, 2013b; *Nakayama and Watanabe*, 2004, 2006, 2008a, 2008b; *Nakayama et al.*, 2006, 2007, 2010, 2012). NICE has been coupled with complex subsystems in irrigation, urban water use, stream junction, and dam/canal in order to develop coupled human and natural systems and to analyze the impact of anthropogenic activity on ecohydrologic change. The unsaturated layer divides the canopy into two layers and soil into three layers in the vertical dimension in the SiB2 (Simple Biosphere model 2) [*Sellers et al.*, 1996]. About the saturated layer, the NICE solves

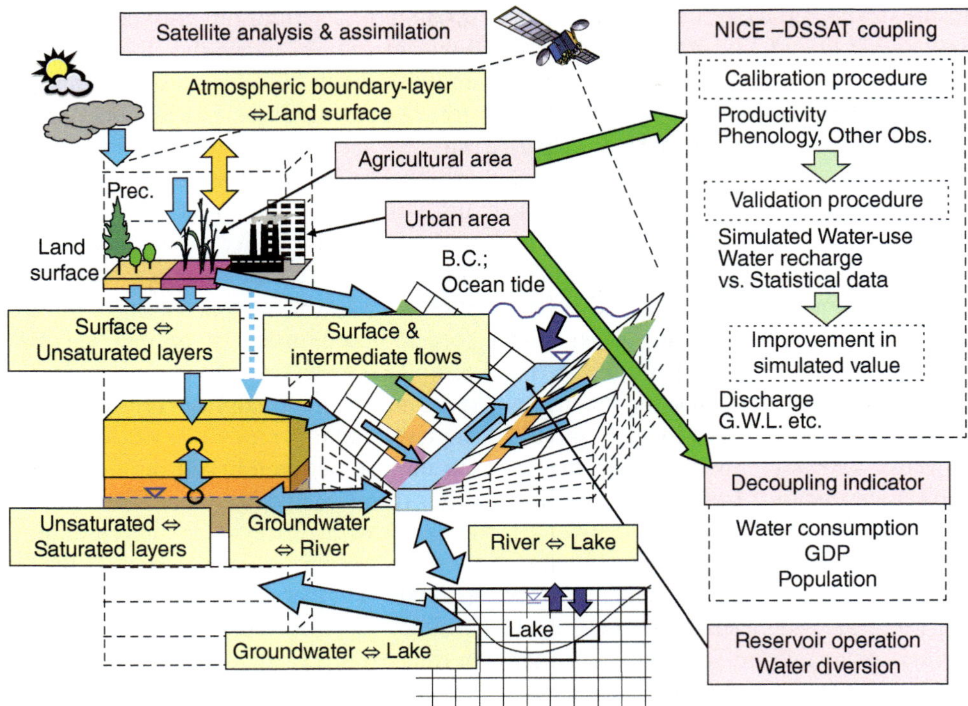

Figure 33.2 National Integrated Catchment-based Eco-Hydrology (NICE) model. NICE is coupled with complex subsystems in irrigation, urban water use, stream junction, and dam/canal.

three-dimensional groundwater flow for both unconfined and confined aquifers. The surface runoff model consists of hillslope and stream network models. The hillslope hydrology can be expressed by the two-layer surface runoff model including freezing/thawing processes. The NICE connects each submodel by considering water/heat fluxes: gradient of hydraulic potentials between the deepest unsaturated layer and the groundwater, effective precipitation, and seepage between river and groundwater [*Nakayama and Watanabe*, 2004]. In an agricultural field, NICE is coupled with DSSAT (Decision Support Systems for Agro-technology Transfer) [*Ritchie et al.*, 1998] to include different functions of representative crops (wheat, maize, soybean, and rice) [*Nakayama et al.*, 2006; *Nakayama and Watanabe*, 2008b]. The model simulates automatically dynamic growth processes and biomass formulation by inputting previous point data for each crop type [*Wang et al.*, 2001; *Liu et al.*, 2002; *Tao et al.*, 2006], spatial distribution of crop types [*Chinese Academy of Sciences*, 1988; *Fang et al.*, 2006], and irrigation area [*Liu*, 1996] in which automatic irrigation mode supplies crop water requirements, assuming that average available water in the top layer falls below soil moisture at field capacity for cultivated fields [*Nakayama*, 2011a, 2011b; *Nakayama et al.*, 2006] and that water level is maintained in irrigation schedule of irrigated water depth for paddy fields [*Nakayama and Watanabe*, 2008b]. To evaluate the role of flood storage ability in Dongting and Poyang

lakes, which drain into the Changjiang through narrow outlets (Figure 33.1), the NICE was combined with lake and stream junction models [*Nakayama and Shankman*, 2013a; *Nakayama and Watanabe*, 2008b]. This energy-based method addresses river discharge across the junction by performing standard step backwater and forwater calculations with continuity and energy equations. Some of the water is transferred through a network of many dams (reservoirs) and canals to main cities and cultivated fields such as the NCP and the Dalian. Because limited data are available on discharge control at most of these dams adjacent to these reservoirs, a constant ratio of inflow to outflow — instead of the more complicated storage-runoff function model — was simply used for the major dams for which no data were available [*Nakayama*, 2011a].

33.2.2. Model Input Data, Statistical and Satellite Analyses

In 6-h reanalyzed data of downward radiation, precipitation, atmospheric pressure, air temperature, air humidity, and wind speed, FPAR (fraction of photosynthetically active radiation), and LAI (leaf area index) were input into the model after interpolation of ECMWF (European Centre for Medium-Range Weather Forecasts) in inverse proportion to the distance back-calculated in each grid. Because the ECMWF precipitation had the least reliability and underestimated observed peak values, rain gauge daily precipitation collected at 4271 meteorological stations were used to correct the ECMWF value [*Nakayama,* 2011a; *Nakayama et al.,* 2010]. Mean elevation was calculated by using a global digital elevation model (DEM; GTOPO30) [*U.S. Geological Survey,* 1996]. Digital land cover data were categorized. Forests, grasses, bushes, and shrubs cover mountainous and hilly areas in the upper regions. Paddy fields are widely distributed along the Changjiang mainstream, whereas cultivated fields are mainly in the Yellow River (Figure 33.1). Vegetation class and soil texture were categorized and digitized by using the Vegetation and Soil Maps of China (1:4,000,000) [*Chinese Academy of Sciences,* 1988]. The geological structures were divided into four types on the basis of hydraulic conductivity, the specific storage of porous material, and specific yield by scanning and digitizing the geological material [*Geological Atlas of China,* 2002] and core-sampling data.

The Organisation for Economic Co-operation and Development (OECD) developed procedures to decouple environmental pressures from economic growth [*OECD,* 2002]. Decoupling can be either absolute or relative values. Absolute decoupling occurs when the environmental variable is stable or decreasing while the economic driving force is growing. Relative decoupling occurs when the environmental variable is growing, but not as quickly as the economic variable. A decoupling indicator is calculated from the ratio of pressure to driving force at the end to the value at the start of a given time period. In the author's previous research [*Nakayama et al.,* 2010], the decoupling indicator procedure was applied to identify impacts of driving force such as social activities and its pressure such as water consumption and economic growth on the environment in urban areas. There are various indicators of driving force, such as population, standard of living, and urbanization, in addition to the gross domestic product (GDP). The main objective of this statistical analysis is to evaluate the linkage between urban development, water resource management, and environmental degradation. Because the study area such as Dalian is a highly industrialized city in China, it is preferable to use the GDP as a typical driving force as a first step. In an analysis of the impact of water abstraction from the Biliu River on the environmental degradation in the catchment, the pressure is the normalized difference vegetation index (NDVI) as an index of environmental degradation, and the driving force is water supply from the Biliu River. Details are described in *Nakayama et al.* [2010].

After the NDVI was calculated from NOAA/AVHRR (Advanced Very High Resolution Radiometer) satellite data collected over East Asia during 1982–1999 downloading the original data from *WebPaNDA* [2008], the time-integrated NDVI (TINDVI) was calculated from satellite NDVI images with a resolution of 6 km by using the nearest-neighbor method. This value is defined as area under curve described by NDVI with time because dry biomass is linearly related to TINDVI and crop growth rate is related to NDVI [*Yang et al.,* 1998]. The previous research showed that simulated dry biomass, which agreed well with the observed value from previous field experiments [*Nakayama et al.,* 2006], has a linear relationship to TINDVI with a high correlation (r^2 is greater than 0.950) [*Nakayama,* 2011b]. The result also implied that maximum growth rate generally occurs when NDVI reaches a plateau stage. This relation is useful for estimating the potential rate of primary production in the basin. These input data are fine enough to estimate ecohydrological process in the scale of a few hundreds to thousands of kilometers in this study.

33.2.3. Model Modification for TGD and SNWTP

Water level at the TGD would be operated between 145.0 m (flood season) and 175.0 m (normal season) above sea level; above see level depending on optimal flood control water-level schemes on conditions that the spill discharge should be limited to less than 55,000 m³/s for the minimization of flood risk in the downstream and larger than 35,000 m³/s for the maximization of navigation benefits [*Liu et al.,* 2008]. The author's previous studies [*Nakayama and Shankman,* 2013a, 2013b) modified the NICE to use one-dimensional water budget and the reservoir operation by applying the relation between water level and storage volume for the estimation of released discharge (Figure 33.2). The model showed estimated discharge at the reservoir during the late summer for a flood event did not greatly change after the dam construction, which implies the dam's inability to have a substantial effect on the discharge during severe floods in the same way as previous research [*Shankman and Liang,* 2003]. For prediction of hydrologic change in normal flood, monthly discharge change [*Yang et al.,* 2006; *Li et al.,* 2007; *Dai et al.,* 2010; *Yi et al.,* 2010] was directly input into the model [*Nakayama and Shankman,* 2013a, 2013b]. The model also showed flood water level would become less extreme, and the range at which the water level increases would be dampened downstream after the filling of the TGD due to increased evaporation and water usage [*Yi et al.,* 2010). The middle route of the SNWTP diverts water from the Danjiangkou Dam at the upstream of the Hanjiang River to northern China (Figure 33.1). The previous research also modified the NICE to input directly water transfer schemes of 9.5×10^9 m³ and 13.0×10^9 m³ per year in the model by considering monthly runoff [*Li et al.,* 2007; *Nakayama and Shankman,* 2013a, 2013b].

33.2.4. Boundary Conditions and Running the Simulation

At the upstream boundaries, the reflecting condition on the hydraulic head was used, assuming that there is no inflow from the mountains in the opposite direction [*Chen et al.*, 2005; *Nakayama*, 2011b; *Nakayama and Watanabe*, 2008b]. At the eastern sea boundary, constant head was set at 0 m. In river grids decided by digital river network from 1 50,000 and 1 100,000 topographic maps, inflows or outflows from the riverbeds were simulated at each time step depending on the difference in the hydraulic heads of groundwater and river.

The simulation areas are 530 km wide by 840 km long in the NCP [*Nakayama*, 2011a; *Nakayama et al.*, 2006], 60 km wide by 110 km long in the Dalian [*Nakayama et al.*, 2010], and 3000 km wide by 2000 km long covering most of the Changjiang River and the Yellow River basins [*Nakayama*, 2011b, 2012c; *Nakayama and Shankman*, 2013a, 2013b; *Nakayama and Watanabe*, 2008b] in the Albers (WGS, 1984; World Geodetic System 1984) coordinates (Figure 33.1). These areas were discretized with a grid spacing of 200 m to 10 km in the horizontal direction and into 20 layers with a weighting factor of 1.1 (finer at the upper layers) in the vertical direction. The upper layer was set at 2 m depth, and the 20th layer was defined as an elevation of −500 m from the sea surface. Simulations were performed with a time step of $\Delta t = 1\,h - 6\,h$ for 1987–1988 and 1998–1999 after 1 year of warmup period until equilibrium, and parameters were estimated by a comparison of simulated steady state value with that published in previous studies. Previous observed data of river discharge [*Changjiang Water Conservancy Committee*, 1987, 1988, 1998, 1999; *Yellow River Conservancy Committee*, 1987, 1988), soil moisture [*Entin et al.*, 2000; *Robock et al.*, 2000], and groundwater level [*China Institute for Geo-Environmental Monitoring*, 2003] were used for calibration and validation in addition to values published in the literature [*Clapp and Hornberger*, 1978; *Rawls et al.*, 1982] in the same way as the author's previous work [*Nakayama*, 2011a, 2011b; *Nakayama and Watanabe*, 2008b; *Nakayama and Shankman*, 2013a, 2013b; *Nakayama et al.*, 2010].

33.3. RESULTS

33.3.1. Water Resources Degradation Due to Human Intervention in Semiarid Regions

NICE simulated the change in the groundwater level over the previous half century in the NCP [reprint from *Nakayama*, 2011a] (Figure 33.3). The model reproduced excellently the observed groundwater levels [*Shimada*, 2000] during the same period and showed that the water table has decreased year by year because the simulation includes the process of irrigation, industry, and domestic water uses in the extension of only irrigation [*Nakayama et al.*, 2006]. In particular, the model reproduces better cone depressions around the bigger cities (Cangzhou, Hengshui, Baoding, etc.) owing to excessive groundwater use in the urban area [*Nakayama*, 2011a; *Nakayama et al.*, 2006]. Hydraulic gradients are steeper around the Taihang and Yanshan mountains, and are relatively flat in the plain area. Because the groundwater level is higher in the mountainous areas, groundwater constantly flows into the plain, and the mountainous area plays an important role as a recharge region. Generally, the groundwater level is drawn down from March to June–July mainly as a result of winter wheat production and then starts to recover as the rainy season starts [*Nakayama et al.*, 2006]. The simulated groundwater levels reproduce actual levels very accurately in both the mountainous and plain areas.

33.3.2. Vegetation Degradation Affected by Economic Growth in Water-Scarce Urban Areas

The economy has grown faster than water consumption in Dalian, which shows a pronounced relative decoupling between economic growth and environmental pressure in Dalian during 1992–2007 [reprint from *Nakayama et al.*, 2010] (Figure 33.4). The consumption of water derived from the Biliu River shows a relative decoupling from the GDP of the urban area (Figure 33.4a). Though the Dalian government has increased the water price several times to control water consumption, the decoupling indicator values declined, were unstable, and were negative twice in 8 years (1998–1999 and 2004–2005). The value dropped from 0.43 in 1993–1994 to −0.37 in 1998–1999, became positive at 0.33 again in 2003–2004, and then declined again. These trends show that the environmental pressure of water consumption has increased with economic growth in urban areas, although relative decoupling predominated, especially when the decoupling indicator value was negative. The impact of water withdrawal from the Biliu Reservoir on environmental degradation of NDVI downstream is shown in Figure 33.4b. The decoupling indicator value gradually decreased from 0.75 (relative decoupling) in 1988–1989 to −0.25 (no decoupling) in 2006–2007, although the periods of 1985–1988 take irregular values due to the start of dam operation and incorrect data. This previous result [*Nakayama et al.*, 2010] shows increasing environmental degradation with increasing abstraction of water from the Biliu River. These results also indicate that the environmental degradation in the Biliu River catchment will grow more serious with economic growth in the coming years.

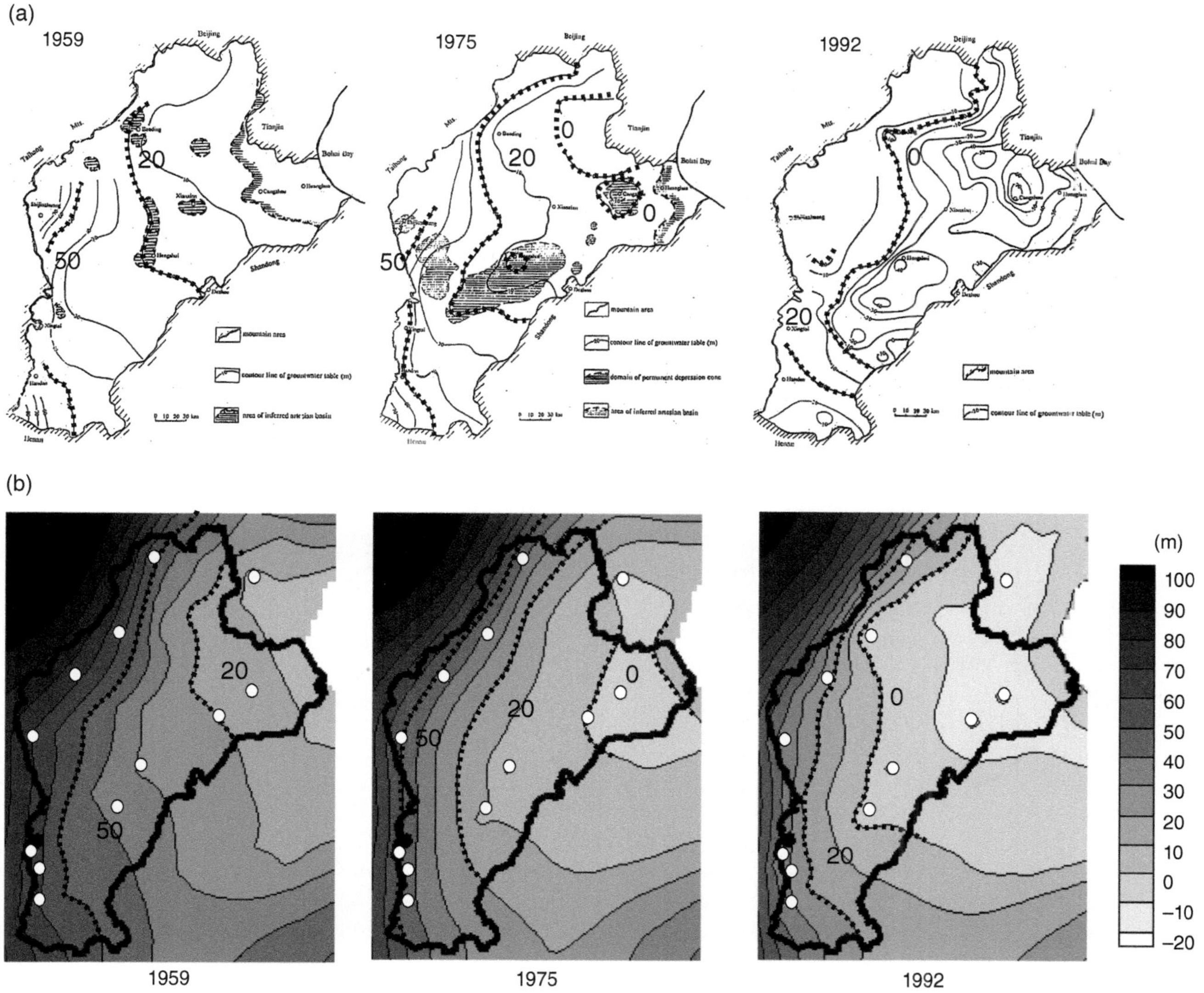

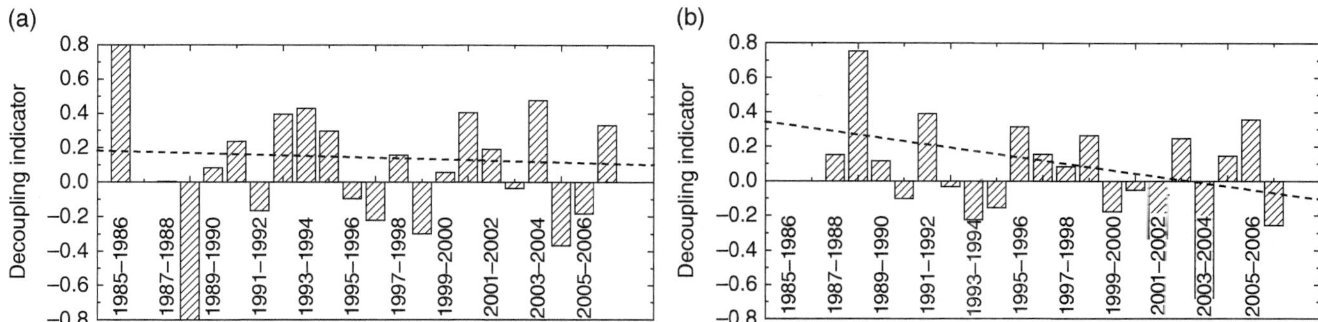

Figure 33.3 Groundwater degradation in the NCP from 1959 to 1992 [reprint from *Nakayama*, 2011a]: (a) observed value [*Shimada*, 2000] and (b) simulated result with irrigation and urban water uses. Dotted lines show the groundwater levels 0 m, 20 m, and 50 m above sea level. In (b) circles show the main cities with predominant water use.

Figure 33.4 Annual trends of decoupling indicators between (a) water consumption and GDP in the urban area of Dalian city and (b) NDVI and water consumption [reprint from *Nakayama et al.*, 2010]. Dashed lines are least-squares regression lines estimated from annual decoupling indicators (bars).

33.3.3. Impact of the TGD and SNWTP on Changjiang Downstream in Humid Regions

The river discharges around lake regions in Changjiang River area are very sensitive and go forward and backward depending on the hydroclimatic conditions around the junction (Figure 33.1), which is closely related to the surface - groundwater interactions and the flood storage ability of the lake [*Nakayama and Watanabe*, 2008b]. Though the flood risk around Poyang Lake decreases moderately with controlled discharge decreasing during July – September, in particular, in the case that controlled discharge is equal to 40,000 m³/s, the model predicted that the TGD might increase the lake water level and flood occurrence during May – June regardless of controlled discharge in exchange for the benefit (flood decrease) in summer flood season in the same way as previous studies [*Shankman and Liang*, 2003] [reprint from *Nakayama and Shankman*, 2013b] (Figure 33.5a). The discharge and water level during the severe 1998 flood would not greatly change without limitation of spill discharge after the dam construction as suggested by *Shankman and Liang* [2003]. The simulated results also show flood risk around the lake might be further accelerated through the constant increase in water level about 1–2 m if the hydrologic change would cause downstream channel aggradation by sediment deposition at 22 years after the completion of the TGD [*Yang et al.*, 2006; *Chen et al.*, 2008; *Xu and Milliman*, 2009],

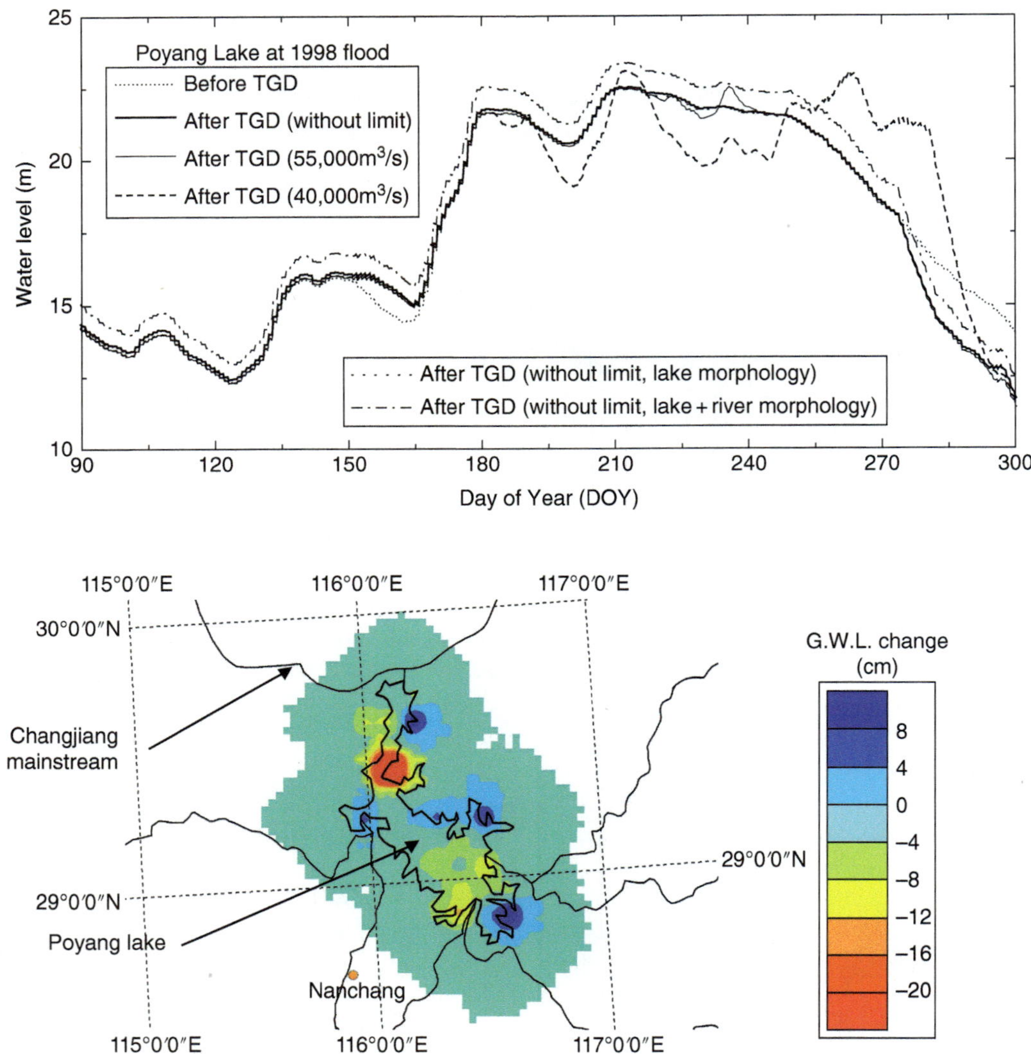

Figure 33.5 Impact of the TGD and SNWTP on hydrologic change around Poyang Lake [reprint from *Nakayama and Shankman*, 2013b]: (a) lake water level during severe flood event and (b) groundwater level difference between the average level before and after their constructions. In (a), the spill discharge at the TGD was also changed by considering navigation benefits and flood risk in the downstream. In (b) the SNWTP scenario is the water transfer scheme of 13.0 × 10⁹ m³ per year [*Li et al.*, 2007].

although the sediment deposition in the lake [*Dai et al.*, 2005; *de Leeuw et al.*, 2009] might not affect so greatly the hydrologic change in the long-term periods. This previous result [*Nakayama and Shankman*, 2013b] indicates that it is important to manage not only the flood discharge but also sediment loads at the whole basin during the TGD operation and the SNWTP in addition to permanent removal of some levees and opening of others during severe floods. Though the effect of the SNWTP is not so remarkable as that of the TGD, the hydrologic change would occur in spring and early summer around the lake region in accordance with the water supply period of rice planting [*U.S. Department of Agriculture*, 1994] and the probability of water shortage would be high, in particular, around the Hanjiang River basin, due to the sensitivity to climate change [*Li et al.*, 2007]. The groundwater seepage to lakes might play a more important role along the mid-lower reaches in a normal year [reprint from *Nakayama and Shankman*, 2013b] (Figure 33.5b), which is supported by the previous research that it pays about 30% of hydrologic budget in maintaining streamflow in the drought season there [*Dai et al.*, 2010]. In particular, the average level would change greatly around the inlet of major tributaries into the lake and around the junction of the mainstream after the TGD and SNWTP constructions, which implies the complex river-lake-groundwater interactions there [*Nakayama and Shankman*, 2013b].

33.3.4. Satellite Analyses for Effect of Irrigation on Ecosystem Change on Continental Scale

The mean TINDVI gradients during 1982–1999 in various field crops (wheat, maize, and rice) at four stations were compared with trends of crop yields in the previous research of *Tao et al.* [2006] [*reprint from Nakayama*, 2012c] (Figure 33.6a). The correlation of both values is relatively good ($r^2 = 0.986$) and the TINDVI gradient has a linear relation to the yield trend. The spatial pattern of the mean TINDVI gradient in agricultural fields shows a generally increasing tendency, particularly in the Yellow River downstream and the NCP [reprint from *Nakayama*, 2012c] (Figure 33.6b), which is closely related to the increasing tendency for winter wheat production in the downstream area [*USDA*, 1994], increase in irrigation water use [*Yang et al.*, 2004] and chemical fertilizer, changes in crop varieties, improvements in technology such as agricultural machines, and other agronomic changes. The situation is a little different in the Changjiang River, where it shows a generally decreasing tendency, particularly in the mid-lower reaches and around the lakes. This is caused mainly by an increase in lake reclamation, levee construction, and the resultant relative decrease in rice productivity in the lower reaches [*USDA*, 1994; *Shankman and Liang*, 2003; *Zhao et al.*, 2005;

Nakayama and Watanabe, 2008b]. The decrease in the TINDVI gradient near the Bohai Sea, the East China Sea, and the Taihang Mountains was due to several effects including groundwater degradation, seawater intrusion, and rapid urbanization in the areas surrounding bigger cities [*Brown and Halweil*, 1998]. Generally, this previous result [*Nakayama*, 2012c] shows that the satellite analysis is an effective tool to grasp ecohydrological process on a continental scale and suggests that the increase in irrigation water use is one of the reasons for the increase in crop production [*Yang et al.*, 2004].

33.4. DISCUSSIONS

This chapter summarized comprehensively the author's various studies about different land covers such as natural, irrigated, and urban areas in China [*Nakayama*, 2011a, 2011b, 2012c; *Nakayama and Shankman*, 2013a, 2013b; *Nakayama and Watanabe*, 2008b; *Nakayama et al.*, 2010] to transcend each study's objective and to synthesize the impact of anthropogenic activity on ecohydrological changes at the basin scale there for adaptation strategy and effective water resources management in the future. Environmental pollution is recently becoming intertwined with degradations of water environment, imbalance of hydrologic cycle, thermal environment, and contamination, for example. Because water stress has intensified water use conflicts between various sectors, in particular, in China, we need an improvement in cumulative diagnosing threat framework for effective policy making and global water security. An integrated approach combining sophisticated process-based model, satellite image, and statistical analysis in this study showed a close relationship among water resource, economic growth, and ecosystem degradation in the environment. Though water resources are vital for human activity, its overexploitation causes the serious hydrologic change such as drying-out the river, groundwater drawdown, and seawater intrusion (Figure 33.3). The results indicated a close relationship between water resource and economic growth, which also causes various ecosystem degradations and its serious burden on the environment (Figure 33.4). However, effective management of water resource is also powerful for decision making on sustainable development and adaptation to climate change and urbanization on a global scale (Figure 33.5). Satellite analysis is also valuable for estimating complex ecohydrological changes and its relation to ecosystem degradation there by coupling with a sophisticated process-based model (Figure 33.6).

Further evaluation of the water-heat-mass cycle in China is also necessary relative to national security and international contributions [*Nakayama*, 2011a, 2011b; *Nakayama and Watanabe*, 2008b; *Nakayama et al.*, 2006, 2010]. The Changjiang, Yellow, and Huaihe rivers are among the "seven big rivers" of China. Although it is

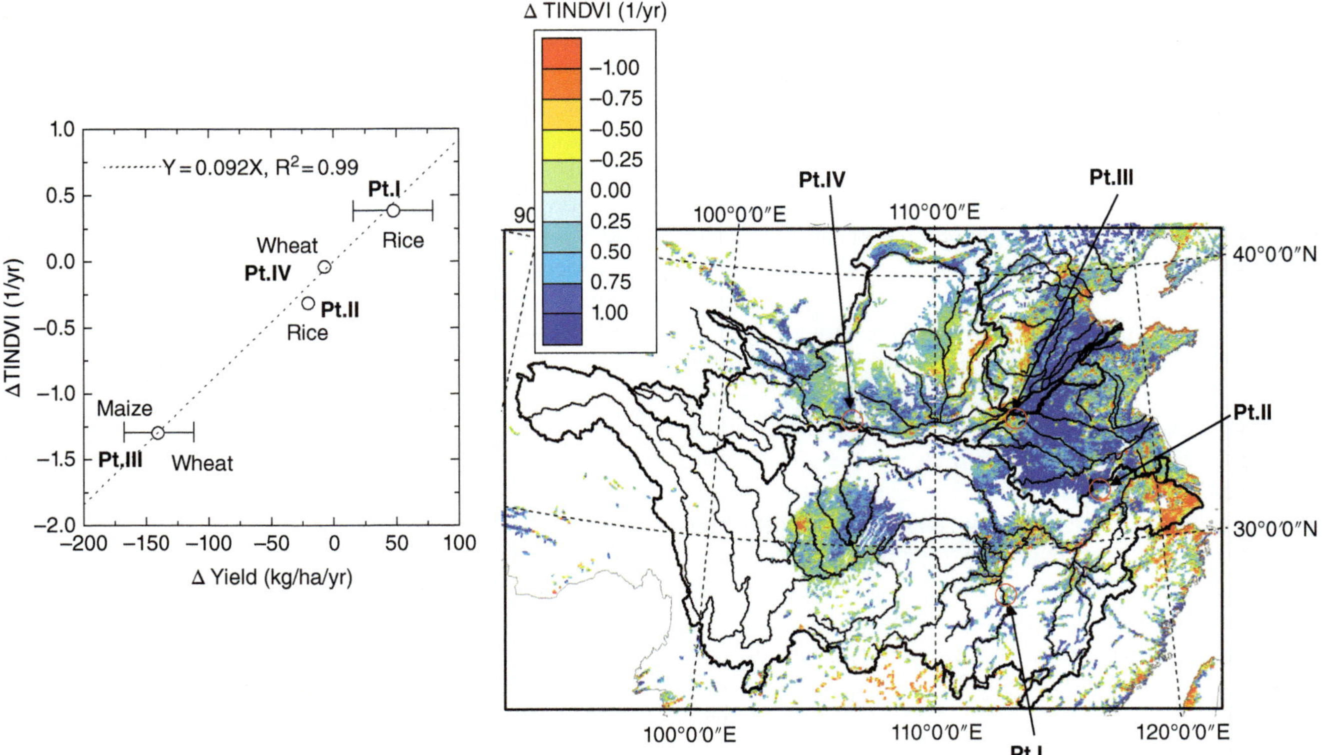

Figure 33.6 TINDVI over the agricultural fields by NOAA/AVHRR satellite images in the agricultural fields [reprint from *Nakayama*, 2012c]: (a) comparison of TINDVI with trend in observed yields at four stations (Pt. I: Changsha, Pt. II: Hefei; Pt. III: Zhengzhou; Pt. IV: Tianshui) [*Tao et al.*, 2006] and (b) spatial patterns in mean TINDVI gradient (per year) during 1982–1999. In (b) black squire line is the border of simulation area in the Albers coordinates in the same way as Figure 33.1.

estimated that the TGD and SNWTP would remove these imbalances as far as possible (Figure 33.5), there is an urgent need to clarify comprehensively the relations of these two extremes affected by human activity, develop a process-based model, and present an integrated assessment system for appropriate water resource management to achieve sustainable development [*Nakayama*, 2012c; *Nakayama and Shankman*, 2013a, 2013b]. Therefore, accurate evaluation of the optimum amount of transferred water, and socioeconomic and environmental consequences are essential in their basins. Future installments of this study will evaluate water-heat-mass dynamics using the NICE model to simulate the changes caused by human activity and to aid accurate estimation on the amount of water to be transferred between basins. This assessment system will also play an important role in the improvement in biogeochemical activity in spatial and temporal hot spots [*Frei et al.*, 2012] and boundless biogeochemical cycle along terrestrial-aquatic continuum for global environmental change [*Cole et al.*, 2007; *Battin et al.*, 2009].

It is also important to research the mutual interaction and feedback of hydrogeomorphology and vegetation dynamics to further explain the positive feedback between geomorphology and eco-hydrology in the natural and irrigated areas. This is necessary for the effectiveness of the river/lake restoration project implemented in Dongting and Poyang lake regions [*Zhao et al.*, 2005; *Nakayama and Watanabe*, 2008b; *Nakayama and Shankman*, 2013a] and the importance of the process-based model in assessing the connection between hydrologic change and vegetation succession [*Nakayama*, 2012b, 2013]. Recent research suggests serious concerns against extrapolation of experimental results at a small scale to entire landscapes and the necessity to bridge the gap between ecosystems at various scales [*Deegan et al.* 2012]. From this point of view, a future research will be more powerful to reevaluate the ecosystem as an extension of the "metabolic theory of ecology" [*Brown et al.*, 2004] from the perspective of a meta-ecosystem analysis by considering multiscaled aspects between global-regional-micro (genetic) levels (Figure 33.7) in the same way as the "river continuum concept" [*Vannote et al.*, 1980]. It is heuristically powerful to identify spatial coupling of local ecosystems including energy, materials, and organisms across ecosystem boundaries. In addition, it is necessary to assess the impact of water-heat-material cycle changes through human activities

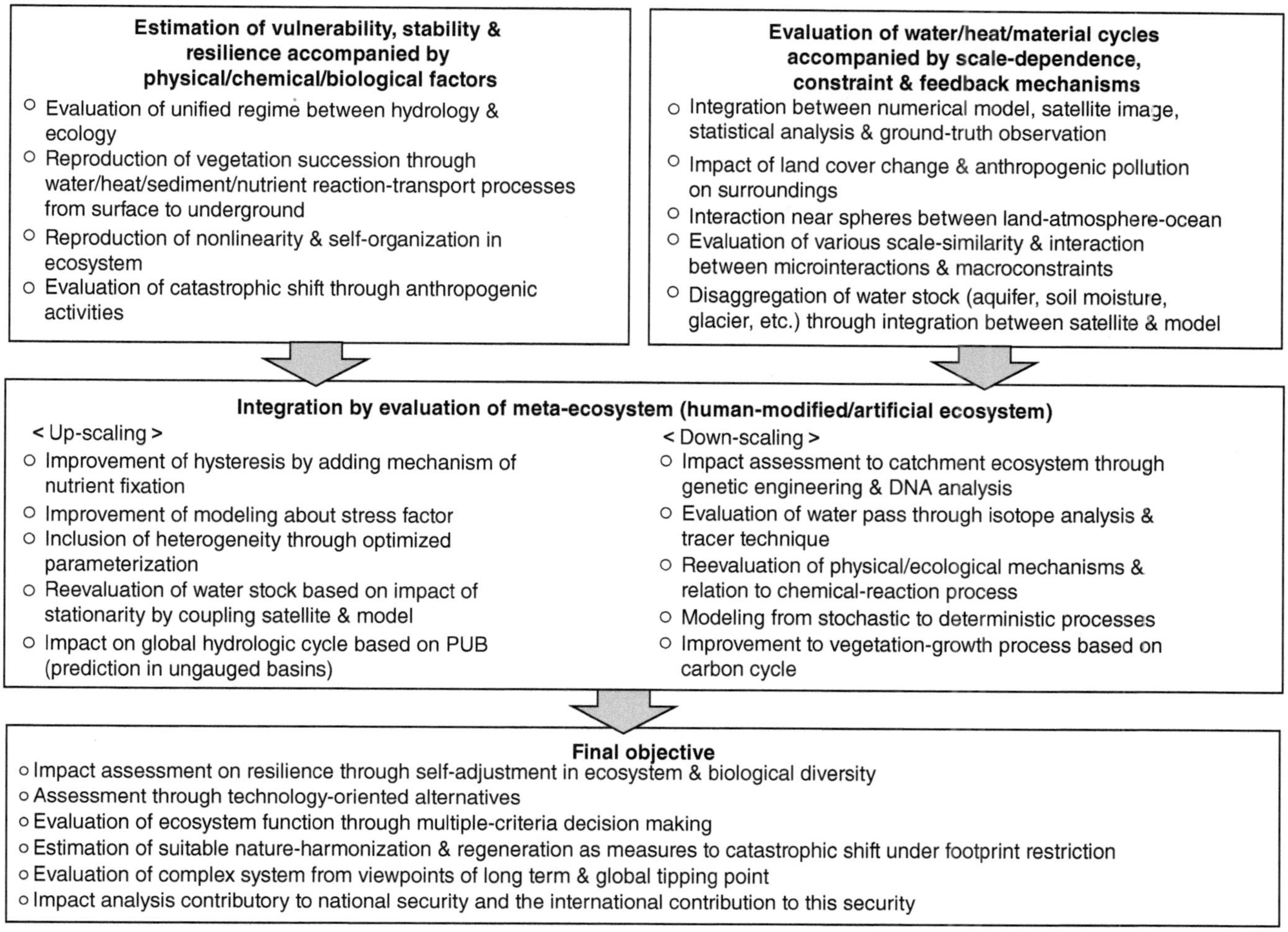

<table>
<tr>
<td>

Estimation of vulnerability, stability & resilience accompanied by physical/chemical/biological factors

- Evaluation of unified regime between hydrology & ecology
- Reproduction of vegetation succession through water/heat/sediment/nutrient reaction-transport processes from surface to underground
- Reproduction of nonlinearity & self-organization in ecosystem
- Evaluation of catastrophic shift through anthropogenic activities

</td>
<td>

Evaluation of water/heat/material cycles accompanied by scale-dependence, constraint & feedback mechanisms

- Integration between numerical model, satellite image, statistical analysis & ground-truth observation
- Impact of land cover change & anthropogenic pollution on surroundings
- Interaction near spheres between land-atmosphere-ocean
- Evaluation of various scale-similarity & interaction between microinteractions & macroconstraints
- Disaggregation of water stock (aquifer, soil moisture, glacier, etc.) through integration between satellite & model

</td>
</tr>
</table>

Integration by evaluation of meta-ecosystem (human-modified/artificial ecosystem)

< Up-scaling >
- Improvement of hysteresis by adding mechanism of nutrient fixation
- Improvement of modeling about stress factor
- Inclusion of heterogeneity through optimized parameterization
- Reevaluation of water stock based on impact of stationarity by coupling satellite & model
- Impact on global hydrologic cycle based on PUB (prediction in ungauged basins)

< Down-scaling >
- Impact assessment to catchment ecosystem through genetic engineering & DNA analysis
- Evaluation of water pass through isotope analysis & tracer technique
- Reevaluation of physical/ecological mechanisms & relation to chemical-reaction process
- Modeling from stochastic to deterministic processes
- Improvement to vegetation-growth process based on carbon cycle

Final objective
- Impact assessment on resilience through self-adjustment in ecosystem & biological diversity
- Assessment through technology-oriented alternatives
- Evaluation of ecosystem function through multiple-criteria decision making
- Estimation of suitable nature-harmonization & regeneration as measures to catastrophic shift under footprint restriction
- Evaluation of complex system from viewpoints of long term & global tipping point
- Impact analysis contributory to national security and the international contribution to this security

Figure 33.7 Integrated procedures through meta-ecosystem analysis.

on ecosystem function, analysis environmental impact, and limiting factor accompanied by the ecosystem change. Moreover, a quantitative estimation is important for more effective measures and available technologies as alternative scenarios for the final objective of sustainable development under the constraint of global security in addition to environmental degradation. It is a future research to develop integrated assessment system flexibly available to various scaled ecosystems consisting of urban areas, forests, grasslands, cultivated fields, paddy fields, wetlands, rivers, lakes, groundwater, oceans, and atmosphere on the basis of integrations between numerical simulation, satellite data, statistical analysis, and ground-truth observation in multiple scales, as described below (Figure 33.7):

1. Construction of assessment models in multiple scales, including ecosystem dynamism in regional scales, impact assessment of urban activities on regional water-heat-material cycles, quantification of potential capacity in the ecosystem, numerical simulation in the scenario of technology for environmental improvement, and integration of modeling and political scenarios through an evaluation of sustainability.

2. Impact assessment of various stress factors on a multiscaled ecosystem, development of a modeling system including the multiscaled aspect between global-regional-micro levels and a feedback mechanism between hydrology and ecology and evaluation of vulnerability in ecosystem functions.

3. Impact assessment to contribute to global security relative to items 1 and 2, up-scaling and down-scaling procedures focusing on urban, agricultural, and natural areas, reevaluation of natural energy and environmental resources, and objective-oriented research combined with political scenarios and decision making on the basis of multiple criteria.

33.5. CONCLUSION

In this study, the author's various studies about different land covers such as natural, irrigated, and urban areas in China were reviewed and summarized in order to synthesize the impact of anthropogenic activity on ecohydrological changes in the catchment or basin scales there for adaptation strategy and effective water resources management in the future. The process-based NICE model was coupled with complex subsystems in irrigation, stream junction, reservoir operation, and water transfer to evaluate cause and effect of uneven water resource in the entire Changjiang and Yellow river basins. An integrated approach combining this sophisticated model, satellite image, and statistical analysis showed a broadly close relationship among water resource, economic growth, and ecosystem degradation in the environment. The model included different growth processes of representative crops and predicted hydrologic change after TGD and SNWTP to estimate whether dilemmas among water stress, crop productivity, and ecosystem degradation would diminish in the basins. Further, the heterogeneous pattern of TINDVI gradient in agricultural fields helped to estimate uneven crop yield and its relation to water availability. Though water resources are vital for human activity, its overexploitation sometimes causes serious hydrologic change such as drying-out a river, groundwater fall, and seawater intrusion. These summarized results indicated comprehensively a close relationship between water resource and economic growth, which also causes various ecosystem degradations and its serious burden on the environment. However, effective management of water resource such as further refinement of TGD and SNWTP is also powerful for decision making and adaptation strategy for sustainable development. These results are very important to clarify the effect of human activity on ecosystem degradation caused by the ecohydrological changes beyond separate previous results.

ACKNOWLEDGMENTS

Some of the simulations in this study were made on an NEC SX-8 supercomputer at the Center for Global Environment Research (CGER), NIES. Dr. D. Shankman, Rhodes College, USA, Dr. M. Watanabe, Keio University, Japan, Dr. K. Xu, National Institute for Environmental Studies, Japan, and Dr Y. Yang, Chinese Academy of Sciences, China, gave helpful comments and discussions about the study area.

REFERENCES

Alley, W. M., R. W. Healy, J. W. LaBaugh, and T. E. Reilly (2002), Flow and storage in groundwater systems, *Science*, *296*, 1985–1990.

Barnaby, W. (2009), Do nations go to war over water? *Nature*, *458*, 282–283.

Baron, J. S., N. L. Poff, P. L. Angermeier, C. N. Dahm, P. H. Gleick, N. G. Hairston, R. B. Jackson, C. A. Johnston, B. D. Richter, and A. D. Steinman (2002), Meeting ecological and societal needs for freshwater, *Ecol. Appl.*, *12*(5), 1247–1260.

Battin, T. J., S. Luyssaert, L. A. Kaplan, A. K. Aufdenkampe, A. Richter, and L. J. Tranvik (2009), The boundless carbon cycle, *Nature Geosci. 2*, 598–600.

Brown, L. R., and B. Halweil (1998), China's water shortage could shake world food security, *World Watch*, *11*(4), 10–18.

Brown, J. H., J. F. Gillooly, A. P. Allen, V. M. Savage, and G. B. West (2004), Toward a metabolic theory of ecology, *Ecology*, *85*, 1771–1789.

Changjiang Water Conservancy Committee (1987), Annual report of Changjiang water and sediment, interior report of the committee (in Chinese).

Changjiang Water Conservancy Committee (1988), Annual report of Changjiang water and sediment, interior report of the committee (in Chinese).

Changjiang Water Conservancy Committee (1998), Annual report of Changjiang water and sediment, interior report of the committee (in Chinese).

Changjiang Water Conservancy Committee (1999), Annual report of Changjiang water and sediment, interior report of the committee (in Chinese).

Chen, J. Y., C. Y. Tang, Y. J. Shen, Y. Sakura, A. Kondoh, and J. Shimada (2003), Use of water balance calculation and tritium to examine the dropdown of groundwater table in the piedmont of the North China Plain (NCP), *Environ. Geol. 44*, 564–571.

Chen, Z. Y., Z. L. Nie, Z. J. Zhang, J. X. Qi, and Y. J. Nan (2005), Isotopes and sustainability of ground water resources, North China Plain, *Ground Water*, *43*(4), 485–493.

Chen, X., Y. Yan, R. Fu, X. Dou, and E. Zhang (2008), Sediment transport from the Yangtze River, China, into the sea over the Post-Three Gorges Dam Period: A discussion, *Quatern. Int.*, *186*, 55–64, doi:10.1016/j.quaint.2007.10.003.

China Institute for Geo-Environmental Monitoring (2003), China Geological Environment Infonet, Database of groundwater observation in the People's Republic of China, available at http://www.cigem.gov.cn.

Chinese Academy of Sciences (1988), Administrative division coding system of the People's Republic of China, Chinese Academy of Sciences, Beijing.

Clapp, R. B., and G. M. Hornberger (1978), Empirical equations for some soil hydraulic properties, *Water Resourc. Res.*, *14*, 601–604.

Cole, J. J., et al. (2007), Plumbing the global carbon cycle: Integrating inland waters into the terrestrial carbon budget, *Ecosystems*, *10*, 171–184, doi:10.1007/s10021-006-9013-8.

Dai, S. B., S. L. Yang, J. Zhu, A. Gao, and P. Li (2005), The role of Lake Dongting in regulating the sediment budget of the Yangtze River, *Hydrol. Earth Syst. Sci.*, *9*(6), 692–698.

Dai, Z., J. Du, A. Chu, J. Li, J. Chen, and X. Zhang (2010), Groundwater discharge to the Changjiang River, China, during the drought season of 2006: Effects of the extreme drought and the impoundment of the Three Gorges Dam, *Hydrogeol. J.*, *18*, 359–369.

Deegan, L. A., D. S. Johnson, R. S. Warren, B. J. Peterson, J. W. Fleeger, S. Fagherazzi, and W. M. Wollheim (2012), Coastal eutrophication as a driver of salt marsh loss, *Nature*, *490*, 388–394, doi:10.1038/nature11533.

de Leeuw, J., D. Shankman, G. Wu, W. F. de Boer, J. Burnham, Q. He, and H. Yesou (2009), Strategic assessment of the magnitude and impacts of sand mining in Poyang Lake, China, *Regional Environ. Change*, *10*, 95–102, doi:10.1007/s10113-009-0096-6.

D'Odorico, P., F. Laio, and L. Ridolfi (2010), Does globalization of water reduce societal resilience to drought? *Geophys. Res. Lett.*, *37*, L13403, doi:10.1029/2010GL043167.

Entin, J. K., A. Robock, K. Y. Vinnikov, S. E. Hollinger, S. Liu, A. Namkhai (2000), Temporal and spatial scales of observed soil moisture variations in the extratropics, *J. Geophys. Res.*, *105*(D9), 11,865–11,877.

Fang, W., H. Imura, F. Shi (2006), Wheat irrigation water requirement variability (2001–2030) in the Yellow River Basin under HADCM3 GCM scenarios, *Jpn. J. Environ. Sci.*, *19*(1), 3–14.

Frei, S., K. H. Knorr, S. Peiffer, J. H. Fleckenstein (2012), Surface micro-topography causes hot spots of biogeochemical activity in wetland system: A virtual modeling experiment, *J. Geophys. Res.*, *117*, G00N12, doi:10.1029/2012JG002012.

Geological Atlas of China (2002), Geological Publisher, Beijing (in Chinese).

Gleick, P. H., and M. Palaniappan (2010), Peak water limits to freshwater withdrawal and use, *Proc. Nat. Acad. Sci. USA*, *107*(25), 11,155–11,162, doi:10.1073/pnas.1004812107.

Hoekstra, A. Y., and P. Q. Hung (2005), Globalisation of water resources: International virtual water flows in relation to crop trade, *Global Environ. Change*, *15*, 45–56, doi:10.1016/j.gloenvcha.2004.06.004.

Jackson, R. B., S. R. Carpenter, C. N. Dahm, D. M. McKnight, R. J. Naiman, S. L. Postel, and S. W. Running (2001), Water in a changing world, *Ecol. Appl.*, *11*(4), 1027–1045.

Li, Q., Z. Zou, Z. Xia, J. Guo, and Y. Liu (2007), Impacts of human activities on the flow regime of the Yangtze River, in *Changes in Water Resources Systems: Methodologies to Maintain Water Security and Ensure Integrated Management*, edited by N. van de Giesen, et al., IAHS Publ. *315*, pp. 266–275 IAHS Press, Wallingford, UK.

Liu, J. Y. (1996), *Macro-Scale Survey and Dynamic Study of Natural Resources and Environment of China by Remote Sensing*, Chinese Science and Technology, Beijing (in Chinese).

Liu, C., and J. Xia (2004), Water problems and hydrological research in the Yellow River and the Huai and Hai River basins of China, *Hydrol. Process.*, *18*, 2197–2210, doi:10.1002/hyp.5524.

Liu, C., X. Zhang, Y. Zhang (2002), Determination of daily evapotranspiration of winter wheat and corn by large-scale weighting lysimeter and micro-lysimeter, *Agric. Forest Meteorol.*, *111*, 109–120.

Liu, P., S. Guo, and W. Li (2008) Optimal design of seasonal flood control water levels for the Three Gorges Reservoir, in *Hydrological Sciences for Managing Water Resources in the Asian Developing World*, edited by X. Chen, et al., IAHS Pub. *319*, pp. 270–277. IAHS Press, Wallingford, UK.

Ma, J., A. Y. Hoekstra, H. Wang, A. K. Chapagain, and D. Wang (2006), Virtual versus real water transfers within China, *Philos. Trans. Roy. Soc. B*, *361*, 835–842.

Milly, P. C. D., K. A. Dunne, and A. V. Vecchia (2005), Global pattern of trends in streamflow and water availability in a changing climate, *Nature*, *438*, 347–350, doi:10.1038/nature04312.

Milly, P. C. D., J. Betancourt, M. Falkenmark, R. M. Hirsch, Z. W. Kundzewicz, D. P. Lettenmaier, and R. J. Stouffer (2008), Stationarity is dead: Whither water management? *Science*, *319*, 573–574, doi:10.1126/science.1151915.

Nakayama, T. (2008a), Factors controlling vegetation succession in Kushiro Mire, *Ecol. Model.*, *215*, 225–236, doi:10.1016/j.ecolmodel.2008.02.017.

Nakayama, T. (2008b), Shrinkage of shrub forest and recovery of mire ecosystem by river restoration in northern Japan, *Forest Ecol. Manag.*, *256*, 1927–1938, doi:10.1016/j.foreco.2008.07.017.

Nakayama, T. (2009), Simulation of ecosystem degradation and its application for effective policy-making in regional scale, in *River pollution Research Progress*, edited by M. N. Gallo and M. H. Ferrari, pp. 1–89, Nova Science Pub., New York.

Nakayama, T. (2010), Simulation of hydrologic and geomorphic changes affecting a shrinking mire, *River Res. Appl.*, *26*(3), 305–321, doi:10.1002/rra.1253.

Nakayama, T. (2011a), Simulation of complicated and diverse water system accompanied by human intervention in the North China Plain, *Hydrol. Process.*, *25*, 2679–2693, doi:10.1002/hyp.8009.

Nakayama, T. (2011b), Simulation of the effect of irrigation on the hydrologic cycle in the highly cultivated Yellow River Basin, *Agric. Forest Meteorol.*, *151*, 314–327, doi:10.1016/j.agrformet.2010.11.006.

Nakayama, T. (2012a), Visualization of missing role of hydrothermal interactions in Japanese megalopolis for win-win solution, *Water Sci. Technol.*, *66*(2), 409–414, doi:10.2166/wst.2012.205.

Nakayama, T. (2012b), Feedback and regime shift of mire ecosystem in northern Japan, *Hydrol. Process.*, *26*(16), 2455–2469, doi:10.1002/hyp.9347.

Nakayama, T. (2012c), Impact of anthropogenic activity on eco-hydrological process in continental scales, *Procedia Environ. Sci.*, *13*, 87–94, doi:10.1016/j.proenv.2012.01.008.

Nakayama, T. (2013), For improvement in understanding eco-hydrological processes in mire, *Ecohydrol. Hydrobiol.*, *13*, 62–72, doi:10.1016/j.ecohyd.2013.03.004.

Nakayama, T., and T. Fujita (2010), Cooling effect of water-holding pavements made of new materials on water and heat budgets in urban areas, *Landscape and Urban Planning*, *96*, 57–67, doi:10.1016/j.landurbplan.2010.02.003.

Nakayama, T., and S. Hashimoto (2011), Analysis of the ability of water resources to reduce the urban heat island in the Tokyo megalopolis, *Environ. Pollut.*, *159*, 2164–2173, doi:10.1016/j.envpol.2010.11.016.

Nakayama, T., and D. Shankman (2013a), Impact of the Three-Gorges Dam and water transfer project on Changjiang floods, *Global Planet. Change*, *100*, 38–50, doi:10.1016/j.gloplacha.2012.10.004.

Nakayama, T., and D. Shankman (2013b), Evaluation of uneven water resource and relation between anthropogenic water withdrawal and ecosystem degradation in Changjiang and Yellow River basins, *Hydrol. Process.*, *27*(23), 3350–3362. doi:10.1002/hyp.9835.

Nakayama, T., and M. Watanabe (2004), Simulation of drying phenomena associated with vegetation change caused by invasion of alder (Alnus japonica) in Kushiro Mire, *Water Resourc. Res.*, *40*(8), W08402, doi:10.1029/2004WR003174.

Nakayama, T., and M. Watanabe (2006), Simulation of spring snowmelt runoff by considering micro-topography and phase changes in soil layer, *Hydrol. Earth Syst. Sci. Disc.*, *3*, 2101–2144.

Nakayama, T., and M. Watanabe (2008a), Missing role of groundwater in water and nutrient cycles in the shallow eutrophic Lake Kasumigaura, Japan, *Hydrol. Process.*, *22*, 1150–1172, doi:10.1002/hyp.6684.

Nakayama, T., and M. Watanabe (2008b), Role of flood storage ability of lakes in the Changjiang River catchment, *Global Planet. Change*, *63*, 9–22, doi:10.1016/j.gloplacha.2008.04.002.

Nakayama, T., Y. Yang, M. Watanabe, and X. Zhang (2006), Simulation of groundwater dynamics in North China Plain by coupled hydrology and agricultural models, *Hydrol. Process.*, *20*(16), 3441–3466, doi:10.1002/hyp.6142.

Nakayama, T., M. Watanabe, K. Tanji, and T. Morioka (2007), Effect of underground urban structures on eutrophic coastal environment, *Sci. Total Environ.*, *373*(1), 270–288, doi:10.1016/j.scitotenv.2006.11.033.

Nakayama, T., Y. Sun, and Y. Geng (2010), Simulation of water resource and its relation to urban activity in Dalian City, Northern China, *Global Planet. Change*, *73*, 172–185, doi:10.1016/j.gloplacha.2010.06.001.

Nakayama, T., S. Hashimoto, and H. Hamano (2012), Multi-scaled analysis of hydrothermal dynamics in Japanese megalopolis by using integrated approach, *Hydrol. Process.*, *26*(16), 2431–2444, doi:10.1002/hyp.9290.

Organisation for Economic Co-operation, and Development (OECD) (2002), Indicators to measure decoupling of environmental pressure from economic growth, available at http://www.olis.oecd.org/olis/2002doc.nsf/LinkTo/sg-sd(2002)1-final.

Rawls, W. J., D. L. Brakensiek, and K. E. Saxton (1982), Estimation of soil water properties, *Trans. ASAE*, *25*, 1316–1320.

Ren, L., M. Wang, C. Li, and W. Zhang (2002), Impacts of human activity on river runoff in the northern area of China, *J. Hydrol.*, *261*, 204–217.

Rich, V. (1983), Yangtze to cross Yellow River, *Nature*, *305*, 568.

Ritchie, J. T., U. Singh, D. C. Godwin, and W. T. Bowen (1998), Cereal growth, development and yield, in *Understanding Options for Agricultural Production*, edited by G.Y. Tsuji, G. Hoogenboom, and P.K. Thornton, pp. 79–98, Kluwer, Dordrecht, The Netherlands.

Robock, A., Y. V. Konstantin, S. Govindarajalu, K. E. Jared, E. H. Steven, A. S. L. Suxia, and A. Namkhai (2000), The global soil moisture data bank, *Bull. Am. Meteorol. Soc.*, *81*, 1281–1299. http://climate.envsci.rutgers.edu/soil_moisture/.

Sellers, P. J., D. A. Randall, G. J. Collatz, J. A. Berry, C. B. Field, D. A. Dazlich, C. Zhang, G. D. Collelo, and L. Bounoua (1996), A revised land surface prameterization (SiB2) for atmospheric GCMs. Part I : Model formulation, *J. Climate*, *9*, 676–705.

Shankman, D., and Q. Liang (2003), Landscape changes and increasing flood frequency in China's Poyang Lake region, *Prof. Geographer*, *55*(4), 434–445.

Shimada, J. (2000), Proposals for the groundwater preservation toward 21st century through the view point of hydrological cycle, *J. Jpn. Associ. Hydrol. Sci.*, *30*, 63–72 (in Japanese).

Tao, F., M. Yokozawa, Y. Xu, Y. Hayashi, and Z. Zhang (2006), Climate changes and trends in phenology and yields of field crops in China, 1981–2000, *Agric. Forest Meteorol.*, *138*, 82–92.

U.S. Department of Agriculture (USDA) (1994), Major World Crop Areas and Climatic Profiles, World Agricultural Outlook Board, USDA, Agricultural Handbook No. 664, available at http://www.usda.gov/oce/weather/pubs/Other/MWCACP/MajorWorldCropAreas.pdf.

U.S. Geological Survey (USGS) (1996), GTOPO30 Global 30 Arc Second Elevation Data Set, USGS, available at http://www1.gsi.go.jp/geowww/globalmap-gsi/gtopo30/gtopo30.html.

Vannote, R. L., G. W. Minshall, K. W. Cummings, J. R. Sedell, and C. E. Cushing (1980), The river continuum concept, *Can. J. Fish. Aquatic Sci.*, *37*, 130–137.

Varis, O., and P. Vakkilainen (2001), China's 8 challenges to water resources management in the first quarter of the 21st century, *Geomorphology*, *41*, 93–104.

Vorosmarty, C. J., et al. (2010), Global threats to human water security and river biodiversity, *Nature*, *467*, 555–561, doi:10.1038/nature09440.

Wada, Y., L. P. H. van Beek, C. M. van Kempen, J. W. T. M. Reckman, S. Vasak, and M. F. P. Bierkens (2010), Global depletion of groundwater resources, *Geophys. Res. Lett.*, *37*, L20402, doi:10.1029/2010GL044571.

Wang, H., L. Zhang, W. R. Dawes, and C. Liu (2001), Improving water use efficiency of irrigated crops in the North China Plain—Measurements and modeling, *Agric. Forest Meteorol.*, *48*, 151–167.

WebPaNDA (2008), NOAA AVHRR 10-days composite database over East Asia, Univ. of Tokyo, available at http://webpanda.iis.u-tokyo.ac.jp/.

Xu, K., and J. D. Milliman (2009), Seasonal variations of sediment discharge from the Yangtze River before and after impoundment of the Three Gorges dam, *Geomorphology*, *104*, 276–283, doi:10.1016/j.geomorph.2008.09.004.

Yang, H., and A. Zehnder (2001), China's regional water scarcity and implications for grain supply and trade, *Environ. Plan. A*, *33*, 79–95.

Yang, L., B. K. Wylie, L. L. Tieszen, and B. C. Reed (1998), An analysis of relationships among climate forcing and time-integrated NDVI of grasslands over the U.S. Northern and Central Great Plains, *Remote Sen. Environ.*, *65*, 25–37.

Yang, D., C. Li, H. Hu, Z. Lei, S. Yang, T. Kusuda, T. Koike, and K. Musiake (2004), Analysis of water resources variability in the Yellow River of China during the last half century using historical data, *Water Resourc. Res.*, *40*, W06502, doi:10.1029/2003WR002763.

Yang, Z., H. Wang, Y. Saito, J. D. Milliman, K. Xu, S. Qiao, and G. Shi (2006), Dam impacts on the Changjiang (Yangtze) River sediment discharge to the sea: The past 55 years and after the Three Gorges Dam, *Water Resourc. Res.*, *42*, W04407, doi:10.1029/2005WR003970.

Yellow River Conservancy Committee (1987), Annual report of river discharge at Yellow River, Interior report of the committee (in Chinese).

Yellow River Conservancy Committee (1988), Annual report of river discharge at Yellow River, Interior report of the committee (in Chinese).

Yi, Y., Z. Wang, and Z. Yang (2010), Impact of the Gezhouba and Three Gorges Dams on habitat suitability of carps in the Yangtze River, *J. Hydrol.*, *387*, 283–291, doi:10.1016/j.jhydrol.2010.04.018.

Zhao, S., J. Fang, S. Miao, B. Gu, S. Tao, C. Peng, and Z. Tang (2005), The 7-decade degradation of a large freshwater lake in central Yangtze River, China, *Enviro. Sci. Technol.*, *39*, 431–436.

INDEX

Page references followed by f denote figures. Page references followed by t denote tables.

Remote Sensing of the Terrestrial Water Cycle, Geophysical Monograph 206. First Edition. Edited by Venkat Lakshmi.
© 2015 American Geophysical Union. Published 2015 by John Wiley & Sons, Inc.